Peter M.A. Sloot   C.J. Kenneth Tan
Jack J. Dongarra   Alfons G. Hoekstra (Eds.)

Computational Science – ICCS 2002

International Conference
Amsterdam, The Netherlands, April 21-24, 2002
Proceedings, Part I

Springer

Series Editors

Gerhard Goos, Karlsruhe University, Germany
Juris Hartmanis, Cornell University, NY, USA
Jan van Leeuwen, Utrecht University, The Netherlands

Volume Editors

Peter M.A. Sloot
Alfons G. Hoekstra
University of Amsterdam, Faculty of Science, Section Computational Science
Kruislaan 403, 1098 SJ Amsterdam, The Netherlands
E-mail: {sloot,alfons}@science.uva.nl

C.J. Kenneth Tan
University of Western Ontario, Western Science Center, SHARCNET
London, Ontario, Canada N6A 5B7
E-mail: cjtan@acm.org

Jack J. Dongarra
University of Tennessee, Computer Science Department
Innovative Computing Laboratory
1122 Volunteer Blvd, Knoxville, TN 37996-3450, USA
E-mail: dongarra@cs.utk.edu

Cataloging-in-Publication Data applied for

Die Deutsche Bibliothek - CIP-Einheitsaufnahme

Computational science : international conference ; proceedings / ICCS 2002,
Amsterdam, The Netherlands, April 21 - 24, 2002. Peter M. A. Sloot (ed.). -
Berlin ; Heidelberg ; New York ; Barcelona ; Hong Kong ; London ; Milan ;
Paris ; Tokyo : Springer
Pt. 1 . - (2002)
  (Lecture notes in computer science ; Vol. 2329)
  ISBN 3-540-43591-3

CR Subject Classification (1998): D, F, G, H, I, J, C.2-3
ISSN 0302-9743
ISBN 3-540-43591-3 Springer-Verlag Berlin Heidelberg New York

Springer-Verlag Berlin Heidelberg New York
a member of BertelsmannSpringer Science+Business Media GmbH

http://www.springer.de

© Springer-Verlag Berlin Heidelberg 2002
Printed in Germany

Typesetting: Camera-ready by author, data conversion by PTP-Berlin, Stefan Sossna e.K.
Printed on acid-free paper   SPIN: 10846759   06/3142   5 4 3 2 1 0

Preface

Computational Science is the scientific discipline that aims at the development and understanding of new computational methods and techniques to model and simulate complex systems.

The area of application includes natural systems – such as biology, environmental and geo-sciences, physics, and chemistry – and synthetic systems such as electronics and financial and economic systems. The discipline is a bridge between 'classical' computer science – logic, complexity, architecture, algorithms – mathematics, and the use of computers in the aforementioned areas.

The relevance for society stems from the numerous challenges that exist in the various science and engineering disciplines, which can be tackled by advances made in this field. For instance new models and methods to study environmental issues like the quality of air, water, and soil, and weather and climate predictions through simulations, as well as the simulation-supported development of cars, airplanes, and medical and transport systems etc.

Paraphrasing R. Kenway (R.D. Kenway, Contemporary Physics. 1994): 'There is an important message to scientists, politicians, and industrialists: in the future science, the best industrial design and manufacture, the greatest medical progress, and the most accurate environmental monitoring and forecasting will be done by countries that most rapidly exploit the full potential of *computational science*'.

Nowadays we have access to high-end computer architectures and a large range of computing environments, mainly as a consequence of the enormous stimulus from the various international programs on advanced computing, e.g. HPCC (USA), HPCN (Europe), Real-World Computing (Japan), and ASCI (USA: Advanced Strategie Computing Initiative). The sequel to this, known as 'grid-systems' and 'grid-computing', will boost the computer, processing, and storage power even further. Today's supercomputing application may be tomorrow's desktop computing application.

The societal and industrial pulls have given a significant impulse to the rewriting of existing models and software. This has resulted among other things in a big 'clean-up' of often outdated software and new programming paradigms and verification techniques. With this make-up of arrears the road is paved for the study of real complex systems through computer simulations, and large scale problems that have long been intractable can now be tackled. However, the development of complexity reducing algorithms, numerical algorithms for large data sets, formal methods and associated modeling, as well as representation (i.e. visualization) techniques are still in their infancy. Deep understanding of the approaches required to model and simulate problems with increasing complexity and to efficiently exploit high performance computational techniques is still a big scientific challenge.

The International Conference on Computational Science (ICCS) series of conferences was started in May 2001 in San Francisco. The success of that meeting motivated the organization of the meeting held in Amsterdam from April 21–24, 2002.

These three volumes (Lecture Notes in Computer Science volumes 2329, 2330, and 2321) contain the proceedings of the ICCS 2002 meeting. The volumes consist of over 350 – peer reviewed – contributed and invited papers presented at the conference in the Science and Technology Center Watergraafsmeer (WTCW), in Amsterdam. The papers presented reflect the aims of the program committee to bring together major role players in the emerging field of computational science.

The conference was organized by The University of Amsterdam, Section Computational Science (http://www.science.uva.nl/research/scs/), SHARCNET, Canada (http://www.sharcnet.com), and the Innovative Computing Laboratory at The University of Tennessee.

The conference included 22 workshops, 7 keynote addresses, and over 350 contributed papers selected for oral presentation. Each paper was refereed by at least two referees.

We are deeply indebted to the members of the program committee, the workshop organizers, and all those in the community who helped us to organize a successful conference. Special thanks go to Alexander Bogdanov, Jerzy Wasniewski, and Marian Bubak for their help in the final phases of the review process. The invaluable administrative support of Manfred Stienstra, Alain Dankers, and Erik Hitipeuw is also acknowledged. Lodewijk Bos and his team were responsible for the local logistics and as always did a great job.

ICCS 2002 would not have been possible without the support of our sponsors: The University of Amsterdam, The Netherlands; Power Computing and Communication BV, The Netherlands; Elsevier Science Publishers, The Netherlands; Springer-Verlag, Germany; HPCN Foundation, The Netherlands; National Supercomputer Facilities (NCF), The Netherlands; Sun Microsystems, Inc., USA; SHARCNET, Canada; The Department of Computer Science, University of Calgary, Canada; and The School of Computer Science, The Queens University, Belfast, UK.

Amsterdam, April 2002   Peter M.A. Sloot,
                                                          Scientific Chair 2002,

                                                     on behalf of the co-editors:
                                                            C.J. Kenneth Tan
                                                            Jack J. Dongarra
                                                          Alfons G. Hoekstra

Organization

The 2002 International Conference on Computational Science was organized jointly by The University of Amsterdam, Section Computational Science, SHARCNET, Canada, and the University of Tennessee, Department of Computer Science.

Conference Chairs

Peter M.A. Sloot, Scientific and Overall Chair ICCS 2002 (University of Amsterdam, The Netherlands)
C.J. Kenneth Tan (SHARCNET, Canada)
Jack J. Dongarra (University of Tennessee, Knoxville, USA)

Workshops Organizing Chair

Alfons G. Hoekstra (University of Amsterdam, The Netherlands)

International Steering Committee

Vassil N. Alexandrov (University of Reading, UK)
J. A. Rod Blais (University of Calgary, Canada)
Alexander V. Bogdanov (Institute for High Performance Computing and Data Bases, Russia)
Marian Bubak (AGH, Poland)
Geoffrey Fox (Florida State University, USA)
Marina L. Gavrilova (University of Calgary, Canada)
Bob Hertzberger (University of Amsterdam, The Netherlands)
Anthony Hey (University of Southampton, UK)
Benjoe A. Juliano (California State University at Chico, USA)
James S. Pascoe (University of Reading, UK)
Rene S. Renner (California State University at Chico, USA)
Kokichi Sugihara (University of Tokyo, Japan)
Jerzy Wasniewski (Danish Computing Center for Research and Education, Denmark)
Albert Zomaya (University of Western Australia, Australia)

Local Organizing Committee

Alfons Hoekstra (University of Amsterdam, The Netherlands)
Alexander V. Bogdanov (Institute for High Performance Computing and Data Bases, Russia)
Marian Bubak (AGH, Poland)
Jerzy Wasniewski (Danish Computing Center for Research and Education, Denmark)

Local Advisory Committee

Patrick Aerts (National Computing Facilities (NCF), The Netherlands Organization for Scientific Research (NWO), The Netherlands
Jos Engelen (NIKHEF, The Netherlands)
Daan Frenkel (Amolf, The Netherlands)
Walter Hoogland (University of Amsterdam, The Netherlands)
Anwar Osseyran (SARA, The Netherlands)
Rik Maes (Faculty of Economics, University of Amsterdam, The Netherlands)
Gerard van Oortmerssen (CWI, The Netherlands)

Program Committee

Vassil N. Alexandrov (University of Reading, UK)
Hamid Arabnia (University of Georgia, USA)
J. A. Rod Blais (University of Calgary, Canada)
Alexander V. Bogdanov (Institute for High Performance Computing and Data Bases, Russia)
Marian Bubak (AGH, Poland)
Toni Cortes (University of Catalonia, Barcelona, Spain)
Brian J. d'Auriol (University of Texas at El Paso, USA)
Clint Dawson (University of Texas at Austin, USA)
Geoffrey Fox (Florida State University, USA)
Marina L. Gavrilova (University of Calgary, Canada)
James Glimm (SUNY Stony Brook, USA)
Paul Gray (University of Northern Iowa, USA)
Piet Hemker (CWI, The Netherlands)
Bob Hertzberger (University of Amsterdam, The Netherlands)
Chris Johnson (University of Utah, USA)
Dieter Kranzlmüller (Johannes Kepler University of Linz, Austria)
Antonio Lagana (University of Perugia, Italy)
Michael Mascagni (Florida State University, USA)
Jiri Nedoma (Academy of Sciences of the Czech Republic, Czech Republic)
Roman Neruda (Academy of Sciences of the Czech Republic, Czech Republic)
Jose M. Laginha M. Palma (University of Porto, Portugal)

James Pascoe (University of Reading, UK)
Ron Perrott (The Queen's University of Belfast, UK)
Andy Pimentel (The University of Amsterdam, The Netherlands)
William R. Pulleyblank (IBM T. J. Watson Research Center, USA)
Rene S. Renner (California State University at Chico, USA)
Laura A. Salter (University of New Mexico, USA)
Dale Shires (Army Research Laboratory, USA)
Vaidy Sunderam (Emory University, USA)
Jesus Vigo-Aguiar (University of Salamanca, Spain)
Koichi Wada (University of Tsukuba, Japan)
Jerzy Wasniewski (Danish Computing Center for Research and Education, Denmark)
Roy Williams (California Institute of Technology, USA)
Elena Zudilova (Corning Scientific, Russia)

Workshop Organizers

Computer Graphics and Geometric Modeling
 Andres Iglesias (University of Cantabria, Spain)
Modern Numerical Algorithms
 Jerzy Wasniewski (Danish Computing Center for Research and Education, Denmark)
Network Support and Services for Computational Grids
 C. Pham (University of Lyon, France)
 N. Rao (Oak Ridge National Labs, USA)
Stochastic Computation: From Parallel Random Number Generators to Monte Carlo Simulation and Applications
 Vasil Alexandrov (University of Reading, UK)
 Michael Mascagni (Florida State University, USA)
Global and Collaborative Computing
 James Pascoe (The University of Reading, UK)
 Peter Kacsuk (MTA SZTAKI, Hungary)
 Vassil Alexandrov (The Unviversity of Reading, UK)
 Vaidy Sunderam (Emory University, USA)
 Roger Loader (The University of Reading, UK)
Climate Systems Modeling
 J. Taylor (Argonne National Laboratory, USA)
Parallel Computational Mechanics for Complex Systems
 Mark Cross (University of Greenwich, UK)
Tools for Program Development and Analysis
 Dieter Kranzlmüller (Joh. Kepler University of Linz, Austria)
 Jens Volkert (Joh. Kepler University of Linz, Austria)
3G Medicine
 Andy Marsh (VMW Solutions Ltd, UK)
 Andreas Lymberis (European Commission, Belgium)
 Ad Emmen (Genias Benelux bv, The Netherlands)

Automatic Differentiation and Applications
> H. Martin Buecker (Aachen University of Technology, Germany)
> Christian H. Bischof (Aachen University of Technology, Germany)

Computational Geometry and Applications
> Marina Gavrilova (University of Calgary, Canada)

Computing in Medicine
> Hans Reiber (Leiden University Medical Center, The Netherlands)
> Rosemary Renaut (Arizona State University, USA)

High Performance Computing in Particle Accelerator Science and Technology
> Andreas Adelmann (Paul Scherrer Institute, Switzerland)
> Robert D. Ryne (Lawrence Berkeley National Laboratory, USA)

Geometric Numerical Algorithms: Theoretical Aspects and Applications
> Nicoletta Del Buono (University of Bari, Italy)
> Tiziano Politi (Politecnico-Bari, Italy)

Soft Computing: Systems and Applications
> Renee Renner (California State University, USA)

PDE Software
> Hans Petter Langtangen (University of Oslo, Norway)
> Christoph Pflaum (University of Würzburg, Germany)
> Ulrich Ruede (University of Erlangen-Nürnberg, Germany)
> Stefan Turek (University of Dortmund, Germany)

Numerical Models in Geomechanics
> R. Blaheta (Academy of Science, Czech Republic)
> J. Nedoma (Academy of Science, Czech Republic)

Education in Computational Sciences
> Rosie Renaut (Arizona State University, USA)

Computational Chemistry and Molecular Dynamics
> Antonio Lagana (University of Perugia, Italy)

Geocomputation and Evolutionary Computation
> Yong Xue (CAS, UK)
> Narayana Jayaram (University of North London, UK)

Modeling and Simulation in Supercomputing and Telecommunications
> Youngsong Mun (Korea)

Determinism, Randomness, Irreversibility, and Predictability
> Guenri E. Norman (Russian Academy of Sciences, Russia)
> Alexander V. Bogdanov (Institute of High Performance Computing and Information Systems, Russia)
> Harald A. Pasch (University of Vienna, Austria)
> Konstantin Korotenko (Shirshov Institute of Oceanology, Russia)

Sponsoring Organizations

The University of Amsterdam, The Netherlands
Power Computing and Communication BV, The Netherlands
Elsevier Science Publishers, The Netherlands
Springer-Verlag, Germany
HPCN Foundation, The Netherlands
National Supercomputer Facilities (NCF), The Netherlands
Sun Microsystems, Inc., USA
SHARCNET, Canada
Department of Computer Science, University of Calgary, Canada
School of Computer Science, The Queens University, Belfast, UK.

Local Organization and Logistics

Lodewijk Bos, MC-Consultancy
Jeanine Mulders, Registration Office, LGCE
Alain Dankers, University of Amsterdam
Manfred Stienstra, University of Amsterdam

Table of Contents, Part I

Computer Science – Computer Systems Models

Scientific Computing – Stochastic Algorithms

Complex Systems Applications 2

Computer Science – Networks

Scientific Computing – Domain Decomposition

Complex Systems Applications 3

Computer Science – Code Optimization

Methods for Complex Systems Simulation

Grid and Applications

Problem Solving Environment 1

Data Mining

Computer Science – Scheduling and Load Balancing

Problem Solving Environment 2

Problem Solving Environments 3

Computational Fluid Dynamics 2

Complex Systems Applications 4

Scientific Computing – Computational Methods 2

Scientific Computing – Computational Methods 3

Table of Contents, Part II

Modern Numerical Algorithms

Network Support and Services for Computational Grids

Stochastic Computation: From Parallel Random Number Generators to Monte Carlo Simulation and Applications

Global and Collaborative Computing

Climate Systems Modelling

Parallel Computational Mechanics for Complex Systems

Tools for Program Development and Analysis

Automatic Differentiation and Applications

Table of Contents, Part III

Computing in Medicine

High Performance Computing
in Particle Accelerator Science and Technology

Geometric Numerical Algorithms: Theoretical Aspects and Applications

Soft Computing: Systems and Applications

PDE Software

Numerical Models in Geomechanics

Education in Computational Sciences

Computational Chemistry and Molecular Dynamics

Geocomputation and Evolutionary Computation

Modeling and Simulation in Supercomputing and Telecommunications

Determinism, Randomness, Irreversibility, and Predictability

Keynote Papers

The UK e-Science Core Programme and the Grid

Tony Hey[1] and Anne E. Trefethen[2]

[1]Director UK e-Science Core Programme EPSRC, Polaris House, North Star Avenue
Swindon SN2 1ET, UK,
Tony.Hey@epsrc.ac.uk
[2] Deputy Director UK e-Science Core Programme EPSRC, Polaris House, North Star
Avenue Swindon SN2 1ET, UK,
Anne.Trefethen@epsrc.ac.uk

Abstract. e-Science encompasses computational science and adds the new dimension
of the routine sharing of distributed computational, data, instruments and specialist
resources. This paper describes the £120M UK 'e-Science' initiative, focusing on the
'Core Programme'. The majority of the funding is for large-scale e-Science pilot
projects in many areas of science and engineering. The pilot e-Science projects that
have been announced to date are briefly described. These projects span a range of
disciplines from particle physics and astronomy to engineering and healthcare, and
illustrate the breadth of the UK e-Science Programme. The requirements of these
projects and the recent directions in developments of open source middleware have
provided direction for the roadmap of development within the Core Programme. The
middleware infrastructure needed to support e-Science projects must permit routine
the sharing of any remote resources as well as supporting effective collaboration
between groups of scientists. Such an infrastructure is commonly referred to as the
Grid. The focus of the paper is on the Core Programme that has been created in order
to advance the development of robust and generic Grid middleware in collaboration
with industry. The key elements of the Core Programme will be outlined including
details of a UK e-Science Grid.

1. Introduction

e-Science in many ways seems a natural extension to computational science. It is of
course much more than this. Dr John Taylor, Director General of Research Councils
in the UK Office of Science and Technology (OST), introduced the term e-Science.
Taylor saw that many areas of science are becoming increasingly reliant on new ways
of collaborative, multidisciplinary working. The term e-Science is intended to capture
this new mode of working [1]:

P.M.A. Sloot et al. (Eds.): ICCS 2002, LNCS 2329, pp. 3–21, 2002.
© Springer-Verlag Berlin Heidelberg 2002

> *'e-Science is about global collaboration in key areas of science and the next generation of infrastructure that will enable it.'*

The infrastructure to enable this science revolution is generally referred to as the Grid [2]. The use of the term *Grid* to describe this middleware infrastructure resonates with the idea of a future in which computing resources, compute cycles and storage, as well as expensive scientific facilities and software, can be accessed on demand like the electric power utilities of today. These 'e-Utility' ideas are also reminiscent with the recent trend of the Web community towards a model of 'Web services' advertised by brokers and consumed by applications. We will illustrate toward the end of this paper how these two models are likely to intersect.

In the next section we outline the general structure of the UK e-Science programme. Following on from that we describe the UK funded e-Science Pilot projects and discuss the likely requirements from these projects. In Section 3 we describe the UK e-Science Core Programme in some detail.. We conclude with some remarks about the evolution of grid middleware architecture and the roadmap for contributions in the UK.

2. The UK e-Science Programme

Under the UK Government's Spending Review in 2000, the Office of Science and Technology (OST) was allocated £98M to establish a 3-year e-Science R&D Programme. This e-Science initiative spans all the Research Councils - the Biotechnology and Biological Sciences Research Council (BBSRC), the Council for the Central Laboratory of the Research Councils (CCLRC), the Engineering and Physical Sciences Research Council (EPSRC), the Economic Social Research Council (ESRC), the Medical Research Council (MRC), the Natural Environment Research Council (NERC) and the Particle Physics and Astronomy Research Council (PPARC). A specific allocation was made to each Research Council (see Figure 1), with PPARC being allocated the lion's share (£26M) so that they can begin putting in place the infrastructure necessary to support the LHC experiments that are projected to come on stream in 2005. The Central Laboratories at Daresbury and Rutherford (CLRC) have been allocated £5M specifically to 'grid-enable' their experimental facilities. The sum of £10M has been specifically allocated towards the procurement of a new national Teraflop computing system. The remainder, some £15M, is designated as the e-Science 'Core Programme'. This sum is augmented by an allocation of £20M from the Department of Trade and Industry making a total of £35M for the Core Programme.

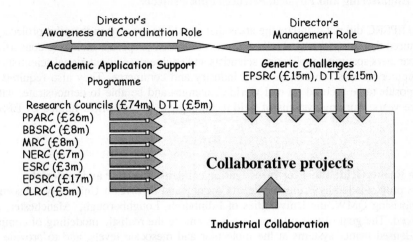

Fig. 1. *Structure and funding for UK e-Science Programme*

As is common in DTI programmes, the DTI contribution of £20M requires a matching contribution from industry. It is also expected that there will be industrial contributions to the individual Research Council e-Science pilot projects making a total industrial commitment to the e-Science programme of well over £20M. The goal of the Core Programme is to support the e-Science pilot projects of the different Research Councils and work with industry in developing robust, 'industrial strength' generic grid middleware. Requirements and lessons learnt in the different e-Science applications will inform the development of more stable and function grid middleware that can assist the e-Science experiments and be of relevance to industry and commerce.

3. The UK e-Science Pilot Projects

As noted above, within the UK e-Science programme each research council will fund a number of pilot projects in their own application areas. The research councils are each taking a slightly different approach in the selection of pilot projects, which has the consequence that whilst some research councils have committed all their e-Science funds, others are only partially through that process. We describe below those projects that have been funded so far.

3.1 Engineering and Physical Sciences Pilot Projects

The EPSRC did not mandate the areas that should be covered in the pilot projects but required that the proposal should be consortia based proposals including consortia of researchers spanning application scientists and engineers, computational scientists and computer scientists and those from industry and commerce. They also required the proposals to be focused on real world challenges and be able to demonstrate, within three years, working systems of Grid technologies. The six pilots funded by EPSRC are:

The RealityGrid; a tool for investigating condensed matter and materials
This project is led by Professor Peter Coveney and involves a University consortium comprising QMW, the Universities of Edinburgh, Loughborough, Manchester, and Oxford. The goal of this pilot project is to enable the realistic modelling of complex condensed matter systems at the molecular and mesocale levels, and to provide the setting for the discovery of new materials. Integration of high performance computing and visualisation facilities is critical to this pilot, providing a synthetic environment for modelling the problem that will be compared and integrated with experimental data. The RealityGrid involves the active collaboration of industry: AVS, SGI and Fujitsu are collaborating on the underlying computational and visualisation issues, Schlumberger and the Edward Jenner Institute for Vaccine Research will provide end-user scientific applications to evaluate and test the environment and tools produced by the project.

Comb-e-Chem - Structure-Property Mapping: Combinatorial Chemistry and the Grid
The Comb-e-Chem project is concerned with the synthesis of new compounds by combinatorial methods. It is a collaboration between the Universities of Southampton and Bristol, led by Dr Jeremy Frey. The university consortium is working together with Roche Discovery, Welwyn, Pfizer, and IBM. Combinatorial methods provide new opportunities for the generation of large amounts of original chemical knowledge. To this end an extensive range of primary data needs to be accumulated, integrated and relationships modelled for maximum effectiveness. The project intends to develop an integrated platform that combines existing structure and property data sources within a grid-based information-and knowledge-sharing environment. The first requirement for this platform is to support new data collection, including process as well as product data, based on integration with electronic lab and e-logbook facilities. The next step is to integrate data generation on demand via grid-based quantum and simulation modelling to augment the experimental data. For the environment to be usable by the community at large, it will be necessary to develop interfaces that provide a unified view of resources, with transparent access to data retrieval, online modelling, and design of experiments to populate new regions of scientific interest. The service-based grid-computing infrastructure required will extend to devices in the laboratory as well as databases and computational resources.

Distributed Aircraft Maintenance Environment: DAME
The collaborating universities on this pilot are York, Oxford, Sheffield, and Leeds, and the project is led by Professor Jim Austin. The project aims to build a grid-based distributed diagnostics system for aircraft engines. The pilot is in collaboration with Rolls Royce and is motivated by the needs of Rolls Royce and its information system partner Data Systems and Solutions. The project will address performance issues such as large-scale data management with real time demands. The main deliverables from the project will be a generic Distributed Diagnostics Grid application; an Aero-gas turbine Application Demonstrator for the maintenance for aircraft engines; and techniques for distributed data mining and diagnostics. Distributed diagnostics is a generic problem that is fundamental in many fields such as medical, transport and manufacturing and it is hoped that the lessons learned and tools created in this project will be suitable for application in those areas.

Grid Enabled Optimisation and DesIgn Search for Engineering (GEODISE)
GEODISE is a collaboration between the Universities of Southampton, Oxford and Manchester, together with BAE Systems, Rolls Royce, Fluent. The goal of this pilot is to provide grid-based seamless access to an intelligent knowledge repository, a state-of-the-art collection of optimisation and search tools, industrial strength analysis codes, and distributed computing and data resources.

Engineering design search and optimisation is the process whereby engineering modelling and analysis are exploited to yield improved designs. In the next 2-5 years intelligent search tools will become a vital component of all engineering design systems and will steer the user through the process of setting up, executing and post-processing design, search and optimisation activities. Such systems typically require large-scale distributed simulations to be coupled with tools to describe and modify designs using information from a knowledge base. These tools are usually physically distributed and under the control of multiple elements in the supply chain. Whilst evaluation of a single design may require the analysis of gigabytes of data, to improve the process of design can require assimilation of terabytes of distributed data. Achieving the latter goal will lead to the development of intelligent search tools. The application area of focus is that of computational fluid dynamics (CFD) which has clear relevance to the industrial partners.

DiscoveryNet: An e-Science Testbed for High Throughput Informatics
The DiscoveryNet pilot has a slightly different aim than those of the other projects, in that it is focused on high throughput. It aims to design, develop and implement an advanced infrastructure to support real-time processing, interpretation, integration, visualization and mining of massive amounts of time critical data generated by high throughput devices. The project will cover new technology devices and technology including biochips in biology, high throughput screening technology in biochemistry and combinatorial chemistry, high throughput sensors in energy and environmental

science, remote sensing and geology. A number of application studies are included in the pilot - analysis of Protein Folding Chips and SNP Chips using LFII technology, protein-based fluorescent micro array data, air sensing data, renewable energy data, and geohazard prediction data. The collaboration on this pilot is between groups at Imperial College London, lead by Dr Yike Guo, and industrial partners Infosense Ltd, Deltadot Ltd, and Rvco Inc.

myGrid: Directly Supporting the e-Scientist

This pilot has one of the larger consortiums comprising the Universities of Manchester, Southampton, Nottingham, Newcastle, and Sheffield together with the European Bioinformatics Insitute. The goal of myGrid is to design, develop and demonstrate higher level functionalities over an existing Grid infrastructure that support scientists in making use of complex distributed resources. An e-Scientist's workbench will be developed in this pilot project. The workbench, not unlike that of Comb-e-Chem, aims to support: the scientific process of experimental investigation, evidence accumulation and result assimilation; the scientist's use of the community's information; and scientific collaboration, allowing dynamic groupings to tackle emergent research problems.

A novel feature of the proposed workbench is provision for personalisation facilities relating to resource selection, data management and process enactment. The myGrid design and development activity will be driven by applications in bioinformatics. myGrid will develop two application environments, one that supports the analysis of functional genomic data, and another that supports the annotation of a pattern database. Both of these tasks require explicit representation and enactment of scientific processes, and have challenging performance requirements. The industrial collaborators on this project are GSK, AstraZeneca, IBM and SUN.

3.2 Particle-Physics and Astronomy Pilot Projects

The Particle Physicists and Astronomers are communities that perhaps have the most easily recognizable requirement for e-Science. The world-wide particle physics community is planning an exciting new series of experiments to be carried out on the new 'Large Hadron Collider' (LHC) experimental facility under construction at CERN in Geneva. The goal is to find signs of the Higgs boson, key to the generation of mass for both the vector bosons and the fermions of the Standard Model of the weak and electromagnetic interactions. These LHC experiments are on a scale never before seen in physics, with each experiment involving a collaboration of over 100 institutions and over 1000 physicists from Europe, USA and Japan. When operational in 2005, the LHC will generate petabytes of experimental data per year, for each experiment. The physicists need to put in place an LHC Grid infrastructure that will permit the transport and data mining of such distributed data sets. in which the particle physicists are working to build a Grid that will support these needs. Similarly the astronomers have the goal of providing uniform access to a federated, distributed repository of astronomical data spanning all wavelengths from radio waves to X rays. At present, astronomical data using different wavelengths are taken with different telescopes and stored in a wide variety of formats. Their goal is to create something like a 'data warehouse' for astronomical data that will enable new types of studies to be performed. Again, the astronomers are considering building a Grid infrastructure to support these Virtual Observatories. The two projects being funded by PPARC are:

GridPP

GridPP is a collaboration of Particle Physicists and Computer Scientists from the UK and CERN. The goal of GridPP is to build a UK Grid and deliver Grid middleware and hardware infrastructure to enable testing of a prototype of the Grid for the Large Hadron Collider (LHC) project at CERN, as described above. The GridPP project is designed to integrate with the existing Particle Physics programme within the UK, thus enabling early deployment and full testing of Grid technology. GridPP is very closely integrated with the EU DataGrid project [3] and, EU DataTag [4], and related to the USA projects NSF GriPhyN [5], DOE PPDataGrid [6], and NSF iVDGL [7]

AstroGrid

The AstroGrid project has been described early in terms of the requirements within this community. The project is collaboration between astronomers and computer scientists at the universities of Edinburgh, Leicester, Cambridge, Queens Belfast, UCL and Manchester, together with RAL. The goal of AstroGrid is to build a grid infrastructure that will allow a 'Virtual Observatory', unifying the interfaces to astronomy databases and providing remote access as well as assimilation of data. The Virtual Observatory is a truly global problem and AstroGrid will be the UK contribution to the global Virtual Observatory collaborating closely with the US NVO [8] project and the European AVO [9] project.

3.3 Biotechnology and Biological Sciences Research Council

Biological science is now one of the major areas of focus of scientific research worldwide. This effort is largely interdisciplinary teams working on specific problems of significant biological interest that underpin UK industry: pharmaceuticals, healthcare, biotechnology and agriculture. BBSRC has made a significant investment in bioinformatics research in recent years to establish a strong bioinformatics research community and have built on this investment with the e-Science funding focusing on post genomics, structural biology/structural genomics and generic technology development. The two pilots funded are described below:

A GRID Database for Biomolecular Simulations
This project is led by Prof Mark Sansom of Oxford University with collaborators at Southampton University, Birkbeck College, York University, Nottingham University and the University of Birmingham. The aims of the project include establishing a consortium for biomolecular simulations based on distributed, shared data structures; to establish a biomolecular simulation database; to establish metadata and tools to provide interrogation and datamining of federated datasets. This GRID approach with associated metadata facilities will facilitate access to simulation results by non-experts eg in the structural biology and genomics communities.

A distributed pipeline for structure-based proteome annotation using GRID technology
This project aims to use GRID technology to provide a structure-based annotation of the proteins in the major genomes (proteomes) and will develop the prototype for a national/distributed proteome annotation GRID. The annotation will primarily employ sophisticated homology and fold recognition methods to assign protein structures to the proteomes and generate 3D models. The consortium plan to establish local databases with structural and function annotations; to disseminate to the biological community the proteome annotation via a single web-based distributed annotation system (DAS); to share computing power transparently between sites using GLOBUS; and to use the developed system for comparison of alternative approaches for annotation and thereby identify methodological improvements. The consortium comprises University College London, the European Bioinformatics Institute, and is led by Prof Mike Sternberg of Imperial College.

3.4 Medical Research Council Pilot Projects

The Medical Research Council has recently funded the projects described below and anticipates funding further pilots in the near future. MRC like other councils required multidisciplinary consortia but restricted the problem areas to cancer or brain sciences for this call. The four proposals funded to date are described below.

Co-ordination, integration and distribution of sequence and structural family data
This project is led by Dr A Bateman of the European Bioinformatics Institute, together with the Sanger Centre, the MRC Centre for Protein Engineering and University College London. The goal of the project is to bridge the gap between sequence and structural information to improve access and utility of this data for biologists. Using Grid technologies the consortium will co-ordinate and integrate information from five existing databases (SCOP, CATH, EMSD, PFAM, and Interpro) allowing researchers from the UK and around the world to increase the pace of their basic research. They plan to facilitate genome annotation by extending the functional annotations associated with these integrated resources.

Bioinformatics, Molecular Evolution and Molecular Population Genetics.
Professor J Hein at Oxford University plans to employ Statistical Alignment.to improve comparative versions of problems such as gene finding, protein and RNA secondary prediction. His project also includes the application of coalescent theory to variation in the human genome. This would focus on finding computational accelerations to present methods and would be applied to data generated by the Sanger Centre. He will also study models of protein structure evolution and pathogen sequence evolution. Sequences of pathogens are being determined on a huge scale world wide and issues such as intrapatient evolution, origin of pathogens, clonality of pathogens and much more can be addressed by such data using the proper methods.

Artificial Neural Networks in Individual Cancer Management
The objective of this project is to develop and train a system of artificial neuronal networks (ANNs) providing information on prognosis, staging and optimal (multidisciplinary) management in patients with (a) breast, (b) upper GI and (c) colorectal cancers. In the first part of the study, the neuronal net system will be developed based on an existing system used by the particle physicists in Manchester by working in cooperation with academic clinical oncologists and a research nurse in Dundee using three national patient databases that are available for the project. The patient data used for development and training of the ANNs will be obtained by random retrieval from the three databases. In the second part of the project, validation of the ANNs for Cancer Management will be evaluated (in terms of staging, predicted optimal treatment and prognosis) with the observed management and outcome of the patients. The project is a collaboration between the universities of Dundee and Manchester and is led by A Cuschieri of the former.

Biology of Ageing E-Science Integration and Simulation System (BASIS)
Prof T Kirkwood of the University of Newcastle is employing e-Science techniques to the issue of ageing, which is a complex biological process involving interactions between a number of biochemical mechanisms whose effects occur at the molecular, cell, tissue and whole organism levels. The project will look to providing integration of data and hypotheses from diverse sources and to develop an innovative biomathematical and bioinformatics system.

3.5 Natural Environment Research Council

NERC has a ten year vision for the development of coupled Earth System models that improve both the understanding of the complex interactions within and between the biosphere, the geosphere, the atmosphere and the hydrosphere and the ability to confidently predict and resolve environmental problems. e-Science and Grid technologies clearly play an important role in advancements in this area. NERC has funded 3 pilots projects to date and plan to fund a similar number in a few months time.

Climateprediction.com: Distributed computing for global climate research

The climate*prediction*.com project aims to perform a probabilistic forecast of climate change, by exploring parameter-space in a coupled AOGCM (the UM) as a Monte Carlo simulation. The model will be distributed to businesses and individuals to run on their PCs, using up idle cycles. The runs will perform a pre-packaged 100-year simulation 1950-2050. Individual forecasts will be weighted by the quality of fit to observations (1950-2000), yielding a probabilistic forecast independent of expert judgment. Initial runs with a "slab" (thermodynamic ocean) version will help prioritise parameter perturbations. The project will have the challenging problem of tera-scale, distributed data archival and analysis. The project lead is Myles Allen at Oxford University who has collaborators at RAL and Open University.

GRID for Environmental Systems Diagnostics and Visualisation
The aim of this project is to develop a grid for ocean diagnostics with high-resolution marine data from the NERC OCCAM model and data from the Met Office Forecasting Ocean-Atmosphere Model. A prototype of the system can be seen at www.nerc-essc.ac.uk/las. Integrated in this environment will be sophisticated remote visualisation tools to allow the visualisation of large amounts of environmental data created in oceanic and atmospheric science. The project includes environmental research through the exploration of the thermohaline circulation and the calculation of water transformation budgets from the models and data assimilation techniques; and new visualisation techniques for unstructured grid fluid models.

The project is being led by Keith Haines of the University of Reading and includes collaborations with the Oceanographic centre at Southampton University, Imperial College, Manchester University, CLRC, the Meteorological Office, SGI, AVS and BMT Ltd.

Environment from the molecular level: an e-Science project for modelling the atomistic processes involved in environmental issues
The consortium in this project will develop are focusing on the processes at an atomistic level for modeling environmental problems. The e-Science challenge is to scale up the length and times scales of the simulation techniques from the molecular level, through the mesoscopic scale and towards the length and times scales of human experiences. The project will develop a collaboratory toolkit for archiving and mining of simulation output and visulaistaion of critical events together with metadata creation for the simulation of processes at the molecular level. The driving environmental applications are transport of pollutants and containment of radioactivity. This is a large consortium involving individuals from University College London, Bath University, Reading University, the Royal Institution and CLRC Daresbury.

3.6 Concluding Remarks

EPSRC and PPARC were the first to fund pilot projects and have almost completed the process of collecting requirements and beginning to provide direction for development of middleware. The other projects have not yet had time to complete this process. It is clear, however, that an overwhelming theme throughout the pilots is the need to access, manage and mine large amounts of disparate datasets. The datasets are, for some applications such as GridPP, largely in flat files, whilst for others, such as the bioinformatics, are in a variety of databases. The present generation of Grid middleware does not provide the tools required for these data-centric applications. This has been acknowledged and in the later section on directions for the Grid we will provide some suggestions of the UK can play a leading role in addressing this issue.

4. The UK e-Science Core Programme

4.1 Structure of the Core Programme

As we have explained, the goal of the e-science Core Programme is to identify the generic middleware requirements arising from the e-Science pilot projects. In collaboration with scientists, computer scientists and industry, we wish to develop a framework that will promote the emergence of robust, industrial strength Grid middleware that will not only underpin individual application areas but also be of relevance to industry and commerce. The Core Programme has been structured around six key elements:

1. Implementation of a National e-Science Grid based on a network of e-Science Centres
2. Promotion of Generic Grid Middleware Development
3. Interdisciplinary Research Collaboration (IRC) Grid Projects
4. Establishment of a support structure for e-Science Pilot Projects
5. Support for involvement in International activities
6. Support for e-Science networking requirements

We briefly discuss each of these activities below.

4.2 The UK e-Science Grid and the e-Science Centres

Nine e-Science Centres have been established at the locations shown on the map of the UK in Figure 2.

A National e-Science Centre has been established in Edinburgh, managed jointly by Glasgow and Edinburgh Universities. Eight other regional centres have been established – in Belfast, Cardiff, Manchester, Newcastle, Oxford, Cambridge, London (Imperial College) and Southampton - giving coverage of most of the UK.

Manchester currently operates the UK's national Supercomputer service. The Centres have three key roles:

1. to allocate substantial computing and data resources and run a standard set of Grid middleware to form the basis for the construction of the UK e-Science Grid testbed;
2. to generate a portfolio of collaborative industrial Grid middleware and tools projects;
3. to disseminate information and experience of the Grid within their local region.

The Centres have a pre-allocated budget for industrial collaborative Grid projects of £1M (£3M for the National Centre) requiring matching funds in cash or kind from industry.

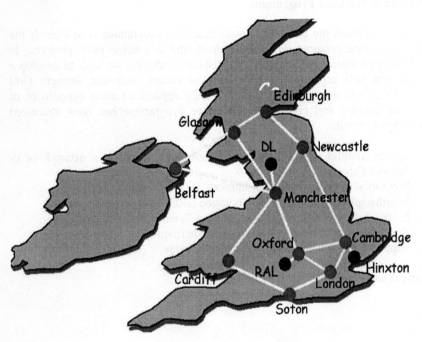

Fig. 2. The UK e-Science Grid

Figure 2 also shows the Rutherford and Daresbury Laboratory sites of CLRC. These national laboratories are key sites of the UK e-Science Grid. The Hinxton site near Cambridge is also shown in Figure 2: Hinxton hosts the European Bioinformatics Institute (EBI), the Sanger Centre and an MRC Institute. This constitutes one of the major centres of genomic data in the world. It is therefore important that this site is linked to the e-Science Grid with sufficient bandwidth to support the e-Science bioinformatics projects.

The National Centre in Edinburgh has also been funded to establish an 'e-Science Institute'. This Institute is organizing a series of multidisciplinary research seminars covering a wide range of topics, with scientists and experts from all over the world. The seminars have many different formats, from a one-day workshop to a month-long 'e-Science Festival', planned for summer 2002. Their brief is to make the Institute an internationally known centre for stimulating intellectual debate on all aspects of e-Science.

In addition, AccessGrid nodes have been established in each Centre to aid collaboration both within and outside the UK (Figure 3). The AccessGrid system was developed at Argonne National Laboratory in the USA and makes use of MBONE and Multicast technologies to provide a more natural video-conferencing experience between multiple sites that allows direct integration of Grid simulations and visualisation [10]. This system allows easy interaction between the Centres and will be used to experiment with innovative ways of working and teaching.

Fig. 3. An AccessGrid Session at the NorthWest e-Science Centre

4.3 Promotion of Grid Middleware Development

In order for the UK programme to be successful we must provide projects with a common starting point and educate a core team of people with the necessary experience in building and running a Grid. The starting point of middleware has been chosen as Globus [11], Condor [12] and SRB [13]. This choice has been based upon the NASA Information Power Grid (IPG) [14] which is based on this set of middleware and has therefore proven them to some extent.   A difference between the IPG and the UK e-Science Grid is that the former consists of a homogeneous set of compute resoures whereas the centres are contributing a heterogeneous collection of resources to this Grid, including supercomputer and cluster computing systems as well as diverse data storage systems.   Hence we will be stretching the capabilities of the starting middleware.   An Engineering Task Force (ETF) has been created in order to ensure that the centers are able to overcome any obstructions this might create. The UK e-Science Grid will certainly be a good test of the basic Globus infrastructure and the digital certificate based security system.

The Core Programme is in discussions with major IT companies such as IBM, Sun, Oracle and Microsoft, as well as with the Globus, Condor and SRB teams concerning the future development of Grid middleware. In this respect, it is encouraging that both IBM and Sun have given strong endorsements to working with the Globus team to take forward the production of improved and robust Grid middleware. The software that will emerge will offer considerably more functionality than the present Grid middleware and will also be produced to industrial quality. In a recent interview, Irving Wladawsky-Berger made the commitment that 'all of our systems will be enabled to work with the grid, and all of our middleware will integrate with the software' [15]. He also saw that there would be 'early adopter' industries such as petro-chemical, pharmaceutical and engineering design with the grid making significant impact on more general commerce and industry by 2003 or 2004. This is certainly what we are seeing in the UK e-Science Grid projects described above. A recent paper [16] co-authored by the Globus team and individuals from IBM sets the direction for this new evolution of grid middleware. The recent paradigm is referred to as the 'Open Grid Software Architecture' (OGSA) and describes the drawing together of two threads of recent development, namely webservices and grid technologies.

In the Core Programme, a total of £11M has been allocated for collaborative industrial Grid middleware projects via the individual e-Science Centres. An additional £5M (plus a matching industrial contribution) is available through an 'Open Call' with no deadlines. A major task for the Core Programme is the capture of requirements for the Grid infrastructure from each of the e-Science pilot projects. These include computational, data storage and networking requirements as well as the desired Grid middleware functionality. In order that the projects do not dissipate their energies by fruitless re-explorations of common ground, the Core Programme has commissioned a number of reports on the present state of Grid middleware, which aare currently available at the National Centre website (www.nesc.ac.uk). In addition, a Grid DataBase Task Force (DBTF), led by Norman Paton from Manchester, has been set up to examine the question of Grid middleware interfaces to Relational DataBase Management Systems and the federation of different data sources. Their preliminary ideas have been discussed with the Globus team and with IBM, Oracle and Microsoft. The DBTF has both a short term remit - to look at developing an interface with some minimal useful functionality as soon as possible - and a longer term remit – to look at research issues beyond flat files and relational data. The task force has produced a paper on the initial interface definitions for discussion at the next Global Grid Forum (GGF4) [17].

In addition, a Grid Architecture Task Force has been set up, led by Malcolm Atkinson, Director of the National e-Science Centre in Edinburgh, to look at overall architectural directions for the Grid. They are tasked with producing a 'UK Grid Road Map' for Grid middleware development. Again, the ATF is tasked with identifying some specific short term goals as well as identifying longer term research issues. Initial ideas from the DBTF point towards the implementation of a database interface in terms of a 'Grid Services' model along the lines of Web Services and this fits very well with OGSA, leading to a Grid middleware designed as a 'Service Oriented Architecture' with Grid services consumed by higher level applications. We expect to see the eventual emergence of agreed 'Open Grid Services' protocols, with specified interfaces to operating systems and RDBMS below this layer and offering a set of Grid services to applications above. It is important that at least a subset of standard protocols that go beyond the present Globus model are agreed as quickly as possible in order to ensure that the Grid middleware development in the application projects can proceed effectively. We intend to collaborate with the Globus team and assist in taking forward the open source implementation of these standards.

4.4 Interdisciplinary Research Collaboration Grid Projects

The EPSRC in the UK has funded three, six-year, computer science (CS) oriented, Interdisciplinary Research Collaborations (IRCs). These are major projects that fund key CS research groups from a number of universities to undertake long-term research in three important areas. The Equator project, led by Tom Rodden from Nottingham, is concerned with technological innovation in physical and digital life [19]. The Advanced Knowledge Technologies project (AKT) led by Nigel Shadbolt from Southampton is concerned with the management of the knowledge life cycle [20]. Lastly, the DIRC project, led by Cliff Jones from Newcastle and Ian Sommerville from Lancaster, is concerned with the dependability of computer-based systems. A fourth IRC, jointly funded by EPSRC and the MRC, is the MIAS project led by Mike Brady from Oxford. This application focussed IRC is concerned with translating data from medical images and signals into clinical information of use to the medical profession.

These IRCs were selected after an open competitive bidding process and represent a unique pool of expertise in these three key software technologies and in the important multidisciplinary application area of medical informatics. For this reason the Core Programme is funding projects with each of these IRCs to enable them to consider the implications of Grid technology on their research directions. In addition, we are funding two collaborative projects combining the software technologies of the Equator and AKT IRCs with the MIAS application project. In effect these projects constitute a sort of 'Grand Challenge' pilot project in 'e-Healthcare'.

As a concrete example of the potential of Grid technologies for e-Healthcare, consider the problem that Mike Brady and his group at Oxford are investigating in the MIAS IRC. They are performing sophisticated image processing techniques on mammogram and ultrasound scans for breast cancer tumours. In order to locate the position of possible tumours with maximum accuracy it is necessary for them to construct a Finite Element Model of the breast using an accurate representation of the properties of breast tissue. The informatics challenge for the Grid middleware is to deliver accurate information on the position of any tumour within the breast to the surgeon in or near the operating theatre. Clearly there are many issues concerning security and privacy of data to be resolved but the successful resolution of such issues must be a high priority not only for healthcare in the UK but also world-wide. This also illustrates another aspect of the Grid. Hospitals do not particularly wish to be in the business of running major computing centres to do the required modelling and analysis required by modern medical technologies. They would much rather be in the position of being able to purchase the required resource as a 'Grid service' on a pay-as-you-go basis. This illustrates the way in which Grid Services middleware can encourage the emergence of new 'Application Service Provider' businesses.

4.5 Support for e-Science Projects

In order to provide support for e-Science application projects, a Grid Support Centre has been established, jointly funded by the Core Programme and PPARC (www.grid-support.ac.uk). The Centre provides support for the UK 'Grid Starter Kit' which initially consists of the Globus Toolkit, Condor and the Storage Resource Broker middleware. The support team is available to answer questions and resolve problems for Grid application developers across the UK on a 9 to 5 basis. They also test and evaluate new middleware software and control future upgrades to the Grid Starter Kit. Further roles of the center include to help educate the systems support staff at the UK e-Science Centres and this is being executed through the ETF, and to provide a certificate authority for the UK.

4.6 International Collaboration

It is important to ensure that the UK e-Science community are actively communicating and collaborating with the international community. It is therefore desirable to encourage the development of an informed UK community on Grid technologies and provide funding for them to play an active role in the development of internationally agreed Grid protocols at the Global Grid Forum. We have therefore funded a 'GridNet' network project which has a substantial travel budget for attendance of UK experts at relevant standards bodies – the Global Grid Forum, the IETF and W3C, for example.

The UK programme is also concerned to create meaningful links to international efforts represented by projects such as the EU DataGrid and the US iVDGL projects.

We are therefore funding Grid fellowships for young computer scientists from the UK to participate in these projects. The National e-Science Centre is also tasked with establishing working agreements with major international centres such as Argonne National Laboratory, San Diego Supercomputing Center and NCSA in the USA. We are also looking to establish other international links and joint programmes.

4.7 Networking

The UK e-Science application projects will rely on the UK universities network SuperJANET4 for delivering the necessary bandwidth. The backbone bandwidth of SuperJANET4 is now 2.5 Gbps and there is funding in place to upgrade this to 10 Gbps by mid 2002 In order to gather network requirements from the UK e-Science project - in terms of acceptable latencies and necessary end-to-end bandwidth – the Grid Network Team (GNT) has been established. Their short-term remit is to identify bottlenecks and potential bottlenecks as well as look to longer term Quality of Service issues. A joint project with UKERNA and CISCO will be looking at such traffic engineering issues and another project will be looking at the question of bandwidth scheduling with the EU DataGrid project. Both of these network R&D projects are jointly funded by the Core Programme and PPARC.

4.8 Demonstrator Projects

The Core Programme has also funded a number of 'Grid Demonstrator' projects. The idea of the demonstrators is to have short-term projects that can use the present technology to illustrate the potential of the Grid in different areas. We have tried to select demonstrators across a range of applications. Examples include a dynamic brain atlas, a medical imaging project using VR, a robotic telescope project, automated chemical data capture and climate prediction. The demonstrators are being used to illustrate what e-Science and the Grid has to offer, now and in the future. They are suitable for a variety of audience, scientific, non-scientific, academic and industrial. They have also provided a 'bleeding' edge review of the near term requirements for middleware development.

6. Conclusions

There are many challenges to be overcome before we can realize the vision of e-Science and the Grid described above. These are not only technical issues such as scalability, dependability, interoperability, fault tolerance, resource management, performance and security but also more people-centric relating to collaboration and the sharing of resources and data.

As we have noted the UK pilot projects all demand an infrastructure and tools for data management and mining. This implies interfaces for federating database and techniques for metadata generation along side other data issues. Also as noted, OGSA takes the ideas of web services and extends them to provide support for grid services. The data services within OGSA are yet to be tackled. We have groups in the UK who are already taking a lead on these issues and by working with industry; applications and the Globus team are setting directions for discussion. This is clearly an area where the UK can and will make valuable contributions to the evolving Grid middleware.

Still there are many other challenges yet to be addressed. Two important areas not yet mentioned are security and scientific data curation. For the Grid vision to succeed in the industrial and commercial sector, the middleware must be secure against attack. In this respect, the present interaction of Globus with firewalls leaves much to be desired. Another issue concerns the long-term curation and preservation of the scientific data along with its associated metadata annotations. A three-year programme is clearly unable to provide a solution to such a long-term problem but it is one that must be addressed in the near future. Our present e-Science funding has enabled the UK to make a start on these problems and for the UK to play a full part in the development of the international Grid community. If the Grid middleware succeeds in making the transition from e-Science to commerce and industry in a few years, there will clearly be many new application areas to explore.

Acknowledgements

The authors thank Juri Papay for valuable assistance in preparing this paper and Jim Fleming and Ray Browne for their help in constructing the Core Programme and to our colleagues in other research councils who have provided the information regarding research council pilot projects. We also thank Paul Messina for encouragement and insight and Ian Foster and Carl Kesselman for their constructive engagement with the UK programme.

References

[1] e-Science definition: http://www.clrc.ac.uk
[2] Foster, I. and Kesselman, C. (eds.). *The Grid: Blueprint for a New ComputingInfrastructure*. Morgan Kaufmann, 1999.
[3] EU DataGrid: http://www.datagrid.cnr.it/, http://www.cern.ch/grid/
[4] EU DataTag: http://www.datatag.org
[5] GriPhyn: http://www.griphyn.org/
[6] PPDataGrid: http://www.ppdg.net/
[7] iVDGL: http://www.ivdgl.org/
[8] US NVO: http://www.srl.caltech.edu/nvo/
[9] EU AVO: http://www.astro-opticon.org/archives.html

[10] AccessGrid: http://www-fp.mcs.anl.gov/fl/accessgrid/
[11] Foster I. and Kesselman C., *Globus: A Metacomputing Infrastructure Toolki*t, International Journal of Supercomputer Applications, 11(2): 115-128, 1997.
[12] M. Litzkow, M. Livny and M. Mutka, 'Condor– A Hunter of Idle Workstations', Proceedings of the 8th International Conference of Distributed Computing Systems, pages 104-111, June 1988.
[13] Storage Resource Broker: http://www.npaci.edu/DICE/SRB
[14] NASA Information Power Grid: http://www.nas.nasa.gov/About/IPG/ipg.html
[15] IBM position: http://www.ibm.com/news/us/2001/08/15.html
[16] Foster, I., Kesselman, C., Nick, J. and Tuecke, S., *The Physiology of the Grid: Open Grid Services Architecture for Distributed Systems Integration*, to be presented at GGF4, Feb. 2002.
[17] Paton, N.W., Atkinson, M.P., Dialani, V., Pearson, D., Storey, T. and Watson, P., *Database Acccess and Integration Services on the Grid*, UK DBTF working paper, Jan. 2002
[18] Watson, P., Databases and the Grid. Technical Report CS-TR-755, University of Newcastle, 2001.
[19] Equator Project: http://www.equator.ac.uk/projects/DigitalCare.html
[20] Advanced Knowledge Technologies: http://www.ecs.soton.ac.uk/~nrs/projects.html
[21] De Roure, D., Jennings, N. and Shadbolt, N., *Research Agenda for the Semantic Grid: a Future e-Science Infrastructure*, Report for the UK e-Science Core Programme Directorate and NeSC TR: 2002-1, December 2001, www.semanticgrid.org. [1] http://www.e-science.clrc.ac.uk/

Community Grids

Geoffrey Fox[1,2,5], Ozgur Balsoy[1,3], Shrideep Pallickara[1], Ahmet Uyar[1,4]
Dennis Gannon[2], and Aleksander Slominski[2]

[1]Community Grid Computing Laboratory, Indiana University
gcf@indiana.edu, spallick@indiana.edu
[2]Computer Science Department, Indiana University
gannon@cs.indiana.edu, aslom@cs.indiana.edu
[3]Computer Science Department, Florida State University
ozgur@csit.fsu.edu
[4]EECS Department, Syracuse University
auyar@syr.edu
[5]School of Informatics and Physics Department, Indiana University

Abstract. We describe Community Grids built around Integration of technologies from the peer-to-peer and Grid fields. We focus on the implications of Web Service ideas built around powerful event services using uniform XML interfaces. We go through collaborative systems in detail showing how one can build an environment that can use either P2P approaches like JXTA or more conventional client-server models.

1. Introduction

The Grid [1-5] has made dramatic progress recently with impressive technology and several large important applications initiated in high-energy physics [6,7], earth science [8,9] and other areas [29,30]. At the same time, there have been equally impressive advances in broadly deployed Internet technology. We can cite the dramatic growth in the use of XML, the "disruptive" impact of peer-to-peer (P2P) approaches [10,11] and the more orderly but still widespread adoption of a universal Web Service approach to Web based applications [12-14]. We have discussed this recently [15,16] with an emphasis on programming environments for the Grid [17]. In particular we described the important opportunities opened up by using Web service ideas as the basis of a component model for scientific computing [18-22]. This builds on the DoE Common Component Architecture (CCA). This paper also discussed the implications for portals and computational science applications on the Grid. In the following, we look at other facets of the integration of Grid with P2P and Web Service technology. In particular we discuss the overall architecture in section 2 with attention to the implications of adopting a powerful event service as a key building block [23-24]. Web services are discussed in section 3 with special focus on the possibility of building science and Engineering as a Web Service – what can be

P.M.A. Sloot et al. (Eds.): ICCS 2002, LNCS 2329, pp. 22–38, 2002.

termed e-Science. P2P technologies are very relevant for collaboration [25,26] and we discuss this in section 4; an area addressed for the Grid [31], including a seminal paper by Foster and collaborators [27] addressing broad support for communities. Section 5 gives more detail on our proposed event model, which integrates both P2P and more traditional models – in particular, that of the commercial Java Message Service [28].

2. Architecture

We view the "world" as built of three categories of distributed system components: raw resources, clients and servers shown in fig. 1. These describe the different roles of machines in our distributed system. Clients provide user interfaces; raw resources

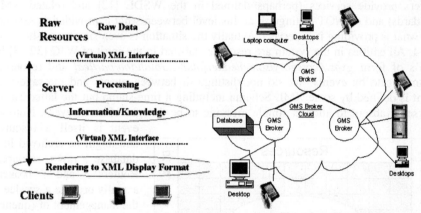

Fig. 1: XML-based Architecture

Fig. 2: Distributed Raw Resources, Servers, and Clients

provide "raw data" either from simulations or data sources; servers map between clients and raw resources and are specified by two XML specified interfaces; that between raw resource and server and that between client and server. Actually exploding the "server" layer inside fig. 1 finds an interlinked set of servers each of which linkage is itself described by XML interfaces. Note that the three functions can be thought of as roles and a given computer can have one, two or all of these three roles. Our architecture then should be termed as a three-role model rather than the more traditional three-tier model used in many current systems. We need to view our system in this way because in a peer-to-peer (P2P) system [11,23], one does not see the clear identification of machine and roles found in a classic Grid application involving say a workstation client going through some middleware to clearly identified back-end supercomputers.

The components of our system of whatever role are linked by message passing infrastructure shown in fig. 2. This we also term the event-bus and it has significant features, which we will elaborate later. We assume that all messages will be defined in XML and the message infrastructure – called GMS (Grid Message Service) in fig.

2 – can support the publish-subscribe mechanism. Messages are queued by GMS from "publishers" and then clients subscribe to them. XML tag values are used to define the "topics" or "properties" that label the queues. We can simplify and abstract the system as shown in figs. 3 and 4. We have divided what is normally called Middleware into two. There are routers and/or brokers whose function is to distribute messages between the raw resources, clients and servers of the system. We consider that the servers provide services (perhaps defined in the WSDL [12] and related XML standards) and do NOT distinguish at this level between what is provided (a service) and what is providing it (a server). Actually the situation is even simpler as shown in fig. 4. All entities in the system are resources labeled in the spirit of W3C [32,33] by URI's of form *gxos://category/someheader/blah/…/blah/foo/bar/leaf* and resources communicate by events. We do not distinguish between events and messages; an event is defined by some XML Schema including a time-stamp but the latter can of course be absent to allow a simple message to be thought of as an event. Note an event is itself a resource and might be archived in a database raw resource. Routers and brokers actually provide a service – the management of (queued events) and so these can themselves be considered as the servers corresponding to the event or message service. Note that in fig. 1, we call the XML Interfaces "virtual".

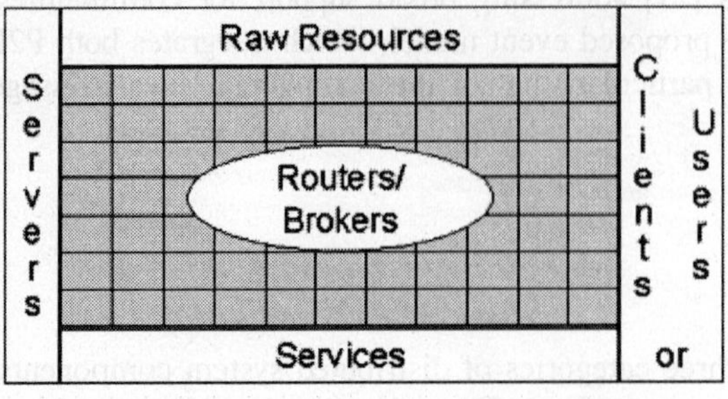

Fig. 3: One View of System Components

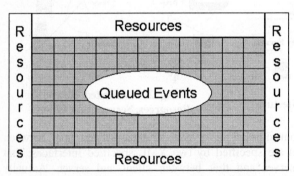

Fig. 4: Simplest View of System Components

This signifies that the interface is logically defined by an XML Schema but could in fact be implemented differently. As a trivial example, one might use a different syntax with say *<sender>meoryou</sender>* replaced by *sender:meoryou* which is an easier to parse but less powerful notation. Such simpler syntax seems a good idea for "flat" Schemas that can be mapped into it. Less trivially, we could define a linear algebra web service in WSDL but compile it into method calls to a Scalapack routine for high performance implementation. This compilation step would replace the XML SOAP based messaging [34] with serialized method arguments of the default remote invocation of this service by the natural in memory stack based use of pointers to binary representations of the arguments.

In the next four subsections we summarize some features and issues for the four components of the architecture events/messages, clients/users, servers/services and raw resources.

2.1 Event and Message Subsystem

We discuss the event or message service further in Sec. 5 but we elaborate first on our choice of this as an essential feature. We see several interesting developments in this area where we can give four examples: there is SOAP messaging [34]; the JXTA peer-to-peer protocols [10]; the commercial JMS message service [28]; and finally a growing interest in SIP [35] and its use in instant messenger standards [36]. All these approaches define messaging principles but not always at the same level of the OSI stack; further they have features that sometimes can be compared but often they make implicit architecture and implementation assumptions that hamper interoperability and functionality. We suggest breaking such frameworks into subsystem capabilities describing common core primitives. This will allow us to compose them into flexible systems, which support a range of functionality without major change in application interfaces. Here SOAP defines a message structure and is already a "core primitive" as described above; it is "only" XML but as discussed above, a message specified in XML could be "compiled to other forms such as RMI either for higher performance or "just" because the message was linking two Java programs. In some of our work, we use publish-subscribe messaging mechanisms but

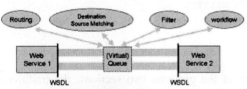

Fig 5: Communication Model showing Sub-services of Event Service

of course this is often unnecessary and indeed occurs unacceptable overhead. However it does appears useful to define an event architecture such as that of fig. 5, allowing communication channels between Web services which can either be direct or pass through some mechanism allowing various services on the events. These could be low-level such as routing between known source and destination or the higher-level publish-subscribe mechanism that identifies the destinations for a given published event. Some routing mechanisms in peer-to-peer systems in fact use dynamic routing mechanisms that merge these high and low level approaches to communication. We use the term virtual queue in fig. 5 because again we can in many cases preprocess (or "compile") away the queue and transmit messages directly. As an example, consider an audio-video conferencing web service. It would use a simple publish/subscribe mechanism to advertise the availability of some video feed. A client interested in receiving the video would negotiate (using the SIP protocol perhaps) the transmission details. The video could either be sent directly from publisher to subscriber; alternatively from publisher to web service and then from web service to subscriber; as a third option, we could send from the web service to the client but passing through a filter that converted one codec into another if required. In the last case, the location of the filter would be negotiated based on computer/network performance issues – it might also involve proprietary software only available at special locations. The choice and details of these three different video transport and filtering strategies would be chosen at the initial negotiation and one would at this stage "compile" a generic interface to its chosen form. One could of course allow dynamic "run-time compilation" when the event processing strategy needs to change during a particular stream. This scenario is not meant to be innovative but rather to illustrate the purpose of our architecture building blocks in a homely example. Web services are particularly attractive due to their support of interoperability, which allows the choices described. One could argue that the complexity of fig. 5 is unnecessary as its "luxury" features are an unacceptable overhead. However as the performance of networks and computers increase, this "luxurious" approach can be used more and more broadly. For instance, we have shown that JMS (the Java Message Service which is a simple but highly robust commercial publish-subscribe mechanism) can be used to support real-time synchronous collaboration. Note this application only requires latencies of milliseconds and not the microseconds needed by say MPI for parallel computing. Thus JMS with a message-processing overhead of a millisecond in good implementations can be used here. As explained in Sec. 5, we have developed a system that allows general XML-based selection on our message queues and this generalizes the simple topic and property model of JMS. The XML selection can specify that the message be passed though a general Web service, and so the "subscription" mechanism supports both event selection and filtering of the messages. In the collaboration application, this mechanism allows the same event service to support multiple clients – say a hand-held device and a high-end workstation – which would need different views (versions) of an event.

2.2 Clients

In the "pure view" of the architecture of the previous two sections, the traditional workstation (desktop) client has at least two roles – rendering and providing services. There has been a trend away from sophisticated clients (Java Applets and complex JavaScript) towards server based dynamic pages using technologies like ASP and JSP (Active and Java Server Pages). This reflects both the increased performance of networks and the more robust modular systems that can be built in the server-based model. There are several good XML standards for rendering – XHTML [37] and SVG [38] define traditional browser and 2D vector graphics respectively. These are the XML display formats of fig. 1; their support of the W3C Document Object Model [39] (DOM) allows both more portable client-side animation and shared export collaboration systems described in sec. 4. We do not expect this to replace server side control of "macroscopic" dynamic pages but built in animation of SVG (which can be considered a portable text version of Flash technology) and scripted event control will allow important client side animation. XML standards on the client should also allow universal access and customization for hand-held devices -- possibly using the WML [40] standard for PDA's and VoiceXML [41] for (Cell)-phones. The 3D graphics standard X3D [42] is likely to be another important XML rendering standard.

2.3 Servers and Services

Servers are the most important feature of our community grid architecture. They "host" all application and system services ranging in principle from Microsoft word through a 1024 node parallel simulation. They have multiple input and output ports defined by (virtual) XML Interfaces. There has been substantial work on many system Services in both the Grid and broader communities. We find Object registration, lookup and persistence; security, fault tolerance, information and collaboration. There are also the set of services common in computing environments such as Job submission, transparent login, accounting, performance, file access, parameter specification, monitoring, and visualization. An online education system using this architecture would have curriculum authoring, course scheduling and delivery, registration, testing, grading, homework submission, knowledge discovery, assessment, and learning paths as some of the services. We see current typically monolithic systems being broken up into small Web services and this will enable easier delivery of capabilities customized to particular communities. As one makes the basic services "smaller", flexibility increases but typically performance suffers as communication overhead increases. Here efficient "compilation" techniques will be important to use the optimal implementation of communication between the ports (see fig. 5) of linked web services. Static "compilation" can be supplemented by
dynamic choice of communication mechanism/stub that will use the optimal solution (based on many criteria such as bandwidth, security, etc.). Looking at peer-to-peer technology, we see important issues as to the appropriate implementation infrastructure. Is it to be based on a relatively large servers or a horde of smaller peers.

2.4 Raw Resources

The XML interfaces exhibited by servers represented "knowledge" or processed data and have typically "universal" form. The raw resources – databases, sensors, and supercomputers – also use XML interfaces but these can reflect nitty gritty detail. One sees approaches like JDBC (databases), SLE (spacecraft sensors [43]), HDF (scientific data) and MathML. The first two are true interfaces, the last two "raw" data format.

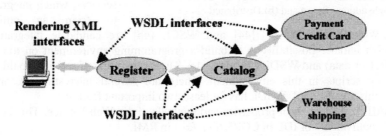

Fig. 6. An Online Shopping system with component Web Services

3. Web Services

Web Services are being developed actively by many major companies (Ariba, IBM, Microsoft, Oracle, Sun) with the idea typified in fig. 6 of componentizing Business to Business and Business to Customers applications.

We suggest that a similar approach is useful in both Grid system services shown in table 1 but also more generally to develop „Science as a Web Service" – one could term the latter e-Science.

Table 1. Basic Grid services

Security Services	Authorization, authentication, privacy
Scheduling	Advance reservations, resource co-scheduling
Data Services	Data object name-space management, file staging, data stream management, caching
User Services	Trouble tickets, problem resolution
App Services	Application tracking, performance analysis
Monitoring Service	Keep-alive meta-services

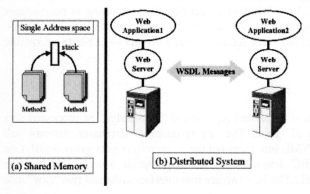

Fig. 7. Schematic of two bindings of a Web Service (a) Single address Space and (b) Distributed

We see WSDL [12-14] – the Web Services Definition Language – as a well thought through proposal. It is incomplete in some ways and more research is needed to decide how best to enhance it in such areas as the integration of multiple services together to form composite systems. Figure 6 shows 4 component Web services, which integrate to form an online shopping Web Service. WSFL [44] and WSCL [45] are candidate integration standards but another possibility is to build a programming environment on top of basic XML (for data) and WSDL (for methods). Then integration of services could be specified by scripts in this environment. There are several interesting research projects in this area [49,50]. WSDL has (at least) two important features:

1) An XML specification of properties and methods of the Web Service. This is an XML „equivalent" of IDL in CORBA or Java in RMI.
2) A distinction between the abstract application interface and its realization gotten by binding to particular transport (such as HTTP) and message (such as SOAP) formats.

The result is a model for Services with ports communicating by messages with a general XML specified structure. WSDL allows multiple bindings of a given interface and so supports the "compilation model" described in section 2. As an extreme example, fig. 7 illustrates that one could use WSDL to specify purely local methods and allow implementations that are either distributed or confined within a given address space. This carefully defined model can be integrated with the ideas of DoE's common component architecture (CCA) project [18-20] and this leads to an interesting "Grid programming model for high performance applications". Substantial research is needed into the optimization ("compiling") of Web services but we seem to

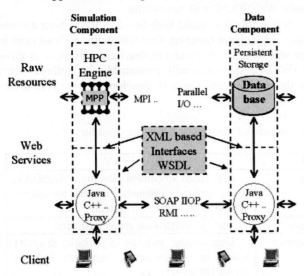

Fig. 8. An approach to integrating high performance and commodity systems

finally have a promising approach to integrating the best approaches of the high performance computing and commodity software communities. In fig. 8, we show how one can potentially define (composite) Web services with both commodity (SOAP, IIOP, RMI) messaging and high performance parallel MPI standards. For instance the Gateway [46-48] approach to integrating distributed object and high performance modules, builds a wrapper for the latter and implements a very loose coupling between a backend "parallel raw resource" and its proxy in the commodity layer. WSDL allows a tighter integration and this could lead to better commodity interfaces to high performance services. In the same vein as fig. 7, some variant of WSDL could provide a portable framework for linking shared and distributed memory parallel programming models. A WSDL formulation of MPI could involve two types of ports; firstly high performance with "native bindings" where basic data transfer would implemented; however one could use "commodity ports" for the less compute intensive MPI calls and for resetting some of the overall MPI parameters.

It is very important to note that Web Services are and should be composable. However as long as composed service can be exposed through WSDL it does not matter how they were composed. Therefore all those composition mechanisms are really interchangeable by means of WSDL - this is layer of abstraction that adds up to the robustness of Web Services technology. There are other Web Service technologies -- UDDI [51] and WSIL [52] are two approaches for registering and lookup of such services. This is a critical service but current approach seems incomplete. The matching of syntax of Web Service port interfaces is addressed but not their semantics. Further the field of XML meta-data registering and look up is much broader than Web Services. It seems likely that future versions of UDDI should be built on terms of more general XML Object infrastructure for searching. Possibly developments like the Semantic Web [53,54] will be relevant.

We expect there to be many major efforts to build Web Services throughout a broad range of academic and commercial problem domains. One will define web services in a hierarchical fashion. There will be a set of broadly defined services such as those defined in table 2. These (WSDL) standards will presumably be defined either through a body like W3C or through de facto standards developed by the commodity industry.

Table 2. General Collaboration, Planning and Knowledge Grid Services

People Collaboration	Access Grid - Desktop AV
Resource Collaboration	P2P based document Sharing, WebDAV, News groups, channels, instant messenger, whiteboards, annotation systems
Decision Making Services	Surveys, consensus, group mediation
Knowledge Discovery Service	Data mining, indexes (myGoogle: directory based or unstructured), metadata indexes, digital library services
Workflow Services	Support flow of information (approval) through some process, secure authentication of this flow. Planning and documentation
Authoring Services	Multi-fragment pages, Charts, Multimedia
Universal Access	From PDA/Phone to disabilities

Table 3. Science and Engineering Generic Services

Authoring and Rendering	Storage Rendering and Authoring of Mathematics, scientific whiteboards, nD (n=2,3) support, GIS, Virtual worlds
Multidisciplinary Services	Optimization (NEOS), image processing, netsolve, ninf, Matlab as a collaborative Grid Service
Education Services	Authoring, curriculum specification, assessment and evaluation, self paced learning (from K-12 to Lifelong)

Although critical for e-Science, one will build Science and engineering Services on the top of those in table 2. e-Science itself will define its own generic services as in table 3 and then refine them into areas like research (table 4) and education. Further hierarchical services would be developed on the basis of particular disciplines or perhaps in terms of approaches such as theory and experiment.

Table 4. Science and Engineering Research

Portal Services	Job control/submission, scheduling, visualization, parameter specification
Legacy Code Support	Wrapping, application Integration, version control, monitoring
Scientific Data Services	High Performance, special formats, virtual data as in Griphyn, scientific journal publication, Geographical Information Systems
Research Support Services	Scientific notebook/whiteboard, brainstorming, seminars, theorem proving
Experiment Support	Virtual Control Rooms (accelerator to satellite), Data analysis, virtual instruments, sensors (Satellites to field work to wireless to video to medical instruments (Telemedicine Grid Service)
Outreach	Multi-cultural customization, multi-level presentations

4. Collaboration

One of the general services introduced in the earlier sections was collaboration. This is the capability for geographically distributed users to share information and work together on a single problem. The basic distributed object and Web Service model described in this paper allows one to develop a powerful collaborative model. We originally built a collaborative system TangoInteractive [55,56], which was in fact designed for Command and Control operations, which is the military equivalent of crisis management. It was later evolved to address scientific collaboration and distance education [57,58]. Our new system Garnet has been built from scratch around the model of section 2. In particular Garnet views all collaboration as mediated by the universal event brokering and distribution system described in sections 2.1 and 5.

One of the attractive features of the web and distributed objects is the natural support for asynchronous collaboration. One can post a web page or host a Web Service and then others can access it on their own time. Asynchronous collaboration as enabled by basic web infrastructure, must be supplemented by synchronous or real-time interactions between community members. The field of synchronous collaboration is very active at the moment and we can identify several important areas:
(1) Basic Interactive tools including Text chat, Instant Messenger and White boards
(2) Shared resources including shared documents (e.g. PowerPoint presentation,), as well shared visualization, maps, or data streaming from sensor.

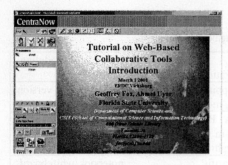

Fig. 9. Typical Shared Document System from Centra commercial collaboration system

(3) Audio-video conferencing illustrated by both commercial systems and the recent high-end Access Grid from Argonne [59] shown in fig. 5.

There are several commercial tools that support (1) and (2) – Interwise, Centra, Placeware and WebEx are best known [60-63]. They look to the user similar to the screen in fig. 9 – a shared document window surrounded by windows and control panels supporting the collaborative function. All clients are presented the same or a similar view and this is ensured by an event service that transmits messages whenever an object is updated. There are several ways objects can be shared:

Shared Display: The master system brings up an application and the system shares the bitmap defining display window of this application [64]. This approach has the advantage that essentially all applications can be shared and the application does not need any modification. The disadvantage is that faithful sharing of dynamic windows can be CPU intensive (on the client holding the frame-buffer). If the display changes rapidly, it may not be possible to accurately track this and further the network traffic could be excessive, as this application requires relatively large messages to record the object changes

Native Shared Object: Here one changes the object to be shared so that it generates messages defining its state changes. These messages are received by collaborating clients and used to maintain consistency between the shared object's representations on the different machines. In some cases this is essentially impossible, as one has no access to the code or data-structures defining the object. In general developing a native shared object is a time consuming and difficult process. It is an approach used if you can both access the relevant code and if the shared display option has the problems alluded to earlier. Usually this approach produces much smaller messages and lower network traffic than shared display – this or some variant of it (see below) can be the only viable approach if some clients have poor network connectivity. This approach has been developed commercially by Groove Networks using COM objects. It appears interesting to look at this model with Web services as the underlying object model.

Shared Export: This applies the above approach but chooses a client form that can be used by several applications. Development of this client is still hard but worth the cost if useable in many applications. For example one could export applications to the Web and build a general shared web browser, which in its simplest form just shares the defining URL of the page. The effort in building a shared browser can be amortized over many applications. We have built quite complex systems around this concept – these systems track frames, changes in HTML forms, JSP (Java Server Page) and other events. Note the characteristic of this approach – the required sharing bandwidth is very low but one now needs each client to use the shared URL and access common (or set of mirrored) servers. The need for each client to access servers to fetch the object can lead to substantial bandwidth requirements, which are addressed by the static shared archive model described below. Other natural shared export models are PDF, SVG [38], X3D [42], Java3D or whatever formats ones scientific visualization system uses.

Static Shared Archive: This is an important special case of shared export that can be used when one knows ahead of time what objects are to be shared, and all that changes in the presentation is the choice of object and not the state within the object. The system downloads copies of the objects to participating clients (these could be URL's, PowerPoint foils or Word documents). Sharing requires synchronous notification as to which of the objects to view. This is the least flexible approach but gives in real-time, the highest quality with negligible real-time network bandwidth. This approach requires substantially more bandwidth for the archive download – for example, exporting a PowerPoint foil to JPEG or Windows Meta File (WMF) format increases the total size but can be done as we described before the real-time session.

It can be noted that in all four approaches, sharing objects does not require identical representations on all the collaborating systems. Even for shared display, one can choose to resize images on some machines – this we do for a palmtop device with a low-resolution screen sharing a display from a desktop.

Real-time collaborative systems can be used as a tool in Science in different modes:

(a) Traditional scientific interactions – seminars, brainstorming, conferences – but done at a distance. Here the easiest to implement are structured sessions such as seminars.

(b) Interactions driven by events (such as earthquakes, unexpected results in a phsics experiment on-line data system, need to respond to error-condition in a sensor) that require collaborative scientific interactions, which must be at a distance to respond to a non-planned event in a timely fashion. Note this type of use suggests the importance of collaborating with diverse clients – a key expert may be needed in a session but he or she may only have access through a PDA.

We are developing in Garnet a shared SVG browser in the shared export model. The new SVG standard has some very attractive features [38]. It is a 2D vector graphics standard, which allows hyperlinked 2D canvases with a full range of graphics support – Adobe Illustrator supports it well. SVG is a natural export format for 2D maps on which one can overlay simulations and sensor data. As well as its use in 2D scientific visualization, SVG is a natural framework for high quality educational material – we have built a filter that automates the PowerPoint to SVG conversion. The work on SVG can viewed as a special case of a shared W3C DOM [39] environment and it can be extended to sharing any browser (such as XHTML [37]) supporting this document object model.

We mentioned audio-video conferencing earlier in this section where we have used a variety of commercial and research tools with the Access Grid [59] as high-end capability. We are investigating using the Web Service ideas of the previous sections to build a Audio Video Conferencing Web Service with clients using publish-subscribe metaphor to stream audio-video data to the port of the web service that integrates the different systems using the H323 and SIP standards. More generally we expect shared Web Services to be an attractive framework for future work in Garnet.

5. Event Service

There are some important new developments in collaboration that come from the peer-to-peer (P2P) networking field[32]. Traditional systems such as TangoInteractive and our current Garnet environment [26] have rather structured ways of forming communities and controlling them with centralized servers. The P2P approach [23] exemplified by Napster, Gnutella and JXTA [10] uses search techniques with "waves of agents" establishing communities and finding resources As described in section 2, Peer-to-Peer Grids [16] should be built around Web services whose collaborative capabilities support the P2P metaphors.

As one approach to this, we are generalizing the design of the Garnet collaboration system described in the previous section. Currently this uses a central publish-subscribe server for coordinating the collaboration with the current implementation built around the commercial JMS (Java Message Service) [28] system. This has proved very successful, with JMS allowing the integration of real-time and asynchronous collaboration with a more flexible implementation than the custom Java Server used in our previous TangoInteractive environment.

Originally we realized that Garnet's requirements for a publish/subscribe model were rather different than that for which JMS was developed. Thus we designed some extensions, which we have prototyped in Narada [25,66,67] – a system first described in the PhD thesis of Pallickara [68]. Narada was designed to support the following capabilities

- The matching of Published messages with subscribers is based on the comparison of XML based publisher topics or advertisements (in a JXTA parlance) with XML based subscriber profiles.
- The matching involves software agents and not just SQL-like property comparisons at the server as used by JMS.
- Narada servers form a distributed network with servers created and terminated as needed to get high performance fault tolerant delivery.

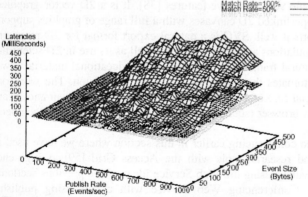

The Narada server network was illustrated in fig. 2 where each cluster of clients instantiates a Narada server. The servers communicate with each other while peer-to-peer methods are used within a client subgroup. Fig. 10 illustrates some results from our

Fig. 10. Message Transit times (labeled latencies) for Narada Event Infrastructure as a function of Event size and publish rate for three different subscription rates

initial research where we studied the message delivery time as a function of load. The results are from a system with the event brokering service supported by 22 server processes and with the "measuring" subscriber 10 hops away from publisher. The three matching values correspond to the percentages of subscribing clients to which messages are delivered. We found that the distributed network scaled well with adequate latency (a few milliseconds) unless the system became saturated. On the average, the time per hop between brokers was about 2 milliseconds. We expect this to decrease by a factor of about three in an "optimized production system". Nevertheless, our current pure Java system has adequate performance for several applications. The distributed cluster architecture allows Narada to support large heterogeneous client configurations that scale to arbitrary size. Now we are evolving these ideas to explicitly include both P2P and Web service ideas. We have already extended Narada so that it can function in a mode equivalent to JMS and are currently comparing it with our commercial JMS (from SonicMQ [69]) when used in Garnet and other circumstances. Soon we will integrate Narada with JXTA so that it can interpolate between "central server" (JMS) and distributed P2P (JXTA) models. Web services designed to use the event model of section 2 will then be able to run "seamlessly" with the same API to either classic 3-tier (client-server) or distributed P2P architectures.

Garnet supports PDA access and for this uses a modification – HHMS (Hand Held Message Service) – of the event service. Currently a conventional client subscribes to events of interest to the PDA on the JMS server. This special client acts as an adaptor and exchanges HHMS messages with one or more PDA's. It seems that PDA's and Cell-phones are a especially good example on the opposite end of the scale to clusters and supercomputers. They also require specialized protocols because of performance and size considerations but here because they are slow and constrained devices. In these cases an optimized protocol is not a luxury, it is requirement and a pure XML approach does not perform well enough. However we can still describe what is to be done in XML/WSDL and then translate it into binary (by an adaptor as described above). This is another example of run-time compilation.

We emphasis that it is unlikely that there will be a single event service standard and in this case using XML for events and messages will prove very important in gaining interoperability on the Grid. As long as there are enough similarities between the event systems, the XML specified messages can be automatically transformed by use of an event system adapter that can run as web service and allow for seamless integration of event services as part of middle tier (perhaps this is a filter in Fig. 5).

References

1. The Grid Forum http://www.gridforum.org
2. GridForum Grid Computing Environment working group (http://www.computingportals.org) and survey of existing grid portal projects. http:www.computingportals.org/cbp.html
3. 27 Papers on Grid Computing Environments http://aspen.ucs.indiana.edu/gce/index.html
4. "The Grid: Blueprint for a New Computing Infrastructure", Ian Foster and Carl Kesselman (Eds.), Morgan-Kaufman, 1998. See especially D. Gannon, and A. Grimshaw, "Object-Based Approaches", pp. 205-236, of this book.

36   G. Fox et al.

5. Globus Grid Project http://www.globus.org
6. GriPhyN Particle Physics Grid Project Site, http://www.griphyn.org/
7. International Virtual Data Grid Laboratory at http://www.ivdgl.org/
8. NEES Earthquake Engineering Grid, http://www.neesgrid.org/
9. SCEC Earthquake Science Grid http://www.scec.org
10. Sun Microsystems JXTA Peer to Peer technology. http://www.jxta.org.
11. "Peer-To-Peer: Harnessing the Benefits of a Disruptive Technology", edited by Andy Oram, O'Reilly Press March 2001.
12. Web Services Description Language (WSDL) 1.1 http://www.w3.org/TR/wsdl.
13. Definition of Web Services and Components http://www.stencilgroup.com/ideas_scope_200106wsdefined.html#whatare
14. Presentation on Web Services by Francesco Curbera of IBM at DoE Components Workshop July 23-25, 2001. Livermore, California. http://www.llnl.gov/CASC/workshops/ components_2001/viewgraphs/FranciscoCurbera.ppt
15. Dennis Gannon, Randall Bramley, Geoffrey Fox, Shava Smallen, Al Rossi, Rachana Ananthakrishnan, Felipe Bertrand, Ken Chiu, Matt Farrellee, Madhu Govindaraju, Sriram Krishnan, Lavanya Ramakrishnan, Yogesh Simmhan, Alek Slominski, Yu Ma, Caroline Olariu, Nicolas Rey-Cenvaz; "Programming the Grid: Distributed Software Components, P2P and Grid Web Services for Scientific Applications" GRID 2001, 2nd International Workshop on Grid Computing, Denver November 12 2001.
16. Geoffrey Fox. and Dennis Gannon, "Computational Grids," *Computing in Science & Engineering*, vol. 3, no. 4, July 2001
17. GrADS Testbed: Grid application development software project. http://hipersoft.cs.rice.edu.
18. R. Armstrong, D. Gannon, A. Geist, K. Keahey, S. Kohn, L. Mcinnes, S. Parker, and B. Smolinski, "Toward a Common Component Architecture for High Performance Scientific Computing," *High Performance Distributed Computing Conference*, 1999.   See http://z.ca.sandia.gov/~cca-forum.
19. B. A. Allan, R. C. Armstrong, A. P. Wolfe, J. Ray, D. E. Bernholdt and J. A. Kohl, "The CCA Core Specification In a Distributed Memory SPMD Framework," to be published in *Concurrency and Computation : Practice and Experience*
20. Talk by CCA (Common Component architecture) lead Rob Armstrong of Sandia at LLNL July 23-25 2001 meeting on Software Components http://www.llnl.gov/CASC/workshops/components_2001/viewgraphs/RobArmstrong.ppt
21. R. Bramley, K. Chiu, S. Diwan, D. Gannon, M. Govindaraju, N. Mukhi, B. Temko, M. Yechuri,   "A Component Based Services Architecture for Building Distributed Applications," *Proceedings of HPDC*, 2000.
22. S. Krishnan, R. Bramley, D. Gannon, M. Govindaraju, R. Indurkar, A. Slominski, B. Temko, R. Alkire, T. Drews, E. Webb, and J. Alameda, "The XCAT Science Portal," Proceedings of SC2001, 2001.
23. Geoffrey Fox, "Peer-to-Peer Networks," *Computing in Science & Engineering*, vol. 3, no. 3, May2001.
24. Geoffrey Fox, Marlon Pierce et al., *Grid Services for Earthquake Science*, to be published in Concurrency and Computation: Practice and Experience in ACES Special Issue, Spring 2002. http://aspen.ucs.indiana.edu/gemmauisummer2001/resources/gemandit7.doc
25. Geoffrey Fox and Shrideep Pallickara, An Event Service to Support Grid Computational Environments, to be published in Concurrency and Computation: Practice and Experience, Special Issue on Grid Computing Environments.
26. Fox, G. Report on Architecture and Implementation of a Collaborative Computing and Education Portal. http://aspen.csit.fsu.edu/collabtools/updatejuly01/erdcgarnet.pdf. 2001.

27. Ian Foster, Carl Kesselman, Steven Tuecke, The Anatomy of the Grid: Enabling Scalable Virtual Organizations http://www.globus.org/research/papers/anatomy.pdf
28. Sun Microsystems. Java Message Service. http://java.sun.com/products/jms
29. W. Johnston, D. Gannon, B. Nitzberg, A. Woo, B. Thigpen, L. Tanner, "Computing and Data Grids for Science and Engineering," Proceedings of SC2000.
30. DoE Fusion Grid at http://www.fusiongrid.org
31. V. Mann and M. Parashar, "Middleware Support for Global Access to Integrated Computational Collaboratories", Proc. of the 10th IEEE symposium on High Performance Distributed Computing (HPDC-10), San Francisco, CA, August 2001.
32. W3C and IETF Standards for Universal Resource Identifiers URI's: http://www.w3.org/Addressing/
33. Resource Description Framework (RDF). http://www.w3.org/TR/REC-rdf-syntax
34. XML based messaging and protocol specifications SOAP. http://www.w3.org/2000/xp/.
35. SIP Session Initiation Protocol standard http://www.ietf.org/html.charters/sip-charter.html
36. SIP for Instant Messaging and Presence Leveraging Extensions (simple) http://www.ietf.org/html.charters/simple-charter.html
37. XHTML Browser Standard in XML http://www.w3.org/TR/xhtml1/
38. W3C Scalable Vector Graphics Standard SVG. http://www.w3.org/Graphics/SVG
39. W3C Document Object Model DOM http://www.w3.org/DOM/
40. Wireless application Protocol (includes WML) http://www.wapforum.org/
41. Voice Extensible Markup Language http://www.voicexml.org
42. X3D 3D Graphics standard http://www.web3d.org/x3d.html
43. Space Link Extension SLE developed by The Consultative Committee for Space Data Systems http://www.ccsds.org
44. Frank Laymann (IBM), Web services Flow Language WSFL, http://www-4.ibm.com/software/solutions/webservices/pdf/WSFL.pdf
45. Harumi Kuno, Mike Lemon, Alan Karp and Dorothea Beringer, (WSCL – Web services Conversational Language), "Conversations + Interfaces == Business logic" , http://www.hpl.hp.com/techreports/2001/HPL-2001-127.html
46. Gateway Computational Portal, http://www.gatewayportal.org;
47. Marlon. E. Pierce, Choonhan Youn, Geoffrey C. Fox, The Gateway Computational Web Portal http://aspen.ucs.indiana.edu/gce/C543pierce/c543gateway.pdf, to be published in Concurrency and Computation: Practice and Experience in Grid Computing environments Special Issue 2002;
48. Fox, G., Haupt, T., Akarsu E., Kalinichenko, A., Kim, K., Sheethalnath, P., and Youn, C. The Gateway System: Uniform Web Based Access to Remote Resources. 1999. *ACM Java Grande Conference.*
49. IPG Group at NASA Ames, CRADLE Workflow project.
50. Mehrotra, P and colleagues. Arcade Computational Portal. http://www.cs.odu.edu/~ppvm
51. Universal Description, Discovery and Integration Project UDDI, http://www.uddi.org/
52. Peter Brittenham, An Overview of the Web Services Inspection Language (WSIL), http://www-106.ibm.com/developerworks/webservices/library/ws-wsilover/
53. Semantic Web from W3C to describe self organizing Intelligence from enhanced web resources. http://www.w3.org/2001/sw/
54. Berners-Lee, T., Hendler, J., and Lassila, O., "The Semantic Web," Scientific American, May2001.
55. Beca, L., Cheng, G., Fox, G., Jurga, T., Olszewski, K., Podgorny, M., Sokolowski, P., and Walczak, K., "Java Enabling Collaborative Education Health Care and Computing," Concurrency: Practice and Experience, vol. 9, no. 6, pp. 521-533, May1997.
56. Beca, L., Cheng, G., Fox, G., Jurga, T., Olszewski, K., Podgorny, M., and Walczak, K. Web Technologies for Collaborative Visualization and Simulation. 1997. Proceedings of the Eighth SIAM Conference on Parallel Processing for Scientific Computing.

57. Fox, G. and Podgorny, M. Real Time Training and Integration of Simulation and Planning using the TANGO Interactive Collaborative System. 1998. International Test and Evaluation Workshop on High Performance Computing.
58. Fox, G., Scavo, T., Bernholdt, D., Markowski, R., McCracken, N., Podgorny, M., Mitra, D., and Malluhi, Q. Synchronous Learning at a Distance: Experiences with TANGO Interactive. 1998. Proceedings of Supercomputing 98 Conference.
59. Argonne National Laboratory. Access Grid. http://www.mcs.anl.gov/fl/accessgrid.
60. Interwise Enterprise Communication Platform http://www.interwise.com/
61. Centra Collaboration Environment. http://www.centra.com.
62. Placeware Collaboration Environment. http://www.placeware.com.
63. WebEx Collaboration Environment. http://www.webex.com.
64. Virtual Network Computing System (VNC). http://www.uk.research.att.com/vnc.
65. openp2p P2P Web Site from O'Reilly http://www.openp2p.com.
66. Geoffrey C. Fox and Shrideep Pallickara , A Scalable Durable Grid Event Service. under review IEEE Internet Computing. Special Issue on Usability and the World Wide Web.
67. Geoffrey C. Fox and Shrideep Pallickara , An Approach to High Performance Distributed Web Brokering   ACM Ubiquity Volume2 Issue 38. November 2001.
68. Pallickara, S., "A Grid Event Service." PhD Syracuse University, 2001.
69. SonicMQ JMS Server http://www.sonicsoftware.com/

Conference Papers

A Conceptual Model for Surveillance Video Content and Event-Based Indexing and Retrieval

Farhi Marir, Kamel Zerzour , Karim Ouazzane and Yong Xue

Knowledge Management Group (KMG), School of Informatics and Multimedia
Technology, University of North London, Holloway Road, London N7 8DB, UK
{f.marir, k.zerzour, k.ouazzane,Y.xue}@unl.ac.uk

Abstract. This paper addresses the need for a semantic video-object approach for efficient storage and manipulation of video data to respond to the needs of several classes of potential applications when efficient management and deductions over voluminous data are involved. We present the VIGILANT conceptual model for content and event-based retrieval of video images and clips using automatic annotation and indexing of contents and events representing the extracted features and recognised objects in the images captured by a video camera in a car park environment. The underlying video-object model combines Object-Oriented modelling (OO) techniques and Description Logics (DLs) Knowledge representation. The OO technique models the static aspects of video clips and instances and their indexes will be stored in an Object-Oriented Database. The DLs model will extend the OO model to cater for the inherent dynamic content descriptions of the video, as events tend to spread over a sequence of frames.

1. Introduction and Related Work

Recently, video surveillance industry, particularly in the UK, has experienced growth rates of over 10 % year. CCTV camera systems have been installed on company premises, stores of city and around councils. The total annual UK market is believed to be valued at £2billions. The bulk of this expenditure is spent on hardware i.e. cameras, recording equipment and control centres. Currently, the overwhelming bulk of video systems involve multiplexing CCTV channels onto single 24-hour videotapes. The extensive range of commercial and public safety Content-Based Image/Video Retrieval applications includes law enforcement agencies, intruder alarms, collation of shopping behaviours, automatic event logging for surveillance and car park administration. It also include art galleries and museum management, architectural and engineering design, interior design, remote sensing and management of earth resources, geographic information systems, medical imaging, scientific database management systems, weather forecasting, retailing, fabric and fashion design, trademark and copyright database management.

The problem of providing suitable models for indexing and effectively retrieving videos based on their contents and events has taken three main directions. Database research concentrated on providing models to mostly handle static and aspects of data video (with little or no support for the dynamic aspects of video), and classification is process indices are usually performed offline. From the Image Processing point of

P.M.A. Sloot et al. (Eds.): ICCS 2002, LNCS 2329, pp. 41–50, 2002.
© Springer-Verlag Berlin Heidelberg 2002

view, video and image features are created automatically during image (frame) capture. These features usually include motion vectors and, in most cases an object-recognition module is employed to depict object attributes such as identity, colour and texture. The need for a synergy between database and computer vision approaches to handle video data is inevitable. This collaborative approach involves database modelling, Vision/Image Processing, Artificial Intelligence and Knowledge Base Systems and other research areas [1].

The Database indexing realm includes the works carried by [4] who developed an EER video model that captures video data structure (sequence, scene, and shot) and supports thematic indexing based on manually inserted annotations (Person, Location, Event). They also provided a set of operations to model video data. [6] investigated the appropriateness of the existing object-oriented modelling for Multimedia data modelling. He recognised that there is a need for more modelling tool to handle video data as conventional object modelling suffer from three main limitations when dealing with video images and scenes. The video data is raw and its use is independent from why it is created, the video description is incremental and dynamic, and finally the overlapping of meaningful video scenes. Content-based video indexing and retrieval architecture was proposed in [13]. This architecture extends a database management system (DBMS) with capabilities to parse, index and retrieve or browse video clips. Semantic indexing was attained through a frame-based knowledge base (similar to a record in traditional databases) in the shape of a restricted tree form. The database also includes textual descriptions of all frames present in the knowledge base.

With regards to Image Processing, objects may also be classified on their motion by extracting trajectories, which are used with other primitives (colour, shape) as keys to index moving objects [2]. Description Logics is also used to build Terminology Servers to handle video data content in a bottom up approach [3]. DLs proved beneficial in terms of dynamic, incremental classification, knowledge extensibility and evolution, precise and imprecise querying abilities. Recently, standard stochastic context-free grammars (SCFG) have been employed to contextually label events in an outdoor car park monitoring [5]. Objects and events are incrementally detected using a confidence factor with events being detected on the basis of Spatial Locations when objects tend to enter or leave the scene. But no storage issues were addressed. [10] modelled object interaction by Behaviour and Situation agents using Bayesian Networks. The system applies textual descriptions for dynamic activities in the 3D world by exploiting two levels of description: Object levels where each detected object is assigned a behaviour agent, and Interaction level using situation agents. [9] developed a system to detect abandoned objects in unattended railway stations. The system is able to reveal to the operator the presence of abandoned objects in the waiting room. It also has the capability to index video sequences based on an event detection module, which classifies objects into abandoned, person, and structural change (e.g. chair changes position) or lighting effect, with the aid of a Multilevel Perception Neural Network.

2. The VIGILANT Video Object Model

VIGILANT is an end-to-end intelligent semantic video object model for efficient storage, indexing and content / event-based retrieval of *(real-time)* surveillance video in a car park environment. The proposed system diagnoses the problems of content extraction, (semantic) labelling and efficient retrieval of video content. The system is intended to automatically extract moving objects, identify their properties and label the events they participate in (e.g. a car entering the car park).

* Capture: A suitable means for capturing surveillance video footage with associated semantic labelling of events happening in the car park had to be implemented. An unattended camera was set up to record events (e.g. a car entering the car park) and a frame grabber is needed to process this information and make it available for the next stage: indexing and content extraction.
* Segmentation and Content Extraction: The most desirable segmentation of surveillance video material is clearly separation into individual meaningful events (a person getting off the car) with participating objects (a car and a person). Semantic video segmentation is a highly challenging task and usually segmentation based on simpler breaks of the video (e.g. a camera break) are opted for [12]. This involves examining the structure of the video and using statistical measures (e.g. colour histograms) [13]; cuts are determined from scene to scene. The VIGILANT projects attempts to exploit the frame-to-frame object tracking as a means of providing shot cuts, whereby each new object appearing in the scene gives rise to a potential event which ends when the tracking of that particular object finishes (e.g. object leaves the camera view). At each frame various statistical features (colour, position, time, speed) are extracted from object at each frame and the content (especially identities of objects and their activities) is incrementally built as objects are tracked from frame to frame using DLs knowledge base (events) and Neural Networks (objects).
* Content / Event Indexing: Different indexing schemes have been researched [1] and these are classified into various abstraction levels depending on the types of features being extracted. Conventional database indexing tools have proved inadequate for video labelling due to the dynamic and incremental nature of video content [6]. The VIGILANT system will couple Object Oriented Database (OODB) tools with Description Logics (DLs). The former will be employed to model the static aspects of video (contextually independent content such as object identity and maybe colour) using concepts of classes and relationships. The latter will deal with dynamic contents that spread over a sequence of frames (object activities) by means of concepts and roles. Various issues will need to be addressed at this stage primarily concept tractability, DLs reasoning tools and the ability of the DLs knowledge base to access secondary storage.
* Retrieval: The way video clips are retrieved is strongly tied to the way in which they are indexed. Current retrieval practice relies heavily on syntactic primitive features such as colour, texture, shape and spatial arrangement, and little work has been done into indexing (and hence retrieval) based on semantic contextual features (events) [1]. To respond to this need, this work proposes to research and

implement queries for the second and third level of abstraction. We proposed to develop a real-time Video-Object Query Language (Video-QQL) which will uses the semantic content, indexes and annotations generated by the Index Module. The Video-OQL module will not be limited to the first level of abstraction i.e. (primitive features) but respond to complex queries such as "display video clips showing red cars entering the car park between 11am and 2pm". (a full description of the VIGILANT's proposed UML modelled can be found in [14]).

3. VIGILANT's Architecture

The complete architecture of VGILANT is shown in Figure 1 with the parts to be addressed in this paper in **bold**. The Video-Indexing Module parses the VIGILANT test file, which encodes the extracted features and recognised objects, to either populate the video-object database (if the object and events are well recognised) or to populate the description logics knowledge base (if not yet known). Then indices are generated in each video-object to reference a frame or a sequence of frames (clips). The Schema Manager Module will implement a mechanism and channels of communication between the video-object and description logics (DL) schemas to reflect the dynamic changes of the video content. The Real-time Video-OQL Module will provide visual facilities, through the VIGILANT user interface, for the user to access and browse retrieved images or video clips of a particular scene using the indices and content of the video already stored in the video-object database and DL knowledge base.

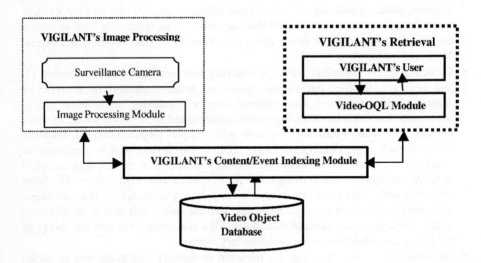

Fig. 1. VIGILANT's Proposed Architecture

4. VIGILANT Content and Event Indexing

The Digital image Research Centre (DIRC) at Kingston University has implemented an event-detection and tracking algorithm that detects and encodes visual events captured by a security camera [7, 8]. The object and the event streams are parsed automatically to recover the visual bounds and pixels representing moving objects (cars, persons, etc.) in the park

The output of this process will be used by the Video-Index Module to either populate the video-object database (if the object or the events are well recognised) or to enrich the DL knowledge base (if the objects or the events are not yet known). As soon as an instance of a video object or event is created and stored in the video-object database or the DL knowledge base, the Video-Index will generate automatically a logical pointer to the digital file containing the video shot (an image or a video clip). To eliminate redundant video indices in the database, the index-generating process will only be triggered when the video-Index Module detects a change of content or events in the car park.

Figure 2 summarises the work of the collaborating three schemas (video-object, description logics and the schema manage).

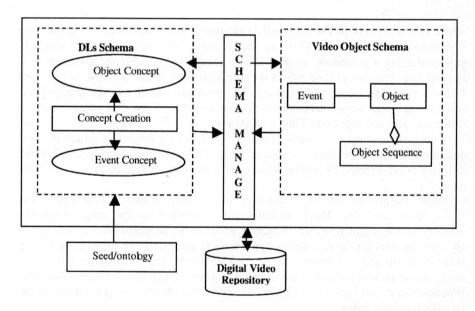

Fig. 2. The VIGILANT's Proposed Content/Event Indexing Module shows the collaborative work between both the DLs schema and the Video Object schemas through the Schema manager

4.1 The Schema Manager and The Video-Object Schema

Video content is difficult to model within a structured data model such as relational or object oriented mainly due to the dynamic changes of this content. The proposed solution is to develop a dynamic descriptive video object model reflecting the evolution of the content as camera shots is taken. It uses object-oriented (OO) modelling to present the hierarchical structure (static) of the video content enhanced with Description Logics (DLs) to model the incremental aspect of that video content (dynamic).

The VIGILANT's both Video Object and Schema Manager's schemas are described in [14], we here further describe the DLs schema.

4.2 The Description Logics Schema

The *DLkb* domain knowledge represents concepts held by objects in the class, and is defined by the Equation

$$DLkb(C) = <F[fn,method], R[rn,rolebvr], CM[qc,val]> \tag{1}$$

where C is a concept in the knowledge base representing an object of a given class (moving objects or events). *Fn* is an object feature (e.g. colour, type,...) which is extracted using a procedure *method*. *R* is a relationship connecting a feature to a concept (e.g. has_colour) and *role* is the role's semantic behaviour (e.g. has_colour). *CM* is a condition mapper that converts the condition value(s) of the query into a certain data *value* whose data type is the same as that of the data values of the *F* (afternoon gets converted into Time > 12:00 am).

Two basic concepts are needed for the VIGILANT's project to reflect the two directions of the undertaken research (content and event based retrieval). Content indexing is represented by *moving object concepts* (car, person) and events by *event concepts*.

Object features include object identity, colour, speed, position within a particular frame, speed and size. These attributes are available from the image-processing module, except object types which require utilising the Semantic Neural Net. When an object appears for the first time a **hypothesis** is made about its identity, then using other syntactic object features (colour, height, width, and area) the hypothesis is strengthened until at the end of the tracking process the hypothesis becomes a **thesis**. Whenever an object is processed a new version of that object concept is created in the DLs domain knowledge.

Event concepts are created in a similar manner. The first version of the object concept calls for an event concept to be created with the following features:
(1)A temporary event type, which is then strengthened as the tracking, continues over time. (2)The event start time is the first object's time. (3) The end time is the last object version's time, and (4)An object sequence containing all the versions of one object (or many) involved in that particular event.

Equations (2) and (3) show a formal definition of a moving-object and an event concept, while in figures 4 and 5 are samples of the definition of the concept Event in both OO and DLs environments

<MovingObject>
  Feature{[Type, Ext_Type],[Colour, Ext_Colour], ...}   **(2)**
  Role{[Type, Has_Type], [Colour, Has_Colour],...}
  Condition_Mapper{ [Red,[120,24,167], [fast, 90km/h],...}

<Event>
  Feature{[Type, Ext_Type],[StartTime, Ext_STime], ...}   **(3)**
  Role{[Type, Has_Type], [STime, Has_Has_STime],...}
  Condition_Mapper{ [Type,[Car_Entering,Person_Leaving,...],}

```
Class: VideoShot{

StartTime, EndTime, BackgroundClour, NumOfObjects, etc;

Car("door open", "window close", etc);

Person("walking", "getting in Car", etc);

//Event part- Relationships between objects

Person,Car ("get in", "get out", ect);

Car, CarPark("Enter", "leave", etc);

}//End of videoShot Class
```

Fig. 3. Example of typical Class concept Definition of an Event

(defind-concept VideoShot(AND

  (define-concept Colour(aset red green yellow ...)(ALL StartsAt Num)
  (ALL EndsAt Num)(ALL HasNumberOfObjects Num)
  ;Composition part
  (ALL ContainsComp VideoShotComp)
  (define-concept State (aset CLODE OPEN)
  (define-primitive-concept Door CarComp)
  (define-primitive-concept Window CarComp) etc...
  ;Event part
  (ALL ContainsEvent Event)
  (define-concept EventType (aset GetIn GetOut Enters Leaves ...))
  (define-concept GetIn (ALL Event (ALL HasEventType(aset GetIn))
                      (ALL HasFirstInteractor Person)
                      (ALL HasSecondInteractor Car))) etc

Fig. 4. Example of typical DLs concept Definition of an Event

5. VIGILANT's Content and Event Based Retrieval

While Content-based image retrieval (CBIR) systems currently operate effectively at the lowest of these levels, a variety of domains, including crime prevention, medicine, architecture, fashion and particularly video surveillance demand higher levels of retrieval.

To respond to this urgent need, this work proposes to research and implement queries for the second and third level of abstraction. For this, it is proposed to develop a real-time Video-Object Query Language (Video-QQL) which will uses the semantic content, indexes and annotation generated by the Real-time Index Module. The Real-time Video-OQL module will not be limited to the first level of abstraction i.e. (primitive features) but respond to complex queries such as "display all the video clips showing red cars entering the car park between 11am and 2pm". This could be expressed in OQL-Video as:

```
SELECT VideoObject = VideoClip

FROM VideoClip.Cars AND VideoClip.CarPArk

WHERE VideoClip.Event.EventType = "Car Entering Car Park" AND

    VideoClip.Time = IN(11am, 3pm) AND

    Cars.Colour = "Red"
```

To achieve such complex queries, the Video-OQL will combine the facilities provided by the object database Object Query Language (OQL) and the description logics (DL) descriptive language. This combination of Object database and DL queries will also widen the power of Video-OQL to deal with exact queries as in database query languages as well as with incomplete and imprecise queries.

In the VIGILANT project, similar to [11], a query Q is modelled as:

$$Q = dm(rf[ac, c]) \qquad (4)$$

where dm is the domain or the set of classes to be retrieved as a result of the query Q, rf is the retrieval function, ac is the attribute(s) for which the condition(s) is/are specified, and c is the condition value(s) of the attribute(s).an example query is "find video clips of cars". This is expressed as:

```
ObjectSequence(find_equal[VideoTape.VideoShot.Frame.Object.Type,Car])
```

This is regarded as a primitive query and more sophisticated queries can be built by combining primitive queries. Consider the following query statement: "find video clips of red cars entering the car park between 11:00 am and 12:00 am", this can be expressed in our model as:

```
ObjectSequence(

    equal_type[videoTape.VideoShot.Frame.Object.Type,Car]&&

    equal_color[videoTape.VideoShot.Frame.Object.Color,Red] &&

    equal_event[videoTape.VideoShot.Event.Type,Entering] &&

    time[videoTape.VideoShot.Frame.Object.StartTime,11am&12am]
```

6. Conclusions

The VIGILANT's proposed model couples Description Logics Domain Knowledge schemas which deal with the static content with Object-Oriented schemas which handle the evolution of video content. Challenges associated with this scheme include DLs inability to handle large volumes of data. This will be addressed through (1) enhancing the DL inference engine with facilities to access secondary storage (where both the concepts that compose the DL schema and the instances derived from the schema might be stored) and also. (2) Improving the performance problems with the reclassification and recognition of instances whenever conceptual schemas or instances are changed. Another equally important challenge will be the mapping of both OO and DLs schemas.

Although the OO and DL models seem to complement each other, this work will be facing two main challenges. The first is mapping between DL and OO models or schemas. Tight coupling is preferred in this system to allow on-demand traffic of instances between the video object database and the DLs knowledge base on demand. Fully recognised objects and events are transferred to the video object database permanently (persistent objects and events). Unknown objects (and events) reside in the knowledge base until further information about their content is acquired. If still unrecognised, a decision has to be made whether to transfer them to the video object database or leave them in DLs knowledge base. The second challenge is addressing the limitation of DL in accessing external storage when the number of DL concepts (class) individuals (objects) grows and requires external storage.

7. References

1. J.P. Eakins, M.E. Graham, "Content-Based Image Retrieval, Report to JISC Technology Applications Program", 1999.
2. E. Sahouria, "Video Indexing Based on Object Motion. Image and Video Retrieval for National Security Applications: An Approach Based on Multiple Content Codebooks", The MITRE Corporation, Mclean, Virginia, 1997, USA.

3. C. A. Goble, C. Haul and S. Bechhofer, "Describing and Classifying Multimedia using the Description Logics GRAIL", SPIE Conference on Storage and Retrieval of Still Image and Video IV, San Jose, CA, 1996.
4. R. Hjelsvold, R.Midtstraum and O. Sandsta, "A Temporal Foundation of Video Databases", Proceedings of the International Workshop on Temporal Databases, 1995.
5. Y. Ivanov, C. Stauffer, A. Bobick and W. E. L. Grimsom, "Video Surveillance Interactions", Second IEEE International Workshop on Visual Surveillance, 1999.
6. E. Oomoto, and K. Tanaka, "OVID: Design and Implementation of a Video-Object Database System". IEEE Transaction on Knowledge and Data Engineering, Vol. 4, No 4, 1994.
7. J. Orwell, P. Massey, P. Romagnino, D. Greenhill, and G. A. Jones, "A Multi-Agent Framework for Visual Surveillance", Proceedings of the 10th International conference on Image Analysis and Processing, 1996.
8. J. Orwell, P. Romagnino, and G. A. Jones, "Multi- Camera Colour Tracking", Proceedings of the second IEEE International Workshop on Visual Surveillance, 1999.
9. C. S. Regazzoni, and E.Stringa, "Content-Based Retrieval and Real-Time Detection from Video Sequences Acquired by Surveillance Systems", IEEE International conference on image Processing ICIP98, 3(3), 1999.
10. P. Romagnino, T. Tan, and K. Baker, "Agent Oriented Annotation in Model Based Visual Surveillance", In ICCV, pp 857-862, 1998
11. A. Yoshitaka, S. Kishida, M. Hirakawa and T. Ichikawa, "Knowledge Assisted Content-Based Retrieval for Multimedia Databases", IEEE '94.
12. P. Aigrain, H. Zhang , and D. Petkovic, "Content-based Representation and Retrieval of Visual Media: A state-of-the-art Review", Multimedia Tools and Applications, 3(3), pp 79-202, 1996.
13. S.W. Smoliar and H. Zhang, "Content-Based Video Indexing and Retrieval", IEEE Multimedia 1(2), pp 62-72, 1994.
14. K. Zerzour, F. Marir and J. Orwell, A Descriptive Object Database for Automatic Content-Based Indexing and Retrieval of Surveillance Video Images and Clips, ICIMADE'01 International Conference on Intelligent Multimedia and Distance Education, Fargo, ND, USA, June 1-3

Comparison of Overlap Detection Techniques

Krisztián Monostori[1], Raphael Finkel[2], Arkady Zaslavsky[1], Gábor Hodász[3],
Máté Pataki[3]

[1] School of Computer Science and Software Engineering, Monash University, 900
Dandenong Rd, Caulfield East, VIC 3145, Australia
{Krisztian.Monostori,Arkady.Zaslavsky}@csse.monash.edu.au
[2] Computer Science, University of Kentucky, 773 Anderson Hall, Lexington, KY 40506-
0046, USA
raphael@cs.uky.edu
[3] Department of Automation and Applied Informatics,
Budapest University of Technology and Economic Sciences
1111 Budapest, Goldmann György tér 3. IV.em.433., Hungary,
s8043hod@hszk.bme.hu, pm205@hszk.bme.hu

Abstract. Easy access to the World Wide Web has raised concerns about
copyright issues and plagiarism. It is easy to copy someone else's work and
submit it as someone's own. This problem has been targeted by many systems,
which use very similar approaches. These approaches are compared in this
paper and suggestions are made when different strategies are more applicable
than others. Some alternative approaches are proposed that perform better than
previously presented methods. These previous methods share two common
stages: chunking of documents and selection of representative chunks. We
study both stages and also propose alternatives that are better in terms of
accuracy and space requirement. The applications of these methods are not
limited to plagiarism detection but may target other copy-detection problems.
We also propose a third stage to be applied in the comparison that uses suffix
trees and suffix vectors to identify the overlapping chunks.

1 Introduction

Digital libraries and semi-structured text collections provide vast amounts of
digitised information online. Preventing these documents from unauthorised copying
and redistribution is a hard and challenging task, which often results in avoiding
putting valuable documents online [5]. The problem may be targeted in two
fundamentally different ways: copy prevention tries to hinder the redistribution of
documents by distributing information on a separate disk, using special hardware or
active documents [6] while copy detection allows for free distribution of documents
and tries to find partial or full copies of documents.

One of the most current areas of copy detection applications is detecting
plagiarism. With the enormous growth of the information available on the Internet,
students have a handy tool for writing research papers. With the numerous search

P.M.A. Sloot et al. (Eds.): ICCS 2002, LNCS 2329, pp. 51–60, 2002.
© Springer-Verlag Berlin Heidelberg 2002

engines available, students can easily find relevant articles and papers for their research. But this tool is a two-edged sword. These documents are available in electronic form, which makes plagiarism feasible by cut-and-paste or drag-and-drop operations.

Academic organisations as well as research institutions are looking for a powerful tool for detecting plagiarism. There are well-known reported cases of plagiarised assignment papers [1,2]. Garcia-Molina [5] reports on a mysterious Mr. X who has submitted papers to different conferences that were absolutely not his work. Such a tool is also useful for bona fide researchers and students who want to be sure not to be accused of plagiarism before submitting their papers.

Copy-detection mechanisms may be used in many other scenarios. Nowadays books are published with companion CDs containing the full-text version of the book, which is very useful for searching purposes but bears the risk that the book may be posted on Web sites or newsgroups. A copy-detection tool could use the full-text version of the book and try to locate near replicas of the document on the Internet. Search-engines could also benefit from copy-detection methods: when search results are presented to the user they could be used as a filter on documents, thus preventing users from e.g. reading the same document from different mirror sites. There are various other application areas where copy-detection mechanisms could be used, but they are not listed here because of space limitations. We believe that copy-prevention mechanisms are not the right solution because they impede bona fide researchers while protecting against cheating and law infringement.

In this paper, we compare existing copy-detection mechanisms, which, we have found, have a lot in common. We suggest different application areas for different methods. The most common two-stage approach is discussed in the following section. Section 3 discusses different alternative approaches to split the document into chunks based on our test results. Section 4 presents alternative approaches to select a representative set of chunks, and in Section 5 we summarize our results and look at future work.

2 Overview of Copy-Detection Methods

There are commercial copy-detection systems on the market [10, 4], whose algorithms are not published for obvious reasons. Research prototype systems include the SCAM (Stanford Copy Analysis Method) system [5], the Koala system [7], and the "shingling approach" of [3].

The common approach of these prototype systems to the copy-detection problem can be summarized in five steps:
1. Partition each file into contiguous chunks of tokens.
2. Retain a relatively small number of representative chunks.
3. Digest each retained chunk into a short byte string. We call the set of byte strings derived from a single file its **signature** or **fingerprint**.
4. Store the resulting byte strings in a hash table along with identifying information.

5. If two files share byte strings in their signatures, they are **related**. The closeness of relation is the proportion of shared byte strings.

We defer the discussion of the first two stages to Sections 3 and 4, and we discuss steps 3,4, and 5 in the following subsections.

2.1   Digesting

We could store the chunks themselves, but we choose not to do so for two reasons. First, chunks may be quite large, and we wish to limit the amount of storage required. Second, chunks contain the intellectual property of the author of the file that we are processing. We prefer not to store such property in order to reduce fears that our tools can themselves be used to promote plagiarism or that the database can be used for breaches of confidentiality and privacy.

Instead of storing the chunks, we reduce them by applying a digesting tool. We use the MD5 algorithm [11], which converts arbitrary byte streams to 128-bit numbers. Storing 128 bits would waste too much storage without offering any significant advantage over shorter representations. Of course, the more hex bits we store, the more accurate our system will be. Storing fewer bytes means that two chunks may produce the same digest value even when they are different. These undesired cases are called false positives. In Section 3 where we discuss different chunking strategies, we analyse the effect of different digest sizes on the accuracy of the system. Here we also note that in our system we use an extra stage when we compare documents. This extra stage uses exact string matching techniques to find the actual overlapping chunks. Thus some false positives are not of great concern, because they are eliminated in this stage. We refer to this extra stage as the MDR approach [8, 9].

Of course, other systems use other hashing schemes to create the digest representation of a given chunk, and it would be interesting to compare the effect of different hashing schemes on the number of false positives, but that comparison is beyond the scope of this paper.

2.2   Storing Hash Values

Hash values generated using the methods described above need to be stored in some kind of a database. We may choose to store the hash values in a general-purpose database management system, or we can develop a special-purpose system tailored to store the hash values and their postings efficiently. In our system we have used Perl [12] hash data structures to store hash values.

2.3   Comparison

Different similarity measures can be defined for different application areas. Here we list the three most common measures. Let us denote the digest of file F by $d(F)$.

- Asymmetric similarity $$a(F,G) = \frac{|d(F) \cap d(G)|}{|d(F)|}$$

- Symmetric similarity $$s(F,G) = \frac{|d(F) \cap d(G)|}{|d(F)| + |d(G)|}$$

- Global similarity $$g(F) = \frac{|d(F) \cap (\cup_G d(G))|}{|d(F)|}$$

Of course, the size of documents has a significant effect on the results. For example if document F is much smaller than document G, and document F is entirely contained in document G, then the $a(F,G)$ value will be much higher than the $s(F,G)$ value. The global similarity measure is used for defining the similarity between a given document and a set of documents.

3   Chunking

In this section we analyse different chunking methods proposed in different prototype systems. Subsection 3.1 discusses existing techniques while subsection 3.2 analyses our test results and proposes a new technique.

3.1   Existing Chunking Methods

Before we compare chunking methods, we have to decide what is the smallest unit of text that we consider overlap. One extreme is one letter, but in this case, almost all English language documents would have 100% overlap with other documents, because they share the same alphabet. The other extreme is considering the whole document as one unit. This method could not identify partial overlap between documents.

Prototype systems and our experiments show that the length of chunks should be somewhere between 40 and 60 characters for plagiarism-detection purposes. The Koala system [7] uses 20 consonants, which translates into 30-45 characters; the "shingling approach" [3] considers 10 consecutive words, which is approximately 50-60 characters; the SCAM system [6] analyses the effect of different chunk sizes: one word, five words, ten words, sentences. Our exact-comparison algorithm (MDR approach) used in the extra stage of the comparison process uses 60 characters as a threshold.

The selection of chunking strategy has a significant effect on the accuracy of the system. In the ideal case, we would index all possible α-length chunks of a document. In a document of length l there are $l\text{-}\alpha\text{-}1$ possible α-length chunks [7]. A suffix tree is a data structure that stores all possible suffixes of a given string, thus all possible α-length chunks. Suffix trees are used in the MDR approach.

We could consider non-overlapping chunks of α-length character sequences. The problem with this approach is that adding an extra word or character to the document would shift all boundaries, so two documents differing only in a word would produce two totally different sets of chunks.

The "shingling approach" of Broder et al. [3] uses words as the building blocks of chunks. Every ten-word chunk is considered. As an example, consider the following sentence: "copy-detection methods use some kind of a hash-function to reduce space requirements".

Ten-word chunks generated from this "document": "copy-detection methods use some kind of a hash-function to reduce", "methods use some kind of a hash-function to reduce space", and "use some kind of a hash-function to reduce space requirements".

They call these ten-word chunks shingles. To store every ten-word chunk requires too much space, so Broder et al. [3] also consider a strategy to select representative chunks of a document. Selection algorithms are discussed in the next section.

Garcia-Molina et al. [6] compare different chunking strategies. At the finest granularity, word chunking is considered with some enhancements, such as eliminating stopwords. Another possibility is to use sentences as chunks to be indexed. The problem with sentences is that sentence boundaries are not always easy to detect. They tested their algorithm on informal texts, such as discussion-group documents, because of their informality they often lack punctuation. Sentence chunking also suffers when identifying partial sentence overlap. We have run sentence chunking on our document set of RFC documents and found that the average chunk length using only the full stop as a termination symbol is about 13 words; if we consider other punctuations as chunk boundaries the average is 3.5 words. The problem with sentence chunking is that it does not allow for tuning. By tuning we mean that different problem areas might require different chunk lengths.

To overcome the boundary-shifting problem of non-overlapping chunks, Garcia-Molina et al. [6] introduced hashed-breakpoint chunking. Instead of strict positional criteria to determine chunk boundaries, such as every 10th word is a chunk boundary, they calculate a hash value on each word and whenever this hash value modulo k is 0, it is taken as a chunk boundary. They studied the performance of their prototype system with $k=10$ and $k=5$. The expected length of a chunk with this hashed-breakpoint chunking is k words. As discussed earlier, $k=10$ is closer to the empirically defined threshold value of 40-60 characters. The problem with this approach is that the chunk sizes may differ. We can easily have one-word chunks and also 15-20 word chunks. See the next subsection about this problem.

3.2 Chunking Strategy Tests

The most promising chunking strategy is hashed-breakpoint chunking. It avoids the shifting problem without the need to store overlapping chunks. In this subsection we analyse the hashed-breakpoint strategy in details.

It is true that the expected chunk length is k in case of k-hashed breakpoint chunking, but if a common word happens to be a chunk boundary, the average chunk length may be much smaller than the expected average.

We have used two document sets for comparison. The first set is the set of RFC (Request for Comment) documents; the second set comprises different Hungarian translations of the Bible. We have chosen these two sets because we expect some overlap within these document sets for obvious reasons.

Figure 1 shows the average chunk length in the function of the k value. We can see that the higher the k value, the higher the chance is of greater deviation from the expected result. This behaviour can be explained by the uncertainty of this chunking strategy.

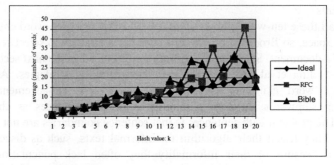

Fig 1. Average chunk length

We have picked nine pairs of documents from the Bible translations and compared them with different hash values. The results shown in *Figure 2* reflect the uncertainty of this chunking method.

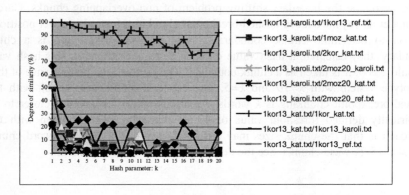

Fig 2. Overlap percentage

The chosen pairs are different translations of the same part of the Bible. The correlation between *Figures 1* and *2* is obvious. For example, the first pair has a peak at $k=16$ following a low at $k=14$ and $k=15$. In *Figure 1*, we can see that at $k=14$ and $k=15$ the average chunk length is higher than expected; if we have longer chunks, the chance for overlap is smaller. On the contrary, at $k=16$ we have a low in *Figure 1*, which means that we have shorter chunks, so the chance for overlap is higher.

In *Figure 2* we can also see that the chosen k value has a significant effect on the amount of overlap detected. We propose to use more than one k value for comparison, and we can either choose the average of the reported overlaps or the maximum/minimum values depending on the actual application. This strategy eliminates the sensitivity of the algorithm to different k values. *Figure 3* shows the results of this approach. We aggregated the number of chunks reported by different k values and calculated the result based on the total number of chunks.

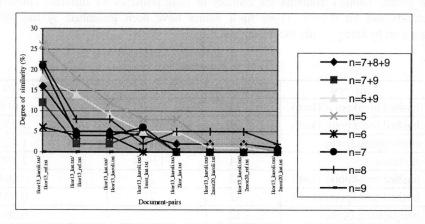

Fig 3. The effect of aggregate k values

Of course, in our applications false positives are more desirable than false negatives, because false positives can be eliminated in our extra stage of exact comparison, while we will never report documents missed by the first stage.

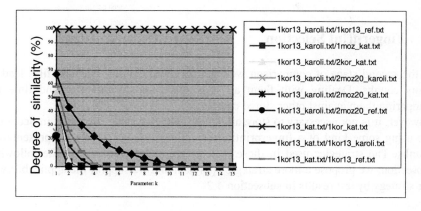

Fig. 4. Overlapping chunks

We have also conducted a test on different chunk sizes. In this test we have used overlapping chunks of different sizes because of their better predictability. The results are shown in *Figure 4*. As we expected longer chunks result in less detected overlap. Based on these experiences, we can conclude that for finding documents similar in style, we can choose a k value of 2 or 3, which would find similar phrases. For plagiarism detection, $k=8$ or $k=9$ seems to be a good parameter value, while for detecting longer overlaps, we can choose values greater than 10.

False positives are a problem common to every approach. We use hashing to identify chunks, and we may have identical hash values when the original chunks do not match. *Table 1* contains the number of false positives for different chunking methods and bit depths. These hash values have been generated by the MD5 algorithm by keeping only the left-most k bits.

Table 1. False positives

Method	bit-depth	false positives	false positive (%)
hashed breakpoint (k=6)	24	8434	1.6868
hashed breakpoint (k=9)	24	6790	1.3580
overlapping (k=6)	24	7115	1.4230
sentence	24	13954	2.7908
hashed breakpoint (k=6)	32	23	0.0046
hashed breakpoint (k=9)	32	21	0.0042
overlapping (k=6)	32	26	0.0052
sentence	32	15	0.0030

The tests were carried out on 500,000 chunks. The results show a dramatic decrease in the number of false positives when we move from 24 bits to 32 bits. We suggest using 32 bits, not only because the number of false positives is less than 0.01%, but it is also an easier data width to handle in today's computers.

4   Fingerprint Selection

In the second stage of comparison we could store all chunks, but long files lead to many chunks. Dealing with them all uses space for storing them and time for comparing them against other stored chunks.

However, it is not necessary to store all chunks. Other systems discussed in Section 2 are using some kind of a fingerprinting or culling strategy to select representative chunks. The strategies used in those systems are mainly random. In the following subsection, we propose a more strategic approach and we support the applicability of our strategy by test results in subsection 4.2.

4.1 Selection strategies

A short chunk is not very representative of a text. The fact that two files share a short chunk does not lead us to suspect that they share ancestry. In contrast, very long chunks are very representative, but unless a plagiariser is quite lazy, it is unlikely that a copy will retain a long section of text.

We therefore discard the longest and the shortest chunks. We wish to retain similar chunks for any file. We have experimented with two culling methods. Let n be the number of chunks, m the median chunk size (measured in tokens), s the standard deviation of chunk size, b a constant, and L the length of an arbitrary chunk.

Sqrt. Retain $\lceil \sqrt{n} \rceil$ chunks whose lengths L are closest to m.

Variance. Retain those chunks such that $|L-m| \le bs$. Increase b, if necessary, until at least $\sqrt{n}$ chunks are selected. We start with $b=0.1$.

4.2 Test Results

We have tested our fingerprinting method on the RFC data set. We have found that the Sqrt method does not store enough chunks, thus the Variance method is preferable. In *Table 2* we show the results of these tests based on some RFC documents with known overlap. The table shows asymmetric similarity, that is RFC 1 compared to RFC 2 does not necessarily provides the same result as RFC 2 compared to RFC 1. We consider the MDR method as accurate because it is based on exact matching. SE is our signature extraction method while OV is the overlapping chunk method.

Table 2 shows that our SE method tends to underestimate large overlaps while overestimating small overlaps (overlap is given as percentage). The overlapping-chunks method seems to provide more accurate results but at a much higher storage cost. Overestimation is not a problem in our system because we use the MDR approach as a final filter, which correctly identifies overlapping chunks.

Table 2. Asymmetric similarities

RFC 1	RFC 2	MDR 1	MDR 2	SE 1	SE 2	OV 1	OV 2
1596	1604	99	99	91	92	94	94
2264	2274	99	99	96	95	94	94
1138	1148	96	95	93	92	91	89
1065	1155	96	91	71	68	84	79
1084	1395	86	84	58	64	79	75
1600	1410	72	77	52	48	58	61
2497	2394	19	17	33	27	16	15
2422	2276	18	3	23	6	15	2
2392	2541	16	12	27	17	13	10

5   Conclusion and Future Work

In this paper we have analysed different chunking strategies used in copy-detection systems. We know of no other studies that have analysed these systems from this aspect. We have made suggestions on how to select chunking strategies and also touched on the problem of fingerprint selection, which is also crucial because of space-efficiency. Our SE algorithm provides very close results to the exact comparison and it overestimates small overlaps, which is a more desirable behaviour than underestimating, because false positives can be found by applying an extra filter on the results. Of course, too many false positives are not desirable, either. In the future we will also study the effects of different hash functions on false positives. We plan to run our tests on more document sets and we are also developing an online system where these methods are available for copy-detection applications (more specifically plagiarism detection).

6   References

1. Argetsinger A. Technology exposes cheating at U-Va. *The Washington Post*, May 8, 2001.
2. Benjaminson A. Internet offers new path to plagiarism, UC-Berkeley officials say. *Daily Californian*, October 6, 1999.
3. Broder A.Z., Glassman S.C., Manasse M.S. Syntatic Clustering of the Web. *Sixth International Web Conference*, Santa Clara, California USA. URL http://decweb.ethz.ch/WWW6/Technical/Paper205/paper205.html
4. EVE Plagiarism Detection System. *URL http://www.canexus.com*, 2000
5. Garcia-Molina H., Shivakumar N. The SCAM Approach To Copy Detection in Digital Libraries. *D-lib Magazine*, November, 1995.
6. Garcia-Molina H., Shivakumar N. Building a Scalable and Accurate Copy Detection Mechanism. *Proceedings of 1st ACM International Conference on Digital Libraries (DL'96)* March, Bethesda Maryland, 1996.
7. Heintze N. Scalable Document Fingerprinting. *Proceedings of the Second USENIX Workshop on Electronic Commerce*, Oakland, California, 18-21 November, 1996. URL http://www.cs.cmu.edu/afs/cs/user/nch/www/koala/main.html
8. Monostori K., Zaslavsky A., Schmidt H. MatchDetectReveal: Finding Overlapping and Similar Digital Documents. *Information Resources Management Association International Conference* (IRMA2000), 21-24 May, 2000 at Anchorage Hilton Hotel, Anchorage, Alaska, USA. pp 955-957, 2000.
9. Monostori K., Zaslavsky A., Schmidt H. Parallel Overlap and Similarity Detection in Semi-Structured Document Collections. *Proceedings of 6th Annual Australasian Conference on Parallel And Real-Time Systems (PART '99)*, Melbourne, Australia, 1999. pp 92-103, 1999.
10. Plagiarism.org, the Internet plagiarism detection service for authors & education. *URL http://www.plagiarism.org* , 1999.
11. Rivest R. L.. RFC 1321: The MD5 Message-Digest Algorithm. *Internet Activities Board*, April 1992.
12. Wall L. and Schwartz R. L. Programming Perl. *O'Reilly & Associates*, Inc., 981 Chestnut Street, Newton, MA 02164, USA, 1992.

Using a Passage Retrieval System to Support Question Answering Process[1]

Fernando Llopis, José Luis Vicedo and Antonio Ferrández

Departamento de Lenguajes y Sistemas Informáticos
University of Alicante
Alicante, Spain
{llopis, vicedo, antonio}@dlsi.ua.es

Abstract. Previous works in Information Retrieval show that using pieces of text obtain better results than using the whole document as the basic unit to compare with the user's query. This kind of IR systems is usually called *Passage Retrieval (PR)*. This paper discusses the use of our PR system in the question answering process (QA). Our main objective is to examine if a PR system provide a better support for QA than Information Retrieval systems based in the whole document..

1 Introduction

Given a user's query, Information Retrieval (IR) systems return a set of documents, which are sorted by the probability of containing the required information. Since IR systems return whole documents, there is an additional work for the user, who has to read the whole document to search for the required information. However, when searching for specific information, this last user's task can be carried out by Question Answering systems (QA), which are tools that usually work on the output of an IR system, and try to return the precise answer to the user's query.

IR systems work on measuring the similarity between each document and the query by means of several formulas that typically use the frequency of query terms in the documents. This way of measuring means that larger documents could have a greater chance at being considered relevant, because of the higher number of terms that could coincide with that of the query.

In order to solve this problem, some IR systems measure the similarity in accordance with the relevance of the pieces of adjoining text that form the documents, where these pieces of text are called passages. These kinds of IR systems, which are usually called

---

[1] This paper has been partially supported by the Spanish Government (CICYT) project number TIC2000-0664-C02-02

P.M.A. Sloot et al. (Eds.): ICCS 2002, LNCS 2329, pp. 61–69, 2002.

Passage Retrieval (PR), allows that the similarity measure is not affected by the size of the document. PR systems return the precise piece of text where it is supposed to find the answer to the query, a fact that is especially important when large documents are returned. PR systems are more complex than IR, since the number of textual units to compare is higher (each document is formed by several passages) and the number of tasks they accomplish is higher too (above all when passage splitting is performed after processing each query as it is proposed in [7]). Nevertheless, the complexity added to the system is rewarded with a significant increase in the performance of the system. For example, in [1] the improvement reaches a 20%, and in [7] it does a 50%.

As well as obtaining better precision than document retrieval systems, PR return the most relevant paragraphs to a query, information that could be used as input by a QA system. This fact allows reducing QA systems computational complexity (they work on small fragments of text instead of documents) and improves QA's precision, since the passage list retrieved is more relevant to the query than the results of a document retrieval system.

The PR system presented in this paper is called IR-n. It defines a novel passage selection model, which forms the passages from sentences in the document. IR-n has been used in the last Cross-Language Evaluation Forum (CLEF-2001) and in the last Text REtrieval Conference (TREC-2001) for the Question Answering track. Part of the better performance obtained in TREC-2001 with reference to TREC-2000, is due to IR-n.

The following section presents PR and QA backgrounds. Section 3 describes the architecture of our proposed PR system. In section 4, we give a detailed account of the tests, experiments and results obtained. Finally, we present the conclusions of this work.

2 Backgrounds in Question Answering and Passage Retrieval

The Computational Linguistic community has shown a recent interest in QA, and it comes after developing Information Extraction systems, which have been evaluated in Message Understanding Conferences (MUC). Specifically, the interest was shown when in TREC-8, there appears a new track on QA that tries to benefit from large-scale evaluation, that was previously carried out on IR systems, in previous TREC conferences.

If a QA system wants to successfully obtain a user's request, it needs to understand both texts and questions to a minimum level. That is to say, it has to carry on many of the typical steps on natural language analysis: lexical, syntactical and semantic. This analysis takes much more time than a statistical analysis that is carried out in IR. Besides, as QA has to work with as much text as IR, and the user needs the answer in a limited period of time, it is usual that an IR system processes the query and after, the QA system will continue with its output. In this way, the time of analysis is greatly decreased.

Some of the best QA systems are the following: [2],[3],[10],[5] After studying these systems, it seems agreeable the following general architecture, that is formed by four modules, where document retrieval module is accomplished by using IR technology:

- Question Analysis.
- Document Retrieval.
- Passage Selection.
- Answer Extraction.

Different IR models are used at the Document Retrieval stage in QA systems however, the best results are obtained by systems that apply PR techniques for this task.

The main differences between different PR systems are the way that they select the passages, that is to say, what they consider a passage and its size. According to the taxonomy proposed in [1], the following PR systems can be found: discourse-based model, semantic model and window model. The first one uses the structural properties of the documents, such as sentences or paragraphs (e.g. the one proposed in [9], [12]) in order to define the passages. The second one divides each document into semantic pieces, according to the different topics in the document (e.g. those in [4]). The last one uses windows of a fixed size (usually a number of terms) to form the passages [1], [6].

It would seem coherent that discourse-based models are more effective since they use the structure of the document itself. However, the problem with them is that the results could depend on the writing style of the document's author. On the other hand, window models have the advantage that they are simpler to use, since the passages have a previously known size, whereas the remaining models have to bear in mind the variable size of each passage. Nevertheless, discourse-based and semantic models have the advantage that they return logical and coherent fragments of the document, which is important if these IR systems are used for other applications such as Question Answering.

The passage extraction model that we are proposing allows us to benefit from the advantages of discourse-based models since logical information units of the text, such as sentences, form the passages. Moreover, another novel proposal in our PR system is the relevance measure, which unlike other discourse-based models, is not calculated from the number of passage terms, but from the fixed number of passage sentences. This fact, allows a simpler calculation of this measure unlike other discourse-based or semantic models. Although we are using a fixed number of sentences for each passage, we consider that our proposal differs from the window models since our passages do not have a fixed size (i.e. a fixed number of words) because we are using variable length sentences.

3 System Overview

In this section, we describe the architecture of the proposed PR system, namely IR-n, focusing on its two main modules: the indexation and the document extraction modules.

3.1 Indexation Module

The main aim of this module is to generate the dictionaries that contain all the required information for the document-extraction module. It requires the following information for each term:

- The number of documents that contain the term.
- For each document:
  - The number of times that the term appears in the document.
  - The position of each term in the document: the number of sentence and position in the sentence.

Where we consider as terms, the stems produced by the "Porter stemmer" on those words that do not appear in a list of stop-words, list that is similar to those used in IR systems. For the query, the terms are also extracted in the same way, that is to say, their stems and positions in the query one query word that do not appear in the list of stop-words.

3.2 Document Extraction Module

This module extracts the documents according to it's similarity with the user's query. The scheme in this process is as follows:

1. Query terms are sorted according to the number of documents in which they appear, where the terms that appear in fewer documents are processed firstly.
2. The documents that contain some query term are extracted.
3. The following similarity measure is calculated for each passage p with the query q:

$$\text{Similarity\_measure}(p, q) = \sum_{t \in p \wedge q} W_{p,t} * W_{q,t}$$

Where:

$W_{p,t} = \log_e(f_{p,t} + 1)$.
$f_{p,t}$ is the number of times that the term t appears in the passage p.
$W_{q,t} = \log_e(f_{q,t} + 1) * idf$.
$f_{q,t}$ is the number of times that the term t appears in the query q.
$idf = \log_e(N / f_t + 1)$.
N is the number of documents in the collection.
f_t is the number of documents that contain the term t.

4. Each document is assigned the highest similarity measure from its passages.
5. The documents are sorted by their similarity measure.
6. The documents are presented according to their similarity measure.

As it will be noted, the similarity measure is similar to the cosine measure presented in [11]. The only difference is that the size of each passage (the number of terms) is not used to normalise the results. This difference makes the calculation simpler than other discourse-based PR systems or IR systems, since the normalization is accomplished according to a fixed number of sentences per passage. Another important detail to notice is that we are using N as the number of documents in the collection, instead of the number of passages. That is because in [7] it is not considered relevant for the final results.

The optimum number of sentences to consider per passage is experimentally obtained. It can depend on the genre of the documents, or even on the type of the query as it is suggested in [6]. We have experimentally considered a fixed number of 20 sentences for the collection of documents with which we worked [8]. Table 1 presents the experiment where the 20 sentences per passage obtained the best results.

Table 1. Precision results obtained on *Los Angeles Times* collection with different number of sentences per passage.

Recall	Precision IR-n					
	5 Sent.	10 Sent.	15 Sent.	20 Sent.	25 Sent.	30 Sent.
0.00	0.6378	0.6508	0.6950	*0.7343*	0.6759	0.6823
0.10	0.5253	0.5490	0.5441	*0.5516*	0.5287	0.5269
0.20	0.4204	0.4583	0.4696	*0.4891*	0.4566	0.4431
0.30	0.3372	0.3694	0.3848	*0.3964*	0.3522	0.3591
0.40	0.2751	0.3017	0.2992	*0.2970*	0.2766	0.2827
0.50	0.2564	0.2837	0.2678	*0.2633*	0.2466	0.2515
0.60	0.1836	0.1934	0.1809	*0.1880*	0.1949	0.1882
0.70	0.1496	0.1597	0.1517	*0.1498*	0.1517	0.1517
0.80	0.1213	0.1201	0.1218	*0.1254*	0.1229	0.1279
0.90	0.0844	0.0878	0.0909	*0.0880*	0.0874	0.0904
1.00	0.0728	0.0722	0.0785	*0.0755*	0.0721	0.0711

Table 2. Experiments with a different number of overlapping sentences

Recall	IR-n with 1 overlap.	IR-n with 5 overlap.	IR-n 10 overlap.
0.00	0.7729	0.7211	0.7244
0.10	0.7299	0.6707	0.6541
0.20	0.6770	0.6072	0.6143
0.30	0.5835	0.5173	0.5225
0.40	0.4832	0.4144	0.4215
0.50	0.4284	0.3704	0.3758
0.60	0.3115	0.2743	0.2759
0.70	0.2546	0.2252	0.2240
0.80	0.2176	0.1914	0.1918
0.90	0.1748	0.1504	0.1485
1.00	0.1046	0.0890	0.0886
Medium	0.4150	0.3635	0.3648

As said, the proposed PR system can be classified into discourse-based models since it is using variable-sized passages that are based on a fixed number of sentences (but with a different number of terms per passage). The passages overlap each other, that is to say, let us suppose that the size of the passage is N sentences, then the first passage will be formed by the sentences from 1 to N, the second one from 2 to N+1, and so on. We have decided to overlap just one sentence based on the following experiment, where several numbers of overlapping sentences have been tested. In this experiment, Table 2, it will observed that only one overlapping sentence obtained the best results.

4 Evaluation

This section presents the experiment proposed for evaluating our approach and the results obtained. The experiment has been run on the TREC-9 QA Track question set and document collection.

4.1 Data Collection

TREC-9 question test set is made up by 682 questions with answers included in the document collection. The document set consists of 978,952 documents from the TIPSTER and TREC following collections: AP Newswire, Wall Street Journal, San Jose Mercury News, Financial Times, Los Angeles Times, Foreign Broadcast Information Service.

4.2 Experiment

In order to evaluate our proposal we decided to compare the quality of the information retrieved by our system with the ranked list retrieved by the ATT information retrieval system.

First, ATT IR system was used for retrieving the first 1000 relevant documents for each question. Second, our approach was applied over these 1000 retrieved documents in order to re-rank the list obtained by ATT system for each question. This process was repeated for different passage lengths. These lengths were of 5, 10, 15, 20, 25 and 30 sentences.

The final step was to compare the list retrieved by ATT system and the four lists obtained by our system. Table 3 shows the precision obtained in each test. The precision measure is defined as the number of answers included at 5, 10, 20, 30, 40, 50, 100 and 200 first documents or paragraphs retrieved in each list.

4.3 Results Obtained

Table 3 shows the number of correct answers included into the top N documents of each list for the 681 test questions. The first column indicates the number N of first ranked documents selected for measuring the precision of each list. The second column shows the results obtained by ATT list. The remaining columns show the precision obtained by our system for passages of 5, 10, 15, 20, 25 an 30 sentences.

It can be observed that our proposal obtains better results than the original model, being 20 sentences the optimum passage length for this task.

Table 3. Obtained results on *TREC* collection and 681 TREC queries.

Answer included	ATT system	IR-n system					
		5 Sent.	10 Sent.	15 Sent.	20 Sent.	25 Sent	30 Sent.
At 5 docs	442	444	464	488	508	532	531
At 10 docs	479	505	531	549	561	570	572
At 20 docs	517	551	573	584	595	599	599
At 30 docs	539	575	596	600	612	617	618
At 50 docs	570	599	611	623	624	637	637
At 100 docs	595	614	631	640	644	650	646
At 200 docs	613	634	643	648	654	655	655

5 Conclusions and Future Works

In this paper, a passage retrieval system for QA purposes has been presented. This model can be included in the discourse-based models since it is using the sentences as the logical unit to divide the document into passages. The passages are formed by a fixed number of sentences. As results show, this approach performs better than a good IR system for QA tasks. Results seem better if we take into account that our passage approach returns only a small piece of the whole document. This fact enhances our results because the amount of text that a QA system has to deal with is much smaller than using a whole document. This way, our approach achieves a reduction in QA processing time without affecting system performance.

Several areas of future work have appeared while analysing results. First, we have to investigate the application of question expansion techniques that allow increasing system precision. Second, we have to measure the performance of our system by applying directly our paragraph retrieval method over the whole document collection instead of using it for reordering a given list. And thirdly, it would be very useful for QA systems to develop a passage validation measure to estimate the minimum number of passages that a QA system has to process to be sure that it contains the answer to the question. All these strategies need to be investigated and tested.

References

[1]   Callan, J. *Passage-Level Evidence in Document Retrieval*. In Proceedings of the 17 th Annual ACM SIGIR Conference on Research and Development in Information Retrieval, Dublin, Ireland, 1994, pp. 302-310.

[2]   Clarke, C.; Derek, C.; Kisman, I. and Lynam, T. R. Question Answering by Passage Selection (MultiText Experiments for TREC9). In Nineth Text REtrieval Conference, 2000.

[3]   *Harabagiu, S.; Moldovan, D.; Pasca, M.; Mihalcea, R.; Surdeanu, M.; Bunescu, R.; Gîrju, R.; Rus, V. and Morarescu, P. FALCON: Boosting Knowledge for Answer Engines. In Nineth Text REtrieval Conference, 2000.*

[4]   Hearst, M. and Plaunt, C. *Subtopic structuring for full-length document access*. Proceedings of the Sixteenth Annual International ACM SIGIR Conference on Research and Development in Information Retrieval, June 1993, Pittsburgh, PA, pp 59-68

[5]   *Ittycheriah, A.; Franz, M.; Zu, W. and Ratnaparkhi, A. IBM's Statistical Question Answering System. In Nineth Text REtrieval Conference, 2000.*

[6]   Kaskiel, M. and Zobel, J. *Passage Retrieval Revisited* SIGIR '97: Proceedings of the 20th Annual International ACM July, 1997, Philadelphia, PA, USA, pp 27-31

[7]   KaszKiel, M. and Zobel, J. *Effective Ranking with Arbitrary Passages*. Journal of the American Society for Information Science, Vol 52, No. 4, February 2001, pp 344-364.

[8]   Llopis, F. and Vicedo, J. *Ir-n system, a passage retrieval system at CLEF 2001* Working Notes for the Clef 2001 Darmstdt, Germany , pp 115-120

[9]   Namba, I *Fujitsu Laboratories TREC9 Report*. Proceedings of the Tenth Text REtrieval Conference, TREC-10. Gaithersburg,USA. November 2001, pp 203-208

[10]   Prager, J.; Brown, E.; Radev, D. and Czuba, K. One Search Engine or Two for QuestionAnswering. In *Nineth Text REtrieval Conference*, 2000.

[11]   Salton G. *Automatic Text Processing: The Transformation, Analysis, and Retrieval of Information by Computer,* Addison Wesley Publishing, New York. 1989

[12]   Salton, G.; Allan, J. Buckley *Approaches to passage retrieval in full text information systems.* In R Korfhage, E Rasmussen & P Willet (Eds.) Prodeedings of the 16 th annual international ACM-SIGIR conference on research and development in information retrieval. Pittsburgh PA, pp 49-58

XML Design Patterns Used in the EnterTheGrid Portal

Ad Emmen

Genias Benelux bv, James Stewartstraat 248, NL-1325 JN Almere,
The Netherlands,
emmen@genias.nl

Abstract. EnterTheGrid is the largest independent Grid portal in existence. It has been build entirely on XML-technology, from the database up to the user interface. In this paper we describe several XML design patterns, that were developed for the site. A hybrid XML data model is used as the basis to describe the items of the site. A search engine is developed using XSLT/XPath and it is describer how information from different HPCN sources can be combined.

1 Introduction

EnterTheGrid [1] is a portal on the Internet that provides a single access point to anyone interested in The Grid. Currently, the three main parts of EnterTheGrid are the catalogue, the EnterTheGrid/Primeur HPCN and Grid news services - on the web and by e-mail, and XML based resources (web services) like the TOP500 in XML. The server currently attracts some 30.000 different visitors each month.

Everybody is using XML these days at least so it seems, but when one looks a little bit deeper, it is often just a report that is produced in XML out of a relational database, or some configuration file written in XML instead of property/value pairs.

Important as these applications are, they only scratch the surface of what is possible with current XML-technology. Two paradigms prohibit the view over the landscape of new possibilities:
1. A database is a relational database with an SQL interface
2. The only good programming model is an object oriented one, except when you do math then it is ok to use Fortan, or C.

A popular public domain based approach is for instance a combination of PHP [2] and MySQL [2]. Also the main commercial solutions implement this approach, with proprietary relational database systems. Also emerging e-market places [4] often follow this approach.

P.M.A. Sloot et al. (Eds.): ICCS 2002, LNCS 2329, pp. 70–77, 2002.
© Springer-Verlag Berlin Heidelberg 2002

During the past years, relational database management systems are dominant. The difference between the relational database itself and the database management system has been blurred, and people tend to see it as one thing. Hence they associate a relational database with the possibility to handle large quantities of data in a secure and easy to use way.

Fact is that XML-documents are basically hierarchical and do not lend themselves in general to be stored effeciently in a relational database. Most of the time what people need, however, is a reliable XML-database *management system*. Once you make the paradigm shift, one sees that relational databases are much too restricted to store general XML-documents. Unfortunately, although there is a lot of development in the area, native XML-database management systems they cannot match relational database management systems yet.

The other paradigm is that the only good programming language is an object oriented one. As a result, most people do not notice that XSLT (Extended Stylesheet Language - Transformations) [5], is in fact a specialized functional programming language that can convert any set of XML-documents in a new different set. With the Xpath [5] standard for addressing any item in an XML-document, it can be used to build powerful search engines that can match SQL based ones.

Do not throw away your relational database, because there are application fields where it is superior to anything else, do not throw away you Java or C++, because there are application fields where you cannot do without, but when you start a new XML application, do not take for granted it has to be build on top of a relational database and has to be programmed in an object oriented programming language.

2 Basic Design

EnterTheGrid is a community management system [6]. There is a news section, there is a calendar, organisation profiles, etc. A typical information item, such as a news article, is somewhere in the middle between a data and document oriented type. It does not have the regular structure of a data oriented type, because, for instance, the body part of the article can exhibit a complex structure with paragraphs, tables and lists. On the other hand, they do not have the rich complexity of a real document oriented type like a manual or full report.

Hence we did design a document type definition that has a limited number of elements in a fixed structure. For parts, like the body of the article or a description of a company, we allow the use of XHTML type of structure constructs. This proves to be very powerful, as it allows to develop generic tools around this definition.

Each article, organisation profile, etc, is stored as a separate entity (an XML external parsed entity). An index (an XML document) is created that indeed includes all relevant items as references to these XML entities, into an XML document, creating one logical view of all items, the same as if they were stored in one physical XML-document. This base index can then be used to create sub indices to speed up processing. This separation between physical and logical structure of an XML document, inherited from SGML, is seldom used, but provides in fact a powerful storage mechanism.

Because each article, organisation profile, etc, is a physical different entity, they can also be updated easily, independent from each other. For this purpose we did choose to create a web based editor that only needs an HTML browser to operate. There is no Java or other programming executable needed on the client side. There is some Javascript to ease editing, but one can do without. There are other approaches possible. For instance, creating an editor as a Java applet, using a proprietary language from one of the browser vendors or using Javascript. One even create a stand-alone editor that has to be downloaded. However, all of these approaches have the disadvantage that they do not work all off the time. Because we want authors having access from any place and any type of computer, they could even be using it only once, the HTML-only client approach is the best.

The advantage of this approach is that everything can be done on the server. The disadvantage is that everything has to be done on the server. In our case, the bulk of the editor is written in XSLT, with a little Java to get it started. It then runs in a standard Servlet engine. The advantage of doing all processing on the server is the control on what is happening. There are never surprises because of the client software not working as expected.

On the down side the problem is that the user who edits a contribution has to wait for the server to respond, which in practice is not a big problem. The main problem that arises is that choices and sub-choice dependencies (for instance subsections that differ for each section), can only be updated after contact with the server. Another disadvantage of this approach is, although the user is guided as much as possible, it still is possible in some fields to enter text that is not well-formed XML. This then has to be reported back by the server.

The editor itself is fully configurable, because of the use of XML. There are several XML configuration files that steer the process. Each time the user gets a new editor page, or editor page update, it is automatically created from this configuration files. There are two important ones.

The first configuration file describes which part of a document may be edited by a user, and how they should be called for him. So for instance when a news article in a specific publication has a title, called "Kop" and a lead part, called "Tekst", but no large body part or reference section, then the user gets only two fields with these names. He does not see how they are actually coded in XML, which is vmp:title and "vmp:lead" respectively. In the field "Tekst" a complex content structure can be entered. When some time later the user wants to edit for instance an organisation profile, he gets the fields an labels belonging to an organisation profile.

The community manager maintainer can change he configuration file, and hence what a user is allowed to change or enter in an item. Different groups of users, for instance, editors and authors, can have different editors. The layout of the editor, i.e. where the fields and labels, size, colours, and images, are determined by an XSL-FO [5] file. XSL-FO, the formatting part of the Extended Stylesheet Language is an excellent format to describe the layout of 2-D objects, be it for paper or for screens. Most people use it as an intermediate language to specify printed material in for instance PDF format. The idea is that we have to "wait" for browsers that are XSL-FO aware before using them for display on a screen. We, however, use it as a specification language for defining the automatically generated editor's layout. Hence, the page that is send to the user consists of XHTML and CSS. It is automatically composed from the item to be edited (if it exists) in XML, the configuration file for what has to be displayed and the XSL-FO definition of the lay-out. In fact there are a number of other general applicable XML files that may be used. For instance for formatting dates, or translations to other languages.

The publishing system contains further a publishing module, a search module and an administration module. The search module is the most novel and interesting one to describe.

3 Implementing an XSLT/XPath Search Engine

Because all the content is stored as XML, we cannot use SQL as a search language. We could use a standard text based search-engine approach that is used on a lot of web sites. Although they can give very good results in general cases, they cannot use the structure that is inherent in the XML-content. To express this we need an XML aware search approach. Although there is work done on specifying an Xquery language[5], the XSLT/Xpath standards, already provide enough searching and location possibilities for a good implementation. Hence, our search module is build using XSLT/Xpath technology. Take for instance the XML fragment:

```
<vmp:article>
<vmp:title>Jansen en Jansen</vmp:title>
<vmp:author>Piet<vmp:author>
</vmp:article>

<vmp:article>
<vmp:title>Pietersen en Pietersen</vmp:title>
<vmp:author>Jan<vmp:author>
</vmp:article>
```

Then we can display the article title of the article that has "Jan" as author with:

```
<xsl:template match="vmp:article">
<xsl:if select="contains(vmp:author, 'Jan')">
<xsl:value-of select="vmp:title" />
</xsl:if>
</xsl:template>
```

To generalize this, we have to use a variable containing the search string and pass this to the template. If we want to make it case insensitive we can use the XSLT translate() function to translate both the element content and the search string to the same case. To search for more texts in different parts of the article, we can use for instance:

```
<xsl:if select="contains(vmp:author, 'Jan')
               and contains(vmp:title, 'Piet')">
```

When we use variables as search strings, for instance called for instance $author and $title, this would be:

```
<xsl:if select="contains(vmp:author, $author)
               and contains(vmp:title, $title)">
```

When a search string is empty, the result of the contains() function is "true" which is what we need for the complete if-statement to work correctly. When a user inadvertently enters a space in the search-field, or adds a space or return character at the end, the contains() function will return "false". To get the desired result, one can strip these characters with the normalize-space() function available in XSLT.

Because the general XML structure of each item is the same, extending this to a full search engine is not too difficult. For the displaying the search query to the user, we, again use the XML configuration file that maps each element name to the label too be used.

4 Combining Information from Different Sources - The TOP500 and "Overview of Recent Supercomputers"

Currently, most of the World Wide Web consists of HTML web pages. In these pages, the original content description is lost. No longer is an address an address, but it is marked up as a paragraph or pargraph, perhaps with some bold typeface added. One of the promises of the web services hype is that these new services will solve the problem for you. In fact, that is an easy promise: once you put your information in an XML-document and make that accessible through on a URI, you can access the information in it using XSLT/XPath and combine it with other sources into a new XML-document, as we will show next.

There are many sources of information on supercomputers available. One of the most important ones is the TOP500 list of most powerful supercomputers in the world. It contains data on the position in the TOP500, the performance and where the machine is located. Another source is the *Overview of recent supercomputers* [7] that contains a description of the supercomputer systems available, with system parameters, and benchmark information. With XML-technology we have combined information from the two sources to generate a new document, containing a description of each architecture, extended with a sub list of the TOP500 on that machine. The new document is generated with XSL-Transformation-stylesheets, that leave the original sources untouched. When there is a new version of for instance the TOP500, running the stylesheet is enough to generate a new, updated version of the assembled information document.

We produced an XML-version of the TOP500 [8] and showed that, by using stylesheets, this can be used to select and combine information from several lists easily. Although this is already very useful, combining information with information from other sources can make the TOP500 even more valuable. The before mentioned report *Overview of Recent Supercomputers* by Aad van der Steen e.a is such an information source. The report describes all the supercomputer systems currently available in detail. For each system it lists, for instance, the operating systems, compilers, and parameters like clock cycle, peak performance, memory size and interconnect speed. Also it contains information on benchmarks.

The previous editions of the overview report were available on paper as the primary source and html version on the web derived from the LaTex input source.

Looking up and comparing information in the Overview report was simply a way of scanning through the document with the traditional table-of-contents as guidance. Combing information from the report with that from the TOP500 was a tedious process. Let us take as an example that you want to have a list of all the TOP500 sites that belong to a certain system architecture from the Overview report. You go to the TOP500 web site. Create a sublist for system A, print it out and stick it into the Overview report. Than do the same for system B, C and so one. I guess only a handful of people went through this process.

With XML technology, that becomes a simple task. But before we describe how to do it, let us recall what XML is. With XML you can not only write information in a document, but also describe what the information is about. When you are writing a book, you can tell what a chapter is, what a section, etc. When you write about the TOP500 list, you can tell which item is the TOP500 position and which is the address of the site that has the supercomputer. Each information element is contained by XML-tags that look like HTML tags. The difference is that you can define your own tags to exactly describe the information in your document.

When you have XML documents, you can use XSLT (Extensible Stylesheet Language Transformations) to extract information from one or more documents and combine that into a new one. You can specify in detail what information you need. The new document can be another XML document, an HTML document, or a text. When you apply a formatter to it, you can generate RTF (word), PDF, LaTex, a database record or whatever format you need.

XSLT is a powerful tool. You can for instance extract information like: "Give me the country name of the last system in the TOP500 list, that according to the Overview report could have an HPF compiler on it." I do not know what you can do with this information, but you can ask it.

Although this looks a bit like a data base query, it is important to realise that we are locating information in documents, not in a relation database.

With the TOP500 already available in XML, the first task was to make the Overview report available in XML. The source text of the report is written in LaTeX. Although LaTex gives a good control over the page-lay-out, it does not provide us with much clues as what type of information is described. Most of the important data on machines is hidden in tables. Hence we decide to first turn that information into

XML designing a new Document Type Definition suited for description of supercomputing systems as applied in the report.

This is an excerpt showing how a description of a supercomputer model looks like.

```
<ors:model>
<ors:name>VPP5000</ors:name>
<ors:clock-cycle unit="nsec">3.3
</ors:clock-cycle>
<ors:processor-performance
unit="gflops">9.6</ors:processor-performance>
<ors:peak-performance unit="tflops">1.22</ors:peak-
performance>
<ors:memory-node unit="gbyte">16</ors:memory-node>
<ors:memory-maximal unit="tbyte">2</ors:memory-maximal>
<ors:number-of-processors>
   <ors:min>4</ors:min>
   <ors:max>128</ors:max>
</ors:number-of-processors>
<ors:communication-bandwidth>
   <ors:point-to-point unit="gbitpsec">1.6</ors:point-
to-point>
   <ors:processor-memory
unit="gbitpsec">38.4</ors:processor-memory>
</ors:communication-bandwidth>
</ors:model>
```

Note that from the names of the tags, one can easily deduce what the information means. We defined an attribute "unit" to indicate what a number is representing.

When you define your own tags, you can choose to make them as short as possible, or as descriptive as possible. Although in the latter case documents tend to become more lengthy, I choose nevertheless for that one: it makes the documents much more self explanatory.

Next we create an XSLT stylesheet. It takes the XML document of the systems from the Overview report, and for each system matches it with manufacturer information in the TOP500 list. The lines that do this are:

```
<xsl:template match="ors:system">
...
<xsl:for-each
   select="document('top500-199911.xml')
           /top500:list/top500:site/top500:manufacturer">
...
```

The output document, also XML, has a traditional report type of structure, with chapters, sections, and tables. For this we have another stylesheet that we apply to reports and that can turn it for instance into an HTML document. We have also made this available over the Web [9].

5 Conclusion

We have showed that with standard XML and XSLT, it is possible to build an EnterTheGrid portal with the same functionality as with a traditional relational database system approach. However, in addition, it is possible to exploit the XML structure of the stored information in an XML-aware search engine that delivers better results.

We have combined the information from several HPC-resources translated into XML into a "web service" where a report is enriched with information from the TOP500. This would have been very difficult without an XML-based approached.

The methods described form basic XML design patterns that can also be deployed in other applications.

6 References

1. http://enterthegrid.com and http://www.hoise.com/primeur, last visited, January 2002.
2. PHP website, http://www.php.net, last accessed, January 2002.
3. MySQL website, http://www.mysql.org, last accessed, January 2002.
4. Kaplan, J.O, Sawhney, M, E-hubs: The new B2B Market places. Harvard Business review, May-June 2000, 97-103, 2000.
5. The World Wide Web Consortium style initiative (XSL, XSLT, Xpath), http://www.w3c.org/style, last visited January 2002.
6. Netgain, expanding markets through virtual communities, John Hagel II, Arthur G. Armstrong, Harvard Business School Press, ISBN 0-87584-759-5, 1997.
7. Van der Steen, A.J..: Overview of recent supercomputers, tenth revised edition, Stichting NCF, ISBN 90-72910-03-6 also available:
   http://www.phys.uu.nl/~steen/web00/overview00.html,
8. Meuer, H.W., Strohmaier, E., Dongarra, J.J, and Simon, H.D. TOP500 Supercomputer sites, University of Mannheim, 2001, XML version available on
   http://www.hoise.com/primeur/analysis/top500, last accessed, November 2001,
9. Van der Steen, A.J..: Overview of recent supercomputers, tenth revised edition, machine descriptions in XML, enriched with TOP500 data,
   http://www.hoise.com/vmp/examples/ors/index.html, last accessed, November 2001.

Modeling Metadata-Enabled Information Retrieval

Manuel J. Fernández-Iglesias, Judith S. Rodríguez, Luis Anido, Juan Santos,
Manuel Caeiro, and Martin Llamas

Grupo de Ingeniería de Sistemas Telemáticos.
Departamento de Ingeniería Telemática.
Universidade de Vigo, Spain
manolo@det.uvigo.es,
WWW:http://www-gist.det.uvigo.es/

Abstract. We introduce a proposal to theoretically characterize Information Retrieval (IR) supporting metadata. The proposed model has its foundation in a classical approach to IR, namely vector models. These models are simple and implementations are fast, their term-weighting approach improve retrieval performance, allow partial matching, and support document ranking. The proposed characterization includes document and query representations, support for typical IR-related activities like stemming, stoplist application or dictionary transformations, and a framework for similarity calculation and document ranking. The classical vector model is integrated as a particular case in the new proposal.

1 Introduction

Metadata are traditionally defined as *data about data, information about information*. Metadata facilitate querying over documents by providing semantic tags. Interpreted this way, they could be introduced in the IR domain to improve the relevance of the documents retrieved as a response to a user query. IR systems may be provided with this higher level, structured information about document contents for users to query the system not only about existing document data, but also about these higher level descriptions. This poses the question of how to utilize the provided semantic tags.

Besides, due to the Web's inherent redundancy, lack of structure and the high percentage of unstable data, IR systems for the Web (e.g. search engines) is and will be a hot research topic in the next years. Although some promising results are already available [3] [4], as typically happens in other emerging fields metadata-based IR is far from being stable.

In this line, we feel that a theoretical characterization of metadata-enabled IR will contribute to a solid foundation of the field, which will facilitate the development of tools and methods to analyze existing systems and new proposals, and eventually improve the performance and features of IR systems. This characterization should be abstract enough to be fully independent from metadata models.

P.M.A. Sloot et al. (Eds.): ICCS 2002, LNCS 2329, pp. 78–87, 2002.

In this paper we propose a framework to represent metadata-enabled Information Retrieval that supports relevance calculation. Our proposal is based on the classical vector model for information retrieval [7]. Despite its simplicity, it is generally accepted that this model is either superior or at least as good as known alternatives (c.f., for example, [2], pages 30 & 34).

2   Basic Concepts and Notation

$\mathbb{N}$ and $\mathbb{R}$ represent respectively the set of natural numbers and the set of real numbers. A matrix over $\mathbb{R}$ is a rectangular array of real numbers. These real numbers are named *matrix components*. The symbol $\mathcal{M}_{m \times n}(\mathbb{R})$ denotes the collection of all $m \times n$ matrices over $\mathbb{R}$. The symbol $\mathcal{M}(\mathbb{R})$ denotes the collection of all matrices over $\mathbb{R}$.

Matrices are denoted by capital letters, and $A = [a_{ij}]$ means that the element in the i-th row and j-th column of the matrix A equals a_{ij}. Operations on matrices and particular types of matrices are defined and denoted as usual (see for example [8]). $A^T = [a_{ji}]$ represents the transpose of matrix A. This notation is extended to vectors (i.e. row vector x^T is the transpose of x).

We define the *unit vector* $u_j^n = (u_i) \in \mathbb{R}^n$ as an n-dimensional vector such that

$$u_i = \begin{cases} 1 \ j = i \\ 0 \ j \neq i \ 1 \leq i \leq n \end{cases}$$

We define $u^n = \sum_{i=1}^{n} u_i^n$. We omit the superscript n if the dimension of these vectors is clear from the context.

Matrices are associated with particular types of functions called linear transformations. We associate to $M \in \mathcal{M}_{m \times n}(\mathbb{R})$ the function $T_M : \mathbb{R}^n \to \mathbb{R}^m$ defined by $T_M(x) = Mx$ for all $x \in \mathbb{R}^n$.

A subset $V \subset \mathbb{R}^n$ is called a *subspace of* $\mathbb{R}^n$ if $0 \in V$, and V is closed under vector addition and scalar multiplication. Vectors $x_1, \ldots, x_n$ belonging to a subspace $V \subset \mathbb{R}^n$ are said to form a basis of V if every vector in V is a linear combination of the x_i and the x_i are linearly independent. Every vector in V is expressible as a linear combination of the x_i. Vectors u_i^n, $i = 1, \ldots n$ form a basis for $\mathbb{R}^n$. We name this basis the *canonical basis* of $\mathbb{R}^n$.

3   The Classical Vector Model for Information Retrieval

Basically, an Information Retrieval model (IR model) defines a document and query representation and formalizes users' information needs, that is, it states how documents should be stored and managed, and which information about documents should be kept to efficiently fetch the most relevant documents as a response to a user query. An IR model also defines how and from which information the relevance of a given document should be estimated. In our context, classical IR models are those that do not support metadata information.

Definition 1 (IR model). *Given a collection of documents, an Information Retrieval model is a quadruple* $\mathcal{IR} = (D, Q, \mathcal{F}, R(q_i, d_j))$ *where*

1. *D is the set of document representations.*
2. *Q is the set of queries (i.e. representations of users' information needs).*
3. *$\mathcal{F}$ is a framework for modeling document representations, queries and their relationships.*
4. *$R(q_i, d_j)$ is a ranking function that returns a real number for every query $q_i \in Q$ and document representation $d_j \in D$. This function defines an ordering among the documents with respect to the query q_i.*

We have chosen the vector model as our foundation because it is simple and implementations are fast, its term-weighting approach improves retrieval performance, allows partial matching, and permits document ranking. The vector model assigns positive real numbers (*weights*) to index terms in queries and documents. Weights will be eventually used to compute the degree of similarity between documents stored in the IR system and user queries. By sorting the retrieved documents according to their similarity, the vector model [2] supports partial (query) matching.

Definition 2 (Vector Model). *Let t be the number of index terms in the system and k_i a generic index term. Let d_j be a generic document, and q a user query. For the vector model, we associate a positive real number $w_{i,j}$ to each pair (k_i, d_j) called* weight of term k_i in document d_j. *We also associate a positive real number $w_{i,q}$ to each pair (k_i, q) called* weight of term k_i in query q.
    The document vector for d_j is defined as $\boldsymbol{d}_j^T = (w_{1,j}, \ldots w_{t,j})$, and the query vector for q as $\boldsymbol{q}^T = (w_{1,q}, \ldots w_{t,q})$

Queries trigger the calculation of a similarity function $sim(\boldsymbol{q}, \boldsymbol{d}_j)$ that tries to estimate the degree of similarity between the query and the documents in the collection. This similarity is estimated from a measure of the distance between the two vectors, and is typically based on the scalar product (e.g. the cosine of the angle between $\boldsymbol{d}_j$ and $\boldsymbol{q}$). The results from the query, a set of relevant documents, is ranked according to their similarity and presented to the user.

The $w_{i,j}$ ($w_{i,q}$) are defined from a set of statistic functions whose domain is $\mathcal{IR}$ and return a real number. We can use statistics computed for single terms in the database (e.g. $idf_i = \log(N/n_i)$, where N is the number of documents in the system, and n_i the number of documents where term i appears, describes the discrimination power of term i), computed por single documents (e.g. the number of distinct terms in document j) or global statistics (e.g. N).

The best known term-weighting schemes for the vector model (*tf-idf*) are given by (variations of):

$$w_{i,j} = \frac{n_{ij}}{max_l\{n_{lj}\}} \times idf_i$$

where n_{ij} is the frequency of index term k_i in document d_j[2].

4   The Metadata Matrix Model for Information Retrieval

We have to define a representation for documents and queries, a framework to define the relationships among documents, queries and representations, and a ranking function. We start from a collection of t metadata documents. In our context, metadata documents are documents composed by a set of nodes, where each node has a tag and a content.

Definition 3 (Metadata document). *Let T and M be two sets known respectively as* term dictionary *and* tag dictionary. *A metadata document is composed by a set of pairs (m, B_m), where m is a tag from M and B_m a list of terms from T.*

In the definition above, B_m is the list of terms bound to tag m. This structure results, for example, from the parsing of an XML document (DOM tree [6]). However, different representations may be produced from a single XML document, depending on the model selected to label nodes or the value equality chosen.

```
Document 1                        Document 2
<List>                            <List>
 <Title>things to do </Title>      <Item><Abstract>do</Abstract></Item>
 <Item>read</Item>                  <Item>write</Item>
 <Item>write</Item>                </List>
 <Item>read</Item>
</List>


(List.Title, things to do)        (List.Item.Abstract, do)
(List.Item, read write read)      (List.Item, write)
```

Fig. 1. Example metadata documents

Example 1. In figure 1 appear two XML documents belonging to a simple collection. In a given application they may be converted into the metadata documents appearing below the corresponding XML documents in the figure. In this example, tags correspond to *Label Path Expressions* [1].

4.1   Metadata Documents in a Matrix Model

In our proposal, documents and queries will be represented as matrices:

Definition 4 (Metadata Matrix Model). *Let t be the size of the term dictionary and k_i a generic index term. Let m be the size of the tag dictionary and l_j a generic tag. Let d_k be a generic document, and q a user query. For the matrix model, we associate a positive real number $w_{i,j,k}$ to each triplet (k_i, l_j, d_k)*

called weight of term k_i in document d_k when bound to tag l_j. *We also associate a positive real number* $w_{i,j,q}$ *to each triplet* (k_i, l_j, q) *called* weight of term k_i in query q when bound to tag l_j.

The document matrix for d_k *is defined as* $D_k = [w_{i,j,k}]$, *and the query matrix for* q *as* $Q = [w_{i,j,q}]$.

From the classical point of view, this model is based on a $t \times m$-dimensional real subspace, where both queries and documents are elements from $\mathcal{M}_{t \times m}(\mathbb{R})$. For a given document matrix $D_k = (d_{k,1}, \ldots, d_{k,m})$, each column is a vector $d_{k,j}$ containing the weights of the terms bound to tag j in the corresponding document . The same applies to user queries: a query matrix $Q = (q_1, \ldots, q_m)$ is composed by m vectors $q_j^T = (w_{1,j,q}, \ldots w_{t,j,q})$, and the $w_{i,j,q}$ represent how much the user is interested in documents containing term i bound to tag j.

We can see that the proposed matrix model extends the classical vector model: the vector model is a matrix model where $m = 1$.

Example 2. At the bottom of figure 1 appear two metadata documents belonging to a simple collection (c.f. example 1). Let us assume that $w_{i,j,k} = n_{ijk}$, where n_{ijk} is the raw frequency of term i in document k for tag j. Some examples about the information managed for this IR system are presented below:

- Dictionaries: $T = \{things, to, do, read, write\}$; $t = 5$;
  $M = \{List.Title, List.Item, List.Item.Abstract\}$; $m = 3$
- Example queries: $q_a = \{$Retrieve docs. containing *read* at tag *List.Item* $\}$;
  $q_b = \{$Retrieve docs. containing *write* or *do* in any tag at or below *List.Item* $\}$
- Matrices: $D_1 = ((1, 1, 1, 0, 0)^T, (0, 0, 0, 2, 1)^T, (0, 0, 0, 0, 0)^T)$
  $D_2 = ((0, 0, 0, 0, 0)^T, (0, 0, 0, 0, 1)^T, (0, 0, 1, 0, 0)^T)$
  $Q_a = ((0, 0, 0, 0, 0)^T, (0, 0, 0, 1, 0)^T, (0, 0, 0, 0, 0)^T)$
  $Q_b = ((0, 0, 0, 0, 0)^T, (0, 0, 1, 0, 1)^T, (0, 0, 1, 0, 1)^T)$

4.2   Documents as Linear Transformations

Document representations are $t \times m$ matrices. Matrix algebra states that these matrices can be seen as the representation of a linear transformation between two subspaces. These subspaces will be defined as the *tag subspace* and *term subspace*.

Definition 5 (Tag subspace, content profile). *Let m be the number of distinct metadata tags in the collection. We define $V_l \subset \mathbb{R}^m$ as the subspace whose vectors define content profiles. The i-th coordinate of a vector $v \in V_l$ defines the weight of tag l_i in the corresponding content profile. Null coordinates in v represent tags not participating in the profile.*

A canonical base for V_l is composed by the set of content profiles $u_j^m, 1 \le j \le m$. Each u_j^m represents a profile with a single tag. This canonical base represents the tag dictionary.

Content profiles represent the relative weights of metadata tags in a given context. For a given collection, they define how relevant is a given tag or set of tags with respect to the others, and characterize the tag dictionary.

Definition 6 (Term subspace, content suit). *Let t be the number of distinct index terms in the collection. We define $V_k \subset \mathbb{R}^t$ as the subspace whose vectors define content suits. The i-th coordinate of a vector $v \in V_k$ defines the weight of term k_i in the corresponding content suit. Null coordinates in v represent terms not participating in the suit.*

A canonical base for V_k is composed by the set of content suits u_j^t, $1 \le j \le t$. Each u_j^t represents a content suit with a single term. This canonical base represents the term dictionary.

Content suits represent sets of index terms and their relative weight for a given context. They characterize the term dictionary for a collection. Then, we can associate a linear transformation $T_{D_k} : V_l \to V_k$ to a document representation $D_k \in \mathcal{M}_{t \times m}(\mathbb{R})$. This transformation assigns content suits to content profiles, that is, given a content profile, the linear transformation associated to a document representation D_k returns the content suit corresponding to that content profile for that document.

As stated in definition 4, a document matrix defines the weights of terms when bound to a given tag. Given a content profile defining the relative relevance of tags, a document matrix defines the corresponding content according to that content profile.

Additionally, for a given collection content profiles can be used to obtain new representations of documents where metadata information is *filtered out* according to the profiles. As a consequence, this interpretation of document matrices as linear transformations formalizes the generation of classical vector models from matrix ones.

Example 3. Let us consider matrix D_1 in example 2 (c.f. also figure 1). This matrix corresponds to the linear transformation $D_1 x = y$ where $x \in V_l$ and $y \in V_k$. Content profile $x^T = (1\ 0\ 0)$ defines a context where we are only interested in terms bound to tag *List.Title*. The corresponding content suit is $y^T = (1\ 1\ 1\ 0\ 0)$, which represents the raw frequencies of the terms bound to that tag in the original document.

For content profile $x^T = (1\ 1\ 1)$ (i. e. a profile where all tags are equally relevant), we get $y^T = (1\ 1\ 1\ 2\ 1)$, that is, a content suit having all terms in the original document. Term weights in this suit correspond to the raw frequencies of index terms in the original document.

Note that $x^T = (1\ 0\ 0)$, when applied to all documents in the collection, will generate a classical vector model where documents will only retain information bound to tag *List.Title*, whereas $x^T = (1\ 1\ 1)$ will generate a classical vector model where documents have the same index terms and all structure provided by metadata is lost.

4.3   Queries and Ranking as Linear Transformations

Content profiles and content suits also apply to queries. Query matrices can also be considered linear transformations from the profile subspace to the suit subspace. In this case, a content profile represents at what extent users are interested in terms bound to a given set of tags, whereas content suits represent the relative relevance of a set of index terms for a given query.

On the other side, in section 3 we defined a ranking function that returns a real number for each query $q_i \in Q$ and document representation $d_j \in D$. This function defines an ordering among the documents w.r.t. the query q_i. Besides, queries can be interpreted as functions that take as a parameter a query matrix and return a ranking vector. This idea can also be formalized using subspaces and linear transformations:

Definition 7 (Ranking subspace, ranking). *Let N be the size of the collection. We define $V_r \subset \mathbb{R}^N$ as the subspace where vectors in V_r define rankings. The i-th coordinate of a vector $v \in V_r$ defines the relative position of document d_i in a ranking of documents. Null coordinates in v represent documents not participating in the ranking.*

A canonical base for V_r is composed by the set of rankings u_j^N, $1 \leq j \leq N$. Each u_j^N represents a ranking with a single document.

Example 4. For the collection in example 2, ranking $v^T = (1, 0)$ represents a ranking for query q_a whereas $v^T = (0.1714, 0.5)$ is a ranking for query q_b (c. f. section 5)

5   Relevance analysis in a matrix model

Relevance analysis will based on the same principles as the classical vector model. In our case, queries trigger the calculation of a similarity function $sim(Q, D_j)$ to estimate the degree of similarity between a query and the documents in the collection.

Classical vector-based similarity functions are now calculated from document and query matrices. Similarity is based on a measure of the distance between a document matrix and a query matrix in the corresponding $t \times m$ subspace. Retrieved documents are ranked according to the corresponding results.

The statistics available to compute similarities can now be enriched with new ones that take into account metadata information, like statistics computed for single tags (e.g. the number of documents containing tag j), computed for single tags in a given document (e.g. the number of terms bound to tag j in document k), or computed for single tags for a given term (e.g. $idf_{ij} = \log(N_j/n_{ij})$ represents the discrimination power of term i for contents bound to tag j). Obviously, other combinations are possible. As far as we know, the evaluation of useful metadata-dependent statistics is an open problem.

Existing results for the vector model can be translated to the matrix model if we note that queries and documents are elements from a $t \times m$ subspace.

Similarity functions based on the scalar product of vectors are typically used in the classical vector model. For example

$$sim(q, d) = \frac{< q, d >}{|q||d|} = \frac{q_1 d_1 + \ldots + q_n d_n}{|q||d|}$$

measures the similarity of document d with respect to query q as the cosine of the angle between d and q.

For the matrix model, we can also define a similarity function having the properties of a scalar product:

$$sim(Q, D) = \frac{< Q, D >}{|Q||D|} = \frac{tr(Q^T D)}{|Q||D|} = \frac{tr(Q^D T)}{|Q||D|} = \frac{\sum_i \sum_j q_{ij} d_{ij}}{|Q||D|}$$

where tr represents the trace of a matrix. In the new scenery, this approach has further advantages due to the properties of the scalar product combined with those of the product of matrices.

Proposition 1. *Let D and Q be respectivelty a document matrix and a query matrix. Let T_T be a transformation in the term subspace and T_M a transformation in the tag subspace. Then*

$$< Q, DT_T > = < QT_T^T, D > \; ; \; < Q, T_M D > = < T_M^T Q, D >$$
$$< Q, T_{M1} T_{M2} D T_{T1} T_{T2} > = < T_{M2}^T T_{M1}^T Q T_{T2}^T T_{T1}^T, D >$$

The proof is based on the algebraic manipulation of the expressions above.

Property 1 establishes a relation between transformations on queries and documents insofar similarity calculation is concerned, which opens the door to the methodical characterization of transformations and their influence on the computed similarity in a metadata-enabled world. Besides, as the vector model is a particular case of the matrix model, available results for the former can be systematically adapted to the latter.

Example 5. Let us take the collection in figure 1. Transforming documents using $u^T = (1\ 1\ 1)$ generates classical vector models where all metadata information is lost. For each index term k_i, the resulting vector $d_k = D_k u$ has as weights $w_{i,k} = \sum_j w_{i,j,k}$.

Assuming in this example that weights correspond to raw frequencies, all text bound to any tag will be equally relevant. After this transformation being applied to all documents, we get a classical IR vector model where metadata-specific information will not be considered for relevance calculation. Vectors for the transformed model are

$$d_1^T = (1, 1, 1, 2, 1), \; d_2^T = (0, 0, 1, 0, 1), \; q_a^T = (0, 0, 0, 1, 0), \; q_b^T = (0, 0, 2, 0, 2)$$

To calculate the similarity between documents and queries we can now apply known results for this classical model. Let $sim(d_j, q) = \cos(\angle w_j w_q)$, where

$w_q = q$, and $w_{ij} = n_{ij}/n_i$, n_i being the number of documents where term i appears.

For query $q = q_b$ we have $n_{things} = n_{to} = n_{read} = 1$, $n_{do} = n_{write} = 2$, and therefore $n_i^T = (1\ 1\ 2\ 1\ 2)$. Then, $w_1^T = (1\ 1\ 0.5\ 2\ 0.5)$, $w_2^T = (0\ 0\ 0.5\ 0\ 0.5)$, and the corresponding values for the similarity function are $sim(d_1, q_b) = 0.277$, $sim(d_2, q_b) = 1$.

We conclude that d_2 is more relevant to the query q_b. Note that all metadata information was lost, and d_2 is composed only by the terms in the query.

Example 6. A unit vector u_j generates classical IR systems whose documents contain only the information bound to tag j. For a given tag dictionary T of size m, we can generate m classical IR systems projecting the original IR system using u_j, $j = 1 \ldots m$. Then, m similarity results can be calculated for a given query. We have to select a procedure to combine these values into a single one for ranking purposes. For this example, the procedure selected is based on the similarity estimation for the extended boolean model[2]. We will assume that the query string is composed by a set of *or*-ed subqueries, each bound to a tag.

For the extended boolean model $sim_{bool-ext} = \frac{1}{m'}\sqrt{\sum_{m=1}^{m'} sim_m^2}$, where m' is the number of non-null similarities. For the query $q = q_b$ we have:

Projection	d_1^T	q^T	$sim_i(d_1, q)$
u_1 *List.Title*	$(1\ 1\ 1\ 0\ 0)$	$(0\ 0\ 0\ 0\ 0)$	0
u_2 *List.Item*	$(0\ 0\ 0\ 2\ 1)$	$(0\ 0\ 1\ 0\ 1)$	0.1714
u_3 *List.Item.Abstract*	$(0\ 0\ 0\ 0\ 0)$	$(0\ 0\ 1\ 0\ 1)$	0
Document 1: $sim_t(d_1, q) = 0.1714$			
Projection	d_2^T	q^T	$sim_i(d_2, q)$
u_1 *List.Title*	$(0\ 0\ 0\ 0\ 0)$	$(0\ 0\ 0\ 0\ 0)$	0
u_2 *List.Item*	$(0\ 0\ 0\ 0\ 1)$	$(0\ 0\ 1\ 0\ 1)$	0.707
u_3 *List.Item.Abstract*	$(0\ 0\ 1\ 0\ 0)$	$(0\ 0\ 1\ 0\ 1)$	0.707
Document 2: $sim_t(d_2, q) = 0.5$			

If we compare the results above with those from example 5, we see that the relevance of both documents to the query q_b decreases. This is due to the role played by metadata. We see that the term *do* is not bound to tag *List.Item* in the first document. For the second document we see that, although both query terms are relevant to the query, they are bound to different tags.

6   Concluding remarks

The need for efficient information retrieval an management tools for the Web, and the introduction of advanced markup, determined the evolution of Information Retrieval techniques to take into account metadata. As a consequence, research is necessary to study the real contribution of metadata to the performance of IR systems. A suitable theoretical framework to formally characterize the different aspects of this problem may be helpful.

In this paper we have introduced a matrix-based characterization for meta-data-based IR where documents, user queries, and document and term transformations are modeled as matrices. This proposal is independent of the metadata model as long as metadata documents can be represented as a collection of tag-value pairs. Therefore, it can be easily applied to several metadata and tag (meta)language proposals, like XML or XML-derived languages. Furthermore, it seamlessly integrates previous results from classical IR.

The proposal can also be seen as an extension of the vector model for IR. It is complete in the sense that it provides a document and query representation framework and a ranking function. We hope that this characterization will contribute to construct a solid foundation for metadata-enabled IR.

Presently, we are using the proposal discussed in this paper to evaluate different relevance calculation methods for metadata IR using DelfosnetX[5]. Indeed, the first aim of DelfosnetX was to provide a workbench to validate our proposal, so it was designed to easily test the performance of different configurations for the matrix model. The system automatically fetches and (re)calculates a comprehensive set of statistics to be used to compute and test different similarity functions and relevance analysis methods. This approach also allows to study the relative performance of classical and metadata-oriented IR systems. An Application Programmer Interface (API) is also provided to easily customize DelfosnetX for particular applications.

References

1. S. Abiteboul, D. Quass, J. McHugh, J. Widom, and J. Wiener. The lorel query langauge for semistructured data. *International Journal of Digital Libraries*, 1(1):68 – 88, 1997.
2. R. Baeza-Yates and B. Ribeiro-Neto. *Modern Information Retrieval.* Addison-Wesley, 1999.
3. F. J. Burkowski. An algebra for hierarchically organized text-dominated databases. *Information Processing & Management*, 28(3):333–348, 1992.
4. C. L. A. Clarke, G. V. Cormack, and F. J. Burkowski. An algebra for structured text search and a framework for its implementation. *The Computer Journal*, 38(1):43–56, 1995.
5. M. Fernández, P. Pavón, J. Rodríguez, L. Anido, and M. Llamas. Delfosnetx: A workbench for XML-based information retrieval systems. In *Procs. of the 7th International Symposium of String Processing and Information Retrieval*, pages 87–95. IEEE Comp. Soc. Press, 2000.
6. A. L. Hors, P. L. Hégaret, L. Wood, G. Nicol, J. Robie, M. Champion, and S. Byrne, editors. *Document Object Model Level 2 Core Specification.* W3 Consortium, 1998. W3C Recommendation.
7. G. Salton and M. E. Lesk. Computer evaluation of indexing and text processing. *Journal of the ACM*, 15(1):8–36, 1968.
8. H. Schneider and G. P. Barker. *Matrices and Linear Algebra.* Dover Books on Advanced Mathematics. Dover Pubns, 2nd edition, 1989.

Spontaneous Branching in a Polyp Oriented Model of Stony Coral Growth

Roeland Merks, Alfons Hoekstra, Jaap Kaandorp, and Peter Sloot

University of Amsterdam, Faculty of Science,
Section of Computational Science,
Kruislaan 403, 1098 SJ Amsterdam, The Netherlands
{roel,alfons,jaapk,sloot}@science.uva.nl,
WWW home page: http://www.science.uva.nl/research/scs/

Abstract. A three-dimensional model of diffusion limited coral growth
is introduced. As opposed to previous models, in this model we take a
"polyp oriented" approach. Here, coral morphogenesis is the result of the
collective behaviour of the individual coral polyps. In the polyp oriented
model, branching occurs spontaneously, as opposed to previous models
in which an explicit rule was responsible for branching. We discuss the
mechanism of branching in our model. Also, the effects of polyp spacing
on the coral morphology are studied.

1   Introduction

Stony corals are colonial reef-building organisms, building an external calcium
skeleton. The individual coral polyps reside in skeletal structures called corallites,
and form a layer of living tissue surrounding the skeleton. *Skeletal growth* occurs
as the polyps deposit new skeleton on top of the old skeleton (reviewed in [1]). In
this process, the corallites move outwards, leaving behind a trail of coral skeleton.
As the coral surface expands due to skeletal growth, the corallites move away
from each other and divide, filling up the resulting space, a process called *tissue
growth* (reviewed in [2]).

This seemingly simple growth process generates a wide range of colony morphologies; examples are spherical, thinly branched and plate-like forms. These
morphologies are species specific, but also show high variability within one
species. This intraspecific variability is caused by environmental parameters,
such as light availability and the amount of water flow. In previous model studies, the effect of these environmental parameters on coral morphology has been
assessed (light availability, [3, 4] and flow conditions, [5, 6]). In the present study,
we concentrate on morphologic differences between species; we keep the environmental parameters constant, and vary the properties of the individual polyps
instead.

Our current work is based on the three-dimensional radiate accretive growth
model [6, 7]. In this model, coral growth is simulated by the accretion of discrete
layers that are represented by a triangular mesh. The "coral" grows in a simulated fluid [6], in which nutrients are dispersed by advection and diffusion. The

P.M.A. Sloot et al. (Eds.): ICCS 2002, LNCS 2329, pp. 88–96, 2002.

local thickness of the accreted layer depends on the flux of nutrients to the coral surface, which is a nutrient sink.

In this paper, we use a "polyp-based model", in which we model coral growth starting from the individual polyps. Coral growth is seen as the collective result of a process taking place in individual polyps. Each polyp takes up resources, deposits skeleton, buds off new polyps and dies. With this "polyp-based" approach, we have been able to make a number of simplifications relative to the model of Kaandorp et al. [6]. The most important of these is the omission of the socalled *curvature rule*. In the previous models this rule was necessary to generate branching growth forms. Here, we show that branching occurs without the *curvature rule* if we model the resource fluxes for each corallite invidually. In this way, a curvature effect comes out naturally. Relative to a plain surface, at a convex surface the corallites are fanned out and less of the resources near them are taken up by their competitors. At a concave surface polyps point towards each other and interfere strongly.

We concentrate exclusively on skeletal growth. In this paper we assume that the main rate limiting factor for skeletal growth is the availability of diffusing resources such as organic food or calcium ions. The rate of tissue growth is entirely dictated by the rate of skeletal growth, as it generates space in which new corallites are inserted.

In the remainder of this paper, we first present the "polyp oriented" model, and discuss some of the differences with the previous models. Then, we discuss the branching mechanism as it occurs in our model. After that, a parameter study is presented. We show a correlation between the spacing of the model polyps and the branch size in the final morphology.

2   The Model

We model the growth of the coral skeleton, which takes place at the living surface of the coral colony, using a triangular mesh. Diffusion of resources and other chemicals is modelled using the moment propagation method [8–10].

In short, a simulation step proceeds as follows. We start with the dispersion of resources. Since we use a grid based method, we first map the triangular mesh onto a three-dimensional grid using a voxelization method [11]. This grid-based representation of the mesh is then filled to generate a solid obstacle. Resources are dispersed from the top plane of the simulation box, and absorbed by the "polyps"[1]. As soon as the influx of "resource" at the top plane balances the absorption of resources by the "polyps" (that is, if the resource field is stable), the resource flux ϕ_i at each "polyp" p_i is measured. Now, we move back to the triangular mesh representation for an *accretive growth* step. On top of the previous layer, a new layer of skeleton is built, whose local thickness l_i depends on the growth function $g(\phi_i)$. In the next sections these steps will be described in more detail.

---

[1] Throughout this text, the name of an entity e will be put in quotes if it is used to denote "model of e".

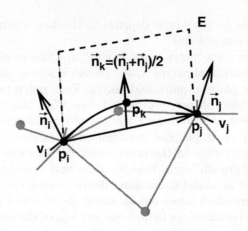

Fig. 1. Interpolation of the coral surface for "polyp" insertion. A third order polynomial f is constructed, running tangential to the coral surface at p_i and p_j. The new "polyp" p_k is constructed at $p_k = (\frac{1}{2}, f(\frac{1}{2}), 0)$, relative to an orthonomal basis B where $p_i = (0,0,0)$, $p_j = (1,0,0)$ and $z = 0$ at plane E

2.1   Accretive Growth

As in the previous models of radiate accretive growth, the coral surface is described by a triangular mesh. In the present model, we model the individual coral polyps as the vertices p_i of the triangular mesh. In contrast to the previous models, the triangles do not have a biological meaning. Each "polyp" p_i consists of a coordinate x_i, y_i, z_i in three-dimensional space and of one or more scalars n_i. In the present model, we use a single scalar n_i which contains the resource influx ϕ_i.

In each growth step, a longitudinal element l_i is placed on top of each "polyp" p_i along the mean normal vector of the surrounding triangles. The length of l_i is determined by the growth function g. In the present work, this is a linear function of the resource flux ϕ_i. The "polyp" p_i is then moved to the end of the longitudinal element l_i, as if it pushes itself upwards by excreting calcium skeleton. In this process, the values of the scalars n_i are retained.

After an accretive growth step, the length of the link Λ_{ij} between polyp p_i and p_j may have changed, depending on the curvature of the coral surface. A "polyp" is inserted or removed accordingly. If the distance $|\Lambda_{ij}|$ between two "polyps" p_i and p_j drops below the fusion threshold θ_{FUSE}, the two polyps fuse. A new, fused "polyp" p_k is placed at the middle of the interpolypal link, and the value of the scalars n_k are set equal to the mean $n_k = (n_i + n_j)/2$. Then, the "polyps" p_i and p_j are removed.

If the distance between two "polyps" p_i and p_j rises above the insertion threshold θ_{INS}, a new polyp is inserted. The position of this polyp is determined using a third order interpolation of the coral surface (see fig. 1). As for "polyp" fusion, the values of the scalars is set to mean of the parent "polyps", $n_k = (n_i + n_j)/2$. Note that in case of "polyp" insertion, the parent "polyps" are retained in

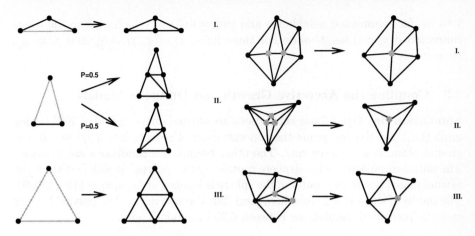

Fig. 2. Triangle insertion (left) and fusion (right) rules. Gray links are divided. Gray nodes are fused

the simulation. The insertion and fusion threshold are set relative to the initial mean distance between "polyps", as $\theta_{\text{FUSE}} = \frac{3}{4}\langle|\Lambda_{ij}|\rangle$ and $\theta_{\text{INS}} = \frac{3}{2}\langle|\Lambda_{ij}|\rangle$, where $\langle|\Lambda_{ij}|\rangle$ is the mean initial length of the interpolypal links.

As "polyps" are inserted and deleted from the simulation, new interpolypal connections are created and old ones are removed. In this way triangles are inserted and deleted from the mesh, according to the the triangle insertion and deletion rules. These rules are summarized in figure 2.

2.2   Resource Diffusion

After the completion of an accretive growth step, the diffusion and the absorption of the limiting resource is simulated. We solve the diffusion equations on a cubic lattice. For this, the triangular mesh describing the surface of the coral colony is mapped onto the cubic lattice, for which we use the triangle voxelization method by Huang et al [11]. The resulting hollow "shell" is filled using a fast, heuristic three-dimensional seed fill algorithm.

The diffusion equation is solved using the moment propagation method [8–10]. This method is used in conjunction with the lattice Boltzmann method (reviewed in [12]), and has been shown to solve the advection-diffusion equation [9]. At each lattice node, a scalar quantity $P(x,t)$ is redistributed over the neighbouring nodes, according to the probability f_i that a fluid particle moves with velocity c_i after collision,

$$P(x, t+1) = \sum_i \frac{f_i(x - c_i)P(x - c_i, t)}{\rho(x - c_i)}. \tag{1}$$

For a stationary fluid the velocity probabilities f_i are identical for each lattice site, and depend only on the length of the lattice link. In the 19-velocity lattice we use, there are three types of links: one velocity 0 link, six links of velocity

1 to the face-connected neighbours and twelve links of velocity $\sqrt{2}$ to the edge-connected diagonal neighbours. For these links, $f_i = \frac{1}{3}$, $f_i = \frac{1}{18}$ and $f_i = \frac{1}{36}$, respectively.

2.3 Coupling the Accretive Growth and Diffusion Models

Simulation Set Up All our simulations are carried out on a grid of 200^3 lattice units (l.u.). At the top plane the concentration of resources is fixed at 1.0, the ground plane is a resource sink. The other boundaries conditions are periodic. The initial condition is a hemisphere containing 81 "polyps" positioned at on the ground plane, where the center of the sphere is located at $(x, y, z) = (100, 100, 0)$. The radius of the initial hemisphere (and thus the spacing of the "polyps") varies over the presented simulations between 6.25 l.u. and 12.5 l.u.

Resource Absorption At each time step of the diffusion algorithm, the "polyps" absorb resource, thus acting as resource sinks. The position of each "polyp" in the simulation is mapped onto the cubic lattice. Since the "polyps" are part of the coral surface, the mapped positions have been marked as "solid" in the voxelisation algorithm. Therefore, the absorption takes place at a small distance l_{reach} from the "polyp" position, normal to the coral surface. Throughout the simulations presented in this paper $l_{\text{reach}} = 3$ l.u., thus ensuring that resources are absorbed at least from the first node marked as "fluid".

The diffusion algoritm is iterated until the influx of resources from the source plane balances the total outflux of resource into the "coral polyps" and the ground plane. We measure the resource influx ϕ_i for each "polyp" p_i individually.

As the "coral" grows towards the top plane, the resoure fluxes ϕ_i increase. In order to fix the growth rate, the resource absorption is normalised against the maximum resource flux, $\phi_i \rightarrow \phi_i/\phi_{\text{max}}$. The rationale behind this is that we fix the thickness of the boundary layer between the growing coral colony and the rest of the sea water, in which the nutrient concentration is assumed constant and constantly mixed.

3 Results

3.1 Coral Branching

In figure 3 [2] a possible outcome after 84 iterations of the diffusion limited accretive growth process is shown. The "coral" is grown upon an initial hemisphere of radius 6.25 l.u., giving a mean "interpolyp" distance of 1.9 l.u. Note that the growth form is strongly branched, although no special assumptions were made to get branching growth.

In previous models of accretive coral growth [6,13], an explicit *curvature rule* was necessary to enforce branching and branch splitting. This rule, simulating

---

[2] VRML files and MPEG movies of the objects shown in the figures have been made available at http://www.science.uva.nl/~roel/ICCS2002

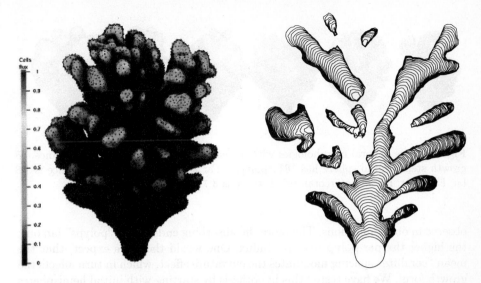

Fig. 3. Simulated branching coral after 84 growth steps. The size of the cubic grid is 200^3. The coral contains 18006 "polyps", which are shown as black dots. The mean "interpolypal" distance is 1.9 l.u. Left panel: outside view, color scale: normalized resource flux. Right panel: cross section, in which the successive growth layers are visible

the amount of contact with the environment, increases the growth rate at places of high curvature. If the curvature drops below some positive threshold value, growth is completely suppressed. This procedure leads to a branching instability.

Here we show that branching occurs spontaneously if the resource fluxes are resolved for the "polyps" individually. In our model, there is no explicit *curvature rule*. Yet, branching is explained by a similar process. Absorption takes place at a distance l_{reach} from the skeleton. This corresponds to real coral polyps, in which the mouth and tentacles are located at a small distance from the coral surface. At a convex location, the "polyps" fan out, thus getting better access to the resources than the "polyps" located at a flat or concave part of the "coral". At concave sites the resources are quickly depleted, because many "polyps" absorb resources from the same location and the resource replenishment is poor. Since the local deposition rate depends linearly on the resource flux, the local skeletal growth rate correlates positively with the local curvature which promotes branching. The precise correspondence between the curvature of the "coral" surface, the flux of resources into the "polyps" and the skeletal growth rate is the subject of ongoing research.

3.2   Branch Size Depends on "Polyp" Spacing

In section 3.1 we have shown that a curvature effect, caused by the "polyps" fanning out at convex sites, is responsible for the branching growth forms we

Fig. 4. Range of coral morphologies with increasing mean "intercorallite" distance. 85 growth steps. Initial sphere has 181 "polyps". Far left: "intercorallite" distance is 1.8 l.u. Far right: mean "intercorallite" distance is 3.7 l.u.

observe in our simulations. The more the absorbing ends of the "polyps" fan out, the higher the per polyp resource influx. One would therefore expect, that the mean "corallite" spacing modulates the curvature effect, which in turn affects the growth form. We have tested this hypothesis by starting with initial hemispheres of increasing radius, each of them having 181 "polyps". The initial spacing of the "polyps" dictates the mean "intercorallite" distance in the growth form, because the insertion and fusion threshold θ_{INS} and θ_{FUSE} are set relative to the intitial mean "intercorallite" distance.

In figure 4 a range of growth forms obtained after accretive growth steps is shown. At the far left, the mean "intercorallite" distance is 1.9 l.u., increasing to 3.7 l.u. at the far right. With increasingly wide "corallite" spacing, the branches become thicker and less numerous. Also, the branching split less quickly; the branch grows for a longer time before splitting up. The far right "coral" still branches like the "coral" at the left, though, which can be seen from figure 5. In this experiment, we have continued the growth of the "coral" with a mean "intercorallite" distance of 3.7 l.u. until we obtained 150 growth layers.

4   Conclusion and Discussion

Branching: the Polyp Competition Hypothesis In the polyp oriented model, branching occurs as an emergent property. This finding contrasts previous models of branching coral growth, where an explicit curvature rule was introduced to enforce branching [6, 7, 13].

Our model suggests a biological explanation for branching growth in stony corals. At convex sites, the polyps fan out, thus getting better access to the diffusing resources. At concave sites, the polyps point towards each other, thus interfering in the uptake of resources. In this way, a curvature effect comes out as a natural consequence of the competition between the polyps to take up resources from the water.

As the mean intercorallite distance increases (fig. 4), the coral branches less strongly. This observation fits well within the "polyp competition hypothesis"

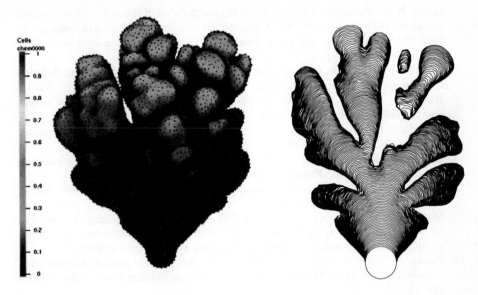

Cells
chem0000

1
0.9
0.8
0.7
0.6
0.5
0.4
0.3
0.2
0.1
0

Fig. 5. Coral with initial "intercorallite" distance of 3.7 l.u. (fig. 4, far right) with 150 growth steps

that we proposed above. The competition for resources becomes less strong for larger mean "intercorallite" distances.

Polyp Oriented Modelling We have developed a polyp oriented approach to modelling coral growth. In this polyp oriented approach, branching occurs as a natural consequence of the competition for resources between the "polyps" at a curved surface. The polyp based approach will be useful in understanding how genetic information drives the morphogenesis of coral colonies. Genetics affects coral morphology indirectly, by setting the properties of the individual coral polyps. We can only come to an understanding of the genetic guidance of coral morphogenesis, through a thorough understanding of how the properties of individual polyps shape the coral colony. In the study of corallite spacing we have made a first step in this direction.

References

1. Buddemeier, R., Kinzie III, R.: Coral growth. Oceanography and Marine Biology: An Annual Review **14** (1976) 183–225
2. Harrison, P., Wallace, C.: Reproduction, dispersal and recruitment of scleractinian corals. In Dubinsky, Z., ed.: Coral Reefs. Ecosystems of the world. Volume 25. Elsevier Science Publishers B.V., Amsterdam (1990) 133–207
3. Graus, R.R., Macintyre, I.G.: Light control of growth form in colonial reef corals: computer simulation. Science (1976)

4. Graus, R.R., Macintyre, I.G.: Variation in growth forms of the reef coral *montastrea annularis* (Ellis and Solander): A quantitative evaluation of growth response to light distribution using computer simulations. Smithsonian Contributions to the Marine Sciences (1982)
5. Kaandorp, J.A., Lowe, C.P., Frenkel, D., Sloot, P.M.A.: Effect of nutrient diffusion and flow on coral morphology. Phys. Rev. Lett. **77** (1996) 2328–2331
6. Kaandorp, J.A., Sloot, P.M.A.: Morphological models of radiate accretive growth and the influence of hydrodynamics. J. Theor. Biol. **209** (2001) 257–274
7. Kaandorp, J.A.: Analysis and synthesis of radiate accretive growth in 3 dimensions. J. Theor. Biol. **175** (1995) 39–55
8. Lowe, C.P., Frenkel, D.: The super long-time decay of velocity fluctuations in a two-dimensional fluid. Physica A **220** (1995) 251–260
9. Warren, P.B.: Electroviscous transport problems via lattice-Boltzmann. Int. J. Mod. Phys. C **8** (1997) 889–898
10. Merks, R.M.H., Hoekstra, A.G., Sloot, P.M.A.: The moment propagation method for advection-diffusion in the lattice Boltzmann method: validation and Péclet number limits. Submitted to Journal of Computational Physics (2001)
11. Huang, J., Yagel, R., Filippov, V., Kurzion, Y.: An accurate method for voxelizing polygon meshes. In: 1998 ACE/IEEE Symposium on Volume Visualization. (1998) 119–126
12. Chen, S., Doolen, G.D.: Lattice Boltzmann method for fluid flows. Annu. Rev. Fluid Mech. **30** (1998) 329–364
13. Kaandorp, J.A., Sloot, P.M.A.: Parallel simulation of accretive growth and form in three dimensions. Biosystems. **44** (1997) 181–192

Local Minimization Paradigm in Numerical Modelling of Foraminiferal Shells

Paweł Topa[1], Jarosław Tyszka[2]

[1] Institute of Computer Sciences, University of Mining and Metallurgy, al. Mickiewicza 30, 30-059 Kraków, Poland
email: topa@uci.agh.edu.pl
[2] Institute of Geological Sciences, Polish Academy of Sciences, Cracow Research Centre, ul. Senacka 1, 31-002 Kraków, Poland
email: ndtyszka@cyf-kr.edu.pl

Abstract. We present a new approach to modelling of foraminiferal shells. Previous models referred to fixed reference axes and neglected apertures, which play a crucial role in morphogenesis of shells. Our 2D preliminary model applies the moving reference system based on introducing of apertures and minimization of the local communication path (LCP). LCP defines the position of every final aperture. A formal description of simple analytical methods with some elements of randomness is given in this paper. Selected examples of simulated shells are figured.

1 Introduction

The emergence of forms in the growth process of foraminiferal shells is the essential problem in ontogenesis of these creatures. Foraminifera are single-celled organisms (protozoans) that construct shells. They inhabit all marine and marginal marine environments from very shallow to the deep ocean floor. Depending on the group, the shell may be made of organic compounds, sand grains and other particles cemented (agglutinated) together, or secreted from crystalline calcium carbonates. Foraminifera are a unique group of autonomous unicellulars reaching a size class (typically 0.1 mm to 1 mm in size, up to 20 cm) which is characteristic of small metacellular organisms [1]. Foraminiferal shells occur in an enormous variety of shapes (Fig.1). The majority of foraminifera are built of chambers, which are cavities containing the protoplasm surrounded (enveloped) by a firm wall [2]. The shape of a shell results from growth processes and depends on a chamber form, location and the type of aperture as well as the final chamber arrangement (Fig.1).

Modelling of foraminifera started very early with the classical work of Berger [3]. Nevertheless, so far simple regular morphologies (e.g. planispiral, helicoidal, uncoiled) that do not express complexity of foraminiferal shell patterns have only been simulated. It is therefore necessary to find a new alternative approach for further progress in theoretical morphospace of these organisms. The general aim is to gain a better understanding of foraminiferal shell morphology and its

P.M.A. Sloot et al. (Eds.): ICCS 2002, LNCS 2329, pp. 97–106, 2002.

incredible variability, and thus, to find essential geometric growth rules. Systematics of foraminifera is based on shell morphology (shell composition and microstructure, chamber form and arrangement, aperture type etc.). Theoretical shell morphology can help to verify some taxonomic rules. This study briefly summarizes previous foraminiferal models and presents a new approach to model basic morphologies.

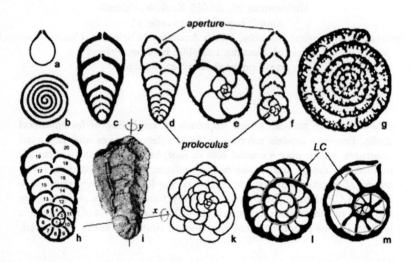

Fig. 1. Cross sections of foraminiferal shells; (a) *Fissurina*, unilocular shell; (b) *Spirillina*, non-septate bi-chambered; (c) *Pseudonodosaria*, multilocular uniserial shell; (d) *Bolivina*, biserial; (e) *Globorotalia* low helicoidal; (f) *Ammobaculites*, planispiral becoming uniserial; (g) *Reticulinella*, complex foraminifer with multiple apertures (not expressed here); (h, i) *Heterantyx*, planispiral switching to biserial; first 12 chambers rotate along the 'x' axis, then (chambers 13-20) along the 'y' axis, SEM view; (k) *Planorbulina* low helicoidal to random coiling; (l) *Miscellanea*, planispiral with the shortest global communication path (LC) via foramina (m) *Lenticulina*, planispiral shell with the longest global communication path (LC). Not to scale, actual size range from 0.1 to 1.3 mm.

2 Overview of foraminiferal modelling

Foraminiferal shells can be divided into 3 informal groups corresponding to the trend of increasing complexity (Fig. 1): (i) unilocular shells (Fig. 1a); (ii) multilocular shells (multichambered forms) (Fig. 1b-m); (iii) complex shells (multilocular shells with chambers divided into smaller chamberlets and/or having complex wall structure) (Fig. 1g).

Multilocular foraminifera enlarge in discrete growth processes of serial chamber additions successively united into the shell during ontogeny (Fig. 1c-m). The first usually globular chamber is called proloculus. The opening on every last chamber, through which pseudopodia (rhizopodia) extrude, is termed the aperture. Adjacent chambers are separated by septa but connected by foramina (Fig. 1c-m). Another type of foraminiferal morphology is produced by tubular chambers (Fig. 1b), which show a different growth pattern closely related to accretive growth of mollusks and/or branching system of plants (see [4], [5] for overview).

Over three decades ago, Berger [3] in his pioneering work already created the first theoretical morphospace of foraminifera. The theoretical morphospace has three parameters which define simple step-by-step rotation of a circle with a certain overlap and expansion of circle radius. This model simulates isometric growth (all three parameters are held constant through ontogeny) and is confined to planispiral and trochospiral shells composed of spherical chambers (Fig. 1e,l).

Signes et al. [6] designed a similar three dimensional theoretical morphogenetic model with two basic assumptions concerning foraminiferal growth: the shape of the chambers in the shell remains constant with growth, and the volume of each new chamber increases in a constant proportion to the pre-existing volume of the shell. This model produces isometric growth with coiling in a fixed-reference frame, which is similar to Raup's [7] and Berger's [3] models. Webb and Swan [8] partly extended this classical theoretical morphospace to uncoiled morphologies based on the different definition of the angle between successive chambers, which was not referred to the center of the shell.

Some other authors also simulated allometry of foraminiferal shells. Brasier [9] who produced a working morphospace model of foraminiferal form using four parameters, which actually correspond to Berger's parameters expanded by the degree of extension of growth along coiling axis and the degree of chamber compression and overlap. This last parameter includes allometry into the system. Another important difference is that chambers are not directly rotated but translated. An interesting model was presented by De Renzi ([10], [11]) who simulated allometric growth of some larger foraminifera based on the logistic model applied in polar coordinates.

3   New approach

All the mentioned simulations use theoretical axes, which have no morphogenetic or physiological meaning. Chambers (circles or spheres) are only rotated and translated along these artificial axes, which are fixed and serve as a reference line for growth process. Therefore, these models can simulate plane planispiral, trochospiral (helicospiral) or uniserial chamber arrangement, but cannot simulate more complex patterns. For instance, they cannot model gradual or abrupt changes of coiling axis that cause different chamber arrangements (Fig. 1f, h, i, k).

It seems to be clear that it is necessary to give up the fixed-reference frame based on theoretical coiling axes in favour of moving-reference system. In general, the moving reference model is based on simple principles of motion and step-wise growth [12]. At each growth step, the aperture migrates from its present position to a new position, according to locally defined rules [12]. Such models have been developed for simulation of ammonites. Okamoto [13] proposed a tube model for all types of shell coiling, including heteromorph forms with abrupt changes of coiling pattern. His approach integrates accretional growth of the aperture (opening of the shell) without defining any fixed coordinate system. Similar moving-reference frame is used in simulating radiate accretive growth of marine sessile organisms, such as corals and sponges, where growth axis is associated with the local maximum of growth (e.g. [14]). A comparable approach is also used in simulation of plant growth [15].

In order to define a moving-reference system for modelling of foraminifers, it is also reasonable to use apertures, which seem to be essential for location of a new growing chamber. Analyses of different modes of chamber growth in foraminifera suggest that the position of the aperture controls local chamber arrangement (see Fig. 1c-m). The problem is that although the processes associated with chamber formation are relatively well understood, we still know very little of how foraminiferal apertures are formed in nature.

Hottinger [16] discerned, the foraminifer must devise methods to shorten distances between the first and last compartments of its shell. Brasier [17] also analysed the energetics of protoplasmic pathways through the organism and concluded that foraminifers show the trend towards minimizing distance from the back of the first chamber (proloculus) to the most proximal aperture in the final chamber (Fig. 1l). He standardized this cumulative distance and named it MinLOC (minimum line of communication).

This 'rule' seems to be valid for many foraminiferal architectures, but not for all of them. *Lenticulina* and other coiled lagenids are curious exceptions to this rule because the foramina are located at the outer margin of the chamber [16] that creates the longest possible global line of communication (Fig. 1m). Nevertheless, local distances between adjacent foramina are in fact minimal. Thus, it is a paradox that the shortest distance between adjacent chambers creates the longest global line of communication between the first and last chambers. The conclusion is that the "the growth programme" of a foraminifer does not directly control cumulative arrangement of chambers. It does control formation of every chamber itself and optimisation (minimization) of the local distance between the last and the new-formed aperture of a new chamber. This minimization of the local communication path (LCP) seems to be more general (even if it is not the only rule) and should help to define simulated apertures.

4   Model

The model, we present in this paper, is significantly simplified. We have limited our study to a 2D case only. The modelling in 3D is more complicated and not necessary at this stage of work. The chambers are represented by regular circles. Although, real chambers can have very diverse shapes, the shells with spheroidal chambers are relatively frequently occurring. Additionally, we can apply simple analytical methods to compute the aperture location.

The modelling of development of foraminifera consists of discrete steps in which successive chambers are added to the forming shell. The position of the newly created chamber depends on the aperture location of the previous chamber and so called "vector of growth" (see Fig. 2). "Vector of growth" is attached at the aperture of the chamber c_i and pointed at the center of chamber c_{i+1}. Thus, in fact the aperture represents the moving reference point and the vector indicates the current direction of growth. The aperture of the new chamber is calculated according to the minimization principle. The distance between the apertures of the two successive chambers must be as short as possible. The apertures cannot be enclosed within any other already existing chamber. The most exterior aperture must be connected with the first chamber (proloculus) by a line of communication running through all the previous apertures (foramina). Fig. 3 presents a few hypothetical successive steps of simulation of ontogenetic shell development.

For convenience, calculations are made in circular co-ordinates. Instead of

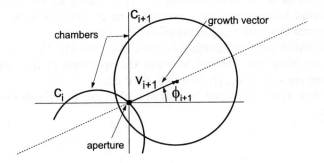

Fig. 2. Vector of growth

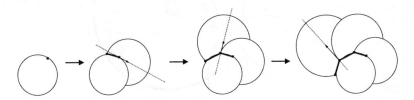

Fig. 3. Modelling the development of foraminferal shell

euclidian x,y coordinators we use the length of vector $|v_{i+1}|$ and angle ϕ_{i+1} (see Fig. 2).

We define the two parameters, which control the development of shell:

1. GF — chamber growth ratio: the ratio between the sizes (radius) of two successive chambers,
2. V (Φ, R_v) — vector of growth (Φ and R_v stand for it polar coordinates).

All the previous models assume invariance of their parameters during the modelling of the shell development. In our model, the foregoing parameters do not have to be constant and they can fluctuate within the ranges, we specify. The chamber growth ratio and growth vector are chosen at random within the given range. In such a way we can model the natural susceptibility of some species to forming a specific shell pattern. Specifying the exact values instead of the ranges of fluctuations, let us partly track models delivered from the fixed reference system.

Our algorithm works as follows:
Size and position of the aperture of the first chamber are set arbitrarily as the starting parameters of the simulation. In loop, the following operations are performed:

1. the position of the new chamber is calculated:
   – the radius of the new chamber (r_{i+1}) fluctuate within the given range adjacent to the mean value estimated on real foraminifera.
   – the growth vector v_{i+1} is calculated at random.
2. the position of the new aperture is calculated (see Fig 4):
   – we minimize the distance between the apertures of the two successive chambers - $min(|u_{i+1}|)$ (see appendix A),
   – the new aperture cannot be placed within any previously created chamber.

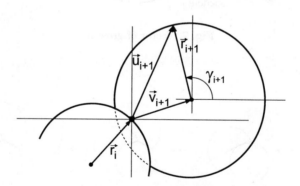

Fig. 4. Calculating the aperture location

The area, in which the aperture cannot be located, is determined by calculating the points, in which the circle of the new chamber is crossing over all the previous chambers.

The simplifications we have made, let us solve the minimization problem in an analytical way. Detailed description of the used mathematical methods can be found in Appendix A.

5 Results

Fig 5 presents the forms generated by using our models. The three first pictures have been supplemented by the numbering of successive chambers and the communication line connecting its apertures. This line usually crosses every previous chamber, which is another simplification, not affecting our general model.

Simulated shells show either relatively stable growth pattern (Fig. 5a, b, d),

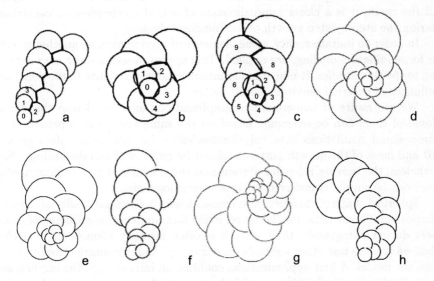

Fig. 5. Shells generated by computer; (a,f) biserial form; (b,d) spiral form; (c, e) initially spiral, after 7 chambers changes to biserial; (g,h) alternately biserial and spiral.

or abrubtly change the mode of coiling (Fig. 5c, e, g, h). Most of the simulated shell forms do have real counterparts (compare Fig 1). Some forms mimic abnormal shells usually related to environmental stress (Fig. 5f, h). Another set of generated shells is very chaotic and flexible to growth switch-overs (Fig. 5g). These shells may resemble some real attached (adherent) foraminifera, as well as some irregular (in 3D) agglutinated foraminifera.

We observe that ranges of fluctuations of the growth vector $V(\Phi, R_v)$ are crucial for generating forms. The narrow range of Φ and longer R_v results in elongated, biserial forms (Fig. 5a,f), while wider range of Φ and short R_v lead to formulation of spiral shells (Fig. 5b,d). Other combinations of these parameters usually follows to mixed forms (Fig. 5c,e,g,h).

6   Conclusions

A new approach to modeling of foraminiferal shells has been presented. Previous models referred to fixed axes and neglected apertures [3], [6]. Our 2D simple model applies the moving reference system which has already been described to simulate heteromorphic shells of ammonites [13], [12]. This system is based on introducing of apertures as reference points, which in reality and in our model are responsible for emplacement of every new chamber. Minimization paradigm of the local communication path (LCP) helps to define an aperture in every newly added chamber. LCP rule derives from previous studies [16], [17] suggesting global shortening of the distance between the first and the last chamber via internal foramina and an external aperture. This rule seems to be based on local optimization during formation of a new chamber. Even if it does not explain all the cases, it is a closer approximation of actual morphogenetic constraints during the step-by-step growth of foraminiferal shells.

In order to imitate reality, some elements of randomness, as another novelty in foraminiferal modeling, are applied. This approach seems to be very promising for further studies. It may mimic random genetic variability (mutations) and influence of external (environmental) factors.

We are aware of numerous oversimplifications introduced into the model. Some of them can be overcome based on the same analytical approach. Two-dimensional simulations have only limited value, because foraminifera grow in 3D and most of the growth patterns cannot be reduced to two dimensions. Nevertheless, the moving reference system and the minimization paradigm applied here can be simply transferred to three-dimensional space.

Spherical shape of chambers also tremendously reduces variability of chamber shapes in different taxa. Furthermore, many foraminifers change shape of chambers during ontogenesis, their growth is often strongly allometric. It is clear that we have to test other methods (under study) to incorporate chamber shape into the model. A first approximation could be an introduction the implicit surfaces (blobs, metaballs)[18] to model chambers with irregular shapes. An overall form of the whole specimen directly depends on cumulative succession of chambers. Single apertures per chamber represent another simplification of real shell variability. Multiple apertures will be essential for future attempts to simulate complex foraminiferal shells.

Our model does not require high performance computing. An analytical method of optimization makes that the most difficult calculations are made on paper (see Appendix A). Moving our method to 3D with simple spherical chambers would complicate the paper calculations, but the computational cost will be kept on the same level. Thanks to this, we can implement presented model in Java language as an applet (with Java 3D as a graphic library) and published it on WWW page for broader audience of researchers interested in morphogenetic studies. Models with chambers represented by implicit surfaces, will require advanced numerical optimization methods and they will be more demanding for computational power.

7   Acknowledgments

We are grateful to Drs Sheila J Stubbles, Witold Alda, Mariusz Paszkowski and two anonymous reviewers for their contribution to this paper.

A   Distance minimization

We minimize the length of the vector $u_{i+1} = v_{i+1} + r_{i+1}$ (see Fig. 4).

$$|u_{i+1}| = \sqrt{(xv_{i+1} + xr_{i+1})^2 + (yv_{i+1} + yr_{i+1})^2} =$$
$$\sqrt{xv_{i+1}^2 + 2xv_{i+1}xr_{i+1} + xr_{i+1}^2 + yv_{i+1}^2 + 2yv_{i+1}yr_{i+1} + yr_{i+1}^2} \qquad (1)$$

Let's make the following assumption:
The coordinators of vector v_{i+1} are known therefore we can simplify our formula by introducing variables $a = xv_{i+1}$ and $b = yv_{i+1}$:
Lets express vector r_{i+1} in polar coordinators:

$$xr_{i+1} = r_{i+1}\cos\gamma_{i+1}, \quad yr_{i+1} = r_{i+1}\sin\gamma_{i+1}$$

The radius r_{i+1} is also known, so we can introduce variable r: $r_{i+1} = r$.
The angle γ_{i+1} must be calculated.
Now, the formula 1 has form:

$$\sqrt{a^2 + 2ar\cos\gamma_{i+1} + r^2\cos\gamma_{i+1}^2 + b^2 + 2br\sin\gamma_{i+1} + r^2\sin\gamma_{i+1}} \qquad (2)$$

We reduce this formula to the form:

$$\sqrt{c + 2r(a\cos\gamma_{i+1} + a\sin\gamma_{i+1})}, \quad where \quad c = a^2 + b^2 + r^2 \qquad (3)$$

We can treat this formula as a function $f(\gamma_{i+1})$. The first derivative of the function is:

$$f'(\gamma_{i+1}) = \frac{2r(b\cos\gamma_{i+1} - a\sin\gamma_{i+1})}{2\sqrt{c + 2r(a\cos\gamma_{i+1} + b\sin\gamma_{i+1})}} \qquad (4)$$

The left side of the equation must be equated to zero.

$$\frac{2r(b\cos\gamma_{i+1} - a\sin\gamma_{i+1})}{2\sqrt{c + 2r(a\cos\gamma_{i+1} + b\sin\gamma_{i+1})}} = 0$$

The γ_{i+1} value must preserve the following conditions:

$$2r(b\cos\gamma_{i+1} - a\sin\gamma_{i+1}) = 0 \qquad (5)$$

$$2\sqrt{c + 2r(a\cos\gamma_{i+1} + b\sin\gamma_{i+1})} \neq 0 \qquad (6)$$

Solving the equation 5 we obtain the points in which function $f(\gamma_{i+1})$ achieve extremal values:

$$\gamma_{i+1} = \arctan\frac{b}{a} + k\pi, \quad where \quad k = 0, 1, 2... \qquad (7)$$

Returning to the original nomenclature of variables, the γ_{i+1} can be calculated:

$$\gamma_{i+1} = \arctan \frac{yv_{i+1}}{xv_{i+1}} + k\pi, \quad where \quad k = 0, 1, 2... \qquad (8)$$

$\gamma_{i+1} \in \langle 0, 2\pi \rangle$. To calculate the angle in which the function achieve minima, we examine the points: $\arctan \frac{yv_{i+1}}{xv_{i+1}}$ and $\arctan \frac{yv_{i+1}}{xv_{i+1}} + \pi$.

References

[1] Hottinger L., Functional morphology of benthic foraminiferal shells, envelopes of cells beyond measure, *Micropaleontology*, 46:supplement no. 1, pp. 57-86, 2000.

[2] Lipps J.H., *Fossil Prokaryotes and Protists*, Blackwell, Boston, 1993.

[3] Berger W.H., Planktonic foraminifera: basic morphology and ecologic implications., *Journal of Paleont.*, 6(43):1369-1383, 1969.

[4] Fowler D.R., Meinhardt H., Prusinkiewicz P., Modeling seashells, *Proceedings of SIGGRAPH'92. Computer Graphics*, 26(2):379-387, 1992.

[5] McGhee Jr., G.R., *Theoretical Morphology. The Concept and its Application. Perspectives in Paleobiology and Earth History*, Columbia University Press, New York, 1999.

[6] Signes M., Bijma J., Hemleben C., Ott R., A model for planktic foraminiferal shell growth, *Paleobiology*, 19:71-91, 1993.

[7] Raup D.M., Geometric analysis of shell coiling: General problems, *Journal of Paleontology*, 40:1178-1190, 1966.

[8] Webb L.P., Swan A.R.C, Estimation of parameters of foraminiferal test geometry by image analysis, *Paleontology*, 39:471-475, 1996.

[9] Brasier, *Microfossils*, Allen & Unwin, London 1980.

[10] De Renzi M., Shell coiling in some larger foraminifera: general comments and problems, *Paleobiology*, 14(4):387-400, 1988.

[11] De Renzi M., Theoretical morphology of logistic coiling exemplified by tests of genus *Alveolina* (larger foraminifera), *Neues Jahrbuch für Geologie und Paläontologie. Abhandlungen*, 195:241-251, 1995.

[12] Ackerly S.C, Kinematics of accretionary shell growth, with examples from brachiopods and molluscus *Paleobiology*, 15:147-164, 1989.

[13] Okamoto T., Analysis of heteromorphy ammonoids by differential geometry, *Palaeontology*,31:35-52, 1988.

[14] Kaandorp J.A., A formal description of radiate accretive growth *J. theor. Biol.*, 166:149-161, 1994.

[15] Prusinkiewicz P., Lindenmayer A., *The Algorithmic Beauty of Plants*, Springer-Verlag, New York, 1990.

[16] Hottinger L., Comparative Anatomy of Elementary Shell Structures in Selected Larger Foraminifera, *In: Hedley, R.H. and Adams, C.G. (eds.) Foraminifera*, 3:203-266, 1978.

[17] Brasier, Foraminiferid architectural history; a review using the MinLOC and PI methods, *Journal of Micropalaeontology*, 1:95-105, 1982.

[18] Opalach A., Maddock S., Introduction to Modelling and Animation using Implicit Surfaces, Overview of Implicit Surfaces. Course Notes No 3, pp 1.1-1.13, 1995.

Using PDES to Simulate Individual-Oriented Models in Ecology: A Case Study[1]

Remo Suppi, Pere Munt, and Emilio Luque

Dept. of Computer Science, University Autonoma of Barcelona,
08193, Bellaterra, Spain
Remo.Suppi@uab.es Pere.Munt@campus.uab.es Emilio.Luque@uab.es

Abstract. The present work outlines the results of the Parallel Discrete Event Simulation (PDES) utilization for solving an individual based model: Fish Schools. This type of model cannot be solved through analytical methods, thus simulation techniques are necessary. The greater problem presented by this type of model is the high computing capacity necessary for solving middle-high size problems (hundreds to thousands of individuals). PDES is presented as a useful and low cost tool for individual-oriented models simulation, since it allows a scalable and efficient solution in accordance with the problem to be simulated.

1. Introduction

One of the main drawbacks to the use in ecology of individual-oriented models is the necessary computing capacity for their simulation. This type of model chooses the individual as basic element of the system. The ecosystem is described by dynamic and static individuals properties. The behaviour of an individual can differ from the behaviour of other individuals of the same or other species. This type of model cannot be solved in analytical form and it is therefore necessary to use simulation techniques in obtaining the ecosystem's dynamical behaviour. For complex systems (thousands of individuals), it is necessary to use advanced simulation techniques and parallel-distributed computing systems to give an efficient response to this type of problems. PDES (parallel and distributed event simulation) is a useful tool (and indispensable in some instances) for providing response to complex problems within an acceptable time. There are two main objectives to PDES: to reduce the simulation execution time, and to increase the potential dimension of the problem to be simulated. This paper demonstrates the use of PDES in solving a type of individual-oriented model: Fish Schools. The next section is a summary of the individual-orient model characteristics. Section 3 shows the implementation of the model on a distributed simulator. Section 4 presents the experimental framework and sections 5 and 6 the conclusions and references, respectively.

---

[1] This work has been supported by the CICYT under contract TIC98-0433

P.M.A. Sloot et al. (Eds.): ICCS 2002, LNCS 2329, pp. 107–116, 2002.

2. Individual-oriented models

There are considerable references to individual-oriented models (IoM) in the litera-ture [3,7,6]. The model definition is based on recognizing each individual as an autonomous agent that acts according to a set of biological rules.

One of the most representative applications of IoM is used to describe the move-ment of given species (schools, flocks, herds, etc) [5,6]. The IoM utilization allows us to determine the movement of a species group by using the movement of each mem-ber. The Fish Schools is an IoM application for the movement of fish species. [8,6,10]. From the observation, was discovered that fish can describe very complex figures in their movement, but that these figures are governed by three basic postu-lates from the point of view of the individual:

- To avoid collisions
- Speed coupling
- To obtain a position in the centre of the group.

These rules express both the individual's need for survival and its instinct for pro-tection (the need to escape from predators). Each fish in the model is represented as a point in a three-dimensional space with an associated speed. And each fish changes position and speed simultaneously after a certain period Δt. The actions that the model describes for each fish are:

1- Each fish chooses up to X neighbour fish (X=4 seems sufficient for most schools), which will be those nearest and with direct vision.

2- Each fish reacts in accordance with the direction and distance of each neighbour. Three influence radios and three possible reactions are estab-lished. The final reaction will be the average of the reactions experimented on each neighbour.

   a. If the neighbour is found within the smaller radio, the fish will carry out an "opposed to address" movement -repulsion action- (to avoid collisions).

   b. If the neighbour is within a second influence radio, the fish will adopt the same direction as the neighbour.

   c. If the neighbour is within a third radio, the fish will move towards it.

3- Each fish calculates its new position according to the new direction.

This generates a very simple model, but one that allows very complex behaviour to be described (an implementation with applets in 2D can be found in [4]). As a coun-terpart, very high computing power is necessary, since the complexity algorithm is of $O(N^2)$, where N is the number of fish (each fish attempts to find the neighbour fish by inspecting all other fish in the school).

Each fish Fi is defined by its position $\vec{p_i}$ and velocity $\vec{v_i}$, and chooses its potential neighbours by watching concentric zones of incremental radio until finding X fish. To calculate the distance between two fish, the Euclidean distance is used:

$$Dist(p_a, p_b) = \sqrt{(p_{ax} - p_{bx})^2 + (p_{ay} - p_{by})^2 + (p_{az} - p_{bz})^2} \tag{1}$$

The potential neighbours are chosen by using the algorithm of front priority. This algorithm calculates all the angles formed by $\overrightarrow{v_i}$ and $\overrightarrow{p_i} - \overrightarrow{p_j}$ (being the collision angle between p_i and p_j); the X neighbours with the smallest angle are chosen. Figure 1 shows the neighbour selection of p_1 using the front priority algorithm. p_2 is the neighbour selected, since the angle v_1-v_{21} (v_2, (p_2-p_1)) is less than v_1-v_{31} (v_3, (p_3-p_1)).

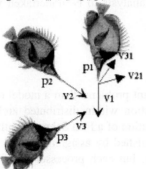

Fig. 1. Neighbour selection

Each F_i, once they have selected the X neighbours, must then determine the reaction (rotation of the v_i) to each F_j. β_{ij} will be the F_i reaction with respect to F_j expressed in spherical coordinates. Each fish F_j can be found within one of three possible areas of influence with respect to F_i (A: near, B: middle, C: far):

- If $Dist(F_j, F_i) \leq A$, F_i has a repulsion reaction with respect to F_j and $\beta_{ij} = (1, \pi, \pi)$
- If $A < Dist(F_j, F_i) \leq B$, F_i adopts a parallel position with respect to F_j and $\beta_{ij} = (1, \theta_j, \varphi_j)$
- If $B < Dist(F_j, F_i) \leq C$, F_i is guided toward F_j and $\beta_{ij} = (1, \theta_i - \theta_j, \varphi_i - \varphi_j)$

Finally, reaction β is obtained (mean value for all β_{ij}) and v_i is rotated according to β. Object-Oriented methodology and UML techniques [1] were used for the model implementation and the analysis phase, respectively. [5]

3. PDES simulation

As model of the parallel discrete event simulation (PDES), a set of logical processes (LP) managing a distributed event lists was considered. These processes interact exclusively by exchanging time-stamped messages. The PDES mechanisms can be divided in two categories: *conservative* and *optimistic*. *Conservative* approaches use synchronization to avoid causality errors. In these algorithms, the events are processed when it is certain that the execution order is correct. On the other hand, in *optimistic* algorithms, each LP processes the events as soon as they are available and this execution, in some instances, can mean causality errors. Nevertheless, the algorithm has detection mechanisms to avoid these errors and to recover the causality [11,12,9].

The fish school simulator was built on the base of a PDES simulation kernel developed at the UAB (written in C++). This kernel is designed to be executed in Unix stations and PVM. The union of the simulation kernel and the model classes is accomplished through the inheritance of the kernel classes and virtual functions implementation. This simulation kernel uses an STL library for data structure management. [9,2]. Based on the UAB kernel, two simulator versions were developed: sequential and distributed. The distributed version will be used to make performance analysis using different simulation PDES algorithms. The sequential version will allow speedup and performance analysis to be undertaken with respect to the distributed version.

3.1 Model distribution

One of the most important problems than a model of these characteristics outlines is the individual's distribution within distributed architecture computing elements. Lorek [8] outlines a distribution of a Fish School simulator for real time visualization. The distribution is accomplished by assigning a static partition of the fish group to simulate in each processor, but each processor has the information on the fish assigned in other processors.

Our solution is based on the problem division in a set of logical processes (LP), which will be executed in the different processors. For each LP, an initial partition of the problem (number of fish) is assigned and this quantity will change dynamically during the simulation. The LPs have a physical zone of the problem (cubes in fig. 2) to simulate (Spatially Explicit Simulation) and the fish movement will imply migrations between the LPs.

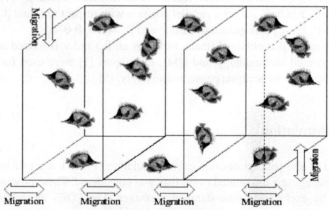

Fig. 2. Model distribution

The present model implements two types of messages between LPs: neighbour petition within a given zone and migration. To reduce the information quantity that the LPs should share, a point-to-point petition was designed (under demand): only the neighbours of a specific zone are requested. If the LP simulated area is a cube, only the neighbours that share a cube side will be requested (see fig. 2).

4. Experimental Studies

The experimental framework was developed on a cluster of machines executing Linux SuSE 6.4 and interconnected by Fast Ethernet. The tools used for the development were: Pvm 3.4.3, Xpvm 1.1.5, Gcc 2.95 and Ddd 3.2 (Date Display Debugger). Figure 3 a,b,c shows three types of animation frames obtained from the simulator traces.

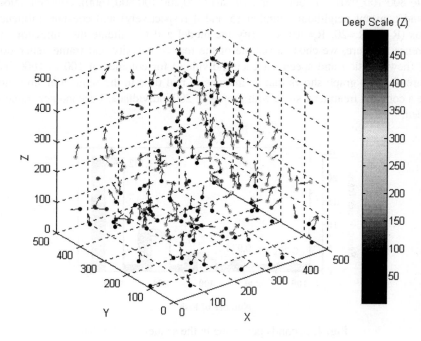

Fig. 3.a 3-D Fish Visualization

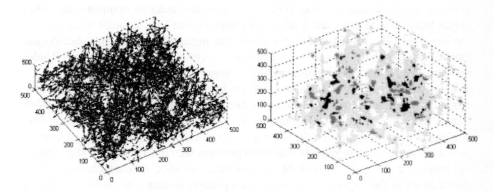

Fig. 3.b, c. Visualization of trajectory & density of fish
(10 frame animations)

Figure 3.a shows a distribution of 300 fish. The point and colours indicate their position and depth (with respect to the colour scale), respectively, and the arrows indicate their direction and speed. Figure 3.b shows a path graph for 300 fish during a 100-frame simulation, and figure 3.c shows fish density in a given zone (red=high density, green=low density).

The first step in the development process was the design and development of the sequential simulator version. This version was analysed with different territory sizes (up to 800:100:100), number of individuals (100,200,400,800,1600), constant velocity and constant neighbours number (5 and 4 respectively) and constant influence radios (R_1=5, R_2=20, R_3=30) to verify the model and to validate the simulator. As reference measure, we chose a sample of the real complexity and frame generation time (new position and speed for each individual) for colonies of 100 to 1600 fish (figure 4). This graph shows that animations in real time in the sequential simulator have a complex treatment, since with groups of 200 individuals, 1 second per frame is needed.

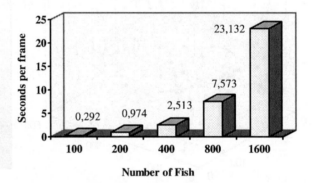

Fig. 4. Seconds per frame in the sequential simulator

After the sequential version, the following step was the development of distributed versions. The first objective was to build the simulator under an optimistic algorithm (Time Warp) and to solve the load balance between the LPs, in order to avoid synchronizations. The results under this simulation method are acceptable, but do not have a homogeneous behaviour for different fish colonies (as was expected). Under the Time Warp algorithm (optimistic), the model shows acceptable behaviour for small quantities of fish (<100 individual) on large territories (800x100x100) where migration events are few and therefore the need for synchronization is minimal. With a more realistic evaluation of fish quantities (>100 individual), the simulation model was unable to advance due to the great quantity of external events produced both by migration and neighbour request. The main problem with the Time Warp algorithm is the rollback chains [11]. In an IoM of these characteristics, rollback chains are produced easily in the migration and neighbours petition actions. The problem is due to the fact that the fish do not have a regular movement and often come and go on the same point (see figure 3.b). This behaviour generates a very high quantity of migrations on the LPs shared cube face. Such activity means that synchronization between the two LPs is very difficult.

The second step was the utilization of conservative distributed simulation algorithms to control the synchronization events. Our objective was to avoid the 'out of order' events processing to remove the synchronization events. Each LP_j will have to execute the following steps (in each iteration):

1. To generate the iteration results $T_{j,i}$ (state + migrations).
2. To wait until all the $LP_{k\,k\neq j}$ sharing information with LP_j finish their iteration $T_{k,i}$.
3. To process the external events.
4. To generate the iteration $T_{j,i+1}$

The worse case is when an LP does not know with which LP shares information. In this case, a global waiting time will be necessary and therefore the global simulation time progression (GVT) will be conditioned by the speed of the slowest LP. This situation does not exist when the experiments are accomplished on homogeneous machines and similar computing charges. Figure 5 shows the time per frame for a conservative distributed simulation on 1,2,4,6,8 processors. The simulation characteristics are: territory=800:100:100; individual=100,200,400,800,1600; speed=5; neighbour number=4; influence radio= 5:20:30.

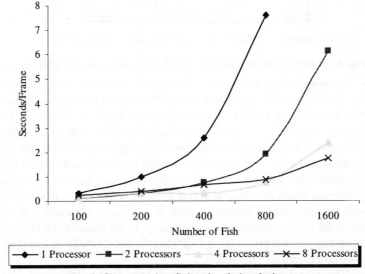

Fig. 5. Conservative fish school simulation

From figures 5 and 6, the following conclusions can be extracted:

- The model scales well: as a rule, an increase in processor number facilitates a reduction in frame time to values below 1 second.
- The frame generation time cannot be reduced without the limit increasing processor number. Figure 6 shows that the tendency line for 1600 fish is asymptotically to 0,15s (decreasing this time would require acting on the communications model). For visualizations in real time, approximately 4 fps for 400 fish and 4 processors are obtained.

- The model to be simulated must be analysed carefully, there are situations in which adding processors does not bring about benefits. This case is present in the current model (fig. 4): it is not necessary to use 8 processors to simulate below 800 individuals (the cube generates a high communication granularity for these cases).
- The results obtained are excellent, but results above the linearity require explanation.

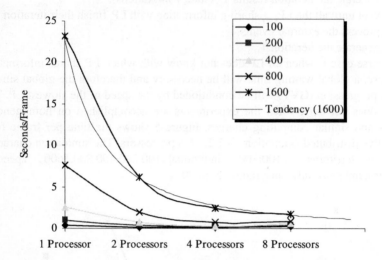

Fig. 6. Improvement & tendency line for the simulation of 1600 individuals

Figure 7 shows speedup with respect to the sequential version and for different fish colonies. This figure confirms that at least 100 fish per processor are required in order to obtain acceptable speedups. A comment with respect to this profit is necessary. Parallel programs use linear speedup as maximum profit, but figure 7 show that the profit obtained is superior in given points. This singularity is explained by the model complexity calculation. The IoM Fish Schools has complexity $O(N^2)$ [8], since the data structure that sustains the individuals does not have the implicit location (neighbours are found searching in N fish positions). The partition generation (section 3.1) adds this information implicitly: each process must accomplish the neighbour's searches in the current partition and, eventually, in the shared cube faces. In this way, the complexity for two processes is $O(2*(N/2)^2)$, and $2*(N/2)^2$ is less than N^2. In this sense, two comments are required:

    a)   The model distribution generates a considerable additional advantage that must be taken into account in the PDES speedup calculation.

    b)   The model series was not implemented with complexity $O(k*(N/k)^2)$, because the same model proposed by Lorek & Sonnenschein was used [8]. In the Lorek model, the local territory notion does not exist. Each processor has global information of the whole territory and seeks neighbours in N fish positions, only obtaining good speedups with low and unrealistic granularity (speedup=16 with 10 fish per processor).

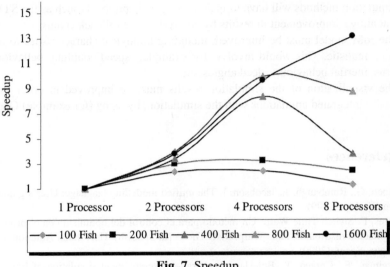

Fig. 7. Speedup

5. Conclusions

The ecological systems simulation is a field that requires considerable computing power for realistic models, and parallelism is a very useful tool in solving this type of problem. The present study shows very good results for IoM models and efficient solutions for such problems by using PDES. The operative prototypes were built using a PDES kernel developed at the UAB and are executed on a low-cost distributed architecture based on Linux.

The IoM Fish Schools model was developed using conservative and optimistic PDES algorithms. The results are better in the first type of protocols, due to the model characteristics. The main conclusions that can be extracted are:

- Model scalability is possible and acceptable. The division of the problem to be simulated into partitions (cubes) increases the speedup with respect to the existing parallel versions [8]. For a fixed size problem (number of fish) there is an optimal number of processor maximizing the speedup. Increasing the processors number should be accompanied of a problem size growing in order to keep a higher speedup.
- Performance is good with respect to large data animations, but there is a limit essentially imposed both by the communications model and the architecture. A reduction in this time would mean changing the communications model (for example using MPI) and modifying the network communications technology (Gigabit Ethernet).

Future work is guided towards:

- The need to include within the study an increase in individual and processor numbers in order to verify model scalability.
- Simulation methods will have to include optimistic protocols such as the STW [11] that allows improvement in results by controlling the rollback chains.
- The IoM model must be improved, including biological characteristics to make it more realistic. This would involve, for example: speed coupling, behaviour patterns, inertial behaviours, dead angles, etc.
- The visualización of the simulation results must be improved in order to allow totally integrated animations with the simulation, by using (for example) OpenGL.

6. References

1. Booch, G., Rumbaugh. J., Jacobson, I. The unified modelling language: User's guide. Adisson-Wesley. (1999)
2. Cores, F. Switch Time Warp: Un método para el control del optimismo en el protocolo de simulacion distribuïda Time Warp. MsC Thesis (in Spanish). Universitat Autònoma de Barcelona. Spain. (2000)
3. Deelman, E., Coraco, T., Boleslaw, K. Parallel discrete event simulation of lime disease. Pacific Biocomputing Conference. (1996). 191-202.
4. ECOTOOLS: High level tools for modelling and simulation of individual-oriented ecological models. http://www.offis.uni-oldenburg.de/projekte/ecotools (1999)
5. Fishwick, P., Sanderson, J.G., Wolf, W. A multimodeling basis for across-trophic-level ecosystem modelling. Trans. SCS. Vol. 15. No. 2. (1998) 76-89.
6. Huth, A., Wissel, C. The simulation of movement of fish schools. Journal of Theoretical Biology 156. (1992) 365-385.
7. Kreft, J. Booth, G, Wimpenny, W. BacSim, a simulator for individual-based modelling of bacterial colony growth. Microbiology 144. (1998) 3275-3287.
8. Lorek, H, Sonnenschein, M. Using parallel computers to simulate individual oriented models: a case study. European Simulation Multiconference (ESM). (1995) 526-531.
9. Munt, P. Simulació distribuïda en PVM: Implementació dels algorismes TimeWarp i Switch Time Warp. Graduate Thesis (in Catalan). Universitat Autònoma de Barcelona. Spain. (1999)
10. Proctor, G., Winter, C. Information flocking, data visualisation in Virtual Worlds using emergent behaviours. Proc. Int. Conf. of Virtual Worlds Vol. 1434. Springer-Verlag. (1998) 168-176.
11. Suppi, R., Cores, F, Luque, E.. Improving optimistic PDES in PVM environments. Lecture Notes in Computer Science. Springer-Verlag 1908. (2000). 304-312.
12. Sloot, P. Kaandorp, J., Hoekstra, A., Overeinder, B. Distributed Cellular Automata: Large Scale Simulation of Natural Phenomena. Solutions to Parallel and Distributed Computing Problems: Lessons from Biological Sciences. ISBN: 0-471-35352-3 (2001), 1-46,

In Silico Modelling of the Human Intestinal Microflora

Derk Jan Kamerman and Michael H.F. Wilkinson

Institute for Mathematics and Computing Science,
University of Groningen, P.O. Box 800, 9700 AV Groningen, The Netherlands

Abstract. The ecology of the human intestinal microflora and its interaction with the host are poorly understood. Though more and more data
are being acquired, in part using modern molecular methods, development of a quantitative theory has not kept pace with this development.
This is in part due to the complexity of the system, and to the lack of
simulation environments in which to test what the ecological effect of
a hypothetical mechanism of interaction would be, before resorting to
laboratory experiments. The MIMICS project attempts to address this
through the development of a system for simulation of the intestinal
microflora. In this paper the design, possibilities and relevance of this
simulator including an intestinal wall are discussed.

1 Introduction

This paper introduces an new field of application for computer simulation: the
intestinal microflora and its interaction with the host. The human intestines
harbour a complex microbial ecosystem, part bound to the intestinal wall or
embedded in the mucus layer covering the wall, part inhabiting the lumen, especially of the large intestine. The intestinal microflora form a highly complex
community of an estimated 400 species. However, since the greater part (60-85
%) of the microscopically visible bacteria in faecal content cannot be cultured
[8], the exact number of species, and even their nature and role in the ecosystem are not really known. The importance of the gut ecosystem stems from the
fact that it is considered a first line of defence against invading pathogens [14],
mediated through competition for substrate and wall binding sites, production
of toxins by the resident microflora, etc. However, the precise mechanisms are
poorly understood, particularly because of the culturability problems.

The Model Intestinal Microflora In Computer Simulation (MIMICS) project
aims to provide a flexible means of modelling and simulating the intestinal microflora, and in particular those ecological processes which have consequences
for the host. It will provide a platform for medical researchers to test hypotheses
in silico before commencing with (costly) in vitro or in vivo experiments. The
MIMICS project is inspired by a number of developments in medicine, concurrent with the increased availability of powerful computing facilities. The first
development in medicine which is relevant to this project, and which indeed

P.M.A. Sloot et al. (Eds.): ICCS 2002, LNCS 2329, pp. 117–126, 2002.

initiated it, is the rapid and alarming increase in antibiotics resistance [3]. In line with this there is a growing realization of the importance of the bacterial microflora inhabiting the human intestines in maintaining the hosts defenses, and in performing a large number of other important functions.

A separate development is the increased use of computer simulation techniques in many fields of medicine, both for teaching and research purposes. Though this trend has been clearly visible in fields such as surgery [6], in the field of the intestinal microflora this trend is much less pronounced, and only fairly recently have there been any significant developments [1].

The interest in modelling is growing in this particular community due to the increased volume of data on the intestinal ecology due to modern detection methods such as DNA probes [11]. The earliest computer models of the intestinal microflora date from the early eighties [4], but subsequently very little work has been done on modelling the intestinal microflora using computers and mathematics [2, 7]; for a review see [1]. The MIMICS project is the first, and so far unique attempt to model the human intestinal microflora by means of parallel high performance computers [1, 16].

This paper describes the main tool: a large scale cellular automaton which can simulate both metabolic and transport processes in the human intestine including the intestinal wall. The software runs on the Cray J932 supercomputer of the Centre for High Performance Computing.

2   Modelling the ecosystem

2.1   General approach

In the intestinal ecosystem various processes take place e.g. flow, diffusion, peristalsis, metabolic processes, interactions between bacteria, and immunological processes. Given the complexity of this system we have to restrict ourselves to what Levins calls *strategic* modelling [10] aimed at qualitative understanding of the types of behaviour of a class of systems. This contrasts with what Levins calls *tactical* modelling aimed at quantitative prediction of the behaviour of a specific system e.g. a patient. Via this strategic modelling, using in silico simulation, we would like to obtain a better understanding of the ecosystem, and to propose ideas and suggestions for further in vivo, in vitro, and in silico studies regarding the mechanisms in and properties of the ecosystem. We aim to find out what behaviour can be found in theory, which part of the ecosystem is crucial for which kind of behaviour, and which parameters are influential. Furthermore, we aim to test consistencies of hypotheses, and to calibrate our model with in vivo observations.

Because of the many unknowns we have to keep our model simple, and it does not make sense to elaborate on certain knowns. For instance, as long as peristalsis is not implemented, it is not reasonable to model the flow in accordance to the sophisticated Navier Stokes equations. There is simply no gain in accuracy. Furthermore we want a flexible organization of the software, allowing easy insertion of extra types of metabolism, and allowing easy exchange of one way of modelling metabolism for another. This is why we computationally separate metabolism from flow and diffusion, in different subroutines.

2.2   General model of the ecosystem

The most prominent entity in the ecosystem of the human intestine is the intestine itself. We model it as a straight tube of varying diameter. From stomach to anus we distinguish: small intestine (length $4.98m$, diameter $0.03m$), caecum (length $0.12m$, diameter $0.09m$), and colon (length $0.90m$, diameter $0.06m$). Axially we divide the intestine into 100 parts of equal length, and radially we subdivide each part into 10 equi-volume conaxial cylinder-like volume elements. The geometry is shown in fig. 1.

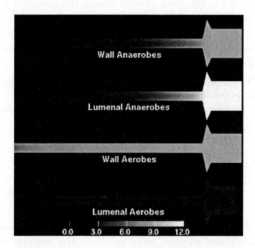

Fig. 1. The model intestinal geometry: four cross-sections of the intestine showing the concentrations of four bacterial species (in ^{10} log number bacteria / ml).

The entities we distinguish next to the intestine are: food (one kind so far), oxygen, and various types of bacteria. The bacteria are subdivided based on their metabolism, e.g. anaerobes (do not use oxygen to grow) and aerobes (need oxygen to grow), and based on their ability to attach to the intestinal wall i.e. wall attaching bacteria and strictly lumenal bacteria.

The processes we distinguish in the ecosystem are metabolism and transport processes. The transport processes are transport into and out of the ecosystem and transport within the ecosystem. Transport into and out of the ecosystem includes: influx of food, oxygen and bacteria from the stomach, transport of food and oxygen through the intestinal wall, and efflux through the anus. Transport within the ecosystem consists of bulk motion modelled as laminar flow and turbulent and diffusive mixing modelled as diffusion with a rather large diffusion coefficient.

In each time step, for which we use a value of about 5 minutes, we simulate metabolism, laminar flow, and diffusion consecutively. We thus separate computationally what in reality are simultaneously occurring processes.

3   Metabolism in more detail

In the metabolism subroutine we not only compute bacterial growth by metabolic activity and the corresponding decrease in food and oxygen concentrations, but also the increase and decrease of bacterium concentrations due to adhesion to and detachment from the intestinal wall. We do this for each of the volume elements separately, hence the computations can be done in parallel very well.

There is a large volume of literature devoted to modelling bacterial growth (see [9] for an overview). We have chosen one of the simplest but quite realistic models: the interacting Monod model, which is based on Michaelis-Menten-type kinetics [5]. The differential equations we use are solved by means of a fourth-order Runge-Kutta method, which can however easily be replaced by some other differential equation solver.

For each species or strain of bacteria k with lumenal concentration $[X]_k(t)$ ($mol\ C\ m^{-3}$), the increase in bacterial concentration due to the metabolism of that bacterium in the lumen, denoted by $\mu_{l,k}(t)$ ($mol\ C\ m^{-3}s^{-1}$), is modelled by the following differential equation

$$\mu_{l,k}(t) = \left((\mu_{O_2,k} \frac{[O_2](t)}{K_{R,O_2,k}+[O_2](t)} + \mu_{an,k}) \frac{[S](t)}{K_{S,k}+[S](t)} \right.$$
$$\left. -\kappa_{O_2,k} \frac{[O_2](t)}{K_{T,O_2,k}+[O_2](t)} - \mu_{b,k} \right) [X]_k(t). \quad (1)$$

In this equation, $[S](t)$ and $[O_2](t)$ represent the concentrations of food (or substrate) ($mol\ C\ m^{-3}$) and oxygen ($mol\ m^{-3}$) in the volume element, as is experienced by the bacterium, i.e., the partial volume effect of bacteria is taken into account. This equation states that growth is based on the metabolism (first term), is reduced by killing due to toxicity of oxygen (second term) and by maintenance energy costs $\mu_{b,k}$. For the parameters in this equation and the following see table 1.

The aerobic bacteria have no anaerobic metabolism ($\mu_{an,k} = 0$), and have zero maximum oxygen kill rate $\kappa_{O_2,k}$. The anaerobic bacteria have an anaerobic metabolism (therefore $\mu_{an,k} > 0$) which is inhibited in the presence of oxygen ($\mu_{O_2,k} < 0$). On top of that the anaerobic bacteria we model suffer from losses due to the toxicity of oxygen ($\kappa_{O_2,k} > 0$). A more comprehensive discussion of the above Monod based equation can be found in [15].

Next to the distinction in anaerobes and aerobes, we have a distinction in strictly lumenal and wall attaching bacteria. In each volume element along the wall we model a wall population next to a lumenal population of the latter. It is essential that we give the bacteria which can attach to the wall a penalty for this extra ability. If not, it can easily be shown that these bacteria would outcompete the strict lumenals under all circumstances, which is both uninteresting

Table 1. Parameters describing a bacterial metabolism, after [5]

Symbol	Meaning	aerobes		anaerobes		Units
		lumenal	wall	lumenal	wall	
μ_{O_2}	maximum specific aerobic growth rate	$6.6 \cdot 10^{-4}$	$6.0 \cdot 10^{-4}$	$-1.1 \cdot 10^{-4}$	$-1.0 \cdot 10^{-4}$	s^{-1}
$\mu_{O_2,w}$	as above, when wall bound	0	$5.4 \cdot 10^{-4}$	0	$-0.9 \cdot 10^{-4}$	s^{-1}
μ_{an}	maximum specific anaerobic growth rate	0	0	$1.1 \cdot 10^{-4}$	$1.0 \cdot 10^{-4}$	s^{-1}
$\mu_{an,w}$	as above, when wall bound	0	0	0	$0.9 \cdot 10^{-4}$	s^{-1}
μ_b	maintenance costs	$1.0 \cdot 10^{-6}$	$1.0 \cdot 10^{-6}$	$1.0 \cdot 10^{-6}$	$1.0 \cdot 10^{-6}$	s^{-1}
K_S	food uptake saturation constant	$2.0 \cdot 10^{+1}$	$2.0 \cdot 10^{+1}$	$2.0 \cdot 10^{+1}$	$2.0 \cdot 10^{+1}$	$mol\ C\ m^{-3}$
K_R	respiratory oxygen uptake rate constant	$1.0 \cdot 10^{-2}$	$1.0 \cdot 10^{-2}$	$1.0 \cdot 10^{-3}$	$1.0 \cdot 10^{-3}$	$mol\ O_2\ m^{-3}$
K_T	toxic oxygen uptake rate constant	$1.0 \cdot 10^{-2}$	$1.0 \cdot 10^{-2}$	$1.0 \cdot 10^{-3}$	$1.0 \cdot 10^{-3}$	$mol\ O_2\ m^{-3}$
κ_{O_2}	maximum oxygen kill rate	0.0	0.0	$1.0 \cdot 10^{-6}$	$1.0 \cdot 10^{-6}$	s^{-1}
α_{O_2}	yield of aerobic metabolism	1.0	1.0	1.0	1.0	
α_{an}	yield of anaerobic metabolism	1.0	1.0	1.0	1.0	
α_κ	fraction of oxygen killed bacteria returned as food	1.0	1.0	0.5	0.5	
β_μ	maximum respiratory oxygen uptake rate	$1.5 \cdot 10^{-4}$	$1.5 \cdot 10^{-4}$	$1.0 \cdot 10^{-7}$	$1.0 \cdot 10^{-7}$	s^{-1}
β_κ	maximum oxygen uptake rate due to toxic effect	0.0	0.0	$1.0 \cdot 10^{-7}$	$1.0 \cdot 10^{-7}$	s^{-1}
p_b	probability to attach to the wall	0	0.9	0	0.9	
R_{det}	detachment rate	0	$3.0 \cdot 10^{-6}$	0	$3.0 \cdot 10^{-6}$	s^{-1}
R_{adh}	adhesion rate	0	$3.0 \cdot 10^{-5}$	0	$3.0 \cdot 10^{-5}$	s^{-1}

and unrealistic. We penalize the bacterium with this wall attaching ability by giving it a smaller maximum specific growth rate, if only for the extra DNA it has to reproduce. Additionally, we penalize the bacterium while it is attached to the wall by giving it a smaller maximum specific growth rate again, because it has to express genes it otherwise would not, and it has a reduced effective surface area available for food and oxygen uptake.

The bacteria attached to the wall produce offspring according to (1), just as their lumenal counterparts do. Attached to the wall however, the growth rate $\mu_{w,k}(t)$ ($mol\ C\ m^{-2}s^{-1}$) is lower, resulting in

$$\mu_{w,k}(t) = \left(\left(\mu_{O_2,w,k} \frac{[O_2](t)}{K_{R,O_2,k}+[O_2](t)} + \mu_{an,w,k} \right) \frac{[S](t)}{K_{S,k}+[S](t)} \right.$$
$$\left. -\kappa_{O_2,k} \frac{[O_2](t)}{K_{T,O_2,k}+[O_2](t)} - \mu_{b,k} \right) [W]_k(t).$$
(2)

In this equation $[W]_k(t)$ is the concentration ($mol\ C\ m^{-2}$) of bacteria of strain k attached to the wall, and the w in $\mu_{O_2,w,k}$ and $\mu_{an,w,k}$ denotes the reduction of the maximum specific growth rate of the bacterium in its wall attached state.

Some modifications are required however, in which we follow [7]. First of all, only part of this offspring will immediately attach to the wall, the rest will migrate into the lumen. A fraction $\eta_w(t)$ of the offspring of the wall attached bacteria emerges at a free place at the wall. The value of $\eta_w(t)$ in the volume element at hand depends on the number of bacteria already attached to the wall, and on the maximum number of bacteria that can attach in that place $\eta_w(t) = 1 - \frac{\sum_k [W]_k(t)}{[W]_{max}}$. Each daughter cell of a bacterium attached to the wall, emerging at a free place, is assigned a probability $p_{b,k}$ to attach (or bind) to the wall. The rest of the offspring migrates into the lumen along the wall. This results in the following two equations: the increase in lumenal concentration of bacterium k due to growth of the wall attached ones

$$\frac{d[X]_k}{dt}(t)|metab_{W_k} = \max\{0, \frac{A}{V}(1 - p_{b,k} * \eta_w(t)) * \mu_{w,k}(t)\}, \tag{3}$$

and the increase in concentration at the wall of bacterium k due to growth of wall attached bacteria

$$\frac{d[W]_k}{dt}(t)|metab = \mu_{w,k}(t) - \frac{V}{A}\frac{d[X]_k}{dt}(t)|metab_{W_k}. \tag{4}$$

In these equations an extra provision is taken in case the growth of the wall attached bacteria would become negative, in that case migration of negative amounts of offspring into the lumen must be prohibited. Furthermore, V is the volume of the volume element and A is the area of the intestinal wall in or along that volume element. Equations (1) and (3) combine to the differential equation for the bacterial concentration in a volume element due to bacterial growth,

$$\frac{d[X]_k}{dt}(t)|metab = \mu_{l,k}(t) + \frac{d[X]_k}{dt}(t)|metab_{W_k}, \tag{5}$$

in which $\frac{d[X]_k}{dt}(t)|metab_{W_k} = 0$ if the volume element is not along the intestinal wall.

Next to migration of offspring of wall attached bacteria, there is migration of lumenal bacteria to the wall (adhesion), and migration of wall attached bacteria into the lumen (detachment).

$$\frac{d[W]_k}{dt}(t)|det = -R_{det,k}[W]_k(t) \tag{6}$$

$$\frac{d[X]_k}{dt}(t)|det = \frac{A}{V}R_{det,k}[W]_k(t) \tag{7}$$

$$\frac{d[W]_k}{dt}(t)|adh = \eta_w(t)R_{adh,k} * c * [X]_k(t) \tag{8}$$

$$\frac{d[X]_k}{dt}(t)|adh = -\frac{A}{V}\eta_w(t)R_{adh,k} * c * [X]_k(t) \tag{9}$$

The detachment rate R_{det} and the adhesion rate R_{adh} are both measured in s^{-1}, and c denotes a constant on the order of the radius of a bacterium, for the time being taken to be $0.5 * 10^{-6}(m)$.

Bacterial growth, modelled in (1) and (2) is associated with uptake and release of oxygen and food. The next two equations balance the uptake and release of oxygen and food, by both the lumenal population and the wall population of the strains of bacteria present.

$$
\frac{d[O_2]}{dt}(t) = -\sum_k \left(\beta_{\mu,k} \frac{[O_2](t)}{K_{R,O_2,k}+[O_2](t)} \frac{[S](t)}{K_{S,k}+[S](t)} + \beta_{\kappa,k} \frac{[O_2](t)}{K_{T,O_2,k}+[O_2](t)} \right)[X]_k(t)
$$

$$
-\frac{A}{V}\sum_k \left(\beta_{\mu,w,k} \frac{[O_2](t)}{K_{R,O_2,k}+[O_2](t)} \frac{[S](t)}{K_{S,k}+[S](t)} + \beta_{\kappa,k} \frac{[O_2](t)}{K_{T,O_2,k}+[O_2](t)} \right)[W]_k(t)
$$

$$(10)$$

$$
\frac{d[S]}{dt}(t) = \sum_k \left(-(\frac{\mu_{O_2,k}}{\alpha_{O_2,k}} \frac{[O_2](t)}{K_{R,O_2,k}+[O_2](t)} + \frac{\mu_{an,k}}{\alpha_{an,k}}) \frac{[S](t)}{K_{S,k}+[S](t)} \right.
$$

$$
\left. + \alpha_{\kappa,k}\kappa_{O_2,k} \frac{[O_2](t)}{K_{T,O_2,k}+[O_2](t)} \right)[X]_k(t)
$$

$$
+\frac{A}{V}\sum_k \left(-(\frac{\mu_{O_2,w,k}}{\alpha_{O_2,k}} \frac{[O_2](t)}{K_{R,O_2,k}+[O_2](t)} + \frac{\mu_{an,w,k}}{\alpha_{an,k}}) \frac{[S](t)}{K_{S,k}+[S](t)} \right.
$$

$$
\left. + \alpha_{\kappa,k}\kappa_{O_2,k} \frac{[O_2](t)}{K_{T,O_2,k}+[O_2](t)} \right)[W]_k(t)
$$

$$(11)$$

4 Flow and diffusion

We model both flow and diffusion by difference equations. In contrast to metabolism, transport and diffusion take place not only within but also between volume elements. In modelling diffusion, we take a short enough time step to ensure that only diffusion between immediate neighbours need be considered. Difference equations can lead to numerical instabilities. In order to avoid those instabilities, we divide each volume element into four subdivisions, each bordering on a single subdivision of one of its neighbours, then analytically obtain the concentrations after diffusion in just each pair of bordering subdivisions of neighbouring volume elements, and after that we obtain the final concentrations in each volume element by taking the mean over the concentrations in the corresponding 4 constituent subdivisions.

We thus model flow and diffusion rather crudely, but the essential features (local mixing and overall flow) are preserved. The exact model used plus discussion can be found on the internet [15].

5 Simulation results

The first 90 days after colonization of a sterile intestine were simulated, with and without the possibility of attachment to the wall. The results with binding are shown in figures 1 and 2. The four bacterial species simulated have metabolic parameters as in table 1. Further parameters such as flow rate, oxygen and food influx, etc. were set as in [16]

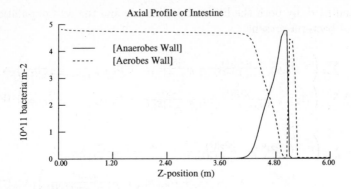

Fig. 2. Concentrations of anaerobic and aerobic bacteria attached to wall.

6   Discussion

In describing the results we will focus on those that give an impression of the tenability of modelling the human intestinal microflora in silico.

In our simulations we find that in initially sterile and oxygen rich bowels, as in newborns, first aerobic bacteria will prosper thereby reducing the oxygen concentration, and thus clearing the way for the anaerobic bacteria in the large intestine. The anaerobes, which initially cannot colonize at all due to the initial oxygen richness of the lumenal environment, will then prosper in the large intestine, reducing the oxygen concentration even further, making growth largely impossible for aerobic bacteria, and reducing the food concentration substantially. These findings are in line with in vivo studies, e.g. [12]. In the small intestine however, the anaerobes do not manage to colonize substantially due to the high flow rate of the bowel contents and the corresponding oxygen richness of the environment [16].

It is interesting that here we find a division of the ecosystem into two areas, an area where anaerobes are in the majority (large intestine), and an area where the aerobes are in the majority (small intestine). It is remarkable, that the concentrations of anaerobes ($3*10^{11}$ $bacteria\ g^{-1}$) and aerobes ($5*10^8$ $bacteria\ g^{-1}$) in the large intestine, are in good agreement with what is found in vivo [8]. By contrast, though the numbers of bacteria in the small intestine is within the observed range, the ratio of anaerobes to aerobes is not. In our simulation we do not find many anaerobes though they are observed in vivo. The literature concerning these data is scarce however, and the accuracy is not very high. This difference in available data stems from the fact that measurements from the large intestine are obtained from faecal samples; samples from the small intestine require biopsies [8].

We find almost no bacterial growth in the colon. Most of the growth takes place in the caecum, which might be due to the light digestibility of the food we use. In the colon almost all of the growth which does occur, occurs along the intestinal wall. In vivo, Poulsen et al. [11] also observed a growing population along the intestinal wall, combined with a static one in the lumen of the colon of mice.

We find aerobes attached to the wall throughout the small intestine using the carrying capacity of the wall almost to the utmost. In addition, we find that when wall attachment is possible, the aerobes in the lumen of the small intestine are almost exclusively wall attaching bacteria, despite their slower growth. Otherwise, the aerobes in the lumen are almost exclusively strictly lumenal. This suggests wall attachment is an asset for aerobic bacteria.

Another thing we find in our simulations is that in contrast to the small intestine we find almost no bacteria attached to the wall in the distal colon. We do however find bacteria attached to the wall in the proximal colon. Note that these differences are not due to the wall model itself. In vivo, wall attachment of bacteria is found in the small intestine [8]. By contrast, in the large intestine of mice, Poulsen found bacteria in the mucus covering the wall, but not actually attached [11]. In our model the differences are due to different limits on growth. In the small intestine growth in the lumen is limited by both oxygen influx and dilution by the flow. Growth in the distal colon however, insofar existing, is limited by food and oxygen but not by dilution. Therefore bacteria in the small intestine that attach to the wall avoid dilution and thus a restriction on growth, finding a new restriction in the available space at the wall. Bacteria in the distal colon however do not escape a restriction on growth by attaching to the wall, but would become more restricted in their growth because of their penalty for being attached to the wall.

7   Concluding remarks

This modelling is a start. In due course more aspects will be modelled. As knowledge proceeds more accurate modelling can be employed, and as we want to uncover deeper laying facts more accurate modelling must be employed. So in time our model will grow in size and quality and more computing power will be needed.

In the near future we would like to add facultative bacteria, a mucus layer, a better geometry, different kinds of food, etcetera. In the not so near future we might integrate our software with in silico models of the immune system which are being built by others [13].

Next to biological adaptations we would like to make the software available for the community, therefore we would like to improve the interface with the user, but also to port the software from our shared memory vector machine to a cluster, making it available for a broader public.

Despite the simplicity of the modelling system used, several salient features of the intestinal microflora can be reproduced. Though tactical modelling may always be beyond our possibilities, this study shows strategic modelling is feasible.

References

1. H. Boureau, L. Hartmann, T. Karjalainen, I. Rowland, and M. H. F. Wilkinson. Models to study colonisation and colonisation resistance. *Microbial Ecology in Health and Disease*, 12 suppl. 2:247–258, 2000.
2. M. E. Coleman, D. W. Dreesen, and R. G. Wiegert. A simulation of microbial competition in the human colonic ecosystem. *Apllied and Environmental Microbiology*, 62:3632–3639, 1996.
3. J. Davies. Bacteria on the rampage. *Nature*, 383:219–220, 1996.
4. R. Freter, H. Brickner, J. Fekete, M. M. Vickerman, and K. V. Carey. Survival and implantation of escherichia coli in the intestinal tract. *Infection and Immunity*, 39:686–703, 1983.
5. J. Gerritse, F. Schut, and J. C. Gottschal. Modelling of mixed chemostat cultures of an anaerobic bacterium Comamonas testosteroni, and an anaerobic bacterium Veilonella alcalescencs: comparison with experimental data. *Applied and Environmental Microbiology*, 58:1466–1476, 1992.
6. S.B. Issenberg, W. C. McGaghie, I. R. Hart, J. W. Mayer, J. M. Felner, E. R. Petrusa, R. A. Waugh, D. D. Brown, R. R. Safford, I. H. Gessner, D. L. Gordon, and G. A. Ewy. Simulation technology for health care professional skills training and assessment. *JAMA*, 282:861–866, 1999.
7. D. E. Kirschner and M. J. Blaser. The dynamics of helicobacter pylori infection in the human stomach. *Journal of Theoretical Biology*, 176:281–290, 1995.
8. B. Kleessen, E. Bezirtzoglou, and J. Mättö. Culture-based knowledge on biodiversity, development and stability of human gastrointestinal microflora. *Microbial Ecology in Health and Disease*, 12 suppl. 2:54–63, 2000.
9. A. L. Koch. The monod model and its alternatives. In A. L. Koch, J. A. Robinson, and G. A. Milliken, editors, *Mathematical Modeling in Microbial Ecology*, pages 62–93. Chapman & Hall, New York, 1998.
10. D. Levins. *Evolution in a Changing Environment.* Princeton University Press, Princeton, 1968.
11. L. K. Poulsen, F. Lan, C. S. Kristensen, P. Hobolth, S. Molin, and K. A. Krogfelt. Spatial distribution of Escherichia coli in the mouse large intestine inferred from rRNA in situ hybridization. *Infection and Immunity*, 62:5191–5194, 1994.
12. R. W. Schaedler, R. Dubos, and R. Costello. Association of germfree mice with bacteria isolated from normal mice. *J. Exp. Med.*, 122:77–82, 1963.
13. H. B. Sieburg, J. A. McCutchan, O. Clay, L. Caballero, and J. J. Ostlund. Simulation of hiv-infection in artificial immune system. *Physica D*, 45:208–228, 1990.
14. D. van der Waaij, J. M. Berghuis-De Vries, and J. E. C. Lekkerkerk-Van der Wees. Colonization resistance of the digestive tract in conventional and antibiotic treated mice. *Journal of Hygiene*, 69:405–411, 1971.
15. M. H. F. Wilkinson. The mimics cellular automaton program design and performance testing. MIMICS Technical Report 1, Centre for High Performance Computing, University of Groningen, 1997. Available via http://www.cs.rug.nl/~michael/ under downloadable publications as pdf-file.
16. M. H. F. Wilkinson. Nonlinear dynamics, chaos-theory, and the "sciences of complexity": their relevance to the study of the interaction between host and microflora. In P. J. Heidt, V. Rusch, and D. van der Waaij, editors, *Old Herborn University Seminar Monograph 10: New Antimicrobial Strategies*, pages 111–130, Herborn-Dill, Germany, 1997. Herborn Litterae.

A Mesoscopic Approach to Modelling Immunological Memory

Yongle Liu and Heather J. Ruskin

School of Computer Applications
Dublin City University, Dublin 9, Ireland

Abstract. In recent years, the study of immune response behaviour through mathematical and computational models has been the focus of considerable efforts. We propose a mesoscopic model to combine the most useful features of the microscopic and macroscopic approaches, which have been the alternatives to date. Cellular automata and Monte Carlo simulation are used to describe the humoral and T-cell mediated immune response, where the nature of the response induced depends on the polarization of T_H1 and T_H2 cells. Memory immunity is introduced to our model, so that we can simulate primary and secondary immune response. The high affinity between memory B-cells and antibodies contributes to a quick and intense response to repeated infection. The experiments on $P_{affB-AB}$ and $P_{affMB-AB}$, which control antibody production, explore the different roles of B-cell and memory B-cell in immune response stages. The duration of immunological memory is also studied.

Keywords: Cellular automata; Monte Carlo; Humoral; T-cell mediated; Immune response; Immunisation; Mesoscopic.

1 Introduction

A number of mathematical and computational models have been developed over recent years to describe the human immune system and its response to threat. These efforts can be grouped principally in terms of their focus on microscopic or macroscopic features respectively. Microscopic models thus provide detailed information on the immune system, such as T-cell polarisation, (where naive CD4 T-cells differentiate into T_H1 and T_H2 cells) [1-2], vaccine complexity [1], repertoire descriptors, (where receptors of T and B-cells, epitope and peptide of antigen and antibody are described by bit-strings [3]), and antigen-specific clonal expansion, (where only one kind of T-cell is likely to be specific to a particular antigen) [4, 5]. The microscopic model IMMSIM is a well-known and much studied example [4, 5]. It includes seven entities or cell types, with bit-strings used to simulate the key-lock interactions between the elements of the immune system. In comparison to the detailed microscopic approach, other cellular automata-based or similar models have until now used less cell types or simpler inter-cell interactions [9-16], whereas ordinary differential equations (ODE)-based [6-8] or continuous models are essentially macroscopic in outlook. The aim has been to study the overall disease evolution, such as global aspects of the viral infection,

P.M.A. Sloot et al. (Eds.): ICCS 2002, LNCS 2329, pp. 127–136, 2002.

overall symptoms, and related statistics. Thus, use of reduced and simplified interactions and system parameters permits an overview of the entire population of cells and molecules. Mannion et al. [9, 10], proposed models of cellular automata (Monte Carlo) type for the T-cell mediated response to HIV. Four cell types, two sets of interactions, and two global parameters, P_{mut} and P_{mob}, which built on earlier models with few cell types [12, 13], are involved in this MC model.

In contrast, we propose a mesoscopic model here, which seeks to bridge microscopic and macroscopic features. The cell types and global parameters in the original model [9,10] are retained, and an increasing number of cells of different types are used to more densely populate the host space. This approach thus increases the detail on the immune space, with all cell populations interacting on an individual cell-by-cell basis, rather than as a well-mixed cell population described by ODEs [6-8]. Furthermore, Instead of using the combination of scalars for immune cells and antigen and allowing several of them on a single lattice site (like IMMSIM [4, 5]), we choose Booleans for all cell states. We simplify the cell populations, permitting a maximum of one cell of each type per site, which allows us to simulate on a larger-scale (computationally expensive for a detailed microscopic model), but compromises on highly local detail (at a given site). However, population dynamics of immune cells and overall disease evolution is obtained from the whole lattice, which is large scale. Hence, we describe an intermediate approach between microscopic and macroscopic.

In particular, we include *both* humoral and T-cell mediated immune response in what follows. Thus, the mesoscopic model attempts to describe T-cell polarizations, and to characterise primary and secondary humoral response.

2   Model

2.1   The entities

The immune system is extremely complex with many different cell types, proteins and molecules involved in its efficient operation. It is impossible to take into account all factors in a computational model, so we limit the cell types to eight key ones in the humoral and T-cell mediated immune response. These cells are not precisely those of a former 8-cell model proposed by Pandey [17], since some different entities are included and some omitted. Included here are Macrophage (M), T_H1 cell (T1), T_H2 cell (T2), Cytotoxic T-cell (CT), Memory T-cell (MT), B-cell (B), Memory B-cell (MB), Antibody (AB) and Antigen (AG). As well as the key entities described, receptors, molecules and other signals are also involved in immune response operation, which control e.g. the proliferation and elimination of the eight on which we focus. In our model, we do not consider these factors in terms of separate units, but present their function through parameters, which control cell-cell interactions.

2.2   Updating cells on the host systemic spatial lattice

The human immune system is represented by a simple cubic lattice, where each site has six neighbouring sites, with square cross-section. The lattice has linear dimension L, with L^3 sites, and periodic boundary conditions, so that edges of the lattice are wrapped. Any site can be occupied by any one of the eight cell types, but there is *at most one* cell of each cell type at a site. The cellular state is described by a Boolean variable, where "true" is used to indicate the high concentration of a given cell type at one site and "false" the low concentration.

A Monte Carlo method is used, which stochastically updates cellular states, following interactions, with one update of the whole lattice equal to one Monte Carlo step (MCS): (asynchronous updating). A site is randomly selected to be updated and this selection is repeated L^3 times for one MCS. Around 1000 MCS are used to monitor the disease evolution.

In updating cell types at any one site, a number of features, such as growth, inter-cell interactions and death must be considered in order to mimic real biological processes. States of all cells *at a given site* are simultaneously updated. At the beginning of an update, the information is saved in temporary states. All states evolve from these temporary ones, so that results of updating all cell types are independent of the sequence used.

2.3   The growth

Every cell type at a site at time $t+1$ depends on the states of its six neighbouring sites at time t. In Equation (1), if any cell type at a given site has a neighbouring site, occupied by the same cell type with state = "true" at time t, then the state of the given cell type will evolve to "true" at time $t+1$.

$$S'_{ic}(x,y,z,t+1) = S_{ic}(x,y,z,t).or.S_{ic}(x+1,y,z,t)$$
$$.or.S_{ic}(x-1,y,z,t).or.S_{ic}(x,y+1,z,t).or.S_{ic}(x,y-1,z,t) \qquad (1)$$
$$.or.S_{ic}(x,y,z+1,t).or.S_{ic}(x,y,z-1,t)$$

where S' is the intermediate state of a given cell; ic is the cell type and x, y and z are the coordinates of the site. Equation (1) is the same as that of our original model [9, 10], but cells may not always proliferate successfully. In every inter-site interaction of any one cell type, a probability $(P_{prolrate}(ic))$ is assigned to successful growth. Different values of $P_{prolrate}(ic)$ reflect the fact that every cell type has different growth rate to proliferate in real immune system. After growth, the cellular states are temporary as before, and are used as a basis for updating the cellular states in inter-cell interactions. These involve different cell types at any one site, i.e. account for affinities and triggers between cell types.

2.4   Interactions

The following set of intra-site inter-cell interactions are used to describe the immune-response to antigens. All cellular states at time $t+1$ are evolved from the temporary states (S'_{ic}) after growth.

$$
\begin{align}
M(t+1) &= M'.or.AG'.and.[not[M'.and.AG']] \tag{2.1}\\
T1(t+1) &= [T1'.or.[AG'.and.M']].and.[not[T1'.and.AG']] \tag{2.2}\\
T2(t+1) &= [T2'.or.[AG'.and.M']].and.[not[T2'.and.AG']] \tag{2.3}\\
CT(t+1) &= CT'.or.[AG'.and.M'.and.T1'] \tag{2.4}\\
B(t+1) &= B'.or.[T2'.and.AG'] \tag{2.5}\\
AB(t+1) &= AB'.or.[[B'.or.MB'].and.AG'] \tag{2.6}\\
AG(t+1) &= AG'.and.[not[CT'.or.AB']] \tag{2.7}
\end{align}
$$
(2)

Our model thus attempts to describe in more detail the immune system be-haviour through simple interaction equations, limited though these still are in terms of the biological reality. All the interactions above are taken to be stochas-tic, and the success or failure of every interaction is controlled by independent probabilities. Values of all parameters are set between "0" and "1", with a proba-bility close to "1" implying that the specific inter-cell interaction is highly likely. The mechanism for cooperative interactions is described below.

In Equation (2.1), the presence of antigen, but no macrophage at one site, implies that the antigen can induce the growth of a macrophage with a probabil-ity $P_{affAG-M}$. If both antigen and macrophage are present at the same site, the antigen can kill the macrophage with a probability $P_{infectM}$. In Equation (2.2) and (2.3), the presence of an antigen and a macrophage together can activate growth and differentiation of T-cells with a probability P_{affTH}. A proportion (P_{ropT1}) of the newly generated T-cells differentiate into T_H1 T-cells, and the others $(1-P_{ropT1})$ into T_H2 T-cells. If no macrophage presents to the antigen, it becomes free and can kill T_H1 and T_H2 cells with a probability $P_{infectTH}$. Equation (2.4) shows that with probability $P_{affT1-CT}$, a cytotoxic T-cell grows when an antigen, a macrophage and a T_H1 T-cell are all present at that site. Further, the presence of a T_H2 T-cell and an antigen can induce the growth of a B-cell with a probability $P_{affT2-B}$ (Equation (2.5)). Antibody secretion is described in Equation (2.6). When the antigen presents at the same site, a B-cell secretes antibody with a probability $P_{affB-AB}$; a memory B-cell can se-crete antibody with a higher probability $P_{affMB-AB}$ than that for the B-cell, and the antibodies spread to all six neighbouring sites. Equation (2.7) describes the elimination of antigen. A cytotoxic T-cell (with a probability $P_{killCT-AG}$)or an antibody cell (with a probability $P_{killAB-AG}$) will kill an antigen which is present at the same site.

2.5   Death of all cell types

One MCS simulates activity in a unit period of time equivalent to a time period in the immune-response, which is equivalent to the smallest half-life of the eight entities. Consequently some cells may die naturally in this period. All the cells in the lattice are allowed to experience a natural death process with probability P_{death}. This death process is assumed (somewhat naively) to occur after cellular growth and the inter-cell interactions have taken place, ensuring that a new generation of cells are produced. New born cells and original cells are taken to have the same probability of natural death, (again a simplification) with no account taken of degradation of cell function in our model. This treatment of new born cells and original cells as effectively memory-less is assumed to have little effect on the evolution of the total populations, which are the major focus for study, but this is clearly a limitation on the model as a whole. We have based the probability of cell death in a unit of time on the biological half-life data, where *half-life* is defined to be the time required for half the number of a given cell type to be eliminated. Then

$$P_{death} = e^{-\frac{(ln2) \times \sigma}{\tau}}$$ (3)

where P_{death} is the death rate of the cell type, τ is half-life, σ is the period that one MCS represents in the real immune system.

2.6 The cellular mobility

In the immune system, realistically, all cell types are mobile. For example, T-cells and B-cells circulate continuously from the bloodstream to the lymphoid tissues and back to blood, and macrophages bring antigens from blood to lymphoid organs. The mobility of these cells increases the opportunity to interact with other cells. At the end of each MCS, all cells are permitted to move to their neighbouring sites with probability P_{mob}, which is non-directional i.e., one cell can move to any one of its six neighbouring sites with equal probability. Directional mobility, in which one cell can move in only selected axial directions e.g. positive only, may have some relevance, for example in chemotaxis. The mobility algorithm remains as described in the original model [9, 10] with the motivation for movement of a given cell based on the attraction for it to interact with other cell types. For example, a cytotoxic T-cell, antibody or memory B-cell randomly move to a neighbouring site occupied by an antigen to kill it, whereas, an antigen randomly moves to a neighbouring site occupied by a macrophages, T_H1 or T_H2 cell to infect them. The T_H1 cell, T_H2 cell or B-cell randomly move to any one of their six neighbouring sites, and do not require further special conditions.

3 Results

In order to investigate the immunological memory, we have included memory B-cells in our model, which have higher affinity to antibodies than B-cells. When B-cells die in the death process, some of them become memory B-cells. The parameter $P_{transRateMB}$ is the rate that decides what proportion of those dying B-cells can differentiate into memory B-cells. $P_{transRateMB}$ is set to 0.1 in the simulations.

   We show a typical simulation of immunisation in Fig. 1 for selected entities only. At the start of the simulation, the system has no immune cells, and only low density of antigen (1%) uniformly distributed on a S.C. lattice of size 50 ×

50×50. The antigen induces a quick and intense immune response, in which both humoral and T-cell mediated entities are involved. Antigen levels peak then decrease, because the immune response suppresses further growth. Finally, the antigen is eliminated, and the primary immune-response is complete. After the primary immune-response, all immune cells except memory B-cells gradually decrease then disappear, but memory B-cells remain due to their long half-life. The immune system is thus returned to the healthy state, but retains memory of the antigen. If the same dose of antigen is again distributed randomly into the lattice, the secondary immune-response is activated. This is characterised by faster and more effective response to antigen [18]. A large amount of antibodies are produced very quickly, so the density of antigens is limited to a low level, which is far below that of primary immune response. The density of B-cells is also lower than in the primary response. This suggest secondary immune response is far more intense and effective than the primary, and that the antigen causes less damage to the immune system. This behaviour is similar to the behaviour of a real immune system.

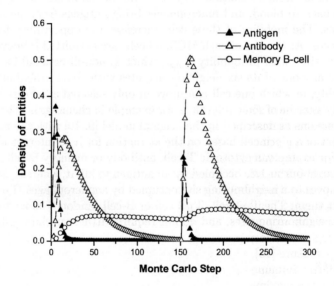

Fig. 1. Immunological memory to repeated infection

In our model, the ability of B-cells and memory B-cells to secrete antibody are controlled by parameters $P_{affB-AB}$ and $P_{affMB-AB}$, and the humoral immune response is predominantly driven by these two parameters. In Fig. 2, we look at the effect of $P_{affB-AB}$ and $P_{affMB-AB}$ on the antigen population in primary and secondary immune response. We fix the value of $P_{affMB-AB}$, and increase the value of $P_{affB-AB}$ from 0.3 to 1.0 in Fig. 2a. The selection of other parameters involved in this humoral immune response are shown in Table 1. We vary the parameters individually to investigate how sensitive results are to parameter values chosen. $P_{affAG-M}$, P_{affTH} and $P_{killAB-AG}$, which control the growth of immune cells, just affect the growth speed of these cells, and have little influence on values of peak densities. The period of an immune cell's half-life and $P_{prolRate-AG}$ decide the peak densities of these immune cells and antigen respectively. With low density of macrophage, $T_H 2$ cell or B-cell, there are not enough antibodies produced, so antigens can not to be eliminated from the immune system. If $P_{prolRate-AG}$ is high, antigens have a high growth rate, and survive the immune response.

Parameter	Value	Parameter	Value
$P_{transRateMB}$	0.13	$P_{prolRate-AG}$	0.2
$P_{affAG-M}$	1.0	Half-life of M	10(days)
P_{affTH}	1.0	Half-life of T2	10(days)
P_{T2-B}	1.0	Half-life of AB	20(days)
$P_{killAB-AG}$	1.0	Half-life of MB	400(days)

Table 1. Parameter selections in Fig.2.

With the increase of $P_{affB-AB}$, the peak antigen population increases in primary response, but remains at the same level in the secondary, which suggests that the parameter principally influences the former, and is negligible for the latter. In Fig. 2b, the value of $P_{affB-AB}$ is fixed, and the value of $P_{affMB-AB}$ is changed from 0.3 to 1.0. Little difference in primary immune response is observed, but peak antigen density increases, when $P_{affMB-AB}$ is increased for the secondary stage. This phenomenon suggests that $P_{affMB-AB}$ drives the secondary immune response, but has little effect on the primary.

Immunological memory is sustained by long-lived memory B-cells, induced by the original exposure [18], so that memory B-cells provide long-term protection after the primary immune-response is over. In Fig.3, we investigate results from our model for the maximum antigen density in secondary immunity (D_{maxAG}) vs. the time interval between antigen injections(Δt), where an expression of the form

$$D_{maxAG} \propto \Delta t^{\beta} \tag{4}$$

clearly applies. If the antigen injection interval is short, a more intense secondary response is induced, and the antigen population is limited to a low level. If the interval is very long, the secondary immune response is not as effective as before, and maximum antigen density achieves the same level as that in primary immune response (dotted line Fig.3). Results thus support the fact that immunological

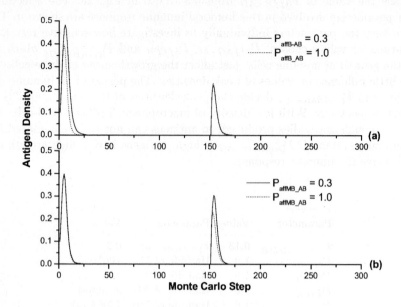

Fig. 2. $P_{affMB-AB}$ is fixed to 1.0 (a) and $P_{affB-AB}$ is fixed to 1.0 (b); Lattice size is $30 \times 30 \times 30$; Ten samples are generated for each value of $P_{affB-AB}$ and $P_{affB-AB}$, and the average results are displayed.

memory gradually fades after exposure to initial antigen attack. The immunological memory is related to half-life of memory B-cells and the transfer rate of B to memory B-cells. The larger the percentage of memory B-cells, the longer the duration of relative immunity.

4   Conclusion

In conclusion, we studied immunological memory in this paper through a mesoscopic model incorporating key cell types and stochastic interactions with T_H1 T_H2 differentiation. Main features of the immunological memory were captured, but details of cell population dynamics are not presented. In repeated infection, we obtained a more rapid and intense secondary response compared to that of the original exposure, so that antigen population is suppressed to a low level. A preliminary analysis of the influence of probabilities assigned to antibody secretion of B and memory B-cells demonstrated that first infection immune response is driven predominantly by B-cells while memory B-cells drive repeated infection. Immunological memory (based on memory B-cells) diminishes with time, so that after a very long period, the antigen invasion is treated as a first infection because of the loss of memory to the antigen. This is reflected in our model (Fig.3) with $D_{maxAG} \propto \Delta t^\beta$. The long half-life of memory B-cells and large transfer rate from B to memory B-cells sustain long immunological memory.

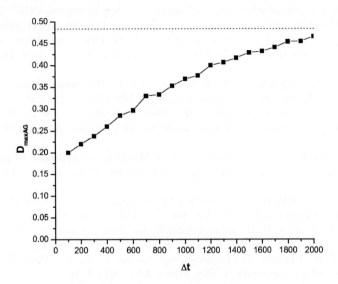

Fig. 3. Decline in immunological memory. Maximum antigen density in primary immune response (dotted line).

Our model attempted an intermediate approach between microscopic and macroscopic, and results of this paper show that some features of immune response could be reproduced, notably immunological memory. Currently, we are also studying T-cell mediated immune response specific to HIV, and cooperation of the humoral and the T-cell mediated reactions, as well as chemotaxis (directional mobility) from a mesoscopic viewpoint.

References

1. Kohler, B., Puzone, R., Seiden, P.E., Celada,F.: A systematic approach to vaccine complexity using an automaton model of the cellular and humoral immune system I. Viral characteristics and polarized response. Vaccine **19** (2001), 862–876
2. Ortega, N.R.S., Pinheiro, C.F.S., Tome, T., Felicio, J.R.D.: Critical behavior of a probabilistic cellular automaton describing a biological system. Physica A **255** (1998), 189–200
3. Lagreca, M.C., Almeida, R.M.C., Santos, R.M.Z.: A dynamical model for the immune repertoire. Physica A **289** (2001), 191–207
4. Celada, F., Seiden, P.E.: A computer model of cellular interactions in the immune system. Immunology Today **13** (1992), 56–62
5. Bezzi, M., Celada, F., Ruffo, S., Seiden, P.E.: The transition between immune and disease states in a cellular automaton model of clonal immune response. Physica A **245** (1997), 145–163

6. Boer, R.J.D., Oprea, M., Antia, R., Murali-krishna, K., Ahmed, R., Perelson, A.S.: Recruitment times, proliferation, and apoptosis rates during the $CD8^+$ T-cell Response to lymphocytic choriomeningitis virus. J. Virology. **75** (2001), 10663–10669

7. Boer, R.J.D., Segel, L.A., Perelson, A.S.: Pattern formation in one- and two-dimensional shape-space models of the immune system. J. Theor. Biol. **155** (1992), 295–333

8. Hershberg, U., Louzoun, Y., Atlan, H., Solomon, S.: HIV time hierarchy: winning the war while loosing all the battles. Physica A **289** (2001), 178–190

9. Mannion, R., Ruskin, H., Pandey, R.B.: Effect of mutation helper T-cells and viral population: A computer simulation model for HIV. Theor. Biosci. **119** (2000a), 10–19

10. Mannion, R., Ruskin, H., Pandey, R.B.: A Monte Carlo approach to population dynamics of cells in a HIV immune response model. Theor. Biosci. **119** (2000b), 145–155

11. Kougias, C.F., Schulte, J.: Simulating the immune response to the HIV-1 virus with cellular automata. J. Stat. Phys. **60**, Nos. 1/2, (1990), 263–273.

12. Stauffer, D. Pandey, R. B.: Immunologically motivated simulations of cellular automata. Computers in Physics, **6**, No. 4, JUL/AUG (1992), 404–410

13. Pandey, R.B., Stauffer, D.: Immune response via interacting three dimensional network of cellular automata. J. Phy. France **50** (1989), 1–10

14. Pandey, R.B., Stauffer, D.: Metastability with probabilistic cellular automata in an HIV infection. J. Stat. Phys. **61**, Nos. 1/2, (1990), 235–240

15. Mielke, A., Pandey, R.B.: A computer simulation study of cell population in a fuzzy interaction model for mutating HIV. Physica A **251** (1998), 430–438

16. Pandey, R.B.: Computer simulation of a cellular automata model for the immune response in a retrovirus system. J. Stat. Phys., **54**, Nos. 3/4, (1989), 997–1010

17. Pandey, R.B.: Growth and decay of a cellular population in a multicell immune network. J. Phys. A: Math. Gen. **23** (1990), 4321–4331

18. Janeway, C.A., Travers, J.P., Walport, M., Capra, J.D.: Immunobiology. Garlan Publishing (1999).

A New Method for Ordering Binary States Probabilities in Reliability and Risk Analysis*

Luis González

Department of Mathematics,
University of Las Palmas de Gran Canaria,
Edificio de Informática y Matemáticas,
35017 Las Palmas de Gran Canaria, Spain
luisglez@dma.ulpgc.es

Abstract. A new method is proposed to select the binary n-tuples of 0s and 1s by decreasing order of there occurrence probabilities in stochastic Boolean models. This method can be applied to evaluate fault trees in Reliability Engineering and Risk Analysis, as well as to many other problems described by a stochastic Boolean structure. The selecting criterion is exclusively based on the positions of 0s and 1s in the binary n-tuples. In this way, the computational cost in sorting algorithms is drastically reduced, because the proposed criterion is independent of the probabilities of the Boolean variables. Every step, the algorithm extends the set of selected binary states, obtaining the binary n-tuples just by adding 0 (1) at the end of all the (some) previously selected binary $n - 1$-tuples.

1 Introduction

In many applications of Probability Calculus and Operations Research, the basic variables are Boolean variables with a prefixed occurrence probability. This happens, for example, in Reliability Theory and Risk Analysis, where the technical system is described by a Boolean function Φ depending on the basic components of the system. This Boolean function is often represented by a fault tree, a picture describing the superposition of component faults to create system faults [2], [9]. Then, assuming that: $\Phi = 1$ if the system fails, the system unavailability, usually called the top event probability, is evaluated by computing the probability $\Pr\{\Phi = 1\}$.

Many methods have been developed to evaluate the top event probability in Fault Tree Analysis. These methods are usually classified into two great groups: deterministic methods and probabilistic methods [3]. The first ones provide exact lower and upper bounds on top event probability [1], [7], [8]. The second ones provide an estimation of the system unavailability by using sampling and variance reduction techniques [6]. Both, deterministic and probabilistic methods, can also be subdivided into two classes: direct methods and indirect methods. The direct methods do not require any qualitative analysis of the Boolean function Φ [6], [7], [8]. The indirect methods are based on a previous analysis of the

---

* Partially supported by MCYT, Spain. Grant contract: REN2001-0925-C03-02/CLI

P.M.A. Sloot et al. (Eds.): ICCS 2002, LNCS 2329, pp. 137–146, 2002.

fault tree logic: minimal cut sets, minimal paths, common cause failure analysis, etc. [1], [5].

It is very simple to describe an elemental direct method to evaluate $\Pr\{\Phi = 1\}$ prefixing the maximum admissible error (difference between upper and lower bounds). This method is based on the canonical normal forms of the Boolean function Φ and it can be summarised as follows [4], [9].

The probability of a Boolean function Φ taking value 1 can be estimated using the canonical normal forms of Φ. More precisely, we can obtain lower and upper bounds L, U on system unavailability by taking any subsets S_1, S_0 of the sets of binary n-tuples for which $\Phi = 1$, $\Phi = 0$, respectively:

$$S_1 \subseteq \{u \in \{0,1\}^n \ / \ \Phi(u) = 1\} \quad ; \quad S_0 \subseteq \{u \in \{0,1\}^n \ / \ \Phi(u) = 0\}.$$

$$L = \sum_{u \in S_1} \Pr\{u\} \leq \Pr\{\Phi = 1\} \leq 1 - \sum_{u \in S_0} \Pr\{u\} = U. \tag{1}$$

where, assuming that the Boolean variables are mutually independent, we have:

$$\forall \, u = (u_1, u_2, \ldots, u_n) \in \{0,1\}^n : \quad \Pr\{u\} = \prod_{i=1}^{n} p_i^{u_i} (1 - p_i)^{1-u_i}. \tag{2}$$

that is, $\Pr\{u\}$ is the product of factors $\begin{cases} p_i, & \text{if} \ \ u_i = 1 \\ 1 - p_i, & \text{if} \ \ u_i = 0 \end{cases}$.

Then, for any set of selected binary n-tuples, the maximum error in the estimation will be:

$$U - L = 1 - \sum_{u \in S_0 \, \cup \, S_1} \Pr\{u\}. \tag{3}$$

Therefore, the accuracy in the estimation of the top event probability improves at the same time as the total sum $\sum_{u \in S_0 \, \cup \, S_1} \Pr\{u\}$ of the probabilities of all the selected elementary states increases, and this sum is completely independent of the fault tree logic.

Of course, the main question is how to select the minor number of elementary states, with probabilities as large as possible, in order to minimise the computational cost. The first answer to this question: ordering n-tuples attending to their probabilities is not valid, because of the exponential nature of the problem and the high computational costs in sorting algorithms. Then, it is essential to propose an efficient criterion that allows the identification of elementary states with probabilities as large as possible, in order to obtain a good approximation of the top event probability using as few elementary states as possible. For this purpose, we have established a characterisation theorem [4], which provides us the above mentioned criterion.

In Sect. 2, we state this theorem and a few related explanations for a better understanding of the new results we propose in this paper. Beginning with our new approaches, in Sect. 3 we establish a classification of the n-tuples of 0s and 1s into three groups: Top, Bottom and Jumping n-tuples. The Top n-tuples correspond to the binary states with largest occurrence probabilities. In Sect. 4, we present a simple and elegant algorithm to generate all the Top n-tuples. Next, we prove that this algorithm selects the binary states by decreasing order of their occurrence probabilities. Finally, in Sect. 5 we present our conclusions.

2 Previous Results

In [4] we have established the following theorem that *a priori* assures us that the occurrence probability of a given elementary state is always greater (lower) than another one, for any values of basic probabilities:

Theorem 1. *An intrinsic order criterion for binary n-tuples probabilities.*
Let $x_1, \ldots, x_n$ be n Boolean variables (corresponding to n basic components of a system), with basic probabilities $p_i = \Pr\{x_i = 1\}$ $(1 \leq i \leq n)$ verifying:

$$0 < p_1 \leq \ldots \leq p_n \leq \frac{1}{2}. \tag{4}$$

Then, the probability of the n-tuple $(u_1, \ldots, u_n) \in \{0,1\}^n$ is intrinsically greater than or equal to the probability of the n-tuple $(v_1, \ldots, v_n) \in \{0,1\}^n$ (that is, with independence on probabilities p_i) if and only if the matrix

$$\begin{pmatrix} u_1 \ldots u_n \\ v_1 \ldots v_n \end{pmatrix}. \tag{5}$$

has not any $\binom{1}{0}$ column, or for each $\binom{1}{0}$ column there exists (at least) one corresponding preceding $\binom{0}{1}$ column (Condition IOC).

Remark 1. In the following we assume that basic probabilities p_i always satisfy condition (4). Note that this hypothesis is not restrictive because, if for some i: $p_i > \frac{1}{2}$, we only need to consider the Boolean variable $\overline{x_i}$, instead of x_i.

Remark 2. The $\binom{0}{1}$ column preceding to each $\binom{1}{0}$ column is not necessarily allocated at the immediately previous position.

Remark 3. The term *corresponding*, used in Theorem 1, has the following meaning: for each two $\binom{1}{0}$ columns in matrix (5), there must exist (at least) two different $\binom{0}{1}$ columns preceding to each other.

The partial order relation established by Theorem 1 between the n-tuples of $\{0,1\}^n$ is called intrinsic order, because it only depends on the positions of 0s and 1s in the n-tuples, but not on the basic probabilities p_i. In the following we shall note with the symbol $\preceq$ this order relation:

Definition 1. *For all* $u = (u_1, \ldots, u_n), v = (v_1, \ldots, v_n) \in \{0,1\}^n$:

$$(u_1, \ldots, u_n) \succeq (v_1, \ldots, v_n) \Leftrightarrow \Pr\{u\} \geq \Pr\{v\}; \quad \textit{for all } p_i \ \textit{s.t.:} \quad (4) \Leftrightarrow$$
$$\begin{pmatrix} u_1 & \cdots & u_n \\ v_1 & \cdots & v_n \end{pmatrix} \textit{ satisfies IOC.}$$

Example 1. $(0,0,1,1) \succeq (1,1,0,0)$ because $\begin{pmatrix} 0 & 0 & 1 & 1 \\ 1 & 1 & 0 & 0 \end{pmatrix}$ satisfies *IOC* (Remark 2):

$$\Pr\{(0,0,1,1)\} \geq \Pr\{(1,1,0,0)\}, \textit{ for all } p_1 \leq p_2 \leq p_3 \leq p_4 \leq \frac{1}{2}. \quad (6)$$

Example 2. $(0,1,1,1) \not\succeq, \not\preceq (1,1,0,0)$ because neither $\begin{pmatrix} 0 & 1 & 1 & 1 \\ 1 & 1 & 0 & 0 \end{pmatrix}$ nor $\begin{pmatrix} 1 & 1 & 0 & 0 \\ 0 & 1 & 1 & 1 \end{pmatrix}$ satisfy *IOC* (Remark 3). Therefore, depending on p_i:

$$\Pr\{(0,1,1,1)\} \leq, \geq \Pr\{(1,1,0,0)\}. \quad (7)$$

For instance, if for all $i = 1, 2, 3, 4$:

$$p_i = 10^{-1}i \Rightarrow \Pr\{(0,1,1,1)\} = 2.1 \cdot 10^{-2} > \Pr\{(1,1,0,0)\} = 8.4 \cdot 10^{-3}. \quad (8)$$

$$p_i = 10^{-2}i \Rightarrow \Pr\{(0,1,1,1)\} = 2 \cdot 10^{-5} < \Pr\{(1,1,0,0)\} = 1.8 \cdot 10^{-4}. \quad (9)$$

Now, we present the graph of the intrinsic order for $n = 4$.

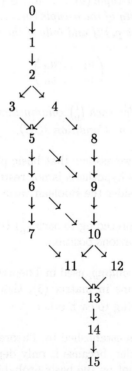

Fig. 1. Graph of the partially ordered set $\left(\{0,1\}^4, \preceq \right)$

The 4-tuples are given using decimal numbering and the relation $a \succeq b$ is noted by $a \longrightarrow b$. To finish Sect. 2, let us mention that intrinsic order implies lexicographic (truth table) order, i.e.: If $(u_1, \dots, u_n) \succeq (v_1, \dots, v_n)$ then, $(u_1, \dots, u_n)$ is placed at previous position to $(v_1, \dots, v_n)$ in the truth table [4]. The converse is false, and this is the most simple counter-example $(n = 3)$: $(0, 1, 1) \not\succeq (1, 0, 0)$ (Remark 3).

3   Top, Bottom and Jumping n-Tuples

The subgraph of $\left(\{0, 1\}^4, \preceq \right)$ corresponding to $n = 3$ (Fig. 1), shows that there exist only two binary 3-tuples non-comparable by intrinsic order: $3 \not\succeq 4$ and $3 \not\preceq 4$. Therefore, when we write the eight binary 3-tuples by decreasing order of their occurrence probabilities, only two cases are possible:

$$
\begin{array}{cc}
0 & 0 \\
1 & 1 \\
2 & 2 \\
3 & 4 \\
\cdots & \cdots \\
4 & 3 \\
5 & 5 \\
6 & 6 \\
7 & 7 \\
\end{array}
$$

Fig. 2. Binary 3-tuples by decreasing order of their occurrence probabilities

The left column corresponds to the case $\Pr\{(0, 1, 1)\} \geq \Pr\{(1, 0, 0)\}$, and the right column corresponds to the case $\Pr\{(0, 1, 1)\} \leq \Pr\{(1, 0, 0)\}$. Anyway, Fig. 2 shows that for the two possible cases: the 3-tuples 0, 1 and 2 are among the four first ones; the 3-tuples 5, 6 and 7 are among the four last ones and the 3-tuples 3 and 4 can be allocated at both positions (depending on basic probabilities p_i). This fact suggests us the following definition.

Definition 2. *Let the 2^n binary n-tuples be ordered by decreasing order of their occurrence probabilities. Then:*
(i) *The binary n-tuple $(u_1, \dots, u_n)$ is called Top if it is always among the 2^{n-1} first n-tuples (for any basic probabilities p_i).*
(ii) *The binary n-tuple $(u_1, \dots, u_n)$ is called Bottom if it is always among the 2^{n-1} last n-tuples (for any basic probabilities p_i).*
(iii) *The binary n-tuple $(u_1, \dots, u_n)$ is called Jumping if it is neither Top nor Bottom.*

Denoting:

$$
\begin{aligned}
\mathcal{T}(n) &= \{(u_1, \dots, u_n) \in \{0, 1\}^n \, / \, (u_1, \dots, u_n) \text{ is Top}\} \\
\mathcal{B}(n) &= \{(u_1, \dots, u_n) \in \{0, 1\}^n \, / \, (u_1, \dots, u_n) \text{ is Bottom}\} \\
\mathcal{J}(n) &= \{(u_1, \dots, u_n) \in \{0, 1\}^n \, / \, (u_1, \dots, u_n) \text{ is Jumping}\}.
\end{aligned}
\tag{10}
$$

Definition 2 leads to the following partition of the set of n-tuples of 0s and 1s:

$$\{0,1\}^n = \mathcal{T}(n) \cup \mathcal{J}(n) \cup \mathcal{B}(n). \tag{11}$$

Now, we characterise the Top, Bottom and Jumping n-tuples, by a simple and elegant criterion. For this purpose, we need to state the following definition.

Definition 3. *For all* $(u_1, \ldots, u_n) \in \{0,1\}^n$, *the complementary of* $(u_1, \ldots, u_n)$ *is the* n-*tuple:*

$$(u_1, \ldots, u_n)^c = (1 - u_1, \ldots, 1 - u_n). \tag{12}$$

Obviously, the sum of two complementary n-tuples is always $(1, \overset{n}{\ldots}, 1) \equiv 2^n - 1$. Then, we can observe in Fig. 2 that the complementary 3-tuples are always allocated at symmetric positions with respect to the medium line. Indeed, this fact is an immediate consequence of the following result.

Lemma 1. *For all* $u, v \in \{0,1\}^n$:

$$\Pr\{u\} \geq \Pr\{v\} \Leftrightarrow \Pr\{u^c\} \leq \Pr\{v^c\}. \tag{13}$$

Proof. For all $u = (u_1, \ldots, u_n), v = (v_1, \ldots, v_n) \in \{0,1\}^n$, from (2) we get:

$$\Pr\{u\} \geq \Pr\{v\} \Leftrightarrow \prod_{i=1}^n p_i^{u_i} (1 - p_i)^{1-u_i} \geq \prod_{i=1}^n p_i^{v_i} (1 - p_i)^{1-v_i} \Leftrightarrow$$

$$\prod_{i=1}^n p_i^{(1-v_i)-(1-u_i)} (1 - p_i)^{v_i-u_i} \geq 1 \Leftrightarrow$$

$$\prod_{i=1}^n p_i^{1-u_i} (1 - p_i)^{1-(1-u_i)} \leq \prod_{i=1}^n p_i^{1-v_i} (1 - p_i)^{1-(1-v_i)} \Leftrightarrow \Pr\{u^c\} \leq \Pr\{v^c\} \quad \square$$

From Lemma 1 we obtain the following nice characterisation of Top, Bottom and Jumping n-tuples.

Theorem 2. *For all* $(u_1, \ldots, u_n) \in \{0,1\}^n$:
(i) $(u_1, \ldots, u_n) \in \mathcal{T}(n)$ *iff* $(u_1, \ldots, u_n)$ *has not any 1, or for each 1 there exists (at least) one corresponding preceding 0 (Condition T).*
(ii) $(u_1, \ldots, u_n) \in \mathcal{B}(n)$ *iff* $(u_1, \ldots, u_n)$ *has not any 0, or for each 0 there exists (at least) one corresponding preceding 1 (Condition B).*
(iii) $(u_1, \ldots, u_n) \in \mathcal{J}(n)$ *iff* $(u_1, \ldots, u_n)$ *has at least one 1 without its corresponding preceding 0, and at least one 0 without its corresponding preceding 1 (Condition J).*

Remark 4. The term *corresponding*, used in Theorem 2, has the same meaning that the one explained in Remark 3 for Theorem 1: Condition T (B) requires that for each two 1s (0s) in $(u_1, \ldots, u_n)$ there must exist (at least) two different 0s (1s) preceding to each other.

Proof (of Theorem 2). (i) Using Lemma 1, we have:

$$(u_1, \ldots, u_n) \in \mathcal{T}(n) \Leftrightarrow (u_1, \ldots, u_n)^c \in \mathcal{B}(n) \Leftrightarrow$$

$$(u_1, \ldots, u_n) \succeq (u_1, \ldots, u_n)^c \Leftrightarrow$$

$$\begin{pmatrix} u_1 & \cdots & u_n \\ 1 - u_1 & \cdots & 1 - u_n \end{pmatrix} \text{ satisfies } IOC \Leftrightarrow (u_1, \ldots, u_n) \text{ satisfies } T.$$

(ii) Using (i), we have:

$$(u_1, \ldots, u_n) \in \mathcal{B}(n) \Leftrightarrow (u_1, \ldots, u_n)^c \in \mathcal{T}(n) \Leftrightarrow$$

$$(u_1, \ldots, u_n)^c \text{ satisfies } T \Leftrightarrow (u_1, \ldots, u_n) \text{ satisfies } B.$$

(iii) Using (i) and (ii), we have:

$$(u_1, \ldots, u_n) \in \mathcal{J}(n) \Leftrightarrow (u_1, \ldots, u_n) \notin \mathcal{T}(n) \cup \mathcal{B}(n) \Leftrightarrow$$

$$(u_1, \ldots, u_n) \text{ does not satisfy } T \text{ and does not satisfy } B \ \square$$

Example 3. For $n = 1, 2, 3$ we obtain:

$$\mathcal{T}(1) = \{(0)\} \ ; \ \mathcal{J}(1) = \emptyset \ ; \ \mathcal{B}(1) = \{(1)\}. \tag{14}$$

$$\mathcal{T}(2) = \{(0,0), (0,1)\} \ ; \ \mathcal{J}(2) = \emptyset \ ; \ \mathcal{B}(2) = \{(1,0), (1,1)\}. \tag{15}$$

$$\mathcal{T}(3) = \{(0,0,0), (0,0,1), (0,1,0)\} \ ; \ \mathcal{B}(3) = \{(1,0,1), (1,1,0), (1,1,1)\}$$
$$\mathcal{J}(3) = \{(0,1,1), (1,0,0)\}. \tag{16}$$

In the following, we restrict to the Top n-tuples, since they correspond to the binary states with largest probabilities. The following theorem states three elementary necessary conditions for Top n-tuples.

Theorem 3. *For all $u = (u_1, \ldots, u_n) \in \mathcal{T}(n)$:*
(i) $u_1 = 0$
(ii) The number of 0s of u is greater than or equal to its number of 1s.
(iii) $(u_1, \ldots, u_k) \in \mathcal{T}(k)$; $\forall k = 1, 2, \ldots, n$

Proof. Using Theorem 2-(i) the proof is straightforward $\square$

Identifying all the binary states with the same decimal (lexicographic) numbering, for any number of components, i.e.:

$$(u_1, \ldots, u_n) \equiv (0, u_1, \ldots, u_n) \equiv (0, 0, u_1, \ldots, u_n) \equiv \ldots \tag{17}$$

the following theorem states that the sets of Top n-tuples can be disposed in an increasing chain for inclusion.

Theorem 4. $\forall n \in N: \ \mathcal{T}(1) \subset \mathcal{T}(2) \subset \ldots \subset \mathcal{T}(n) \subset \mathcal{T}(n+1) \subset \ldots$

Proof. Using Theorem 2-(i), we have:

$$\forall n \in N: \ (u_1, \ldots, u_n) \in \mathcal{T}(n) \Rightarrow (0, u_1, \ldots, u_n) \in \mathcal{T}(n+1) \ \square$$

Now, we can present in Sect. 4 an efficient algorithm to generate all the Top n-tuples.

4   Generation of Top n-Tuples

The simple description T of Top n-tuples (Theorem 2-(i)) allows us to present a simple algorithm for their generation (illustrated in Fig. 3). Starting from the only Top 1-tuple (see subgraph $n = 1$ in Fig.1, or Example 3):

$$T(1) = \{(0)\}. \tag{18}$$

we only need adding, in an adequate way, 0 (1) at the end of all (some) Top $n - 1$-tuples, to obtain all the Top n-tuples. More precisely:

Theorem 5. *For all $n \geq 2$, we have:*
 (i) If $n - 1$ is odd, then, adding 0 at the end of all the Top $n - 1$-tuples and adding 1 at the end of all the Top $n - 1$-tuples, we obtain all the Top n-tuples.
 (ii) If $n - 1$ is even, then, adding 0 at the end of all the Top $n - 1$-tuples and adding 1 at the end of the Top $n - 1$-tuples for which the number of 0s is greater than the number of 1s, we obtain all the Top n-tuples.

Proof. First, we prove that the addition of 0s and 1s, in the described way, always generates different Top n-tuples. Taking into account the characterisation (T) of the Top binary states given in Theorem 2-(i), we deduce:
- On one hand, for both cases (i) and (ii), the addition of one 0 at the end of any Top $n - 1$-tuple obviously leads to a Top n-tuple.
- On the other hand, if $n - 1$ is odd (case (i)) then, all the Top $n - 1$-tuples have at least one 0 which is not preceding to any corresponding 1. The same happens, when $n - 1$ is even, for those Top $n - 1$-tuples for which the number of 0s is greater than the number of 1s (case (ii)). Then, for both cases, the addition of one 1 at the end of the described Top $n - 1$-tuples also leads to a Top n-tuple.
- Obviously, all the obtained Top n-tuples are different because, if they are generated by the same $n - 1$-tuple then, they differ in the nth component.

Second, we must prove that all the Top n-tuples can be generated by the described addition algorithm. For both cases (i) and (ii), we have: If $(u_1, \ldots, u_n) \in T(n)$, then suppressing the last component u_n $(u_n = 0, 1)$ we get $(u_1, \ldots, u_{n-1})$. Theorem 3-(iii) guarantees that $(u_1, \ldots, u_{n-1}) \in T(n-1)$ and, obviously, this is the Top $n - 1$-tuple which generates $(u_1, \ldots, u_n)$ $\square$

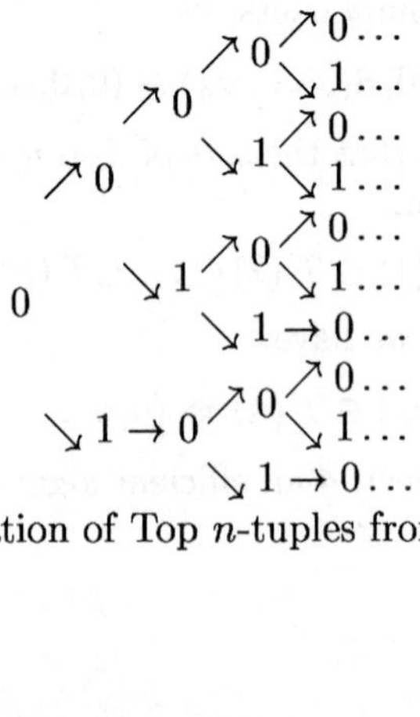

Fig. 3. Generation of Top n-tuples from $T(1) = \{(0)\}$

The number of rows in the nth column $(n = 1, 2, 3, 4, 5, \dots)$ is the number of Top n-tuples, which can be written following the arrows from the first column (0) to each row of the nth column. Obviously, all Top n-tuples (for each natural number n) begin with 0 (Theorem 3-(i)) and they are generated following lexicographic order.

Example 4. For $n = 5$ we have:

$$\mathcal{T}(5) = \left\{ \begin{array}{l} (0,0,0,0,0), (0,0,0,0,1), (0,0,0,1,0), (0,0,0,1,1), (0,0,1,0,0), \\ (0,0,1,0,1), (0,0,1,1,0), (0,1,0,0,0), (0,1,0,0,1), (0,1,0,1,0) \end{array} \right\}.$$

According to the identification (17), next theorem is to justify that the above algorithm selects the binary states by decreasing order of their occurrence probabilities. More precisely:

Theorem 6. *The occurrence probability of any Top n-tuple generated from the Top $n-1$-tuple $(u_1, \dots, u_{n-1})$, by the algorithm described in Theorem 5, is less than or equal to $\Pr\{(0, u_1, \dots, u_{n-1})\}$ (for any basic probabilities verifying (4)).*

Proof. Indeed, the assertion stated by this theorem is more general. We shall prove that, according to the identification (17), the addition of 0 or 1 at the end of any $n-1$-tuple, always produces n-tuples with minor occurrence probabilities (for any basic probabilities p_i such that (4)), i.e. $\forall (u_1, \dots, u_{n-1}) \in \{0,1\}^{n-1}$:

$$(0, u_1, \dots, u_{n-1}) \succeq (u_1, \dots, u_{n-1}, 0) \succeq (u_1, \dots, u_{n-1}, 1). \qquad (19)$$

Using the matrix description IOC of the intrinsic order (Theorem 1) the second inequality is obvious. To prove the first one, let us suppose that matrix

$$\begin{pmatrix} 0 & u_1 & \dots & u_{n-2} & u_{n-1} \\ u_1 & u_2 & \dots & u_{n-1} & 0 \end{pmatrix}. \qquad (20)$$

does not satisfy IOC. Then, matrix (20) contains (at least) one $\binom{1}{0}$ column without its corresponding preceding $\binom{0}{1}$ column. For a homogeneous notation we define $u_n = 0$, and let $\binom{u_{m-1}}{u_m}$ be the left-most $\binom{1}{0}$ column in matrix (20) without its corresponding preceding $\binom{0}{1}$ column. Then, attending to the particular pattern of matrix (20), we have:

$$u_{m-1} = 1 \Rightarrow u_{m-2} = 1 \Rightarrow \dots \Rightarrow u_1 = 1.$$

and thus, $\binom{u_{m-1}}{u_m}$ has its corresponding preceding $\binom{0}{1}$ column: the first column of matrix (20). But this contradicts the hypothesis on $\binom{u_{m-1}}{u_m}$ $\square$

5 Conclusions

We have justified the convenience of selecting binary n-tuples with occurrence probabilities as large as possible in Reliability Theory, Risk Analysis and in many other application areas with stochastic Boolean structures. Using an intrinsic order criterion (IOC) for binary n-tuples probabilities [4], we have classified the binary n-tuples of 0s and 1s into three groups: Top, Bottom and Jumping n-tuples. We have characterised these three types of n-tuples, attending to the relative positions of their 0s and 1s. The Top n-tuples correspond to the binary states with largest occurrence probabilities and we have constructed a simple algorithm for their generation. Starting from the only Top 1-tuple (0), this algorithm obtains all the Top n-tuples just by adding 0 (1) at the end of all the (some) previous Top $n-1$-tuples. The algorithm selects the binary states respecting lexicographic order, and by decreasing order of their occurrence probabilities. The main advantage is that the proposed method is exclusively based on the positions of 0s and 1s in the binary n-tuples, but it selects the binary states with the intrinsically largest occurrence probabilities, that is, for any (basic) probabilities of the Boolean variables.

References

1. Coudert, O., Madre, J.C.: Metaprime: An Interactive Fault-Tree Analyzer. IEEE Trans. Reliability **43**(1) (1994) 121-127
2. Fussell, J.B.: A Formal Methodology for Fault-Tree Construction. Nuclear Ebg. And Design **52** (1973) 337-360
3. Galván, B.J., García, D., González, L.: Un Método Determinista para la Evaluación de Funciones Booleanas Estocásticas: Análisis Comparativo. In: Font, J., Jorba, A. (eds.): Proc. XIV CEDYA -IV CMA. Barcelona (1995) 257-258
4. González, L., García, D., Galván, B.J.: An Intrinsic Order Criterion to Evaluate Large, Complex Fault Trees. IEEE Trans. Reliability, to appear.
5. Kumamoto, H., Henley, E.: Top-Down Algorithm for Obtaining Prime Implicants Sets on Non Coherent Fault Trees. IEEE Trans. Reliability **27**(4) (1978) 242-249
6. Kumamoto, H., Tanaka, T., Inoue, K., Henley, E.: Dagger Sampling Montecarlo for System Unavailability Evaluation. IEEE Trans. Reliability **29**(2) (1980) 122-125
7. Page, L.B., Perry, J.E.: Direct Evaluation Algorithms for Fault-Tree Probabilities. Computers in Chemical Engineering **15**(3) (1991) 157-169
8. Rai, S.: A Direct Approach to Obtain Tighter Bounds for Large Fault Trees with Repeated Events. In: Evans Associates (eds.): Proc. Annual. Rel.& Maint. Symposium, Vol. 23. North Carolina (1994) 475-480
9. Schneeweiss, W.G.: Boolean Functions with Engineering Applications and Computer Programs. Springer-Verlag, Berlin Heidelberg New York (1989)

Reliability Evaluation Using Monte Carlo Simulation and Support Vector Machine

C.M. Rocco Sanseverino and J.A. Moreno

Universidad Central, Facultad de Ingeniería, Apartado 47937, Caracas 1040A, Venezuela
e-mail: {rocco,jose}@neurona.ciens.ucv.ve

Abstract. In this paper we propose the use of Support Vector Machine (SVM) to evaluate system reliability. The main idea is to develop an estimation algorithm by training a SVM on a restricted data set, replacing the system performance model evaluation by a simpler calculation. The proposed approach is illustrated by an example. System reliability is properly emulated by training a SVM with a small amount of information.

1 Introduction

Reliability evaluation of real engineering systems is often performed using simulation tools. Indeed, reliability indices of a system can be seen as the *expected value* of a test function applied to a system state $\mathbf{X}_i$ (vector representing the state of each element) in order to assess whether that specific configuration corresponds to an operating or failed state [1].

For example, in a *s-t* network, to assess if a selected state $\mathbf{X}_i$ corresponds to an operating or failed state, we need to determine its connectivity, which requires knowledge of the cut sets or path sets of the system [2] or to use a depth-first procedure [3].

In other cases, such as telecommunication networks and other real systems, the success of the network requires that a given state is capable of transporting a required flow. To evaluate a state, for example, the max-flow min-cut algorithm [4-5] can be used.

In general, to determine the state of the system (operating or failed) as a function of the state of its components, it is necessary to evaluate a function (one or more) that is called System Function [6] or Structure Function (SF) [7].

In a Monte Carlo simulation, system reliability is evaluated by generating several systems states and evaluating the SF. Since a large number of SF evaluations are required, a fast, approximated algorithm substitutes its evaluation.

There are several approaches that have been used to address the definition of these approximated algorithms [8-12]. In this work, an empirical model built by training a Support Vector Machine (SVM) is presented. SVM provides a novel approach to the

P.M.A. Sloot et al. (Eds.): ICCS 2002, LNCS 2329, pp. 147–155, 2002.

two-category classification problem (operating or failed)[13]. Nowadays, SVM has reached a high level of maturity, with algorithms that are simple, easy to implement, faster and with good performance [14].

The organization of the paper is as follow: Section 2 contains an overview of Monte Carlo approach. The SVM technique is presented in Section 3. Finally, section 4 presents the proposed approach and the results of an example.

2 Monte Carlo Approach

Monte Carlo is one of the methods used to evaluate system reliability. The basic idea is to estimate reliability by a relatively simple sampling plan that requires little information about the system under study [15].

For example, a system state depends on the combination of all component states and each component state can be determined by sampling the probability that the component appears in that state [5].

If it is assumed that each component has two states (failure and success) and that component failures are independent events, then the state of the ith component (x_i) can be evaluated using its failure probability Pf_i and a random number U_i, distributed uniformly between [0,1], as [5]:

$$X_i = \begin{cases} 1 & (\text{succes state}) \text{ if } U_i \geq Pf_i \\ -1 & (\text{failure state}) \text{ if } 0 \leq U_i < Pf_i \end{cases}$$

The jth-state of the system containing NC components is expressed by the vector $X_j = (x_1, x_2, \ldots, x_{NC})$

In general, to evaluate if X_j is an operating or failure state, a SF has to be defined and evaluated at X_j. The SF should have the form:

$$SF(X_j) = \begin{cases} 1 & \text{if system is operational in this state} \\ -1 & \text{if system is failed in this state} \end{cases}$$

In the Monte Carlo approach the conceptual scheme for reliability evaluation consist on [1]:

1. Select a system state X_j
2. *Calculate the System Function for the selected state X_j*
3. Update the estimate of the expected value of $SF(X_j)$ ($E(SF(X_j))$)
4. Calculate the uncertainty of the estimate
5. If the uncertainty is acceptable (within a target tolerance), stop; otherwise return to step 1)

Other sample techniques exist which are more effective [15]. But, in general, all Monte Carlo Methods require SF evaluations. As previously mentioned, the SF depends on the type of reliability evaluation required. In this paper, the emulation of the SF using a SVM is considered.

3   Support Vector Machine

Support Vector Machines provide a new approach to the two-category classification problem [13].

SVMs have been successfully applied to a number of applications ranging from particle identification, face identification and text categorization to engine detection, bioinformatics and data base marketing. The approach is systematic and properly motivated by statistical learning theory [16].

SVM is an estimation algorithm ("learning machine") based on [13]:
- Parameter estimation procedure ("Training") from a data set
- Computation of the function value ("Testing")
- Generalization accuracy ("Performance")

Training involves optimization of a convex cost function: there are no local minima to complicate the learning process. Testing is based on the model evaluation using the most informative patterns in the data (the support vectors). Performance is based on error rate determination as test set size tends to infinity [17].

Suppose $\mathbf{X}_i$ is a system state and y_i is the result of applying the SF to $\mathbf{X}_i$: $y_i = SF(\mathbf{X}_i)$.

Consider a set of N training data points $\{(\mathbf{X}_1,y_1), \ldots (\mathbf{X}_N,y_N)\}$. The main idea is to obtain a hyperplane that separates failed from non-failed in this space, that is, to construct the hyperplane H: $y = \mathbf{w} \cdot \mathbf{X}\text{-}b = 0$ and two hyperplanes parallel to it:

$$H_1: y = \mathbf{w} \cdot \mathbf{X}\text{-}b = +1 \text{ and}$$
$$H_2: y = \mathbf{w} \cdot \mathbf{X}\text{-}b = -1$$

with the condition, that there are no data points between H_1 and H_2, and the distance between H_1 and H_2 (the margin) is maximized. Figure 1 shows the situation [18].

The quantities $\mathbf{w}$ and b are the parameters that control the function and are referred as the weight vector and bias [16].

The problem can be formulated as:

$$\text{Min } \tfrac{1}{2} \mathbf{w}^T\mathbf{w}$$
$$\mathbf{w},b$$
$$\text{s.t} \quad y_i(\mathbf{w} \cdot \mathbf{X}\text{-}b) \geq 1$$

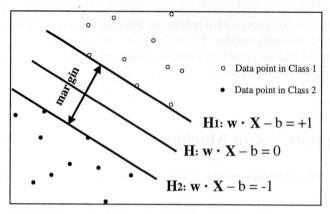

Fig. 1. Decision hyperplanes generated by a linear SVM [18]

This is a convex, quadratic programming problem in (**w**, b), in a convex set. Once a SVM has been trained, it is simple to determine on which side of the decision boundary a given test pattern $\mathbf{X}^*$ lies and assign the corresponding class label, using sgn ($\mathbf{w} \cdot \mathbf{X}^* + b$).

When the maximal margin hyperplane is found, only those points which lie closest to the hyperplane have $\alpha_i > 0$ and these points are the support vectors, that is, the critical elements of the training set. All other points have $\alpha_i = 0$. This means that if all other training points were removed and training was repeated, the same separating hyperplane would be found [13]. In figure 2, the points a, b, c, d and e are examples of support vectors [18].

Small problems can be solved by any general-purpose optimization package that solves linearly constrained convex quadratic programs. For larger problems, a range of existing techniques can be used [16].

If the surface separating the two classes is not linear, the data points can be transformed to another high dimensional feature space where the problem is linearly separable. Figure 3 is an example of such transformation [16].

The algorithm that finds a separating hyperplane in the feature space can be obtained in terms of vector in input space and a transformation function $\Phi(\cdot)$. It is not necessary to be explicit about the transformation $\Phi(\cdot)$ as long as it is known that a *kernel function* $\mathbf{K}(\mathbf{X}_i, \mathbf{X}_j)$ is equivalent to a dot product in some other high dimensional feature space [13,16-19].

There are many kernel functions that can be used this way, for example [13,16]:

$\mathbf{K}(\mathbf{X}_i, \mathbf{X}_j) = e^{-\|Xi-Xj\|^2/2\sigma^2}$ the Gaussian radial basis function kernel

$\mathbf{K}(\mathbf{X}_i, \mathbf{X}_j) = (\mathbf{X}_i \cdot \mathbf{X}_j + m)^p$ the polynomial kernel

The Mercer's theorem is applied to determine if a function can be used as kernel function [19].

With a suitable kernel, SVM can separate in the feature space the data that in the original input space was non-separable. This property means that we can obtain nonlinear algorithms by using proven methods to handle linearly separable data sets [19].

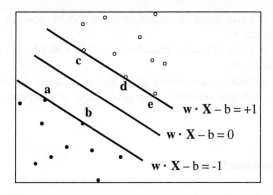

Fig. 2. Example of support vectors [18]

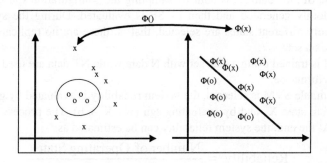

Fig. 3. A non-linear separating region transformed in to a linear one [16]

The choice of the kernel is a limitation of the support vector approach. Some work has been done on limiting kernels using prior knowledge [19]. However, it has been noticed that when different kernel functions are used in SVM, they empirically lead to very similar classification accuracy [19]. In these cases, the SVM with lower complexity should be selected.

The performance of a binary classifier is measured using sensitivity, specificity and accuracy [20]:

$$\text{sensitivity} = \frac{TP}{TP + FN}$$

$$\text{specificity} = \frac{TN}{TN + FP}$$

$$\text{accuracy} = \frac{TP + TN}{TP + TN + FP + FN}$$

where:
TP=Number of True Positive classified cases (SVM correctly classifies)
TN=Number of True Negative classified cases (SVM correctly classifies)

FP= Number of False Positive classified cases (SVM labels a case as positive while it is a negative)

FN= Number of False Negative classified cases (SVM labels a case as negative while it is a positive)

For reliability evaluation, *sensitivity* gives the percentage of correctly classified operational states and the *specificity* the percentage of correctly classified failed states.

4 Proposed Approach

In order to apply SVM in Monte Carlo reliability evaluations, a data set using the system state vector $\mathbf{X}_i$ and $y_i = SF(\mathbf{X}_i)$ is built to train the SVM. Since a SVM is going to replace the SF, the data set is built by sampling the configuration space. A system state is randomly generated and then its SF is evaluated. During the system state generation only different states are selected, that is, there are no replicated states in the training data set.

The SVM is trained using a data set with N data while NT data are used to evaluate the SVM performance.

Once a suitable SVM is selected, the system reliability is evaluated by generating a random system state $\mathbf{X}^*_i$ and by evaluating sgn $(\mathbf{w} \cdot \mathbf{X}^*_i + b)$. The process is repeated NM times. At the end, the system reliability can be estimated as:

$$\text{Reliability} = \frac{\text{Number of Operating States}}{\text{NM}}$$

4.1 Example

Figure 4 shows the system to be evaluated [21]. Each link has reliability r_i and capacity of 100 units. The goal is to evaluate the reliability between the source node *s* and the terminal node *t*. A system failure is defined when the flow at terminal node is less than 200 unit. Using a pure Monte Carlo approach and a max-flow min-cut algorithm as the SF, the estimated reliability is 0.93794.

In this case there are 2^{21} possible combinations. The state space is randomly sampled and a training data set with 500 different states is generated. Different kernels were tried and it was found that the best SVM has a second order polynomial kernel, with only 90 support vectors.

Once the SVM is trained, it is used to evaluate the system reliability. Table 1 shows the size of the test data set, the system reliability based on the number of operating states obtained evaluating the SF, the system reliability using SVM and the relative error:

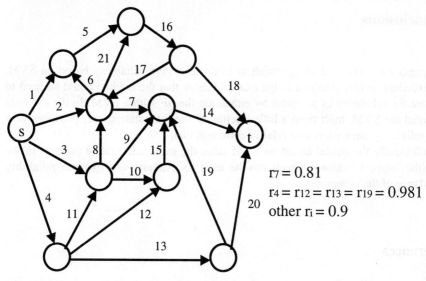

Fig. 4. Network for example 4.1 [21]

$r_7 = 0.81$

$r_4 = r_{12} = r_{13} = r_{19} = 0.981$

other $r_i = 0.9$

$$\text{Relative Error } (\%) = \frac{(\text{SF Evaluation - SVM Evaluation})}{\text{SF Evaluation}} \times 100$$

System reliability obtained using SF or SVM are very close. It is interesting to note that the system reliability is properly emulated using only a little fraction of the state space (90 support vectors/2^{21} states = 0.0043 %).

The complexity of a SVM is $O(NC \cdot NSV)$ [13], where NC is the number of components in the system and NSV is the number of support vector, while the complexity of the max flow algorithm is $O(N^3)$ [4], where N is the number of nodes. An additional speed up factor can be obtained, using the virtual support vector method or the reduced set method. Description of those techniques can be found in [13].

Table 1. SF and SVM Reliability results

Testing data set size	SF Reliability	SVM Reliability	Relative Error (%)
1000	0.9390	0.9480	-0,96%
5000	0.9390	0.9410	-0,21%
10000	0.9367	0.9382	-0,16%
20000	0.9382	0.9398	-0,17%

5 Conclusions

This paper has presented an approach to evaluate system reliability based on SVM. The excellent results obtained in the example show that the method could be used to evaluate the reliability of a system by emulating the SF with a SVM. In the example presented the SVM, built from a little fraction of the total state space, produces very close reliability estimation with relative error less than 1 %

Additionally the model based on SVM takes the most informative patterns in the data (the support vectors) which can be used to evaluate approximate reliability importance of the components.

References

1. Pereira M.V.F, Pinto L.M.V.G.: "A New Computational Tool for Composite Reliability Evaluation", *IEEE Power System Engineering Society Summer Meeting*, 1991, 91SM443-2
2. Billinton, R. Allan R.N: *Reliability Evaluation of Engineering Systems, Concepts and Techniques*. Second Edition. Plenum Press. 1992
3. Reingold E., Nievergelt J., Deo N.: *Combinatorial Algorithms: Theory and Practice*, Prentice Hall, New Jersey, 1977
4. Papadimitriou C. H., Steiglitz K.: *Combinatorial Optimization: Algorithms and Complexity*, Prentice Hall, New Jersey, 1982
5. Billinton, R. Li W.: *Reliability Assessment of Electric Power System Using Monte Carlo Methods*. Plenum Press. 1994
6. Dubi A.: "Modeling of Realistic System with the Monte Carlo Method: A Unified System Engineering Approach", *Proc. Annual Reliability and Maintainability Symposium*, Tutorial Notes, 2001
7. Pohl E.A., Mykyta E.F.: "Simulation Modeling for Reliability Analysis", *Proc. Annual Reliability and Maintainability Symposium*, Tutorial Notes, 2000
8. Marseguerra M., Masine R., Zio E., Cojazzi G.: "Using Neural Networks as Nonlinear Model Interpolators in Variance Decomposition-Based Sensitivity Analysis", *Third International Symposium on Sensitivity Analysis of Model Output*, Madrid, June 2001.
9. Merrill H., Schweppe F.C., "Energy Strategic Planning for Electric Utilities Part I and Part II, Smarte Methodology", *IEEE Transactions on Power Apparatus and Systems*, Vol. PAS - 101, No. 2, February 1982
10. Sapio B.: "SEARCH (Scenario Evaluation and Analysis through Repeated Cross- impact Handling): A new method for scenario analysis with an application to the Videotel service in Italy", *International Journal of Forecasting*, (11) 113-131, 1995
11. Mukerji R., Lovell B.: "Creating Data Bases for Power Systems Planning Using High Order Linear Interpolation", *IEEE Transactions on Power Systems*, Vol.3, No.4, November 1988
12. Rahman S., Shrestha G.: "Analysis of Inconsistent Data in Power Planning", *IEEE Transactions on Power Systems*, Vol.6, No.1, February 1991

13. Burges C.: "A tutorial on Support Vector Machines for Patter Recognition", http://www.kernel-machines.org/
14. Platt J.:"Fast Training of Support Vector Machines using Sequential Minimal Optimization", http://www.research.microsoft.com/~jplatt
15. Fishman G.: "A Comparison of Four Monte Carlo Methods for Estimating the Probability of s-t Connectedness", *IEEE Transaction on Reliability*, Vol. R-35, No. 2, June 1986
16. Cristianini N., Shawe-Taylor J.: *"An introduction to Support Vector Machines"*, Cambridge University Press, 2000
17. Campbell C.: "Kernel Methods: A survey of Current Techniques", http://www.kernel-machines.org/
18. http://www.ics.uci.edu/~xge/svm/
19. Campbell C.: "An Introduction to Kernel Methods", In R.J. Howlett and L.C. Jain, editors, *Radial Basis Function Networks: Design and Applications*, page 31. Springer Verlag, Berlin, 2000
20. Veropoulos K., Campbell C., Cristianini N.: "Controlling the Sensitivity of Support Vector Machines", *Proceedings of the International Joint Conference on Artificial Intelligence, Stockholm, Sweden, 1999 (IJCAI99)*, Workshop ML3, p. 55-60.
21. Yoo Y.B., Deo N.: "A Comparison of Algorithm for Terminal-Pair Reliability", *IEEE Transaction on Reliability*, Vol. 37, No. 2, June 1988

On Models for Time-Sensitive Interactive Computing

Merik Meriste [1] and Leo Motus[2]

[1] University of Tartu, Estonia
merik.meriste@ut.ee
[2] Tallinn Technical University, Estonia,
leo.motus@dcc.ttu.ee

Abstract. Agent-based paradigm is increasingly applied to building computing systems where conventionally latent requirements, e.g. time-sensitivity of data and event validity, and/or truth-values of predicates, timeliness of communication, and others become essential for correct functioning of systems. In many such cases empirical demonstration of expected behaviour is not sufficient, formal verification of certain properties becomes desirable. This assumes that interaction-based models of computing are to be enhanced by introducing sufficiently sophisticated time. Such enhancement is not too simple since time has been abstracted away from models of computing during the evolution of computer science. This paper introduces some preliminary ideas for developing time-sensitive interaction-based models of computations.

1 Introduction

The interaction-based models of computations have emerged from three independent research domains. Computer science and software engineering have reached the interaction-based models when developing methods for formal program verification (Milner (1980), Milner (1999)), and developing theoretical foundations for object-oriented programs and for programming in the large (Wegner (1997)). Interaction-based models of computations have attracted attention in distributed artificial intelligence due to the evolution of agent and multi-agent paradigm (see for instance Ferber (1999) and Wooldridge (2000b)). Many applications of real-time software are known for high dependency requirements. This has lead to studies that, departing from specific timing requirements of real-time software have lead to interaction-based models of computation (see, for instance Quirk (1977), Motus (1995)), applicable from the specification stage of system development. The latter models are not quite comparable with the conventional computer science models, because they focus on timing analysis in a component-based system.

Interaction-based models of computing exceed algorithmic models in formal power (see for details the notion of persistent Turing machines in Wegner (1997), Wegner and Goldin (1999)). For instance, persistent Turing machine can model any discrete-state computing agent that sequentially interacts with its environment. As a rule those models describe indefinitely on-going computations on a set of persistent Turing machines, whereas the interaction may depend on the pre-history of computations. Typical practical examples are multi-agent systems, real-time systems (e.g. systems

P.M.A. Sloot et al. (Eds.): ICCS 2002, LNCS 2329, pp. 156–165, 2002.
© Springer-Verlag Berlin Heidelberg 2002

where computer is directly interacting with real-world processes – influencing thus behaviour of real-world processes and/or is being controlled by these processes, human-machine collaborative decision-making systems, and others.

Need for time-sensitive models of computations stems from specific problems in application domains. Computer control systems, increasing number of communication systems, multi-media, many distributed artificial intelligence and agents' applications naturally lead to time-constraint, concurrent software that assumes time-sensitive models of computations for its verification. Conventional computer science and software engineering have (unfortunately) successfully eliminated time (except as a trivial parameter for ordering events) from theories and tools related to program development. Most of the conventional theoretical thinking in computer science is based on an (not very realistic) assumption of completely known causal relations. Natural consequence of this assumption is that interest to time arises only at the physical design and/or implementation stages of software (e.g. scheduling) – and even then it is based on trivial performance type measures of execution time and execution frequency. Timeliness of interactions and validity of used and/or produced data is of no interest and cannot be evaluated by the existing models of computations.

Incomplete knowledge of causal relations is usual in time-constraint concurrent software and also in many agent-based applications. Software operation in such systems is, to a large extent, based on estimates, beliefs and/or deliberate approximations of causal relations – just to enable the use of conventional computing methods. Quite often causal relations are approximated by time constraints – just for the sake of economic considerations, or to enable the use of existing (too complex) knowledge, or to avoid paradoxes. For example a paradoxical situation may occur in continuous time when a causal relation becomes questionable if the time interval between the cause and effect reduces infinitely.

Another unavoidable feature of real-time and agent-based software is its indefinitely on-going nature of execution and persistency of interactions' influence on the future behaviour. Wegner and Goldin (1999) state that models of computations must be adequately enriched in order to model persistence of computing agent state and on-going interaction with a possibly incomputable environment. An example from conventional computer science illustrates what may happen if the models are not properly expressive and analysable – consider a non-terminating program as a model of indefinitely on-going program. Because of too abstract notion (i.e. non-termination) that hides the intrinsic features of the on-going computations, the whole topic of non-termination was left out of the mainstream theory for a long time (as not easily tractable and therefore not interesting). Just remember that termination is often a precondition to verifiability of a program. This is a subjective obstacle to verification of such programs. The objective obstacle is that the usually applied first-order predicate calculus cannot describe the functioning of an interaction-based non-terminating program – one needs higher order predicate calculus for that (see for instance, a claim in Wegner and Goldin (1999), an example in Lorents et alii (1986)). A statement from Goldin and Keil (2001) partly explains the essence why higher-order logic is needed – no sequence of preordained steps (or series of interactions) can model the multiple-stream encounters between multi-interacting evolved agents and their environments.

Serious confusion also rises when one tries to explain the essence of a non-terminating program to the end-user (who is not a specialist in computer science). Besides, the notion of "non-terminating program" is an approximate description that hides inner details of the program. More precise, and also more acceptable for non-computer specialists is the description of a "non-terminating program" as a set of interacting, repeatedly activated, terminating programs (Quirk (1997)). This concept easily enables to introduce sufficiently sophisticated time and naturally leads to formal verification of timing properties in the early life-cycle stages – e.g. specification and logical design (Lorents et alii (1986) and Motus (1995)).

Conceptually closer to the traditional notion of "non-terminating program" is the departing point for timed process algebra studied by Caspi and Halbwachs (1986)) as an approach to timing analysis of software. This approach encountered serious theoretical problems due to use of continuous time that has hindered its practical application.

Interaction-based computing is an input/output stream processing view on computations as opposed to algorithmic computations where an input string is processed, output string is generated and then the algorithm terminates. The input/output stream processing emphasises the persistent nature of interactive computing. Characteristic to interactive computing is the phenomenon of emergent behaviour (Simon, 1970) that can be observed in multi-agent systems, real-time systems, human organizations, and in many other applications. In such systems agent's learning implies adaptation and, adaptation in turn implies interaction. Learning is an example of history-dependent behaviour that cannot be adequately modelled within an algorithmic (i.e. without history) model of computations neither in an order-of-events setting (Goldin and Keil, 2001).

Besides the emergent behaviour, interactive computations have the following features that are not present in algorithmic computations: (1) stream input-output, (2) interleaving of inputs and outputs during computations, (3) history-dependent behaviour of the computing agents. Causal relations built into the computing system define admissible permutations in interleaving input and output streams. When the causal relations are not completely known one cannot define precisely which of the occurring permutations are admissible. Another difficulty is related to exceptional permutations potentially occurring when one implements forced (Wegner and Goldin 1999 use "true") concurrency that occurs in the case of multi-stream interaction machines or forced (and too high) frequency of input stream for sequential interaction machines. Such phenomena may occur, for instance, as disturbances even in algorithmic computations on multi-processors with pipeline or matrix architecture.

Such anomalies, including those potentially caused by incomplete knowledge of causal relations, can be avoided by introducing suitable time constraints. However, this assumes that instead of an abstract notion of non-terminating program is considered a more precise inner structure – e.g. persistent Turing machine (suggested by Wegner and Goldin (1999)), or a set of interacting, countable number of times activated, terminating programs (suggested by Quirk (1977), Motus (1995)).

This paper focuses on preparatory aspects of developing time-sensitive interaction-based model of computations – selection of suitable candidates for comparative study of interaction-based models of computations, and on discussing the sufficient complexity of time to be used in the model of computations. Most of the future work is planned to be theoretical. However, occasional experimental checks of the theory are important. In addition to the existing tool that implements an interaction-based model of computations (for timing analysis of systems behaviour), the authors have started design and implementation of another "by-product" – a test-bed for assessment and analysis of development and evolution of multi-agent systems. The testing tools and experiments will be considered in a separate paper.

2   Interactive Models of Computations

There is no widely accepted model of interactive computations for the time being. Instead, several concepts and models have been suggested. For instance, the calculus of communicating systems and pi-calculus (Milner 1980, Milner 1999), Q-model (Motus and Rodd 1994); temporal logic of reactive and concurrent systems (Manna and Pnueli, 1991), logic of rational agents (Wooldridge, 2000a), input/output automata (Lunch and Tuttle, 1989), abstract state machines (Gurevich, 1993), attributed automata (Meriste and Penjam 1995), interaction machines (Wegner 1997).

In the pragmatic world of real-time systems the mainstream research is still trying to patch holes in the algorithmic approach. Models of interactive computing are attracting attention slowly, in spite of the successful start in the 1980-es and some interesting practical results obtained recently. For instance, consider a paper on application of a time-sensitive model of interactive computations (Naks and Motus (2001)), and suggestion for a sufficiently sophisticated time-model to be applied for RT UML project (Motus and Naks (2001)).

Respective studies with models of interactive computations in the agents' community and object-oriented community have progressed faster, see for instance, Milner (1999), Wegner (1997), Wegner and Goldin (1999), Wooldridge (2000a), Jennings (2001). Respecting the computer science traditions, however, time has been abstracted away from the models of interactive computations resulting from the latter communities.

Modelling the stream-based behaviour of a computing system has been the key problem in those communities. Behaviour usually denotes a set of I/O streams that obey certain pre-defined ordering discipline. In many cases the ordering discipline are defined by setting only a few constraints on certain interactions. This enables to cope with potentially non-countable number of different behaviours generated by a persistent interactive machine (with countable number of interactions). It is suggested that by changing mathematical framework from induction to co-induction, from least fixed point to greatest fixed point, the problem of model expressiveness can be resolved (see for details Wegner and Goldin (1999), Goldin and Keil (2001).

The analysing capability of models is also important, although somewhat neglected at this development phase. In many cases the analysing capability can be reduced to the question of defining admissible behaviours, distinguishing equivalent sets of behaviours from each other, and demonstration that only admissible behaviours are generated (or that no prohibited behaviour is generated). Behaviour is defined by order imposed on elements in the I/O stream, usually by precedence relation. Concurrent occurrence of stream elements is often demonstrated by interleaving streams.

Difficulties may appear when considering the case of incompletely known causal relations between the agents (components, etc), or the case when causal relations are completely known, but the use of this knowledge is computationally too expensive (e.g. takes more computing power than one has). This usually implies that I/O stream elements cannot be ordered *a priori*. On the outskirts of computer science computing systems seldom function fully alone, independently and autonomously (i.e. normally they interact directly with the surrounding world). In such cases the admissible behaviours are often dynamically defined by responses from a natural (or artificial) world processes that are (at least partially) out of the control of the computing system. Remember, for instance, the case of emergent behaviours (Goldin and Keil (2001)).

In many applications not only the logical order of stream elements is important for defining behaviours. In time-critical applications, for instance, the length of time interval between the respective elements of the I/O stream may be decisive for behaviour classification. This implies the introduction of a metric time in addition to logical time (ordering). One metric time per system may not be sufficient. In multi agent systems and in real-time systems each agent (component, subsystem) may have its own time system, imposed by its own immediate environment. Those separate time systems are often quite different from each other and can only approximately be projected to a single system timeline. Nevertheless, for analysing properties of inter-agent (inter-component) interactions, one needs all those different time systems – approximate system's time may not be sufficient for analysing salient time properties of interactions (Motus (1995).

The way ahead. The authors of this paper believe that interaction based computing systems are increasingly applied to time-sensitive applications. Such applications provide essentially incomplete information about their (potentially dynamically changing) intrinsic details and interactions – e.g. goal functions, characteristics of components, inner structure, influence from the environment, etc. Formal analysing tools are not yet available, although their development is supported by OMG initiatives in developing real-time extensions to UML, and ongoing research in applying agent-based paradigm (e.g. distributed computer control systems, mobile robotics, banking). Nevertheless, the everyday practice is still relying on "trial and error" method.

To partially suppress a new round of "trial and error" activities it makes sense to try and merge the favourably advanced theoretical ideas from all the three involved source domains – computer science, real-time systems, and multi-agent systems – and develop a time-sensitive model of interactive computations.

Based on the above discussion the major question (in the context of this paper) in models of interactive computations seems to be "how to improve their analysing capabilities in the case of incomplete information, emerging behaviours and quantitative time-sensitive behaviours". Previous theoretical and experimental experience in the domains of real-time systems and time-critical reasoning lead the authors to the following

Working hypothesis: *Introduction of reasonably sophisticated time system into the description of model of interactive computations enables to reduce indeterminacy of behaviours generated by the model by approximating incompletely known causal relations, catering for quantitative time constraints, and ordering interactions in time even on multi-stream interaction machines.*

Hence, co-inductive definitions permit to consider the space of all computational processes as a well-defined set, even if the input streams are generated dynamically (Goldin and Keil (2001)). The sufficiently sophisticated time system adds the ability to introduce adequate ordering disciplines – required for analysis of system properties. It is stressed that time system can also cater for forced concurrency created (pure computational) problems in multiprocessors with pipeline and/or matrix architectures – by introducing additional ordering constraints.

2.1   Criteria for Qualifying as a Candidate for Time-Sensitive Model of Interactive Computations

The first problem is to fix (preliminary) criteria that limit the number of candidate models to be considered in more details for developing time-sensitive model of interactive computations. The qualified models are then studied further to understand their specific and common properties, one or two candidates will eventually be chosen.

The first criterion emphasises that essential requirements of multi-agents and real-time systems are to be met by candidate models. It seems reasonable to depart from the characteristic properties for multi-stream interaction machines stated by Wegner and Goldin (1999):

- Capability to handle true concurrency (called forced concurrency in Motus and Rodd (1994))
- Higher than first order-logic is required to describe their functioning.

By pure coincidence the same features characterise the Q-model and related weak second order predicate calculus discussed in Motus and Rodd (1994) and Lorents et alii (1986).

The second criterion facilitates comparison of candidate models by partition them into subgroups. During the preliminary analysis the candidates are subjectively classified based on the (at least seemingly) prevailing methodology, e.g.:

- Process algebras represented, for instance, by CCS and PI-calculus (Milner (1980) and Milner (1999)); history transformer suggested by Caspi and Halbwachs (1986) is also a good candidate
- State machines, represented, for instance, by interaction machines by Wegner (1997), attributed automata (Meriste and Penjam (1995), statecharts (Harel (1987))

- Logical framework based models, represented by temporal logic (Manna and Pnueli (1991), logic of rational agents' Wooldridge (2000a), weak second order predicate calculus with time (Lorents et alii (1986)).
- Other (pragmatic) models represented, for instance, by Wooldridge (2000b), Harel (2001), Motus and Rodd (1994), and others.

Thorough research of the qualified models will hopefully reveal one or two suitable candidates for further elaboration to a time-sensitive model of interactive computations by introducing the required system of time-features. The resulting time-sensitive model is to be used as one of the models for describing agents and systems of agents in a concurrently developed test-bed – the experimental side of this theoretical study. The respective methods developed for formal analysis of systems properties form the basis of the assessment tool to be developed and used in the test-bed.

3   On concepts of Time as Required in Time-Sensitive Multi-agent Systems

Different authors have used the notions of "time", "time properties", "timing analysis", and "time models" in a variety of contexts and meanings. Some comments follow in order to reduce potential confusion and to explain the essence of the time as used in models of computations. A large set of disparate views on the role of time in computing is partly the reason for the excessive variety of terms, beliefs and understandings. For instance, implementation stage (e.g. scheduling theory) needs different time notion than specification and design stages. In the latter stages the time constraints are derived from the environmental and user requirements and imposed on components and their interaction. In the implementation stage (scheduling) all those time constraints are projected on one single time-line and checked for satisfaction – if the system is schedulable. If not a new error and trial iteration is started. This is where sophisticated time and formal timing analysis is of real use (see, pre-run-time scheduling (Xu and Parnas, (2000)); timing analysis at specification stage, (Motus and Rodd (1994))).

It has been demonstrated in Motus and Rodd (1994) that a conventionally used single time variable per system can be sufficient only for assessing performance type of timing properties (usually the focus of scheduling methods). Instead, a rather sophisticated time model should be used in order to handle also the truly salient types of timing properties – time-correctness of events and data, and time correctness of interactions between parts of an agent, agents, coalitions of agents, etc.

A system is compiled from components (reused software pieces, COTS, agents, subsystems, etc) whose dynamic properties are important for reaching the goal of the component, and the system as a whole. A component is loosely connected to the other components (by exchanging messages), and may manage its own connections with the environment. Due to the remarkable potential autonomy of components each component may have its one, independent of the other components, time system. Usual members of a component's time system are, for instance:

- thermodynamic time for accounting the overall progress of the component and its environment,
- reversible time for potential handling of faults and exceptional cases inside of the component, and
- a set of relative times for describing and analysing interactions with the other components.

At the first glance, such time model looks extremely sophisticated. In reality it boils down to intuitively well understandable for a non-computer specialist, and easily specifiable model that enables automatic verification of many timing properties (see, Motus and Rodd (1994), Motus (1995)) and simulation sessions on an automatically generated prototype already at the specification stage. Automatic verification means that the user is saved from theoretical nuances required for statement theorems about common timing problems – e.g. non-violation of imposed timing constraints during an interaction. Application-specific timing properties still need specific handling (e.g. statement and proof of theorems) as usual.

4 Conclusions

The number of time-sensitive applications of interactive computing is increasing fast. At the same time practical development is, especially at the early development stages, still largely based on trial and error method. This paper discusses premises for developing a time-sensitive model of interactive computations that would enhance theoretical understanding of what goes on at the early stages of systems development and assist in early detection and elimination of inconsistencies – including timing inconsistencies.

The paper discusses the necessity of time in computing systems and states a working hypothesis that reasonably sophisticated time system could substantially assist in reducing indeterminacy introduced by incompletely known causal relations. Time could also facilitate the handling of constraints imposed on computational processes and their interactions by direct communication of the computing system with its dynamically changing environment.

The paper also sketches a way towards systematic development of the time-sensitive model of interactive computations, starting form the variety of existing models and also considering the experience obtained in formal timing analysis of real-time software.

It is acknowledged that true stable progress in developing a theoretical model can be guaranteed only by merging the theoretical study with experiments. Therefore, the authors suggest that, as a part of future activities, building a test-bed is to be started. Such a test-bed would serve as a discovery system for agent-related knowledge, in addition to being just an assessment environment for theoretical results. For instance, the test-bed would enable (based on the model that is to be developed) explicit experimental study of time sensitivity of parts of agents (related to communication, learning, and adaptation), and agents' interactions.

Acknowledgment

The partial financial support provided by ETF grant 4860 and grants nr 0140237s98, 0250556s98 from Estonian Ministry of Education is acknowledged.

References

Caspi P. and Halbwachs N. (1986) "A Functional Model for Describing and Reasoning about Time Behaviour of Computing Systems", *Acta Informatica*, **Vol.22**, 595-627

Ferber J. (1999) "Multi-Agent Systems", Addison-Wesley, 509 pp

Goldin D. and Keil D. (2001) "Interaction, Evolution, and Intelligence", Congress on Evolutionary Computation, Korea, May, 2001. http://www.cs.umb.edu/~dqg/papers/cec01.doc

Gurevich Y. (1993). Evolving algebras 1993: Lipari guide. *In* Borger, Ed., *Specification and validation methods,* 1995, pp. 231-243.

Harel D. (1987) "State-charts: a visual formalism for complex systems", Science of Computer Programming, **Vol.8** (1987), p.231

Harel D. (2001) "From play-in scenarios to code: an achievable dream" Computer, **Vol.34**, No.1, 53-60

Jennings N.R. (2001) "An agent-based approach for building complex software systems" *Communications of the ACM*, **Vol.44**, No.4, pp.35-41

Lorents P., Motus L., Tekko J. (1986) "A language and a calculus for distributed computer control systems description and analysis", Proc. on Software for Computer Control, Pergamon/Elsevier, 159-166

Lynch N. A. and Tuttle. M. R. (1989) An introduction to input/output automata. *CWI Quarterly* 2(3) (September 1989), pp. 219-246.

Manna Z. and Pnueli A. (1991) "The temporal logic of Reactive and Concurrent Systems", Springer Verlag

Meriste M. and Penjam J. (1995) "Attributed Models of Computing". Proc. of the Estonian Academy of Sciences. Engineering, **Vol.1**, No.2, pp.139-157

Milner R.A. (1980) "A Calculus of Communicating Systems" LNCS, **Vol.92**, Springer Verlag, 171p

Milner R.A. (1993) "Elements of Interaction". Turing Award Lecture, *Communications of the ACM*, **Vol.36**, No.1, pp.78-89

Milner R. (1999) "*Communicating and Mobile Systems: The PI-calculus*", Cambridge University Press, 161p

Motus L. and Naks T. (2001) "Time models as used in Q-model and suggested for RT UML". Proc. 5th World Multi-Conference on Systemics, Cybernetics and Informatics, **vol. XI**, 467-472

Motus L (1995) "Timing problems and their handling at system integration", in S.G.Tsafestas and H.B.Verbruggen (eds) *Artificial Intelligence in Industrial Decision Making, Control and Automation*, Kluwer Publ., pp.67-88

Motus L. and Rodd M.G. (1994) "*Timing analysis of real-time software*", Pergamon/Elsevier

Naks T. and Motus L. (2001) "Handling Timing in a Time-critical Reasoning System – a case study". *Annual Reviews in Control*, **vol.25**, 157-168

Quirk W. and Gilbert (1977) "The formal specification of the requirements of complex real-time systems", AERE, Harwell, rep. No. 8602

Wegner P. (1995) "Interaction as a Basis for Empirical Computer Science", ACM Computing Surveys, **Vol.27**, No.5, pp.80-91

Wegner P. (1997) "Why Interaction is More Powerful than Algorithms", *Comm. of ACM* **Vol.40**, No.5, pp.80-91

Wegner P. and Goldin D. (1999) "Co-inductive Models of Finite Computing Agents", *Electronic Notes in Theoretical Computer Science,***Vol.19**, www.elsevier.nl/locate/entcs

Wooldridge M. (2000a) "Reasoning about rational agents" MIT Press, 227 p

Wooldridge M. (2000b) "On the Sources of Complexity in Agent Design", *Applied Artificial Intelligence*, **Vol.14**, No.7, 623-644

Xu J. and Parnas D.L. (2000) "Priority Scheduling versus Pre-run-time Scheduling", The International Journal of Time-critical Computing Systems, **vol.18**, no.1, 7-23

Induction of Decision Multi-trees Using Levin Search*

C. Ferri-Ramírez   J. Hernández-Orallo   M.J. Ramírez-Quintana

DSIC, UPV, Camino de Vera s/n, 46020 Valencia, Spain.
{cferri,jorallo,mramirez}@dsic.upv.es

Abstract. In this paper, we present a method for generating very expressive decision trees over a functional logic language. The generation of the tree follows a short-to-long search which is guided by the MDL principle. Once a solution is found, the construction of the tree goes on in order to obtain more solutions ordered as well by description length. The result is a multi-tree which is populated taking into consideration computational resources according to a Levin search. Some experiments show that the method pays off in practice.

Keywords: Machine Learning, Decision-tree Induction, Inductive Logic Programming (ILP), Levin search, Minimum Description Length (MDL).

1 Introduction

In a classification problem, the goal is to produce a model that will predict the class of future examples with high accuracy from a set of training examples (cases). Each case is described by a vector of attribute values, which represents a mapping from attribute values to classes. The attributes can be continuous or discrete, whereas we consider that the class can have only discrete values. One of the most known classification methods is based on the construction of decision trees. In a decision tree, each node contains a test on an attribute, each branch from a node represents a possible outcome of the test, and each leaf contains a class prediction. A decision tree is usually induced by recursively replacing leaves by test nodes, starting at the root. Classic decision-tree learners such as CART [2], ID3 [16], C4.5 [18] or FOIL [17] have given good results; however, they do not have enough flexibility to trading result quality with processing cost.

In this paper, we present an algorithm for the induction of decision trees which is able to obtain more than one solution, looking for the best one or combining them in order to improve the accuracy. To do this, once a node has been selected to be split (an AND node) the other possible splits at this point (OR nodes) are suspended until a new solution is required. In this way, the search space is an AND/OR tree [12] which is traversed producing an increasing number of solutions for increasing provided time. Since each new solution is

* This work has been partially supported by CICYT under grant TIC 2001-2705-C03-01, Generalitat Valenciana under grant GV00-092-14, and Acción Integrada hispano-italiana HI2000-0161

built following the construction of a complete tree, our method differs from other approaches such as the boosting method [6, 19] which induces a new decision tree for each solution. The result is a multi-tree rather than a tree. We perform a greedy search for each solution, but once the first solution is found the following ones can be obtained taking into consideration a limited computation time. Therefore, our algorithm can be considered *anytime* in a certain way [3].

We focus on functional logic languages as representation languages. In the Functional Logic Programming (FLP) paradigm [7], conditional programs are sets of rules and, hence, they can also be represented as trees. The context of this work is the generation of a conditional FLP program (a hypothesis) from examples. This allows us to include the type information of the function profile in the split criterion. On the other hand, since a function f is defined by equations of the form $f(X_1, \ldots, X_n) = Y$, we consider the function result Y as another attribute to be tested. All this extends the kind of tests performed in classical decision-tree induction approaches.

Our method uses a heuristic based on the Minimum Description Length (MDL) principle [21]. Hence, the decision tree is built in a short-to-long way. The MDL principle has been previously used in the induction of decision trees but just within the post-pruning phase [11, 20]. Also, the MDL principle has been used as a stopping criterion (*pre-pruning*) [17], as a measure for globally evaluating discretisations of continuous attributes [15], and for restructuring decision trees [14]. In our approach, the MDL principle is used at the generation phase which is justified because other quality criteria based on discrimination such as the *information gain* [18], the *information gain ratio* [18] or the *Gini heuristic* [2] are not useful for functions that have a recursive definition or that use concepts of the background knowledge. Another reason is that the guidance of the search by the MDL principle ensures a better use of computational resources following a Levin search [9]. We use the MDL principle as split criterion, as stopping criterion and also as solution tree selection criterion. In this way, we present a uniform framework based on the same measure for constructing the tree, selecting the split, selecting second-best trees to explore and selecting or combining hypotheses.

The paper is organised as follows. Section 2 presents our method for constructing decission trees defining new criteria. Section 3 illustrates the construction of decision multi-trees and presents a way of combining the obtained solutions in order to give more precise predictions. Some experimental results are shown in Section 4. Finally, Section 5 concludes the paper.

2   Decision Trees by Levin search

For the construction of a decision tree we consider 9 kinds of partitions [5], including the comparison of an attribute to a variable, the inclusion of constructor based types, the split over real numbers, the equality between variables, the introduction of a function from the background knowledge and recursive calls.

With these kinds of partitions the adaptation of classical split criteria, such as those used by C4.5 / FOIL or CART, would not be suitable, because these measurements are devised to reward partitions which correctly discriminate the class of the result, be it a predicate or a function. However, this may be misleading for recursive functions where this recursive call appears directly on the right hand side of a rule. As we have stated in the introduction, one proper way to order the search space is by the description length of the hypothesis. By definition, a top-down construction of a decision tree is short-to-long, since it adds conditions and after a partition is made the tree is larger to describe. However, this is not sufficient. The idea is to devise a split criterion such that partitions that presumably lead to shorter trees should be selected first.

There exists a suitable paradigm for guiding the construction of the tree: the MDL principle. If we assume $P(h) = 2^{-K(h)}$ where $K(\cdot)$ is the descriptional (Kolmogorov) complexity of a hypothesis h, and $P(E|h) = 2^{-K(E|h)}$ with E being the evidence, we can obtain the so-called maximum a posteriori (MAP) hypothesis as follows [10]:

$$h_{MAP} = argmax_{h \in H} P(h|E) = argmin_{h \in H} (K(h) + K(E|h))$$

This last expression is the MDL principle, which means that the best hypothesis is the one which minimises the sum of the description of the hypothesis and the description of the evidence using the hypothesis.

Initially, when there is only the root of the tree, the hypothesis is empty and the length of its description $(K(\emptyset))$ is almost zero, while the description of the data $(K(E|\emptyset))$ is large. At the end of the construction of the decision tree, the description of the hypothesis $K(h)$ may be large, and each branch constitutes a rule of the program, while the description of the data by using the tree, i.e. $K(E|h)$, will have been reduced considerably. If the resulting tree is good, the term $K(h) + K(E|h)$ is smaller than initially.

The way to construct the tree is to select those partitions (whose description will swell the $K(h)$ part) so that the $K(E|h)$ is reduced considerably. In other words, the best partition will be the one which minimises the term $K(h)+K(E|h)$ after the partition. Once a tree is finished, we will explore second-best splits in accordance with space-time resources. Therefore, *we populate the tree up to a limit number of nodes or time.* The result of this process behaves as a Levin search since solutions are found in a short-to-long fashion. The Levin search guarantees the optimal order of computational complexity [9] or, more precisely, it is the fastest method for inverting functions save for a large multiplicative constant [10]. In what follows, we will introduce an approximation for $K(E|h)$ and for $K(h)$.

Notation

Let S be a set of *sorts*[1], also called *types*. An S-sorted signature Σ is an $S^* \times S$-sorted family $\langle \Sigma_{w,s} | w \in S^*, s \in S \rangle$. $f \in \Sigma_{w,s}$ is a function symbol of arity w and type s; the arity of a function symbol expresses which data sorts it

---

[1] A sort is a name for a set of objects.

expects to see as input and in what order, and the s expresses the type of data it returns. Also, we consider Σ as the disjoint union $\Sigma = \mathcal{C} \uplus \mathcal{F}$ of symbols $c \in \mathcal{C}$, called *constructors*, and symbols $f \in \mathcal{F}$, called *defined functions*. Let $\mathcal{X}$ be a countably infinite set of *variables*. Then $\mathcal{T}(\Sigma, \mathcal{X})$ denotes the set of *terms* built from Σ and $\mathcal{X}$, and $\mathcal{T}(\mathcal{C}, \mathcal{X})$ is the set of constructor terms. The set of variables occurring in a term t is denoted $Var(t)$. A term t is a *ground term* if $Var(t) = \emptyset$. Given a term $f(t_1, \ldots, t_n)$ where $f \in \Sigma_{(s_1, \ldots, s_n), s}$, then $Type(t_i) = s_i$.

Estimate for K(h) and K(E|h)

Given a node ν of a decision tree, we denote its set of conditions by C_ν. The Boolean function $leaf(\nu)$ returns true if ν is a leaf of the tree; $\pi(\nu)$ denotes the partition in ν; $child_i(\nu)$ represents the i-th child of ν and $range(\nu)$ is the number of children of ν. Let us denote with E_ν the set of examples which are consistent with the conditions C_ν. The open variables OV_ν of ν are the variables that do not appear in the lhs of an equality of a condition of C_ν. Let us denote with OVR_ν the set of variables of real type in OV_ν. $OVNR_\nu = OV_\nu - OVR_\nu$.

A *predictive* MDL criterion just describes the class of the examples. For instance, if a leaf is assigned a class c_j then the examples that fall into this leaf need not be described. Only exceptions need be described. However, the construction of our algorithm is guided by *descriptive* MDL, i.e., the examples must be described completely, including not only the class but also the arguments. For instance, if a leaf is assigned a class c_j then the examples that fall into this leaf need be described (except the class). Additionally, exceptions need be described completely, including the class.

Given a partition P, its information can be obtained as[2]:

$$InfoPart(P, \nu) = \log 10 + InfoP(P, \nu)$$

The first term of the above formula is used to select the partition from the 9 possible partitions (the tenth option corresponds to no split, i.e. a leaf). Note that leaves also have $InfoPart$, which is equal to $\log 10$, because the node cannot be exploited further. $InfoP(P, \nu)$ denotes the cost in bits of each partition. Details about the cost of the partitions can be found in [5]. Thus, $K(h)$ can be estimated as follows:

$$K(h) \approx InfoHead(h) + \sum_{\nu \in nodes(h)} InfoPart(\pi(\nu), \nu) \qquad (1)$$

where $InfoHead(h)$ captures the information which is required to code the profile of the function to be learned (which must include the arity and types of the function).

The estimate for $K(E|h)$ is based on the construction of tables. Given a node ν, a table T_i is constructed for each different type τ_i of the set $OVNR_\nu$. The table contains an entry for each different term of type τ_i which appeared in the evidence. With $|T_i|$, we denote the number of elements in table T_i. Each term is denoted by $term_{i,j}$, with j ranging from 1 to $|T_i|$. The information required for each entry in the table, $Info(term_{i,j})$, is defined in the following way:

---

[2] All logarithms in this paper are binary logarithms.

- for finite discrete types, $Info(term_{i,j}) = \log n$, with n being the number of possible values of the type.
- for constructor-based types, $Info(term_{i,j})$ is defined as the cost in bits of selecting the appropriate constructors (from the possible set of constructors applicable at each moment) to describe the term.

From here, we can define the information which is required to describe the table as:

$$InfoTable(T_i) = |T_i| + \sum_{j=1..|T_i|} Info(term_{i,j})$$

Note that a table is constructed for each different type, not for each different non-real open variable $(OVNR)$.

Next we have to give a definition for the information required to give values to all the open variables in order to describe an example. Given an example e from E_ν, we have to code the substitution for non-real variables (just referring to the position in its corresponding table) and the substitution for real (continuous) variables in a different way. Let us denote the argument k of the lhs of example e of real type as $RealValue_k(e)$. We consider the cost of a real number (denoted as $infoR$) as a constant.

Thus, we can define the information which is required to code an example, given the node ν and using the tables, as:

$$Info(e|\nu) = \sum_{k \in OVNR_\nu} \log |T_{Type(k)}| + \sum_{k \in OVR_\nu} (1 + InfoR)$$

Note that this part is the same for all the examples.

Finally, we can define the cost of coding the whole set of examples that fall under ν, i.e. E_ν, as:

$$Info(E_\nu|\nu) = |E_\nu| + \sum_i InfoTable(T_i) + \sum_{e \in E_\nu} Info(e|\nu)$$

The first term codes the number of examples that will be described. With this, we have an estimate for the second term of the definition of the MDL principle, $K(E|h)$. In this way,

$$K(E|h) \approx \sum_{\nu \in leaves(h)} Info(E_\nu|\nu) \qquad (2)$$

Information of the Tree

Now, we introduce the information for the whole tree. Given a tree T with root node ν, the cost of describing T is defined as:

$$InfoTree(E,T) = InfoHead(T) + InfoN(E,\nu)$$

where $InfoN(E,\nu)$ can be obtained recursively in the following way:

$$InfoN(E,\nu) = \begin{cases} Info(E_\nu|\nu) & \text{if } leaf(\nu) \\ InfoPart(\pi(\nu)) + \sum_{i=1..range(\nu)} InfoN(E, child_i(\nu)) & \text{otherwise} \end{cases}$$

Initially, the tree with just one node only has information about the function profile. Since this node is a leaf, $InfoTree(E,T) = InfoHead(T) + Info(E_\nu|\nu)$.

When the tree is being constructed, any exploited node ν (which is still a leaf) has an approximate value for the information, given by $Info(E_\nu|\nu)$, independently of the possible partitions that there could be underneath. However, when this node is exploited, then the value is substituted by the sum of the information of the partition and the information required for the children subtrees.

Since the first approximations are useful when populating and pruning the tree, we will denote the $InfoN$ of a node up to depth d by $InfoN_d(E, \nu)$. Obviously $InfoN_\infty(E, \nu) = InfoN(E, \nu)$ whereas $InfoN_0(E, \nu) = Info(E_\nu|\nu)$.

3   Constructing the Multi-Tree

In this section, we define the construction of decision multi-trees. At this time, we can establish the way in which one tree is build from the root $f(X_1, X_2, ..., X_{n-1}) = X_n$, which is an open node. First of all, the *node selection criterion* chooses the node that is explored first. From all the open nodes, we select the node with less $InfoN_0(E, \nu)$, i.e. with less $Info(E_\nu|\nu)$. This criterion has little relevance, since all the open nodes must be explored sooner or later. Secondly, the *split selection criterion* is much more important, since it selects between the many possible partitions. The $InfoN_1(\nu)$ of the node ν is determined for every possible split. The split with less $InfoN_1$ is selected. Its children are new open nodes. The other partitions are preserved as suspended nodes.

When pruning is not activated (when data is assumed to be noise-free) the stopping criterion is easy to determine. A node is closed when the class is consistent with all the examples that fall into that node (i.e. E_ν). When pruning is activated, the stopping criterion is given by the pruning criterion that we describe next.

The MDL criterion has been used for pre-pruning and post-pruning decision trees. Usually, a predictive MDL criterion has been applied for this purpose [20]. Note that exceptions are much costlier in the case of predictive MDL criterion, because there is a great difference from regular examples (no extra bits are needed) and exceptions (the arguments and class need be coded). This means that the predictive MDL criterion would prune too late in many cases. For this reason, we will also use a descriptive criterion for pruning.

More concretely, the descriptive MDL criterion is used for the *pre-pruning criterion* in the following way. A node is pruned when the description at level $n + 1$ is greater than at level n. More formally, let us suppose a node ν, then the tree should be pre-pruned when:

$$InfoN_0(E_\nu|\nu) < InfoPart(\pi(\nu)) + \sum_{i=1..range(\nu)} InfoN_0(E, child_i(\nu))$$

Now that we can have non-closed nodes which can be pruned and we need an extra bit for every node to tell whether a node is pruned or not. In the case a node is pruned, the node is assigned the most common class under that node. This has to be coded as well, with a cost $\log nC$ where nC is the number of classes. Note that, in the above formula $InfoN_0$ does not need to code the class for all the examples which are consistent with their most common class. For exceptions, however, this class has to be coded as usual, taking into account all the possible values appearing in the exceptions.

The generation of the tree stops when all nodes are closed or pruned.

As we have stated in the introduction, once a tree is concluded, new trees can be generated in order to construct a multi-tree. To do this, we have to establish a new criterion, a *tree selection criterion*. Consider the set of possible splits at depth 1 of a node ν, where σ_1 is the best split and σ_k is the best split that has not yet been exploited. Let us denote the node with split σ_1 as ν_1. Let us denote the node with split σ_k as ν_k. We define the 'rival ratio' $\rho(\nu) = InfoN_1(\nu_1)/InfoN_1(\nu_k)$. Once a tree has been completely constructed, the next tree can be explored, beginning with the split with the greatest rival ratio. This next tree has to be fully completed before selecting another tree.

Finally, when different solutions are obtained we have to select one of them. We have considered two *solution selection criteria*: the MDL principle $(K(h) + K(E|h))$ or Occam's Razor $(K(h))$.

The use of a multi-tree ensures that, when the problem is small or there is time enough, many solutions will be generated. The first idea is to select the best solution according to the solution selection criterion. As we have commented on in the introduction, each solution of the tree can be expressed as a functional logic program, which provides a comprehensible model.

Another option is to combine hypothesis in order to improve the accuracy, however this represents the loose of a comprehensible model. The method used for combining solutions is propagating upwards a vector of the probabilities of nodes and they are combined whenever an OR-node is found.

4   Experiments

The method presented in the previous section has been implemented in a machine learning system (named FLIP2), which is able to induce problems from arbitrary evidence using a functional logic language as representation language. This system and examples is publicly available in [4]. We have performed different experiments using this system which show that our multi-tree approach pays off in practice.

All the examples induced were extracted from the UCI repository [13] and from the Clementine system [8] sample examples, and are well-known by the machine learning community. Some of them contain noisy data and real arguments. We have split the data sets in two similar-sized parts, using one part as the train set and the other one as the test set.

Table 1 contains the results of the experiments: the accuracy and the number of rules of the solution program depending on the number of hypotheses induced (*Numtree*). The experiments were executed on a Pentium III 800 Mhz with 175 MB of memory. The experiments demonstrate that the system is able to induce programs from a complex evidence (i.e. large number of examples and many parameters). The increasing of *Numtree* allows to get shorter theories (i.e. more comprehensible), without an important worsening of accuracy.

Table 1. Accuracy and number of rules generated in the learning of some classification problems.

Numtree	1		10		100		1000	
	Rules	Accuracy	Rules	Accuracy	Rules	Accuracy	Rules	Accuracy
cars	126	85.53	126	85.53	101	85.65	69	84.03
house-votes	71	86.70	71	86.70	53	93.11	49	89.90
tic-tac-toe	346	65.55	297	70.35	263	75.99	252	74.94
nursery	471	91.34	467	91.37	408	91.77	364	92.37
monks1	17	94.90	17	94.90	7	100	7	100
monks2	100	69.90	97	69.90	89	79.16	61	79.62
monks3	35	88.42	35	88.42	28	87.26	22	88.19
drugs	134	92.09	132	92.90	131	92.72	129	93.00
tae	41	57.33	40	60.00	38	60.00	37	61.33

The accuracy obtained by using the combination of hypotheses is detailed in Table 2. In most cases the use of the combination produces an improvement of the result. However, the raise of the accuracy is not linear w.r.t. *Numtree*. It could be interesting to determine automatically the optimal *Numtree* depending on the features of the problem.

Table 2. Accuracy obtained in the learning of some classification problems using the combination of hypotheses.

Numtree	1	10	100	1000
Example	Accuracy	Accuracy	Accuracy	Accuracy
cars	85.53	85.53	90.16	92.12
house-votes	86.69	88.99	92.66	92.20
tic-tac-toe	65.55	82.46	83.71	85.39
nursery	91.34	91.45	92.98	93.98
monks1	94.90	94.90	100	91.90
monks2	69.90	69.91	79.17	79.63
monks3	88.43	90.05	92.59	86.80
drugs	92.09	92.45	93.09	93.09
tae	57.33	57.33	57.33	58.67

An experimental comparison of accuracy with some well known ML systems is illustrated in Table 3. The results for FLIP2 are obtained with *Numtree* = 1000 and using solution combination. *C5.0* is a decision tree learner, *Rules* is a rule induction algorithm[3]. The two are part of the data-mining package Clementine 5.2 from SPSS[8]. *C4.5* [18] represents the results of FLIP2 using the splitting criterion of *C4.5* generating just one solution. As can be seen in the tables, the accuracy of FLIP2 is superior or similar to the other systems in general depending on the problem.

Another technique that permits to adapt the learning process to the amount of resources available is *Boosting* [19]. This mechanism has been incorporated successfully to *C5.0*. Figure 1 shows how *Boosting* increases constantly the resources required (time and memory) depending on the number of iterations of the algorithm.The reason is because each run does not use the information generated in previous iterations. On the contrary, FLIP2 needs more resources initially, every step it reuses the trees generated previously, thus the increasing of resources required is slower.

---

[3] Note that *Rules* is not able to deal with examples with noisy classes like *tae*.

Table 3. Accuracy comparison between some ML systems.

Example	FLIP2	C5.0	Rules	C4.5
cars	92.12	88.54	85.88	90.39
house-votes	92.20	94.50	94.5	94.04
tic-tac-toe	85.39	80.38	77.45	78.70
nursery	93.98	95.99	95.73	95.67
monks1	91.90	87.90	100	78.00
monks2	79.63	65.05	65.74	69.90
monks3	86.80	97.22	94.44	92.12
drugs	93.09	97.27	95.05	90.18
tae	58.67	54.67	-	38.67

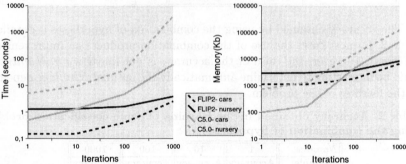

Fig. 1. Time and memory required by FLIP2 and _C5.0_ with _Boosting_ depending on the number of iterations.

5   Conclusions and future work

Much recent work in the area of machine learning has been devoted to the combination of results given by different learners and by several iterations of the same learner [1, 19]. In the latter case, the algorithm is re-run from scratch with different samples from the training data, giving a bank of hypotheses, from which the best solution can be selected or a voting can be made to give a combined solution.

In this paper, we have presented an algorithm that has been designed to give multiple solutions in an efficient way, re-using parts of other solutions and maintaining the same common structure for all the solutions. We have called this structure a multi-tree. The accuracy of the first solution given by our algorithm is comparable to other systems, such as C5.0. However, the multi-tree can be further populated in order to improve this accuracy. Moreover, a new tree is not generated each time from scratch, but just some new branches are explored. This makes that once a first solution has been found the time which is required to produce the next n trees increases in a sublinear way.

This behaviour is similar to an anytime algorithm, as we have stated, which increases the quality of the solution with increasing time. This quality is improved for the best solution (which can be expressed as a comprehensible functional logic program) or as a combination of the multiple solutions (which is less comprehensible but usually improves further the accuracy). The anytime character of our algorithm makes it very suitable for data-mining applications.

As future work we plan to include the complete set of partitions, in order to induce recursive problems and to address problems with deep structure. Finally, more experimental work with this kind of problems must also be performed.

References

1. E. Bauer and R. Kohavi. An empirical comparison of voting classification algorithms: Bagging, boosting and variants. *Machine Learning*, 36:105–139, 1999.
2. Leo Breiman, J. H. Friedman, R. A. Olshen, and C. J. Stone. *Classification and Regression Trees*. Wadsworth Publishing Company, 1984.
3. T. Dean and M. Boddy. An analysis of time-dependent planning. In *Proc. of the 7th National Conference on Artificial Intelligence*, pages 49–54, 1988.
4. C. Ferri, J. Hernández, and M.J. Ramírez. The FLIP system homepage. http://www.dsic.upv.es/~flip/, 2000.
5. C. Ferri, J. Hernández, and M.J. Ramírez. Learning MDL-guided Decision Trees for Constructor-Based Languages. In *WIP track of 11th Int. Conf. on Inductive Logic Progr,ILP01*, pages 39–50, 2001.
6. Y. Freund and R.E. Schapire. Experiments with a new boosting algorithm. In *Proc. of the 13th Int. Conf. on Machine Learning (ICML'1996)*, pages 148–156. Morgan Kaufmann, 1996.
7. M. Hanus. The Integration of Functions into Logic Programming: From Theory to Practice. *Journal of Logic Programming*, 19-20:583–628, 1994.
8. SPSS Inc. Clementine homepage. http://www.spss.com/clementine/.
9. L.A. Levin. Universal Search Problems. *Problems Inform. Transmission*, 9:265–266, 1973.
10. M. Li and P. Vitányi. *An Introduction to Kolmogorov Complexity and its Applications*. 2nd Ed. Springer-Verlag, 1997.
11. M. Mehta, J. Rissanen, and R. Agrawal. MDL-Based Decision Tree Pruning. In *Proc. of the 1st Int. Conf. on Knowledge Discovery and Data Mining (KDD'95)*, pages 216–221, 1995.
12. N.J. Nilsson. *Artficial Intelligence: a new synthesis*. Morgan Kaufmann, 1998.
13. University of California. UCI Machine Learning Repository Content Summary. http://www.ics.uci.edu/~mlearn/MLSummary.html.
14. N.C. Berkman P.E. Utgoff and J.A. Clouse. Decision tree induction based on efficient tree restructuring. *Machine Learning*, 29(1):5–44, 1997.
15. B. Pfahringer. Compression-based discretization of continuous attributes. In *Proc. 12th International Conference on Machine Learning*, pages 456–463. Morgan Kaufmann, 1995.
16. J. R. Quinlan. Induction of Decision Trees. In *Read. in Machine Learning*. M. Kaufmann, 1990.
17. J. R. Quinlan. Learning Logical Definitions from Relations. *M.L.J*, 5(3):239–266, 1990.
18. J. R. Quinlan. *C4.5: Programs for Machine Learning*. Morgan Kaufmann, 1993.
19. J. R. Quinlan. Bagging, Boosting, and C4.5. In *Proc. of the 13th Nat. Conf. on A.I. and the Eighth Innovative Applications of A.I. Conf.*, pages 725–730. AAAI Press / MIT Press, 1996.
20. J. R. Quinlan and R. L. Rivest. Inferring Decision Trees Using The Minimum Description Length Principle. *Information and Computation*, 80:227–248, 1989.
21. J. Rissanen. Modelling by shortest data description. *Automatica*, 14:465–471, 1978.

A Versatile Simulation Model for Hierarchical Treecodes

P.F. Spinnato, G.D. van Albada, and P.M.A. Sloot

Section Computational Science
Faculty of Science
University of Amsterdam
Kruislaan 403, 1098 SJ Amsterdam, The Netherlands
{piero, dick, sloot}@science.uva.nl
http://www.science.uva.nl/research/scs/index.html

Abstract. We present here a performance model which simulates different versions of the hierarchical treecode on different computer architectures, including hybrid architectures, where a parallel distributed general purpose host is connected to special purpose devices that accelerate specific compute-intense tasks. We focus on the inverse square force computation task, and study the interaction of the treecode with hybrid architectures including the GRAPE boards specialised in the gravity force computation. We validate the accuracy and versatility of our model by simulating existing configurations reported in the literature, and use it to forecast the performance of other architectures, in order to assess the optimal hardware-software configuration.

1 Introduction

The problem of computing the force interactions in a system of N mutually interacting particles is one of the important challenges to Computational Science. The gravitational interactions among stars in a stellar globular cluster, or the electrostatic interactions among ions in a chemical solution are typical instances of such a problem, generally known as the N-body problem. As interactions are long ranged, the N-body problem scales as $\mathcal{O}(N^2)$, leading to an unsustainable computational load for realistic values of N. Special software and hardware tools have been developed in order to make the simulation of realistic N-body systems feasible.

Two of the most important contributions to this research have been the introduction of the hierarchical treecodes [1], and the realisation of the GRAPE class of Special Purpose Devices (SPDs) [2]. Such approaches aim at reducing the time spent in computing the forces. The treecode reaches this goal by computing partial forces on a given particle from a truncated multipole expansion of groups of particles, instead of adding each single force contribution. Groups become larger and larger as their distance from the given particle increases. This technique allows to decrease the computing time of the force evaluation to

P.M.A. Sloot et al. (Eds.): ICCS 2002, LNCS 2329, pp. 176–185, 2002.

$\mathcal{O}(N \log N)$, at the cost of a reduced accuracy, due to the truncated multipole expansion. The highest multipole order is usually the quadrupole.

The GRAPE is a very fast SPD, containing an array of pipelines, each one hardwiring the operations needed to compute a gravitational (or electrostatic) interaction. The latest development in the GRAPE series, GRAPE-6, will contain more than 18 000 hierarchically organised pipelines in its final configuration, for an aggregated peak performance of about 100 TeraFlop/s. A computing environment able to take advantage of both approaches could provide a further boost to N-body computation. Research in the direction of merging these two approaches has been carried out for about a decade [3, 4], but the attained performance has been limited by several factors, the most important being related to the difficulty of using GRAPEs efficiently in a distributed architecture. When selected nodes of a distributed architecture are GRAPE hosts, and GRAPEs are used to compute forces also for particles stored in remote nodes, a huge amount of communication is necessary to perform this task, leading to a very high communication overhead.

In this paper, we explore the possibility of using a parallel host in order to improve the performance of the system. Efficient parallelisation of the treecode is not straightforward, and much efforts have been devoted to this task [5, 6]. Coupling a parallel host with a set of GRAPE boards introduces a further complication, as said above, which makes it difficult to estimate the optimal configuration of the resulting hybrid architecture. In order to understand the complex interplay of the parallel algorithm, the parallel general purpose host, and the multiple board SPD, we developed a performance model, by means of which we can predict the performance of real systems [7, 8]. Using this tool, the task of designing the optimal computing environment for the integration of N-body systems can be made less difficult. A detailed discussion on our performance modelling approach is given in [7].

In the following sections, we present our model, the hardware, and the software components of the modelled environment. Then we show the modelling of configurations reported in the literature, which validate our model. Finally we show the results concerning the model of a configuration that integrate GRAPE boards in a distributed parallel architecture.

2   The Performance Model

2.1   Modelling Approach

Our modelling approach can be sketched as a so-called Y-chart, represented in fig. 1. We define an *application model*, where each task of the software application is described in terms of the operations performed, and the workload that such operations produce. Workload is expressed as a function of the application parameters $\mathcal{P} = \{\pi_1, \pi_2, \dots\}$. Our application is here the treecode; a table containing the modelling expressions of each task of the treecode could not be included here because of page limitations. See [9] for an extended report which includes the table.

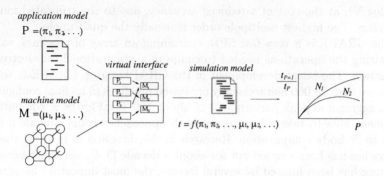

Fig. 1. Performance modelling process. The output of the simulation model, t, is the modelled execution time. The rightmost sketch represents a speedup graph (P is the number of processors), showing the behaviour produced by two different values assigned to a certain parameter N. The meaning of the other symbols is explained in the text.

In the *machine model*, each architecture resource is specified in terms of the time spent in accomplishing the task it is designed to perform, as a function of the machine parameters $\mathcal{M} = \{\mu_1, \mu_2, \dots\}$. Such quantities are basic performance parameters, such as clock frequency and communication bandwidth.

A *virtual interface* maps each task of the application model to the appropriate machine resource; possible resource contention is managed and resolved by the performance modelling environment. The result of the task mapping is the *simulation model*, which is the actual simulator of the whole system. It simulates an instantiation of the treecode, running on an instantiation of a computer architecture. According to the input parameters, it can simulate different treecode implementations, running on different architectures, including SPDs. The simulation model outputs the execution time of the application, the utilisation of the various hardware components, and other performance measures.

Our performance model is implemented in PAMELA (PerformAnce ModEling LAnguage) [10]. PAMELA is a C-style procedure-oriented simulation language in which a number of operators model the basic features of a set of concurrent processes. A detailed overview on PAMELA is given in [10].

2.2   Model Components

Software Application   The hierarchical treecode [1] is one of the most popular numerical methods for particle simulation involving long range interactions. It is widely used in the Computational Astrophysics community to simulate systems like single galaxies or clusters of galaxies. It reduces the computational complexity of the N-body problem from $\mathcal{O}(N^2)$ to $\mathcal{O}(N \log N)$, trading higher speed with lower accuracy. We give now a short description of the main treecode procedures. A simple pseudo-code sketching the basic tasks of the treecode is given in fig. 2.

The treecode approach for computing forces on a given particle ξ is to group particles in larger and larger cells as their distance from ξ increases, and compute force contributions from such cells using truncated multipole expansions. The grouping is realised by inserting the particles one by one into the initially empty simulation cube. Each time two particles are into the same cube, such cube is divided into eight 'child' cubes, whose linear size in one half of their parent's. This proce-

```
t = 0
while (t < t_end)
   build tree
   for each particle
      traverse tree to
      compute forces
   integrate orbits
```

Fig. 2. Pseudocode sketching the basic treecode tasks.

dure is repeated until the two particles find themselves into different cubes. Hierarchically connecting such cubical cells according to their parental relation leads to a hierarchical tree data structure. When force on particle ξ is computed, the tree is traversed looking for cells that satisfy an appropriate criterion (see [11] for a detailed overview on acceptance criterions). By applying this procedure recursively starting from the tree root, i.e. the cell containing the whole system, all the cells satisfying the acceptance criterion are found.

The original treecode algorithm has been modified in several ways, in order to improve performance. The tree traversal phase has been optimised by performing a single traversal for a group of nearby particles [12], whereas the original algorithm performs a tree traversal for each particle. This drastically reduces the number of tree traversals, and allows concurrent force computation on vector machines. It is also ideal when the treecode is used with GRAPE, because each pipeline of the array contained in a GRAPE board can compute force on a different particle simultaneously. The drawback of this technique is an increase in memory usage, because for each group an interaction list containing all the cells interacting with the group must be written and stored in memory.

The use of interaction lists is also useful for parallelisation on distributed systems. The possibility of decoupling tree traversal and force computation through interaction list compilation, allows the implementation of latency hiding algorithms for the retrieval of cell information stored in a remote processor memory [6]. We will refer to this version of the parallel treecode as HOT (HOT is the acronym of Hashed Oct-Tree, which is the name given to the code by its authors).

Another modification consists in computing force only on a small fraction of the N particles at each code iteration, a criterion originally introduced in the direct N-body code (*cf.* [13]). In this case, each particle is assigned an individual time-step, and at each iteration only those particles having an update time below a certain time are selected for force evaluation [5]. In this code, a different approach for remote interactions computation is also implemented: data of the selected particles are sent to the remote processors, interactions are computed remotely, and results are received back. A further modification consists in rebuilding the local tree less frequently than at every iteration. This version

```
t = 0
while (t < t_end)
    if code is HOT
        build local tree
        exchange data to build global tree
        for each group
            build interaction list
            (communications needed for remote data retrieval)
        for each group
            for each particle in group
                compute forces
    if code is GDT
        if it is time to rebuild tree
            build local tree
        for each selected particle
            traverse local tree to compute local forces
        send particles to remote nodes
        receive particles from remote nodes
        compute force on remote particles
        send forces to remote nodes
        receive forces from remote nodes
        integrate orbits
```

Fig. 3. Pseudocode sketching the generic parallel treecode tasks. HOT and GDT are the two versions of the treecode modelled in this work. Tasks involving communication are highlighted.

will be referred as GDT, which is a short for GADGET, the name given to the code by its authors. In fig. 3 we give a pseudo-code representation of the generic algorithm that our model simulates.

Application Model Many other versions of the treecode have been proposed, implementing different tools and techniques. A recent report on that is given in [5]. Our performance model has been designed to reproduce the behaviour of state-of-the-art parallel treecodes, running on distributed architectures, and able to make use of dedicated hardware. For a synoptic table of the modelling expressions of the application tasks described above, see [9].

Computer Architecture The parallel system simulated in our machine model is a generic distributed multicomputer, where given nodes can be connected to one or more SPDs. When SPDs are present, the appropriate task is executed on them. The application model needs no modification in this case. According to an input parameter which tells whether SPDs are present, the mapping interface chooses the routine that maps the task to the SPD, or to the general purpose processor. Since we are interested in SPDs dedicated to the gravity force computation, the machine model of the SPD reproduces the GRAPE activity, and its communication with the host. The modelling of the fairly complicated data exchange machinery between GRAPE and its host is discussed in [8].

The hardware characteristics of the simulated multicomputer are parameterised by two constants, τ_p and τ_n. The first quantity, τ_p, accounts for the processor speed, its value being the amount of floating points operations per nanoseconds; τ_n accounts for the network speed, and its value is the transfer rate in $\mu s/B$. In the execution model, each computation-related function (those containing a cp constant in the modelling expressions reported in the table included in [9]) will be multiplied by τ_p, each communication-related function (those containing a cm constant) will be multiplied by τ_n.

3 Results

3.1 Model Validation

We present here the result of running our simulation model with modelling parameters such that performance measurements reported in the literature are reproduced. We show for each case the scaling with the total particle number N of each task of the code, compared with the corresponding real system timings, as reported by the measurements authors. Finally we present a plot comparing the total compute time for a code iteration of each configuration.

HOT on Touchstone Delta The Touchstone Delta was a one-of-a-kind machine installed at Caltech in the early nineties. It consisted of 512 i860 computing nodes running at 40 MHz, and connected by a 20 MB/s network. The performance measurements reported in [6] are based on a run using the whole 512 nodes system, and consist in a timing breakdown of a code iteration taken during the early stage of evolution of a cosmological simulation, when the particle distribution is close to uniform. The total number of particles is $N = 8.8 \cdot 10^6$. Implementation limitations prevented our performance model to simulate 512 concurrent processes, so that we limited our simulation to 32 processes, and scaled down 16-fold the measured compute time reported in [6]. Since the communication overhead for that run was just $\simeq 6\%$, we assumed a linear scalability of the code. Page limitations prevented us from presenting here a figure showing the timing breakdown of our simulation. The figure is included in [9].

The figure shows that the force computation task and the tree traversal are the most expensive tasks. The relative computational weight of each task is qualitatively well reproduced by our model. Quantitatively, a certain discrepancy between our model and the real system timings originates from an over-estimation of the tree traversal and the force computation tasks, which also results in over-estimating the total time, as shown in fig. 5. Since the main goal of our present model is versatility with respect to both hardware and software modelling, and qualitative accuracy, a quantitative discrepancy in a single case does not limit the general validity of the model.

GDT on T3E Page limitations forced us to cut this section. The main conclusions here are that the trend in the measurements suggests a saturation in the

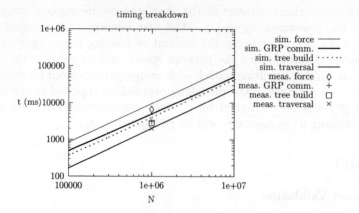

Fig. 4. Timings of the sequential treecode tasks. The real system timings are also reported. The hardware architecture consists of a Compaq 500 MHz workstation connected to a GRAPE-5.

attained performance, arguably due to an increasing load imbalance. This trend is not visible in our model results, because load imbalance is not modelled. In the overall, our model results match the real system timings. See [9] for a full description of this case.

Sequential Treecode on GRAPE-5 Here we simulate the configuration described in [14]. In that case, a modified treecode is used to simulate a system containing about one million particles, and groups of $\simeq$ 2000 particles share the same interaction list. This code is run on a Compaq workstation with a 500 MHz Alpha 21264 processor, connected to a GRAPE-5 board containing 96 virtual pipelines,[1] each one able to compute a force interaction in 75 nanoseconds. Estimating a force interaction as 30 flops, the aggregate performance of a GRAPE-5 board is 38.4 Gflop/s. Fig. 4 shows the results of our simulation model, compared with the real system timings, as reported in [14].

In this case, the force computation task is performed by the GRAPE. An important fraction of the total timing is taken by the communication between the host and the GRAPE. The decrease of importance of the tree traversal task, due to the particle grouping technique, is clearly observable.

Cases Comparison We compare here the three cases presented above. We show in fig. 5 a plot of the time taken by a code iteration versus N, as obtained

---

[1] A GRAPE-5 board contains in fact 16 physical pipelines, each one running at 80 MHz, which is 6 times the speed of the board bus. The board 'sees' 16*6 = 96 logical pipelines, running at 80/6 MHz. Appropriate hardwiring manages the data exchange between the pipelines and the board.

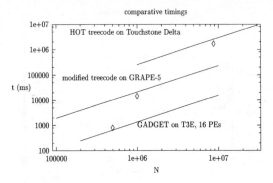

Fig. 5. Timings of a code iteration for the three simulated configurations.

from our simulation model, compared with the real system measurements. The value for the HOT code on the Touchstone Delta is 16 times greater than the value reported in [6], in order to scale their 512 processor run to our 32 processor simulation. The converse scaling would have resulted in simulation values overlapping the values for the GRAPE case.

The simulation values match well the real system measurements. This allows us to use our model to forecast the behaviour of other configurations with a good degree of confidence. Results from such simulations are given in the next section.

3.2   Model Forecasts

We explore in this section the possibility of using a hybrid architecture consisting of a distributed general purpose system, where single nodes host zero or more GRAPE boards. We span the two-dimensional parameter space defined by the two quantities P, the number of nodes, and G, the number of GRAPEs. We assign to those quantities values as follows: $P \in \{1, 2, 4, 8, 12, 16, 20, 24\}$, $G \in \{0, 1, 2, 4, 8, 12, 16\}$. We simulate the same software configuration as described in the previous section with respect to the case related to the sequential treecode on GRAPE-5. The SPD we simulate in this case is the GRAPE-4 [2], whose performance per board is 30 Gflop/s, comparable to GRAPE-5's. It provides a higher accuracy with respect to GRAPE-5, and is used in fields as Globular Cluster dynamics on Planetesimal evolution, where high computing precision is required. The general purpose nodes are assumed to perform a floating point operation in 2 ns, and the communication network is assumed to have a 100 MB/s throughput.

When one of the nodes of the distributed system is a GRAPE host, forces on particles that it stores can be computed on the GRAPE hosted by it. Forces on particles residing on nodes that do not host GRAPEs can be computed on remote GRAPEs, provided that both particle positions and particle interaction

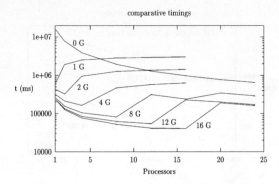

Fig. 6. Timings for a system with P processors and G GRAPEs. Comparison with a system without GRAPES (marked as 0 G) is also provided.

lists be sent to the node hosting the GRAPE. This implies a very large communication traffic. With our simulation we try to evaluate the relevance of this communication overhead.

Fig. 6 shows our results. It is clear that, as long as all nodes are connected to one (or more) GRAPE, a remarkable performance benefit appears. For comparison, we also provide the timing of a system without GRAPEs. When no all nodes are GRAPE hosts, the very large communication overhead due to sending particle and interaction list data is disruptive for performance. This result suggests that the communication task needs a very careful analysis, in order to design an efficient parallel treecode for hybrid architectures. Here we assumed that a 'non graped' node sends all its data to a single 'graped' node. We discuss this point further in the next section. The plot in fig. 6 also features an oscillatory behaviour, particularly evident in the case with 8 GRAPEs. The local minima (i.e. better performances) correspond to configurations where P is an exact multiple of G. In this case the computational load on the GRAPEs is perfectly balanced, whereas in the other cases some GRAPE bears a higher computational load from remote data.

4   Conclusions and Future Work

The efficient integration of hierarchical treecodes and hybrid architectures, consisting of distributed parallel systems powered by GRAPE SPDs, could lead to a very high performance environment for the solution of the N-body problem. In order to assess the optimal configuration, we realised a performance model, with which we can explore the parameter space of this problem. We validated our model by simulating existing configurations and comparing our results to real system measurements. We used our model to evaluate the performance of a hybrid architecture, and highlighted that an efficient implementation of such architecture is made difficult by an intrinsic high communication overhead. Issues like latency hiding, or partial redistribution of work to remove load imbalance, could help to solve this problem, and will be the object of further research. The model would also benefit from an accurate parameterisation of load imbalance. An experimental configuration realised according to those guidelines will allow us to fine-tune our model, which in turn will give a feedback to improve the actual configuration, in the direction of finding the optimal interaction of software, parallel host, and SPD.

Acknowledgements We acknowledge Arjan van Gemund for having made the PAMELA package available to us. We also acknowledge Jun Makino for the GRAPE boards that he kindly made at our disposal.

References

1. Barnes, J.E., Hut, P.: A hierarchical $O(N \log N)$ force-calculation algorithm. Nature **324** (1986) 446–449
2. Makino, J., Taiji, M., Ebisuzaki, T., Sugimoto, D.: GRAPE-4: a massively parallel special-purpose computer for collisional N-body simulations. Astrophysical Journal **480** (1997) 432–446
3. Athanassoula, E., Bosma, A., Lambert, J.C., Makino, J.: Performance and accuracy of a GRAPE-3 system for collisionless N-body simulations. Montly Notices of the Royal Astronomical Society **293** (1998) 369–380
4. Makino, J.: Treecode with a special-purpose processor. Publications of the Astronomical Society of Japan **43** (1991) 621–638
5. Springel, V., Yoshida, N., White, S.D.M.: GADGET: a code for collisionless and gasdynamical cosmological simulations. New Astronomy **6** (2001) 79–117
6. Warren, M.S., Salmon, J.K.: A parallel hashed oct-tree N-body algorithm. In: Proceedings of the Supercomputing '93 conference, ACM press (1993) 12–21
7. Spinnato, P.F., van Albada, G.D., Sloot, P.M.A.: Performance modelling of distributed hybrid architectures. IEEE Transactions on Parallel and Distributed Systems **submitted** (2001) Also available at: http://www.science.uva.nl/~piero.
8. Spinnato, P.F., van Albada, G.D., Sloot, P.M.A.: Performance prediction of N-body simulations on a hybrid architecture. Computer Physics Communications **139** (2001) 34–44
9. Spinnato, P.F., van Albada, G.D., Sloot, P.M.A.: A versatile simulation model for hierarchical treecodes. Technical report CS-2002-01, Dept. of Computer Science, University of Amsterdam (2002) Also available at our group's web-site.
10. van Gemund, A.: Performance prediction of parallel processing systems: The PAMELA methodology. In: Proceedings of seventh ACM International Conference on Supercomputing, ACM press (1993)
11. Salmon, J.K., Warren, M.S.: Skeletons from the treecode closet. Journal of Computational Physics **111** (1994) 136–155
12. Barnes, J.E.: A modified tree code: Don't laugh, it runs. Journal of Computational Physics **87** (1990) 161
13. Aarseth, S.J.: From NBODY1 to NBODY6: The growth of an industry. Publications of the Astronomical Society of the Pacific **111** (1999) 1333–1346
14. Kawai, A., Fukushige, T., Makino, J., Makoto, T.: GRAPE-5: A special-purpose computer for N-body simulations. Publications of the Astronomical Society of Japan **52** (2000) 659–676

Computational Processes in Iterative Stochastic Control Design

Semoushin I. V.[1] and Gorokhov O. Yu.[1]

Ulyanovsk State University, L. Tolstoy Str. 42, 432700 Ulyanovsk, Russia

Abstract. In this article we analyse computational processes and investigate characteristics of iterative stochastic algorithms for control and identification based on the auxiliary performance index (API) approach. A workable API is the one which (1) depends on available values only, (2) has the same domain as the original performance index (OPI) has, and (3) achieves its minimum at the same point in the domain as the OPI does. In the article we show that the proposed method for replacing the OPI by an API is applicable for identification the discrete-time steady-state Kalman filter included into the closed-loop stochastic control systems.

1 Introduction

In order to obtain a good model for control under the real conditions of uncertainty, it has been shown that iterative design procedures are adequate [1]. One iteration involves direct identification of plant model from experimental data and replacement the old model by the new one. Thus the question of identifiability is central in the procedure. This question is twofold: theoretical identifiability and practical identifiability. Basic concepts of theoretical identifiability for plants described by linear difference equations with constant coefficients are set forth in [2]. Numerical methods of parameter identification have also received much consideration, e.g., [3], [4]. A review and unification of linear identifiability concepts can be found in [5].

System identification combines system modelling with parameter estimation. In-depth coverage of system identification in its first unified and mathematically rigorous exposition, including the new general theory of Minimum Prediction Error (MPE) identification methods, is given by Caines [6]. Consistency and other basic qualitative properties of parameter estimators have been thoroughly analyzed, e.g., for a broad class of MPE methods in [6] and [7].

In order to relate the contribution of this paper to the existing control systems literature, we emphasize the fact that in classical MPE methods [6] the set of models $\mathfrak{M}(\theta)$ with a vector parameter θ is constructed to predict the given, i.e., *available* observation process y, for example, $y = \begin{bmatrix} z \\ u \end{bmatrix}$ as for closed-loop control systems with the applied control input u and observed output z. By virtue of the fact, the prediction error $e(t_i)$ in classical MPE methods is available to specify the criterion functions. In our paper, we consider stochastic control systems, however the prediction error we have is not available because it is defined as

P.M.A. Sloot et al. (Eds.): ICCS 2002, LNCS 2329, pp. 186–195, 2002.

difference between the output of $\mathfrak{M}(\theta)$ and the optimal (steady-state) estimator of state vector x. Hence the object for which we construct a set of $\mathfrak{M}(\theta)$ is not given, i.e. is *not available*, being the optimal (steady-state) Kalman filter. It is a sort of *hidden object* to be identified from the observed data y. This object, intended to be used in the system feedback, is hidden in the given data source, and we must "extract" it from there.

We do it by constructing an auxiliary error-like process $\varepsilon(t_i)$ such that the relation holds:

$$J_\varepsilon(\theta) \overset{def}{=} \mathrm{E}\{\|\varepsilon(t_i)\|^2\} = \mathrm{E}\{\|e(t_i)\|^2\} + \mathrm{Const} \tag{1}$$

where $\mathrm{Const} > 0$ denotes a quantity independent of $\mathfrak{M}(\theta)$. Auxiliary Performance Index $J_\varepsilon(\theta)$, when minimized instead of Original Performance Index $J_e(\theta) \overset{def}{=} \mathrm{E}\{\|e(t_i)\|^2\}$, theoretically gives the indirect way to estimate the true (i.e., optimal) parameter θ° because the set of $\mathfrak{M}(\theta)$ contains $\mathfrak{M}(\theta^\circ)$ [8], [9].

In this paper, by doing a number of computational experiments, we consider the practicality of this approach as applied to filters that are included into the closed-loop stochastic control systems described by the following equations

$$\begin{aligned}
x(t_{i+1}) &= \Phi x(t_i) + \Psi u(t_i) + \Gamma w(t_i) \\
z(t_i) &= H x(t_i) + v(t_i) \\
u(t_i) &= -G_0 \tilde{x}(t_i^+)
\end{aligned} \tag{2}$$

where $x(t_i) \in \mathbf{R}^n$ is the state vector, $z(t_i) \in \mathbf{R}^m$ is the measurement vector, $u(t_i) \in \mathbf{R}^r$ is the control input calculated, for lack of optimal filtered estimate $\hat{x}(t_i^+)$ and optimal regulator matrix G_r, through their suboptimal values $\tilde{x}(t_i^+)$ and G_0, and $\{w(t_0), w(t_1), \ldots\}$ and $\{v(t_1), v(t_2), \ldots\}$ are zero-mean independent sequences of independent and identically distributed random vectors $w(t_i) \in \mathbf{R}^q$ and $v(t_i) \in \mathbf{R}^m$ with covariances Q and R respectively.

2   General Algorithm

In this section we present the general computational algorithm for adaptive estimation of parameters that are subject to change. We assume that the uncertainity can reside in matrices Φ, Γ and covariances Q and R. The model state estimates under conditions of uncertainty are obtained by *the feedback suboptimal filter*

$$\begin{aligned}
\tilde{x}(t_{i+1}^-) &= \Phi_0 \tilde{x}(t_i^+) + \Psi u(t_i) \\
\tilde{x}(t_i^+) &= \tilde{x}(t_i^-) + K_0 \nu(t_i) \\
\nu(t_i) &= z(t_i) - H\tilde{x}(t_i^-)
\end{aligned} \tag{3}$$

where control $u(t_i)$ is calculated by a suboptimal control law $u(t_i) = -G_0 \tilde{x}(t_i^+)$. The gain K_0 is replaced by the result of each iteration (the whole identification process) and G_0 is recalculated after each iteration according to LQG control theory. The initial values for K_0 and G_0 are set as some nominal values which are chosen a'priori to satisfy the stability conditions. *The adaptive model is* appended to this closed-loop system and started with initial state taken from the suboptimal filter (3). It has the following form

$$
\begin{aligned}
\tilde{g}(x_{i+1}) &= A\hat{g}(t_i) + Bu(t_i) \\
\hat{g}(t_i) &= \tilde{g}(t_i) + D\eta(t_i) \\
\eta(t_i) &= z(t_i) - H_*\tilde{g}(t_i)
\end{aligned}
\tag{4}
$$

where $A = T\Phi T^{-1}, B = T\Psi, H_* = HT^{-1}$ and T is the observability matrix defined in [8], [9]. Let us assume that the collective parameter θ represents the set of adjustable parameters in the model (the Kalman gain D and parameters from matrix A) indexed accordingly.

Denote the stackable vector of $\eta(t_j), t_j \in [t_{i-s+1}, t_i]$ as $H(t_{i-s+1}, t_i)$ where s is the maximal partial observability index, *the model error* between the adaptive and suboptimal models then can be written in the following form

$$
\varepsilon(t_i) = \mathcal{N}(D)H(t_{i-s+1}, t_i)
\tag{5}
$$

where $\mathcal{N}(D)$ is the structure transformation of adaptive model gain D as defined in [8], [9].

The sensitivity model that reflects the influence of the adjustable parameters on the model error (5) and in fact is the partial derivatives of vector $\varepsilon(t_i)$ wrt. vector θ, is defined by two types of recursions according to the placement of adjustable parameter. Let μ denote the sensitivity model state vector, then for parameters θ_j of Kalman gain D we have

$$
\begin{aligned}
\tilde{\mu}_j(t_i) &= A\hat{\mu}_j(t_{i-1}) \\
\hat{\mu}_j(t_i) &= (I - DH_*)\tilde{\mu}_j(t_i) + \frac{\partial D}{\partial \theta_j}\eta(t_i)
\end{aligned}
\tag{6}
$$

For the adjustable parameters θ_j of transition matrix A the recursions take form

$$
\begin{aligned}
\tilde{\mu}_j(t_i) &= \frac{\partial A}{\partial \theta_j}\hat{g}(t_{i-1}) + A\hat{\mu}_j(t_{i-1}) \\
\hat{\mu}_j(t_i) &= (I - DH_*)\tilde{\mu}_j(t_i)
\end{aligned}
\tag{7}
$$

Both recursions start with initial values $\hat{\mu}_j(t_0) = 0$ for each θ_j. Let vector $\xi_j(t_i)$ be equal to $-H_*\tilde{\mu}_j(t_i)$. The history for vectors $\xi_j(t_i)$ and $H(t_{i-s+1}, t_i)$ should be accumulated during the iterations (3)-(7) till s last values are recalculated.

The sensitivity matrix $\mathcal{S}(t_i)$ is computed as follows

$$
\mathcal{S}(t_i) = \left(\frac{\partial \varepsilon(t_i)}{\partial \theta_j}\right) = \left(\frac{\partial \mathcal{N}(D)}{\partial \theta_j}H(t_{i-s+1}, t_i) + \mathcal{N}(D)\frac{\partial H(t_{i-s+1}, t_i)}{\partial \theta_j}\right)
\tag{8}
$$

where $\frac{\partial H(t_{i-s+1}, t_i)}{\partial \theta_j}$ is the stackable vector $\xi_j(t_k), t_k \in [t_{i-s+1}, t_i]$.

The gradient model then is defined as the production of the transposed sensitivity matrix $\mathcal{S}(t_i)$ with $\varepsilon(t_i)$

$$
\begin{aligned}
\mathcal{G}(t_i) &= \mathcal{S}^T(t_i)\varepsilon(t_i) \\
\hat{\mathcal{G}}(t_i) &= \beta\hat{\mathcal{G}}(t_{i-1}) + (1 - \beta)\mathcal{G}(t_i)
\end{aligned}
\tag{9}
$$

where β is the exponential smooth factor.

Three types of adaptation procedures can be used to generate the new estimates for adjustable parameters. Simple stochastic approximation procedure (SSAP)

$$\pi(t_i) = \hat{\theta}(t_i) - \lambda(t_{i+1})\hat{\mathcal{G}}(t_i), \quad \lambda(t_{i+1}) = 1/(t_{i+1} + 1) \tag{10}$$

(Here and below $\pi(t_i)$ denotes a trial value for $\hat{\theta}(t_{i+1})$). The suboptimal adaptation procedure (SAP) is defined for each adjustable parameter θ_j through recursion

$$p_j(t_{i+1}) = p_j(t_i) + \|\tfrac{\partial\varepsilon(t_i)}{\partial\theta_j}\|^2$$
$$\pi(t_i) = \hat{\theta}(t_i) - \text{diag}(p_j(t_{i+1}))\hat{\mathcal{G}}(t_i), \tag{11}$$

The optimal adaptation procedure (OAP) is given by

$$P(t_{i+1}) = P(t_i) + \mathcal{S}^T(t_i)\mathcal{S}(t_i)$$
$$P(t_{i+1})\Delta\hat{\theta}(t_i) = -\hat{\mathcal{G}}(t_i) \tag{12}$$
$$\pi(t_i) = \hat{\theta}(t_i) + \Delta\hat{\theta}(t_i)$$

The stability condition

$$\rho\left[(I - DH_*)A\right] < 1 \tag{13}$$

for estimate $\pi(t_i)$ should be checked and if new estimates satisfy (13), $\hat{\theta}(t_{i+1}) = \pi(t_i)$.

3   Numerical Experiments

We start from the statement of the examples to demonstrate the features of auxiliary performance index (API) algorithm considered in the section 2. The subsection 3.2 is dedicated to the analysis and investigation of the adaptive model estimates behaviour.

3.1   Numerical Examples

We consider on the following examples:

E1 *Controlled plant.* The first order controller model that follows the dynamics of the reference variable and is described by the equations

$$x(t_{i+1}) = \Phi x(t_i) + \Psi u(t_i) + w_d(t_i)$$
$$z(t_i) = x(t_i) + v_d(t_i) \tag{14}$$

where $\Phi = 0.82$, $\Psi = 0.18$. Covariances Q_d of the zero-mean white Gaussian noise $w_d(t_i)$ that is equal to $Q_d = 0.084Q$ (Q is the parameter of the experiment) and R_d of $v_d(t_i)$ are subject to change their values. The reference model which state should be tracked is given by

$$x_r(t_{i+1}) = \Phi_r x_r(t_i) + w_{dr}(t_i)$$
$$z_r(t_i) = x_r(t_i) + v_{dr}(t_i) \tag{15}$$

where $\Phi_r = 0.61$ and covariances of the noises Q_{dr} and R_{dr} may change their values. Two Kalman gains of controller and reference model should be independently estimated.

E2 *Second order model* with possible changes in the covariances of the measurement and state noises is

$$x(t_{i+1}) = \begin{bmatrix} 0 & 1 \\ f_1 & f_2 \end{bmatrix} x(t_i) + \begin{bmatrix} 0 \\ \beta \end{bmatrix} u(t_i) + \begin{bmatrix} 0 \\ \alpha \end{bmatrix} w_d(t_i)$$

$$z(t_i) = Hx(t_i) + v_d(t_i)$$

(16)

where $\alpha = 0.4$, $\beta = 1.0$, $f_1 = -0.8$ and $f_2 = 0.1$. The measurement matrix H is $\begin{bmatrix} 0, 1 \end{bmatrix}$. Kalman gain vector of dimension 2 should be estimated in the adaptation process.

E3 *Second order model* the same as in E2, with changes in the parameters f_1, f_2 of transition matrix $\varPhi$ and covariances of noises. The dimension of the vector θ of all adjustable parameters is 4 ($\hat{f}_1, \hat{f}_2$ and adaptive Kalman gain).

In the examples the optimal stochastic LQG control $u(t_i)$ is calculated from the independent Kalman estimates for the states of corresponding models.

The experiment is started under the initial conditions and then, after the Kalman estimates become stable, switching in the covariances of the noises or the transition matrix (in example **E3**) occurs. The adaptive estimator is started to identify the new Kalman filter parameters to be used to complete the current iteration step.

3.2   Properties of Algorithm

For the analysis of the estimation process behaviour two numerical characteristics are taken into account: the integral percent error (IPE) and the average normalized error (ANE) of parameters estimates. The influence of the signal-noise ratio, dimension of the vector of adjustable parameters, stability conditions on the calculation process characteristics are summarized in this subsection.

Adjustable Parameter Dimension. The influence of the dimension of adjustable parameter that may change its value and should be identified during adaptive process has shown on the graphs Fig. 1 and Fig. 2. On the Fig. 1 the covariance Q_d of the state noise is changed from the nominal value of 0.01 to 0.1, the Fig. 2 represents the case when both covariances Q_d and Q_{dr} are changed from 0.01 to 0.1. The speed and quality of the estimation process is decreased as the number of parameters that change values increases as depicted on the Fig. 1.

Adaptation Procedure. The influence of the adaptation procedure choice on the ANE characteristic of algorithm has shown on the Fig. 3 and the Fig. 4. It can be noted that the selection of OAP in the considered experiments increases the estimation power of the algorithm. The compromise between the number of calculations and the estimation quality is SAP which used in experiments.

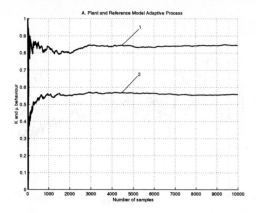

Fig. 1. Example E1. Adaptive estimation process (1-plant model, 2-reference model). The changes in the covariance of plant model.

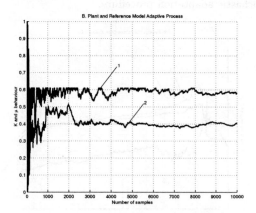

Fig. 2. Example E1. Adaptive estimation process (1-plant model, 2-reference model). The changes in the covariances of plant and reference models.

Signal-Noise Ratio. The IPE characteristic process for different levels of signal-noise ratio for examples E2 and E3 depicted on the Fig. 5 and Fig. 6 shows that the impact on the estimation quality of the algorithm differs in both considered examples. In A-graph of Fig. 5 as the SNR increases from 0.01 to 10, the better performance in the sense of IPE can be observed. When the uncertainty resides also in the transition matrix Φ the ability of Kalman filter to estimate state of the object becomes crucial. As the noise component in the state equation grows the integral estimation error reaches the unimprovable lower bound (Fig. 6).

Stability Properties. The effect of the unimprovable lower IPE bound is determined also by the stability properties of the object. The contribution of the

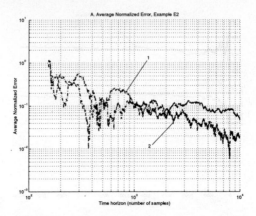

Fig. 3. Example E2. Average normalized error for different adaptation procedures: 1-SAP, suboptimal adaptation procedure, 2-OAP, optimal adaptation procedure, 3-SSAP, simple stochastic adaptation procedure.

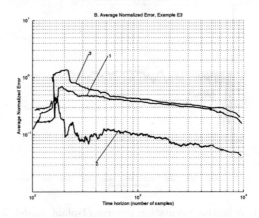

Fig. 4. Example E3. Average normalized error for different adaptation procedures: 1-SAP, suboptimal adaptation procedure, 2-OAP, optimal adaptation procedure, 3-SSAP, simple stochastic adaptation procedure.

dynamic part of the state equation depends on the placement of the object eigenvalues with respect to unit circle on the complex plane. At the fixed level of SNR as the eigenvalues moving to the coordinate origin the quality of obtained estimates decreasing as depicted at the Fig. 7. The influence is more distinct in the case of example E3 again as in the previous subsection.

Initial Values. The results of test runs of the algorithm for different initial values of estimated parameter are shown at Fig. 9 and Fig. 10 at levels SNR 0.1 and 10.0. At the level of 0.1 the algorithm is more sensitive to the initial value than for 10.0 as it is represented at Fig. 10.

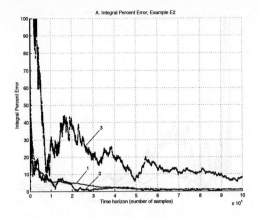

Fig. 5. Example E2. Integral percent error for different level of SNR: 1 - $SNR = 10$, 2 - $SNR = 0.1$, 3 - $SNR = 0.01$, 4 - $SNR = 0.001$.

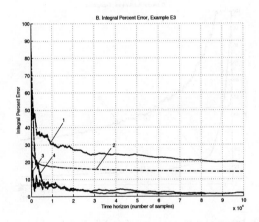

Fig. 6. Example E3. Integral percent error for different level of SNR: 1 - $SNR = 10$, 2 - $SNR = 0.1$, 3 - $SNR = 0.01$, 4 - $SNR = 0.001$.

4   Conclusions

The results reported in the paper show the applicability of the proposed API method to the iterative stochastic control design for the descrete-time closed-loop stochastic control systems. The numerical experiments determined the influence of different factors (the object stability properties, the signal noise ratio, the choice of the initial estimate value) on the algorithm robustness and performance. Some additional experimental effects revealed during the numerical simulation to be theoretically justified.

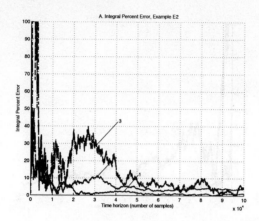

Fig. 7. Example E2. Integral percent error for different level of stability of the object: 1 - $\rho = 0.1$, 2 - $\rho = 0.2$, 3 - $\rho = 0.5$.

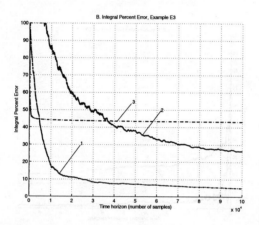

Fig. 8. Example E3. Integral percent error for different level of stability of the object: 1 - $\rho = 0.1$, 2 - $\rho = 0.2$, 3 - $\rho = 0.5$.

References

1. R. R. Schrama. "Accurate identification for control: the necessity of an iterative scheme". *IEEE Transactions on Automatic Control*, Vol. 37(7), pp. 991–994 (1992).
2. Eykhoff, P. (1974); System Identification; Wiley (pp. 684)
3. Saridis, G.N. (1977); Self-Organizing Control of Stochastic Systems; Marcel Dekker, Inc. (pp. 400)
4. Tsypkin, Ya. Z., E.D. Avedyan and O.V. Gulinsky (1981); On Convergence of the Recursive Identification Algorithms; IEEE Trans. Aut. Control, Vol. AC-26, No. 5 (pp. 1009–1017)
5. Nguyen, V. V. and E.F. Wood (1982); Review and Unification of Linear Identifiability Concepts; SIAM Review, Vol. 24 (pp. 34–51)

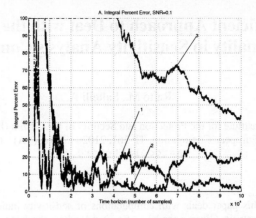

Fig. 9. Example E2. The influence of initial value of estimates for the $SNR = 0.1$. 1 - θ_0, 2 - $5\theta_0$, 3 - $10\theta_0$.

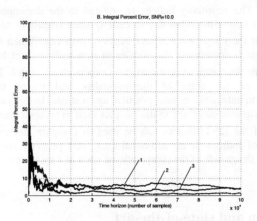

Fig. 10. Example E2. The influence of initial value of estimates for the $SNR = 10.0$. 1 - θ_0, 2 - $5\theta_0$, 3 - $10\theta_0$.

6. Caines, P.E. (1988); Linear Stochastic Systems; Wiley (pp. 874)
7. Ljung, L. (1978); Convergence Analysis of Parametric Identification Methods; IEEE Trans. Aut. Control, Vol. AC-23, No. 5 (pp. 770–783)
8. Semoushin, I. V. (1987); Active Methods of Adaptation and Fault Detection for Stochastic Control Systems; D.Sc. Dissertation; Leningrad Institute of Aircraft Equipment Engineering (pp. 426) [*in Russian*]
9. Semoushin, I. V. and J. V. Tsyganova (2000); Indirect Error Control for Adaptive Filtering; Proc. 3rd European Conference on Numerical Mathematics and Advanced Applications, ENUMATH-99, (eds. P. Neittaanmaki, T. Tiihonen and P. Tarvainen), World Scientific, Singapore (pp. 333-340)

An Efficient Approach to Deal with the Curse of Dimensionality in Sensitivity Analysis Computations

Ratto M., Saltelli A.

Institute for the Protection and Security of the Citizen (IPSC)
JRC, European Commission, TP 361, 21020 ISPRA (VA) - ITALY
marco.ratto@jrc.it

Abstract. This paper deals with computations of sensitivity indices in global sensitivity analysis. Given a model $y=f(x_1,...,x_k)$, where the k input factors x_i's are uncorrelated with one another, one can see y as the realisation of a stochastic process obtained by sampling each of the x_i's from its marginal distribution. The sensitivity indices are related to the decomposition of the variance of y into terms either due to each x_i taken singularly, as well as into terms due to the cooperative effects of more than one. When the complete decomposition is considered, the number of sensitivity indices to compute is (2^k-1), making the computational cost grow exponentially with k. This has been referred to as the curse of dimensionality and makes the complete decomposition unfeasible in most practical applications. In this paper we show that the information contained in the samples used to compute suitably defined subsets A of the (2^k-1) indices can be used to compute the complementary subsets A^* of indices, at no additional cost. This property allows reducing significantly the growth of the computational costs as k increases.

1 Introduction and state-of-the-art

Global sensitivity analysis aims to quantify the relative importance of input variables or factors in determining the value of an assigned output variable y, defined as

$$y = f(x_1,...,x_k) \tag{1}$$

A recent review of applications of this discipline can be found in [1-2]. Global sensitivity analysis is based on the variance decomposition scheme proposed by Sobol' [3], whereby the total unconditional variance of y can be decomposed as:

$$V(y) = \sum_i V_i + \sum_i \sum_{j>i} V_{ij} + ... + V_{12...k} \tag{2}$$

where

$$V_i = V\left(E\left(Y|x_i\right)\right) \tag{3}$$

P.M.A. Sloot et al. (Eds.): ICCS 2002, LNCS 2329, pp. 196–205, 2002.

$$V_{ij} = V\left(E\left(Y\middle|x_i, x_j\right)\right) - V_i - V_j \tag{4}$$

and so on. The development in (2) contains k terms of the first order V_i, $k(k-1)/2$ terms of the second order V_{ij} and so on, till the last term of order k, for a total of (2^k-1) terms. The V_{ij} terms are the second order (or two-way) terms, analogous to the second order effects described in experimental design textbooks (see e.g. [4]). The V_{ij} terms capture that part of the effect of x_i and x_j that is not described by the first order terms. If one divides the terms in equation (2) by the unconditional variance V, one obtains the sensitivity indices of increasing order S_i, S_{ij} and so on. The so defined sensitivity indices are nicely scaled in [0, 1]. Formula (2) has a long history, and various authors have proposed different versions of it. A discussion can be found in [5-7]. Sobol's version of formula (2) is based on a decomposition of the function f itself into terms of increasing dimensionality, i.e.:

$$f(x_1, x_2, ..., x_k) = f_0 + \sum_i f_i + \sum_i \sum_{j>i} f_{ij} + ... + f_{12...k} \tag{5}$$

where each term is function only of the factors in its index, i.e. $f_i=f_i(x_i)$, $f_{ij}=f_{ij}(x_i,x_j)$ and so on. Decompositions (2, 5) are unique provided that the input factors are independent and that each individual term (5) is square integrable and the integral over any of its own variables is zero [3].

Considering a generic term of cardinality c in the development (2), the set $\mathbf{u} = (x_{i_1}, x_{i_2}, ..., x_{i_c})$ of the factors in the generic term and the remaining set $\mathbf{v} = (x_{l_1}, x_{l_2}, ..., x_{l_{k-c}})$ can be defined. The evaluation of the term requires the computation a multidimensional integral of the following form

$$V\left(E\left(y\middle|\mathbf{u}\right)\right) = \int E^2\left(y\middle|\mathbf{u} = \tilde{\mathbf{u}}\right) p(\tilde{\mathbf{u}}) d\tilde{\mathbf{u}} - E^2(y) \tag{6}$$

where $p(\mathbf{u})$ denotes the joint probability density function of factors in the set $\mathbf{u}$, which, upon orthogonality, can be expressed as

$$p(\mathbf{u}) d\mathbf{u} = \prod_{j=i_1}^{i_c} \left(p_j\left(x_j\right) dx_j\right) \tag{7}$$

Equation (6) is computationally impractical. In a Monte Carlo frame, it implies a double loop: the inner one to compute the conditional expectation of y conditioned to the subset $\mathbf{u}$ (integral over the complementary $\mathbf{v}$ subspace) and the outer to compute the integral over the $\mathbf{u}$ subspace. The integral in (6) has been rewritten by Ishigami and Homma [8], Sobol' [3], as:

$$\int E^2\left(y\middle|\mathbf{u} = \tilde{\mathbf{u}}\right) p(\tilde{\mathbf{u}}) d\tilde{\mathbf{u}} = \int \left\{\int f(\tilde{\mathbf{u}}, \mathbf{v}) p(\mathbf{v}) d\mathbf{v}\right\}^2 p(\tilde{\mathbf{u}}) d\tilde{\mathbf{u}} =$$

$$\int \left\{\int f(\tilde{\mathbf{u}}, \mathbf{v}) f(\tilde{\mathbf{u}}, \mathbf{v}') p(\mathbf{v}) d\mathbf{v} \, p(\mathbf{v}') d\mathbf{v}'\right\} p(\tilde{\mathbf{u}}) d\tilde{\mathbf{u}} = \tag{8}$$

$$\int f(\mathbf{u}, \mathbf{v}) f(\mathbf{u}, \mathbf{v}') p(\mathbf{v}) d\mathbf{v} \, p(\mathbf{v}') d\mathbf{v}' \, p(\mathbf{u}) d\mathbf{u}$$

The expedient of using the additional integration variable primed, allows us to realise that the integral in (8) is the expectation value of the function F of a set of $(2k-c)$ factors:

$$F(\mathbf{u}, \mathbf{v}, \mathbf{v}') = f(\mathbf{u}, \mathbf{v}) f(\mathbf{u}, \mathbf{v}') =$$
$$f(x_{i_1}, x_{i_2}, ..., x_{i_c}, x_{l_1}, x_{l_2}, ..., x_{l_{k-c}}) f(x_{i_1}, x_{i_2}, ..., x_{i_c}, x'_{l_1}, x'_{l_2}, ..., x'_{l_{k-c}}) = \qquad (9)$$
$$f(x_1, x_2, ... x_k) f(x_{i_1}, x_{i_2}, ..., x_{i_c}, x'_{l_1}, x'_{l_2}, ..., x'_{l_{k-c}})$$

The integral (8) can be hence computed using a single Monte Carlo loop. The Monte Carlo procedure that follows was proposed by Saltelli et al. [9].

Two input sample matrices $\mathbf{M}_1$ and $\mathbf{M}_2$ are generated:

$$\mathbf{M}_1 = \begin{matrix} x_{11} & x_{12} & \cdots & x_{1k} \\ x_{21} & x_{22} & \cdots & x_{2k} \\ \cdots & & & \\ x_{n1} & x_{n2} & \cdots & x_{nk} \end{matrix} , \qquad \mathbf{M}_2 = \begin{matrix} x'_{11} & x'_{12} & \cdots & x'_{1k} \\ x'_{21} & x'_{22} & \cdots & x'_{2k} \\ \cdots & & & \\ x'_{n1} & x'_{n2} & \cdots & x'_{nk} \end{matrix} \qquad (10)$$

where n is the sample size used for the Monte Carlo estimate. In order to estimate the sensitivity measure for a generic subset $\mathbf{u}$, i.e.

$$S_{\mathbf{u}} = V(E(y|\mathbf{u}))/V(y) \qquad (11)$$

$$V(E(y|\mathbf{u})) = U_{\mathbf{u}} - E^2(y) \qquad (12)$$

$$U_{\mathbf{u}} = \iint \int f(\mathbf{u}, \mathbf{v}) f(\mathbf{u}, \mathbf{v}') d\mathbf{u} d\mathbf{v} d\mathbf{v}' \qquad (13)$$

we need an estimate for both $E^2(y)$ and $U_{\mathbf{u}}$. The former can be either obtained from values of y computed on the sample in $\mathbf{M}_1$ or $\mathbf{M}_2$. $U_{\mathbf{u}}$ can be obtained from values of y computed on the matrix $\mathbf{N}_{\mathbf{u}}$, whose columns are taken from matrix $\mathbf{M}_1$ for factors in subset $\mathbf{u}$ and from matrix $\mathbf{M}_2$ for factors in subset $\mathbf{v}$ (i.e. $\mathbf{v}'$):

$$\mathbf{N}_{\mathbf{u}}[1, ..., n; i_1, ..., i_c] = \begin{matrix} x_{1i_1} & \cdots & x_{1i_c} \\ x_{2i_1} & \cdots & x_{2i_c} \\ \cdots & \cdots & \cdots \\ x_{ni_1} & \cdots & x_{ni_c} \end{matrix} \quad \mathbf{N}_{\mathbf{u}}[1, ..., n; l_1, ..., l_{k-c}] = \begin{matrix} x'_{1l_1} & \cdots & x'_{1l_{k-c}} \\ x'_{2l_1} & \cdots & x'_{2l_{k-c}} \\ \cdots & \cdots & \cdots \\ x'_{nl_1} & \cdots & x'_{nl_{k-c}} \end{matrix} \qquad (14)$$

i.e. by:

$$\hat{U}_{\mathbf{u}} = \frac{1}{n-1} \sum_{r=1}^{n} f(x_{r1}, x_{r2}, ..., x_{rk}) f(x_{ri_1}, x_{ri_2}, ..., x_{ri_c}, x'_{rl_1}, x'_{rl_2}, ..., x'_{rl_{k-c}}) \qquad (15)$$

If one thinks of matrix $\mathbf{M}_1$ as the "sample" matrix, and of $\mathbf{M}_2$ as the "re-sample" matrix, then $\hat{U}_{\mathbf{u}}$ is obtained from products of values of f computed from the sample matrix times values of f computed from $\mathbf{N}_{\mathbf{u}}$, i.e. a matrix where all factors $\mathbf{v}$ are re-sampled. This procedure allows estimating all terms but the last one, quantifying the k order interaction effect[1]. The latter can be estimated by difference, subtracting from the unconditional variance the remaining (2^k-2) terms. Error estimates for $\hat{U}_\mathbf{u}$'s are discussed in the original reference of Sobol'. A bootstrap based alternative is discussed in [5].

So, a complete analysis of all effects of increasing order for the k input factors can be obtained by computing all terms in (2), but these are as many as (2^k-1). This problem has been referred to by Rabitz et al. [7] as "the curse of dimensionality". The computational cost of estimating all effects in (2) is in fact as high as $n(2^k-1)$, where again n is the sample size used to estimate one individual effect[2].

2   A More Efficient Approach

Consider to have the matrices $\mathbf{M}_1$ and $\mathbf{M}_2$, and to have run the model for the parameter values of matrix $\mathbf{M}_1$ and for a subset m of matrices $(\mathbf{N}_{\mathbf{u}1}, \ldots, \mathbf{N}_{\mathbf{u}m})$ of the whole set of (2^k-1) $\mathbf{N}_{\mathbf{u}i}$ matrices to be considered. According to the standard procedure described in the previous section, m indices can be computed by applying the estimator (15), by multiplying one at a time the vectors of n model runs for matrices $(\mathbf{N}_{\mathbf{u}1}, \ldots, \mathbf{N}_{\mathbf{u}m})$, that we define $[f(\mathbf{N}_{\mathbf{u}1}), \ldots, f(\mathbf{N}_{\mathbf{u}m})]$, by the vector of n model runs for parameter values $\mathbf{M}_1$, that we define $f(\mathbf{M}_1)$. However, nothing impedes to apply the same estimator (15) to all pair-wise combinations of the set of m vectors of model evaluations $[f(\mathbf{N}_{\mathbf{u}1}), \ldots, f(\mathbf{N}_{\mathbf{u}m})]$. By doing so, another set of $m(m-1)/2$ of $\hat{U}_\mathbf{u}$ estimates, *with repetitions*, can be obtained. Such an additional set has no additional computational cost, and so can be used for making the computation of the whole set of (2^k-1) indices more efficient. This idea has been proposed by Saltelli [10], who demonstrated that, once first order effects and total effects have been computed, at the cost of $n(k+2)$ model runs, double estimates can be obtained for the effects of all groups of two factors taken together and all groups of $(k-2)$ factors taken together, at no additional cost[3].

Some formalism is now needed, in order to allow a correct problem formulation for the efficient approach. The matrices $\mathbf{M}_1$, $\mathbf{M}_2$ and $\mathbf{N}_{\mathbf{u}i}$ can be represented by k-dimensional Boolean arrays as follows:

$$a_{(1,\ldots,k)} = [0,0,\ldots,0] \text{ for } \mathbf{M}_1 ; a_0 = [1,1,\ldots,1] \text{ for } \mathbf{M}_2$$

$$a_{(i_1,\ldots,i_{c_i})}[j] = \begin{cases} 0 \text{ if } j \in [i_1,\ldots,i_{c_i}] \\ 1 \text{ if } j \in [l_1,\ldots,l_{k-c_i}] \end{cases} \text{ for } \mathbf{N}_{\mathbf{u}i} \tag{16}$$

---

[1] It is easy to see that for the k order term, the matrix $\mathbf{N}_\mathbf{u}$ equals the sample matrix and the estimator (15) would simply give the unconditional variance.

[2] $n(2^k-2)$ is needed to compute all effects but the k order, and n more to compute $\hat{E}(y)$, $V(y)$.

[3] Total indices are defined as $S_{Ti}=1-S_{\sim i}$, where $\sim i$ means all factors except i, and quantifies the effect of i alone (main effect) plus the interaction terms with all other factors [1,3,10].

where c_i is the cardinality of the element $\mathbf{u}_i$ of the decomposition (2) and $(l_1,\ldots,l_{k-c_i})$ is., as usual, the array of indices of the complementary set of factors $\mathbf{v}$.

For example, for $k=3$, the whole set of arrays is defined by:

$$
\begin{array}{lll}
& a_1 = (0,1,1) & a_{12} = (0,0,1) \\
a_{123} = (0,0,0) & a_2 = (1,0,1) & a_{13} = (0,1,0) \\
a_0 = (1,1,1) & a_3 = (1,1,0) & a_{23} = (1,0,0)
\end{array}
\tag{17}
$$

and the estimates $\hat{U}_{\mathbf{u}}$ and their Boolean representations s_i can be expressed by:

$$
\begin{aligned}
\hat{U}_1 &= 1/(n-1)f(\mathbf{M}_1)\cdot f(\mathbf{N}_1); \quad s_1 = (1,0,0) = (0,0,0)\otimes(0,1,1) \\
\hat{U}_2 &= 1/(n-1)f(\mathbf{M}_1)\cdot f(\mathbf{N}_2); \quad s_1 = (0,1,0) = (0,0,0)\otimes(1,0,1) \\
\hat{U}_3 &= 1/(n-1)f(\mathbf{M}_1)\cdot f(\mathbf{N}_3); \quad s_1 = (0,0,1) = (0,0,0)\otimes(1,1,0) \\
\hat{U}_{12} &= 1/(n-1)f(\mathbf{M}_1)f(\mathbf{N}_{12}); \quad s_{12} = (1,1,0) = (0,0,0)\otimes(0,0,1) \\
\hat{U}_{13} &= 1/(n-1)f(\mathbf{M}_1)f(\mathbf{N}_{13}); \quad s_{13} = (1,0,1) = (0,0,0)\otimes(0,1,0) \\
\hat{U}_{23} &= 1/(n-1)f(\mathbf{M}_1)f(\mathbf{N}_{23}); \quad s_{23} = (0,1,1) = (0,0,0)\otimes(1,0,0)
\end{aligned}
\tag{18}
$$

where $\otimes$ is a bit-wise Boolean operator which gives 1 if the elements in the array are equal and 0 if they are different. The $\otimes$ operator allows to know which estimate $\hat{U}_{\mathbf{u}}$ can be obtained by taking pair-wise combinations of the arrays of Monte Carlo runs $[f(\mathbf{N}_{\mathbf{u}1}), \ldots, f(\mathbf{N}_{\mathbf{u}m})]$. For example:

$$
a_{13}\otimes a_3 = (0,1,0)\otimes(1,1,0) = (0,1,1) = s_{23} \Rightarrow 1/(n-1)f(\mathbf{N}_{13})\cdot f(\mathbf{N}_3) = \hat{U}_{23}
\tag{19}
$$

With this formalism, the various array elements a_i can also be interpreted as co-ordinates of the vertices of a k-dimensional unit hypercube, while the sensitivity indices can be represented by edges and hyper-diagonals joining couples of vertices a_i. With this interpretation the standard procedure consists of drawing all edges and hyper-diagonals starting from the origin a_{123}.

Another interpretation of arrays a_i is that they can be seen as the k-bit binary representations of the sequence of natural numbers $[0, 1, \ldots, 2^k-1]\in \aleph$. This can be useful for a synthetic representation of arrays a_i trough integer numbers.

Theorem 1

Let m^* be the minimum dimension of the subset A of *opportunely* chosen elements $[a_{\mathbf{u}1},\ldots,a_{\mathbf{u}m^*}]$, for which, combining pair-wise the elements of A through the $\otimes$ operator, the whole set of s_i elements can be obtained, then

$$
m_{MIN} = \mathrm{int}^+\left[\left(1+\sqrt{1+8(2^k-2)}\right)/2\right]\leq m^* < 2^k-2
\tag{20}
$$

where $\mathrm{int}^+(z)$ is the smallest integer larger the real number z.

Proof
First let us demonstrate the right-hand side inequality. The cost of the standard procedure is as high as $n(2^k-1)^2$. However, for efficiency purposes, $\hat{E}(y)$ and $V(y)$ can be estimated also from any of the block matrices used for the (2^k-2) $\hat{U}_\mathbf{u}$ estimates, making the maximum cost to be as high as $n(2^k-2)$. Moreover, the minimum number m^* is surely *smaller* than (2^k-2), because by combining an arbitrary pair of arrays $[f(\mathbf{N_{u1}}), ..., f(\mathbf{N_{um}})]$, $m>1$, an additional estimate $\hat{U}_{\mathbf{u}(m+1)}$ can be computed at zero cost, reducing the total cost *at least* by n.

The left-hand side can be demonstrated by introducing the constraint that the number of pair-wise combinations of the m^* elements must be at least as high as (2^k-2), i.e.:

$$m^*(m^*-1)/2 \geq 2^k - 2 \tag{21}$$

which gives the minimum constraint for m^*.

Remarks

1. Inequality (21) would become and equation if each combination gives a different index, i.e. no repeated estimates are obtained in the optimal a_i subset.
2. Theorem 1 shows that the order of the growth rate of the computational cost of the full decomposition can be halved, since the cost is reduced from $o(2^k)$ to $o(2^{(k+1)/2})$. This makes the complete decomposition feasible at least for 'not too high' values of k (e.g. $k \leq 11$, see next sections). On the other hand the exponential form is unaltered. This implies that for large values of k, the curse of dimensionality problem remains impracticable.
3. Theorem 1 does not tell us anything about the choice of the m^* elements allowing the efficient approach. This will be the solution of an optimisation problem, which we have not yet solved for the general case, but only for some values of k. In section 3 we will show a geometrical solution of the optimisation problem for $k=3$, 4. In section 4 we will show some experimental results obtained from a simple (but not optimal!) algorithm for the selection of suitable elements, which allows appreciating the degree of simplification offered by the proposed approach for larger k values.

2.1   Geometrical Solution of the Optimisation Problem for $k=3, 4$

For the case of three factors, formula (20) gives that $m_{MIN}=4$. The complete decomposition in this case reads:

$$V(y) = V_1 + V_2 + V_3 + V_{12} + V_{13} + V_{23} + V_{123} \tag{22}$$

Let us consider the 3D unit-cube representing our optimisation problem. The three main effects can be represented by choosing three face diagonals taken on three distinct non-parallel faces. The three second order effects are represented by choosing three non-parallel edges. Our optimisation problem can be seen as identifying the minimum set of vertices, which allow drawing three non-parallel edges and three non-parallel faces. It is easy to verify that it is sufficient to consider the 4 vertices shown in Fig. 1 in order to draw all required edges and diagonals. Since in this case $m^*=m_{MIN}$, the optimisation problem for $k=3$ is solved.

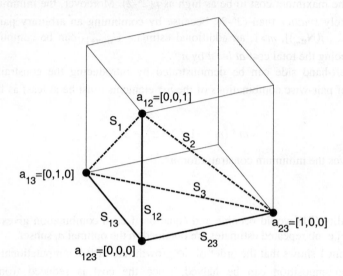

Fig. 1. Graphical interpretation of the efficient approach for $k=3$. Solid lines are second order effects [edges]; dashed lines are first order effects [face diagonals].

Table 1. Results of the application of the operator $\otimes$ to all pair-wise combinations of optimal subset of elements a_i for $k=4$.

Bits ↔ ℵ		a_{1234}	a_{123}	a_{124}	a_{134}	a_{234}	a_0
$[0,0,0,0] \leftrightarrow 0$	a_{1234}	$\hat{V}(y)$					
$[0,0,0,1] \leftrightarrow 1$	a_{123}	$\hat{U}_{123}$	$\hat{V}(y)$				
$[0,0,1,0] \leftrightarrow 2$	a_{124}	$\hat{U}_{124}$	$\hat{U}_{12}$	$\hat{V}(y)$			
$[0,1,0,0] \leftrightarrow 4$	a_{134}	$\hat{U}_{134}$	$\hat{U}_{13}$	$\hat{U}_{14}$	$\hat{V}(y)$		
$[1,0,0,0] \leftrightarrow 8$	a_{234}	$\hat{U}_{234}$	$\hat{U}_{23}$	$\hat{U}_{24}$	$\hat{U}_{34}$	$\hat{V}(y)$	
$[1,1,1,1] \leftrightarrow 15$	a_0	$\hat{E}^2(y)$	$\hat{U}_4$	$\hat{U}_3$	$\hat{U}_2$	$\hat{U}_1$	$\hat{V}(y)$

In the case of $k=4$, formula (20) gives that $m_{MIN}=6$. By extending the geometrical solution found for $k=3$, we took as initial trial subset of vertices the origin a_{1234} and its 4 adjacent vertices: a_{234}, a_{134}, a_{124}, a_{123}. Finally by adding the vertex opposite to the origin a_0, it can be easily verified from Table 1 that the whole set of 15 indices can be computed from these 6 elements. Again, since our estimated m^* equals m_{MIN}, the solution find is the optimal solution for the efficient approach. In the first column of Table 1, the 4-bit binary representation of a_i and the natural numbers corresponding to such a binary array are also shown. A final remark regards the cross-combination of two complementary elements (see a_{1234} and a_0 in Table 1). In such a case, the result of the operator $\otimes$ gives the squared expected value, since we are making the scalar product of two independent random vectors. On the other hand, applying the operator $\otimes$ to repeated elements (diagonal elements of the matrix), we get estimates of the unconditional variance of y.

2.2 Some Experimental Results for $k>4$

For $k>4$, we were not able to find similar geometrical considerations to find the optimal subset of vertices in the k dimensional hyper-cube. The problem should be solved with an optimisation algorithm. Such an optimisation problem can be a formidable task. Consider the easiest case of $k=5$. The total number of elements among which the optimal subset has to be found is as large as $2^5 = 32$, while $m_{MIN}=9$. If no optimisation algorithm is applied, but a complete search of combinations is performed, the combinatory study requires the evaluation of 32 elements grouped by 9, i.e. 28048800 possible combinations. Further examples of costs for the full search procedure are shown in Table 2. It can be seen that, for $k>5$, a procedure based on a full search of combinations is clearly unaffordable, at least on a workstation or PC.

Table 2. Cost of a plain search of combinations for some values of k.

k	m_{MIN}	Elements (2^k)	Combinations for $m^*=m_{MIN}$	Combinations for $m^*=m_{MIN}+1$	Combinations for $m^*=m_{MIN}+2$
5	9	32	28048800	64512240	-
6	12	64	$3.3\ 10^{12}$	$1.3\ 10^{13}$	$4.8\ 10^{13}$
7	17	128	$6.1\ 10^{20}$	$3.8\ 10^{21}$	$2.2\ 10^{22}$

So, to find possible suitable combinations for $k>4$, we applied a simple and relatively quick algorithm, which is not an optimisation algorithm but gives sufficiently small values of m^* to appreciate the advantage of the new approach. The algorithm takes an initial trial subset of $k+1$ elements, given by the origin $a_{1,...,k}$ and the k adjacent vertices, as we did for $k=3, 4$. Then it adds one element at a time in such a way that the additional number of indices computable introducing the new element is maximum. The algorithm stops when the elements a_i selected are sufficient to compute the whole set of (2^k-1) indices. In the case of $k=5$, the algorithm gave a number of elements $m^*=10$. In such a case, we verified with a complete search of combinations that none of the 28048800 combinations of 9 elements allows the computation of the whole set of indices, so that also for $k=5$ the optimisation problem was solved.

The results of this procedure for $k=3, ..., 11$ are shown in Table 3. The advantages of the new approach are clear. If the standard procedure begins to be hard to implement for $k=7$, with the new approach, we can afford a complete analysis at least for $k=11$. Moreover, even if the 'true' values of $m*$ were smaller than the estimates $\hat{m}*$ shown in Table 3, the subsets of a_i elements that we found were sufficiently small to bridge most of the gap between the cost of the standard procedure and the limit $m(MIN)$. In fact, given the latter inferior limit, the further improvement, which may be given by the true optimisation, will not be as large as the one presented here.

In Table 4 we show the subsets of a_i elements represented by arrays of natural numbers in the interval $[0, 2^k-1]$. By taking the k-bit binary representation of such arrays, we get the a_i elements allowing the estimation of all terms in the decomposition (2).

Table 3. Comparison of the computational cost of the standard and the new procedure for some values of k.

k	Number of indices (2^k-1)	Cost of the standard procedure	Estimated cost of the new procedure $\hat{m}*$	Minimum limit for cost [formula (20)]
3	7	$7\,n$	$4\,n^4$	$4\,n$
4	15	$15\,n$	$6\,n^4$	$6\,n$
5	31	$31\,n$	$10\,n^4$	$9\,n$
6	63	$63\,n$	$14\,n$	$12\,n$
7	127	$127\,n$	$20\,n$	$17\,n$
8	255	$255\,n$	$29\,n$	$24\,n$
9	511	$511\,n$	$42\,n$	$33\,n$
10	1023	$1023\,n$	$67\,n$	$46\,n$
11	2047	$2047n$	$98\,n$	$65\,n$

Table 4. Arrays of natural numbers representing the optimal a_i conbinations detected by the simplified algorithm for some values of k.

k	Optimal combinations for the new approach
3	[0 1 2 4]
4	[0 1 2 4 8 15]
5	[0 1 2 4 8 15 16 17 18 19]
6	[0 1 2 3 4 5 6 8 16 24 32 40 48 63]
7	[0 1 2 4 8 15 16 22 28 32 44 51 57 64 77 85 94 106 107 112]

3 Conclusions

In this paper we presented some efficient procedures for numerical experiments aimed at sensitivity analysis of model output. We have focused here on the computation of sensitivity indices that are based on decomposing the variance of the target function in a quantitative fashion. The approach presented aimed to identify to what extent it is

---

[4] In this case the optimisation problem has been solved, so the estimate is correct.

possible to fight the so-called "curse of dimensionality", that hinders the use of quantitative sensitivity analysis for computationally expensive models.

The approach allows reducing by a factor 2 the growth rate of computational effort required for the complete decomposition of the unconditional variance of a model output. It consists of using the information contained in the Monte Carlo samples applied for the computation of a given subset of terms, to compute the remaining set of terms, at no additional cost. The complete solution of the optimisation problem is still in progress and we managed to treat it in a simplified way only for $k<12$.

On the base of this incomplete results, we can state that even assuming for n the value of 1000, the new procedure can make the full decomposition affordable for models whose cost per run is in the range from milliseconds or lower to some seconds. If k does not exceed 6, the model cost per run can increase to some minutes. For models whose execution is in the tenths of minutes to a day range, quantitative methods are not applicable and efficient qualitative methods such as that of Morris [11] should be used (see [12] for a review).

References

1. Saltelli, A., Tarantola, S., Campolongo, F.: Sensitivity analysis as an ingredient of modelling. Statistical Science 15 (2000) 377-395.
2. Saltelli, A., Chan, K., Scott, M. (eds.): Sensitivity Analysis. Wiley Series in Probability and Statistics. John Wiley and Sons, Chichester (2000).
3. Sobol', I. M.: Sensitivity estimates for nonlinear mathematical models. Matematicheskoe Modelirovanie 2 (1990) 112-118, translated as: Sobol', I. M.: Sensitivity analysis for nonlinear mathematical models. Mathematical Modelling & Computational Experiment 1 (1993) 407-414.
4. Box, G. E. P., Hunter, W. G., Hunter, J. S.: Statistics for experimenters. John Wiley and Sons, New York (1978).
5. Archer, G., Saltelli, A., Sobol', I. M.: Sensitivity measures, ANOVA like techniques and the use of bootstrap. Journal of Statistical Computation and Simulation 58 (1997) 99-120.
6. Rabitz, H., Ali\c{s}, Ö. F., Shorter, J., Shim, K.: Efficient input-output representations. Computer Physics Communications 117 (1999) 11-20.
7. Rabitz, H., Ali\c{s}, Ö. F.: Managing the Tyranny of Parameters in Mathematical Modelling of Physical Systems. In: Saltelli, A. Chan K., Scott M. (eds.): Sensitivity Analysis. Wiley Series in Probability and Statistics. John Wiley and Sons, Chichester (2000) 199-223
8. Ishigami, T., Homma, T.: An importance quantification technique in uncertainty analysis for computer models. Proceedings of the ISUMA '90, First International Symposium on Uncertainty Modelling and Analysis, University of Maryland, December 3-6 (1990).
9. Saltelli A., Andres, T. H., Homma, T.: Some new techniques in sensitivity analysis of model output. Computational Statistics and Data Analysis 15 (1993) 211-238
10. Saltelli, A.: Making best use of model evaluations to compute sensitivity indices. Submitted to Computer Physics Communications (2002).
11. Morris, M. D.: Factorial Sampling Plans for Preliminary Computational Experiments. Technometrics 33 (1991) 161-174
12. Campolongo, F., Saltelli, A.: Design of experiment. In: Saltelli, A. Chan K., Scott M. (eds.): Sensitivity Analysis. Wiley Series in Probability and Statistics. John Wiley and Sons, Chichester (2000) 51-63.

Birge and Qi Method for Three-Stage Stochastic Programs Using IPM

Georg Ch. Pflug and Ladislav Halada

Department of Statistics and Decision Support Systems, Vienna University,
Universitaesstr. 5, A-1010 Wien, Austria,
e-mail:georg.pflug@univie.ac.at

Abstract. One approach how to solve a linear optimization is based on interior point method. This method requires the solution of large linear system equations. A special matrix factorization techniques that exploit the structure of the constraint matrix has been suggested for its computation. The method of Birge and Qi has been reported as efficient, stable and accurate for two-stage stochastic programs. In this report we present a generalization of this method for three-stage stochastic programs.

1 Introduction

Solving the deterministic equivalent formulation of two-stage stochastic programs using interior point method requires the solution of linear systems of the form

$$(ADA^t)\, dy = b. \tag{1}$$

Solving of this problem requires more then $90 - 95\%$ of total programming time [1]. Birge and Holmes [2] compared different methods for the solution of this system. They found that the factorization technique based on the work of Birge and Qi (BQ) [3] is more efficient and stable than other methods. They also suggested BQ for parallel computation. A parallel version of BQ for two-stage stochastic programs was also implemented on an Intel iPSC/860 hypercube and a Connection Machine CM-5 with nearly perfect speedup [4]. According to our knowledge, this method has not been used for a three-stage stochastic program so far. The aim of this report is analysis of the BQ method and a suggestion how the BQ method can be used for three-stage stochastic programming. Sect. 2 briefly describes the BQ matrix decomposition. An application of the BQ to a three-stage stochastic model is given in Sect. 3.

2 BQ Factorization for Two-Stage Model

For expositional clarity, we start with a two-stage model which has been resolved in [2]. Let $A^{(2)}$ be a constraint matrix of the two-stage stochastic model and $D^{(2)}$

P.M.A. Sloot et al. (Eds.): ICCS 2002, LNCS 2329, pp. 206–215, 2002.

a positive definite and diagonal matrix

$$
A^{(2)} = \begin{pmatrix} A_0 & & & & \\ T_1 & A_1 & & & \\ T_2 & & A_2 & & \\ \vdots & & & \ddots & \\ T_k & & & & A_k \end{pmatrix}, \quad D^{(2)} = \begin{pmatrix} D_0 & & & & \\ & D_1 & & & \\ & & D_2 & & \\ & & & \ddots & \\ & & & & D_k \end{pmatrix}, \quad (2)
$$

where A_i are $m_i \times n_i$ and D_i are positive definite diagonal $n_i \times n_i$ matrices, $i = 0, 1, 2, \ldots, k$. T_i are $m_i \times n_0$ matrices, $i = 1, 2, \ldots, k$. We assume that A_i have full row rank and $m_i \leq n_i$, $i = 0, 1, 2, \ldots, k$. For wit sake of completness, let us denote $R^{(2)} = A^{(2)} D^{(2)} (A^{(2)})^t$. Then $R^{(2)}$ can be expressed as sum of a block-diagonal matrix and product of matrices

$$
R^{(2)} = \mathcal{R}^{(2)} + U^{(2)} \left[\mathcal{D}^{(2)} (V^{(2)})^t \right] = \mathcal{R}^{(2)} + U^{(2)} \left[(W^{(2)})^t \right], \quad (3)
$$

where

$$
\mathcal{R}^{(2)} = Diag(I_{m_0}, A_1 D_1 A_1^t, \ldots, A_k D_k A_k^t) = Diag(I_{m_0}, R_1, \ldots, R_k) \quad (4)
$$

and

$$
U^{(2)} \mathcal{D}^{(2)} (V^{(2)})^t = \begin{pmatrix} A_0 \, I_{m_0} \\ T_1 \\ T_2 \\ \vdots \\ T_k \end{pmatrix} \begin{pmatrix} D_0 & \\ & I_{m_0} \end{pmatrix} \begin{pmatrix} A_0^t & T_1^t \, T_2^t \, \ldots \, T_k^t \\ -I_{m_0} & \end{pmatrix}. \quad (5)
$$

Note $\mathcal{R}^{(2)}$ is the diagonal matrix with positive definite matrices on the diagonal entries. Now, if we need the inverse of $R^{(2)}$, we can use the Sherman-Morrison-Woodbury formula. It holds [5]

$$
(R^{(2)})^{-1} = (\mathcal{R}^{(2)})^{-1} - (\mathcal{R}^{(2)})^{-1} U^{(2)} (G^{(2)})^{-1} (V^{(2)})^t (\mathcal{R}^{(2)})^{-1}, \quad (6)
$$

if and only if both $\mathcal{R}^{(2)}$ and $G^{(2)}$ are nonsingular, where

$$
(G^{(2)})^{-1} = [I_{n_0+m_0} + (W^{(2)})^t (\mathcal{R}^{(2)})^{-1} U^{(2)}]^{-1} \mathcal{D}^{(2)}. \quad (7)
$$

In the matrix form

$$
G^{(2)} = \begin{pmatrix} D_0^{-1} + A_0^t A_0 + \sum_{i=1}^{k} T_i^t R_i^{-1} T_i \, A_0^t \\ -A_0 & 0 \end{pmatrix} = \begin{pmatrix} \hat{G}^{(2)} & A_0^t \\ -A_0 & 0 \end{pmatrix}. \quad (8)
$$

It has been proved [2] that $\hat{G}^{(2)}$ is positive definite and symmetric matrix and $G^{(2)}$ is nonsingular. Hence, the conditions for the validity of (6) are fulfilled.

Thus, we can rewrite the solution of the system $R^{(2)}\,dy^{(2)} = b$ by the relation (6) as follows: $dy^{(2)} = p^{(2)} - s^{(2)}$, where

$$R^{(2)}\,p^{(2)} = b \tag{9}$$
$$G^{(2)}\,q^{(2)} = (V^{(2)})^t\,p^{(2)} \tag{10}$$
$$R^{(2)}\,s^{(2)} = U^{(2)}\,q^{(2)}. \tag{11}$$

More precisely,

$$R^{(2)}\,p^{(2)} = \begin{pmatrix} I_{m_0} & & & \\ & R_1 & & \\ & & \ddots & \\ & & & R_k \end{pmatrix} \begin{pmatrix} p_0^{(2)} \\ p_1^{(2)} \\ \vdots \\ p_k^{(2)} \end{pmatrix} = \begin{pmatrix} b_0 \\ b_1 \\ \vdots \\ b_k \end{pmatrix}. \tag{12}$$

Hence, the vector $p^{(2)}$ can be computed component-wise by solving sub-block systems

$$p_0^{(2)} = b_0$$
$$R_i\,p_i^{(2)} = b_i, \qquad i = 1, 2, ...k. \tag{13}$$

We can proceed similarly in the case of computation of the vector $s^{(2)}$. It gives

$$s_0^{(2)} = A_0\,q_1^{(2)} + q_2^{(2)}$$
$$R_i\,s_i^{(2)} = T_i\,q_1^{(2)}, \qquad i = 1, 2, ...k. \tag{14}$$

Because $R_i = A_i D_i A_i^t, \quad i = 1, 2, ...k$ is symmetric positive definite matrix, its Cholesky decomposition can be used for the solution of both (13) and (14). The solution of (10) can be found by exploiting the matrix structure $G^{(2)}$. We have

$$G^{(2)}\,q^{(2)} = \begin{pmatrix} \hat{G}^{(2)} & A_0^t \\ -A_0 & 0 \end{pmatrix} \begin{pmatrix} q_1^{(2)} \\ q_2^{(2)} \end{pmatrix} = \begin{pmatrix} \hat{v}_1^{(2)} \\ \hat{v}_2^{(2)} \end{pmatrix}, \tag{15}$$

where

$$\begin{pmatrix} \hat{v}_1^{(2)} \\ \hat{v}_2^{(2)} \end{pmatrix} = (V^{(2)})^t\,p^{(2)} = \begin{pmatrix} A_0^t p_0^{(2)} + \sum_{i=1}^k T_i^t p_i^{(2)} \\ -p_0^{(2)} \end{pmatrix}. \tag{16}$$

The vectors $q_i^{(2)}$ and $\hat{v}_i^{(2)}$, i=1,2 have the size corresponding to the matrix structure $G^{(2)}$. By an elimination process, for (15) we get

$$[\,A_0(\hat{G}^{(2)})^{-1}A_0^t\,]\,q_2^{(2)} = A_0\,(\hat{G}^{(2)})^{-1}\,\hat{v}_1^{(2)} + \hat{v}_2^{(2)} \tag{17}$$
$$\hat{G}^{(2)}\,q_1^{(2)} = \hat{v}_1^{(2)} - A_0^t\,q_2^{(2)}.$$

Because both matrices $A_0(\hat{G}^{(2)})^{-1}A_0^t$ and $\hat{G}^{(2)}$ are symmetric positive definite, one can use their Cholesky decomposition for solving these systems. The procedure for sequential computing of the vector $dy^{(2)}$ by (9)-(11) has been named

Findy in [2]. The parallel version of Findy has been formulated in [4]. Formally, parameters of this procedure are $Findy(\mathcal{R}^{(2)}, A_0, D_0, T_1, \ldots, T_k, b, dy^{(2)})$. In [4] this procedure has been implemented on a distributed-memory multiple-instruction multiple-date (MIMD) message-passing parallel computers, an Intel iPSC/860 hypercube and a Connection Machine CM-5. Results are reported with the solution of the linear systems arising when solving stochastic programs with $98,304$ scenarios, which correspond to deterministic equivalent linear programs with up to $1,966,090$ constraints and $13,762,630$ variables. From the timing data presented in this paper it is evident that the speed-up and the efficiency is most influenced by the percentage of time spent in communication and also by the ratio m_i/n_i.

3   BQ Method Applied to Three-Stage Stochastic Model

Let $A^{(3)}$ be here a constraint matrix of a three-stage stochastic model

$$
A^{(3)} = \begin{pmatrix}
A_0 & & & \\
T_{10} & A_{10} & & \\
 & T_{11} & A_{11} & \\
 & T_{12} & & A_{12} \\
 & T_{13} & & & A_{13} \\
T_{20} & & & & & A_{20} \\
 & & & & & T_{21} & A_{21} \\
 & & & & & T_{22} & & A_{22} \\
 & & & & & T_{23} & & & A_{23}
\end{pmatrix} = \begin{pmatrix}
A_0 & & \\
T_1^{(3)} & A_1^{(2)} & \\
T_2^{(3)} & & A_2^{(2)}
\end{pmatrix}, \quad (18)
$$

where A_0 is $m_0 \times n_0$ and A_{ij} are $m_{ij} \times n_{ij}$ matrices. T_{ij} has the size conformable to the matrices A_0 and A_{ij}. Let $D^{(3)}$ be diagonal and positive definite matrix

$$
D^{(3)} = Diag(D_0; D_{10}, \ldots, D_{13}; D_{20}, \ldots, D_{23}) = Diag(D_0, D_1^{(2)}, D_2^{(2)}), \quad (19)
$$

where D_0 and D_{ij} are diagonal $n_0 \times n_0$ and $n_{ij} \times n_{ij}$ matrices with positive entries, $i = 1, 2; \ j = 0, 1, 2, 3..$ Again, we assume that A_0 and A_{ij} have full row rank and $m_0 \le n_0, \ m_{ij} \le n_{ij}, \ i = 1, 2; \ j = 0, 1, 2, 3$.

The matrix $R^{(3)} = A^{(3)} D^{(3)} (A^{(3)})^t$ in the denotation of the largesized blocks matrix $A^{(3)}$ can be decomposed as follows:

$$
R^{(3)} = \mathcal{R}^{(3)} + U^{(3)} [\, \mathcal{D}^{(3)} \, (V^{(3)})^t \,] = \mathcal{R}^{(3)} + U^{(3)} [\, (W^{(3)})^t \,], \quad (20)
$$

where

$$
\mathcal{R}^{(3)} = Diag(I_{m_0}, A_1^{(2)} D_1^{(2)} (A_1^{(2)})^t, A_2^{(2)} D_2^{(2)} (A_2^{(2)})^t) = Diag(I_{m_0}, R_1^{(2)}, R_2^{(2)})
$$

$$
(21)
$$

and

$$U^{(3)} \mathcal{D}^{(3)} (V^{(3)})^t = \begin{pmatrix} A_0 & I_{m_0} \\ T_1^{(3)} & \\ T_2^{(3)} & \end{pmatrix} \begin{pmatrix} \mathcal{D}_0 & \\ & I_{m_0} \end{pmatrix} \begin{pmatrix} A_0^t & (T_1^{(3)})^t & (T_2^{(3)})^t \\ -I_{m_0} & & \end{pmatrix}. \quad (22)$$

By the Sherman-Morrison-Woodbury formula we obtain again

$$(R^{(3)})^{-1} = (\mathcal{R}^{(3)})^{-1} - (\mathcal{R}^{(3)})^{-1} U^{(3)} (G^{(3)})^{-1} (V^{(3)})^t (\mathcal{R}^{(3)})^{-1}, \quad (23)$$

if and only if $\mathcal{R}^{(3)}$ and $G^{(3)}$ are nonsingular. Here

$$(G^{(3)})^{-1} = [\, I_{n_0+m_0} + \mathcal{D}^{(3)} (V^{(3)})^t (\mathcal{R}^{(3)})^{-1} U^{(3)}]^{-1} \mathcal{D}^{(3)}$$
$$G^{(3)} = (\mathcal{D}^{(3)})^{-1} + (V^{(3)})^t (\mathcal{R}^{(3)})^{-1} U^{(3)}. \quad (24)$$

In the matrix form:

$$G^{(3)} = \begin{pmatrix} \mathcal{D}_0^{-1} + A_0^t A_0 + \sum_{i=1}^{2}(T_i^{(3)})^t(R_i^{(2)})^{-1}T_i^{(3)} & A_0^t \\ -A_0 & 0 \end{pmatrix} = \begin{pmatrix} \hat{G}^{(3)} & A_0^t \\ -A_0 & 0 \end{pmatrix}. \quad (25)$$

The validity of (23) follows from the same reason as for the two-stage model. Thus, the solution of $R^{(3)} \, dy^{(3)} = b^{(3)}$ can be expressed by the inversion as $dy^{(3)} = p^{(3)} - s^{(3)}$ while

$$\mathcal{R}^{(3)} p^{(3)} = b^{(3)}, \quad (26)$$
$$G^{(3)} q^{(3)} = (V^{(3)})^t p^{(3)}, \quad (27)$$
$$\mathcal{R}^{(3)} s^{(3)} = U^{(3)} q^{(3)}. \quad (28)$$

The equations (26)-(28) represent the decomposition of the original problem into three sub-problems. An advantage of such decomposition is that $\mathcal{R}^{(3)}$ is the block-diagonal matrix amenable to further decomposition.

3.1   Solving the Equation $\mathcal{R}^{(3)} p^{(3)} = b^{(3)}$

It is easy to see from the equation

$$\mathcal{R}^{(3)} p^{(3)} = \begin{pmatrix} I_{m_0} & & \\ & R_1^{(2)} & \\ & & R_2^{(2)} \end{pmatrix} \begin{pmatrix} p_0^{(3)} \\ p_1^{(3)} \\ p_2^{(3)} \end{pmatrix} = \begin{pmatrix} b_0^{(3)} \\ b_1^{(3)} \\ b_2^{(3)} \end{pmatrix} \quad (29)$$

that this system represents the following independent systems

$$p_0^{(3)} = b_0^{(3)} \quad (30)$$
$$R_i^{(2)} p_i^{(3)} = b_i^{(3)}, \quad i = 1, 2, \quad (31)$$

where $R_i^{(2)} = A_i^{(2)} D_i^{(2)} (A_i^{(2)})^t$, $i = 1, 2$ represent the matrix of the two-stage model problem, which has been described in Sect.2. Therefore, (31) can be solved by the procedure Findy or Parallel Findy, depending on the number of processors. Its input parameters are readable from the entries of matrix $A_i^{(2)}$, $D_i^{(2)}$ defined in (18)-(19). The right-hand side and the solution vector are $b_i^{(3)}$ and $p_i^{(3)}$, $i = 1, 2$, respectively. It is clear that in our case the parameters are

$$Findy(\mathcal{R}_i^{(2)}, \; A_{i0}, \; D_{i0}, T_{i1}, \ldots, T_{i3}, b_i^{(3)}, p_i^{(3)}), \quad i = 1, 2$$

where $\mathcal{R}_i^{(2)}$ is the diagonal matrix in the decomposition $R_i^{(2)}$, i.e.

$$R_i^{(2)} = \mathcal{R}_i^{(2)} + U_i^{(2)} (W_i^{(2)})^t \quad i = 1, 2 \tag{32}$$

and

$$\mathcal{R}_i^{(2)} = Diag(I_{m_{i0}}, \; R_{i1}, R_{i2}, R_{i3}), \quad R_{ij} = A_{ij} D_{ij} A_{ij}^t, \quad i = 1, 2 \quad j = 1, 2, 3$$

$$U_i^{(2)} = \begin{pmatrix} A_{i0} \; I_{m_{i0}} \\ T_{i1} \\ T_{i2} \\ T_{i3} \end{pmatrix}, \quad (W_i^{(2)})^t = \begin{pmatrix} D_{i0} & \\ & I_{m_{i0}} \end{pmatrix} \begin{pmatrix} A_{i0}^t & T_{i1}^t \; T_{i2}^t \; T_{i3}^t \\ -I_{m_{i0}} & \end{pmatrix}. \tag{33}$$

If this procedure is applied for the given values, having in the mind the relations (30)-(31) we are able to compose the vector $p^{(3)}$.

3.2 Solving the Equation $G^{(3)} q^{(3)} = (V^{(3)})^t p^{(3)}$

Solving of this equation requires to have the entries of the right-hand side vector and the sub-block matrix $\hat{G}^{(3)}$ available. With this aim we denote the elements of the vector $(V^{(3)})^t p^{(3)}$ as $(\hat{v}_1^{(3)}, \hat{v}_2^{(3)})^t$. Then we have

$$\begin{pmatrix} \hat{v}_1^{(3)} \\ \hat{v}_2^{(3)} \end{pmatrix} = \begin{pmatrix} A_0^t p_0^{(3)} + \sum_{i=1}^{2} T_{i0}^t p_{i0}^{(3)} \\ -p_0^{(3)} \end{pmatrix}, \tag{34}$$

where $p_{i0}^{(3)}$, $i = 1, 2$ is the vector of the first m_{i0}-elements of $p_i^{(3)}$. We know from (25) that

$$\hat{G}^{(3)} = D_0^{-1} + A_0^t A_0 + \sum_{i=1}^{2} (T_i^{(3)})^t (R_i^{(2)})^{-1} T_i^{(3)}. \tag{35}$$

For the relatively complicated expression $(T_i^{(3)})^t (R_i^{(2)})^{-1} T_i^{(3)}$ we can prove that

$$(T_i^{(3)})^t (R_i^{(2)})^{-1} T_i^{(3)} = T_{i0}^t (\hat{T}_{i0} - T_{i0}), \quad i = 1, 2 \tag{36}$$

where $\hat{T}_{i0}$ is the solution of the equation

$$[A_{i0} (\hat{G}_i^{(2)})^{(-1)} A_{i0}^t] \hat{T}_{i0} = T_{i0}, \quad i = 1, 2. \tag{37}$$

Really, according to (6) we have

$$(T_i^{(3)})^t (R_i^{(2)})^{-1} T_i^{(3)} =$$

$$(T_i^{(3)})^t [(R_i^{(2)})^{-1} - (R_i^{(2)})^{-1} U_i^{(2)} (G_i^{(2)})^{-1} (V_i^{(2)})^t (R_i^{(2)})^{-1}] T_i^{(3)} =$$

$$(T_i^{(3)})^t (R_i^{(2)})^{-1} T_i^{(3)} - (T_i^{(3)})^t (R_i^{(2)})^{-1} U_i^{(2)} (G_i^{(2)})^{-1} (V_i^{(2)})^t (R_i^{(2)})^{-1} T_i^{(3)}.$$

It holds

$$(T_i^{(3)})^t (R_i^{(2)})^{-1} T_i^{(3)} = T_{i0}^t T_{i0}, \quad (T_i^{(3)})^t (R_i^{(2)})^{-1} U_i^{(2)} = (T_{i0}^t A_{i0}, T_{i0}^t), \tag{38}$$

$$(V_i^{(2)})^t (R_i^{(2)})^{-1} T_i^{(3)} = \begin{pmatrix} A_{i0}^t T_{i0} \\ -T_{i0} \end{pmatrix}. \tag{39}$$

Therefore

$$(T_i^{(3)})^t (R_i^{(2)})^{-1} T_i^{(3)} = T_{i0}^t T_{i0} - (T_{i0}^t A_{i0}, T_{i0}^t) \begin{pmatrix} \hat{G}_i^{(2)} & A_{i0}^t \\ -A_{i0} & 0 \end{pmatrix}^{-1} \begin{pmatrix} A_{i0}^t T_{i0} \\ -T_{i0} \end{pmatrix}. \tag{40}$$

Now let

$$\begin{pmatrix} K & L \\ M & N \end{pmatrix} = \begin{pmatrix} \hat{G}_i^{(2)} & A_{i0}^t \\ -A_{i0} & 0 \end{pmatrix}^{-1}. \tag{41}$$

According to [6]

$$N = [A_{i0}(\hat{G}_i^{(2)})^{-1} A_{i0}^t]^{-1} \tag{42}$$

$$L = -(\hat{G}_i^{(2)})^{-1} A_{i0}^t [A_{i0}(\hat{G}_i^{(2)})^{-1} A_{i0}^t]^{-1} \tag{43}$$

$$M = [A_{i0}(\hat{G}_i^{(2)})^{-1} A_{i0}^t]^{-1} A_{i0}(\hat{G}_i^{(2)})^{-1} \tag{44}$$

$$K = (\hat{G}_i^{(2)})^{-1} - (\hat{G}_i^{(2)})^{-1} A_{i0}^t [A_{i0}(\hat{G}_i^{(2)})^{-1} A_{i0}^t]^{-1} A_{i0}(\hat{G}_i^{(2)})^{-1}. \tag{45}$$

Now, if we use (42)-(45) in (40) we obtain

$$(T_i^{(3)})^t (R_i^{(2)})^{(-1)} T_i^{(3)} = T_{i0}^t [A_{i0}(\hat{G}_i^{(2)})^{-1} A_{i0}^t]^{-1} T_{i0} - T_{i0}^t T_{i0}. \tag{46}$$

Thus,

$$\hat{G}^{(3)} = D_0^{-1} + A_0^t A_0 + \sum_{i=1}^{2} T_{i0}^t (\hat{T}_{i0} - T_{i0}). \tag{47}$$

We remember that the Cholesky decomposition of matrix $A_{i0}(\hat{G}_i^{(2)})^{-1} A_{i0}^t$ has been performed during the procedure Findy applied on matrix $R_i^{(2)}$, $i = 1, 2$. Thus, this decomposition is available already, only triangular solver is used for the computation of $\hat{T}_{i0}$, $i = 1, 2$ in this step. Having the values of $\hat{G}^{(3)}$ and $(\hat{v}_1^{(3)}, \hat{v}_2^{(3)})^t$ we can solve the system

$$\begin{pmatrix} \hat{G}^{(3)} & A_0^t \\ -A_0 & 0 \end{pmatrix} \begin{pmatrix} q_1^{(3)} \\ q_2^{(3)} \end{pmatrix} = \begin{pmatrix} \hat{v}_1^{(3)} \\ \hat{v}_2^{(3)} \end{pmatrix}. \tag{48}$$

The standard elimination process applied on this system yields

$$[(A_0 \, (\hat{G}^{(3)})^{-1} \, A_0^t] \, q_2^{(3)} = A_0 \, (\hat{G}^{(3)})^{-1} \, \hat{v}_1^{(3)} + \hat{v}_2^{(3)} \tag{49}$$

$$\hat{G}^{(3)} \, q_1^{(3)} = \hat{v}_1^{(3)} - A_0^t \, q_2^{(3)}. \tag{50}$$

Thus, to solve (48) the following procedure is required:

PROCEDURE $\text{Updy}(\hat{G}^{(3)}, \, A_0, \, \hat{v}_1^{(3)}, \hat{v}_2^{(3)})$
(a) Form the Cholesky decomposition of $\hat{G}^{(3)}$
(b) Solve the systems $\hat{G}^{(3)} B_0 = A_0^t$
(c) Form the Cholesky decomposition of $A_0 B_0$
(d) Solve the systems (49) and (50).

3.3   Solving the Equation $\mathcal{R}^{(3)} s^{(3)} = U^{(3)} q^{(3)}$

The system $\mathcal{R}^{(3)} \, s^{(3)} = U^{(3)} q^{(3)}$ could be solved in a similar way as in Sect. 3.1. The right-hand side equals

$$U^{(3)} q^{(3)} = \begin{pmatrix} A_0 & I_{m_0} \\ T_1^{(3)} & 0 \\ T_2^{(3)} & 0 \end{pmatrix} \begin{pmatrix} q_1^{(3)} \\ q_2^{(3)} \end{pmatrix} = \begin{pmatrix} A_0 \, q_1^{(3)} + q_2^{(3)} \\ T_1^{(3)} \, q_1^{(3)} \\ T_2^{(3)} \, q_1^{(3)} \end{pmatrix}. \tag{51}$$

Thus, the system has the form

$$\mathcal{R}^{(3)} s^{(3)} = \begin{pmatrix} I_{m_0} & & \\ & R_1^{(2)} & \\ & & R_2^{(2)} \end{pmatrix} \begin{pmatrix} s_0^{(3)} \\ s_1^{(3)} \\ s_2^{(3)} \end{pmatrix} = \begin{pmatrix} A_0 \, q_1^{(3)} + q_2^{(3)} \\ T_1^{(3)} \, q_1^{(3)} \\ T_2^{(3)} \, q_1^{(3)} \end{pmatrix}, \tag{52}$$

from which we obtain independent equations

$$s_0^{(3)} = A_0 \, q_1^{(3)} + q_2^{(3)} \tag{53}$$

$$R_i^{(2)} s_i^{(3)} = T_i^{(3)} \, q_1^{(3)}, \; i = 1, 2. \tag{54}$$

The last two equations are again solvable by the procedure Findy as in Sect. 3.1 with the right-hand side $T_i^{(3)} q_1^{(3)}$, $i = 1, 2$. But, owing to the structure of this vector, where only the first m_{i0}- entries are nonzero, we suggest for its computation a modification the already mentioned procedure Findy. The solution $s_i^{(3)}$ of (54) can be expressed as $s_i^{(3)} = \hat{p}_i^{(3)} - \hat{s}_i^{(3)}$, where $\hat{p}_i^{(3)}$ and $\hat{s}_i^{(3)}$, $i = 1, 2$ fulfil

$$\mathcal{R}_i^{(2)} \, \hat{p}_i^{(3)} = T_i^{(3)} \, q_1^{(3)} \tag{55}$$

$$G_i^{(2)} \, \hat{q}_i^{(3)} = (V_i^{(2)})^t \, \hat{p}_i^{(3)} \tag{56}$$

$$\mathcal{R}_i^{(2)} \, \hat{s}_i^{(3)} = U_i^{(2)} \, \hat{q}_i^{(3)}. \tag{57}$$

In the matrix form

$$\mathcal{R}_i^{(2)}\,\hat{p}_i^{(3)} = \begin{pmatrix} I_{m_{i0}} & & & \\ & R_{i1} & & \\ & & R_{i2} & \\ & & & R_{i3} \end{pmatrix} \begin{pmatrix} \hat{p}_{i0}^{(3)} \\ \hat{p}_{i1}^{(3)} \\ \hat{p}_{i2}^{(3)} \\ \hat{p}_{i3}^{(3)} \end{pmatrix} = \begin{pmatrix} T_{i0}\,q_1^{(3)} \\ 0 \\ 0 \\ 0 \end{pmatrix}, \tag{58}$$

from which we have immediately

$$\hat{p}_{i0}^{(3)} = T_{i0}q_1^{(3)}, \quad i = 1, 2 \tag{59}$$

$$\hat{p}_{ij}^{(3)} = 0, \quad i = 1, 2 \quad j = 1, 2, 3. \tag{60}$$

To find the solution of (56) means to solve the following matrix equation

$$\begin{pmatrix} \hat{G}_i^{(2)} & A_{i0}^t \\ -A_{i0} & 0 \end{pmatrix} \begin{pmatrix} \hat{q}_{i1}^{(3)} \\ \hat{q}_{i2}^{(3)} \end{pmatrix} = \begin{pmatrix} A_{i0}^t\,\hat{p}_{i0}^{(3)} \\ -\hat{p}_{i0}^{(3)} \end{pmatrix}. \tag{61}$$

The last equation (57) represents the system

$$\mathcal{R}_i^{(2)}\hat{s}_i^{(3)} = \begin{pmatrix} I_{m_{i0}} & & & \\ & R_{i1} & & \\ & & R_{i2} & \\ & & & R_{i3} \end{pmatrix} \begin{pmatrix} \hat{s}_{i0}^{(3)} \\ \hat{s}_{i1}^{(3)} \\ \hat{s}_{i2}^{(3)} \\ \hat{s}_{i3}^{(3)} \end{pmatrix} = \begin{pmatrix} A_{i0}\hat{q}_{i1}^{(3)} + \hat{q}_{i2}^{(3)} \\ T_{i1}\hat{q}_{i1}^{(3)} \\ T_{i2}\hat{q}_{i1}^{(3)} \\ T_{i3}\hat{q}_{i1}^{(3)} \end{pmatrix}, \tag{62}$$

from which we have

$$\hat{s}_{i0}^{(3)} = A_{i0}\,\hat{q}_{i1}^{(3)} + \hat{q}_{i2}^{(3)}, \quad i = 1, 2 \tag{63}$$

$$R_{ij}\,\hat{s}_{ij}^{(3)} = T_{ij}\,\hat{q}_{i1}^{(3)}, \quad i = 1, 2 \quad j = 1, 2, 3. \tag{64}$$

With the vector $\hat{s}_{ij}^{(3)}$ available, we have the result

$$s_i^{(3)} = \hat{p}_i^{(3)} - \hat{s}_i^{(3)} = \begin{pmatrix} \hat{p}_{i0}^{(3)} - \hat{s}_{i0}^{(3)} \\ -\hat{s}_{i1}^{(3)} \\ -\hat{s}_{i2}^{(3)} \\ -\hat{s}_{i3}^{(3)} \end{pmatrix}, \quad i = 1, 2. \tag{65}$$

Finally, the computing of $s_i^{(3)}$, $i = 1, 2$ consist of the following steps:

PROCEDURE Findysparse($\mathcal{R}_i^{(2)}, A_{i0}, T_{i1}, T_{i2}, T_{i3}, T_i^{(3)}q_1^{(3)}, s_i^{(3)}$)
1. Set $\hat{p}_{i0}^{(3)} = T_{i0}q_1^{(3)}$, $i = 1, 2$
2. Solve the system (61)
3. (a) Set $\hat{s}_{i0}^{(3)} = A_{i0}\hat{q}_{i1}^{(3)} + \hat{q}_{i2}^{(3)}$, $i = 1, 2$
    (b) Solve $R_{ij}\hat{s}_{ij}^{(3)} = T_{ij}\hat{q}_{i1}^{(3)}$, $i = 1, 2; j = 1, 2, 3$
4. Set $s_i^{(3)} = \hat{p}_i^{(3)} - \hat{s}_i^{(3)}$ $i = 1, 2$.

Note that the Cholesky decomposition of the system matrices is available in step 2 and 3(b). These decomposition has been computed by the procedure Findy. In the end, the result of a three-stage stochastic model problem equals $dy^{(3)} = p^{(3)} - s^{(3)}$. This difference is obtained by the following computational process :

1. Call Findy $(\mathcal{R}_i^{(2)}, A_{i0}, D_{i0}, T_{i1}, \ldots, T_{i3}, b_i^{(3)}, p_i^{(3)})$, $i = 1, 2$,
2. Call Updy$(\hat{G}^{(3)}, A_0, \hat{v}_1^{(3)}, \hat{v}_2^{(3)})$
3. Call Findysparse$(\mathcal{R}_i^{(2)}, A_{i0}, T_{i1}, T_{i2}, T_{i3}, T_i^{(3)} q_1^{(3)}, s_i^{(3)})$ i=1,2
4. Form $dy^{(3)}$ as difference $p^{(3)} - s^{(3)}$.

The procedures in steps 1 and 3 are independent and can be computed at the same time. Step 2 represents a binding of existing two-stage models and enables to calculate $s_0^{(3)}$ and the parameter $q_i^{(3)}$, i=1,2 for Findysparse(). Roughly, the process may be symbolically written as:

$$dy^{(3)} = \begin{pmatrix} b_0 \\ Findy(\mathcal{R}_1^{(2)}, ..., p_1^{(3)}) \\ Findy(\mathcal{R}_2^{(2)}, ..., p_2^{(3)}) \end{pmatrix} - \begin{pmatrix} A_0 q_1^{(3)} + q_2^{(3)} \\ Findysparse(\mathcal{R}_1^{(2)}, ..., s_1^{(3)}) \\ Findysparse(\mathcal{R}_2^{(2)}, ..., s_2^{(3)}) \end{pmatrix}. \qquad (66)$$

4 Conclusion

The aim of our paper has been to use the BQ factorization technique for three-stage stochastic program in a framework of an interior point method. As we can see, this technique leads to the solution of independent subproblems. Moreover, these subproblems are again scalable into smaller linear system of equations. The whole process contains a serial coordination step, but the range of a sequential computation is not critical for large-scale stochastic program. A parallel implementation of this algorithm within an interior point code and an extension to the multistage model will be topics of our future work.

References

1. Vladimirou, H. Zenios, S.A.: Scalable parallel computations for large-scale stochastic programming,Annals of Oper. Res. 90,1999, pp.87-129.
2. Birge, J.R., Holmes, D.F.: Efficient solution of two-stage stochastic linear programs using interior point methods, Comput. Optim. Appl., 1, 1992,pp. 245-276.
3. Birge, J.R., Qi, L.: Computing block-angular Karmarkar projections with applications to stochastic programming, Management Sci., 34,1988, pp.1472-1479.
4. Jessup, E.R.,Yang, D., Zenios, S.A.: Parallel factorization of structured matrices arising in stochastic programming, SIAM J.Optimization, Vol. 4., No.4,1994, pp. 833-846.
5. Golub,G.H.,Van Loan, Ch.F.: Matrix Computations, The Johns Hopkins University Press, 1996.
6. A.Ralston : A first course in numerical analysis, McGraw-Hill, inc., New York, 1965.

Multivariate Stochastic Models of Metocean Fields: Computational Aspects and Applications

A.V. Boukhanovsky

State Oceanographic Institute, St. Petersburg branch, Russia
Institute for High Performance Computing and Data Bases
120 Fontanka ,St. Petersburg, 198005, Russia
E-mail: avb@fn.csa.ru

1. Introduction

Metocean data fields (atmospheric pressure, wind speed, sea surface and air temperature, sea waves etc.) are multivariate and multidimensional, i.e. have a complex spatial and temporal variability. Only 10-20 years back the environmental databases wholly consisted of the time series (ship observation, sea and coastal monitoring stations, automatic probes and buoys, satellites) in fixed points of spatial regions. For processing of such data the different kinds of software are developed, e.g. [11]. They interprets the information in terms of random values (RV) or time series (TS) models only.

Development of environmental hydrodynamical models and use them for data assimilation and reanalysis [14], has allowed to create global information arrays of metocean fields in points of a regular spatial-temporal grid. So, the results of meteorological fields reanalysis may be used as source for hydrodynamic simulation of metocean (generally, hydrologic) fields, e.g., sea waves fields [8], water temperature and salinity fields [12] et al. This way allow to obtain the ensemble of metocean data fields in a regular grid points with certain temporal step. Hence, for processing and generalization of such data the model of a nonstationary inhomogeneous spatial-temporal random field (RF) must be considers.

Due to the high dimension of data in set of grid points, the multivariate statistical analysis (MSA) are applied to their processing. The goal of MSA is solution of three global problems - reduction of dimensionality (RD), detection of dependence (DD), and detection of heterogeneity (DH) of the random data [1]. Its operates by the canonical variables (principal components, canonical correlations, factor loadings), includes regression, discriminant analyses, and analysis of variance and covariance, classification, clustering and multidimensional scaling.

The main complexity of direct application of classical MSA to analysis of metocean spatial–temporal fields is connected with the fact, that all the procedures are developed for model of multivariate RV only [1,2,15,19]. Some of them are generalized for model of multivariate time series [5]. But there are two problems in the generalization of these procedures to RF model:

- *Various physical nature of metocean fields.* Hence, the various mathematical abstract objects may be used for their statistical description. E.g. the field of

P.M.A. Sloot et al. (Eds.): ICCS 2002, LNCS 2329, pp. 216–225, 2002.

atmospheric pressure is scalar function, and the simplest moments (mathematical expectation and variance) are scalar values too. The wind speed field is Euclidean (geometric) vector function. The mathematical expectation of such kind of data is Euclidean vector too, and variance is dyadic tensor. The field of water parameters (temperature, salinity, oxygen) is the affine (algebraic) vector function with affine vector of mathematical expectation and matrix of variance. Hence, it is necessary previously to introduce the basic algebraic and statistical operation for each kind of abstract objects. General discussion about this is in the paper [6].

- *Requirements to computational aspects of MSA RF procedures.* For traditional MSA RV with sufficiently small number of variables, the methods of linear algebra are developed well [10]. But for random fields of high dimension, especially in a small spatial and temporal grids, some of results by these methods became ill–conditioned due to strong spatial and temporal connectivity of data. Hence, it is necessary to develop special computational tools for MSA RF on the base of functional analysis in infinite–dimensional spaces of functions [3].

The goals of this paper are:

- To demonstrate the general approach for MSA RF in arbitrary functional space (scalar, Euclidean or affine vector) considering three main problems: RD, DD, DH.
- To synthesize the stochastic models on the base of MSA RV result for further ensemble simulation for investigation of non–observable rare environmental events.

2. General Approach

Let us consider the infinite–dimensional space of functions $\mathbf{H}$ with the operations of addition $(\mathbf{f}+\mathbf{g}) \in \mathbf{H}$, multiplication $(\mathbf{f} \circ \mathbf{g}) \in \mathbf{H}$, and scalar product $(\mathbf{g},\mathbf{f}) \in \mathbf{R}$ of elements $\mathbf{f}$, $\mathbf{g}$. Concept of scalar product is obvious only to scalar values. For Euclidean and affine vectors it generalizes concept of scalar product both in discrete space and in continuous space. If we consider the functional spaces of metocean events as Hilbert [3], in each of them any element η can be presented as an infinite converging series on some system of basic elements (scalar, Euclidean of affine functions) $\{\varphi_k\}$:

$$\eta = \sum_k c_k \phi_k .$$

(1)

The back transformation

$$c_k = (\eta, \phi_k)$$

(2)

defines an isomorphism between functional space $\mathbf{H}$ and discrete space $\mathbf{C}$, $c_k \in \mathbf{C}$.

Due to linearity of (1,2), the mathematical expectation and variance of η may be expressed from the moments of coefficients c_k:

$$m_\eta = \sum_k m_k \phi_k , \qquad (3)$$

$$K_\eta = \sum_k \sum_p k_{kp} [\phi_k \circ \phi_p] . \qquad (4)$$

Here m_k – scalar mathematical expectations of c_k, k_{kp} – covariances between c_k, c_p.

Full system of orthogonal functions $\{\phi_k\}$ may be obtained by means of Gram–Shmidt [3] orthogonalization procedure in accordance with of equation

$$(\phi_k, \phi_s) = \delta_{ks} N_k , \qquad (5)$$

where N_k is a norm. Even for Euclidean or affine vector fields it is complicate to obtain such functions in convenient form.

Let us consider optimal basis $\{\varphi_k\}$ from all $\{\phi_k\} \in H$. Following [17], this basis is generates by the equation

$$(K_\eta, \varphi_k) = D_k \varphi_k \qquad (6)$$

in the certain functional space H. System $\{\varphi_k\}$ is orthogonal, but coefficients c_k are independent random variables with variances D_k. For all other choices of $\{\phi_k\}$ coefficients c_k are correlated due to (4).

The main advantage of model (1) is passage from random field η (scalar, Euclidean or affine vector multivariate function) to set of scalar coefficients or scalar time series $\{c_k\}$. Hence, for different kinds of data the model of RV is valid for $\{c_k\}$. Below let us consider some applications of this approach for analysis and synthesis of various metocean fields.

3. Applications to analysis

3.1. Reduction of dimensionality [6]. Let us consider the scalar fields of sea level pressure (SLP) $\zeta(\vec{r}, t)$ and wind speed (VS) at the level 10 (m) $\vec{V}(\vec{r}, t)$ by the reanalysis [13] data array. Here $\vec{r} = (x, y)$ are the spatial coordinates, t is time. Traditionally for RD the expansion on empirical orthogonal basis (EOF [14]) are uses. For scalar fields of SLP the EOFs are the eigenfunctions of correlation kernel, and equation (6) became to Fredholm II integral equation [17]:

$$\int_{\langle R \rangle} K(\vec{r}, \vec{r}_1) \varphi_k(\vec{r}_1) d\vec{r}_1 = \lambda \varphi(\vec{r}) . \qquad (7)$$

Spectrum $\{\lambda_k\}$ of such kernel is discrete. Application of quadrature methods for (7) solution results in matrix representation without avoiding multicollinearity of the problem. Therefore let us use projective (variational) methods [11] for obtaining orthogonal basis. The idea of this methods is using of expansion (1) for representation of each EOF in (7). If we substitute (3) in (7), the equation (7) transform to algebraic eigenvalue problem. Occasionally such method allows to obtain analytical solution for some types of modeling representations of autocorrelation function $K_\eta(\bullet)$ of non-homogeneous field [5].

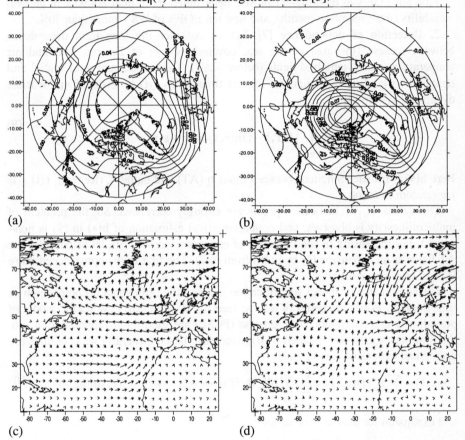

(a)   (b)

(c)   (d)

Fig. 1. First EOFs of monthly–mean SLP (North Hemisphere) and VS (North Atlantic). (a,c)–January, (b,d)–July

When we generalize (1) on Euclidean vector field $\vec{V}(\vec{r},t)$ in accordance with the rules of vector sums and dyadic (external) products, the equation (6) transforms in two homogeneous Fredholm II equations:

$$\int K_{uu}(\vec{r}_1,\vec{r}_2)\varphi(\vec{r}_2)d\vec{r}_2 + \int K_{uv}(\vec{r}_1,\vec{r}_2)\psi(\vec{r}_2)d\vec{r}_2 = \lambda\varphi(\vec{r}_1), \tag{8}$$

$$\int K_{vu}(\vec{r}_1,\vec{r}_2)\varphi(\vec{r}_2)d\vec{r}_2 + \int K_{vv}(\vec{r}_1,\vec{r}_2)\psi(\vec{r}_2)d\vec{r}_2 = \lambda\psi(\vec{r}_1),$$

with respect to components $\vec{\Psi} = (\phi, \psi)$. Here $K_{\bullet\bullet}$ – correlation functions between components of Euclidean vector. Using of (8) allows to simplify the interpretation of vector EOFs. E.g., in fig. 1 the first EOFs of monthly–mean SLP (a,b) and VS (c,d) are shown for January and July. It is clearly seen the great zones of wind speed variability near Iceland and Azores.

First ten EOF's of monthly–mean SLP determined more than 80–90% of general variability (from month to month), and first ten EOF's of WS – more than 70%.

2. Detection of dependence [7]. Let us consider the long–term dependence between iceness (the area of the sea, covered by ice) of Barents Sea, and air temperature spatial–temporal fields (below – AT). Take in account joint spatial and temporal variability, we define the iceness time series $\zeta(t)$ in terms of multivariate dynamical system [4]:

$$\zeta(t) = \int\limits_{<R>} \int\limits_{0}^{\infty} h(\vec{r}, t - \tau)\eta(\vec{r}, \tau)d\tau d\vec{r} + \varepsilon(t). \tag{9}$$

Here $h(\bullet)$ is transfer functions between input η (AT) and output ζ (iceness), $\varepsilon(t)$ – is lag random function.

The values of AT in nearest points are strongly correlated. So, the problem of multicollinearity is observes, and nonparametrical estimation of $h(\bullet)$ in (9) is non–correct. For increasing of conditionality of equations, born by (9), let us consider the AT fields as an expansion (1) by EOFs from (6,7). The first EOF approximated more 90% of total variability.

The second step of model simplification is using the orthogonal decomposition in time domain. While the series of iceness (see fig. 2) are clearly cyclic, the model of periodically correlated stochastic process (PCSP) are used for it representation. So, the time series may be expanded by trigonometric basis. Generally we obtaining bi–orthogonal expansion for AT fields

$$\eta(\vec{r}, t) = \sum_{n} a_n(t)\psi_n(\vec{r}) = \sum_{n}\sum_{m} b_{nm}^{(c)}\psi_n(\vec{r})\cos(\omega_m t) + b_{nm}^{(s)}\psi_n(\vec{r})\sin(\omega_m t) \tag{10}$$

and expansion for iceness time series

$$\zeta(t) = \sum_{m} \beta_m^{(c)}\cos(\omega_m t) + \beta_m^{(s)}\sin(\omega_m t). \tag{11}$$

This model allows to transfer from model of correlated RFs to model of depended RV $\left\{b_k^{(s)}, b_j^{(c)}, \beta_m^{(s)}, \beta_n^{(c)}\right\}$, so the procedure of dependence detecting is simplified. The joint correlation matrix $K\left[b_k^{(s)}, b_j^{(c)}, \beta_m^{(s)}, \beta_n^{(c)}\right]$ is rarefied, but between several coefficients the dependence is high. If we calculate the canonical correlations [15] between $\left\{b_k^{(s)}, b_j^{(c)}\right\}$ and $\left\{\beta_m^{(s)}, \beta_n^{(c)}\right\}$, we obtain $\lambda_1 = 0.8$ and $\lambda_2 = 0.6$, so, the general dependence between considered factors is sufficiently high. In the fig. 2 the time series of iceness from 1960 to 1980 are shown by observations and model simulation by (9–11). It is clearly seen the good agreement between to graphs.

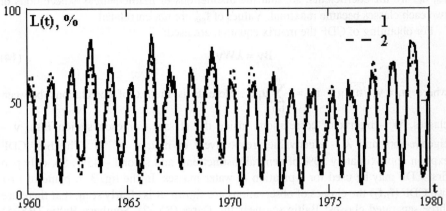

Fig. 2. Time series of iceness (%) of the Barents sea. 1 – observations, 2 – simulation (9–11).

3. Detection of heterogeneity [8]. Let us consider the example of typization of water mass of Baltic sea on the base of long–term observations on four International monitoring stations BY–2, BY–5, BY–15, BY–28, see fig. 3a. The water temperature **T**, salinity **S**, conditional density ρ and dissolved oxygen concentration **O₂** are jointly used for typization. All the data obtained each month from 1960 to 1980, in standard depths: z=0(10)100(25)… meters.

The generalization of assimilated data in terms of mathematical expectations and variances of initial data demands to operate with the grid function of mathematical expectation in 8832 points, and correlation matrix with more than $39 \cdot 10^6$ independent elements.

For decreasing of dimensionality the bi–ortogonal expansion of (1) on the basis $\{\varphi_k(z),\psi_s(t)\}$ for each oceanographic value (T,S,ρ,O_2) is considered

$$\zeta(z,t) = \sum_k a_k(t)\varphi_k(z) = \sum_k \sum_s b_{ks}\varphi_k(z)\psi_s(t) = \sum_s \eta_s(z)\psi_s(t). \qquad (12)$$

In paper [8] shown, that the spatial basis $\varphi_k(z)$ is Chebyshev polynomials, and temporal one $\psi_s(t)$ – is trigonometric. So, the expression (12) allows to transfer from model of vertical–inhomogeneous RF $\zeta(z,t)$ to system of RV $\{b_{ks}\}$ for T,S,ρ,O_2 independently. Due to the physical dependence between water mass parameters, the values $\{b_{ks}\}$ are correlated. Let us use its for typization in terms of discriminant variables. One of general procedures of discriminant analysis is obtaining of canonical discriminant functions (KDF) f_{pm}, as follows linear combination [15]:

$$f_{pm} = u_0 + \sum_k \sum_s u_{ks} b_{ks}^{(pm)} \qquad (13)$$

Here $b_{ks}^{(pm)}$ are coefficients of expansion (12) for monitoring station p in a year m, u_0, u_{ks} are the coefficients, so, that the distinctions of mathematical expectations of two each classes became maximal. Values of f_{pm} are not correlated.

For obtaining of CDF the matrix equation are used:

$$Bv = \lambda Wv \qquad (14)$$

where $u_i = v_i\sqrt{n-g}$, $u_0 = -\sum_{i=1}^{n}u_ix_i$. Here B is matrix of correlations between classes, W – matrix of correlations inside the classes (for each station BY), v – eigenvectors and λ – eigenvalues respectively of matrix $W^{-1}B$. The first CDF explain approximately 50% of general distinctions, and second – 45%. So, only two first CDF may be used for typization of water masses. In the fig. 3 in plane of first two CDF (f_1,f_2) the classes of observations are shown. It is clearly seen, that there are three separated classes: Baltic sea near the Darss (BY–2), Southern Baltic (BY–5) and Central Baltic (BY–15,28).

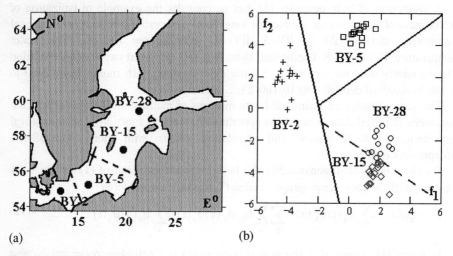

(a)   (b)

Fig. 3. Fig. 3. Typization of water masses of Baltic sea. (a) – Positions of International monitoring stations (b) – Data of four monitoring stations in first two CDF (f_1,f_2) plane.

4.   Application to ensemble simulation and extreme analysis

Representation (1) allows to synthesize the model ensemble of metocean fields on the base of MSA RF results by means of stochastic models. These models are used for interval estimation, testing of hypotheses about latent properties of phenomena and investigation of rare non–observable situations, e.g. extreme values of metocean fields once T years.

For identification of stochastic model let us consider (1) in terms of factor analysis [1]:

$$\eta = \sum_k c_k \phi_k + \varepsilon.$$ (15)

Here c_k are the factors, basis elements ϕ_k – became the factor loadings, and ε is the specific factor. If all the factors obtained by means (6), this method called as "principal factor method" for scalar variables.

Model (15) allows generate the model ensemble by means of Monte–Carlo approach on the base of variances of coefficients c_k, and specific factor ε [20]. For example, in the fig. 4 the directional annual WS extremes in North Atlantic (55N,30W) are shown. These values are obtained by means of direct estimation on simulated ensemble [19] by the model, based on EOF's from fig.1. It is seen, that strongest direction is West (28 m/s), and weakest – is North–East.

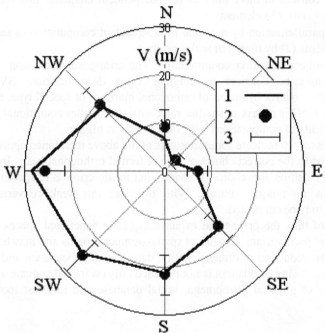

Fig. 4. Estimation of annual WS directional extremes in North Atlantic (55N,30E). 1 – simulation, 2 – estimation on reanalysis data, 3 – 95% confidence intervals

In the same fig. the values of annual extremes on reanalysis [13] data are shown too. The 95% confidence intervals covered both simulated and reanalysis estimates. Thus, it shown the good agreement between stochastic model and data.

5.   Discussion and conclusions

The above presented examples shown, that the using of orthogonal expansion (1) for representation of multivariate and multidimensional spatial–temporal metocean fields allows to obtain some advantages in comparison with classical matrix methods of MSA RV, e.g.:

- Simplification of probabilistic model (from non–scalar RF to system of scalar RV);
- Possibility to obtain the analytical solutions – a step for complex mathematical investigation of physical phenomena;
- Avoiding of quantization problem (methods of interpolation between greed points and objective analysis are not required);
- Decreasing of data dimensionality (in example 3, the initial data correlation matrix consists of more than $39 \cdot 10^6$ independent elements, and reduced matrix in (14) – only 153 elements);
- 100% parallelization by data due to independent computation of each values of coefficients (2) by means of scalar product;
- Improvement of matrix conditionality. The orthogonal expansion (1) in matrix terms may be consider as singular value decomposition (SVD) [10] of correlation matrix by means of orthogonal matrixes of special type. So, the result of such SVD is matrix of smaller rank with the smaller conditional number. So, the stability of computations with such matrix is higher.

It is necessary to indicate some shortcuts of the above mentioned approach. One of them is based in the correct choice of type of formal orthogonal basis. In some cases, as temporal rhythms, this choice is obvious, but for description of spatial patterns the problem has no unique solution. With this fact the weak convergence of the expansions may be connected.

Despite of this, the orthogonal expansions (1) in functional spaces remains the power tool of multivariate statistics of spatial–temporal fields and may be used for all the problems: reduction of dimensionality, detection of dependence and detection of heterogeneity. Using of this tool is not be based only on formal statistic approach, but the specifics of physical phenomena, initial database and hindcast tools must take into account.

Acknowledgement

This research is supported by INTAS grant Open 1999 N666.

References

1.   Anderson T.W. An introduction to multivariate statistical analysis. John Wiley, NY, 1948.

2.   Bartlett M.S. Multivariate analysis. J. Roy. Stat. Soc. Suppl. 9(B), 1947, 176–197.
3.   Balacrishnan, Applied functional analysis. NewYork, John Wiley, 1980
4.   Bendat J.S., Piersol A.G. Random data. Analysis and measurement procedures. John Wiley, NY, 1989.
5.   Brilinger D. Time series. Data analysis and theory. Holt, Renehart and Winston, Inc., New York, 1975.
6.   Boukhanovsky A.V., Degtyarev A.B., Rozhkov V.A. Peculiarities of computer simulation and statistical representation of time–spatial metocean fields. LNCS #2073, Springer–Verlag, 2001, pp.463–472.
7.   Boukhanovsky A.V., Ivanov N.E., Makarova A.V. Probabilistic analysis of spatial–temporal metocean fields. Proc. of Reg. Conf "Hydrodinamical methods of weather foreacast and climate investigation", St.Petersburg, Russia, 19–21 June, 2001 (in press, in Russian).
8.   Boukhanovsky A.V., Kokh A.O., Rozhkov V.A., Savchuk O.P., Shaer I.S. Statistical analysis of water masses of Baltic sea. Proc. of GOIN, St.Petersburg, Gymiz, Russia, 2001 (in press, in Russian)
9.   Cardone V.J., Cox A.T., Swail V.R. Specification of global wave climate: is this the final answer ?// 6th Int. Workshop on Wave Hindcasting and Forecasting. Monterey, California, November 6-10, p.211-223
10. Golube G.H., Van Loan C.F. Matrix computations (2nd ed.)., London, John Hopkins University Press, 1989.
11. Gould S.H. Variational methods for eigenvalue problems. University of Toronto Press, 1957.
12. Hamilton G.D. Processing of marine data. JCOMM Technical Report, WMO/TD #150, 1986.
13. Hansen I.S. Long–term 3–D modelling of stratification and nutrient cycling in the Baltic sea. Proc. of III BASYS Annual Science Conf., September 20–22, 1999, pp. 31–39.
14. Hotelling H. Relations between two sets of variables. Biometrika, 28, 1936, pp. 321–377.
15. Johnson R.A., Wichern D.W. Applied multivariate statistical analysis. Prentice-Hall Internalional, Inc., London, 1992, 642 pp.
16. Kalnay E., M. Kanamitsu, R. Kistler, W. Collins, D. Deaven, L. Gandin, M. Iredell, S. Saha, G. White, J. Woollen, Y. Zhu, A. Leetmaa, R. Reynolds, M. Chelliah, W. Ebisuzaki, W.Higgins, J. Janowiak, K. C. Mo, C. Ropelewski, J. Wang, R. Jenne, D. Joseph. The NCEP/NCAR 40-Year Reanalysis Project. Bulletin of the American Meteorological Society, №3, March, 1996.
17. Loeve M. Fonctions aleatories de second odre. C.R. Acad. Sci. 220, 1945.
18. Lopatoukhin L.J., Rozhkov V.A., Ryabinin V.E., Swail V.R., Boukhanovsky A.V., Degtyarev A.B. Estimation of extreme wave heights. JCOMM Technical Report, WMO/TD #1041, 2000.
19. Mardia K.V. Kent J.T., Bibby J.M. Multivariate analysis. London, Academia Press Inc., 1979
20. Ogorodnikov V.A., Prigarin S.M. Numerical modelling of random processes and fields: algorithms and applications. VSP, Utrecht, the Netherlands, 1996, 240 p.

Simulation of Gender Artificial Society: Multi-agent Models of Subject-Object Interactions

Julia Frolova[1] and Victor Korobitsin[1]

Omsk State University, Mathematical Modelling Chair,
55A, Mira pr., 644077 Omsk, Russia
{frolova, korobits}@univer.omsk.su
http://www.univer.omsk.su/socsys

Abstract. In the paper is presented an approach to simulation of gender interaction in artificial society. We are studying the process of the male's and female's interaction based on subject-object relations. On the one hand, the females radiate the charm pulses. Responding to these pulses, the males choose a female. On the other hand, the males accumulate a resource and females choose a male, having the maximum resource wealth. The computer simulation were realized on $\mathcal{SWARM}$. We reveal three types of agent interaction. These types is characterized by a subject-object interaction coefficient, defining the stability of agent interaction.

1 Introduction

The aim of research is a studying the behaviour of males and females and their differences, we research not only a man but also a society as a whole.

We consider a gender system in society. It differs from population by inclusion of gender structure. The existence of two types of agent (male and female) is needed under construction of artificial society model. We intend to add the gender structure in artificial society model for its adequacy to real society. Gender System is a collection of social relations between males and females that are defined their interpersonal interactions. The system of individual's connection has the informational structure, presented as a communication model of society.

Computer Simulation has the huge possibilities for understanding the society: the research of individuals, individual interactions, family, social group and organization, ethnoses, and social systems [2]. The computer models of gender systems help to develop the sociological approaches, reproducing and playing in real relations. Replacing the real gender relations by the computer model we come from the notion about gender as a special type of intrigue [3]. Both in intrigue and in discourse the source structure has a form of collection of simple sentences, connected between itself by various logical relations. The elements of discourse and intrigue are the events, participants and "non-events". The non-events include: the circumstances accompanying the events, the background

P.M.A. Sloot et al. (Eds.): ICCS 2002, LNCS 2329, pp. 226–235, 2002.

explaining the events, the estimation of event participants, the information correlating discourse (or intrigue) with events.

Considering the gender system we are studying the process of the male's and female's interaction based on subject-object relations [4]. Each individual can be presented and as subject and as object of interaction. Defining the individual state, we describe the respective act of subject and the change of object state in artificial society.

Defining the gender as an intrigue allows easy to go to constructing the model of gender interaction in artificial society. This is because, model is the playing simulation, reproducing the parameters a close to intrigue (see Table 1).

Table 1. Gender as Intrigue vs Computer Model

Gender as Intrigue	Computer Model
• participants	• agents
• circumstances accompanying the events	• rules
• events	• agent interactions
• background explaining the events	• environment
• estimation of event participants	• interacting the agents with environment
• information correlating the intrigue with events	• change into environment

2 Description of interaction process between males and females

The communication process between males and females has a collection of main elements of interaction. The problem of the computer model is an attempt to select they.

Under gender contact, the people are interacting on relations of confidence, understanding, supports, respects, and imitations. But they are in condition of the fights with the individuals of the same sex. The interaction between the different sexes is built on delight and pulses for males and on material and emotional support for females. Assigning the even chance for agents of artificial society and limiting the environment for they, we create the competition for agents of same sex. For satisfying its need, the agent must overtake an adversary. Then he will has got the necessary interaction.

The males and females are the equivalent individuals but they have the various gender particularities.

On the one hand, the females have an unique device for influence on the male behaviour. Herewith the males have own system of perception. For example, the influence device is a beauty, glamour and etc. In this instance the female is defined as a subject of action and the male is an object of influence in gender relations. This definition is a foundation of the first presented model.

On the other hand, the males have the resource and females are consuming it. In this instance the female is defined as an object and the male is a subject of action in gender relations. This definition is a foundation of the second presented model.

The result of interaction process is characterized by *a Gender Interaction Coefficient*, given by

$$\mathcal{G} = \frac{C_f}{F} : \frac{C_m}{M},$$

where F, M are numbers of females and males, respectively. C_f, C_m are numbers of interacting females and males, respectively. This numbers are shown how much agents participate in the interaction process. Thus the Gender Interaction Coefficient is given by the relation of the interaction proportions of females and males. If $\mathcal{G}$ equals 1.0 then we can observe the interaction process with gender equality in artificial society. When $\mathcal{G}$ less than 1.0 then the interaction majority is formed by the males. If $\mathcal{G}$ more than 1.0 then the males participate in the interaction with lesser activity.

These models were realized on multi-agent modelling system $\mathcal{SWARM}$ [1]. This system allow to program the artificial society, defining the respective parameters, rules of behaviour and interaction for agents and environment. Designed object-oriented $\mathcal{SWARM}$ libraries allows to demonstrate the agents' behaviour on computer display in real time and to get necessary characteristics by graphs and diagrams.

3   The first model

The pulses of female charm are the power of action for attraction between males and females. The attraction is an important component of interpersonal attractiveness. It is very involved with a need of interaction one with other. The feminine influence possesses the different degree of intensity, defining power of influence. The male is presented in models as a creature, encircled with all sides by radiated pulses. But he can make choice, realizing the possibility of the individual liberty of movement. The field of gender interactions is formed by means of made choice. It is defined as a superposition of two personal spaces.

The personal space of any individual is characterized by the communication potential, reflecting the condition of the contact process. The communication potential is feature of the personal possibilities, defining the quality of the individual contact. Its level depends on a sociability, nature and toughness of contacts, installed with other people. We may consider the interpersonal attraction as a condition and result of compatibility of two persons under interaction.

3.1   Constructing model of pulse influence

We apply the multi-agent modelling method for constructing model of pulse influence. Hereinafter we will describe the environment, agents, rule of movement and interaction for agents.

The Environment. The model describes the interaction of males and females inside the limited space (environment). We define the environment as an interaction field. Let the field be a square-wave area on plane. The interaction field wraps around from right to left (that is, were you to walk off the screen to the right, you would reappear at the left) and from top to bottom, forming a doughnut - technically, torus. We denote the model attributes:

T is the maximum simulated time for the model;

X, Y are the horizontal and vertical field dimensions, respectively.

Also, we denote the model variable:

t is a time $(0 \leq t \leq T)$.

The Agents. We define the male as a m-agent, the female as a f-agent. At every time each agent has a position given by an ordered pair (x, y) of horizontal and vertical lattice coordinates, respectively $(x \in [0, X], y \in [0, Y])$. Two agents are not allowed to occupy the same position. We shall investigate a random distribution of M m-agents and F f-agents in environment.

All agents have the own personal space and communication potential. The communication potential value is a variable state. Define it as $H_f(t)$ and $H_m(t)$ for f-agent and m-agent, respectively.

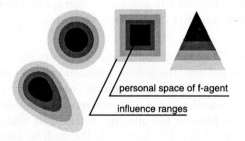

Fig. 1. An influence ranges

Each f-agent is source of pulse radiation and one has the influence range on m-agent. This area forms by radiation of charm pulses with various intensity level I_f. They can be the some forms included the personal space of f-agent. The maximum value of intensity achieves in this space and reduces with distribution on field of interaction. The intensity value is a variable state depended on communication potential level. The influence of f-agent on m-agent is reduced when this m-agent come away from source of pulse radiation. The interaction field is created through superposition of influence ranges of several f-agents. The example of areas is shown in figure 1.

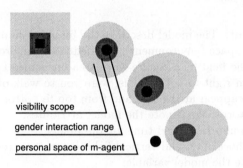

visibility scope

gender interaction range

personal space of m-agent

Fig. 2. An interaction ranges

Each m-agent has the individual system of perception. He look over the field of possible interaction within a some area defined as a visibility scope. This area can be the some form and not include the personal space of m-agent. Also, the gender interaction range is defined inside the visibility scope. The process of interaction between m-agent and f-agent can only occur f-agent within gender interaction range of m-agent. The example of ranges is shown in figure 2.

The movement rule. The m-agents are also given a movement rule. From interaction field within the visibility scope of m-agent, find the nearest unoccupied position of maximum intensity of influence, go there.

The contact rule. The interaction process of agents occurs according to contact rule. If f-agent fall into the gender interaction range of m-agent then the interaction process is realizing. Under interaction process, the communication potential is being increased and the influence intensity is being decreased. In the absence of interaction process, the influence intensity is being increased and the communication potential is being decreased. For each m-agent

$$H_m(t+1) = H_m(t) + \alpha N_m - \beta,$$

where α is an increase rate of communication potential; β is a decrease rate of one under interaction absence; N_m is a number of f-agents interacting with this m-agent. The value N_m is defined by the number of f-agents inhered in the interaction range of this m-agent. For each f-agent

$$H_f(t+1) = H_f(t) + \alpha N_f - \beta,$$

where N_f is a number of m-agents interacting with this f-agent.

The influence intensity is changed by the following rule

$$I_f = \frac{I_0}{H_f + 1},$$

where I_0 is a maximum value of intensity.

3.2 Computer simulation of interaction process based on the pulse influence

We shall investigate a random distribution of 53 f-agents - the color is black - and 47 m-agents - the color is white, as shown in figure 3, on the left.

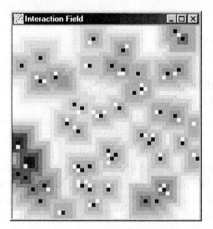 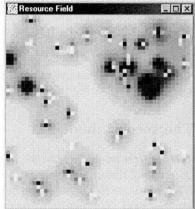

Fig. 3. Interaction field and Resource field

Environment consists of 2500 locations arranged on a 50×50 lattice with the influence intensity level at every site initially. The intensity score is highest at the peaks in the field around the cell occupied by f-agent - where the color is most black - and falls off in a series of terraces. The intensity scores range from some maximum - here $I_0 = 100$ - at the peaks to zero.

The value α (increase rate of communication potential) is set 0.015 and the value β (decrease rate of communication potential) is set 0.01. The visibility scope of m-agent, influence range of f-agent, and gender interaction range of m-agent have the square forms with linear sizes 11, 17, and 3, respectively.

Describe the simulation process in dynamics. The m-agent is moving on the field according to movement rule, and fall into the influence range of some f-agent. The m-agent go to the cell with maximum value of influence intensity. When the f-agent fall into the gender interaction range of m-agent then the interaction process is realized. Under interaction process the agent characteristics are changed according to the contact rule. The result of interaction series is a decrease of influence level of f-agent. In consequence of which the m-agent continues the search of f-agent with high influence level. And the process of interactions is being replayed.

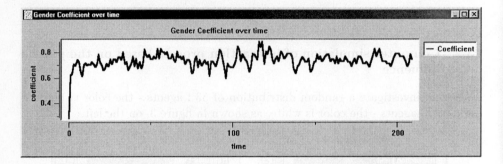

Fig. 4. Gender Interaction Coefficient value dynamics

On the figure 4 is presented the graph of Gender Interaction Coefficient value dynamics. $\mathcal{G}$ is fluctuating in interval $[0.6, 0.9]$.

4   The second model

Similarly the first model, we use the multi-agent modelling method for constructing the model of resource consumption. We also denote the model attributes and variables, and lattice. The interaction of m-agent and f-agent is realized within the resource field. The field also has the form of torus. The m-agent accumulate the resource. The resource field is formed according to resource wealth of m-agent. The f-agent has a need in consumption of resource. Therefore he is moving for searching the m-agent, having maximum resource wealth.

4.1   Constructing model of resource consumption

The Environment. In this model we define the environment as a resource field. Each (x, y) cell is characterized by the attributes:

ε is a rate of resource distribution;

$\beta_0(x, y)$ is a rate of resource depletion at cell (x, y);

$\gamma_0(x, y)$ is a resource loss at cell (x, y)

and the state variables:

$u(x, y, t)$ is a level of resource at cell (x, y) at time t;

$p_m(x, y, t)$ is a occupancy status of cell (x, y) by m-agent at time t. This function has a value equaled 1.0 in case the cell is occupied by m-agent and 0.0 in other case;

$p_f(x, y, t)$ is a occupancy status of cell (x, y) by f-agent at time t. This function has a value equaled 1.0 in case the cell is occupied by f-agent and 0.0 in other case.

The Agents. Each m-agent is a source of resource, which change with time. He has the initial resource capacity. The initial resource forms the resource field at first time. The m-agent is characterized by the attributes:

β_m is a rate of the resource accumulation;

γ_m is a level of resource loss. It is set from interval $[\gamma_{min}, \gamma_{max}]$;

ζ_m is an additional resource bonus;

s_m is an initial level of resource. It is set from interval $[s_{min}, s_{max}]$;

(x_m, y_m) is a field cell occupied by the m-agent

and the state variable:

$w(t)$ is a amount of resource wealth at time t. Its value equals $u(x_m, y_m, t))$.

Each f-agent is characterized by the attributes:

v is a visibility scope;

β_f is a rate of the resource consumption

and the state variable:

(x_f, y_f) is a field cell occupied by the f-agent.

The movement rule. The f-agents are given a movement rule, which is a kind of gradient search algorithm. From resource field within the visibility scope of f-agent, find the nearest unoccupied position of maximum level of resource, go there and consume the resource.

The rule of resource change on field. Under influence of agent's behaviour, the resource level in cell (x, y) is being changed at time. The resource level is changed in the following events:

- if this cell occupied by m-agent then resource level is being increased at a rate β_m;
- if this cell occupied by f-agent then resource level is being decreased at a rate β_f;
- if this cell unoccupied then resource level is being decreased at a rate β_0.

Besides the change of resource level, the resource is being distributed in the field at a rate ε.

Finally, the resource field dynamics is defined by the equation

$$u^{t+\tau} = u^t + \tau \left(\varepsilon \Delta u^t + \beta u^t - \gamma (u^t)^2 + \zeta \right)$$

where $u^t = u(x, y, t)$,

$$\Delta u^t = \frac{u^t_{x-1,y} + u^t_{x,y-1} - 4u^t_{x,y} + u^t_{x+1,y} + u^t_{x,y+1}}{h^2},$$

$$u(x, y, 0) = s_m p_m(x, y, 0),$$

$$\beta(x, y, t) = \beta_m p_m(x, y, t) - \beta_f p_f(x, y, t) - \beta_0,$$

$$\gamma(x, y, t) = \gamma_0 + (\gamma_m - \gamma_0) p_m(x, y, t),$$

$$\zeta(x, y, t) = \zeta_m p_m(x, y, t).$$

The constant h is a size of lattice cell and τ is a time teak.

4.2   Computer simulation of interaction process based on the resource consumption

We shall investigate a random distribution of 53 f-agents - the color is white - and 47 m-agents - the color is black, as shown in figure 3, on the right.

Environment consists of 2500 locations arranged on a 50×50 lattice with the resource level at every site initially. The resource score is highest at the peaks in the field around the cell occupied by m-agent - where the color is most black - and falls off in a series of terraces.

Describe the simulation process in dynamics. The f-agent is moving on the field according to movement rule. Every time he consumes the resource in the cell occupied by himself. The f-agent go to the cell with maximum value of resource. The m-agent is accumulating the resource. It is being distributed on the resource field according to the rule of resource change on field. When the f-agent occupies the nearest cell around m-agent then the interaction process is realized.

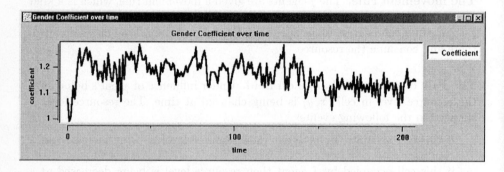

Fig. 5. Gender Interaction Coefficient value dynamics

On the figure 5 is presented the graph of Gender Interaction Coefficient value dynamics. $\mathcal{G}$ is fluctuating in interval $[1, 1.3]$.

5   Simulation results

5.1   The interaction types

Under the computer simulation with various initial data, we reveal three types of agent interaction. These types is characterized by *a subject-object interaction coefficient* $\mathcal{K}$. There are three types of interaction:

- first type - stable. It is realized when rate of resource accumulation approximately equals the rate of resource consumption ($\mathcal{K} \approx 1$);
- second type - unstable ($\mathcal{K} \ll 1$);
- third type - multi-contact ($\mathcal{K} \gg 1$).

Under first interaction type, the contact process is a stable. In this case, the one subject is interacting with one object at long time. The unstable interaction is the contact process, realizing at short time. Under third interaction type, the one subject can interact with some objects at long time. In the first model the coefficient is defined by $\mathcal{K} = \beta/\alpha$, and in the second model — $\mathcal{K} = \beta_m/\beta_f$.

5.2   The Gender Interaction Coefficient

Although the agents are constantly moving and number of interaction is changed in artificial society, but the main characteristic of interaction process is staying on the fixed level. Such characteristic in our models is the Gender Interaction Coefficient. From the observe of coefficient value dynamics (see figures 4, 5) we have got the following results:

- in the first model, $\mathcal{G}$ is fluctuating about 0.75. In this case the males have more activity than females for searching the interaction partner;
- in the second model, $\mathcal{G}$ is fluctuating about 1.15. Here the females play the leader role on the amount of interaction.

6   Conclusions

Analyzing the results of investigation, we can make the following conclusions:

- the particularity of gender interaction modelling is the consideration of two agents' type with various sex. Therefore the gender relations differ from the interpersonal interactions by the competition of agents into group of its sex and the search of interaction partner, having opposite sex;
- defining the process of female's and male's interaction on base of subject-object relations influence on the contact intensity. The agent is more active in the contact process when he is the interaction object;
- considering the gender as intrigue, we can say that the gender equality is a possible, if the female or male will acting not only as an object but also as a subject of interaction.

The research enables to define the influence of gender structure on society dynamics. It gives the possibility for studying the interconnection and the controlling factors of socium.

References

1. Epstein, J.M., Axtell. R.: Growing Artificial Societies: Social Science from the Bottom Up. Brookings Institution Press, Washington (1996)
2. Guts, A.K., Korobitsin, V.V., Laptev, A.A., Pautova, L.A., Frolova, J.V.: Mathematical Models of Social Systems. Omsk State University Press, Omsk (2000)
3. Haleeva, I.I.: Gender as Cognition Intrigue. Moscow State Linguistic University Press, Moscow (2000)
4. Parsons, T.: Social Structure and Personality. Free Press, New York (1964)

Memory Functioning in Psychopathology

Roseli S. Wedemann[1], Raul Donangelo[2],
Luís Alfredo V. de Carvalho[3], and Isa H. Martins[1]

[1] Instituto de Matemática e Estatística
Universidade do Estado do Rio de Janeiro
Rua S. Francisco Xavier, 524
20550-013 Rio de Janeiro, RJ, Brazil
{roseli, isa}@ime.uerj.br
[2] Instituto de Física
Universidade Federal do Rio de Janeiro
Caixa Postal 68528
21945-970 Rio de Janeiro, RJ, Brazil
donangel@if.ufrj.br
[3] Programa de Sistemas, COPPE - LUPA, UniCarioca
Universidade Federal do Rio de Janeiro
Caixa Postal 68511
21945-970 Rio de Janeiro, RJ, Brazil
alfredo@cos.ufrj.br

Abstract. The mental pathology known as neurosis, its aetiology, and
the development of symptoms are described in terms of their relation to
memory function. We propose, based on a neural network model, that
neurotic behavior may be understood as an associative memory process
in the brain, and that the linguistic, symbolic associative process involved
in psychoanalytic working-through can be mapped onto a corresponding
process of reconfiguration of the neural network. The model is illustrated
through a computer simulation implementation. We relate the sensitivity
to temperature and the adaptive capabilities of our model, with the sensitivity of cortical map modulation to the catecholamines (norepinephrine
and dopamine). The signal-to-noise ratio regulated by these substances
influence thought associativity, suggesting a continuum from psychotic
functioning through to normal and neurotic behavior and creativity.

1 Introduction

We propose a schematic functional model for some concepts associated with
neurotic mental processes as described by Sigmund Freud and further developed
on by Jacques Lacan [1–7]. Our description is based on the current view that the
brain is a cognitive system composed of neurons, interconnected by a network
of synapses, that cooperate locally among themselves to process information
in a distributed fashion. Mental states thus appear as the result of the global
cooperation of the distributed neural cell activity in the brain. We also consider
that the emergence of a global state of the neural network of the brain generates
a bodily response which we will call an *act*.

P.M.A. Sloot et al. (Eds.): ICCS 2002, LNCS 2329, pp. 236–245, 2002.
© Springer-Verlag Berlin Heidelberg 2002

Based on the view of the brain as a parallel and distributed processing system [8, 9], we assume that human memory is encoded in the architecture of the neural net of the brain. By this we mean that we record information by reconfiguring the topology of our neural net, *i.e.* the set of *active* neurons and synapses that interconnect them to each other, along with the intensities and durations of these connections [10]. We will often refer to this reconfiguration process as *learning*.

Once a memory trace has been stored, whenever a stimulus excites the neural net to a certain energy state S, the net will stabilize into a local minimum energy state S_m, in which the circuit corresponding to the stored memory trace most similar to the stimulus becomes active, and, as a result generating the respective output. This is referred to as an *associative memory mechanism* [11, 8].

Since there is no clear consensus on the relevance of quantum effects in the macroscopic phenomena that underly molecular and cellular activity in the brain, we therefore take a classical approximation of these phenomena. In particular, we disregard the possibility of the occurrence of nondeterministic events in brain activity caused by quantum effects. We thus attribute any unpredictability in mental behavior to the sensitivity of the non-linear, complex neural networks to initial states and to internal and external parameters, which cannot be determined exactly.

Finally, we assume that each brain state (global state of the neural network) represents only one mental state. That is equivalent to affirming, in linguistic terms, that each symbol is associated to only one significance (meaning), so that we have a one-to-one functional mapping. We suggest that the symbol is represented physiologically by a minimal energy state of the neural net configuration, which encodes the memorized symbolic information.

The remainder of the paper is organized as follows. In the next section, we introduce concepts associated with neurotic phenomena as described by Freud, which we model in sections 3 and 4. In section 5 we describe the simulation experiments and their results, and draw our conclusions and perspectives for further work.

2   Basic Concepts Associated with Neuroses

As a result of the assumptions we have made in the introduction, each repressed or traumatic memory trace stored in our brain is associated, as other non-traumatic memories, to one of the possible minimum energy states of our biological neural net.

It is one of the early findings of psychoanalytic research regarding the *transference neuroses*, that traumatic and repressed memories are knowledge which is present in the subject but which is inaccessible to him, *i.e.* momentarily or permanently inaccessible to the subject's conscience. It is therefore considered *unconscious* knowledge [12, 13, 1]. They arise from events which give the mind a stimulus too powerful to be dealt with in the normal way, and thus result in permanent disturbances.

Freud observed that his neurotic patients systematically repeated symptoms in the form of ideas and impulses. This tendency to systematically repeat symptoms was called by Freud a *compulsion to repeat* [2]. He related the compulsion to repeat to repressed or traumatic memory traces. The original cause of the repression is given by a conflict associated with libidinal fixation and frustration. The neurosis installs itself in the form of " a contention between wishful impulses, or, . . . a psychical conflict. One part of the personality champions certain wishes while another part opposes them and fends them off." [1].

The result of clinical experience in psycho-analysis since late nineteenth century has revealed that patients with strong neurotic symptoms have been able to obtain relief and cure of painful symptoms through a mechanism called *working-through* [3]. The procedure aims at developing knowledge regarding the causes of symptoms, by accessing unconscious memories. Psycho-analysis aims not only at constructing conscious knowledge of the repressed, unconscious material, but beyond that and most importantly, at understanding and changing the way in which the analysand obtains satisfaction [2, 6]. For this purpose, it is fundamental that the analyst positions himself in the sessions in a manner that compels the analysand to a new outcome. Lacan emphasizes the creative nature of transference [14]. A mere repetition, such as that which takes place in our everyday life, simply reinforces the traumas and repressions.

3   Functional Model for the Neuroses

We propose that the neuroses manifest themselves as an associative memory process. An associative memory is a mechanism where the network returns a given stored pattern when it is shown another input pattern sufficiently similar to the stored one [11, 8]. Although neural networks have been proposed to help diagnose neuroses and schizophrenia [15], we are not aware of any previous effort to model neuroses with a distributed processing, neural network approach.

We propose in our model that the compulsion to repeat neurotic symptoms can be described by supposing that a neurotic symptom is acted when the subject is presented with a stimulus which resembles, at least partially, a repressed or traumatic memory trace. The stimulus causes a stabilization of the neural net onto a minimal energy state corresponding to the original memory trace, which in turn generates a neurotic response (an act).

In neurotic behavior associated with a stimulus, the act is not a result of the stimulus as a new situation but a response to the original repressed memory trace. The original repression can be accounted for by a mechanism which inhibits the formation of certain synaptic connections. The inhibition may be externally imposed, for example by cultural stimulation or the relation with the parents, and internalized so that the subject inhibitively stimulates the regions associated with the memory traces, not allowing the establishment of certain synaptic connections.

We thus map the linguistic, symbolic associative process involved in psycho-analytic working-through into a corresponding process of reinforcing synapses among memory traces in the brain. These connections should involve declarative memory, leading to at least partial transformation of the repressed memory to consciousness. This has a relation to the importance of language in psychoanalytic sessions and the idea that unconscious memories are those that cannot be expressed.

We propose that as the analysand symbolically elaborates the manifestations of unconscious material and creates new editions of old conflicts through transference in psychoanalytic sessions, he is reconfiguring the topology of the neural net in his brain by actually creating new connections and reinforcing older ones, among the subnetworks that store the repressed memory traces. The network topology which results from this reconfiguration process will stabilize onto new energy minima associated with new conscious or unconscious responses or acts.

In our model, it is clear why repetition in psychoanalysis is specially important. Neuroscience has established that memory traces are established by repeatedly reinforcing, through stimulation, the appropriate synaptic connections. This is exactly the learning process in a neural network and accounts for the long durations of psychoanalytic processes. Much time may be needed to overcome resistances in order to access and interpret repressed material, and even more to repeat and reconfigure the net in a learning process.

Our model differentiates, from the physiological point of view, psychoanalytic working-through from the neurotransmitting drug therapy, characteristic of psychiatric treatment. Drugs change network response by having a more global effect over the net. This global effect cannot account for the selective fine-tuning process achieved by reconfiguring individual synapses through psychoanalytic working -through.

4   Computational Model

Memory functioning is modeled by a Boltzmann machine [11, 16] We consider N nodes, which are connected symmetrically by weights $w_{ij} = w_{ji}$. The states of the units S_i, take output values in $\{-1, 1\}$. Because of the symmetry of the connections, there is an energy functional

$$H(\{S_i\}) = -\frac{1}{2} \sum_{ij} w_{ij} S_i S_j \, , \qquad (1)$$

which allows us to define the Boltzmann distribution function for network states

$$P(\{S_i\}) = \exp\left[-\frac{H(\{S_i\})}{T}\right] \Big/ \sum_{\{S_i\}} \exp\left[-\frac{H(\{S_i\})}{T}\right] \, , \qquad (2)$$

where T is the network temperature parameter. Pattern retrieval on the net is achieved by a standard simulated annealing process, in which the network temperature T is gradually lowered by a factor α.

In our simulations, initially, we take random connection weights w_{ij}. We also consider that the network is divided into two weakly linked subsets, representing the conscious and unconscious parts of the memory. This is done by multiplying the connections between the two subsets by a number less than one.

Once the network is initialized, we find the stored patterns by presenting many random patterns to the Boltzmann machine, with an annealing schedule α that permits stabilizing onto the many local minimum states of the network energy function. These initially stored patterns, associated as they are to two weakly linked subnetworks, represent the neurotic memory states. One verifies that when initializing the "conscious" nodes of the memory into one of these states and the "unconscious" nodes on a random configuration, the net rapidly evolves into one of the neurotic memory states. In [17], the authors propose a neurocomputational model to describe how the original memory traces are formed in cortical maps.

In order to simulate the working-through process, one should stimulate the net by means of a change in a randomly chosen node n_i belonging to the "unconscious" section of a neurotic memory pattern. This stimulus is then presented to the network and, if the Boltzmann machine retrieves a pattern with conscious configuration different than that of the neurotic pattern, we interpret this as a new conscious association, and enhance all weights from n_i to the changed nodes in the conscious cluster. The increment values are given by

$$\Delta w_{ij} = \beta * S_i * S_j * w_{max} , \tag{3}$$

where β is the learning parameter chosen in $(0,1)$, and w_{max} the maximum absolute initial synaptic strengths. We note that new knowledge is learned only when the stimulus from the analyst is not similar to the neurotic memory trace.

This procedure must be repeated for various reinforcement iterations in an adaptive learning procedure, and also repeat each set of reinforcement iterations for various initial annealing temperature values. The new set of synaptic weights will define a new network configuration.

5   Simulation Model Illustration

For a network with $N = 32$ nodes such that $N_{unc} = 16$ of them belong to the unconscious subset, the model depends on the parameters listed in table 1, which also gives the corresponding values for our simulation experiment. The memory configurations before working-through are shown in table 2. After the learning process, one obtains a new set of patterns shown in table 3. In both tables, minimum energy values are listed in the last column. Many repetitions, are needed in the learning process for the observation of new network states.

Table 4 shows the associativity capability of the network before working-through, i.e. of a neurotic network, as a function of initial annealing temperature, by presenting a single random pattern to the Boltzmann machine and varying the temperature . The second column lists the energy value associated with the pattern onto which the machine stabilizes for each temperature. With a small annealing schedule, we see in table 4 that as temperature is varied in the model, new memory configurations are attainable from a given initial pattern. In our

Table 1. Parameter set for illustrative simulation experiment.

N	32
number of nodes in each subset, N_{unc}	16
initial temperature for finding global minimum with simulated annealing	1.0
initial temperature for seeking local minima with simulated annealing	0.05
lowest temperature for simulated annealing	0.001
α for finding global minimum	0.95
α for finding local minima	0.50
number of iterations per temperature in simulated annealing	20
number of trial patterns for initial configuration pattern detection	10000
parameter for initializing intercluster synapses	0.5
β	0.3
number of iterations for reinforcement learning	1000

initial experiments, some simulations required a temperature increase to allow association of stimulus from the analyst from a neurotic state to a new conscious configuration.

In neural network modeling, temperature is inspired from the fact that real neurons fire with variable strength, and there are delays in synapses, random fluctuations from the release of neurotransmitters, and so on. These are effects that we can loosely think of as noise [11, 8, 10]. So temperature in Boltzmann machines controls noise. In our model, temperature, *i.e.* noise, allows associativity among memory configurations, lowering synaptic inhibition, in an analogy with the idea that freely talking analytic sessions, and stimulation from the analyst lower resistances and allow greater associativity.

It is also possible to relate our simulation experiments with the model described in [17], where the signal-to-noise ratio at neuronal level, induced by the catecholamines (dopamine, norepinephrine, etc.), alter the way the brain performs associations of ideas. An increase of dopamine (hyperdopaminergy) promotes an enhancement of signal and therefore of the signal-to-noise ratio. Since noise is responsible for more free association of ideas, its reduction weakens associativity resulting in fixation to certain ideas. On the other hand, a decrease in dopamine increases noise which permits a broader association of ideas. Certainly, excessive noise may disorganize thought by permitting the association of distantly correlated ideas and breaking the logic of thought. This suggests a phenomenologic continuum from low signal-to-noise ratio (low dopamine level) corresponding to logically disorganized thought, passing through a slightly low dopamine level associated with creative thought and arriving at high signal-to-noise ratios, responsible for fixations to ideas and delusions.

We can thus relate temperature and noise in our Boltzmann machine to associativity of thought as in [17]. Very high temperatures, allow the production of logically disorganized thought because they allow associations of distant, usually uncorrelated ideas as in the low signal-to-noise ratio characteristic of low dopamine levels. However, at low temperatures, the Boltzmann machine inhibits

Table 2. Memorized patterns for initial network configuration.

Conscious	Unconscious	Energy
0 1 1 0 1 1 1 1 0 1 0 0 1 0 1 1 0 1 1 0 0 1 0 1 1 0 0 0 1 1 0 0		-216.2
0 1 1 1 1 1 1 1 0 1 1 0 1 0 1 1 0 0 1 0 1 1 0 1 1 0 0 0 1 1 1 0		-208.2
0 1 1 0 1 1 1 1 0 1 0 0 1 0 1 1 0 1 1 0 1 1 0 1 0 0 0 0 1 1 1 0		-215.0
0 1 1 0 1 1 1 1 0 1 0 0 1 0 1 1 0 0 1 0 0 1 0 1 1 0 0 0 1 1 1 0		-215.6
0 1 1 0 1 1 1 1 0 1 0 0 1 0 1 1 0 0 1 0 0 1 0 1 1 0 0 0 1 1 0 1		-215.9
0 1 1 0 1 1 1 0 0 0 1 0 1 0 1 1 0 0 0 0 0 0 1 0 1 1 0 1 0 0 0 1		-167.8
1 1 1 0 1 1 1 1 0 1 0 0 1 0 1 1 0 1 0 0 1 1 0 0 1 0 0 0 1 1 0 0		-208.2
0 1 1 1 1 1 1 1 0 1 1 0 1 0 1 1 0 1 1 1 1 1 0 0 1 0 1 0 1 1 0 0		-200.9
0 1 1 1 1 1 1 0 0 0 1 0 1 0 1 1 1 0 1 0 0 0 1 0 1 0 0 1 0 0 0 0		-169.1
0 1 1 1 1 1 1 0 0 0 1 0 1 0 1 1 1 0 1 0 0 0 1 0 1 1 0 1 0 0 0 1		-171.4
0 1 1 1 1 1 1 1 0 1 1 0 1 0 1 1 0 0 1 0 0 1 0 0 1 0 0 1 1 1 0 0		-198.1
1 0 0 1 0 1 1 0 1 0 1 0 0 0 1 1 0 1 0 1 1 0 1 0 0 1 0 1 0 0 0 0		-128.9
1 0 0 1 0 1 1 0 1 0 1 0 1 0 1 0 1 1 1 0 1 0 1 0 1 0 0 0 0 1 0 0 1 0		-133.9
0 1 0 1 0 1 1 0 1 0 1 0 1 1 0 1 0 1 1 1 1 1 0 1 1 0 0 0 0 0 0 0 0		-105.0

the associative process becoming fixated in a local miminum of the energy function, equivalent to a hyperdopaminergic state characteristic of delusions and inflexible thought.

It is also known that norepinephrine levels control synaptic learning. Low or high norepinephrine levels corresponding to much or little noise respectively, inhibit the establishment of synapses and learning. Psychoanalysis promotes new thought associations and learning, and this cannot be carried through if norepinephrine and dopamine levels are too high or too low. We suggest that psychoanalytic working-through explores the neurophysiologic potentialities of the subject, inducing network reconfiguration. When these potentialities are limited by chemical alterations, such as in strong psychotic cases, working-through should be limited or even impossible.

The model is in agreement with psychoanalytic experience that working-through is a slow process, where the individual slowly elaborates knowledge by re-associating weakly connected memory traces and new experiences. This self-reconfiguration process, which we represent in the model by a change in network connectivity, will correspond to new outcomes in the subjects life history. Repetition is an important component of this process and our model illustrates this by the repetitive adaptive reinforcement learning involved in the simulation of working-through. Although biologically plausible and in accordance with aspects of clinical experience described by psycho-analysis, the model is very schematic and far from explaining the complexities of mental processes. Although still in lack of experimental verification, it seams to be a good metaphorical view of the basic concepts of neurotic behavior described by Freud, to which we have referred.

We are now proceeding to systematically study the parameter dependency of the model. If possible, we will try an interpretation of these parameter dependencies as associated to memory functioning in the brain and psychic apparatus functioning.

Acknowledgements

We are grateful for fruitful discussions regarding basic concepts of psychoanalytic theory, with the psychoanalyst Professor Angela Bernardes of the Escola Brasileira de Psicanálise of Rio de Janeiro and the Department of Psychology of the Federal Fluminense University, psychoanalyst Dr. Nicolau Maluf and Eduardo Aguilar, both graduate students in neuroscience of the Federal University of Rio de Janeiro. We also thank the members of the Escola Brasileira de Psicanálise of Rio de Janeiro for their warm reception of the author R.S. Wedemann in some of their seminars. The initial discussions regarding language and logic of the unconscious with Professor Ricardo Kubrusly of the Institute of Mathematics of the Federal University of Rio de Janeiro were also enlightening. We acknowledge partial financial support from the Conselho Nacional de Desenvolvimento Científico e Tecnológico (CNPq).

Table 3. Memorized patterns after the simulated working-through process.

Conscious	Unconscious	Energy
1 1 0 1 1 1 1 1 0 1 1 0 1 0 1 1 0 0 1 1 1 1 1 0 0 0 0 1 1 1 1 1		-428.0
0 1 0 0 1 1 1 1 0 1 0 0 1 1 1 1 0 1 0 1 1 1 0 0 1 0 1 0 1 1 0 0		-198.4
0 1 1 0 1 1 1 1 0 1 0 0 1 0 1 1 0 1 0 0 0 0 0 1 1 1 0 1 1 1 0 1		-220.4
0 1 0 0 1 1 1 1 0 1 0 0 1 1 1 1 0 1 1 1 0 1 0 0 1 0 1 0 1 1 0 0		-198.9

References

1. Freud, S.: *Introductory Lectures on Psycho-Analysis*. Standard Edition . W. W. Norton and Company, 1966. First German edition in 1917.
2. Freud, S.: *Beyond the Pleasure Principle*. Standard Edition . The Hogarth Press, London, 1974. First German edition in 1920.
3. Freud, S.: *Remembering, Repeating and Working-Through*. Standard Edition. Volume XII, The Hogarth Press, London, 1953-74. First German edition in 1914.
4. Freud, S.: *The Ego and the Id*. Standard Edition . W. W. Norton and Company, 1960. First German edition in 1923.
5. Lacan, J.: *O Seminário, Livro 2: O Eu na Teoria de Freud e na Técnica da Psicanálise*. Jorge Zahar Editor, Brazil, 1985.
6. Bernardes, A.C.: *Elaboration of Knowledge in Psychoanalysis, a Treatment of the Impossible*. Doctoral thesis. Programa de Pós-graduação em Teoria Psicanalítica, UFRJ, Rio de Janeiro, Brazil, 2000 (in Portuguese).

7. Forrester, J.: *The Seductions of Psychoanalysis, Freud, Lacan and Derrida*. Cambridge University Press, 1990.
8. Rumelhart, D.E., McCleland, J.L. (ed.): *Parallel Distributed Processing: Explorations in the Microstructure of Cognition*. 2 Volumes . The MIT Press, Cambridge, MA, 1986.
9. Varela, F.J., Thompson, E., Rosch, E.: *The Embodied Mind*. The MIT Press, Cambridge, MA, 1997.
10. Kandel, E.R., Schwartz, J.H., Jessel, T.M.: *Principles of Neural Science*. MacGraw Hill, 2000.
11. Hertz, J.A., Krogh, A., Palmer, R.G. (ed.): *Introduction to the Theory of Neural Computation*. Lecture Notes, Vol. I, Santa Fe Institute, Studies in the Science of Complexity, 1991.
12. Janet, P.: *Les actes inconscients et la memoire*. Rev. Philosoph., **13**, 238. (319), 1888.
13. Breuer, J., Freud, S.: *On the Psychical Mechanism of Hysterical Phenomena: Preliminary Communication*. Collected Papers 1, 24; Standard Edition, London 1924-50. First German edition in 1893.
14. Lacan, J.: *O Seminário, Livro 8: A Transferência*. Jorge Zahar Editor, Brazil, 1992.
15. Zou Y.Z., Shen, Y.C., Shu, L.A. et al. : *Artificial neural network to assist psychiatric diagnosis*. Brit. J. Psychiat. 169: (1) 64-67 Jul. 1996.
16. Barbosa, W.C.: *Massively Parallel Models of Computation*. Ellis Horwood Limited, 1993.
17. Mendes, D.Q., Carvalho, L.A.V.: *Creativity and Delusions: The Dopaminergic Modulation of Cortical Maps*. Submitted to the Journal of Theoretical Medicine in 2001.

Table 4. Associativity capability of the neurotic network.

Temperature	Energy
0.001	-208.2
0.04	-216.2
0.08	-216.2
0.10	-215.0
0.14	-216.2
0.18	-215.6
0.20	-216.2
0.24	-215.0
0.28	-216.2
0.30	-215.9
0.34	-215.9
0.38	-216.2
0.40	-216.2
0.44	-216.2
0.48	-215.9
0.50	-215.9
0.54	-216.2
0.58	-216.2
0.60	-216.2
0.64	-216.2
0.68	-216.2
0.70	-216.2
0.74	-215.9
0.78	-215.0
0.80	-216.2
0.84	-216.2
0.88	-216.2
0.90	-216.2
0.94	-215.6
0.98	-215.6
0.99	-215.9

Investigating e-Market Evolution

John Debenham
University of Technology, Sydney
debenham@it.uts.edu.au

Abstract. A market is in equilibrium if there is no opportunity for risk-free, or low-risk, profit. The majority of real markets are not in equilibrium thus presenting the opportunity for novel forms of transactions to take advantage of such risk-free, or low-risk, profits. The introduction of such novel forms of transaction is an instance of market evolution. A project is investigating the market evolutionary process in a particular electronic market that has been constructed in an on-going collaborative research project between a university and a software house. The way in which actors (buyers, sellers and others) use the market will be influenced by the information available to them, including information drawn from outside the immediate market environment. In this experiment, data mining and filtering techniques are used to distil both individual signals drawn from the markets and signals from the Internet into meaningful advice for the actors. The goal of this experiment is first to learn how actors will use the advice available to them, and second how the market will evolve through entrepreneurial intervention. In this electronic market a multiagent process management system is used to manage all market transactions including those that drive the market evolutionary process.

1   Introduction

A market is in equilibrium if there is no opportunity for risk-free, or low-risk, profit. The majority of real markets, such as the stock market, are not in equilibrium thus presenting the opportunity for transactions to take advantage of such risk-free, or low-risk, profits. Such transactions may be of an innovative, novel form. For example, the practice of corporate asset-stripping—although now common-place in many countries—would have been such a novel transaction when the practice commenced. The introduction of such novel forms of transaction is an instance of market evolution. The project described here aims to derive fundamental insight into how e-markets evolve. To achieve this it addresses the problem of identifying timely information for e-markets with their rapid, pervasive and massive flows of data. This information is distilled from individual signals in the markets themselves and from signals observed on the unreliable, information-overloaded Internet. Distributed, concurrent, time-constrained data mining methods are managed using intelligent business process management technology to extract timely, reliable information from this unreliable environment. The perturbation of market equilibrium through entrepreneurial action is the essence of market evolution. Entrepreneurship relies both on intuition and information discovery. The term 'entrepreneur' is used here in its technical sense [1].

P.M.A. Sloot et al. (Eds.): ICCS 2002, LNCS 2329, pp. 246–255, 2002.

An electronic market has been constructed in an on-going collaborative research project between a university and a software house. This electronic market forms a subset of the system described here; it is called the *basic system*. The goal of this subset is to identify timely information for traders in an e-market. The traders are the buyers and sellers. This basic system does not address the question of market evolution. The basic system is constructed in two parts: the e-market and the actors' assistant. The e-market has been constructed by Bullant Australasia Pty Ltd—an Australian software house with a strong interest in business-to-business (B2B) e-business [www.bullant.com]. The e-market is part of their on-going research effort in this area. It has been constructed using Bullant's proprietary software development tools. The e-market was designed by the author. The actors' assistant is being constructed in the Faculty of Information Technology at the University of Technology, Sydney. It is funded by two Australian Research Council Grants; one awarded to the author, and one awarded to Dr Simeon Simoff.

One feature of the whole project is that every transaction is treated as a business process and is managed by a process management system. In other words, the process management system makes the whole thing work. The process management system is based on a robust multiagent architecture. The use of multiagent systems is justified first by the distributed nature of e-business, and second by the critical nature of the transactions involved. The environment may be unreliable due to the unreliability of the network and components in it, or due to the unreliability of players—for example, a seller may simply renege on a deal.

2   Actor classes

For some while there has been optimism in the role of agents in electronic commerce. "During this next-generation of agent-mediated electronic commerce,.... Agents will strategically form and reform coalitions to bid on contracts and leverage economies of scale...... It is in this third-generation of agent-mediated electronic commerce where companies will be at their most agile and markets will approach perfect efficiency." [2]. There is a wealth of material, developed principally by micro-economists, on the behaviour of rational economic agents. The value of that work in describing the behaviour of human agents is limited in part by the inability of humans to necessarily behave in an (economically) rational way, particularly when their (computational) resources are limited. That work provides a firm foundation for describing the behaviour of rational, intelligent software agents whose resource bounds are known, but more work has to be done [3]. Further, new market mechanisms that may be particularly well-suited to markets populated by software agents is now an established area of research [4] [5]. Most electronic business to date has centred on on-line exchanges in which a single issue, usually price, is negotiated through the application of traditional auction-based market mechanisms. Systems for multi-issue negotiation are also being developed [6], also IBM's Silkroad project [7]. The efficient management of multi-issue negotiation towards a possible solution when new issues may be introduced as the negotiation progresses remains a complex problem [8].

Given the optimism in the future of agents in electronic commerce and the body of theoretical work describing the behaviour of rational agents, it is perhaps surprising that the basic structure of the emerging e-business world is far from clear. The majority of Internet e-exchanges are floundering, and it appears that few will survive [9]. There are indications that exchanges may even charge a negative commission to gain business and so too market intelligence [op. cit.]. For example, the Knight Trading Group currently pays on-line brokers for their orders. The rationale for negative commissions is discussed in [28]. One reason for the recent failure of e-exchanges is that the process of competitive bidding to obtain the lowest possible price is not compatible with the development of buyer-seller relations. The preoccupation with a single issue, namely price, can overshadow other attributes such as quality, reliability, availability and customisation. A second reason for the failure Internet e-exchanges is that they deliver little benefit to the seller—few suppliers want to engage in a ruthless bidding war [op. cit.]. The future of electronic commerce must include the negotiation of complex transactions and the development of long-term relationships between buyer and seller as well as the e-exchanges. Support for these complex transactions and relationships is provided here by *solution providers*.

A considerable amount of work has been published on the comparative virtues of open market e-exchanges and solution providers that facilitate direct negotiation. For example, [10] argues that for privately informed traders the 'weak' trader types will systematically migrate from direct negotiations to competitive open markets. Also, for example, see [11] who compare the virtues of auctions and negotiation. Those results are derived in a supply/demand-bounded world into which signals may flow. These signals may be received by one or more of the agents in that world, and so may cause those agents to revise their valuation of the matter at hand.

3   The e-market

The construction of experimental e-markets is an active area of research. For example, [12] describes work done at IBM's Institute for Advanced Commerce. There are two functional components in the *basic e-market*: the e-exchange and a solution provider. The *solution provider* is 'minimal' and simply provides a conduit between buyer and seller through which long term contracts are negotiated. The *solution provider* in its present form does not give third-party support to the negotiation process.

An e-exchange is created for a fixed duration. An *e-exchange* is a virtual space in which a variety of market-type *activities* can take place at specified times. The time is determined by the e-exchange *clock*. Each activity is advertised on a notice *board* which shows the start and stop time for that activity as well as what the activity is and the *regulations* that apply to players who wish to participate in it. A human player works though a PC (or similar) by interacting with a *user agent* which communicates with a *proxy agent* or a solution provider situated in the e-market. The inter-agent communication is discussed in Sec 3. The user agents may be 'dumb', or 'smart' being programmed by the user to make decisions. Each activity has an *activity manager* that ensures that the regulations of that activity are complied with.

When an e-exchange is created, a specification is made of the e-exchange *rules*. These rules will state who is permitted to enter the e-exchange and the roles that they are permitted to play in the e-exchange. These rules are enforced by an *e-exchange manager*. For example, can any player create a sale activity (which could be some sort of auction), or, can any player enter the e-exchange by offering some service, such as advice on what to buy, or by offering 'package deals' of goods derived from different suppliers? A high-level view of the e-market is shown in Fig. 1.

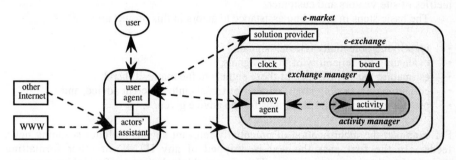

Fig. 1. High-level model of the e-market and user

The activities in the basic e-market are limited to opportunities to buy and sell goods. The regulations for this limited class of activities are called *market mechanisms* [5]. The subject of a negotiation is a *good*, buyers make *bids*, sellers make *asks*. Designing market mechanisms is an active area of research. For example, see optimal auctions [4]. One important feature of a mechanism is the 'optimal' strategy that a player should use, and whether that strategy is "truth revealing" [11].

4   The actors' assistant

In 'real' problems, a decision to use an e-exchange or a solution provider will be made on the basis of general knowledge that is external to the e-market place. Having decided to use either an e-exchange or a solution provider, the negotiation strategy used will also depend on general knowledge. Such general knowledge will typically be broadly based and beyond the capacity of modern AI systems whilst remaining within reasonable cost bounds.

E-markets reside on the Internet alongside the vast resources of the World Wide Web. In the experiments described here, the general knowledge available is restricted to that which can be gleaned from the e-markets themselves and that which can be extracted from the Internet in general—including the World Wide Web. The actors' assistant is a workbench that provides a suite of tools to *assist* a buyer or seller in the e-market. The actors' assistant does *not* attempt to replace buyers and sellers. For example, there is no attempt to automate 'speculation' in any sense. Web-mining tools assist the players in the market to make informed decisions. One of the issues in operating in an e-market place is coping with the rapidly-changing signals in it. These signals include: product and assortment attributes (if the site offers multiple products), promotions shown, visit attributes (sequences within the site, counts, click-streams) and business agent attributes. Combinations of these signals may be vital information to an actor. A new generation of data analysis and supporting techniques—collectively labelled as *data mining* methods—are now applied to stock market analysis, predictions and other financial and market analysis applications [13]. The application of data mining methods in e-business to date has predominantly been within the B2C framework, where data is mined at an on-line business site, resulting in the derivation of various behavioural metrics of site visitors and customers.

The basic steps in providing assistance to actors in this project are:

- identifying potentially relevant signals;
- evaluating the reliability of those signals;
- estimating the significance of those signals to the matter at hand;
- combining a (possibly large) number of signals into coherent advice, and
- providing digestible explanations for the advice given.

For example, the identification of potentially relevant signals includes scanning news feeds. In the first pass the text is stripped of any HTML or other formatting commends—this is *preprocessing*. Then keyword matches are performed by *scanners*. Provided the number of 'hits' has not been too large, these two steps alone produce useful information. Bots that have been built into the system include: News Hub [www.newshub.com], NewsTrawler [www.newstrawler.com] and CompanySleuth [www.companysleuth.com]. Following the prepossessing and scanning steps, an *assessment* is made of the overall reliability of the source. At present this is simply a measure of the overall reputation that the source has for accuracy. In addition, *watchers* detect changes to material on the Web. Here URLyWarning [www.urlywarning.com] and other watcher bots are used to identify pages in which designated information *may* have changed; they may be used to trigger a detailed search of other pages.

The estimation of the significance of a signal to a matter at hand is complicated by the fact that one person may place more faith in the relevance of a particular signal than others. So this estimation can only be performed on a personal basis. This work does *not*, for example, attempt to use a signal to predict whether the US dollar will rise against the UK pound. What it *does* attempt to do is to predict the value that an actor will place on a signal [14]. So the feedback here is provided by the user in the form of a rating of the material used. A five point scale runs from 'totally useless' to 'very useful'. Having identified the signals that a user has faith in, "classical" data mining methods [15] are then applied to combine these signals into succinct advice again using a five point scale. This feedback is used to 'tweak' the weights in Bayesian networks and as feedback to neural networks [16]. Bayesian networks are preferred when some confidence can be placed in a set of initial values for the weights. The system is able to raise an alarm automatically and quickly when a pre-specified compound event occurs such as: four members of the board of our principal supplier "Good Co" have resigned, the share price has dropped unexpectedly and there are rumours that our previous supplier "Bad Co" is taking over "Good Co".

The actors' assistant integrates two different approaches in data mining — the data driven and the hypothesis-driven approach. In the data-driven approach the assistant is just "absorbing" the information discovered by the scanners. It only specifies broad parameters to constrain the material scanned. For example, in the text analysis of the news files a text miner observes the frequencies of word occurrences and co-occurrences that appear to be relevant to a keyword such as 'steel prices'. The result of this process is an initial representative vocabulary for that news file. In the hypothesis-driven approach, the actors' assistant specifies precisely what it is looking for, for example, it formulates a hypothesis that a fall in the price of steel is likely within a month. The combination of data-driven and hypothesis driven approaches aims to provide a mechanism for meeting tight time constraints. Managing and synchronising the actors' assistant is handled by process management plans in the user agents. For example, a request is made for the best information on the Sydney Steel Co to be delivered by 4.00pm. This request triggers a business process. Things can go wrong with this process, for example a server may be down, in which case the process management plans activate less-preferred but nevertheless useful ways of obtaining the required information by the required time.

5   Process management

Fig 1. may give the false impression that all the process management system does is to support communication between the user agents and their corresponding proxy agents. All transactions are managed as business processes, including a simple 'buy order', and a complex request for information placed with an actor's assistant. Building e-business process management systems is business process reengineering on a massive scale, it often named *industry process reengineering* [17]. This can lead to considerable problems unless there is an agreed basis for transacting business. The majority of market transactions are constrained by time ("I need it before Tuesday"), or more complex constraints ("I only need the engine if I also have a chassis and as long as the total cost is less than..). The majority of transactions are *critical* in that they must be dealt with and can't be forgotten or mislaid. Or at least it is an awful nuisance if they are. So this means that a system for managing them is required that can handle complex constraints and that attempts to prevent process failure.

E-market processes will typically be *goal-directed* in the sense that it may be known *what* goals have to be achieved, but not necessarily *how* to achieve those goals today. A goal-directed process may be modelled as a (possibly conditional) sequence of goals. Alternatively a process may be *emergent* in the sense that the person who triggers the process may not have any particular goal in mind and may be on a 'fishing expedition' [18]. There has been little work on the management of emergent processes [19]. There a multiagent process management system is described that is based on a three-layer, BDI, hybrid architecture. That system 'works with' the user as emergent processes unfold. It also manages goal-directed processes in a fairly conventional way using single-entry quadruple-exit plans that give almost-failure-proof operation. Those plans can represent constraints of the type referred to above, and so it is a candidate for managing the operation of the system described in Sec. 2.

Multiagent technology is an attractive basis for industry process re-engineering [20] [21]. A multiagent system consists of autonomous components that negotiate with one another. The scalability issue of industry process reengineering is "solved"—in theory—by establishing a common understanding for inter-agent communication and interaction. Standard XML-based ontologies will enable data to be communicated freely [22] but much work has yet to be done on standards for communicating expertise [23]. Results in ontological analysis and engineering [24] [23] is a potential source for formal communication languages which supports information exchange between the actors in an e-market place. Systems such as CommerceNet's Eco [www.commerce.net] and Rosettanet [www.rosettanet.org] are attempting to establish common languages and frameworks for business transactions and negotiations. Specifying an agent interaction protocol is complex as it in effect specifies the common understanding of the basis on which the whole system will operate.

A variety of architectures have been described for autonomous agents. A fundamental distinction is the extent to which an architecture exhibits deliberative (feed forward, planning) reasoning and reactive (feed back) reasoning. If an agent architecture combines these two forms of reasoning it is a *hybrid architecture*. One well reported class of hybrid architectures is the three-layer, BDI agent architectures. One member of this class is the INTERRAP architecture [25], which has its origins in the work of [26]. A multiagent system to manage "goal-driven" processes is described in [19]. In that system each human user is assisted by an agent which is based on a generic three-layer, BDI hybrid agent architecture similar to the INTERRAP architecture. That system has been extended to support emergent processes and so to support and the full range of industry processes. That conceptual architecture is adapted slightly for use here; see Fig 2(a). Each agent receives messages from other agents (and, if it is a personal agent, from its user) in its message area. The world beliefs are derived from reading messages, observing the e-market and from the World Wide Web (as accessed by an actor's assistant).

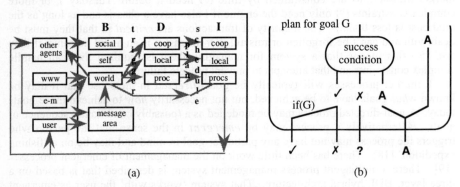

(a)   (b)

Fig 2. (a) conceptual architecture, (b) the four plan exits

Deliberative reasoning is effected by the non-deterministic procedure: "on the basis of current *beliefs*—identify the current *options*, on the basis of current options and existing commitments—select the current commitments (called the agent's *goals* or *desires*), for each newly-committed goal choose a *plan* for that goal, from the selected plans choose a consistent set of things to do next (called the agent's *intentions*)". A *plan* for a goal is a conditional sequence of sub-goals that may include iterative or recursive structures. If the current options do not include a current commitment then that commitment is dropped. In outline, the reactive reasoning mechanism employs triggers that observe the agent's beliefs and are 'hot wired' back to the procedural intentions. If those triggers fire then they take precedence over the agent's deliberative reasoning. The environment is intrinsically unreliable. In particular plans can not necessarily be relied upon to achieve their goal. So at the end of every plan there is a *success condition* which tests whether that plan's goal has been achieved; see Fig 2(b). That success condition is itself a procedure which can succeed (✓), fail (✗) or be aborted (**A**). So this leads to each plan having four possible exits: success (✓), failure (✗), aborted (**A**) and unknown (**?**). In practice these four exists do not necessarily have to lead to different sub-goals, and so the growth in the size of plan with depth is not quite as bad as could be expected.

KQML (Knowledge Query and Manipulation Language) is used for inter-agent communication [27]. Each process agent has a *message area*. If agent A wishes to tell something to agent B then it does so by posting a message to agent B's message area. Each agent has a *message manager* whose role is to look after that agent's message area. Each message contains an instruction for the message manager.

The first step in the design process for a multi-agent system is determining the system organisation. In the application described here, the system organisation consists of one process agent for each (human) user. There are no other agents in the system.

The system operates in an environment whose performance and reliability will be unreliable and unpredictable. Further, choices may have to be made that balance reliability with cost. To apply the deliberative reasoning procedure requires a mechanism for *identifying* options, for *selecting* goals, for *choosing* plans and for *scheduling* intentions. A plan may perform well or badly. The process management system takes account of the "process knowledge" and the "performance knowledge". *Process knowledge* is the wisdom that has been accumulated, particularly that which is relevant to the process instance at hand. *Performance knowledge* is knowledge of how effective agents, people, systems, methods and plans are at achieving various things. A plan's *performance* is defined in terms of: the likelihood that the plan will succeed, the expected cost and time to execute the plan, the expected value added to the process by the plan, or some combination of these measures. If each agent knows how well the choices that it has made have performed in the past then it can be expected to make decisions reasonably well as long as plan performance remains reasonably stable. One mechanism for achieving this form of adaptivity is reinforcement learning. An alternative approach based on probability is described in [19]. In addition, an agent may know things about the system environment, and may have some idea of the *reason why* one choice lead to failure. An agent's belief in these reasons may result from communication with other agents. Such beliefs may be used to revise the "historical" estimates to give an *informed* estimate of plan performance that takes into account the *reasons why* a plan behaved the way that it did [op. cit.].

6   Conclusion

One of the innovations in this project is the development of a coherent environment for e-market places, a comprehensive set of actor classes and the use of a powerful multiagent process management system to make the whole thing work. The use of a powerful business process management system to drive all the electronic market transactions unifies the whole market operation. The development of computational models of the basic market transactions, deploying those models in the e-market place, and including them as part of the building blocks for creating a complete e-market place provides a practical instrument for continued research and development in electronic markets.

Acknowledgment

The work described herein was completed whilst the author was a visitor at the CSIRO Joint Research Centre for Advanced Systems Engineering, Macquarie University, Sydney. The contribution to the work by the members of that Centre is gratefully acknowledged.

References

[1]   Israel M. Kirzner Entrepreneurial Discovery and the Competitive Market Process: An Austrian Approach" Journal of Economic Literature XXXV (March) 1997 60-85.
[2]   R. Guttman, A. Moukas, and P. Maes. Agent-mediated Electronic Commerce: A Survey. Knowledge Engineering Review, June 1998.
[3]   Moshe Tennenholtz. Electronic Commerce: From Economic and Game-Theoretic Models to Working Protocols. Invited paper. Proceedings Sixteenth International Joint Conference on Artificial Intelligence, IJCAI'99, Stockholm, Sweden.
[4]   Milgrom, P. Auction Theory for Privatization. Cambridge Univ Press (2001).
[5]   Bichler, M. The Future of E-Commerce: Multi-Dimensional Market Mechanisms. Cambridge University Press (2001).
[6]   Sandholm, T. Agents in Electronic Commerce: Component Technologies for Automated Negotiation and Coalition Formation. Autonomous Agents and Multi-Agent Systems, 3(1), 73-96.
[7]   Ströbel, M. Design of Roles and Protocols for Electronic Negotiations. Electronic Commerce Research Journal, Special Issue on Market Design 2001.
[8]   Peyman Faratin. Automated Service Negotiation Between Autonomous Computational Agents. PhD dissertation, University of London (Dec 2000).
[9]   R. Wise & D. Morrison. Beyond the Exchange; The Future of B2B. Harvard Business review Nov-Dec 2000, pp86-96.
[10]   Neeman, Z. & Vulkan, N. Markets Versus Negotiations. The Hebrew University of Jerusalem Discussion Paper 239. (February 2001).
[11]   Bulow, J. & Klemperer, P. Auctions Versus Negotiations. American Economic Review, 1996.
[12]   Kumar, M. & Feldman, S.I. Business Negotiations on the Internet. Proceedings INET'98 Internet Summit, Geneva, July 21-24, 1998.

[13]   B. Kovalerchuk & E. Vityaev. Data Mining in Finance: Advances in Relational and Hybrid Methods. Kluwer, 2000.

[14]   J. Han, L.V.S. Lakshmanan & R.T. Ng. Constraint-based multidimensional data mining. IEEE Computer, 8, 46-50, 1999.

[15]   Han, J. & Kamber, M. Data Mining: Concepts and Techniques. Morgan Kaufmann (2000).

[16]   Chen, Z. Computational Intelligence for Decision Support. CRC Press, Boca Raton, 2000.

[17]   Feldman, S. Technology Trends and Drivers and a Vision of the Future of e-business. Proceedings 4th International Enterprise Distributed Object Computing Conference, September 25-28, 2000, Makuhari, Japan.

[18]   Fischer, L. (Ed). Workflow Handbook 2001. Future Strategies, 2000.

[19]   Debenham, J.K.. Supporting knowledge-driven processes in a multiagent process management system. Proceedings Twentieth International Conference on Knowledge Based Systems and Applied Artificial Intelligence, ES'2000: Research and Development in Intelligent Systems XVII, Cambridge UK, December 2000, pp273-286.

[20]   Jain, A.K., Aparicio, M. and Singh, M.P. "Agents for Process Coherence in Virtual Enterprises" in Communications of the ACM, Volume 42, No 3, March 1999, pp62—69.

[21]   Jennings, N.R., Faratin, P., Norman, T.J., O'Brien, P. & Odgers, B. Autonomous Agents for Business Process Management. Int. Journal of Applied Artificial Intelligence 14 (2) 145—189, 2000.

[22]   Robert Skinstad, R. "Business process integration through XML". In proceedings XML Europe 2000, Paris, 12-16 June 2000.

[23]   Guarino N., Masolo C., and Vetere G., OntoSeek: Content-Based Access to the Web, IEEE Intelligent Systems 14(3), May/June 1999, pp. 70-80

[24] · Uschold, M. and Gruninger, M.: 1996, Ontologies: principles, methods and applications. Knowledge Engineering Review, 11(2), 1996.

[25]   Müller, J.P. "The Design of Intelligent Agents" Springer-Verlag, 1996.

[26]   Rao, A.S. and Georgeff, M.P. "BDI Agents: From Theory to Practice", in proceedings First International Conference on Multi-Agent Systems (ICMAS-95), San Francisco, USA, pp 312—319.

[27]   Finin, F. Labrou, Y., and Mayfield, J. "KQML as an agent communication language." In Jeff Bradshaw (Ed.) Software Agents. MIT Press (1997).

[28]   Kaplan, Steven and Sawhney, Mohanbir. E-Hubs: The New B2B Marketplace. Harvard Business Review 78 May-June 2000 97-103.

[29]   Shane, Scott. Prior knowledge and the discovery of entrepreneurial opportunities. Organization Science 11 (July-August), 2000, 448-469.

Markets as Global Scheduling Mechanisms: The Current State

Junko Nakai

AMTI at NASA Ames Research Center, Mail Stop 258-6, Moffett Field, CA
94305-1000
nakai@nas.nasa.gov

Abstract. Viewing the computers as "suppliers" and the users as "consumers" of computing services, markets for computing services/resources have been examined as one of the most promising mechanisms for global scheduling. We first establish how economics can contribute to scheduling. We further define the criterion for a scheme to qualify as an application of economics. Many studies to date have claimed to have applied economics to scheduling. If their scheduling mechanisms do not utilize economics, contrary to their claims, their favorable results do not contribute to the assertion that markets provide the best framework for global scheduling. For any of the schemes examined, we could not reach the conclusion that it makes full use of economics.

1 Scheduling and Economics

A distributed computing system brings about opportunities for enhancing the efficiency in computing, and hence, for increasing the value of existing facilities for computation. One of the new administrative features necessitated for that purpose is global scheduling. The core problem we face in scheduling computing jobs is exceedingly similar to that in the economy. If there were an unlimited amount of computing resources, jobs would be executed as soon as they are submitted, eliminating the need for scheduling. As there is a limit to the availability of resources, scheduling of computing jobs becomes a problem of allocating limited resources. We may envision a situation where the users of computers are required to give up the limited resources they are endowed with in exchange for access to the computing resources. Further, if the endowment could be used for more than one item or occasion, which result in outcomes differing in importance, then the allocation of computing resources is precisely the economic problem in its most fundamental form. Modern economics is recognized as the science that studies human behavior as a relationship between scarce means which have alternative uses, as asserted by Robbins [17].[1]

---

[1] The most basic activity in an economy is an exchange of goods and services. Whenever there is an exchange of goods or services, a price is established, which is simply the rate of exchange of goods or services involved. Thus, when we refer to one of the three, an economy, an exchange, and a price, the other two necessarily exist. Markets are forums for exchanges of goods and services, where exchanges are voluntarily

P.M.A. Sloot et al. (Eds.): ICCS 2002, LNCS 2329, pp. 256–265, 2002.
© Springer-Verlag Berlin Heidelberg 2002

When the so-called pricing of computing resources started to attract increasing attention in the late 1960s, it was quickly acknowledged that pricing can be seen as a kind of scheduling mechanism and that both concern allocation of resources [16].[2] The advent of networks of computers did not alienate scheduling from economics, and scheduling for distributed computer systems has also been recognized as an activity under resource management [2, 12]. Markets for computing resources have been embraced by many as the key components in the operation of distributed systems [1, 3, 4, 5, 10, 11, 14, 15, 18, 19, 20, 21, 22, 23].

1.1   The Objective of an Economy

An additional insight to the economic problem, as defined by Robbins, was provided by Hayek [6]. He made it clear that the problem was how to allocate resources so that they are used in the best way possible. Inasmuch as the best uses are known only to individuals, the economic problem becomes "a problem of the utilization of knowledge which is not given to anyone in its totality" [6]. By pursuing the "best" use of each resource, our attempt to allocate resources becomes one of maximizing private values associated with resource use. Certainly, this is also the problem posed to the group of users who are to share a set of computing facilities. The computing capabilities should be shared, and for that reason, information on the requirements for each job, including what resources and the value of its successful execution, is required. Unfortunately, this information is usually available only to the user. In fact, utilization of users' private information has been recognized as one of the advantages of pricing over other types of scheduling [16].

We now discuss the importance of scarcity of resources, including budget, and that of the existence of alternative uses of resources in the formation of value itself. A variety of needs, which cannot all be fulfilled and each of which leads to an outcome that is different in importance, is what brings the economic problem into its existence. If resources are obtainable whenever desired and/or their uses result in outcomes identical in value, there would be no question of allocating the resources for their best use or to maximize value from use.

We distinguish two types of scarcity. It may take the straightforward form of a finite limit to the amount available, an amount smaller than is required to fulfill all needs, in which case the multiplicity of possible uses of the resource, with outcomes that are distinguishable in importance, leads to the economic problem. If a limited amount of a resource is available during a particular time period (which cannot be used any other time) and that resource is designated for only one use, the economic problem does not exist because there is a unique possibility: allocation of the entire resource for that single use. Scarcity may

---

initiated. Markets do not prevail in all types of economies, and economies are not synonymous with markets.

[2] Pricing stepped into the limelight because the most common default scheduling mechanism, first-come, first-served, was often combined with additional rules, indicating that it alone was not meeting needs [16].

take another form: resources having dual effects, each of which is associated with a positive or a negative value, depending on the amount allocated. The second type of scarcity also serves the function provided by alternative uses, which vary in importance of attainable outcomes, in the first kind. Both the existence of alternative uses with different, resultant utility levels (in the first kind of scarcity), and the possible negative effects of resource use (in the second type of scarcity) make it necessary for resource users to evaluate and compare the values of various allocations.

1.2   Application of the Economics to Scheduling

We have laid out above the problem an economy faces, and clarified the motivation behind economic activities. The existence of a motivation to meet an objective is a necessary condition for the attainment of the objective. In the following sections, we analyze various pricing schemes for computing resources, with this necessary condition as a yardstick for determining whether a scheme is an application of economics. When the participants in the scheme maximize utility (or equivalently, value), given constraints, by choosing among needs that cannot be simultaneously fulfilled and are different in achievable utility, the economic motivation exists, and therefore, the necessary condition is satisfied for calling the scheme an application of economics to scheduling. If a computing-service provider is directly involved in the process of determining service allocation (i.e., the provider interacts with users in the process), the criterion must be satisfied not only by the users but also by the provider.

2   Markets and Computer Networks: from early 1980s to present

A computer network consists not only of multiple users, but also of multiple service providers, appearing much more complex [19] and closer to a market as we know from everyday life [1, 5, 10, 11, 15, 21]. The seeming ease with which markets allocate resources, albeit its complexity [1, 5, 11, 15, 19], and the solid existence of theoretical microeconomics (particularly, the general equilibrium theory) [5, 11, 15, 19], have captured the imagination of researchers in the field of networks. We examine below the representative market models for distributed computing systems (the Contract Net Protocol [18], the Enterprise [14], a model by Kurose and Simha [11], Agoric System [15], Spawn [10, 21], WALRAS [3, 22, 23], Mariposa [19, 20], and the Grid Architecture for Computational Economy [1]),[3] focusing on the ones that make an extensive and direct use of the general equilibrium theory: the model by Kurose and Simha, and WALRAS.

---

[3] In citing such models, those for database systems that assume light traffic, are often included, e.g., Mariposa.

2.1 Model by Kurose and Simha

The paper by Kurose and Simha [11] is one of the oft-cited papers on distributed computer systems with relation to economics. We briefly discuss the general equilibrium theory, which their model draws on, with attention to the so-called tâtonnement process and the problematic aspects of the process. While the second type of scarcity existed for the nodes in the model, they were not utility maximizers, disqualifying the model as an application of economics. The utility of the resource allocator, called auctioneer, was undefined.

General Equilibrium Theory and Resource-Directed Approach. Kurose and Simha implemented one of what they call the two basic microeconomic approaches, the price-oriented and the resource-oriented approaches, which are better known as a tâtonnement process (for reaching an equilibrium) with pure price adjustment and that with pure quantity adjustment, respectively. Hence, the model applies the general equilibrium theory in economics, a theory that concerns price and quantity determination in equilibrium initiated by Léon Walras in the late 19th century. The economy under consideration in the general equilibrium theory is one in which the number of agents in the system is large enough so that any one of them cannot affect prices by acting alone, and agents do not collaborate.[4]

The tâtonnement process with pure quantity adjustment in a pure exchange economy proceeds as follows: The auctioneer informs the utility-maximizing user-agents of their entitled quantities of resources and the agents report back the marginal utilities at those quantities. The auctioneer changes the allocation of resources so that more inputs are allocated to the agents with higher marginal utilities. The process continues until an equilibrium in price and quantity is attained. This is the approach preferred and adopted by the authors, because all interim allocations are feasible, unlike the price-oriented approach. We note that the process does not guarantee convergence to a unique equilibrium, even if one exists, without further restrictions on the economy.

Another attractive feature of the resource-directed approach was reported: "When analytic formulas are used to compute performance, successive iterations of the algorithm result in resource allocations of strictly increasing systemwide utility."[5] Putting aside the issue of incentive compatibility, we may say that optimization by local agents led to an optimal solution for the entire system

---

[4] In other words, all participants in the economy, producers and consumers, are price-takers.

[5] Heal [7] concluded that in the tâtonnement process with pure quantity adjustment more information exchange would be required than in the process with pure price adjustment. Hurwicz [9] pointed out that the total information would be of higher dimension in the process with pure quantity adjustment compared to the process with pure price adjustment, but also added that whether the difference was significant was "somewhat controversial."

because the global utility was set equal to the sum of utilities of local agents.[6] The iterations in resource allocation conform with the economic motivation (or, are economically feasible) in the adopted framework, only under the condition that honest reporting of utility levels is ensured.

The two advantages described above, feasibility (in a discrete process) and monotonicity, were labeled two desirable properties of tâtonnement [13]/gradient [9]-based processes for reaching an equilibrium, by Malinvaud.[7] Kurose and Simha proposed algorithms which were based on a tâtonnement process with pure quantity adjustment. Theoretical investigation of their most basic algorithm by Heal [7] had shown that it indeed exhibits both properties. For the process to function, however, one condition has to be met: the central authority's knowledge of an initial allocation that is feasible, which is the cost of obtaining the desirable properties according to Hurwicz [9].

Optimal File Allocation. The distributed system chosen for investigation was a network of nodes, which were assumed capable of communicating with any other in the network. The problem was how to allocate files optimally.[8] The cost of communication to each node was defined to be the average delay in the transmission of messages. The cost of access delay was defined to be the expected time in access delay. In turn, the sum of the cost of communication and the cost of access delay was called the expected cost of access to the file source at a node. Therefore, allocating all files at one node may reduce the cost of communication, but only by increasing the cost of access delay, because that node must handle all inquiries in the system. The expected cost of access to the entire network was the sum of the expected cost at each node. The optimal file allocation was taken to be the allocation that minimizes the expected cost of access for the whole network.[9]

The performance of three decentralized algorithms was examined. They all employed gradient processes; each node computed the first derivative of the utility function (i.e., marginal utility) and/or the second derivative, evaluated at a specific point, and sent that information to the central node (or alternatively, to all other nodes). The paper concluded that all algorithms had the following desirable properties: feasibility in all iterations, strict monotonicity, and fast convergence.

---

[6] Consequently, the global objective function was a function only of local objective functions increasing in all of its arguments, and unresponsive to the names of the local agents; local optimization coincided with global optimization.

[7] He was concerned about the possibility of slow or "disorganized" convergence to an equilibrium, which implied that the existence of an equilibrium and convergence to one through a tâtonnement process, by themselves, do not guarantee practicality of the process as one in economic planning.

[8] Each node had a local look-up table, which provided information on the file fragment locations so that a request not met locally could be sent to an appropriate node.

[9] In order to cast the problem as a utility maximization problem, the utility was set equal to the negative of the expected cost.

Problems in General Equilibrium Theory One of the problems in the general equilibrium theory is that the tâtonnement process cannot do without an auctioneer; decentralized economic planning does not truly qualify as a decentralized system with features such as lack of a single point of failure, as envisaged by many of the researchers in the field of global scheduling. Heal's model [7], on which the model by Kurose and Simha is based, was decentralized only in the sense that information was collected from the local agents. The mechanisms employed by Kurose and Simha for adjustment of the system towards an equilibrium are informationally decentralized [9], but that is not equivalent to decentralized decision-making.[10] Indeed, the processes do not concern local decision-making, just as Heal's model does not.

Another problem in the general equilibrium theory is also carried over to the model by Heal, as well as to that by Kurose and Simha: incentive incompatibility.[11] Note that individual nodes could have increased the final utility by falsely reporting the levels of marginal utility that were above the actual levels. However, the local agents in the investigated network acted so as to fulfill the global goal, by forgoing the opportunity to increase their own utilities. Moreover, the utility of auctioneers is always ignored in the general equilibrium theory, and so it is in the models by Heal, Kurose and Simha.

The above discussion also serves as an analysis of whether the resource allocation scheme was driven by economic considerations of the agents in the system. The model relied on nodes' balancing the benefits and the costs of owning a file fragment; the second type of scarcity (and hence, the existence of alternative uses in the broad sense) was present. The fact that the nodes reported their marginal utilities honestly to the central node (or, to all other nodes), in face of feasible cheating and attainment of higher utility, together with the fact that honesty did not factor into the utility defined, indicate that the nodes were not utility-maximizing agents; the first part of the necessary condition to be an economic application was not met. The utility of the auctioneer, an active participant in the resource allocation process, was left undefined. The study neither validates nor invalidates the appropriateness of creating markets for computing resources in distributed systems.

2.2   WALRAS

WALRAS, like the model by Kurose and Simha, is an application of the general equilibrium theory to scheduling [22]. It implemented the tâtonnement process with pure price adjustment. WALRAS is incentive incompatible, as any model based on the general equilibrium theory with a finite number of agents would

---

[10] An informationally decentralized process is one which has informational requirements that are no greater than those for a perfectly competitive process [8].

[11] Although reporting the derivatives of the pertinent functions to the central authority is one of the standard elements in the tâtonnement processes, its incentive compatibility has been established only when the number of traders is infinite in a pure exchange economy, if no forced, initial redistribution of endowments is allowed [9].

be (if there exists neither production nor the possibility of redistributing initial endowments); it is unsuitable to be called an economic application.

General Equilibrium Theory and WALRAS. In the standard tâtonnement process with pure price adjustment, there is an auctioneer who informs agents of the prices, and the agents report back the amounts of goods they demand (or more precisely, demand for goods over and above the amount endowed) at those prices. Such reports are called bids. The auctioneer calculates the new prices, according to the predetermined rules and the amounts of demand reported, and the process repeats until the prices no longer need to be adjusted.

WALRAS differed from the standard tâtonnement process in that demand functions were reported by the agents (instead of a point on a demand function) and that the auctioneer dealt with each good separately [3, 22]. Moreover, not all agents reported their demands for all goods in each time period [3]. Random draws, which were independent across time and agents, determined which bids were submitted. For the unselected combinations of agents and goods, the bids from the previous period were used. The advantage of their asynchronous bidding was small price oscillations [3, 23]. Cheng and Wellman [3] favored their approach over other processes for attaining an equilibrium in the general equilibrium theory, since they saw fewer opportunities for strategic interactions and no trade took place until an equilibrium was achieved (no resource was allocated based on intermediate results, which were by definition not global optima and may have been irreversible if implemented). The utility function of the resource users, $u(x)$, was of constant elasticity of substitution: $u(x) = \left(\sum_{j=1}^{k} \alpha_j (x^j)^\rho \right)^{1/\rho}$, where α_j's were randomly generated coefficients from a uniform distribution, x^j was the amount of good j, and ρ was fixed at 0.5 (for the main simulation). Therefore, the resulting excess-demand functions, for each agent and for the entire economy, had the property of gross substitutability [3]. The existence of an equilibrium was guaranteed by the preferences implied from the utility functions (which were continuous, strictly convex, and locally nonsatiated) and non-negative total excess-demand (as Cheng and Wellman implicitly demonstrated with experiments).[12] Moreover, gross substitutability ensured the uniqueness of equilibrium and convergence to that equilibrium point on any price path. While the adaptive learning behavior of the auctioneers justified the rules of WALRAS, users remained simple price-takers who reported their excess-demand functions honestly. The users would not have reported the true demand functions if they acted so as to maximize utilities, as is evidenced by the utility function shown above.

Convergence and Other Problems. We concentrate here on the most comprehensive results given in "The WALRAS Algorithm" [3]. An examination of 100 randomly generated economies, with five or seven agents of utility as described above (where j is equal to 5), showed that the median behavior of the

---

[12] See Figure 1 in "The WALRAS Algorithm" [3].

system was a rapid convergence to the equilibrium at the beginning, and leveling off at a small, positive amount of total excess-demand after 150 iterations.[13] When values other than 0.5 for the substitution coefficient were adopted, convergence was not seen even after 5,000 iterations in some cases.

The feasibility of the proposed algorithm and singularity of equilibrium are obtained only under restricted circumstances [3]; we need restrictions on the agents' utility functions. Gross substitutability of the aggregate excess-demand function is a sufficient condition, and the authors reported that they could not find a class of utility functions that are not grossly substitutable and yet converge to an equilibrium. Whatever the necessary conditions may be for convergence, the utility functions employed must represent the preferences of users. WALRAS explored the case where every agent had an identical utility function. How it would serve a system of users with various utility functions and how it would compare with other scheduling mechanisms are yet to be seen.

2.3   Other Models

The information exchange process in the Contract Net Protocol [18] does not appear too different from that in a non-distributed computing system, if evaluated according to the characterization provided by the author of the protocol. Many of the details necessary for implementation were left unspecified, including agents' utility; we are unable to conclude that the protocol is an economic application.

The Enterprise system [14] is a fleshed-out version of the Contract Net Protocol, which connected personal workstations, using a local area network. Although favorable simulation results were reported, we cannot attribute them wholly to the scheme's general features. Neither can we conclude that the scheme is an application of economics, because the utility of the service providers in the scheme was undefined. There was no mention of user budget, without which we cannot evaluate whether the scarcity and alternative-use condition for users was satisfied.

Spawn [10, 21] is a resource allocation mechanism for a network of heterogeneous workstations whose agents are sellers (i.e., owners of workstations, who are not using them at any given moment) and buyers of CPU time. The special feature of Spawn is its spawning process or dividing a task into subtasks. The agents' utilities (that of both buyers and sellers of computing resources) were not defined; it was impossible to confirm that the scheme satisfied the necessary condition to be an economic application.

Mariposa is a distributed database and storage system for non-uniform, multi-administrator, wide area networks [19, 20]. Some of the advantages of markets, which are claimed also as those of Mariposa, do not hold unconditionally, and the practicality of the objectives chosen for the artificial agents is

---

[13] Equilibria reached through tâtonnement processes in a framework as the one adopted by WALRAS are Pareto-optimal, which are often interpreted as desirable allocations [23].

in question. Some, but not all, of the agents' objectives (and thus, utilities of some agents) were proposed. We could not conclude that the scheme satisfied the necessary condition for being an economic application.

An agoric system is an intellectual exploration as to what economics may be able to do for distributed computing systems. Without more information than has been provided in the paper [15], we cannot draw a conclusion as to whether the system is an application of economics.

The Grid Architecture for Computational Economy [1] aims at incorporating an economic model into a grid system with existing middleware, such as Globus and Legion. Although utilities for service providers and users were suggested, there was an implication that no alternative uses for user budget existed; the scheme does not qualify as an economic application.

3   Future Direction

We could not conclude that the criterion for an economic application was satisfied by any of the models examined. Moreover, there was no comparison of their performances with those using other scheduling mechanisms. Hence, no support was provided for establishing that economics is a necessary component for superior performance of distributed computing systems.

Many of the studies examined justified their use of economics based on the desirable results given by the general equilibrium theory, which is more narrowly focused than the whole discipline of economics. How relevant is the general equilibrium theory to scheduling? More generally, does the theory capture what drives the economy to behave well as a system? Conversely, are markets capable of producing the favorable outcomes as the general equilibrium theory implies (and subsequently, asserted by the architects of market-based scheduling)? We believe that many of these questions can be answered through a careful reading of the general equilibrium theory and its related fields.

References

1. Buyya, R., Abramson, D., Giddy, J.: A Case for Economy Grid Architecture for Service Oriented Grid Computing. Presented at the 10th Heterogeneous Computing Workshop, San Francisco, April 23, 2001
2. Casavant, T. L., Kuhl, J. G.: A Taxonomy of Scheduling in General-Purpose Distributed Computing Systems. IEEE Trans. Soft. Eng. 14 (1988) 141-154
3. Cheng, J. Q., Wellman, M. P.: The WALRAS Algorithm: A Convergent Distributed Implementation of General Equilibrium Outcomes. Computational Econ. 12 (1998) 1-24
4. Clearwater, S. H.: Why Market-Based Control? In: Clearwater, S. H. (ed.): Market-Based Control World Scientific Publishing, Singapore (1996)
5. Ferguson, D. F., Nikolaou, C., Sairamesh, J. Yemeni, Y.: Economic Models for Allocating Resources in Computer Systems. In: Clearwater, S. H. (ed.): Market-Based Control. World Scientific Publishing, Singapore (1996)

6. Hayek, F. A.: The Use of Knowledge in Society. Amer. Econ. Rev. **35** (1945) 519-530
7. Heal, G.: Planning without Prices. Rev. Econ. Stud. **36** (1960) 347-362
8. Hurwicz, L.: On Informationally Decentralized Systems. In: McGuire, C. B. Radner, R. (eds.): Decision and Organization North-Holland Publishing, New York (1972)
9. Hurwicz, L.: The Design Mechanisms for Resource Allocation. Amer. Econ. Rev. **63** (1973) 1-30
10. Huberman, B. A., Hogg, T.: Distributed Computation as an Economic System. J. Econ. Persp. **9** (1995) 141-152
11. Kurose, J. F., Simha, R.: A Microeconomic Approach to Optimal Resource Allocation in Distributed Computer Systems. IEEE Trans. Comp. **38** (1989) 705-717
12. MacKie-Mason, J. K., Varian, H. R.: Pricing the Internet. In: Kahin, B., Keller, J. (eds.): Public Access to the Internet. MIT Press, Cambridge, Massachusetts (1995)
13. Malinvaud, E.: Decentralized Procedures for Planning. In: Malinvaud, E., Bacharach, M. O. L. (eds.): Activity Analysis in the Theory of Growth and Planning. Macmillan, London (1967)
14. Malone, T. W., Fikes, R. E., Grant, K. R., Howard, M. T.: Enterprise: A Marketlike Task Scheduler for Distributed Computing Environments. I:; Huberman, B. A. (ed.): The Ecology of Computation. Elsevier Science Publishers, North-Holland (1988)
15. Miller, M. S., Drexler, K. E.: Markets and Computation: Agoric Open Systems. http://www.agorics.com/agoricpapers.html, 7 July 2000
16. Nielsen, N. R.: The Allocation of Computer Resources—Is Pricing the Answer? Comm. ACM. **13** (1970) 467-474
17. Robbins, L. C.: An Essay on the Nature and Significance of Economic Science. Macmillan, London (1984, originally published in 1932)
18. Smith, R. G.: The Contract Net Protocol: High-Level Communication and Control in a Distributed Problem Solver. IEEE Trans. Comp. **c-29** (1980) 1104-1113
19. Stonebraker, M., Devine, R., Kornacker, M. Litwin, W., Pfeffer, A., Sah, A., Staelin, C.: An Economic Paradigm for Query Processing and Data Migration in Mariposa. http://sunsite.berkeley.edu/Dienst/UI/2.0/Describe/ncstrl.-ucb/S2K-94-49
20. Stonebraker, M., Aoki, P. M. Litwin, W., Pfeffer, A., Sah, A., Staelin, C., Yu, A.: Mariposa: a Wide-Area Distributed Database System VLDB J. **5** (1996) 48-63
21. Waldspurger, C. A., Hogg, T., Huberman, B. A. Kephart, J. O., Stornetta, S. Spawn: A Distributed Computational Economy. IEEE Trans. Soft. Eng. **18** (1992) 103-117
22. Wellman, M. P.: Market-Oriented Programming Environment and its Application to Distributed Multicommodity Flow Problems. J. Art. Intel. Res. **1** (1993) 1-23
23. Wellman, M. P.: Market-Oriented Programming: Some Early Lessons. In: Clearwater, S. H. (ed.): Market-Based Control. World Scientific Publishing, SIngapore (1996)

Numerical Simulations of Combined Effects of Terrain Orography and Thermal Stratification on Pollutant Distribution in a Town Valley

S. Kenjereš, K. Hanjalić, and G. Krstović

Department of Applied Physics, Delft University of Technology
Lorentzweg 1, 2628 CJ Delft, The Netherlands
kenjeres@ws.tn.tudelft.nl; hanjalic@ws.tn.tudelft.nl,
WWW home page: http://www.ws.tn.tudelft.nl

Abstract. Combined effects of terrain orography and thermal stratification on the dispersion of pollutants in a mountainous town valley over a diurnal cycle are numerically simulated by a time-dependent Reynolds-averaged Navier-Stokes (T-RANS) approach. The T-RANS model was incorporated into a finite volume NS solver for three-dimensional non-orthogonal domains, using Cartesian vector and tensor components and collocated variable arrangement. Prior to the full scale simulations, the T-RANS approach was validated in test situations where the effects of thermal stratification and terrain orography are separated, showing good agreement with the available experimental and simulation data. The full scale simulations were performed in a realistic orography over two diurnal cycles for two cases of the initial thermal stratification, both with a prescribed time and space variation of ground temperature and pollutant emission - reflecting the daily activities in the town. The results confirmed that T-RANS approach can serve as a powerful tool for predicting local environments at micro and meso scales.

1 Introduction

Most urban areas are continuous sources of heat and pollution as a result of a variety of human activities, e.g. industrial processes, transportation, agriculture, etc. In addition, a high percentage of urban areas is covered with concrete and asphalt which store and reflect incoming radiation causing a significantly warmer surface-layer air than that of their natural surroundings. As a result, the urban areas form a kind of local heat islands in the surrounding countryside, Stull [9]. Urban city landscape with tall buildings and streets of different sizes together with surrounding terrain orography create very complex local geometry. These local boundary conditions together with an imposed initial temperature distribution in atmosphere (thermal stratification) form a complex interactions mechanism between heat transfer and corresponding atmospheric pollutant emission. The prediction of this complex interaction mechanism is of vital importance for estimating possible toxic pollutant distribution that may pose a risk to human health. It is also the major prerequisite for optimum control of air quality:

P.M.A. Sloot et al. (Eds.): ICCS 2002, LNCS 2329, pp. 266–275, 2002.

future city planning and optimum location of industrial zones, design of city transportation system, control of traffic and industrial activities during critical meteorological periods, etc. Current practice relies on semi-empirical methods and simple integral modelling of pollutant dispersion with prescribed wind conditions, whereas situations at micro and meso scales dominated by buoyancy are usually beyond the reach of such models. Large eddy simulations (LES) is a possible option, but a hybrid LES/RANS approach ('ultra' CFD problems; Hunt [2]) seems a more viable option that can provide, under lower costs, required detailed insights into above mentioned complex phenomena.

In this paper, we propose the transient Reynolds-averaged Navier-Stokes (T-RANS) approach as potentially efficient, numerically robust and physically accurate method for simulation of combined effects of terrain orography and thermal stratification on pollutant dispersion. These effects will be analysed by performing simulations of a realistic environmental problem of a medium-size valley town with significant residential and industrial pollution. The critical periods are the winter cloudy windless days when the lower atmosphere in the valley is capped with an inversion layer preventing any convection through it. The air movement and the pollutant dispersion are solely generated by the day ground heating in surroundings and urban areas. While any realistic conditions can be imposed, we consider at present - as a part of a preliminary study - an idealised situation with space and time sinusoidal variation of temperature and concentration, both imitating two diurnal cycles. In order to accommodate a very complex terrain orography, a finite-volume Navier-Stokes solver for three-dimensional flows in structured non-orthogonal geometries, based on Cartesian vector and tensorial components and collocated variable arrangement, was applied.

Prior to the full scale simulations that include both effects - thermal stratification and terrain orography - the T-RANS approach was validated in cases of unsteady turbulent penetrative convection of unstable mixed layer and classical Rayleigh-Bénard convection over flat and wavy surfaces of different topology and over a range of Ra numbers.

2   The Time-Dependent RANS (T-RANS): equations and subscale models

This approach can be regarded as Very Large Eddy Simulations (VLES) in which the stochastic motion is modelled using a $\langle k \rangle - \langle \varepsilon \rangle - \langle \overline{\theta^2} \rangle$ Algebraic Stress/Flux/Concentration (ASM/AFM/ACM) single-point closure models as the "subscale model", where $\langle \rangle$ denoted the time-resolved motion. The turbulent stress tensor, $\tau_{ij} = \langle u_i u_j \rangle$, heat flux vector, $\tau_{\theta i} = \langle \theta u_i \rangle$ and concentration flux vector, $\tau_{ci} = \langle c u_i \rangle$, were derived by truncation of the modelled RANS parent differential transport equations by assuming weak equilibrium, i.e. $(D/Dt - \mathcal{D})\overline{\phi u_i} = 0$, but retaining all major flux production terms (all treated as time-dependent). In contrast to Large Eddy Simulation (LES), the contribution of both modes to the turbulent fluctuations are of the same order of magnitude.

Environmental fluid flows are described by standard conservation laws for mass, momentum, energy and concentration. For the resolves ('filtered') motion, equations can be written in the essentially same form as for the LES:

$$\frac{\partial \langle U_i \rangle}{\partial t} + \langle U_j \rangle \frac{\partial \langle U_i \rangle}{\partial x_j} = \frac{\partial}{\partial x_j} \left(\nu \frac{\partial \langle U_i \rangle}{\partial x_j} - \tau_{ij} \right)$$
$$- \frac{1}{\rho} \frac{(\langle P \rangle - P_{ref})}{\partial x_i} + \beta g_i (\langle T \rangle - T_{ref}) \tag{1}$$

$$\frac{\partial \langle T \rangle}{\partial t} + \langle U_j \rangle \frac{\partial \langle T \rangle}{\partial x_j} = \frac{\partial}{\partial x_j} \left(\frac{\nu}{Pr} \frac{\partial \langle T \rangle}{\partial x_j} - \tau_{\theta j} \right) \tag{2}$$

$$\frac{\partial \langle C \rangle}{\partial t} + \langle U_j \rangle \frac{\partial \langle C \rangle}{\partial x_j} = \frac{\partial}{\partial x_j} \left(\frac{\nu}{Sc} \frac{\partial \langle C \rangle}{\partial x_j} - \tau_{cj} \right) \tag{3}$$

where $\langle \rangle$ stands for resolved ensemble-averaged quantities and $\tau_{ij}, \tau_{\theta j}$ and τ_{cj} represent contributions due to unresolved scales to momentum, temperature and concentration equation respectively, which were provided by the subscale model. In the present work which is still at the preliminary stage, the adopted 'subscale' expression are given as follows. For turbulent stresses we applied eddy viscosity expression:

$$\tau_{ij} = -\nu_t \left(\frac{\partial \langle U_i \rangle}{\partial x_j} + \frac{\partial \langle U_j \rangle}{\partial x_i} \right) + \frac{2}{3} \langle k \rangle \delta_{ij} \tag{4}$$

The turbulence heat and concentration fluxes are expressed by AFM counterparts:

$$\tau_{\theta i} = -C_\phi \frac{\langle k \rangle}{\langle \varepsilon \rangle} \left[\tau_{ij} \frac{\partial \langle T \rangle}{\partial x_j} + \xi \tau_{\theta j} \frac{\partial \langle U_i \rangle}{\partial x_j} + \eta \beta g_i \langle \theta^2 \rangle \right] \tag{5}$$

$$\tau_{ci} = -C_\phi \frac{\langle k \rangle}{\langle \varepsilon \rangle} \left[\tau_{ij} \frac{\partial \langle C \rangle}{\partial x_j} + \xi \tau_{cj} \frac{\partial \langle U_i \rangle}{\partial x_j} \right] \tag{6}$$

The closure of the expressions for subscale quantities is achieved by solving the equations for turbulence kinetic energy $\langle k \rangle$, its dissipation rate $\langle \varepsilon \rangle$ and temperature variance $\langle \theta^2 \rangle$, resulting in three-equation model $\langle k \rangle - \langle \varepsilon \rangle - \langle \theta^2 \rangle$, Kenjereš and Hanjalić [4]:

$$\frac{D \langle k \rangle}{Dt} = \mathcal{D}_k + P_k + G_k - \langle \varepsilon \rangle$$

$$\frac{D \langle \varepsilon \rangle}{Dt} = \mathcal{D}_\varepsilon + P_{\varepsilon 1} + P_{\varepsilon 2} + G_\varepsilon - Y$$

$$\frac{D \langle \theta^2 \rangle}{Dt} = \mathcal{D}_\theta + P_\theta - \langle \varepsilon_\theta \rangle \tag{7}$$

3   Results

3.1   Unsteady turbulent penetrative convection of unstable mixed layer

The ability of the proposed Algebraic-Stress-Flux-Concentration subscale Model (ASM/AFM/ACM) to reproduce correctly the flows with a strong thermal strat-

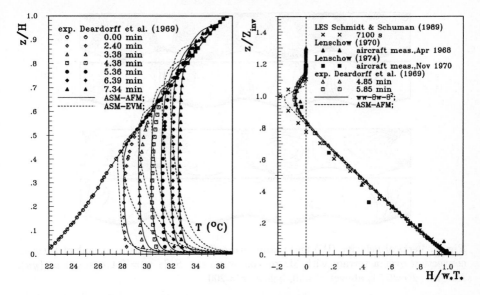

Fig. 1. Time evolution of vertical temperature profiles in a mixed layer heated from below -left, Normalised vertical heat flux in a mixed layer heated from below- right: comparison between experiments and simulations.

ification was first tested on unsteady turbulent penetrative convection of an unstable mixed layer. Due to heating from bellow, the initially stable environment in the near-ground region becomes unstable and interactions between stable and unstable regions occur. With further intensification of the heat transfer in the vertical direction, the location of the interface between the two regions moves up with time, causing the mixed turbulent ground layer to grow into a stable region. Another feature of the phenomenon is the entrainment of overlaying non-turbulent fluid into mixed layer causing very steep gradients at the interface as well as a reversal of the sign of the heat flux as a result of the compensating cooling of the fluid in the stable region, Nieuwstadt *et al.* [7]. These features explain why turbulent penetrative convection represents a very challenging test case for turbulence models, Mahrt [6]. Fig. 1a shows the time evolution of temperature profiles obtained with ASM/AFM model. The experimental data (Deardorff *et al.*[1]; Willis and Deardorff [10]) are very well reproduced in the middle as well as at the upper edge of the mixed layer. A further simplification of turbulent heat flux to anisotropic eddy viscosity model (denoted with EVM), which basically excludes temperature variance from consideration, resulted in serious deterioration of the predicted results. This confirms that the AFM representation of turbulent heat flux is the minimum level of modelling which can provide close agreement with experimental temperature profiles. The predicted vertical heat flux suitably scaled with the inversion height and the product of the buoyancy velocity ($w_* = (\beta g Q_s Z_{inv})^{1/3}$) and temperature ($T_* = Q_s / w_*$) is shown in Fig. 1b. Here the solid line represents result obtained with the full second-order closure

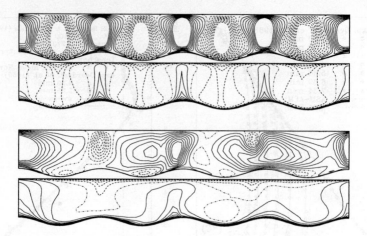

Fig. 2. Vertical velocity $\langle W \rangle$ and temperature $\langle T \rangle$ distributions in turbulent thermal convection over a terrain with 3D regular waviness, $S_B(x,y) = 0.1\ cos(x\pi)\ cos(y\pi)$, $Ra=10^7$, $Pr=0.71$; above- $\tau^*=50$, below- $\tau^*=200$.

for both momentum and temperature equations. As seen, the ASM/AFM result shows good agreement with several sets of laboratory and field measurements, and with LES data of Schmidt and Schumann [8].

3.2   Transient Rayleigh-Bénard convection over horizontal wavy walls

The potential of the T-RANS approach to capture the effects of bottom wall topology was verified against the DNS and LES results of Krettenauer and Schumann [5] for turbulent convection over wavy terrain - both under identical conditions. Their main finding was that the total heat transfer was only slightly affected by terrain orography compared to flat bottom wall situation. The gross feature of the flow statistics, such as profiles of turbulence fluxes and variance were not very sensitive to the variation of the bottom wall topology. On the other hand, they reported that the motion structure persisted considerably longer over the wavy terrain than over flat surfaces. While the Krettenauer and Schumann [5] were interested mainly in the *final* turbulence statistics under steady conditions, we extended our study to time evolution of turbulence quantities and large coherent structures. The effect of the horizontal wavy wall on the mean properties $\langle W, T \rangle$ is presented in Fig. 2. At the initial stage of heating, $\tau^* = \sqrt{\beta g \Delta T / H} = 50$, all quantities show a regular flow pattern determined by the wall configuration. The plumes raise from the surface peaks and sink into the surface valleys, portraying 25 characteristic locations. At $\tau^*=200$ the initial organisation of the flow cannot be observed anymore. Thermal plumes occupy a significantly larger space and not only the regions close to the bottom surface peaks, as found in the initial phase of flow development, Hanjalić and Kenjereš [3]. It can be concluded that the T-RANS approach was able to capture impor-

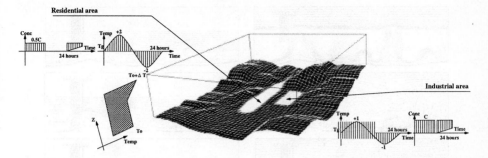

Fig. 3. Specification of imposed boundary and initial conditions for a middle-size town located in a mountain area; Two distinct areas (industrial and residential) are represented by different pollution emission ($C = 100$ and 50% respectively) and different heat source intensities ($T_g = \pm 1$ and ± 2 respectively); Initial thermal stratification is also presented.

tant phenomena of time and spatial evolution of flow and turbulence structures which reflect specific horizontal surface topologies.

3.3   Air circulation and pollutant dispersion in a town valley

As an illustration of the potential of the T-RANS approach for predicting atmospheric thermal convection and pollutant transport in a real, full scale situation over a mountainous conurbation, we considered diurnal variation of air movement and pollutant dispersion over a medium-sized town situated in a mountain valley, with distinct residential and industrial zones. These two zones are represented by different pollution emission ($C=50\%$ and 100% respectively) and different heat source intensities ($T_g=\pm 2$ and ± 1 respectively) from surrounding areas, over two diurnal cycles, Fig. 3. Simulated domain covers an area of $12 \times 10 \times 2.5$ km, which was represented by an averaged mesh size of 100 m in each direction. Two consecutive diurnal cycles were simulated (0-24h, day (I) and day (II)) with a time step of 2.5min. Wall functions were applied for the ground plane. At the top boundary, we prescribed constant temperature and assumed symmetry boundary condition for the velocity. The side boundaries were artificially extended and treated as symmetries for all variables. Two different situation with respect to imposed thermal stratification were analysed. The imposed vertical profile of potential temperature of dry air is uniform in lower level and with identical linear distribution ($\Delta T = 4$) in upper level. The base of the inversion layer (the switch from uniform to linear temperature) is located at $z/H=2/3$ (≈ 1600m from the valley deepest point) for the first case and at $z/H=1/3$ ($\approx 800m$) for the second case. The domain height (H) and the characteristic initial temperature gradients (or surface fluxes) give very high values of Rayleigh number, i.e. $\mathcal{O}(10^{17})$.

Since the experimental measurements were not available for quantitative validation of presented simulations, no direct verification is possible. Instead, we

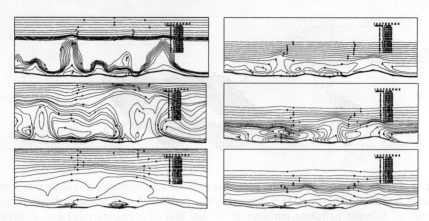

Fig. 4. Potential temperature distributions at characteristic location where both residential and industrial zone are present, for different time instants, $\tau=1$ p.m., 6 p.m. (day I), 10 a.m. (day II): left- weak stratification, right- strong stratification.

present some results that illustrate features of interest to environmental studies. The time evolution of the potential temperature distribution in a characteristic vertical plane that crosses both the residential and industrial zones, for two different stratifications, is shown in Fig. 4. In the initial stage of heating, strong thermal plumes appear, originating both from locally increased rate of heating (for industrial and residential zones) as well as from the hill orography. As time progresses, strong mixing in the lower layer is observed. At the beginning of the new cycle, due to nocturnal cooling, upper inversion layer is again moved towards the ground. As seen, stronger stratification causes significant damping of the thermal plume and transport activities, Fig. 5.

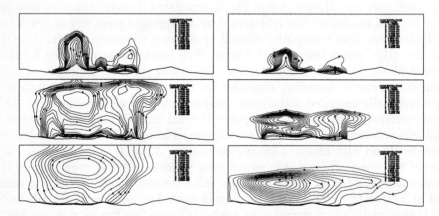

Fig. 5. Concentration distributions at characteristic location where both residential and industrial zone are present, for different time instants, $\tau=1$ p.m., 6 p.m. (day I), 10 a.m. (day II): left- weak stratification, right- strong stratification.

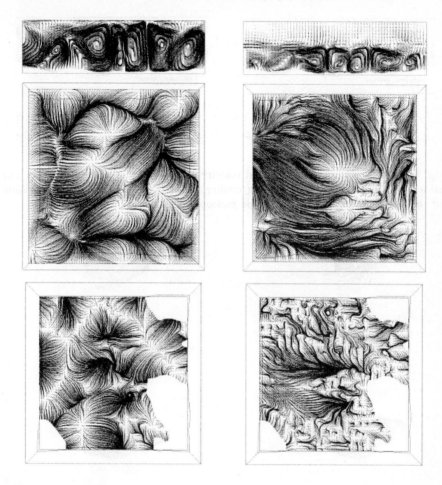

Fig. 6. Instantaneous trajectories in characteristic vertical plane (over residential and industrial zones) and two horizontal planes: first- at the base of inversion layer, second- 50m from ground; τ=6 p.m., day II, case with weak stratification: left- weak stratification, right- strong stratification.

Same conclusion can be drawn from Fig. 6 where instantaneous trajectories in one vertical (residential+industrial zones) and two horizontal (at inversion layer location and at 50m above the ground) planes. Since the capping inversion is closer to ground in the second case, vertical motion is significantly suppressed and the number of distinct thermal plumes is significantly larger in plane at 50m from ground than for the weak stratification conditions. Numerical simulation also revealed interesting phenomena of up/down-slope inertial motions few hours after offset of heating/cooling of ground, as shown in Fig. 7. Finally, the three-dimensional concentration fields, represented by a characteristic concentration isosurface, are presented in Fig. 8. As expected, stronger stratification

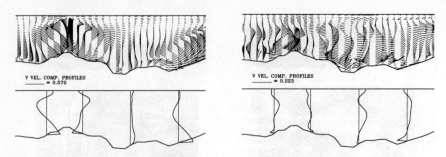

Fig. 7. Velocity vectors and horizontal velocity component profiles indicating an inertial motion, 2hrs after onset of heating/cooling, day (II), case with weak stratification: left- up-slope motion, right- down-slope motion.

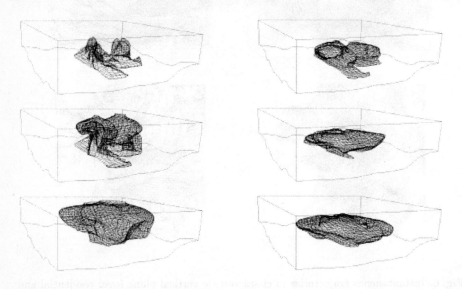

Fig. 8. Time evolution of concentration ($\langle C \rangle$=0.01), $\tau = 1$ p.m., 6 p.m., day (I), 10 p.m., day (II) respectively; left- weak stratification, right- strong stratification.

reduces convective transport resulting in an increase of pollutant concentrations over residential and industrial zones.

4   Conclusions

Numerical simulations of combined effects of terrain orography and thermal stratification on pollutant dispersion in a town valley were performed using the time-dependent Reynolds-averaged Navier-Stokes method (T-RANS). The approach can be regarded as very large eddy simulation, with a single-point closure playing the role of the 'subscale' model. In comparison with the conventional LES, the model of the unresolved motion (here a reduced algebraic $\langle \theta u_i, c u_i, k - \varepsilon - \theta^2 \rangle$ model) covers a much larger part of turbulence spectrum whereas the large deterministic structure was fully resolved. Validation of the proposed T-RANS approach was performed for situations where the effects of thermal stratification and terrain orography were separated, and for which a good experimental and numerical database exist. Unsteady turbulent penetrative convection of unstable mixed layer and convection over horizontal wavy walls demonstrated a very good predictive potential of the proposed approach. The full scale simulations of pollutant dispersion in a town valley with distinct residential and industrial zones under differently imposed thermal stratification, portrayed qualitatively very reasonable results and at the same time confirmed numerical efficiency and robustness of the proposed approach. We believe that the T-RANS can be used as a potentially powerful and efficient tool for prediction of local environments.

References

1. Deardorff, J. W. and Willis, G. E. and Lilly, D. K.: Laboratory investigation of non-steady penetrative convection. J. Fluid Mech., **35**, (1969) 7–31,
2. Hunt, J.C.R.: UltraCFD for Computing Very Complex Flows. ERCOFTAC Bulletin, No.45, (2000) 22–23
3. Hanjalić, K. and Kenjereš, S.: T-RANS simulations of deterministic eddy structure in flows driven by thermal buoyancy and Lorentz force. Flow Turbulence and Combustion, Vol.66, (2001) 427–451
4. Kenjereš, S. and Hanjalić, K.: Transient analysis of Rayleigh-Bénard convection with a RANS model. Int. J. Heat and Fluid Flow, **20** (1999) 329–340
5. Krettenauer, K. and Schumann, U.: Numerical simulation of turbulent convection over wavy terrain. J. Fluid Mech., **237** (1992) 261–299
6. Mahrt, L.: Stratified Atmospheric Boundary Layers and Breakdown of Models. Theoret. Comput. Fluid Dynamics, **11** (1998) 263–279
7. Nieuwstadt, F. T. M. and Mason, P. J. and Moeng, C. H. and Schumann, U.: Large-eddy simulation of the convective boundary layer: A comparison of four computer codes. Proceedings of the 8th Turbulent Shear Flow Symposium, Munich, Germany (1991) 1.4.1–1.4.6
8. Schmidt, H. and Schumann, U.: Coherent structure of the convective boundary layer derived from large-eddy simulations. J. Fluid Mech., **200** (1989) 511–562
9. Stull, R. B.: An Introduction to Boundary Layer Meteorology. Kluwer Academic Publishers (1988)
10. Willis, G. E. and Deardorff, J. W.: Laboratory Model of Unstable Planetary Boundary Layer. J. Atmos. Sci., **31** (1974) 1297–1307

The Differentiated Call Processing Based on the Simple Priority-Scheduling Algorithm in SIP6*

Chinchol Kim, Byounguk Choi, Keecheon Kim, Sunyoung Han

Department of Computer Science and Engineering, Konkuk University,
1, Hwayangdong, Kwangin-gu, Seoul, 143-701, Korea
{bredkim, buchoi, kckim, syhan}@konkuk.ac.kr

Abstract. We proposed and implemented a differentiated call processing mechanism in SIP6 (Session Initiation Protocol based on IPv6) implemented by this paper. In order to satisfy the service quality in VoIPv6, call quality and call setup quality should be provided in the Internet. End-to-end QoS for call quality and call setup quality in the VoIP Server like to gatekeeper/gateway, SIP Proxy Server, and SIP Redirect Server are essential. In order to support the end-to-end QoS, RSVP, and DiffServ proposed by IETF have been used. Nevertheless, call setup quality for call processing in VoIP Server has not been supported in the current VoIP systems. To solve this problem, we proposed a concept of differentiated call processing mechanism in SIP6 Server. For differentiating services, we use the Flow Label field of the IPv6 header and the predefined service levels. In this paper, we presented the design and implementation of SIP6. We explain the simple priority-scheduling algorithm that is applied to support QoS call processing. We also demonstrated the better performance by supporting the differentiated call processing service to satisfy the user's required QoS.

1 Introduction

H.323 and SIP protocol are used for call control in the VoIP technology [9][10]. H.323 has gained the reputation because of its stability and performance through the long time research. H.323 has extensible and flexible because it permits the backward compatibility, E.164 numbers, URLs, TAs, e-mail address, H.323 IDs, and mobile UIMs. But, it has more problem in terms of complexity than SIP [9]. SIP only handles function to establish and control the sessions and has a simple structure providing a good mobility and interaction with other protocols (HTTP, RTSP, and SMTP). As a result, the current Internet society has made the SIP as a standard in many fields such as 3GPP, VoIP, Messaging Systems and ALL-IP system (next generation mobile system). However, H.323 and SIP is do not support the QoS for call processing.

---

* This work is supported by the grant No. R01-2001-00349 from the Basic Research Program of the Korea Science and Engineering Foundation.

P.M.A. Sloot et al. (Eds.): ICCS 2002, LNCS 2329, pp. 276–285, 2002.

VoIP is needs a QoS support for call setup and voice quality. It also requires a security association among peers for authentication. Those requirements can resolved with the IPv6 techniques of Flow Label, AH, and ESP header.

Currently, in order to satisfy the VoIPv6 service quality, QoS for call and call setup quality must be supported in the Internet. In order to support these QoS, the followings are required.

- End-to-end QoS for call quality over the network
- QoS for call setup quality for call processing in VoIP Server during the session establishment.

In order to satisfy the end-to-end QoS, RSVP and DiffServ proposed by IETF have been used. RSVP supports resource reservation mechanism in a local environment and DiffServ supports the differentiated service in the Internet backbone. Nevertheless, In establishing a session, QoS features to meet the call setup quality in VoIP Server cannot be supported by using RSVP and DiffServ. To solve these problems, we propose a concept of differentiated call processing mechanism that uses a simple priority-scheduling algorithm in SIP6. A service level to satisfy the user's required QoS is categorized in to High Quality, Medium Quality, and Normal Quality. A service level is marked in the Flow Label field of the IPv6 header. In this paper, we explain the design and implementation of SIP6 supporting QoS call processing and also demonstrate the better performance resulted from supporting the differentiated call processing service to satisfy the user's required QoS.

This paper presents the system design in section 2. The implementation result is explained in section 3. And the performance analysis is shown in section 4. Finally, we describe the conclusion and future work in section 5.

2   System Design

The SIP6 (Session Initiation Protocol based on IPv6) proposed in this paper is a SIP protocol supporting a differentiated call processing mechanism based on the IPv6 protocol. This system consists of SIP6 Server and User Agent. In order to establish a session, two members exchange information through the SIP6 Server.

Fig. 1 shows the system components and the following explains each component.

- *SIP6 Daemon* contains Proxy Server, Redirect Server, Location Server, and Registrar Server. It refers to a configuration file. QoS Processor receives all IPv6 packets and applies a differentiated call processing for satisfying the user's required QoS.
- *Web Registrar Server* is responsible for registering the information of user location; it is associated with a user database through the web.
- *User Database* is responsible for adding, deleting, and updating the user information.

- **SIP6 UA** is a SIP6 User Agent. This is divided into two parts: SIP6 User Agent Client for requesting a session and SIP6 User Agent Server for accepting a session. When SIP6 UA Client requests a session and registers the information of user location, SIP6 UA Client marks up the service level for the user's required QoS through the QoS marker.

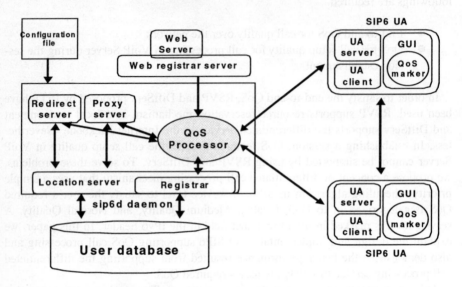

Fig. 1. SIP6 System Components

2.1 SIP6 Server Design

SIP6 Server provides the services for user registration, management, user location, call-forward, and call-redirect. It is consisted of QoS Processor and SIP6 Server Module.

2.1.1 QoS Processor

QoS Processor applies a differentiated call processing in the session establishment. It defines three service levels (High Quality, Medium Quality, Normal Quality). If QoS Processor receives a packet, Classify Processor classifies the packets according to the service level. Priority-scheduling Processor processes a packet by using a differentiated call processing mechanism. Message Parsing Processor and Method Processor parses the messages and call the SIP6 Server Module. Fig. 2 shows the schedule for processing messages in the QoS Processor.

The Classify Processor is an UDP receiver that receives the SIP6 messages. Those received messages are divided into the Request and Response message types. If the received message is a Request message, the differentiated call processing mechanism is applied. If the received message is a Response message, this message is sent to

Message Parsing Processor. In the case of Request message, the messages are classified and stored at the High, Medium or Normal Quality Buffer according to the service level specified in the Flow Label field of the IPv6 packet. The service level is defined as follows.

- High Quality has the highest level priority and requires a fast session establishment by using the High Quality Buffer.
- Medium Quality has the middle level priority and uses the Medium Quality Buffer.
- Normal Quality has the lowest level priority and uses the Normal Quality Buffer.

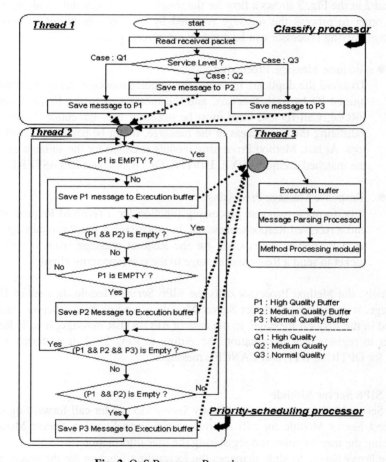

Fig. 2. QoS Processor Procedure

The Priority-scheduling Processor continually examines the messages stored in the multi buffer (High, Medium, Normal Quality Buffer) and applies the differentiated service according to the service levels. It can dynamically process the three buffers at a time when a message arrives. Priority-scheduling processor is based on a simple priority-scheduling algorithm. The simple priority-scheduling algorithm operates as follows.

- In the first level priority buffer, this processor directly calls the Message Parsing Processor to process a message.
- After processing the message, it looks for the buffer of next high priority.
- While processing a message in the lower priority buffer, it can be interrupted and handle a message with the higher priority.

By this mechanism, the differentiated call processing service can be supported. Thread 2 in the Fig. 2 shows a flow for the simple priority-scheduling algorithm.

In order to process the message received by Priority-scheduling Processor, the Message Parsing Processor can be divided into several parts.

- Request Message Processing
  To avoid the duplicate message, the Request message can be mapped into a unique ID by hash function, and the message with the Request-URI, TO, FROM, Call-ID and Cseq must be stored in the "Execution Buffer". After validating the uniqueness of the message, it can be parsed to detect the errors. At last, Method Processing module is called for the suitable service for the matched methods (BYE, INVITE, OPTIONS, and REGISTER).

- Response Message Processing
  It performs a response processing function for a received Request message and a received Response message. In processing a Response message, it sets up a Response Status Code for the Request message. And it generates a socket to send a Response message to the corresponding client

Finally, the Method Processor calls the SIP6 Server Module. In case of INVITE message, it calls Proxy/Redirect Server Module according to the Server Action included in the Request message. In the case of REGISTER message, it calls Registrar Server to register user information. According to RFC 2543, it performs a proper work for OPTION, BYE and CANCEL messages.

2.1.2 SIP6 Server Module

SIP6 Server Module is consisted of Proxy Server Module for call forwarding service, Redirect Server Module for call redirect service, and Location Server Module for servicing the user location and registering the user information.

The Proxy Server Module manages the forwarding function for the received calls. Depending on the received methods, it invites a session by forwarding INVITE method after getting user location information from Location Server Module. A transmitted message sets the service timeout during which Proxy Server processes the message. When the timeout occurs, it will cancel the request and send Response message with 487 Status Code. If it is an ACK, ACK method is sent to the destination client.

The Redirect Server Module processes the redirect function for the received calls. When it receives INVITE method, it initializes the connection header for the Response message. After setting the service timeout, it gets information of user location from Location Server. If it receives the address as a domain, this module will send URI (User ID, Host, Port number) to the client who has requested INVITE after mapping the IP address through a DNS resolving function of the system.

The Location Server Module is responsible for enrolling the user information and tracking the user location. It includes the Registrar to receive the user enrollment, User Input Table to manage the user ID authorized for accessing the system, User Identity Table to give the multiple IDs per user (So there is a unique user ID used in the session establishment in this table), and User Contact Table to manage the user's current location.

User information from the Registrar can be transformed to the unique user ID that is in the User Identity Table and registered into the User Contact Table. The registered data are used to serve the request of user location from the Proxy and Redirect Server. The user entry information consists of User ID, SIP6 UA port number, URL, Server Action, and Service Level.

2.2 SIP6 UA Design

SIP6 UA uses the IPv6 UDP socket interface. This module has GUI for the user interface, UA Client Module and UA Server Module. GUI includes QoS Marker for setting the service levels and interfaces for user registration, session invitation. SIP6 UA Server and Client Module perform the function to send and receive calls in order to establish sessions.

2.3 Call Processing Procedure and Message

The SIPv6 suggested in this paper follows the call processing procedure and message formats specified in IETF RFC 2543.

3   Implementation

3.1   SIP6 Implementation

SIP6 Server is implemented with C language on Linux Kernel 2.2.x supporting the IPv6. The programs are designed to compose the Proxy, Redirect, Location, and Registrar Server. It is executed by daemon process named SIP6d.

When we start a SIP6 Server, it creates a thread that can open an IPv6 UDP socket and it receives an INVITE Request message. When SIP6d receives a message, the message is classified by checking the service level specified in the Flow Label field of IPv6 header. SIP6d also creates another thread to process the classified Request message sequentially according to the service level proposed by this paper. Finally, According to the method type and Server Action, the thread selectively calls Proxy, Redirect, or Registrar Server Module.

3.2 QoS Processor Implementation

We use the Flow Label field of the IPv6 header to specify the service levels in applying the differentiated call processing mechanism. Currently, the Flow Label field is composed of 24-bit to define the QoS. But it has not been used yet. In the Fig. 3, we defined the code values for service levels.

```
/* qos.h  Header File */

...

/* define the High Quality Code Value */

#define    HighQuqlity              1

/* define the Medium Quality Code Value */

#define    MediumQuality            2

/* define the Normal Quality Code Value */

#define    NormalQuality            0

...
```

Fig. 3. Code value for service levels defined in qos.h header file

The IPv6 socket structure (struct sockaddr_in6) includes the variable for the Flow Label field in the IPv6 header. In the implementation, the sin6_flowinfo variable in sockaddr_in6 structure is used to specify the code value for the service level.

```
struct  sockaddr_in6 {
        uint8_t              sin6_len;
        sa_family_t          sin6_family;
        in_port_t            sin6_port;
        uint32_t             sin6_flowinfo;
        struct               in6_addr      sin6_addr;
```

When a session is established, SIP6 UA specifies a service level in sin6_flowinfo field of socket_in6 structure. If SIP6 Server receives an INVITE Request message, it checks a service level in sin6_flowinfo field and dynamically processes the INVITE message according to the service level specified by the user. It is based on the simple priority-scheduling algorithm proposed by this paper.

4 Performance Analysis

In order to analyze the performance of differentiated call processing in the SIP6 Server, we implemented a thread test program. We used a Linux system for the SIP6 Server and a windows system for the test program on the IPv6 Testbed in Konkuk University.

We analyzed and compared the performance for call processing time in SIP6 and SIP from Columbia University. Test scenario is as following.

- The test program concurrently sends the INVITE and REGISTER message increasingly of the same number step by step according to the service level to SIP6 Server. In the case of the SIP Server of Columbia University, differentiated service scheme is not applied.
- The test program measures the average of message processing time in establishing a session in the SIP6 and SIP Server.

Fig. 4 presents the test results of average time to process REGISTER message in the SIP6 Registrar Server and SIP Registrar Server. Fig. 5 and Fig. 6 present the test results of average time for processing INVITE messages in the SIP6 and SIP Proxy Server and Redirect Server.

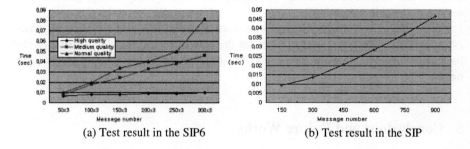

(a) Test result in the SIP6   (b) Test result in the SIP

Fig. 4. The average time for processing REGISTER messages in Registrar Server

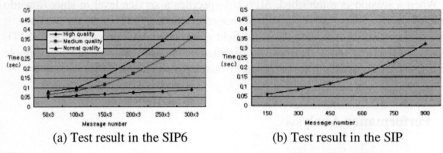

(a) Test result in the SIP6   (b) Test result in the SIP

Fig. 5. The average time for processing INVITE messages in Redirect Server

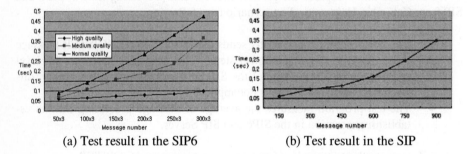

(a) Test result in the SIP6   (b) Test result in the SIP

Fig. 6. The average time for processing INVITE messages in Proxy Server

As shown in these results, SIP6 proposed by this paper supports a differentiated call processing service to satisfy the user's required QoS. As INVITE and REGISTER messages increases, the average time in processing messages in the SIP6 Server displays a remarkable difference according to the service levels. Particularly, INVITE and REGISTER message specifying the High Quality always gets the fast service in the call processing.

The results of performance analysis demonstrates that the differentiated call processing service gives better performance for the service requiring a fast session establishment when the network traffic and the requests for session establishment are heavy.

5   Conclusions and Future Works

In this paper, we proposed a differentiated call processing mechanism in the SIP6 based on IPv6. Because we implemented an IPv6 in this system, it can use the QoS and the security features of the IPv6. We applied the QoS based on the differentiated service using the simple priority-scheduling algorithm for call processing. We presented that the performance improved.

In the future, we are going to add an authentication service by using AH (Authentication Header), ESP (Encapsulation Security Payload) header of IPv6.

References

1. Handley, H.Schulzrine, E.Schooler, and J.Rosenberg. : SIP: session initiation protocol, RFC 2543 (1999)
2. A.Johnston, S.Donovan, et al. : SIP Telephony Call Flow Examples, Internet Draft (2000)
3. R.Gilligan, S.Thomson, J.Bound, W.Stevens. : Basic Socket Interface Extensions for IPv6, RFC 2553 (1999)
4. R. Hinden, S. Deering. : Internet Protocol version 6(IPv6) specification, RFC 2460 (1998)
5. K. Nichols, S. Blake, F. Baker, D. Black. : Definition of the Differentiated Services Field (DS Field) in the IPv4 and IPv6 Headers, RFC 2474 (1998)
6. P. Ferquson and G. Huston. : Quality of Service, John Wiley & Sons (1998)
7. S. Donovan, J. Rosenberg. : The SIP Session Timer, Internet Draft (2001)
8. ITU-T Recommendation H.323 (1996)
9. I. Dalgic and H. Fang. : Comparison of H.323 and SIP for IP Telephony Signaling, Photonics East, Proceeding of SPIE'99, Boston, Massachusetts (1999)
10. Henning Schulzrinne and Jonathan Rosenberg. : Signaling for Internet Telephony (1998)
11. S.Blake, M.Carlson, E.Davies, Z.wang, and W.Weiss. : An architecture for differentiated service, Request for Comments (Proposed Standard) 2475, Internet Engineering Task Force (1998)
12. W.Almesberger, T.Ferrari. and J.-Y.Le Boudec. : SRP: a scalable resource reservation protocol for the internet, Tech. Rep. SSC/1998/009, EPFL, Lausanne, Switzerland (1998)
13. P.P.Pan and H.Schulzrinne. : YESSIR: A simple reservation mechanism for the Internet, in Proc. International Workshop on Network and Operating System Support for Digital Audio and Video (NOSSDAV), (Cambridge, England) (1998)
14. Y. Bernet, R. Yavatkar, P.Ford, F.Baker, L. Zhang, M. Speer, and R. Braden. : Interoperation of RSVP/Intserv and diffserv networks, Internet Draft (1999)

A Fuzzy Approach
for the Network Congestion Problem

Giuseppe Di Fatta, Giuseppe Lo Re, Alfonso Urso

CERE, Centro di studio sulle Reti di Elaboratori, C.N.R.,
viale delle Scienze, 90128 Palermo, Italy
{ difatta, lore, urso }@cere.pa.cnr.it

Abstract. In the recent years, the unpredictable growth of the Internet has moreover pointed out the congestion problem, one of the problems that historically have affected the network. This paper deals with the design and the evaluation of a congestion control algorithm which adopts a Fuzzy Controller. The analogy between Proportional Integral (PI) regulators and Fuzzy controllers is discussed and a method to determine the scaling factors of the Fuzzy controller is presented. It is shown that the Fuzzy controller outperforms the PI under traffic conditions which are different from those related to the operating point considered in the design.

Keywords: *AQM, Congestion Control, Fuzzy Control.*

1 Introduction

In the recent years, the unpredictable growth of the Internet has moreover pointed out the congestion problem, one of the problems that historically have affected the network. The network congestion phenomenon is induced when the amount of data injected in the network is larger than the amount of the data that can be delivered to destination. Two different approaches, that can be considered complementary parts of a single main strategy, can be adopted to solve the above problem. The approach which historically has represented the beginning of network congestion control is a so called end-to-end approach; in this approach when data sources infer congestion occurrences from packet losses, they properly reduce their transmit rate. This is, for instance, the approach adopted by the Transmission Control Protocol. The effectiveness of a control system where sources are responsible for congestion control is based essentially on the fact that all, or at least most of the agents running through the network respect its rules. In this environment non compliant flows can obtain larger bandwidth against the ones which correctly obey the control laws. The end-to-end peers detect congestion level by inferring it from packet losses. A packet loss could mean that one of the intermediate routers does not have enough memory space to host it before its retrasmission on the appropriate link towards the destination. The simplest and also most deployed policy adopted by a router to

P.M.A. Sloot et al. (Eds.): ICCS 2002, LNCS 2329, pp. 286–295, 2002.

manage its queues, is a First Come First Served policy which is implemented by means of a First In First Out queue management. Such a policy, known as Drop Tail for router queue management, presents several disadvantages such as, the higher delays suffered by packets when they go through longer queues. To eliminate the Drop Tail disadvantages and to anticipate the source answers to incipient congestion situations, authors in [1] proposed the adoption of a Random Early Detection (RED) policy. RED is an active policy of queue management which involves the dropping, or marking, of packets when the queue average length ranges between a minimum and a maximum threshold. The probability of packet dropping/marking is obtained from the average queue length accordingly to a linear law. In the last years, the active queue management policies have been object of a large interest in the scientific networking community and several proposals have been presented to find more effective control policies than RED. Among these we will refer to REM (Random Exponential Marking) [6] e PI (Proportional Integral) [9]. Nevertheless, these AQM policies suffer the disadvantage that they are unable to maintain their performances as the number of TCP flows increases. In order to avoid such a problem, it is possible to use a congestion control algorithm based upon a fuzzy logic controller. In fact, in the cases of controlling high-order nonlinear systems, fuzzy logic controllers often produce results better than those obtained by using classical control techniques. Accordingly, many research efforts have been carried out in the development of fuzzy logic controllers [10],[11],[12],[13]. On the other hand, the design of a fuzzy logic controller is not straightforward, because of the heuristic involved with control rules and membership functions; moreover, the tuning of the parameters of a fuzzy logic controller, as scaling factors, membership functions and control rules is a very complex task. Currently there are not many simple methods available for the design of the fuzzy knowledge base and for the tuning of a fuzzy logic controller. Therefore, the designers have to devise a fuzzy knowledge base by heuristic methods, employing experience and, accordingly, the parameters of a fuzzy control system are tuned repeatedly by a trial and error method. This leads to a well-known fact that the design of a fuzzy logic controller is more difficult than the design of a conventional controller. In this paper, in order to obtain a method for the tuning of a fuzzy controller, the analogy between Proportional Integral (PI)regulators and fuzzy controllers is discussed and a method to determine the scaling factors of the fuzzy controller is presented. The remainder of the paper is structured in the following way. In section 2 we will examine more deeply RED, REM and PI. Section 3 deals with the design of a Fuzzy controller and how its project is related to the PI one. In section 4 we present the set of experiments we carried out and discuss the results. Section 5 is devoted to conclusions and to considerations for possible future investigation researches.

2   Active Queue Management Policies

The name AQM (Active Queue Management) indicates those policies of router queue management that allow: a) the detection of network congestion, b) the notification of such occurrences to the hosts on the network borders, c) the adoption of a suitable control policy. With reference to the congestion notification, two different approaches can be followed. The first one involves the setting of a bit called ECN (Early Congestion Notification) in a sample of the packets flowing through the router. In turn, the destination will transmit such information to the source piggybacking it into the acknowledgement message. The second approach, used for those protocols that are not able to manage the ECN bit, involves a more drastic action, i.e. a probabilistic dropping of packets with the aim to induce a reaction of the flow sources. The current Transmission Control Protocol of the Internet is not able to manage the ECN bit and as a consequence only the second solution may be adopted by the Internet routers. The first AQM policy proposed for the Internet has been RED that, as previously mentioned, involves the dropping of packets accordingly to a probability law which ranges linearly when the average queue length is varying between a minimum and a maximum threshold. RED calculates the average queue length by assigning different weights to old value and current measure. This means the adoption of a low pass filter to reduce the high frequency variation of the instantaneous queue. This behavior is a precise design choice of the authors to overcome the oscillations which may be induced by isolated bursts. In the last years several objections have been raised against RED, among which the difficulty of setting proper RED parameters according to network conditions, and the dependence of the queue length in steady state from the number of flows.

Namely, a growth of the flow number involves that of the average queue length, which, in turn, could exceed the maximum threshold, and all packets would be dropped. However, it is not possible to increase too much the maximum threshold, because this would mean higher queuing delays. On the other hand, if the maximum threshold is set to a low value, this would mean a bad usage of the link because of severe buffer oscillations. From these considerations follows that it is very difficult to find out the right trade-off, and it is not possible to tune RED to achieve both high link utilization and low delay and packet losses.

Several proposals have been presented in the recent years which introduce RED improvements [2] [3] [4]. Recently different approaches than RED have been proposed; among these two proposals REM and PI which, although obtained independently and following full different theoretical approaches, seem to represent, accordingly to their authors, the same solution [9]. REM is the result of a linear modelling of the problem and of its resolution in its dual form. Differently from RED, the REM solution differentiates between the congestion measure of each router and the dropping probability. In the REM model the authors introduce a measure called price that eliminates the dependence of the dropping probability from the current value of the queue size. The derived algorithm uses the current queue size and the difference from a desired value to calculate the dropping probability accordingly to an exponential law. Such a feature owns the additivity property, so thus a source can calculate the price of the whole path using the knowledge of the total number of packets dropped on the path.

The PI controller for AQM uses classical control system techniques to design well suited control law for the router queue management. In particular, a non-linear dynamic model for TCP/AQM has been developed in [7]. Once the model is linearized around an operating point, a stable PI linear controller is designed in order to satisfy the project specifications. The authors in [9] show that in the PI controller the Proportional part is equivalent to RED when the input low-pass filter is removed. The usage of a proportional controller leads to a lower time of response but also to lower stability margins; moreover, the proportional controller has a steady state regulation error, where such an error is defined as the difference between the steady state output queue and the reference value. In order to overcome the above disadvantages, the integral term is added which has the characteristic to give steady state error equal to zero and to give higher stability margins.

3 Fuzzy Controller

The AQM policies described in the above section suffer the disadvantage that they are unable to maintain performance, in terms of speed of response, as the number of TCP flows increases. From a control point of view, with reference to the PI controller, this disadvantage is essentially due to the fact that the high frequency gain of the open loop transfer function is fixed because the controller design is carried out considering a particular value of the TCP flows. Namely, when the load increases the high frequency gain decreases and the system bandwidth becomes lower which implies a slower system in terms of rise and settling time. To overcome this disadvantage we use a fuzzy logic controller. In recent years, fuzzy logic controllers, especially Fuzzy Proportional-Integral (FPI) controllers have been widely used for processes control owing to their heuristic nature associated with simplicity and effectiveness for both linear and nonlinear systems [16], [17]. In fact, for single-input single-output systems, fuzzy logic controllers can be essentially seen as PI type associated with nonlinear gain. Because of the nonlinear property of control gain, FPI controllers can achieve better system performance than the conventional PI controllers. On the other hand, due to the existence of non linearity, it is usually difficult to conduct theoretical analyses to explain why FPI controllers can achieve better performance. Consequently, it is useful to explore the nonlinear control properties of FPI controllers to improve the closed-loop performance. Moreover, a method which allows to obtain the parameters of the fuzzy controllers is needed.Systematic methods for the determination of FPI scale factors have been developed in [14] [15] [18], which take advantage from the analogy of FPI and conventional PI controllers . More precisely, a set of relationships between the scale factors and the gains of PI controller are obtained from such an analogy. Then, the gains of PI controller are determined so as to satisfy, for instance, requirements on bandwidth of the control loop. Finally, scale factors are computed from the gains of PI controller. As already said, fuzzy controllers taken into account in this paper are the PI-type fuzzy controllers, as depicted in figure 1, where all quantities are considered at the generic discrete instant kT_s, with T_s the sampling period, $e = q - q_r$ is

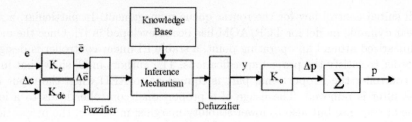

Fig. 1. *Structure of FPI controller*

the error on the controlled variable q (queue size), $\Delta e = e\,(kT_s) - e\,((k-1)\,T_s)$ is the variation of the error, Δp is the increment of the control variable p (probability of packet marking/dropping) and K_e, K_{de}, and K_o are scaling factors to be determined. It should be noticed that these fuzzy controllers have an intrinsic PI action. As a consequence, the steady-state behavior in the operating point cannot be different in the two controllers. Once, the relationships between the parameters of the fuzzy controller and those of the standard PI one have to be determined; then a method of synthesis of the standard controller has to be chosen. In this paper, with the aim to synthesize the standard PI controller, we adopt the dynamic model of TCP behaviour using fluid-flow and stochastic differential equation analysis developed by Hollot et al.[7]. Furthermore, the synthesis of the PI controller is carried out following the guidelines to design stable controllers given in [9].

Finally, we can derive the relationships between the gains of the PI controller and the scale factors of the fuzzy one. Let us suppose that a set of rules and membership function on normalized axes have been assigned, the fuzzy controller generates a non linear surface of the form:

$$y = f\,(\bar{e}, \Delta\bar{e})\,, \tag{1}$$

whereas the relation $p - y$ is that of a discretized integrator of gain K_o. The corresponding standard PI regulator can be seen as a plane whose equation, using the trapezoidal integration method and with reference to e, Δe and Δp axes, is:

$$\Delta p = K_I T_s e + \left(K_P + \frac{K_I T_s}{2}\right) \Delta e. \tag{2}$$

Moreover, it is possible to approximate the surface (1) with the plane:

$$y = K'_e \bar{e} + K'_{de} \Delta\bar{e}, \tag{3}$$

where K'_e and K'_{de} are obtained by means of minimization of the following index:

$$J = \int \int_{-\epsilon}^{\epsilon} \left(K'_e \bar{e} + K'_{de} \Delta\bar{e} - f\,(\bar{e}, \Delta\bar{e})\right)^2 d\bar{e}\,d\Delta\bar{e}, \tag{4}$$

where the parameter ϵ defines a well suited interval around the operating point.

A least square solution for the minimization of the above index it is possible by discretization of the integral in (4). Finally, the output Δp of the PI controller (2) is made equal to the output of linear approximation (3) scaled by K_o. Therefore, the final analogy relationships can be given as follows:

$$\begin{cases} K_P = K_o K'_{de} K_{de} - \dfrac{K_I T_s}{2} \\ K_I = \dfrac{K_o K'_e K_e}{T_s} \end{cases} \tag{5}$$

Given K_P and K_I from the sinthesys of the PI controller, the equations (5) provide a degree of freedom to determine the parameters K_e, K_{de} and K_o of the fuzzy controller. We use this degree of freedom to set the parameter K_o in order to fix the maximum allowed variation of the control variable Δp.

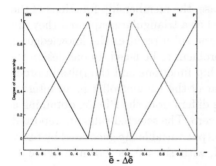 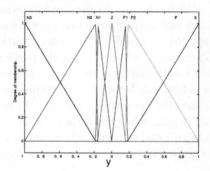

Fig. 2. *Fuzzy Membership Functions of Input and Output Variables.*

de\e	MN	N	Z	P	MP
MN	N3	N3	N3	Z	P1
N	N3	N2	N1	Z	P2
Z	N3	N1	Z	P1	P3
P	N2	Z	P1	P2	P3
MP	N1	Z	P3	P3	P3

Table 1. Fuzzy Rules

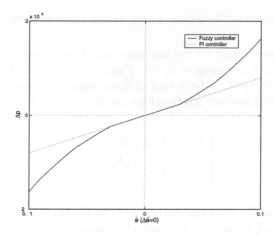

Fig. 3. *The comparison between Fuzzy and PI output around* $\bar{e} = 0$ *and with* $\Delta\bar{e} = 0$.

With reference to the fuzzy knowledge base, the membership functions of the input and output fuzzy variables are adopted as a triangular shape and they are depicted in figure 2. The determination of the rules requires the knowledge or experience of experts about the particular problem to be faced. In our design we select both the peak values of the membership functions and the rules in order to obtain a behaviour which reproduces that of the PI controller in an interval around the operating point ($\bar{e} = 0$), whereas differs from the PI one, obtaining a higher gain, when the error is different from zero. The selected rules are reported in table 1 and the comparison between the output variable Δp generated by fuzzy controller and that one generated by the PI, with $\Delta\bar{e} = 0$, is shown in figure 3.

4   Experiments and Performance Evaluation

We have implemented the fuzzy controller as an active queue manager under the well known $ns2$ simulator [19]. In this environment we also implemented the PI controller according to the pseudo-code reported in [9].

The network topology adopted in the experiments is reported in the figure 4. The network load is generated by FTP and HTTP sources. All the flows are conveyed in the bottleneck link with 15 Mbps bandwidth capacity between the AQM router R_0 and the router R_1. The Router R_2, R_3, R_4, R_5 are introduced to measure the different FTP and HTTP input and output traffic in the bottleneck link. HTTP flows are short lived flows with a bursty behaviour and can not be easily controlled by the congestion control mechanism. We adopted them as noise traffic. FTP flows, with their intensive data transfers, represent the traffic load to be controlled in the fluid flow dynamic model.

The propagation delay of the paths between sources and destinations is uniformily distributed in the range 160-240 msec. The maximum queue length in

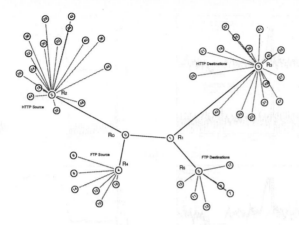

Fig. 4. *The experimental topology*

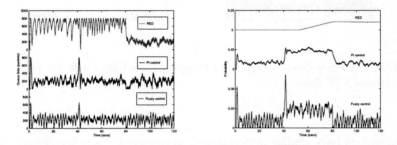

Fig. 5. *RED, PI and Fuzzy Queue Lengths and Dropping Probabilities with 60 additional FTP flows.*

the AQM router R_0 is 800 packets. The parameters of the PI and Fuzzy controllers are determined for operating conditions where the number of flows is 60 and the desidered value of the queue length is 200 packets.

In all the experiments we adopted 60 FTP flows and 180 HTTP flows which last the whole simulation time (from 0 to 120 seconds) and additional FTP flows which transmit in the time interval between 40 and 80 seconds. We carried out three different experiments, where the additional number of FTP flows are respectively 60 (figure 5), 240 (figure 6.a), and 480 (figure 6.b).

Charts in the left side of figures show the queue lengths obtained by AQM controllers, whereas charts in the right side show the corresponding values of the dropping probability. Figures 5 confirms that PI outperforms RED, as reported in [9] and shows also the same behaviour of the Fuzzy and PI controllers. The comparison between Fuzzy and PI will be the subject of the remainder of this section. Figures 5 and 6 show that in steady traffic conditions both the PI and Fuzzy controllers are able to mantain the queue length at the reference value (200 packets). In operating conditions near to the design ones (figure 5) both

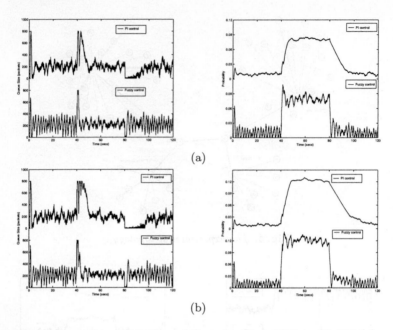

Fig. 6. *PI and Fuzzy Queue Lengths and Dropping Probabilities with 240 (a), and 480 (b) additional FTP flows.*

the controllers have similar performances, as expected. The effectiveness of the Fuzzy controller becomes evident when a greater increment of the load brings the operating point far away from the one considered in the design (figures 6.a and 6.b). Specifically, in such cases the Fuzzy controller shows a lower time of response than the PI one. Furthermore, when the additional load stops at 80 seconds, the PI controller brings the queue length to zero for a longer time than the Fuzzy one, thus producing a longer underutilization of the link. This phenomenon is confirmed by the faster response of the Fuzzy control variable (the dropping probability) shown in the right charts. Such a better behaviour is due to the characteristic of the Fuzzy controller to change its closed loop gain under different operating conditions.

5   Conclusions

In this paper we developed a Fuzzy controller for active queue management. The design has been carried out in analogy with a PI controller and the experiments showed that it outperforms the PI one in situations of functioning different than those related to the operating point of the design. Such a better behaviour is due to the characteristic of the Fuzzy controller to change its closed loop gain under different operating conditions. Further experiments are required to better exploit and show such a characteristic.

References

1. S. Floyd and V. Jacobson, "Random Early Detection Gateway for Congestion Avoidance", IEEE/ACM Transactions on Networking, Vol. 1, N. 4, Aug. 1993, pp. 397 - 413.
2. W. Feng, D. Kandlur, D. Saha, K. Shin, A Self-Configuring RED Gateway, INFO-COM '99, March 1999
3. T. J. Ott, T. V. Lakshman, and L. H. Wong, SRED: Stabilized RED, Proceedings IEEE INFOCOM '99, New York, March 1999.
4. David D. Clark and Wenjia Fang, Explicit Allocation of Best Effort Packet Delivery Service, ACM Transactions on Networking, August,1998.
5. S.H. Low and D. E. Lapsley, "Optimization Flow Control, I: Basic Algorithm and Convergence", IEEE/ACM Transactions on Networking, Vol. 7, N. 6, Dec. 1999, pp. 861 - 874.
6. S. Athuraliya, V. Li, S.H. Low and K. Yin, "REM: Active Queue Management", IEEE Network, Vol. 15, N. 3, May/June 2001 pp. 48 - 53.
7. C. Hollot, V. Misra, D. Towsley, W. Gong, "Fluid-based Analysis of a Network of AQM Routers Supporting TCP Flows with an Application to RED", Proc. of ACM SIGCOMM 2000 Aug. 2000, pp. 151 - 160.
8. C. Hollot, V. Misra, D. Towsley, W. Gong, "A Control Theoretic Analysis of RED", Proc. of IEEE INFOCOM Apr. 2001.
9. C. Hollot, V. Misra, D. Towsley, W. Gong, "On Designing Improved Controllers for AQM Routers Supporting TCP Flows", Proc. of IEEE INFOCOM Apr. 2001.
10. E. H. Mamdani, "Application of Fuzzy Algorithms for Control of Simple Dynamic Plant", Proc. Inst. Elect. Eng., vol. 121, no. 12, pp. 1585-1588, 1974.
11. D. A. Rutherford, G. C. Bloore, "The Implementation of Fuzzy Algorithms for Control", Proc. IEEE, vol 34, no. 4, pp. 572-573, 1976.
12. C. C. Lee, "Fuzzy Logic in Control Systems: Fuzzy Logic Controller -Part I," IEEE Trans. Syst., Man, Cybern., vol. 20, pp. 404-435, 1990.
13. M. Maeda, S. Murakami, "A Self-tuning Fuzzy Controller", Fuzzy Sets Syst., vol. 51, pp. 29-40, 1992.
14. Jian-Xin Xu, Chang-Chieh Hang, Chen Liu, "Parallel Structure and Tuning of a Fuzzy PID Controller", Automatica 36 (2000) pp. 673-684.
15. Misir D., Malki H. A., Chen G., "Design and Analysis of a Fuzzy Proprortional-Integral-Derivative Controller", Fuzzy Sets and Systems 79 (1996) pp. 297-314.
16. Li W., "Design of a Hybrid Fuzzy Logic Proportional plus Conventional Integral-Derivative Controller", IEEE Trans. on Fuzzy Systems 6 (4) (1998) pp. 449 - 463.
17. Malki H. A., Li H. D., Chen G. R., "New Design and Stability Analysis of Fuzzy Proprortional-Derivative Control System", IEEE Trans. on Fuzzy Systems 2 (4) (1994) pp. 245 - 254.
18. Alonge F., D'Ippolito F., Raimondi F. M., Urso A., "Method for Designing PI-Type Fuzzy Controllers for Induction Motor Drives", IEE Proceedings on Control Theory and Applications. vol. 148, n. 1, January 2001, p. 61-69
19. ns-2, network simulator (ver. 2), http://www.isi.edu/nsnam/ns/

Performance Evaluation of Fast Ethernet, Giganet and Myrinet on a Cluster

Marcelo Lobosco, Vítor Santos Costa, and Claudio L. de Amorim

Programa de Engenharia de Sistemas e Computação, COPPE, UFRJ
Centro de Tecnologia, Cidade Universitária, Rio de Janeiro, Brazil
{lobosco, vitor, amorim}@cos.ufrj.br

Abstract. This paper evaluates the performance of three popular technologies used to interconnect machines on clusters: Fast Ethernet, Myrinet and Giganet. To achieve this purpose, we used the NAS Parallel Benchmarks. Surprisingly, for the LU application, the performance of Fast Ethernet was better than Myrinet. We also evaluate the performance gains provided by VIA, a user lever communication protocol, when compared with TCP/IP, a traditional, stacked-based communication protocol. The impacts caused by the use of Remote DMA Write are also evaluated. The results show that Fast Ethernet, when combined with a high performance communication protocol, such as VIA, has a good cost-benefit ratio, and can be a good choice to connect machines on a small cluster environment where bandwidth is not crucial for applications.

1 Introduction

In the last few years, we have seen a continuous improvement in the performance of networks, both in reduced latency and in increased bandwidth. These improvements have motivated interest in the development of applications that can take advantage of parallelism in clusters of standard workstations.

Fast Ethernet, Giganet [1] and Myrinet [2] are three popular interconnection technologies used to build cluster systems. Fast Ethernet is a cheap LAN technology that can deliver 100Mbps bandwidth, while maintaining the original Ethernet's transmission protocol, CSMA/CD. TCP/IP is the most popular communication protocol for Fast Ethernet, although other protocols can be used, such as VIA [3]. TCP/IP is a robust protocol set developed to connect a number of different networks designed by different vendors into a network of networks. However, the reliability provided by TCP/IP has a price in communication overhead. To ensure reliable data transfer, protocol stack implementations like TCP/IP usually require data to be copied several times among layers and that communicating nodes exchange numerous protocol-related messages during the course of data transmission and reception. The number of protocol layers that are traversed, data copies, context switches and timers directly contributes to the software overhead. Also, the multiple copies of the data that must be maintained by the sender node and intermediate nodes until receipt of the data-packet is acknowledged contributes to reduce memory resources and further slows down transmission.

P.M.A. Sloot et al. (Eds.): ICCS 2002, LNCS 2329, pp. 296–305, 2002.

Giganet and Myrinet are more expensive technologies that provide low latency, high bandwidth, end-to-end communication between nodes in a cluster. Giganet provides both a switch and a host interface. The switch is based on a proprietary implementation of ATM. Giganet's host interface is based on a hardware implementation of VIA. The Virtual Interface Architecture (VIA) is a user-level memory-mapped communication architecture that aims at achieving low latency and high bandwidth communication within clusters of servers and workstations. The main idea is to remove the kernel from the critical path of communication. The operating system is called just to control and setup the communication. Data is transferred directly to the network by the sending process and is read directly from the network by the receiving process. Even though VIA allows applications to bypass the operating system for message passing, VIA works with the operating system to protect memory so that applications use only the memory allocated to them. VIA supports two types of data transfers: Send-Receive, that is similar to the traditional message-passing model, and Remote Direct Memory Access (RDMA), where the source and destination buffers are specified by the sender, and no receiver is required. Two RDMA operations, RDMA Write and RDMA Read, respectively write and read remote data.

Myrinet is the most popular high-speed interconnect used to build clusters. Myrinet also provides both a switch and a host interface. Myrinet packets may be of any length, and thus can encapsulate other types of packets, including IP packets, without an adaptation layer. Each packet is identified by type, so that Myrinet, like Fast-Ethernet, can carry packets of many types or protocols concurrently. Thus, Myrinet supports several software interfaces. The GM communication system is the most popular communication protocol for Myrinet. It provides reliable, ordered delivery between communication endpoints, called ports. This model is connectionless in that there is no need for client software to establish a connection with a remote port in order to communicate with it. GM also provides memory protected network access. Message order is preserved only for messages of the same priority, from the same sending port, and directed to the same receiving port. Messages with differing priority never block each other.

This paper studies the impacts of these three popular cluster interconnection technologies on application performance, since previous works pointed out that interconnection technology directly impacts the performance of parallel applications [4]. To achieve this purpose, we used the NAS Parallel Benchmark (NPB) [5] to measure the performance of a cluster when using each of the interconnection networks presented previously. The main contributions of our work are a) the comparative study of three popular interconnections technologies for clusters of workstations; b) the evaluation of the performance gains provided by VIA, a user lever communication protocol, when compared with TCP/IP, a traditional, stacked-based communication protocol; c) the evaluation of the impacts caused by the use of Remote DMA Write on VIA and d) an explanation for the poor performance of the LU benchmark on Myrinet. The results show that Fast Ethernet, when combined with a high performance communication protocol, such as VIA, has a good cost-benefit ratio, and can be a good choice when connecting machines on a small cluster environment where bandwidth is not crucial for applications. This paper is organized as follows. Section 2 presents the applications used in the study and their results. Section 3 concludes the work.

2   Performance Evaluation

Our experiments were performed on a cluster of 16 SMP PCs. Each PC contains two 650 MHZ Pentium III processors. For the results presented in this paper, we used just one processor on each node. Each processor has a 256 Kb L2 cache and each node has 512 Mb of main memory. All nodes run Linux 2.2.14-5.0. Table 1 shows a summary of the interconnect specifications used in the performance evaluation. To run VIA on Fast Ethernet, we use the NESRC's M-VIA version 1.0 [6], a software implementation of VIA for Linux.

Table 1. Summary of interconnect specifications

	Fast Ethernet	Giganet	Myrinet
Switch	Micronet SP624C	cLAN 5300	M2L-SW16
Network Card	Intel EtherExpress Pro 10/100	cLAN NIC	M2L-PCI64B
Link Speed	100Mbps	1.25Gbps	1.28Gbps
Topology	Single Switch	Thin Tree	Full-Crossbar
Protocol	TCP/IP and VIA	VIA	GM
Middleware	MPICH1.2.0 and MVICH1-a5	MVICH1-a5	MPICH1.2..8

Figure 1 shows the latency and bandwidth of TCP/IP and M-VIA on the Intel eepro100 card, VIA on Giganet and GM on Myrinet. The figures show that M-VIA's latency is 70% of the TCP/IP's. Giganet's and Myrinet's latency is an order of magnitude smaller. Giganet's latency is smaller until 28 bytes; after this Myrinet's latency is smaller. The bandwidth to send 31487 bytes is 10.5 MB/s on Fast Ethernet with TCP/IP, 11.2 MB/s on Fast Ethernet with M-VIA, 98.28 MB/s on Giganet and 108.44 MB/s on Myrinet.

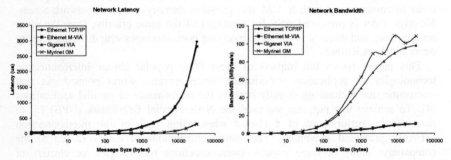

Fig. 1. Network Latency and Bandwidth

The NAS Parallel Benchmarks (NPB) are widely used to evaluate the performance of parallel machines [5]. The set consists of five kernels – EP, FT, MG, CG and IS – and three applications – BT, LU and SP – derived from computational fluid dynamics (CFD) codes. Each benchmark has five classes: S, W, A, B and C. Problems size grows from class S to class C. Our experiments used class B. We did not run FT and BT in our experiments. The sequential version of BT did not run. We could not compile FT in our experiments because it requires a Fortran 90 compiler (we used GNU g77 version 2.91.66 to compile the source codes). All NAS benchmarks use

MPI [7]. We used MPICH 1.2.0 in our experiments for TCP/IP, MPICH 1.2.8 for GM and MVICH 1-alpha 5 [8] for VIA. The implementation of MVICH is at the early stages, so it is neither completely stable nor optimized for performance. We expect MVICH results presented here to improve in future versions. We ran the applications with and without RDMA for the tests with VIA (both Fast Ethernet and Giganet). This option is enabled at compilation.

Table 2. Sequential times of applications

Program	Program Size	Seq. Time (s)	Std. Dev. (%)
IS	2^{25}, Iterations = 10	69.03	0.71
MG	256x256x256, Iterations = 20	342	0.52
CG	75,000, Iterations = 75	2,346	0.01
SP	102x102x102, Iterations = 400	6,783	0.46
EP	2^{31}, Iterations = 10	1,537	0.08
LU	102x102x102, Iterations = 250	7,553	0.30

The sequential execution times for the applications are presented in Table 2. Each application was run 5 times; the times presented are an average of these values. We also present the standard deviation of the times. All execution times are non-trivial, with LU and SP having the longest running-times. IS and MG have much shorter running-times, but still take more than a minute.

Table 3. Messages and data at 16 processors

Program	Messages	Transfers (MB)	Medium Size (KB/m)	Bw per CPU (MB/s)
IS	5,420	1,281.19	242.05	16.18
MG	42,776	667.25	15.97	1.74
CG	220,800	13,390.54	62.1	5.03
SP	153,600	14,504.17	96.69	1.82
EP	90	0.003	0.03	0.00
LU	1,212,060	3,525.70	2.97	0.45

Table 3 shows the total amount of data and messages sent by the applications, as well as medium message size (in kilobytes per message) and average bandwidth per processor, when running on 16 nodes. We can observe that the benchmarks have very different characteristics. EP sends the smallest number of messages, while LU sends the larger amount of messages. Although it takes the least time to run, IS sends the largest messages, hence requiring very high bandwidth, above the maximum provided by Fast Ethernet. SP is a long running-time application and also sends the largest volume of data, so it has moderate average bandwidth. The least bandwidth is used by EP, which only sends 90 messages in more than a thousand seconds. Table 4 presents a more detailed panorama of the message sizes sent by each application. SP and EP are opposites in that SP sends a lot of large message and EP few small messages. IS sends the same number of small and large messages. MG has a relative uniform distribution. Last, LU mostly sends medium-sized messages, and also some larger messages.

Table 5 shows the application's speedups for 16 processors. Speedup curves are presented in Figure 2. To understand the performance of the benchmarks, we used both the log and trace options available at the Multiprocessing Environment library.

Table 4. Message numbers by size

Size	IS	MG	CG	SP	EP	LU
$x < 10^1$	2,560	760	124,800	0	60	0
$10^1 \le x < 10^2$	0	8,960	2,400	0	30	60
$10^2 \le x < 10^3$	0	9,920	0	0	0	3×10^5
$10^3 \le x < 10^4$	300	11,520	0	0	0	9×10^5
$10^4 \le x < 10^5$	0	9,664	0	57,600	0	0
$x \ge 10^5$	2,560	1,952	93,600	96,000	0	12,000

Figure 3 presents the execution time breakdown of the benchmarks. We include only four communications calls (MPI_Send, MPI_Recv, MPI_Isend and MPI_Irecv) and two synchronizations calls (MPI_Wait and MPI_Waitall). All other MPI calls are constructed through the combination of these six primitives calls. We show breakdowns for the Myrinet and Giganet configurations. The breakdowns are quite similar, except for LU. Note that computation time dominates the breakdown. IS has very significant send and waitall times, and LU has very significant recv time. The dominance of computation time indicates we can expect good speedups, as it is indeed the case.

Table 5. Speedups – 16 processors

Application	MPICH		MVICH			
	Ethernet	Myrinet	Ethernet	Eth RDMA	Giganet	Gig RDMA
IS	1.43	11.85	NA	NA	NA	13.95
MG	10.95	14.28	12.36	NA	13.92	14.03
CG	7.70	14.11	9.72	NA	13.36	13.59
SP	NA	13.57	11.93	NA	13.4	13.66
EP	15.92	16.01	16.00	16.00	16.02	16.02
LU	12.76	7.22	15.16	NA	15.48	15.66

IS. Integer Sort (IS) kernel uses bucket sort to rank an unsorted sequence of keys. IS sends a total of 5,420 messages; 2,560 messages are smaller than 10 bytes and 2,560 are bigger than 10^5 bytes. IS requires a total of 16.18 MB/s of bandwidth per CPU. IS running on Myrinet spends 55% of the time on communication, while the version running on Giganet with RDMA spends 48% of time on communication. So, this is a communication bound application and just the bandwidth required per CPU by IS explains the poor performance of Fast Ethernet. We found that the difference between Giganet and Myrinet stems from the time spent in the recv and waitall primitives. For 16 nodes, Giganet spends 0.78s and 1.55s, respectively, in recv and waitall, while Myrinet spends 0.90s and 2.26s (Figure 3a). This seems to suggest better performance from the Giganet switch. VIA is effective in improving performance of Fast Ethernet. For 8 nodes, the speedup of the M-VIA version is 3.66, against 1.02 of the TCP/IP version (figure 2a). We believe this result to be quite good, considering the average bandwidth required by the application. Unfortunately, we could not make M-VIA run with 16 nodes, due to an excessive number of messages. M-VIA with RDMA version only runs with 2 nodes, so we cannot evaluate its effectiveness in improving performance.

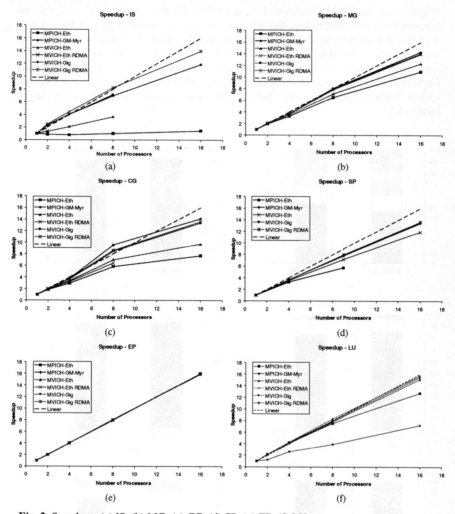

Fig. 2. Speedups: (a) IS, (b) MG, (c) CG, (d) SP, (e) EP, (f) LU

MG. MG uses a multigrid method to compute the solution of the three-dimensional scalar Poisson equation. MG sends a total of 42,776 messages. Despite of sending more messages than IS, the medium message size of MG is smaller and the benchmark runs for longer. This is reflected in the average bandwidth required of only 1.74 MB/s, smaller than the bandwidth required by IS. As shown in Figure 3b, MG is a computation bound application, spending less than 10% of the time doing communication on both Myrinet and Giganet. Fast Ethernet presented a good speedup on 16 nodes: 10.95 for TCP/IP and 12.36 for VIA. The speedups of Myrinet and Giganet are quite similar: Myrinet achieved a speedup of 14.28, against 14.03 of Giganet with the RDMA support. Without RDMA, the speedup of Giganet is 13.93

(Figure 2b). So RDMA support does not offer, for this application, a significant improvement in performance. Indeed, for 8 nodes, the performance of the version with RDMA support on Fast Ethernet was slightly worst: 7.10 against 7.15 for the version without RDMA. For 16 nodes on Fast Ethernet, the version with RDMA support has broken. The reason why Myrinet has a slightly better performance than Giganet is the time spent in the send primitive: Myrinet spends 1.4s, while Giganet spends 1.55s (Figure 3b). We believe MG may benefit from Myrinet's larger bandwidth.

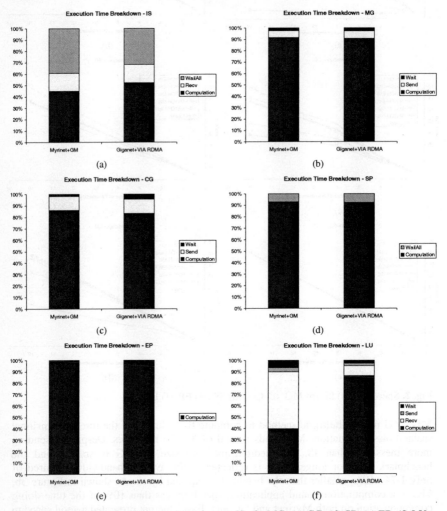

Fig. 3. Execution time breakdown - 16 nodes: (a) IS, (b) MG, (c) CG, (d) SP, (e) EP, (f) LU

CG. In the CG kernel, a conjugate gradient method is used to find an estimate of the largest eigenvalue of a large, symmetric positive definite sparse matrix with a random pattern of nonzeros. CG sends a total of 220,800 messages; 124,800 of which are smaller than 9 bytes and 93,600 bigger than 10^5. In spite of the large number of messages sent, CG is a computation bound application: 16% of the time is spend on communication on Giganet and 14% on Myrinet. The bandwidth required by this benchmark is 5.03 MB/s. Myrinet achieves the better speedup: 14.11 on 16 nodes. Giganet achieves a speedup of 13.59 with the RDMA support and 13.36 without RDMA. TCP/IP over Fast Ethernet achieved a speedup of 7.7. Substituting TCP/IP by M-VIA makes the speedup grow up to a very reasonable 9.72 (Figure 2c). Again, M-VIA with RDMA was broken for 16 nodes. For 8 nodes, the version with RDMA has worst performance (6.42) than the version without RDMA (7.00). This time, the time spent in the wait primitive by Myrinet, 3.28s, is smaller than the time Giganet spends on it, 7.5s. Giganet also spends more time in the send primitive (21.05s) than Myrinet (19.7s), which contributed to the worst performance of Giganet on this benchmark (Figure 3c). Again, we believe Myrinet may benefit from larger bandwidth.

SP. SP solves 3 uncoupled systems of non-diagonally dominant, scalar, pentadiagonal equations using the multi-partition algorithm. In the multi-partition algorithm, each processor is responsible for several disjoint sub-blocks of points of the grid, which are called cells. The information from a cell is not sent to the next processor until all linear equations of the cell have been solved, which keeps the granularity of communications large (57,600 messages sent by SP are between 10^4 and 10^5-1 bytes, and 96,000 are bigger than 10^5 bytes). As CG and MG, SP is also a computation bound application, spending less than 8% on communication. The bandwidth required by SP is 1.82 MB/s. SP requires a square number of processors. Myrinet and Giganet have similar performance. Myrinet achieved a speedup of 13.57 on 16 nodes, and Giganet achieved a speedup of 13.66 when running with RDMA support and 13.4 without RDMA. We were surprised to find that TCP/IP over Fast Ethernet was broken for 16 nodes. M-VIA over Fast Ethernet achieved a speedup of 11.93, reasonably close to the faster networks (Figure 2d).

EP. In the Embarrassingly Parallel (EP) kernel, each processor independently generates pseudorandom numbers in batches and uses these to compute pairs of normally distributed numbers. This kernel requires virtually no inter-process communication; the communication is just needed when the tallies of all processors are combined at the end of computation. All systems achieved linear speedups in this application (only TCP/IP was slightly slower – see Figure 2e).

LU. The Lower-Upper (LU) diagonal kernel uses a symmetric successive over-relaxation (SSOR) procedure to solve a regular sparse, block (5x5) lower and upper triangular system of equations in 3 dimensions. The SSOR procedure proceeds on diagonals, which progressively sweep from one corner on the third dimension to the opposite corner of the same dimension. Communication of partition boundary data occurs after completion of computation on all diagonals that contact an adjacent partition. This results in a very large number (1.2×10^6) of very small messages (about 2 Kb). The bandwidth required by LU is a little: 0.45 MB/s. Surprisingly, the performance of Myrinet is terrible in this benchmark. It achieves a speedup of 7.22 on 16 nodes, while TCP/IP over Fast Ethernet achieves a speedup of 12.76. M-VIA is effective in improving the performance of LU on Fast Ethernet, achieving a speedup of 15.16. Giganet has the better performance for LU: 15.66 for the version with RDMA, and 15.48 for the version without RDMA (Figure 2f). While Giganet spends 13.5% of the time sending messages, Myrinet spends almost 40%. The send, recv and wait primitives are responsible for the poor performance of LU on Myrinet. While Myrinet spends 522s, 39s and 67s on send, recv and wait, respectively, Giganet spends 41s, 11s and 11s (Figure 3f). Previous work [9] also pointed this poor performance of LU on Myrinet. They also used MPICH-GM on Myrinet, but used MPI/Pro on Giganet. The work attributed the poor performance of LU on Myrinet to the pooling-based approach adopted by MPICH-GM (MPI/Pro adopts a interrupt-based approach). But MVICH also adopts the pooling-based approach, and its performance is good, which indicates that the theory presented in [9] is not correct. Instead, we suggest that the problem may stem from the way GM allocates memory for small blocks. GM has a cache for small blocks. We believe that cache management for the very many small blocks created by the application may be significantly hurting performance.

Analysis. The results show that M-VIA was effective in improving the performance of all benchmarks on Fast Ethernet, when compared with the version that uses TCP/IP. The performance on Ethernet with VIA is 3.6 times better than the version with TCP/IP for IS; 1.12 times better for MG; 1.26 times better for CG; 1.25 times better for SP; and 1.18 times better for LU. The biggest difference between VIA and TCP/IP appeared in the application that has the biggest bandwidth requirement, IS, which suggests that the gains provided by VIA are related to the bandwidth required by the application: the bigger the bandwidth required, the bigger the performance difference between VIA and TCP/IP. The RDMA support provided by VIA on Giganet was effective in improving the performance of IS: it runs 1.14 times faster than the version without RDMA. For all other applications, the RDMA support contributes for a small improvement in performance (less than 2%). On Ethernet, the version with RDMA support was quite unstable due to sequence mismatch messages. This occurs because the NIC has dropped packets under heavy load: the interrupt service routine on the receiving node cannot process the packets as fast as the NIC receives them, causing the NIC to run out of buffer space. Because M-VIA supports only the unreliable delivery mode of VIA, these events do not cause a disconnect and the MVICH library gets different data than what it is expecting. The behavior of MVICH in this case is to abort. It is also interesting to note that on Ethernet, the benchmarks with RDMA support performed 8% worst on CG. In IS, for 2 nodes, the performance of the version with RDMA was 6% better than the version without RDMA. Recall that for this benchmark, the configurations with more than 4 nodes have broken. In all other benchmarks, the performance was equal. The results of both Giganet and Ethernet indicate that the RDMA support is only effective when the message size is huge. As could be expected, Ethernet, even with VIA support, performed worst than the best execution time, either Giganet's or Myrinet's, for all benchmarks. The only exception was EP, where Ethernet performed as well as Myrinet and Giganet. This happened because EP's almost does not require communication. For IS, Ethernet had its worst performance (124% slower than Giganet) due to the high bandwidth required by this application. The better performance, after EP, is LU: only 3% slower than Giganet. Not surprisingly, LU is the application with the smaller bandwidth requirement after EP. Other performances are 15% worst for MG and SP and 45% for CG. The performance on LU and MG suggests that Ethernet is a good choice to connect machines on a small cluster environment where bandwidth is not crucial for applications, since it has a good cost-benefit ratio. Ethernet is 10 times cheaper than Giganet and Myrinet, and its performance is 2 times slower for the worst application, IS. If performance is most important, a good rule of thumb would be to choose Fast Ethernet with TCP/IP if your application requires very small bandwidth, and to use a faster protocol such as M-VIA on the same hardware if your application requires up to 5 MB/s. Giganet performed better than Myrinet for IS (18%) and LU (116%); had a similar performance for EP and SP; and performed slightly worst for MG (2%) and CG (4%). These results show the influence of the MPI implementation over performance. The raw communication numbers presented in Figure 1 could suggest that Myrinet would have the best performance, indicating that the implementation of MPICH-GM does not take full advantage of the performance of Myrinet.

3   Summary and Conclusions

This paper evaluated the impact of three popular cluster interconnection technologies, namely Fast Ethernet, Giganet and Myrinet, over the performance of the NAS Parallel Benchmarks (NPB). The results show that Fast Ethernet, when combined with a high performance communication protocol, such as VIA, has a good cost-benefit ratio, and can be a good choice to connect machines on a small cluster environment where bandwidth is not crucial for applications. We also evaluated the performance gains provided by VIA, when compared with TCP/IP, and found that VIA is quite effective in improving the performance of applications on Fast Ethernet. The RDMA support provided by VIA was evaluated, and we conclude that it is only effective when the messages exchanged by the applications are huge. Last, our results showed Giganet performing better than Myrinet on the NPB. We found the main difference in LU, where Myrinet performance is poor due to the MPI implementation for this interconnect technology, and not due to the pooling-based approach adopted by the MPI implementation, as a previous work has pointed out.

References

1. Giganet, Inc. http://www.emulex.com/
2. Myricom, Inc. http://www.myri.com/
3. Compaq, Intel, and Microsoft. VIA Specification 1.0. Available at http://www.viarch.org.
4. Wong, F, et alli. Architectural Requirements and Scalability of the NAS Parallel Benchmarks. SuperComputing'99, Nov 1999.
5. Bailey, D, et alli. The NAS Parallel Benchmarks. Tech. Report 103863, NASA, July 1993.
6. NERSC. M-VIA. http://www.nersc.gov/research/FTG/via/.
7. Message Passing Interface Forum. http://www.mpi-forum.org/.
8. NERSC. MVICH. http://www.nersc.gov/research/FTG/mvich
9. Hsieh, J, et alli. Architectural and Performance Evaluation of GigaNet and Myrinet Interconnects on Clusters of Small-Scale SMP Servers. SuperComputing'00, Nov 2000.

Basic Operations on a Partitioned Optical Passive Stars Network with Large Group Size [*]

Amitava Datta and Subbiah Soundaralakshmi

Department of Computer Science & Software Engineering
The University of Western Australia
Perth, WA 6009
Australia
email : {datta,laxmi}@cs.uwa.edu.au

Abstract. In a *Partitioned Optical Passive Stars (POPS)* network, $n = dg$ processors are divided into g groups of d processors each and such a POPS network is denoted by $POPS(d, g)$. There is an optical passive star (OPS) coupler between every pair of groups. Each OPS coupler can receive an optical signal from any one of its source nodes and broadcast the signal to all the destination nodes. The time needed to perform this receive and broadcast is referred to as a time *slot* and the complexity of an algorithm using the POPS network is measured in terms of number of slots it uses. Since a $POPS(d, g)$ requires g^2 couplers, it is unlikely that in a practical system the number of couplers will be more than the number of processors. In other words, in most practical systems, the group size d will be greater than the number of groups g, i.e., $d > \sqrt{n} > g$. Hence, it is important to design fast algorithms for basic operations on such POPS networks with large group sizes. We present several fast algorithms for basic arithmetic operations on $POPS(d, g)$s such that $d > \sqrt{n} > g$. Our algorithms require significantly less number of slots for these operations compared to the best known algorithms for these problems designed by Sahni [8].

keywords: optical computing, partitioned optical passive stars network, arithmetic operations

1   Introduction

The *Partitioned Optical Passive Stars (POPS)* network was proposed in [3–5, 7] as a fast optical interconnection network for a multiprocessor system. The POPS network uses multiple optical passive star (OPS) couplers to construct a flexible interconnection topology. Each OPS coupler can receive an optical signal from any one of its source nodes and broadcast the received signal to all of its destination nodes. The time needed to perform this receive and broadcast is

[*] This research is partially supported by a University of Western Australia research grant.

referred to as a *slot*. A POPS network is divided into g groups of d processors each. Hence, the total number of processors in the network is $n = dg$. We denote such a POPS network as a $POPS(d, g)$ network.

Berthomé and Ferreira [1] have shown that POPS networks can be modeled by directed stack-complete graphs with loops. This is used to obtain optimal embedding of rings and de Bruijn graphs into POPS networks. Berthomé *et al.* [2] have also shown how to embed tori on POPS networks. Gravenstreter and Melhem [4] have shown the embedding of rings and tori into POPS networks.

Sahni [8] has shown simulations of hypercubes and mesh-connected computers using POPS networks. Sahni [8] has also presented algorithms for several fundamental operations like data sum, prefix sum, rank, adjacent sum, consecutive sum, concentrate, distribute and generalize. Though it is possible to solve these problems by using existing algorithms on hypercubes, the algorithms presented by Sahni [8] improve upon the complexities of the simulated hypercube algorithms. In another paper, Sahni [9] has presented fast algorithms for matrix multiplication, data permutations and BPC permutations on the POPS network. One of the main results in the paper by Sahni [8] is the simulation of an SIMD hypercube by a $POPS(d, g)$ network. Sahni [8] has shown that an n processor $POPS(d, g)$ can simulate every move of an n processor SIMD hypercube using one slot when $d = 1$ and $2\lceil d/g \rceil$ slots when $d > 1$. It only makes sense to design specific algorithms for a $POPS(d, g)$ network for the case when $d > 1$. For the case $d = 1$, the POPS network is a completely connected network and it is easier to simulate the corresponding hypercube algorithm. Moreover, any algorithm designed for a $POPS(d, g)$ network should perform better than the corresponding simulated algorithm on the hypercube.

Most of the algorithms designed by Sahni have different complexities for different group sizes. He usually expresses the complexity of these algorithms for three cases, **(i)** $d = g = \sqrt{n}$, **(ii)** $d < \sqrt{n} < g$ and **(iii)** $d > \sqrt{n} > g$. Gravenstreter and Melhem [4] mention that for most practical systems, the number of couplers will be less than the number of processors. Indeed, for a $POPS(d, g)$, with $dg = n$, if the number of groups $g > \sqrt{n}$, we need $g^2 > n$ couplers to connect such a network. It is unlikely that a practical system will be built with such a high number of couplers. One would expect that the number of couplers in most practical systems will be significantly smaller than the number of processors. In other words, the group size $d > \sqrt{n}$ in a practical system and hence, $g < \sqrt{n}$. We feel that cases **(i)** and **(ii)** are unrealistic in most practical systems since in these cases the number of OPS couplers g^2 is equal to or greater than n, the number of processors. Hence, our main motivation is to design fast algorithms for $POPS(d, g)$ networks such that $d > \sqrt{n} > g$.

In this paper, we present fast algorithms for two fundamental arithmetic operations, data sum and prefix sum, for POPS networks with large group size and hence, small number of groups. Our algorithms improve upon the complexities obtained by Sahni for these problems when $d > \sqrt{n} > g$. We present the comparison of the algorithms by Sahni [8] and our algorithms in Table 1.

Problem	Sahni's [8] algorithm	Our algorithm
Data sum	$\lceil \frac{d}{g} \rceil \log n$	$\frac{d}{g} + 2 \log g - 1$
Prefix sum	$\frac{2d}{g}(1 + \log g) + \log d + 1$	$\frac{2d}{g} + 4 \log g + 6$

Table 1. The comparison between our algorithms and Sahni's [8] algorithms. All complexities are in terms of number of slots. In this paper, all logarithms are to the base 2.

The rest of this paper is organized as follows. In Section 2, we discuss some details of the POPS network. We present our data sum algorithm in Section 3. Finally, we present the prefix sum algorithm in Section 4.

2   The POPS network

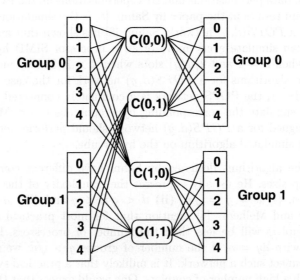

Fig. 1. A 10-processor computer connected via a $POPS(5,2)$ network.

A $POPS(d,g)$ network partitions the n processors into g groups of size d each and optical passive stars (OPS) couplers are used to connect such a network when every pair of groups is connected by a coupler. Hence, overall g^2 couplers are needed. Each processor must have g optical transmitters, one transmitter each for transmitting to the g OPSs for which it is a source node. Also, each processors should have g optical receivers, for receiving data from each of the g couplers. Each OPS in a $POPS(d,g)$ network has degree d. In one slot, each OPS can receive a message from any one of its source nodes and broadcast the message to all of its destination nodes. However, in one slot, a processor can receive a message from only one of the OPSs for which it is a destination node. Melhem *et al.* [7] observe that faster all-to-all broadcasts can be implemented by allowing a processor to receive different messages from different OPSs in the same slot. However, in this paper, we will assume that only one message can be received by a processor in one slot. A ten-processor computer connected via a $POPS(5,2)$ network is shown in Figure 1.

The g groups in a POPS network are numbered from 1 to g. A pair of groups is connected by an optical passive star (OPS) coupler. For coupler $c(i,j)$, the source nodes are the processors in group j and the destination nodes are the processors in group i, $1 \leq i,j \leq g$. Note that we have shown each group twice in Figure 1 to indicate that processors in each group are both receivers and senders.

The most important advantage of a POPS network is that its diameter is 1. A message can be sent from processor i to processor j ($i \neq j$) in one slot. We use the notation $group(i)$ to indicate the group that processor i is in. To send a message from processor i to processor j, processor i uses the coupler $c(group(j), group(i))$. Processor i first sends the message to $c(group(j), group(i))$ and then $c(group(j), group(i))$ sends the message to processor j. Similarly, one-to-all broadcasts also can be implemented in one slot.

We refer to the following result obtained by Sahni [8].

Theorem 1. *[8] An n processor $POPS(d,g)$ can simulate every move of an n processor SIMD hypercube using one slot when $d = 1$ and using $2\lceil d/g \rceil$ slots when $d > 1$.*

One consequence of Theorem 1 is that we can simulate an existing hypercube algorithm on a POPS network.

In our algorithms, we use the following results obtained by Sahni [8].

Lemma 1. *[8] If n data items are distributed one per processor in a $POPS(g,g)$ with $n = g^2$ processors, their sum can be computed in $\lceil \frac{g}{g} \rceil \log n = 2 \log g$ slots.*

Lemma 2. *[8] If n data items are distributed one per processor in a $POPS(d,g)$ with $n = gd$ processors, with $1 < d \leq g$, their prefix sum can be computed in $3 + \log n + \log d$ slots.*

3   Data sum

Initially n data items are distributed among n processors, one item per processor in a $POPS(d,g)$. The purpose of the *data sum* operation is to compute the sum of these n data items and bring the sum in the first processor of the first group. We consider the case when $d > \sqrt{n} > g$.

Our algorithm is based on the following strategy. Our main aim is to reduce the number of data items rapidly so that there are only g data items in each group. Since there are g groups, we can use Sahni's algorithm [8] i as stated in Lemma 1 at this point to compute the data sum of these g^2 remaining elements.

To start with, we divide the processors in each group into two categories.

- The first g processors in each group i are called *receivers*. The set of g receivers in the i-th group, $1 \leq i \leq g$, is denoted by R_i and the j-th receiver in the i-th group is denoted by $r_{i,j}$, $1 \leq j \leq g$.
- The $(d - g)$ processors other than the receivers in each group are called the *senders*. The set of senders in the m-th group, $1 \leq m \leq g$, is denoted by S_m.

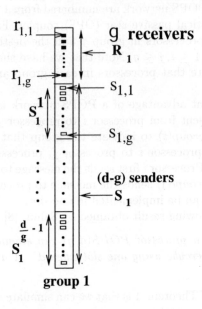

Fig. 2. The *Receiver* and *Sender* sets for group 1.

Consider the i-th group, $1 \leq i \leq g$. First, we divide the set $S_i (1 \leq i \leq g)$ of $d - g$ senders in the group $i(\ 1 \leq i \leq g)$ into $\frac{d}{g} - 1$ subsets, each consisting of g processors and we denote them by $S_i^1, \ldots, S_i^{\frac{d}{g}-1}$. Note that there are g senders in each subset S_i^j, $1 \leq j \leq \frac{d}{g} - 1$. We denote the senders in the subset S_i^j of S_i by $s_{i,(j-1)g+1}, s_{i,(j-1)g+2}, \ldots, s_{i,jg}(\ 1 \leq j \leq (\frac{d}{g} - 1))$. The structure of the first group is shown in Figure 2.

In each stage, we perform the following computation in each group $i(1 \leq i \leq g)$ in parallel. At stage $k(\ 1 \leq k \leq \frac{d}{g} - 1)$, the participating subsets are $S_1^k, S_2^k, \ldots, S_g^k$ from group 1, group 2, ..., group g respectively. For the subset S_j^k in group j, the senders $s_{j,(k-1)g+1}, \ldots, s_{j,kg}$ send their data to the receivers $r_{1,j}$ in group 1, $r_{2,j}$ in group 2 ..., $r_{g,j}$ in group g respectively. The couplers used for this data movement are $c(*, j)$ where '*' indicates all the values from 1 to g. This computation takes one slot. This data movement is performed in parallel by all the g subsets $S_i^k(1 \leq i \leq g)$. Hence, the computation at stage k takes one slot. Further, g^2 data items are moved in stage k and g^2 different couplers are used since each subset S_i^k of S_i $(1 \leq i \leq g)$ in group i uses the g couplers connected to group i. Note that each receiver receives only one data item. The computation is the same for each stage except that the participating subsets are different. Since there are $\frac{d}{g} - 1$ stages, the total number of slots is $\frac{d}{g} - 1$. The data movement at Stage 1 is illustrated in Figure 3.

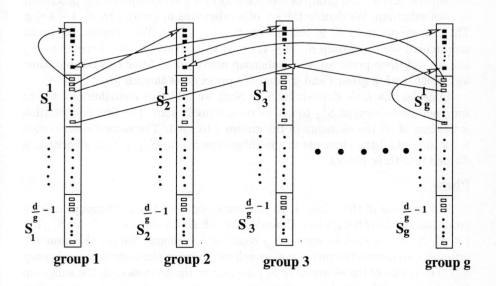

Fig. 3. At stage 1, the participating subsets are $S_1^1, S_2^1, \ldots, S_g^1$. The data movement is shown for subsets S_1^1 and S_g^1.

Once the receivers receive the data items, they add the received data with their existing data. In the k-th stage $(1 \leq k \leq \frac{d}{g}-1)$, the g senders $s_{i,(k-1)g+1}, \ldots,$ $s_{i,kg}$ in S_i^k send their data items$(1 \leq i \leq g)$ to i-th receivers $r_{1,i}, \ldots, r_{g,i}$ of the groups $1, 2, \ldots, g$ respectively. Each receiver add the data item that it receives at this stage to its accumulated sum. Note that after stage k, each group has $d - kg$ senders holding data items.

After $\frac{d}{g}-1$ stages, all the senders have sent their data to the receivers. Hence, each group has now g receivers holding their accumulated sums. Our next task is to add the accumulated sums in these g^2 receivers and bring this sum to the first processor in the first group. We do this using Sahni's [8] algorithm in Lemma 1. This computation takes $2 \log g$ slots. Hence, the sum of all the data items in the $n = dg$ processors is computed after $\frac{d}{g} + 2 \log g - 1$ slots. We have shown the following lemma.

Lemma 3. *The sum of n data items initially given in a $POPS(d,g)$ with one data item per processor, $dg = n$ and $d > \sqrt{n} > g$, can be computed in $\frac{d}{g} + 2 \log g - 1$ slots.*

4 Prefix sum

In this section, we present an algorithm on a $POPS(d, g)$ for computing prefix sum in $\frac{2d}{g} + 4 \log g + 6$ slots when $d > \sqrt{n} > g$. The strategy behind our algorithm is the following.

We first divide each group of d processors into g subgroups with $\frac{d}{g}$ processors in each subgroup. We denote the set of g subgroups in group i by $S_i, 1 \leq i \leq g$. The j-th subgroup in S_i is denoted by $S_{i,j}, 1 \leq j \leq g$. We compute the prefix sum in each such subgroup $S_{i,j} (1 \leq i, j \leq g)$ with respect to the elements in $S_{i,j}$ and we call these prefix sums as subgroup prefix sums. After this computation, we are left with g groups and g subgroup prefix sums in each group.

Consider the k-th element in $S_{i,j}$. Next we add two contributions to the subgroup prefix sum of $S_{i,j}^k$ to get its overall prefix sum. The first contribution is the sum of all the elements in the groups 1 to $i - 1$. The second contribution is the sum of all the elements in the subgroups $S_{i,1}$ to $S_{i,j-1}$. Our algorithm is divided into three phases.

Phase 1.

The purpose of this phase is to distribute copies of the g subgroups of each group among the other groups. Consider the i-th group and its set $S_i = \{S_{i,j} | 1 \leq j \leq g\}$. Note that each subgroup $S_{i,j}$ contains $\frac{d}{g}$ elements. We use the same notation $S_{i,j}$ to denote the processors as well as the input elements in the subgroup $S_{i,j}$. The copies of the elements in $S_{i,j}$ are sent to the processors in the subgroup $S_{j,i}$ of S_j in group j, one element per processor, by using the coupler $c(j, i)$. This data movement is done for all $S_{i,j}, 1 \leq i \leq g, 1 \leq j \leq g$, in parallel. Note that there is no coupler conflict as for any two couplers, both of their indices do not match. In other words, we are transmitting data from g^2 subgroups and we have g^2 couplers and hence, there is no coupler conflict.

This data transmission is performed in $\frac{d}{g}$ slots since there are these many elements in each subgroup $S_{i,j}$ and a coupler can receive one data item in each slot.

Phase 2.

We compute the subgroup prefix sums in this phase. Consider the subgroup $S_{i,j}$ of S_i, the j-th subgroup in the i-th group. We denote the $\frac{d}{g}$ elements in this subgroup as $S_{i,j}^k, 1 \leq k \leq \frac{d}{g}$. Recall that copies of all the elements in $S_{i,j}$ has been sent to the subgroup $S_{j,i}$ in Phase 1.

Now, the processors in $S_{j,i}$ broadcast the elements of $S_{i,j}$ to the processors in $S_{i,j}$ in $\frac{d}{g}$ slots. This broadcast is done by the processors in $S_{j,i}$ one by one.

- In the first slot, $S_{j,i}^1$ broadcasts the element $S_{i,j}^1$ to the processors $S_{i,j}^k, 2 \leq k \leq \frac{d}{g}$. These processors add $S_{i,j}^1$ with the input they hold. In general, in the m-th slot, $1 \leq m \leq \frac{d}{g}$, the processor $S_{j,i}^m$ broadcasts the element $S_{i,j}^m$ to the processors $S_{i,j}^k, m + 1 \leq k \leq \frac{d}{g}$.
- This computation is done by all the subgroups in all the groups in parallel,

i.e., in the subgroups $S_{i,j}, 1 \leq i \leq g, 1 \leq j \leq g$. Note that the coupler used to do the broadcasts from processors in $S_{j,i}$ to processors in $S_{i,j}$ is $c(i,j)$. Hence, all the subgroups can communicate with the help of the g^2 couplers and there is no coupler conflict as each coupler $c(i,j)$ will have its two indices i and j different from all other couplers. Further, each process broadcasts at most one element and receives at most one element in each slot. The total number of slots required in this phase is $\frac{d}{g}$.

Note that at the end of the $\frac{d}{g}$ slots, all the processors in $S_{i,j}$ have computed the prefix sum of its element with respect to the elements in $S_{i,j}$ We call these prefix sums as subgroup prefix sums and the subgroup prefix sum of the element $S_{i,j}^k, 1 \leq i, j \leq g$ is denoted by $SubPrefix(S_{i,j}^k)$.

end of Phase 2

For the computation in the next phase, we are interested in the prefix sums $SubPrefix(S_{i,j}^{\frac{d}{g}}), 1 \leq i, j \leq g$. Note that $S_{i,j}^{\frac{d}{g}} = \sum_{m=1}^{\frac{d}{g}} S_{i,j}^m$, i.e., the last element $S_{i,j}^{\frac{d}{g}}$ of the subgroup $S_{i,j}$ holds the sum of all the elements in it. Hence, the sum $\sum_{j=1}^g SubPrefix(S_{i,j}^{\frac{d}{g}})$ gives the sum of all the elements in the i-th group and we call this the *group-sum* for group i and denote it by GS_i.

Consider the element $S_{i,j}^k (1 \leq k \leq \frac{d}{g})$, i.e., the k-th element of the j-th subgroup $S_{i,j}$ in group i. We denote the actual prefix sum of $S_{i,j}^k$ w.r.t all the elements in $POPS(d,g)$ by $prefix(S_{i,j}^k)$. The computation of $prefix(S_{i,j}^k)$ is performed as follows. After the computation in Phase 2, we know $SubPrefix(S_{i,j}^k)$. We add the following two quantities to $SubPrefix(S_{i,j}^k)$ to compute $prefix(S_{i,j}^k)$.

(i). The first quantity is $\sum_{m=1}^{i-1} GS_m$, i.e., the contribution of all the groups before the i-th group.

(ii). The second quantity is $\sum_{p=1}^{j-1} SubPrefix(S_{i,p}^{\frac{d}{g}})$, i.e., the contribution of all the subgroups before the j-th subgroup in group i. We discuss the computation of these two contributions **(i)** and **(ii)** in Phase 3.

Phase 3.

After the computation in Phase 2, each of the g groups has g subgroup prefix sums. Hence, for each group i, we can compute the prefix sums of the subgroup prefix sums $SubPrefix(S_{i,1}^{\frac{d}{g}}), SubPrefix(S_{i,2}^{\frac{d}{g}}), \ldots, SubPrefix(S_{i,j}^{\frac{d}{g}})$. by considering our $POPS(d,g)$ as a $POPS(g,g)$ and using Sahni's algorithm [8] as described in Lemma 2. In this case, $d = g$ and $n = g^2$ and hence, the number of slots required is $3 + \log n + \log d = 3 + \log(g^2) + \log g = 3 + 3 \log g$.

Note that the quantity $GS_i, 1 \leq i \leq g$, will be available in the i-th group after this computation. We now compute the prefix sums of the GS_i's by using our $POPS(d,g)$ as a $POPS(1,g)$ and by using Sahni's algorithm for simulating a hypercube algorithm on a $POPS(1,g)$. This computation takes $\log g$ slots.

Now, the last processor in each group $i, 1 \leq i \leq g$ has got the prefix sums $\sum_{j=1}^i GS_j$. This processor computes $\sum_{j=1}^{i-1} GS_j = \sum_{j=1}^i GS_j - GS_i$ and broadcasts this quantity to all the processors in group $i, 1 \leq i \leq g$ by using the coupler $c(i,i)$. All processors add this quantity to their current prefix sum. This computation takes one slot.

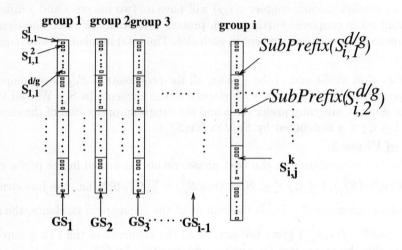

Fig. 4. Illustration for the computation in Phase 3. $prefix(S_{i,j}^k)$ is computed by adding $\sum_{m=1}^{i-1} GS_m$ and $\sum_{p=1}^{j-1} SubPrefix(S_{i,p}^{\frac{d}{g}})$ to $SubPrefix(S_{i,j}^k)$.

We have to compute the final contribution in the prefix sum of each element due to the contribution of the subgroup prefix sums within its own group. Recall that we already know the subgroup prefix sums due to the computation in the first part of this phase. Each group i has g subgroup prefix sums $SubPrefix(S_{i,1}^{\frac{d}{g}})$, $\sum_{p=1}^{2} SubPrefix(S_{i,p}^{\frac{d}{g}})$, ..., $\sum_{p=1}^{j} SubPrefix(S_{i,p}^{\frac{d}{g}})$, ..., $\sum_{p=1}^{g} SubPrefix(S_{i,p}^{\frac{d}{g}})$.

We need to broadcast the subgroup prefix sum $\sum_{p=1}^{j-1} SubPrefix(S_{i,p}^{\frac{d}{g}})$ to all the processors in $S_{i,j}$ and these processors should update their existing sum by adding this quantity. If we use the coupler $c(i,i)$ to do this broadcast, then we will need g slots. Instead, we do this broadcast for all the groups in two slots in the following way.

• We specify $g - 1$ processors from each of the g groups as receivers. We denote the $g - 1$ receivers for group j, $1 \le j \le g$ by $r_{j,k}, 1 \le k \le (g - 1)$. The processors $S_{i,j}^{\frac{d}{g}}(1 \le j \le (g - 1))$ in group i holding the $(g - 1)$ sub-prefix sums $\sum_{p=1}^{j} SubPrefix(S_{i,p}^{\frac{d}{g}})$ send these sub prefix sums to the i-th receivers in each group. For example, $SubPrefix(S_{i,1}^{\frac{d}{g}})$ is sent to the receiver $r_{1,i}$, $\sum_{p=1}^{2} SubPrefix(S_{i,p}^{\frac{d}{g}})$ is sent to the receiver $r_{2,i}$ and in general, the sum $\sum_{p=1}^{m} SubPrefix(S_{i,p}^{\frac{d}{g}})$, $1 \le m \le (g - 1)$ is sent to the receiver $r_{m,i}$, for all

$1 \leq i \leq g$. This data movement takes one slot. Processors in group i use the couplers $c(*, i)$, where '*' means all indices $1, \ldots, g$. Hence, no two couplers used in this data transfer, will have both indices same and in the same order. This shows that there is no coupler conflict.

- Next, the j-th receiver $r_{k,j}, 1 \leq k \leq (g-1)$ in group k broadcasts the sum $\sum_{p=1}^{k} SubPrefix(S_{i,p}^{\frac{d}{g}})$ to all the processors in the subgroup $S_{j,k+1}$ using the coupler $c(j, k)$. This broadcast is done by all the receivers $r_{k,j}, 1 \leq k \leq g, 1 \leq j \leq (g-1)$. There is no coupler conflict as the the coupler indices depend on the corresponding receiver index and no two receivers have matching indices.

- Finally, each processor in each of the subgroups updates its prefix sum by adding the received quantity with its existing prefix sum.

The total slots required for the computation in this phase is $4 \log g + 6$. This concludes the computation of prefix sum. Hence, we have the following lemma.

Lemma 4. *The prefix sum of n data items initially given in the $n = dg$ processors of a $POPS(d, g)$ with $d > \sqrt{n} > g$, can be computed in $\frac{2d}{g} + 4 \log g + 6$ slots.*

References

1. P. Berthomé and A. Ferreira, "Improved embeddings in POPS networks through stack-graph models", *Proc. Third International Workshop on Massively Parallel Processing Using Optical Interconnections*, pp. 130-135, 1996.
2. P. Berthomé, J. Cohen and A. Ferreira, "Embedding tori in partitioned optical passive stars networks", *Proc. Fourth International Colloquium on Structural Information and Communication Complexity (Sirocco '97)*, pp. 40-52, 1997.
3. D. Chiarulli, S. Levitan, R. Melhem, J. Teza and G. Gravenstreter, "Partitioned optical passive star (POPS) multiprocessor interconnection networks with distributed control", *Journal of Lightwave Technology*, **14** (7), pp. 1901-1612, 1996.
4. G. Gravenstreter and R. Melhem, "Realizing common communication patterns in partitioned optical passive stars (POPS) networks", *IEEE Trans. Computers*, **47** (9), pp. 998-1013, 1998.
5. G. Gravenstreter, R. Melhem, D. Chiarulli, S. Levitan and J. Teza, "The partitioned optical passive stars (POPS) topology", *Proc. Ninth International Parallel Processing Symposium*, IEEE Computer Society, pp. 4-10, 1995.
6. V. Prasanna Kumar and V. Krishnan, "Efficient template matching on SIMD arrays", *Proc. 1987 International Conference on Parallel Processing*, pp. 765-771, 1987.
7. R. Melhem, G. Gravenstreter, D. Chiarulli and S. Levitan, "The communication capabilities of partitioned optical passive star networks", *Parallel Computing Using Optical Interconnections*, K. Li, Y.Pan and S. Zheng (Eds), Kluwer Academic Publishers, pp. 77-98, 1998.
8. S. Sahni, "The partitioned optical passive stars network : Simulations and fundamental operations", *IEEE Trans. Parallel and Distributed Systems*, **11** (7), pp. 739-748, 2000.
9. S. Sahni, "Matrix multiplication and data routing using a partitioned optical passive stars network", *IEEE Trans. Parallel and Distributed Systems*, **11** (7), pp. 720-728, 2000.

3D Mesh Generation for the Results of Anisotropic Etch Simulation

Elena V. Zudilova, Maxim O Borisov

Integration Technologies Department
Corning Scientific Center
4 Birzhevaya Linia, St. Petersburg 199034, RUSSIA
tel.: +7-812-32292060   fax: +7-812-3292061
e-mail: {ZudilovaEV; BorisovMO}@corning.com
http://www.corning.com

Abstract. The paper is devoted to the development of 3D mesh generator that provides the interface between etch simulation tool of IntelliSuite™ CAD for MEMS™ and its analyses compounds. Paper gives a brief introduction to IntelliSuite and its anisotropic etch simulator. The rest of the paper is devoted to the algorithm of 3D mesh generation based on special requirements to Finite Element Mesh. This algorithm can be divided into two main stages: manual 2D meshing of the upper surface and automatic 3D mesh generation for simplified object obtained from the previous stage. Paper contains the detailed description of both stages of the algorithm.

1. Introduction

The paper is devoted to the development of mesh generation algorithm for 3D object which geometry is defined by the unstructured set of triangular and quadrangular polygons that form its surface.

The algorithm has been developed for building the interface between anisotropic etch simulator of IntelliSuite™ CAD for MEMS™ and its performance analysis components. This interface will allow to use 3D model of MEMS (Micro Electro Mechanical Systems) generated during etch simulation process for further mechanical, electrostatic, electromagnetic and other types of analyses provided by IntelliSuite, unfortunately, not available today.

To solve this task a stand-alone application has been developing. It permits a user to simplify manually the shape of initial 3D geometry of the object by smoothing slopes of etching pits via 2D surface meshing. Then this simplified 3D object is processed and meshed automatically by 3D mesh generator.

The introduction to IntelliSuite and its mesh requirements are given in section 2 of the paper. Section 3 is devoted to the first stage of the algorithm - manual 2D surface meshing. Section 4 contains the description of the second stage - automatic 3D mesh generation.

P.M.A. Sloot et al. (Eds.): ICCS 2002, LNCS 2329, pp. 316–323, 2002.

2. IntelliSuite Anisotropic Etch Simulation

IntelliSuite CAD for MEMS provides the ability to simulate with high accuracy different classes of MEMS devices induced mechanically, electrostatically and electromagnetically and then to obtain the graphical presentation of the appearance of each simulated device. It is an integrated software complex which assists designers in optimizing MEMS devices by providing them access to manufacturing databases and by allowing them to model the entire device manufacturing sequence, to simulate behavior and to see obtained results visually without having to enter a manufacturing facility. [1, 4]

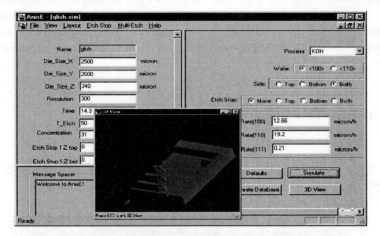

Fig. 1 IntelliSuite example of anisotropic etch simulation results

Etch simulation tool of IntelliSuite AnisE® permits to generate 3D model for anisotropic etching of silicon. With AnisE user can layout a microstructure, view 3D representation, access information about the etch rates of different etchants and simulate automatically the etching under different time, temperature, and concentration parameters [6]. Fig. 1 contains a screen-shot of AnisE GUI with a sample of 3D visualization.

3D geometry of MEMS device generated by AnisE can not be processed further by performance analyses components of IntelliSuite. It occurred because the geometry of MEMS device generated by AnisE, as a set of triangles and quadrangles defining its surface, does not satisfy Finite Element Mesh [5] requirements on which IntelliSuite is based.

A stand-alone application has been developing for building interface between AnisE and performance analyses of IntelliSuite. This component generates solid mesh from initial MEMS device geometry of AnisE in accordance with FEM requirements of IntelliSuite. These requirements are as follows [4, 7]:

- Mesh elements are hexahedrons;
- All edges are uninterrupted;
- Limits for edges ratio (1:1; 1:10);

- Lower and higher angle bounds ($\alpha_{min} = 30°$; $\alpha_{max} = 120°$);
- Optimum number of finite meshed elements.

All mesh requirements listed above are determined by ABAQUS on which IntelliSuite simulation routines are based.

The algorithm of 3D mesh generation is divided into two main stages - manual 2D surface meshing and automatic solid meshing.

First, the upper surface of the object is to be meshed by user. As a result, the new surface formed by quadrilateral polygons that satisfy angle and ratio requirements is generated. User also has the possibility to simplify the geometry of initial object by smoothing slopes or deleting insignificant pits. So at the end of the following stage a new simplified 3D object is generated. Its geometry is defined by the meshed upper surface, lower and front facets. Then this new geometry is processed by automatic 3D mesh generator and final solid mesh that satisfies all the requirements above is generated. The detailed description of each stage of the algorithm is provided in the next two sections of the paper.

3. Manual Surface Meshing

2D meshing of the upper surface of the object is conducted manually. Fig.2 represents the screen-shot of the graphical editor intended for this purpose.

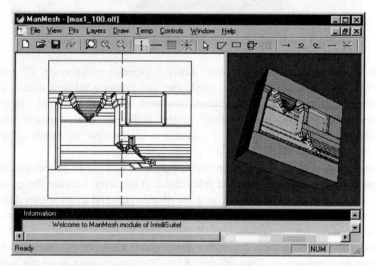

Fig. 2 Current GUI of 2D layer dependent editor (Win-platform: MFC, Vtk 3.1.2)

2D surface meshing and smoothing slopes of the object is conducted through 2D layer dependent editor [9] - the upper left zone on Fig. 2. Any projection of the object can be selected to be active for meshing [1, 3]. User can work with a concrete etching pit or with the object as a whole. At the beginning of surface meshing user detects the significant layers on the vertical or horizontal section of the object (Fig. 3).

User can draw mesh elements - generic quadrilateral polygons and rectangles. As soon as a new mesh element is added, the automatic checker runs. This checker verifies overlapping of elements, whether angles and edges ratios satisfy mesh requirements or not. If the obtained quadrangle satisfies all mesh requirements, it becomes colored as it is shown on Fig. 4.

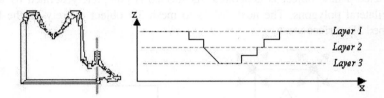

Fig. 3 Sample of definition of significant layers

The upper right zone on Fig. 2 contains interactive 3D viewer that represents geometry of MEMS device. As soon as any type of editing is conducted, viewer updates its screen. Special routine for etching pits recognition has been developed. It gives a user a possibility to delete insignificant pits from the object. This routine scans the active surface of the object and selects all connected faces that form a separate pit. As a result etching pits on 3D view are colored differently.

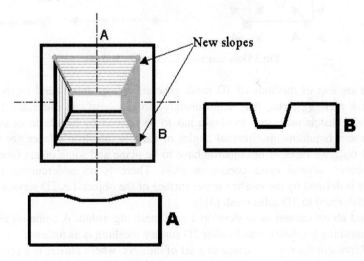

Fig. 4 Sample of smoothing slopes of a simple etching pit

The main goal of working with 2D layout editor is to get a new 3D object which upper surface will be meshed by quadrilateral polygons, convex and unstructured.

Special routine checks whether all final edges are uninterrupted or not. If it is true then the manual part of meshing is completed.

4. Automatic Generation of Solid Mesh

The next part of the algorithm is aimed to the generation of 3D mesh by processing data obtained from the previous stage. When the manual part of the algorithm is completed a new object is generated. Its surface is completely defined by a set of quadrilateral polygons. The next task is to mesh 3D object entirely on the base of obtained 2D surface mesh.

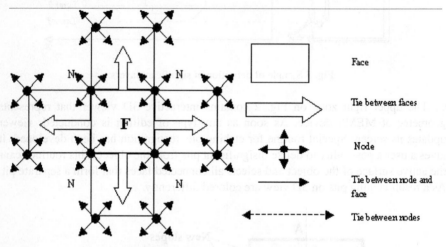

Fig.5 Data structure: faces, nodes and ties

There are lots of methods of 3D mesh generation, e.g., tetrahedral mesh [8], 20-node brick mesh [2], etc., that, unfortunately, do not satisfy in our case. The final solid model that is necessary to obtain has to be formed by hexahedrons satisfying several mesh requirements. Internal angles of each hexahedron's faces are between α_{min} and α_{max}, all faces of hexahedron have to be plane and some others (see section 2). Moreover, several extra conditions exist. There is no undercutting [6]. The geometry is defined by the meshed upper surface of the object, i.e. 2D surface mesh is to be transformed to 3D solid mesh [3].

All said above caused us to develop a new mesh algorithm. A common algorithm of solid meshing for AnisE results after 2D surface meshing is as follows.

(1) Represent the object surface as a set of clusters, where *cluster* is a set of faces, nodes and ties between them (see Fig. 5); so each face knows its neighbors and their ties (sizes, angles, etc.).

(2) Build the adjacency matrix for all faces, so that each face knows the shortest way to space, where *space* is the empty region that surrounds the surface of an object.

CYCLE (until the whole object is meshed):
(3) Surface analysis - finding the cluster with the slope of face plane less than α_{min} or the cluster that completely defines a hexahedron (see Fig. 6);

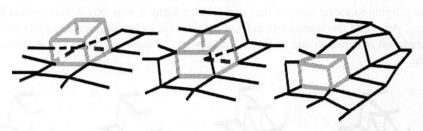

Fig. 6 Samples of clusters defining hexahedrons completely

(4) If there are no such faces, then cut the object vertically by planes formed from surface quadrangles.
(5) Build a hexahedron from the current cluster if this cluster completely defines hexahedron (Fig. 6):
- If obtained hexahedron satisfies mesh requirements, then add missing faces if necessary and remove this hexahedron from the object with replacing obtained emptiness by appropriate faces;
- If obtained hexahedron does not satisfy requirements, then divide a cluster into several ones trying to locate all the changes inside.
(6) If cluster does not define hexahedron completely, then build missing faces with tendency to horizontal planes.
(7) Return to step 3.

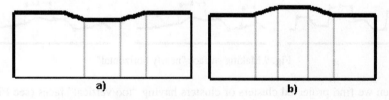

Fig. 7 Sample of 3D meshing: faces are horizontal or nearly horizontal

After data restructuring surface can be analyzed as a set of clusters. One of the approaches of how it can be done is shown on Fig. 7. If all angles between each face and the vertical axis Z is greater than α_{min}, then the object should be cut vertically by planes defined by 2D surface mesh elements (quadrangles) obtained from the previous stage.

If several faces of a cluster are "too vertical" (the angle between the face and axis Z is less than α_{min}), then it can be tried to build a hexahedron from this cluster.

The definition of a hexahedron depends on the number of ties in cluster and spatial location of faces. A cluster can completely define a hexahedron, e.g. if a cluster has 8 nodes and only 24 node-to-node ties.

In the case when a cluster does not completely define a hexahedron and several faces are not "nearly horizontal" (see Fig. 7), one another approach can be used. We can generate all missing parts, faces or edges, from 2D surface mesh elements with tendency to optimizing (construction of plane intersections, point-wise construction, etc.), trying to locate them so that angles being formed with axis Z were greater than α_{min}. After that this completely defined polygon has to be cut from the object and the emptiness obtained should be filled with appropriate faces and nodes as it is shown on Fig. 8.

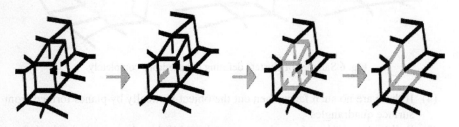

Fig. 8 Finding a cluster that defines a hexahedrons and removing it

Some clusters may completely define elements but obtained hexahedrons do not satisfy requirements (facets are not plane or angles are not correct). It can happen because of the characteristics of faces and their spatial location. Thus this completely defined cluster should be divided into several clusters that form correct hexahedrons. It is desirable that all changes should be localized inside initial cluster. New generated hexahedrons should be removed from the object as it was described above.

Fig. 9 Making surface "nearly horizontal"

Then we find projected clusters or clusters having "too vertical" faces (see Fig. 9), construct hexahedrons and cut them off till we get "nearly horizontal" surface or mesh the object as a whole.

5. Conclusion

The uniqueness of the algorithm described in the paper consists in the combination of manual simplification of 3D geometry with automatic mesh generation. It will give a user the ability to control the process of generating mesh. The following algorithm being implemented now is the first step of the long work aimed to the development of the interface between anisotropic etch simulator of IntelliSuite CAD for MEMS and its performance analyses components. The final goal of this work is to develop automatic 3D Finite Element Mesh generator that can be corrected by user if necessary [9].

These algorithms both elaborated and projected can be also used for solution of various mesh generation tasks when 3D object is defined by a set of triangles and/or quadrangles and mesh requirements are similar to those listed above.

Adding to IntelliSuite a new component based on the approach represented in the paper will permit to serve better users needs. IntelliSuite users will be able to go on processing anisotropic etch simulation results via electrostatic, mechanic, electromagnetic and other types of analyses within a uniform design environment that they can not do today.

References

1. Bogdanov A.V., Stankova E.N., Zudilova E.V.: Visualization Environment for 3D Modeling of Numerical Simulation Results, Proceedings of the 1st SGI Users' Conference, pp. 487-494 (2000).
2. Dhondt G.D.: Unstructured 20-node brick element meshing, Computer Aided Design N 33, pp. 233-249 (2001).
3. Farin G.: Curves and Surfaces for Computer Aided Geometric Design: A Practical Guide, Academic Press, Boston (1993).
4. He Y., Marchetti J., Maseeh F.: MEMS Computer-aided Design. Proc. of the 1997 European Design &Test Conference and Exhibition Microfabrication (1997).
5. Joun M.S., Lee M.C.: Quadrilateral Finite-Element Generation for Mesh Quality Control for Metal Forming Simulation, Int. J. Num. Meth. Engng., 40(21), pp.4059-4075 (1997).
6. Marchetti J., He Y., Than O., Akkaraju S.: Process Development for Bulk Silicon Etching using Cellular Automata Simulation Techniques. Proc. of SPIE's 1998 Symposium on Micromachining and Microfabrication, Micromachined Devices and Components, Santa Clara, CA (1998).
7. Maza S., Noel F., Leon J.C.: Generation of quadrilateral meshes on free-form surfaces, Computers and Structures, 71, pp.505-524 (1999).
8. O'Rourke J.: Computational Geometry in C (second edition), Cambridge University Press, 377 p. (1998)
9. Upson C., Faulhaber J., Kamins D.: The Application Visualization System: A Computational Environment for Scientific Visualization, IEE Computer Graphics and Applications. 9(4), pp. 30-42 (1989).

A Fractional Splitting Algorithm for Non–overlapping Domain Decomposition

Daoud S. Daoud and D. S. Subasi

Department of Mathematics, Eastern Mediterranean Univ., Famaqusta, North
Cyprus - Mersin 10 Turkey
daoud.daoud@emu.edu.tr

Abstract. In this paper we study the convergence of the non overlapping domain decomposition for solving large linear system arising from semi discretization of two dimensional initial value problem with homogeneous boundary conditions, and solved by implicit time stepping using first and, two alternatives of second order FS-methods. The interface values along the artificial boundary condition line are found using explicit forward Euler's method for the first order FS-method, and for the second order FS-method to use extrapolation procedure for each spatial variable individually. The solution by the non overlapping domain decomposition with FS-method is applicable to problems that requires the solution on non uniform meshes for each spatial variables which will unable us to use different time stepping over different subdomains, and with the possibility of extension to three dimensional problem.

1   Introduction and Notation

In domain decomposition the interface boundary conditions for one or two dimensional heat equations have been studied by [DDD1,BLR1,Me1] and [Me2]. In two dimensional problem its generally required iterative solution methods, with good preconditioner, such as preconditioning conjugate gradient method [DDD1,Me1,Me2], to solve the subproblems.

In literature one of the methods received great attention in solving two or three dimensional parabolic equation is the Fractional splitting method, FS-method, or dimensional splitting method [H1,GM1,KL1,S2]. The FS method was firstly introduced by Godunov, 1959 [Go1], and now its one of the classical methods [GM1,MG1,Ya1].

Our aim in this work is to present simpler criteria for defining interface boundary condition for two dimensional parabolic problem, which could be easily generalized to three dimensional problem. The approach is by incorporating interface boundary conditions defined for one dimensional problem with the one dimensional parabolic equation accomplished from FS-method along each spatial variable. The presented form of combination will reduce the cost of the algorithm and the storage requirements. This combination of the FS with the domain decomposition method has been recently considered by Gaiffe, et.al, [GGL1], in order to provide

P.M.A. Sloot et al. (Eds.): ICCS 2002, LNCS 2329, pp. 324–334, 2002.

a time discretization scheme to be different over each sub domains. In section 2, we present the model problem definition, the FS-method classifications, and their appropriate definition for the artificial boundary at the interface line for the 2-dimensional heat equation. In section 3, we presented the theoretical aspects related to the error bound, stability and the convergence for the non overlapping FS-method of different order. The numerical experiment which support the convergence analysis, the comparison between the first and second order FS-method, and the relation with the number of subdomains are presented in section 4.

2 Preliminaries

2.1 The Model Problem

In this work we consider the domain decomposition solution of linear two dimensional parabolic problem on the unit rectangular domain $\Omega = (0,1) \times (0,1)$ with $0 \leq t \leq T$, given by

$$
\begin{aligned}
\frac{\partial u}{\partial t} &= -\frac{\partial^2 u}{\partial x^2} - \frac{\partial^2 u}{\partial y^2} && on \ \Omega \\
B.C. : u(x,y,t) &= 0 && on \ \Gamma = \partial\Omega, \ and \\
I.C. : u(x,y,0) &= u_0(x,y) \ on \ \Omega \cup \Gamma
\end{aligned}
\tag{1}
$$

Let $h = \frac{1}{N+1}$, be the mesh spacing for the spatial variables x and y, and let Ω_h be the spatial grid domain defined by $\Omega_h = \{(x,y)| \ x_i = ih, \ y_j = jh, 1 \leq i,j \leq N\}$.
The grid functions $u_{i,j}^m$ representing the discrete solution over Ω_h, represented by vectors in $\mathbf{R}^{N \times N}$, with natural row ordering for the grid functions at time t_m.
The spatial operator $-\frac{\partial^2}{\partial x^2} - \frac{\partial^2}{\partial y^2}$, is approximated by the standard second order finite difference discretization (central difference) over Ω_h , and after assembling the unknowns it will leads to a semi discrete system of first order differential equations [LP1].

$$
\frac{du}{dt} = -Au,
\tag{2}
$$

where A is the global matrix of coefficients presented with 5 non zero diagonal entries. The exact solution of (1) satisfies the two term recurrence relation involving the matrix exponential function of the matrix A given by

$$
u(t + \delta t) = e^{-\delta t A} u(t),
\tag{3}
$$

where δt is the time step.
Due to the existence of the constituent splitting of A as $A = A_1 + A_2$, which they (i.e. A_1 and A_2) are of tridiagonal entries. Then $e^A \simeq e^{A_1 + A_2}$ will be considered in terms of e^{A_1} and e^{A_2} as induced by the fractional step method FS [MPR1,Ya1,GM1,MG1].
The fractional splitting method FS due to Yanenko [Ya1], is based on simple splitting of the spatial operator along the spatial variables.

The first step tow ardssolving the model problem (1) is to consider the FS method, whic hcan b e expressed as follo ws:

$$\frac{1}{2}u_t = u_{xx} \quad \text{ov er } [t_m, t_m + \tfrac{\delta t}{2}], \quad \text{and}$$
$$\frac{1}{2}u_t = u_{yy} \quad \text{ov er } [t_m + \tfrac{\delta t}{2}, t_{m+1}]. \tag{4}$$

The solution for (4), is b y theuse of the semi discretization, with resp ect to the spatial v ariable, suc h that the recursiv e solution for u_1, and u_2 over the time in terval $[t_m, t_{m+1}]$, is giv en by

$$u(t_m + \tfrac{\delta t}{2}) = e^{-\delta t A_1}u(t_m),$$
$$u(t_m + \delta t) = e^{-\delta t A_2}u(t_m + \tfrac{\delta t}{2}), \tag{5}$$

and the solution of (5), will b e appro ximatedising P ade'appro ximation for $e^{-\delta t A_i}$.

The FS methods are classified according to the order of the exp onen tial appro ximationeither as a locallysecond order $i.e.$ $O(\delta t^2)$ represen tedb y (1,0) pade' appro ximationor equiv alen tlythe bac kw ardEuler's scheme giv en b y:

$$e^{-\delta t A_i} \simeq (I + \delta t A_i)^{-1} + O(\delta t^2), \tag{6}$$

or as a locallythird order $i.e.$ $O(\delta t^3)$ whic h represen ted b y (1,1) P ade' pro ximationor equiv alen tly the CranNicolson(C.N.)-scheme [H1,MPR1], giv en b y

$$e^{-\delta t A_i} \simeq (I + \frac{\delta t}{2} A_i)^{-1}(I - \frac{\delta t}{2} A_i) + O(\delta t^3). \tag{7}$$

The in troducedclassification is consequeted from the order of the p ow er series represen tation of $e^{\delta t A_i}$, and also from the commutativit y of A_1 and A_2 e.g. [H1 ,MPR1]. F or the first order FS-method (3) is rewritten as follo ws;

$$e^{-\delta t A} \simeq e^{-\delta t A_1}e^{-\delta t A_2}. \tag{8}$$

In a non commutativ ecase, the highest p ossible order to consider for $e^{\delta t A_i}$, is the first order [H1,KL1,MPR1,S1,Sw1], and to achiev e a higher order p ow er series for$e^{\delta t A}$, Strang [S1] presen t analgorithm to pro vide a third order appro ximationfor $e^{\delta t A}$ achiev edb y the follo wing splitting of A g iven b y

$$e^{-\delta t A} \simeq e^{-\frac{\delta t}{2} A_1}e^{-\delta t A_2}e^{-\frac{\delta t}{2} A_1}, \tag{9}$$

when, a locally, thirdorder appro ximationsfor $e^{-\frac{\delta t}{2} A_1}$ and $e^{-\delta t A_2}$ are considered [MPR1].

A further prop osalto pro vide a second order presen tationb y means of po wer series of $e^{-\delta t A_i}$, is due to Sw aynd[Sw1], giv en b y

$$e^{-\delta t A} = \frac{1}{2} \left[e^{-\delta t A_1}e^{-\delta t A_2} + e^{-\delta t A_2}e^{-\delta t A_1} \right]. \tag{10}$$

3   Interface Boundary Condition

3.1   First Order FS-Method

F or the first order FS method w e willconsider the definition of the in terface b oundary conditions b y the F orw ardEuler's appro ximationdue to Dawson, et, al [DDD1] for each spatial variable, along the interface line is given by the following algorithm:

Algorithm 1:
1. Let $\bar{x}(= ph)$ for $1 < p < N$, be an interface point for the spatial variable x, such that the domain Ω is then split into $\Omega_{\bar{x},1} = \{(x_i, y)|0 < x_i < \bar{x}, 0 < y < 1\}$, and $\Omega_{\bar{x},2} = \{(x_i, y)|\bar{x} < x_i < 1, 0 < y < 1\}$. At $t = t_m + \frac{1}{2}\delta t$, the interface boundary condition along the interface line $(\bar{x}, y_j)$, $j = 1, ..., N$, is given by

$$u_{\bar{x},y_j}^{m+\frac{1}{2}} = u_{p,j}^{m+\frac{1}{2}} = ru_{p-1,j}^m + (1 - 2r)u_{p,j}^m + ru_{p+1,j}^m, \quad j = 1, ..., N,$$

where $r = \frac{\delta t}{H_x}$, and H_x is the integral multiple of the mesh spacing h_x with respect to the spatial variable x. The intermediate solution $u^{m+\frac{1}{2}}$ is then given by solving $(I + \delta t A_1)u^{m+\frac{1}{2}} = u^m$, over $\Omega_{\bar{x},1}$ and $\Omega_{\bar{x},2}$ respectively using the backward Euler's scheme.

2. Let $\bar{y}(= qh)$ for $1 < q < N$, be an interface point for the spatial variable y, such that the domain Ω is then split as follows: $\Omega_{\bar{y},1} = \{(x, y_i)|0 < x < 1, 0 < y < \bar{y}\}$, and $\Omega_{\bar{y},2} = \{(x, y_i)|0 < x < 1, \bar{y} < y < 1\}$. At $t = t_{m+1}$ the interface boundary conditions along the interface line $(x_i, \bar{y})$ for $i = 1, ..., N$, is given by

$$u_{x_i,\bar{y}}^{m+1} = u_{i,q}^{m+1} = ru_{i,q+1}^{m+\frac{1}{2}} + (1 - 2r)u_{i,q}^{m+\frac{1}{2}} + ru_{i,q-1}^{m+\frac{1}{2}}, \quad i = 1, ..., N,$$

where $r = \frac{\delta t}{H_y^2}$ and H_y is the integral multiple of the mesh spacing h_y, with respect to the spatial variable y.
Then solve for u^{m+1} the system $(I + \delta t A_2)u^{m+1} = u^{m+\frac{1}{2}}$, over $\Omega_{\bar{y},1}$ and $\Omega_{\bar{y},2}$.

In our analysis we will consider the maximum norm and define the max norm of the error at the interface boundary points by

$$\|\tilde{E}_{x_p}^m\|_j = max_j|U_{p,j}^m - u_{p,j}^m|, \text{ and } \|\tilde{E}_{y_p}^m\|_i = max_i|U_{i,q}^m - u_{i,q}^m|,$$

where x_p and y_p are the interface points for x and y respectively, and $\tilde{E}^m = max\{\tilde{E}_{x_p}^m, \tilde{E}_{y_p}^m\}$.

Lemma 1. *If A_1 and A_2 are symmetric and positive definite matrices, then the first order-FS method is stable in maximum norm independent of h and δt.*

The following theorem due to Mathew, et al [MPR1], concerns the truncation error of the first order-FS method.

Theorem 1. *The first order-FS method approximate solution W^{m+1} of $(I + \delta t A)u^{m+1} = u^m$ solves*

$$(I + \delta t A)W^{m+1} + (\delta t^2 \sum_{1 \leq k_1, k_2 \leq 2} A_{k_1} A_{k_2})W^{m+1} = W^m.$$

The local truncation error T_{FS1} of the first orderFS method has the terms $T_{FS1} = T + (\delta t^2 \sum_{1 \leq k_1, k_2 \leq 2} A_{k_1} A_{k_2})u^{m+1}$ in addition to the terms in the local truncation error T when exact solvers are used. Here u^{m+1} is the exact solution of the parabolic equation (1).

Proof. For the proof see [MPR1].

From the incorporation of the interface boundary conditions given by Algorithm 1, with the FS method, the error bound estimates for the overall error $E^{m+1} = u^{m+1} - W^{m+1}$ at time $t = t_{m+1}$, is given by the following theorem.

Theorem 2. *If A_1, and A_2 are symmetric and positive semi definite matrices, then the error bound for algorithm 1, satisfies the relation $||E^{m+1}|| \leq ||E^m|| + O(\delta t^2)$, and hence algorithm 1, is stable.*

The error at (x_i, y_j, t_m) due to the utilization of Dawson's algorithm in the first order FS-method subject to non-overlapping subdomains is given by.

Theorem 3. *If $\frac{1}{2}|\frac{\partial^2 u}{\partial t^2}|$, $\frac{1}{12}|\frac{\partial^4 u}{\partial x^4}|$, and $\frac{1}{12}|\frac{\partial^4 u}{\partial y^4}|$ are bounded by constant C on $\Omega \times T$, and $\delta t \leq max\{\frac{H_y^2}{2}, \frac{H_x^2}{2}\}$, then*

$$\max_{i,j,m} |u_{i,j}^m - W_{i,j}^m| \leq \frac{C^2}{16}[h_x^2 + h_y^2 + H_x^3 + H_y^3 + \delta t^2],$$

where H_x, and H_y are he integral multiplies of h_x, and h_y respectively.

Proof. Consider the solution method given in (5) for each spatial variable over time interval of length $\frac{\delta t}{2}$, with non overlapping two subdomains, then;

$$\max_{i,j,m} |u_{i,j}^m - W_{i,j}^m| \leq (\max_{j,m} |u(x_i, y_j, t^m) - W(x_i, y_j, t^m)) \atop (\max_{i,m} |u(x_i, y_j, t^m) - W(x_i, y_j, t^m)|). \quad (11)$$

For the heat equation defined by $E_{i,j}^{m+1}$, with the homogeneous boundary conditions, and the artificial boundary conditions at the interface line (Algorithm 1), then for the spatial variable x, we have

$$\max_{i,m} |u(x_i, y_j, t^m) - W(x_i, y_j, t^m)| \leq C[\frac{1}{8}(\frac{\delta t}{2} + h_x^2) + \frac{1}{4}H_x(H_x^2 + \frac{\delta t}{2})]. \quad (12)$$

Similarly for the spatial variable y,

$$\max_{j,m} |u(x_i, y_j, t^m) - W(x_i, y_j, t^m)| \leq C\left[\frac{1}{8}(\frac{\delta t}{2} + h_y^2) + \frac{1}{4}H_y(H_y^2 + \frac{\delta t}{2})\right]. \quad (13)$$

By substituting (12), and (13), in (11),

$$\max_{i,j,m} |u_{i,j}^m - W_{i,j}^m| \leq C[\frac{1}{8}(\frac{\delta t}{2} + h_x^2) + \frac{1}{4}H_x(H_x^2 + \frac{\delta t}{2})] \atop C\left[\frac{1}{8}(\frac{\delta t}{2} + h_y^2) + \frac{1}{4}H_y(H_y^2 + \frac{\delta t}{2})\right]. \quad (14)$$

After performing the desired product for the terms in the last equation and summing up the terms of the same order the desired results is obtained given as follows,

$$\max_{i,j,m} |u_{i,j}^m - W_{i,j}^m| \leq \frac{C^2}{16}[h_x^2 + h_y^2 + H_x^3 + H_y^3 + \delta t^2]. \quad (15)$$

3.2   Interface Boundary Condition- Second Order FS-Method

The globally second order FS-method can be achieved using Strang's splitting for A, $A = \frac{1}{2}A_1 + A_2 + \frac{1}{2}A_1$, such that

$$e^{-\delta t A} \simeq e^{-\frac{\delta t}{2}A_1} e^{-\delta t A_2} e^{-\frac{\delta t}{2}A_1}, \tag{16}$$

[KL1,MPR1,S1,S2].
The exponential term $e^{-\delta t B}$, $(\delta t B = \frac{\delta t}{2}A_1$, or $\delta t A_2)$ in (16), are approximated by $(1,1)$ Pade' approximation of locally third order $O(\delta t^3)$, denoted by $R_{1,1}(\delta t B)$, given by

$$e^{-\delta t B} \simeq R_{1,1}(-\delta t B) = (I + \frac{\delta t}{2}B)^{-1}(I - \frac{\delta t}{2}B) + O(\delta t^3), \tag{17}$$

or equivalently the C.N. scheme [H1,MPR1].
The solution $u(t + \delta t)$ of (3), is evaluated by using an intermediate solution vectors, internal estimation vectors, w and v, as follows;

$$u(t + \delta t) = e^{-\frac{\delta t}{2}A_1} w, \text{ where } w = e^{-\delta t A_2} v, \text{ and} v = e^{-\frac{\delta t}{2}A_1} u. \tag{18}$$

In addition to the above scheme (18), Swayne [Sw1] presented a further type of second order approximation to $e^{-\delta t A}$ given by :

$$e^{-\delta t A} = \frac{1}{2}[e^{-\delta t A_1} e^{-\delta t A_2} + e^{-\delta t A_2} e^{-\delta t A_1}], \tag{19}$$

such that the solution $u(t + \delta t)$ of (3), is given by;

$$u(t + \delta t) = \frac{1}{2}[u^{(1)}(t + \delta t) + u^{(2)}(t + \delta t)]. \tag{20}$$

To provide an interface boundary condition u_I^m, for each spatial variables, with second order accurate in time $i.e.$ with leading error term $O(\delta t^3)$ we considered an extrapolation for $u^{m-\gamma\delta t}$ and $u^{m-\beta\delta t}$, for $0 < \gamma \leq 1$ and $1 < \beta \leq 2$, given by

$$(\beta - \gamma)u_I^m = \beta u^{m-\gamma\delta t} - \gamma u^{m-\beta\delta t} + O(\delta t^3), \tag{21}$$

for $m \geq 2$, $\gamma \leq 1$, and $\beta = 2$, then (21) is given by

$$u_I^m = 2u^{m-1} - u^{m-2} + O(\delta t^3),$$

where $u_I^m = u_{p,j}^m$ or $u_I^m = u_{i,q}^m$ are the interface boundary conditions for the spatial variables x or y respectively. The above approach has also a flexibility to be used even for estimating the interface boundary condition for a higher space dimension, and its a stable scheme for positive δt [LM1]. The domain decomposition algorithm for (1), by the second order FS defined by Strang's splitting method with the boundary condition defined by (21) is given by the following algorithm:

Algorithm 2:
Let u^1 be the solution from the second order FS method or any second order method in time e.g. C.N., and u^0 be the initial condition of (1), then for $m \geq 2$, and with the interface boundary condition given by (21) according to each spatial variable.

1. Let $(\bar{x}, y_j)$, $\bar{x} = ph$, $1 < p < N$, $j = 1, ..., N$ be interface points for the spatial variable x over $[t_m, t_{m+\frac{1}{2}}]$, and solve for v, $v = e^{-\frac{\delta t}{2}A_1}u$.

2. Let $(x_j, \bar{y})$, $\bar{y} = qh$, $1 < q < N$, $i = 1, ..., N$ be interface points for the spatial variable y over $[t_m, t_m + \delta t]$, and solve for w, $w = e^{-\delta t A_2}v$.

3. Solve for u, $u = e^{-\frac{\delta t}{2}A_1}w$ over$[t_m + \frac{\delta t}{2}, t_m + \delta t]$, with the interface boundary conditions given by (21) with respect to the spatial variable x, i.e. at the interface points $(\bar{x}, y_j)$.

The domain decomposition algorithm for (1), by the second order FS defined using Swayne's splitting method with the boundary condition defined by (21) is given by the following algorithm:

Algorithm 3:

Let u^1 be the solution from the second order FS method or any second order method in time e.g. C.N., and u^0 be the initial condition of (1), then for $m \geq 2$,

1. Let $(\bar{x}, y_j)$, $\bar{x} = ph$, $1 < p < N$, $j = 1, ..., N$ be interface points for the spatial variable x over $[t_{m-1}, t_m]$, with interface boundary conditions given by (21), at t_m, and solve for u^*, $u^* = e^{-\delta t A_1}u$.

2. Let $(x_i, \bar{y})$, $\bar{y} = qh$, $1 < q < N$, $i = 1, ..., N$ be interface points for the spatial variable y over $[t_{m-1}, t_m]$ with interface boundary conditions given by (21), at t_m, and solve for w, $w = e^{-\delta t A_2}u^*$, $(u^{(1)} = w)$.

3. Let $(x_i, \bar{y})$, $\bar{y} = qh$, $1 < q < N$, $i = 1, ..., N$ be interface points for the spatial variable y over $[t_{m-1}, t_m]$, with interface boundary conditions given by (21), at t_m, and solve for $\tilde{u}$, $\tilde{u} = e^{-\delta t A_2}u$.

4. Let $(\bar{x}, y_j)$, $\bar{x} = ph$, $1 < p < N, j = 1, ..., N$ be interface points for the spatial variable x over $[t_{m-1}, t_m]$ with interface boundary conditions given by (21), at t_m, and solve for v, $v = e^{-\delta t A_2}\tilde{u}$, $(u^{(2)} = v)$.
   Then $u(t + \delta t) = \frac{1}{2}(u^{(1)} + u^{(2)})$.

The stability of the second order FS method defined by Strang's splitting [MPR1], and the Swayne's proposal governed by the stability of the rational approximation $R_{1,1}(\delta t A_i)$ for the exponent of the matrices A_1, and A_2, which is due to the stability of the C.N.-Scheme, [H1,MPR1,Sw1]. The following theorem concerns the truncation error for the second order FS-splitting.

Theorem 4. *The second order-FS method approximates solution W^{m+1} of*

$$R_{1,1}(\delta t A)u^{m+1} = u^m,$$

by Strang's Splitting, solves

$$R_{1,1}(\tfrac{\delta t}{2}A_1)R_{1,1}(\delta t A_2)R_{1,1}(\tfrac{\delta t}{2}A_1)W^{m+1} + (\delta t^3 \sum_{j,1 \leq k_1,k_2,k_3 \leq 2} \tau_j A_{k_1}A_{k_2}A_{k_3})W^{m+1} = W^m,$$

and by Swayne's proposal solves

$$\tfrac{1}{2}(R_{1,1}(\delta t A_1)R_{1,1}(\delta t A_2) + R_{1,1}(\delta t A_2)R_{1,1}(\delta t A_1))W^{m+1} +$$
$$(\delta t^3 \sum\nolimits_{j,1\leq k_1,k_2,k_3\leq 2} \rho_j A_{k_1} A_{k_2} A_{k_3})W^{m+1} = W^m.$$

The local truncation error T_{FS} of the second order FS-method has the following terms in addition to the terms in the local error T when exact solvers are used,

$$T_{FS}(strang) = T + (\delta t^3 \sum_{j,1\leq k_1,k_2,k_3\leq 2} \tau_j A_{k_1} A_{k_2} A_{k_3})u^{m+1},$$

and

$$T_{FS}(swayne) = T + (\delta t^3 \sum_{j,1\leq k_1,k_2,k_3\leq 2} \rho_j A_{k_1} A_{k_2} A_{k_3})u^{m+1},$$

corresponding to Strang's Splitting and Swayne's proposal, respectively. Here u^{m+1} is the exact solution of the parabolic equation (21).

Corollary 1. *The maximum norm of the $O(\delta t^3)$ term, in $(e^{\delta t A} - e^{\frac{\delta t}{2} A_1} e^{\delta t A_2} e^{\frac{\delta t}{2} A_1})$ by Strang's splitting is given by;*

$$\frac{O(\delta t^3)}{6}(30\|A_1\|^2\|A_2\| + 20\|A_1\|\|A_2\|^2),$$

and the maximum norm of the $O(\delta t^3)$, term $(e^{\delta t A} - \tfrac{1}{2}(e^{\delta t A_1}e^{\delta t A_2} + e^{\delta t A_2}e^{\delta t A_1}))$, by Swayne's proposal is given by;

$$O(\delta t^3)(2\|A_1\|^2\|A_2\| + 2\|A_2\|^2\|A_1\|)$$

4   Numerical Experiments and Conclusion

The proposed algorithms 1, 2, and 3 are examined to solve the following two dimensional heat equation

$$\frac{\partial u}{\partial t} = \frac{\partial^2 u}{\partial x^2} + \frac{\partial^2 u}{\partial y^2}, 0 \leq x,y \leq 1,\ t > 0 \qquad (22)$$

with boundary conditions $u(0,y,t) = u(1,y,t) = 0$, $0 \leq y \leq 1$, and $u(x,0,t) = u(x,1,t) = 0$, $0 \leq x \leq$ $t > 0$, and initial conditions $u(x,y,0) = \sin(\pi x)\sin(\pi y)$, $0 \leq x,y \leq 1$. The standard central difference approximation was used for the discretization for each spatial variable x, and y, on uniform grid with $h = \frac{1}{100}$, and $\frac{1}{150}$, for the spatial variables x, and y, and the discretized problem is then solved over time intervals $t = 0.1,\ 0.2,\ 0.5$, and $t = 1$. The solution is then discussed according to different number of subdomains.
Figure 1 plots the maximum error by algorithms 1, 2, and 3, for 2, and 15 subdomains by mesh spacing $h = \frac{1}{150}$, over $t = [0,1]$.
The figure indicated that the error by the second order FS-algorithms, algorithms 2, and 3 has less error than the first order, and that is due to the order of splitting as referred in theorems 3, 4, and also due to

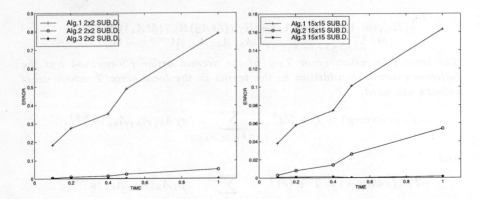

Fig. 1. The error by the FS-algorithms 1, 2, and 3, on the left for 2 subdomains and on the right for 15 subdomains, for $h = \frac{1}{150}$, over the time interval $[0,1]$.

the error at the artificial boundary at the interface predicted by the forward Euler's scheme which posses a restriction on the δt as given in theorem 3, not like the case in the prediction of the interface for the second order FS-method. Also, the error by algorithm 3 produce the solution with less error than the solution by algorithm 2 which is due to corollary 1 of theorem 4. Figure 2, shows the plots of maximum error by the algorithms 1 and 2, respectively over different time intervals $t = 0.1$, 0.2, 0.4, 0.5, and 1, solved for different number of sub domains, we concluded that the algorithm of the first order FS-method depends on the number of the subdomains but on the contrary the second order FS-method does not depend on the number of the sub domains.   The

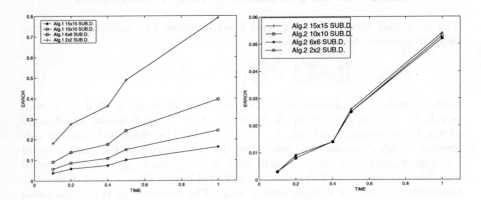

Fig. 2. The effect of the number of subdomains on the error of the first order FS-algorithm 1 on the left for and on the right for FS-algorithm 2 for $h = \frac{1}{150}$, over the time interval $[0,1]$.

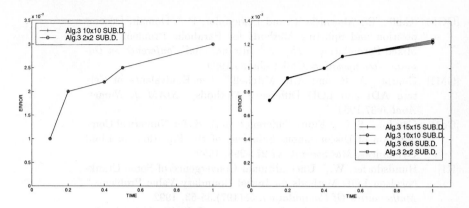

Fig. 3. The second order FS-algorithm 3, is independent on the number of the subdomains, on the right for $h = \frac{1}{150}$ and on the left for $h = \frac{1}{100}$ over the time interval $[0,1]$.

deterioration of the accuracy(error) in the domain splitting algorithm for larger time interval is mainly due to the truncation error in the discretization and the interface prediction of the boundary values at the artificial boundary of the interface. A similar conclusion is concluded by algorithm 3 as illustrated in figure 3 for different mesh spacing. For both types of splitting by algorithms 2, and 3 the numerical results demonstrates that the error from the solution of the model problem does not depend on the number of the sub domains, not like the case in the first order FS-method algorithm 1, as shown in fig(2).

The main features of the presented work are the following points:

1. The algorithm provide a solution of the original problem on subdomains and thus allows a refinement in time variable, and in xy-plane, according to each space variable independently and then solve over each subdomain for each space variable with independent mesh spacing. The algorithm is also extendable to three dimensional problem.

2. The second order FS- method, by Swayne's algorithm, by itself is a completely parallel algorithm called *parallel alternating scheme* [H1], according to each term in (19), and provides a further parallelism enhanced by the solution of the subproblem over a non overlapping subdomains according to each spatial variable.

References

[BLR1] Blum, H.,Lisky, S. and Ranacher, R.," A domain Splitting Algorithm for Parabolic Problems," *Computing*, 49:11-23, 1992.

[DDD1] Dawson,C.N.,Du, Q., and Dupont, D.F." A finite Difference Domain Decomposition Algorithm for Numerical Solution of the Heat Equation," *Mathematics of Computation*, 57:63-71, 1991.

[GGL1] Gaiffe, S., Glowinski,R. and Lemonnier, P. " Domain Decomposition and Splitting Methods for Parabolic Problems via a Mixed Formulation", *The* 12th *International Conference on Domain Cecomposition,* Chiba -Japan,1999.

[GM1] Gourlay, A. R. and A.R. Mitchell," The Equivalence of Certain ADI and LOD Difference Methods," *SIAM J. Numer. Anal.*,6:37,1969.

[Go1] Goudnov, S. K.," Finite Difference Methods for Numerical Computation of Discontinuous Solutions of the Equations of Fluid Dynamics", *Mat.Sbornik,* 47:271-295, 1959.

[H1]   Hundsdorfer, W.," Unconditional Convergence of Some Crank-Nicolson LOD Methods for Initial Boundary Value Problems," *Mathematics Of Computation,* 58(197):35-53, 1992.

[KL1] Khan, L.A., and P.L. Liu," An Operator-Splitting Algorithm for the Three-Dimensional Diffusion Equation," *Numerical methods for Partial Differential Equations* 11:617-624, 1995.

[LM1] Lawson, J. D. and Morris J, LI, " The Extrapolation of First Order Methods for Parabolic Partial Differential Equations. I," *SIAM J. Numer. Anal.* 15:1212-1224, 1978 .

[LP1] Lapidus, L., and Pinder, G.F.: " Numerical Solution of Partial Differential Equations in Science and Engineering", *John Wiley,* 1982.

[Me1] Meurant, G.A.," A Domain Decomposition Method for Parabolic Problems," *Applied Numerical Mathematics,* 8:427-441,1991.

[Me2] Meurant,G.A.," Numerical Experiments with Domain Decomposition Method for Parabolic Problems on Parallel Computers," 4th *"Int. Sypm. on Domain Decomposition methods for Partial Differential Equations,"* ed. Glowinski,R.,Kunetsov,Y.A., Meurant,G.A., ,Periaus, J., and Widlund,O.W., SIAM, PA., 394-408,1991.

[MG1] Mitchell, A.R. and Griffiths, D.F.," The Fnite Difference Methods in Partial Differential Equations",*John Wiley,* 1980.

[MPR1] Mathew, T.P., Polyakov, P.L., Russo,G., and Wang, J.," Domain Decomposition Operator Splittings for the Solution of Parabolic Equations," *SIAM J.Sci.Comput.* 19:912-932, 1998.

[S1]   Strang, G.," Accurate Partial Difference Methods I: Linear Cauchy Problems," *Arch. Rational Mech. Anal.*, 12:392-402, 1963.

[S2]   Strang,G.," On the construction and Comparison of Difference Schemes," *SIAM J. Numer. Anal.*, 5:506-517, 1968.

[Sw1] Swayne, D.A.," Time Dependent Boundary and Interior Forcing in Locally One Dimensional Schemes," *SIAM J.Sci.Stat.Comput,* 8:755-767, 1987.

[Ya1] Yanenko, N.N.," The method of Fractional Steps; The Solution of Problems of Mathematical Physics in Several Variables", *Springer Verlag,* 1971.

Tetrahedral Mesh Generation for Environmental Problems over Complex Terrains*

Rafael Montenegro, Gustavo Montero, José María Escobar,
Eduardo Rodríguez, and José María González-Yuste**

University Institute of Intelligent Systems and Numerical Applications in Engineering,
University of Las Palmas de Gran Canaria,
Edificio Instituto Polivalente, Campus Universitario de Tafira,
35017 Las Palmas de Gran Canaria, Spain
{rafa, gustavo}@dma.ulpgc.es, escobar@cic.teleco.ulpgc.es,
barrera@dma.ulpgc.es, josem@sinf.ulpgc.es

Abstract. In the finite element simulation of environmental processes that occur in a three-dimensional domain defined over an irregular terrain, a mesh generator capable of adapting itself to the topographic characteristics is essential. The present study develops a code for generating a tetrahedral mesh from an "optimal" node distribution in the domain. The main ideas for the construction of the initial mesh combine the use of a refinement/derefinement algorithm for two-dimensional domains and a tetrahedral mesh generator algorithm based on Delaunay triangulation. Moreover, we propose a procedure to optimise the resulting mesh. A function to define the vertical distance between nodes distributed in the domain is also analysed. Finally, these techniques are applied to the construction of meshes adapted to the topography of the southern section of La Palma (Canary Islands).

1 Introduction

The problem in question presents certain difficulties due to the irregularity of the terrain surface. Here we construct a tetrahedral mesh that respects the orography of the terrain with a given precision. To do so, we only have digital terrain information. Furthermore, it is essential for the mesh to adapt to the geometrical terrain characteristics. In other words, node density must be high enough to fix the orography by using a linear piecewise interpolation. Our domain is limited in its lower part by the terrain and in its upper part by a horizontal plane placed at a height at which the magnitudes under study may be considered steady. The lateral walls are formed by four vertical planes. The generated mesh could be used for numerical simulation of natural processes, such as wind field adjustment [9], fire propagation [8] and atmospheric pollution. These phenomena have the main effect on the proximities of the terrain surface. Thus node density increases in these areas accordingly.

---

* Partially supported by MCYT, Spain. Grant contract: REN2001-0925-C03-02/CLI
** The authors acknowledge Dr. David Shea for editorial assistance

P.M.A. Sloot et al. (Eds.): ICCS 2002, LNCS 2329, pp. 335–344, 2002.

To construct the Delaunay triangulation, we must define a set of points within the domain and on its boundary. These nodes will be precisely the vertices of the tetrahedra that comprise the mesh. Point generation on our domain will be done over several layers, real or fictitious, defined from the terrain up to the upper boundary, i.e. the top of the domain. Specifically, we propose the construction of a regular triangulation of this upper boundary. Now, the refinement/derefinement algorithm [3, 11] is applied over this regular mesh to define an adaptive node distribution of the layer corresponding to the surface of the terrain. These process foundations are summarised in Sect. 2. Once the node distribution is defined on the terrain and the upper boundary, we begin to distribute the nodes located between both layers. A vertical spacing function, studied in Sect. 3, is involved in this process.

The node distribution in the domain will be the input to a three-dimensional mesh generator based on Delaunay triangulation [2]. To avoid conforming problems between mesh and orography, the tetrahedral mesh will be designed with the aid of an auxiliary parallelepiped. Section 4 is concerned with both the definition of the set of points in the real domain and its transformation to the auxiliary parallelepiped where the mesh is constructed. Next, the points are placed by the appropriate inverse transformation in their real position, keeping the mesh topology. This process may give rise to mesh tangling that will have to be solved subsequently. We should, then, apply a mesh optimisation to improve the quality of the elements in the resulting mesh. The details of the triangulation process are developed in Sect. 5; those related to the mesh optimisation process are presented in Sect. 6. Numerical experiments are shown in Sect. 7, and, finally, we offer some concluding remarks.

2   Adaptive Discretization of the Terrain Surface

The three-dimensional mesh generation process starts by fixing the nodes placed on the terrain surface. Their distribution must be adapted to the orography to minimise the number of required nodes. First, we construct a sequence of nested meshes $T = \{\tau_1 < \tau_2 < ... < \tau_m\}$ from a regular triangulation τ_1 of the rectangular area under consideration. The τ_j level is obtained by previous level τ_{j-1} using the 4-T Rivara algorithm [12]. All triangles of the τ_{j-1} level are divided in four sub-triangles by introducing a new node in the centres of each edge and connecting the node introduced on the longest side with the opposite vertex and with the other two introduced nodes. Thus, new nodes, edges and elements named *proper* of level j appear in the τ_j level. The number of levels m of the sequence is determined by the degree of discretization of the terrain digitalisation. In other words, the diameter of the triangulation must be approximately the spatial step of the digitalisation. In this way we ensure that the mesh is capable of obtaining all the topographic information by an interpolation of the actual heights on the mesh nodes. Finally, a new sequence $T' = \{\tau_1 < \tau_2' < ... < \tau_{m'}'\}$, $m' \leq m$, is constructed by applying the derefinement algorithm; details may be seen in [3, 11]. In this step we present the derefinement parameter ε that fixes the precision with which we intend to approximate the terrain topography. The difference in absolute value between the resulting heights at any point of the mesh $\tau'_{m'}$ and its corresponding real height will be less than ε.

This resulting two-dimensional mesh $\tau'_{m'}$ may be modified when constructing Delaunay triangulation in the three-dimensional domain, as its node position is the only information we use. We are also interested in storing the level in which every node is proper so as to proceed to the node generation inside the domain. This will be used in the proposed vertical spacing strategies.

3   Vertical Spacing Function

As stated above, we are interested in generating a set of points with higher density in the area close to the terrain. Thus, every node is to be placed in accordance with the following function

$$z_i = a\,i^\alpha + b \ . \tag{1}$$

so that when the exponent $\alpha \geq 1$ increases, it provides a greater concentration of points near the terrain surface. The z_i height corresponds to the ith inserted point, in such a way that for $i = 0$ the height of the terrain is obtained, and for $i = n$, the height of the last introduced point. This last height must coincide with the altitude h of the upper plane that bounds the domain. In these conditions the number of points defined over the vertical is $n + 1$ and (1) becomes

$$z_i = \frac{h - z_0}{n^\alpha} i^\alpha + z_0 \ ; \quad i = 0, 1, 2, ..., n \ . \tag{2}$$

It is sometimes appropriate to define the height of a point in terms of the previous one, thus avoiding the need for storing the value of z_0

$$z_i = z_{i-1} + \frac{h - z_{i-1}}{n^\alpha - (i-1)^\alpha} \left[i^\alpha - (i-1)^\alpha \right] \ ; \quad i = 1, 2, ..., n \ . \tag{3}$$

In (2) or (3), once the values of α and n are fixed, the points to insert are completely defined. Nevertheless, to maintain acceptable minimum quality of the generated mesh, the distance between the first inserted point ($i = 1$) and the surface of the terrain could be fixed. This will reduce to one, either α or n, the number of degrees of freedom. Consider the value of the distance d as a determined one, such that $d = z_1 - z_0$. Using (2),

$$d = z_1 - z_0 = \frac{h - z_0}{n^\alpha} \ . \tag{4}$$

If we fix α and set free the value of n, from (4) we obtain

$$n = \left(\frac{h - z_0}{d} \right)^{1/\alpha} \ . \tag{5}$$

Nevertheless, in practice, n will be approximated to the closest integer number. Conversely, if we fix the value of n and set α free, we get

$$\alpha = \frac{\log \frac{h-z_0}{d}}{\log n} . \tag{6}$$

In both cases, given one of the parameters, the other may be calculated by expressions (5) or (6), respectively. In this way, the point distribution on the vertical respects the distance d between z_1 and z_0. Moreover, if the distance between the last two introduced points is fixed, that is, $D = z_n - z_{n-1}$, then the α and n parameters are perfectly defined. Let us assume that α is defined by (6). For $i = n - 1$, (2) could be expressed as

$$z_{n-1} = \frac{h-z_0}{n^\alpha} (n-1)^\alpha + z_0 . \tag{7}$$

and thus, by using (6),

$$\frac{\log (n-1)}{\log n} = \frac{\log \frac{h-z_0-D}{d}}{\log \frac{h-z_0}{d}} . \tag{8}$$

From the characteristics which define the mesh, we may affirm *a priori* that $h - z_0 > D \geq d > 0$. Thus, the value of n will be bounded such that, $2 \leq n \leq \frac{h-z_0}{d}$, and the value of α cannot be less than 1. Moreover, to introduce at least one intermediate point between the terrain surface and the upper boundary of the domain, we must verify that $d + D \leq h - z_0$. If we call $k = \frac{\log \frac{h-z_0-D}{d}}{\log \frac{h-z_0}{d}}$, it can be easily proved that $0 \leq k < 1$. So, (8) yields

$$n = 1 + n^k . \tag{9}$$

If we name $g(x) = 1 + x^k$, it can be demonstrated that $g(x)$ is contractive in $\left[2, \frac{h-z_0}{d}\right]$ with Lipschitz constant $C = \frac{1}{2^{1-k}}$, and it is also bounded by

$$2 \leq g(x) \leq 1 + \left(\frac{h-z_0}{d}\right)^k \leq \frac{h-z_0}{d} . \tag{10}$$

In view of the fixed point theorem, we can ensure that (9) has a unique solution which can be obtained numerically, for example, by the fixed point method, as this converges for any initial approximation chosen in the interval $\left[2, \frac{h-z_0}{d}\right]$. Nevertheless, the solution will not generally have integer values. Consequently, if its value is approximated to the closest integer number, the imposed condition with distance D will not exactly hold, but approximately.

4   Determination of the Set of Points

The point generation will be carried out in three stages. In the first, we define a regular two-dimensional mesh τ_1 for the upper boundary of the domain with the required density of points. Second, the mesh τ_1 will be globally refined and subsequently derefined to obtain a two-dimensional mesh $\tau'_{m'}$ capable of fitting itself to the topography of the terrain. This last mesh defines the appropriate node distribution over the terrain surface. Next, we generate the set of points distributed between the upper boundary and the terrain surface. In order to do this, some points will be placed over the vertical of each node P of the terrain mesh $\tau'_{m'}$, attending to the vertical spacing function and to level j ($1 \leq j \leq m'$) where P is proper. The vertical spacing function will be determined by the strategy used to define the following parameters: the topographic height z_0 of P; the altitude h of the upper boundary; the maximum possible number of points $n+1$ in the vertical of P, including both P and the corresponding upper boundary point, if there is one; the degree of the spacing function α; the distance between the two first generated points $d = z_1 - z_0$; and the distance between the two last generated points $D = z_n - z_{n-1}$. Thus, the height of the ith point generated over the vertical of P is given by (2) for $i = 1, 2, ..., n-1$.

Regardless of the defined vertical spacing function, we shall use level j where P is proper to determine the definitive number of points generated over the vertical of P excluding the terrain and the upper boundary. We shall discriminate among the following cases:

1. If $j = 1$, that is, if node P is proper of the initial mesh τ_1, nodes are generated from (2) for $i = 1, 2, ..., n-1$.

2. If $2 \leq j \leq m' - 1$, we generate nodes for $i = 1, 2, ..., min(m' - j, n - 1)$.

3. If $j = m'$, that is, node P is proper of the finest level $\tau'_{m'}$, then any new node is generated.

This process has its justification, as mesh $\tau'_{m'}$ corresponds to the finest level of the sequence of nested meshes $T' = \{\tau_1 < \tau'_2 < ... < \tau'_{m'}\}$, obtained by the refinement/derefinement algorithm. Thus the number of introduced points decreases smoothly with altitude, and they are also efficiently distributed in order to build the three-dimensional mesh in the domain.

We set out a particular strategy where values of α and n are automatically determined for every point P of $\tau'_{m'}$, according to the size of the elements closest to the terrain and to the upper boundary of the domain. First, the value of d for each point P is established as the average of the side lengths of the triangles that share P in the mesh $\tau'_{m'}$. A unique value of D is then fixed according to the desired distance between the last point that would be theoretically generated over the different verticals and the upper boundary. This distance is directly determined according to the size of the elements of the regular mesh τ_1. Once d and D are obtained, for every point P of $\tau'_{m'}$, their corresponding value of n is calculated by solving (9). Finally, the vertical spacing function is determined when obtaining the value of α by (6). This strategy approximately respects both the required distances between the terrain surface and the first layer and the imposed distance between the last virtual layer and the upper boundary.

5   Three-dimensional Mesh Generation

Once the set of points has been defined, it will be necessary to build a three-dimensional mesh able to connect the points in an appropriate way and which conforms with the domain boundary, i.e., a mesh that respects every established boundary.

Although Delaunay triangulation is suitable to generate finite element meshes with a high regularity degree for a given set of points, this does not occur in the problem of conformity with the boundary, as it generates a mesh of the convex hull of the set of points. It may be thus impossible to recover the domain boundary from the faces and edges generated by the triangulation. To avoid this, we have two different sorts of techniques: *conforming Delaunay triangulation* [10] and *constrained Delaunay triangulation* [5]. The first alternative is inadequate for our purpose, as we wish the resulting mesh to contain certain predetermined points. Moreover, given the terrain surface complexity, this strategy would imply a high computational cost. The second alternative could provide another solution, but it requires quite complex algorithms to recover the domain boundary.

To build the three-dimensional Delaunay triangulation of the domain points, we start by resetting them in an auxiliary parallelepiped, so that every point of the terrain surface is on the original coordinates x, y, but at an altitude equal to the minimum terrain height, z_{min}. In the upper plane of the parallelepiped we set the nodes of level τ_1 of the mesh sequence that defines the terrain surface at altitude h. Generally, the remaining points also keep their coordinates x, y, but their heights are obtained by replacing their corresponding z_0 by z_{min} in (2). The triangulation of this set of points is done using a variant of Watson incremental algorithm [2] that effectively solves the problems derived from the round-off errors made when working with floating coma numbers.

Once the triangulation is built in the parallelepiped, the final mesh is obtained by re-establishing its original heights. This latter process can be understood as a compression of the global mesh defined in the parallelepiped, such that its lowest plane becomes the terrain surface. In this way, conformity is ensured.

Sometimes when re-establishing the position of each point to its real height, poor quality, or even *inverted* elements may occur. For inverted elements, their volume V_e, evaluated as the Jacobian determinant $|J_e|$ associated with the map from reference tetrahedron to the physical one e, becomes negative. For this reason, we need a procedure to untangle and smooth the resulting mesh, as analysed in Sect. 6.

We must also take into account the possibility of getting a high quality mesh by smoothing algorithms, based on movements of nodes around their initial positions, depends on the *topological quality* of the mesh. It is understood that this quality is high when every *node valence*, i.e., the number of nodes connected to it, approaches the valence corresponding to a regular mesh formed by *quasi-equilateral* tetrahedra.

Our domain mesh keeps the topological quality of the triangulation obtained in the parallelepiped and an appropriate smoothing would thus lead to high quality meshes.

6   Mesh Optimisation

The most accepted techniques for improving valid triangulation quality are based upon local smoothing. In short, these techniques locate the new positions that the mesh nodes must hold so that they optimise a certain objective function based upon a quality measurement of the tetrahedra connected to the adjustable or free node. The objective functions are generally useful for improving the quality of a valid mesh. They do not work properly, however, in the case of inverted elements, since they show singularity when the tetrahedra volumes change their sign. To avoid this problem we can proceed as in [4], where an optimisation method consisting of two stages is proposed. In the first, the possible inverted elements are untangled by an algorithm that maximises the negative Jacobian determinants corresponding to the inverted elements. In the second, the resulting mesh from the first stage is smoothed. We propose here an alternative to this procedure in which the untangling and smoothing are performed in the same stage. To do this, we shall use a modification of the objective function proposed in [1]. Thus, let $N(v)$ be the set of the s tetrahedra attached to free node v, and $\mathbf{r} = (x, y, z)$ be its position vector. Hence, the function to minimise is given by

$$F(\mathbf{r}) = \sum_{e=1}^{s} f_e(\mathbf{r}) = \sum_{e=1}^{s} \frac{\sum_{i=1}^{6}(l_i^e)^2}{V_e^{2/3}} . \tag{11}$$

where f_e is the objective function associated to tetrahedron e, l_i^e $(i = 1, ..., 6)$ are the edge lengths of the tetrahedron e and V_e its volume. If $N(v)$ is a valid sub-mesh, then the minimisation of F originates positions of v for which the local mesh quality improves [1]. Nevertheless, F is not bounded when the volume of any tetrahedron of $N(v)$ is null. Moreover, we cannot use F if there are inverted tetrahedra. Thus, if $N(v)$ contains any inverted or zero volume elements, it will be impossible to find the relative minimum by conventional procedures, such as steepest descent, conjugate gradient, etc. To remedy this situation, we have modified function f_e in such a way that the new objective function is nearly identical to F in the minimum proximity, but being defined and regular in all $\mathbb{R}^3$. We substitute V_e in (11) by the increasing function

$$h(V_e) = \frac{1}{2}(V_e + \sqrt{V_e^2 + 4\delta^2}) . \tag{12}$$

such that $\forall V_e \in \mathbb{R}$, $h(V_e) > 0$, being the parameter $\delta = h(0)$. In this way, the new objective function here proposed is given by

$$\Phi(\mathbf{r}) = \sum_{e=1}^{s} \phi_e(\mathbf{r}) = \sum_{e=1}^{s} \frac{\sum_{i=1}^{6}(l_i^e)^2}{[h(V_e)]^{2/3}} . \tag{13}$$

The asymptotic behaviour of $h(V_e)$, that is, $h(V_e) \approx V_e$ when $V_e \to +\infty$, will make function f_e and its corresponding modified version ϕ_e as close as required, for a value of δ small enough and positive values of V_e. On the other hand, when $V_e \to -\infty$, then $h(V_e) \to 0$. For the *most* inverted tetrahedra we shall then have a value of ϕ_e further from the minimum than for the *less* inverted ones. Moreover, with the proposed objective function Φ, the problems of F for tetrahedra with values close to zero are avoided. Due to the introduction of parameter δ, the singularity of f_e disappears in ϕ_e. As smaller values of δ are chosen, function ϕ_e behaves much like f_e. As a result of these properties, we may conclude that the positions of v that minimise objective functions F and Φ are nearly identical. Nevertheless, contrary to what happens to F, it is possible to find the minimum of Φ from any initial position of the free node. In particular, we can start from positions for which $N(v)$ is not a valid sub-mesh. Therefore, by using the modified objective function Φ, we can untangle the mesh and, at the same time, improve its quality. The value of δ is selected in terms of point v under consideration, making it as small as possible and in such a way that the evaluation of the minimum of Φ does not present any computational problem. Finally, we would state that the steepest descent method has been the one used to calculate the minimum of the objective function.

7   Numerical Experiments

As a practical application of the mesh generator, we have considered a rectangular area in the south of La Palma island of $45.6 \times 31.2 \ km$, where extreme heights vary from 0 to 2279 m. The upper boundary of the domain has been placed at $h = 9 \ km$. To define the topography we used a digitalisation of the area where heights were defined over a grid with a spacing step of 200 m in directions x and y. Starting from a uniform mesh τ_1 of the rectangular area with an element size of about $2 \times 2 \ km$, four global refinements were made using Rivara 4-T algorithm [12]. Once the data were interpolated on this refined mesh, we applied the derefinement algorithm developed in [3, 11] with a derefinement parameter of $\varepsilon = 40 \ m$. Thus, the adapted mesh approximates the terrain surface with an error less than that value. The node distribution of τ_1 is the one considered on the upper boundary of the domain.

The result obtained is shown in Fig. 1, fixing as the only parameter distance $D = 1.5 \ km$. In this case, the mesh has 57193 tetrahedra and 11841 nodes, with a maximum valence of 26. The node distribution obtained with this strategy has such a quality that it is hardly modified after five steps of the optimisation process, although there is initial tangling that is nevertheless efficiently solved; see Fig. 2, where $q(e)$ is the quality measure proposed in [4] for tetrahedron e. In fact, to avoid inverted tetrahedra, the technique proposed in Sect. 6 has been efficiently applied. Moreover, the worst quality measure of the optimised mesh tetrahedra is about 0.2.

We note that the number of parameters necessary to define the resulting mesh is quite low, as well as the computational cost. In fact, the complexity of 2-D refinement/derefinement algorithm is linear [11]. Besides, in experimental results we have approximately obtained a linear complexity in function of the

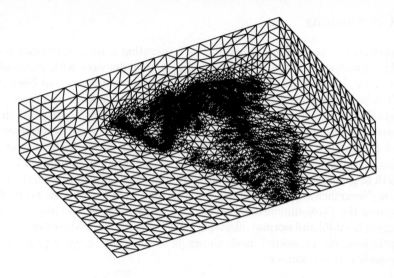

Fig. 1. Resulting mesh after five steps of the optimisation process

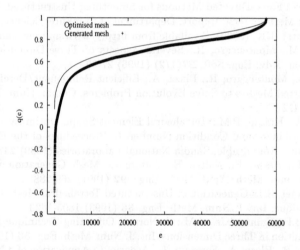

Fig. 2. Quality curves of the generated mesh and the resulting mesh after five steps of the optimisation process. Function $q(e)$ is a quality measure for tetrahedron e

number of points for our algorithm of 3-D Delaunay triangulation [2]. In the present application only a few seconds of CPU time on a Pentium III were necesary to construct the mesh before its optimisation. Finally, the complexity of each step of the mesh optimisation process is also linear. In practice we have found acceptable quality meshes appling a limited number of steps of this latter algorithm.

8   Conclusions

We have established the main aspects for generating a three-dimensional mesh capable of adapting to the topography of a rectangular area with minimal user intervention. In short, an efficient and adaptive point generation has been stated which is well distributed in the domain under study, because it preserves the topographic information of the terrain with a decreasing density as altitude increases. Points are generated using refinement/derefinement techniques in 2-D and the vertical spacing function here introduced. Next, with the aid of an auxiliary parallelepiped, a proceeding based on Delaunay triangulation has been set forth to generate the mesh automatically, assuring conformity with the terrain surface. Nevertheless, the obtained point distribution could also be useful in generating the three-dimensional mesh with other classical techniques, such as advancing front [6] and normal offsetting [7]. Finally, the procedure here proposed for optimising the generated mesh allows us to solve the tangling problems and mesh quality at the same time.

References

1. Djidjev, H.N.: Force-Directed Methods for Smoothing Unstructured Triangular and Tetrahedral Meshes. Tech. Report, Department of Computer Science, Univ. of Warwick, Coventry, UK, (2000). Available from http://www.andrew.cmu.edu/user
2. Escobar, J.M., Montenegro, R.: Several Aspects of Three-Dimensional Delaunay Triangulation. Adv. Eng. Soft. **27**(1/2) (1996) 27-39
3. Ferragut, L., Montenegro, R., Plaza, A.: Efficient Refinement/Derefinement Algorithm of Nested Meshes to Solve Evolution Problems. Comm. Num. Meth. Eng. **10** (1994) 403-412
4. Freitag, L.A., Knupp, P.M.: Tetrahedral Element Shape Optimization Via the Jacobian Determinant and Condition Number. In Proceedings of the Eighth International Meshing Roundtable. Sandia National Laboratories (1999) 247-258
5. George, P.L., Hecht, F., Saltel, E.: Automatic Mesh Generation with Specified Boundary. Comp. Meth. Appl. Mech. Eng. **92** (1991) 269-288
6. Jin, H., Tanner, R.I.: Generation of Unstructured Tetrahedral Meshes by Advancing Front Technique. Int. J. Num. Meth. Eng. **36** (1993) 1805-1823
7. Johnston, B.P., Sullivan Jr., J.M.: A Normal Offsetting Technique for Automatic Mesh Generation in Three Dimensions. Int. J. Num. Meth. Eng. **36** (1993) 1717-1734
8. Montenegro, R., Plaza, A., Ferragut, L., Asensio, I.: Application of a Nonlinear Evolution Model to Fire Propagation. Nonlinear Anal., Th., Meth. App. **30**(5) (1997) 2873-2882
9. Montero, G., Montenegro, R., Escobar, J.M.: A 3-D Diagnostic Model for Wind Field Adjustment. J. Wind Eng. Ind. Aer. **74-76** (1998) 249-261
10. Murphy, M., Mount, D.M., Gable, C.W.: A Point-Placement Strategy for Conforming Delaunay Tetrahedralization. In Symposium on Discrete Algorithms (2000) 67-74
11. Plaza, A., Montenegro, R., Ferragut, F.: An Improved Derefinement Algorithm of Nested Meshes. Adv. Eng. Soft. **27**(1/2) (1996) 51-57
12. Rivara, M.C.: A Grid Generator Based on 4-Triangles Conforming. Mesh-Refinement Algorithms. Int. J. Num. Meth. Eng. **24** (1987) 1343-1354

Domain Decomposition and Multigrid Methods for Obstacle Problems

Xue–Cheng Tai *

Department of Mathematics, University of Bergen, Johannes Brunsgate 12, 5007, Bergen, Norway.
tai@mi.uib.no,
http://www.mi.uib.no/~tai

Abstract. Uniform linear mesh independent convergence rate estimate is given for some proposed algorithms for variational inequalities in connection with domain decomposition and multigrid methods. The algorithms are proposed for general space decompositions and thus can also be applied to estimate convergence rate for classical block relaxation methods. Numerical examples which support the theoretical predictions are presented.

1 Some subspace correction algorithms

Consider the nonlinear convex minimization problem

$$\min_{v \in K} F(v), \quad K \subset V , \tag{1}$$

where F is a convex functional over a reflexive Banach space V and $K \subset V$ is a nonempty closed convex subset. In order to solve the mimization problem efficiently, we shall decompose V and K into a sum of subspaces and subsets of smaller sizes respectively as in [4] [7]. More precisely, we decompose

$$V = \sum_{i=1}^{m} V_i, \quad K = \sum_{i=1}^{m} K_i, \quad K_i \subset V_i \subset V , \tag{2}$$

where V_i are subspaces and K_i are convex subsets. We use two constants C_1 and C_2 to measure the quality of the decompositions. First, we assume that there exits a constant $C_1 > 0$ and this constant is fixed once the decomposition (2) is fixed. With such a $C_1 > 0$, it is assumed that any $u, v \in K$ can be decomposed into a sum of $u_i, v_i \in K_i$ and the decompositions satisfy

$$u = \sum_{i=1}^{m} u_i , \quad v = \sum_{i=1}^{m} v_i, \quad \text{and} \quad \left(\sum_{i=1}^{m} \|u_i - v_i\|^2 \right)^{\frac{1}{2}} \le C_1 \|u - v\| . \tag{3}$$

* This work was partially supported by the Norwegian Research Council under projects 128224/431 and SEP-115837/431.

P.M.A. Sloot et al. (Eds.): ICCS 2002, LNCS 2329, pp. 345–352, 2002.

For given $u, v \in K$, the decompositions u_i, v_i satisfying (3) may not be unique. We also need to assume that there is a $C_2 > 0$ such that for any $w_i \in V, \hat{v}_i \in V_i$, $\tilde{v}_j \in V_j$ it is true that

$$\sum_{i=1}^{m} \sum_{j=1}^{m} |\langle F'(w_{ij} + \hat{v}_i) - F'(w_{ij}), \tilde{v}_j \rangle| \le C_2 \left(\sum_{i=1}^{m} \|\hat{v}_i\|^2 \right)^{\frac{1}{2}} \left(\sum_{j=1}^{m} \|\tilde{v}_j\|^2 \right)^{\frac{1}{2}}, \quad (4)$$

In the above, F' is the Gâteaux differential of F and $\langle \cdot, \cdot \rangle$ is the duality pairing between V and its dual space V', i.e. the value of a linear function at an element of V. We also assume that there exists a constant $\kappa > 0$ such that

$$\langle F'(v_1) - F'(v_2), v_1 - v_2 \rangle \ge \kappa \|v_1 - v_2\|_V^2, \quad \forall w, v \in V . \quad (5)$$

Under the assumption (5), problem (1) has a unique solution. For some nonlinear problems, the constant κ may depend on v_1 and v_2. For a given approximate solution $u \in K$, we shall find a better solution w using one of the following two algorithms.

Algorithm 1 *Choose a relaxation parameter $\alpha \in (0, 1/m]$ and decompose u into a sum of $u_i \in K_i$ satisfying (3). Find $\hat{w}_i \in K_i$ in parallel for $i = 1, 2, \cdots, m$ such that*

$$\hat{w}_i = \arg \min_{v_i \in K_i} G(v_i) \quad with \quad G(v_i) = F\left(\sum_{j=1, j \ne i}^{m} u_j + v_i \right). \quad (6)$$

Set $w_i = (1 - \alpha)u_i + \alpha \hat{w}_i$ and $w = (1 - \alpha)u + \alpha \sum_{i=1}^{m} \hat{w}_i$.

Algorithm 2 *Choose a relaxation parameter $\alpha \in (0, 1]$ and decompose u into a sum of $u_i \in K_i$ satisfying (3). Find $\hat{w}_i \in K_i$ sequentially for $i = 1, 2, \cdots, m$ such that*

$$\hat{w}_i = \arg \min_{v_i \in K_i} G(v_i) \quad with \quad G(v_i) = F\left(\sum_{j<i} w_j + v_i + \sum_{j>i} u_j \right) \quad (7)$$

where $w_j = (1 - \alpha)u_j + \alpha \hat{w}_j$, $j = 1, 2, \cdots i - 1$. Set $w = (1 - \alpha)u + \alpha \sum_{i=1}^{m} \hat{w}_i$.

Denote u^* the unique solution of (1), the following convergence estimate is correct for Algorithms 1 and 2 (see Tai [6]):

Theorem 1. *Assuming that the space decomposition satisfies (3), (4) and that the functional F satisfies (5). Then for Algorithms 1 and 2, we have*

$$\frac{F(w) - F(u^*)}{F(u) - F(u^*)} \le 1 - \frac{\alpha}{(\sqrt{1 + C^*} + \sqrt{C^*})^2}, \quad C^* = \left(C_2 + \frac{[C_1 C_2]^2}{2\kappa} \right) \frac{2}{\kappa}. \quad (8)$$

2   Some Applications

We apply the algorithms for the following obstacle problem:

$$\text{Find } u \in K, \quad \text{such that} \quad a(u, v - u) \geq f(v - u), \quad \forall v \in K, \tag{9}$$

with $a(v, w) = \int_\Omega \nabla v \cdot \nabla w \, dx$, $K = \{v \in H_0^1(\Omega) | \ v(x) \geq \psi(x) \text{ a.e. in } \Omega\}$. It is well known that the above problem is equivalent to the minimization problem (1) assuming that $f(v)$ is a linear functional on $H_0^1(\Omega)$. For the obstacle problem (9), the minimization space $V = H_0^1(\Omega)$. Correspondingly, we have $\kappa = 1$ for assumption (5). The finite element method shall be used to solve (9). It shall be shown that domain decomposition and multigrid methods satisfy the conditions (3) and (4). For simplicity of the presentation, it will be assumed that $\psi = 0$.

2.1   Overlapping domain decomposition methods

For the domain Ω, we first divide it into a quasi-uniform coarse mesh partitions $\mathcal{T}_H = \{\Omega_i\}_{i=1}^M$ with a mesh size H. The coarse mesh elements are also called subdomains later. We further divide each Ω_i into smaller simplices with diameter of order h. We assume that the resulting finite element partition $\mathcal{T}_h$ form a shape regular finite element subdivision of Ω. We call this the fine mesh or the h-level subdivision of Ω with the mesh parameter h. We denote by $S_H \subset W_0^{1,\infty}(\Omega)$ and $S_h \subset W_0^{1,\infty}(\Omega)$ the continuous, piecewise linear finite element spaces over the H-level and h-level subdivisions of Ω respectively. For each Ω_i, we consider an enlarged subdomain Ω_i^δ consisting of elements $\tau \in \mathcal{T}_h$ with $distance(\tau, \Omega_i) \leq \delta$. The union of Ω_i^δ covers $\bar{\Omega}$ with overlaps of size δ. For the overlapping subdomains, assume that there exist m colors such that each subdomain Ω_i^δ can be marked with one color, and the subdomains with the same color will not intersect with each other. Let Ω_i^c be the union of the subdomains with the ith color, and $V_i = \{v \in S_h | \ v(x) = 0, \ x \notin \Omega_i^c\}$, $i = 1, 2, \cdots, m$. By denoting the subspaces $V_0 = S_H$, $V = S_h$, we find that

$$a). \quad V = \sum_{i=1}^m V_i \quad \text{and} \quad b). \quad V = V_0 + \sum_{i=1}^m V_i. \tag{10}$$

Note that the summation index is now from 0 to m instead of from 1 to m when the coarse mesh is added. For the constraint set K, we define

$$K_0 = \{v \in V_0 | \ v \geq 0\}, \quad \text{and} \quad K_i = \{v \in V_i | \ v \geq 0\}, \ i = 1, 2, \cdots, m. \tag{11}$$

Under the condition that $\psi = 0$, it is easy to see that (2) is correct both with or without the coarse mesh. When the coarse mesh is added, the summation index is from 0 to m. Let $\{\theta_i\}_{i=1}^m$ be a partition of unity with respect to $\{\Omega_i^c\}_{i=1}^m$, i.e. $\theta_i \in V_i$, $\theta_i \geq 0$ and $\sum_{i=1}^m \theta_i = 1$. It can be chosen so that

$$|\nabla\theta_i| \leq C/\delta, \quad \theta_i(x) = \begin{cases} 1 & \text{if } x \in \tau, \ \text{distance } (\tau, \partial\Omega_i^c) \geq \delta \text{ and } \tau \subset \Omega_i^c, \\ 0 & \text{on } \overline{\Omega \backslash \Omega_i^c}. \end{cases} \tag{12}$$

Later in this paper, we use I_h as the linear Lagrangian interpolation operator which uses the function values at the h-level nodes. In addition, we aslo need a nonlinear interpolation operator $I_H^\ominus : S_h \mapsto S_H$. Assume that $\{x_0^i\}_{i=1}^{n_0}$ are all the interior nodes for $\mathcal{T}_H$ and let ω_i be the support for the nodal basis function of the coarse mesh at x_0^i. The nodal values for $I_H^\ominus v$ for any $v \in S_h$ is defined as $(I_H^\ominus v)(x_0^i) = \min_{x \in \omega_i} v(x)$, c.f [6]. This operator satisfies

$$I_H^\ominus v \leq v, \ \forall v \in S_h, \quad \text{and} \quad I_H^\ominus v \geq 0, \ \forall v \geq 0, v \in S_h. \tag{13}$$

Moreover, it has the following monotonicity property

$$I_{h_1}^\ominus v \leq I_{h_2}^\ominus v, \quad \forall h_1 \geq h_2 \geq h, \quad \forall v \in S_h. \tag{14}$$

As $I_H^\ominus v$ equals v at least at one point in ω_i, it is thus true that for any $u, v \in S_h$

$$\|I_H^\ominus u - I_H^\ominus v - (u - v)\|_0 \leq c_d H |u - v|_1, \quad \|I_H^\ominus v - v\|_0 \leq c_d H |v|_1, \tag{15}$$

where d indicates the dimension of the physical domain Ω, i.e. $\Omega \subset R^d$, and

$$c_d = \begin{cases} C & \text{if } d = 1; \\ C \left(1 + \left|\log \frac{H}{h}\right|^{\frac{1}{2}}\right) & \text{if } d = 2, \\ C \left(\frac{H}{h}\right)^{\frac{1}{2}} & \text{if } d = 3, \end{cases}$$

With C being a generic constant independent of the mesh parameters. See Tai [6] for a detailed proof.

2.2   Decompositions with or without the coarse mesh

If we use the overlapping domain decomposition without the coarse mesh, i.e. we use decomposition (10.a), then we will get some domain decomposition algorithms which are essentially the block-relaxation methods. Even in the case $V = R^n$, the analysis of the convergence rate for a general convex functional $F : R^n \mapsto R$ and a general convex set $K \subset R^n$ is not a trivial matter, see [2] [3]. A linear convergence rate has been proved in [1] [5] for the overlapping domain decomposition without the coarse mesh. However, all the proofs require that the computed solutions converge to the true solution monotonically. Numerical evidence shows that linear convergence is true even if the computed solutions are not monotonically increasing or decreasing. In the following, we shall use our theory to prove this fact.

For any given $u, v \in S_h$, we decompose u, v as

$$u = \sum_{i=1}^m u_i, \quad v = \sum_{i=1}^m v_i, \quad u_i = I_h(\theta_i u), \quad v_i = I_h(\theta_i v). \tag{16}$$

In case that $u, v \geq 0$, it is true that $u_i, v_i \geq 0$. In addition,

$$\sum_{i=1}^m \|u_i - v_i\|_1^2 \leq C \left(1 + \frac{1}{\delta^2}\right) \|u - v\|_1^2,$$

which shows that $C_1 \leq C(1 + \delta^{-1})$. It is known that $C_2 \leq \sqrt{m}$ with m being the number of colors. From Theorem 1, the following rate is obtained without requiring that the computed solutions increase or decrease monotonically:

$$\frac{F(w) - F(u^*)}{F(u) - F(u^*)} \leq 1 - \frac{\alpha}{1 + C(1 + \delta^{-2})}.$$

For Algorithm 2, we can take $\alpha = 1$.

Numerical experiments and the convergence analysis for the two-level domain decomposition method, i.e. overlapping domain decomposition with a coarse mesh, seems still missing in the literature. To apply our algorithms and theory, we decompose any $u \in K$ as

$$u = u_0 + \sum_{i=1}^{m} u_i, \quad u_0 = I_H^{\ominus} u, \ u_i = I_h(\theta_i(u - u_0)). \tag{17}$$

From (13) and the fact that $u \geq 0$, it is true that $0 \leq u_0 \leq u$ and so $u_i \geq 0$, $i = 1, 2, \cdots, m$, which indicates that $u_0 \in K_0$ and $u_i \in K_i$, $i = 1, 2, \cdots, m$. The decomposition for any $v \in K$ shall be done in the same way. It follows from (15) that $\|u_0 - v_0\|_1 \leq C\|u - v\|_1$. Note that $u_i - v_i = I_h(\theta_i(u - v - I_H^{\ominus} u + I_H^{\ominus} v))$. Using estimate (15) and a proof similar to those for the unconstrained cases, c.f. [7], [8], it can be proven that $\|u_i - v_i\|_1^2 \leq c_d \left(1 + \frac{H}{\delta}\right) \|u - v\|_1^2$. Thus

$$\left(\|u_0 - v_0\|_1^2 + \sum_{i=1}^{m} \|u_i - v_i\|_1^2\right)^{\frac{1}{2}} \leq C(m)c_d \left(1 + \left(\frac{H}{\delta}\right)^{\frac{1}{2}}\right) \|u - v\|_1.$$

The estimate for C_2 is known, c.f. [7], [8]. Thus, for the two-level domain decomposition method, we have $C_1 = C(m)c_d \left(1 + \frac{\sqrt{H}}{\sqrt{\delta}}\right)$, $C_2 = C(m)$, where $C(m)$ is a constant only depending on m, but not on the mesh parameters and the number of subdomains. An application of Theorem 1 will show that the following convergence rate estimate is correct:

$$\frac{F(w) - F(u^*)}{F(u) - F(u^*)} \leq 1 - \frac{\alpha}{1 + c_d^2(1 + H\delta^{-1})}.$$

2.3   Multigrid decomposition

Multigrid methods can be regarded as a repeated use of the two-level method. We assume that the finite element partition $\mathcal{T}_h$ is constructed by a successive refinement process. More precisely, $\mathcal{T}_h = \mathcal{T}_{h_J}$ for some $J > 1$, and $\mathcal{T}_{h_j}$ for $j \leq J$ is a nested sequence of quasi-uniform finite element partitions, see [6], [7], [8]. We further assume that there is a constant $\gamma < 1$, independent of j, such that h_j is proportional to γ^{2j}. Corresponding to each finite element partition $\mathcal{T}_{h_j}$, let $\{x_j^k\}_{k=1}^{n_j}$ be the set of all the interior nodes. Denoted by $\{\phi_j^i\}_{i=1}^{n_j}$ the nodal basis functions satisfying $\phi_j^i(x_j^k) = \delta_{ik}$. We then define a one dimensional subspace $V_j^i = \text{span}(\phi_j^i)$. Letting $V = \mathcal{M}_J$, we have the following trivial space decomposition:

$$V = \sum_{j=1}^{J} \sum_{i=1}^{n_j} V_j^i. \tag{18}$$

Each subspace V_j^i is a one dimensional subspace. For any $v \geq 0$ and $j \leq J - 1$, define $v_j = I_{h_j}^{\ominus} v - I_{h_{j-1}}^{\ominus} v \in \mathcal{M}_j$. Let $v_J = v - I_{h_{J-1}}^{\ominus} v \in \mathcal{M}_J$. A further decomposition of v_j is given by $v_j = \sum_{i=1}^{n_j} v_j^i$ with $v_j^i = v_j(x_j^i)\phi_j^i$. It is easy to see that

$$v = \sum_{j=1}^{J} v_j = \sum_{j=1}^{J} \sum_{i=1}^{n_j} v_j^i.$$

For any $u \geq 0$, it shall be decomposed in the same way, i.e.

$$u = \sum_{j=1}^{J} \sum_{i=1}^{n_j} u_j^i, \; u_j^i = u_j(x_j^i)\phi_j^i, \; u_j = I_{h_j}^{\ominus} u - I_{h_{j-1}}^{\ominus} u, \; j < J; \; u_J = u - I_{h_{J-1}}^{\ominus} u.$$

$$\tag{19}$$

It follows from (13) and (14) that $u_j^i, v_j^i \geq 0$ for all $u, v \geq 0$, i.e. $u_j^i, v_j^i \in K_j^i = \{v \in V_j^i : v \geq 0\}$ under the condition that $\psi = 0$. Define

$$\tilde{c}_d = \begin{cases} C, & \text{if } d = 1; \\ C(1 + |\log h|^{\frac{1}{2}}), & \text{if } d = 2; \\ Ch^{-\frac{1}{2}}, & \text{if } d = 3. \end{cases}$$

The following estimate can be obtained using approximation properties (15) (see [6]):

$$\sum_{j=1}^{J} \sum_{i=1}^{n_j} \|u_j^i - v_j^i\|_1^2 \leq C \sum_{j=1}^{J} h_j^{-2} \|u_j - v_j\|_0^2 \leq \tilde{c}_d^2 \sum_{j=1}^{J} h_j^{-2} h_{j-1}^2 \, |u - v|_1^2 \leq \tilde{c}_d^2 \gamma^{-2} J \, |u - v|_1^2,$$

which proves that

$$C_1 \cong \tilde{c}_d \gamma^{-1} J^{\frac{1}{2}} \cong \tilde{c}_d \gamma^{-1} |\log h|^{\frac{1}{2}}.$$

The estimate for C_2 is known, i.e. $C_2 = C(1 - \gamma^d)^{-1}$, see Tai and Xu [8]. Thus for the multigrid method, the error reduction factor for the algorithms is

$$\frac{F(w) - F(u^*)}{F(u) - F(u^*)} \leq 1 - \frac{\alpha}{1 + \tilde{c}_d^2 \gamma^{-2} J}.$$

2.4   Numerical experiments

We shall test our algorithms for the obstacle problem (9) with $\Omega = [-2, 2] \times [-2, 2]$, $f = 0$ and $\psi(x, y) = \sqrt{x^2 + y^2}$ when $x^2 + y^2 \leq 1$ and $\psi(x, y) = -1$ elsewhere. This problem has an analytical solution [6]. Note that the continuous obstacle function ψ is not even in $H^1(\Omega)$. Even for such a difficult problem,

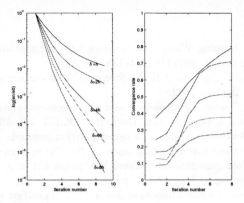

Fig. 1. Convergence for the two-level method for decomposition (17) with different overlaps, $h = 4/128$, and $H = 4/8$.

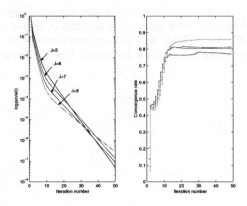

Fig. 2. Convergence for the multigrid method

uniform linear convergence has been observed in our experiments. In the implementations, the non-zero obstacle can be shifted to the right hand side.

Figure 1 shows the convergence rate for Algorithm 2 with different overlapping sizes for decomposition (17). Figure 2 shows the convergence rate for Algorithm 2 with the multigrid method for decomposition (19) and J indicates the number of levels. In the figures en is the H^1-error between the computed solution and the true finite element solution and $e0$ is the initial error. $\log(en/e0)$ is used for one of the subfigures. The convergence rate is faster in the beginning and then approaches a constant after some iterations.

References

1. Lori Badea and Junping Wang. An additive Schwarz method for variational inequalities. *Math. Comp.*, 69(232):1341–1354, 2000.
2. R. W. Cottle, J. S. Pang, and R. E. Stone. *The linear complementary problem.* Academic Press, Boston, 1992.
3. Z.-Q. Luo and P. Tseng. Error bounds and convergence analysis of feasible descent methods: A general approach. *Ann. Oper. Res.*, 46:157–178, 1993.
4. X.-C. Tai. Parallel function and space decomposition methods. In P. Neittaanmäki, editor, *Finite element methods, fifty years of the Courant element, Lecture notes in pure and applied mathematics*, volume 164, pages 421–432. Marcel Dekker inc., 1994. Available online at http://www.mi.uib.no/~tai.
5. X.-C. Tai. Convergence rate analysis of domain decomposition methods for obstacle problems. *East-West J. Numer. Math.*, 9(3):233–252, 2001. Available online at http://www.mi.uib.no/~tai.
6. X.-C. Tai. Rate of convergence for some constraint decomposition methods for nonlinear variational inequalities. Technical Report 150, Department of Mathematics, University of Bergen, November, 2000. Available online at http://www.mi.uib.no/~tai.
7. X.-C. Tai and P. Tseng. Convergence rate analysis of an asynchronous space decomposition method for convex minimization. *Math. Comput.*, 2001.
8. X-C. Tai and J.-C. Xu. Global and uniform convergence of subspace correction methods for some convex optimization problems. *Math. Comput.*, 71:105–124, 2001.

Domain Decomposition Coupled with Delaunay Mesh Generation

Tomasz Jurczyk[1] and Barbara Głut[1]

University of Mining and Metallurgy, Cracow, Poland,
{jurczyk,glut}@uci.agh.edu.pl

Abstract. This paper presents a domain decomposition method for finite element meshes created using the Delaunay criterion. The decomposition is performed at the intermediate stage of the generation process and is successively refined as the number of nodes in the mesh increases. The efficiency of the algorithm and the quality of the partitioning is evaluated for different partition heuristics and various finite element meshes.

1 Introduction

The mesh partitioning problem is to partition the vertices (or elements) of a mesh in k roughly equal partitions, such that the communication overhead during the parallel simulation is minimized. The quality of the partition is a function of several criteria, depending upon the computer architecture and the problem being computed. In most cases, the cut-size of the partition (the number of edges connecting vertices in different clusters) is assumed as the main quality coefficient. The graph partitioning problem is NP-complete (e.g. [1]), regardless of the number of partitions. However, there have been developed many algorithms calculating reasonably good partitionings[2, 3].

Recently a class of multilevel algorithms[4, 5] has emerged as a highly effective method for computing a partitioning of a graph. The basic structure of a multilevel partitioning algorithm consist of three phases: coarsening, initial partitioning and refinement. In the first phase the multilevel algorithm repeatedly reduces the size of the graph by joining pairs of adjacent nodes. At each step the *maximum matching* is found by applying some strategy of edge collapsing, the graph is reduced and the weights of nodes and edges are updated. The second phase partitions the reduced graph into required number of clusters. Since the number of nodes is small, the partitioning can be effectively computed. The only requirement for the partitioning algorithm is to be able to handle graphs with weighted nodes and edges. In the third step the graph is repeatedly uncoarsened until it reaches the original size. Because of the weighted nodes and edges, the balanced partition of the smaller graph corresponds to a balanced partition of the larger graph with the same cut size. At each step of uncoarsening the graph can be further improved by local refinement heuristics.

In this article we present a similar method applied to the decomposition of finite element meshes generated with the Delaunay triangulation. In the process

P.M.A. Sloot et al. (Eds.): ICCS 2002, LNCS 2329, pp. 353–360, 2002.

of triangulation using the Delaunay criterion the mesh is created by an incremental refinement of initial simple triangulation. First phase consists of successive insertion of all boundary nodes. Then, inner nodes are added until the quality measure (e.g. size or geometric quality) for all triangles in the mesh is satisfactory. Inner nodes are inserted in the currently worst areas of the triangulation, which results in the uniform refinement of the mesh. In the presented method, instead of coarsening the complete mesh, an intermediate mesh is selected as a coarse graph and is initially partitioned into the required number of clusters. The process of the further construction of the mesh is similar to the uncoarsening stage in the multilevel method. Each newly inserted node becomes a part of some cluster, depending upon the partition numbers of adjacent nodes. Periodically, after a specified number of nodes are included into the mesh, a smoothing process is started. The smoothing method should be fast and improve the quality of the partitioning locally.

The remainder of the paper is organized as follows. Section 2 defines the graph partitioning problem and presents shortly the most important heuristics for both the partitioning and smoothing task. Section 3 presents the description of the domain decomposition during the Delaunay triangulation. In section 4 various parameters of the presented method are evaluated.

2   Graph Partitioning

Given a graph $G = (V, E)$ with nodes V ($n = |V|$ is the number of nodes) and undirected edges E, let $\pi : V \rightarrow \{0, 1, \ldots k - 1\}$ be a *partition* of a graph G, that distributes the nodes among k clusters $V_0, V_1, \ldots V_{k-1}$. The *balance* of π is defined as follows: $bal(\pi) := \max |V_i|; 0 \leq i < k - \min |V_j|; 0 \leq j < k$; and the *cut-size*: $cut(\pi) := |\{\{v, w\} \in E; \pi(v) \neq \pi(w)\}|$.

2.1   Decomposition Algorithms

There are several different criteria a heuristic can focus on. Many methods are very powerful with regard to one criterion, but don't satisfy others. The following description summarizes the most popular schemes.

The *greedy* strategy[6] starts with one cluster holding the whole graph, others being empty. Then, nodes are repeatedly moved to the successive underweighted clusters in such way, that each movement minimally increases the cut size. If the number of clusters $k > 2$ both direct approach and recursive bisection can be applied.

The *layer* method[7] starts with a random initial node and repeatedly identifies successive layers of adjacent nodes. After all nodes are visited, the graph is accordingly split with the first cluster containing $\frac{n}{2}$ nodes from successive layers.

In the simple *geometric* bisection the longest expansion of any dimension is determined and the nodes are split according to their coordinate in this dimension. The *inertial* method[8] considers nodes as mass points and cuts the graph with the line (plane) orthogonal to the principle axis of this collection of mass points. In the *shortest cut* strategy several lines (planes) containing the mass center of the graph are created. All edges of the graph are inspected, the *cut size* for each of these lines (planes) is evaluated, and the line (plane) with the smallest number of crossing edges is selected.

The *spectral* bisection heuristic[7, 8] is based on algebraic graph theory. The *Laplacian matrix* is constructed on the basis of the adjacency information of the graph and the eigenvector $\bar{y}$ of the second smallest eigenvalue of this matrix is used to bisect the graph. This method gives good results, unfortunately the time and space required to compute accurately the eigenvalues and eigenvectors of the *Laplacian matrix* is very high.

The *multilevel* class[4, 5]) of graph partitioning methods employs the reduction of large graphs both to shorten the partitioning time and to improve the solution quality. First the graph is successively reduced by collapsing vertices and edges, the smaller graph is partitioned and finally uncoarsened level by level with addition of local refinement.

2.2   Refinement Algorithms

The initial (global) partitioning can be further improved by applying some local smoothing method, which concentrates on the refining of the current partitioning in order to obtain a lower *cut size* (or to improve others desirable properties).

The *Kernighan-Lin* heuristic[9, 10] is the most frequently used local refinement method. Originally, it uses a sequence of node pair exchanges to determine the changes in the initial partitioning. After the given number of steps is executed, the history of changes is checked and the step with the best incremental gain is selected. The whole procedure continues until there is no improvement. Several improvements and generalization were proposed since the original publication of this algorithm such as moving single nodes, limiting the maximum number of moved nodes in a single pass, using cost metrics other than cut-size, refining k-way partition.

3   Simultaneous Mesh Generation and Decomposition

The preparation stage of the parallel numerical simulation of physical processes using finite element method consists usually of several phases: definition of the process domain, generation of the finite element mesh, decomposition of this mesh in the given number of clusters and mapping these clusters to the processor-nodes of the parallel machine. In the approach presented in this paper, the mesh generation procedure (based on the Delaunay property) is coupled with the mesh decomposition. By running the partitioning procedure at an early phase of the mesh construction, both time complexity and the quality of the overall process can be improved. It should be noted, that addition of the decomposition procedures doesn't influence the geometrical structure of the final finite element mesh.

The domain decomposition can be incorporated into the mesh generation scheme as follows:

[First stage]
  Create simple mesh (two triangles) containing all boundary nodes.
  Insert all boundary nodes into the mesh
    (respecting the Delaunay property).
[Second stage]
  Evaluate the quality coefficients for all triangles.
  While the worst triangle is not good enough:
    insert a new node inside this triangle,
    retriangulate the mesh locally.
    (*) if(node_count == N1) initial decomposition
    (*) else if(node_count == N1 + k*N2) decomposition smoothing

The initial decomposition is performed when the number of nodes (both boundary and inner ones) reaches N1. After this decomposition the smoothing procedure is called in regular intervals (every N2 number of inserted nodes).

In Fig. 1 there are presented successive steps of the mesh generation process using the Delaunay property. Figure 1a shows the mesh containing boundary nodes only. Figures 1b,c present intermediate mesh with a number of inner nodes already inserted and finally the complete mesh.

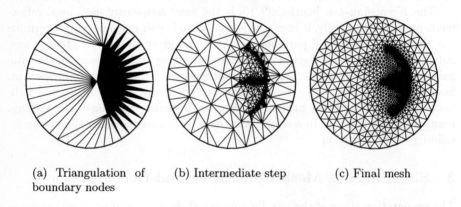

(a) Triangulation of   (b) Intermediate step   (c) Final mesh
boundary nodes

Fig. 1. Successive steps of Delaunay mesh generation

3.1   Initial Decomposition

The decomposition of the mesh can take place at any time during the second phase of the Delaunay triangulation algorithm, while the mesh is being refined by insertion of inner nodes. At the beginning of this phase the mesh may substantially differ from the final mesh. As the refinement of the mesh progresses, the mesh structure stabilizes. The initial partitioning of the mesh should be performed when the overall variance of the mesh density is reasonably uniform. The "resemblance" of the intermediate mesh to the final form is increasing with each inserted inner node. Unfortunately, so does the time required for the partitioning process.

The choice of the step when the initial partitioning is run (the parameter N1) influences noticeably the final quality of the mesh decomposition. Partitioning of the mesh at an early stage allows to concentrate on the main characteristics of the mesh structure. Unfortunately, if the final mesh density is variegated, splitting the mesh too soon can result in highly unbalanced clusters.

The selection of the partitioning heuristic is arbitrary (e.g. any method described in the section 2.1 can be used). The obtained partitioning, however, should be of good quality (with connected clusters), so that any further smoothing would adjust the boundaries of clusters locally only.

3.2 Decomposition Smoothing

After the initial partitioning of the mesh, the generation process is continued. As the number of inserted nodes increases, the structure of the mesh can alter and an adequate adjustment of the decomposition separators should be performed. Additionally, finer mesh has more degrees of freedom, and the quality of decomposition can be further improved.

In order to meet these needs, a smoothing method is periodically executed (every N2 inserted nodes and when the mesh generation process ends). A class of local refinement algorithms that tend to produce good results are those based on the Kernighan-Lin algorithm. In the variation used in this work the vertices are incrementally swapped among partitions to reduce the edge-cut and balance of the partitioning.

The selected smoothing method influences greatly both the quality of the obtained decomposition and the running time of the whole process. Both frequency of smoothing and the complexity of the method itself should be carefully selected. The complexity aspect is especially important as the number of partitions increases, which makes the complexity of this method much higher.

3.3 Introduction of Inner Nodes

During the continued process of the generation of a preliminary partitioned mesh, all newly inserted nodes must be assigned to the proper cluster. The decision is based on the nodes adjacent to the node in question. If all these nodes belong to a single cluster (Fig. 2a) the new node is assigned to the same cluster. However, if there are more available clusters (Fig. 2b,c) several criteria can be considered:

- *cut-size* – cluster with the most number of edges incident to this node should be selected.
- *balance* – prefers the smallest of incident clusters.
- other quality properties (e.g. avoiding nodes adjacent to several different clusters).

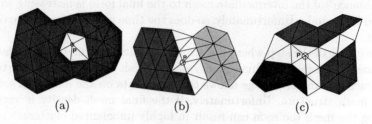

(a)   (b)   (c)

Fig. 2. Identification of the inner nodes

4 Results

The described method was tested on several domains with various control spaces, which resulted in generation of meshes with different shape and the variation of the density of elements. All meshes had roughly equal number of nodes and were partitioned into eight clusters. In [11] these meshes are presented along with a detailed analysis (including several aspects of the decomposition quality) of the results obtained with different partitioning algorithms.

In this work we tested the efficiency of the proposed algorithm and the influence of different values of parameters N1 and N2 (along with other parameters of the decomposition process). The results were evaluated using several quality coefficients such as cut-size, balance, boundary balance, local cut-size, connectivity, multiconnected nodes, and running time.

Table 1 presents the results obtained for a three selected meshes (Fig. 3) containing about 8000 nodes partitioned into 8 clusters. The cut-size was chosen as a representative coefficient of the decomposition quality. The Roman numbers denote partitioning algorithms used to perform the initial decomposition: (I) direct greedy, (II) recursive greedy, (III) recursive layer, (IV) recursive inertial, and (V) recursive shortest-cut. In each case the partitioning was further improved by a Kernighan-Lin smoothing method.

Table 1. Cut-size of meshes decomposed into 8 clusters with different methods (P0 denotes the decomposition of the final mesh, P1 – simultaneous mesh generation and decomposition with the parameters N1=1000, N2=1000, and P2 – with the parameters N1=1500, N2=500)

	cheese			mountain			gear		
	P0	P1	P2	P0	P1	P2	P0	P1	P2
I	659	769	596	804	868	834	599	805	685
II	632	647	610	819	808	795	553	595	650
III	542	572	513	915	782	674	519	855	590
IV	588	607	701	920	717	711	501	625	525
V	471	613	522	699	694	662	404	455	432

Coefficients presented in table 1 visualize the typical characteristic of obtained results. Depending upon the various parameters of the overall mesh generation and decomposition process, different variants gain prevalence. In the discussed example, the results were in most cases better when the value of parameter N1 (denoting the step at which the initial partitioning is performed) was increased from 1000 to 1500, and the frequency of smoothing (N2) was decreased from 1000 to 500. For different meshes the quality of decomposition calculated during the mesh generation as compared to the partitioning of the complete mesh has no regular tendency. E.g. the partitioning during the generation of the mesh (P2) is usually better than the partitioning of the final mesh (P0) in case of the mesh mountain, and consistently worse for the mesh gear.

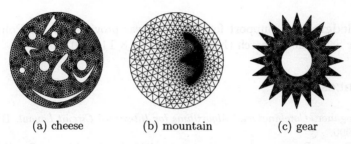

(a) cheese   (b) mountain   (c) gear

Fig. 3. Tested meshes

Fig. 4 shows the intermediate phases and the final result of the simultaneous mesh generation and partitioning. As can be seen, the final partitioning is much alike the initial one. This property is very advantageous, especially with respect to the task of a parallel mesh generation, when the partitions should be preliminary distributed to the proper processors of a parallel machine and any further movement of these nodes results in the degradation of the overall performance.

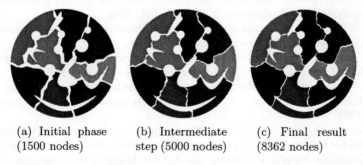

(a) Initial phase   (b) Intermediate   (c) Final result
(1500 nodes)   step (5000 nodes)   (8362 nodes)

Fig. 4. Example of generation and decomposition process (N1=1500, N2=500).

5   Summary

In our work we tested and compared the efficiency and quality of the decomposition of various algorithms applied to variegated finite element meshes. The experiments with the presented approach of combined mesh generation and partitioning have shown that these algorithms work quite well for many different meshes. The main requirements needed for a successful employment of this scheme are a good, global initial partitioning algorithm and fast locally smoothing method.

The future research will concentrate on improving the mesh refinement procedure in order to make it more efficient in terms of both the decomposition quality and the running time. The aspects of the presented approach connected with the problem of a parallel mesh generation will also be addressed.

Acknowledgments. Support for this work was provided by the Polish Committee for Scientific Research (KBN) Grant No. 8 T11F 01216.

References

1. T. Lengauer. *Combinatorial Algorithms for Integrated Circuit Layout.* B.G. Teubner, 1990.
2. R.A. Davey. *Decomposition of Unstructured Meshes for Efficient Parallel Computation.* PhD thesis, University of Edinburgh, 1997.
3. K. Schlogel, G. Karypis, and V. Kumar. Graph partitioning for high-performance scientific simulations. In J. Dongarra and et. al., editors, *CRPC Parallel Computing Handbook.* Morgan Kaufmann, 2000.
4. B. Hendrickson and R. Leland. A multilevel algorithm for partitioning graphs. Technical Report SAND93-0074, Sandia National Laboratories, Albuquerque, New Mexico 87185, 1993.
5. G. Karypis and V. Kumar. Multilevel k-way partitioning scheme for irregular graphs. *Journal of Parallel and Distributed Computing*, 48(1):96–129, 1998.
6. C. Farhat. A simple and efficient automatic FEM domain decomposer. *Computers and Structures*, 28(5):579–602, 1988.
7. H.D. Simon. Partitioning of unstructured problems for parallel processing. *Computing Systems in Engineering*, 2(2/3):135–148, 1991.
8. R.D. Williams. Performance of dynamic load balancing algorithms for unstructured mesh calculations. *Concurrency*, 3:457–481, 1991.
9. B.W. Kernighan and S. Lin. An effective heuristic procedure for partitioning graphs. *The Bell Systems Technical Journal*, pages 291–308, Feb. 1970.
10. C.M. Fiduccia and R.M. Mattheyses. A linear-time heuristic for improving network partitions. In *Proc. of the 19th IEEE Design Automation Conference*, pages 175–181, 1982.
11. T. Jurczyk, B. Głut, and J. Kitowski. An empirical comparison of decomposition algorithms for complex finite element meshes. Presented at Fourth International Conference on Parallel Processing and Applied Mathematics, Nałęczów, 2001.

Accuracy of 2D Pulsatile Flow in the Lattice Boltzmann BGK Method

A. M. Artoli, A. G. Hoekstra, and P. M. A. Sloot

Section Computational Science, Informatics Institute, Faculty of Science
University of Amsterdam, Kruislaan 403, 1098 SJ Amsterdam, The Netherlands
{artoli, alfons, sloot}@science.uva.nl
http://www.science.uva.nl/research/scs/

Abstract. We present detailed analysis of the accuracy of the lattice Boltzmann BGK (LBGK) method in simulating oscillatory flow in a two dimensional channel. Flow parameters such as velocity and shear stress have been compared to the analytic solutions. Effects of different boundary conditions on the accuracy have been tested. When the flow is driven by a body force, we have observed a shift in time between the theory and the simulation, being about +0.5, which when taken into account enhances the accuracy at least one order of magnitude.

1 Introduction

Recently, it has been demonstrated that the lattice-Boltzmann method (LBM) is useful in simulation of time-dependent fluid flow problems [1]. The method is attracting more attention for its inherent parallelism and straightforward implementation [2]. It is well known that, for steady flow, LBM is second order accurate in the velocity fields and the stress tensor when second order boundary conditions are applied [3],[4], [5]. Although the accuracy has been studied extensively for steady flow, studies on the accuracy of the LBM method for time dependent flows are quite rare [6].

Pulsatile flow characteristics are quite important in haemodynamics: Cardiovascular diseases are considered as the main causes of death in the world [7]. It is reasonably believed that the shear stress plays a dominant role in locality, diagnosis and treatment of such diseases [8]. With LBM, it is possible to compute the local components of the stress tensor without a need to estimate the velocity gradients from the velocity profiles [9], [5]. In addition, the equation of state defines the pressure as a linear function of the density which makes it quite simple to obtain the pressure from the density gradient. All these advantages makes the LBM suitable for simulating time-dependent flows.

In this article, we present the capability and investigate the accuracy of the lattice Boltzmann BGK (LBGK) model in recovering analytic solutions for oscillatory two dimensional (2D) channel flow in a range of Womersley and Reynolds numbers. We will first shortly review the model, then present the simulation results and conclude with a discussion.

P.M.A. Sloot et al. (Eds.): ICCS 2002, LNCS 2329, pp. 361–370, 2002.
© Springer-Verlag Berlin Heidelberg 2002

2   The Lattice Boltzmann BGK Method

Although the LBGK model historically evolved from the lattice gas automata, it can be derived directly from the simplified Boltzmann BGK equation [10] which mainly describes rarefied gases in the Boltzmann gas limit. This approximation works well for incompressible and isothermal flows. The method ensures isotropy and Galilean invariance. It is called LBGK since it simplifies the collision operator of the Boltzmann equation via the single particle relaxation time approximation proposed by Bhatnagar, Gross and Krook in 1954 [11]. The evolution equation of the distribution functions is [10]

$$\frac{\partial f}{\partial t} + \boldsymbol{\xi} \cdot \nabla f = -\frac{1}{\lambda}(f - g), \tag{1}$$

where $f \equiv f(x, \boldsymbol{\xi}, t)$ is the single particle distribution function, $\boldsymbol{\xi}$ is the microscopic velocity, λ is the collision relaxation time and g is the Maxwell-Boltzmann distribution function. The numeric solution for f is obtained by discretizing Eq.(1) in the velocity space $\boldsymbol{\xi}$ using a finite set of velocities $\mathbf{e}_i$ without violating the conservation laws of the hydrodynamic moments. This gives [10]

$$\frac{\partial f_i}{\partial t} + \mathbf{e}_i \cdot \nabla f_i = -\frac{1}{\lambda}\left(f_i - f_i^{(eq)}\right), \tag{2}$$

where $f_i(\mathbf{x}, t) \equiv f(\mathbf{x}, \mathbf{e}_i, t)$ and $f_i^{(eq)}(\mathbf{x}, t) = f^{(0)}(\mathbf{x}, \mathbf{e}_i, t)$ are the distribution function and the equilibrium distribution function of $\mathbf{e}_i$, respectively. The equilibrium distribution function, which is the Taylor expansion of the Maxwellian distribution, takes the following form [12] in the limit of low Mach number:

$$f_i^{(eq)} = w_i \left(\rho + \frac{3}{v^2}\mathbf{e}_i \cdot \mathbf{u} + \frac{9}{2v^4}(\mathbf{e}_i \cdot \mathbf{u})^2 + \frac{3}{2v^2}\mathbf{u} \cdot \mathbf{u}\right), \tag{3}$$

where w_i is a weighting factor, $v = \delta_x/\delta_t$ is the lattice speed, and δ_x and δ_t are the lattice spacing and the time step, respectively. The values of the weighting factor and the discrete velocities depend on the used LBM model and can be found in literature (see e.g. ref.[12]). In this study, we have used the improved incompressible D2Q9i model [12], which has three types of particles on each node; a rest particle, four particles moving along x and y principal directions with speeds $|\mathbf{e}_i| = \pm 1$, and four particles moving along diagonal directions with speeds $|\mathbf{e}_i| = \sqrt{2}$. The hydrodynamic density, ρ, and the macroscopic velocity, $\mathbf{u}$, are determined in terms of the particle distribution functions from

$$\rho = \sum_i f_i = \sum_i f_i^{(eq)} \tag{4}$$

and

$$\rho\mathbf{u} = \sum_i \mathbf{e}_i f_i = \sum_i \mathbf{e}_i f_i^{(eq)}. \tag{5}$$

Equation (2) is then discretized in space and time into the well-known lattice BGK equation [6]

$$f_i(\mathbf{x} + \mathbf{e}_i \delta_t, \mathbf{e}_i, t + \delta_t) - f_i(\mathbf{x}, \mathbf{e}_i, t) = -\frac{1}{\tau}[f_i(\mathbf{x}, \mathbf{e}_i, t) - f_i^{(0)}(\mathbf{x}, \mathbf{e}_i, t)] \quad (6)$$

where $\tau = \frac{\lambda}{\delta_t}$ is the dimensionless relaxation time. Taylor expansion of Eq. (6) up to $O(\delta_t^2)$ and application of the multi-scale Chapman-Enskog technique [13] by expanding f_i about $f_i^{(0)}$, we can derive the evolution equations and the momentum flux tensor up-to second order in the Knudsen number. The Navier-Stokes equation can be derived from this equation to yield

$$\partial_t \mathbf{u} + (\mathbf{u} \cdot \nabla)\mathbf{u} = -\frac{1}{\rho}\nabla p + \nu \nabla^2 \mathbf{u} \quad (7)$$

where $p = \rho c_s^2$ is the scalar pressure and ν is the kinematic viscosity of the lattice Boltzmann model, given by

$$\nu = c_s^2 \delta_t (\tau - \frac{1}{2}) \quad (8)$$

where c_s is the speed of sound. Also, the stress tensor is

$$\sigma_{\alpha\beta} = -\rho c_s^2 \delta_{\alpha\beta} - \left(1 - \frac{1}{2\tau}\right) \sum_{i=0} f_i^{(1)} e_{i\alpha} e_{i\beta}. \quad (9)$$

which can be computed directly during the collision process [5], without a need to compute the derivatives of the velocity fields. This extensively enhances the lattice Boltzmann BGK method, as other CFD methods are more elaborate and estimate the stress tensor components from the simulated velocity field. With this LBM model, the flow fields and the stress tensor components are accurate up to the second order in the Knudsen number. We emphasize that, with compressible LBM models, when the forcing term is time-dependent, compressibility effects may arise and using an incompressible model is a must [6]. However, since the D2Q9i model has already been tested for steady flows, for which it was proposed, we test it here for unsteady flows.

3   Simulations

We have conducted a number of 2D simulations for time dependent flow in a channel. Various boundary conditions have been tested: the Bounce-back on the nodes, the body force, periodic boundaries, non slip, and inlet and outlet boundary conditions. For all simulations described below, unless otherwise specified, the flow is assumed to be laminar, the Mach number is low, the driving force is the body force, the Reynolds number is defined as $Re = \frac{L^3 G}{8\nu^2}$ [3], the Womersley number is defined as $\alpha = \frac{L}{2}\sqrt{\frac{2\pi}{\nu T}}$ [14] and the Strouhal number is defined as

$St = \frac{\alpha^2}{\pi Re}$ [7], where L is the channel width, $\nu = \frac{2\tau - 1}{6}$ is the kinematic viscosity, G is the body force, T is the sampling period and $\omega = \frac{2\pi}{T}$ is the angular frequency. The density is ρ and the pressure gradient is sinusoidal with amplitude A which is normalized to 1 by dividing by the magnitude of the body force when comparing simulation results to analytic solutions.

3.1   Oscillatory channel flow

We have studied the flow in an infinite 2D channel due to an oscillatory pressure gradient $\frac{\partial P}{\partial x} = A sin(\omega t)$, where A is a constant. The pressure gradient is implemented by applying an equivalent body force G. The analytic solution for the velocity in this case is given by[14]

$$u(y,t) = Re \left[-\frac{A}{\rho \omega} e^{-i\omega t} \left(1 - \frac{cosh\left[\sqrt{b}(y - L/2)\right]}{cosh\left[\sqrt{b}L/2\right]} \right) \right] \tag{10}$$

where ρ is the fluid density, L is the width of the channel, and $b = -i\omega/\nu$. To check the accuracy of the LBGK model, we have performed a number of simulations. At first, we have used the bounce-back on the nodes which is quite simple and is known to be of second order accuracy[3]. The driving force for the flow is the body force G. Periodic boundary conditions are used for the inlet and the outlet boundaries. Both the Reynolds and the Womersley numbers were kept fixed by fixing the distance L between the two plates and varying the relaxation parameter τ, the period T and the body force G. The error in velocity at each time step is defined by

$$Ev = \frac{\sum_{i=1}^{n} |u_{th} - u_{lb}|}{\sum_{i=1}^{n} u_{th}} \tag{11}$$

where u_{th} is the analytic solution for the horizontal velocity, u_{lb} is the velocity obtained from the LBGK simulations, T is the period and n is the number of lattice nodes representing the width of the channel. The overall error , Eav , is averaged over the period T. The relaxation time ranges from $\tau = 0.6$ to $\tau = 3$, the body force ranges from $G = 25 \times 10^{-5}$ to $G = 0.04$ and the sampling period lies in the range $500 - 20$, giving corresponding values of $0.2 - 5$ for δt, with $\delta_t = 1$ corresponding to the case where $\tau = 1$. The system was initialized by setting the velocity to zero everywhere in the system. The convergence criterion is attained by comparing simulation results from two successive periods and the stop criterion is when this difference is less than 10^{-7}.

The agreement between the simulation and analytic solutions is quite good, as it is shown by the dashed lines in Fig. 1 which shows the velocity profiles for $\alpha = 4.34$ for $t = 0.75T$. However, there seems to be a small shift in time between the simulation and the theory. This shift is a function of time and τ. We have found that if we assume that the theory lags the simulation with a half time step, i.e $t_{lb} = t_{th} + 0.5$, the error reduces at least one order of magnitude for all τ values. Figure 1 shows a typical simulation result compared to the analytic solution with and without shifting the time coordinate. We have used this observation to compare the simulation results with the shifted analytic solution which leads to excellent agreement for all values of time, as shown in Fig. 2. The error behavior is shown in Fig. 3, from which it can be seen that the error is minimum at $\tau = 1$, as expected, since there is no slip velocity at this specific case. An error of the order of the round-off error could be reached for the special case when $\tau = 1$, when the bounce-back on the nodes is used [3], and assuming the 0.5 time shift.

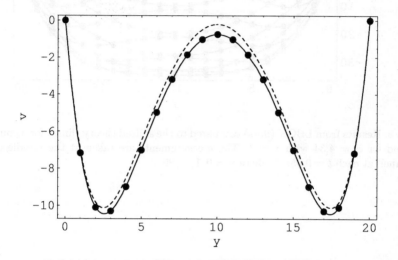

Fig. 1. Velocity profiles at $t = 0.75T$ with $\tau = 1$ and $\alpha = 4.34$ in a 2D oscillatory channel flow. The dots are the LBGK results. The dashed line is the analytic solution and the solid line is the analytic solution with a shift of 0.5 in time.

This shift in time has been observed before by Matthaeus [15]. It may be attributed to the way the driving force is imposed in the simulation and the used LBM model. We also believe that the way in which time coordinates are discretized may have some effects on this shift. For the other cases, when $\tau \neq 1$, the effect of the slip velocity dominates. Up to now, the slip velocity has analytic formulae for steady channel flow, but not yet for unsteady flow.

3.2   Non slip boundaries

In order to remove errors arising from the slip velocity, we have conducted similar simulations with a no-slip boundary condition [16] at the walls and periodic boundary conditions at the inflow and the outflow boundaries. The body force that corresponds to a desired Reynolds number drives the flow. The Strouhal

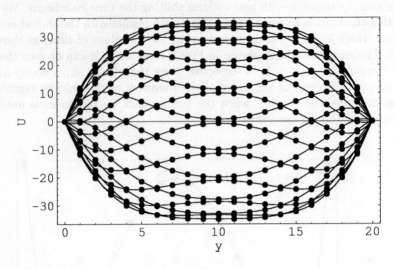

Fig. 2. Results from LBGK (dots) compared to the shifted theory (lines) for a complete period for $\alpha = 4.34$ with $\tau = 1$. The measurements are taken at the middle of the channel, at each $t = 0.05nT$ where $n = 0, 1, ..., 20$.

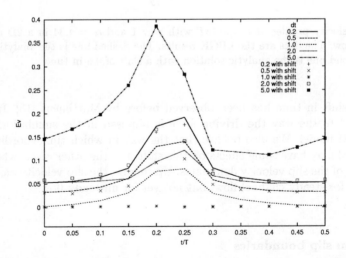

Fig. 3. Error behavior as a function of δ_t without (lines) and with (points) time-shift correction $\alpha = 4.34$ in a 2D oscillatory tube flow

number is kept constant by keeping fixed both the Reynolds and the Womersley numbers and looking at the accuracy in time. This is done by fixing the width L and assigning the corresponding values for the sampling period T, the body force G, and the relaxation parameter τ. In this way, δ_t will change. The results are shown in Fig.4 which shows the error behavior as a function of δ_t. From this figure, we clearly see that the LBM is of first order in time (slope = 0.9). The error is again decreased when the half-time step correction is used, specially at $\tau = 1$, as shown in the same graph. From this experiment, we conclude that the shift in time doesn't depend on the used boundary condition.

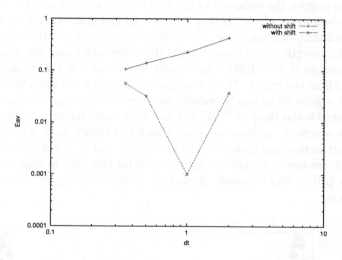

Fig. 4. Overall Relative Error of velocity vs δ_t for $\alpha = 15.53$ and $Re = 10$ in a 2D oscillatory channel flow. The slope of the straight line is 0.90. The dashed line is the error with reference to the shifted in time analytic solution.

3.3   Influence of the Reynolds number

We have conducted another set of simulations to see the influence of the Reynolds number on the error in the flow fields. Here, the relaxation parameter is kept fixed at the value $\tau = 1$, the width of the channel is varied to achieve higher Reynols numbers and the period is changed analogously to keep the Womersley parameter constant. The length of the channel is $5L$. In summary, we change the Reynolds number Re, the body force G, the width L and the period T. Simulations for Reynolds numbers in the range $1 \to 200$ at $\alpha = 15.533$ were performed. Figure 5 shows comparisons of numerical and analytical solutions of the velocity profile for $Re = 200$ at $t = 0.2T$. Similar agreements between theory and simulations has been observed for the whole period at different Reynolds numbers (data not shown). When compared to the analytic solutions, with and without shift in time, the error decreases from 0.0085 to 0.00024 for $Re = 200$.

It can be seen that the difference between the two analytic solutions (the original and the shifted) becomes less as the Reynolds number increases. This suggests that the shift is inversely proportional to the applied body force which may have direct influence on it. It is therefore necessary to conduct another set of simulations in which the body force has no influence.

3.4   Inlet and Outlet Pressure Boundary Conditions

In order to remove the influence of the body force, we have conducted another set of simulations in which the flow in a 2D channel is driven with a sinusoidal pressure gradient of magnitude $A = 0.001/L_x$ where L_x is the length of the channel. The length of the channel is 10 times the width and the period of the driving pressure is $T = 1000$. The density at the inlet is $1 + A sin(\omega t)$ and is set to be 1.0 at the outlet. The convergence criterion is attained by comparing simulation results from two successive periods and the stop criterion is when this difference is less than 10^{-7}. All the flow fields were initialized from zero. We have again observed excellent agreement with the theory, as it is shown in Fig. 6 . The shift in time has diminished in magnitude, but it is still there. From this experiment, we conclude that the main reason for this shift in time is due to an artifact in LBGK and is mainly attributed to incorrect implementation of the forcing term.

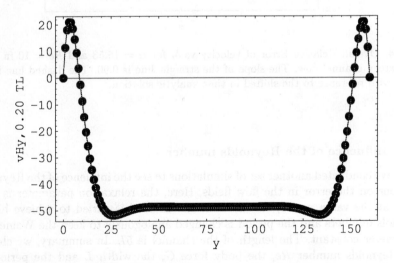

Fig. 5. Velocity profiles obtained from LBGK simulations (dots) for $\alpha = 15.53$ and $Re = 200$ in a 2D oscillatory channel flow, showing excellent agreement with the analytic solutions (lines). The effect of time shift is not observable.

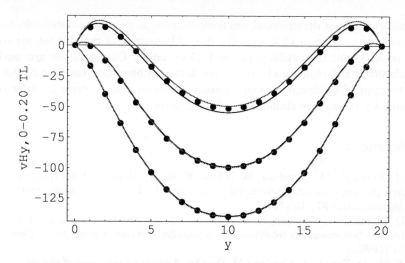

Fig. 6. Velocity profiles obtained from LBGK simulations (dots) for $\alpha = 4.00017$ in a 2D oscillatory channel flow when inflow and outflow boundary conditions are used. Selected simulation times are shown. The shift in time (lines) has little effect but still there.

3.5   Discussion and Conclusion

We have shown that the lattice Boltzmann LBGK model can be used to simulate time dependent flows in 2D within acceptable accuracy if suitable simulation parameters and accurate boundary conditions are used. We have conducted a number of 2D simulations for time-dependent flow in a channel with different boundary conditions. We have observed a shift in time and have shown that the lattice Boltzmann BGK model is more accurate when a half time step correction is added to the time coordinates. We have investigated the time shift association with the used boundary conditions, and have found that it is always present for the cases we have studied. The effects of the Womersley, the Reynolds and the Strouhal numbers have also been studied. We also have looked at the accuracy in time for time-dependent flows and have shown that it is of first order, in agreement with the theory[10].

We have also measured the shear stress from the non-equilibrium parts of the distribution functions and have found good agreement with the analyticly derived stresses. However, it has been reported that LBGK is thermodynamically inconsistent and the forcing term leads to incorrect energy balance equation when the acceleration is not constant in space [17]. Therefore, it is argued that, it is better to use a general LBM model rather then the LBGK to overcome problems arising from artifacts in the LBGK model, since it is not suitable for dense gases. The modified Lattice Boltzmann method that is derived from the Enskog equation in which the H theorems can be proved and the forcing term recovers correct energy balance equations [17]. However, on the other hand, the LBGK is simple and could work properly if being used cautiously. Currently, we have performed a number of three dimensional simulations using a quasi-incompressible model. The error is more than 12 % with the simple bounce-back rule due mainly to the slip velocity and the artifacts in the LBKG model. Currently, we are working in enhancing the model and using more accurate boundary conditions. All this will be applied to simulations of a human pressure systolic cycle in order to be suitable for blood flow simulations of large arteries.

References

1. M. Krafczyk, M. Cerrolaza, M. Schulz, E. Rank, Analysis of 3D transient blood flow passing through an artificial aortic valve by Lattice-Boltzmann methods, *J. Biomechanics* **31**, 453(1998).
2. D. Kandhai, A. Koponen, A. G. Hoekstra, M. Kataja, J. Timonen and P.M.Sloot, Lattice-Boltzmann hydrodynamics on parallel systems, *Comp. Phys. Com.* **111**, 14 (1998).
3. X. He, Q. Zou, L. S. Luo, and M. Dembo. Analytical solutions of simple flow and analysis of non-slip boundary conditions for the lattice Boltzmann BGK model, *J. of Stat. Phys.*, **87**, 115 (1996).
4. S. Hou, Q. Zou, S. Chen, G. Doolen and A. C. Cogley. Simulation of cavity flow by the lattice-Boltzmann method, *J.Comp. Phys.* **118**, 118 (1995).
5. A. M. Artoli, D. Kandhai, H. C. J. Hoefsloot, A. G. Hoekstra, and P. M. A. Sloot, Accuracy of the stress tensor in Lattice Boltzmann BGK Simulations, submitted to *J. Comp. Phys.*
6. Xiaoyi He and Li-Shi Luo, Lattice Boltzmann Model for the Incompressible Navier-Stokes Equation, *J. Stat. Phys.* **88** 927 (1997).
7. Y. C. Fung, *Biomechanics:Circulation, 2nd edn.*, Springer, 192 (1997).
8. D. A. McDonald, *Blood flow in Arteries, 2nd edn.*, The Camelot Press, (1974).
9. A. J. C. Ladd, Numerical simulations of particulate suspensions via a discretized Boltzmann equation. Part I. Theoretical foundation, *J. Fluid Mech.* **271** 285 (1994).
10. X. He and L. Luo, Theory of the lattice Boltzmann method: from the Boltzmann equation to the lattice Boltzmann equation, *Phys. Rev. E*, **56**, 6811 (1997).
11. P. L. Bhatnagar, E. P. Gross, and M. Krook, A model for collision processes in gases. I. Small amplitude processes in charged and neutral one-component system, *Phys. Rev. A*, **94**, 511 (1954).
12. Q. Zou, S. Hou, S.Chen and G.D. Doolen. An improved incompressible lattice Boltzmann model for time independent flows, *J. Stat. Phys.*, October(1995).
13. B. Chopard and M. Droz, *Cellular Automata modeling of Physical Systems*, Cambridge University Press (1998).
14. C. Pozrikidis, Introduction to Theoretical and Computational Fluid Dynamics, OUP, (1997)
15. W. H. Mattheus, Bartol Research Institute, University of Delware, personal communication, International Conference on Discrete Simulation of Fluid dynamics, Cargese, July(2001).
16. Q. Zou and X. He, On pressure and velocity boundary conditions for the lattice Boltzmann BGK model, *Phys. Fluids* **9**, 1591 (1997).
17. Li-Shi Luo, Theory of the lattice Boltzmann method: Lattice Boltzmann models for non-ideal gases, *Phys.Rev. E* **62**, 4982(2000).

Towards a Microscopic Traffic Simulation of All of Switzerland

Bryan Raney[1], Andreas Voellmy[1], Nurhan Cetin[1], Milenko Vrtic[2], and Kai Nagel[1]

[1] Dept. of Computer Science, ETH Zentrum IFW B27.1,
CH-8092 Zürich, Switzerland
[2] Inst. for Transportation Planning IVT, ETH Hönggerberg HIL F32.3,
CH-8093 Zürich, Switzerland

Abstract Multi-agent transportation simulations are rule-based. The fact that such simulations do not vectorize means that the recent move to distributed computing architectures results in an explosion of computing capabilities of multi-agent simulations. This paper describes the general modules which are necessary for transportation planning simulations, reports the status of an implementation of such a simulation for all of Switzerland, and gives computational performance numbers.

Keywords: Traffic Simulation, Transportation Planning, TRANSIMS, parallel computing

1 Introduction

Human transportation has physical, engineering, and socio-economic aspects. This last aspect means that any simulation of human transportation systems will include elements of adaptation, learning, and individual planning. In terms of computerization, these aspects can be much better described by rules which are applied to individual entities than by equations which are applied to aggregated fields. This means that a rule-based multi-agent simulation is a promising method for transportation simulations (and for socio-economic simulations in general). By a "multi-agent" simulation we mean a microscopic simulation that models the behavior of each traveler, or *agent*, within the transportation system as an individual, rather than aggregating their behavior in some way. These agents are intelligent, which means that they have strategic, long-term goals. They also have internal representations of the world around them which they use to reach these goals. Adding the term "rule-based" indicates that the behavior of the agents is determined by sets of rules instead of equations. Thus, a rule-based multi-agent simulation of a transportation system will apply to each agent individually. This means such a simulation differs significantly from a microscopic simulation of, say, molecular dynamics, because unlike molecules, two "traveler" particles (agents) in identical situations within a transportation simulation will in general make different decisions.

Such rule-based multi-agent simulations run well on current workstations and they can be distributed on parallel computers of the type "networks of coupled workstations."

P.M.A. Sloot et al. (Eds.): ICCS 2002, LNCS 2329, pp. 371–380, 2002.

Since these simulations do not vectorize, this means that the jump in computational capability over the last decade has had a greater impact on the performance of multi-agent simulations than for, say, computational fluid-dynamics, which also worked well in vector-based computational environments. In practical terms, this means that we are now able to run microscopic simulations of large metropolitan regions with more than 10 million travelers. These simulations are even fast enough to run them many times in sequence, which is necessary to emulate the day-to-day dynamics of human learning, for example in reaction to congestion.

In order to demonstrate this capability and also in order to gain practical experience with such a simulation system, we are currently implementing a 24-hour microscopic transportation simulation of all of Switzerland. Switzerland has 7.2 million inhabitants. Assuming 3 to 3.5 trips per person per day, this will result in about 20–25 million trips. This number includes pedestrian trips (like walking to lunch), trips by public transit, freight traffic, etc. The number of car trips on a typical weekday in Switzerland is currently about 5 million (see [1] for where the data comes from). The goal of this study is twofold:

– Investigate if it is possible to make TRANSIMS realistic enough to be useful for such a scenario, and how difficult this is.
– Investigate the computational challenges and how they can be overcome.

This paper gives a short report on the current status. Section 2 describes the simulation modules and how they were used for the purposes of this study. Section 3 describes the input data, i.e. the underlying network and the demand generation. Besides "normal" demand, we also describe one where 50 000 travelers travel from random starting points within Switzerland to the Ticino, which is the southern part of Switzerland. We use this second scenario as a plausibility test for routing and feedback. This is followed by Sect. 4, which describes some results, and by a Summary.

2   Simulation Modules

Traffic simulations for transportation planning typically consist of the following modules (Fig. 1):

– **Population generation**. Demographic data is disaggregated so that one obtains individual households and individual household members, with certain characteristics, such as a street address, car ownership, or household income [2]. – This module is not used for our current investigations but will be used in future.
– **Activities generation**. For each individual, a set of activities (home, going shopping, going to work, etc.) and activity locations for a day is generated [3, 4]. – This module is not used in our current investigations but will be used in future.
– **Modal and route choice**. For each individual, modes and routes are generated that connect activities at different locations [5].
– **Traffic micro-simulation**. Up to here, all individuals have made *plans* about their behavior. The traffic micro-simulation executes all those plans simultaneously [6]. In particular, we now obtain the result of *interactions* between the plans – for example congestion.

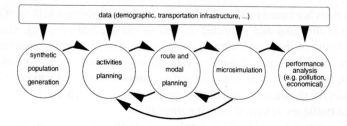

data (demographic, transportation infrastructure, ...)

| synthetic population generation | activities planning | route and modal planning | microsimulation | performance analysis (e.g. pollution, economical) |

Figure 1. TRANSIMS modules

- **Feedback**. In addition, such an approach needs to make the modules consistent with each other. For example, plans depend on congestion, but congestion depends on plans. A widely accepted method to resolve this is systematic relaxation [7] – that is, make preliminary plans, run the traffic micro-simulation, adapt the plans, run the traffic micro-simulation again, etc., until consistency between modules is reached. The method is somewhat similar to a standard relaxation technique in numerical analysis.

This modularization has in fact been used for a long time; the main difference is that it is now feasible to make all modules completely microscopic, i.e. each traveler is individually represented in all modules.

2.1 Routing

Travelers/vehicles need to compute the sequence of links that they are taking through the network. A typical way to obtain such paths is to use a shortest path Dijkstra algorithm [5]. This algorithm uses as input the individual link travel times plus the starting and ending point of a trip, and generates as output the fastest path.

It is relatively straightforward to make the costs (link travel times) time dependent, meaning that the algorithm can include the effect that congestion is time-dependent: Trips starting at one time of the day will encounter different delay patterns than trips starting at another time of the day. Link travel times are fed back from the micro-simulation in 15-min time bins, and the router finds the fastest route based on these 15-min time bins. Apart from relatively small and essential technical details, the implementation of such an algorithm is fairly standard [5]. It is possible to include public transportation into the routing [8]; in our current work, we look at car traffic only.

2.2 Micro-Simulation

We use two different micro-simulations, one being the micro-simulation of the TRAN-SIMS [9] project and the other one being a so-called queue micro-simulation that we also use for computational performance testing. The TRANSIMS micro-simulation is a complex package with many rules and details. In order to speed up the computation, the driving rules are based on the cellular automaton (CA) method [10] with additional rules for lane changing and protected as well as unprotected turns [6]. The result is, within the limits of the capabilities of a CA, a virtual reality traffic simulation. Note that besides the usual traffic dynamics, vehicles also follow routes as specified above. This means, for example, that vehicles need to change lanes in order to be in one of the allowed lanes for the desired turning movement. Vehicles which fail to do this because of too much traffic are removed from the simulation.

The queue simulation [11, 12] is simpler in its traffic dynamics capabilities. Streets are essentially represented as FIFO (first-in first-out) queues, with the additional restrictions that (1) vehicles have to remain for a certain time on the link, corresponding to free speed travel time; and that (2) there is a link storage capacity and once that is exhausted, no more vehicles can enter the link.

A major advantage of the queue simulation, besides its simplicity, is that it can run directly off the data typically available for transportation planning purposes. For the more complicated TRANSIMS micro-simulation, a lot of data conversion and additional assumptions have to be made.

2.3   Feedback

As mentioned above, plans (such as routes) and congestion need to be made consistent. This is achieved via a relaxation technique:

1. Initially, the system generates a set of routes based on free speed travel times.
2. The traffic simulation is run with these routes.
3. 10% of the population is gets new routes, which are based on the link travel times of the last traffic simulation.
4. This cycle (i.e. steps (2) and (3)) is run for 50 times; earlier investigations have shown that this is more than enough to reach relaxation [13].

Note that this implies that routes are fixed during the traffic simulation and can only be changed between iteration runs. Work is under way to improve this situation, i.e. to allow online re-planning.

3   Input Data and Scenarios

The input data consists of two parts: the street network, and the demand.

3.1   The Street Network

The network that is used was originally developed for the Swiss regional planning authority (Bundesamt für Raumentwicklung). It has since been updated, corrected and calibrated by Vrtic at the IVT. The network has the fairly typical number of 10 572 nodes and 28 622 links. Also fairly typical, the major attributes on these links are type, length, speed, and capacity.

As pointed out above, this is enough information for the queue simulation. However, since the TRANSIMS micro-simulation is extremely realistic with respect to details such as number of lanes, turn and merge lanes, lane connectivity across intersections, or traffic signal phases, plausible defaults need to be generated for those elements from the available network files. For example, all intersections are assumed as "no control", which is a TRANSIMS category meaning that the simulation does not make any special provisions in order to deal with traffic stream priorities. The result will depend on the sequence in which the simulation goes through the links. More details can be found in [14].

3.2 The "Gotthard" Scenario

In order to test our set-up, we generated a set of 50 000 trips going to the same destination. Having all trips going to the same destination allows us to check the plausibility of the feedback since all traffic jams on all used routes to the destination should dissolve in parallel. In this scenario, we simulate the traffic resulting from 50 000 vehicles which start between 6am and 7am all over Switzerland and which all go to Lugano, which is in the Ticino, the Italian-speaking part of Switzerland south of the Alps. In order for the vehicles to get there, most of them have to cross the Alps. There are however not many ways to do this, resulting in traffic jams, most notably in the corridor leading towards the Gotthard pass. This scenario has some resemblance with real-world vacation traffic in Switzerland.

3.3 The "Switzerland" Scenario

For a realistic simulation of all of Switzerland, the starting point for demand generation is a 24-hour origin-destination matrix from the Swiss regional planning authority (Bundesamt für Raumentwicklung). For this matrix, the region is divided into 3066 zones. Each matrix entry describes the number of trips from one zone to another during a typical 24-hour workday; trips within zones are not included in the data. The original 24-hour matrix was converted into 24 one-hourly matrices using a three step heuristic which uses departure time probabilities and field data volume counts. These matrices are then converted to individual (disaggregated) trips using another heuristic. The final result is that for each entry in the origin-destination matrix we have a trip which starts in the given time slice, with origin and destination links in the correct geographical area. More details can be found in [14].

In the long run, it is intended to move to activity-based demand generation. Then, as explained above one would start from a synthetic population, and for each population member, one would generate the chain of activities for the whole 24-hour period.

4   Some Results

Figure 2 shows a typical result after 50 iterations with the TRANSIMS micro-simulation for the Gotthard scenario. The figures show the 15-minute aggregated density of the links in the simulated road network, which is calculated for a given link by dividing the number of vehicles seen on that link in a 15-minute time interval by the length of the link (in meters) and the number of traffic lanes the link contains. In all of the figures, the network is drawn as the set of small, connected line segments, re-creating the road-ways as might be seen from an aerial or satellite view of the country. The lane-wise density values are plotted for each link as a 3-dimensional box super-imposed on the 2-dimensional network, with the base of a box lying on top of its corresponding link in the network, and the height above the "ground" set relative to the value of the density. Thus, larger density values are drawn as taller boxes, and smaller values with shorter boxes. Longer links naturally have longer boxes than shorter links. Also, the boxes are color coded, with smaller values tending toward green, middle values tending toward yellow, and larger values tending toward red. In short, the higher the density (the taller/redder the boxes), the more vehicles there were on the link during the 15-minute time period being illustrated. Higher densities imply higher vehicular flow, up to a certain point (the yellow boxes), but any boxes that are orange or red indicate a congested (jammed) link. All times given in the figures are at the end of the 15-minute measurement interval. The Gotthard tunnel is indicated by a circle; the destination in Lugano is indicated by an arrow.

As expected, many routes towards the single destination are equally used. In particular, many longer but uncongested routes are used in the final iteration (shown here) which are initially empty. It turns however out that only a subset of routes towards the final destination is used. This is related to the unrealistic intersection dynamics caused by the "no control" intersections: There are many plausible routes which are at a disadvantage at critical intersections and which are for that reason only used by very few vehicles.

Figure 3 shows a snapshot of the initial run (i.e. without feedback) for the TRAN-SIMS micro-simulation and for the queue micro-simulation, both based on the same set of route plans. The visual similarity of both simulations is confirmed by analysis: Fig. 4 shows a link-by-link comparison between the two simulations for the time from 7am to 7:15am. It is clear that the two simulations are highly correlated in both quantities.

Figure 5 shows a preliminary result of the Switzerland scenario. In particular, this is a result before any feedback iterations were done. As one would expect, there is more traffic near the cities than in the country.

5   Computational Issues

A metropolitan region can consist of 10 million or more inhabitants which causes considerable demands on computational performance. This is made worse by the relaxation iterations. And in contrast to simulations in the natural sciences, traffic particles (= travelers, vehicles) have internal intelligence. As pointed out in the introduction, this internal intelligence translates into rule-based code, which does not vectorize but runs well on modern workstation architectures. This makes traffic simulations ideally suited for clusters of PCs, also called Beowulf clusters. We use domain decomposition, that is, each CPU obtains a patch of the geographical region. Information and vehicles between the patches are exchanged via message passing using MPI (Message Passing Interface).

Table 1 shows computing speeds for different numbers of CPUs for the queue simulation. The simulation scales fairly well for this scenario size and this computing architecture up to about 10 CPUs. The TRANSIMS micro-simulation is somewhat slower,

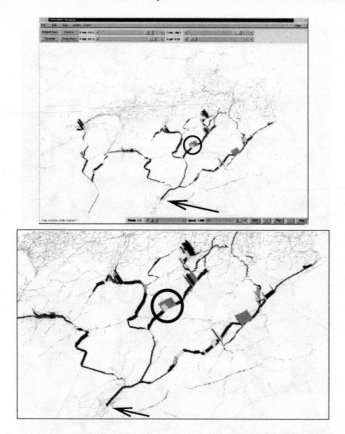

Figure 2. Snapshot at 8:00AM. The circle shows the traffic jam before the Gotthard tunnel. The arrow indicates the destination of all vehicles. TOP: All of Switzerland; BOTTOM: Zoom-In on interesting area

but scales up to a RTR of 50 using about 30 CPUs [15, 16]. In fact, the bottleneck to faster computing speeds is the latency of the Ethernet interface [15, 16], which is about 0.5–1 msec. Since we have in the average six neighbors per domain meaning six message sends per time step, running 100 times faster than real time means that between 30% and 60% of the computing time is used up by message passing. As usual, one could run larger scenarios at the same computational speed when using more CPUs. However, running the same scenarios faster by adding more CPUs would demand a low latency communication network, such as Myrinet, or a supercomputer. Systematic computational speed predictions for different types of computer architectures can be found in Refs. [15, 16].

6 Summary

In terms of travelers and trips, a simulation of all of Switzerland, with more than 10 million trips, is comparable with a simulation of a large metropolitan area, such as London

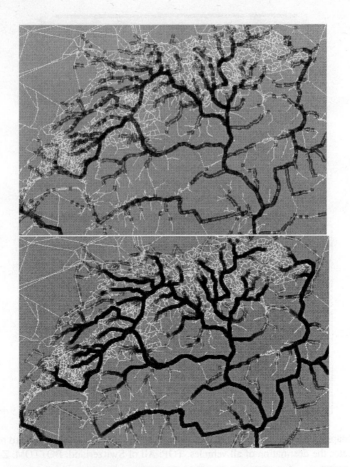

Figure 3. Iteration 0, 7h15; TOP: TRANSIMS; BOTTOM: Queue-Sim.

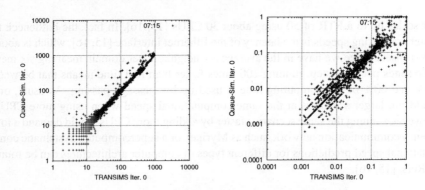

Figure 4. Link-by-link comparison TRANSIMS vs. Queue-Sim (iteration 0; 7h15). LEFT: Throughput. RIGHT: Density. The strong diagonal lines in the density comparison stem from links with very few cars on them: In those cases, they have very low densities which are related by small integer ratios, translating into constant offset on the logarithmic scales.

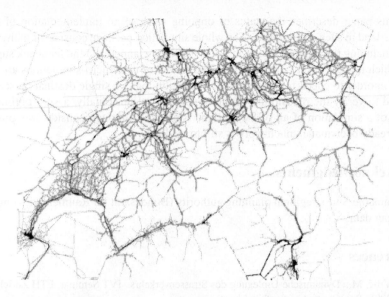

Figure 5. Switzerland at 8AM: Red (very dark gray) areas represent jammed traffic (in the cities); green and blue (mid-grays) represent flowing traffic; yellow (light gray) represents empty or nearly-empty roads. Very preliminary result.

Table 1. Computational performance of the queue micro-simulation on a Beowulf Pentium cluster. The center column gives the number of seconds taken to run the first 3 hours of the Gotthard scenario. The right column gives the real time ratio (RTR), which is how much faster than reality the simulation is. A RTR of 100 means that one simulates 100 seconds of the traffic scenario in one second of wall clock time.

Number of Procs	Time elapsed	real time ratio
1	597	18.09
4	358	30.17
5	261	41.23
8	151	71.14
9	131	82.34
12	123	87.15
17	105	102.27
25	103	104.10
33	115	93.13

or Los Angeles. It is also comparable in size to a molecular dynamics simulation, except that travelers have considerably more "internal intelligence" than molecules, leading to complicated rule-based instead of relatively simple equation-based code. Such multi-agent simulations do not vectorize but run well on distributed workstations, meaning that the computing capabilities for such simulations have virtually exploded over the last decade.

This paper describes the status of ongoing work of an implementation of all of Switzerland in such a simulation. The whole simulation package consists of many modules, including the micro-simulation itself, the route planner, and the feedback supervisor which models day-to-day learning. The results of two micro-simulations are compared in order to check for errors in the simulation logic; a single destination scenario is used to verify the plausibility of the replanning set-up. Finally, a very preliminary result of a simulation of all of Switzerland is shown. Although considerable progress has already been made, much work is still to be done.

7   Acknowledgments

We thank the Swiss regional planning authority (Bundesamt für Raumentwicklung) for the input data.

References

[1] Vrtic, M.: Dynamische Umlegung des Strassenverkehrs. IVT Seminar, ETH Zürich, Dec 2001. See www.ivt.baug.ethz.ch/vortraege

[2] Beckman, R.J., Baggerly, K.A., McKay, M.D.: Creating synthetic base-line populations. Transportation Research Part A – Policy and Practice 30 (1996) 415–429

[3] Vaughn, K., Speckman, P., Pas, E.: Generating household activity-travel patterns (HATPs) for synthetic populations (1997)

[4] Bowman, J.L.: The day activity schedule approach to travel demand analysis. PhD thesis, Massachusetts Institute of Technology, Boston, MA (1998)

[5] Jacob, R.R., Marathe, M.V., Nagel, K.: A computational study of routing algorithms for realistic transportation networks. ACM Journal of Experimental Algorithms 4 (1999)

[6] Nagel, K., Stretz, P., Pieck, M., Leckey, S., Donnelly, R., Barrett, C.L.: TRANSIMS traffic flow characteristics. Los Alamos Unclassified Report (LA-UR) 97-3530 (1997)

[7] : DYNAMIT/MITSIM (1999) Massachusetts Institute of Technology, Cambridge, Massachusetts. See its.mit.edu.

[8] Barrett, C.L., Jacob, R., Marathe, M.V.: Formal language constrained path problems. Los Alamos Unclassified Report (LA-UR) 98-1739, see transims.tsasa.lanl.gov (1997)

[9] TRANSIMS, TRansportation ANalysis and SIMulation System: (since 1992) See transims.tsasa.lanl.gov.

[10] Nagel, K., Schreckenberg, M.: A cellular automaton model for freeway traffic. Journal de Physique I France 2 (1992) 2221

[11] Gawron, C.: An iterative algorithm to determine the dynamic user equilibrium in a traffic simulation model. International Journal of Modern Physics C 9 (1998) 393–407

[12] Simon, P.M., Nagel, K.: Simple queueing model applied to the city of Portland. International Journal of Modern Physics C 10 (1999) 941–960

[13] Rickert, M.: Traffic simulation on distributed memory computers. PhD thesis, University of Cologne, Germany (1998) See www.zpr.uni-koeln.de/~mr/dissertation.

[14] Voellmy, A., Vrtic, M., Raney, B., Axhausen, K., Nagel, K.: Status of a TRANSIMS implementation for Switzerland (in preparation)

[15] Rickert, M., Nagel, K.: Dynamic traffic assignment on parallel computers in TRANSIMS. Future generation computer systems 17 (2001) 637–648

[16] Nagel, K., Rickert, M.: Parallel implementation of the TRANSIMS micro-simulation. Parallel Computing (in press)

Modelling Traffic Flow at an Urban Unsignalised Intersection

H. J. Ruskin and R. Wang

School of Computer Applications, Dublin City University, Dublin 9, Ireland
{hruskin, rwang}@compapp.dcu.ie

Abstract. This paper proposes a new way to study traffic flow at an urban unsignalised intersection, through detailed space considerations, using cellular automata (CA). Heterogeneity and inconsistency are simulated by incorporation of different categories of driver behaviour and reassignment of categories with given probabilities at each time step. The method is able to reproduce many features of urban traffic, for which gap-acceptance models are less appropriate. Capacities of the minor-stream in a TWSC intersection are found to depend on flow rates of major-streams, also changes with *flow rate ratio* (**FRR**= *flow rate of near lane: flow rate of far lane*). Hence flow rates corresponding to each stream must be distinguished. The relationship between the performance of intersections and other traffic flow parameters is also considered. Vehicle movements in this paper relate to *left-side driving*, such as found in UK/Ireland. However, rules are generally applicable.

Keywords: Modelling, cellular automata, unsignalised intersection, capacity, TWSC

1 Introduction

Two types of unsignalised intersections have been the main focus in modelling uncontrolled intersection flow. These are the two-way stop-controlled intersection (**TWSC**) and all-way stop-controlled intersection (AWSC). AWSC and TWSC are typical in North America and UK/ Ireland respectively. We focus on the latter here.

Performance measurements for TWSC have included *capacity,* (the maximum number of vehicles that can pass through an intersection from a given road), queue-length and delay. Both empirical and analytical methods have been used. The former includes Kimber's model [1] and the linear capacity model [2], while the most common analytical method uses the gap-acceptance criterion [3].

Cellular automata (CA) models provide an efficient way to model traffic flow on highway and urban networks, [4-6]. The CA model is designed to describe stochastic interaction between individual vehicles, independently of headway distribution and can be applied to most features of traffic flow, whether or not these can be described by a theoretical distribution. Features modelled may include multi-streams on the major road, heterogeneous vehicles (passenger and heavy vehicles), and intersections with or without flaring.

P.M.A. Sloot et al. (Eds.): ICCS 2002, LNCS 2329, pp. 381–390, 2002.

2   Background

The basic assumption of gap-acceptance models is that the driver will enter the intersection when a safe opportunity or "gap" occurs in the traffic. Gaps are measured in time and correspond to *headway,* (defined as distance divided by speed). Critical gap and follow-up time are the two main parameters, where the *critical gap* is defined as the minimum time interval required for one minor-stream vehicle to enter the intersection. The *follow-up time* is the time span between two departing vehicles, under the condition of continuous queueing.

Gap-acceptance models are, however, unrealistic in assuming that drivers are consistent and homogenous [7, 8]. A *consistent driver* would be expected to behave in the same way in all similar situations, while in a *homogenous population*, all drivers have the same critical gap and are expected to behave uniformly [9]. In any simulation, therefore, driver type may differ and the critical gap for *a particular driver* should be represented by a stochastic distribution, Bottom and Ashworth [10].

In gap acceptance models, estimation of the critical gap has attracted much attention, with use of a mean critical gap also proposed [11-13]. Maximum likelihood estimation of the mean critical gap has been widely accepted [3, 12-15], but the basic assumption is still that all drivers are consistent.

Tian et al. [16] investigated the factors affecting critical gap and follow-up time, concluding that drivers use shorter critical gap at higher flow and delay conditions. Many other factors have also been found to affect critical gap [16-18], so that a critical value, obtained for any given situation, is unlikely to be generally applicable.

Further, gap-acceptance models have failed to consider conflicts between the two major-streams. When right-turning vehicles (for left-side driving) in the major-stream of a narrow road give way to straight-through vehicles from the opposing stream, a queue will form on the major-stream behind the subject vehicle, (i.e. turning-left and going-straight vehicles share the same lane). The headway distributions are affected so that original gap-acceptance criteria no longer apply.

At an unsignalised intersection in an urban network, adjacent intersections with traffic lights will have grouped the vehicles into a queue (or queues) during the red signal phases, and platoons will thus be present, (i.e. a filtering effect). The filtering of traffic flow by traffic signals has a significant impact on capacity and performance [19]. In particular, the gap-acceptance model can be applied *only* when no platoon is present [20]. Otherwise, no minor-stream vehicle can enter the intersection, as the mean headway within a platoon is supposed to be less than the critical gap. If traffic signal cycles are known and co-ordinated, the platoon pattern may be predictable. Otherwise, traditional gap-acceptance is not readily applied [20].

Headway distributions are also affected by traffic lights and in absence of these, platoon formation will occur due to the vehicle speeds. Further, critical gap is not easy to define and implement when several traffic streams are involved [3] and gap-acceptance does not specifically allow for modelling directional flow [16].

A CA model is thus proposed, using analogous but more flexible methodology compared to gap acceptance, (e.g. spatial and temporal details of vehicle interactions can be described). This not only facilitates understanding of the interaction between the drivers, but can also be applied to situations for which headway distributions are insufficient to describe traffic flow. This paper considers combinations of available space on several major/minor streams and extends previous work on single conflicting flows [21].

A CA ring was firstly proposed for unsignalised intersections [5, 22]. All entry roads are "connected" on the ring. The car "on the ring" has priority over any new entry. However, there is no differentiation between the major and minor entry roads and all vehicles have equal priority to move into the ring (intersection), which compromises usual TWSC rules. A further CA model variant for intersections is described [23].

3   Methodology

A two-speed one-dimensional deterministic CA model, [5, 24, 25], is used to simulate interaction between drivers on the intersection *only*. The speed of a vehicle is either 0 or $1(v_{max}=1)$, i.e. the vehicle can move only one cell in a time-step, (1 second for our model). The length of each cell corresponds to "average speed" on given intersection, e.g. length of 1 cell $\cong 13.9$ m for speed of $50km/h$.

While multi-speed models [4] are somewhat similar, these have many features, which are superfluous for urban features, or to representation of driver behaviour [5]. Moreover, vehicle dynamics are often less important than driver interactions in simulating queue formation in urban networks [22].

A multi-speed one-dimensional deterministic CA model [4, 26] is used here to model the traffic flow on the *straight* roads only (intersection excluded). Speed of vehicles is 0, 1 or 2 ($v_{max}=2$), corresponding to speed of 0, $25km/h$ and $50km/h$. Length of 1 cell $\cong 7$ m. The difference between the t

The update rules for each time-step are:
- Vehicle moves v (= 0 or 1:two-speed CA or 0, 1 or 2 : multi-speed CA) cells ahead
- Find the number of empty cells ahead = E
- If $v < E$ and $v < v_{max}$, then v increased to $v+1$
- If $v \geq E$, then $v = E$

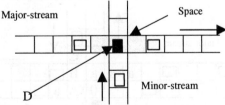

Fig. 1. Two-stream intersections with a rational driver D

A two-stream intersection (Fig. 1) is used to illustrate the driver interaction. Three cells give the minimum theoretical acceptable space for non-interruption of major-stream. Drivers are categorised as *aggressive, rational* and *conservative*.

A driver accepting a 3-cell space as the *minimum acceptable space* (**MAP**) is *rational*. A 2-cell space corresponds to *aggressive* behaviour. The effect is the blocking of the vehicle that has priority by the sub-rank vehicle. *Conservative* behaviour corresponds to *MAP* ≥ *4* cells. Drivers are likely to prefer longer gaps for the more complex decision involved in turning, even though longer gaps are not required theoretically [17]. We assume therefore that most driver behaviour can be classified as rational or conservative. Probabilities associated with driver types sum to 1. Drivers are randomly reassigned to different categories with given probabilities at each time step, prior to checking whether the space meets the MAP. In this way, heterogeneity and inconsistency of driver behaviour are incorporated.

According to the rules of the road, a vehicle from a minor-stream has to obey a stop sign before it can enter an intersection. Our simulation ensures that all vehicles from the minor-stream will stop for at least one time-step (equal to 1 second). For minor-stream vehicles travelling straight-ahead or right-turning, two time-steps delay is allowed, in order to make a decision, (two major-streams are checked). We denote the time required as stop-sign-delay-time (**SSDT**). Thus, the follow-up time for a minor-stream in the simulation will be from 3 to 4 seconds, which agrees with the recommended follow-up time from observed data [16].

The main difference between our CA model and gap-acceptance models, in general, is that the critical gap in the gap-acceptance model and the MAP in our model have different temporal and spatial content, although both provide criteria for a driver to take action. For the gap-acceptance model, where the conflicting flow includes more than two streams, the *gap* is normally defined as *the time taken for two vehicles from conflicting streams to pass through the path of the subject vehicle*. Without distinguishing the direction that each vehicle comes from, the critical gap then has strong temporal meaning but is weak in spatial detail. However, in our model, the space required (in terms of different number of vacant cells required in *each* conflicting stream) is clearly specified so that temporal and spatial details are known precisely for each different movement indifferent streams, (details below), and the driver decision process is thus fully specified. Also, the critical gap is a fixed single value, whereas the MAP is a multi-value distribution corresponding to the distribution of driver behaviour.

3.1 Minor-stream Vehicles: Straight-ahead (*SA*) and Right-turning (*RT*)

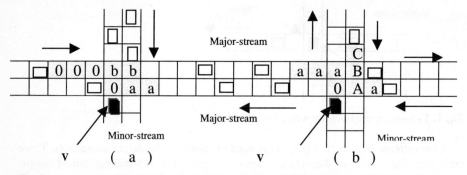

Fig. 2. A *rational* vehicle from a minor-stream moving (a) Straight-ahead (b) Right

For TWSC, conditions for the marked SA vehicle V to move into the intersection are illustrated. A *rational* driver needs to observe the 8 marked cells before s/he can drive into the intersection (Fig. 2a), whereas a *conservative* driver needs to check 10 cells. The marked cells, 0, a, b and c, correspond to: "0" cell is vacant, "a" cell either vacant or occupied by a vehicle that will turn left, "b" cell *not* occupied by a right-turning vehicle and "c" cell either occupied by a right-turning vehicle or vacant.

In Fig. 2b, a SA or LT vehicle from the opposing minor-stream in cell marked "c" also has priority over the RT vehicle V from the given minor-stream according to the rules of the road. However, priorities between minor-stream vehicles might not be distinct [16]. Drivers were observed to enter the intersection on a first-come, first-served basis. The movement of a RT vehicle from a minor-stream does not need to consider opposing vehicles if one of the following conditions is met.

- The first cell in the opposing minor-stream is vacant
- A RT vehicle is the first vehicle in the opposing minor-stream
- The first vehicle in the opposing minor-stream arrives at a stop-line in less than SSDT

3.2 Left-turning (LT) Vehicle from a Minor-stream and RT from a major stream

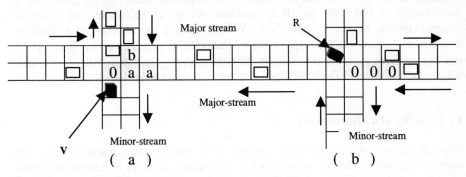

Fig. 3. A *rational* vehicle (a) LT from a minor-stream (b) RT from a major stream

Similar conditions apply to a driver turning left from a minor-stream (MiLT) and right turning from a major-stream (MaRT). A *rational* MiLT driver needs to check 4 marked cells before entering the intersection (Fig. 3a), whereas a MaRT vehicle R needs to check 3 marked cells (Fig. 3b).

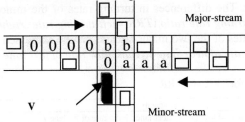

Fig. 4. A rational SA long vehicle from a minor-stream

3.3 Long Vehicles and Modified Intersections

The case for a long vehicle has been considered briefly, based on occupation of two cells (Fig. 4). Crudely, *rational* movement for long vehicle through the intersection requires a check on 10 cells (as for a *conservative* car driver) in the simple model. Preliminary results indicate that long vehicles reduce throughput, as expected, but the impact of distribution assumptions has yet to be investigated. Flared minor-stream increases (e.g. [20]), can also be accommodated in our model, but have not been investigated to date.

4. Model Implementation and Results

Based on assumptions described, we studied performance, (capacity, time delay and queue-length) of a TWSC intersection under different values of traffic flow parameters, such as arrival rate (traffic volume) and turning rate (turning proportion). Experiments were carried out for 36,000 time-steps (equivalent to 10 hours) for a street-length of 100 cells on all approaches. All driver behaviour was assumed *rational* unless otherwise specified. Vehicles arrive according to a Poisson distribution, (where $\lambda \leq 0.5$ (equivalent to 1800vph) in general for free flow). If all arriving vehicles pass the intersection without queueing, the flow rate $\lambda=0.1$, 0.2, 0.3… equivalents are 360vph, 720vph, 1080vph… respectively.

4.1 Capacity of a Minor-street

When a RT or SA vehicle from a minor-street involves two major-streams, the capacity depends on their flow rates and configurations. In order to determine impact of different turning rates and different major-stream combinations, a TWSC intersection is studied, which contains only right-turning and left-turning vehicles in the minor-stream. All major-streams are assumed to have only SA vehicles. The total number of vehicles per hour in major-streams is assumed to be 1440 vph, which is split between the near-lane stream, (vehicles coming from the right), and far-lane stream, (vehicles coming from the left). Both left-turning-rate (**LTR**) and right-turning-rate (**RTR**) are varied. The differences in turning rates of the minor-stream can be expressed in terms of *turning rate ratio* (***TRR** =left-turning rate: right-turning rate*). The difference in flow rates of the two major-streams can be expressed in terms of *flow rate ratio* (***FRR**= flow rate of near-lane stream: flow rate of far- lane stream*).

Table 1. Capacity of Minor-street for TRR and FRR

TRR	Capacity (vph)				
	FRR(=Flow rate of near lane : Flow rate of far lane)				
	1440:0	1080:360	720:720	360:1080	0:1440
1:0	196	397	585	755	900
0.75:0.25	193	363	483	527	415
0.5:0.5	190	331	413	408	286
0.25:0.75	183	308	361	337	217
0:1	177	288	321	286	180

Table 1 indicates that both TRR and FRR affect capacity. TRR has been varied by increasing the number of right-turning vehicles in the minor-street. We find that the capacity of the minor-stream decreases in general when TRR decreases. However, this effect differs as FRR varies. In general, a vehicle manoeuvre, which requires conflicting streams to be crossed, leads to reduction in capacity. This is clearly illustrated, for example, in the final column of the table where far lane flow is heavy and right turning ratio is gradually increased. Similarly, a large percentage of minor stream vehicles joining a busy near-lane will be delayed and so on.

4.2 Capacity of a Major-street

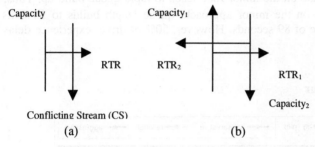

Fig. 5. Traffic configurations of shared lane on the major-streams

A right-turning vehicle in a shared major-street, where right-turning, straight-going and left-turning vehicles are on the one lane, can block SA and LT vehicles behind and in the same stream. Right-turning rates (RTR) of major-streams thus have great impact on capacities of major-streams. Two configurations have been studied (Fig. 5), with the analysis of major-street capacity following that of Chodur [27].

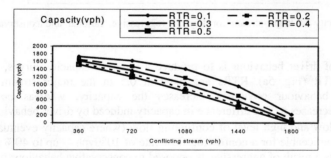

Fig. 6. Capacity of a major-stream as for Fig. 5a for rational driver behaviour

Table 2. Capacities and capacity ratio vs. right turning rate ratio

	RTR₁:RTR₂					
	0.4:0.1	0.3:0.1	0.2:0.1	0.2:0.2	0.2:0.3	0.2:0.4
Cap₁:Cap₂	~1:4	~1:3	~2:1	1:1	~3:2	2:1
Cap₁(vph)	413	541	758	1164	1373	1480
Cap₂(vph)	1659	1616	1508	1164	911	740

Fig. 6 shows unsurprisingly that the capacity of the major-stream declines rapidly with RTR as flow rate of the conflicting major-stream increases (Fig. 5a). Similarly, Table. 2 with RT vehicles in both major-streams yields a similar relation to that found form empirical study by Chodur [27], with $Capacity_1: Capacity_2 = RTR_2: RTR_1$.

4.3 Queue-length and Delay

Minor stream queue-length is found to depend on *degree of saturation, d* (= *flow rate/capacity*) and also on arrival and turning rates of major and minor streams. For example, for *LTR: SAR: RTR = 0.2:0.7:0.1* and *FRR=0.15:0.15* on the major stream even a low arrival rate on the minor road leads to rapid queue build-up. Thus, for *d* =0.9 *(λ≅ 0.13)* say on the minor approach, queue length builds to 34 cells with maximum delay time of 89 seconds. However, 50% of driver experience delay < 18 seconds.

4.4 Driver Behaviour

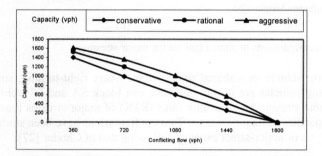

Fig. 7. Capacity of major-stream for Fig. 5a (aggressive, rational and conservative driver behaviour)

The effect of driver behaviour is to produce a series of capacity curves, similar to those for RTR (Fig. 5a). RTRs are fixed at 0.5 in the major stream (Fig. 7). *Aggressive* behaviour obviously increases the capacity, whereas *conservative* behaviour decreases it. The difference in capacity induced by driver behaviour is less-marked for low and high level of conflicting flow (where capacity eventually → 0), but is most noticeable for a conflicting flow rate of 1080vph -- up to 40% difference in capacity as a result of aggressive as opposed to conservative behaviour is observed for conflicting flow rates between 720 to 1440 vph.

Table.3 illustrates effects of different driver behaviour populations on capacity figures. In each scenario, turning rates and arrival rates are fixed. $\lambda_{1,2\ or\ 3} < 0.5$, $\lambda_4=0.8$ >> 0.5 for approach 4 (minor-stream) only.

An approximate linear relationship is observed between capacities and driver behaviour ratio. Hence driver behaviour roughly determines the capacity of an unsignalised TWSC intersection.

Table 3. Minor- stream capacity vs. driver behaviour

Modelled Scenarios	Driver Population (Rational :Conservative)				
	1:0	0.75:0.25	0.5:0.5	0.25:0.75	0:1
Scenario 1	518	492	464	435	406
Scenario 2	412	377	343	308	269
Scenario 3	527	504	482	461	437

In general, our CA model agrees well with the validated results obtained from empirical and simulation studies e.g. [27].

5   Summary

A cellular automata model is used to simulate directly the interactions between drivers at a TWSC intersection using detailed space considerations. Heterogeneity and inconsistency of driver behaviour are also investigated and driver distribution is shown to have noticeable impact on capacity of major and minor streams, where a distribution, biased in favour of conservative driver behaviour, leads to a reduction.

The capacity of the minor-stream is shown to depend not only on the flow rates of major-streams, but also on flow rate ratios. The capacity of a minor-stream decreases when LTR decreases, but is less marked for FRR increases, which depend on increased flow rate of the near-lane.

Lacking real data, the distribution of driver behaviour is arbitrarily decided in the experiments, but the model can be used to investigate various assumptions and conditions of performance for TWSC intersections together with other features of urban traffic for which gap-acceptance models are less applicable.

Acknowledgement

We should like to acknowledge useful discussions with Zong Tian, Texas Transportation Institute, U.S.A.

References

1. Kimber, R. M.: The Traffic Capacity of Roundabouts. Transport and Road Research Laboratory Report 942, TRRL, Crowthorne, England (1980)
2. Brilon, W., Wu N., Bondzio, L.: Unsignalised Intersections in Germany—A State of the Art 1997. In: Kyte M. (ed.): Proceedings of Third International Symposium on Intersections without Traffic Signals. Portland, Oregon, U.S.A., University of Idaho (1997) 61-70
3. Tian, Z., Vandehey, M., Robinson, B. M., Kittelson, W. Kyte, M., Troutbeck, R., Brilon, W., Wu, N.: Implementing the maximum likelihood methodology to measure a driver's critical gap, Transportation Research A, Vol. 33. Elsevier (1999) 187-197
4. Nagel, K., Schreckenberg, M.: A cellular automaton model for freeway traffic. Journal de Physique I (France), 2: 2221 (1992)
5. Chopard, B., Dupuis, A., Luthi, P.: A cellular Automata model for urban traffic and its application to the city of Geneva. In: Wolf D. E. and Schreckenberg M.(eds.):Traffic and Granular Flow '97. World Scientific (1998) 153-168

6. Wahle, J., Neubert, L., Esser J. , Schreckenberg, M.: A cellular automaton traffic flow model for online simulation of traffic. Parallel Computing, Vol. 27 (5). Elsevier (2001) 719-735
7. Robinson, B. W., Tian, Z.: Capacity and level of Service at Unsignalised Intersection: Final Report Volume1, Two-Way-Stop-Controlled Intersections. Access 28th July 2001. http://books.nap.edu/books/nch005/html/R5.html#pagetop (1997)
8. Troutbeck, R. J., Brilon, W.: Unsignalized Intersection Theory, Revised Traffic Flow Theory. http://www-cta.ornl.gov/cta/research/trb/CHAP8.PDF (1997)
9. Plank, A. W., Catchpole, E. A.: A General Capacity Formula for an Uncontrolled Intersection. Traffic Engineering Control. Vol. 25 (60) (1984) 327-329
10. Bottom, C. G., Ashworth R.: Factors affecting the variability of driver gap-acceptance behaviour. Ergonomics, Vol. 21 (1978) 721-734
11. Miller, A. J.: Nine Estimators for Gap-Acceptance Parameters in Traffic Flow and Transportation. In: G. Newell (ed.): Proceedings of the International Symposium on the Theory of Traffic Flow and Transportation, Berkeley, California. Elsevier (1972) 215-235
12. Troutbeck, R. J.: Estimating the Critical Acceptance Gap from Traffic Movements. Physical Infrastructure Centre, Queensland University of Technology. Research Report 92-1 (1992)
13. Brilon, W., Koening, R., Troutbeck, R.: Useful estimation procedures for critical gaps. In: Kyte M., (ed.): Proceedings of Third International Symposium on Intersections without Traffic Signals, Portland, Oregon, U.S.A., University of Idaho (1997)
14. Troutbeck, R. J.: Does gap acceptance theory adequately predict the capacity or a roundabout? In: Proceedings of the 12 ARRB (Australian Road Research Board) conference, Vol. 12, Part 4 (1985) 62-75
15. Hagring, O.: Estimation of critical gaps in two major streams. Transportation Research B, Vol. 34. Elsevier (2000) 293-313
16. Tian, Z., Troutbeck, R., Kyte, M., Brilon, W., Vandehey, M., Kittelson, W., Robinson, B. M.: A further investigation on critical gap and follow-up time Transportation Research Circular E-C108: 4th International Symposium on Highway Capacity (2000) 397-408
17. Harwood, D. W., Mason, J. M., Robert, E. B.: Design policies for sight distance at the stop controlled intersections based on gap acceptance. Transportation Research A, Vol. 33. Elsevier (1999) 199-216
18. Troutbeck, R. J., Kako, S.: Limit priority merge at unsiginalized intersection. Transportation Research A, Vol. 33. Elsevier (1999) 291-304
19. Tracz, M., Gondek, S.: Use of Simulation to Analysis of Impedance Impact at Unsignalised Intersections. In: Brilon, W.(ed): Transportation Research Circular E-C108: 4th International Symposium on Highway Capacity (2000) 471-483
20. Robinson, B. W., Tian, Z., Kittelson, W., Vandehey, M., Kyte, M., Wu, N., Brilon, W., Troutbeck, R. J.: Extension of theoretical capacity models to account for special conditions, Transportation Research A, Vol. 33. Elsevier (1999) 217-236
21. Wang, R., Ruskin, H.J.: Modeling traffic flow at a single-lane urban roundabout. In press, Computer Physics Communications (2002)
22. Chopard, B, Luthi, P. O., Queloz, P.-A.: Cellular Automata Model of Car Traffic in Two-dimensional Street Networks, J. Phys. A, Vol. 29. Elsevier (1996) 2325-2336
23. Esser, J., Schreckenberg, M.: Microscopic Simulation of Urban Traffic Based on Cellular Automata. International Journal of Modern Physics C, Vol. 8, No. 5 (1997) 1025-1026
24. Yakawa, S., Kikuchi, M., Tadaki, S.: Dynamical Phase Transition in One–Dimensional Traffic Flow Model with Blockage, Journal of the Physical Society of Japan, Vol. 63, No. 10 (1994) 3609-3618.
25. Wang R., Ruskin H. J.: Modelling Traffic Flow at a Roundabout in Urban Networks, Working Paper CA-0601, School of Computer Applications, Dublin City University.http://www.compapp.dcu.ie/CA_Working_Papers/wp01.html#0601 (2001)
26 Nagel, K.: Particle Hopping vs. Fluid-dynamic Models for Traffic Flow. In: Wolf D. E. et al. (eds.): Traffic and Granular Flow. World Scientific (1996) 41-56
27. Chodur, J.: Capacity of Unsignalized Urban Junctions. Transportation Research Circular E-C108: 4th International Symposium on Highway Capacity (2000) 357-368

A Discrete Model of Oil Recovery

Germán González-Santos[1] and Cristobal Vargas-Jarillo[2]

[1] Escuela Superior de Física y Matemáticas del IPN, Unidad Profesional Adolfo
López Mateos, México D.F. 07738
gsantos@esfm.ipn.mx
[2] Departamento de Matemáticas del CINVESTAV, A. P. 14-740, México D. F. 07000
cvargas@math.cinvestav.mx

Abstract. We propose the simulation of oil recovery by means of a
molecular type approach. By using a finite set of particles under the
interaction of a Lennard-Jones type potential we simulate the behavior
of a fluid in a porous media, and we show that under certain conditions
the fingering phenomena appears.

1   Introduction

In this work we propose the simulation of oil recovery by means of a molecular
type approach. This means that we consider the materials to be composed of
a finite number of particles, which are approximants for molecules. Porous flow
is studied qualitatively under the assumption that particles of rock, oil and the
flooding flow interact with each other by means of a compensating Lennard-
Jones type potential. We also consider the system to be under the influence of
gravity. We study miscible displacement in an oil reservoir from various sets of
initial data. The velocity and the rate of injection of the ingoing particles proved
to be among the most important parameters that can be adjusted to increase
the rate of production. It is also noted that the fingering phenomenon is readily
detected. This simulation technique has been used in [1-2] and [4] to simulate
several physical systems. Details of this method applied to the study of porous
flow can be founded in [3].

2   Model formulation

Consider a rectangular region R, which is a porous medium. We assume that in
this region we have a resident fluid or oil. We shall introduce a different kind of
fluid which, as a matter of convenience, will be called water, although it is an
aqueous solution which could be a polymeric solution, surfactant solution or a
brine. The physical system consists of $N = N_1 + N_2 + N_3$ particles, $P_1, P_2, ..., P_N$,
with masses $m_1, m_2, ..., m_N$. The particles

$$P_1, P_2, ..., P_{N_1}, \quad \text{Represent rocks,}$$
$$P_{N_1+1}, P_2, ..., P_{N_2}, \quad \text{Oil, and}$$
$$P_{N_2+1}, P_2, ..., P_N \quad \text{Incoming water}$$

P.M.A. Sloot et al. (Eds.): ICCS 2002, LNCS 2329, pp. 391–398, 2002.

For purposes of injection of water and production of oil, two wells are opened, one in the bottom left corner of R, for injection, and other in the diagonally opposite corner for production, see Fig. 1. The variables at time $t = k\Delta t$ are:

$\overline{r}_{i,k}$ Coordinates of the particle,

$r_{i,j,k}$ Distance between the particles P_i and P_j,

$\overline{v}_{i,k}$ Velocity of the particle,

$\overline{a}_{i,k}$ Acceleration of the particle,

$\overline{F}_{i,j,k}$ Local force exerted on P_i by P_j,

$\overline{F}^*_{i,k}$ Local force acting on P_i due to the other particles,

$\overline{f}_{i,k}$ Long range force acting on P_i (like gravity),

$\overline{F}_{i,k}$ Total force on particle P_i,

for $i = 1, 2, ..., N$ and $k = 0, 1, ...$

The local force $\overline{F}_{i,j,k}$ exerted on P_i by P_j is

$$\overline{F}_{i,j,k} = m_i m_j \left[\frac{H_{i,j}}{r_{i,j,k}^{q_{i,j}}} - \frac{G_{i,j}}{r_{i,j,k}^{p_{i,j}}} \right] \frac{\overline{r}_{j,k} - \overline{r}_{i,k}}{r_{i,j,k}}, \tag{1}$$

where the values of the parameters $H_{i,j}$, $G_{i,j}$, $q_{i,j}$ and $p_{i,j}$ depend on the particles which are interacting. The total local force $\overline{F}^*_{i,k}$ acting on particle P_i due to the other particles is given by:

$$\overline{F}^*_{i,k} = \sum_{j=1, j \neq i}^{N} \overline{F}_{i,j,k} \tag{2}$$

Therefore, the total force acting upon the particle P_i is

$$\overline{F}_{i,k} = \overline{F}^*_{i,k} + \overline{f}_{i,k}. \tag{3}$$

The aceleration of P_i is related to the force by Newton's Law

$$\overline{F}_{i,k} = m_i \overline{a}_{i,k}. \tag{4}$$

In general system (4) can not be solved analytically from given initial positions and velocities, therefore it must be solved numerically. For economy, simplicity and relatively numerical stability we use the "leap frog" formulae, which has second-order accuracy in time,

$$\overline{v}_{i,1/2} = \overline{v}_{i,0} + \tfrac{1}{2}\overline{a}_{i,0}\Delta t$$
$$\overline{v}_{i,k+1/2} = \overline{v}_{i,k-1/2} + \tfrac{1}{2}\overline{a}_{i,k}\Delta t$$
$$\overline{r}_{i,k+1} = \overline{r}_{i,k} + \tfrac{1}{2}\overline{v}_{i,k-1/2}\Delta t$$
$$\text{for } i = 1, 2, ..., N, \quad k = 1, 2, ... \ .$$

The number of calculations required to evaluate (1) at each iteration is $O(N^2)$. However this number is much smaller if the potential is truncated for a distance greater than r_c .

3   Boundary Conditions

We assume that the particles of the fluids loose energy when they interact with the walls of the region R, therefore it will be necessary to model the hardness of the wall relative to the reflection of the interacting fluid, and it is done by using the following damping factors acting on the velocity of the reflected particles.

$$\delta_i = 0.4 \; for \; i = N_1 + 1, ..., N_1 + N_2 \;, and$$
$$\delta_i = 0.8 \; for \; i = N_1 + N_2 + 1, ..., N.$$

4   Initial conditions

The rock and oil particles, for two an three dimensions; were set up at the initial time in such a way that they satisfied an equilibrium state, as shown in Fig. 1 and 2.

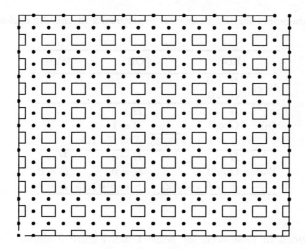

Fig. 1. Initial configuration in two dimensions

5   Numerical results in 2D

Figure 3 shows the system evolution. All the examples were run with time step $\Delta t = E - 5$ on the Sun workstation Ultra 60, the distance between particles of water before going to into the well was $d = 0.5$ and their velocity was $v = 15.0$. The gravity constant was equal to $g = 9.8$. The Lennard-Jones potentential parameters are summarized in table 1.

Figure 4 shows the advancement of water for different times, the shaded area is the region which has been traveled only by water, this means that not oil

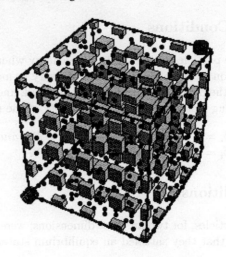

Fig. 2. Initial configuration in three dimensions

Table 1. Parameters for the numerical experiments in two dimensions. In this case $F = 0.5$

	Rock	Oil	Water
Rock	$H = 0$ $G = 0$		
Oil	$H = 1$ $G = 3$ $E = F * \sqrt{13/36}$	$H = 1$ $G = 1$ $E = F * 1.3$	
Water	$H = 1.5$ $G = 0$ $E = F * 13/36$	$H = 1$ $G = 0$ $E = F * 1.15$	$H = 1$ $G = 0$ $E = F$

particle has been in that area for some time. Figure 5 shows the number of particles of oil out and the number of particles of water out versus time. We can see from the graph that for t small, the rate of oil production is higher when v is higher. We can also observe that water comes out of the production well sooner for $v = 100$ than for $v = 15$.

6   Numerical results in 3D

The results in three dimensions are shown in Fig 6. All the examples were run with time step $\Delta t = E - 5$ on a Cluster of PC computers.

The distance between particles of water before going into the well, was $d = 0.5$ and their and velocity was $v = 5.0$. The gravity constant was equal to $g = 9.8$. The Lennard-Jones potential parameters are summarized in table 2.

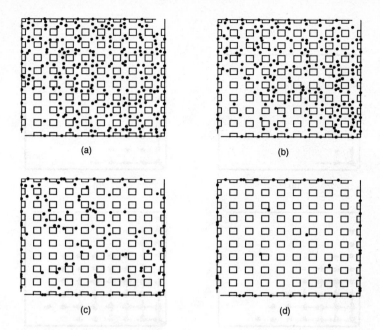

Fig. 3. Evolution of the oil particles. (a) $Time = 0.1$, (b) $Time = 0.8$, (c) $Time = 2.0$, (d) $Time = 4.8$

Figure 6 shows the effect of the oil and water production, when velocity of the water particles is increased. An increment in the velocity of the water particles produces an increment on the oil and water production.

References

1. Greenspan D.: Arithmetic Applied Mathematics. Pergamon, Oxford (1980)
2. Greenspan D., Quasimolecular Modelling, Worl Scientific, Singapore (1991)
3. Vargas-Jarillo C., A Discrete Model for the Recovery of Oil from a Reservoir. Appl. Math. and Comp. 18, 93-118 (1986).
4. Korline M. S., Three Dimensional Computer Simulation of Liquid Drop Evaporation. Comp. and Math. with Appl. 39, 43-52 (2000)

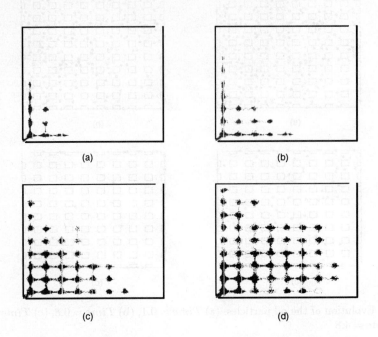

Fig. 4. Advancement of the water for $d = 1, v = 5$ at different times. (a) $Iter = 3E5$, (b) $Iter = 4.25E5$, (c) $Iter = 7.8E5$, and (d) $Iter = 1.2E6$

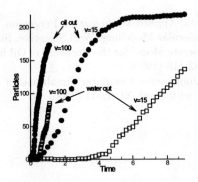

Fig. 5. Comparison of the effect of the velocity of the water particles on the oil and water production

Table 2. Parameters for the numerical experiments in three dimensions. In this case $F = 1.0$.

		Rock	Oil	Water
Rock	$H = 0$ $G = 0$			
Oil	$H = 1$ $G = 3$ $E = F * \sqrt{13/36}$		$H = 1$ $G = 1$ $E = F * 1.3$	
Water	$H = 1.5$ $G = 0$ $E = F * 13/36$		$H = 1$ $G = 0$ $E = F * 1.15$	$H = 1$ $G = 0$ $E = F$

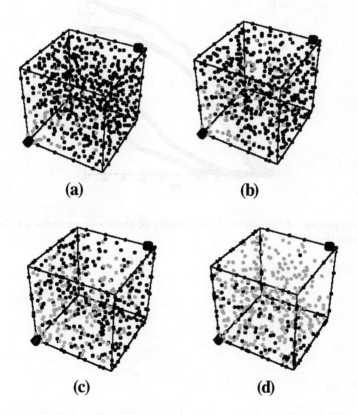

(a)   (b)

(c)   (d)

Fig. 6. Evolution of the oil and water particles at different times. (a)$Iter = 1E6$, (b)$Iter = 2E6$, (c)$Iter = 4E6$, and (d)$Iter = 1E6$

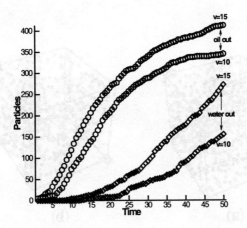

Fig. 7. Comparison of the effect of the velocity of the water particles on the oil and water production

Virtual Phase Dynamics for Constrained Geometries in a Soap Froth

Yu Feng, Heather J. Ruskin and Bao Zhu

School of Computer Applications, Dublin City University, Dublin 9, Ireland
Corresponding email: hruskin@compapp.dcu.ie

Abstract. Soap froths as typical disordered cellular structures, exhibiting spatial and temporal evolution, have been studied through their distributions and topological properties. Recently, persistence has been introduced as a non-topological probe to study froth dynamics at different length scales and to view the froth as a two-phase system. Using a direct simulation method, we have investigated virtual phase dynamics in 2D artificial froths with various initial structures corresponding to controlled disorder. In particular, we examine the special case of a defect ring surrounding a central inclusion in a uniform froth, for different percentages of persistent cells, where this geometry permits comparison with shell-theory. It appears that defect location and pattern of cell inclusion in the virtual phase cause considerable variation in the evolutionary behaviour, leading to non-universal exponents for the phase dynamics. This is probably explained by the fact that the froth is still in the transient period over simulation time-scales, rather than achieving the final stage of persistence. However, distinctive patterns of response can be identified for the different froth regions, despite the limitations on system size.

Keywords: persistence, defect ring, phase dynamics, constrained geometries, transience

1 Introduction

Soap froth as a space-filling cellular network is an intrinsically non-equilibrium system, which has attracted considerable interest in recent years, [1-4]. The froth system evolves to a universal stable state through surface-energy driven diffusion, by means of topological rearrangements in the bubbles (cells). In two-dimensions, these may occur in two ways, namely through side- switching and cell disappearance, (T1 and T2 processes respectively).

The area of an individual cell changes with time during diffusion, according to von Neumann's Law, [5], where cells with number of sides, $n < 6$ will shrink, whilst those with $n > 6$ will grow. Typically, dynamic properties are measured in terms of the side distribution, $f(n)$ and its second moment, μ_2, which achieves a stable value after a certain period of time, i.e. reaches a scaling state. Non-equilibrium behaviour has also been studied, and transient effects have been shown to depend on the nature and amount of disorder in the original structure. In particular, introduction of one or more *defects* (i.e. *topological dislocations*)

P.M.A. Sloot et al. (Eds.): ICCS 2002, LNCS 2329, pp. 399–408, 2002.
© Springer-Verlag Berlin Heidelberg 2002

in an otherwise uniform hexagonal froth affects the speed at which scaling is reached, while a constant μ_2 is not obtained in the special case of a single defect, [6-9].

Additionally, topological order between neighbouring cells has been described by correlation laws, such as that of Aboav-Weaire, [10-12], although more recent studies have indicated that second and third nearest-neighbours are also correlated, i.e. that theories of froth dynamics based on independent bubble approximations are inadequate, [13]. These detailed correlations are based on analysis of the froth as a system of concentric shells, with layers, $j = 1, 2...n$ arranged around a central germ cell, $j = 0$. The froth is said to have the property of shell-structured inflatability if no topological inclusions occur between layers, [14], but most *SSI* froths do not retain the property under system evolution, [15].

While topological measures are a natural choice to describe froth, more general measures can provide a useful basis for comparison with other non-cellular structures. The local decay of persistence, $P(t_0, t) \sim t^{-\theta}$ was first proposed as a new and general probe of non-equilibrium dynamics, [16-18]. Although agreement between theory and experiment on the value of θ is lacking, treating the froth as a two-phase system, (with a *virtual phase* defined by colouring a fraction of the bubbles), permits exploration of the dynamics at different length scales.

Persistence exponents for volume fractions $0 \leq \phi \leq 1$ in a coarsening 2D froth have been investigated both experimentally and computationally, [16-17,19], but no evidence was found for persistence decay in a Voronoi system over simulation time-scales. This, in contrast to experiment, where findings were obtained over extended times. However, for a uniform froth, seeded with one or more defects, some indications of persistent decay have been obtained, [19], although it was not found possible to estimate exponent values with any confidence. Nevertheless, this has suggested that controlling the rate of evolution may be possible for specified geometries in order to observe further persistence effects or, at least, to quantify the dynamics of the virtual phase. In what follows, the focus is principally on the evolution of the virtual phase in a uniform froth, where the geometry observed consists of a ring of defects around a central defect. The methodology is described and results on the dynamics and persistent properties are reported.

2   Methodology

A ring geometry was chosen since, given that only a small percentage of the original cells will contribute to persistence effects, rapid cell suppression is initially required, coupled with the potential for large bubbles to grow. Further, unlimited growth is checked by the ring buffer, so that the froth is not consumed, while the equidistant spacing provides some basis for scaling the observed effects on the froth partition, e.g. [8] and not unlike [20].

The direct simulation method [21-22] employed, provides precise information on individual cell and topological changes as these occur. A distinction is made between *topological* and *diffusive* changes and detailed statistics on cell pressure, area, number of sides etc. are recorded at each time step. Periodic boundary conditions are chosen and equilibration takes place through an iterative procedure. After local relaxation, the froth is analysed for T1/T2 changes. Unlike an intrinsically disordered froth, the uniform froth is stable, (or in mechanical equilibrium), so that evolution must be stimulated. This is achieved by forcing a T1 or T2-formed process, to create a topological dislocation or defect. Clearly, defects are "survivors" (since $n > 6$), and account for a large percentage of the remaining area as the froth evolves [23]. Further, if defects form part of the virtual phase, they will dominate persistent behaviour.

The virtual phase, (Figs. 1 and 2), is defined by colouring a given fraction ϕ of the bubbles at time t_0 and studying persistent properties of this fraction as $t \to t_f$. The number of survivors is crucial as an indicator of the potential for further change. At any time t, quantities of interest include $< A^*(t)/A(0) >$ and $N^*(t)/N(t)$, (namely, the average persistent area ratio, where $A(0)$ is the initial bubble area, and the fraction of coloured bubbles persisting (or persistent region)). The ring geometry network consists in general of a number of defects equi-distant from a single central defect and equi-spaced in relation to each other. (We focus here predominantly on a 12-defect ring, but also consider one with 8 defects). Different choices of cells, for inclusion in the virtual phase of the ring geometry, are illustrated by the concentric shadings around the central defect in Fig. 1. These correspond to volume fraction, $\phi = 0.2$, 0.4, 0.6 respectively (where $\phi = 0.4$ includes the inner ring for $\phi = 0.2$ and $\phi = 0.6$ includes both inner rings). Taking the radius as defined by the number of cell layers crossed from the centre cell to the ring formed by the defects, several ring sizes have also been investigated. Fig. 2 shows the evolution of the basic geometry for given ϕ, where all defects increase in area over time and finally impact upon each other, (Fig. 2 (c)). Systems of size up to 2500 bubbles have been considered for the special ring geometry case, forming a basis for comparison with random and other colouring patterns for the virtual phase dynamics, [19].

3   Results

The evolutionary picture for the ring geometry shows the froth partitioned into the central defect, inner ring (compressed by defects expanding from both the centre and defect ring), defect "buffer" zone and the external region, where *none* of the defects form part of the (shaded) virtual phase. On impact of the defects in the ring, no further evolution is possible and a constrained equilibrium is restored. The behaviour of the overall second moment μ_2 vs t for the ring geometry is similar to that for the single defect case initially, with higher peaks (as several defects suppress smaller bubbles), but with some indication that this rapid rise slows down as defects compete, (for μ_2 around 1.4). Fig. 3 contrasts the ring behaviour with that of a random (Voronoi) froth and the single defect case. Similarly, average bubble area $< A(t) >$ vs t is consistent with the single defect initially, but again diverges slightly, at $t \sim 150$ time steps, as coarsening slows down, (not shown). Large area bubbles are thus still relatively few in number

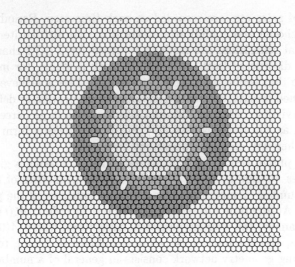

Fig. 1. Hexagonal froth (system size = 2500) with ring defects, with shaded area represen ting coloured volume fraction $\phi = 0.04, 0.2, 0.4$.

in the froth as a whole. In terms of persistence, the average area of persistent regions within a bubble at time t normalised w.r.t. that at t_0, (average area ratio), and the fraction of bubbles containing persistent volume at time t were also considered and are shown in Figs. 4 and 5 respectively for small and large radius ring sizes, ($r_s = 11$, $r_l = 15$) and for ϕ values in the range 0.06-0.4.

In Fig. 4, $< A^*(t)/A(0) >$ vs t, persistent area declines gradually, since the bubbles forming the virtual phase are being squeezed by the central defect and buffer ring. This is observed for both the small and large ring, with the steepest decline in each case corresponding to the value of ϕ for which the virtual phase just exceeds the inner region, (i.e. $\phi = 0.2, 0.4$ for the small and large ring respectively). This is in broad agreement with the results obtained for the single and several defects cases [19], and little difference is shown for colouring one or a few defects of ring.

For the persistent fraction of bubbles, $log N^*(t)/N(t)$ vs. $log < A(t) >$ in Fig. 5 shows broadly similar results to those discussed in [19], with the steepest slope again observed for the virtual phase having volume fraction ϕ corresponding closely to the ring radius. It is very evident that the slope k vs ϕ is highly variable, (comparing also with the case of one and several defects, [19]), and that there is a rapid decline in persistent fraction for some ϕ values, (Fig. 6). Specifically, for those where the radius of the virtual phase is close to or just larger than the defect ring radius. The effect is most marked where the coloured ring has small radius, (see e.g. Fig. 1, $\phi = 0.04$ say), since the coloured bubbles

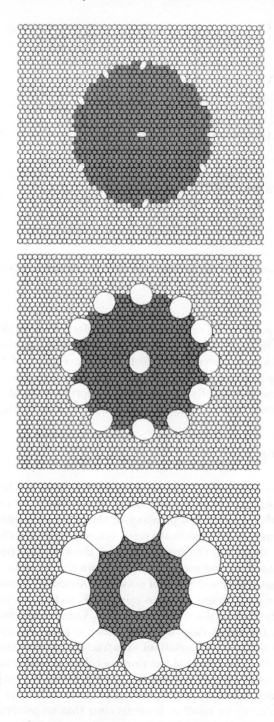

Fig. 2. Evolutionary behaviour of hexagonal froth, (system size = 2500) with ring defects, after time steps of 100, 200, 240

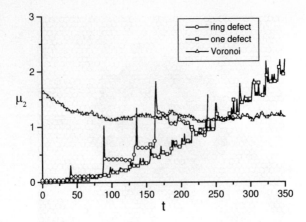

Fig. 3. Second moment, μ_2 vs t for ring defects, Voronoi and one defect case, (system size =2500)

are consumed rapidly by the expanding central defect, whereas the trough in k vs ϕ is both shifted in location, (to higher ϕ) and less marked, as the ring expansion competes with that of the central defect. The convergence of the slope values is emphasised by the comparison with the one and several defects case, where k varies little and is virtually constant in the former case. $\phi \rightarrow 1.0$, persistent fraction $\rightarrow$ constant value = 0, in all cases, as noted also for the Voronoi, [19].

4   Discussion

The virtual phase dynamics in the artificial froth system can be shown to vary enormously for specific colouring and values of ϕ. Clearly, we can bias the evolution to permit survival of few or many of the coloured bubbles. Further, for any complex geometry in a moderate sized system, finite size effects must operate for enclosed regions of the froth and limit the evolution period. Persistent measurements based on ring geometry have shown similar behaviour to that obtained in both hexagonal and Voronoi network. Obviously, increasing the radius of the ring buffer leads to less extreme values of k, so that for very large radius, we are essentially back to the single defect case. Conversely, containment of the virtual phase by the defect rings acts as a buffer to rapid disappearance of cells and mimics slow decay of the persistent area fraction for increasing average area, (Fig. 5). If < 50% of the defects in the ring are coloured, little change is observed in the area ratio or persistent fraction as the large defect has major contribution to those properties. However, as the percentage increases further, large area ratio contributions do occur for small ϕ. It seems clear that no persistent decay of the virtual phase may be observed for these systems, except for specific colouring patterns allied to the ring geometry. For all configurations of the virtual phase

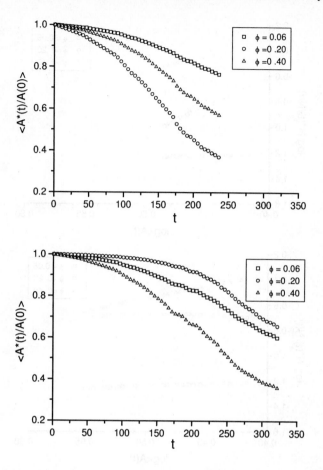

Fig. 4. Persistent area ratio, $< A(t)^*/A(0) >$ (average area of persistent region) vs t, (time steps) for ring defect froth, with $\phi = 0.06$, 0.2 and 0.4. (a) for small ring, (b) for large ring.

and defect geometry, the limiting value, $k=0$, is approached gradually from below as $\phi \rightarrow 1.0$. Extreme values are due to the artificial constraints imposed on the froth evolution and are moderated as these are relaxed. For the constrained system, with the virtual phase contained by the defect ring, clearly

$$k_{max} \propto 1/r \qquad (1)$$

Similar behaviour is observed for a ring geometry with fewer defects and the evolution can be monitored for a slightly longer period. However, time-scales in general are unacceptably short, since the closure of the ring effectively blocks further evolution, and it is evident that true persistence is not observed for the whole system. However, if we consider the ring geometries as a block between

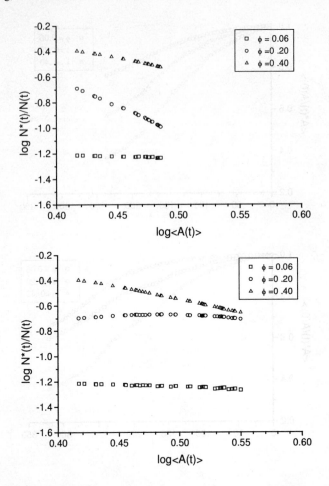

Fig. 5. Log.-log. plot of persistent fraction of bubbles, $N^*(t)/N(t)$ vs. average bubble area for ring defects froth with $\phi = 0.06$, 0.2 and 0.4. (a) and (as in Fig. 4.

the inner and outer froth system, then depending on ring radius, contributions from those zones (in analogy with shell structure analysis) could bring us some new insights on virtual phase dynamics as well as other properties [24].

5   Conclusions

Construction of ring geometries in a 2-D froth has enabled further study of evolutionary properties and virtual phase dynamics. Evolution of the froth can be controlled by constrained geometries to some extent, with μ_2 relatively stable at an earlier stage, which is consistent with a uniform froth with defects. However, true persistence has not been convincingly observed over these simulation times on a froth with ring defects, although various quantities, (the average area ratio

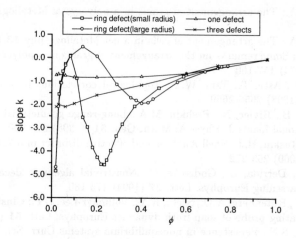

Fig. 6. Comparison of slope k with ϕ for 3 defect case and systems as in Fig. 3.

and fraction of coloured bubbles), have exhibited radically different behaviour for different values of ϕ. These results are in line with those from defect-seeded or randomly defect-seeded hexagonal systems; in particular, it seems that the virtual phase, (as a persistence quantity), is contained by the defect ring buffer, with mimicked decays ($logN^{*}(t)/N(t)$ vs $log < A(t) >$) having non-zero slope k. It is evident that the non-zero decay exponent, a consequence of the artificial geometry, will tend to $\phi \rightarrow 1.0$ as $k \rightarrow 0$, in agreement with earlier findings for a Voronoi-based froth.

References

1. Weaire, D., Rivier, N.: Soap, cells and statistics - random patterns in two dimensions Contemp. Phys. **25** (1984) 59-99
2. Glazier, J.A., Gross, S.P., Stavans, J.: Dynamics of two-dimensional soap froths Phys. Rev. A **36** (1989) 306-312
3. Stavans, J.: The evolution of cellular structures Rep. Prog. in Phys. **56** (1993) 733-789
4. Ruskin, H.J., Feng, Y: Cellular networks: characterising disorder in froths Trends in Stat. Phys. **2** (1998) 131-152
5. von Neumann J.: Grain shapes and other metallurgical applications of topology Metal Interfaces **41** (1952) 108-110
6. Levitan, B.: Evolution of two-dimensional soap froth with a single defect Phys. Rev. Lett. **72** (1994) 4057-4061
7. Ruskin H.J., Feng, Y.: Scaling properties for ordered/disordered 2-D dry froths Physica A **230** (1996) 455-466
8. Ruskin, H.J., Feng, Y.: A note on stages of evolution in a 2D froth Physica A **247** (1997) 153-158
9. Ruskin, H.J., Feng, Y.: The evolution of a two-dimensional soap froth with a single defect J. Phys.: Condens. Matter **7** (1995) L553-L559

10. Aboav, D.A.: The arrangement of grains in a polycrystal Metallography **3** (1970) 383-390
11. Aboav, D.A.: The arrangement of cells in a net Metallography **13** (1980) 43-58
12. Weaire, D.: Some remarks on the arrangement of grains in a polycrystal Metallography **7** (1974) 157-160
13. Szeto, K.Y., Aste, T., Tam, W.Y.: Topological correlations in soap froths Phys. Rev. E **58** (1998) 2656-2659
14. Dubertret, B., Rivier, N., Peshkin, M.A.: Long-range geometrical correlations in two-dimensional foams J. Phys. A: Math. Gen. **31** (1998) 879-900
15. Feng, Y., Ruskin, H.J.: Shell Analysis and effective disorder in a 2D froth. J. Stat. Phys. **99** (2000) 263-272
16. Bray, A.J., Derrida, B., Godreche, C.: Non-trivial algebraic decay in a soluble model of coarsening Europhys. Lett. **27** (1994) 175-180
17. Tam, W.Y., Rutenberg, A.D., Szeto, K.Y., Vollmayr-Lee, B.P.: Cluster persistence: A discriminating probe of soap froth dynamics Europhys. Lett. **51** (2000) 223-229
18. Majumdar, S.N.: Persistence in nonequilibrium systems Curr. Sci. **77** (1999) 370-375
19. Ruskin, H.J., Feng Y., Zhu, B.: Modelling froth dynamics; persistence measures Physica A **293** (2001) 315-323; also Ruskin, H.J., and Feng Y.: Constrained geometries in soap froth dynamics, at conference on Horizons in Complex Systems, Messina, Sicily, Italy, Dec 5th-8th, 2001
20. Aste, T., Boose, D., Rivier, N.: From one cell to the whole froth: a dynamical map Phys. Rev. E. **53** (1996) 6181-91.
21. Weaire, D., Kermode, J.P.: Computer simulation of a two-dimensional soap froth Phil. Mag. B **48** (1983)245-259; 1984. Phil. Mag. B **50** (1984) 379-395
22. Kermode, J.P., Weaire, D.: 2D-FROTH: a program for the investigation of 2-dimensional froths Comp. Phys. Commun. **60** (1990) 75-109
23. Tam, W.Y., Zeitak, R., Szeto, K.Y., Stavans, J.: First-passage exponent in two-dimensional soap froth Phys. Rev. Lett. **78** (1997) 1588-1591
24. Feng, Y., Ruskin, H.J., Zhu, B.: 2002 (in preparation)

A Correction Method for Parallel Loop Execution

Volodymyr Beletskyy

Faculty of Computer Science, Technical University of Szczecin, Zolnierska 49 st.,
71-210 Szczecin, Poland,
Bielecki@man.szczecin.pl

Abstract. A new method of parallel loop execution is presented. Firstly, all the loop iterations are executed in parallel. Then, the ends of pairs of dependent iterations are re-executed. The method requires no conversion of a source loop into an equivalent serial-parallel loop, involving iteration indices to be converted into new ones, that appears to be a typical requirement in the most known loop paralleling approaches such as unimodular and nonunimodular linear transformation methods. Possibilities of serial and parallel correction processes are discussed. Experimental results are considered. As follows from experiments, applying the method can be reasonable for the loops with a small fraction of dependent iterations.

1 Introduction and related work

Since parallel and distributed computing have become increasingly popular, it is significant to develop such compilers that would automatically translate serial programs into effective parallel code. Such code can be run on several processors of a parallel computer or on several nodes of a distributed system.

Numerous methods of loop paralleling have been proposed [1]. However, research in this field is still carried out as the problem of loop paralleling appears extremely complicated.

A new method of parallel loop execution proposed in this paper aims at maximizing the number of iterations that can be processed in parallel for the loops containing the small fraction of dependent iterations. No loop transformations need. First, all the iterations are executed in parallel. Next, dependent iterations need to be corrected. The method requires no conversion of a source loop into an equivalent serial-parallel loop, involving the source iteration indices to be converted into new ones, that appears to be a typical requirement in the most known loop paralleling methods such as unimodular and nonunimodular linear transformation approaches [2], [3], [4], [5].

The method is different from the speculative parallel execution[6] in the following: 1) dependence analysis is fulfilled before loop execution; 2) if cross-iteration dependencies exist, then only the iterations that are the ends of pairs of dependent iterations are re-executed.

The main objective of the paper is to present a concept of the correction method, possibilities of its implementation, experimental results, and future research.

P.M.A. Sloot et al. (Eds.): ICCS 2002, LNCS 2329, pp. 409–418, 2002.

2 Analysis of Data Dependence

Analysis of data dependencies is considered in lots of papers [7], [8], [9]. In this section, we reproduce known knowledge that is necessary to explain a concept of the correction method.

Consider the following generic nested loop

for $I_1 = L_1$, U_1
   for $I_2 = L_2(I_1), U_2(I_1)$
   ...
      for $I_n = L_n(I_1,...,I_{n-1}), U_n(I_1,...,I_{n-1})$
        $H(I_1,...,I_n)$
      end for
   ...
   end for
end for,

where $I_1,...,I_n$ are the iteration indices; L_i and U_i - the lower and upper loop bounds that are linear functions of the iteration indices $I_1,...,I_{i-1}$, implicitly a stride of one is assumed; H is the body of the nested loop. $I = (I_1,...,I_n)'$ is called the iteration vector.

Definition 1. The set $I \subseteq Z^n$ such that

$$I = \{(i_1,...,i_n) \mid L_1 \le i_1 \le U_1,...,L_n(i_1,...,i_{n-1}) \le i_n \le U_n(i_1,...,i_{n-1})\},$$

is an iteration space.

Individual iterations are denoted by tuples of iteration indices.

The dependence relation between two statements constrains the order in which the statements may be executed. There are four types of data dependencies: flow, anti, output, and input dependence. The only true dependence is flow dependence. The other dependencies are the result of reusing the same location of memory and are hence called pseudo dependencies. They can be eliminated by renaming variables as well as by the techniques presented in [6], [10], [11]. From now on, we suppose that the loop has no cross-iteration anti-dependencies. Additionally, we suppose that the loop body has no statements of the form: x=...x...

We often need to know all the pairs of iterations that are dependent and the precise relationship between them, such as the dependence distance vector.

Definition 2. (Dependence Distance Vector) For a pair of iterations $i = (i_1,...,i_n)$ and $j = (j_1,..., j_n)$ such that j is flow dependent on i, the vector $j - i = (j_1 - i_1,..., j_n - i_n)'$ is called the dependence distance vector.

From now on, we write "dependence vector" instead of "dependence distance vector".

To find dependence vectors, equations should be built for each pair of the same named variables ID ($A_1 I + B_1$), ID ($A_2 I + B_2$) that are located in the loop body on both hand sides of assignment statements - the right and the left- or on the left-hand sides only, where A_1, A_2 are matrices, B_1, B_2 are vectors.

The equations mentioned can be written as follows:

$$A_1 *i - A_2 *j = B_2 - B_1, \tag{1}$$
$$K = j - i,$$
$$K > 0.$$

The solution of equations (1), if it exists, can be presented in the following form[12]:

$$K = t_{11} * V_1 + t_{12} * V_2 + ... + t_{1r} * V_r + V_0,$$
$$i = t_{21} * W_1 + t_{22} * W_2 + ... + t_{2s} * W_s + W_0,$$

where: V_0, V_1, ..., V_r, W_0, W_1, ..., W_s are vectors of dimension n with constant elements; t_{1q}, t_{2p}, $q = 1, 2, ..., r$; $p = 1,2,...,s$ are free variables, values of which are arbitrary integer numbers, $0 \le r, s \le n$; to find the boundaries of t_{1q}, t_{2p}, the loop limits should be taken into consideration.

The vector i describes all the iterations that form the beginnings of pairs of dependent iterations, while the vector $j = i + K$ describes the ends of those. For a pair of dependent iterations, the beginning is the iteration that is lexicographically less. To obtain correct results, all the dependent iterations have to be executed   in lexicographical order.

Let us consider the example

for $i_1 = 1, N$
    for $i_2 = 1, N$
       $a(i_1, i_2) = a(i_1 - 2, i_2 - 1)$.

Equations (1) for the loop above are as follows:

$$i_1 - j_1 = -2,$$
$$i_2 - j_2 = -1.$$

The solution to these equations is: $K = (2, 1)'$, $i = t_1 (1, 0)' + t_2 (0, 1)'$, where t_1, t_2 are free variables. For the vector j, we have

$$j = i + K = t_1 (1, 0)' + t_2 (0, 1)' + (2, 1)'.$$

To find the limits of t_1 and t_2, we build inequalities

$$(1,1)' \le j \le (N, N)'   \text{or} \tag{2}$$

$$(1,1)' \le t_1 \ (1, 0)' + t_2 \ (0, 1)' + (2, 1)' \le \ (N, N)'.$$

From (2), we have

$$1 \le t_1 \le N\text{-}2,$$

$$1 \le t_2 \le N\text{-}1.$$

The beginnings of dependent pairs of iterations are as follows: (1,1), (1,2), ..., (1,N), (2,1), (2,2), ..., (2,N), ..., (N,1), (N,2), ..., (N,N), i.e. they belong to the entire iteration space I.

3   Correction Method

The correction method comprises the two steps described below.

1. All the iterations of a loop are executed in parallel applying only old values of variables (the values before the execution of iterations).
2. Given vectors i and K, dependent iterations are determined and the correction of results for these dependent iterations is executed. The correction means a procedure of repeating the execution of the ends of all the pairs of dependent iterations in lexicographical order with regard to new values of variables.

The correction can be executed in various ways: serially or parallel, synchronously or asynchronously, statically or dynamically.

The serial correction means that only one processor fulfils the correction.

The parallel one implies that two or more processors execute the correction simultaneously.

For the synchronous correction at first, a graph representing all the dependent iterations is built. Next, the correction for the iterations of the first layer of the graph is fulfilled, then the correction is executed for the iterations of the second layer and so on, i.e. consecutive layers are scanned serially, while the iterations within each layer are corrected in parallel. If we can correct the iterations represented by the graph independently, i.e. without waiting for the finish of the execution of a particular layer of iterations, then we have the asynchronous correction.

The correction method is considered static if finding dependent iterations and planning the iterations applied for the correction take place before loop execution(at compile-time), else the method is considered as dynamic.

Let us consider the following loop:

```
for i = 3,6
    a(i) = a(i - 3) + 1;
```

and suppose the values of vector elements before loop execution are: a(m) =0, m=1,2,...,6.

When this loop is executed serially, the following values are obtained: a (3) = 1, a(4) = 1, a (5) = 1, a (6) = 2.

Now, let the correction method is to be applied. After the first step(parallel execution of all iterations) we have: a (3) = 1, a (4) = 1, a (5) = 1, and a (6) = 1. The vectors K and i for the loop considered are as follows: K = 3, i = t_1 + 2, $1 \le t_1 \le 4$. Here, we have the only pair of dependent iterations with its beginning in iteration 3 and its end in iteration 6 (j = i + K = 3 + 3 = 6). It means a single correction is only required: the result of iteration 6 is calculated using the new value of iteration 3:

$$A (6) = (3) + 1 = 1 + 1 = 2.$$

The results obtained are the same as those obtained at serial loop execution.

4 Serial Correction

Procedures of the serial correction may be built in different ways. When a loop yields the only pair of vectors i, K, a possible procedure is as follows. From equations (1), the vectors i, K are found. By means of the vector j = i+K, a serial loop is built that executes the correction. The loop is constructed in such a way that tuples of iteration indices yielded with the vector j are increased in lexicographical order. The loop, carrying out the correction, has to execute all the iterations yielded with the vector j, $j \in I$.

Let us consider the example:

```
for i = 1, N
   a(6i - 7) = a(2i + 5);
```

The solution to equations (1) for the loop above is as follows:

$$i = t^*(1) +3,$$
$$K = t^*(2),$$
$$j = i + K = t^* (1) + t^* (2) + 3,$$

where: $1 \le t \le \lfloor N/3 - 3 \rfloor$.
In this case, loop, fulfilling the correction, is as follows:

```
for t = 1, ⌊N/3 - 3⌋
   i = 3t + 3;
   a(6i - 7) = a(2i + 5);
```

or

```
t = 1;
while (i = 3t + 3 ≤ N)
   a(6i - 7) = a(2i + 5); t = t + 1;
```

The example above shows that the correction is required only for every third iteration and applying the correction method might be advantageous.

5 Parallel Correction

Let us now consider the case when the loop 1) yields many dependence vectors K_1, K_2, ..., K_n, but elements   of them are constant; 2) $\{i_1\}=...=\{i_n\}= I$, i.e. each iteration is the beginning of a pair of dependent iterations.

To implement parallel dynamic correction in such a case, the following approach can be applied. Among the vectors K_i, i = 1, ..., n, only such vectors are selected and marked as T_1, T_2, ..., T_m, $m \le n$ that satisfy the condition:

$$i_1 + T_j \ne i_2 + K_i, \quad i=1,2,...,n; \; j=1,2,...,m,$$

where i_1, $i_2 \in I$ and are arbitrary iterations. If for some $K_j \ne T_1$, $l =1,2,...,m$, T_i the condition

$$i_1 + K_j = i_2 + T_i$$

holds, it means that all the ends of pairs of dependent iterations yielded with the vector K_j will be held with the vector T_i as well since for any beginning $i_1 \in I$ of a pair of dependent iterations with the end $i_1 + K_j$, an iteration $i_2 \in I$ exists, which is the beginning of other pair of dependent iterations with the same end. To execute dependent iterations in lexicographical order, according to the approach below, vectors $K_j \ne T_1$, $l=1,2,...,m$ must be skipped.

Let us consider the following example :

```
for i = 1, N
    a(i) = a(i - 2) + a(i - 3);
```

Here, dependence vectors are: $K_1 = 2$, $K_2 = 3$. Only K_1 is the independent vector since it forms all the ends of dependent iterations as the vector K_2 does.

Next, all the iterations of the set $\{J_1\}$ are determined. The set $\{J_1\}$ contains the iterations of layer 1, i.e. the iterations for which the correction should be executed firstly. The set $\{J_1\}$ can be described as follows:

$$\{J_1\}=\{((\{I_{01}+T_1\} \cup \{I_{01}+T_2\} \cup ... \cup \{I_{01}+T_m\})\cap I\},$$

where: $\{I_{01}\}$ is a   set of all the iterations that are the beginnings of pairs of dependent iterations, but they are not the ends of any pair of dependent iterations; $\{I_{01}+T_i\}$ means that to each tuple of the set $\{I_{01}\}$ the tuple T_i is added; $\cup$, $\cap$ are the union and intersection operations on sets.

The set $\{I_{01}\}$ is computed as $\{I_{01}\} = I\backslash(\{J_1\}\cup\{J_2\}\cup...\cup\{J_n\})$, where $J_i = i_i + K_i$ is the solution of equations (1) built for a particular dependence i=1,2,...,n and yielding all the ends of pairs of dependent iterations, "\" is the difference operation on sets.

The sets $\{I_{01} + T_i\}$, i=1,2, ...,m contain only single representatives of identical elements. The correction is executed in parallel for all the iterations of the set $\{J_1\}$.

Farther, we select all the iterations of   layer 2 as follows:

$$\{J_2\} = \{((\{J_1 + T_1\} \cup \{J_1 + T_2\} \cup ... \cup \{J_1 + T_m\}) \cap I\}.$$

All the iterations within layer 2 are executed in parallel and the procedure as above is repeated for next layer, and so on.

The pseudo code of the correction procedure above can be written as follows:

$\{J_1\} = \{((\{I_{01} + T_1\} \cup \{I_{01} + T_2\} \cup ... \cup \{I_{01} + T_m\}) \cap I\}$,   i=1;

WHILE$\{J_i\} \neq \emptyset$

PAR L$\in \{J_i\}$

 {loop body (L);}

 i=i+1;

  $\{J_i\} = \{((\{J_{i-1} + T_1\} \cup \{J_{i-1} + T_2\} \cup ... \cup \{J_{i-1} + T_m\}) \cap I\}$,

where: the instruction WHILE $\{J_i\} \neq \emptyset$ means that as long as the set $\{J_i\}$ contains elements, the loop is executed; PAR L $\in \{J_i\}$ means that each iteration L, belonging to the set $\{J_i\}$, can be executed in parallel.

The procedure above finds all the ends of pairs of dependent iterations. Iterations of each layer are independent, while for dependent iterations on consecutive layers i and i+1, the iterations on layer i are lexicographically less than those on layer i+1. Since layers are scanned serially at the correction, dependent iterations are executed in lexicographical order. It proves the correctness of the proposed approach.

Let us consider the loop:

for $i_1 = 1, N$
 for $i_2 = 1, N$
  $a(i_1, i_2) = a(i_1, i_2 - 1) + a(i_1 - 1, i_2) + a(i_1 - 1, i_2 - 1).$

The dependence vectors for the loop above are: $K_1 = (0, 1)'$, $K_2 = (1, 0)'$, $K_3 = (1, 1)'$. Only the vectors K_1 and K_2 should be chosen for the correction. The set $\{I_{01}\}$ equals $\{(1,1)\}$. The set$\{\{I_{01} + T_1\} = \{(1,2)\}$ and the set $\{I_{01} + T_2\} = \{(2,1)\}$. The loop applied to execute the correction is as follows:

$(J_1\} = \{(1,2)\} \cup \{(2,1)\} = \{(1,2), (2,1)\}$

i=1;

WHILE $\{J_i\} \neq \emptyset$

PAR L $\in \{J_i\}$

 $a(i_1, i_2) = a(i_1, i_2 - 1) + a(i_1 - 1, i_2) + a(i_1 - 1, i_2 - 1);$

 i=i+1;

 $\{J_i\} = \{((\{J_{i-1} + (0,1)\} \cup \{J_{i-1} + (1,0)\}) \cap I\}.$

The number of correction layers for the loop in the last example equals to 2N - 2. For N=3, the iterations to be corrected on each layer are as follows:

1. $\{J_1\}=\{(1,2), (2,1)\}$.
2. $\{J_2\}=\{J_1+(0,1)\}\cup\{J_1+(1,0)\}=\{(1,2)+(0,1),(2,1)+(0,1)\}\cup\{(1,2)+(1,0),(2,1)+(1,0)\} = \{(1,3), (2,2)\}\cup\{(2,2), (3,1)\}=\{(1,3),(2,2),(3,1)\}$.
3. $\{J_3\}=(\{J_2+(0,1)\}\cup\{J_2+(1,0)\})\cap I=(\{(1,3)+(0,1),(2,2)+(0,1),(3,1)+(0,1)\}\cup\{(1,3)+(1,0),(2,2)+(1,0),(3,1)+(1,0)\})\cap I=(\{(1,4),(2,3),(3,2)\}\cup\{(2,3),((3,2),(4,1)\})\cap I=(\{(1,4), (2,3),(3,2), (4,1)\})\cap I =\{(2,3), (3,2)\}$.
4. $\{J_4\}=(\{J_3+(0,1)\}\cup\{J_3+(1,0)\})\cap I=(\{(2,3)+(0,1),(3,2)+(0,1)\}\cup\{(2,3)+(1,0),(3,2)+(1,0)\})\cap I = (\{(2,4), (3,3)\}\cup\{(3,3), (4,2)\})\cap I =(\{(2,4), (3,3), (4,2)\})\cap I=\{(3,3)\}$.

If $\{i_i\} \neq I$, $i=1,2,...,n$, then the correction procedure above should be modified. After the execution of the iterations of layer i, from the set $\{J_i\}$ all the iterations that are not the beginnings of pairs of dependent iterations yielded with J_i must be deleted and the iterations that are the beginnings of pairs of dependent ones given with J_k, $k\neq i$ must be copied to the set $\{J_k\}$.

6   Experiments

To evaluate the efficiency of the correction method, several experiments on Power Challenge supercomputer with three working processors and the IRIX operating system were performed.   IRIX provides support for creating multithread applications by means of the Power C compiler .
   The following C loop

```
for (a1=1; a1<=n; a1++) {                                    (3)
    T[10*a1] = f(T[a1], t);
    T[11*a1] = f(T[a1], t);
    T[13*a1] = f(T[a1], t); }
```

was executed with different implementations of the correction method, where f(T[a1], t) is a function with two parameters T[al] and t. The parameter t defines the time of the execution of each iteration.
   For this loop, there are three dependence vectors $K_1=9t_1$, $K_2=10t_2$, $K_3=12t_3$ stating flow dependencies with the beginnings in the iterations $i_1=t_1$, $i_2=t_2$, and $i_3=t_3$ correspondingly. There are three dependence vectors $K_4=t_4$, $K_5=3t_5$, $K_6=2t_6$ designating output dependencies with the beginnings in the iterations $i_4=10t_4$, $i_5=10t_5$, and $i_6=11t_6$ correspondingly. Applying the correction method for the above loop is permissible since there are no anti-dependencies. The independent vectors are K_1, K_4, and K_6.
   Correction procedures were implemented in accordance with the modified approach presented in Section 5. For the static correction, all the ends of pairs of dependent iterations were calculated end memorized in a container in lexicographical

order at compile-time. At the dynamic correction, the data required were calculated at run-time. At all the cases, the results after serial loop execution and after the corrections were the same.

Figure 1 illustrates how speedup depends on the average execution time of one iteration at the execution of loop (3) with the static serial, dynamic serial, and dynamic parallel corrections.

The average execution time of one iteration is the quotient of the division of the whole time of the loop execution by the number of iterations executed. This time as well as the number of the iterations executed greatly influences speedup at parallel executing the loop on Power Challenge since creating and synchronizing threads with IRIX need some time. If this time is comparable or more than the time of the loop execution, then the efficiency of parallel applications is low enough. It is why we have investigated how speedup depends on the average execution time of one iteration and the number of iterations.

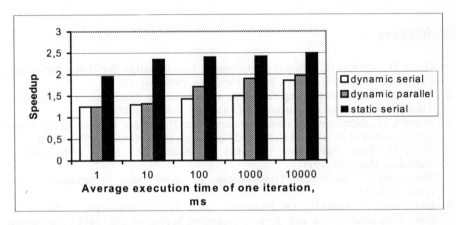

Fig.1. Speedup dependence on the average execution time of one iteration at fulfilling loop (3): 10.000 iterations, 2,6% dependent iterations

On the basis of analysis of the results received, the following conclusions can be made:

1. The top speedup was reached with the static serial correction. The explanation of this fact is that all the date necessary for the static correction are collected at compile-time, while at dynamic correction, the date are generated at run-time. But the static correction requires many memory for its implementation. All the data(dependent iterations) must be stored in memory before the loop execution.

2. As the average execution time of one iteration increases, speedup for each approach investigated increases as well since the percentage of the time required to create and synchronize parallel threads decreases.

It may be expected that if the average execution time of one iteration is large enough and the percentage of dependent iterations is small, applying the correction method can be reasonable.

7 Conclusion

The concept of the correction method, enabling executing loops in parallel, was presented. The method involves two steps: the first step provides the parallel execution of all the iterations of a loop, while the second one fulfils the correction of the ends of pairs of dependent loop iterations.

Further research on the correction method needs. The main work to be done is as follows: determining a scope of effective applying the   method; developing algorithms of the method implementation for different approaches: static/dynamic, serial/parallel, synchronous/asynchronous; comparing the correction method with other known loop paralleling techniques; combining the correction method with other loop paralleling approaches; program implementation of different techniques of the method; examination of the method by means of standard benchmarks and different general-purpose applications.

References

1. Bacon, D., Graham S., Sharp O.:   Compiler Transformation for High – Performance Computing. Computing Surveys, 26 (4) (December, 1994) 345-420
2. Lim A.W., Lam S.: Maximizing Parallelism and Minimizing Synchronization with Affine Transformations. Parallel Computing, 24(3-4) (May, 1998) 445-475
3. Huang C.H., Sadayappan P.: Communication – Free Hyperplane Partitioning of Nested Loops. Journal of Parallel and Distributed Computing, 19(2) (October, 1993) 90-102
4. Wolfe M.: High Performance Compilers for Parallel Computing. Addison -Wesley Publishing Company (1995)
5. Ramanujam J.: Non-Singular Transformations of Nested Loops. In Supercomputing 92 (1992) 214-223
6. Rauchwerger L., Padua D.: The Privatizing Doall Test: A Run -Time Technique for Doall Loop Identification and Array Privatization. In Proc. of the 1994 International Conf. on Supercomputing (July, 1994) 33-43
7. Banerjee U.: Dependence Analysis for Supercomputing. Kluwer Academic Publishers (1988)
8. Li Z., Yew P., Zhu C.: An Efficient Data Dependence Analysis for Parallelizing Compilers. IEEE Trans. Parallel Distributed Systems, 1(1) (1990) 26-34
9. Wolfe M. Tseng Chau-Wen: The Power Test for Data Dependence. IEEE Trans. Parallel Distributed Systems, 3(5) (1992) 591-601
10. Feautrier P.: Array Expansion. Conference Proceedings on International Conference on Supercomputing, ACM (1988) 429-441
11. Brandes T.: The Importance of Direct Dependencies for Automatic Parallelization. Conference Proceedings on International Conference on Supercomputing, ACM (1988) 407- 417
12. Gregory R.I., Krishnamurthy E.V.: Methods and Applications of Error-Free Computation. Springer-Verlag (1984)

A Case Study for Automatic Code Generation on a Coupled Ocean–Atmosphere Model

P. van der Mark[1,*], R. van Engelen[2,**], K. Gallivan[2,**], and W. Dewar[3]

[1] Leiden Institute of Advanced Computer Science,Leiden University, P.O. Box 9512,
2300 RA Leiden, The Netherlands
pmark@liacs.nl
[2] Department of Computer Science, Florida State University, Tallahassee,
FL 32306-4530, USA
{engelen,gallivan}@cs.fsu.edu
[3] Department of Oceanography, Florida State University, Tallahassee,
FL 32306-4320, USA
dewar@ocean.fsu.edu

Abstract. Traditional design and implementation of large atmospheric models is a difficult, tedious and erroneous task. With the CTADEL project we propose a new method of code generation, where the designer describes the model in an abstract high-level specification language which is translated into highly optimized Fortran code. In this paper we show the abilities of this method on a coupled ocean–atmosphere model, in which we have to deal with multi-resolution domains and different time-steps. We, briefly, describe a new concept in compiler design, the use of templates for code generation, to elevate the burden of choosing architecture optimized numerical routines.

1   Introduction

Simulation of atmospheric models is a computationally intensive task, where developers of those models have to make a trade-off between numerical accuracy and computational resources. First, approximations are made in the formulation of the physical processes in a mathematical model by discarding certain details that are assumed to have little impact on the overall outcome. Second, approximations are made by the discretization of the model and its numerical solving methods and the resolution of the discrete grid.

Increasing processing power and the development of new high-performance architectures such as large networks of workstations (NOW) (for example the Beowulf project [9]) have led to opportunities for developers of numerical models to change the focus on more numerical detail, finer resolution or larger computational domains. This requires a significant programming effort to implement the changes in the simulation application since a simple plug-and-play development

---

* Supported in part by the FSU Cornerstone Program for Centers of Excellence.
** Supported in part by NSF grant CCR-0105422.

P.M.A. Sloot et al. (Eds.): ICCS 2002, LNCS 2329, pp. 419–428, 2002.

paradigm with software components for scientific applications does not yet exist. This can lead to huge amounts of undocumented code, for which new versions have to be developed with every new computer architecture emerging.

In most cases, however, a model can be described on an abstract level, for example mathematical equations. One of the aims of the CTADEL project is to enable developers to formulate the problem on a high level, refraining from architecture specific dependencies, while producing highly efficient codes for a variety of target platforms [13]. In this paper we show the possibilities of CTADEL on a coupled ocean–atmosphere model, a quasi-geostrophic climate dynamics model [2]. Coupled models impose certain difficulties on their implementation; often these models use multiple resolutions on the computational grids and separate time steps. This also assesses certain constraints on the interaction between the different parts of the model and the parallelization of the model.

For specifications in CTADEL the ATmospheric MOdeling Language (AT-MOL) [14] was developed in close collaboration with meteorologists at the Royal Netherlands Meteorological Institute (KNMI) to ensure ease of use, concise notation, and the adaptation of common notational conventions. The high-level constructs in ATMOL are *declarative* and *side-effect free* which is required for the application of transformations to translate and optimize the intermediate stages of the model and its code. ATMOL is *strict* and requires the typing of objects before they are used. This helps to pinpoint problems with the specification at an early stage before code synthesis takes place. ATMOL supports both high-level as well as low-level language constructs such as Fortran-like programming statements which are used to implement and optimize the target numerical code.

This paper is organized as follows, in section 2 we will explain some of the basic theory behind the original quasi-geostrophic model, while we discuss our specification of the model in section 4 and the CTADEL system in section 3. Results from experiments conducted with the produced codes by CTADEL and the original hand-written code are shown in section 5 and we draw some conclusions in section 7.

2   Quasi-Geostrophic Dynamics

In this section we will explain the quasi-geostrophic dynamics model some more. For the sake of understandability and readability we only briefly touch the theory and interested readers are referred to [2]. The model describes a mid-latitude coupled climate model, which is used in an attempt to understand how the ocean climatology is modified by atmospheric coupling. At present, the best comprehensive coupled climate models run at resolutions far coarser than those needed to model the inertial recirculation. The model presented here is an attempt to explicitly include eddies in general, and inertial recirculation in particular, within the framework of an idealized climate setting. The basic model consists of a quasi-geostrophic channel atmosphere coupled to a simple, rectangular quasi-geostrophic ocean. Heat and momentum exchanges between the ocean and the atmosphere are mediated via mixed layer models and the system is driven by steady, latitudinally dependent incident solar radiation.

Solar radiation is the basic force behind the global climate. About 30% is reflected back to space while the remaining part is absorbed mostly at bottom interface, whether it be land or water. Heat transfers to the atmosphere occur either through sensible or latent fluxes, or long wave radiative surface emissions to which the atmosphere is almost totally opaque.

The model is based on a classical β-plane mid-latitude representation and a two layered version of the quasi-geostrophic (QG) equations. For the most part, classical QG models are adiabatic, i.e. thermodynamics are neglected [4]. If the standard QG scaling is used and the usual Rossby number expansion is employed, the momentum equations yield the non-dimensional vorticity equation

$$(\frac{\partial}{\partial t} + J(p_i, .))(\nabla^2 p_i + \beta y) = w_{i1z} + HF, \tag{1}$$

where J denotes the usual Jacobian operator, p_i is the layer pressure, y is the meridional coordinate, HF denotes the horizontal frictional effects and all variables are non-dimensional.

The final dimensional forms of the QG equations for both ocean layers can now be extracted, for example the first layers reads,

$$\frac{d}{dt}q_1 = A_{1h}\nabla^6 p_1 + \frac{f_0}{H_1}(w_{ek} - E(-H_1 - h_{1+}))$$
$$q_1 = \frac{\nabla^2 p_1}{f_0} + \beta y - \frac{f_0}{g'H_1}(p1 - p2), \tag{2}$$

where F_1 denotes the effects of the upper and lower layer horizontal frictional processes. Here $\nabla^6 p_1$ is used as closure for for the latter and super slip boundary conditions (i.e. $\nabla^2 p_i = \nabla^4 p_i = 0$).

3   Ctadel

Many attempts have been made to solve scientific or physical models numerically using computers. Today, a large collection of libraries, tools, and Problem Solving Environments (PSEs) have been developed for these problems. The computational kernels of most PSEs consist of a large library containing routines for several numerical solution methods. These routines form the templates for the resulting code. The power of this library determines the numerical knowledge of such a system.

A different approach to those so-called library-based PSEs consists of a collection of tools that generate code based on a problem specification without the use of a library. The numerical knowledge is determined by the expressiveness of the problem specification language and the underlying translation techniques. CTADEL belongs to this class of PSEs. Besides, a hardware description of the target platform is included in CTADEL's problem specification. This makes it possible to produce efficient codes for different types of architectures.

The CTADEL system provides an automated means of generating specific high performance scientific codes. These codes are optimized for a number of different architectures, like serial, vector, or shared virtual memory and distributed memory parallel computer architectures. One of the key elements of this system is the usage of algebraic transformation techniques and powerful methods for global common subexpression elimination. These techniques ensure the generation of efficient high performance codes.

For a more detailed description of CTADEL see [11].

4   Specification and Implementation

In this section we describe the specification of the quasi-geostrophic model using the ATMOL specification language for CTADEL and one of the main problems with this specification, the separate domain resolutions.

4.1   Spatial resolutions

As seen in section 2 both the ocean and atmosphere model have their own spatial domain resolution, which leads to several problems when combining both models into a mixed model. For example, when exchanging data between the two models, a spatial interpolation has to take place. Another problem arises with differentiation of a function, CTADEL can't assume a default grid size but has to determine this from the context. For example, one would normally write a differentiation like,

```
coordinates := [dx,dy].
Q = diff(P,y).
```

which would be translated into

```
      DO 10 j = 1,m-1
        DO 20 i = 1,n
    Q(i,j) = (P(i,j-1) - P(i,j)) / dy
20      CONTINUE
10 CONTINUE
```

Since we work with two different grid sizes, this is not conceivable. One possibility is to hard-code the distance between two grid-points into the specification, which is not desirable. Therefor we couple the specification of a domain to it's spatial grid sizes. For example, the definition of a domain could look like,

```
domain_a := i=1..X_a by j=1..Y_a by k=1..Z_a.
domain_a'coordinates := [dxa,dya].
```

If we define a variable with this domain CTADEL knows which distances to choose from. A problem arising with this implementation is when two variables with different domains are combined in one equation. Since this has been avoided in the original model, we did not investigated this problem.

4.2 Implementation

For the sake of separation of concern we divided the original model in three clearly distinguishable sub-parts,

1. Initialization, in this initial data for the model is calculated for the ocean and atmosphere models. We discuss this step some more in subsection 4.3.
2. Ocean step, this sub-part performs one time step for the ocean model, which we will explain in subsection 4.4.
3. Atmosphere step, analogue with the ocean part, this sub-part performs one time for the atmosphere model. This part highly resembles the ocean step and we therefor don't discuss this part.

A possible extension to the specification could be a high-grain parallelization of steps two and three. In figure 4.2 we see the current sequential execution of the model, while figure 4.2 a possible parallel implementation is shown. Unfortunately data dependencies did not allow for a trivial parallelization; for example it has to be taken into account when data has to be exchanged between the models. This would require a redesign of the original model, which is beyond the focus of this work.

Fig. 1. Sequential execution of the ocean and atmosphere steps

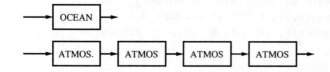

Fig. 2. Parallel execution of the ocean and atmosphere steps

In the implementation of the original model, several numerical library calls where made. For example, for finding the solution to a tridiagonal matrix a tridiag solver was called. In [10] we examined the use of calling external library calls for implicit equations, while for this specification we explore the use of templates even more [14] The use of templates makes it possible to offer the designer of the model a number of standard numerical solution methods, but leaves the actual implementation to CTADEL. The user can therefor specify the usage of a method while CTADEL can pick an appropriate and optimal implementation for the target architecture. For example when a Fourier analysis is needed the user can specify this like

```
V = Fourier(Q),
```

without bothering about the actual implementation; CTADEL decides if a library call should be made to an existing optimized library or fill in a the code itself. In the current implementation CTADEL makes use of default numerical functions, like spline2 from [7].

4.3   Initialization

In the first phase of the model, a number of general input parameters and initial data are read in, after which model specific data is calculated from these parameters. The original implementation of the model also could perform special tasks in this initialization phase, like some sort of crash recovery from an aborted previous run. We did not implement this in our specification.

An example equation belonging to this first step is equation (2) which reads in the CTADEL specification like,

```
qcomp(P,DX,DY,H,GP,NL) := (C + sign *(f_zero/(GP * H))*(DPO)
          where C=(DX*(lapl P)/f_zero + beta * Y)
          where sign=(1 if k>1 \\ -1 otherwise)
          where (DPO=(P-(P@(k=k+1)))) if k<NL \\
                ((P@(k=k-1))-P) otherwise)
          where Y=(j-1)*DY
      ).
CalcO(P,DX,DY, H,GP) :=
        ( zq(P,DY, H,GP,nlo) if borderj_O \\
          mqo(P,DY, H,GP) if borderi_O \\
          qcomp(P, DX, DY, H, GP,nlo) otherwise
      ).
```

This specification shows two generalized functions (including some boundary conditions) which can be used like,

```
qo = CalcO(po,dxom2, dyo, H_o, g_prime).
```

In the quasi-geostrophic model for the calculation of the barotropic mode, a chsolv routine function is used, which uses a Fourier and a tridiag solver. In our specification we make use of templates to deal with these solvers; in the current specification CTADEL calls an external library function [7] to find a solution. A sample specification for this reads like,

```
tp_ch2 = (ch_solv(boundary_south, aa, ba) where aa=1/dya^2).
tp_ch3 = (ch_solv(boundary_north, aa, ba) where aa=1/dya^2).
```

4.4 Ocean step

In this part of the specification, one ocean time step is calculated. A typical 'dynamics' equation which can be found is the calculation of the auxiliary currents, e.g. wind tendencies. The specification for equation (refeqn:aux) for the ocean dynamics read like,

```
u_ag = (-C*(P1-P2)*dxam2/f_zero
        where P1=((pa@(j=j+1)) if j<nya \\
                  (pa@(j=nya)) otherwise)@(k=1)
        where P2=((pa@(j=j-1)) if j>1 \\
                  (pa@(j=1)) otherwise)@(k=1)
        where C=(0.5 if noborder(nya) \\ 0 otherwise)
        ).
v_ag = (-1/2*((P@(i=i+1))-(P@(i=i-1)))*dxam2/f_zero
          where P=pa@(k=1)) if noborder(nya) \\
                0 otherwise.
```

5 Preliminary results

In this section the code generated by CTADEL for the quasi-geostrophic model is compared with the original hand-written code. Since the original code was not written with efficiency in mind, we can expect a performance difference with the generated code by CTADEL which performs some aggressive optimizations like common subexpression elimination. Standard compiler optimizations like loop optimization [5] where not applied. We ran the code on a scalar type of architecture; a commodity personal computer, running the linux/GNU operating system, with linux kernel version 2.4.10. The used machine was equipped with a 700 Mhz AMD Athlon processor and 128 MByte.

All test-runs used an input grid with variable horizontal and vertical points and two layers per model. For some experiments we used a fixed number of thousand time-steps while we also conducted experiments with diverging time-steps. In order to reduce external influences on these times, we run each experiment a number of times and calculated the average execution time. Because the deviations from these average times are in the order of tenths of percents, we do not include them in the figures.

For the compilation of the programs we used the GNU Fortran compiler, version 2.95.2, with standard optimization turned on (g77 -O2). With each run of the programs we compared the output of the generated code with the reference code. Since numerical solution methods are used over several time-steps, small differences appear in the output of both programs. However, these differences turned out to be relatively small and therefore acceptable.

In figure 3 we run both generated code and the hand-written code with a varying input grid size and a constant number of 1000 time-steps. Since both the atmosphere and the ocean model use different input grid sizes we have scaled them equitable. As we see from both the graphs, the code generated by CTADEL

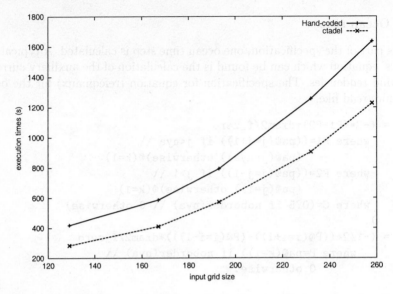

Fig. 3. Execution of both models with a varying input grid size, execution times in seconds

has a performance gain of 30% over the hand-written code. We also see that neither of the codes has a linear execution time with increasing input sizes.

In figure 4 we have examined the execution times of both the codes as a function of the number of time-steps. As one might expect, this is a linear relation. As for figure 3 the automaticly generated code from CTADEL has a performance gain of around 30% over the hand-written code.

These experiments show that automatic generated code can vanquish the hand-written codes for the quasi-geostrophic model. We should, however, point out that the original code was not optimized with respect to performance. Previous studies [10, 12] have shown that generated code can compete with hand-optimized codes.

6   Related work

Not much work has been done in the field of automatic generation of program codes based on a mathematical specification. Therefore programmers tend to 'convert' the problem by hand into a program or use libraries and/or problem solving environments (PSE). Some PSEs exist which produce program codes. For example the MathWorks package is a compiler for Mathlab specifications which generates C and C++ codes. There also exists special purpose libraries for PSEs, which can handle a specific problem. Examples are the REDTEN [3] for Maple and RICCI [6] for Mathematica.

The Falcon project [1] takes a different approach, comparable with CTADEL. It uses an existing high-level array language, MATLAB, as source language and

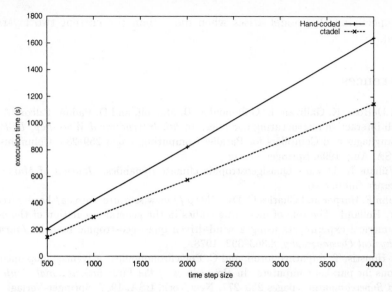

Fig. 4. Execution of both models with varying number of time steps, execution times in seconds

performs static, dynamic, and interactive analysis to generate Fortran 90 programs with directives for parallelism. It includes capabilities for interactive and automatic transformations.

7   Conclusion

In this paper we show a new method of code generation for large scale numerical models. Instead of translating a mathematical model into source code by hand and maintaining hand-optimized versions for several target computer architectures, the designer can specify the model in an abstract specification language which is automaticly compiled into highly-optimized target-dependent Fortran codes. We explored this on a coupled ocean–atmosphere model, in which we have to deal with multi-resolution domains and different time-steps. We also, briefly, addressed the concept of templates in our CTADEL compiler, which can elevate the burden from the model-designer of picking efficient numerical solution methods.

As we can see from the preliminary results in section 5 the generated code by CTADEL is clearly more efficient considering execution time. Since the original hand-written code was not optimized for speed, this is not a big surprise. More important is the ability to generate code for multiple architectures by using CTADEL. For example, it's fairly trivial to automaticly generate codes for a distributed memory architecture with use of the MPI message parsing while this is a tedious task to do with the original hand-written code. Another advantage from specifying the model arises when more than the current two layers are needed.

References

1. L. DeRose, K. Gallivan, E. Gallopoulos, B. Marsolf, and D. Padua. Falcon: A matlab interactive restructuring compiler. In *8th International Workshop, LCPC'95, Languages and Compilers for Parallel Computing*, pages 269–288, Columbus OH, USA, Aug. 1995. Springer Verlag.
2. William K. Dewar. Quasigeostrophic climate dynamics. *Journal of Marine Research*, Submitted.
3. John F. Harper and Charles C. Dyer. http://www.scar.utoronto.ca/ harper/redten/.
4. W. Holland. The role of mesoscale eddies in the general circulation of the ocean: numerical experiments using a wind-driven quasi-geostrophic model. *Journal of Physical Oceanography*, 8:363–392, 1978.
5. E. Houstis, T. Papatheodorou, and C. Polychronopoulos. Advanced loop optimizations for parallel computers. In *Proceedings of the First International Conference on Supercomputing*, pages 255–277, New York, USA, 1987. Springer-Verlag.
6. University of Washington. Department. of Mathematics. http://www.math.washington.edu/~lee/Ricci/.
7. William H. Press, Saul A. Teukolsky, William T. Vetterling, and Brian P. Flannery. *Numerical Recipes in Fortran 77*. Cambridge University Press, 1992.
8. K. Shafer Smith and Geoffrey K. Vallis. The scales and equilibration of midocean eddies: Freely evolving flow. *Journal of Physical Oceanography*, 31:554–571, February 2001.
9. T. Sterling, D. Savarese, D. J. Becker, J. E. Dorband, U. A. Ranawake, and C. V. Packer. BEOWULF: A parallel workstation for scientific computation. In *Proceedings of the 24th International Conference on Parallel Processing*, pages I:11–14, Oconomowoc, WI, 1995.
10. Paul van der Mark, Gerard Cats, and Lex Wolters. Automatic code generation for a turbulence scheme. In *Proceedings of the 15th International Conference of Supercomputing*, pages 252–259, Sorrento, Italy, June 2001. ACM.
11. R.A. van Engelen. *Ctadel: A Generator of Efficient Numerical Codes*. PhD thesis, Universiteit Leiden, 1998.
12. Robert van Engelen, Lex Wolters, and Gerard Cats. Ctadel: A generator of multiplatform high performance codes for pde-based scientific applications. In *Proceedings of the 10th International Conference on Supercomputing*, pages 86–93, Philadelphia, USA, May 1996. ACM.
13. Robert van Engelen, Lex Wolters, and Gerard Cats. Tomorrow's weather forecast: Productive program generation in atmospheric modeling. *IEEE Computational Science and Engineering*, 4(3):22–31, 1997.
14. Robert A. van Engelen. Atmol: A domain-specific language for atmospheric modeling. *Journal of Computing and Information Technology*, 2001.

Data-Flow Oriented Visual Programming Libraries for Scientific Computing

Joseph M. Maubach and Wienand Drenth

Technical University of Eindhoven, Faculty of Mathematics and CS,
Scientific Computing, Postbus 513, NL-5600 MB Eindhoven,
The Netherlands
{maubach, drenth}@win.tue.nl

Abstract. The growing release of scientific computational software does not seem to aid the implementation of complex numerical algorithms. Released libraries lack a common standard interface with regard to for instance finite element, difference or volume discretizations. And, libraries written in standard languages such as FORTRAN or c++ need not even contain the information required for combining different libraries in a safe manner.

This paper introduces a small standard interface, to adorn existing libraries with. The interface aims at the – automated – implementation of complex algorithms for numerics and visualization. First, we derive a requirement list for the interface: it must be identical for different libraries and numerical disciplines, support interpreted, compiled and visual programming, must be implemented using standard tools and languages, and adorn libraries in the absence of source code. Next, we show the benefits of its implementation in a mature (visual) programming environment [1], [2] and [3]), where it adorns both public domain and commercial libraries. The last part of this paper describes the interface itself. For an example, the implementational details are worked out.

1 Introduction

This paper introduces techniques for the construction of software libraries which aid the implementation of complex numerical algorithms. Its focus is on the *unaltered reuse of existing* public domain and commercial *libraries* such as for instance packages from www.netlib.org: The Linear Algebra Package LAPack [4], FFTPack [5], Piece-wise Polynomial Package PPPack [6], Templates [7], etc. This reuse of scientific numerical software is achieved without the definition of a complete across-libraries communication language as in Open-Math [8] and MathML [9]. Instead, we adorn parts of existing standard-lacking numerical libraries with a minimal but powerful interface, ensuring fast communication in a standard computer language. The operations we interface are in general more

P.M.A. Sloot et al. (Eds.): ICCS 2002, LNCS 2329, pp. 429–438, 2002.
© Springer-Verlag Berlin Heidelberg 2002

complex than the basic operations in [10], [11], and our interface is less complex than the one published in [12].

In section 3, we demonstrate how to adorn existing libraries with an interface, which satisfies at least the following so-called *use-requirements*: The interface should be similar across different numerical packages and disciplines, must enable and stimulate all of interpretation/compilation/visual programming, must be generated in the absence of the original source code, must be formulated in a wide-spread computer language, and must be built using a standard tools. Whenever possible, existing de facto fundamental data standards must be supported: matrices in MATLAB [13] and a free clone octave [14], matrices and vectors in Open-Math [8] and LATEX [15] (both matrix and table format), lists in Mathematica [16] and a free clone Maxima [17] various dataset formats of VTK [18], Open Inventor [19], OpenGL [20] and java3d [21].

Each of the proposed use-requirements has its advantages. To mention a few: A similar interface aids the implementation of complex algorithms, interpretation is convenient, complication ensures speed, and visual programming provides application-level overview (see [1], [22]) as well as visualization/solver interaction (see [1] and [3], but also the use-cases presented in [12] and the references cited therein). Building an interface from a library in the absence of source code is a requirement for the reuse of a commercial library ([23], [24], [25], etc.). Using a wide-spread standard fast language turns out to be feasible, so it is a must.

The remainder of this paper is composed as follows. First, section 2 shows how the to be introduced interface can be used to implement complex numerical algorithms in a rather convenient manner. The interface itself is presented in section 3, which also provides an example implementation.

2   Data-Flow oriented Scientific Computing

Because our proposed interface must aid the implementation of complex numerical problems, we first examine the nature of numerical *complex*. We distinguish two categories of *complex* complex numerical problems.

We call a numerical complex problem *global*, if a range of numerical problems for which each a software solution exists, is combined into one. An example would be a transient problem involving multiple conservation equations in a moving domain, where different state variables and equations have to be discretized using different methods such as finite element and volume techniques.

The second category of complex problems is called *local*. An example of this category is a single non-trivial discretization or iterative solver which requires various types of input and output data.

The distinction into local and global complex problems is of course relative, so all techniques used to solve the local ones, solve the global ones – see [1] for detailed information on the global case.

For the remainder of this paper, our example of a complex numerical problem is the leading-order problem for the solidification of an amount of molten material flowing past a relative cold solid wall from [26], based on [27] and [28]. As described in more detail in [29], an axis-symmetric splash of metal moves along the inside wall of a cylinder as in figure 2b. The related x and y coordinate-directions are along and perpendicular to the inside wall. The metal splash is part molten (liquid) and part solid: Its total height is $\phi(x, t)$, and the height of the internal solid/liquid interface is $0 \leq \psi(x, t) \leq \phi(x, t)$. The temperature is denoted by $\theta(x, y, t)$. In different subdomains, the splash's movement is determined using: The conservation of mass:

$$\frac{\partial \phi}{\partial t} + \frac{\partial}{\partial x}(u\phi) = -\frac{\partial \psi}{\partial t}; \tag{1}$$

momentum:

$$\frac{\partial}{\partial t}(u\phi) + \frac{\partial}{\partial x}(u^2\phi) = -u\frac{\partial \psi}{\partial t}, \tag{2}$$

and energy:

$$\frac{\partial \theta}{\partial t} + \frac{\partial}{\partial x}(u\theta) + \frac{\partial}{\partial y}\left(\{(u\psi)_x - u_x y\}\theta - D\frac{\partial \theta}{\partial y}\right) = 0, \tag{3}$$

in combined with initial, Dirichlet, Neumann, Robin, and Stephan conditions. The discretization and related iterative solution algorithm we use in our example program is presented in [29]).

This problem is complex at a global level and also at a local level: ϕ and ψ are represented using splines (PPPack), are visualized super-imposed on a cylinder (VTK, Open Inventor (OI)), and are obtained solving systems of equations (LAPack and SEPRAN). Figure 1 shows part of an implementation, loaded into the NumLab visual programming environment (see [1]). The implementation uses functions (*modules*) from different libraries including LAPack, PP-Pack, SEPRAN, VTK, OI, etc., as well as operating system commands: The mpeg module (hidden inside the group <PostProcessViewer). As most numerical programs, the program in figure 1 can be divided into two parts: (1) The part which reads the input and performs the numerical computations; (2) The part which visualizes the computed results. The latter postprocessing part also writes the output, for instance .mpeg files. Figure 2a shows the automatically built editors which can be used to enter values of arguments which are not connected. Figure 2b depicts a VTK module, which shows the initial shape of the splash of metal, superimposed with graphical operations onto the inside of a cylinder wall. Both so-called viewers offer solid, hidden-line, wire-frame and other representations, as well as interactive 3-d manipulation of the visualized objects.

In order to understand the program related to figure 1 the depicted visible graphical elements must be explained. Each rectangle such as MIteration is a graphical representation of a function, using the interface proposed in section 3. For

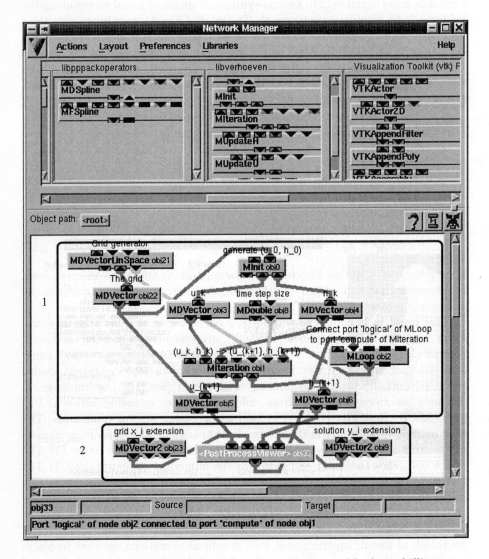

Fig. 1. A program which calculates deformations caused by laser drilling

the sake of convenience, we call this representation also *module*. Within each module, we distinguish bottom and a top area. Modules have either input or output arguments (input/output are not part of a function's interface). Each of its *arguments made available via the interface description*, is mapped to a small colored rectangular area at the top or bottom the module. Using standard data-flow terminology, these small rectangles are called *ports*.

Different builtin argument types are colored in a different manner, all pointers and derived types are colored using a default (green). The function's interface determines the position of each argument port. Next, each port is either connected, or behind the scenes a default value is used. Output is connected to input using the mouse (click on bottom port, drag to top port and release. The connection is established when the data types concur). Default values are altered using a graphical user interface which is generated from the function's interface, and which pops up when selected (clicking on the module reveals a list of options, amongst which is `Edit`). Building complex numerical algorithms with NumLab is convenient because of the libraries LAPack, EigPack, SEPRAN, PPPack, VTK, OI which have been interfaced and because of those which have been added (data-types, data-exchange in MATLAB/Mathematica/Open-Math/LaTeX/VTK/other formats, discretized ordinary and partial differential related operators). All implementations such as LAPack have been interfaced for that part which covers the currently conducted research. This means that about half a million lines of code out of 4 million is interfaced, covering about 1000 modules,

Fig. 2. (a) Left: The user alters a module's default input values using an automatically built editor. (b) Right: The contour $\phi(x,0)$ of the liquid metal on the cylinder wall

3   The construction of data-flow oriented libraries

With the NumLab environment from the previous section kept in mind, we show how to interface existing libraries such that all promised use-requirements are met. Our techniques uses design patterns (inheritance, proxies) such as in [31], [32], [33] and [34].

First, we regard the choice of language for the implementation of the interface. To begin with, because of all toolboxes must offer a similar visual programming interface, a common language must be used. Next, the use-requirement that all toolboxes can also be interpreted, requires the use of a common language which can be interpreted. Not introducing a new language implies using an existing one, and using source-libraries means that the common language must be a kind of union of all source-libraries' languages (Pascal, FORTRAN, c, c++). Thus, the interpreter/compiler must be able to distinguish all related data types.

With the NumLab environment from section 2 kept in mind, we assume that the common language is a *large subset of c++*: That subset of c++ which cint c++ interpreter can interpret – see the manual of [35]. The interpreter cint can interpret most c++ programs, even when templates are used in combination with compiled code.

Now, we introduce the interface structure (implementation) itself. We restrict ourselves to interfaces for fortran libraries because most of the numerical implementation we use are written in fortran. Also c++ code can be adorned with our interface (using proxies instead of direct inheritance from class-level to module-level below), but the description falls outside the scope of this paper.

The first layer is the so-called *source-layer*. It contains (the functions from) the libraries which the user wants to use. Assume we want to use the compiled function f(z, x), in combination with documentation which describes that both the *input argument* x (only read from) and the *output argument* z (only written into). For the sake of a simple presentation we assume that both arguments are documented to be of type X. The source of the documentation (include file, reference manual, etc.) is irrelevant. For languages such as c++, input and output can be deduced when the keyword const is used in a proper manner. However, using this keyword at the right moment is difficult and often forgotten, so even for c++ program extra documentation is recommended in principle. Other languages such a FORTRAN lack the const feature and absolutely require related documentation.

The second and middle layer is called *class-layer*. It contains one c++ class for each function in the bottom layer. The class has a signature which is derived from the source. Assuming we know which variable is input and which one output, the class which interfaces the source f() is called class cf, and defined as follows

```
class cf
{
public:
  cf();
  cf(X &z, const X x);
  int callback(X &z, const X x);
  void print(ostream &) const;
};
  ostream &operator(ostream &, const cf &);
```

The second constructor uses the call-back on the function `callback()` which returns value `Ok` if all went fine during its call-back to `f()` , and returns value `NotOk` if the call to `f()` somehow failed. If for instance `f()` was a higher level LAPack function, its success or failure could be detected inspecting the output argument with name `info`. The `print()` and output methods `<<`

The the third and last layer, is called *module-layer*. This layer contains the modules. Each module is a class, which delegates all operations to class-level, and which in addition adorns the class-level with a few standard data-flow operations. The module related to class `cf` is named `mf`:

```
class mf : public cf, public mbase
{
 public:
  mf();
  mf(X &z, const X x);
  integer callback(X *z, const X *x);
  void print(ostream &) const;

 public:
  int update();
  const X *get_x();
  void set_z(const X *);
  X* get_z();
  void set_z(X *);

 private:
  const X *x;
  X *z;
};
 ostream &operator(ostream &, const cf &);
```

The network manager calls the `set_*()` and `get_*()` operations when making a connection, or when using the editor to edit non-connected default values of an `mf` instance. The network manager calls `update()` whenever an `mf` module must be executed, and of course the standard implementation of `update()` is:

```
int update()
  { return (z && x && callback(*z, *x) == Ok) ? Continue : Halt;
```

The mbase module is a module which implements no functionality, but is required for a different purpose outside the scope of this paper. The values Halt and Continue are returned to the network manager, which halts the network when the module returns Halt. Deviations from this rule have not been observed for numerical libraries. Next to the module, the module level also requires extra descriptions related to data-flow properties: each argument must be at the top or bottom for data-flow purposes and each argument must be read or write in data-flow sense. This information, in combination whether specific not-connected ports are required (see section 2) is added in an ascii file mf.mh, containing the meta information:

```
module mf : module mbase
{
input:
 write "info x" "const X*" (set_x, get_x);
output:
 write "info z" "X*" (set_z, get_z):
};
```

As can be observed variables read from and written into can both be at the top and at the bottom of the module. This turns out to be required in practice. What seems even more strange but is correct: Both the *read from* (input) and *written into* (output) arguments are logged as write in the . mh file. The reason is that the module mf needs both a pointer to an X module representing x as well as a pointer to an X module representing z in order to perform its function. The names of the arguments x and z are not specified, because these arguments are accessed using the set_*() and get_*() operations.

Next, in the meta information file the user can include default values for specific data types such as X. The default value for all pointer types X * is value zero. The default value for an other type X is the value returned using the constructor X(). All default values can be overwritten on a per module basis. By default, all pointer typed ports must be connected. Lifting this status and thus using the default value is done adding the keyword optional at the end of the line before the semi-colon. For each variable, an info string is given (shown in the editor and as help balloon text).

Due to space limitations, a wide range of available sub-problem solutions can not be mentioned. Among the most interesting are the automatic generation of the interfaces of libraries, a discussion of the merits of different possible interface solutions, and the various technical ideas/constructions and limitations encountered when attaching c++ interfaces to FORTRAN libraries.

3.1   Conclusions

This paper presents an interface for numerical software which consists of two small interfaces on top of the original unaltered source. The first small c++

interface ensures that different sources use common c++ datatypes. This facilitates the writing of implementations which combine different sources for numerical and visualization purposes. The second small c++ interface ensures that the sources can be used in a visual programming environment, and also facilitates the automated construction of (numerical) algorithms. The two interfaces are small with respect to the size of the original sources and eliminate the difference between network- and script-representations of implementations.

References

1. Maubach, J., Telea, A.: The NUMLAB Numerical Laboratory for Computation and Visualisation. Computing and Visualization in Science, accepted (2001)
2. Telea, A.: Visualisation and Simulation with Object-Oriented Networks. Ph.D. Thesis. Technical University of Eindhoven, The Netherlands (2001)
3. Telea, A., Wijk, J.J. van: VISSION: An Object Oriented Dataflow System for Simulation and Visualization. Proceedings of IEEE VisSym '99, eds. E. Groeller, W. Ribarsky, Springer (1999) 95–104
4. Anderson, E., Bai, Z., Bischof, C., et al.: LAPACK user's guide. SIAM Philadelphia (1995)
5. Swarztrauber, P.N.: Fortran subprograms for the fast Fourier transform of periodic and other symmetric sequences. http://www.netlib.org/fftpack/index.html
6. Boor, C. de: Piece Wise Polynomial Package.
   http://www.netlib.org/pppack/index.html
7. Barret, R., Berry, M., Dongarra, J., Pozzo, R.: Templates for the Solution of Linear Systems, 2nd Edition. SIAM Philadelphia (1995)
8. Caprotti, O. and Cohen, A.M.: On the role of OpenMath in interactive mathematical documents. J. Symbolic Computation, special issue on the integration of computer algebra and deduction systems, to appear
9. WC3: Mathematical Markup Language (MathML[tm]) 1.01 Specification.
   http://www.w3.org/1999/07/REC-MathML-19990707
10. Kodosky, J., MacCrisken, J., Rymar, G.: IEEE Workshop on Visual Languages, (1991) 34–39
11. Auguston, M., Delgado, A.: IEEE Proceedings on Visual Languages, (1997) 152–159
12. Wydaeghe, B.: Component Composition Based on Composition Patterns and Usage Scenarios. Ph.D. Thesis. Vrije Universiteit Brussel, Belgium (2001)
13. Matlab: Matlab Reference Guidex. The Math Works Incorporated (1992)
14. Eaton, J.W.: GNU Octave, a High-Level Interactive Language for Numerical Computations. http://www.octave.org (1997)
15. Lamport, L.:
    LaTeX User's guide & Reference Manual. Addison-Wesley Publishing Company (1986)
16. Wolfram, S.: The Mathematica Book 4-th edition. Cambridge University Press (1999)
17. Shelter, W.: Maxima for Symbolic Computation.
    http://maxima.sourceforge.net
18. Schroeder, W., Martin, K., Lorensen, B.: The Visualization Toolkit: An Object-Oriented Approach to 3D Graphics. Prentice Hall (1995)

19. Wernecke, J.: The Inventor Mentor: Programming Object-Oriented 3D Graphics with Open Inventor. Addison-Wesley (1993)
20. Jackie, N., Davis, T., Woo, M.: OpenGL Programming Guide. Addison-Wesley (1993)
21. The Java 3D Application Programming Interface.
    http://java.sun.com/products/java-media/3D/
22. Upson, C., Faulhaber, T., Kamins, D., Laidlaw, D., Schlegel, D., Vroom, J., Gurwitz, R., Dam, A., van: The Application Visualization System: A Computational Environment for Scientific Visualization. In: IEEE Computer Graphics and Applications (1989) 30–42. http://www.avs.com
23. NAG: FORTRAN Library, Introductory Guide, Mark 14. Numerical Analysis Group Limited and Inc. (1990)
24. IMSL: FORTRAN Subroutines for Mathematical Applications, User's Manual. IMSL (1987)
25. Segal, G.: The SEPRAN Finite Element Package.
    http://ta.twi.tudelft.nl/sepran/sepran.html
26. Smith, W.R.: Models for solidification and splashing in laser percussion drilling. Rana Report 01-02, Deparment of Mathematics and Computing Science, Technical University of Eindhoven, The Netherlands (2001)
27. Whitham, G.B.: Linear and Nonlinear Waves. John Wiley, New York (1974)
28. Elliott, C.M., Ockendon, J.R.: Mathematical Models in the Applied Sciences. Cambridge University Press, New York (1997)
29. Verhoeven, J.C.J.: Numerical Methods for Solidification Problems. Internal report in preparation, Technical University of Eindhoven, The Netherlands (2001)
30. Harten, A.: High resolution schemes for hyperbolic conservation laws. J. Comput. Phys. **49** (1983) 357–393
31. Rumbaugh, J., Blaha, M., Premerlani, W., Eddy, F., Lorensen, W.: Object-Oriented Modelling and Design. Prentice-Hall (1991)
32. Booch, G.: Object-Oriented Analysis and Design. Benjamin/Cummings, Redwood City, CA, second edition (1994)
33. Budd, T.: An Introduction to Object-Oriented Programming. Addison-Wesley (1997)
34. Meyer, B.: Object-oriented software construction. Prentice Hall (1997)
35. Brun, R., Goto, M., Rademakers, F.: The CINT C/C++ Interpreter.
    http://root.cern.ch/root/CintInterpreter.html

One Dilemma – Different Points of View

Ivanna Ferdinandova

International Doctorate in Economic Analysis, Departament d'Economia i d'Història Econòmica, Universitat Autònoma de Barcelona

Abstract. We build a simulation model imitating the structure of human reasoning in order to study how people face a Repeated Prisoner's Dilemma game. The results are ranged starting from individual learning in which case the worst result -defection- is obtained, passing through a partial imitation, where individuals could end up in cooperation or defection, and reaching the other extreme of social learning, where mutual cooperation can be obtained. The influence of some particular strategies on the attainment of cooperation is also considered.

1 Introduction

Probably the most difficult problems are those that seem simple, and the most studied those faced by many people. A problem belonging to both groups is the so called Prisoner's Dilemma. The fable, which made it famous, is a story about two suspects in a crime who are put in a separated cells and told the following rules: if both of them confess, each will be sentenced to three years in prison, if only of them confesses, he will be free and the other one will be send to jail for four months, if neither of them confesses they will be send to prison only for one month. What makes the problem interesting is the conflict between individual interest to defect (D) and confess, compared to the collective interest of cooperating (C) with the opponent. If the game is played only once then both suspects have no common interest and they will defect. Solving the repeated problem becomes a challenge. On the one hand, the solution of the classical Game Theory to a Finitely Repeated Prisoner's Dilemma is defection in all periods, but on the other, experiments reveal a significant level of cooperation. This contradiction raises the question of whether there exist some conditions that promote cooperation.

In particular this work focuses on the learning process, and analyzes how different models of learning, can influence the final outcome.

We build an ACE model of the repeated game in which, players' learning is modeled using an explicit evolutionary process - genetic algorithms. The pioneer in simulating Repeated Prisoner's Dilemma (RPD) was Axelrod [3], who applied genetic algorithms to evolve RPD strategies against a constant environment. His work was continued by many economists and computer scientists' papers. Among them is the paper of Miller [7], who studies the evolution of cooperation, starting from random population of strategies, which evolve in a changing environment. The authors mentioned above model learning at population level, i.e. each player

P.M.A. Sloot et al. (Eds.): ICCS 2002, LNCS 2329, pp. 439–448, 2002.
© Springer-Verlag Berlin Heidelberg 2002

learns from the best individuals in the population, assumption that is difficult to justify in the context of PD. This model of learning implicitly means that a player imitates the other players strategies, even from players he has never played with, which is equivalent to removing the assumption of no communication and changing the essence of the dilemma. With this scenario cooperation is easily obtained.

Another possibility is to construct the other extreme of individual learning, where each individual learns only from his own experience and this becomes a reason for mutual defection.

The distinction between individual and social learning was studied in the work of Vriend [9], where he analyzed Cournot oligopoly and found that with social learning the results converge to the Walrasian output level, whereas with individual learning they converge to the Cournot - Nash outcome.

Our results move in the same directions as those of Vriend, i.e. social learning leads to socially optimal outcome, and individual to the egoistic one.

Having constructed these models we can notice that both assumptions are too extreme. People do not only learn from the others, neither do they learn only from proper experience. A scenario, which combines both assumptions, is more realistic. Here arises the problem of how to construct it. If we use the classical crossover for creating new strategies, half of the experiments end up defecting and the other half cooperating. But again we face a similar problem - how can a player imitate a strategy that he does not know. This problem can be avoided by using a different type of crossover, where each player is trying to understand and imitate to same extent (without copying the strategy) the behavior of his opponent (Vilà [8]). This reduces the number of cooperative outcomes and confirms that the type of crossover is not irrelevant to the outcome, in other words the way the information is exchanged is important.

Under this scenario the elements that determine the outcome are the type of strategies the players have at the beginning and how fast is the process of evolution. If the evolution process is fast it is more likely that players end up defecting, but if it is slow they will experiment with more strategies and could obtain cooperation.

To analyze the importance of players' strategies we introduce players who always play a given strategy. The effects they induce are different. A player who always either defects or cooperates will induce defection, but one who plays a variation of Tit-For-Tat (TFT) (cooperates when the opponent is observed to cooperate and defects when the opponent has defected) makes his opponent cooperate.

The rest of the paper is organized as follows. Section 2 presents the model. Section 3 is dedicated to the results and the last Section 4 summarizes the results and concludes.

2   Model

The purpose of our work is to distinguish between three models of learning - social, individual and imitation. We start by building a basic framework which can be easily adapted to one of the three scenarios.

Two individuals are about to play a RPD. When facing the problem for the first time, each of them is assumed to have K randomly generated strategies.[1]

Let S_i^k be the k strategy of an individual i ($k = 1,..K$, $i = 1,2$). For the purposes of our model we need an abstraction of the process by which the player implements this strategy. This role will be played by a machine called finite automata. A finite automata is a system which responds to discrete inputs (in our case the actions of the opponent) with discrete outputs (player's own actions). The type of finite automata adopted here is a Moore machine[2].

Having constructed the sets of strategies for both players they are ready to play. Each player chooses one of his strategies to play r times PD. When the repeated game is over the strategy S_i^k receives payoff Π_i^k being the sum of the payoffs in each single game. We assume that the next time players meet they will experiment with another strategy.

Once the players have tried all their strategies, they analyze the results obtained. A strategy that has relatively high payoff will be kept in the memory as a good one, and used in the future, and the strategies that performed bad will be replaced with a combination of the existing strategies. A strategy that performed well in one trail is not necessarily good, it just appeared to be good against a given strategy of the opponent. For example, a strategy that always defects will be a very good choice if played against a strategy that always cooperates, the first one will have an average payoff of five and the second of zero. But the first strategy may be a bad choice against TFT.

The process explained above is evolutionary and will be modeled using Genetic Algorithms. In our context the population will be each player's strategies and the environment they play in - his opponent's strategies.

Using this framework and changing only the process of formation of the new strategies we are able to construct the three scenarios.

The first one resembles the model of Miller [7] and almost coincides with the classical framework in this problem. In Miller's model, each agent has only one strategy and the strategies of all agents evolve together i.e. there is only one population evolving. We use the same structure but our agents have more strategies. The strategies of both players evolve together, forming one population in which good strategies will be kept and bad replaced with a combination of the current strategies. This can be the case when players talk and discuss different plans. The result Miller obtained in the case of perfect information is that at the end all individuals have the same strategy and their behaviour converges to the cooperative result.

---

[1] A strategy is a plan of how to behave in all possible circumstances.
[2] For more detailed description of a Moore machine see Miller [7] or Vilà [8].

Since the assumptions of the social learning scenario contradict with the essence of the dilemma, another possibility is to construct the other extreme - individual learning, where each player learns only from his own experience. Players stay at different rooms and they do not try to interpret each other's behaviour, but just adapt. Technically this is a two population model based on Vilà [8], where a repeated discrete principal-agent game is analyzed. The difference between this model and the social learning is that now each player can use only his own strategies. The process of learning is internal and is based only on proper experience.

The two scenarios, as described above are good as benchmark models, but too extreme. A better option is the imitation scenario, which combines the assumptions of the previous two. Players are not allowed to communicated, but they do make an attempt, to understand each other. It starts like the individual learning scenario until the moment when each player has his strategies with the same structure. After this point each of them imitates the strategy of his opponent. If we use the classical method for creating new strategies (single cut crossover[3]) we are facing the same problem as in the social learning. In order to avoid it we use a different procedure for generating new strategies i.e. different type of crossover due to Vilà [8]. Our player can not observe the strategies of his opponent, but he observes the history of the repeated game, which includes both players actions. Let h_i be the history of player's i actions. Then assume that player 1 asks himself - what would I do if I were him and somebody played against me in the way I have played against him. Since he does not know the strategy of his opponent he can only imagine playing with his strategy, from his opponent's position a history of random length and comparing his action after this history with the real action of the opponent. If they do not coincide player one replaces his action with his partner's action.

All the variations of the original framework are summarized in the following table:

The structure of the model:

1. Generate K strategies for individual i, i=1,2;

2. Randomly match each strategy of one individual with one from the other;

3. Repeat until all strategies are played once. Each strategy receives payoff Π_i^k i=1,2. k=1,..K;

4. Form new strategies

    4.1 Social Learning (as in Miller [7])

    4.2.Individual Learning

        a) include the best K/2 strategies of each individual in his new group of strategies;

---

[3] With single cut crossover you combine two strategies by cutting them in two parts and interchanging the second parts.

b) select two strategies of an individual i to be parents. The probability of strategy S_i^k of being selected is:

$$P(S_i^k) = \frac{\Pi_i^k}{\sum_k \Pi_i^k} \qquad (1)$$

c) create K/2 new strategies for each individual using the crossover and then apply the mutation;

4.3.Imitation - steps a, b and c remain the same until both individuals have their strategies with the same structure. After that step b is changed with

b') one of the strategies selected belongs to the individual himself and the other is the best strategy of his opponent.

5. Repeat steps 1÷4 R times, where R is the number of repetitions.

---

3   The Results

3.1   Parameter Values

The simulations were performed under the following conditions:

Number of individuals	2
Number of strategies of each individual	4
Number of rounds	300
Number of repetitions	10 000
Probability of mutation	0.001
Overlapping generations	1/2
Crossover type	partial imitation or single cut
Length of bit string	60

$$(2)$$

	C	D
Payoff structure		
C	3 , 3	0 , 5
D	5 , 0	1 , 1

The results are robust to changes in most of the parameters (number of strategies, length of bit string and payoff structure), but the choice of some of them requires some discussion. One of them is the number of strategies. Usually the size of the population chosen in similar models is higher, but having in mind that in our context the population consists of strategies, it is difficult to assume that people have 50 or 100 strategies. We have run simulations with 50 or 100 strategies, but the results obtained were not different. The number of rounds and the number of repetitions, were chosen to guarantee convergence. The probability of mutation determines, among other things, how adaptive the players are. The higher is the probability of mutation, the easier it will be for one player to adapt to the changes of the behavior of the other. But if it becomes too high they will modify their behavior too often. The value chosen is standard.

3.2   Results

Our results under the social learning scenario coincides with the one obtained in the literature i.e. cooperation. The outcome is independent of the type of crossover used. A typical evolution of the average payoffs of both players is depicted in Figure 1.

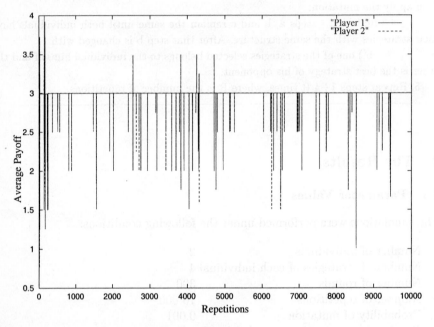

Fig. 1. In the case of *Social Learning* players cooperate and receive average payoff of three.

The same stability of the result is obtained also in the individual learning case but the outcome, as Figure 2 shows, is defection.

One possible explanation is that, when the evolution is common, the strategies that the players will have in the next period come out of the same process. Therefore it is very likely that they will be similar for both players and they will move together. Hence they have to choose between cooperate or not, and the first outcome is clearly better.

The imitation scenario has two possible results depending on the type of crossover assumed. If we use the traditional crossover (single cut) players can end up defecting or cooperating depending on how fast is the learning process and what type of strategy do they have before the imitation begins. This difference can be seen in the two graphs at Figure 3. Intuitively if players learns slowly the search field becomes bigger and the possibility of cooperation increases. But at the same time even if they learn fast but at least one of them uses a strategy

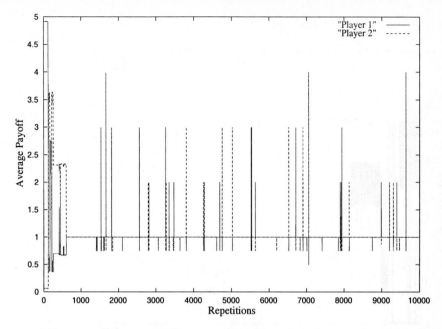

Fig. 2. In the scenario of *Individual Learning* the only possible result is defection, which gives both players an average payoff of one.

that induces cooperation this is enough to make cooperation the only possible outcome. In order to check the influence of different strategies we introduce a player who always plays the same strategy. The three possible players are a player who cooperates, player who defects and players who plays TFT. If the first two are introduced they induce defection, but the third one is able to "teach" his opponent to cooperate. One possible explanation for the experiments, which converge to defection is that it is very difficult for a strategy like TFT to appear in a random environment.

If instead of adopting the classical crossover we choose the one that does not violate the assumptions of no communication the only possible result as depicted in Figure 4 is defection.

4   Concluding remarks

The results of the three scenarios we have constructed suggest that one should be very careful when deciding which one to choose.

Special attention should be given to the assumption of no communication, which we found to be very influential. Intuitively if people cannot talk this lowers the confidence and decreases the probability of cooperation. Going back to the initial structure of the dilemma, if the two suspects can discuss, they will be able to find plans or strategies, which guarantee that neither will confess. Making

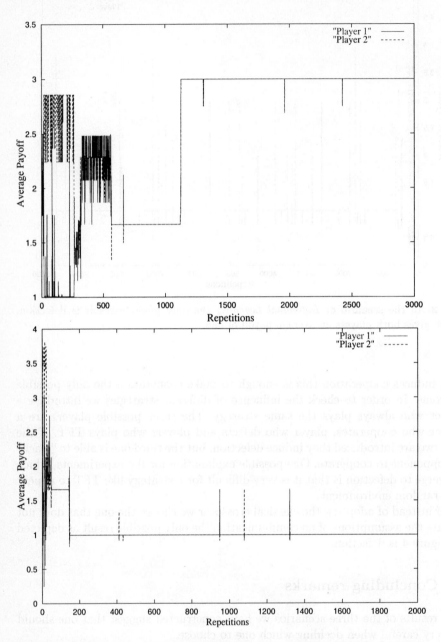

Fig. 3. *Imitation scenario* The left graph depicts a case in which the learning process is slow and the result is cooperation. Exactly the opposite can be seen at the right graph, where the players end up defecting

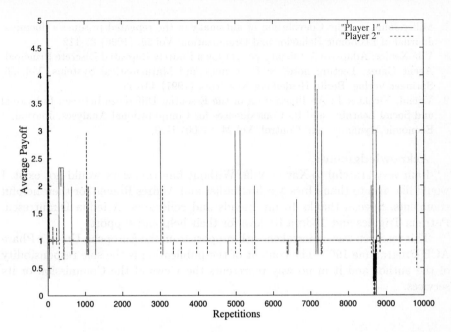

Fig. 4. *Imitation scenario with partial imitation crossover.* The result is defection.

the learning process personal, complicates it and independently on the possible attempts of reasoning about the other person's behavior, the cooperation fails. The imitation of the opponent's behavior leads to cooperation only if one of the agents plays TFT.

Summarizing, we have found two conditions, that promote cooperation. On the one hand, that is the possibility of communication and on the other the influence of a TFT player.

References

1. Arthur, Brian W.: Complexity in Economic Theory: Inductive Reasoning and Bounded Rationality, American Economic Review, Vol.84,(1994) 406-411
2. Axelrod R.: The Evolution of Cooperation, Basic Books, New York (1984)
3. Axelrod R.: The complexity of cooperation: Agent based models of competition and collaboration, Priceton University Press (1997)
4. Goldberg David E.: Genetic Algorithms in Search, Optimization and Machine Learning, Addison Wesley (1989)
5. Holland J.H: Adaptation in Natural and Artificial Systems, University of Michigan Press (1975)
6. Holland J.H., Holyoak K. J., Nisbett R.E., Thagard P. R.: Induction: Process of Inference, Learning and Discovery, MIT Press (1989)

7. Miller,John H.: The Coevolution of automata in the repeated prisoner's dilemma, Journal of Economic Behavior and Organization, Vol.29, (1996) 87-112
8. Vilà Xavier: Adaptive Artificial Agents Play a Finitely Repeated Discrete Principal-Agent Game. Lecture notes in Economics and Mathematical Systems, Vol.456, Springer-Verlag, Berlin Heidelberg New York (1997) 437-56
9. Vriend, Nicolaas J.: An Illustration of the Essential Difference between Individual and Social Learning and Its Consequences for Computational Analyses, Journal of Economic Dynamics and Control, Vol.24, (2000) 1-19

Acknowledgments

I am very grateful to Xavier Vilà. Without him this work would not exist. I would like also to thank Inés Macho Stadler and Andrés Romeu for their helpful comments. Special thanks to my friends and colleagues: Ariadna Dumitrescu, Patricia Trigales and Todora Radeva for their help and support.

This research was undertaken with support from the European Union's Phare ACE Programme 1997. The content of the publication is the sole responsibility of the author and it in no way represents the views of the Commission or its services.

Business Agent

I.-Heng Meng[1], Wei-Pang Yang[1], Wen-Chih Chen[2], Lu-Ping Chang[2]

[1] Institute of Computer and Information Science.,
National Chiao-Tung University, Hsinchu, Taiwan, R.O.C
ihmeng@iii.org.tw
[2] Advanced e-Commerce Technology Lab., Institute for Information Industry, ROC
wjchen@iii.org.tw, clp@iii.org.tw

Abstract. The Internet and World Wide Web represent an increasingly important channel for retail commerce as well as business transactions. However, there are almost 5 billion pages or sites on the Internet and WWW. There lacks an integrated mediator business agent which is around the internet to connect between suppliers and users. Therefore, an intelligent business broking agent between supply and demand is needed for using and sharing the information efficiently and effectively. In this paper we proposed a new business agent architecture. The Business Spy Agent (BSA) captures the supply and demand information from e-commence sites automatically. Supply and Demand Analysis Mechanism (SDAM) uses NLP technologies to extract product and trading information. Domain ontology is used to classify between supply and demand and to divide them into subsets. A benefit model is proposed to handle the pairing between sub-supply and sub-demand. And a divide-and-conquer algorithm is proposed to handle the whole matching between supply and demand. In the negotiation process, the architecture uses NLP and template to produce negotiation text to handle negotiation through mail. Finally, Hidden Business Mining Mechanism (HBMM) adopts data mining technology to achieve hidden business mining. The architecture covers the four process of buying behavior including support and need identification, Product brokering, Merchant brokering, Negotiation.

1. Introduction

WWW had become the successful Platform for message exchange. Users can search and exchange messages through the Internet and WWW. Therefore, more and more trading have been done on the internet and WWW. Recently, there are almost 5 billion pages or sites on the Internet and WWW. Because of drowning in information, users or companies wasted lots times on searching and browsing huge number of sites to get the information of products and satisfy the demands. There lacks a integrated mediator business agent in sites which is around the internet, to connect the suppliers and users. Therefore, an intelligent business broking agent between supply and demand is needed for using and sharing the information efficiently and effectively.

P.M.A. Sloot et al. (Eds.): ICCS 2002, LNCS 2329, pp. 449–457, 2002.

2. Related work

Up to now, most of business agents in the Internet support the services which like electronic market. In these service sites, the supporters can use the "Posting System" which is supported by the service site to publish the products information. Users can retrieve the products information through retrieval system provided by sites. Some sites provide the electronic prices quotation system which let users require price of products on sites. The XML technology is used for information exchanging makes the buying process electrolyzed. The sites including http://www.personalogic.com, http://www.firefly.com, http://jango.excite.com, http://www.onsale.com, http://www.ebary.com/aw, http://auction.eecs.imich.edu, http://ecommerce.media.mit.edu/tete-a-tete/ and http://bf.cstar.ac.com/bfare are famous. These business agents usually provide electronic marketing sites which contain all business transactions. In other words, from products information to buying behaviors are all in this sites and can't across the sites. The sites can't help users to find the products and help supplier to find the hidden business. When the user has demands, he can use the category or search interface provided by site to retrieve the products. After finding the products, user can do compare the price, purchase the products, and entering payment information. Many descriptive theories and models sought to capture buying behavior, including the Andreasen model, the Bettman information-processing model, Nicosia model. All share six similar fundamental stages of the buying process [6]:
(1)Supply and Demand Identification. (2)Product Brokering. (3)Merchant Brokering. (4)Negotiation. (5)Purchase and Delivery. (6)Production Service and Evaluation.

The Purchase and Delivery process and Production Service and Evaluation process can't be completed by business agent. But others process can be done by business agent technologies. Most of the business agents didn't handle the Need Identification and only the T@T handles the processes including Product Brokering, Merchant Brokering and Negotiation[4]. We hope business agent is more intelligence. It can identify the real need of user through the huge product information in internet. Moreover, it can help user handle the processes including the Supply and Demand Identification process, the Product Brokering process, the Merchant Brokering process and the Negotiation process.

For the users, business agent could help user search needed products, compare price, handle negotiation and purchase. The things user must handle by self are payment and get the products. For supplier, business agent can help supplier find the hidden business. In this paper we propose a new architecture to construct business agent which use the NLP(Natural Language Process) and knowledge retrieval technologies to match supply and demand intelligently.

3. Business Agent

3.1 System architecture

In the architecture, business agent contain 5 mechanism as shown in figure 1: (1)Business Spy Agent (BSA) — The supply and demand information is captured in electronic market site automatically. The supply and demand information could be analyzed in next mechanism to construct supply and demand list. (2)Supply and Demand Analysis Mechanism (SDAM)— Using NLP to analyze the information which is aggregated by Business Spy Agent and construct the supply and demand list. (3)Supply and Demand Clustering Mechanism (SDCM)— Using the clustering algorithm to cluster the information which is analyzed by previous mechanism. The clustering algorithm is based on the domain ontology [1]. (4) Matching and Negotiation Mechanism (MNM)— uses a benefit model and a matching algorithm to match supply and demand. (5) Hidden Business Mining Mechanism (HBMM) — Using the data mining algorithm to mine the hidden business from business trading.

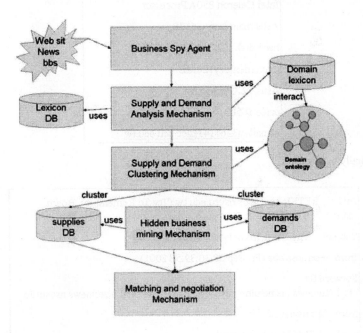

Fig. 1. System architecture

3.2 Business spy agent

There are many activities when a trading is made in internet. First, the supply (for suppler) and demand (for buyer) messages are posted. Then user waits for someone

452   I.-H. Meng et al.

response destructively or does more posting and search activities. It is time-consuming processing to finish the business trading. In our architecture, the BSA (business spy agent) is used to capture the supply and demand from internet [3]. In order to complete the job, the knowledge base is used to store the famous electronic markets' URL (http://www.bid.com.tw), BBS (bbs.ntu.edu.tw) and news, etc. These sites have many supply and demand information which can be captured by business spy agent.

3.3 Supply and Demand Analysis Mechanism

The SDAM analyzes the information which is captured by BSA, and filters the redundancy and useless information. After filtering, the useful supplies and useful needs information are formatted. For example, the supply and demand are some simply free word sentences in Bulletin Board System (BBS). As shown in figure 2 and figure 3:

```
Sell   Pentium 586 computer
Intel Celeron 850A Processor
Asus main-board 6178
hard-disk IBM 40GB
main memory128MB
Asus 52X CD-ROM
price $6500
e-mail: u3261143@cc.ncu.edu.tw
```

Fig. 2. Supply information

```
Author: RETER.bbs@bbs.nsysu.edu.tw (Try To Remember), Board:
market
Subject:   buy PCMCIA network card
from : Formosa bbs (Fri Sep 28 10:39:58 2001)
Forward Site:
cis_nctu!news.cis.nctu!netnews.csie.nctu!news.civil.ncku!news.nsysu!Fo
from : bbs.nsysu.edu.tw
wanted PCMCIA network card
Plz reply with prices and product information
```

Fig. 3. Demand information

The SDAM uses the NLP technology to analyze the information and construct useful supply information" and "demand information". The analysis processing contains three steps:

(1) Supply and demand detection: The purpose of this step is to separate the supply and demand by using corpus which contain concept of trading. For example, "sell", "buy", "wanted", "ask for"… etc. Using NLP technology, the supply and demand information can be separated by this way.

(2) Product information detection: After separate supply or demand information, the product information is detected. In this paper, we focus on computer hardware peripherals. We construct the domain ontology of computer hardware peripheral then using the ontology and NLP technology to extract the production information. The domain ontology and production can used in SDCM latter to achieve matching.

(3) Transaction information detection: The transaction information is detected in this step. The information contains delivery time, place, contact information … etc.

3.4 Supply and Demand Clustering Mechanism

After the supply and demand information is detected through SDAM, the information can be clustered by SDCM. The purpose of clustering is to speed up matching supply and demand. The supply and demand belong to the same cluster are matched to finish supply and demand matching. In this paper, we use the ontology based clustering algorithm to cluster supply and demand. Each node in ontology considered as a cluster, Supply information and demand information are mapped to some nodes on ontology as shown in figure 4.

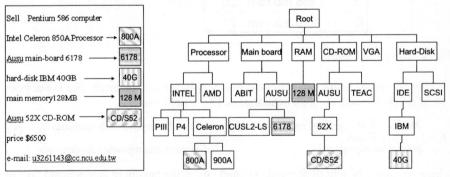

Fig. 4. Supply and demand mapping to nodes on ontology

The other advance of the clustering algorithm is that supplies and demands can be divided into sub-supplies and sub-demands. For example, the demand for computer can be divided to sub-demands including processor, main-board, hard-disk, main-memory…etc. When matching supply and demand, these sub-demands can be fitted by some sub-supplies and the whole matching can be achieved.

3.5 Matching and Negotiation Mechanism

After clustering, supplies and demands are matched in this step. We describe the three types of supplies and demands matching and propose a Divide and Conquer Matching algorithm to achieve the work.

Type1: In this type just on supply can satisfy demand (one to one matching) as shown in figure 5. This type is the simplest type of matching and we just match the supply and demand. Then we can do the negotiation.

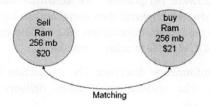

Sell
Ram
256 mb
$20

buy
Ram
256 mb
$21

Matching

Fig. 5. One to one matching

Type2: In this type many supplies can satisfy demand (many to one matching) as shown in figure 6. The best supply is selected to match for demand.

Seller	Item	price
A	256 mb ram	20
B	256 mb ram	21
C	256 mb ram	19.5
D	256 mb ram	20
E	256 mb ram	22

buy
Ram
256 mb
$21

Matching

Fig. 6. Many to one matching

The benefit model which can select the best supply is based on price. The lower price of total sub-demand, the high benefit gets. The benefit model is:

$$B_i = \begin{cases} Min(\sum_{j \subseteq sub-demand\, of\, i} B_{ij} \text{ , price of supply pair of } D_i) & \text{, if } D_i \text{ can decompose to sub-demands } D_{ij} \\ \text{price of supply pair of } D_i & \text{, if } D_i \text{ can't decompose to sub-demands} \end{cases} \quad (1)$$

where B_i is the benefit of demand D_i and B_{ij} is the benefit of sub-demand D_{ij} of demand D_i.

Type3: Different supplies must be combined to satisfy demand in this type as show in figure7. Demand must be decomposed to sub-demands. For each sub-demand, a supply is matched to satisfy it or decompose it to more sub-demands. We propose a Divide and Conquer Matching algorithm to achieve the goal. The algorithm is shown in figure 8.

seller	item	specification
B	Memory	256 mb
A	CPU	P-3 800 hz
B	HD	30 G

Matching

Item	specification
CPU	P-3 700 above
Memory	256 mb
HD	30 G above

Fig. 7. Different supplies must be combined to satisfy demand

Function Match(D_0)
    For demand D_0 do begin
        If D_0 can be divided into sub-demands according to domain ontology then
            Divide D_0 into sub-demands $\{ D_{01}, D_{02}, ..., D_{0k} \}$
            $B_0 = 0$
            For each sub-demands D_{0k} do begin
                $B_{0k} = $ Match(D_{0k})
                $B_0 = B_0 + B_{0k}$
            End For
        End If
        If D_0 can be matched with supply without divided D_0 then
            $C_0 = $ price of matching supply with D_0
        Else
            $C_0 = 0$
        End If
    End For
    If $C_0 <> 0$ then
        Return Min(B_0, C_0)
    Else
        Return B_0
    End If
End Function

Fig. 8. Divide and conquer algorithm for supply and demand matching

After supplies and demands are matched, the negotiation [2] must be handled. In our architecture, we provide hiding negotiation mechanism and use auto-mail technology to achieve negotiation. The supply and demand information in internet (WWW, BBS, News… etc.) contain some static information (product, price, time, place … etc). The buy and sell templates are constructed previously. The negotiation mail is constructed by using NLP and select suitable template. Then the negotiation mechanism sends negotiation mails to suppler and user for achieving negotiation. The buying template is showed in figure 9.

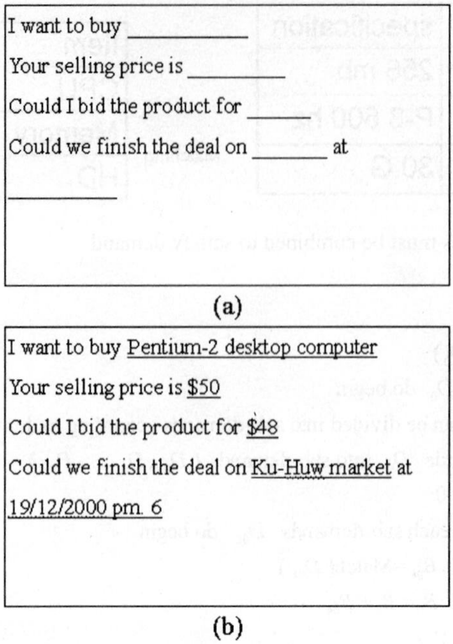

```
I want to buy _____
Your selling price is _____
Could I bid the product for _____
Could we finish the deal on _____ at


_____
```

(a)

```
I want to buy Pentium-2 desktop computer
Your selling price is $50
Could I bid the product for $48
Could we finish the deal on Ku-Huw market at
19/12/2000 pm 6
```

(b)

Fig. 9. (a) buying template (b) buying negotiation text

3.6 Hidden business mining (HBM) mechanism

In previous section, the negotiation is processed when there are demands. Sometimes, supply products need wait long time for sell or can't sell forever. In our architecture, The HBM (Hidden Business Mining) mechanism is used. The HBMM uses data mining technology to mining the hidden business and doing promotion. The association rules algorithm is used to achieve the goal [7]. The HBM mines the association rules by using history transaction log. If there is a rule "buy printer => buy printer paper", then the system retrieves the uses which had buy printer without printer paper. And the system does promotion of printer paper. The HBMM makes the process of business more complete.

4 Conclusion and Future work

In this paper, we propose a business agent architecture which covers the four process of buying behavior (Support and Demand Identification, Product Brokering, Merchant Brokering, Negotiation). The BSA captures the supply and demand information from e-commence sites automatically. SDAM uses NLP technologies to extract product and transaction information. Domain ontology is used to classify supply and demand and divide them into subsets. A benefit model is proposed to handle the matching between sub-supply and sub-demand. And a divide and conquer algorithm is proposed to handle the whole matching between supply and demand. In the negotiation process, the architecture uses NLP and template to product negotiation text and mailed through mail mechanism to handle negotiation. Finally, HBMM adopts data mining technology to achieve hidden business mining.

Acknowledgements
This research was supported by the   III Innovative and Prospective Technologies Project   of Institute for Information Industry and sponsored by MOEA , ROC

5 References

1. Bergamaschi, S., et al., "An Intelligent Approach to Information Integration," Guarino, N. (ed.) Formal Ontology in Information Systems. IOS Press 1998.
2. Chavez, A., Dreilinger, D., Guttman, R., and Maes, P. A real-life exper-iment in creating an agent marketplace. In Proceedings of the Second Inter-national Conference on the Practical Application of Intelligent Agents and Multi-Agent Technology PAAM'97 (London, U.K., Apr.). Practical Application Company, London, 1997.
3. Doorenbos, R., Etzioni, O., and Weld, D. A scalable comparison-shop-ping agent for the World-Wide Web. In Proceedings of the First Interna-tional Conference on Autonomous Agents Agents'97 (Marina del Rey, Calif., Feb. 5–8). ACM Press, N.Y., 1997, pp. 39–48
4. Guttman, R., and Maes, P. Agent-mediated integrative negotiation for retail electronic commerce. In Proceedings of the Workshop on Agent-Medi-ated Electronic Trading AMET'98 (Minneapolis, May 1998).
5. Moukas, A., Guttman, R., and Maes, P. Agent-mediated electronic com-merce: An MIT Media Laboratory perspective. In Proceedings of the Inter-national Conference on Electronic Commerce (Seoul, Korea, Apr. 6–9). ICEC, Seoul, 1998, pp. 9–15.
6. Pattie Maes, Robert H. Guttman, AND Alexandros G. Moukas. Ahents That Buy and Sell. COMMUNICATIONS OF THE ACM March 1999/vol. 42, No. 3.
7. R. Srikant, Q.Vu, and R. Agrawal. Mining association rules with item constraints. In Proc. 1997 Int. Conf. Knowledge Discovery and Data Mining (KDD'97), Newport Beach, CA, Aug.1997.

On the Use of Longitudinal Data Techniques for Modeling the Behavior of a Complex System

Xaro Benavent [1], Francisco Vegara[1], Juan Domingo[1], and Guillermo Ayala[2]

[1] Instituto de Robótica, Univ. de Valencia,
Polígono de la Coma, s/n, Aptdo. 22085,
46071-Paterna (Spain),
{Xaro.Benavent,Francisco.Vegara,Juan.Domingo}@uv.es,
[2] Dpto. de Estadística e Investigación Operativa, Univ. de Valencia
Dr. Moliner, 50.
46100-Burjasot (Spain),
Guillermo.Ayala@uv.es

Abstract. This work presents the use of longitudinal data analysis techniques to fit the accelerations of a real car in terms of some previous throttle pedal measurements and of the current time. Different repetitions of the same driving maneuvers have been observed in a real car, which constitute the data used to learn the model. The natural statistical framework to analyze these data is to consider it as a particular case of longitudinal data.

Different fits are given and tested as a first step in order to explain the relationship between variables describing the control of the car by the driver and the final variables describing the movement of the vehicle.

Results show that the approach can be valid in those cases in which a temporal implicit dependency can be assumed and in which several realizations of the experiment in similar conditions are available; in such cases an analytical model of the system can be obtained which has the ability to generalize, i.e. to show a robust behavior when faced to input data not used in the model construction phase.

1   Introduction

Many natural phenomena or artificial devices can be modeled as input/output systems, i.e. as entities that take signals from their environment by means of direct acquisition from sensors or by explicit data introduction and that generate outputs that change the state of the system itself and/or of its environment. The problem of modeling such systems with a digital computer consists on obtaining a computing device connectable to sensors if appropriate, or at least an algorithm, that accounts for the (a priori unknown) relationship between inputs and outputs.

In general terms, two main approaches can be used to model an unknown system: the symbolic and the non-symbolic way. The first one use knowledge of the internal behavior of the system that is expressed in algorithmic or mathematical terms; in the case of dynamic systems, differential equations are the

P.M.A. Sloot et al. (Eds.): ICCS 2002, LNCS 2329, pp. 458–467, 2002.
© Springer-Verlag Berlin Heidelberg 2002

most common formulation. The second way of getting a model is the recollection of data taken along the time at regular time intervals or at predetermined instants; this approach can be divided again into parametric and non-parametric models, depending on whether the functional form of the relationship between inputs and outputs is explicitly stated or not. Linear models are an example of parametric models, since the functional form of the relationship between inputs and outputs is explicitly stated by the experimenter [4].

The approach we propose consists on the estimation of a parametric model for the system, which takes elements from both of the aforementioned possibilities: it needs data collected from the real system but it generates an algebraic expression relating inputs and outputs. The use of parametric models is a standard way to work in statistics: the experimenter proposes a reasonable model for the dependency between inputs and outputs, taking into account the statistical variability, which involves the input variables and some unknown parameters and then optimal values (under a given criterion) for these parameters are estimated. Models with increasing complexity or with special assumptions can be proposed if the data come from a physical phenomenon that generates values along the time (time series) and also if the data represent several realizations of the same stochastic process (i.e. several experiments). When both conditions are met, the appropriate theoretical framework to deal with that case is the analysis of longitudinal data, as described in [3]. This work presents an application of such formulation to the restricted modeling of a real car, considering the usual controls of the car (throttle and brake pedals, handwheel, etc.) as inputs and the accelerations measured by appropriate sensing devices on board of the car as outputs. The final purpose of this model is its usage embedded into the software to control a driving simulator; the simulator contains a cockpit resembling that of a real car, including a handwheel, throttle and brake pedals and gearbox lever. All these controls are appropriately sensorized and they are intended to be the inputs of the model. The simulator has been provided, too with solid state accelerometers that allow the measurement of the acceleration that the driver feels. The outputs of the model should be such accelerations, and the platform will have to be moved so as to produce at least scaled versions of them.

This work is organized as follows: section 2 describes the system and the problem specifying the inputs and outputs; section 3 makes explicit the assumed model and the formulation used and shows how the model has been applied to our problem; section 4 details the experiments and results, and finally section 5 states the conclusions and the proposed future work.

2   Description of the system

Driving simulators are becoming popular as a way to evaluate different security and mechanical issues of real cars, and also to test the psychophysical behavior of typical drivers and to evaluate their reactions in different driving situations [6]. Movement is usually reproduced by means of mechanical platforms connected to electrical or electro-neumatic actuators. To make the simulation platform behave in a way similar to a real car, a model of the car itself is needed; this means to know its behavior in terms of the generated movement and acceleration when the driver interacts with the controls (throttle, brake pedal, handwheel, etc.) in the usual way. Up to now, the prevalent approach has been to model the behavior of the vehicle by means of Newton's laws of mechanics, taking into account the known forces and torques ([5]). Differential equations can be programmed so that their output is the state of movement of the vehicle at each moment, which we will have to reproduce by moving the platform appropriately. The set of differential equations use to be quite complex, several simplifications are normally assumed and the knowledge of several parameters is needed, amongst others suspension stiffness, friction coefficient between the vehicle and the floor and others not easily measurable.

On the contrary, our solution starts by experimenting several typical driving maneuvers in a real car that has been sensorized. This means that several sensors and recording devices have been connected to the main controls to get a record of the input values, and also a triaxial accelerometer was installed on board of the car, so that the accelerations along the x, y and z axis of a given coordinate system are measured and recorded. They are the outputs and the model will have to calculate them from the current and former values of the inputs. In this case is clear that the measured acceleration will depend not only on the current readings of the controls, but also on the past history of the car, which has determined its present state, and also on the time elapsed from the beginning of the maneuver. Due to this, the outputs can be considered as time series. It is common in the context of time series the use of ARMA (Autoregressive Moving Average) or ARIMA (Autoregressive Integrated Moving Average) models [9], but this is not appropriate in our case, since the model we intend to determine has to be obtained from several experiments performed in different conditions (throttle pressed up to different levels, etc.) in order to be able to generalize, so it is said, to account for all the different situations that were used to learn, and also for similar ones, since the behavior of the driver when the model is working on the simulator will be similar, but not identical in each occasion. All these reasons have motivated the choice of models based on the formulation of longitudinal data, since it has theoretical properties appropriate to accomplish these requirements.

3   Linear models for longitudinal data

Longitudinal data consist on several continuous or categorical responses taken from one or more experimental units (different repetitions of the same experiment performed independently). The analysis of longitudinal data is closely related with that of time series but it presents two main differences. First, the different time series are considered as a sample of a population. Second, the interest in the correlation structure of longitudinal data is usually minor, but covariance must be adjusted in the process of data analysis to ensure valid inferences on the structure of the mean of the response.

The main subject of the paper is to approximate the observed acceleration of a car from the previous throttle pedal lectures and the current time. All data are observed at the time instants $(t_1, \ldots, t_n)$. Let y_j and x_j be the acceleration and p explanatory variables (some previous throttle pedal lectures, the time and different functions of them) at the time t_j. The simplest model and the classical option is to assume that

$$y_j = \sum_{k=1}^{p} x_{jk}\beta_k + \epsilon_j \qquad (1)$$

where $\boldsymbol{\beta} = (\beta_1, \ldots, \beta_p)$ are unknown parameters and ϵ_j, the experimental error, a normal (or Gaussian) random variable with zero mean and variance σ^2. It is denoted usually as $\epsilon_j \sim N(0, \sigma^2)$ (from now on, $\sim$ means *distributed as*). Furthermore, the different ϵ_j's are assumed independent and identically distributed with variance σ^2. Based on this assumptions, linear model theory permit us to estimate the unknown parameters and to predict the accelerations. However, this approach can not be applied to our case since the different experimental errors are not independent. Figure 1 shows the autocorrelation function of the errors plotted against the time lag when the time and three previous throttle pedal lectures are used to fit the model. Real data of a straight-line acceleration maneuver was used for this example.

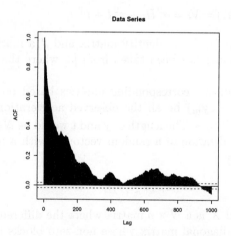

Fig. 1. Autocorrelation function of the residuals observed for a straight-line acceleration maneuver with linear fit (see text for details)

This serial correlation must be taken into account to estimate the parameters $\boldsymbol{\beta}$. We must be aware that have longitudinal data. Different individuals (the different proofs of the same maneuver) are measured repeatedly through time. The natural experimental unit with longitudinal data is the vector $\mathbf{y} = (y_1, \ldots, y_n)$. It is assumed that the same operation is repeated m times, and $\mathbf{y}_i = (y_{i1}, \ldots, y_{in_i})$

are the accelerations observed in the i-th proof at the times $\mathbf{t}_i = (t_{i1}, \ldots, t_{in_i})$ where n_i (which is the total number of times) can be possibly different between different proofs. It is assumed independence between different repetitions but no within a given maneuver observed. This basic hypothesis has to be assumed by the simpler linear model given in equation 1.

It will be assumed that the different $\mathbf{y}_i$'s can be considered as independent realizations of a random vector $\mathbf{Y}_i$ with a multivariate normal (or Gaussian) distribution i.e.

$$\mathbf{Y}_i = X_i \boldsymbol{\beta}' + \boldsymbol{\epsilon}_i \tag{2}$$

being $\boldsymbol{\epsilon} = (\epsilon_{i1}, \ldots, \epsilon_{in_i})$ and

$$\epsilon_{ij} = U_i + W_i(t_{ij}) + Z_{ij}, \tag{3}$$

where U_i is a normal random variable (independent for different i's) with zero mean and variance ν^2; Z_{ij} is another normal random variable with zero mean and variance τ^2 (independent for different i's and j's) and finally W_i is a stationary Gaussian process such that $W_i(t_{ij}) \sim N(0, \sigma^2)$ and whose autocorrelation function is ρ. Two different autocorrelation functions will be used in this work: $\rho(u) = \exp\{-\phi \mid u \mid\}$ and the Gaussian $\rho(u) = \exp\{-\phi u^2\}$. For different i's, it will be assumed that the different realizations of the Gaussian process W_i are independent. Under this model, it can be easily verified that

$$var(Y_i) = V_i = \sigma^2 H_i + \tau^2 I + \nu^2 J, \tag{4}$$

being $H_i(j, k) = \rho(\mid t_{ij} - t_{ik} \mid)$, I the identity matrix and J a matrix of ones. The notation used in the paper has been taken from [3] where the reader can find more details.

The proposed model and the corresponding analysis based on it uses all data jointly. Let $\mathbf{y} = (\mathbf{y}_1, \ldots, \mathbf{y}_m)$ be all the observed accelerations and $\mathbf{t} = (\mathbf{t}_1, \ldots, \mathbf{t}_m)$, the whole set of times. The length of $\mathbf{y}$ and $\mathbf{t}$ would be $N = \sum_{i=1}^{N} n_i$. It is assumed that $\mathbf{y}$ is a realization of a random vector $\mathbf{Y}$ with a multivariate normal distribution given by

$$Y \sim N(X\boldsymbol{\beta}, V(t, \theta)), \tag{5}$$

where $\theta = (\sigma^2, \phi, \tau^2, \nu^2)$ and X is a $N \times p$ matrix where the different X_i's have been stacked. V is a block-diagonal matrix whose non-zero blocks are the V_i's previously considered i.e.

$$\mathbf{Y} = X\boldsymbol{\beta} + \boldsymbol{\epsilon}. \tag{6}$$

The log-likelihood of the observed data $\mathbf{y}$ is then

$$\mathbf{L}(\boldsymbol{\beta}, \theta) = -\frac{1}{2}\{nm \log(\sigma^2) + \sum_{i=1}^{m} \log(\mid V_i \mid)\} + \frac{1}{\sigma^2}(\mathbf{y} - X\boldsymbol{\beta}')'V^{-1}(\mathbf{y} - X\boldsymbol{\beta}'). \tag{7}$$

The parameters $(\boldsymbol{\beta}, \theta)$ will be estimated by using the maximum likelihood estimators (MLE) i.e. the values that give the maximum of the likelihood given by the former equation. The software package *Oswald*, a library of S-PLUS, has been used.[1]

The global behavior of a car can be seen as a juxtaposition of different behaviors depending on the maneuver: acceleration, braking, steering, etc. In this work we will deal exclusively with the straight-line acceleration maneuver, but the behavior of the model could be extended to other driving maneuver by identifying them in the same way as done with the aforementioned maneuver. This was the approach adopted in [1] to build a complete car simulator.

The inputs to the system are the signals that have been obtained from the most important controls of the car: throttle position, brake pedal force, angle of the handwheel and a time vector that will be generated from the beginning of the maneuver. The outputs of the system are the accelerations in the three axes X, Y and Z depending on the maneuver we want to learn. One of the inputs to the system, the time vector, does not refer to any of the controls of the car. This input has been fed into the system because it is important to know if we have pushed the brake pedal or the throttle pedal up to any position (for example, a 50% of its final position) at the beginning of the development of the maneuver or later; indeed, the acceleration of the car will be different in each case.

After doing several tests to determine how many and which previous instants contained relevant information for the dynamic car's modeling, it was decided to use 50 and 100 previous samples from the current instants, for each of the input signals. Given our sampling time, which is $T = 0.01s$, this means half and one second before present.

Since identification was done for each of the driving maneuvers separately, not necessarily all available input had to be used for each of them. For example, straight line acceleration maneuver does not need the brake pedal and handwheel angle, since they are both null in this case. This type of heuristic knowledge may help in the reduction of the dimensionality of the input space.

In order to model the straight-line acceleration maneuver, we need data which are sufficiently representative of the general behavior of the maneuver. In this case we had available data from different runs done for various conditions: throttle pedal pushed at 10%, 30% and 75% of its total allowed run, which are obviously different. Following the notation used by [7] where the concept of *experimental unit* represents the repetition of each of the tests, each experimental unit contains one of the three available data banks for the acceleration maneuver, each of which refers to each of the three different final positions of the throttle pedal.

4   Results

Several experiments have been performed by using the models of equations 2 and 3 with different sets of explanatory variables and assuming different autocorrelation structure. The results are shown in table 1. This table displays the

---

[1] *Oswald* is a copyright ©1997 David M. Smith and Lancaster University

MLE of the parameters $\hat{\nu}^2, \hat{\tau}^2$ and $\hat{\phi}^2$, and the maximum log-likelihood reached. The usual S-PLUS notation for the formulae is used (see [8]).

Table 1. Results with the different models fitted showing the value of the maximum log-likelihood (column headed L) besides the MLE: $\hat{\nu}^2$, $\hat{\sigma}^2$, $\hat{\tau}^2$ and $\hat{\phi}^2$

Num.	Model	$\hat{\nu}^2$	$\hat{\sigma}^2$	$\hat{\tau}^2$	$\hat{\phi}^2$	L
1	$y \sim x_1 + x_2 + x_3$ $\rho(u) = exp(-\phi \cdot |u|^2)$	$9.1e^{-9}$	0.223	0.0033	0.0385	-770.9
2	$y \sim x_1 + x_2 + x_3$ $\rho(u) = exp(-\phi \cdot |u|)$	0.517	0.659	$3.79e^{-10}$	0.009	-729.1
3	$y \sim x_1 * x_2 * x_3$ $\rho(u) = exp(-\phi \cdot |u|)$	0.822	0.938	$34.59e^{-11}$	0.0058	-700.8
4	$y \sim poly(x_1, 2) + poly(x_2, 2)$ $+poly(x_3, 2)$ $\rho(u) = exp(-\phi \cdot |u|)$	0.584	0.711	$9.76e^{-11}$	0.0078	-709.1
5	$y \sim poly(x_1, 2) + poly(x_2, 2)$ $+poly(x_3, 2)$ $\rho(u) = exp(-\phi \cdot |u|^2)$	$3.11e^{-8}$	0.2197	0.0032	0.0386	-766.8
6	$y \sim poly(x_1, 2) + poly(x_2, 2)$ $+poly(x_3, 2) + poly(time, 2)$ $\rho(u) = exp(-\phi \cdot |u|)$	0.133	0.263	$3.49e^{-10}$	0.021	-701.9
7	$y \sim poly(x_1, 2) + poly(x_2, 2)$ $+poly(x_3, 2) + tex(t1) + tex(t2)$ $+tex(t3)$ $\rho(u) = exp(-\phi \cdot |u|^2)$	0.089	0.0845	0.0031	0.0501	-703.7
8	$y \sim poly(x_1, 2) + poly(x_2, 2)$ $+poly(x_3, 2) + tex(t1) + tex(t2)$ $+tex(t3)$ $\rho(u) = exp(-\phi \cdot |u|)$	0.042	0.186	$3.94e^{-10}$	0.0298	-697.7

For instance, models 1 and 2 (first and second rows of the table) are denoted by $y \sim x_1 + x_2 + x_3$, which means that the longitudinal data model uses as explanatory variables three throttle pedal lectures in the current and two former instants. The second row indicates the auto-correlation function used. The third model denoted by $y \sim x_1 * x_2 * x_3$ uses as explanatory variables the original variables and all the cross-products of the variables x_1, x_2 and x_3 i.e. $x_1 x_2$, $x_1 x_3$ and so on. The expression $poly(x_1, 2)$ means that the original, x_1 and x_1^2 have been used as explanatory variables. $tex(t1)$ denotes a vector in which time increases linearly; $tex(t2)$ is a vector in which time increases quadratically in the first time interval, up to a certain time t_0 and $tex(t3)$ is a vector in which time increases linearly only from t_0 up to the end of the maneuver. In all the cases t_0 was chosen as 50 sampling periods.

Let us look at the model 4 with more detail. This model uses polynomials of order up to 2 for each of the explanatory variables. The likelihood observed is -709.1 i.e. it is the fifth better fit from this point of view. Note that the parameters estimated are based on three experimental units. The plots in figure 2 show the obtained adjusted data that the model gives for each experimental unit overlaid to the real data. It is clear that the results are not good, since acceleration goes out of the desired range for the real outputs (there are even negative accelerations), but the generalization is acceptable. Nevertheless, the result would be unusable in a simulator since negative accelerations would be opposite to the type of sensation that is expected by the driver.

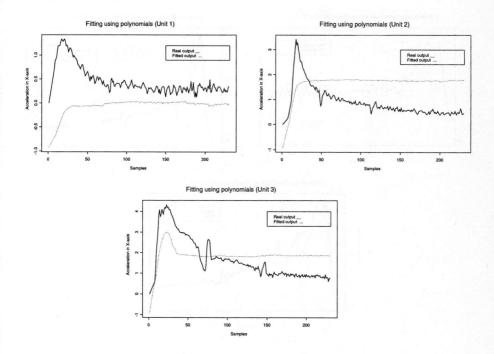

Fig. 2. Comparison between real and fitted values for experimental units 1, 2 and 3 respectively. A model with polynomials of order 2 (experiment 4) has been used.

Model 6 is the first one in which time is used explicitly. Experiments 7 and 8 use the tex function with the two user-defined vectors of time, t_1 and t_2, explained above. These vectors have been created in this way so that a quadratic fitting can be done for the first part of the plot and a linear fitting for the rest of the maneuver. If we observe the behavior of the acceleration in this type of maneuver (output to be fitted) it is clear that the creation of these two vectors is a sensible option. Table 1 shows the obtained results for this type of models; it can be seen that they provide better fits that every previous model giving a maximum log-likelihood of -697.7. Plots in figure 3 show the comparison between the real signal and fitted data. Results are now clearly better; they could be used in our practical case, though not with a completely real sensation. The model could be further refined to get better results by changing the variables in the correlation function and testing other combinations in the formula for the linear model.

Obviously, more explanatory variables and more complex relations provide a better fit. However, they remains two open questions: the variables selection (how many previous pedal lectures and in which times to observe) and a deeper study of the functional relation between the explanatory variables and the observed acceleration.

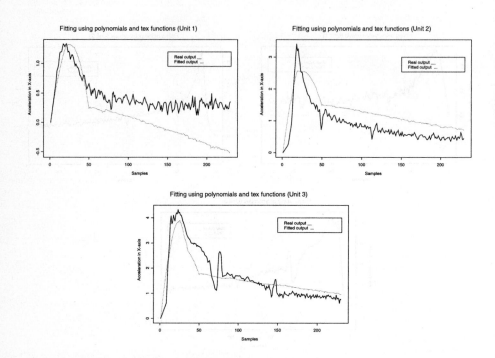

Fig. 3. Comparison between real and fitted values using the formula 8 of table 1 for experimental units 1, 2 and 3 respectively.

5   Conclusions

An important objective of this work was to determine to what extent longitudinal data models can be considered as a valid alternative for the modeling of the dynamic behavior of a real system, or at least of a restricted part of it. The performed experiments shows that, in our case, longitudinal data analysis can be a feasible approach only for sufficiently sophisticated models. Explicit usage of time as an input introduces valuable information, but this can only be done if the initial instant of time (the instant in which the data start to arrive) is known. Moreover, controlling the specific type of dependency of the output with respect to time, and making it different for different time intervals of the experiment increases the goodness of the fit, as can be seen in the second experiment, in which this has been done. On the contrary, the choice of the autocorrelation function of the errors (exponential or Gaussian) does not appear to have a decisive influence, at least for this case.

With respect to the adequacy of the model for its intended purposes, the fit obtained in the second experiment can be considered as sufficient for its use in our driving simulator, given the limited ability of people to perceive absolute values of accelerations or differences of them. Other dynamic systems may require a better fit, but this could probably be achieved by using models involving more complex dependencies between inputs and output. On the other hand, the generalization capabilities of the model are also appropriate for this case, and this is the main reason to choose longitudinal data with preference to simpler linear models for problems involving the control of dynamic systems whose behavior is similar, but strictly different for each trial.

An important practical aspect is the detection of the point in time in which the dependency of the output with time changes, for instance from linear to quadratic, and which roughly corresponds in our case to a change in the automatic gearbox. This could be done by using a different model involving information such as engine speed and remains as a future work.

References

1. X. Benavent. *Modelizacion y control de sistemas no lineales utilizando redes neuronales y modelos lineales. Aplicacion al control de una plataforma de simulacion*. PhD thesis, Universitat de Valencia, 2001.
2. J. P. Chrstos and P. A. Grygier. Experimental testing of a 1994 Ford Taurus for NADSdyna validation. Technical Report SAE Paper 970563, Society of Automotive Engineers, Inc., 1997.
3. P.J. Diggle, K.-Y. Liang, and S.L. Zeger. *Analysis of Longitudinal Data*. Oxford Science Publications, 1994.
4. N.R. Draper and H. Smith. *Applied Regression analysis*. Princeton University Press, 1998.
5. T. D. Gillespie. *Fundamentals of Vehicle Dynamics*. Society of Automotive Engineers, Inc., 1992.
6. S. Nordmark. Driving simulators trends and experiences. In *Driving Simulation Conference. Real Time System 94*, Paris. France, January 1994.
7. D.M. Smith and P.J. Diggle. Oswald: Object-oriented software for the analysis of longitudinal data in S. Technical Report MA94/95, Lancaster University Department of Mathematics and Statistics, 1994.
8. W.N. Venables and B.D. Ripley. *Modern Applied Statistics with S-PLUS*. Springer, third edition edition, 1999.
9. P.J. Brockwell and R.A. Davis. *Time Series: Theory and Methods*. Springer Series in Statistics, Springer, 2nd Edition, 1991.

Problem of Inconsistent and Contradictory Judgements in Pairwise Comparison Method in Sense of AHP

Miroslaw Kwiesielewicz [1] and Ewa van Uden [2]

[1] Technical University of Gdansk
Faculty of Electrical and Control Engineering
Narutowicza 11/12, Gdansk, POLAND, 80-952
Phone: +48 58 3472124, Fax: +48 58 3471802
mkwies@ely.pg.gda.pl

[2] Technical University of Gdansk
Faculty of Electrical and Control Engineering
Narutowicza 11/12, Gdansk, POLAND, 80-952
Phone: +48 58 3471225, Fax: +48 58 3471802
cotynato@ely.pg.gda.pl

Abstract. The aim of this paper is to show the relationship between inconsistent and contradictory matrices of data obtained as a result of the pairwise comparison of factors in the sense of the Analytic Hierarchy Process. The consistency check is performed to ensure that judgements are neither random nor illogical. This paper shows that even if a matrix will pass a consistency test successfully, it can be contradictory. Moreover an algorithm of checking contradictions is proposed.

1 Introduction

The methodology of AHP decomposes a complex decision problem into elemental issues to create a hierarchical model. The decision variables (factors) are situated at the lowest level. The elements with similar nature are grouped at the same interim levels and the overall goal at the highest level. At each level of the hierarchy, paired comparison judgements are obtained among stimuli. After normalisation of the results from each level, the aggregation is made.

The main aim of the pairwise comparison method in the AHP is to make a ranking of n given factors or alternatives. To compare the factors often a scale $\{1/9, 1/8, \ldots, 1/2, 1, 2, \ldots, 8, 9\}$ introduced by Saaty [6] is used. The numbers from the scale above are displayed in the matrix of judgement $R = \lfloor r_{ij} \rfloor_{nxn}$ expressing a relative significance of the factors F_i and F_j.

Matrix R is said to be reciprocal if $r_{ij} = 1/r_{ji}$ for all $i, j = 1, 2, \ldots, n$ and $r_{ij} > 0$.

The matrix R is said to be consistent if $r_{ij} r_{jk} = r_{ik}$ for all $i, j, k = 1, 2, \ldots, n$.

There are some reasons of having inconsistent data, namely the mistakes of a person providing his or her opinions or using 9-point semantic scale [5]. To show this phenomenon assume the following exemplary judgement matrix:

P.M.A. Sloot et al. (Eds.): ICCS 2002, LNCS 2329, pp. 468–473, 2002.
© Springer-Verlag Berlin Heidelberg 2002

$$R = \begin{bmatrix} 1 & 3 & 5 \\ 1/3 & 1 & - \\ 1/5 & - & 1 \end{bmatrix}.$$

To keep the consistency the numbers $R(2,3) = 5/3$ and $R(3,2) = 3/5$ should be placed in the empty entries respectively. These numbers however do not belong to assumed scale! To avoid such a situation, one can use a scale containing only the multiple of one number e.g. 2. A scale constructed in this way would take a form: $\{...1/8, 1/4, 1/2, 1, 2, 4, 8...\}$. A quantity of numbers appearing on the scale would depend on complexity of the problem. A lot of authors discuss the problems of scale e.g. Finan and Hurley [3]. A reasonable requirement is to make inconsistencies as few as possible. The most often approach is to calculate a consistency index according to the formula [6,7]:

$$C.I. = \frac{\lambda_{max} - n}{n - 1} \qquad (1)$$

where λ_{max} denotes the maximal eigenvalue of matrix R. When matrix R is consistent then $\lambda_{max} = n$ and $C.I. = 0$.

It's obvious that $C.I.$ should be as close as possible to zero, but till now it is not said precisely what values are permissible. The standard procedure allows to accept a set of judgements when $C.I.$ is not bigger than $1/10$ of the mean consistency index of randomly generated matrices. The mean consistency for indices matrices of size 3x3 to 9x9, proposed by Saaty and Vargas [7] are shown in Table 1.

Table 1. Mean consistency index for randomly generated matrices.

Matrix size (n)	Mean consistency C.I.
3	0.5381
4	0.8832
5	1.1045
6	1.2525
7	1.3334
8	1.4217
9	1.4457

Karpetrovic and Rosenbloom [4] give some examples when the comparisons are neither random nor illogical but they fail the consistency test.

This paper separates a subset of so-called contradictory matrices from inconsistent matrices and shows that such matrices may pass the consistency test and data may be acceptable. At the end an algorithm for checking contradictions is presented and some numerical examples are included.

2   Basic Concepts

Let us introduce the following definition of contradictory judgement matrix, based on the transitivity rule.

Definition 1

The matrix R is called contradictory if there exists $i, j, k : 1, 2, \ldots, n$ such that any of the detailed below cases hold:

$$r_{ij} > 1 \ \wedge \ r_{ik} < 1 \ \wedge \ r_{jk} > 1 \qquad (2)$$

$$r_{ij} < 1 \ \wedge \ r_{ik} > 1 \ \wedge \ r_{jk} < 1 \qquad (3)$$

$$r_{ij} = 1 \ \wedge \ r_{ik} > 1 \ \wedge \ r_{jk} < 1 \qquad (4)$$

$$r_{ij} = 1 \ \wedge \ r_{ik} < 1 \ \wedge \ r_{jk} > 1 \qquad (5)$$

$$r_{ij} = 1 \ \wedge \ r_{ik} = 1 \ \wedge \ r_{jk} < 1 \qquad (6)$$

$$r_{ij} = 1 \ \wedge \ r_{ik} = 1 \ \wedge \ r_{jk} > 1. \qquad (7)$$

Denoting: $\gg$ - "better", $\ll$ -"worst", $=$ - "equal", it is equivalent to contradictory judgements concerning factors F_i, F_j, F_k :

$$F_i \gg F_j \ and \ F_i \ll F_k \ and \ F_k \ll F_j \qquad (8)$$

$$F_i \ll F_j \ and \ F_i \gg F_k \ and \ F_k \ll F_j \qquad (9)$$

$$F_i = F_j \ and \ F_i \gg F_k \ and \ F_k \ll F_j \qquad (10)$$

$$F_i = F_j \ and \ F_i \ll F_k \ and \ F_k \gg F_j \qquad (11)$$

$$F_i = F_j \ and \ F_i = F_k \ and \ F_k \ll F_j \qquad (12)$$

$$F_i = F_j \ and \ F_i = F_k \ and \ F_k \gg F_j. \qquad (13)$$

It's easy to notice that in this case a transitivity rule does not hold.

An example of a matrix with acceptable *C.I.*, but which is contradictory is:

$$R = \begin{bmatrix} 1 & 2 & 1/2 & 2 & 1/2 & 2 & 1/2 & 2 \\ 1/2 & 1 & 4 & 1 & 1/4 & 1 & 1/4 & 1 \\ 2 & 1/4 & 1 & 4 & 1 & 4 & 1 & 4 \\ 1/2 & 1 & 1/4 & 1 & 1/4 & 1 & 1/4 & 1 \\ 2 & 4 & 1 & 4 & 1 & 4 & 1 & 4 \\ 1/2 & 1 & 1/4 & 1 & 1/4 & 1 & 1/4 & 1 \\ 2 & 4 & 1 & 4 & 1 & 4 & 1 & 4 \\ 1/2 & 1 & 1/4 & 1 & 1/4 & 1 & 1/4 & 1 \end{bmatrix}$$

The consistency index for this matrix is equal to $C.I. = 0.07252$. Moreover, note that when the entries $R(2,3)$ and $R(3,2)$ are replaced, the fully consistent matrix in the sense of the consistency index $C.I.$ (1) is obtained. Analysing this matrix one may notice that $R(1,2) = 2$, $R(1,3) = 1/2$. So, one may say that factor F_2 is better then F_1, factor F_1 is better then F_3 and it would be natural to expect that F_2 were better than F_3, but $R(2,3) = 4$, what means that F_3 is better than F_2.

Studying this example another remark may be made. Intuitively one could expect that contradictory is the extreme form of inconsistency, but it is not true. This example shows that a matrix may be very close to perfect consistency, but it's possible to find some contradicts in the judgements. Graphically the structure of comparison matrices may be presented as shown in Fig.1.

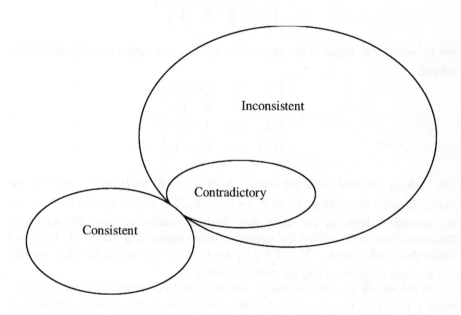

Fig. 1. Structure of the set of comparison matrice

The last task undertook by authors is to create an effective algorithm testing the contradictory. The proposal may be formulated in the following way:

FOR ($i, j, k = 1$ AND $i \neq j \neq k$) TO n DO
IF $\log r_{ij} \cdot \log r_{ik} \leq 0$ AND $\log r_{ik} \cdot \log r_{jk} < 0$
THEN STOP - CONTRADICTORY
ELSE

IF $\log r_{ij} = 0$ AND $\log r_{ik} = 0$ AND $\log r_{jk} \neq 0$

THEN STOP - CONTRADICTORY

ELSE

OK.

3 Calculation example

Let us assume that an expert intended to create the following matrix of judgements comparing four factors F_1, F_2, F_3, F_4:

$$A = \begin{bmatrix} 1 & 4 & 8 & 2 \\ 1/4 & 1 & 2 & 1/8 \\ 1/8 & 1/2 & 1 & 1/9 \\ 1/2 & 8 & 9 & 1 \end{bmatrix},$$

but by mistake he replaced the entries a_{13} and a_{31}. In such a case his matrix has a form:

$$B = \begin{bmatrix} 1 & 4 & 1/8 & 2 \\ 1/4 & 1 & 2 & 1/8 \\ 8 & 1/2 & 1 & 1/9 \\ 1/2 & 8 & 9 & 1 \end{bmatrix}.$$

The ranking obtained from the matrix A is $F_1 \gg F_4 \gg F_2 \gg F_3$, whereas the matrix B gives the ranking: $F_4 \gg F_1 \gg F_3 \gg F_2$. It is easy to see that the "cost" of his mistake is high, in the sense that obtained ranking is quite different from the correct one. If he applied the algorithm from chapter 2 to matrix B, he would notice that $\log b_{12} \cdot \log b_{13} < 0$ and $\log b_{13} \cdot \log b_{23} < 0$, what would be a hint for him that he made a mistake filling the entries of matrix B.

In the matrix B one can observe the situation from definition 1, formulae (2), namely $b_{12} > 1 \land b_{13} < 1 \land b_{23} > 1$. The consequence of the opinions expressed in matrix B is that factor F_1 is better than F_2, the factor F_2 is better than F_3, but the factor F_1 is worst than F_3, what obviously gives a contradiction. The algorithm created in the previous chapter allows not only to state but also localising the mistakes that can appear in judgements.

4 Conclusions

In this paper a new approach to consistency, inconsistency and contradictory of judgement matrices obtained as a result of the pairwise comparisons in the AHP was introduced. The method can be easily applied for the case with many experts - it is logical to check their judgements independently and to exclude these from them, who provided contradictory opinions.

The approach presented can be extended for fuzzy approaches to the problem considered, [1,2], i.e. for the fuzzy case, contradictory in the opinion of authors, should be checked at least for the modal values.

It is necessary to stress that the consistency test is not sufficient for statement of a fact whether a set of judgements should be accepted or rejected. Therefore an additional contradictory test should be made. Without such a test, it is possible to obtain absolutely arbitrary ranking as it was shown in calculation example. Moreover, the algorithm allows localising the contradictions in judgements, so they can be easily corrected.

References

1. Buckley, J.J.: Fuzzy Hierarchical Analysis. Fuzzy Sets and Systems, Vol. 17 (1985) 233-247.
2. Buckley, J.J.: Fuzzy Hierarchical Analysis Revisited. Proc. Eight International Fuzzy Systems Association World Congress, 1999 August 17-20, Taipei, Taiwan ROC.
3. Finan , J.S., Hurley, W.J.: Transitive Calibration of the AHP Verbal Scale. European Journal of Operational Research, Vol. 112 (1999) 367-372.
4. Karpetrovic, S., Rosenbloom, E.S.: A Quality Control Approach to Consistency Paradoxes in AHP. European Journal of Operational Research, Vol. 119 (1999) 704-718.
5. Murphy, C.K.: Limits on the Analytic Hierarchy Process from its Consistency Index. European Journal of Operational Research, Vol. 65 (1993) 138-139.
6. Saaty T.L. The Analytic Hierarchy Process, Mac Grew Hill (1980)
7. Saaty, T., Vargas, L.: Comparison of Eigenvalue, Logarithmic Least Squares and Least Squares Methods in Estimating Ratios. Mathematical Modelling, Vol. 5 (1984) 309-324.

An Integration Platform for Metacomputing Applications

Toan Nguyen and Christine Plumejeaud

INRIA Rhône-Alpes
655, Avenue de l'Europe
Montbonnot, F-38334 Saint Ismier Cedex
Toan.Nguyen@inrialpes.fr

Abstract. Simulation and optimisation applications involve a large variety of codes that result in high CPU loads on existing computer systems. Advances in both hardware and software, including massively parallel computers, PC-clusters and parallel programming languages somewhat alleviate current performance penalties. It is the claim of this paper that formal specification techniques, together with distributed integration platforms, provide a sound and efficient support for high performance distributed computing in metacomputing environments. Formal specifications provide rigourous and provable approaches for complex applications definition and configuration, while distributed integration platforms provide standardised deployment and execution environments for coupling heterogeneous codes in problem-solving environments supporting multi-discipline applications.

1. Introduction

Simulation and optimisation applications, e.g., as used for digital mockups in aerospace design, and for numerical propulsion systems simulation [5], raise important challenges to the computer science community [9, 12]. They require powerful and sophisticated computing environments, e.g., parallel computers, PC-clusters. They also require efficient communication software, e.g., message passing protocols. Although these problems are not new, they have given rise to specific solutions that have become de-facto standard, e.g., MPI and PVM.

Simultaneously, computer science has provided interesting solutions to the problems of:

– formal specification of distributed and communicating systems
– standardized exchange and development environments for distributed applications

This paper explores the design and implementation of an integration platform which is a computerised environment dedicated to the *specification, configuration, deployment and execution* of collaborative multi-discipline applications. Such applications may

P.M.A. Sloot et al. (Eds.): ICCS 2002, LNCS 2329, pp. 474–483, 2002.

invoke a variety of software components from various disciplines that are connected together to collaborate on common projects.

It details the development of an integration platform called *CAST* (an acronym for " *Collaborative Applications Specification Tool* ") allowing the execution of test-cases for aerodynamic design in the aerospace industry. The platform is based on an high-level graphic user interface (Figure 1) allowing the formal specification of optimisation applications, based on the work of Milner's process communication algebra SCCS [8]. It is deployed on a network of workstations (NOW) and PC-clusters connected by a high-speed network called VTHD [13] and communicating through a CORBA object request broker [10]. This was developed in part for a european Esprit HPCN project, called " DECISION ": *Integrated Optimisation Strategies for Increased Engineering Design Complexity* [3].

The paper is organised as follows. Section 2 is an overview distributed integration platforms. It details the *CAST* integration platform dedicated to distributed multi-discipline applications, including example test-cases. Section 3 is a conclusion.

2. Distributed Integration Platforms

Research in operating systems has emphasised the advantages of software components for distributed computing, e.g., EJB (Enterprise Java Beans), CCM (Corba Components Model) [1]. They allow the definition, deployment and execution of distributed applications formed by various heterogenous software modules which were not necessarily designed to cooperate. Further, they can reside on heterogeneous platforms, provided that they communicate through a common hardware or software medium.

Hence, the idiosyncracies of the data and module distribution is left to the software developers. This is in contrast with dedicated standards that are used for parallel programming, e.g., MPI. Here, the developers must code inside the modules the grouping and spawning of the execution processes. The actual application code is therefore mixed with distribution and parallelisation statements. This is not the case with environments where modularity and transparent distribution is supported, e.g., CORBA. Here, the users' code can be used as is, without modifications.

A particular test-case will illustrate the following sections. It consists in the shape optimisation for an airfoil in stationary aerodynamic conditions [6]. The goal is to reduce the shock-wave induced drag (Figure 2). Two algorithms are involved. One, which computes the airflow around the airfoil, the other which optimises the airfoil's shape. The entry data are the airfoil shape (RAE2612), its angle of attack (2 degrees), the airflow speed (Mach 0.84) and the number of optmisation steps (100). This problem requires several CPU hours on an entry level SUN Sparc workstation running the Solaris 2.6 operating system.

2.1 Principles

CAST is an integration platform designed to fulfill the requirements of distributed simulation and optimisation applications in complex engineering design projects.

Example applications are industrial projects in the electronics and aerospace industry, in particular concerning multi-discipline optimisation, e.g., coupled aerodynamics, structure, acoustics and electromagnetics optimisation.

The target user population includes project managers and engineers that are experts in their particular domain, i.e., electromagnetics, aerodynamics. They are well aware of computer technology, but not necessarily experts in formal process specifications, theoretical computational mathematics, nor object-oriented, CORBA and components programming.

This constrains the integration platform to provide a high-level user interface, with no exotic idiosyncrasies related to process algebras and distributed computing.

The fundamentals of the platform are therefore:
- to rely on a sound theoretical background for process specifications
- transparent and widely accepted standards for distributed computing
- transparent use of a large variety of software
- transparent use of distributed and heterogeneous computer platforms
- hign-level user interface for ease of use by end-users, that are experts in non computer science areas of interest.

The platform supports the definition, configuration, deployment and execution of multi-discipline applications distributed over NOW, LAN, WAN and PC-clusters.

They are formed by collaborating modules which may run on heterogeneous computers. The modules may be implemented using various programming languages and they dynamically exchange data.

They are synchronised by user-specified plans which include sequences, embedded and interleaved loops, and fork operators.   These synchronisation operators are components of a process algebra which is primarily based on Milner's SCCS algebra [8].

For instance, the formal specification of the HBCGA (Hybrid Binary Coded Genetic Algorithm) wing shape optimisation example detailed in Section 2 produces the following SCCS formula:

```
InitBCGA:InitHybrid:BGGA:(TRUE:(END)+FALSE:(FUN:(TRUE:(
HYBRID:(TRUE:(=>InitHybrid)+FALSE:(=>FUN)))+FALSE:(=>BG
GA))))
```

where BCGA, FUN, HYBRID and the various INIT modules are connected by the loop operator, depicted by '=>', the sequence operator ' :' and the choice operator '+'. Other SCCS operators are available, including the synchronisation of tasks '&' and the parallelisation operator '/'. The icons on the left-hand menu are used to build the application on the screen by the use: they represent the SCCS operators. The tasks are defined by the users on menu driven screens that are not depicted here. Note that this example involves two embedded loops and two interleaved loops.

The icons on the top menu line are tools to invoke various operations: opening of a file, undo/redo, save to file, etc. Predefined applications may be loaded and modified on-line or included in new applications, i.e., existing applications may be saved and reuused later to form new ones.
The SCCS formula is produced automatically by *CAST 2.0* from the application's graphic specification. It is used internally to generate the adequate distributed execution structure.

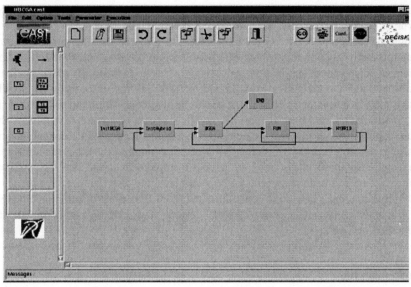

Fig. 1. *CAST: the user inerface*

The integration platform provides a high-level icon-based interface to the users which can invoke simulation, verification, configuration, deployment and execution tools. The result of the above HBCGA example is depicted by Figure 2.

Hardware configurations are uniformly supported, i.e., workstations, PC-clusters and computers connected through NOW, LAN, WAN are transparently supported through the definition of execution locations for the user modules. This is done by specifiying the CORBA name servers for the various modules. Currently, no dynamic migration of modules is supported   by CAST. The end-user has no choice concerning the execution location of the modules. Execution of the modules are invoked by dynamic request to the ORB: the name server includes the name of the methods implementing the required codes in the wrapper object definitions for any particular module.

2.2 Specification of Complex Applications

The benefit gained from using process algebras to define the user applications is twofold:
- it constrains the application developers to specify rigorously the modules interactions and scheduling, without programming first the modules inputs/outputs, thus leaving implementations considerations to a later phase in the application design
- it produces formal specifications which can be processed by automatic verification and reduction algorithms, thus ensuring further computational properties, e.g., fairness, deadlocks freeness, accessibility of the modules, etc.

Although the last point is not yet implemented, it is foreseen that the introduction of verification and reduction algorithms will permit a significant gain in the integration of complex distributed applications involving a large number of modules. This provides a sound and theoretical background on which to rely in order to eventually formally specify, prove and verify the application processes. For example, electronic chips co-simulation involves dozens of modules which cooperate to simulate the various aspects of electronics: electrical, thermal, clocks, etc. Managing and controlling the interactions of these modules may require intricate relationships. Proving that that these interactions are properly implemented and that the simplest simulator has indeed been designed can be a cumbersome task that may benefit from model validation and verification tools inherited from formal verification techniques. This is common use in electronic chip design, and can be straightforwardly implemented in CAST.

2.3 Implementation

A demonstrator was developed during a first six months phase (April-October 1998).This preliminary Corba-compliant integration platform was demonstrated during the second DECISION project review in Sophia-Antipolis, October 1998. It relied on the *ILU* object request broker (ORB) from Xerox [4] and the *OLAN* distributed application configuration environment [1].

ILU has interesting features, including interfaces with a variety of programming languages, e.g., C, C++, Lisp, Java, Python. But inherent limitations (e.g., no interface repositories, no multi-threading) and performance penalties lead however to the replacement of these software.

The replacement by the *Orbacus* ORB [10] from Object-Oriented Concepts, Inc., a fully Corba-compliant ORB, started in November 1998. *Orbacus* permits a seamless integration with the C++ and Java languages. This made possible the fast implementation of the Corba version of *CAST* on Unix workstations.

The prototype was demonstrated in April 1999. A contributing factor was the C++ language chosen for the implementation of *CAST*. Based on these grounds, the upgrade to an extensive Corba-compliant integration platform, named *CAST 2.0*, was therefore straightforward.

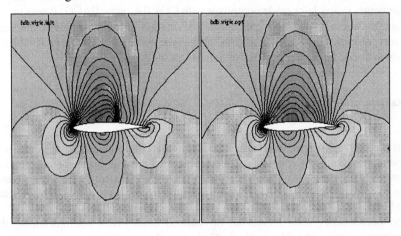

Fig. 2. *Wing shape optimisation: induced drag reduction in high-speed cruise configuration*

Extensions to the platform have since been implemented to support the dynamic plug-in of solvers and optimisers to the platform. This uses the full dynamic creation and invocation capabilties of the CORBA object-request broker.

2.4 Integration of Parallel Codes

A later development phase started in 2000 to support parallel codes and execution on remote PC-clusters over wide-area networks. By the end of year 2000, this was operational concerning the execution of parallel codes on PC-clusters. The users can now define distributed simulation and optimisation applications that involve sequential and parallel codes that cooperate. The codes run on parallel computers, NOW and PC-clusters that are distributed over wide area networks.

Interfacing existing codes that involve MPI statements with CORBA was a challenge. Various work has already been carried out concerning the integration of parallel codes in distributed computing environments and was of invaluable help for this matter. Our implementation uses the notion of " parallel CORBA objects " developed in PACO by the *PARIS* project at IRISA [11]. Basically, a parallel CORBA object is a collection of similar objects running in parallel on a PC-cluster for example. They are managed by a particular server object which interfaces with the remote clients. The parallelism is therefore transparent to the end user. This provides for the seamless integration of parallel codes, written in Fortran using MPI for example, up to the interface of the code with the parallel CORBA objects. PACO is itself built on the MICO object-request broker [7]. MICO has therefore become the new ORB layer used by *CAST* since September 2000.

The 2D wing shape optimisation example (Figure 2) presented in Section 2 was run on a PC-cluster using a parallel version of the genetic algorithm written in Fortran, using MPI statements. It was later made compliant with CORBA using PACO. It was run as a set of servers and a *CAST* client on the network. In the last implementation, there are several servers, one which is dedicated to the genetic optimisation algorithm, and the others which are dedicated to the CFD and mesh calculations.

2.5 Metacomputing

The next step, which started in January 2001, involved the execution of simulation applications concerning aerodynamic optimisation of airfoils in a metacomputing environment, including workstations and PC-clusters distributed over several locations at INRIA Rennes, INRIA Sophia-Antipolis near Nice and INRIA Rhône-Alpes in Grenoble all connected by a high-speed gigabits/sec network (Figure 5).

Based on performance tests conducted on various deployment configurations, the *CAST* software runs on the PC-cluster in Grenoble, the parallel genetic optimisation algorithm runs on the PC-cluster in Sophia-Antipolis, and several instances of the CFD solver run on the PC-cluster in Rennes. CAST is a CORBA client for the genetic algorithm server, which acts in turn as a client for the CFD solvers (Figure 3).

The PC-cluster in Grenoble includes 225 Bi-Pentium III 733 MHz processors connected by a 100 Mbits/sec FastEthernet network, running Linux (Mandrake 7.0). The PC-cluster in Rennes includes 20 bi-Pentium III, 500 MHz, and bi-Pentium II, 450 MHz., connected by a FastEthernet 100 Mbits/sec network, running Linux (Debian 2.2). The PC-cluster in Sophia-Antipolis includes 33 bi-Pentium III processors (19 running at 933 MHz and 14 at 500 MHz), running Linux 2.2.

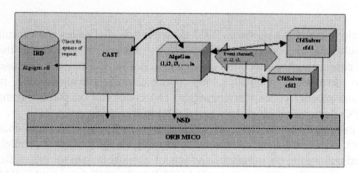

Fig. 3. Clients and servers for the optimisation application.

Several instances of the CFD solver run in parallel to account for the candidate solutions generated by the genetic algorithm. Each CFD solver is allocated several processors to account for the domain decomposition of the area around the airfoil, which is statically divided into four sub-domains (Figure 4). One processor is allocated to each of the sub-domains. The airflow around the wing shape is computed each time the airfoil is modified by the optimisation algorithm, i.e., for each optimisation loop. Each optimisation loop generates several candidate solutions, which are then evaluated in parallel, i.e., the mesh and the flow is computed for each optimisation loop and for each solution generated by the optimisation algorithm.

In this example, the CFD solvers are the most time-consuming tasks. It is therefore of most importance to duplicate and parallelize the mesh generation and computation of the airflow around the wing shape. This is why there are several instances of the solver executing in parallel for each candidate solution, in fact four of them for one wing shape instance. There are in turn several candidate solutions that are produced by the genetic optimisation algorithm that are evaluated in parallel. The processor allocation is done statically: a requested number of processors is queried on each PC-cluster prior to the execution of the application. It is then allocated to the specific tasks to execute. Various performance tests have been conducted for this particular application. The PC-clusters are connected by the high-speed gigabits/sec VTHD network [13] (Figure 5).

3. Conclusion

This paper presents a software integration platform called *CAST* that combines both techniques of formal specifications and distributed object-oriented development, for the specification, configuration, deployment and execution of complex engineering applications in a unified and operational way.

The platform was developed on Unix workstations and communicate through a commercial CORBA object request broker. It was successfully tested, showing blazing performance, against complex optimisation applications in a distributed environment involving several remotely located PC-clusters at INRIA, connected to a gigabits/sec high-speed network called VTHD.

The target user population includes project managers and engineers that are expert in their particular domain, e.g., aerodynamics, electromagnetics.

The application tasks and their interactions are transparently mapped to distributed software components. They are implemented using object-oriented programming techniques. The actual user modules may include simultaneously non object-oriented code, e.g., Fortran subroutines, as well fully object-oriented software, e.g., C++. In this case, *CAST* takes automatically into account the diversity of the interfaces and parameters for the various modules.

The *CAST* integration platform also offers a high-level graphics interface for the specification of complex multi-discipline applications, thus making the formal algebraic and theoretical background on which it is based, as well as the technicalities of distributed computing, totally transparent to the end-users.

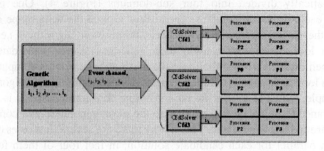

Fig. 4. *Interaction between the genetic algorithm and the CFD solvers.*

Also, an interesting feature in *CAST* is that it supports the coupling of CORBA and non-CORBA modules simultaneously. This provides for a smooth transition from existing application software and mathematical libraries to state-of-the-art integration platforms and metacomputing environments. It is also expected that this will permit the connection to third party environments that support widely accepted communication standards and exchange protocols, e.g., CORBA.

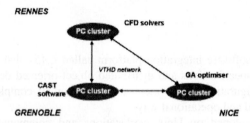

Fig. 5. The optimisation application in the metacomputing environment.

Although metacomputing has focused much attention and efforts in the last few years, large scale applications, including CPU and graphics intensive applications that require large amounts of computing ressources are still demanding easy to use high-performance problem-solving environments [2]. Examples of such   applications include the co-simulation of electronic chips in the *CADNET* project, numerical propulsion systems simulation [5] and multidiscipline aircraft simulation, as for the Digital Dynamic Aircraft approach in the *AMAZE* project. Indeed, the design environments required now by aircraft manufacturers will include CFD, CSM, acoustics and electromagnetics simulation and optimisation, ranging from the initial CAD outlines up to the dynamic behavior of the airframe and systems under operating constraints. This will put an unprecedented load on hardware and software resources in problem-solving environments.

Still lacking specific features, e.g., load balancing, guaranteed QoS and security, the *CAST* integration platform provides the functionalities that make it a good candidate for high-performance metacomputing environments dedicated to multi-discipline simulation applications.

The software documentation, including the software specifications and the *CAST* user's manual, is available on-line at: http://www.inrialpes.fr/sinus/cast. Further information on the project is available at: http://www.inrialpes.fr/sinus.

Acknowledgments

The authors wish to thank Alain Dervieux and Jean-Antoine Desideri from the *SINUS* project at INRIA Sophia-Antipolis, as well as Jacques Periaux from Dassault-Aviation (Pôle Scientifique), for their strong support and helpful advice.
Part of this project is funded by INRIA through ARC " Couplage " and by the French RNRT program, through the *VTHD* project. It was also partly supported by the EEC *Esprit* Program, through the HPCN " *DECISION* " project: " *Integrated Optimisation Strategies for Increased Engineering Design Complexity* ".

References

1. Balter, R., et al. Architecturing and configuring distributed applications with OLAN. Proc. IFIP Int'l. Conf. Distributed Systems platforms. Middleware '98. Lake District. September 1998.
2. Barnard, S., et al. Large-scale distributed computational fluid dynamics on the Information Power Grid using Globus. Proc. NASA HPCC/CAS Workshop. NASA Ames Research Center. February 2000.
3. J. Blachon, T. Nguyen " DECISION prototype integration platform: CAST CORBA prototype ". Soft-IT Workshop. Invited lecture. " Next generation of interoperable simulation environments based on CORBA ", INRIA-ONERA-Simulog. June 1999.
4. XEROX ILU Reference manual. 1998.
   ftp://ftp.parc.xerox.com/pub/ilu/2.0a14/manual_html
5. Lopez, I.,et al. NPSS on NASA's IPG: using Corba and Globus to coordinate multidisciplinary aeroscience applications. Proc. NASA HPCC/CAS Workshop. NASA Ames Research Center. February 2000.
6. Marco N., Lanteri S. A two-level parallelization strategy for genetic algorithms applied to shape optimum design. Research Report no. 3463. INRIA. July 1998.
7. http://www.mico.org
8. Milner R. Calculi for synchrony and asynchrony. Theoretical computer science. Vol. 25, no. 3. July 1983.
9. Nguyen G.T., Plumejeaud C. Integration of multidiscipline applications in metacomputing environments. Systèmes, Réseaux et Calculateurs Parallèles. Hermès Ed. To appear 2002.
10. Object-Oriented Concepts, Inc. ORBacus for C++ and Java. Version 3.1.1. Object-Oriented Concepts, Inc. 1998.
11. Th. Priol. Projet PARIS " Programmation des systèmes parallèles et distribués pour le simulation numérique distribuée ". INRIA. May 1999.
12. Sang, J., Kim, C., Lopez, I. "Developing Corba-based distributed scientific applications from legacy Fortran programs" Proc. NASA HPCC/CAS Workshop. NASA Ames Research Center. February 2000.
13. http://www.telecom.gouv.fr/rnrt/projets/pres_d74_ap99.htm

Large-Scale Scientific Irregular Computing on Clusters and Grids

Peter Brezany[1], Marian Bubak[2,3], Maciej Malawski[2], and Katarzyna Zając[2]

[1]Institute for Software Science, University of Vienna,
Liechtensteinstrasse 22, A-1090 Vienna, Austria
[2]Institute of Computer Science, AGH, al. Mickiewicza 30, 30-059 Kraków, Poland
[3]Academic Computer Centre – CYFRONET, Nawojki 11, 30-950 Kraków, Poland
brezany@par.univie.ac.at, {bubak | malawski | kzajac}@uci.agh.edu.pl

Abstract. Data sets involved in many scientific applications are often too massive to fit into main memory of even the most powerful computers and therefore they must reside on disk, and thus communication between internal and external memory, and not actual computation time, becomes the bottleneck in the computation. The most challenging are scientific and engineering applications that involve irregular (unstructured) computing phases. This paper discusses an integrated approach, namely ideas, techniques, concepts and software architecture for implementing such data intensive applications on computational clusters and the computational Grid, an emerging computing infrastructure. The experimental performance results achieved on a cluster of PCs are included.
Keywords: Grid, clusters, irregular problems, out-of-core computing, runtime library.

1 Introduction

In the past few years, *clusters of workstations and PCs* have emerged as an important and rapidly expanding approach to parallel computing [7]. These systems often employ commodity processors and networking hardware as well as open source operating systems and communication libraries to deliver supercomputer performance at the lowest possible price. They also provide a large data storage capacity which allows efficient implementations of large-scale applications which are input and output (I/O) intensive. In order to effectively exploit the power of clusters, appropriate programming environments and development tools are a must [6].

There are many applications whose computational requirements exceed the resources available at even the largest supercomputing centres. For these applications, *computational grids* [12] offer the promise of access to increased computational capacity through the simultaneous and coordinated use of geographically separated large-scale computers linked by networks. Through this technique, the "size" of an application can be extended beyond the resources available at a particular location - indeed, potentially beyond the capacity of any individual computing site anywhere [17, 19]. The fundamental challenge is to make Grid

P.M.A. Sloot et al. (Eds.): ICCS 2002, LNCS 2329, pp. 484–493, 2002.

systems widely available and easy to use for a wide range of applications. This is crucial for accelerating their transition into fully operational environments.

This paper addresses challenging problems associated with development of an important class of scientific and engineering applications, called *irregular (unstructured) applications*, on clusters and Grid systems. In such applications, access patterns to major data arrays are only known at runtime, which requires runtime preprocessing and analysis in order to determine the data access patterns and consequently, to find what data must be communicated and where it is located. An example of an irregular code is shown in Fig. 1; here, a and b are called *data arrays* and x and y are called *index arrays*.

```
for (i = 0; i < n; i++)
    a[x[i]] = b[y[i]];
```

Fig. 1. An example of an irregular loop.

Real irregular applications typically involve large data structures. Runtime preprocessing provided for these applications results in construction of additional large data structures which increase the memory usage of the program substantially. Consequently, a parallel program may quickly run out of memory. Therefore, some data structures must be stored on disks and fetched during the execution of the program. Such applications and data structures are called *out-of-core (OOC) applications* and *OOC data structures*, respectively. On the other hand, for *in-core (IC)* programs it is possible to store all their data in the main memory. The performance of an OOC program strongly depends on how fast the processors the program runs on can access data from disks.

OOC problems can be handled in three different ways: (1) virtual memory (VM) which allows the in-core program version to be run on larger data sets, (2) specific OOC techniques which explicitly interface file I/O and focus on its optimization, and (3) the whole large scale irregular application is ported to the computational Grid. Although VM is an elegant solution (it provides the programming comfort and ensures the program correctness), it has been observed [1] that the performance of scientific applications that rely on virtual memory is generally poor due to frequent paging in and out of data. In [3] we presented the library lip which supports the development of OOC irregular applications on distributed-memory systems. So far, to our best knowledge, there has been no research effort devoted to the implementation of irregular applications on computational Grids.

This paper presents an integrated approach to the implementation of large-scale irregular applications on clusters and Grids. The approach is based on a new library called G-lip we are developing at present. Section 2 discusses the parallelization method we apply to irregular in-core and OOC applications and execution models for cluster and Grid environments. Section 3 deals with the design and implementation of the G-lip library. Preliminary performance results are discussed in Section 4, and we briefly conclude in Section 5.

2   Parallelization Strategy and Execution Models

The standard strategy for processing parallel loops with irregular accesses for distributed-memory systems [22], generates two code phases, called the *inspector* and the *executor*. The inspector analyzes the data access pattern at runtime. From the access pattern, the array distribution, and the work distribution of the loop, communication schedules (CSs) can be derived. The schedules are used in the executor to actually perform the required communication for the loop and the computation specified by the loop. Optionally, a dynamic *partitioner* can be applied to the loop [2, 21] (See Fig. 2(a)). In this approach, it is assumed that all data structures reside in main memory; therefore, it is called *in-core parallelization strategy*.

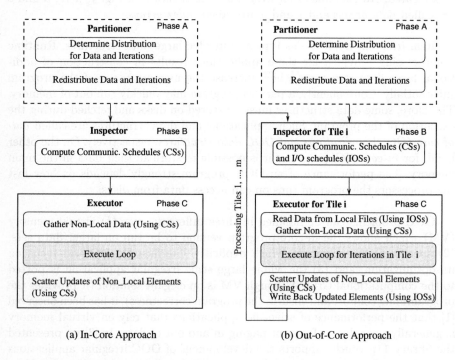

(a) In-Core Approach   (b) Out-of-Core Approach

Fig. 2. Computing phases of processing of loops with irregular data access.

The *out-of-core parallelization strategy* is a natural extension of the in-core approach. The set of iterations assigned to each processor by the work distributor is split into a number of *iteration tiles*. Sections of arrays referenced in each iteration tile, called *data tiles*, are small enough to fit in local memory of that processor. The inspector-executor strategy is then applied to each iteration tile (See Fig. 2(b)). Besides the communication schedules (CSs), the inspector also computes the *I/O schedules* which describe the required I/O operations. The optional partitioning phase involves the use of a disk-oriented partitioner which uses OOC data (for example, the node coordinates of the mesh) to determine a new data and loop iteration distribution.

2.1   Execution Model for Clusters

A typical abstract cluster architecture[1] is shown in Fig. 3. Data and index arrays can be either persistently stored in a (mass-) storage system attached to the front-end node and dynamically distributed across disks of the compute nodes in the computation initialization phase, or can be already available on the compute node disks as the product of previous computation task. The optional partitioning phase can optimize this data distribution.

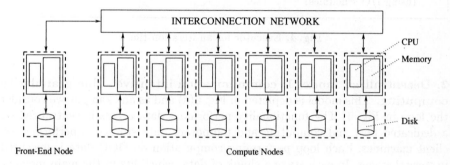

Fig. 3. The cluster model.

For each iteration tile, the inspector computes the computation and I/O schedules sections of index arrays. Once preprocessing is completed, one may carry out the necessary memory allocation, I/O, communication, and computation, following the plan established by the inspector. Fig. 4 outlines the steps involved.

2.2   Execution Models for the Grid

We consider the following four computing scenarios: a centralized model, two decentralized models (with fine and coarse granularity computing) and a distributed supercomputing model.

1. Centralized model. Functionality performing data access (i.e. via *Replica Manager* - see the next Section) and OOC computing is centralized on a dedicated host, which may be a massively parallel computing and storage system, freeing client machines for other tasks such as data visualization and data analysis. This concept is also supported by NetSolve [8].

---

[1] A node of a real cluster may include several processors and a set of disks.

- Before execution of the tile
  1. Data to be used on this processor or sent to remote processors is read from local files (using I/O schedules)
  2. Send/Receive calls transport non-local data (using communication schedules)
  3. Data is written into buffers
- Execution of the tile is carried out
  - non-local operand data is accessed from buffers
  - non-local data assignments go to buffers
- After execution of the tile
  1. Non-local data to be stored in remote files is read from buffers
  2. Send/receive calls transport non-local data (using communication schedules)
  3. Data is written back into local files for longer term storage (using I/O schedules)

Fig. 4. Executor for an iteration tile.

2. Decentralized model 1: communication intensive - fine granularity computing. This model is depicted in Fig. 5(a) and as an example, we consider the loop in Fig. 1. The functionality performing data access is centralized on a dedicated host and computing and temporary data storage is performed on client machines. Each loop performing computation on OOC data is executed in several stages. In each stage a chunk of data, which fits in the main memory of the client machine, is fetched from the remote host, new data is computed, and stored on the remote host, if necessary. The server and client cooperate together in a pipeline mode. The server can prepare array sections in advance; the client may inform it about its I/O needs in advanced in the form of so called I/O requirements graph. This model is provided for clients with small storage resources. The server can use *Replica Manager* service in order to access data (see the next Section).

3. Decentralized model 2: less communication intensive - coarse granularity computing. This model is depicted in Fig. 5(b). Functionality performing data access is centralized on a dedicated host and computing and temporary data storage is performed on client machines. Each loop performing computation on OOC data is executed at the client/server level in several stages. In each stage a chunk of data, which fits in the disk memory of the client machine, is fetched from the remote host, new data is computed, and stored on the remote host, if necessary. At the client machine level, the execution of each stage is split into several sub-stages. In each sub-stage a chunk of data which fits in the main memory of the client machine is fetched from the disk memory of this machine, new data is then computed, and stored in the disk memory, if necessary.

4. Distributed supercomputing model. This model is considered for applications whose computational and storage requirements are so demanding that

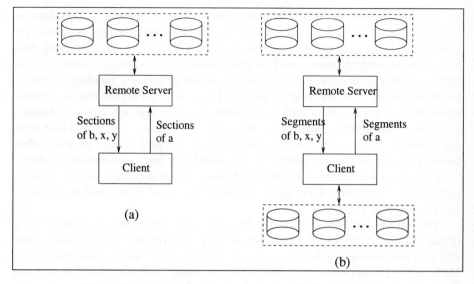

Fig. 5. Execution schemes for irregular OOC on the Grid.

they can be met only by combining multiple high-capacity resources on a computational Grid into a simple, virtual distributed supercomputer. In this case, the execution model is similar to that discussed in the cluster context.

3   Design and Implementation of the G-lip Library

3.1   G-lip library

The library G-lip we are developing has to support both compute and data intensive parts of applications on clusters and the Grid.

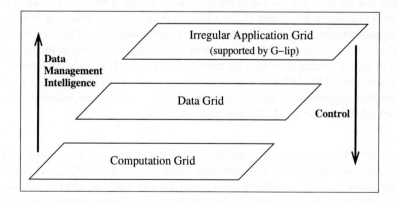

Fig. 6. A 3-layer Grid architecture.

Recently the term *"Data Grid"* has come into use to describe a specialization and extension of the computational Grid for data intensive Grid applications [11], and several Data Grid projects have been started [18]. They aim to develop data management architectures based on two fundamental services: storage system and metadata management. These services allow to create a number of components whose services are used by G-lip. There are two representative components: replica management and replica selection. The role of a *Replica Manager* is to create (or delete) copies of file instances, or *replicas*, within specified storage systems. Often, a replica is created because the new storage location offers better performance or availability for accesses to or from a particular location.

From the Grid architectural point of view, G-lip is considered as a component of the Application Grid layer developed on top of Computation and Data Grid layers (see Fig. 6).

The implementation of the G-lip library is partially based on re-engineering of the lip library [3], which was based on MPI and MPI-IO. The communication and I/O mechanisms of G-lip are based on the Globus Toolkit [10] component MPICH-G. To support all the four Grid execution models outlined in the last Section, G-lip supports both in-core and out-of-core irregular computations.

3.2   Java implementation of lip

The lip library was originally written in C language. As Java is becoming increasingly popular language in distributed and parallel computing [14], there was a need to provide Java bindings to lip library in order to achieve interoperability of the library. The approach to creating Java bindings was to use Java Native Interface (JNI) [15] as a tool for creating Java wrappers to lip functions. JNI enables native code to use Java objects, call Java methods, access Java arrays and take full advantage of Java specific features such as exceptions handling and threads. All this can be accessed by calling appropriate functions from JNI API. During this work it was realized that interface creation process by using bare JNI becomes time consuming and subject to errors. As a solution of this problem the Java to Native Interface Extension called *Janet* was created [16, 4, 5]. *Janet* [13] is the extension of Java language that allows inserting native (i.e., C) language statements directly into Java source code files. The files with such extended Java syntax are input for *Janet* translator that creates automatically Java and native source files with all required JNI calls. It frees programmer from calling low level JNI API and makes interface creation more clear and flexible.

The Java implementation of the lip library contains the following classes: LIP, Schedule, Datamap, IOBufmap and auxiliary.

Class LIP   This class involves methods and fields for lip setup, index translation and communication. The methods Localize() and OOC_localize() are used for index translation. The former translates globally numbered indices into locally numbered counterparts. It also updates the communication schedule so it could be used to transmit the non-local data. The latter function transforms the indices which point to the data residing on disk into indices pointing to the memory data buffer. The results of mapping memory contents into disk is stored in IO mapping structure. Two communication methods Gather() and Scatter() perform collective communication among nodes before and after the computational phase. They support numerous communication patterns and data copy/update schemes and use the schedule object to obtain information about the source and destination of the data.

Class Schedule This class includes group of methods used to create and manipulate a schedule objects that store patterns used for communication.

Class Datamap This class contains function for partitioning, data distribution and redistribution. At present, lip supports coordinate-based partitioner using Hilbert's curve [20] to determine what part of data should reside on the same node. The library supports both in-core and the OOC version of the partitioner. A partitioner is activated using the create_hilbert_distribution() call. The resulting data mapping is stored in a object called Datamap. The remapping method remap_ooc() can be used to redistribute the data.

Class IOBufmap This class includes methods to manipulate I/O buffer mapping objects which store information about the mapping of the memory buffer onto the corresponding disk file. There are methods, e.g. IOBufmap_get_datatype(), to obtain the MPI derived data types according to the information stored in the I/O buffer mapping object which, in turn, may be used to perform data movement by MPI-IO functions.

Other classes and interfaces Java bindings to lip provide additional classes to encapsulate additional features. Interface DataDistribution can represent various distributions handled by lip, class IndicesDatatypes serves as a parameter to IOBufmap.getDatatype function, encapsulating both needed MPI datatypes, interface Indexer represents index array with index layout, class ContIntIndexer implements Indexer interface for integer array with contiguous index layout and class LipException maintains error handling in lip library.

4   Performance Results

To test the performance of Java bindings to lip library we used a generic irregular loop with data located out–of–core. The code is shown in Fig. 7. The data array consisted of 113400 cells with double precision values. The experiments were carried out on a cluster of 6 LINUX workstations connected by 100Mb Fast Ethernet with Intel Celeron 600 MHz processor, 64 MB RAM. For out–of–core data exchange each node was using the local disk drive. The results presented in Fig. 8 show the scalability for both C and Java version of the program.

```
for(i=0;i<edge_counter;i++)
{
    y[i] = -x[ perm[i] ];
}
for(i=0;i<edge_counter;i++)
{
    x[ perm[i] ] += y[i];
}
```

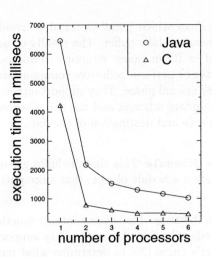

Fig. 7. Irregular loops used for measurements of lip performance.

Fig. 8. The comparison of performance of C and Java OOC lip computation.

At present, we are putting on the Grid an irregular kernel code [9] derived from the industrial application PAM-CRASH and clusters located in Kraków and Vienna connected by the Internet are used as a testbed Grid architecture.

5   Summary

This paper proposes ideas and execution models for solving irregular and OOC problems on clusters and in Grid environments. It presents the G-lip library and the Java extension to lip, which was created to achieve interoperability of the libraries. In the near future we will extend these libraries by optimization modules, which will minimize data transfer between Grid memory hierarchies by advanced caching techniques. Further, we will perform performance measurements in a real Grid environment using industry code kernels.

References

1. Bershad, B., Black, D., DeWitt, D., Gibson, G., Li, K., Peterson, L., and Snir, M.: Operating System Support for High-Performance Parallel I/O Systems. Technical Report CCSF-4, Scalable I/O Initiative (1994)
2. Brezany, P., et al.: Automatic Parallelization of the AVL FIRE Benchmark for a Distributed-Memory System. In: Dongarra, J. et al. (eds.): Proceedings of PARA95. LNCS Vol. 1041 (1996) 50-60
3. Brezany, P., Bubak, M., Malawski, M., and Zając, K.: Advanced Library Support for Irregular and Out-of-Core Parallel Computing. In: Hertzberger, B., Hoekstra,

B., Williams, R. (eds.): Proc. Int. Conf. High Performance Computing and Networking, Amsterdam, June 25-27, 2001, LNCS Vol. 2110, Springer-Verlag (2001) 435–444

4. Bubak, M., Kurzyniec, D., Luszczek, P.: A Versatile Support for Binding Native Code to Java. In: Bubak, M., Afsarmanesh, H., Williams, R., Hertzberger, B. (eds.): Proc. Int. Conf. HPCN Amsterdam, May 2000, LNCS Vol. 1823. Springer-Verlag (2000) 373-384

5. Bubak, M., Kurzyniec D., Luszczek, P.: Creating Java to Native Code Interfaces with Janet Extension. in: Proceedings of 1st SGI Users Conference, Cracow, Poland, October 2000, ACC Cyfronet UMM (2000) 283–294

6. Carns, P.H. et al.: PVFS: A Parallel File System for Linux Clusters. In: Proc. of the Externe Linux Track: 4th Annual Linux Showcase and Conference Oct. 2000

7. Buyya, R. (ed.): High Performance Cluster Computing, Vol. 1 and 2. Prentice Hall, New Jersey (2000)

8. Arnold, D., Agrawal, S., Blackford, S., Dongarra, J., Miller, M., Sagi, K., Shi, Z. and Vadhiyar, S.: Users' Guide to NetSolve V1.4 Tech. Report CS-01-467 University of Tenessee, Knoxville (2001)

9. HPF+ Project. Technical Reports, http://www.par.univie.ac.at/project/hpf+/deliverables.html

10. Foster, I. and Kesselman, C.: Globus: A Metacomputing Infrastructure Toolkit. International Journal of Supercomputer Applications, 11(2) (1997) 115–128

11. Chervenak, A., Foster, I., Kesselman, C., Salisbury, C., and Tuecke, S.: The Data Grid: Towards an Architecture for the Distributed Management and Analysis of Large Scientific Datasets. To be published in the Journal of Network and Computer Applications, 2001

12. Foster, I., Kesselman, C., and Tuecke, S.: The Anatomy of the Grid: Enabling Scalable Virtual Organizations. Intl. J. Supercomputer Applications, 15(3) (2001)

13. Janet homepage: http://www.icsr.agh.edu.pl/janet

14. Java Grande Forum. http://www.javagrande.org

15. JavaSoft: Java Native Interface. http://java.sun.com/products/jdk/1.2/docs/guide/jni/

16. Kurzyniec, D.: Creating Java to Native Code Interfaces with Janet Extension. M. Sc. Thesis, Department of Computer Science at University of Mining and Metallurgy. Cracow, Poland, (August 2000)

17. Messina, P.: Distributed Supercomputing Applications. In: I. Foster and C. Kesselman (eds.), The Grid. Blueprint for a New Computing Infrastructure, Morgan Kaufmann Publ., (1999)

18. Oldfield, R.: Summary of Existing and Developing Data Grids. White paper for the Remote Data Access group of the Global Grid Forum 1, Amsterdam, (March 2001)

19. Oldfield, R. and Kotz, D.: Armada: A parallel File System for Computational Grids. In Proceedings of the IEEE/ACM International Symposium on Cluster Computing and the Grid, Brisbane, Australia, (May 2001) 194–201

20. Ou, C.-W., Ranka, S.: SPRINT: Scalable Partitioning, Refinement, and INcremental partitioning Techniques. http://citeseer.nj.nec.com/169110.html

21. Ponnusamy, R. et al.: CHAOS Runtime Library. Techn. Report, University of Maryland, (May 1994)

22. Saltz, J., Crowley, K., Mirchandaney, R. and Berryman, H.: Run-time Scheduling and Execution of Loops on Message Passing Machines. Journal of Parallel and Distributed Computing, 8(2) (1990) 303–312

High Level Trigger System for the LHC ALICE Experiment

H. Helstrup[1], J. Lien[1], V. Lindenstruth[2], D. Röhrich[3], B. Skaali[4],
T. Steinbeck[2], K. Ullaland[3], A. Vestbø[3], and A. Wiebalck[2]
for the ALICE Collaboration

[1] Bergen College, P.O. Box 7030,
5020 Bergen, Norway
[2] Kirchhoff Institute for Physics, University of Heidelberg, Schröderstrasse 90,
69120 Heidelberg, Germany
[3] Department of Physics, University of Bergen, Allegaten 55,
5007 Bergen, Norway
[4] Department of Physics, University of Oslo, P.O. Box 1048 Blindern,
0316 Oslo, Norway

Abstract. The ALICE experiment at the Large Hadron Collider (LHC) will produce a data size of up to 75 MByte/event at an event rate of up to 200 Hz resulting in a data rate of ~15 GByte/s. Online processing of the data is necessary in order to select interesting events or sub-events (high-level trigger), or to compress data efficiently be modeling techniques. Both require a fast parallel pattern recognition. Processing this data at a bandwidth of 10-20 GByte/s requires a massive parallel computing system. One possible solution to process the detector data at such rates is a farm of clustered SMP-nodes based on off-the-shelf PCs, and connected by a high bandwidth, low latency network.

1 Introduction

The ALICE experiment [1] at the upcoming Large Hadron Collider (LHC) at CERN will investigate Pb-Pb collisions at a center of mass energy of about 5.5 TeV per nucleon pair. The main detector is a 3-dimensional tracking device, the Time Projection Chamber (TPC), which is specially suited for handling large multiplicities. The detector is readout by 570,000 ADC-channels, producing a data size of ~75 MByte/event [3]. The event rate is limited by the bandwidth of the permanent storage system. Without any further reduction or compression ALICE will be able to take events up to 20 Hz. Higher rates are possible by either selecting interesting events and sub-events (high-level trigger), or compressing the data efficiently by modeling techniques. Both high-level triggering and data compression requires pattern recognition to be performed online. In order to process the detector information of 10-20 GByte/s a massive parallel computing system (high-level trigger system) is needed. The construction of such a system includes development of both hardware and software optimized for a high throughput data analysis and compression.

P.M.A. Sloot et al. (Eds.): ICCS 2002, LNCS 2329, pp. 494–502, 2002.

2   Functionality

From a trigger point of view the detectors in ALICE can be divided into two categories: fast and slow. Fast detectors provide information for the trigger system at every LHC crossing. Decisions at trigger 0,1 and 2 are made using information from these detectors. The slow detectors (such as the TPC) are tracking drift detectors and need longer time span after the collisions to deliver their data. Their slowness is compensated for by the detailed information they provide. The ALICE High-Level Trigger (HLT) system is intended to take advantage of this information in order to reduce the data rate as far as possible to have reasonable taping costs. The data is then recorded onto an archive-quality medium for subsequent offline analysis.

A key component of the HLT system is the ability to process the raw data performing track pattern recognition in realtime. About 20,000 particles per interaction each produce about 150 clusters in the detectors. These signals has to be read out, processed, recognized and grouped into track segments. Based on the detector information – clusters and tracks – data reduction can be done in different ways:

- Generation and application of a software trigger capable of selecting interesting events from the input data stream ("High-Level Trigger")

- Reduction in the size of the event data by selecting sub-events

- Reduction in the size of the event data by compression techniques

2.1   Event rate reduction

The ALICE TPC detector will be able to run at a rate up to 200 Hz for heavy ion collisions, and at up to 1 kHz for p-p collisions [3]. In order to increment the statistical significance of rare processes, dedicated triggers can select candidate events or sub-events. By analyzing tracking information from different detectors online, selective or partial readout of the relevant detectors can be performed, thus reducing the event rate. The tasks of such a high-level trigger are (sub)event selections based upon the online reconstructed track parameters of the particles, e.g. to select events which contains e^+-e^- candidates coming from a quarkonium decay. In the case of low multiplicity events from p-p collisions the online pattern recognition system can be used to remove pile-up events from the data stream.

2.2   Realtime data compression

Data compression techniques can be divided into two major categories: lossy and lossless. Lossy compression concedes a certain loss of accuracy in exchange for greatly increased compression. Lossless compression consists of those techniques guaranteed to generate an exact duplicate of the input data stream after a compress/expand cycle.

General lossless or slightly lossy methods like entropy and vector quantizers can compress tracking detector data only by factor 2-3 [2]. The best compression method is to find a good model for the raw data and to transform the data into an efficient representation. By online pattern recognition and a compressed data representation an event size reduction by a factor of 15 can be achieved [3]. The information is stored as model parameters and (small) deviations from the model. The results are coded using Huffman and vector quantization algorithms. All correlations in the data have to be incorporated into the model. The precise knowledge about the detector performance, i.e. analog noise of the detector and the quantization noise, is necessary to create a minimum information model of the data. Realtime pattern recognition and feature extraction are the input to the model.

3   Online pattern recognition

Both high-level triggering and data modeling require a fast parallel pattern recognition. The data modeling scheme is based on the fact that the information content of the TPC are tracks, which can be represented by models of clusters and track parameters. Fig. 1 shows a thin slice of the TPC detector, clusters are aligned along the trajectories (helices). The local track model is a helix; the knowledge of the track parameters helps to describe the shape of the clusters in a simple model [5] [6]. The pattern recognition reconstructs clusters and associates them with local track segments. Note that track recognition at this time can be redundant, i.e. clusters can belong to more than one track and track segments can overlap. Once the pattern recognition is completed, the track can be represented by helix parameters.

3.1   Cluster finder and track follower

In the STAR [7] experiment the online tracking is divided into two steps: [8] Cluster finding and track finding. A cluster finder searches for local maxima in the raw ADC-data. If an isolated cluster is found, the centroid position in time and pad direction is calculated. In the case of overlapping clusters simple unfolding procedures separate charge distributions. Due to the missing track model there are no cluster models available, therefore the centroid determination is biased. The list of space points is given to the track finder, which combines the cluster centroids to form track segments.

A simple cluster finder and a track follower which applies conformal mapping [9] in order to speed up fitting routines has been adapted to ALICE TPC data. This method has shown to work efficiently only in the case of low multiplicity Pb-Pb collisions and for p-p collisions, where the amount of overlapping clusters are small.

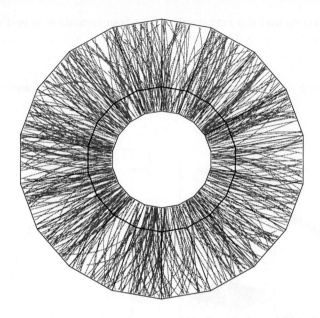

Fig. 1. Simulated Pb-Pb collision at LHC. Only a small fraction of the tracks in the ALICE TPC detector is shown

3.2 Hough transform

For the high cluster densities that are expected for heavy ion collisions at LHC, an alternative is being developed. In this case track segments finding is performed directly on the raw data level, using a special case of the Hough transformation technique.

The Hough transform is a standard tool in image analysis that allows recognition of global patterns in a image space by recognition of local patterns (ideally a point) in a transformed parameter space. The basic idea is to find curves that can be parametrized in a suitable parameter space. In its original form one determines a curve in parameter space for a given signal corresponding to all possible tracks with a given parametric form it could possibly belong to. All such curves belonging to the different signals are drawn in parameter space. This space is then discretized and entries are histogrammed – one divides parameter space up into boxes and counts the number of curves in each box. If the peaks in the histogram exceeds a given threshold, the corresponding parameter values define a potential track.

In ALICE, the transformation is done assuming the tracks follow circles in the transversal plane. In order to simplify the transformation, the detector volume is divided into sub-volumes in pseudorapidity, η. If one restricts the analysis to tracks starting from a common interaction point (assumed to be the vertex), the track model is characterized by only two parameters; the emission angle with the beam axis ϕ_0 and the curvature κ. The transformation is performed using the equation:

$$\kappa = \frac{2}{R} sin(\phi - \phi_0) \tag{1}$$

where R and ϕ are the polar coordinates of the point to be transformed. Each

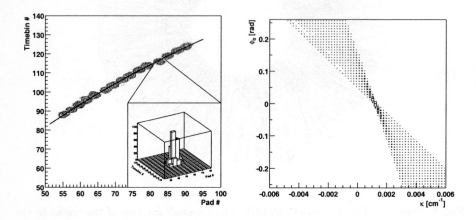

Fig. 2. On the left is shown clusters and associated track segments over 15 successive padrows in the TPC. The right figure shows a corresponding histogram resulting from a Hough transform applied on the signals produced by the track. Each pixel (padrow,timebin, pad) on the track is transformed into a quantised curve in parameter space (κ, ϕ_0), and the corresponding bins are incremented with the ADC-value of the pixels. The intersection of these curves is located at the parameters of the track

active pixel (ADC-value above threshold) transforms into a sinusoidal line extending over the whole ϕ_0-range in the parameter space, see Fig. 2. The corresponding bins in the histogram are incremented with the ADC-value of the transformed pixel. The super-position of all these point transforms produce a maxima at the circle parameter of the track. The track recognition is now done by searching for local maxima in the parameter space. Once the track parameters are known, cluster finding on the raw ADC-data can be performed by a straight forward unfolding of the clusters.

3.3   FPGA implementation

In order to minimize the CPU-load on the HLT processors, as much as possible of the pattern recognition will be implemented in dedicated hardware located on the PCI receiver cards in the first layer of the HLT system (see Fig. 5). Fast isolated cluster finding and the Hough transformation are simple algorithms that will be implemented into one or several FPGAs.

Fig. 3 shows a sketch of the architecture of the PCI receiver cards. One possible implementation scenario is the existing ALTERA Excalibur series of FPGAs, which include a choice of an ARM or MIPS embedded processor. The design will allow for most of the pattern recognition steps to be implemented in the FPGA. The internal SRAM can be extended by external high-speed SRAM if necessary. All additional logic, including the PCI interface, can also be implemented in the FPGA.

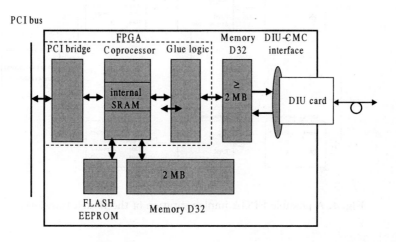

Fig. 3. PCI Receiver Card. The DIU is the interface to the optical detector link

Fig. 4 shows a block diagram of a possible implementation of the Hough transformation. In order to gain speed the raw input data will be processed time multiplexed and in parallel. The input is 10 bit raw ADC data coming from the front-end electronics of the detector. Every nonzero pixel is sorted according to its pseudorapidity, η, which is provided using a look-up table stored in the internal memory of the FPGA. In addition, the polar coordinates are provided using a combination of look-up tables and calculation algorithms. The actual transformation is done in a address calculator which calculates the addresses of counters which is incremented with a 8 bit representation of the ADC value of the pixel. Each counter corresponds to a bin in the histogram in (κ,ϕ_0)-space. There are maximum 128 histograms on each card, and the size of each histogram is $64\times64\times12$ bits.

The peaks in each of the histograms will be found by means of firmware algorithms on the FPGA, and the resulting list of track candidates will be shipped over PCI to the host for further processing. Alternatively the histograms may also be shipped to the host, and the peak finding performed on the CPU.

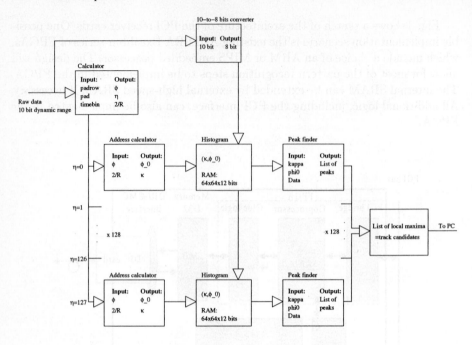

Fig. 4. A possible FPGA implementation of the Hough transform

4 Architecture

The ALICE HLT system will be located in the data flow after the front-end electronics of the detector and before the event-building network. A farm of clustered
SMP-nodes, based on off-the-shelf PCs and connected with a high bandwidth,
low latency network provide the necessary computing power for online pattern
recognition and data compression. The system nodes are interfaced to the front-
end electronics of the detector via their internal PCI-bus.

The hierarchy of the farm has to be adapted to both the parallelism in the
data flow and to the complexity of the pattern recognition scheme. Fig. 5 shows
a sketch of the foreseen hierarchy of the system. The TPC detector consists of
36 sectors, each sector being divided into 6 sub-sectors. The data from each
sub-sector are transferred via an optical fiber from the detector into 216 receiver
nodes of the PC farm. A hierarchical network interconnects all the receiver processors. Each sector is processed in parallel, results are then merged in a higher
level. The first layer of nodes receive the data from the detector and performs the
pre-processing task, i.e. cluster and track seed finding on the sub-sector level. The
next two levels of nodes exploit the local neighborhood: track segment finding on
sector level. Finally all local results are collected from the sectors and combined
on a global level: track segment merging and final track fitting. The total farm
will consist of 500-1000 nodes, partially equipped with FPGA co-processors.

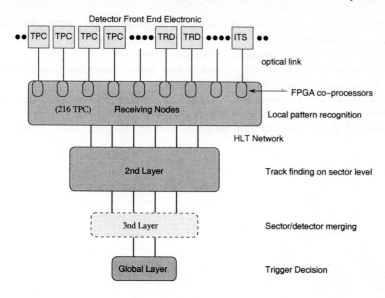

Fig. 5. Architecture of the HLT-system. The data from each sub-detector are transferred via optical fiber to the receiver processors. Results of the local processing on the receiver nodes are merged at a higher level. At the final global level all the processed data are merged, and the trigger decision can be made

The farm has to be fault tolerant and scalable. Possible network technology candidates are amongst others SCI [4] and Gigabit-Ethernet. The operating system will be Linux, using a generic message passing API for communication with the exception of data transfers, which use directly remote DMA.

5   Summary

The High-Level Trigger system for the ALICE experiment at the LHC accelerator has to process data at a rate of 10-20 GByte/s. The architecture of a PC cluster and data analysis software which fulfills this requirement are currently under study. The information from the online pattern recognition can be used for data compression, to derive physical quantities (e.g. momentum, impact parameters etc.) for High-Level Trigger decisions and for online monitoring.

References

1. ALICE collaboration, *Technical Proposal*, CERN/LHCC/95-71 (1995).
2. J. Berger et al., *TPC Data Compression*, Subm. to Nucl. Instrum. Methods Phys. Res., A
3. ALICE collaboration, *Technical Design Report of the Time Projection Chamber*, CERN/LHCC 2000-001 (2000).

4. SCI (Scalable Coherent Interface) Standard Compliant, ANSI/IEEE 1596-1992.
5. H. Appelshäuser, PhD thesis, University of Frankfurt 1996.
6. J. Günter, PhD thesis, University of Frankfurt 1998.
7. STAR collaboration, *STAR Conceptual Design Report*, Lawrence Berkeley Laboratory, University of California, PUB-5347 (1992).
8. C. Adler et al., *The Proposed Level-3 Trigger System for STAR*, IEEE Transactions on Nuclear Science, Vol. 47, No.2 (2000).
9. P. Yepes, *A Fast Track Pattern Recognition*, Nucl. Instrum. Meth. **A380** (1996) 582.

The Gateway Computational Web Portal: Developing Web Services for High Performance Computing

Marlon Pierce[1], Choonhan Youn[2], and Geoffrey Fox[3]

[1] Community Grids Laboratory, Indiana University, Bloomington IN, 47405-7000, USA,
pierceme@asc.hpc.mil,
WWW home page: http://www.gatewayportal.org
[2] Department of Electrical Engineering and Computer Science, Syracuse University, Syracuse NY 13244, USA
cyoun@indiana.edu,
[3] Departments of Computer Science and Physics, School of Informatics, Community Grids Laboratory, Indiana University, Bloomington, IN 47405-7000, USA
gcf@indiana.edu

Abstract. We describe the Gateway computational web portal, which follows a traditional three-tiered approach to portal design. Gateway provides a simplified, ubiquitously available user interface to high performance computing and related resources. This approach, while successful for straightforward applications, has limitations that make it difficult to support loosely federated, interoperable web portal systems. We examine the emerging standards in the so-called web services approach to business-to-business electronic commerce for possible solutions to these shortcomings and outline topics of research in the emerging area of computational grid web services.

1 Overview: Computational Grids and Web Services

As computational grid technologies such as Globus[1] make the transition from research projects into mainstream, production-quality software, one of the key challenges that will be faced by computing centers is simplifying the use of the Grid for users, making contact with the large pool of scientists and engineers who are using computational tools on PCs and who are unfamiliar with high performance computing.

Grid computing environments[2, 3] seek to address this issue, typically through the use of browser-based web portals. Most computational portals have adopted a multi-tiered architecture, adapted from commercial web portals. The tier definitions vary from portal to portal, but in general fall into three main categories, analogous to the three components of the model-view-control design pattern: a front-end that is responsible for the generation, delivery, and display of the user interface, a back end consisting of data, resources, and specific implementations of services, and a middle tier that acts as a control layer, handling user requests

P.M.A. Sloot et al. (Eds.): ICCS 2002, LNCS 2329, pp. 503–512, 2002.

and brokering the interactions through a general layer to specific back end implementations of services. Each of these layers can consist of several sub-layers.

Most computational portals implement a common set of services, including job submission, file transfer, and job monitoring. Given the amount of overlap in these projects, it is quite natural to propose that these portals should work together so that we can avoid duplicating the effort it takes to develop complicated services and deploy them at specific sites. The challenge to interoperability is that portals are built using different programming languages and technologies. We believe the key to portal interoperability is the adoption of lightweight, XML-based standards for remote resources, web service descriptions, and communication protocols. These are by their nature independent of the programming languages and software tools used by different groups to implement their clients and services. We thus can build distributed object systems without relying on the universal use of heavyweight solutions such as CORBA[4], although these will certainly be included in the larger web services framework.

In this paper we describe our work in developing and deploying the Gateway Computational Web Portal. We consider lessons learned about the limitations of our current approach and examine how these limitations can be overcome by adopting a web services approach. We describe specific requirements for high performance computing web services and outline our future research work in these areas.

2   The Design of the Gateway Computational Web Portal

Computational web portals are designed to provide high-level user interfaces to high performance computing resources. These resources may be part of a computational grid, or they may be stand-alone resources. As argued in [5], the real value of computational portals is not just that they simplify access to grid services for users but also that they provide a coarse-grained means of federating grid and non-grid resources. Also, because they are web-based, portals provide arguably the correct architecture for integrating the highly specialized grid technologies (suitable for development at national laboratories and universities) with commercial software applications such as databases and collaborative technologies.

We have developed and are deploying a computational portal, Gateway [6, 7, 5], for the Aeronautical Systems Center and Army Research Laboratory Major Shared Resource Centers, two high performance computing centers sponsored by the US Department of Defense's High Performance Computing Modernization Program. The primary focus of this project is to assist PC-based researchers in the use of the HPC resources available from the centers. In addition, as we have described elsewhere [8], this approach also has potential applications in education.

We note here for clarity that computational web portals are simply special cases of general web portal systems. We concentrate on HPC services because this is the scope of the project, not for any technical reasons. Computational portals can (and should) take advantage of more typical commercial portal services such as address books, calendars, and information resources. Another interesting area of work is the combination of computational portals with collaborative portal tools, as described in Ref. [8].

2.1   Gateway XML Descriptors

XML metadata descriptions serve as the interface to host machines, applications, and available services, describing what codes and machines are available to the user, how they are accessed and used, and what services can be used to interact with them. We store this information in a static data record on the web server and use it to both generate dynamic content for the user and to generate back end requests. For example, a description of an application includes the number of input files required to run it, so the same display code can be used to generate input forms for different codes.

Application Description By application, we refer specifically to third party scientific and engineering codes. All of these have common characteristics for running on a command line, so in our application description we seek to capture this information in an XML data record. For a particular application, we need to capture at least the following to run it:

1. The number of input files the code takes.
2. The number of input parameters the code takes.
3. The number of output files the code generates.
4. The input/output style the code uses.
5. Available host machines (see below).

By input and output files, we refer specifically to data files. Parameters are anything else that needs to be passed to the code on the command line, such as the number of nodes to use in a parallel application. This is highly code specific. I/O style includes standard Unix redirects or C-style command line arguments.

An example of an XML description is given in [8]. We note here that we have found it useful to adopt a shallow tree structure with general tag names. The information in the XML tree consists of name/value pairs, and we use general code for extracting this information from the data record. Thus, we do not make the pretense that our description above is complete. We can extend the XML description later with additional name/value pairs without having to rewrite any code that handles the XML.

HPC Description We have developed a description mechanism for HPC systems using the same viewpoint as our Application Description: we want a data record that contains a minimal amount of information, but which can be extended as needed, since we cannot anticipate all requirements of all codes and hosts from the beginning. The software that handles this code must of course be general in order to handle future description extensions.

Let us now examine the minimal contents of a Host Descriptor. This would include the following:

1. The DNS name of the host.
2. The type of queuing system it uses.
3. The full path of the executable on the host.
4. The working or scratch directory on the host.
5. The full path to the queue submission executable.

Again, an example of this can be seen in [8]. Note we have adopted as a convention that the Application Descriptor is the root of the tree, and the host machines are sub-nodes.

Service Description   As described in 2.2, Gateway provides a number of generic services. These are implemented using WebFlow (Java- and CORBA-based middleware). However, the services themselves should be independent of the implementation. Thus all computational portals could potentially use the same interface description for a particular set of services, and any particular portal could radically redesign its middleware without changing the service interface.

An example of one of Gateway's Service Descriptors is given in [8]. For a single implementation, this is somewhat redundant, since for WebFlow we must translate this into IDL suitable for dynamic CORBA. Now that a standard for service descriptions (the Web Services Description Language, or WSDL[13]) is available from the larger web community, we will adopt this in future development work.

2.2   Gateway Architecture

For Gateway, we identified the following as the core services that we must implement in the portal:

1. Secure identification and authorization over an extended session;
2. Information services for accessing descriptions of available host computers, applications, and users;
3. Job submission and monitoring;
4. File transfer;
5. Remote file access and manipulation;
6. Session archiving and editing.

These services constitute what we term a system portal: it is not tied to specific back end applications, instead implementing these services in a generic fashion that is application neutral. Applications from specific computational areas such as chemistry or structural mechanics can be added to the portal in a well-defined fashion through the XML descriptors.

Gateway's implementation of these services is schematically illustrated in Figure 1. Clients, typically browsers but also custom applications for file browsing, contact a web server over a secure HTTP connection. Security details are highly site-specific. The user interface is implemented using JavaServer Pages (for display) combined with specially written Java components maintained by the web server's servlet engine. These latter represent a set of objects local to the server. These components can handle specific server-side tasks but can also act as proxies for our distributed object software, WebFlow. WebFlow components can be distributed amongst different, geographically distributed computers. WebFlow components in turn act as proxies to back end resources, including stand alone HPC resources plus also computing grids and batch visualization resources. These back end resources are reached via remote shell operations. We are in the process of developing specific components for Globus using the Java CoG Kit[12]. Thus WebFlow provides a single entry point to distributed resources. For a more comprehensive description Gateway, please see [7, 5].

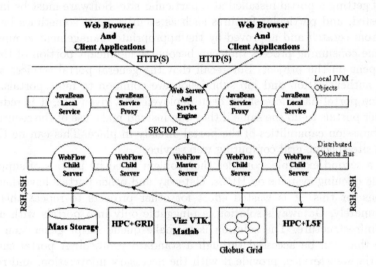

Fig. 1. The Gateway computational portal is implemented in a multi-tiered architecture. SECIOP is a wire protocol for secure CORBA. RSH and SSH are secure remote shell commands.

The services and general architectural approach described above are not unique to Gateway. Most other portal projects implement their versions of these services as well. An extensive list of such projects is given in Refs. [2, 3]. In particular, the HotPage/GridPort [9], Mississippi Computational Web Portal [10], and NASA Information Power Grid [11] provide additional examples of standard computational web portals. As we will describe below, these services thus should be viewed as the atomic services of a computing grid web services system in the next generation of portal systems

2.3   The Need for Interoperability

There are some limitations to the traditional portal approach. Primarily, it is concerned with accessing heterogeneous back end resources through a particular middle tier software implementation. Thus as designed now, most portal projects are not interoperable as they are tied to their particular implementation of the control layer. Yet interoperability is desirable both from the point of view of good software design and the realities of deployment. Many different groups have undertaken portal projects. Even within the DOD's Modernization Program there have been several such projects. Each portal tends to carve out a fiefdom for itself and none of these portals should expect to become the single solution for the entire program. Instead, it is in their interest to work together in order to take advantage of each project's particular strengths and avoid duplication of effort.

To see how this is an advantage, consider the difficulties (technical and political) of getting a portal installed at a particular site. Software must be installed and tested, and particular features such as security must sometimes be developed from scratch and approved by the appropriate management groups. This is a time consuming process that can become a significant portion of the total time spent on the project. But recall that the general portal services such as secure authentication and job submission are common to most portals. Thus, once one portal has successfully been deployed at a given site, it is redundant for other portals to do the same if they can instead make use of the security and job submission capabilities of the portal already in place. This can be thought of as a site-specific grid computing web service.

As a second example, consider the problem faced by portals of supporting multiple queuing systems on HPCs. Gateway and other portals have developed solutions for this. It is wasted effort for other projects to repeat this work. Unfortunately, Gateway's solution is pluggable only into portals with a Java-based infrastructure. The solution is to make queue script generation into a service that can be accessed through a standard protocol. A portal can then access the web service, provide it with the necessary information, and retrieve the generated script for its own use. This is an example of a function-specific grid computing web service.

3   Towards a Grid Web Service Portal Architecture

Given the potential advantages of interoperability, the proper architecture must now be determined. For this we can take our cue from the design goals of the underlying Grid software[14]: portal groups need to agree to communication protocols. Once this is done, they don't need to make further agreements about control code APIs and service implementations so long as the protocols are supported. The web services model [15, 16] has the backing of companies such as IBM[17] and Microsoft[18] as the standard model for business-to-business commerce. Let us now consider the aspects of this model for high performance computing.

Distributed Object Model The point of view that we take is that all back end resources should be considered as objects. These may be well-defined software objects (such as CORBA objects) but may also be hardware, applications, user descriptions, online documents, simulation results, and so forth. As described below, these objects will be loosely coupled through protocols rather than particular APIs.

Resource and Service Descriptions The generalized view of resources as objects requires that we describe the meta-data associated with the object and provide a means of locating and using it. XML is appropriate for the first and we can assign a Uniform Resource Identifier (URI) for the latter. We must likewise provide an XML description of middle-tier services (see Section 2.2). WSDL is an appropriate descriptor language. Note that this defines objects in a language independent way, allowing them to be cast into the appropriate language later, using for example Castor[20] to create Java objects. Likewise, this distributed object system is protocol-independent. The object or service must specify how it is accessed, and can provide multiple protocols.

Resource and Service Discovery Once we have described our object, it must be placed in an XML repository that can be searched by clients. Clients can thus find the resource they are looking for and how to access it. Commercial web services tend to use UDDI[19], but other technologies such as LDAP servers may also be used. Castor provides a particularly powerful way of converting between Java components and XML documents and when combined with an XML database, can serve as a powerful object repository specialized for Java applications [8].

Service Binding Following the discovery phase, the client must bind to the remote service. WSDL supports bindings to services using different mechanisms (including SOAP and CORBA). Thereafter, interactions can be viewed as traditional client-server interactions.

There is nothing particularly new about this architecture. Client-server remote procedure calls have been implemented in numerous ways, as has the service repository. WebFlow, as just one example, implements a specialized naming service for looking up servers and their service modules. The difference here is that the repository is designed to offer a standard XML description of the services using WSDL. The service description itself is independent of wire protocols and Remote Procedure Call (RPC) mechanisms.

There still is the problem that the client must implement the appropriate RPC stubs to access the server. Given the number of protocols and RPC mechanisms available, there is an advantage in the various grid computing environment projects coming to an agreement on a specific RPC protocol. Guided by the surge of development in support of SOAP[21] for web services and its compatibility with HTTP, we suggest this should be evaluated as the appropriate *lingua franca* for interoperable portals. Legacy and alternative RPC mechanisms (such as used by WebFlow) can be reached through bridges.

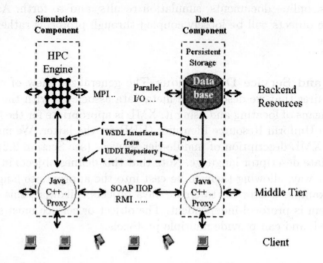

Fig. 2. The extended multi-tiered architecture for web services will be used to support interoperability between portals.

The architecture for this kind of web services system is illustrated in Figure 2. The basic point is that common interfaces can be used to link different multi-tiered components together. In the shown instance we have two distinct services (and HPC service and a database service) perhaps implemented by different groups. Clients can access these services by first contacting the service description repository. Essentially, the local object sub-tier and the distributed object sub-tier of the control layer in Figure 1 become decoupled. The local object service proxy must consult the service repository and find a service to use. It then binds to that service with the protocol and mechanism prescribed in the service description.

The above constitutes the basis for web services. We now identify some important extensions that will need to be made.

Security for high performance computing services needs consideration. First, the very fact that a particular web service is available from a certain site may be sensitive information. Thus the service repository must have access and authorization controls placed upon it in order to allow clients to only see services that they are authorized to use. Furthermore, the services themselves must be secure, as they typically cannot be used anonymously. Specific security extensions required by the service description include

1. Authentication mechanisms for client and/or server.
2. Means of identification (certificate, public or private key, session ticket).
3. Requirements for privacy and data integrity.

Within a limited realm (a Globus Grid or interoperating Kerberos realms) this can be greatly simplified through existing single sign-ons capabilities.

A second important consideration for HPC services is the relationship between job composition, workflow, and events. The simpler web services can be thought of as atomic components that can be used to compose complicated services. For example, a job can be scheduled and its output filtered and transfered to a visualization service for analysis. This entire task can be considered to be a single, composite job. Thus we need to define an XML dialect for handling the above type of workflow, together with appropriate software tools for manipulating these workflow documents and executing their content. We specifically see this as an agent that executes the workflow commands, contacting the service repository at each stage and reacting appropriately to event messages from the back end. Job composition is not new: WebFlow has this capability, for example. The difference now is that we must define workflow in an implementation-independent way (again in XML) so that different groups can provide different pieces of the combined service.

Workflow for grids is a complicated undertaking because of the numerous types of error conditions that can occur. These include failures to reach the desired service provider at any particular stage of the workflow either because of network problems or machine failure. Likewise, even if the resource is reached and the service is executed, failure conditions exist within the application. For example, the file may be corrupt and generate incorrect output, so there is no need to attempt the next stage of the workflow procedure. We can thus see that any workflow description language must be coupled with reasonably rich event and error description language.

4   Acknowledgments

Development of the Gateway Computational Web Portal was funded by the Department of Defense High Performance Computing Modernization Program through Programming Environment and Training.

References

1. The Globus Project. http://www.globus.org
2. Grid Computing Environments. http://www.computingportals.org.
3. Grid Computing Environments Special Issue.
   Concurrency and Computation: Practice and Experience.
   Preprints available from http://aspen.ucs.indiana.edu/gce/index.html.
4. Common Object Request Broker Architecture. http://www.corba.org
5. Pierce, M., Fox, G., and Youn, C.: The Gateway computational Web Portal. Accepted for publication in Concurrency and Computation: Practice and Experience.
6. Gateway Home Page. http://www.gatewayportal.org.
7. Fox, G., Haupt, T., Akarsu, E., Kalinichenko, A., Kim, K., Sheethalnath, P., Youn, C.: The Gateway System: Uniform Web Based Access to Remote Resources. ACM Java Grande Conference (1999).

8. Fox, G., Ko, S.-H., Pierce, M., Balsoy, O., Kim, J., Lee, S., Kim, K., Oh, S., Rao, X., Varank, M., Bulut, H., Gunduz, G., Qiu, X., Pallickara, S., Uyar, A., Youn, C.: Grid Services for Earthquake Science. Accepted for publication in Concurrency and Computation: Practice and Experience.
9. NPACI HotPage. https://hotpage.npaci.edu.
10. Mississippi Computational Web Portal. http://www.erc.msstate.edu/labs/mssl/mcwp/index.html
11. Information Power Grid. http://www.ipg.nasa.gov
12. von Laszewski, G., Foster, I., Gawor, J., Lane, P.: A Java Commodity Grid Kit. Concurrency and Computation: Practice and Experience, **13** (2001) 645-662.
13. WSDL: Web Services Framework. http://www.w3c.org/TR/wsdl.
14. Foster, I., Kesselman, C., Tuecke, S.: The Anatomy of the Grid: Enabling Scalable Virtual Organizations. Intl. J. Supercomputer Applications, **15** 2001.
15. Saganich, A.: Java and Web Services. http://www.onjava.com/pub/a/onjava/2001/08/07/webservices.html
16. Curbera, F.: Web Services Overview. http://www.computingportals.org/GGF2/WebServicesOverview.ppt
17. IBM Web Services Zone. http://www.ibm.com/developerworks/webservices
18. Microsoft Developer Network. http://msdn.microsoft.com
19. UDDI: Universal Descriptors and Discovery Framework. http://www.uddi.org.
20. The Castor Project. http://castor.exolab.org

Evolutionary Optimization Techniques on Computational Grids[*]

Baker Abdalhaq, Ana Cortés, Tomás Margalef and Emilio Luque
Departament d'Informàtica, E.T.S.E, Universitat Autònoma de Barcelona, 08193-
Bellaterra (Barcelona) Spain

baker@aows10.uab.es
{ana.cortes,tomas.margalef,emilio.luque}@uab.es

Abstract. Optimization of complex objective functions such as environmental models is a compute-intensive task, difficult to achieve by classical optimization techniques. Evolutionary techniques such as genetic algorithms present themselves as the best alternative to solving this problem. We present a friendly optimization framework for complex objective function on a computational grid platform, which allows easy incorporation of new optimization strategies. This framework was developed using the MW library and the Condor system. The framework architecture is described, and a case study of a forest-fire propagation simulator is then analyzed.

1 Introduction

Models are increasingly being used to make predictions in many fields of environmental science Typically, simulation (a computer program that represents a given model) exhibits difficulties in exactly imitating the real behavior of the simulated system. Basically, this difficulty lies in two main considerations. On the one hand, environmental models frequently require information on a high number of input variables about which it is not usually possible to provide accurate values for all. Therefore, the results provided by the corresponding simulator usually deviate from the system's real behavior. On the other hand, models can be wrongly designed becoming the model itself the main cause of erroneous results.

Generally, model/simulator analysis is mainly focused on one of these two sources of errors. Whichever analysis goal is chosen, the degree of complexity in model analysis, which can involve model validation, verification and calibration, resulted in particularly difficult tasks in terms of computing power.
Since distributed computing systems (also called metacomputers or grid computing environments) have, over the last decade, increased as a great source of computing

* This work was supported by the CICYT under contract TIC 98-0433 and partially supported by the DGICYT (Spain).

P.M.A. Sloot et al. (Eds.): ICCS 2002, LNCS 2329, pp. 513–522, 2002.
© Springer-Verlag Berlin Heidelberg 2002

power, relevant optimization works were addressed to these environments [1][2][3]. All these works are focused on mathematical optimization techniques, which can be applied when some degree of knowledge about the function to be optimized is provided. However, there is still considerable work to do in the area of non-mathematical approaches that are more suitable for optimization processes dealing with functions about which no information is provided. Environmental models/simulators are a good example of this kind of functions. Usually, environmental simulators are provided as black-boxes, and the only known information about them consists of how they should be provided with input information, and the nature of the results that they generate. Under this assumption, mathematical optimization techniques are destined to failure, and non-mathematical schemes such as Genetic Algorithms, Simulating Annealing, Tabu Search and so on, have arisen as the best candidates to approach such problems.

The aim of this work is to provide a framework for the optimization of complex functions (considered as black-box functions) that take advantage of the computing power provided by a grid-computing environment. The BBOF (Black-Box Optimization Framework) system has been developed using the master-worker programming paradigm and uses Condor as a distributed resource management.

The rest of the paper is organized as follows. In section 2, the problem statement is reported. The architecture of the proposed optimization framework is described in section 3. Implementation details are described in section 4. Section 5 shows how BBOF has been applied to a forest-fire propagation simulator and, finally, section 6 presents the main conclusions.

2   Problem Statement

Formal optimization is associated with the specification of a mathematical objective function (called L) and a collection of parameters that should be adjusted (tuned) to optimize the objective function. This set of parameters is represented by a vector referred to as θ. Consequently, one can formulate an optimization problem as follows:

$$\text{Find } \theta^* \text{ that solves } \min_{\theta \in S} L(\theta) \tag{1}$$

where $L : R^p \to R^1$ represents some *objective function* to be minimized (or maximized). Furthermore, θ represents the vector of adjustable parameters (where θ^* is a particular setting of θ), and $S \subseteq R^p$ represents a constrain set defining the allowable values for the θ parameters. Put simply, the optimization problem deals with the aim of defining a process to find a setting for the parameter vector θ, which provides the *best* value (minimum or maximum) for the *objective function L*. This search is carried out according to certain restrictions of the values that each parameter can take.

The whole range of possibilities that can be explored in obtaining the optimization goal is called the *search space,* which *is* referred to as *S*.

As we mentioned in the previous section, we are interested in complex model optimization regardless how the model itself works. Under this assumption, the underlying model/simulator is identified as a complex black-box function about which no information is provided. However, there is the possibility of measuring the quality of the results provided by the simulator for any input vector (θ). Consequently, the objective function (L) involves both executing the simulator and the quality function simultaneously, being the final value provided by the quality function, the value to be minimized or maximized.

Typically, the way to solve equation (1) consists of applying an existent optimization technique to an initial guess for θ in order to obtain a new set of input parameters closer to the optimal solution (θ*). However, applying any optimization technique once alone never leads to a good solution. Therefore, the same process is repeated again by starting with a new or nearly guess.

Beside the strictly mathematical definition of the optimization problem, there are some extra issues that should be considered when dealing with geographically dispersed CPU´s as source of computing power. Some of these features include the dynamic availability of the machines, communication delays between processors, heterogeneity system and scheduling problems. Despite of all these drawbacks, these platforms are well suited to large-scale computations needed by environmental model optimization problems.

3 Distributed optimization framework

As was commented above, optimization techniques typically obtain progressive improvements in the original guess of the vector θ by consecutive executions of the optimization process. A great improvement in this way of proceeding is to consider, not only one guess at a time, but a wide set of guesses and, based on the results obtained for all of these, to automatically generate a new set of guesses and to re-evaluate the objective function for them all, and so on. For this reason, our optimization framework works in an iterative fashion, where it moves step-by-step from an initial set of guesses for θ to a final value that is expected to be closer to the true θ * (optimal vector of parameters) than the initial guesses. This goal is achieved because, at each iteration of this process, a preset optimization technique is applied to generate a new set of guesses that should be better than the previous one.

This iterative scheme is the core of the proposed framework, which is called the Black-Box Optimization Framework (BBOF). BBOF has been implemented in a plug&play fashion. On the one hand, the optimized function can be any system (complex simulator) that has a vector of parameters as input, and provides one or more values as output (a quality function should also be provided to determine the goodness of the results). On the other hand, any optimization technique ranging from evolutionary techniques, such as genetic algorithms or simulating annealing through strictly mathematical schemes, can easily be incorporated.

Due to its characteristics, the proposed framework fits well into the master-worker programming paradigm working in an iterative scheme. An iterative master-worker application consists of two entities: a master and multiple workers. The master is responsible for decomposing the problem into small task (and distribute these tasks among a farm of worker processes), as well as for gathering the partial results in order to produce the final result of the computation. The worker processes receive a message from the master with the next task, process the task, and send the results to the master. The master process may carry out certain computations while tasks of a given batch are being completed. After that, a new batch of tasks is assigned to the master, and this process is repeated several times until completion of the problem (after K cycles or iterations). Figure 1.a schematically shows how master and workers processes interact during one iteration of the iterative process described above, when 3 workers are considered and the number of tasks to be executed by the workers is 8.

If we analyze the above-described behavior, we can easily match each element with the main components of BBOF. In particular, since the evaluation of the black-box and quality functions for each guess of θ are independent of each other, they can be identified as the work done by the workers. Consequently, the responsibility for collecting all the results from the different workers and for generating the next set of guesses by applying a given optimization technique will be concentrated on the master process. Figure 1.b graphically illustrates how this matching is undertaken.

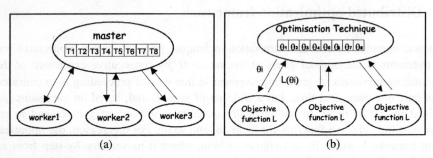

Fig 1. Master and worker process interaction (a) and how these processes are matched to the BBOF elements (b).

Once the architecture of BBOF has been described, we now describe the manner in which the BBOF was implemented.

4 Implementation issues

We should bear in mind that our initial goal was to develop an optimization framework for solving complex problems on a computational grid. For this purpose, we have used Condor as a resource management on a grid-computing environment [4] and the MW library [5] so as to easily implement our master-worker scheme. In the following section, we briefly outline certain basic ideas on Condor and MW.

4.1 Condor

Condor is a software system that runs on a cluster of workstations in order to harness wasted CPU cycles. Condor was first developed for Unix systems, and can currently be executed on a wide range of machines. A Condor pool consists of any number of machines, possibly heterogeneous, which are connected by a network. One machine, the central manager, keeps track of all the resources and jobs in the pool. Condor allows High Throughput Computing (HTC) to be obtained, that is, the obtention of large amounts of processing capacity sustained over long time periods. The resources, i.e. hardware, middleware and software, are large, dynamic and heterogeneous.

4.2 MW

MW is a set of C++ abstract-based classes that must be re-implemented in order to fit the particular application. These classes hide difficult metacomputing issues, allowing rapid development of sophisticated scientific computing applications. Basically, there are three abstract base classes to be re-implemented. The *MWDriver* class correspondents to the master process and contains the control center for distributing tasks to workers. The *MWTask* class describes inputs and outputs –data and results – which are associated with a single unit of work. The *MWWorker* class contains the code required to initialize a worker process and to execute any tasks sent to it by the master. BBOF's implementation is based on these MW classes. As mentioned in section 3, BBOF has two main parts: the objective function (black-box and quality functions) and the optimization technique. In particular, the objective function (L) corresponds to the *MWWorker* class, whereas the optimization technique has been integrated as the *MWDriver* class. Finally, a particular instance of the *MWTask* class directly matches with a given input parameter vector. Concerning communication, there are MW versions that perform communications by using PVM, the file system and sockets. In this work, MW was used with PVM [6], where MW workers are independent jobs spawned as PVM programs. The original MW does not consider multiple iterations of the process depicted in figure 1.a. However, our optimization framework is based on an iterative way of working. For this purpose, we have developed our optimization framework based on an extended version of MW [7], which allows MW to iterate a predetermined number of times.

5   Case study: Forest-fire propagation

Forest fire propagation is considered a challenging problem in the area of simulation, due to the complexity of its physical model, the need for great amount of computation and the difficulties of providing accurate inputs to the model. Uncertainties in the input variables needed by the fire propagation models (temperature, wind, moisture, vegetation features, topographical aspects...) can have a substantial impact on result errors and must be considered. For this reason, optimization methodologies to adjust the set of input parameters of a given model would be provided in order to obtain results as close as possible to real values.

In particular, we have applied the proposed optimization framework to a fire propagation simulator called ISStest [8], which, from an initial fire front (set of points), generates the position of the fire line after a given period of time. As has just been mentioned, forest fire propagation simulators deal with a wide set of input parameters. In the experimental study reported below, we only focus on finding the proper values of the wind input parameter, which is defined as a vector composed of two components: wind speed (w_s) and wind direction (w_d).

Obviously, the expected behavior of the ISStest is that the new fire line provided after a certain preset time exactly matches the real situation of the fire line once that time has passed. Therefore, our objective function L, in this case, consists of executing once ISStest for a given configuration of input values, and evaluating the distance between the real fire line and the fire line obtained by simulation. For this purpose, the Hausdorff distance [9] was chosen. The Hausdorff distance measures the degree of mismatch between two sets of points by measuring the distance of a point from one set that is farthest away from any point of the other, and vice versa. Formally, the directed Hausdorff distance h between two set of points M and I at a specific point in I is:

$$h(M,I) = \max_{m \in M} \left(\min_{i \in I} (distance\ (m,i)) \right) \tag{2}$$

Thus, the Hausdorff distance H is defined as: $H(M,I) = max(h(M,I), h(I,M))$. In this case, our objective resides in finding the global minimum of this function, given that our aim is that the simulated and real fire lines should be the same, and that the difference between the two should therefore be 0.

The experiments carried out considered that wind speed and wind direction remain fixed during the fire-spread simulation process. The real fire line, which was used as a reference during the optimization process, was obtained in a synthetic manner. In other words, we fixed known values for w_s and w_d parameters and, subsequently, the ISStest simulator was executed three times with a setting of the simulation time equal to15 minutes for each execution. This presupposes that the fire propagation was spent 45 minutes. At each execution, the output fire line obtained was used as the initial fire line for the next execution. Finally, the obtained fire line was stored and treated as the real fire line. Obviously, the wind speed and wind direction used during this process (which were 15km/h and 180^0 respectively) were dismissed once all this process had finished. Since for each ISStest execution we have two parameters (wind speed and wind direction), and bearing in mind that we executed the fire simulator three times, the global vector that should be guessed by the optimization technique consists of 6 elements ($\theta = \begin{pmatrix} w_{s1} & w_{d1} & w_{s2} & w_{d2} & w_{s3} & w_{d3} \end{pmatrix}$).

We now describe the experimental study carried out using the forest fire propagation simulator commented above.

5.1 Experimental platform

The experiments reported below were executed on a Linux cluster composed of 21 PC's with Intel Celeron processor 433 MHz, each one having 32 MB RAM and connected with a Fast Ether Net 100 Mb. All the machines are configured so as to use NFS (Network File System) and the Condor system; additionally, PVM are installed on them all.

5.2 Implemented optimization techniques

In this experimental study, we considered two different optimization techniques. Genetic Algorithms (GA) were chosen as a relevant non-mathematical technique within the branch of evolutionary strategies, whereas, on behalf of mathematical optimization techniques, we have adopted a statistical/parabolic method. In the following section, we will outline the main features of each one of these techniques.

Genetic Algorithm
GAs are characterized by emulating natural evolution [10]. Therefore, under the GA scenario, in which one refers to a vector of parameters as chromosome, a gene is identified with one parameter and the population is the set of vectors (chromosome) used for one iteration of the algorithm. The way to obtain the new population consists of applying certain operators (called transmission operators) to the starting set of vectors. These operators (Elitism, Selection, Crossover and Mutation) are in charge of defining the changes to be done on the set of initial vector guesses in order to generate an improved set of vectors.

Statistical/parabolic method
As we previously mentioned, mathematical techniques are supposed to work incorrectly for the kind of problem that we are approaching. However, we implemented a statistical/parabolic method in order to compare the results provided by this scheme with those obtained with the evolutionary techniques described above. This particular

method was chosen once the behavior of the objective function had been outlined for a very simple case. In particular, we considered the simplest case in which the input vector is only composed of two parameters (wind speed and wind direction). This is the case in which the ISStest is executed only once for a fixed period of time, and for which the wind does not change during the whole simulation. Figure 2 shows the value of the objective function when wind speed is fixed to the corresponding real value and the free parameter (wind direction) varies within its corresponding search space (from 0° to 360°). Since a similar behavior was observed for wind direction, a good approximation to the obtained shape seemed to be the parabolic one.

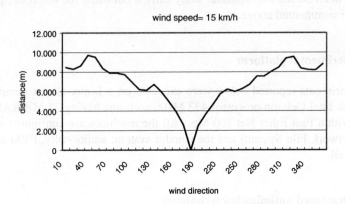

Fig.2 Objective function cross-section at optimal wind speed (15 km/h).

The formal expression to describe this kind of functions corresponds to equation 2, where f is the approximation of the objective function L.

$$f(X) = \beta_0 + \beta_{11} x_1 + \beta_{12} x_1^2 + \dots + \beta_{1n} x_1^n \qquad (3)$$
$$+ \dots + \beta_{p1} x_p + \beta_{p2} x_p^2 \dots + \beta_{pn} x_p^n$$

In order to find the optimum of this function using calculus, we need to derive it for each variable x_i and to find the zero of the derivative. The formula describing this derivation form and the equation we should solve is the following:

$$\frac{\partial f(X)}{\partial x_i} = \beta_{i1} + 2\beta_{i2} x_i + \dots + n\beta_{i2} x_i^{n-1} = 0$$

$$\forall i = 1 \dots p \qquad (4)$$

This approach can help us, if and only if, the objective function can be approximated with good precision to some function that is calculus friendly.

5.3 Experimental results

In this section, we will describe the main results of a preliminary set of experiments, which were performed with the aim of comparing both mathematical optimization techniques (parabolic method) against a new generation of strategies such as genetic algorithms. As we commented, our optimization framework was implemented by using an iterative master-worker programming scheme. In the case of the mathematical approach, the optimization process was iterated once, and the number of vectors evaluated (task to be assigned to the workers) were 200. In contrast to this, when the genetic algorithm was applied, the number of iterations of the whole optimization process were 20, whereas the number of vectors to be evaluated at each iteration were 10. Therefore, the total number of times that the objective function was evaluated is the same in both cases. Furthermore, since the initial set of guesses for both strategies was obtained in a random way, the global optimization process was performed four times; the mean values for all the results (Hausdorff distance in m.) is shown in table 1.

Table 1. Comparison between all the algorithms for homogeneous wind field

Algorithm	Genetic	Statistical/parabolic
Distance (m)	11	443
Evaluations	200	200

As we can observe, GA on average reaches a distance equal to 11m, which is one order of magnitude less than the result obtained in the case of the mathematical approach. The mathematical technique suffers both from deviation from the optimal due to the model and the rare shape of the objective function.

6   Conclusions

In this work, we have described a friendly framework for optimizing black-box functions called BBOF (Black-Boxes Optimization Framework). This framework was developed using the MW library and the Condor system. We applied BBOF to a forest-fire propagation simulator (ISStest) including two optimization techniques: the Genetic Algorithm and the Statistical/parabolic method.

A basic set of experiments were performed and the results denote the difficulty exhibited by classical optimizers in minimized complex objective function such as the one studied. In contrast to this, evolutionary optimization techniques such as genetic algorithms provide substantial improvements in results. These preliminary results have encouraged us to continue our study in this area, as they confirm our expectations.

References

1. Czyzyk J., Mesnier M.P., More J.J.: The Network-Enabled Optimization Systems (NEOS) Server. Preprint MCS-P615-1096, Argonne National Laboratory, Argone, Illinois, (1997).
2. Ferris M., Mesnier M., More J.: NEOS and Condor: Solving optimization problems over the Internet, Preprint ANL/MCS-P708-0398, available at http://www-unix.mcs.anl.gov/metaneos, (March 1998)
3. Linderoth J. and Wright S.J: Computational Grids for Stochastic Programming, Optimization. Technical Report 01-01, UW-Madison, Wisconsin-USA,(October 2001)
4. Livny M. and Raman R.: High Throughput Resource Management, Computational Grids: The Future of High-Performance Distributed Computing. Edited by Ian Foster and Carl Kesselman, published by Morgan Kaufmann (1999)
5. Goux J.-P., Kulkarni S., Linderoth J., Yoder M.: An enabling framework for master-worker applications on the computational grid. Proceedings of the Ninth IEEE Symposium on High Performance Distributed Computing (HPDC9), Pittsburgh, Pennsylvania, (August 2000) 43–50
6. Geist A., Beguelin A., Dongarra J., Jiang W., Manchek R. and Sunderam V.: PVM: Parellel Virtual Machine A User's Guide and Tutorial for Networked Parallel Computing. MIT Press (1994)
7. Heymann E., Senar M.A., Luque E. and Livny M.: Evaluation of an Adaptive Scheduling Strategy for Master-Worker Applications on Clusters of Workstations. Proceedings of the 7th International Conference on High Performance Computing (HiPC'2000), LNCS series, vol. 1971 (2000) 214–227
8. Jorba J., Margalef T., Luque E., J. Campos da Silva Andre, D. X Viegas: Parallel Approah to the Simulation Of Forest Fire Propagation. Proc. 13 Internationales Symposium "Informatik fur den Umweltshutz" der Gesellshaft Fur Informatik (GI). Magdeburg (1999)
9. Reiher E., Said F., Li Y. and Suen C.Y.: Map Symbol Recognition Using Directed Hausdorff Distance and a Neural Network Classifier. Proceedings of International Congress of Photogrammetry and Remote Sensing, Vol. XXXI, Part B3, Vienna, (July 1996) 680–685
10.Thomas Baeck, Ulrich Hammel, and Hans-Paul Schwefel: Evolutionary Computation: Comments on the History and Current State. IEEE Transactions on Evolutionary Computation, Vol. 1, num.1 (April 1997) 3–17

Eclipse and Ellipse: PSEs for EHL Solutions using IRIS Explorer and SCIRun

Christopher Goodyer and Martin Berzins

Computational PDEs Unit, School of Computing, University of Leeds, LS2 9JT, UK
ceg@comp.leeds.ac.uk, martin@comp.leeds.ac.uk
http://www.comp.leeds.ac.uk/cpde/

Abstract. The development of Problem Solving Environments (PSEs) makes it possible to gain extra insight into the solution of numerical problems by integrating the numerical solver and solution visualisation into one package. In this paper we consider building a PSE using IRIS Explorer and SCIRun. The differences in these two PSEs are contrasted and assessed. The problem chosen is the numerically demanding one of elastohydrodynamic lubrication. The usefulness of these packages for present and future use is discussed.

1 Introduction

The field of scientific computing is concerned with using numerical methods to solve real world problems in fields such as engineering, chemistry, fluid flow or biology, which are typically defined by a series of partial differential equations. In solving these problems the ability to use high quality visualisation techniques allows the user to better understand the results generated, to identify any points of interest, or potential difficulties and to obtain greater insight into the solution to a problem more quickly.

This paper will investigate the use of *Problem Solving Environments* (PSEs) to solve a demanding numerical problem in computational engineering. PSEs combine several important stages for the generation of numerical results into one body, thus having synchronous computation and visualisation. There are three ways in which even basic PSEs are advantageous. These are that the input parameters can all be set, or adjusted at run time; the numerical solver is only one part of the PSE and hence it can be possible to change solution methods, if appropriate; and finally the visualisation is an innate component of the package, and results can be visualised and studied as the calculation proceeds. Computational steering gives the PSE another advantage over traditional solution methods because this allows the test problem and/or the solution methods, to be updated during the calculation. The user, thus, "closes the loop" of the interactive visually-driven solution procedure [1].

The PSE construction in this paper has been done in order to compare, contrast and assess the usefulness and the ease of implementation of a challenging engineering problem, in IRIS Explorer [2] and SCIRun [1], two different software packages designed for building PSEs.

P.M.A. Sloot et al. (Eds.): ICCS 2002, LNCS 2329, pp. 523–532, 2002.
© Springer-Verlag Berlin Heidelberg 2002

The numerical problem selected for integration into a PSE is that of *elasto-hydrodynamic lubrication* (EHL) in, for example, journal bearings or gears. This mechanical engineering problem requires sophisticated numerical techniques to be applied in order to obtain solutions quickly. An engineer, for example, may want to establish solution profiles for a particular lubricant under certain operating conditions. With a PSE these could be quickly tuned to give the desired results, before tackling, say, a more demanding transient problem. The numerical code for solving EHL problems used in these PSEs is described in detail in [3], and the required changes to its structure will be set out below. Examples of where this extra insight occurs will be given.

The EHL problem will be described briefly in Section 2 which includes the equation system to be solved, and an outline of the numerical techniques used in the code. Section 3 considers the two PSEs developed. After some general issues have been covered, in Section 3.1 details of the implementation of the EHL PSE into IRIS Explorer, known as ECLIPSE (ElastohydrodynamiC Lubrication Interactive Problem Solving Environment) are given. In Section 3.2 SCIRun is discussed in a similar manner, with the construction of the PSE, known inside SCIRun as ELLIPSE, detailed. In Section 4 the differences between the two systems discussed, both conceptually and in terms of their structure and usefulness as frameworks for building PSEs. Finally, in Section 5 some conclusions are drawn about future development of PSEs bearing in mind the likely needs, in reference to large problem sizes, parallelism and grid-based computations.

2   The Numerical Problem

Elastohydrodynamic lubrication occurs in journal bearings and gears, where, in the presence of a lubricant, at the point of contact there is a very large pressure exerted on a very small area, often up to 3 G Pa. This causes the shape of the contacting surfaces to deform and flatten out at the centre of the contact. There are also significant changes in the behaviour of the lubricant in this area.

A typical solution profile is shown in Figure 1a for the pressure profile across a point contact, such as will be considered in this paper. This is the equivalent situation of a ball bearing travelling along a lubricated plane. With oil flow from left to right it can be seen that in the inlet region there is a very low pressure, in the centre of the contact there is a very high pressure, with an even higher pressure ridge around the back of the contact. Finally, in the outflow a cavitation region is formed where bubbles of air have entered the lubricant, and the pressure is assumed to be ambient there. The corresponding surface geometry profile for the bearing has the undeformed parabolic shape of the contact flattened in the high pressure area. The centreline solution is shown in Figure 1b.

The history of the field is detailed out in papers such as [4]; much information about the numerical techniques currently used to obtain fast, stable solutions is given in both [5] and [3], the latter of which describes in great detail the precise methods used in the code employed in this work.

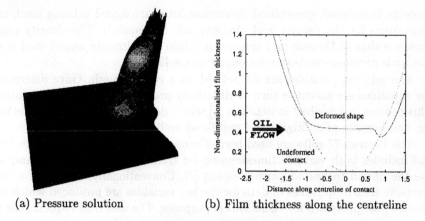

(a) Pressure solution   (b) Film thickness along the centreline

Fig. 1. Typical solutions across an EHL point contact

The EHL system solved depends on many physical parameters concerning both the physical nature of the contacts, the properties of the lubricant used, the loading and the rotation speeds of the surfaces. The solution variables which must be solved for are the pressure profile P, across the domain, the surface geometry H, the viscosity $\bar{\eta}$ and the density $\bar{\rho}$. The full equation system is described in [5] and the appropriate references for the derivations are given therein. The pressure distribution is described by the Reynolds Equation:

$$\frac{\partial}{\partial X}\left(\frac{\bar{\rho}H^3}{\bar{\eta}\lambda}\frac{\partial P}{\partial X}\right) + \frac{\partial}{\partial Y}\left(\frac{\bar{\rho}H^3}{\bar{\eta}\lambda}\frac{\partial P}{\partial Y}\right) - \frac{u_s(T)}{u_s(0)}\frac{\partial(\bar{\rho}H)}{\partial X} - \frac{\partial(\bar{\rho}H)}{\partial T} = 0, \quad (1)$$

where u_s is the sum of the surface speeds in the X-direction at non-dimensional time T, λ is a non-dimensional constant, and X and Y are the non-dimensional coordinate directions, The standard non-dimensionalisation means that the contact has unit Hertzian radius, and that the maximum Hertzian pressure is represented by $P = 1$. The boundary conditions for pressure are such that $P = 0$. For the outflow boundary, once the lubricant has passed through the centre of the contact it will form a free boundary, the *cavitation boundary*, beyond which there is no lubricant. The non-dimensional film thickness, H, is given by:

$$H(X,Y) = H_{00} + \frac{X^2}{2} + \frac{Y^2}{2} + \frac{2}{\pi^2}\int_{-\infty}^{\infty}\int_{-\infty}^{\infty}\frac{P(X',Y')\,\mathrm{d}X'\mathrm{d}Y'}{\sqrt{(X-X')^2+(Y-Y')^2}}, \quad (2)$$

where H_{00} is the central offset film thickness, which defines the relative positions of the surfaces if no deformation was to occur. The two parabolic terms represent the undeformed shape of the surface, and the integral defines the deformation of the surface due to the pressure distribution across the entire domain. The conservation law for the applied force (the Force Balance Equation) is given by:

$$\int_{-\infty}^{\infty}\int_{-\infty}^{\infty}P(X,Y)\,\mathrm{d}X\mathrm{d}Y = \frac{2\pi}{3}. \quad (3)$$

Since an isothermal, generalised Newtonian lubricant model is being used, only expressions for the density and viscosity will be required. The density model chosen is that of Dowson and Higginson, whilst the viscosity model used is the Roelands pressure-viscosity relation, again see [5].

The solution variables are discretised on a regular mesh. Once discretised, the equations are solved in turn to iteratively produce more accurate solutions. This process is described in detail in [5] where detailed descriptions of the need for, and benefits of multigrid and multilevel multi-integration are given.

The Fortran 77 software used here, *Carmehl*, is described by Goodyer in [3] and includes both variable timestepping for transient calculations [6] and also has the option to use adaptive meshing [7]. Conventionally, once execution is complete then output files of data for the key variables are produced which may then be post processed for visualisation purposes. The user may request that the output solution is saved for continuation purposes on a future run.

3 Problem Solving Environments

This section describes the implementation of the Carmehl code into a PSE form suitable for both IRIS Explorer and SCIRun. The differences between the two products, and their implications are described later in Section 4.

IRIS Explorer and SCIRun have several common ideas: both use a visual programming system where individual *modules* are attached together by a *pipeline* structure, representing the dataflow paths. Each module may have several inputs, either from other modules or from widgets on the control panel of the module, and each represents a separate task which must be performed on the input data. Each module usually produces a new output *dataset*, which is then passed to the next module, or modules, downstream until the results are visualised. Since the datatypes required for visualisation are not the same as those used for numerical calculations conversion modules must be used.

3.1 ECLIPSE in IRIS Explorer

In implementating the EHL code as a module in IRIS Explorer [2] it is possible to build on earlier work employing IRIS Explorer for the development of PSEs, such as Wright *et al.* [8]. IRIS Explorer is marketed by NAG as a "advanced visual programming environment" for "developing customised visualisation applications"[1]. In IRIS Explorer a shared memory arena is used and the pipeline of modules is called a *map*. Although the data can be imagined travelling through the map by the wires, in reality it is only passing pointers to structures of known types in the shared memory arena at the end of each module's execution cycle.

A map in IRIS Explorer executes, normally, by a data set either being read in or generated and then control passes to the next module (or modules) downstream in the map. These in turn execute, provided they have all their required

[1] http://www.nag.co.uk/

inputs and control passes again. If a required input is missing then the module will wait until it is received before executing. If a parallel computer is used, then simultaneous module firings will be done on separate processors. This is because IRIS Explorer starts each module as an entirely separate process in the computer. It will be seen how this has both positive and negative consequences.

The Carmehl code has been implemented as one module containing the entirety of the numerical solver. The module's control panel is used to set the dimensions of the computational domain, the mesh refinement level, along with the total number of iterations required on each execution of the module. Other problem specific properties can also be defined on this control panel including information concerning transient problems, along with parameters governing surface features. The actual non-dimensional parameters governing the case in question may also be set here, or, through the addition of further modules, as shown in Figure 2, the actual operating conditions for the case defined upstream in the map, will be displayed. Once the module has completed execution, the datasets of the pressure and film thickness are sent down the map for visualisation.

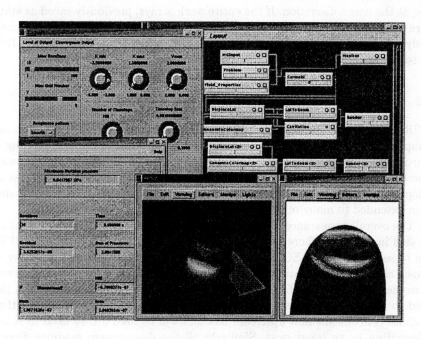

Fig. 2. ECLIPSE running in IRIS Explorer

It is through the addition of extra input modules that steering levels may be abstracted. Shown in Figure 2 are three different input modules which define the physical conditions of the contact, the parameters defining the lubricant, and parameters used to set the number of iterations in the multilevel schemes used. Another input module, not shown here, is that for grid adaptation.

ECLIPSE has been developed from the original Carmehl Fortran code by adding an interface routine written in C. The generation of all the IRIS Explorer data structures and communication is done through the *Application Programming Interface* (API) which is well documented for both C and Fortran. The design of the module control panel is usually done through the Module Builder which allows the widgets to be positioned through a visual interface, rather than by writing code. The Module Builder will also generate the necessary wrapper codes for complete control of the module's firing pattern and communication of data, and these require no alteration by a developer.

Computational steering is implemented in IRIS Explorer using the looping mechanisms provided. Rather than saving results to disk at the end of a run, the work arrays inside Carmehl can be declared as static and hence the previous results are automatically available for use on the next run. A solution used in this manner may provide a good initial estimate for a differently loaded case, or be interpolated for a change of domain size.

The use of the Hyperscribe module [9] would allow another layer of steering to be included. This module stores datasets or variables on disk for future usage, at the user's discretion. If the entire work arrays, previously saved as static, were stored based on the problem's input characteristics then a suite of previously calculated solutions could be created for future invocations of ECLIPSE on separate occasions, or even by other users.

3.2   ELLIPSE in SCIRun

SCIRun has been developed by the SCI group at the University of Utah as a computational workbench for visual programming. Although a longstanding research environment, it has only recently been released as open source software. The discussion below is based on the Version 1.2.0 release[2]. SCIRun was developed originally for calculations in computational medicine [10] but has since been extended to many other applications.

The overall appearance of SCIRun is similar to that of IRIS Explorer, as can be seen in Figure 3 where the implementation of the EHL problem, ELLIPSE, can be seen working. In SCIRun when modules are connected together, they are known as a *network*. The module firing algorithm in SCIRun probes the network from the desired point of firing so that all modules have all the information they need to run, before then sending the information downstream and firing those modules. This means that upstream modules will be fired if they need to supply information to an input port. Similarly all the downstream modules directly affected by the firing will be made aware that new data will be coming.

SCIRun is a multi-threaded program, and hence a single process, with (at least) one thread for each launched module. Therefore every module can have access to all same data without the use of shared memory. This has the advantage that there is more memory available for the generation of datasets to pass between modules, and the disadvantage that any operating system limits on the

---

[2] SCIRun is available from http://www.sci.utah.edu/

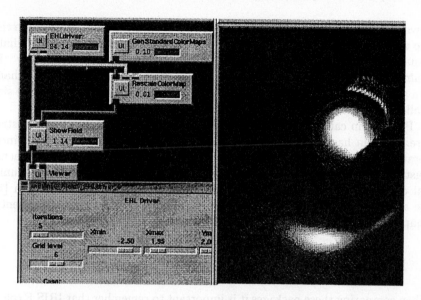

Fig. 3. ELLIPSE running in SCIRun

memory available to a single process apply to the entirety of SCIRun, meaning that calculation and visualisation are all included in the same maximum space allocation defined by the system. It also means that any variables declared as static in one invocation of a module will be the same as used in other invocations, since the operating system cannot differentiate between the two.

SCIRun is written in C++ and requires that at least the driver routine of any contributed module is too. This also means that any of the calls to other SCIRun objects, such as datatypes, needs to be done a SCIRun class. By passing the relevant class pointers to the Fortran, it is possible to return to the class function in order to interact with the SCIRun interface and features.

In Version 1.2.0 all the module control panels, called *UIs* in SCIRun, are written in Tcl and must be programmed by hand. This clearly limits the ease of redesigning the panels, and requires more code to be written to handle the interface between widgets and program variables.

The datatype in SCIRun used to construct meshes for ELLIPSE is *Trisurf*. This is a structure for a surface made up of tesselating triangles. First the list of coordinates of the nodes in the mesh are specified, followed by the connectivities of the points to form the triangles. As yet, SCIRun has no standard modules for manipulating the generated surfaces and so the three dimensional perspective, such as shown in Figure 3, must be included in the creation of the mesh. In order to get the colourmap distributed across the surface then solution values must be stored at each node, again as the mesh is generated.

Since SCIRun is written as a single threaded process it has added flexibility with regard to the rewiring of modules during execution. For the EHL problem, when a transient case is run, the output datasets are prepared and released

down the pipeline for visualisation at the end of each timestep. With more than
one solution variable being solved for, there is obviously a choice as to what is
visualised at anytime. In SCIRun these changes can be made 'on the fly'. For
example if the pressure solution was being visualised, then it is possible to change
to a surface geometry plot between timesteps. This is an important feature since
it allows the user to learn and experiment interactively.

Parallelism can be easily achieved on SCIRun thanks to its threaded structure. SCIRun has its own implementation of threads that can be easily incorporated into a user's code. The use of threads means a shared memory machine
must be used, but within these constraints the parallel performance for numerical calculations is very good. Next generation packages, such as Uintah [11],
use a combination of MPI and threads to achieve massively parallel scientific
computations on terascale computing platforms.

4   Evaluation

When comparing these packages it is important to remember that IRIS Explorer
is a longstanding commercial package supported by NAG whilst the first publi-
cally available version of SCIRun was only released last year. The development
histories of the two packages are obviously different. IRIS Explorer has a large
number of standard modules for reading in various formats of data files, manipulation of datasets, and for the visualisation of this data, with the source
of (practically) every module now being part of the distribution. SCIRun was
developed in a problem driven way, is completely open source and the number
and variety of modules will grow in the coming years.

These different backgrounds are most tellingly reflected in the ease of use of
the environments. For a novice use wanting to visualise output data using IRIS
Explorer it is a relatively simple process. By addition of a further few modules
it is possible to create very intricate output pictures. In SCIRun there is a steep
learning curve at present to be able to visualise data, especially with the smaller
number of visualisation manipulation modules provided. This manipulation must
be done in the initial generation of each mesh and is therefore an additional
computational expense for the main module.

In the previous section it was seen how IRIS Explorer and SCIRun have
very different paradigms for operation: IRIS Explorer launches each module as
a separate process whilst SCIRun is a single, multithreaded process. This has
both positive and negative aspects. The advantage of threading is that it is very
simple to get data transfer from one module to another. It was seen how SCIRun
is more flexible in the rewiring of modules during execution. In IRIS Explorer
new connections can only be made after the module has finished executing rather
than after each timestep. This is different to the work of Walkley et al. [12] where
the numerical calculation is less demanding, hence when the control panel is re-
read every few timesteps, new connections are registered.

In SCIRun all the modules are loaded as shared libraries. This means that for
module developers, coding changes require recompilation of the relevant library,

and currently (though not in future releases) the reloading of SCIRun. In IRIS Explorer only the module which has changed needs to be reloaded using a trivial 'Replace' operation which remembers the data connections of the module.

The design features for constructing a new module in IRIS Explorer benefit from the visual design tools of the Module Builder which allows easy placement of widgets on the control panel and makes the interaction between input variables and those in the driver code very simple. In SCIRun each variable must be captured by the user from the panel, since the module wrapper is generated when the module is first created rather than at compilation, as in IRIS Explorer. This also means that the module wrapper is usually hidden from the developer.

In terms of solving the EHL problem it has been seen that both software packages efficiently handle the PSE structure, achieving similar visualisations by slightly different methods. Using the PSE has been tremendously beneficial in quickly being able to understand complex datasets and see the influence of single parameters. By increasing the regularity of dataset output it is possible to watch the numerical solver converge on the solution. Added insight into the problem was gained using IRIS Explorer's visualisation modules to overlay the surface geometry colour map on the 3D pressure mesh, showing the relationship between pressure and film thickness in a distinctive and hitherto unseen way.

SCIRun benefits from having parallelism at its heart, meaning that incorporating it into an individual module can be accomplished in a relatively straightforward manner. Parallelism in IRIS Explorer has mainly only been done by using the module as a front end to lauching parallel calculations outside of the environment, often on a different machine. Remote processing combined with collaborative visualisation is another area where IRIS Explorer currently takes the lead. In the companion paper [12] Walkley *et al.* show how a similar computational steering IRIS Explorer PSE is first developed for interaction in transient calculations, and then run collaboratively over a network using COVISA [13].

Support for computational steering is central to both packages. Looping of modules is relatively simple to implement in IRIS Explorer through simply wiring relevant modules together, whereas in SCIRun the dataflow mechanism means that care must be taken to ensure that no module is waiting for itself to fire.

5   Conclusions and Future Work

The overall conclusion is that both IRIS Explorer and SCIRun provide good ways to generate PSE environments for problems, such as the EHL problem considered here. An example of how the use of the PSE has enabled extra insight into the problem has been explained.

SCIRun is clearly still in the early stages of its life at Version 1.2.0 whereas IRIS Explorer, now at Version 5.0 is not necessarily too far ahead. Experience suggests that IRIS Explorer's functionality can be recreated in SCIRun, although may involve more programming and a deeper understanding of the software.

It is clear that for the construction of PSEs in the coming years both codes still have work to be done. Parallelism will be very important for increasing problem sizes. This issue has been successfully addressed in the Uintah PSE already developed in Utah [11]. The issues behind visualisation of significantly larger datasets remain to be fully resolved in IRIS Explorer which can create intermediary copies of the datasets for each module manipulating them.

Acknowledgements

This work was funded by an EPSRC ROPA grant. The Carmehl code has been developed over many years of collaboration with Laurence Scales at Shell Global Solutions. Many thanks are due to Jason Wood for technical assistance with IRIS Explorer and to Chris Moulding for help with SCIRun.

References

1. Parker, S.G., Johnson, C.R.: SCIRun: A scientific programming environment for computational steering. In Meuer, H.W., ed.: Proceedings of Supercomputer '95, New York, Springer-Verlag (1995)
2. Walton, J.P.R.B.: Now you see it – interactive visualisation of large datasets. In Brebbia, C.A., Power, H., eds.: Applications of Supercomputers in Engineering III. Computatational Mechanics Publications / Elsevier Applied Science (1993) 139
3. Goodyer, C.E.: Adaptive Numerical Methods for Elastohydrodynamic Lubrication. PhD thesis, University of Leeds, Leeds, England (2001)
4. Dowson, D., Ehret, P.: Past, present and future studies in elastohydrodynamics. Proceedings of the Institution of Mechanical Engineers Part J, Journal of Engineering Tribology **213** (1999) 317–333
5. Venner, C.H., Lubrecht, A.A.: Multilevel Methods in Lubrication. Elsevier (2000)
6. Goodyer, C.E., Fairlie, R., Berzins, M., Scales, L.E.: Adaptive techniques for elastohydrodynamic lubrication solvers. In Dalmaz et al., G., ed.: Tribology Research: From Model Experiment to Industrial Problem, Proceedings of the 27th Leeds-Lyon Symposium on Tribology, Elsevier (2001)
7. Goodyer, C.E., Fairlie, R., Berzins, M., Scales, L.E.: Adaptive mesh methods for elastohydrodynamic lubrication. In: ECCOMAS CFD, Institute of Mathematics and its Applications (2001) ISBN 0-905-091-12-4.
8. Wright, H., Brodlie, K.W., David, T.: Navigating high-dimensional spaces to support design steering. In: VIS 2000, IEEE (2000) 291–296
9. Wright, H., Walton, J.P.R.B.: HyperScribe: A data management facility for the dataflow visualisation pipeline. Technical Report IETR/4, NAG (1996)
10. Johnson, C.R., Parker, S.G.: Applications in computational medicine using SCIRun: A computational steering programming environment. In Meuer, H.W., ed.: Proceedings of Supercomputer '95, New York, Springer-Verlag (1995) 2–19
11. de St. Germain, D., McCorquodale, J., Parker, S., Johnson, C.R.: Uintah: A massively parallel problem solving environment. In: Ninth IEEE International Symposium on High Performance and Distributed Computing. (2000)
12. Walkley, M.A., Wood, J., Brodlie, K.W.: A collaborative problem solving environment in IRIS Explorer. Submitted to: Proceedings of the International Conference on Computational Science 2002. (2002)
13. Wood, J., Wright, H., Brodlie, K.W.: Collaborative visualization. In: Proceedings of IEEE Visualization 97. (1997) 253–259

Parallel Newton-Krylov-Schwarz Method for Solving the Anisotropic Bidomain Equations from the Excitation of the Heart Model

Maria Murillo and Xiao-Chuan Cai

Computer Science Department

University of Colorado

Boulder, CO 80302

(murillo, cai)@colorado.edu

Abstract

We present our preliminary results from applying the Newton-Krylov-Schwarz method for the simulation of the electrical activity of the heart in two dimensions. We use the bidomain nonlinear equations , using a fully implicit time discretization scheme, and solving the resulting large system of equations with a Newton based algorithm at each step. We incorporate anisotropy into our model, and compare the results of using various conductivity ratios between the intra-cellular and extra-cellular domains. We also compare our results with previous work that has been done using less precise techniques, such as explicit and semi-implicit schemes, and less detailed models, such as the Monodomain model, and isotropic as well as quasi-isotropic Bidomain models. The results are obtained using PETSc of the Argonne National Laboratory.

1 Introduction

The rapid increase on heart failure related deaths and other problems prompted the need for a better understanding of all processes involved in the heart system, plus the immediate need to determine mechanisms to either prevent or alleviate these problems. Researchers are now trying to extract more and better data, and to determine better and more complex models which can explain the data. In recent issues of SIAM News, Quarteroni [] has presented a very concise discussion of the current state with respect to modeling the blood circulation in the

P.M.A. Sloot et al. (Eds.): ICCS 2002, LNCS 2329, pp. 533–542, 2002.

heart. In these two articles, Quarteroni also mentions the importance of having a more comprehensive model of the heart system, including the fluid dynamics, the structure, and the electrical signals that activate the process, as well as the various biochemical reactions that are believed to be responsible for many of these interactions. Of all these important components, we concentrate on this paper on the treatment of the electrical activity of the heart. The most recent, and advanced model explains the electrical mechanisms in the heart system by a bidomain model, where an intracellular medium is separated from an extracellular medium by a membrane, and it is across this membrane that the critical processes take place. Once we have a representative model of this electrical activity, we are faced with the challenge of working with this model in a discrete and numerical approach; as Quarteroni mentions, even though great progress has been made in this area, we still need to move forward and develop and effectively implement more accurate techniques that preserve most of the details and accuracy of the given models. Among such techniques, the fully implicit methods are proven to be very accurate, although costly. Incorporating domain decomposition techniques along with these accurate methods, allows us to take advantage of today's parallel supercomputing powers making the solution of the associated systems feasible in time and space. Here we present the preliminary results of incorporating various degrees of anisotropy into this Bidomain model, and solving the set of time-dependent nonlinear equations with a fully implicit Newton-Krylov-Schwarz scheme.

2 The Model

2.1 Bidomain Model

We consider here the Bidomain model which includes three distinct areas within the heart tissue: the intracellular domain, the extracellular domain, and a membrane which separates the two previous ones. Originally, each one of these sections was s considered to be homogeneous, with all its properties being just averages.

This so-called bidomain model is the most accepted today, with slight variations to its complexity. The electrical activity of the heart is believed to be originated by changes in potential across these areas, which in turn, generate waveforms that propagate through the different cells [] . At the intracellular level, it is thought that enzyme activity is largely responsible for the generation of electrical fields at various phase boundaries within the cell. In the transmembrane, an action potential exists due to the effect of changes in ionic permeability and subsequent voltage differences. The extracellular electric field is represented by a waveform which apparently is directly correlated to the actions taking place at the membrane as well as the intracellular area; and, it appears to correspond to the outward and inward flows generated by changes in potential. The bidomain equations [?] can then be expressed as a reaction- diffusion system which consists of a nonlinear parabolic equation in V coupled with an elliptic equation in the extracellular potential V_e

$$\begin{cases} \chi(C_m \frac{\partial V}{\partial t}) + I_{ion} - I_{app} = -\nabla(D_e \nabla V_e) \\ \nabla(D_i \nabla V) = -\nabla(D \nabla V_e) \end{cases},$$

where χ and C_m are constants, I_{ion} is the ionic current, and I_{app} is a current applied to the system. The matrices D refer to the conductivity tensors.

2.2 The Ionic Current

The bidomain equations are used to represent the propagation of the waveform generated by changes in potential. This is uncorrelated to the model being used to represent the internal ionic currents. Some work is being done in trying to understand these currents, how they generate and propagate. DiFrancesco and Noble [5] presented a very complex system with about a dozen of coupled nonlinear equations, and about 50 parameters.

One attempt to simplify even more the expression for this ionic current term was developed by using the Fitzhugh-Nagumo equations [], where only two variables are considered. One other model yet more simplistic than the previous, was proposed by Walton and Fozzard [?]. This model expresses the ionic current as dependent only on V. Although very simplified and tractable, this model is considered to still retain the main characteristics of the cardiac activity without including the smaller details. The ionic current is then expressed as:

$$I_{ion} = AV \left(1 - \frac{V}{V_t}\right) \left(1 - \frac{V}{V_p}\right),$$

where A is a scaling factor, and V_t and V_p are the potential threshold (where activation occurs) and the plateau value, respectively. For simplicity, this latter model for the Ionic current is the one we will be using for our experiments.

2.3 Anisotropy

An important issue to determine an appropriate model for the cardiac activity is the conductivity tensor D . Depending on the structure of this matrix, we are incorporating various degrees of anisotropy to the system. The simplest one would be a diagonal matrix, which indicates that conductivity varies only along the coordinate system. A more complex system requires the inclusion of a full matrix where the conductivity varies in every direction. An associated element is the fact that in the bidomain model, we have different degrees of anisotropy in the intracellular and extracellular media. The anisotropy comes from the fact that the heart is made up of fibers, and conductance is more readily in the direction of these fibers than on the transverse direction.

In this work, we assume our model is two-dimensional and the fiber is oriented in the direction of the axes; this implies we have the following structure for the conductivity tensors in the intracellular and extracellular tissue portions, D_i and D_e respectively:

$$D_i = \begin{bmatrix} \sigma_l^i & \\ & \sigma_t^i \end{bmatrix},$$

$$D_e = \begin{bmatrix} \sigma_l^e & \\ & \sigma_t^e \end{bmatrix},$$

where the superscripts ^{e} and ^{i} refer to the mentioned extra and intracellular domains, and the subscripts $_l$ and $_t$ refer to the longitudinal and transverse directions where we are considering the existence of anisotropy.

Greater simplification can come from the anisotropy of the model; if we consider this two-dimensional case (for simplicity) and assume different conductivity values in the longitudinal as well as transversal direction only, but also assuming that the value of these anisotropy ratios is the same for both the intracellular, and the extracellular domain, then we refer to this as the quasi-isotropic model.

The bidomain model then simplifies when we assume the electrical potential on the interstitial region is zero (Monodomain model), or when the conductivity tensors are just a multiple constant of each other (isotropic or quasi-isotropic Bidomain model) [6].

2.4 Boundary Conditions

For the bidomain model, it is commonly assumed that there is no flux across the boundaries [1]. The reason being that since the potential V seems to have constant values before and behind the exciting front, we assume it is not affected by the outside medium [?].

In particular, for our simplified model, this boundary condition can be expressed as

$$\begin{cases} n^T D_i \nabla V = 0 \\ n^T D \nabla V_e = 0 \end{cases}$$

3 Numerical Approach

There have been numerous attempts to numerically solve for the heart excitation problem. There are three major approaches: explicit, semi-implicit and implicit methods. In the case of the bidomain equations, more emphasis has been placed on the explicit and semi-implicit methods, since these methods are more easily tractable when implemented on a computer. However, there are many limitations associated with these methods, such as stability problems, step size limits in time, and others.

Hooke et al. [6] presented a general version of algebraic transformations that have been applied to the bidomain equations. The explicit Euler's method was the choice for the time integration. In Veronese and Othmer [14] , a hybrid Alternating Direction Implicit (ADI) scheme is used for the nonlinear parabolic equation.

Bogar [1] extended the use of a semi-implicit method for the bidomain model, which has previously been used only for the Monodomain case. The approach is to split the nonlinear term into two parts: one is resolved implicitly (the resulting linear portion), and the other is solved explicitly.

Because of the complexities of implicit methods, very few attempts have been made in using this type of methods. Stalidis et al. [13] used an implicit finite difference method, but applied to the Monodomain model, and similarly in Chudin et al. [4]. Pormann [11] incorporates a similar approach, but using a matrix splitting procedure. Pennacchio and Simoncini [10] introduce the use of an algebraic semi-implicit form which is applied to a system with some degree of anisotropy. Lines [8] uses a similar operator splitting technique, also semi-implicit.

As mentioned above, the existing methods suffer from severe limitations because they use either explicit or semi-implicit approaches for numerically solving the problem at hand. In our case, we propose to use a fully implicit method which does not imply the cost and penalties of conventional fully-implicit methods; it is the Newton-Krylov-Schwarz method based on the Domain Decomposition technique.

3.1 Discretization

In order to numerically solve the system of time-dependent nonlinear partial differential equations which represent the Bidomain Model, we first need to discretise them. We choose to use a fully implicit finite difference scheme. We use a centered, finite difference 5-point stencil for the space variables, and backward Euler for the time variable.

3.2 Newton Method

Once we have the discretized version, we need to put our system of nonlinear equations into the form $F(u) = 0$ for each time step, where

$$\begin{cases} F = (f_1(v), f_2(v), ..., f_M(v)) \\ v = (v_{11}^n, v_{12}^n, ..., v_{mm}^n) \end{cases}$$

This allows us to use Newton's method to solve it [2]. v here represents our state vector (in the case of the heart excitation process, we can assume v represents the values of the potential at each of the points of our domain).

Some particular characteristics of the system of equations for the given problem need to be mentioned: first, the Jacobian matrix is very sparse; this makes the problem a lot more tractable, in terms of both storage and computing time. Second, the type of equations makes it relatively simple to compute the Jacobian elements directly from our given function values in $F(v)$.

At each Newton step, we need to solve the following Linear system of equations:

$$J(v^k)\Delta v_k = -F(v^k)$$

In our particular case, we choose to use the Newton-Krylov technique where each of the individual iterations is solved inexactly, but still having the overall process retain the convergence properties. The idea is to have fewer iterations at each step, specially at the beginning steps where we probably are still far from the solution. As the iterates get closer to the solution, then it might be more necessary, and more productive to have more iterations at each of these linear solver steps.

3.3 Krylov Method

In particular, we choose the Newton-Krylov method for solving (inexactly) the Linear system at each iteration step. When our Jacobian matrix J is s.p.d. , we can compute the iterates by using the Conjugate Gradient method. In our particular research, we need to resort to the use of the GMRES technique in cases where our Jacobian matrix is no longer symmetric; for example, when the bidomain model is allowed to have some complexity (as some degree of anisotropy and inhomogeneity). We improve the efficiency and robustness by means of a Preconditioning technique; in particular, we use the Additive Schwarz method as preconditioner [7]. This preconditioning method is chosen due to its high convergence rate, and equally important, due to its extremely good parallel performance; based on the Domain Decomposition criteria, this method tries to fully utilize the parallel capabilities of a given system to its fullest potential.

3.4 Additive Schwarz Preconditioner

The Schwarz preconditioning is a Domain Decomposition method [] which exhibits ideal features. First, the idea is to subdivide our domain into smaller subdomains, and to solve simpler and smaller problems in each of these subdomains, and then combine these solutions in order to obtain the solution to our more general and complicated problem. Second, parallelism plays a very important factor in these type of decomposition techniques since if the problem is properly subdivided we could fully utilize the parallel capabilities of our machines by working on all subdomains in parallel.

We then solve the problem at each subdomain, and only update the general solution once all the subdomains have been solved for. By using the Domain Decomposition method, we extract from our system of equations only the portion that corresponds to each subdomain (or processor), and compute in parallel

$$M_i = L_i U_i,$$

in the sense of an incomplete factorization.

At the subdomain level, we resort to an inexact solver such as the ILU (Incomplete LU decomposition) [12] technique since it can reduce considerably the time and required storage of each iteration. Once we solve this subdomain problem using the ILU technique, the new iterate is found by combining the solutions at each of the subdomains.

3.5 Algorithm

In summary, we present here the basic steps for the algorithm we use in solving the given bidomain equations:

- Iterate through time:
  1. Using Newton method to solve $F(v) = 0$
  2. At each Newton step:
     - Solve the linear system : $J(v_k)\delta u = -F(v_k)$ with Additive Schwarz combined with Conjugate Gradient (or GMRES)
     - Use ILU to solve for the subdomain problem in Schwarz preconditioning
     - Update Jacobian and the Function
  3. Go to next Newton step (or stop)
- Go to next time step (or stop)

4 Computational Results

The experiments were carried out to implement and solve for the Bidomain equations, in the case that various degrees of anisotropy are included. We first define our two-dimensional domain as a square representing a thin slab of the heart tissue, with dimensions 1 cm by 1 cm. The uniform grid is made up of 64 nodes on each direction, which gives a total of 8192 equations, since there are two degrees of freedom per node (one for the transmembrane potential, and one for the extracellular potential). We maintain this size to be comparable to the experiments performed in [10], and in [1]. The simulation was initiated by applying an initial current of $0.8 A/cm^3$ for a duration of 0.5 msec. as used in [10], and only to a portion of our domain (a small semi-circle around the center of the left edge).

Other parameters were chosen from the literature [10], and are summarized below. Similarly to the author, we have restricted our experiments to one instant of time; therefore, we assume that the information from previous time steps is available.

Table 1. Parameters for Ionic current and others

$\chi = 1000\ cm^{-1}$	$A = 0.04\ mS/cm^2$	$c_m = 0.8\ \mu F/cm^2$
$I_{app} = 0.8\ A/cm^3$	$V_p = 100\ mV$	$V_t = 10\ mV$

For the anisotropy, we assume conductivity coefficients in our matrix D to be along and across the tissue fibers; we used three different set of values which indicate various degrees of anisotropy, and different ratios between the conductivity values in the two domains. We summarize here the values taken from [10], [8], and [1].

Table 2. Conductivity values and Anisotropy ratios

1. Pennacchio [10]

$\sigma_l^i = 2.0\ mS/cm$	$\sigma_l^e = 2.5\ mS/cm$	$ratio = 0.8$
$\sigma_t^i = 0.416\ mS/cm$	$\sigma_t^e = 1.25\ mS/cm$	$ratio = 0.33$

2. Lines [8]

$\sigma_l^i = 3.0\ mS/cm$	$\sigma_l^e = 2.0\ mS/cm$	$ratio = 1.5$
$\sigma_t^i = 0.31525\ mS/cm$	$\sigma_t^e = 1.3514\ mS/cm$	$ratio = 0.23$

3. Bogar [1]

$\sigma_l^i = 4.0\ mS/cm$	$\sigma_l^e = 4.0\ mS/cm$	$ratio = 1.0$
$\sigma_t^i = 1.0\ mS/cm$	$\sigma_t^e = 4.0\ mS/cm$	$ratio = 0.25$

We have previously presented results where we use the Newton-Krylov-Schwarz approach for the simplest cases, the Monodomain case in [3], and for the quasi-Isotropic case in [9], where the ratios are equal, and therefore the equations simplify and become basically the Monodomain equation.

The results that we have obtained are summarized below, and they indicate that the various degrees of anisotropy can affect slightly the rate of convergence of the linear as well as the nonlinear equations we are solving at each Newton step. In general, we have not put major emphasis on getting the linear iterations to totally converge since in our Krylov method, we allow for some approximation. The key portion is to obtain convergence in the nonlinear portion of the method, and we see that we are obtaining such convergence without having to execute a large number of nonlinear iterations.

Table 3. Number of iterations (linear and nonlinear)

$Case$	$Average Linear Iterations$	$Average Nonlinear Iterations$
1	100	7
2	100	10
3	100	20

5 Conclusion

The preliminary results of applying the proposed Newton-Krylov-Schwarz algorithm to solve the more complex model of the electrical activity of the heart, indicate that this method seem capable of dealing with more complicated structure in our model, by incorporating anisotropy and applying the technique to the truly Bidomain model. These results extend from our previous work on the simpler models, and considering the parallel capabilities of this Domain Decomposition technique, major progress is expected when the technique is implemented and extended to larger problems.

References

[1] K. Bogar. *A Semi-Implicit Time Scheme for the Study of Action Potential Propagation in the Bidomain Model.* PhD thesis, University of Utah, 1999.

[2] R. Burden and J. Faires. *Numerical Analysis.* Prindle, Weber and Schmidt, 1985.

[3] X. Cai and M. Murillo. Parallel performance of a newton-krylov-schwarz solver in a ed numerical simulation of the excitation process in the heart. *Eight SIAM Conference on Parallel Processing for Scientific Computing*, page 25, 1997.

[4] E. Chudin et al. Wave propagation in cardiac tissue and effects of intracellular calcium dynamics (computer simulation study). *Progress in Biophysics and Molecular Biology*, 69:225–236, 1998.

[5] D. DiFrancesco and D. Noble. A model of cardiac electrical activity incorporating ionic pumps and concentration changes. *Philosophical Transactions of the Royal Society of London B*, 307:353–398, 1985.

[6] P. Grottum, A. Tveito, and D. Zych. Heart - numerical simulation of the excitation process in the human heart, 1995. http://www.oslo.sintef.no/adv/33/3340/diffpack.

[7] N. Hooke, C. Henriquez, P. Lanzkron, and D. Rose. Linear algebraic transformations of the bidomain equations: Implications for numerical methods. *Mathematical Biosciences*, 120:127–145, 1994.

[8] D. Keyes, Y. Saad, and D. Truhlar. *Domain-Based Parallelism and Problem Decomposition Methods in Computational Science and Engineering*. SIAM, 1995.

[9] G. T. Lines. *Simulating the Electrical Activity of the Heart*. PhD thesis, University of Oslo, 1999.

[10] M. Murillo and X. Cai. Parallel algorithm and software for solving time-dependent nonlinear bidomain equations. *First SIAM Conference on Computational Science and Engineering*, page 69, 2000.

[11] M. Pennacchio and V. Simoncini. Efficient algebraic solution of reaction-diffusion sysems for the cardiac excitation process. *preprint*, pages 1–19, 1999.

[12] J. Pormann. *A Modular Simulation System for the Bidomain Equations*. PhD thesis, Duke University, 1999.

[13] Y. Saad. *Iterative Methods for Sparse Linear Systems*. PWS Kent, 1995.

[14] G. Stalidis et al. Application of a 3-d ischemic heart model derived from mri data to the simulation of the electrical activity of the heart. *Computers in Cardiology*, pages 329–332, 1996.

[15] S. Veronese and H. Othmer. A computational study of wave propagation in a model for anisotropic cardiac ventricular tissue. *Lecture Notes in Computer Science*, 919:248–253, 1995.

Parallel Flood Modeling Systems*

L. Hluchy, V. D. Tran, J. Astalos, M. Dobrucky, G. T. Nguyen

Institute of Informatics, Slovak Academy of Sciences
Dubravska cesta 9, 842 37 Bratislava, Slovakia
viet.ui@savba.sk

D. Froehlich

Parsons Brinckerhoff Quade and Douglas, Inc
909 Aviation Parkway, Suite 1500
Morrisville, North Carolina 27560 USA
Froehlich@pbworld.com

Abstract. Flood modeling is a complex problem that requires cooperation of many scientists in different areas. In this paper, the architecture and results of the ANFAS (Data Fusion for Flood Analysis and Decision Support) project is presented. This paper also focuses on the parallel numerical solutions of flood modeling module that are the most computational-intensive part of the whole ANFAS architecture.

1 Introduction

Over the past few years, floods have caused widespread damages throughout the world. Most continents were heavily threatened. Therefore, modeling and simulation of floods in order to forecast and to make necessary prevention is very important. There are several flood modeling software systems like MIKE [1], FLOWAV [2], however, they are suitable for small problems only. It is the motivation of the ANFAS (Data Fusion for Flood Analysis and Decision Support) project [16] that is supported by European Commission within IST Fifth Framework Programme.

The ANFAS project will:
- use data from the most advanced acquisition technology; in particular remote sensing imagery (optical, radar, interferometry radar) will be incorporated into a conventional Geographical Information System (GIS) database;

---

* This work is supported by EU 5FP ANFAS IST-1999-11676 RTD and the Slovak Scientific Grant Agency within Research Project No. 2/7186/20

P.M.A. Sloot et al. (Eds.): ICCS 2002, LNCS 2329, pp. 543–551, 2002.

544   L. Hluchy et al.

- develop advanced data processing tools for scene modeling and flood simulation; computer vision and scientific computing will be used together in order to take into account information extracted from real images for the calculation of parameters for the simulation model;
- have strong end-users involvement through the definition of the objectives, the conception of the system, the evaluation of the simulation results; it ensures that the system answers to "terrain needs" and are well adapted in practical use;
- involve industrial partners for the realization and design of the final system;
- implement a software modular, i.e. one composed of easily interchangeable blocks.

The functions of ANFAS system will be:
- to perform flood simulation: given a scenario, water flow propagation is simulated; the user can interact with the scene that models the real site in order to add/remove dikes or other man-made;
- to assess flood damage: flood assessment can be done using either the simulation results or the remote sensed images of a real flood event;
- at a prospective level, to analyze floodplain morphology including riverbed and subsurface properties changes due to repetitive floods for a given site.

The output of the system will be:
- visualization: consisting of the bi-dimensional maps of the extent of the flood; if simulation is performed, time sequence maps retracing the propagation of the simulated flood will be provided;
- direct impact assessment, evaluation of the whole affected area; regional and statistical analyses over the affected area; simple socio-economic assessment of damage directly caused by flood.

2   ANFAS Architecture

The ANFAS system is based on a 3-tier architecture (**Fig. 1**):
- the ANFAS client,
- the ANFAS core server,
- the servers – or wrappers – allowing to "connect" (possibly in remote access) external features such as the database, the models and possibly additional processes (e.g. computer vision ones).

Short description of each component follows:
- Database Server
  - GIS ArcView: here all data that represent the sites are stored;
  - GIS access component: responsible for GIS management.
- Modeling server
  - Numerical models such as FESWMS Flo2DH
  - Model access component: it pilots the numerical models
- ANFAS Core Server
  - Application server: it is the bridge between the client and the ANFAS core server

- Data manager: it handles the user sessions (i.e. set of data), especially for what concern scenarios (recording, loading) and collections
- File exchange component: it is responsible for file transfers between servers
- User management component: it manages all end users allowed to perform flood simulations
- Map server component: responsible for the manipulation of geographic data
- Model manager: it drives the numerical models
- Co-operative, explanatory component: it brings functions for the impact assessment.

- ANFAS Client
  - Modelling preparation: this component is responsible for the phase of data sets preparation
  - Modelling activation and follow-up: it permits to follow-up the execution of the numerical models
  - Results exploitation: it brings functions for manipulating the results of a model

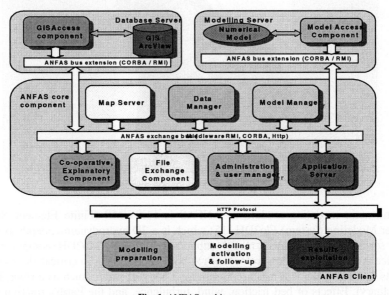

Fig. 1 ANFAS architecture

3   Modeling and Simulation at Vah River Pilot Site

Vah river is the biggest river in Slovakia, therefore it was chosen as a pilot site for the ANFAS project. The following input data are required for modeling and simulation:

- Topographical data: There are several sources of input topographical data: orthophotomaps, river cross-sections, LIDAR (Light Detection and Ranging) data. GIS (Geographical Information System) is used to manage and combine these data, and to product a final map for simulation (e.g. orthophotomaps are better for identifying objects, LIDAR cannot measure river depths).

- Roughness conditions: These data cannot be directly obtained, but are derived from orthophotomaps.
- Hydrological data: These data are obtained from monitoring of past flood events. These data can be used as boundary conditions as well as for calibration and validation of models.

Creating and calibrating models is a challenging process that can be done only by experts in hydrology. SMS (Surface-water Modeling System) [3] provides a convenient graphical interface for pre- and post-processing that can help the experts to speedup the modeling process. **Fig. 2** shows LIDAR data and orthophotomaps in SMS graphical interface.

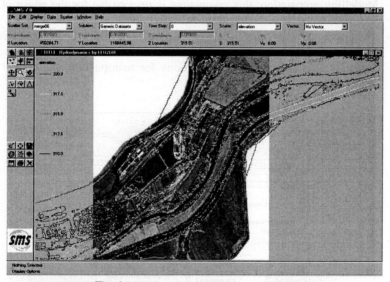

Fig. 2 LIDAR and orthophotomaps in SMS

The main modeling module in ANFAS is FESWMS (Finite Element Surface-Water Modeling System) Flo2DH [15] which is a 2D hydrodynamic, depth averaged, free surface, finite element model supported by SMS. Flo2DH computes water surface elevations and flow velocities for both super- and sub-critical flow at nodal points in a finite element mesh representing a body of water (such as a river, harbor, or estuary). Effects of bed friction, wind, turbulence, and the Earth's rotation can be taken into account. In addition, dynamic flow conditions caused by inflow hydrographs, tidal cycles, and storm surges can be accurately modeled. Since Flo2DH was initially developed to analyze water flow at highway crossings, it can model flow through bridges, culverts, gated openings, and drop inlet spillways, as well as over dams, weirs, and highway embankments. Flow through bridges and culverts, and over highway embankments can be modeled as either 1D or 2D flow. Flo2DH is ideal for modeling complex flow conditions at single- and multiple-bridge and culvert roadway crossings, such as the combination of pressurized flow, weir overflow, and submerged weir overflow, which are difficult to evaluate using conventional models. It is especially well-suited at analyzing bridge embankments and support structures to reduce and eliminate scour during flooding.

4   Numerical Solution of Flo2DH

In Flo2DH, the governing system of differential equations that describe the flow of water in rivers, floodplains, estuaries, and other surface-water flow applications are based on the classical concepts of conservation of mass and momentum. Flo2DH uses the Galerkin finite element method (FEM) [18] to solve the governing system of differential equations. Solutions begin by dividing the physical region of interest into sub-regions called elements. Two-dimensional elements can be either triangular or quadrangular in shape, and are defined by a finite number of node points situated along its boundary or in its interior. Approximations of the dependent variables are submitted into the governing equations, which generally will not be satisfied exactly, thus forming residuals, which are weighted over the entire solution region. The weighted residuals, which are defined by equations, are set to zero, and the resulting algebraic equations are solved for the dependent variables. In Galerkin's method, the weighting functions are chosen to be the same as those used to interpolate values of the dependent variables within each element.

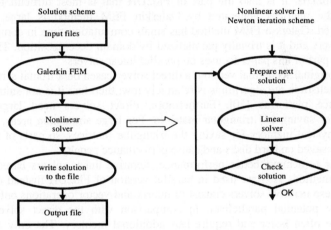

Fig. 3 Numerical scheme of Flo2DH

The weighting process uses integration, which is carried out numerically using Gaussian quadrature on a single element. Repetition of the integration for all elements that comprise a solution region produces a system of nonlinear algebraic equations when the time derivatives are discretized. Because the system of equations is nonlinear, Newton iteration, or its variation, is applied and the resulting system of linear equations is solved using a linear solver. Full-Newton iteration is written as follows:

$$M(a) = b$$
$$J(a_0)(a - a_0) = -(b - M(a_0))$$
$$J(a_0)\ \Delta a_0 = -R(a_0)$$
$$a_1 = a_0 + \omega \Delta a_0\ ,\ \ \omega \in (0,2);$$

generally,
$$J(a_j)\Delta a_j = R(a_j)$$
$$a_{j+1} = a_j + \omega \Delta a_j \quad j = 0,1,2,...n$$

where a_j is a newly computed solution vector, $J(a_j)$ is a Jacobian matrix computed for a_j, $R(a_j)$ is a residual load vector, Δa_i is the incremental solution value. An updated solution is computed by a relaxation factor ω_j. Solving the system of linear equations $J(a_j)\, \Delta a_j = R(a_j)$ is the most time- and memory-consuming part of Flo2DH. The complete numerical scheme of Flo2DH is shown in **Fig. 3**.

5   Parallelizing Flo2DH

Simulation of floods is very computation-expensive. Several days of CPU-time may be needed to simulate large areas. For critical situations, e.g. when a coming flood is simulated in order to predict threatened areas, and to make necessary prevention, long computation times are unacceptable. Therefore, using HPCN platforms to reduce the computational time of flood simulation is imperative [4] [12]. The HPCN version of FESWMS Flo2DH not only reduces computation times but also allows simulation of large-scale problems, and consequently provides more reliable results.

Solving the systems of linear equations is the most time- and memory-consuming part of Flo2DH. It is also the part of Flo2DH that is most difficult to parallelize because the matrices generated by Galerkin FEM method is large, sparse and unstructured. Galerkin FEM method has small computation time in comparison with linear solvers and it is trivially parallelized by domain decomposition. Therefore, the following part of this paper focuses on parallel linear solvers.

In the original sequential version, a direct solver based on a frontal scheme is used. The parallelism in the algorithms is relatively low; therefore it is not suitable for high performance platforms [10]. Furthermore, direct solvers need large additional memory for saving LU triangular matrices. For large simulation areas, there is not enough physical memory for saving the triangular matrices, so parts of the matrices have to be saved on hard disks and cause performance penalty.

In order to achieve better performance, iterative linear solvers based on Krylov subspace methods [19] are used in parallel version of Flo2DH instead of the direct solver. These iterative solvers consist of matrix and vector operations only; thus, they offer large potential parallelism. In comparison with the direct solvers, iterative solvers are often faster and require less additional memory. The only drawback of iterative solvers is that they do not guarantee convergence to a solution. Therefore, preconditioning algorithms [20] are used to improve the convergence of iterative algorithms. Recent studies [5], [8] show that the combination of inexact Newton iteration [7] with Krylov methods and additive Schwarz preconditioners [17] can offer high performance and stable convergence. PETSC [11] library, which is developed in Argonne National Laboratory, is used in implementation of parallel Flo2DH version. It is available for both supercomputers and clusters of workstations.

6   Case Study

Input data are generated by GIS; the data from LIDAR, cross-sections and orthophotomaps are combined. Hydrological data are provided by Vah River Authority which monitors and records all past events in Vah river. Roughness is defined based on image processing procedures of orthophotomaps, which divided the simulated pilot site into areas with the same coverage.

6.1   Hardware Platform

Experiments have been carried out on a cluster of workstations at the Institute of Informatics. The cluster consists of seven computational nodes, each has a Pentium III 550 MHz processor and 384 MB RAM, and of a dual processor master node and 512 MB RAM. All of the nodes are connected by an Ethernet 100Mb/s network. The Linux operating system is used on the cluster. Communication among processes is done via MPI [9].

6.2   Experimental Results

The steady-state simulation converged after 17 Newton iterations. The original Flo2DH with a direct solver takes 2801 seconds for the simulation. The parallel version of Flo2DH takes 404 seconds on a single processor and 109 seconds on 7 processors (Tab. 1). One of the reasons why the parallel version (with iterative solvers) is much faster than original version (with a direct solver) is that the direct solver in the original version needs a lot of additional memory for saving LU triangular matrices. As there is not enough physical memory the matrices, a part of them have to be saved on hard disk, which causes performance penalty. Speedup of the parallel version on our cluster is shown in **Fig. 4.**

Tab. 1 Simulation times (in seconds)

Original version	Parallel version						
	1 processor	2 p.	3 p.	4 p.	5 p.	6 p.	7 p.
2801	404	220	150	136	122	118	109

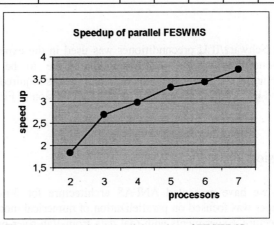

Fig. 4 Speedup of the parallel version of FESWMS

The experiments have revealed that from the implemented Krylov interactive algorithms in PETSTC, Bi-CGSTAB [13] is the fastest and relatively stable, especially when it is used with additive Schwarz/ILU preconditioner. GMRES [6] has the most stable convergence; however, it is much slower than Bi-CGSTAB, and CGS [14] has unstable convergence.

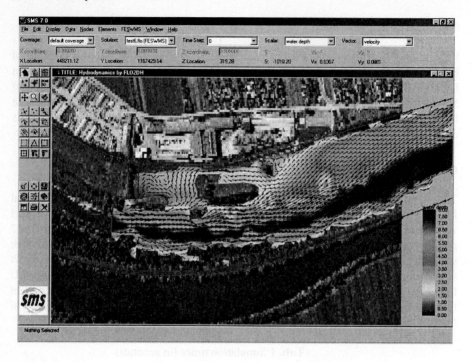

Fig. 5 Simulation results

An additive Schwarz/ILU preconditioner was used in the experiments. However, the preconditioner has several parameters that have to be tuned for better performance. Finding optimal values for these parameters required a large number of experiments and it is a part of future work on parallel version of FESWMS Flo2DH.

7   Conclusion

In this paper, we have presented ANFAS architecture for 3-tier flood modeling system. The paper was focused on parallelization of numerical module Flo2DH of the ANFAS architecture. Experiments with real data from Vah river pilot site show good speedups of our parallel version of Flo2DH. Similar results are also achieved on real data from Loire river pilot site, France. Experiments on larger computer systems (Origin 2800 with 128 processors and Linux clusters with 200 nodes) are planned.

References

1. FLDWAV: 1D flood wave simulation program
   http://hsp.nws.noaa.gov/oh/hrl/rvrmech/rvrmain.htm
2 Mike21: 2D engineering modeling tool for rivers, estuaries and coastal waters
   http://www.dhisoftware.com/mike21/
3. SMS: 2D surface water modeling package
   http://www.bossintl.com/html/sms_overview.html
4. V. D. Tran, L. Hluchy, G. T. Nguyen: Parallel Program Model and Environment,
   ParCo99, Imperial College Press, pp. 697-704, August 1999, TU Delft, The Netherlands.
5. I.S. Duff, H.A. van der Vorst: Developments and Trends in the Parallel Solution of Linear
   Systems, Parallel Computing, Vol 25 (13-14), pp.1931-1970, 1999.
6. Y. Saad, M. H. Schultz: GMRES: A generalized minimal residual algorithm for solving
   nonsymmetric linear systems. SIAM J. Scientific and Statistical Computing, Vol. 7, pp.
   856-869, 1986.
7. R.S. Dembo, S.C. Eisenstat, T. Steihaug: Inexact Newton methods. SIAM J. Numerical
   Analysis, Vol. 19, pp. 400-408, 1982.
8. W.D. Gropp, D.E. Keyes, L.C. McInnes, M.D. Tidriri: Globalized Newton-Krylov-
   Schwarz Algorithms and Software for Parallel Implicit CFD. Int. J. High Performance
   Computing Applications, Vol. 14, pp. 102-136, 2000.
9. MPICH - A Portable Implementation of MPI
   http://www-unix.mcs.anl.gov/mpi/mpich/
10. Selim G. Akl: Parallel Computation Models and Methods, Prentice Hall 1997.
11. PETSc The Portable, Extensible Toolkit for Scientific Computation. http://www-
    fp.mcs.anl.gov/petsc/
12. Pfister G.F.: In Search of Clusters, Second Edition. Prentice Hall PTR, ISBN 0-13-
    899709-8, 1998.
13. H. Vorst: Bi-CGSTAB: A fast and smoothly converging variant of Bi-CG for the
    solution of nonsymmetric linear systems. SIAM J. Scientific and Statistical Computing,
    Vol. 13, pp. 631-644, 1992.
14. P. Sonneveld: CGS, a fast Lanczos-type solver for nonsymmetric linear systems. SIAM
    J. Scientific and Statistical Computing, no. 10, pp. 36-52, 1989.
15. FESWMS - Finite Element Surface Water Modeling.
    http://www.bossintl.com/html/feswms.html
16. ANFAS Data Fusion for Flood Analysis and Decision Support.
    http://www.ercim.org/anfas/
17. X.C. Cai, M. Sarkis: A restricted additive Schwarz preconditioner for general sparse
    linear systems. SIAM J. Scientific Computing Vol. 21, pp. 792-797, 1999.
18. D. Froehlich: Finite element surface-water modeling system: Two-dimensional flow in a
    horizontal plane. User manual.
19. Y. Saad, H. Vorst: Iterative Solution of Linear Systems in the 20-th Century. J. Comp.
    Appl. Math, Vol. 123, pp. 1-33, 2000.
20. K. Ajmani, W.F. Ng, M.S. Liou: Preconditioned conjugate gradient methods for the
    Navier-Stokes equations. J. Computaional Physics, Vol. 110, pp. 68-81, 1994.

Web Based Real Time System for Wavepacket Dynamics

Aleksander Nowiński and Krzysztof Nowiński
ICM Warsaw University
Pawińskiego 5a, 02-106 Warsaw, Poland
{axnow,know}@icm.edu.pl
and
Piotr Bała
Faculty of Mathematics and Computer Science
N. Copernicus University
Chopina 12/18, 87-100 Toruń, Poland,

Abstract. In this paper we describe prototype systemm for web based interface to numerically intense simulations. Presented system solves time-dependent Schroedinger equation using wavepackets. Proposed solution consists from three main parts: client which acts as graphical user interface, server which receives requests form various clients and handles all running jobs. The last part is computing engine which can be treated as external application and in principle can be written in any computer language.

1 Introduction

The history of computer simulations in various areas of physics is relatively long and successful. The increasing rapidly computer power allows nowadays for routine simulations of complicated systems, or allows for advanced theories and models. In the past computers were used for simulations of atomic and molecular systems, for modeling of sets of interacting particles [1], fluid dynamics, quantum chromodynamics and many others [1, 2]. In last ten years available computer power allowed for calculations of wavepacket dynamics obeying time dependent Schroedinger equations, first for one dimensional systems[3] and nowadays for the systems with few degrees of freedom [4].

Developed numerical methods allowed for simulation of quantum evolution of different systems and processes, especially in this cases where classical approximation cannot be used. Numerical simulations in this area are also important because of lack of analytical methods, especially because of lack of analytical solutions of time-dependent Schroedinger equation for nontrivial potentials. The quantum evolution can be predicted analitycally only for very simple potentials (e.g. constant or harmonic), or in very special cases - for example for stationary solutions.

P.M.A. Sloot et al. (Eds.): ICCS 2002, LNCS 2329, pp. 552–561, 2002.

Progress in experimental techniques, as well as new phenomena discovered, implicated interest in simulations of dynamics of quantum systems. The most important examples are photoionization of atoms[5] and molecules[6], dynamics of simple molecular systems[], electron tunneling in superlattices [], proton tunneling in molecules and biomolecules, photo adsorption and desorption[] or, recently, dynamics of Bose Einstein condensates[7, 8].

The wavepacket dynamics based on integration of time-dependent Schroedinger equation, especially in more than one dimensional case can be run only on high performance computers. The existing codes are developed by the research groups and are, in general, not user friendly and have limited visualization capabilities. This is caused by the need of high numerical efficiency of the code, which is, in most cases, run in a batch. One should note, that interpretation of the wavepacket dynamics usually cannot be performed without aid of advanced visualization tools during both simulation setup and analysis. This requires development of coupling mechanism between visualization and simulation codes.

Recent progress in computer technology, especially web based tools makes them good candidate for development of user interfaces to computing programs. Web based interfaces are easy to install, simple to use and can be used in distributed environment. However, even providing powerful user interface, web technology cannot achieve high numerical efficiency required for most simulation problems. This leads us to solutions consisting on web interface for the user and traditional, usually batch type, numerical simulation engine. Web access to high performance resources is being developed significantly in last few years. However most of the development goes to the design and production of general tools which can be used in different situations. Because of this, security is one of the most important and yet not solved issues. The another one is accounting, resource directory services or resource reservation mechanisms. The existing solutions like globus[9], Legion[10] or Unicore[11] address most important issues, Unfortunately, these solutions cannot efficiently create strongly coupled application, especially in the case when advance interactive visualization techniques must be used.

Recently we started effort to create an prototype of distributed computing system with Web interface dedicated to wavepacket dynamics. System is based on computing server, or number of servers, located locally or remotely, eg. in supercomputer centre, which offers computing resources for clients distributed all over the world through internet. In general, user should contact with computing server using web technology, with tools allowing for easy preparation of input data, job submission and steering, and analysis of results.

2 Wavepacket Dynamics

The dynamics of the quantum particle can obtained by solving the time–dependent Schroedinger equation:

$$i\hbar\frac{\partial\Psi(x,t)}{\partial t} = \mathbf{H}\Psi(x,t),\tag{1}$$

with the Hamiltonian:

$$\mathbf{H} = -\frac{\hbar^2}{2m}\Delta + V(x). \tag{2}$$

where m is particle mass.

In general, the time–dependent Schroedinger equation (1) cannot be solved analytically. Recently an effective and powerful numerical methods have been developed and used to investigate different physical problems such as scattering, electron and proton tunneling, photodesorption or photoionization.

As mentioned above, the analytical solution of (1) is not known and numerical methods based on the discrete representation must be used. After a small time step Δt the propagated wavefunction can be obtained by using the time–evolution operator $\mathbf{U}$:

$$\Psi(t + \Delta t) = \mathbf{U}(\Delta t)\Psi(t) = \mathbf{T}e^{-\frac{i}{\hbar}\int_t^{t+\Delta t} \mathbf{H}dt'} \Psi(t), \tag{3}$$

where $\mathbf{T}$ is the time ordering operator. When the Hamiltonian $\mathbf{H}$ is time–independent the time ordering operator can be omitted and the integral can be approximated with the trapezoidal rule:

$$\mathbf{U}(\Delta t) \cong e^{-\frac{i}{\hbar}\mathbf{H}\Delta t}, \tag{4}$$

Once the evolution propagator $\mathbf{U}(\Delta t)$ and its action on the wavefunction are known, the dynamics of the system is obtained and the time–dependent Schroedinger equation (1) is solved.

The different methods has been proposed for the calculation of $\mathbf{U}(\Delta t)$. The most commonly used are: second–order differential scheme [3, 12, 13], split operator method [14–16], Calay method [17, 18], Chebychev polynomial expansion [19, 20] or Lanczos scheme [21–23]. In a particular case an algorithm based on the Chebychev polynomial expansion has been used, which was found to be stable and accurate [13, 24, 25]. The time evolution operator is expanded into Chebychev polynomial series:

$$U(\Delta t) = \sum_{m=1}^{\infty} a_m \Phi_m(-i\mathcal{H}) \cong \sum_{m=1}^{M} a_m \Phi_m(-i\mathcal{H}), \tag{5}$$

where Φ_m are complex Chebychev polynomials and a_m are expansion coefficients.

In numerical applications, the wavefunction $\Psi(t + \Delta t)$ as well as potential for electron motion are represented on a discrete grid. The grid points are usually equally spaced by the Δx, although adaptive grid methods are also possible. The discretization of the wavefunction introduces some limits on the maximal momentum p_{max} represented on the grid:

$$p_{max} = \frac{\pi}{\Delta x}. \tag{6}$$

The momentum limit implies that the energy range of the discretized Hamiltonian is given by the maximal and minimal energy represented on the grid:

$$\Delta E = E_{max} - E_{min},\tag{7}$$

where:

$$E_{min} = V_{min}, \qquad E_{max} = V_{max} + \frac{p_{max}^2}{2m^*},\tag{8}$$

and V_{min} and V_{max} are the minimum and maximum of the potential on the grid.

The potential part of the Hamiltonian can be calculated directly as multiplication of the potential energy and the wavefunction. The most powerful and accurate method of the calculation of the kinetic energy operator is based on the transformation of the wavefunction to the momentum space using the Fourier Transform technique [3]. Since the effect of the operator $\mathbf{H}$ on $\Psi(x,t)$ is calculated using the Fast Fourier Transform (FFT) twice, the described method requires 2M FFT's per time step. The number of FFT's in the Chebychev polynomial method is greater than in other methods of solving the time–dependent Schroedinger equation but longer time steps Δt can be applied and similar computational efficiency is achieved. This method is also found to be most stable and accurate [13, 25].

3   Description of Web Based Numerical Simulation System

System for web access to computer resources for wavepacket dynamics is basically designed as three-part construction - the client, central server (dispositor) and computing server. There can be many clients and many computing servers but only one central server per computing site. This solution is natural, because of the security control at computing site.

3.1   Client

Client is the application that prepares data for computation, sends the request to the central server and collects results. Typically (as in created example) it also offers limited capabilities of result visualization. What is most important, client program is an WWW applet written in Java, so it can be run from any workstation with java capable Web browser. It was tested with Sun Sparc Ultra (Solaris 2.8), Intel PC (Windows 98), Intel PC (RedHat Linux 6.2), Silicon Graphics 02 (Irix 6.5), and no problems were noticed.

The client applet has an visual editor capable to define wave function and any analytic potential on one dimensional computational grid. It also allowed to change some computation parameters, like mesh density, or number of time steps. The client acts as main graphical user interface. Because of required functionality any other existing software could be used. The input of parameters and visualization requirements for wavepacket dynamics cover quite wide range of possible applications and we have decided to spent some effort to develop interface dedicated to such type of simulations. One should note, that client can be also used as user interface for other classes of simulations which are generally based on the mesh. This will however require some additional customization of the client.

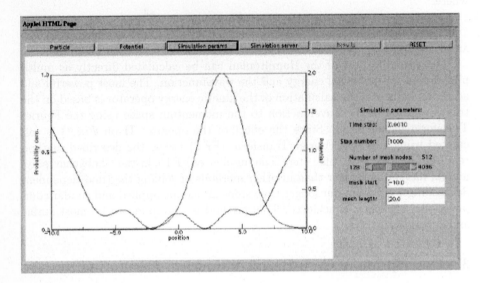

Fig. 1. The Client applet - grid parameters control screen.

3.2   Central Simulation Server

Central server (dispositor) is an application that receives all client requests, and forwards it to proper computation servers. It can provide more than one computation service, so the same dispositor can serve for different remote computation programs, and can be used to provide wide range of services. One computational service is typically a program that can receive remote data and send the result. The program is optimized to gain maximal possible performance and usually is designed to run in batch queue.

The dispositor after receiving request decides which computation server to use (there may be more than one computation server), and forwards the client data to it. Then it returns data back to the client. This model is necessary for Web-based computing, because of security limitations of WWW applets. It also benefits in central management and - what is more important - it hides all the computation from client. Such design simplifies client role: client does not have to know many computation server addresses, but only one - the dispositor address. In case of any system changes for example adding and removing computation servers, client may even not know about it. It also allows to implement basic load

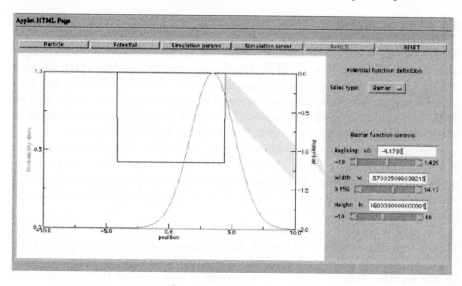

Fig. 2. The Client Applet - change of square well potential parameters.

balancing algorithms, that make system far more effective. At the moment there is no load balancing implemented and computation server is chosen randomly from the list. In the limited access mode, computation dispositor may also have abilities to make client authorization. The last, but not least benefit of the model is that there is only one channel to open in firewall of computing centre which simplifies significantly security management.

3.3 Computation Server

Computation server is rather simple application that receives form dispositor data containing request, and then it runs computation program and finally sends results back to the dispositor. Server is very simple and because of that it's small and fast, not requiring computation power, that must be used for computation program. This allows for server installation on the computer other than computational engine.

At this moment server is separated from computation program, so main numerical code was not changed at all - it *doesn't know* that it's run from the web, not locally. This has strong practical implications: server can be used for any existing computational codes without significant modifications of the numerical engine.

Server has a limit of number of computations that can be run parallelly, to avoid overloading. This this is implemented as system of "tokens", and is also used at the dispositor to determine which machines are free for computations.

3.4 Communication

Network layer is based on standard TCP/IP socket connection and a simple communication protocol over TCP. We decided not to use Java RMI, because in case of running the system on supercomputers, that doesn't have any Java implementation we would have to rewrite the computation server in C, without RMI. It also makes possible to write clients in other languages for example as plugins to data visualization programs or as stand alone applications.

The main goal was to provide user with real-time, interactive simulations of wavepacket dynamics, and therefore existing solutions based on the batch processing cannot be used. Developed computation server is design to start simulations imidiatelly on less loaded machine in the pool, provides user with partial results as soon as they are available.

General scheme of computation is that user prepares his job with client. Then client connects to a dispositor, and requests computation service. Then dispositor decides if there are available computation servers - finds proper token for the service. Then it accepts client's request and receives the job data, and forwards it to the proper server. Computation server receives the data, and uses it to initialize computation (for example writes input data onto disk as files). Then computation server starts to send results as fast as computation program produces it. Dispositor then forwards result to client. This allows for real time visualization of results as soon as they available on the computational server. If the client disconnect job is aborted, and if job has been finished dispositor and computation server disconnects. At this moment there is no possibility of connecting to a running job or reconnection. Client receives results and presents it to the user.

4 Details of System Design

All parts but computation program are written in Java (TM), what makes possible to use whole system on very wide set of machines, without concerning about portability and language implementation details. Effectively it means that we can use cluster of different workstations for different computations, and clients can use in fact any workstation without architecture or operating system limits.

All parts of the system were designed as modules, so in all programs are easily separable modules, responsible for protocol, for request creation and so on. Every module can be easily exchanged and extended, so it's easy to improve system. It was important to make configuration of the system easy and elastic - all servers use an XML configuration files, which provide easy configuration in human-readable form.

The developed system is fully functional and is its applications are not reduced to the wavepacket dynamics. It is capable to serve computations and to perform on line numerical processing such as remote data visualization, access to parts of numerical data and so on. It can be easily adopted for different services. Generally, adding new service to the system requires three steps:

1. Providing computation program,
   Program can be written in any language, in general modifications to the code are not required.
2. Adopting service provider - part of computation server.
   This modifications are small and can be easily performed.
3. Preparing client software
   Client software can be written in any language, however communication with the computing server must be added. Presently this is written in Java and is available in form of library which can be used for different clients.

All this tasks can be easily performed.

5 Conclusions

Developed systems for web based real time wavepacket dynamics proves that it's relatively easy to build small and medium scale system that can be used for serving small and medium scale remote job processing. Presented system is functional and proves that web based real time simulations can be efficiently performed even in the area dominated by the batch processing. The modifications to the existing simulation programs are very limited, in practice, most of existing codes written in Fortran or C can be used as is. Modular approach opens system for other application areas as well as for further development.

It could be used as computation server distributing workstation-scale computations between machines available inside the computing center, as an remote access for custom visualization of parts of the numerical data stored in the center for example for meteograms from weather simulations, or medical images. Nowadays system is used for educational purposes as easy to learn interface to advanced simulations in quantum physics.

Second conclusion is that it's possible to extend the system for classic queuing systems, allowing clients to prepare larger jobs and then disconnect and reconnect to collect results. This, connected with well prepared load-balancing may create nice compact grid application.

Last conclusion is a side effect of project, but we think that it may be most important for future work. The system we used did not involved any modification of the numerical code. We simply took Fortran code, recompiled it for desired architecture and used program as it was - as *black-box* application. It was possible to implement kind of grid without any modifications. What is most important we were able to receive results on line, in time of the running job. This is typical for advanced numerical grid applications, with special network code and carefully prepared for this task (see VisIt, Globus). In our system it was done far simpler way - we connect not to an application but read output files. This approach has three main consequences:

   - very limited interaction with application,
   - no need of any numerical code modification,

- no special skill required from numerical programmer required to prepare new application for the system.

In the present design one has no way to interact (as other grid systems do) except halting the job - but we assume that vast majority of the numerical application are absolutely non interactive, and will not be. What one gets is fact that one can use code as it is. The only recommendation, but not requirement, is that it's better to receive output as an ASCII text or in machine independent format. New developers of the simulation code don't have to learn new skills like network programming, what shall not be underestimated.

Acknowledgements This work is partially supported by European Commission under IST grant 20247. The computations were performed at the ICM, Warsaw University.

References

1. J. W. Eastwood R. W. Hockney. *Computer Simulation Using Particles*. Cambridge University Press, Cambridge, 1987.
2. M. M. Woolfson and G. J. Pert. *An Introduction to Computer Simulation*. Oxford University Press, 1999.
3. D. Kosloff and R. Kosloff. A Fourier method solution for the time dependent Schroedinger equation as a tool in molecular dynamics. *J. Comput. Phys.*, 52:35–53, 1983.
4. D. H. Zhang, Minghui Yang, and S-Y. Lee. Branching ratio in the HD+OH reaction: A full-dimensional quantum dynamics study on a new ab initio potential energy surface. *J. Chem. Phys.*, 114(20), May 2001.
5. J. H. Eberly, R. Grobe, C. K. Law, and Q. Su. Numerical experiments in strong and super-strong fields. In M. Gavrila, editor, *Atoms in string fields*. Academic Press, San Diego, 1992.
6. L. Pesce, Z. Amitay, R. Uberna, S. R. Leone, M. Ratner, and R. Kosloff. Quantum dynamics simulation of the ultrafast photoionization of Li₂. *Journal of Chemical Physics*, 114(3):1259–71, Jan 2001.
7. P. A. Ruprecht, M. J. Holland, K. Burnett, and M. Edwards. Time-dependent solution of the nonlinear Schrodinger equation for Bose-condensed trapped neutral atoms. *Phys. Rev. A*, 51(6):4704–11, 1995.
8. R. Dum, A. Sanpera, K.-A. Suominen, M. Brewczyk, et al. Wave packet dynamics with Bose-Einstein condensates. *Phys. Rev. Lett.*, 80(18):3899–902, 1998.
9. University of Chicago, Chicago. IL USA   The globus project. http://www.globus.org
10. University of Virginia, Charlottesville. VA USA *Legion*. http://legion.virginia.edu
11. Pallas. Germany *Unicore*. http://www.unicore.org
12. R. Kosloff. Time-dependent quantum-mechanical methods for molecular dynamics. *J. Phys. Chem.*, 92(8):2087–2100, April 1988.
13. C. Leforestier, R. H. Bisseling, C. Cerjan, M. D. Feit, R. Freisner, A. Guldberg, A. Hammerich, G. Jolicard, W. Karrlein, H.-D. Meyer, O. Lipkin, O. Roncero, and R. Kosloff. A comparison of different propagation schemes for the time dependent Schrodinger equation. *J. Comput. Phys.*, 94(1):59–80, 1991.

14. M. D. Feit, J. A. Fleck, and A. Steiger. Solution of the Schroedinger equation by a spectral method. *J. Comput. Phys.*, 47:412, 1982.

15. M. D. Feit and J. A. Fleck. Solution of the Schrodinger equation by a spectral method ii. Vibrational energy levels of triatomic molecules. *J. Chem. Phys.*, 78:301–308, 1983.

16. M. D. Feit and J. A. Fleck. Wave packet dynamics and chaos in the Henon-Heiles system. *J. Chem. Phys.*, 80(6):2578–2586, 1984.

17. D. Neuhauser, M. Baer, R. S. Judson, and D. J. Kouri. A time-dependent wave packet approach to atom-diatom reactive collision probabilities: Theory and application to the $H+H_2$ (j=0) system. *Journal of Chemical Physics*, 93(1):312–22, July 1990.

18. D. Neuhauser, M. Baer, R. S. Judson, and D. J. Kouri. The application of time-dependent wavepacket methods to reactive scattering. *Computer Physics Communications*, 63(1-3):460–81, Feb. 1991.

19. U. Peskin, R. Kosloff, and N. Moiseyev. The solution of the time dependent Schrodinger equation by the (t,t') method: The use of global polynomial propagators for time dependent Hamiltonians. *Journal of Chemical Physics*, 100(12):8849–55, June 1994.

20. R. Kosloff. Time dependent approach to femtosecond laser chemistry. In: *Eighteenth International Conference on Physics of Electronic and Atomic Collision (XVIII ICPEAC)*, Aarhus, Denmark, 21-27 July 1993, pages 152–170.

21. G. Torres-Vega. Lanczos method for the numerical propagation of quantum densities in phase space with an application to the kicked harmonic oscillator. *J. Chem. Phys.*, 98(9):7040, 1993.

22. H. Tal-Ezer, R. Kosloff, and C. Cerjan. Low-order polynomial approximation of propagators for the time-dependent Schrodinger equation. *Journal of Computational Physics*, 100(1):179–87, May 1992.

23. R. Kosloff. Propagation methods for quantum molecular dynamics. *Annual Review of Physical Chemistry*, 45:145–152, 1994.

24. H. Tel-Ezer and R. Kosloff. An accurate and efficient scheme for propagating the time dependent schroedinger equation. *J. Chem. Phys.*, 81:3967–3971, 1984.

25. T. N. Truong, J. J. Tanner, P. Bala, J. A. McCammon, D. J. Kouri, B. Lesyng, and D. Hoffman. A comparative study of time dependent quantum mechanical wavepacket evolution methods. *J. Chem. Phys.*, 96:2077–2084, 1992.

The Taylor Center for PCs: Exploring, Graphing and Integrating ODEs with the Ultimate Accuracy

Alexander Gofen

The Smith-Kettlewell Eye Research Institute, 2318 Fillmore St., San Francisco, CA 94102,
USA
galex@ski.org, www.ski.org/gofen

Abstract. A software package intended as a cross-platform integrated
environment for exploring, graphing, real time 2D and 3D "playing" and
integrating ODEs with the Modern Taylor method, is presented. Designed
under Windows in Delphi-6, the package implements a powerful Graphical
User Interface (GUI), and is portable to windowed versions of Linux also. The
downloadable Demo of the package displays a variety of meaningful examples
including such an illustrative as the Problem of Three Bodies and
"Choreography" [8].

1 Introduction

It seems, there were no sophisticated Taylor Solvers designed for PCs since
ATOMFT [2], and definitely not as a cross platform Windows/Linux application
developed in Delphi. Thus, the goal was to develop a new version of such a package,
called the Taylor Center (similarly to "Audio Centers" with radio, cassette, CD
players). It has to take full advantage of:

(1) Hardware breakthroughs such as linear memory and the high speed/memory
characteristics of the modern PCs;

(2) Software breakthroughs such as the new metaphor of modern operating
environments – Graphical User Interface (GUI) and event driven logic, supported by
such developer-friendly and efficient systems as Delphi-6 (for Windows) and Kylex
(for Linux);

(3) Some new features of the Modern Taylor method and its applications.

From the programmatic point of view, any Taylor solver differs from conventional
integrators in that the input – an Initial Value Problem – should be provided not as a
function simply computing the right hand parts, but rather as arithmetic expressions
representing the right hand parts themselves in order to enable *automatic
differentiation*. Typically, these right hand parts must be (automatically) reduced to
the so called canonic equations first [1, 3], but this version of the Taylor solver is built
around an algorithm evaluating postfix sequences obtained while parsing the so called
Linear Elaborated Declarations (in ADA terminology):

P.M.A. Sloot et al. (Eds.): ICCS 2002, LNCS 2329, pp. 562–571, 2002.

$$u_0 = f_0(\{numeric\ values\ only\});$$
$$u_1 = f_1(u_0,\ \{numeric\ values\});$$
$$\dots\dots$$
$$u_n = f_n(u_0,\ u_1,\dots,\ u_{n-1},\ \{numeric\ values\}).$$

(1)

Here each variable first appears in the left hand part before it may be used in the right hand parts of the subsequent equations (just like the principle of declaring constants in Pascal).

To evaluate (1), first we apply a recursive parsing procedure to each line to obtain its postfix sequence, whose operands are either numeric values, or the previously computed variables. The evaluation process of the postfix sequence consists of locating triplets (Operand1, Operand2, Operation $°$) in it, evaluating the elementary equation

$$Result := Operand1\ °\ Operand2,$$

and replacing the triplets with the corresponding Result until the whole sequence reduces to just one term – the final result of the evaluation.

The modern Taylor Method means basically iterative evaluation of recurrent equations similar to (1). Let us consider the standard Initial Value Problem for ODEs:

$$u_k' = f_k(u_1,\dots,\ u_m),\quad u_k|_{t=a} = a_k, \qquad k=1,\dots m$$

(2)

where u_k' means derivatives with respect to t. The equations (2) make it possible to obtain the set of all first order derivatives u_k' by computing the right hand parts. According to the classical Taylor method, these equations make it possible to obtain the set of all N-order derivatives $u_k^{(N)}$ also. To do that, we have to recurrently apply the well known rules of differentiation to both parts of the equations and substitute the already obtained numerical values. As such, the classical Taylor method had not been considered as a numerical method for computers at least for two reasons: applying the formulas for N-th derivative to complex expressions (superposition of several functions in the right hand parts) is both challenging and computationally wasteful due to the exponentially growing volume of calculations in the general case).

On the contrary, the Modern Taylor Method capitalizes on the fact that however complicated the right hand parts are, they are equivalent to a sequence of simple formulas (instructions). Indeed, this sequence of instructions is exactly that resulting from the process of *postfix evaluation* (which is similar to reducing the source system to the so called *Canonical* equations [1, 3]). If the set of the allowed operations consists of the four arithmetic operations (and also raising to constant power, exponent and logarithm), computing the N-th derivative requires only no more than $O(N^2)$ operations and it is easily programmable [6]. Thus, unlike its classical counterpart, the Modern Taylor Method becomes viable and competitive for a computer implementation.

2 How it is implemented

To achieve more flexibility in specifying an Initial Value problem (2), the source data is organized as *Linear Elaborated Declarations* in four sections in the given order: Symbolic constants, Initial values, Auxiliary variables, ODEs (Table 1).

Table 1. How the source data of the Initial Value Problem is organized as *Linear Elaborated Declarations*

Variables and equations	Explanation
$c1 = NumericValue$ $c2 = C2(c1, NumericValues)$	Symbolic constants and functions of them
$u1 = U1(c1, c2,..., NumericValues)$ . . . . $um = Um(c1, c2,..., u1,..., NumericValues)$	Initial values for the Main variables and functions of them
$v1 = V1(c1,..., um, NumericValues)$ $v2 = V2(c1,..., um, v1, NumericValues)$	Auxiliary variables
$u1' = f1(u1,..., um, c1, c2,..., v1, v2,...)$ . . . . $um' = fm(u1,..., um, c1, c2,..., v1, v2,...)$	The source ODEs
	Internal Auxiliary variables

The Symbolic constants and Auxiliary variables make it possible to parameterize Initial Value problems and to optimize computation by eliminating re-calculations of common sub-expressions. For example, in the Three Body Problem the section of *Constants* is:

i15 = -1.5 *{exponent index used in subsequent equations}*
m1 = 1   *{masses}*
. . . .
x1c = 1   *{initial positions of the three bodies}*
. . . . . .
x3c = cos(240)
. . . .

The equations in the section *Auxiliary variables* are these:

dx12 = x1 - x2   *{projections of the distances between the bodies}*
. . . .
r12 = (dx12^2 + dy12^2)^i15   *{common sub-expressions of the ODEs}*
r23 = (dx23^2 + dy23^2)^i15
r31 = (dx31^2 + dy31^2)^i15

The ODEs themselves are:

x1' = vx1

. . . .

vy3' = m2*dy23*r23 - m1*dy31*r31

Without the auxiliary variables, the more familiar, but non-optimized equations would look like this (just the last one)

$$v_{y3}' = m_2(y_2\text{-}y_3)((x_2\text{-}x_3)^2 + (y_2\text{-}y_3)^2)^{-3/2} + m_1(y_1\text{-}y_3)((x_3\text{-}x_1)^2 + (y_3\text{-}y_1)^2)^{-3/2}$$

(see [7] for the downloadable Demo). The data from the Table 1 is to be processed in the following three stages:

(1) Like the list of linear elaborated declarations, it should be parsed line by line into the postfix sequence for the subsequent automatic differentiation process;

(2) The differentiation process up to the specified order is performed. The set of coefficients of the Taylor expansion (also called the Analytical elements) for each variable is obtained;

(3) With the obtained Taylor coefficients, the convergence radius R should be estimated, an integration step $h<R$ determined and the Taylor expansion applied to increment the all main variables. Then, the stages 2 and 3 may be repeated as many times as necessary (unless a singularity is reached).

The syntax checking and parsing for the stage 1 is done just as in the case of an arbitrary arithmetic expression [4]. As to the evaluation, here we need a more sophisticated algorithm, capable of evaluating N-order derivatives for results of each operation in the postfix sequence. To achieve that, we will "compile" the postfix sequences into the three-address instructions of an abstract Processor, and create the software implementation of that Processor. This "compiled set of instructions" for a given problem would remain the same even if we need to switch from real numbers to complex numbers, or from one real type to another, including the cases of software emulated very long floating point types.

3 The Evaluating and Differentiating Processor

The *memory* for this Processor is the special dynamic array of auxiliary variables AuxVars:

```
AuxVars : array of array of extended;
```

An element *AuxVars[i]* corresponds to the variable or symbolic constant in line i of Table 1, while *AuxVars[i, j]* means the j-order derivative of that variable (when non-constant).

The format of the instruction of the Processor is this:

A. Gofen

type TInstruction = record
         ty : (cc, cv, vc, vv); {flags: (C)onstant or (V)ariable}
         op : char;   {operation to perform}
         i1, i2, i3 : word   {addresses, i.e. indexes in AuxVars}
     end;
```

For each postfix triplet (*Operand1, Operand2, Operation*) there is an instruction of the type *TInstruction*, such that *Operand1* is in *AuxVars[i1]*, *Operand2* in *AuxVars[i2]*, and the result of performing of the *Operation* is assigned to *AuxVars[i3]*. To compile a "program" evaluating a postfix sequence, we should scan the postfix sequence from left to right until a first triplet (*Operand1, Operand2, Operation*) is encountered. So, the postfix sequence is encoded in a set of instructions, preserved for the subsequent massive computations of the corresponding arithmetic expression.

Actually, this Processor is designed to do not only simple evaluations like

$$Result := Operand1 \circ Operand2,$$

but also to perform the iterative differentiation up to any given order

$$Result^{(N)} := (Operand1 \circ Operand2)^{(N)}, \quad N = 0, 1, 2,...$$

given the corresponding formulas for computing *N*-th derivative of certain functions – for example the Leibnitz formula

$$(uv)^{(N)} = \sum_{i=0}^{N} u^{(i)} v^{(N-i)}$$

and other similar ones [1,3,6] (all the derivatives are normalized meaning $u^{(i)}/i!$). So, with this Processor we can obtain derivatives of any required order in the initial point *a* for all variables (Table 1) representing the source initial value problem.

## 4. Estimation of the Convergence Radius

To estimate the convergence radius, we use the algorithm based on the formula of Cauchy-Hadamard

$$R^{-1} = sup \left| a_n \right|^{1/n}$$

where Taylor coefficients $a_n$ are obtained for each unknown function in the source ODEs. Each component of the solution may have its own radius, thus the procedure selects the minimal among them, which should be considered as a convergence radius for the whole system. Although the formula is exact, the result is heuristic because the
```

formula is applied to only a finite number of terms, with simplified assumptions for obtaining the supremum. As a matter of fact, in order to obtain the right integration step h, we neither can nor even need to know the exact value of the radius: in any case the step value is going to be adjusted to meet the required accuracy.

So, given the Taylor coefficients $a_0, a_1, ..., a_n$, first of all we obtain their floating point exponents $e_0, e_1, ..., e_n$ using the standard procedure *FrExp(x, m, result)* for all non-zero coefficients, and assigning -4951 (the minimal possible type *extended* exponent) to zero coefficients. These are the rough $\log_2$ of the values, so that logarithms of the sequence of the required fractional powers for further processing are these:

$$e_1/1, \ e_2/2, \ ..., \ e_n/n = s_n.$$

The supremum of the sequence is the greatest point of condensation. To find it for this finite sequence (usually 30-40 terms), the following heuristic procedure is applied.

The sequence $\{s_n\}$ is sorted in descending order (suppose, it is already done). A condensation point of the sequence most probably belongs to such a segment $[s_{i+w}; s_i]$, whose length $s_i - s_{i+w}$ is minimal (w is about $n/6$, usually 4-5). Thus, we scan $\{s_n\}$ for the most narrow difference $s_i - s_{i+w}$, $i=1,2,...$ expecting that the greatest condensation point must be there. The i, delivering that minimum, is further considered for obtaining the average s of $s_i, ..., s_{i+w}$, and then (returning from the binary logarithms back to the basic values), the value 2^{-s} is accepted as a heuristic radius of convergence R. As many numeric experiments showed, this very economical procedure delivers quite a reasonable value for R.

5. The Step Size and Accuracy Control

The problem of selecting the optimal order of approximation N and integrating step h (to meet the required accuracy) is vital for any implementation of the Taylor method [1,3]. For example, according to Moore's rule of thumb, it is advised to set N as the number of decimal digits in the floating point mantissa (19 for the 8-byte mantissa in the type *extended*), and to specify h so that the quotient $q=h/R=0.1$. In the Taylor Center the defaults for these values are $N=30$, $q=0.5$. This q affects each of the three modes of the error control.

In the *No error control* mode the value of q is not automatically adjusted and remains that specified by the user. Generally speaking, the smaller the quotient, the higher the accuracy, but more steps would be necessary to cover the given segment (which may compromise the accuracy).

Even with no error control we can estimate the actual accuracy by switching the direction back and reaching the initial point. The column of the differences then will show how far we are away from the initial values.

Another approach to estimate the accuracy in this mode is to perform the integration until the given point twice with different quotients q, say with $q = 0.2$ and $q = 0.05$, and to compare the results.

The following two methods of error control rely on the values of the absolute and/or relative error tolerance entered by the user.

The *Back-step error control* mode changes the algorithm of integration so that each integration step occurs in a loop of step adjustments in the following way.

(1) The step h obtained in point t with the given starting value of $q = q_o$ (shown on the display) is applied, then the algorithm computes the new element in point $t+h$ and the Taylor expansion in this new point with the step value $(-h)$. Ideally that must be equal to the solution in the previous point t.

(2) If the components of the solution fit the specified error tolerance in the table, this step doesn't require any adjustment and it ends the loop. Otherwise...

(3) The step is split (halved), and the process repeats either until the specified accuracy is reached, or the limit (of 30 splitting) is exceeded. The latter means that the specified accuracy cannot be achieved for the given point due to a variety of reasons. (The 30 splits mean that the last attempt was made with the step as little as 2^{-30} of the initial step).

Similarly, the *Half-step error control* also arranges a loop of step adjustments with the goal of achieving the specified accuracy. It works as follows.

(1) The step h obtained with the given starting value of $q = q_o$ is applied, and then the algorithm computes the new element in point $t+h$ to be used as a vector of check values. The step h is split: $h = h1 + h2$.

(2) The values in the same point $t+h$, $h = h1 + h2$, are obtained in two steps and then compared with the check values. If the components of the solution fit the specified error tolerance, this step doesn't require any adjustment and ends the loop. Otherwise...

(3) The step is split (halved), and the process repeats either until the specified accuracy is reached, or the limit (of 30 splitting) is exceeded. The latter means that the specified accuracy cannot be achieved for the given point.

6 Reaching the Ultimate Accuracy

The ultimate accuracy at each *finite* integration step (i.e. the accuracy up to all available digits of the mantissa) is an ambitious, but achievable goal in the framework of the Taylor method and specifically in this package (the user has to specify the relative error tolerance for the desired components say less than 2^{-64}).

Several cases of essentially unstable problems (the two body problem, or the three body Lagrange case producing symmetrical ellipses with low enough eccentricities) were tested with the goal of reaching the ultimate accuracy. About 150-200 steps were required to cover one period, and this highest accuracy was achieved at each step (maybe except one).

The reason while the Taylor method may fail in achieving the highest accuracy is the same that may happen in any numeric computations with floating point numbers: a *subtractive error*. It occurs when a sum or difference of the computed values is zero, or so near zero, that the exponent of the result is N units smaller than either of the operand's exponents (N is near the length of the mantissa, 64 for the type *extended*). Whether the effect of such a dramatic loss of precision is catastrophic for a given computational process depends on the way it propagates farther to the end result.

For the Taylor method, summing is a part of computing almost all derivatives, thus the subtractive error occurs when any of the n-order derivatives or its computed sub-expressions in a current point reaches a near-zero value. The bigger the n, the easier this effect is eliminated by the factor h^n while selecting a smaller step h. In the worst exceptional case when $n=1$ (or the Taylor series' sum is zero), no small h can cure the problem in this point. Still, unless this point is not a singularity for the solution, there is always a simple remedy: we can avoid this exceptional point selecting the previous step a bit smaller. After that, a new finite step would make it posiible to easily jump over this "problem" point (according to the theory of analytical functions, the zero points may be only isolated).

It is worth noting that whichever integrating method is used, the ultimate accuracy at each and every *one* integration step may still not be enough for obtaining all correct digits in the result of the *two or more* integration steps: such is the nature of essentially unstable problems like those mentioned above. For example, for a Lagrange case of the three body problem, in spite of the fact that the accuracy at each individual step was as high as 10^{-19}, after about 150 such steps covering the full period, the accuracy dropped till only 10^{-13}.

Therefore, in order to achieve the best accuracy in multiple step integration, we have to:

(1) Minimize the number of steps needed to cover the required segment by using a method of integration with *finite* steps independent on the required accuracy. Indeed, this method is the Taylor method.

(2) Implement a longer (or of any given length) type of float numbers and arithmetic over them as a software emulation (it exists in several versions of Pascal).

7 Features of the Taylor Center

With this version of the Taylor Center you can:

- Specify and study the Initial Value Problems for virtually any system of ODEs in the standard format with numeric and symbolic constants and parameters;

- Perform numerical integration of Initial Value Problems with the highest possible accuracy while the step of integration remains finite and does not approach zero (instead, the order of approximation is high);

- Perform integration either "blindly", or graphically visualized. It stops either after a given number of steps, or when an independent variable reaches a given value, or when a <u>dependent</u> variable reaches a given value (as explained in the next item);

- Switch integration between several versions of ODEs defining the same trajectory, but with respect to different independent variables. For example, it is possible to switch the integration from that with respect to t to that by x, or by y in order to reach a given end value (or zeros) of a dependent variable;

- Integrate piecewise-analytical ODEs;

- Specify different methods of controlling the accuracy and the step size;

- Specify accuracy for individual components either as an absolute or relative error tolerance, or both;

- Apply the order of approximation as high as you like (30, or 40 or whatever), and get the solution in the form of the set of analytical elements – Taylor expansions covering the required domain;

- Study Taylor expansions and the radius of convergence for the solution at all points of interest up to any high order (with the only limitation that the terms in the series do not exceed the maximum value of about 10^{4932} implied by the 10-byte implementation of the real type extended);

- Graph color curves (trajectories) for any pair of variables of the solution – up to 7 on one screen – either as plane projections, or as 3D stereo images (for triplets of variables) to be watched through anaglyphic (red/blue) glasses;

- "Play" dynamically the near-real time motion along the computed trajectories either as 2D or 3D stereo animation;

- Explore several meaningful examples supplied with the package such as the problem of Two and Three Bodies. Symbolic constants and expressions make it possible to parameterize the equations and initial values and to easily try different initial configurations of special interest.

In particular, the examples include a fascinating case of the so called Choreography for the Three Body motion, an eight-shaped orbit, discovered just recently (2000) by Chenciner and Montgomery [8]. You can "feed" those equations into the Taylor Center, integrate them, draw the curves and play the motion in the real-time mode all in the same place.

The features of the future versions of the product will include everything above plus the following:

- It will come not only as the Taylor Center GUI executable, but also as the separate units, allowing the users to include them directly in their Delphi projects to perform massive computations with ODEs, and also as a DLL to use in other environments;

- It will implement the Merge procedure and a library of ODEs, defining a large variety of commonly used elementary functions. (Presently, the functions which are not in the allowed limited list may be used also – providing that the user declares the ODEs defining them and properly links them with the source;

- It will work for complex numbers, which will make it possible to study the solutions of ODEs in the complex plane in order to locate and explore their singularities;

- The application will be ported to Linux/Kylex;

- The set of the internal differentiation instructions will be translated into the machine code - to reach the highest possible speed for massive computations. (Meanwhile it is an emulator written in Delphi which runs these instructions). Also, it may be translated into instructions in Pascal, C or Fortran to be further compiled and linked with other applications;

- Integrating by a parameter or integrating boundary value problems [6].

References

1. Moore, R. E.: Interval Analysis. Prenitce-Hall, Englewood Cliffs, N.Y. (1966).
2. Chang Y. F., The ATOMFT User Manual. Version 2.50, 976 W. Foothill Blvd . #298, Claremont, CA 91711 (1988)
3. Gofen, A. M.: Taylor Method of Integrating Ordinary Differential Equations: the Problem of Steps and Singularities. Cosmic Research, Vol. 30, No. 6, pp. 581-593 (1992)
4. Gofen, A. M.: Recursion Excursion. Delphi Informant Magazine. Vol. 6, No. 8 (2000)
5. Gofen, A. M.: Recursive Journey to Three Bodies. Delphi Informant Magazine. Vol. 8, No. 3 (2002)
6. Gofen, A. M.: ODEs and Redefining the Concept of Elementary Functions. These Proceedings (2002)
7. Gofen, A. M.: The Taylor Center Demo for PCs.
   http://www.ski.org/rehab/mackeben/gofen/TaylorMethod.htm
8. Chenciner, Montgomery. The N-body problem "Choreography" www.maia.ub.es/dsg

Classification Rules + Time = Temporal Rules[1]

Paul Cotofrei, Kilian Stoffel

Université Neuchâtel,
Pierre-à-Mazel 7, 2000, Neuchâtel, Suisse
paul.cotofrei@unine.ch, kilian.stoffel@unine.ch

Abstract. Due to the wide availability of huge data collection comprising multiple sequences that evolve over time, the process of adapting the classical data-mining techniques, making them capable to work into this new context, becomes today a strong necessity. In [1] we proposed a methodology permitting the application of a classification tree on sequential raw data and the extraction of the rules having a temporal dimension. In this article, we propose a formalism based on temporal first logic-order and we review the main steps of the methodology through this theoretical frame. Finally, we present some solutions for a practical implementation.

1. Introduction

Data mining is defined as an analytic process designed to explore large amounts of (typically business or market related) data, in search for consistent patterns and/or systematic relationships between variables, and then to validate the findings by applying the detected patterns to new subsets of data. Due to the wide availability of huge amounts of data in electronic form, and the imminent need for turning such data into useful information and knowledge for broad applications including market analysis, business management and decision support, data mining has attracted a great deal of attention in information industry in recent years.

In many applications, the data of interest comprises multiple sequences that evolve over time. Examples include financial market data, currency exchange rates, network traffic data, sensor information from robots, signals from biomedical sources like electrocardiographs, demographic data from multiple jurisdictions, etc. Although traditional time series techniques can sometimes produce accurate results, few can provide easy understandable results. However, a drastically increasing number of users with a limited statistical background would like to use these tools. Therefore, it becomes increasingly important to be able to produce results that can be interpreted by domain expert without special statistical training. In the same time, we have a number of tools developed by researchers in the field of artificial intelligence, which produce understandable rules. However, they have to use ad-hoc, domain-specific techniques for "flattering" the time series to a "learner-friendly" representation. These techniques fail to take into account both the special problems and special heuristics applicable to temporal data and so often unreadable concept description are obtained.

---

[1] This work was supported by the Swiss National Science Foundation (grant N° 2100-063 730).

P.M.A. Sloot et al. (Eds.): ICCS 2002, LNCS 2329, pp. 572–581, 2002.

1.1   State of the Art

The main tasks concerning the information extraction from time series database and on which the researchers concentrated their efforts over the last years may be divided into four directions.

1. *Similarity/Pattern Querying.* The main problem addressed by this body of research concerns the measure of similarity between two (sub-) sequences. Different models of similarity were proposed, based on different similarity measures (Euclidean metric [2], [3], envelope metric ($|X - Y| \leq \varepsilon$)). Different methods were developed (window stitching [4], dynamic time warping based matching [5]) to allow matching similar series despite gaps, translation and scaling. For all methods, the quality of the results and the performance of the algorithms are intrinsically tied to the subjective parameter that is the user-specified tolerance ε.

2. *Clustering/Classification.* In this direction, researchers studied optimal algorithms for clustering/classifying sub-sequences of time series into groups/classes of similar sub-sequences. Different techniques were developed, as the Hidden Markov Model [6], Dynamic Bayes Networks [7] or the Recurrent Neural Networks [8]. Recently, the machine learning approach opened new directions. A system for a supervised classification of signals using piecewise polynomial modeling was developed in [9] and a technique for agglomerative clustering of univariate time series representation was studied in [10].

3. *Pattern finding/Prediction.* These methods, concerning the search for periodicity patterns in time series databases, may be divided into two groups: those that search full periodic patterns and those that search partial periodic patterns, which specify the behavior at some but not all points in time. For full periodicity search there is a rich collection of statistic methods, like FFT [11]. For partial periodicity search, different algorithms were developed which explore properties related to partial periodicity, like the a-priori property and the max-sub-pattern-hit-set property [12] or the segment-wise and the point-wise periodicity [13].

4. *Rule extraction.* Besides these, some researches were devoted to the extraction of explicit rules from time series. Inter-transaction association rules, proposed by Lu, [14] are implication rules whose two sides are totally ordered episodes with time-interval restriction on the events. Cyclic association rules were considered in [15], adaptive methods for finding rules whose conditions refer to patterns in time series were described in [16] and a general architecture for classification and extraction of comprehensible rules was proposed in [17].

1.2 Aims and Scopes

The real-world problem we tried to solve and which practically "pushed" us to develop our methodology consisted in the analysis of an insurance company's database. The question was if it is possible, looking to data containing medical information, to find/extract rules of the form "If a person contracted an influenza necessitating more than five days of hospitalization, then with a confidence p the same person will contract, in the next three month, a serious lung-infection". The size of the database (several Gigabytes) made the application of statistical tools very difficult. Therefore, we searched for alternative techniques to solve these kinds of problems. However, most approaches dealing with the information extraction from time series, described above, have mainly two shortcomings.

The first problem is the type of knowledge inferred by the systems, which is very difficult to be understood by a human user. In a wide range of applications, (e.g. almost all decision making processes) it is unacceptable to produce rules that are not understandable for an end user. Therefore, we decided to develop inference methods that produce knowledge that can be represented in form of general Horn clauses, which are at least comprehensible for a moderately sophisticated user. In the fourth approach, (*Rule extraction*), a similar representation is used. However, the rules inferred by these systems have a more restricted form than the rules we are proposing.

The second problem consists in the number of time series considered during the inference process. Almost all methods mentioned above are based on uni-dimensional data, i.e. they are restricted to one time series. The methods we propose are able to handle multi-dimensional data.

To overcome these problems we developed [1] a methodology that integrates techniques developed both in the field of machine learning and in the field of statistics. The machine learning approach is used to extract symbolic knowledge and the statistical approach is used to perform numerical analysis of the raw data. The overall goal consists in developing a series of methods able to extract/generate/ temporal rules, having the following characteristics:

- Contain explicitly a temporal (or at least a sequential) dimension
- Capture the correlation between time series
- Predict/forecast values/shapes/behavior of sequences (denoted events)

In this article, we extended the methodology with a formalism based on first-order temporal logic, which permits an abstract view on temporal rules. The formalism allows also the application of an inference phase in which "general" temporal rules are inferred from "local" rules. The latter are extracted from the sequences of data. Using this strategy, we can guarantee the scalability of our system to handle huge databases, applying standard statistical and machine learning tools.

The rest of the paper is structured as follows. In the next section, the formalism for the representation of events and temporal rules based on temporal logic is proposed. In Section 3, the main steps of the methodology are summary presented but in connection with the proposed formalism. Some specific problems concerning the practical implementation of the training sets and of the temporal rules are treated in the following section. The final section summarizes our work and points to future research.

2. The Formalism of Temporal Rules

Time is ubiquitous in information systems, but the mode of representation/perception varies in function of the purpose of the analysis [18], [19]. Firstly, there is a choice of a *temporal ontology*, which can be based either on *time points* (instants) or on *intervals* (periods). Secondly, time may have a *discrete* or a *continuous* structure. Finally, there is a choice of *linear* vs. *nonlinear* time (e.g. acyclic graph). For our methodology, we chose a temporal domain represented by linearly ordered discrete instants. This option is imposed by the discrete representation of all databases.

DEFINITION 1. *A single-dimensional linearly ordered temporal domain is a structure* $T_P = (T, <)$, *where T is a set of time instants and "$<$" a linear order on T.*

Because we may always find an isomorphism between T and the set of natural numbers, N, the temporal domain will be usually represented as (N, $<$). As consequence, there is an initial time moment (the first time instant), which is 0.

Databases being first-order structures, the first-order logic represents a natural formalism for their description. Consequently, the first-order temporal logic expresses the formalism of temporal databases. For a better rigor and for a cleaner description of the used terms one may consider the following set of definitions.

A *first-order language* L is constructed over an alphabet containing function symbols, predicate symbols, variables, logical connectives, temporal connectives and qualifier symbols. A *constant* is a zero-ary function symbol and a zero-ary predicate is a *proposition symbol*. There is a particular variable, T, representing the time. There are several special binary predicate symbols (=, <, ≤, >, ≥) known as *relational symbols*.

The basic set of logical connectives is $\{\wedge, \neg\}$ from which one may express $\vee$, $\rightarrow$ and $\leftrightarrow$. The basic set of temporal connectives are F ("sometime"), G ("always"), X ("next") and U ("until"). From these connectives, we may derive X_k, where a k positive means "next k" and a k negative means "previous k".

The syntax of L defines terms, atomic formulae and compound formulae. A *term* is either a variable or a function symbol followed by a bracketed n-tuple of terms ($n \geq 0$). A predicate symbol followed by a bracketed n-tuple of terms ($n \geq 0$) is called an *atomic formula*, or *atom*. If the predicate is a relational symbol, the atom is called a *relation*. A *compound formula* or *formula* is either an atom or n atoms ($n \geq 1$) connected by logical (temporal) connectives and/or qualifier symbols.

A Horn clause is a formula of the form $\forall X_1 \forall X_2 \ldots \forall X_s (B_1 \wedge B_2 \cdots \wedge B_m \rightarrow B_{m+1})$ where each B_i is a positive (non-negated) atom and $X_1, \ldots, X_s$ are all the variables occurring in $B_1, \ldots, B_{m+1}$. The atoms $B_i, i = 1 \ldots m$ are also called implication clauses, whereas B_{m+1} is also known as the implicated clause.

DEFINITION 2. *An event (or temporal atom) is an atom formed by the predicate symbol TE followed by a bracketed $(n+1)$-tuple of terms $(n \geq 1)$ $TE(T, t_1, t_2, \ldots, t_n)$. The first term of the tuple is the time variable, T, the second t_1 is a constant representing the name of the event and all others terms are function symbols. A short temporal atom (or the event's head) is the atom $TE(T, t_1)$.*

DEFINITION 3. *A constraint formula for the event $TE(T, t_1, t_2, \ldots t_n)$ is a conjunctive compound formula, $C_1 \wedge C_2 \wedge \cdots \wedge C_k$, where each C_j is a relation implying one of the terms t_i.*

DEFINITION 4. *A temporal rule is a Horn clause of the form*

$$H_1 \wedge \cdots \wedge H_m \rightarrow H_{m+1} \tag{1}$$

where H_{m+1} is a short temporal atom and H_i are constraint formulae, prefixed or not by the temporal connectives $X_{-k}, k \geq 0$. The maximum value of the index k is called the time window of the temporal rule.

The semantics of L is provided by an interpretation I that assigns an appropriate meaning over a domain D to the symbols of L. The set of symbols is divided into two classes, the class of global symbols, having the same interpretation over all time instants (global interpretation) and the class of local symbols, for which the interpretation depends on the time instant (local interpretation). In the framework of temporal databases, the constants and the function symbols are global symbols, and represent real numbers (with an exception, strings for constants representing the names of events), respectively statistical functions. The events, the constraint formulae and the temporal rules are local symbols, which means they are true only at some time instants. We denote such instants by $i \mapsto P$, i.e. at time instant i the formula P is true. Therefore, $i \mapsto TE(T, t_1, \ldots, t_n)$ means that at time i an event with the name t_1 and characterized by the statistical parameters $t_2, \ldots, t_n$ started (the meaning of the word "event" in the framework of time series will be clarified later). A constraint formula is true at time i iff all relations are true at time i. A temporal rule is true at time i iff either $i \mapsto H_{m+1}$ or $i \mapsto \neg(H_1 \wedge \cdots \wedge H_m)$. (Remark: $i \mapsto X_k P$ iff $i+k \mapsto P$).

However, the standard interpretation for a temporal rule is not conforming to the expectation of a final user for two reasons. The first is that a user wants a global interpretation for rules, that is, rules that are "generally" true. The second is tied on the table of true for a formula $p \to q$: even if p and q are false, the implication is still true. In a real application, a user search temporal rules where both the body, $H_1 \wedge \cdots \wedge H_m$, and the head of the rule, H_{m+1}, are simultaneously true.

DEFINITION 5. *A model for L is a structure $M = (\tilde{T}_p, D, I)$ where $\tilde{T}_p$ is a finite temporal domain and D, (respectively I), the domain, respectively the interpretation for L. We denote by N the cardinality of $\tilde{T}_p$.*

DEFINITION 6. *Given L, a model M for L and an infinite temporal domain T_p that include $\tilde{T}_p$, a global interpretation I_G for an n-ary atom P is a function defined on D^n with values in $true \times [0,1]$. I_G assigns to P the boolean value "true" with a confidence equal with the ratio $card(A)/N$, where $A = \{i \in \tilde{T}_p \mid i \mapsto P\}$.*

The global interpretation is naturally extended to formulae, using the usual probability calculus to obtain the confidence ($P(A \cap B) = P(A) + P(B) - P(A \cap B)$). There is only one exception: for temporal rules the confidence is calculated as a ratio between the number of certain applications (time instants where both the body and the head of the rule are true) and the number of potential applications (time instants where only the body of the rule is true). A useful temporal rule is a rule with a confidence greater than 0.5.

DEFINITION 7. *The confidence of a temporal rule $\forall T(H_1 \wedge \cdots \wedge H_m \to H_{m+1})$ is the ratio $card(A)/card(B)$, where $A = \{i \in \tilde{T}_p \mid i \mapsto H_1 \wedge \cdots \wedge H_m \wedge H_{m+1}\}$ and $B = \{i \in \tilde{T}_p \mid i \mapsto H_1 \wedge \cdots \wedge H_m\}$.*

3. The Methodology

The two scientific communities which made important contribution relevant to data analysis (the statisticians and database researchers) chose two different approaches: statisticians concentrated on the continuous aspect of the data and the large majority of statistical models are continuous models, whereas the database community concentrated much more on the discrete aspects, and in consequence, on discrete models. We are adopting here a mixture of these two approaches, which represents a better description of the reality of data and which generally allows us to benefit from the advantages of both approaches. However, the techniques applied by the two communities must be adapted in order to be integrated in our approach.

The two main steps of the methodology of temporal rules extraction are structured in the following way:

1. Transforming sequential raw data into sequences of events: Roughly speaking, an event can be regarded as a labeled sequence of points extracted from the raw data and characterized by a finite set of predefined features. The features describing the different events are extracted using statistical methods.
2. Inferring temporal rules: We apply an inference process, using sets of events extracted from event database as training sets, to obtain several classification trees. Then temporal rules are extracted from these classification trees and a final inference process will generate the set of global temporal rules.

3.1. The Phase One

The procedure that creates an event database from the initial raw data can be divided into two steps: time series discretisation, which extracts the discrete aspect, and global feature calculation, which captures the continuous aspect. This phase totally defines the interpretation of the domain D and of the function symbols from L.

Time series discretisation. Formally, during this step, the constant symbols t_1 (the second term of a temporal atom) receive an interpretation. Practically, all contiguous subsequences of fixed length w are classified in clusters using a similarity measure and these clusters receive a name (a symbol or a string of symbols). By a misuse of language, an event means a subsequence having a particular shape. There are cases when the database contains explicitly the names of events, like in our insurance case, where we identified the names of events by the names of diseases.

All the methods which construct/find class of events depend on a parameter k which controls the degree of discretisation: a bigger k means a bigger number of events and finally, less understandable rules. However, a smaller k means a rough description of the data and finally, simple rules but without significance.

Global feature calculation. During this step, one extracts various features from each sub-sequence as a whole. Typical global features include global maxima, global minima, means and standard deviation of the values of the sequence as well as the value of some specific point of the sequence, such as the value of the first or of the last point. Of course, it is possible that specific events will demand specific features, necessary for their description (e.g. the average value of the gradient for an event representing an increasing behavior). The optimal set of global features is hard to be defined in advance, but as long as these features are simple descriptive statistics, they can be easily added or removed form the process. From the formal model viewpoint, this step assigns an interpretation to the terms $t_2, \ldots, t_n$ of a temporal atom. They are considered as statistical functions, which take continuous or categorical values, depending on the interpretation of the domain D.

3.2. The Phase Two

During the second phase, we create a set of temporal rules inferred from the events database. This database was obtained using the procedures described above. Two important steps can be defined here:

1. Application of a first inference process, using the event database as training database, to obtain one or more *classification trees* and
2. Application of a second inference process using the previously inferred classification to obtain the final set of temporal rules.

Classification trees. There are different approaches for extracting rules from a set of events. Associations Rules, Inductive Logic Programming, Classification Trees are the most popular ones. For our methodology, we selected the *classification tree approach.* It is a powerful tool used to predict memberships of cases or objects in the classes of a categorical dependent variable from their measurements on one or more predictor variables (or attributes). A classification tree is constructed by recursively partitioning a learning sample of data in which the class and the values of the predictor variables for each case are known. Each partition is represented by a node in the tree. A variety of classification tree programs has been developed and we may mention QUEST [20], CART [21], FACT [22] and last, but not least, C4.5 [23]. Our option was the C4.5 like approach. The tree resulting by applying the C4.5 algorithm is constructed to minimize the observed error rate, using equal priors. This criterion seems to be satisfactory in the frame of sequential data and has the advantage to not favor certain events. To create the successive partitions, the C4.5 algorithm uses three forms of tests in each node: a "standard" test on a discrete attribute, with one outcome ($A = x$) for each possible value x of the attribute A, a more complex test, also based on a discrete attribute, in which the possible values are allocated to a variate number of groups with one outcome for each group, and a binary test, for continuos attributes, with outcomes $A \leq Z$ and $A > Z$, where A is the attribute and Z is a threshold value.

The application of a classification tree algorithm is equivalent, as we will detail in the next section, with the choice of a model for the language L and the procedure that extract rules from the classification tree is equivalent with the creation of a global interpretation I_G and with the calculus of the confidence of temporal rules.

Second inference process. Different classification trees, constructed starting with different training sets, generate finally different sets of rules. The process that tries to infer a general temporal rule from a set of rules implying the same class is similar to the pruning strategy for a classification tree. Consider two temporal rules R_1 and R_2 having the same head and obtained from two different classification trees. Formally, this means we have two models, $M_1 = (\widetilde{T}_1, D, I)$ and $M_2 = (\widetilde{T}_2, D, I)$, and the confidence for each rule was calculated on the corresponding model. If we denote by $\{\overline{R}_i\}$ the set of implication clauses of the rule $R_i, i = 1,2$, then from the set of rules

$\{R_1, R_2\}$ on infer a general rule $R_{1,2}$ having a body of the form $(C_1 \wedge ... \wedge C_s)$ where $C_i \in \bigcap_i \{\overline{R_i}\}$. The confidence of the inferred rule is then calculated on the extended model $M = (\widetilde{T}_1 \cup \widetilde{T}_2, D, I)$. If this confidence is at least equal with the minimum confidence for the initial rules, the general rule is kept, if not, it is rejected.

4. Implementation Problems

Before we can start to apply the decision tree algorithms to an event database, an important problem has to be solved first: establishing the training set. An n-tuple in the training set contains $n-1$ values of the predictor variables (or attributes) and one value of the categorical dependent variable, which represents the class.

Each time series from the initial database was "translated", during the first phase, into a sequence of events. As the classification tree algorithm demands, one of these event series must represent the dependent variable. There are two different approaches on how the sequence that represents the classification (the values of the categorical dependent variable) is obtained. In a supervised methodology, an expert gives this sequence (e.g., in our insurance case, a physician). The situation becomes more difficult when there exists no prior knowledge about the possible classifications. Suppose that our database contains a set of time series representing the evolution of stock price for different companies. We are interested in seeing if a given stock value depends on other stock values. As the dependent variable (the stock price) is not categorical, it cannot represent a classification used to create a classification tree. The solution is to use the sequence of the names of events extracted from the continuous time series as the sequence of classes.

Lets suppose we have k time series representing the predictor variables $s_1, s_2, ... s_k$. Each $s_{ij}, i = 1..k, j = 1..n$ is an event. We have also a sequence $s_c = s_{c1}, ... s_{cn}$ representing the classification. The training set will be constructed using a procedure depending on three parameters. The first, t_0, represents a time instant considered as *present time*. Practically, the first tuple contains the class s_{ct_0} and there is no tuple in the training set containing an event that starts after time t_0. The second, t_p, represents a time interval and controls the further back in time class $s_{c(t_0-t_p)}$ included in the training set. Consequently, the number of tuples in the training set is $t_p + 1$. The third parameter, h, controls the influence of the past events $s_{i(t-1)}, ..., s_{i(t-h)}$ on the actual event s_{it}. This parameter (*history*) reflects the idea that the class s_{ct} depends not only on the events at time t, but also on the events started before time t. Finally, each tuple contains $k(h+1)$ events and one class value. The first tuple is $s_{1t_0}, ..., s_{1(t_0-h)}, ..., s_{k(t_0-h)}, s_{ct_0}$ and the last $s_{1(t_0-t_p)}, ..., s_{k(t_0-t_p-h)}, s_{c(t_0-t_p)}$. The set of time instants at which the events included in the training set start may be structured as a finite temporal domain $\widetilde{T}_p$. The cardinality of $\widetilde{T}_p$ is $t_p + h + 1$. Therefore, the selection of a training set is equivalent with the selection of a model for L and, implicitly, of a global interpretation I_G. As a remark, the parameter h has also a side effect on the final temporal rules: because the time window of any tuple is h, the time window for temporal rules cannot exceed h.

Because the training set depends on different parameters, the process of applying the classification tree will consist in creating multiple training sets, by changing these parameters. For each set, the induced classification tree will be "transformed" into a set of temporal rules.

To adopt this particular strategy for the construction of the training set, we made an assumption: the events $s_{ij}, i = 1..k$, j a fixed value, start all at the same time instant. Formally, this is equivalent with $j \mapsto s_{ij}$, for all $i = 1..k$. The same assumption permits us to solve another implementation problem: the time information is not processed during the classification tree construction, (time is not a predictor variable), but the temporal dimension must be captured by the temporal rules. The solution we chose to "encode" the temporal information is to create a map between the index of the attributes (or predictor variables) and the order in time of the events. The $k(h+1)$ attributes are indexed as $\{A_0, A_1, \ldots, A_h, \ldots, A_{2h}, \ldots A_{k(h+1)-1}\}$. Suppose that the set of indexes of the attributes that appear in the body of the rule is $\{i_0, \ldots, i_m\}$. This set is transformed into the set $\{\bar{i}_0, \ldots, \bar{i}_m\}$, where "$\bar{i}$" means "i modulo $(h+1)$". If t_0 represents the time instant when the event in the head of the rule starts, then an event from the rule's body, corresponding to the attribute A_{i_j}, started at time $t_0 - \bar{i}_j$.

5. Conclusions and Further Directions

The methodology we proposed in this article tries to respond to an actual necessity, the need to discover knowledge from databases where the notion of "time" represents an important issue. We proposed to represent this knowledge in the form of general Horn clauses, a more comprehensible form for a final user without sophisticated statistical background. A formalism based on first-order temporal logic tries to give a theoretical support to the main terms "event" and "temporal rule". To obtain the "temporal rules", a discretisation phase that extracts "events" from raw data is applied first, followed by an inference phase, which constructs classification trees from these events. The discrete and continuous characteristics of an "event" allow us to use statistical tools as well as techniques from artificial intelligence on the same data.

To capture the correlation between events over time, a specific procedure for the construction of a training set is proposed. This procedure depends on three parameters. Among others, the so-called *history* controls the time window of the temporal rules. We wish also to emphasis the fact that our methodology represents, at this moment, just an alternative to others procedures and methods for the extraction of knowledge from time series databases. In order to analyze the quality of the final temporal rules, it is necessary to conduct comparative experiments, using different methods, on the same databases. Finally, we are also interested in an open question, which is strictly connected to the temporal characteristics of the data: At what moment in time, a specific rule is no longer applicable?

References

1. Cotofrei, P, Stoffel K.: Rule Extraction from Time Series Databases using Classification Trees. Proc. of 20th I.A.S.T.E.D. Conference on Applied Informatics, Innsbruck, (2002)
2. Agrawal R., Faloutsos C., Swami A.: Efficient Similarity Search In Sequence Databases. Proc. of IVth Conference F.D.O.A, (1993), 69-84
3. Faloutsos C., Ranganathan M., Manolopoulos Y.: Fast Subsequence Matching in Time-Series Databases. Proc. of ACM SIGMOD, (1994), 419-429
4. Faloutsos C., Jagadish H., Mendelzon A., Milo T.: A Signature Technique for Similarity-Based Queries. Proc. of SEQUENCES97, Salerno, IEEE Press, (1997)
5. Yi B., Jagadish H., Faloutsos C.: Efficient Retrieval of Similar Time Sequences Under Time Warping. IEEE Proc. of Int. Conf. on Data Engineering, (1998), 201-208
6. Rabiner L., Juang B.: An introduction to Hidden Markov Models. IEEE Magazine on Accoustics, Speech and Signal Processing, 3, (1986), 4-16
7. Zweig G., Russel S.: Speech recognition with dynamic Bayesian networks. AAAI, (1998), 173-180
8. Bengio Y.: Neural Networks for Speech and Sequence Recognition. International Thompson Publishing Inc., (1996)
9. Mangaranis S.: Supervised Classification with temporal data. PhD. Thesis, Computer Science Department, School of Engineering, Vanderbilt University, (1997)
10. Keogh E., Pazzani M. J.: An Enhanced Representation of time series which allows fast and accurate classification, clustering and relevance feedback. Proc. of KDD, (1998), 239-243
11. Loether H., McTavish D.: Descriptive and Inferential Statistics: An introduction. Ally and Bacon(ed.), (1993)
12. Han J., Gong W., Yin Y.: Mining Segment-Wise Periodic Patterns in Time-Related Databases. Proc. of IVth Conf. on Knowledge Discovery and Data Mining, (1998), 214-218
13. Han J., Dong G., Yin Y.: Efficient Mining of Partial Periodic Patterns in Time Series Database. Proc. of Int. Conf. on Data Engineering, Sydney, (1999), 106-115
14. Lu H., Han J., Feng L.: Stock movement and n-dimensional inter-transaction association rules. Proc. of SIGMOD workshop on Research Issues on Data Mining and Knowledge Discovery, (1998), 12:1-12:7
15. Ozden B., Ramaswamy S., Silberschatz A.: Cyclic association rules. Proc of International Conference on Data Engineering, Orlando, (1998), 412-421
16. Das G., Lin K., Mannila H., Renganathan G, Smyth P.: Rule Discovery from Time Series. Proc. of IVth Int. Conf. on Knowledge Discovery and Data Mining, (1998), 16-22
17. Waleed Kadous M: Learning Comprehensible Descriptions of Multivariate Time Series. Proc. of Int. Conf. on Machine Learning, (1999), 454-463
18. Chomicki J., Toman D.: Temporal Logic in Information systems, BRICS LS-97-1, (1997)
19. Allen Emerson E.: Temporal and Modal Logic. In Handbook of Theoretical Computer Science, ed. Noth-Holland Pub. Co., (1995)
20. Loh W., Shih Y.: Split Selection Methods for Classification Trees, Statistica Sinica, vol. 7, (1997), 815-840
21. Breiman L., Friedman J.H., Olshen R. A., Stone C. J.: Classification and regression trees, Monterey. Wadsworth & Brooks/Cole Advanced Books & Software, (1984)
22. Loh W., Vanichestakul N.: Tree-structured classification via generalized discriminant analysis (with discussion). Journal of American Statistical Association, (1983), 715-728.
23 Quinland J. R.: C4.5: Programs for Machine Learning. Morgan Kauffmann Publishers, San Mateo, California, (1993)

Parametric Optimization in Data Mining Incorporated with GA-Based Search

Ling Tam, David Taniar, and Kate Smith

Monash University, School of Business Systems, Vic 3800, Australia
{Ling.Tan, David.Taniar, Kate.Smith}@infotech.monash.edu.au

Abstract. A number of parameters must be specified for a data-mining algorithm. Default values of these parameters are given and generally accepted as 'good' estimates for any data set. However, data mining models are known to be data dependent, and so are for their parameters. Default values may be good estimates, but they are often not the best parameter values for a particular data set. A tuned set of parameter values is able to produce a data-mining model of better classification and higher prediction accuracy. However parameter search is known to be expensive. This paper investigates GA-based heuristic techniques in a case study of optimizing parameters of back-propagation neural network classifier. Our experiments show that GA-based optimization technique is capable of finding a better set of parameter values than random search. In addition, this paper extends the island-model of Parallel GA (PGA) and proposes a VC-PGA, which communicates globally fittest individuals to local population with reduced communication overhead. Our result shows that GA-based parallel heuristic optimization technique provides a solution to large parametric optimization problems.

1   Introduction

Data mining is a process of extracting useful information from data. To use any data mining algorithm, a set of parameters often needs to be specified. Generally, a set of default values is provided for a given data mining algorithm, and these default parameter values were considered a 'good' estimate for any data set. However, data mining models are known to be data dependent, and so are for their parameter values. Default values may be good estimates for some data sets, but they are often not tuned to a particular data set.

Parametric search scans through parameter space in order to find the best set of parameter values. However, parameter space is often very large, and brute-force search is simply too expensive to be feasible. This paper investigates genetic algorithm based heuristic search techniques to optimize parameters of data mining algorithms. As a case study, back-propagation neural network (NN) will be used for parameter optimization in this paper.

This rest of paper is organized as follows. Section 2 gives brief introduction on data mining and genetic algorithm (GA). Section 3 shows how to apply genetic algorithms to parametric optimization in a neural network classifier. The proposed

P.M.A. Sloot et al. (Eds.): ICCS 2002, LNCS 2329, pp. 582–591, 2002.
© Springer-Verlag Berlin Heidelberg 2002

parallel version of GA-based search will also be presented. Section 4 describes environment and parameters of performance studies and shows results of performance studies, which is then followed by the conclusions in Section 5.

2 Background

In this section, we introduce data mining and its techniques. Then we briefly describe genetic algorithm (GA) and its operators. Finally, we explain reasons why GA-based search is better than random search.

2.1 Data Mining Techniques

Data mining is a process of discovering previously unknown and potential useful patterns and co-relations from data. Many data mining techniques have been proposed in literature, including *classification, association rule, sequential patterns,* and *clustering* [5].

In *classification*, we are given a set of example records, where each record consists of a number of attributes. One of the attributes is called the classifying attribute, which indicates the class to which each example belongs. The objective of classification is to build a model of how to classify a record based on its attributes. A number of classification methods have been proposed in past, including decision trees and neural networks. In *association rule*, the objective is to discover a set of rules that imply certain association relationships among a set of objects, such as "occur together" or "one implies the other". *Sequential patterns* are to find a common sequence of objects occurring across time from a large number of transactions. *Clustering* identifies sparse and dense regions in a large set of multi-dimensional data point.

To use any of these data mining algorithms, typically a set of parameters often needs to be specified. In this paper, we adopt a neural network classifier as a case study for parametric optimization.

2.2   Genetic Algorithm (GA)

GA is a domain-independent global search technique. It works with a set of solutions that are encoded in strings and a fitness function to evaluate these solutions. These strings represent points in search space. In each iteration (or generation), a new set of solutions is created by crossing some of good solutions in current iteration and by occasionally adding new diversity of search [6]. This process is done through three important operators, i.e. *selection, crossover,* and *mutation*. In the following section, we describe each operator in turn.

Selection. Candidate selection is based on evaluating the fitness of candidates. Candidate fitness is calculated though objective function. Fitness values generally need to be scaled for further use. Scaling is important to avoid early convergence caused by dominant effect of a few strong candidates in the beginning, and to differentiate relative fitness of candidates when they have very close fitness values near the end of run [1]. Two basic methods are stochastic selection and deterministic selection [1]. In stochastic approach, selection is based on a probability of candidate fitness in a population, e.g. roulette wheel method. In deterministic approach, the relative fitness of a candidate is mapped to an integer number, which stands for the number of copies it will be selected.

Crossover. Crossover is the most important operator of GA. It explores new search space by recombining existing search elements, which may produce better solutions based on partial and local solutions. In a simple crossover, the operation takes two candidates and recombines them at a random point. Crossover rate is generally set with a high probability.

Mutation. Mutation is an operator to explore new search space by introducing new search elements not found in existing candidates. In a binary string, the operation flips a single bit. Mutation rate is generally set with a very low probability.

By iteratively applying these operators, GA-based search is able to converge on one of global optimal. GA is a powerful technique to solve optimization problems in a large search space, where random walk is not feasible. The strengths of GA-based search are summarized as follows:

- GA is global search method. It starts with a set of initial points rather than a single point. These points are randomly chosen from the whole search space. Search with a set of points helps GA to skip local optima. In addition, crossover operator allows neighborhood search space of current search points to be covered. And mutation operator injects new search points that may not exist in current search.
- GA-based search is directional with help of selection operator. It is directed at search spaces where current good solutions are found. With each evaluation in selection operator, inferior search points are abandoned and search then focuses on more promising points. Search diversity is large at the beginning with a population of very different fitness values, and then it is gradually reduced to a population of similar or same fitness values at the end.
- Implicit parallelism of GA. When a search point is evaluated, all schemas of which it is an instance are actually evaluated [1]. For example, a search point in binary representation of length n is an instance of 2^n schemas. When this point is evaluated, all schemas are simultaneously evaluated. It has been proven a minimum of M^3 schemas are effectively processed each time when the population of size M is evaluated [7].

It is our intention in this paper to explore GA-based search optimization for data mining algorithms, as well as its potential in parallelism.

3   GA-Guided Parameter Optimization in NN

This section describes details of GA-based search in application to parameter optimization in a back-propagation neural network classifier (NN). The reason we choose to use NN in our parameter optimization case study is that (*i*) parameter value of NN is difficult to choose, and default values are not always good estimates; and (*ii*) it has a relatively large parameter space, which is too expensive to enumerate. A number of parameters need to be specified for a back-propagation NN. The choice of single parameter value and the combining effects of different parameter values have an influence over prediction accuracy of the final data mining model. Parameters in a back-propagation NN include learning rate, momentum, the number of hidden layer and hidden node, and epoch; and their overall search space is 2^{34}.

Like any other GA applications, two essential problems have to be addressed before using GA to parameter optimization.

First is string encoding. Many encoding schemes have been proposed, for example, integer coding and gray coding. There is no standard way to choose these schemes and the choice really depends on the nature expression of problems. The order of putting encoded parameter into a string is also important. In principle, closely associated parameters need to be put together to avoid disruption caused by crossover operator. For parameter optimization problem, we use binary encoding. In addition, we assume no knowledge on relationship of parameters, and parameters are encoded in a random order.

The second issue is how to evaluate string. The fitness of string is evaluated though prediction accuracy of NN model based on a data set. In the following section, we illustrate GA and Parallel version of GA (PGA) to be used in our performance study.

3.1   The Existing GA-Based Searches

There are many variations of GA operators. Our purpose is to see the effectiveness of GA-guided parameter optimization, rather than to compare different GA operators. Therefore, we use the simple genetic algorithm (SGA) described in [1]. However, we use elitism, which always include the best search point from the last iteration to the current iteration, to prevent early convergence. The procedure of SGA is described as the following.

Step 1: Initialize population.

Step 2: For a specified number of iteration, do selection, crossover and mutation according to a set of user specified parameters. And then evaluate population individuals for next iteration.

Two major parallel versions of GA (PGA) are reported in literature. *Island model* (or network model) runs an independent GA with a sub-population on each processor, and the best individuals in a sub-population are communicated either to all other sub-populations or to neighboring population [2, 3].

Cellular model (or neighborhood model) runs an individual on each processor, and cross with the best individual among its neighbors [4]. Cellular model can be seen as a massive parallel extension of island model where population is reduced to a single individual on each processor.

One problem of island model is that best individuals in population are not immediately introduced to local sub-populations due to communication overhead. This may cause inferior search in those sub-populations, which are not physically adjacent to the sub-population where the best individual is found. To solve this problem, we propose an extended version of island model, *Virtual Community PGA* (VC-PGA).

3.2   The Proposed PGA-Based Search: Virtual Community Model (VC-PGA)

In virtual community model, each processor hosts a sub-population just like in island model. A *local community* (LC) is formed among neighboring processors, and each local community has a server processor to facilitate exchanges of best individuals within community and with other communities as shown in Figure 1. To facilitate exchange of best local individuals across community, *virtual community* (VC) can be formed and a virtual server processor performs similar tasks as a local community server processor.

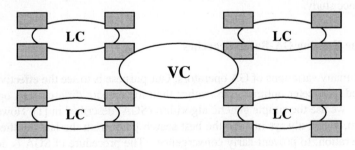

Figure 1. The Proposed VC-PGA Model

Virtual community model gives three advantages over island model: (*i*) Local sub-populations can get globally fittest individuals, (*ii*) The communication overhead is much less expensive, and (*iii*) The evolution surface of solution space is independent from topology of network of workstations. The procedure of VC-PGA is described as follows:

Step 1: Initialize a sub-population of size p on each processor across network. Form a local community with adjacent processors, and randomly choose one processor as its community-server. Form virtual communities by grouping local community-servers. Higher-level of virtual communities can be formed if necessary.

Step 2: On each processor of a local community, run GA for a fixed number of iterations, and send top n fittest of local sub-population to the community-server. At the mean time, compare the fitness of any received individual with the best local fitness, and include it to local population if the received has a higher fitness value.

Step 3: On each community-server, find the best individuals of local community from the received best individuals of community members. Send top n fittest individuals of local community to parent-community server. At the mean time, compare the fitness of any received individual from its virtual community with the best local fitness, and include it to local population if the received has a higher fitness value. Send the best individuals to community members.

4   Performance Evaluation

4.1   Environment and Parameters

Performance study of sequential version GA-based parameter optimization is conducted on a single workstation, while performance of PGA-based parameter optimization is conducted over local area network, which consists of eight homogeneous Pentium workstations. Back-propagation neural network classifier in WEKA [5] data mining library is used to build and evaluate classification models. Fitness value is derived from root mean square error returned by neural network classifier. A small data set, iris.arff (150 records and 4 attributes in each record), is used in experiments. Parameters and their value ranges used in parametric optimization of backpropagation neural network classifier are summarized in Figure 2. Default values in Figure 2 refer to default values used in standard back-propagation Neural Network classifier, and included here for cross reference.

Parameters	Value Range	Bit Length	Default Value
Learning Rate	[0, 1.00]	7	0.3
Momentum	[0, 1.00]	7	0.2
Hidden layer	[1, A] where A is the number of attributes and classifiers	Round($\log_2 A$)	4
Epochs in training	[1, 1000]	10	500
Error threshold in validation testing	[1, 100]	7	30

Figure 2. Parameters in Backpropagation NN Classifier

Elitist deterministic selection is used in both GA and PGA. For GA-related parameters, a high cross over rate and a low mutation rate are used. To study speed-up of PGA, we keep total population size constant, and decrease population size on each workstation while increasing the number of workstations. Parameters in GA and PGA are summarized in Figure 3.

Parameters	GA	PGA
Generation	100	20
Population size	10, 20, 50, 100	15, 20, 30, 60, 120
Cross over rate	0.8	0.8
Mutation rate	0.01	0.01

Figure 3. Parameters in GA and PGA

4.2   Performance Analysis

We compared the performance of GA-based search with random search on a variety of population sizes. The experiments take into consideration of GA overhead by measuring results of random search and GA-based search at end of GA iterations. Performance is measured in terms of average fitness of population, and the best fitness at end of run.

As shown in Figure 4, GA-based search performs considerably better than random search regardless of population size in terms of average fitness. From the figure, we can see the speed of convergence is decreasing while population size increases. It shows that these GA-based searches actually found some good solutions after small number of iterations, and therefore the average fitness of smaller population size is higher than that of larger population size. Another point we want to mention is that at the end of 100 generations, all populations have not converged yet. It is reasonable to use fixed iteration number as a stop criteria than convergence, because it takes a considerable time for a population to converge even although the best solution is already in the population.

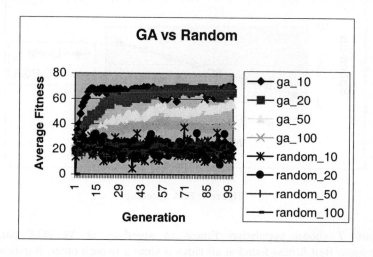

Figure 4. GA vs. Random: Average Fitness

In addition, we compared the best fitness of GA-based search with random search as shown in Figure 5. The best fitness of GA-based search always outperforms random search regardless of population size. However, the out-performance is marginal.

Population Size	GA-based Search	Random Search
10	72.53	71.76
20	73.46	72.29
50	73.89	72.24
100	73.89	72.56

Figure 5. GA vs. Random: Best Fitness

We implemented VC-PGA model on up to 8 stations in a local area network to observe its speed-up. To obtain speed-up, we keep total population size and the number of iterations constant, while using a different cluster size from 1 to 8. For example, population size is 60 on each station in a 2-station cluster, population size 30 on each station in a 4-station cluster and so on. The time for each cluster to complete its job is taken using the total time spent across all stations in a cluster at end of iteration divided by its cluster size. As shown in Figure 6, VC-PGA closely follows a linear speed-up.

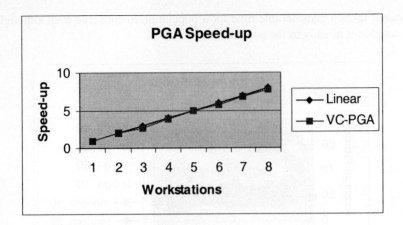

Figure 6. PGA Speed-up

Figure 7 shows population fitness in speed-up of VC-PGA parametric optimization. Best fitness found at all cases is similar to each other. It indicates that search outcomes are comparable among different experiments when more workstations are used to process the same job size. Average population fitness across all sub-populations seems to increase as more nodes are used to work with smaller population size. This is because the results are taken at end of relatively small number of generation where sub-populations have not converged.

Nodes	Population Size	Average Fitness	Best Fitness
1	120	70	77.66264
2	60	71	77.51973
4	30	73.5	77.52943
6	20	73.3	78.20998
8	15	75.3	77.51269

Figure 7. Population Fitness in PGA Speedup

5   Conclusions

Our main objective in this paper is to show that parallel heuristic optimization techniques, like GA can be useful to solve problems where large search space needs to be explored in data mining algorithms. This paper uses simple genetic algorithm (SGA) and a new variant of PGA to optimize parameter space of back-propagation neural network classifier. Our result shows that GA-based parallel heuristic optimization technique provides a solution to large parametric optimization problems.

References

[1] D.E. Goldberg, *Genetic Algorithms in Search, Optimization, and Machine Learning*, Addison-Wesley Publishing Co., 1989.

[2] T. Starkweather, D. Whitley, and K. Mathias, "Optimization using distributed genetic algorithm", *Parallel Problem Solving from Nature*, pp. 176-185, Springer Verlag, 1991.

[3] M. Gorges-Schleuter, "Explicit parallelism of genetic algorithms through population structures", *Parallel Problem Solving from Nature*, pp 150-159, Springer Verlag, 1991.

[4] V. S. Gordon and D. Whitley, "A Machine-Independent Analysis of Parallel Genetic Algorithms", *Complex Systems*, 8:181-214, 1994.

[5] I. H. Witten and E. Frank, *Data Mining: Practical Machine Learning Tools and Techniques with Java Implementations*, Morgan Kaufmann, 2000.

[6] S. M. Sait and Y. Youusef, "Iterative Computer Algorithms with Application in Engineering", *Solving Combinatorial Optimization Problems*, IEEE Computer Society, 1999.

[7] L.B. Booker, D.E. Goldberg, and J.H. Holland, "Classifier Systems and Genetic Algorithms", *Artificial Intelligence*, Vol 40, pp. 235-282, 1989.

Implementing Scalable Parallel Search Algorithms for Data-Intensive Applications

L. Ladányi [1], T. K. Ralphs[*2], and M. J. Saltzman[3]

[1] Department of Mathematical Sciences, IBM T. J. Watson Research Center,
Yorktown Heights, NY 10598, ladanyi@us.ibm.com
[2] Department of Industrial and Systems Engineering, Lehigh University, Bethlehem,
PA 18017, tkralphs@lehigh.edu, www.lehigh.edu/~tkr2
[3] Department of Mathematical Sciences, Clemson University, Clemson, SC 29634,
mjs@clemson.edu, www.math.clemson.edu/~mjs

Abstract. Scalability is a critical issue in the design of parallel software for large-scale search problems. Previous research has not addressed this issue for *data-intensive* applications. We describe the design of a library for parallel search that focuses on efficient data and search tree management for such applications in distributed computing environments.

1 Introduction

This paper describes research in which we are seeking to develop highly scalable algorithms for performing large-scale parallel search in distributed-memory computing environments. This project builds on previous work in which we developed two object-oriented, generic frameworks for implementing parallel algorithms for large-scale discrete optimization problems (DOPs). SYMPHONY (Single- or Multi-Process Optimization over Networks) [12, 10] is a framework written in C and COIN/BCP [13] is a framework written in the same spirit in C++. Because of their generic, object-oriented designs, both are extremely flexible and can be used to solve a wide variety of discrete optimization problems. However, these frameworks have somewhat limited scalability. The goal of this research is to address these scalability issues by designing a more general framework called the Abstract Library for Parallel Search (ALPS).

ALPS is a C++ class library upon which a user can build a wide variety of parallel algorithms for performing tree search. Although the framework is more general than SYMPHONY and COIN/BCP, we will initially be interested in using ALPS to implement algorithms for solving large-scale DOPs. DOPs arise in many important applications, but most are in the complexity class $\mathcal{NP}$-complete so there is little hope of finding provably efficient algorithms [3]. Nevertheless, intelligent search algorithms, such as *branch, constrain, and price* (BCP), have been tremendously successful at tackling these difficult problems.

To support the implementation of parallel BCP algorithms, we have designed two additional C++ class libraries built on top of ALPS. The first, called the

* Funding from NSF grant ACR-0102687 and the IBM Faculty Partnership Program.

P.M.A. Sloot et al. (Eds.): ICCS 2002, LNCS 2329, pp. 592–602, 2002.

Branch, Constrain, and Price Software (BiCePS) library, implements a generic framework for relaxation-based branch and bound. In this library, we make very few assumptions about the nature of the relaxations, i.e., they do not have to be linear programs. The second library, called the BiCePS Linear Integer Solver (BLIS), implements LP-based branch and bound algorithms, including BCP.

The applications we are interested in are extremely *data-intensive*, meaning that they require the maintenance of a vast amount of information about each node in the search tree. However, this data usually does not vary much from parent to child, so we use data structures based on a unique differencing scheme that is memory efficient. This scheme for storing the tree allows us to handle problems much larger than we could otherwise.

A number of techniques for developing scalable parallel branch and bound algorithms have been proposed in the literature [1, 2, 4, 5, 15]. However, we know of no previous work specifically addressing the development of scalable algorithms for data-intensive applications. Standard techniques for parallel branch and bound break down in this setting, primarily because they all depend on the ability to easily shuttle search tree nodes between processors. The data structures we need in order to create efficient storage do not allow this fluid movement. Our design overcomes this difficulty by dividing the search tree into subtrees containing a large number of related search nodes that can be stored together. This requires the design of more sophisticated load balancing schemes that accommodate this storage constraint.

2 Motivation

2.1 Scalability Issues for Parallel Search Algorithms

We will address two primary design issues that are fundamental to the scalability of parallel search algorithms. *Control mechanisms* are the methods by which decisions are made regarding the overall direction of the search, i.e., in what order the search tree nodes should be processed. The design of control mechanisms is closely tied to *load balancing*, the method by which we ensure that all processors have useful work to do. In general, *centralized* control mechanisms allow better decision making, but limit scalability by creating decision-making bottlenecks. Decentralized control mechanisms alleviate these bottlenecks, but reduce the ability to make decision based on accurate global information.

A more important issue for data-intensive algorithms is efficient *data handling*. In these algorithms, there are a huge number of *data objects* (see Sect. 3.2 for a description of these) that are global in nature and must be efficiently generated, manipulated, and stored in order to solve these problems efficiently. The speed with which we can process each node in the search tree depends largely on the number of objects that are *active* in the subproblem. Thus, we attempt to limit the set of active objects in each subproblem to only those that are necessary for the completion of the current calculation. However, this approach requires careful bookkeeping. This bookkeeping becomes more difficult as the number of processors increases.

2.2   Branch and Bound

In order to illustrate the above principles and define some terminology, we first describe the basics of the branch and bound algorithm for optimization. Branch and bound uses a divide and conquer strategy to partition the solution space into *subproblems* and then optimizes individually over each of them. In the *processing* or *bounding* phase, we relax the problem, thereby admitting solutions that are not in the feasible set S. Solving this relaxation yields a lower bound on the value of an optimal solution.[1] If the solution to this relaxation is a member of S, then it is optimal and we are done. Otherwise, we identify n subsets of S, $S_1, \ldots, S_n$, such that $\cup_{i=1}^{n} S_i = S$. Each of these subsets is called a *subproblem*; $S_1, \ldots, S_n$ are also sometimes called the *children* of S. We add the children of S to the list of *candidate subproblems* (those which need processing). This is called *branching*.

To continue the algorithm, we select one of the candidate subproblems, remove it from the list, and process it. There are four possible results. If we find a feasible solution better than the current best, then we replace the current best by the new solution and continue. We may also find that the subproblem has no solutions, in which case we discard, or *fathom* it. Otherwise, we compare the lower bound to upper bound yielded by the current best solution. If it is greater than or equal to our current upper bound, then we may again fathom the subproblem. Finally, if we cannot fathom the subproblem, we are forced to branch and add the children of this subproblem to the list of active candidates. We continue in this way until the list of active subproblems is empty, at which point our current best solution is the optimal one. The order in which the subproblems are processed can have a dramatic effect on the size of the tree. Processing the nodes in *best-first* order, i.e., always choosing the candidate node with the lowest lower bound, minimizes the size of the search tree. However, without a centralized control mechanism, it may be impossible to implement such a strategy.

2.3   Branch, Constrain, and Price

Branch, constrain, and price is a specific implementation of branch and bound that can be used for solving integer programming problems. Early works [6, 7, 11, 16] laid out the basic framework of BCP. Since then, many implementations (including ours) have built on these preliminary ideas. In a typical implementation of BCP, the bounding operation is performed using linear programming. We relax the integrality constraints to obtain a *linear programming (LP) relaxation*. This formulation is augmented with additional, dynamically generated constraints valid for the convex hull of solutions to the original problem. By approximating the convex hull of solutions, we hope to obtain an integer solution. If the number of columns in the constraint matrix is large, these can also be generated dynamically, in a step called *pricing*. When both constraints and variables are generated dynamically throughout the search tree, we obtain the

---

[1] We assume without loss of generality that we wish to minimize the objective function

algorithm known as branch, constrain, and price. Branching is accomplished by imposing additional constraints that subdivide the feasible region. Note that the data objects we referred to in Sect. 2.1 are the constraints and the variables. In large-scale BCP, the number of these can be extremely large.

3   The Library Hierarchy

To make the code easy to maintain, easy to use, and as flexible as possible, we have developed a multi-layered class library, in which the only assumptions made in each layer about the algorithm being implemented are those needed for implementing specific functionality efficiently. By limiting the set of assumptions in this way, we ensure that the libraries we are developing will be useful in a wide variety of settings. To illustrate, we briefly describe the current hierarchy.

3.1   The Abstract Library for Parallel Search

The ALPS layer is a C++ class library containing the base classes needed to implement the parallel search handling, including basic search tree management and load balancing. In the ALPS base classes, there are almost no assumptions made about the algorithm that the user wishes to implement, except that it is based on a tree search. Since we are primarily interested in *data-intensive* applications, the class structure is designed with the implicit assumption that efficient storage of the search tree is paramount and that this storage is accomplished through the compact differencing scheme we alluded to earlier. This means that the implementation must define methods for taking the difference of two nodes and for producing an explicit representation of a node from a compact one. Note that this does not preclude the use of ALPS for applications that are not data-intensive. In that case, the differencing scheme need not be used.

In order to define a search order, ALPS assumes that there is a numerical *quality* associated with each subproblem that is used to order the priority queue of candidates for processing. For instance, the quality measure for branch and bound would most likely be the lower bound in each search tree node. In addition, ALPS has the notion of a *quality threshold*. Any node whose quality falls below this threshold is fathomed. This allows the notion of fathoming to be defined without making any assumption about the underlying algorithm. Again, this quality threshold does not have to be used if it doesn't make sense.

Also associated with a search tree node is its current status. In ALPS, there are only four possible stati indicating whether the subproblem has been processed and what the result was. The possible stati are: `candidate` (available for further processing), `processed` (processed, but not branched or fathomed), `branched`, and `fathomed`. In terms of these stati, processing a node involves computing its quality and converting its status from `candidate` to `processed`. Following processing, the node is either `branched`, producing children, or `fathomed`.

3.2   The Branch, Constrain, and Price Software Library

Primal and Dual Objects. BiCePS is a C++ class library built on top of ALPS, which implements the data handling layer appropriate for a wide variety of relaxation-based branch and bound algorithms. In this layer, we introduce the concept of a BcpsObject, which is the basic building block of a subproblem. The set of these objects is divided into two main types, the *primal objects* and the *dual objects*. Each of the primal objects has a value associated with it that must lie within an interval defined by a given upper and lower bound. One can think of the primal objects as being the *variables*.

In order to sensibly talk about a relaxation-based optimization algorithm, we need the notion of an objective function. The objective function is defined simply as a function of the values of the primal objects. Given the concept of an objective function, we can similarly define the set of dual objects as functions of the values of the primal objects. As with the values of the primal objects, these function values are constrained to lie within certain given bounds. Hence, one can think of the dual object as the *constraints*.

Note that the notion of primal and dual objects is a completely general one. With this definition of variables and constraints, we can easily apply the general theoretical framework of Lagrangian duality. In Lagrangian duality, each constraint has both a *slack* value and the value of a *dual multiplier* or *dual variable* associated with it. Similarly, we can associate a pair of values with each variable, which are the *reduced cost* (the marginal reduction in the objective function value or the partial derivative of the objective function with respect to the variable in question) and the value of the primal variable itself.

Processing. Although the objects separate naturally into primal and dual classes, our goal is to treat these two classes as symmetrically as possible. In fact, in almost all cases, we can treat these two classes using exactly the same methods. In this spirit, we define a subproblem to be comprised of a set of objects, both primal and dual. These objects are global in nature, which means that the same object may be active in multiple subproblems. The set of objects that are active in the subproblem define the current relaxation that can be solved to obtain a bound. Note that this bound is not valid for the original problem unless either all of the primal objects are present or we have proven that all of the missing primal objects can be fixed to value zero.

We assume that processing a subproblem is an iterative procedure in which the list of active objects can be changed in each iteration by *generation* and *deletion*, and individual objects modified through *bound tightening*. To define object generation, we need the concepts of *primal solution* and *dual solution*, each consisting of a list of values associated with the corresponding objects. Object generation then consists of producing variables whose reduced costs are negative and/or constraints whose slacks are negative, given the current primal or dual solutions (these are needed to compute the slacks and reduced costs).

With all this machinery defined, we have the basic framework needed for processing a subproblem. In overview, we begin with a list of primal and dual objects from which we construct the corresponding relaxation, which can be solved to obtain an initial bound for the subproblem. We then begin to iteratively tighten the bound by generating constraints from the resulting primal solution. In addition, we may wish to generate variables. Note that the generation of these objects "loosens" the formulation and hence must be performed strategically. The design also provides for multi-phase approaches in which variable generation is systematically delayed. During each iteration, we may tighten the object bounds and use other logic-based methods to further improve the relaxation.

Branching. To perform branching, we choose a *branching object* consisting of both a list of data objects to be added to the relaxation (possibly only for the purpose of branching) and a list of object bounds to be modified. Any object can have its bounds modified by branching, but the union of the feasible sets contained in all child subproblems must contain the original feasible set in order for the algorithm to be correct.

3.3   The BiCePS Linear Integer Solver Library

BLIS is a concretization of the BiCePS library in which we implement an LP-based relaxation scheme. This simply means that we assume the objective function and all constraints are linear functions of the variables and that the relaxations are linear programs. This allows us to define some of the notions discussed above more concretely. For instance, we can now say that a variable corresponds to a column in an LP relaxation, while a constraint corresponds to a row. Note that the form a variable or a constraint takes in a particular LP relaxation depends on the set of objects that are present. In order to generate the column corresponding to a variable, we must have a list of the active constraints. Conversely, in order to generate the row corresponding to a constraint, we must be given the list of active variables. This distinction between the *representation* and *realization* of an object will be explored further is Sect. 5.

4   Improving Scalability

One of the primary goals of this project is to increase scalability significantly from that of SYMPHONY and COIN/BCP. We have already discussed some issues related to scalability in Sect. 2. As pointed out there, this involves some degree of decentralization. However, the schemes that have appeared in the literature are inadequate for data-intensive applications, or at the very least, would require abandoning our compact data structures. Our new design attempts to reconcile the need for decentralization with our compact storage scheme, as we will describe in the next few sections.

4.1   The Master-Hub-Worker Paradigm

One of the main difficulties with the master-slave paradigm employed in SYMPHONY and COIN/BCP is that the tree manager becomes overburdened with requests for information. Furthermore, most of these requests are *synchronous*, meaning that the sending process is idle while waiting for a reply. Our differencing scheme for storing the search tree also means that the tree manager may have to spend significant time simply *reconstructing* subproblems. This is done by working back up the tree undoing the differencing until an explicit description of the node is obtained.

Our new design employs a master-hub-worker paradigm, in which a layer of "middle management" is inserted between the master process and the worker processes. In this scheme, each hub is responsible for managing a cluster of workers whose size is fixed. As the number of processors increases, we simply add more hubs and more clusters of workers. However, no hub will become overburdened because the number of workers requesting information from it is limited. This scheme, which maintains many of the advantages of global decision-making while moving some of the computational burden from the master process to the hubs, is similar to a scheme implemented by Eckstein in his PICO framework [1].

Each hub is responsible for balancing the load among its workers. Periodically, we must also perform load balancing between the hubs themselves. Note that this load-balancing must be done not only with respect to the *quantity* of work available to each hub, but also the *quality*, i.e., nodes of high-quality must be evenly distributed among the hubs. This is done by maintaining skeleton information about the full search tree in the master process. This skeleton information includes only what is necessary to make load balancing decisions—primarily the quality of each of the subproblems available for processing. With this information, the master is able to match *donor hubs* (those with too many nodes or too high a proportion of high-quality nodes) and *receiver hubs*, who then exchange work appropriately.

4.2   Task Granularity

The most straightforward approach to improving scalability is to increase the task granularity and thereby reduce the number of decisions that need to be made centrally, as well as the amount of data that has to be sent and received. To achieve this, the basic unit of work in our design is an entire *subtree*. This means that each worker is capable of processing an entire subtree autonomously and has access to all of the methods used by the tree manager to manage the tree, including setting up and maintaining its own priority queue of candidate nodes, tracking and maintaining the objects that are active within its subtree, and performing its own processing, branching, and fathoming. Each hub is responsible for tracking a list of subtrees of the current tree that it is responsible for. The hub dispenses new candidate nodes (leaves of one of the subtrees it is responsible for) to the workers as needed and tracks their progress. When a worker receives a new node, it treats this node as the root of a subtree and begins processing that subtree, stopping only when the work is completed or the hub instructs it to stop. Periodically, the worker informs the hub of its progress.

The implications of changing the basic unit of work from a subproblem to a subtree are vast. Although this allows for increased grain size, as well as more compact storage, it does make some parts of the implementation much more difficult. For instance, we must be much more careful about how we perform load balancing in order to try to keep subtrees together. We must also have a way of ensuring that the workers don't go too far down on an unproductive path. In order to achieve this latter goal, each workers must periodically check in with the hub and report the status of the subtree it is working on. The hub can then decide to ask the worker to abandon work on that subtree and send it a new one. An important point, however, is that this decision making is always done in an asynchronous manner. This feature is described next.

4.3   Asynchronous messaging

Another design feature that increases scalability is the elimination of synchronous requests for information. This means that every process must be capable of working completely autonomously until interrupted with a request to perform an action, either by its associated hub (if it is a worker) or by the master process (if it is a hub). In particular, this means that each worker must be capable of acting as an independent sequential solver, as described above. To ensure that the workers are doing useful work, they periodically send an *asynchronous* message to the hub with information about the current state of its subtree. The hub can then decide at its convenience whether to ask the worker to stop working on the subtree and begin work on a new one.

Another important implication of this design is that the workers are not able to assign global identifiers to newly generated objects. In the next section we will explore the consequences of this decision.

5   Data Handling

Besides effective control mechanisms and load balancing procedures, the biggest challenge we face in implementing search algorithms for data-intensive applications is keeping track of the objects that make up the subproblems. Before describing what the issues are, we describe the concept of objects in a little more detail.

5.1   Object Representation

We stated in Sect. 3.2 that an object can be thought of as either a variable or a constraint. However, we did not define exactly what the terms variable or constraint mean as *abstract* concepts. For the moment, consider an LP-based branch and bound algorithm. In this setting, a "constraint" can be thought of as a method for producing a valid row in the current LP relaxation, or, in other words, a method of producing the projection of a given inequality into the domain of the current set of variables. Similarly, a "variable" can be thought of as a method for producing a valid column to be added to the current LP relaxation. This concept can be generalized to other settings by requiring each

object to have an associated method that performs the modifications appropriate
to allow that object to be added to the current relaxation.

Hence, we have the concept that an object's *representation*, which is the way
it is stored as a stand-alone object, is inherently different from its *realization*
in the current relaxation. An object's representation must be defined with respect to both the problem being solved and the form of the relaxation. Within
BiCePS, new categories of objects can be defined by deriving a child class from
the BcpsObject class. This derived class holds the data the user needs to realize
that type of object within the context of a given relaxation and also defines the
method of performing that realization. When sending an object's description to
another process, we need to send only the data itself, and even that should be
sent in as compact a form as possible. Therefore the user must define for each
category of variable and constraint a method for *encoding* the data compactly
into a character array. This form of the object is then used whenever the object
is sent between processes or when it has to be stored for later use. This encoding is also used for another important purpose that is discussed in Sect. 5.2. Of
course, there must also be a corresponding method of decoding as well.

5.2   Tracking Objects Globally

The fact that the algorithms we consider have a notion of global objects is, in
some sense, what makes them so difficult to implement in parallel. In order to
take advantage of the economies of scale that come from having global objects,
we must have global information and central storage. But this is at odds with
our goal of decentralization and asynchronous messaging.

To keep storage requirements at a minimum, we would like to know at the
time we generate an object whether we have previously generated that same
object somewhere else in the tree. If so, we would like to refer to the original
copy of that object and delete the new one. Ideally, we would only need to store
one copy of each object centrally and simply copy it out whenever it is needed
locally.

Instead, we simply ensure that at most one copy of each object is stored
locally within each process. Objects within a process are stored in a hash table,
with the hash value computed from the encoded form. If an object is generated
that already exists, the new object is deleted and replaced with a pointer to
the existing one. Note that we can have separate hash tables for primal and
dual objects, as well as for each object category. The same hashing mechanism
is implemented in the hubs with the additional complexity that when an entire
subtree is sent back from a worker, the pointers to the hashed objects in the
search tree nodes must be reconciled. This is the limit of what can be done
without synchronous messaging.

There is yet one further level of complexity to the storage of objects. In
most cases, we need only keep objects around while they are actually pointed
to, i.e., are active in some subproblem that is still a candidate for processing. If
we don't occasionally "clean house," then the object list will continue to grow
boundlessly, especially in the hubs. To take care of this, we use smart pointers

that are capable of tracking the number of references to an object and deleting the object automatically after there are no more pointers to it.

5.3 Object Pools

Of course, the ideal situation would be to avoid generating the same object twice in the first place. For this purpose, we provide *object pools*. These pools contain sets of the "most effective" objects found in the tree so far so that they may be utilized in other subproblems without having to be regenerated. These objects are stored using a scheme similar to that for storing the active objects, but their inclusion in or deletion from the list is not necessarily tied to whether they are active in any particular subproblem. Instead, it is tied to a rolling average of a quality measure that can be defined by the user. By default, the measure is simply the slack or reduced cost calculated when the object is checked against a given solution to determine the desirability of including that object in the associated subproblem.

6 Conclusions

We have described the main design features of a library hierarchy for implementing large-scale parallel search for data-intensive applications. The issues that must be addressed in order to develop scalable algorithms for such applications are more complex than those for less demanding applications. This project is being developed as open source under the auspices of the Common Optimization Interface for Operations Research (COIN-OR) initiative in order to provide a wide range of users with a powerful and flexible framework for implementing these algorithms. Source code and documentation will be available from the CVS repository at www.coin-or.org.

References

1. Eckstein, J.: Parallel Branch and Bound Algorithms for General Mixed Integer Programming on the CM-5. SIAM Journal on Optimization 4, 794–814, 1994.
2. Eckstein, J.: How Much Communication Does Parallel Branch and Bound Need? INFORMS Journal on Computing 9, 15–29, 1997.
3. Garey, M.R., and Johnson, D.S.: Computers and Intractability: A Guide to the Theory of NP-Completeness. W.H. Freeman and Co., San Francisco, 1979.
4. Gendron, B., and Crainic, T.G.: Parallel Branch and Bound Algorithms: Survey and Synthesis. Operations Research 42, 1042–1066, 1994.
5. Grama, A., and Kumar, V.: Parallel Search Algorithms for Discrete Optimization Problems. ORSA Journal on Computing 7, 365–385, 1995.
6. Grötschel, M., Jünger, M., and Reinelt, G.: A Cutting Plane Algorithm for the Linear Ordering Problem. Operations Research 32, 1195–1220, 1984.
7. Hoffman, K., and Padberg, M.: LP-Based Combinatorial Problem Solving. Annals of Operations Research 4, 145–194, 1985.

8. Kumar, V., and Rao, V.N.: Parallel Depth-first Search. Part II. Analysis. International Journal of Parallel Programming **16**, 501–519, 1987.

9. Kumar, V., and Gupta, A.: Analyzing Scalability of Parallel Algorithms and Architectures. Journal of Parallel and Distributed Computing **22**, 379–391, 1994.

10. Ladányi, L., Ralphs, T.K., and Trotter, L.E.: Branch, Cut, and Price: Sequential and Parallel. In Computational Combinatorial Optimization, D. Naddef and M. Jünger, eds., Springer, Berlin, 223–260, 2001.

11. Padberg, M., and Rinaldi, G.: A Branch-and-Cut Algorithm for the Resolution of Large-Scale Traveling Salesman Problems. SIAM Review **33**, 60–100, 1991.

12. Ralphs, T.K., SYMPHONY Version 2.8 User's Guide. Lehigh University Industrial and Systems Engineering Technical Report 01T-011. Available at www.branchandcut.org/SYMPHONY.

13. Ralphs, T.K. and Ladányi, L.: COIN/BCP User's Guide, 2001. Available at www.coin-or.org.

14. Rao, V.N., and Kumar, V.: Parallel Depth-first Search. Part I. Implementation. International Journal of Parallel Programming **16**, 479–499, 1987.

15. Rushmeier, R., and Nemhauser, G.L.: Experiments with Parallel Branch and Bound Algorithms for the Set Covering Problem. Operations Research Letters **13**, 277–285, 1993.

16. Savelsbergh, M.W.P.: A Branch-and-Price Algorithm for the Generalized Assignment Problem. Operations Research **45**, 831–841, 1997.

Techniques for Estimating the Computation and Communication Costs of Distributed Data Mining·

Shonali Krishnaswamy[1], Arkady Zaslavsky[2], Seng Wai Loke[3],

[1]School of Network Computing, Monash University (Peninsula Campus)
McMahons Rd, Frankston, Victoria –3199, Australia.
Shonali.Krishnaswamy@Infotech.monash.edu.au
[2]School of Computer Science and Software Engineering
900 Dandenong Road, Monash University, Caulfield East, Victoria –3145, Australia.
arkady.zaslavsky@csse.monash.edu.au
[3]School of Computer Science and Information Technology
RMIT University, GPO Box 2476V, Melbourne, Victoria-3001, Australia.
swloke@cs.rmit.edu.au

Abstract. Distributed Data Mining (DDM) is the process of mining distributed and heterogeneous datasets. DDM is widely seen as a means of addressing the scalability issue of mining large data sets. Consequently, there is an emerging focus on optimisation of the DDM process. In this paper we present cost formulae for estimating the communication and computation time for different distributed data mining scenarios.

1 Introduction

Distributed data mining (DDM) addresses the specific issues associated with the application of data mining in distributed computing environments, which are typically characterised by the distribution of users, data, hardware and mining software. DDM is widely seen as a means of addressing the scalability issue of mining large data sets. There are predominantly two architectural models used in the development of DDM systems, namely, *client-server* (CS) and *software agents*. The "agents" category can be further divided on the basis of whether the agents have the ability to migrate in a self-directed manner or not (i.e. whether the agents exhibit the characteristic of *mobility* or not). There is an emerging focus on efficiency and optimisation of response time in distributed data mining [6][9]. A significant issue in the success of an optimisation model is the computation of the cost of different factors in the DDM process. In this paper, we present mathematical cost models that facilitate identification of the different cost components in the DDM process for various strategies such as client-server and mobile agents. These cost models form the basis

---

· THE WORK REPORTED IN THIS PAPER HAS BEEN FUNDED IN PART BY THE CO-OPERATIVE RESEARCH CENTRE PROGRAM THROUGH THE DEPARTMENT OF INDUSTRY, SCIENCE AND TOURISM OF THE COMMONWEALTH GOVERNMENT OF AUSTRALIA.

P.M.A. Sloot et al. (Eds.): ICCS 2002, LNCS 2329, pp. 603–612, 2002.

for apriori estimation of the response time for a given task. In conjunction with the cost models, we have developed a technique for estimating the run time of data mining tasks (which is one of the cost components in the DDM process). The experimental evaluation of this technique establishes its estimation accuracy and validity.

In the following sections of this paper, we present cost models for distributed data mining including cost formulae for estimating the communication time and our technique for apriori estimation of the run time of data mining algorithms. The paper is organised as follows. In section 2 we review related work in the field of distributed data mining with a particular focus on optimisation techniques. Section 3 describes the cost components in different DDM techniques and presents the cost formulae for estimating these components. In section 4, the experimental results of our data mining task run time estimation technique are analysed. Finally, in section 5, we conclude by discussing the current status of our project and the future directions of this work.

2 Related Work

Work on improving performance of DDM systems by using an optimal/cost-efficient strategy has been the focus of [6][9]. IntelliMiner [6] is a client-server distributed data mining system, which focuses on scheduling tasks between distributed processors by computing the cost of executing the task on a given server and then selecting the server with the minimum cost. The cost is computed based on the resources (i.e. number of data sets) needed to perform the task. While this model takes into account the overhead of communication that is increased by having to transfer more datasets it ignores several other cost considerations in the DDM process such as processing cost and size of datasets. In [9], the optimisation is motivated by the consideration that mining a dataset either locally or by moving the entire dataset into a different server is a naïve approach. They use a linear programming approach to partition datasets and allocate the partitions to different servers. The allocation is based on the cost of processing (which is assigned manually in terms dollars) and the cost of transferring data (which is also assigned manually in terms of dollars). The manual assignment of cost requires expert knowledge about different factors that affect the DDM process and quantification of their respective impact, which is a non-trivial task. Thus the computation of the cost is an important question for any optimisation model. In the models discussed, the cost is either assigned manually or computed using a simple technique, which only takes into account the availability of datasets. In the following sections of this paper, we present a technique for computing the cost based on the response time for different mining strategies. We use apriori estimates of the response time for factors such as communication and processing.

3 Cost Components in the Distributed Data Mining Process

In general, optimisation models in DDM attempt to reduce the response time by choosing strategies that facilitate faster communication and/or processing. We propose apriori estimates of the response time as the basis for computing the cost of distributed data mining. In this section we describe the different cost components of the DDM process and present cost formulae for estimating their response times. The response time of a DDM task broadly consists of three components:

1. *Communication:* The communication time is largely dependent on the DDM model and varies depending on whether the task is performed using a client server approach or using mobile agents
2. *Computation.* This is the time taken to perform data mining on the distributed data sets and is a core factor irrespective of the DDM model.
3. *Knowledge Integration.* This is the time taken to integrate the results from the distributed datasets.

The response time for distributed data mining is as follows:

$$T = t_{ddm} + t_{ki} \qquad (1)$$

In (1), T is the response time, t_{ddm} is the time taken to perform mining in a distributed environment and t_{ki} is the time taken to perform knowledge integration. The response time for t_{ddm} in turn can be represented as:

$$t_{ddm} = t_{dm} + t_{com} \qquad (2)$$

In (2), t_{dm} is the time taken to perform data mining and t_{com} is the time involved in the communication. Depending on the model used for distributed data mining (i.e. mobile agents or client server) and the different scenarios within each model, the factors such as communication that determine t_{ddm} will change. This results in a consequent change in the actual cost function that determines t_{ddm}. The modelling and estimation of the knowledge integration (t_{ki}) variable in Eq. (1) is non-trivial as it depends on several unknown factors such as the size and type of results produced by data mining. In fact, one of the primary reasons or rationales for data mining is the discovery of hidden and unknown patterns in the data. Thus, we do not present cost formulae or estimation techniques for the response time of knowledge integration.

3.1 Estimating the Communication Cost

The communication cost in the DDM process varies depending on whether the client-server strategy is followed or the mobile agent model is used.

3.1.1 Mobile Agent Model

This case is characterised by a given distributed data mining task being executed in its entirety using the mobile agent paradigm. The core steps involved include: submission of a task by a user, dispatching of mobile agent (or agents) to the respective data server (or servers), data mining and the return of mobile agent(s) from the data resource(s) with mining results. This model is characterised by a set of mobile agents traversing the relevant data servers to perform mining. In general, this can be expressed as m mobile agents traversing n data sources. There are three possible alternatives within this scenario. The first possibility is $m = n$, where the number of mobile agents is equal to the number of data servers. This implies that one data mining agent is sent to each data source involved in the distributed data mining task. The second option is $m < n$, where the number of mobile agents is less than the number of data servers. The implication of having fewer agents than servers is that some agents may be required to traverse more than one server. We do not consider the third case of $m > n$ since this is in effect equivalent to the case 1 above where there is a mobile agent available per data server. Each of the above alternatives has its own cost function. These cost models are described as follows.

Case 1. Equal number of mobile agents and data servers (m=n).

This is a case where data mining from different distributed data servers is performed in parallel. The algorithm used across the different data servers can be uniform or varied. The system dispatches a mobile agent encapsulating the data mining algorithm (with the relevant parameters) to each of the data servers participating in the distributed data mining activity.

Let n be the number of data servers. Therefore, the number of mobile agents is n (since $m=n$). In order to derive the cost function for the general case involving n data servers and n data mining agents, we first formulate the cost function for the case where there is one data server and one data mining agent.

Let us consider the case where data mining has to be performed at the ith data server (i.e $1 \le i \le n$). The cost function for the response time to perform distributed data mining involving the ith data server is as follows:

$$t_{ddm}=\quad t_{dm}(i) \quad + t_{dmAgent}(AC, i) + t_{resultAgent}(i, AC)$$

The communication terms in the above cost estimate are are $t_{dmAgent}(AC, i)$ and $t_{resultAgent}(i, AC)$. That is,

$$t_{com}= t_{dmAgent}(AC, i) + t_{resultAgent}(i, AC) \tag{3}$$

$t_{dmAgent}(AC, i)$. In our cost model, the representation $t_{mobileAgent}(x, y)$ refers to the time taken by the agent mobileAgent to travel from node x to node y. Therefore $t_{dmAgent}(AC, i)$ is the time taken by the mobile agent dmAgent (which is the agent encapsulating the mining algorithm and the relevant parameters) to travel from the agent centre (AC) to the data server (i). In general, the time taken for a mobile agent to travel depends on the following factors: the size of the agent and the bandwidth between nodes (e.g. in kilobits per second). The travel time is proportional to the size of the agent and is inversely proportional to the bandwidth. This can be expressed as follows:

$$t_{dmAgent}(AC, i) \propto \text{size of dmAgent} \tag{4}$$
$$t_{dmAgent}(AC, i) \propto 1 / \text{bandwidth} \tag{5}$$

From (4) and (5):

$$t_{dmAgent}(AC, i) = (k * \text{size of dmAgent}) / (\text{bandwidth between AC and i})$$

In the above expression for the time taken by the data mining agent to travel from the agent centre to the data server, k is a constant. On adapting the representation used by [8] to model the size of a mobile agent, we express the size of the data mining agent (dmAgent) as:

size of an dmAgent = <dmAgent state, data mining algorithm, input parameters>

$t_{resultAgent}(i, AC)$. This is the time taken for the data mining results to be transferred from the data server (i) to the agent centre (AC). However, estimating this component apriori is not feasible as the size of results obtained from a data mining task is usually unknown.

Since the mining is performed at the distributed locations concurrently, the total communication cost is equal to the time taken by the mobile agent that takes the longest time to travel to its respective remote location. Therefore,

$$t_{com} = \text{max}(t_{dmAgent}(AC, i)), \text{ where } i = 1..n \tag{6}$$

Case 2. Fewer mobile agents than data servers (m<n).

This is the second case, where the number of mobile agents (m) available for distributed data mining is less than the number of data sources (n) participating in a distributed data mining task (i.e. $m < n$). Thus the ith agent (where $1 \le i \le m$) travel to and perform mining at j sites labelled specified in $ds_1, ds_2,...,ds_j$ (where $1 \le j \le n$). We make an assumption that the mining agent returns to the agent centre only after it has accomplished its task in all the respective data sources assigned to it and that results are sent to the central server directly. This allows us to estimate the communication cost more effectively.

The total time taken to mine is therefore the time taken by the agent, which takes the maximum time interval to complete its task. The communication cost estimate for the ith agent's response time is as follows in equation (7):

$$t_{com}(i) = t_{dmAgent}(AC, ds_1) + \sum_{j=2}^{j-1} t_{dmAgent}(j, j+1) + t_{dmAgent}(ds_j, AC)$$

In the above expression, the first term is the time taken by the data mining agent to travel from the agent centre to the first data server in its path. The term involving the summation is the time taken for the agent to travel to the respective data sites within the set assigned to it (excluding the final site). The second last term is the time taken to mine at the last data site in its path. The final term in the expression is the time taken for the agent to travel from the last site on its path to the agent centre.

Since there are m agents operating concurrently, the time t_{com} is the time taken by the agent requiring the longest travel time. Thus,

$$t_{com} = \max(t_{com}(i)), \quad i = 1..m \tag{8}$$

In (8), $t_{com}(i)$ is estimated from equation (7).

3.1.2 Client Server Model

The communication cost formulae for the response time in DDM systems that use the traditional client-server paradigm is presented in this section. Typically, data from distributed sources is brought to the data mining server – a fast, parallel server - and then mined. Let there be n data sites from which data has to be mined. Let s_i be the data set obtained from the ith site (where $1 \le i \le n$). The communication time for DDM for the data set s_i from the ith site is as expressed in equation (9) as follows:

$$t_{com}(i) = t_{dataTransfer}(i, DMS, s_i), \quad 1 \le i \le n \tag{9}$$

The term $t_{dataTransfer}(i, DMS, s_i)$ is the time taken to transfer the data set (s_i) from the ith site to the DDM server (DMS) and is estimated as follows:

$$t_{dataTransfer}(i, DMS, s_i) = \text{size of } s_i / (\text{ bandwidth between i and DMS }) \tag{10}$$

The data transfer can be a significant addition when the data volumes are large and/or the bandwidth is low. In this section, we have presented the cost formulae for the communication component of the DDM process for different mining strategies. We now present our technique for estimating the cost of performing data mining.

3.2 Estimating the Data Mining Cost

In this section, we present application run time estimation techniques as an approach to estimating the time taken to perform data mining. We have developed a novel rough sets based approach to estimating the run time of applications [5]. The motivation for our work in this area comes from application run time estimation techniques proposed by [7]. Application run time prediction algorithms including [1][2][7] operate on the principal that applications that have similar characteristics have similar run times. Thus, a history of applications that executed in the past along with their respective run times is maintained in order to estimate the task run time.

Early work in this area by [1][2] proposed the use of "similarity templates" of application characteristics to identify similar tasks in the history. A similarity template is a set of attributes that are used as the basis for comparing applications in order to determine if they are similar or not. It was proposed by [7] that manual selection of similarity templates was limited and they proposed automated definition and search for templates and used genetic algorithms and greedy search techniques. They were able to obtain improved prediction accuracy using these techniques. For a detailed description of the template identification and search algorithms readers are referred to [7].

We have developed a rough sets based algorithm to address the problem of automatic selection of characteristics that best define similarity to estimate application run times. Rough sets provide an intuitively appropriate theory for identifying good "similarity templates" (or sets of characteristics on the basis of which applications can be compared for similarity). For a detailed explanation of the theoretical soundness of a rough sets based approach and its improved prediction accuracy to identify similarity templates readers are referred to [5]. In this paper, we present a brief overview of our algorithm for use as a means to costing the data mining component in the DDM process.

3.2.1 Rough Sets Algorithm for Estimating t_{dm}

Zdislaw Pawlak introduced the theory of Rough Sets in 1981 as a mathematical tool to deal with uncertainty in data. For a good overview of rough sets concepts, readers are referred to [4]. A data set in rough sets is represented as a table called *Information System*, where each row is an object and each column is an attribute. The attributes are partitioned into condition attributes and decision attributes. The condition attributes determine the decision attribute. The history, as shown in figure 1, is a rough information system, where the objects are the previous applications whose run times (and other properties) have been recorded. The attributes in the information system are the properties about the applications that have been recorded. The decision attribute is the application run time that has been recorded. The other properties that have been recorded constitute the condition attributes. This model of a history intuitively facilitates reasoning about the recorded properties so as to identify the dependency between the recorded attributes and the run time. Thus, it is possible to concretise similarity in terms of the condition attributes that are relevant/significant in determining the decision attribute (i.e. the run time). Thus, the set of attributes that have a strong dependency relation with the run time can form a good similarity template. Having cast the problem of application run time as a rough information system, we now examine the fundamental concepts that are applicable in determining the similarity template.

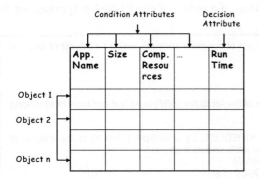

Figure 1: A Task History Modelled as a Rough Information System

Degree of Dependency. Using rough sets it is possible to measure the degree of dependency between two sets of attributes. The measure takes values [0,1] and higher values represent stronger degrees of dependency. It is evident that the problem of identifying a similarity template can be stated as identifying a set of condition attributes in the history that have a strong degree of dependency with the run time. **Significance of Attributes.** The significance of an attribute is the extent by which the attribute alters the degree of dependency between a set of condition and decision attributes. If an attribute is "important" in discerning/determining the decision attribute, then its significance value, which is measured in the range [0,1], will be closer to 1. The similarity template should consist of a set of properties that are important for determining the run time.

Reduct. A reduct consists of the minimal set of condition attributes that have the same discerning power as the entire information system. All superfluous attributes are eliminated from a reduct. According to [4], while it is relatively simple to compute a single reduct, the general solution for finding all reducts is NP-hard. A similarity template should consist of the most important set of attributes that determine the run time without any superfluous attributes. In other words, the similarity template is equivalent to a reduct which has the most significant attributes included.

We now present our rough sets algorithm for identifying similarity templates in figure 2. It is evident that rough sets theory has highly suitable and appropriate constructs for identifying the properties that best define similarity for estimating application run time estimation. Our technique for applying rough sets to identify similarity templates centres round the concept of a reduct. We use a variation of the reduct generation algorithm proposed by [3]. The algorithm proposed by [3] was intended to produce reducts that included user specified attributes. The modified algorithm we use to compute the reduct for use as a similarity template is shown in figure 2. It computes reducts by iteratively adding the most significant attribute to the D-core (note: the *core* of a rough information system is the intersection of all reducts and the *D-core* is the core with respect to the set of decision attributes). An improvement we have included in the algorithm is the identification of any reduct of size $|D\text{-core} + 1|$. From our experiments we found that it is not unusual for the D-core to combine with the a single attribute (typically the most significant attribute) to form a reduct. This iteration in Step 6, is computationally inexpensive as it involves only simple additions and no analysis. Thus, if a reduct of size $|D\text{-core} + 1|$ exists, we find it without further computation.

1. Let A={a_1, a_2,....,a_n}be the set of condition attributes and D be the set of decision attributes.
2. Let C be the D-Core
3. REDUCT = C
4. A1 = A – REDUCT
5. Compute the Significances of the Attributes (SGF) in A1 and sort them in ascending order
6. For i = |A1| to 0
      K(REDUCT, D) =   K(REDUCT, D) + SGF(a_i) /* K(X,Y) is the degree of dependency between attribute sets X and Y */
      If K(REDUCT,D) = K(A,D)
            REDUCT = REDUCT ∪ a_i
            Exit
      End If
      K(REDUCT, D) = K(REDUCT, D) - SGF(a_i)
   End For
7. K(REDUCT, D) = K(REDUCT, D) + SGF($a_{|A1|}$)
8. While K(REDUCT,D) is not equal to K(A,D)
   Do
      REDUCT = REDUCT ∪ a_i (where SGF(a_i) is the highest of the attributes in A1)
      A1 = A1 - a_i
      Compute the degree of dependency K(REDUCT,D)
   End
9. |REDUCT| -> N
10. For i = 0 to N
      If a_i is not in C (that is the original set of attributes of the REDUCT at the start and SGF(a_i) is the least)
            Remove a_i from REDUCT
      End If
      Compute degree of dependency K(REDUCT, D)
      If K(REDUCT, D) not equal to K(A,D)
            REDUCT ∪ a_i -> REDUCT
      End If
11. End For
12. End if

Figure 2: Reduct Generating Algorithm

This process of re-computation of the dependency (to determine if an attribute is significant or not) is expensive as it involves finding equivalence classes and the iteration we initiated was principally an attempt to avoid it. We have also introduced the concept of "incremental evaluation of equivalence classes" to improve the performance of the algorithm. We have presented our rough sets approach to application run time estimation techniques as a way of estimating the cost of processing as it allows prediction of the time taken to perform mining at a given location. Obviously, the success of such a technique is dependent on the prediction accuracy.

4 Experimental Results and Analysis

The viability of using these cost models for DDM optimisation depends on the accuracy of the estimation techniques. Thus, the estimated communication times and predicted data mining task run times must be close to actual run time for these cost formulae to form the basis for effective optimisation. We are implementing mobile agents to perform distributed data mining and do not have experimental results to validate the communication cost models. However, we have implemented our rough sets algorithm and in this section present results from experiments on estimating the run times of data mining tasks. We compiled a history of data mining tasks by running several data mining algorithms on a network of distributed machines running Windows 2000 and Sun OS 5.8 and recording information about the tasks and the environment. We executed several runs of data mining jobs by varying the parameters of the jobs such as the mining algorithm, the data sets, the sizes of the data sets (from 1MB to 20MB) and the machines on which the tasks were run. The algorithms used were from the WEKA package of data mining algorithms [10]. For each data mining job, the following information was recorded in the history: the algorithm, the file name, the file size, the operating system, the version of the operating system, the IP address of the local host on which the job was run, the processor speed, the memory, the start and end times of the job. Currently we record only static information about the machines, however we are implementing a feature to enable recording dynamic information such as memory usage and CPU usage. We used histories with 100 and 150 records and each experimental run consisted of 20 tests. The mean error we recorded was 0.34 minutes and the mean error as a percentage of the mean run time was 8.29. The mean error is less than a minute and the error as a percentage of the actual run times is also very low, which indicates that we obtained very good estimation accuracy for data mining tasks. This good performance accuracy is illustrated in figure 3, which presents the actual and estimated run times from one of our experimental runs.

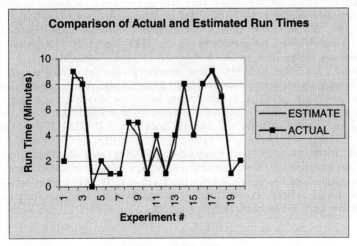

Figure 3: Actual Vs. Estimated Run Times

5 Conclusions and Future Work

This paper has focussed on computing the cost of the distributed data mining process by estimating the response time of the communication and computation components. Accurate costing is vital for optimisation of the DDM process and we address the need for good cost estimation techniques. We have developed cost formulae for estimating the response time of the communication for different strategies (including client server and mobile agent techniques) and a rough sets based application run estimation algorithm for predicting the run times of data mining tasks. We have experimentally validated the prediction accuracy of our rough sets algorithm, which is shown to have low mean errors and high accuracy. We are currently implementing the mobile agent model for DDM so as to experimentally validate the cost formulae for estimating the communication time. Future directions of this work include the development of a DDM optimiser based on the cost models presented in this paper.

References

1. Downey,A,B., (1997), "Predicting Queue Times on Space-Sharing Parallel Computers", *Proc. of the 11th Intl. Parallel Processing Symposium (IPPS)*, Geneva, Switzerland, April.
2. Gibbons,R., (1997), "A Historical Application Profiler for Use by Parallel Schedulers", *LNCS* 1291, Springer-Verlag, pp.58-75.
3. Hu,X., (1995), "Knowledge Discovery in Databases: An Attribute-Oriented Rough sets Approach", PhD Thesis, University of Regina, Canada.
4. Komorowski,J., Pawlak,Z., Polkowski,L., and Skowron, A., (1998), "Rough sets: A Tutorial", in *Rough-Fuzzy Hybridization: A New Trend in Decision Making*, (eds) S.K.Pal and A.Skowron, Springer-Verlag, pp. 3-98.
5. Krishnaswamy,S., Loke,S,W., & Zaslavsky,A, (2002), "Application Run Time Estimation: A Quality of Service Metric for Web-based Data Mining Services", *To Appear in ACM Symposium on Applied Computing (SAC 2002)*, Madrid, March.
6. Parthasarathy,S., and Subramonian,R., (2001), *"An Interactive Resource-Aware Framework for Distributed Data Mining"*, in Newsletter of the IEEE Technical Committee on Distributed Processing, Spring 2001, pp.24-32.
7. Smith,W., Taylor,V., and Foster,I.,(1999), *"Using run-time predictions to estimate queue wait times and improve scheduler performance"*, *LNCS* 1659, Springer-Verlag, pp.202-229.
8. Straßer,M., and Schwehm,M., (1997), "A Performance Model for Mobile Agent Systems", in *Proceedings of the International Conference on Parallel and Distributed Processing Techniques and Applications (PDPTA'97)*, (eds) H. Arabnia, Vol II, CSREA, pp. 1132-1140.
9. Turinsky,A., and Grossman,R., (2000), "A Framework for Finding Distributed Data Mining Strategies that are Intermediate between centralized Strategies and In-place Strategies", *Workshop on Distributed and Parallel Knowledge Discovery at KDD-2000*, Boston, pp.1-7.
10. Witten,I,H., and Eibe,F., (1999), "Data Mining: Practical Machine Learning Tools and Techniques with Java Implementations", Morgan Kauffman.

Distributed Resource Allocation in Ad Hoc Networks

Zhijun Cai and Mi Lu

Department of Electrical Engineering
Texas A&M University, College Station, TX 77843

Abstract. An ad-hoc network is a collection of wireless mobile nodes without the required intervention of any centralized Access Point. With the existence of the hidden terminals and the absence of the central control, the resource allocation is always a very challenging problem in such networks. In this paper, a distributed resource allocation scheme is designed for ad hoc networks, in which each mobile node dynamically searches its Native Index (NI) with negligible communication overhead. By utilizing the NI, the hidden terminal problem can be completely avoided in the resource allocation, and the resource can be efficiently distributed to all the nodes. The proposed resource allocation method can greatly improve the system throughput. Utilizing the method, the broadcast problem can be easily solved and QoS service (Quality of Service) can be efficiently supported.

1 Introduction and Backgrounds

An ad-hoc network is a collection of wireless mobile nodes which has no centralized control unit. Every node has limited transmission range, hence the packets may be relayed over multiple nodes before the destination node is reached. It is widely used in many cases, such as battlefields, search-and-rescue operations, disaster environments. Designing an efficient resource allocation scheme for such networks is always a very challenging problem due to the following reasons.

1. Hidden and exposed terminal problems. Such problems are caused mainly due to the limited transmission radius of the mobile nodes or the unsatisfied wireless environments. If one node is in the range of the receiver, but not the transmitter, the node is termed a *hidden terminal* to the transmitter. Similarly, an *exposed terminal* is a node which is in the range of the transmitter but not the receiver.

2. When nodes communicate, two types of collisions may arise [1]. The first type of collisions, termed the primary collision, means that a node transmitting a packet, cannot be receiving a packet at the same time and this also implies the converse. The second type of collision, termed the secondary collision, means that a node can not receive more than one packets at the same time. In both cases, all packets are rendered useless.

P.M.A. Sloot et al. (Eds.): ICCS 2002, LNCS 2329, pp. 613–622, 2002.

Much work has been done on the resource allocation in ad hoc networks [2–7]. In [2], different kinds of code assignment schemes have been discussed. Two phase algorithms have been devised to assign the codes to the transmitters, receivers and to pair of stations. In [3], an adaptive scheme has been proposed to allocate a portion of the available bandwidth only to those users who are active. The scheme places the nodes in different groups and dynamically divides the total network capacity between the groups. The communication overhead of the existing resource assignment methods is high due to the dynamic network topology, which increases the system offered load. In this paper, a distributed resource allocation scheme is designed for ad hoc networks, in which each mobile node dynamically searches its Native Index (NI) with negligible communication overhead. By utilizing the NI, the hidden terminal problem can be completely avoided in the resource allocation, and the resource can be efficiently distributed to all the nodes.

The rest of this paper is organized as follows. The proposed resource allocation scheme is illustrated in Section 2. Section 3 describes the solution to the broadcast problem and the support of the QoS service. Section 4 gives the performance analysis and the simulation results, and we conclude in Section 5.

2   Proposed Resource Allocation Scheme

2.1   Network Model

In an ad hoc network consisting of N nodes, the packets may be relayed over multiple nodes before the destination node is reached. An example is shown in Figure 1. An identical transmission radius is assumed for all nodes. Each node is given a unique identification number (ID). For a given node A, if the shortest number of hops between A and a particular node is $\leq h$, then that node is an h-hop neighbor of A. Denote $\delta_h(A)$ as the set of all h-hop neighbors of A. For example, in Figure 1, $\delta_1(A) = \{M, B, C\}$. In the following part, the meaning of "neighbors" is identified as *1-hop* neighbors. Define δ_h as the maximum size of $\delta_h(X)$ (for all X; X is a random node in the network). We assume δ_2 is upper-bounded by D.

2.2   NI Distribution

In the proposed resource allocation scheme, each node should obtain its NI (NI $\in \{1, 2, 3, \cdots, D, D+1\}$; the set of $\{1, 2, 3, \cdots, D, D+1\}$ is termed the NI set), and the NI distribution should satisfy the condition that any two nodes within *2-hop* distance should not have the same NI. The NI for the node will be updated when the network topology changes.

In ad hoc networks, each node should identify its neighbors after power-on (known as the neighbor-detection problem), and the general solution to the problem is the periodical beacon signal broadcast, in which each node broadcasts its beacon signal periodically to identify itself after power-on. We next propose a

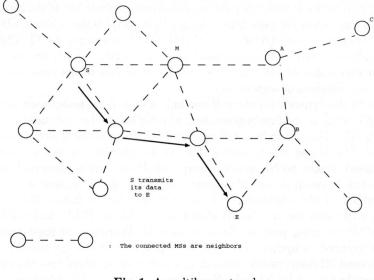

Fig. 1. A multihop network.

ID   NI   MNN   NNIS

Fig. 2. Format of the beacon signal.

method to jointly solve the NI distribution problem with the neighbor-detection problem. The format of the beacon signal is as Figure 2. For the beacon signal of A, the *ID* filed and the *NI* field are A's ID and A's NI respectively. Initially A will set its NI to NULL after power-on. The *MNN* field is the minimum node ID of A's neighbors whose NI is still NULL, while the *NNIS* field is the set of the NIs which have been utilized by A's neighbors. Through the periodical broadcast of the beacon signal, every node will be aware of the NIs, the MNNs and the NNISs of its neighbors. For a given node A, its Available NI Set (ANIS) includes all the remaining NIs that are not occupied by the nodes in $\delta_2(A)$, which is the remaining set subtracted from the NI set by all NNISs of A's neighbors. Every node should maintain its up-to-date ANIS. The NI obtaining process is as follows. If the NI of A is NULL and A acknowledges that its ID is the minimum ID of its *2-hop* neighbors *whose NIs are still NULL*, which means all MNNs of its neighbors are equal to A's ID, A will set its NI to a randomly selected NI from its ANIS. We have the following results.

1. The ANIS of a node will not be empty when the node attempts to obtain its NI, since the number of NIs $(D + 1)$ is greater than the maximum number of *2-hop* neighbors for any node plus 1.

2. Any two nodes within *2-hop* distance can not set their NIs at the same time, since only when *the node* acknowledges that its ID is the minimum ID of all its *2-hop* neighbors *whose NIs are still NULL*, it can set its NI. Therefore, the NIs can be distributed without any conflict (the conflict means the case that two nodes within *2-hop* distance set their NIs to the same one due to the unawareness of each other).

3. Due to the dynamic topology, if one link is broken, no nodes need to update its NI, while if one link is generated, *at most* two nodes *may* need to update their NIs. The reason is as follows. Suppose all nodes have already obtained their NIs. If one link is broken, the NI distribution condition can still be satisfied. Hence no NI update is required. If one link is generated, suppose the link between A and B is generated. Then A will become a new *2-hop* neighbor of B's neighbors. If A and any neighbor of B have the same NI, the node with the smaller ID should set its NI to NULL, and will initiate the NI obtaining process. So does node B. Therefore, at most two nodes are required to update their NIs. The property shows the stability of the proposed NI distribution method. If a new node is added into the network, the node will set its NI to be NULL, and initiate the NI obtaining process.

2.3   Resource Distribution

Two types of collisions may occur during the data communications in ad hoc networks as we have discussed in Section 1. The resource should be assigned to the nodes to avoid the potential collisions.

The resource in the ad hoc network can be represented by the codes, the slots or the frequency channels which corresponds to the CDMA (Code-Division Multiple Access), TDMA (Time-Division Multiple Access) or FDMA (Frequency-Division Multiple Access) respectively. However, only in the TDMA, two types of collisions are possible to be completely avoided. The reason is as follows. Suppose the resource is represented by the codes, and each node obtain an orthogonal code to transmit its data. The code distribution satisfies the condition that any two nodes within *2-hop* distance should not have the same orthogonal code. Utilizing the NI, the code distribution can be simply realized (each NI is an index to an orthogonal code; each node should utilize the code corresponding to its NI). Hence, the secondary collision can be completely avoided. However, the primary collisions still exist since the transmission of the nodes is not scheduled. For example, in the network topology shown in Figure 3, A and B are neighbors while B and C are neighbors. If A and C transmit the data to B simultaneously, no collisions will occur due to the orthogonal codes. However, if A transmits the data to B while B transmits the data to C, B will not obtain the data from A although the codes have been perfectly distributed. Therefore, if the resource has been represented by the codes, the transmission of the nodes should be scheduled to avoid the primary collisions, which is not an easy task to realize in the distributed environments such as ad hoc networks. Similarly, if the resource is represented by the frequency channels, the primary collision can not be avoided either.

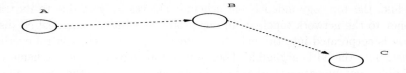

Fig. 3. An example of the resource distribution.

If the resource has been represented by the slots, the secondary collision can be avoided as well as the primary collision. The slot distribution is as follows. The frame is composed of $D+1$ slots. Each NI is an index to a unique slot, and each node should utilize the slot corresponding to its NI to transmit its data. The primary collision will not exist since any two neighbor nodes do not transmit the data in the same data slot. Moreover, the secondary collision will be completely avoided due to the slot distribution condition. In the proposed resource allocation scheme, the network resource is represented by the time slots.

2.4   Dynamic NI Scheme

Since NI is $\in \{1, 2, 3, \cdots, D+1\}$, if the estimated D is smaller, then some nodes may not obtain its NI, which implies these nodes can not have their resource. To address the problem, the dynamic NI scheme can be adopted. If a node does not utilize its NI (which corresponds to its resource) for some time, the node should release its NI, and set its NI to 0 (not NULL). Then other nodes can utilize the released NI. When the node attempts to transmit the data again, the node should set its NI to NULL and initiate the NI obtaining process.

3   The Broadcast Problem and the QoS Service

3.1   Solution to the Broadcast Problem

The broadcast problem is one of the fundamental issues in ad hoc networks, and much work has been done [1, 9, 10]. The existing methods can be divided into two groups, topology-dependent scheme and topology independent scheme. In the topology dependent scheme, global or part of the network topology are utilized to determine the minimum broadcast frame length and transmission schedule, while in the topology independent scheme, the mathematics theories such as the Galois Field theory have been adopted to achieve the efficient solutions with the topology-independent property. The topology-dependent scheme can always produce better solution than the topology independent scheme since the topology information has been utilized to determine the transmission schedule. Hence, if the network topology is fixed or changes slowly, the topology-dependent scheme is the choice. However, in the mobile networks, since the topology-dependent scheme entails significant control overhead to determine the schedule, the topology-independent scheme achieves its gain due to its transparency to the network topology. In our proposed solution, since the NI distribution is corporated into the neighbor detection process, the control overhead to distribute the NI is negligible. Then a new topology-dependent scheme with negligible communication overhead can be proposed. Suppose the time is divided into frames, and each frame is composed of $D+1$ data slots. The entire network is synchronized on the frame and the slot basis. For a given node, say A, suppose its NI is k. Then A should utilize the kth slot to broadcast its data. No conflicts exist when a node broadcasts its data according to the NI distribution condition.

3.2   Designing a MAC (Media Access Control) to Support the QoS

Presently QoS service in ad hoc networks becomes more and more important. For the QoS service, the resource should be allocated on demand (not pre-allocated). The existing on-demand resource allocation methods in ad hoc networks are based on the REQ/ACK dialogue, which is not a collision-free scheme (the REQ and the ACK packets suffer collisions). When the service load of the network is high, due to the collisions, the existing methods perform pretty bad. Next we will propose an efficient MAC to support the QoS. In the proposed method, very little resource will be pre-allocated to every node to avoid the collisions among the resource control packets (the REQ packets and the ACK packets), while the major part of the resource is allocated on-demand.

Time is divided into frames, which is composed of the control subframe and the data subframe. The control subframe is composed of $D+1$ control slots while the data subframe is composed of n_2 data slots. The structure of the frame is illustrated in Figure 4. Each control slot has three fields, the Active Field (AF),

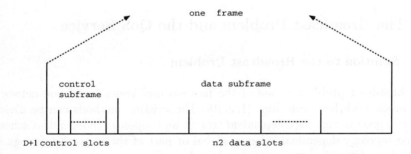

Fig. 4. Frame structure.

the Passive Field (PF) and the Allocation Acknowledgment Field (AAF). A node with its NI as k should utilize the AF of the kth control slot to transmit its control packets. From the NI distribution condition, no collisions will occur for the control packets. For a given node, say A, its control packet should include the resource request information, A's neighbor set, etc.

Resource Status

For every node, in each data slot, it can either transmit or receive the data. For a given node A, its Receiving Slot Set (RSS) includes all the data slots that A utilizes to receive the data, and its Transmitting Slot Set (TSS) includes all the data slots that A utilizes to transmit the data. Its Available Receiving Slot Set (ARSS) includes all the available data slots that A can utilize to receive the data (but not occupied yet). Due to the hidden terminal problem, the ARSS of A includes the slots in which none of A's neighbors including A will transmit the data.

Resource Reservation Process

Suppose A attempts to establish a QoS link to one of its neighbor B. A will transmit a resource request to B in the AF during its control slot. Every neighbor of A will be aware of the request from A. Then all the neighbors of A except B will transmit its RSS while B transmits its ARSS. The transmission strategy is as follows. Since A includes all the IDs of its neighbors in the control packet, every neighbor of A can obtain a sequence number according to the increasing order of all the IDs. For example, suppose A has three neighbors, B, C and D. Assume $ID(B) \leq ID(C) \leq ID(D)$. Then B will obtain the sequence number, 1. The sequence number of C is 2 and D is 3. The PF field is composed of δ_1 mini-fields. Each neighbor of A will transmit its RSS or ARSS in a certain mini-field according to its sequence number. In the above example, B will transmit its ARSS is the $1st$ mini-field, while C transmits its RSS in the $2nd$ mini-field. D will utilize $3rd$ mini-field to transmit its RSS. Considering the definition of δ_1, for a given node, it will receive all the RSSs and the ARSS from its neighbors safely. Based on the received information, A will determine whether and how to allocate the resource for its request. The determination process is as follows. First, A will calculate the available slot set. Suppose the received RSS sets are $RSS_1, RSS_2, RSS_3, ..., RSS_k$, and the received ARSS is $\overline{ARSS}$. Then A will calculate its available slot set as $(\bigcup_{i=1}^{i=k} RSS_i) \cup \overline{(TSS)} \cap (ARSS)$. Here, TSS means the TSS of A. Then it will check whether the available slots are enough for the QoS requirement. If so, it will allocate the required slots for the service. Otherwise it will determine whether to decrease the QoS requirements to utilize the available slots. After that, A will transmit its allocation result in the AAF during the same control slot, and every neighbor of A will receive the information in the AAF. Based on that, the destination node B will update its ARSS and RSS while other neighbor nodes of A update their ARSS. A should also update its own RSS, TSS and the ARSS. Moreover, B will acknowledge the data slots to receive the data from A. The proposed resource reservation process is requester-based (the node which transmits the resource request is termed the requester). That is, if a node attempts to establish the QoS link to another node, it will transmit its request in the AF to query the current local resource information. During the PF, the node will be aware of the current local resource status, and the node can determine how to utilize the available resource to fulfill its QoS requirement. The requester has more flexibility to determine how to utilize the resource. After A has reserved the data slots, it will transmit the data safely in the reserved data slots during every frame. If A finishes its data transmission, A should transmit a resource release message in its control slot, which will update the RSS and the ARSS of A's neighbors including B. A should also update its own TSS, RSS and ARSS.

4   Performance Analysis and Simulation Results

4.1   NI Distribution

Since the NI distribution has been solved together with the neighbor detection problem, the communication overhead is negligible. Due to the dynamic topology, the NI should be updated. The following advantages can be obtained by adoption of the proposed scheme.

1. Local update property, which means the local topology change only affects the local nodes. This is very important to the large mobile networks.
2. Less control overhead. To realize the NI distribution, only the beacon signal is utilized, which is very short ($\approx 2 \log M + \log D + D$ bits).
3. Dynamic NI scheme. If a node does not utilize its NI (which corresponds to its resource) for some time, the node should release its NI, and set its NI to 0 (not NULL).
4. Simplicity. The implementation of the NI distribution is simple since only minor changes have been made to the existing system.

4.2   Network Broadcast

Assume one node, say A, attempts to broadcast a packet to the network. Define the distance between node X and node Y as the minimum hop distance between X and Y, termed $dis(X,Y)$. Suppose $\max(dis(A,X)) = h$ (X is a random node in the network). Then the time cost of the broadcast will be $h \times Frame\_Length$, and N broadcast packets will be transmitted. Note that each node will broadcast the packet once.

4.3   QoS-Supported Resource Allocation

In the proposed scheme, utilizing the $(D+1)$-slot control subframe, every node can transmit its control packet without any collisions, and the network resource can be adaptively allocated to the nodes upon their requests. Due to the collision-free property among the control packets, the system throughput can steadily increased with the system overload. Next we use simulations to show the performance. The number of nodes in the simulated network is 100, and $n_2 = 32$. Assume the network is degree-bounded network with $\delta_1 \leq 8$, and $D = 30$. Define the offered load as the average number of data packets arrived during one data slot for the network and the throughput as the number of data packets transmitted successfully during one data slot. The offered load is uniquely distributed to all nodes. The length of the control slot and the data slot is $64, 2048$ bytes respectively. Figure 5 illustrates the throughput performance. It is seen that the throughput is steadily increased with the system load. Note that the scheme also performs pretty good in light or medium load. Furthermore, we use simulation to

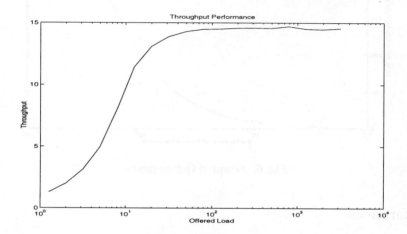

Fig. 5. Throughput performance.

show the network capacity for QoS requests. Assume n QoS requests to reserve the resource arrive at the network, and is uniquely distributed to all the nodes. If the request can not be accepted (not enough resource), the request is failed. Assume all the requests attempt to reserve 2 slots. We illustrate the number of requests that can be accepted utilizing the proposed QoS-supported resource allocation scheme in Figure 6. It is seen that the proposed QoS-Supported scheme can accept most of the QoS requests simultaneously when $n \leq 220$, which represents the network capacity.

5   Conclusions

In this paper, a distributed resource allocation scheme is designed for ad hoc networks, in which each mobile node dynamically searches its NI with negligible communication overhead. By utilizing the NI, the hidden terminal problem can be completely avoided in the resource allocation, and the resource can be efficiently distributed to all the nodes. The proposed resource allocation method can greatly improve the system throughput. Utilizing the method, the broadcast problem can be easily solved and QoS service can be efficiently supported. The performance of the proposed scheme has been analyzed and verified by the simulation results. It is worth while to put continuous efforts in this and related research areas.

(This research was supported by the Texas Advanced Technology Program under Grant No. 000512-0039-1999.)

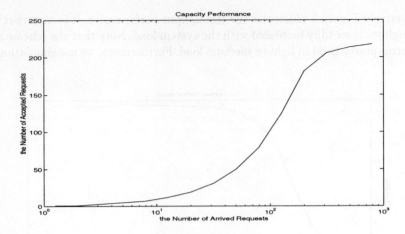

Fig. 6. Accepted QoS requests.

References

1. I. Chlamtac and A. Farago, "Making Transmission Schedules Immune to Topology Changes in Multi-hop Packet Radio Networks," *IEEE Trans. Networking*, vol. 2, no. 1, pp. 23–29, 1994.
2. Limin Hu, "Distributed Code Assignments for CDMA Packet Radio Networks," *IEEE Trans. Networking*, vol. 1, no. 6, pp. 668–677, December 1993.
3. Y. Mohamed, M. K. Gurcan, and L. F. Turner, "A Novel Resource Allocation Scheme For Ad Hoc Radio Local Area Networks ," in *Proc. of VTC'97*, 1997, pp. 1301–1305.
4. Sanjay Lal and E. S. Sousa, "Distributed Resource Allocation For DS-CDMA-based Multimedia Ad Hoc Wireless LANs," *IEEE Journal on Selected Areas in Communications*, vol. 17, pp. 947–967, May 1999.
5. C. Trabelsi, "Access Protocol for Broadband Multimedia Centralized Wireless Local Area Network," in *Proc. of the 2nd IEEE Symposium on Computers and Communications*, 1997, pp. 540–544.
6. A. M. Chou and V. O. K. Li, "Slot Allocation Strategies for TDMA protocols in Multihop Packet Radio Networks," in *Proc. of IEEE INFOCOM'92*, 1992, pp. 710–716.
7. R. Ramaswami and K. K. Parhi, "Distributed Scheduling of Broadcasts in a Radio Network," in *Proc. of IEEE INFOCOM'89*, 1989, pp. 497–504.
8. J. Whitehead, "Distributed Packet Dynamic Resource Allocation (DRA) for Wireless Networks," in *Proc. of the 46th IEEE VTS*, 1996, pp. 111–115.
9. Stefano Basagni, Danilo Bruschi, and Imrich Chlamtac, "A Mobility-Transparent Deterministic Broadcast Mechanism for Ad Hoc Networks," *IEEE Trans. Networking*, vol. 7, no. 6, pp. 799–807, Dec. 1999.
10. I. Chlamtac and O. Weinstein, "The Wave Expansion Approach to Broadcasting in Multihop Radio Networks," *IEEE Trans. Commun.*, vol. 39, pp. 426–433, Mar. 1991.

The Average Diffusion Method for the Load Balancing Problem*

Gregory Karagiorgos and Nikolaos M. Missirlis

Department of Informatics and Telecommunications,
University of Athens,
Panepistimioupolis 157 84, Athens, Greece,
{greg,nmis}@di.uoa.gr

Abstract. This paper proposes the *Average Diffusion* (ADF) method for solving the load balancing problem. It is shown that a sufficient and necessary condition for the ADF method to converge to the uniform distribution of loads is the induced network of processors to be d-regular, connected and not bipartite. Next, we proceed and apply Fourier analysis determining the convergence factor γ in terms of the diffusion parameters c_{ij} (weighted case) when the network of processors is a ring and 2D-torus. It is shown that $c_{ij} = \frac{1}{2}$ and $c_{ij} \in (0, \frac{1}{2})$ when the network is a ring and 2D-torus, respectively, thus solving partially the open problem which concerns the determination of the diffusion parameters c_{ij}.

Keywords: Load balancing, diffusion method, multiple parameters, d-regular graphs, distributed computing.

1 Introduction

1.1 Environmental models: the origin of the problem

Recently a number of numerical models concerning the simulation of environment (weather prediction, air pollution, ocean simulation) have been parallelized resulting in a considerable reduction in time (e.g. see [13] and the references herein). These studies show the suitability of the atmospheric computations for parallelization. These models use a three dimensional grid to simulate the physical processes at the atmosphere. The computations involved in such simulation models are of three types: **"dynamics"**, **"physics"** and **"chemistry"**. Dynamics computations simulate the fluid dynamics at the atmosphere (advection, diffusion) and are carried out on the horizontal domain. Since these computations use explicit numerical schemes to discretize the involved differential equations, they are inherently parallel. Alternatively, the physical and chemistry computations simulate the physical and chemistry processes such as clouds, precipitations, radiative transfer and are carried out on the vertical grid. These

---

* Research is supported by the Greek General Secretary of Research and Technology (*EΠET-II* No. 87) and the National and Kapodistrian University of Athens (No. 70/4/4917).

computations must be carried out for each grid point and do not require any data from its neighbour grid points. As the computations for each grid column are independent, domain decomposition techniques is best to be applied to the horizontal domain. The column computations refer to the physical and chemistry processes which can be subject to significant spatial and temporal variation in the computational load per grid point. As more sophisticated physics and chemistry will be introduced in the above environmental models, these computational load imbalances will tend to govern the parallel performance. Furthermore, on a network of processors, the performance of each processor may differ. To achieve good performance on a parallel computer, it is essential to establish and maintain a balanced work load among the processors. For this reason, it is necessary to calculate the amount of load to be migrated from each processor to its neighbours. Then, it is also necessary to migrate the load based on this calculation. In case of the environmetal models the amount of work on each processor is proportional to the number of grid points on the processor. To our knowledge there is a limited use of distributed load balancing schemes. In [1] we use the diffusion method to solve the load balancing problem in the Regional Atmospheric Modelling System (RAMS) obtaining encouriging results.

1.2   The load balancing problem

The ultimate goal of any load balancing scheme, *static* or *dynamic*, is to improve the perfomance of a parallel application code, using efficiently the parallel or distributed systems. For this purpose the computational work must be well balanced across processors and second the time spent performing interprocessor communication must be small. For many applications it is possible to make a priori estimates of load distribution and such load balancing schemes are called *static*. In other applications the computational requirements vary over time in an unpredictable way and it is no possible to make a priori estimates of load distribution. Such load balancing schemes are called *dynamic*.

Research on dynamic load balancing has focused on suboptimal algorithms that use local information in a distributed memory architecture. These algorithms describe rules for migrating tasks on overutilized processors to underutilized processors in the network. Tradeoffs exist between achieving the goal of completely balancing the load and the communications costs associated with migrating tasks. Every load balancing algorithm decides to give load to the neighbouring processors depending only on local information, i.e. by comparing the load of the underlying processor with the load of the neighbouring processors and thereby trying to reach a *local* equilibrium. After a number of diffusion cycles, which in the worst case is quadratic to the number of processors, a *global* equilibrium of the load will be reached [5].

The parallel and distributed system is modeled as an undirected graph where nodes represent the processors and the edges represent the network connections. Each processor is associated with a real variable which reflect the work load currently running on it. We assume that during the balancing process no load is generated or consumed and the graph does not change. The system is synchronous, homogeneous and the network connections have unbounded capacity. We concentrate only on solving the data flow problem using the ADF method ignoring the migration problem [11].

The original diffusion algorithm has been proposed by Cybenko [9] and, independently, by Boillat [5]. Diffusion type algorithms [9, 5, 17] are some of the most popular ones for the flow problem. The quality of a diffusion algorithm can be measured in terms of number of iterations it requires to reach a balanced state. Recently the Diffusion method was combined with semi-iterative techniques [16] reducing the number of iterations by an order of magnitude [10–12, 14]. In addition, diffusion type methods are also used solving the flow problem for asynchronous distributed systems [3].

This paper is organized as follows. In section 2 we introduce the Average Diffusion (ADF) method. In section 3 we study its convergence analysis and in particular, we find necessary and sufficient conditions for convergence. In section 4 we use Fourier analysis to determine the values of the diffusion parameters c_{ij} and present our results for the ring and the 2D-torus processor graphs. It is shown that $c_{ij} = \frac{1}{2}$ and $c_{ij} \in (0, \frac{1}{2})$ when the network is a ring and 2D-torus, respectively, thus solving partially the open problem which concerns the determination of the diffusion parameters c_{ij}. Finally, our conclusions are discussed in section 5.

2   The Average Diffusion Method

2.1   The method

Let $G = (V, E)$ be a connected, undirected graph with $|V|$ nodes and $|E|$ edges. Let $u_i \in \Re$ be the load of node $v_i \in V$ and $u \in \Re^{|V|}$ be the vector of load values. The average load per processor is

$$\bar{u} = \frac{1}{|V|} \sum_{i=1}^{|V|} u_i.$$

The computation of $\bar{u}$ requires global communication between the processors which is a time consuming process. To avoid the aforementioned problem we restrict the computation between neighbour processors. The load of processor i is computed as the weighted average work load of its direct neighbours as follows:

$$u_i = \frac{1}{\displaystyle\sum_{j \in A(i)} \hat{c}_{ij}} \sum_{j \in A(i)} \hat{c}_{ij} u_j$$

or

$$\sum_{j \in A(i)} \hat{c}_{ij}(u_i - u_j) = 0 \tag{1}$$

where u_i and u_j are the workloads of processors i and j, respectively, $A(i)$ is the set of nearest neighbors and $\hat{c}_{ij}$ are the nonnegative diffusion parameters. In matrix form, (1) is written as

$$L\bar{u} = 0 \qquad (2)$$

where L is the weighted Laplacian matrix [11] of graph G and has the splitting

$$L = D - A,$$

where $D = (d_{ii})$ is the diagonal matrix with $d_{ii} = deg(i)$, $deg(i)$ denotes the degree of i and A is the adjacent matrix of the graph G. For solving the homogeneous linear system (2), we consider the iterative scheme

$$u^{(n+1)} = Bu^{(n)}, \qquad (3)$$

where

$$B = D^{-1}A. \qquad (4)$$

Let $B = (b_{ij})$ with

$$b_{ij} = \begin{cases} c_{ij}, & \text{if } i \neq j \text{ and } j \in A(i) \\ 0, & \text{otherwise} \end{cases}$$

where

$$c_{ij} = \frac{\hat{c}_{ij}}{\displaystyle\sum_{j \in A(i)} \hat{c}_{ij}}. \qquad (5)$$

From (5) it follows that

$$0 < c_{ij} < 1$$

since $\hat{c}_{ij} > 0$. Moreover, for B to be symmetric we must have $c_{ij} = c_{ji}$, which (because of (5)) imposes the condition $A(i) = A(j)$ since $\hat{c}_{ij} = \hat{c}_{ji}$. This means that all the processors must have the same number of neighbors, that is $d_{ii} = d$. In other words, the matrix B is symmetric when the graph is d-regular. In the sequel we develop our analysis under this assumption. The iterative method given by (3), will be referred to as the *Average Diffusion* (ADF) method and the matrix B is its *iteration matrix*. In case of the nonweighted Laplacian L, $\hat{c}_{ij} = 1$ and

$$b_{ij} = \begin{cases} \frac{1}{d} & \text{if } i \neq j \text{ and } j \in A(i) \\ 0 & \text{otherwise.} \end{cases}$$

2.2   The iteration matrix B

For the average work load of ADF to be invariant B must be doubly stochastic [2]. Let us first prove that B is row stochastic, i.e

$$\bar{u} = B\bar{u}, \qquad (6)$$

where $\bar{u}$ is the vector whose every entry is exactly $\frac{\sum_i u_i}{|V|}$, which implies that (6) will hold if $\sum_{j \in A(i)} b_{ij} = 1$. But,

$$\sum_{j \in A(i)} b_{ij} = \sum_{j \in A(i)} c_{ij} = 1.$$

The last equality holds, because of (5). Therefore, B is row stochastic.

Lemma 1 *If the network graph is d-regular, then B is symmetric.*

Proof. For B to be symmetric we must have $B = B^T$, or because of (4) $A = DAD^{-1}$ which holds in case the elements of D are all equal. This is true since the network graph is d-regular. ∎

Finally, B is doubly stochastic since it is symmetric and row stochastic. In the following, we use the properties of B to establish conditions under which the ADF method converges to the uniform load distribution and determine its rate of convergence. Note, that B is also irreducible and nonnegative matrix. The matrix B satisfies the conditions to the *Perron-Frobenius* theorem [2, 16], hence for its eigenvalues μ_i, $i = 1, 2, \ldots, n$, we have

$$\mu_n \leq \mu_{n-1} \leq \ldots \leq \mu_2 < \mu_1 = 1.$$

The last inequality is strict since 1 is a simple eigenvalue. For convergence of ADF we must find conditions under which $\gamma(B) < 1$, where

$$\gamma(B) = \max_{i \neq 1} |\mu_i|$$

which will be referred to as the *convergence factor*.

3   Convergence Analysis

In this section we find necessary and sufficient conditions for the ADF method to converge.

Theorem 1 *The ADF method always converges to the uniform distribution if and only if the induced network is connected and not bipartite.*

Proof. A graph is bipartite if its vertex-set can be partitioned into two parts V_1 and V_2 such that each edge has one vertex in V_1 and one vertex in V_2. If we order the vertices so that those in V_1 come first, then the adjacency matrix of a bipartite graph takes the form [4]

$$A = \begin{bmatrix} 0 & K \\ K^T & 0 \end{bmatrix}, \tag{7}$$

where 0's are used to denote square zero block matrices on the diagonal of B and K is a rectangular nonnegative matrix. Since $B = D^{-1}A$ it is evident that B will also possess the form (7). It is well known that the eigenvalues of a matrix which takes the above form (7) occur in pairs $\pm\mu_i$ [4]. Hence, the eigenvalues of B satisfy

$$-1 = \mu_n < \mu_{n-1} \ldots \le \mu_2 < \mu_1 = 1.$$

Therefore, the ADF method converges to the uniform distribution ($\gamma < 1$) if and only if -1 is not an eigenvalue of B. If the network graph is not bipartite, then -1 is not an eigenvalue of B. ∎

4   Local Fourier analysis

The conventional way to analyze the ADF method is to use matrix analysis [17]. This approach depends on the properties of the resulting diffusion matrix which in turn depends on the topology of the graph. In this section we use an alternative technique to analyze diffusion algorithms. This technique is the Fourier analysis. Assuming that the iterative diffusion methods solve numerically a Partial Differential Equation (e.g. the Convection-Diffusion equation) we can apply Fourier analysis to study its error smoothing effect [6–8, 15]. Fourier analysis applies only to linear constant coefficient PDEs on an infinite domain or with periodic boundary conditions. However, at a heuristic level this approach provides a useful tool for the analysis of more general PDE problems. Following the same idea, we will apply the Fourier analysis approach to the ADF method. When the graph is the ring or the 2D-torus we obtain the same results for the nonweighted case as in [17]. However, our derivation produces results in the weighted case also. In particular, we are able to determine the convergence factor $\gamma(B)$ in terms of the diffusion parameters c_{ij} and study their role in the convergence behaviour of ADF. Next, we will apply this technique for the ring and 2D-torus processor graphs.

4.1   The ring

At a local node the ADF scheme (3) can be written as

$$u_j^{(n+1)} = B_j u_j^{(n)} \tag{8}$$

where $B_j \equiv (c_{j+1}E + c_{j-1}E^{-1})$ is the *local ADF operator* and

$$Eu_j = u_{j+1}, \quad E^{-1}u_j = u_{j-1}$$

are the *forward-shift* and *backward-shift* operators in the x-direction, respectively. Expressing (8) in terms of the error vector $e^{(n)} = u_j^{(n)} - u$ we have

$$e_j^{(n+1)} = B_j e_j^{(n)}, n = 0, 1, 2, \ldots$$

If the error function $e_j^{(n)}$ is the complex sinusoid e^{ikx}, we have

$$B_j e^{ikx} = \mu_j(k)e^{ikx}$$

where

$$\mu_j(k) = c_{j+1}e^{ikh} + c_{j-1}e^{-ikh}, \ h = \frac{1}{N}$$

where N is the number of processors. So, we may view e^{ikx} as an eigenfunction of B_j with eigenvalues $\mu_j(k)$. Furthermore,

$$|\mu_j(k)| = ([(c_{j+1} + c_{j-1})\cos kh]^2 + [(c_{j+1} - c_{j-1})\sin kh]^2)^{1/2}. \tag{9}$$

In the *nonweighted* case $c_j = \frac{1}{d}$, where $d = 2$ for the ring, hence (9) becomes

$$|\mu_j(k)| = cos(kh) \tag{10}$$

where k is selected such that $|\mu_j(k)|$ attains its maximum value less than 1. Letting $k = 2\pi\ell, \ \ell = 0, \pm 1, \pm 2, \ldots, \pm(N-1)$, (10) yields

$$\gamma(B_j) = cos(2\pi h)).$$

Note, that $\gamma(B_j)$ is equal to $\gamma(B)$, where $\gamma(B)$ is determined using matrix analysis [17].

In the weighted case for the operator B_j to be symmetric we must have

$$c_{j+1} = c_{j-1} \tag{11}$$

and (9) yields

$$|\mu_j(k)| = 2c_{j+1}\cos kh.$$

Moreover, for the operator B_j to be row stochastic we must have

$$c_{j-1} + c_{j+1} = 1. \tag{12}$$

From (11) and (12) it follows that

$$c_j = \frac{1}{2}, \ j = 0, 1, 2, \ldots, N-1.$$

Therefore, the diffusion parameters must be equal to $\frac{1}{2}$ in case of the ring network topology and the weighted case coincides with the nonweighted one.

4.2   The 2D-torus

Using a similar approach as in the previous section, we will define the local ADF operator for 2D-torus network graph. Define the x_1-direction, (x_2-direction) *forward-shift* and *backward-shift* operators, E_1 and E_1^{-1} (E_2 and E_2^{-1}), as

$$E_1 u_{ij} = u_{i+1,j}, \ E_1^{-1} u_{ij} = u_{i-1,j},$$
$$E_2 u_{ij} = u_{i,j+1}, \ E_2^{-1} u_{ij} = u_{i,j-1}.$$

Then, the *local ADF operator* for 2D-torus network graph is $B_{ij} \equiv (c_{i+1,j}E_1 + c_{i-1,j}E_1^{-1} + c_{i,j+1}E_2 + c_{i,j-1}E_2^{-1})$ and thus we have

$$B_{ij}e^{i(k_1x_1+k_2x_2)} = \mu_{ij}(k_1,k_2)e^{i(k_1x_1+k_2x_2)},$$

where

$$\mu_{ij}(k_1,k_2) = (c_{i+1,j}e^{ik_1h} + c_{i-1,j}e^{-ik_1h} + c_{i,j+1}e^{ik_2h} + c_{i,j-1}e^{-ik_2h}),$$

$h = \frac{1}{\sqrt{N}}$. Furthermore,

$$|\mu_{ij}(k_1,k_2)| = ([(c_{i+1,j}+c_{i-1,j})\cos k_1h + (c_{i,j+1}+c_{i,j-1})\cos k_2h]^2 +$$
$$[(c_{i+1,j}-c_{i-1,j})\sin k_1h + (c_{i,j+1}-c_{i,j-1})\sin k_2h]^2)^{1/2}. \quad (13)$$

In the *nonweighted* case $c_{ij} = \frac{1}{d}$, where $d = 4$ for the 2D-torus, hence (13) becomes

$$|\mu_{ij}(k_1,k_2)| = \frac{1}{2}(cos(k_1h) + cos(k_2h)) \quad (14)$$

where k_1,k_2 are selected such that $|\mu_{ij}(k_1,k_2)|$ attains its maximum value less than 1. Letting $k_1, k_2 = 2\pi\ell$, $\ell = 0, \pm1, \pm2, \ldots, \pm\sqrt{N} - 1)$, (14) yields

$$\gamma(B_{ij}) = \frac{1}{2}(1 + \cos(2\pi h)).$$

Note, that $\gamma(B_{ij})$ is equal to $\gamma(B)$, where $\gamma(B)$ is determined using matrix analysis [17].

In the weighted case for the operator B_{ij} to be symmetric we must have

$$c_{i+1,j} = c_{i-1,j} \text{ and } c_{i,j+1} = c_{i,j-1} \quad (15)$$

and (13) yields

$$|\mu_{ij}(k_1,k_2)| = 2(c_{i+1,j}\cos k_1h + c_{i,j+1}\cos k_2h).$$

Moreover, for the operator B_{ij} to be row stochastic we must have

$$c_{i+1,j} + c_{i-1,j} + c_{i,j-1} + c_{i,j+1} = 1. \quad (16)$$

Combining (15) and (16) we obtain

$$c_{i+1,j} + c_{i,j+1} = \frac{1}{2} \quad (17)$$

or any combination of c_{ij}s from (15) must have sum equal to $\frac{1}{2}$. From (17) it follows that for $c_{ij} \in (0, \frac{1}{2})$ the ADF method converges and the "optimum" values for c_{ij}, which minimize $\gamma(B_{ij})$, must lie in the above interval. Choosing $c_{ij} = \frac{1}{4}$, the weighted case coincides with the nonweighted one. This was also the case in the ring topology.

5   Conclusions

In this paper we introduced an iterative distributed load balancing scheme, the ADF method. We showed that for the workload to be invariant the graph must be d-regular. Moreover, we found that a necessary and sufficient condition for the convergence of the ADF method is the processor network to be connected and not bipartite. The use of Fourier analysis gave us the ability to find a closed from formula for the eigenvalues of the iteration matrix B in terms of the diffusion parameters c_{ij}. For the ring and 2D-torus the formula coincides with the one given by [17] in the nonweighted case, thus verifying the validity of our approach. In this work we found that the diffusion parameters c_{ij} must be equal to $\frac{1}{2}$ in case of a ring topology. Moreover, c_{ij} must lie in the interval $(0, \frac{1}{2})$ for the 2D-torus, thus solving partially the open problem which concerns the determination of the diffusion parameters c_{ij}. Currently, our research is focused on the determination of the "optimum" values of c_{ij}s for more general graphs ($2d$-regular).

References

1. Balou A., Karagiorgos G., Kontarinis A., Missirlis N. M., *Optimization and parallelization of the RAMS model*, Thechnical Report 3.3, $E\Pi ET\text{-}II$ No. 87, 2001 (in greek).
2. Berman A., Plemmons R.J., *Nonnegative matrices in the mathematical sciences*, Academic Press, 1979.
3. Bertsekas D.P. and Tsitsiklis J.N., *Parallel and distributed computation: Numerical Methods*, Prentice-Hall, 1989.
4. Biggs N. *Algebraic graph theory*, Cambridge University Press, Cambridge, 1974.
5. Boillat J.E., Load balancing and poisson equation in a graph, *Concurrency: Practice and Experience* **2**, 1990, 289-313.
6. Boukas L.A., Missirlis N. M., The parallel local modified SOR for nonsymmetric linear systems, *Intern. J. Computer Math* **68**, 1998, 153-174.
7. Brandt A., Multi-level adaptive solutions to boundary-value problems, *Math. Comput.* **31**, 1977, 333-390.
8. Chan T. F., Kuo C.-C.J., Tong C., Parallel elliptic preconditioners: Fourier analysis and performance on the connection machine, *Computer Physics Communications* **53**, 1989, 237-252.
9. Cybenko G., Dynamic load balancing for distributed memory multi-processors, *J. Parallel Distrib. Comp.* **7**, 1989, 279-301.
10. Diekmann R., Frommer A., Monien B., Efficient schemes for nearest neighbour load balancing, *Parallel Computing* **25**, 1999, 789-812.
11. Hu Y.F, Blake R.J., An improved diffusion algorithm for dynamic load balancing, *Parallel Computing* **25**, 1999, 417-444.
12. Karagiorgos G., Missirlis N.M., Iterative Algorithms for Distributed Load Balancing, In proc. of 4th Intern. Conf. On Principles Of DIstributed Systems, *Special issue of Intern. Journal on Informatics*, 2000, 37-54.
13. Karagiorgos G. and Missirlis N.M., Iterative Load Balancing Schemes for Air Pollution Models", S. M. Margenov, J. Wasniewski, P. Yalamov (Eds.): *Large-Scale Scientific Computing*, Third Intern. Conference, LSSC 2001, **LNCS 2179**, 291-298.

14. Karagiorgos G., Missirlis N.M., Accelerated Diffusion Algorithms for Dynamic Load Balancing, (*submitted*).
15. Kuo C.-C. J., Levy B. C. and Musicus B. R., A local relaxation method for solving elliptic PDE's on mesh-connected arrays, *SIAM j. Sci. Statist. Comput.*, **8**, 1987, 530-573.
16. Varga R., *Matrix iterative analysis*, Prentice-Hall, Englewood Cliffs, NJ, 1962.
17. Xu C.Z. and Lau F.C.M., *Load balancing in parallel computers: Theory and Practice*, Kluwer Academic Publishers, Dordrecht, 1997.

Remote Access and Scheduling for Parallel Applications on Distributed Systems*

M.Tehver**, E.Vainikko***, K.Skaburskas[†] and J.Vedru[‡]

Center of Technological Competence, Tartu University, Estonia

Abstract. Design of WWW-based frontend for distributed computing systems running scheduled tasks in parallel is discussed. A model implementation for DOUG package (Domain Decomposition on Unstructured Grids) is given. The interface consists of input-output forms, task queueing-scheduling system and administrative tools. MySQL database and PHP scripts are used as building blocks for advanced computational environment accessible for groups working together in the Internet across distinct locations. Various implementational issues are discussed including system security and its testing. An approach to program development of the WWW-interfaces for computational packages is presented.

1   Introduction

The widespread development of computer networks has revolutionised the way in which computations are carried out in large scale problems. Distributed systems and dedicated supercomputers are all making use of some sort of communication between fast processors for solving huge computational problems in parallel. At the same time, from the user's point of view, access to the problem solving environments on computers has become more and more straightforward and easy to use. This is due to availability of various types of user interfaces making the environment more easy to understand, learn and use for performing complex tasks. Most of such interfaces assume user's physical presence within the range of local area network. The internet has brought to the existence a new global range of possibilities for user interactions with computer systems. HTML was developed in the beginning mostly for remote information retrieval. But the movement has been gradually towards more two-way communication between the client (the user) and server. Also, public key encryption technologies have made it possible for various new services to appear online on the Internet that need to maintain privacy, authentication and authorisation. Often these services make use of modern database technology in their core.

* This work was supported by Estonian Science Foundation reserch grant #4449.
** Department of Computer Sciences, Tartu University. E-mail: mark@eeter.fi.tartu.ee
*** Department of Mathematical Sciences, University of Bath. E-mail: masev@maths.bath.ac.uk
[†] Institute of Experimental Physics and Technology, Tartu University. E-mail: konstan@ut.ee
[‡] Institute of Material Sciences, Tartu University. E-mail: vedru@ut.ee

P.M.A. Sloot et al. (Eds.): ICCS 2002, LNCS 2329, pp. 633–642, 2002.

In this paper we investigate possibilities of taking use of the global Internet technology for developing a WWW-interface for the packet DOUG (Domain Decomposition on Unstructured Grids [4]) which has been developed for solving large sparse linear systems in parallel. We use publicly available software tools for developing WWW-environment which is capable of more than solely solving large linear systems – the environment is able to store a series of computer experiments' input and output into a database for later access, analysis and visualisation. We discuss the design and components of such systems together with a general guide for building it up on an example of DOUG.

2   Tools for designing interactive web-pages

There are basically two general classes of web based user interfaces. First dynamic web pages with HTML forms. These web pages are dynamically created (on request) at the server and then sent back to the requesting computer. All processing takes place in the server. The second option is to use Java applets. Java applets are small downloadable binary files that are executed on a client computer. This requires special Java runtime support from WWW-browser and uses more resources on the client side. In recent years a lot of research and development has been done for very large distributed computing systems on the Internet. Complex resource managers and scheduling schemes have been developed like Condor[2], CACTUS [1]. The term 'grid computing' is used for such systems [3]. Java is often seen as a primary development language for 'grid components' as it provides a platform-independent secure operating environment. For relatively small workstation clusters that are homogenous and that are assigned only to computational tasks, grid computing is simply overly abstracted. Resource management and basic scheduling are often already implemented in parallel computational packages.

Although Java applets are more powerful, the interface we were planning to implement did not need their additional functionality and we settled on HTML form based interface. Numerous ways exist for creating dynamic web pages. Probably the most common practice is to use Common Gateway Interface (CGI) on Apache web servers. CGI is not tied to any particular programming language although Perl and C are most widely used. Unfortunately CGI is a low-level interface and requires programmer to know many technical details of HTTP protocol and OS. We wanted to exploit higher level tools that had an abstraction level between the CGI interface and grid components. There are numerous such tools. To restrict the choice, we needed the tools to prove following properties:

1. All software had to be free, actively developed and supported.
2. Tools had to be as platform independent as possible.
3. We preferred tools that were easy to set up and maintain.

We chose PHP scripting language and MySQL database system as key components that satisfied all these requirements. Both PHP and MySQL are reliable, with very large user base and good community support.

2.1 Scripting language PHP

PHP is a programming language specially designed for dynamic web pages [7]. The programming is accomplished by using special tags in HTML files (commonly called 'scripts'). On request PHP scripts are automatically processed in a Web server and as a result a regular HTML file (without any special PHP tags) is sent to the client computer. A clear advantage compared to Java applets (which can be viewed as clients in client-server environment) is that both form generation and processing can be coupled together, making changes in user interface easier to handle.

PHP is not the only language designed for WWW-programming. Other well-known languages are Visual Basic Script and JScript within ASP (Active Server Pages). But PHP stands out with excellent support for different Web servers and platform independence. Besides, being developed by open source community, it is completely free to use and supports most widespread database systems.

2.2 MySQL database technology

MySQL is an open source relational database management system that is based on Structured Query Language (SQL). For our goals MySQL database implementation seemed to fit the best. We needed moderate sized databases with fast access while we did not really need the safe transaction support that is present in larger systems. What made MySQL attractive to us was the ease of maintainance and the very good integration with PHP. MySQL has many small aggregate functions that are handy for formatting output for PHP pages. This reduces PHP code size as a lot of processing can done in SQL.

3 Designing web-interface for DOUG

Our aim was to develop an environment for parallel computations that could be used for wide range of problems with sparse matrices. The application of our choice is DOUG (automatic Domain Decomposition on Unstructured Grids) due to its ability of meeting given needs. We needed a computational package that could be run well in small to moderate sized clusters (4-16 computers). It was intended mostly for a relatively small cluster that was connected into an Ethernet or ATM network. DOUG is currently under rapid development with many fascinating features emerging in the near future. The user interface will add even more value to the list of features.

3.1 DOUG overview

DOUG is a black box parallel iterative solver for large sparse linear systems originating from PDEs with some discretisation. The solver is developed at the Department of Mathematical Sciences, Bath University, United Kingdom. The first public release of the code (DOUG 1) [5] came out in 1997. The release of the new version (DOUG 2) with many advanced features is scheduled to June 2002. Here we give a brief overview of both versions and describe input requirements needed to be able to run the code.

DOUG 1. The first version of DOUG was meant for solving finite element discretisations of elliptic partial differential equations on 2D and 3D domains using unstructured grids. The system is assumed to be given in the form of element stiffness matrices resulting in three required input files:

info_file providing information about the number of elements, the number of freedoms, the spatial dimension, the maximum number of freedoms on an element and the number of nodes in the finite element grid.

data_file listing for each element the number of freedoms and freedom numbers.

element_file providing the element stiffness matrix together with the right hand side vector.

After reading in the elemental matrices the elements are divided into p partitions. Domain decomposition method with small overlap (see *e.g.* [8]), namely the additive Schwarz preconditioner is used for parallel iterative solution of the system. Optionally, for better convergence coarse problem is generated automatically by DOUG. For this user has to provide additional information in the following two files:

xyz_file providing spatial coordinates of the nodes in the finite element grid.

freemap_file is needed if the numbering of the freedoms differs from the numbering of the nodes.

In addition, various options have to be passed to the solver. These include the type of preconditioner (1 or 2-level Additive Schwarz method), subsolver type, tolerance, method to use in the outer iterations (PCG, BICGSTAB, PGMRES) and others.

Parallelisation of DOUG is based on the Message Passing Interface (MPI) and is using master-slave setup. The code is written in Fortran 77.

DOUG 2. The new version of DOUG supports several other input formats of linear system. Fully assembled matrices can be used, thus enabling it to solve problems with finite difference discretisation. The files can either be in a standard sparse representation or MATLAB sparse format (.mat-files).

Another new feature of DOUG2 is an ability to solve efficiently blocked systems that arise particularly in linearisation of Navier Stokes equations. The list of new features includes also eigenvalue solver for finding a small number of eigenvalues using a class of inverse iteration methods. Part of the new features is supported in the WWW-interface as well.

3.2   WWW-Interface

Most WWW-based user interfaces are based on forms. User has to fill in basic form elements like text fields or checkboxes. For our purposes these basic elements were important but not sufficent as we needed support for uploading and downloading large datasets. Designing input forms for all input data is not a

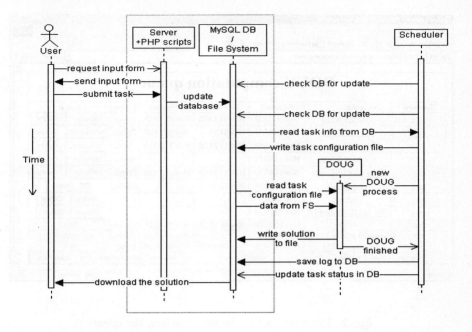

Fig. 1. Dataflow and control sequence diagram in the implementation. (Arrows with bold ends show dataflow. Arrows with regular ends indicate control.)

practical option, as most of the input data comes directly from different programs like MATLAB, Maple or from custom Fortran applications. Fortunately HTTP protocol provides support for both uploading and downloading binary and text files. The interface includes special fields for selection input files that will be uploaded to server on the form submition.

A dataflow and control sequence diagram of our system is presented in Fig.1. The system consists of two parts: input-output interface and task scheduler. MySQL database is used for storage and coordination between parts. It is based on only three entities: User entity describes user accounts, Task entity describes submitted tasks (and includes fields for keeping different date fields that track solution progress) and File entity is used for mapping server side files to tasks (each task may need several input files and produce several output files). Our system also includes an administration module (its functionality is not shown in Fig.1).

Input-output interface design. The central part of our user interface is the task queue. It lists all tasks that have been submitted by showing status (*waiting, running, solved, killed, frozen*), submission date, solving date, description and priority for each task. Queue interface provides direct links (via buttons) to all other scripts of the interface. Queue interface is shown in Fig.2. In the given

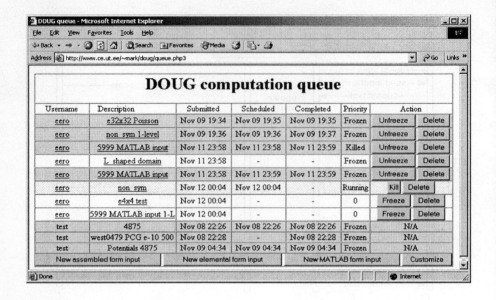

Fig. 2. The main part of the web-interface: the queue.

figure there are eleven tasks in queue, most of which are in *frozen* state (when task is solved, the status is automatically set to *frozen* and *completed* field is set to the current time). One task is currently running and two tasks are waiting. User has option to cancel a running task, or freeze a waiting task. Already solved tasks can be rerun again using the *Unfreeze* button. Note that the three last tasks can not be controlled by the current user as they are submitted by a different user.

While queue interface is completely general to tasks (and did not change after the first version was written), task input and output forms are specific to the computation engine behind the interface and had to be updated several times when new options were added to DOUG. New solving methods were added during the interface development together with new options for displaying the solver output. Currently we have three different input forms for different input data formats (elemental, assembled or MATLAB). Although they are implemented as one script (most options are still common to all formats), several options are specific to each input format.

As forms are generated in the server, client side lacks interactivity. The problem is that task input forms contain many interrelated option dependencies, some options are specific only to a single solution method. To signal this to the user (so that only relevant option combinations could be selected) JavaScript code was embedded into the HTML code. JavaScript is more closely tied to the calling HTML file than it would have been in the case of Java applets. It allows to verify and change forms on client side and is commonly used to make dynamic web pages more user friendly.

While solving a task, DOUG provides some information about solving progress. A lot of this information is not important to end users, though, some of it can be very useful. We extract convergence data from this output and allow users to view graphically the convergence curve while the task is being solved. We use gnuplot utility for graph drawing and redirect its output (which is actually a GIF file) from local file to client computer. So, the user can visually track the solution progress and stop the task solving if convergence is not satisfactory.

We have found the system to be helpful for solving a series of tasks. We have provided support for using input parameters for some previous task as a template for the new task. User can change only relevant parameters and keep the setting of most parameters. For input datasets there is an option for uploading a new dataset or keeping the old. As a result, the user need not upload subsequently the same set of input-files. After submitting the changes, a new task is created and the old one is left unmodified.

Task scheduler. Scheduler is a stand-alone Perl script that searches regularly database for new tasks, dispatches these tasks by launching DOUG when necessary and updates database when the tasks are solved. Although the basic operation is fairly simple, there are many other operations that scheduler performs. For example, user interface has an option to kill a partially solved task. This request is sent via database to the scheduler which kills the corresponding solver process. Another option is to run multiple scheduler processes in parallel. Running several scheduler processes in parallel can give better utilization of resources.

After a task has terminated (or has been killed) it is possible that more than one process is waiting in the queue. We used a simple cost function that is based on the time that the task has spent in queue and a special priority attribute that is associated with each user and assigned by the administrator. The cost function assures the user that all tasks will be chosen within finite time interval and regular users (users with higher priority) are preferred over "guest" users.

We use the UNIX 'crontab' utility to launch scheduler at regular intervals. When run, scheduler first checks that the number of running scheduler processes does not exceed a given limit, if the limit is exceeded, it exits at once. Otherwise it searches for a new task and tries to solve it using DOUG. After the task has been solved, scheduler looks for a new task and exits if none is found. By using 'crontab' utility, we do not need to manually start our system after the computer has been rebooted. As the scheduler script may not be running when a new task is submitted (or can be busy), there is a latency issue. We have chosen a period of 5 minutes between each scheduler process launch (but the frequency can be changed by system administrator.) Thus it is possible that the user has to wait up to five minutes even if system is not busy. So far this has not been a real drawback as most tasks take longer than five minutes to solve.

Tools for system administration, security issues. One of our goals was to use web-interface for all aspects of our system. We use a similar web interface for the administrator of user accounts. This interface lets us create new user accounts, freeze existing accounts, add comments about users and change user priorities (which are used by scheduler as described above). Setting up a new user account takes virtually no time and the account is ready immediately. We have created a username-password system for our user accounts. When an account is created by administrator, the user is given a password that is required for login. All user scripts are protected by username and password checking. We use the HTTP 'cookies' feature for this; after authentication, the user's username and password hash are stored in a cookie. This cookie is kept only for the current browser session and not stored on hard disk. When the browser is closed the cookie will be erased and the user has to log in again next time.

Although not explicitly required by our system, users should use secure HTTPS protocol for login, uploading and downloading. We have not made this requirement mandatory because even if a password is leaked, the most that can be done is downloading solution file or deleting a task. We have not considered that to be a major security problem as input data should still be present in the client computer.

The real security issue is that the system may not expose system resources that are not associated with task solving. Fortunately PHP is quite proven in that respect if proper programming practices are followed. One basic principle of the server-side web-interface security is that all input from the user may never be directly used as arguments with basic OS utilities like *cat*, *ls*, etc. (By substituting user input to arguments, it could be possible to use the system utilities for unintended purposes.) We have carefully avoided such possibilities by checking all input data coming directly from a user before passing it on to command line utilities.

Because our system allows a much smaller set of possible actions and exposes less resources than OS, we think that creating user interface accounts is far safer than creating real usernames in OS. A more detailed log of user actions can be also stored in the server.

3.3   System testing

Our main goal was to get the WWW-interface working for a wide range of problems in different input formats. We start with an example arising from the field of cardiography using finite difference approximations and conclude with some experiments with matrices in the elemental form.

Foucault cardiography problem. An actual scientific problem, the forward problem of Foucault cardiography (FCG) [9], has been chosen as one of the main problems for verifying the work of the system developed. Simulation models of genesis of the FCG-signal involve the calculation of the distribution of the eddy currents induced by electromagnetic field in the human thorax. Combining the finite differences method with the node voltage method of electrical circuits calculation, the problem can be reduced to a system of linear equations.

For the verification of the WWW-interface we used calculation task with reduced dimensions (4875x4875 matrix with 32109 nonzero elements). The system matrix is sparse, symmetric and positive definite.

Generation of the system matrix, righthand side vector and the node coordinates for the forward problem of FCG were perfomed in the MATLAB environment. The matrices, that should be supplied to the solver, were saved in MATLAB .mat-files. So the "New MATLAB form input" form should be chosen from the "DOUG computational queue" (Fig.2) to submit the calculation task.

The submission sequence of a computational task could be as follows. One has to choose an appropriate solver type to be used (CG), the solution accuracy, max. number of iterations, supply files, choose the format of the solution file to be stored (formatted ASCII / binary (MATLAB v5/6 .mat-file)), choose number of processors and write in the description of the task to be identified by in the queue (Fig.2). Then one has to press the "Process" button. PHP script will process the submitted form and the scheduler will start the calculation task. Following the link in "Description" field of the "DOUG computational queue" table (Fig.2), user can get to the "DOUG output" form. From here it is possible to observe visually the convergence of the solution, insert comments to a protocol and, finally, to download the solution file to the local computer.

When the interface is used from a remote network, most operations are performed at once without any noticeable latency. The slowest operation seems to be uploading of input datasets to the server. In our case this takes from several seconds to a few minutes depending on the size of our datasets. Downloading a solved task took up to one minute in our case. The queue interface is displayed almost instantly when the number of tasks in the queue is less than a hundred.

Other testproblems. We have tested the system on various other problems including a set of examples coming with the DOUG 1 code (see [4]). These include Poisson's Equation on a square, L-shaped domain and on a cube. We have tested the system also on non-symmetric problems.

4   Conclusions

There are several key benefits for having a WWW-interface for a computing package:

1. Better utilization of existing resources. By creating a central WWW-server with backend computing cluster, resources for task solving can be more equally divided among users.

2. Faster learning. WWW-based interfaces are become common and users have knowledge using them. Only the important options can be extracted and presented to the end user.

3. Simple progress tracking. Computing packages usually display a lot of information about solving progress. Unfortunately much of it has no use for an ordinary user. WWW-based interfaces can display only relevant information in a suitable format or even graphically.

4. Better security. Because such interfaces are more restricted compared to user accounts, there are also less potential security holes. User actions can be more easily tracked. System resources can be divided in a more complex way than it is possible with normal user accounts.

5. Availabilty and maintainance. WWW-browser is nowadays available on most computers and no additional software is required.

The implementation of the first version of user interface was conducted in one month. This included time for learning how to use the tools, programming and preliminary testing. We see this as a good result which proves that the chosen tools were at sufficent level. We believe that there is a large set of packages that could use similar user interface. The main future direction is to research how to generalize various parts of the user interface for other computational packages.

References

1. Cactus Code, http://www.cactuscode.org.
2. The Condor Project Homepage, http://www.cs.wisc.edu/condor.
3. I.Foster, C.Kesselman, S.Tuecke. The Anatomy of the Grid: Enabling Scalable Virtual Organizations. Intl. J. Supercomputer Applications, 15(3), 2001.
4. M.J.Hagger. Automatic domain decomposition on unstructured grids (DOUG). Advances in Computational Mathematics, 9 (1998) pp. 281-310.
5. M.J.Hagger and L.Stals. Domain Decomposition on Unstructured Grids User Guide v.1.98. Department of Mathematical Sciences, University of Bath, http://www. maths.bath.ac.uk/~parsoft/doug/userguide/userguide.html
6. MySQL Reference Manual for version 4.0.1-alpha, http://www.mysql.com/ documentation/mysql/bychapter/.
7. PHP Manual, http://www.php.net/manual/en/.
8. B.F.Smith, P.E.Bjørstad and W.D.Gropp, Domain Decomposition: Parallel Multilevel Methods for Elliptic Partial Differential Equations. Cambridge University Press, Cambridge, 1996.
9. J.Vedru, K.Skaburskas, S.Malchenko. Solving Simply the Forward Problem for Foucault Cardiography. - Proc. XI Int. Conf. Electrical Bio-impedance, Oslo, June 17-21, 2001. - Oslo, 2001. - pp. 565-568.

Workload Scheduler with Fault Tolerance for MMSC[*]

Juhee Hong, Hocheol Sung, Hoyoung Lee, Keecheon Kim, Sunyoung Han

Department of Computer Science and Engineering , Konkuk University,
1 Hwayangdong, Kwangin-gu, Seoul, 143-701, Korea
{jhhong, bullyboy, hylee, kckim, syhan}@kkucc.konkuk.ac.kr

Abstract. Media server of the MMSC(Multimedia Message Service Center) offers a function that converts sound and image for various types of handsets. MMS Relay requests a media type and format conversion and the media server handles this conversion. When MMS Relay requests a media conversion, a workload scheduler makes the job distributed properly and processed by several media servers. Since the whole system load can be distributed evenly, the performance of MMS system can be improved. The stability and the reliability are provided. In this paper, we propose and implement a workload scheduler in which jobs are processed distributively by monitoring the weights of the jobs in each media server. The weights are assigned to the jobs according to the conversion time. Also, we guarantee a fault tolerance capability of media server to retransmit the jobs fast when fault happens by monitoring the active/idle states and executing the processes.

1 Introduction

With the development of wireless communication and high-speed network, there is a clear need for a more advanced messaging service that allows mobile users to send and receive longer messages with richer contents. In mobile a message service, SMS (Short Message Service) allowing only plain text has been evolved to MMS (Multimedia Messaging Service) that accepts a mixture of different media types including text, image, sound and video. This new messaging service called MMS can be in the future mobile networks such as GPRS and UMTS [1][2][10]. In the multimedia messages, it is possible to perform a format conversion based on the characteristics of the handsets. It is a media server that takes the responsibility of the conversion when a MMS Relay requests a conversion. Since a multimedia message takes longer to convert than text, multiple media servers are needed for faster processing [1][3].

3GPP and WAP Forum specify the necessity of media server for this type conversion, however they don't specify the performance improvement through fast conversion of multimedia message explicitly [2][3]. Therefore, load balancing is necessary

---

[*] This work is supported by NCA(National Computerization Agency) of Korea for Next Generation Internet Infrastructure Development Project 2002.

P.M.A. Sloot et al. (Eds.): ICCS 2002, LNCS 2329, pp. 643–652, 2002.
© Springer-Verlag Berlin Heidelberg 2002

to prevent a overload in a specific media server when multiple media servers convert media formats.

In this paper, we propose and implement a workload scheduler that distributes jobs by monitoring the weight of the jobs in each media server and the scheduler runs the jobs according to weight assigned by the conversion time. The performance and the reliability can be improved in delivery the message. Also, we guarantee fault tolerance capability by retransmitting the job fast when a fault happens by monitoring active/idle states and executing the processes. We show enhanced performance in multiple media servers through our simulation tests.

The rest of this paper is organized as follows. In section 2, we introduce the overall MMSC component and architecture. In section 3, we present a design of workload scheduler supporting load balancing and fault tolerance. In section 4, we present our implementation, and in section 5, we compare and estate test results through the performance evaluation. Finally, section 6 gives our concluding remarks.

2 MMSC(Multimedia Message Service Center)

MMSC is a system for MMS as defined in our testing system based on 3GPP and WAP Forum. MMS is intended to provide a rich set of content to mobile subscribers in their messages, it supports both sending and receiving of such messages and assumes that messaging transmission and retrieval is supported by their handsets [1][3][10]. There are four basic types of equipment within MMSC, these are MMS Relay, MMS Server, Media Server, and Operation console [3][4].

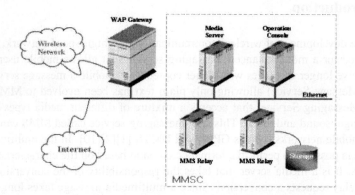

Fig. 1. MMSC(MMSC(Multimedia Message Service Center)

Fig. 1 shows an external configuration of our MMSC. The following explains each elements of Fig. 1.

● MMS Server
   This element is responsible for storing and handling the incoming and outgoing messages.

- MMS Relay

  This element transfers messages between different messaging systems associated with the MMS Server. It should be able to generate a charging data (CDR) when receiving multimedia messages or when delivering multimedia messages to the MMS User Agent or to another MMS environment [1][2].

- User Databases

  This is the system comprised of one or more entities that contain user related information such as subscription and configuration (e.g. user profile, HLR).

- User

  It is a function of application layer to provide the users with the ability to view, compose and handle multimedia messages such as sending, receiving, and deleting.

Depending on the business model, the MMS Server and the MMS Relay may be combined, or distributed across different domains. In practice, MMS system may be integrated in a single physical place and we explain it as a set of components for better understanding.

3   Design of Workload Scheduler

Media server is important and independent server in MMSC, because it performs a format conversion for the multimedia messages based on the characteristics of the handsets [1]. If one MMS Relay and one media server have been consisted separately, routing and scheduling issues between MMS Relay and media server aren't important. But as the necessity of messaging service increases, more than two MMS Relays and media servers are needed. Thus, the routing between MMS Relays and media servers is very important, because it influences the system performance. In order to support this routing function, there must be a monitoring capability in the media servers. Monitoring detects the faults and prevents an overloading problem.

Fig. 2 shows the scheduler suggested in this paper and it performs a load balancing for fast contents conversion in multiple MMS Relays and Media Servers. Because of the distributive processing for a request, it should include fault tolerance to handle the faults happened in a specific Media Server or process.

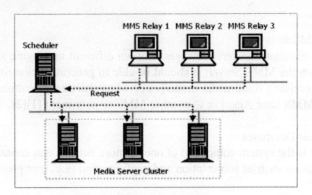

Fig. 2. System topology

We designed a workload scheduler based on the least-connection algorithm because the algorithm directs the network connections to server with the least number of established connections [6]. But the scheme suggested in this paper is based on the sum of the weight of jobs at work in each media server because the connections are within a local network. Weights are assigned for the fast media conversion because it takes different time for each media conversion. These weights are statistical values obtained through our tests for media conversion. Table 1 and 2 show the values.

Media server with the least of the weighted jobs is selected by the scheme below.
If there are n jobs at work in a media server and each job has a weight $Wi(i = 1, \ldots, n)$, the weighted job in that media server is expressed by $\Sigma(1 / W_i)$ $(i = 1, \ldots, n)$.
Therefore, the selected media server has a value of $min\{\Sigma(1 / W_i)\}$ $(i = 1, \ldots, n)$.

Table 1. Statistical value for image format type

(unit : count/second)

Source\Destination	GIF	JPEG	PNG	BMP	WBMP
GIF		5.1	5.0	5.2	4.2
JPEG	3.4		2.5	5.2	1.4
PNG	2.7	5.1		5.2	3.1
BMP	3.0	5.1	2.4		3.5
WBMP	0.2	5.1	5.0	5.2	

Table 2. Statistical value for audio format type

(unit : count/second)

Source\Destination	WAV	MP3	MID
WAV		3.9	2.9
MP3	5.0		2.0
MID	5.1	X	

A scheduler distributes the requests received from MMS Relay using the above scheme. Fig. 3 shows the whole processing diagram of the load balancing. Our system is consisted of two parts, a scheduler and a media server.

- Scheduler
  It is responsible for load balancing and fault tolerance, it creates a worker thread and processes the request.

- Media server
  It performs a media format conversion.

To select a media server with the smallest of weighted jobs, the scheduler retrieves media server list table and finds the states whether it is in active or idle for a candidate media server. This is repeated until the media server is selected to convert the request. When a media server is selected, weighted job field of the list table is increased. Monitoring the conversion state is made known by sending a signal to the media server. We have two signal types, one is a 'complete' signal when process finishes normally and the other is a 'fault' signals when an absurd end of process or other errors happen.

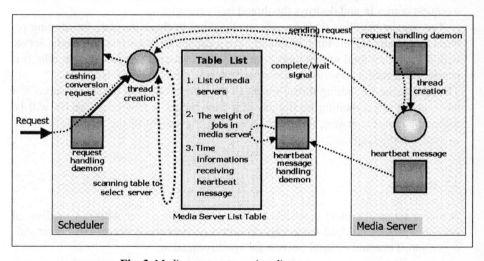

Fig. 3. Media server processing diagram

The current media server receives these signals and knows the conversion state. When a scheduler does not receive a signal in a given time, it regards it as fault and retrieves the next media server. When a complete signal comes to an end, job field in the list table decreases.

Conversion state of media server is reported to the scheduler through the periodic signal. A heartbeat message notifies that the server is active, and the scheduler updates time field of the media server in the list table. Updated time value is calculated from the receiving heartbeat message time plus(+) the given time intervals. If a heartbeat message does not arrive in time, the media server is regarded as fault and retransmits the requests to another server.

4 Implementation

4.1 Modules

The system consists of two parts, a scheduler and a media server. A scheduler forwards the requests to a media server using load balancing and fault tolerance mechanism. It consists of a request-handling module, a media server list table, and a heartbeat message-handling module. Media server appends a heartbeat message-sending module and a complete signal-sending module to monitor the state conversion process.

The request-handling module distributes the conversion requests to media server evenly. It is required to listen a request from MMS Relay and forward appropriately to a media server by the load balancing algorithm. It creates a new thread whenever a request comes in and destroys the thread in its completion.

The request-handling thread needs an algorithm to check the state for sending request to a media server. First, it should have a capability of retrieving a media server with the smallest total weight and checks the state, whether it is in active or idle. If it is selected, a media server will receive and execute the conversion request.

Then the request-handling thread increases the weight of job in the media server list table, and it starts monitoring the job. If a fault occurs, a new media server will be selected again. Otherwise, the thread decreases the weight of the jobs and it is destroyed automatically.

4.2 Structure

A media server list table is a table to retrieve a media server with the smallest of weighted jobs by the load balancing algorithm. It consists of Server ID, Server Address, Job, Time and Critical Section for synchronization.

Server ID is an unique number for identifying each media server and the Server Address is IP address. Job is the weighted value of the job in execution in each media server and Time is the expected time that the media server will be active for some times. The value of the Time is calculated from the incoming time value of heartbeat message added to the predefined interval value. During that time, a scheduler considers that media server is active. Otherwise it is idle. Critical Section is used for locking when the table is being updated.

Table 3. Media server List Table

Server ID	Server Address	Job	Time	Critical Section
Unique identification of each media server	IP address of each media server	The weighted value of the job in execution in each media server	The value of incoming heartbeat + predefined interval value	The variable for synchronization

Table 4. Media server List Table structure

```
typedef struct _ServerEnt {

        int     ServerId;

        SOCKADDR_IN        ServerAddr;

        unsigned int    uWgtOfJob;

        DWORD    dwTime;

        CRITICAL_SECTION        critical_sec;

} ServerEnt;
```

4.3 Fault Tolerance Mechanism

When a media server starts a service, it creates a thread for sending a heartbeat message and forwards it to a scheduler. If a media server succeeds in UDP connection to a scheduler, it starts sending a heartbeat message. Then, the request-handling thread executes UDP listening daemon to receive the message and updates the time field in the media server list table whenever it receives a message. If a heartbeat message does not arrive in time, the media server is regarded as fault, and retransmits the requests to another server.

A media server sends a complete-signal to Process Monitoring Daemon in a scheduler when the job terminates normally, otherwise it sends a fault-signal. We use '1' for a complete-signal and '0' for a fault-signal in this paper.

5 Test Results

We built a test bed in order to compare the performance with the legacy MMS system. So, we prepare five media servers, a scheduler and a client application for playing parts in MMS Relay. We used a Pentium server for media servers and a scheduler. We used windows 2000 compatible OS. Each media server and scheduler was located in a local network. Also, we had to prepare MMS storages for various media files. Each media server connected to MMS storage with a network drive of supporting OS. Therefore, each media server operates as if the source media file for conversion is in the same local drive. And we killed all the unnecessary processes in the media server and scheduler for an accurate test. In order to measures the performance, we have sent between 10 and 100 requests at a time to a media server for message conversion, and then we calculated the average processing completion times. The graph shows the results of this implementation.

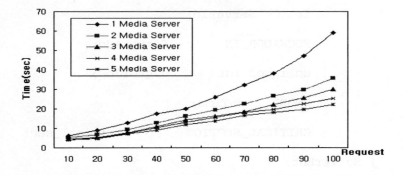

Fig. 4. Using with 1Mbyte wav file

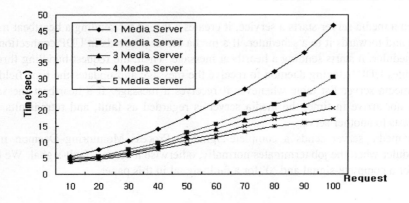

Fig. 5. Using with 500Kbyte wav file

We measured the processing time by increasing the number of media servers in order to compare the performance improvement resulting from our system. As for the various performance tests, we calculated the processing time with changing the file size of different media types. Figures, 4 through 7, show the result of the test. Fig. 4 shows the test result with 1Mbyte audio file. Fig. 5 to Fig. 7 shows the test result with 500Kbyte, 150Kbyte and 50Kbyte respectively. Clearly, the conversion time is different as the number of media servers, and the media server can convert very fast in case of small size of media files. In case of sending 10 requests, there are not different between single media server and five media servers as the graphs show. It spent more time for reaching requests rather than for converting a media file. The increased performance can be seen as the converting requests increased. And the bigger a media file size is, the greater the difference in converting time between a single media server and five media servers. It seems to be obvious that the same result applies to the image files and other audio format files.

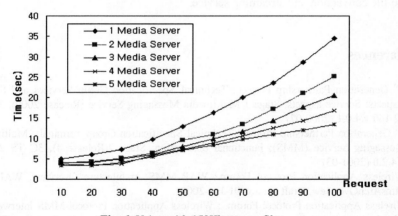

Fig. 6. Using with 150Kbyte wav file

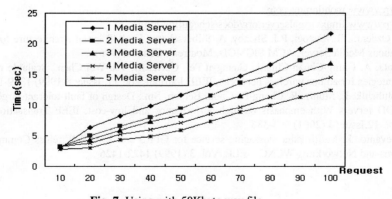

Fig. 7. Using with 50Kbyte wav file

As a result, we found that our system improves the performance dramatically than the legacy MMS system.

6 Conclusions

In this paper, we designed and implemented a workload scheduler by applying a least-connection algorithm in MMSC for MMS defined in our prototype system. The workload scheduler distributes job processing by monitoring the weights of the jobs in each media server to improve the system performance. We also implemented a fault tolerance that is a function of media server of retransmitting the job fast when a fault happens by monitoring the active or idle state of the media server and executing the processes. Test results present clearly that our scheduler performs a fast conversion and provides more stable and reliable system. Future work includes a test for video file conversion and streaming service.

References

1. 3rd Generation Partnership Project.: Technical Specification Group Services and System Aspects; Service aspects; Stage 1 Multimedia Messaging Service (Release 2000), 3G TS 22.140 v.4.0.1 (2000-07)
2. 3rd Generation Partnership Project.: Technical Specification Group Terminals; Multimedia Messaging Service (MMS); Functional description; Stage 2(Release 4), 3G TS 23.140 v.4.2.0 (2001-03)
3. Wireless Application Protocol Forum.: WAP MMS Architecture Overview, WAP-205-MmsArchOverview, Draft version 01-Jun-2000
4. Wireless Application Protocol Forum. : Wireless Application Protocol MMS Interworking with Internet Email Specification, WAP-207-MmsInetInterworking, Draft version 01-Jun-2000
5. http://www.mobilemms.com
6. http://www.linuxvirtualserver.org/docs/scheduling.html
7. B. Ozdenm, R. Rastogi, P.J. Shenoy, A. Silberschatz.: Fault-Tolerant Architecture for Continuous Media Server, ACM SIGMOD, Montreal, Canada(1996)
8. Iwata. A, Ching-Chuan Chiang, Guangyu Pei, Gerla. M, Tsu-Wei Chen.: Scalable routing strategies for ad hoc wireless networks, IEEE Journal, Vol. 17, Issue: 8(1999) 1369-1379
9. Golubchik. L, Muntz. R, Cheng-Fu Chou, Berson. Sm.: Design of fault-tolerant large-scale VOD servers With emphasis on high-performance and low-cost, IEEE Transactions on, Vol. 12,Issue: 4 (2001) 363-386
10. Sevanto. J.: Multimedia messaging service for GPRS and UMTS Wireless Communications and Networking, WCNC. IEEE, Vol. 3 (1999) 1422-1426

A Simulation Environment for Job Scheduling on Distributed Systems

J. Santoso[1,2], G.D. van Albada[2], T. Basaruddin[3], and P.M.A. Sloot[2]

[1] Dept. of Informatics, Bandung Institute of Technology
Jl. Ganesha 10, Bandung 40132 Indonesia
[2] Dept. of Computer Science, University of Amsterdam Kruislaan 403,
1098 SJ Amsterdam The Netherlands
[3] Dept. of Computer Science, University of Indonesia,
Kampus UI Depok, Jakarta Indonesia

Abstract. In this paper we present a simulation environment for the study of hierarchical job scheduling on distributed systems. The environment provides a multi-level mechanism to simulate various types of jobs. An execution model of jobs is implemented to simulate the behaviour of jobs to obtain an accurate performance prediction. For parallel jobs, two execution models have been implemented: one in which the tasks of the job frequently synchronise and effectively run in lock step and a second in which the tasks only synchronise at beginning and end.
The simulator is based on an object approach and on process oriented simulation. Our model supports an unlimited number of workstations, grouped into clusters with their own local resource manager (RM). Work is distributed over these clusters by a global RM. To validate the model, we use two approaches, analysing the main queueing systems and experimenting with real jobs to obtain the actual performance as a reference.

1   Introduction

Workstations are used as computing resources for various applications because of their computing power and cost-effectiveness. Integrating several workstations using a high speed network into a cluster can create a high performance computing environment, and allows users to share expensive resources. Combining several clusters into a wide-area system further increases the potential for of resource sharing. However, an efficient resource management system is needed to utilise the available resources optimally. The main resource management activities are allocating the resources for the applications and scheduling the applications to satisfy predefined performance criteria.

Designing an appropriate scheduling approach for such an environment is an open research question. Scheduling is an NP-hard problem even for homogeneous resources. A good scheduler should be able to satisfy the given constraints, to achieve the scheduling objectives, and to operate with low-overhead. The objectives of the scheduling can be diverse, to obtain a better load sharing, load balancing and to meet user performance expectations. In this case, no optimal schedule can be achieved easily.

P.M.A. Sloot et al. (Eds.): ICCS 2002, LNCS 2329, pp. 653–662, 2002.

The accepted way to study scheduling is through a combination simulation and experimentation. Hence, we have built a simulation environment that allows us to study the scheduling performances of parallel programs in a multi-cluster environment. Simulation models have been widely used to study computer systems and network properties [5, 8]. Damson [6] uses a simulation to predict the performance and scalability properties of a large scale distributed object system. Another example is HASE [1] that is used to simulate computer architectures at multiple levels of abstraction, including hardware and software. The model supports various operation modes, e.g. design, model validation, build simulation, simulate system, and experiments.

Our work differs from the above studies in the following aspect. We do not use our simulation model as the subject of study, but use it to support our study on job scheduling. Our model does not represent a specific system. The level of detail in the simulation is an important parameter that influences the accuracy of the results. Therefore, we decompose the system into several subsystems, and each subsystem can be modelled separately to the required level of detail. We have validated the model by comparing simulation results with the actual performance of PVM jobs and and by comparing with queueing theory.

The goal of our simulation system is to simulate high performance computing (HPC) and high throughput computing (HTC) environments. For HPC we evaluate the efficiency of the system, e.g. on basis of the response time and the slowdown of jobs that are collected when the jobs leave the system. For HTC, the number of finished jobs for a simulation run and total amount of work can be used as a criterion.

This paper is organised as follows. In section 2, our model of distributed system will be described as well as several scheduling strategies used. The functional design of our simulation environment is discussed in section 3, and the implementation is explained in section 4. In section 5, we describe the model validation and in section 6, we come to our conclusions.

2   Conceptual model

2.1   Distributed system

We restrict our model to distributed systems consisting of clusters of workstations(COWs). A distributed system is an aggregation of (possibly heterogeneous) resources, in clusters at geographically distributed sites, e.g an I-WAY system[4]. A cluster is assumed to be an autonomous system with its own policies in managing resources and in scheduling jobs.

The structure of the resource organisation determines the model of its scheduling system. Federated scheduling is used by Legion [10] to exploit remote resources in a wide-area system with the objective to provide high performance and capabilities not available locally. The Globus meta-computing environment [3] implements a scheduling framework that allows for co-allocation, i.e. the co-eval distribution of a job over multiple clusters. In Globus and Legion local cluster managers collaborate to realise global scheduling implicitly. Another approach is an hierarchical organisation [9] that divides the resource management system into global and local levels. The global management system explicitly provides several functionalities that are not available in the local managements system, e.g. global resource monitoring and wide-area scheduling. Our scheduling model is based on this hierarchical architecture.

In our simulation model, we define a COW as a collection of computing resource entities (workstations). Each entity has a set of attributes to indicate the characteristics of the workstation, like processing speed. Each entity has its own CPU scheduler, responsible for allocating processor time to tasks. Each COW has a local scheduler that assigns tasks to workstations in that cluster.

To model the applications running on the distributed system, we consider various synthetic jobs consisting of a number of tasks. We assume a sequential job to have only one task, whereas parallel jobs have at least two tasks. Each task comprises computation and communication phases. All tasks live for the entire duration of the job and represent an equal amount of work.

The synthetic parallel jobs can model a variety of communication and synchronisation patterns. Standard models include the master slave and lock-step SPMD paradigms. In the lock-step SPMD paradigm, the tasks of a parallel job synchronise frequently before resuming the next computation phase. In the master-slave model, the master task sends the work to all slave tasks and collects the output when the computation finishes.

After a job has been created, it is assigned to a cluster by the global scheduler. We will not consider the actual implementation of such a scheduler, but we will assume the existence of an abstract global scheduler. Multiple scheduling criteria can be considered, e.g. clusters are selected for specific applications based on their performance, reliability, connectivity and available resources.

2.2 Hierarchical scheduling

Hierarchical resource management divides the resource management system into several layers. We distinguish a global scheduling layer, a local scheduling layer, and a CPU scheduling layer. The global and local schedulers have different functionalities in several respects. The global scheduler is responsible for assigning arriving jobs to individual clusters by considering the imposed global policies to select the best candidate clusters. The local scheduler in a cluster is responsible for distributing tasks of jobs to a set of resources in its cluster. On those resources, the tasks are scheduled by the CPU scheduler that simulates the operating system scheduler. The local scheduler applies its local policies that need not depend on the external system. For an optimal performance, both global and local schedulers should cooperate and exchange information.

The goal of our scheduler is to optimise one performance measure, while keeping other measures within reasonable bounds. Initially, we will use crude schedulers, like SJF and determine their performance as measured against a large number of criteria. Later, we will experiment with more sophisticated schedulers that consider multiple criteria simultaneously.

Global scheduling The global scheduler is designed for top-level scheduling purposes. Newly arriving jobs will be scheduled by the global scheduler to available clusters. If no cluster is available, the jobs are placed in a global queue. When the resources are available the jobs are dispatched by the global scheduler according to some optimality criteria based on the available information on cluster capacity, load and/or job characteristics.

Local scheduling The local scheduler maintains a single queue to store jobs if they cannot be scheduled immediately. When resources become available, the waiting jobs will be dispatched. Each job consists of a number of tasks that does not vary throughout the lifetime of a job. The local scheduler allocates each task to a resource. By default, the tasks of a parallel job are distributed evenly among the available resources, starting from the least loaded resource. Round-robin scheduling as used by PVM is also modelled. .

The CPU scheduler The runnable tasks on each node are dispatched in a round-robin fashion. As the tasks in a parallel program are assumed to synchronise after every time step, a task becomes runnable again only when all other tasks in the job have advanced to the same point. Gang scheduling can also be simulated. The communications delay is assumed to be a constant value.

3   Functional Design of the Simulation Model

In principle, the simulation model and the modelled system have a similar structure. The components of the simulation model are derived from the components of the system to obtain a representative model and to improve the validity of the model. We have chosen to use an object oriented approach. The correspondence between objects and components allows the object oriented model to mimic the real system. Object oriented models can easily be expanded; a large system can be built incrementally, new object classes can be created from existing classes.

3.1   Interactions and object interfaces

The simulation is based on discrete events and is process oriented. In this approach the management of events is implicit in the management of the processes. We distinguish passive and active objects. The latter are responsible for the control of the internal simulation activities such as event scheduling, time advance and collecting statistics. These objects are associated with an independent thread of control at creation time and can generate events. In our model, there are two classes of active objects, the arrival class and the resource class. When a simulation is started all persistent objects are initialised. Next, the arrival process is activated, Fig. 1. The first task arriving at a resource will activate it. The various objects interact through method invocation and simulation primitives.

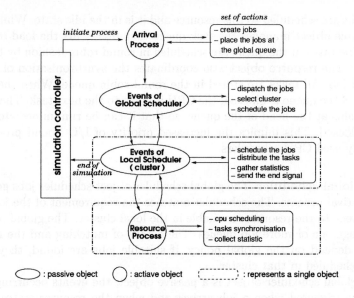

Fig. 1. Simulation process

3.2   Objects and Functionalities

To build our simulation environment, we have defined eight basic classes that represent the actual functionalities of the system.

Class Job and Class Task A *job* object is a passive object. It has a set of attributes describing its characteristics and states. Some attributes are initialised by the *arrival* object when the *job* instance is created; the other attributes are set during its life time in the simulation run, e.g. job response time and start of execution. The information is retrieved when the job leaves the system.

The tasks of a parallel job are also modelled as objects and they are created and initialised by the cluster object when the job starts execution. During the execution the tasks are controlled and scheduled by the resource object.

Class Arrival The *arrival* object generates synthetic jobs of various types until the end of the simulation. Jobs with their attributes are generated individually or in batches according to a stochastic distribution. The generated jobs are placed in the global queue waiting to be scheduled.

Class Resources A resource object represents a computing resource. It has a set of attributes that identifies its characteristics, e.g. speed, maximum load capacity, time quantum length, and synchronisation delay. The values are initialised by the cluster object. A resource object is activated by a cluster object when tasks are scheduled to this resource and it is in the idle state. While active, the resource object manages the task queue that represents the load of the resource. The tasks in the queue are scheduled in round robin fashion by the *CPU scheduler*. The resource object also coordinates the synchronisation of tasks in a parallel job. All tasks are stored in the round-robin queue. When the task at the head of the queue is blocked, the CPU is given to the next task. The blocked task remains at the head of the queue, so that it can be run immediately when it is unblocked. This mimics the increased priority of I/O bound processes in commonly used CPU schedulers.

Class Global scheduler The global scheduler object schedules jobs generated by the arrival process. The scheduler matches the requirement of the job in the global queue to the resources available in the local cluster. The global scheduler can manage one or more job queues. The details of matching and the resulting schedule depend on the global policy. If suitable jobs are found, they can be further scheduled in that cluster.

The global scheduler object is a passive object; the events occurring in this object are triggered when a job arrives and when the resource states change. Scheduling for a new arrival event is performed as follows. When a new job arrives, it is added to the appropriate queue. Next, all clusters are queried about the availability of resources and the job is scheduled, if possible.

Class Cluster A *cluster* is responsible for scheduling jobs, distributing tasks to the resources, and monitoring the state of the resources. All events occurring in this object are triggered by external method invocations from the global scheduler object and the resource objects. Scheduling is initiated when jobs are received from the global scheduler and when resources become available. The cluster maintains one local queue to store jobs that cannot be scheduled immediately. The jobs in this queue are dispatched when resources become available.

Class Job queue The global scheduler and cluster object use *job queue* objects to manage their job queues. Job queue objects have dispatching primitives that allow the global and cluster schedulers to apply various scheduling policies.

Class Task queue A *task queue* object manages the running tasks in the resources. This object has a set of methods to support CPU scheduling, e.g. dispatch the tasks, suspend running tasks, and modify the task state (*ready, blocked, waiting*).

4   Implementation

The simulation environment was developed using the C++SIM simulation package [2]. C++SIM provides an extensive set of simulation and statistical primitives. In a simulator it is important to provide for independent streams of random numbers for each stochastic variable. We added this feature in our system, based on the mtGen generator from Matsumoto [7]. A different random stream can be generated easily by varying the initial seed. The various independent streams are used to generate the random variates of the expected execution time, the job's inter-arrival time, and the job type distribution.

Simulation control and data collection The simulation environment provides a mechanism to collect the outputs within a sub-time-interval in a single long run. This enables us to make replications of runs without re-initialising the simulation parameters. Only the values within a sub-interval are sampled. Sampling periods can be preceded and separated by settling periods in which no statistics are obtained.

Configuring the model The model has a set of configurable parameters. The resource configuration is read from a file during the initialisation. The file describes the number of resources, the characteristics of each resource, and the organisation of those resources. Other parameters are entered as program arguments, e.g. random seed, mean inter-arrival time, scheduling policies and number of simulated jobs. Default values are provided when necessary.

5   Validation

The aim of validation is to ensure that the results of a simulation run can be accepted with enough confidence. It is not possible to prove that the model is absolutely correct. We have used two independent approaches for model validation. In the first approach, we analyse the queueing system at the local and global level for various configurations to verify its consistency with theory and to ensure that the schedulers operate correctly.

In another approach, we validate the model by comparing the simulation results with actual runs of PVM jobs. The objective here is to ascertain that our scheduling models provide an adequate representation for the more complex scheduling on our Unix system. For this purpose, we create the real PVM jobs with the same characteristics as those of the synthetic jobs used in simulation.

5.1   Queueing System : $M/M/4$ and $M/G/1$

For an $M/M/4$ queueing system, we compare the statistics of the job response times computed using the standard theory with those obtained from simulation for various arrival rates with a constant mean execution time. The results are shown in Fig. 2. We have used the following parameter values. The mean execution time $(1/\mu)$ is four unit times; the number of servers is four, the number of jobs for one simulation run is 4000. A single simulation result is obtained by performing 16 simulation runs using different random streams. This allows us to also obtain error estimates and standard deviations of the response times.

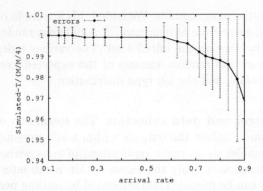

Fig. 2. The response time vs. arrival rate for $M/M/4$. The measured values have been normalised by dividing them by the predicted values from theory. The error bars are derived from the variations between the 16 runs contributing to a single measurement. As the same random streams were used for each arrival rate, the measurements at different arrival rates are not independent.

The results for the $M/M/4$ queueing system show that the simulation results are consistent with theory. From this we may conclude that the code has not made gross scheduling and accounting errors and that the random generators indeed approximate the desired distributions (this had, of course been verified before using more stringent tests).

We also performed a partial test of the treatment of multi-task jobs. As queueing theory considers only sequential jobs, we treat N-task jobs as sequential jobs using N times as much CPU time. Jobs are dispatched using an FCFS policy to a single workstation, and the workstation is allowed to hold one job at a time without preemptive scheduling. As we use a mix of sequential and parallel jobs, this results in an $M/G/1$ scheduling problem, for which the turn-around time can be easily calculated from theory. The simulation results for this configuration (not shown) were also found to be consistent with theory.

5.2   Validation with the PVM jobs

As the goal of our simulation is to describe the behaviour of real jobs in real systems, we have compared the turn-around times predicted by our simulator with those of a batch of simple, real parallel jobs run on a COW. We used PVM jobs (*master-slave and lock-step SPMD models*) with one to four tasks. The main difference between two models is the communication frequency between the tasks: the lock-step SPMD model synchronises at every time step (1 second); master-slave programs are assumed to communicate only at the beginning and the end. In both cases the tasks monitor their own CPU time use in order to ensure consistency with the simulation. Equal numbers of one, two, three, and four-task jobs were generated. The jobs were submitted directly to the workstation with the inter arrival time following an exponential distribution with means of 10, 15 or 20 seconds. We let the PVM round-robin task scheduler distribute the tasks to the workstations. The mean execution time of jobs was 10 seconds. The execution environment consists of four SUN Ultra-5 workstations.

The experimental results are shown in Fig. 3-a and Fig. 3-b. The communication delay of interacting tasks in the actual run has a high variance due to external factors, e.g background load and traffic intensity in the network. When the utilisation of resources is high, these effects are much more obvious, especially for the small jobs (see Fig. 3-b). Master-slave, which has less communications is modelled much more accurately than SPMD.

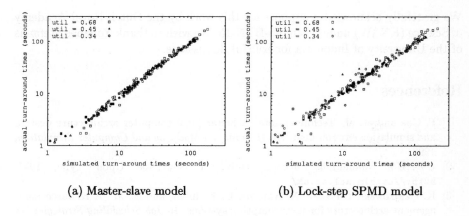

(a) Master-slave model   (b) Lock-step SPMD model

Fig. 3. Performance of simulation and actual run

6   Conclusion

We have built a simulation environment to study job scheduling strategies on a distributed system consisting cluster of workstations. The system is designed and built using an object oriented approach. Our goal is to obtain a better understanding of the interaction of resource elements and scheduling flows. The system incorporates some specific features to provide more flexibility.

- It supports several types of synthetic jobs, including serial and parallel.
- The global scheduler supports a multi queueing system.
- The resource structure and organisation can be configured easily.
- A variety of scheduling policies can be applied.
- A low level job simulation (*CPU scheduler*) is implemented.

With this model, we can study job and task scheduling simultaneously at a low cost, rather than experimenting in the actual system. The behaviour of the running jobs can be observed , which is useful in analysing the performance of the schedule. The individual resource performance can be monitored for dynamic scheduling and load balancing purpose. On the other side, there are simplifications that cause some outputs are not comparable with the actual performance.

- The synthetic jobs are simplified.
- The resources are limited to cluster of workstations.
- The effects of other resources (e.g. I/O, memory cache, network bandwidth, files, and file servers) are not addressed.

Acknowledgements

We gratefully acknowledge support for this work by the Dutch Royal Academy of Science (KNAW) under grant 95-BTN-15. We wish to thank Yohanes Stefanus of the University of Indonesia for fruitful discussions.

References

[1] P. Coe and et. al. Technical note: A hierarchical computer architecture desi gn and simulation environment. *ACM Trans. on Modeling and Computer Simulation*, 1998.

[2] *C++SIM Simulation Package*, Univ. of Newcastle Upon Tyne, 1997. http://cxxsim.ncl.ac.uk/.

[3] K. Czajkowski, I. Foster, N. Karonis, C. K. lman, and et. al. A resource management architecture for metacompting systems. In *Job Scheduling Strategies for Parallel Processing*, volume 1459 of *LNCS*, pages 62–82, Berlin, 1998.

[4] T. A. DeFanti, I. Foster, M. E. Papka, R. Stevens, and T. Kuhfuss. Overview of the I-WAY: Wide area visual supercomputing. *International Journal of Supercomputer Applications and High Performance Computing*, 10(2/3):123–130, 1996.

[5] M. Falkner, M. Devetsikiotis, and I. Lambadaris. Fast simulation of networks of queues with effective and decoupling bandwidths. *ACM Transaction on Modeling and Computer Simulation*, 9(1):45–58, January 1999.

[6] S. Frolund and P. Garg. Design-time sim. of a large-scale distr. object system. *ACM Trans. on Modeling and Computer Simulation*, 8(4):373–400, October 1998.

[7] M. Matsumoto and T. Nishimura. "mersenne twister: A 623-dimensionally equidistributed uniform pseudo-random number generator". In *ACM Transactions on Modeling and Computer Simulation*, volume 8. ACM, January, 1998.

[8] P. Spinnato, G. van Albada, and P. Sloot. Performance prediction of n-body simulations on a hybrid architecture. *Computer Physics Communications*, 139:33–34, 2001.

[9] A. van Halderen, B. Overeinder, P. Sloot, R. van Dantzig, D. Eperma, and M. Livny. Hierarchical resource management in the polder metacomputing initiative. *Parallel Computing*, 24(12/13):1807–1825, November 1998.

[10] J. B. Weissman and A. S. Grimshaw. A federated model for scheduling in wide-area systems. In *Proceedings of the Fifth IEEE International Symposium on High Performance Distributed Computing*, pages 542–550, Syracuse, NY, Aug. 1996.

ICT Environment for Multi-disciplinary Design and Multi-objective Optimisation: A Case Study

W.J. Vankan, R. Maas and M. ten Dam

National Aerospace Laboratory - NLR
ICT Division
Anthony Fokkerweg 2
1059 CM
Amsterdam
The Netherlands
vankan@nlr.nl

Abstract. This paper presents an ICT environment for multi-disciplinary design and multi objective optimisation in which a set of software tools are available for evaluation and approximation of objective functions and for proper control of several optimisation algorithms for multi objective optimisation. The ICT environment provides easy access to the relevant resources in the computer network via a Java based user interface, which can be executed stand-alone or as a Java applet based web client. As an example of application of this environment, some results of a multi objective optimisation study of a blended-wing-body aircraft configuration are shown.

Introduction

Technical product design is hardly ever an exclusively mono-disciplinary exercise. It mostly requires analyses of several different aspects of the product that are often related to different research areas or technical disciplines. For example in aeronautics, the design of an aircraft is determined by many different aspects, of which some important ones are aerodynamic behaviour, structural mechanical properties and flight mechanical and control properties [1]. For each of these aspects there is a specific technical discipline that can deal with highly detailed investigation and analysis of the design in its specific area. Such investigation is usually performed by mono-disciplinary specialists who make use of a variety of dedicated analysis tools. In multi-disciplinary design it is required that these mono-disciplinary investigations effectively contribute to the integrated multi-disciplinary design process [12].

In the EU project MOB [13] the multi-disciplinary design and optimisation of a blended-wing-body (BWB) aircraft configuration is considered. For this purpose an engineering environment for multi-disciplinary design analysis is developed in MOB. This so-called computational design engine (CDE) shall be a flexible system, providing easy access to the partners in the MOB project and allowing them to integrate and interconnect their design and analysis tools. The CDE facilitates the multi-disciplinary design process for the BWB by transparent presentation of the necessary analysis tools, data and documents to the CDE users. The CDE also offers

P.M.A. Sloot et al. (Eds.): ICCS 2002, LNCS 2329, pp. 663–672, 2002.
© Springer-Verlag Berlin Heidelberg 2002

certain facilities for optimisation of the BWB design, based on an efficient but traditional approach.

When considering product design as an optimisation process where the design objectives are optimised subject to certain design constraints, multi-disciplinary design leads to mutually competing objectives and constraints that arise from the different disciplines and must be dealt with simultaneously. Such multi objective optimisation essentially differs from traditional single-objective optimisation. In this case, global or local optima do not exist for all design objectives simultaneously and hence a possibly large set of compromised optimal designs, the so-called Pareto front [8] is searched for. This requires specific search techniques, which can be considered as generalisations of single objective optimisation methods [4].

An ICT (Information and Communication Technology) environment for multi-disciplinary design and multi objective optimisation has been developed in which a set of software tools are available for proper definition and approximation of objective functions and for easy control of several optimisation algorithms for multi objective optimisation. This paper presents an application of this environment to a multi objective design case of a BWB aircraft. The underlying middleware technology that is used to create this ICT environment is briefly addressed. The same underlying middleware is used in the MOB CDE, which enables easy exchange of components between these two environments. The set up of this study, in particular the methods for approximation and multi objective optimisation that have been used, will be briefly described, and some results of the design optimisation are shown.

ICT environment for MDO

Multi-disciplinary design and optimisation (MDO) requires close co-operation of a number of different technical disciplines, in each of which a variety of software tools is available for design and analysis simulations that run on specific computer platforms. The variety of software tools and the associated heterogeneous computer infrastructure and data formats impose certain complexity to close multi-disciplinary co-operation from which the end user in the MDO process should be shielded as much as possible.

Present technologies like CORBA (Common Object Request Broker Architecture), Java and WWW provide possibilities to build end-user oriented, integrated multi-disciplinary design environments. These technologies can be incorporated into middleware systems in order to facilitate interoperability in MDO environments, which are usually operated on distributed heterogeneous computer networks.

SPINEware is such a middleware system that supports the construction and usage of so-called working environments on top of heterogeneous computer networks [2]. A SPINEware-based working environment can be tailored to particular end usage and application areas. A SPINEware working environment provides uniform and network-transparent access to the information, applications, and other resources available from the computer network presented to the user through an intuitive GUI: the SPINEware User Shell. SPINEware also supports web-based access to SPINEware object services via a Java applet implementation of the SPINEware User Shell [10].

In the MOB project SPINEware is being used for the development of the CDE, which is applied to the detailed design and optimisation of the BWB aircraft configuration. This CDE is set up as a distributed environment that is developed from contributions of each of the MOB partners. The main contributions of NLR to the CDE are in the areas of high fidelity aerodynamics, structural mechanics and flight mechanics. Other areas are contributed by other partners. For example, CAD and geometry generation are provided by the University of Delft and low fidelity aerodynamics by the University of Cranfield. Currently, the CDE components from NLR and the universities of Delft and Cranfield are operational and can be used in an integrated multi-disciplinary design process of the BWB (Fig. 1).

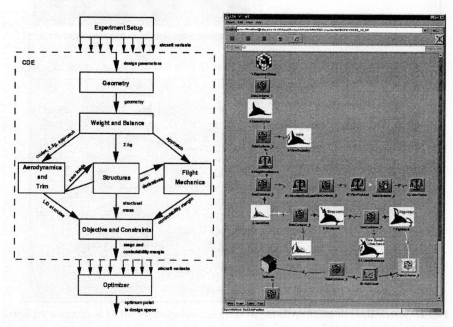

Fig. 1. Schema of the multi-disciplinary BWB design process with the currently operational MOB CDE (left) and its implementation in a SPINEware workflow (right).

At NLR the SPINEware middleware system is also being used to build user-oriented collaborative ICT environments, each dedicated to a certain application area. One such environment has been developed for the field of MDO. A set of software tools required for MDO, multi objective optimisation and approximate modeling have been integrated into this environment, and can be accessed from and executed on different computers of the heterogeneous NLR computer network. The environment is accessible via a standard web browser and comprises facilities for easy tool integration, tool manipulation and tool chaining, job and queue management, and remote operations where CORBA is used for the communication over the network. With this environment a multi objective optimisation study of the MOB BWB aircraft has been performed, where some of the analysis tools of the MOB CDE have been used (Fig. 2).

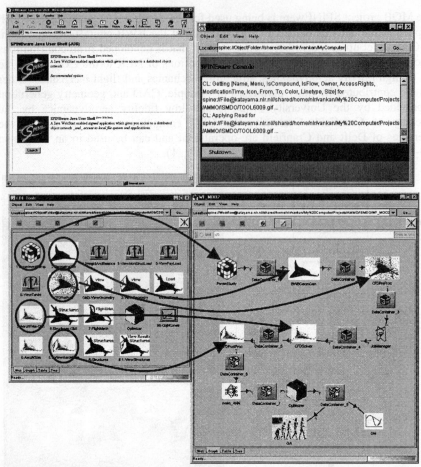

Fig. 2. Screen shots of the SPINEware working environment for MDO as accessed via a web server and using Microsoft Internet Explorer 4.0 as web client. The upper left panel shows the login web page, the upper right shows the SPINEware Console window. In the lower left panel a tool library of the MOB CDE is shown, and the lower right shows the workflow (or tool chain) for the multi objective optimisation of the BWB MDO aircraft, in which some of the CDE tools have been incorporated.

Multi objective optimisation algorithms

In the past decades a strong development of many different optimisation algorithms has taken place. A variety of optimisation methods, ranging from traditional gradient based optimisation methods (GM) to genetic- or evolutionary algorithms (GA), are currently widely available. In general, most GM are typically designed for single objective optimisation, while GA are more suitable for multi objective optimisation. Nevertheless it is desirable to be able to apply both methods to multi objective optimisation problems, because both have certain specific advantages. In general, the advantage of GA is that such algorithms have good global search capabilities, while GM easily get stuck in a local optimum. GM, on the other hand, are generally more efficient than GA in terms of the number of objective functions evaluations that is required for finding an optimum.

The application of both methods to multi objective optimisation problems is not straight-forward, and proper definition of the objective functions and effective control of the optimisation algorithm are required. In aeronautic multi-disciplinary design the evaluation of the objective functions can be computationally very expensive, for example in the case of CFD simulation of the aerodynamic behaviour. Hence a computationally much cheaper approximation model is required for such objective functions because of the large number of objective functions evaluations that is needed in multi objective optimisation [14]. Moreover, in the case of GM the multi objective optimisation process requires flexibly alternating combinations of objective and constraint functions, which can be achieved by specific control of the optimisation algorithms.

The considered MDO environment contains some specific tools for approximate modeling and multi objective optimisation. The approximate modeling tool makes use of artificial neural networks (ANN) as available from the Matlab Neural Network Toolbox [7]. Upon execution, the tool presents a user interface that guides the user through the process of setting up the ANN approximate model (Fig. 3). For multi objective optimisation, a GA and four algorithms based on GM, as available from the Matlab Optimisation Toolbox [6], can be used. The GA tool [11] applies a more or less standard MOGA technique [5], including elitism strategy and options such as niching (on both input and objective) and mating. The multi objective optimisation tool provides an interface to the four different GM based optimisation algorithms (Fig. 3).

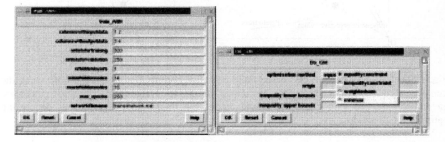

Fig. 3. The interfaces for the ANN and multi objective optimisation tools. Left: the option menu for the training of the ANN. Right: the option menu for specification and control of the GM based multi objective optimisation.

Two out of the four gradient-based methods are presented in this study. The first method is the minimax algorithm, more specific its Matlab implementation *fminimax*. A special strategy is applied by the multi objective optimisation tool in order to obtain the Pareto front instead of the standard single point minimax optimum. In this strategy it is assumed that each objective is positive. Furthermore it is assumed that an initial set of feasible design parameters with their corresponding objective values is available. The Pareto rank-one subset of this initial set is then used as demand values; this rank-one subset is the set of compromised optimal designs, also called the set of non-dominated solutions. The minimax optimisation process will then result, upon convergence of the method, in a point P on the Pareto front [9] (eq. (1), Fig. 4). In formula it reads:

$$P_j = c(p_j^{opt}) \text{ with } p_j^{opt} = \underset{p \in S}{\textbf{arg min}} \, \underset{i}{\textbf{max}} \, \frac{c^i(p)}{c^i(q_j)} \qquad (1)$$

where P is the point on the Pareto front, c is the multi-dimensional objective function, i denotes i-th the component of this objective function, S is the design parameter space, q_j is a design parameter vector corresponding to a rank-one design. If this is repeated for every rank-one point in the design parameter space, a set of points on the Pareto front is obtained.

In the second gradient based method, equality-constraint, the rank-one points of the initial set of feasible design parameters are again the starting point, but here only one objective is minimised while keeping the other fixed (Fig. 4). The implementation that has been chosen is a constrained minimisation using the Matlab function *fmincon*.

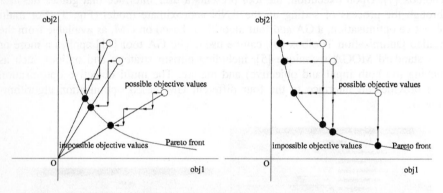

Fig. 4. Illustration of the two GM based multi objective optimisation processes. Left: using the minimax algorithm; right: using the constrained minimisation algorithm. The open circles are the initial objective values and the closed circles the final values. The arrows denote the optimisation processes.

MDO of a blended-wing-body aircraft

In the MOB project a detailed MDO study is conducted on a new BWB aircraft configuration. As a preliminary design analysis, a multi objective optimisation study is applied to some key properties of the BWB in cruise flight: aerodynamic performance (C_{dl}) based on the lift and drag coefficients (C_l and C_d), structural mechanical wing loading (M_l) based on the roll and yaw moments at the centre of mass (M_r and M_l), and flight mechanical unbalance (M_{pA}) based on the pitching moment at the centre of mass (M_p). These properties are represented by the following three objective functions that are minimized:

$$Cdl = \frac{Cd}{Cl}$$

$$Mt = \sqrt{Mr^2 + My^2} \tag{2}$$

$$MpA = |Mp|$$

The BWB design parameters that are varied relative to the BWB reference configuration, are the wing twist, wing sweep, and angle of attack in cruise flight (Fig 5).

Reference BWB   Twist variation   Sweep variation   Angle of attack

Fig. 5. Illustration of the BWB reference configuration, and the design parameters twist, sweep, and angle alpha used in the BWB pre-design study.

The objective function values can be evaluated from simulations of the airflow around the BWB under the cruise flight boundary conditions using a CFD solver for Navier-Stokes equations. However, because these simulations involve large scale and time consuming calculations for CAD geometry re-generation, flow domain discretisation and CFD computation, the ANN tool is used for approximation of the considered BWB properties (C_1, C_d, M_r, M_y and M_p) from which the objectives C_{dl}, M_t and M_{pA} are derived. For this purpose a data set of these properties is generated for the relevant ranges of the design parameters by flow simulations in a limited number of design points. The ANN has been trained using the resulting data set of 342 design points, where the three design parameters are used as ANN inputs and the five BWB properties as ANN outputs. Afterwards the outputs are combined to yield the objectives C_{dl}, M_t and M_{pA}. To train the network the input set is split in separate training, validation and test sets of 250, 50 and 24 design points, respectively [3]. The ANN is a feed-forward network and the optimal number of hidden nodes is nine, which has been automatically determined within a specified range by the ANN tool. Furthermore it has been found that one hidden layer was sufficient and that within the relevant range of design parameters the approximation is acceptable with an error of about 1 %.

Once the ANN is available, the multi objective optimisations can be performed with the different optimisation algorithms. The GA is run with a population of 100 and for 30 generations. The initial population is taken randomly from the 342 input vectors and will contain about 22 rank-one points in the output space. For both the minimax method and the equality-constraint method those inputs of the training set are chosen which have rank-one outputs. There are 55 of those input vectors.

For each individual method, after the optimisation is run, the rank-one points are selected and taken as approximation of the Pareto front. The results are depicted in Fig. 6. In Table 1 the approximate number of function calls, the total number of points resulting from the optimisation, and the number of points in the non-dominated sets are shown.

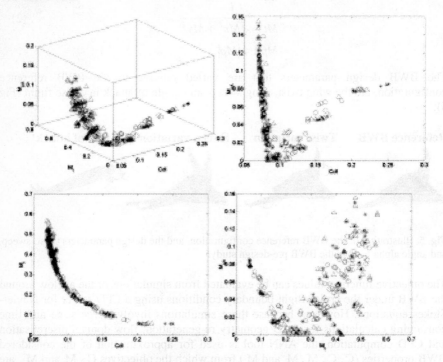

Fig. 6. The results of the different optimisation methods are depicted here. The C_{d1}, M_t and M_{pA} obtained by the GA (o), minimax method (*) and the equality-constraint method (Δ) are depicted in 3D and in 2D projections.

Table 1. Some numbers about the performance of each individual method

Method	GA	minimax	equality-constraint
Total nr of points in final result	100	55	165
Nr of non-dominated points	100	54	113
Number of function calls (about)	4900	2000	6400

The rank-one points for each method can than be compared with the rank-one points of the other methods. Then sub-optimal solutions (i.e. not rank-one) can then be found. In Table 2 the results are summarised. One sees that every method yields some sub-optimal points.

Table 2. The number of non-dominated and dominated points in set #1 when comparing the set with set #2. The numbers left of the slash are the non-dominated points, while those right of the slash are the dominated points.

set #1	set #2		
	GA	minimax	GM
GA	-	96/4	93/7
Minimax	51/3	-	51/3
GM	108/5	100/13	-

Discussion and conclusions

In this paper an ICT environment for multi objective optimisation and multi-disciplinary design has been presented. Special attention has been paid to the functionality for multi objective optimisation and approximate modeling. Several methods for multi objective optimisation, based on GA and GM, and a method for approximate modeling based on ANN, were implemented in this environment. Furthermore, several tools for specific aeronautic design analyses have been adopted from the MDO environment (the CDE) that is developed in the MOB project.

The ICT environment has been applied to a predesign case of the BWB aircraft configuration from the MOB project. In the multi objective optimisation analysis an ANN approximation of the CFD calculations has been made, and an initial comparative study of the use of the GA and two of the GM based methods for MOO has been performed. The results indicate that the different multi objective optimisation methods provide comparable results at comparable computational cost. However, the robustness and global search capabilities of the GA compete with the higher efficiency of the GM due to the smaller size of the required sets of design points. Therefore these methods have complementary functionality.

Acknowledgements
The support of Dr. M. Laban and Dr. J.C. Kok in setting up the CFD simulations of the BWB with NLR's Enflow system is gratefully acknowledged. This study has been partially supported by the EU project MOB, contract number G4RD-CT1999-0172.

References

1. Allwright, S.: Multi-discipline Optimisation in Preliminary Design of Commercial Transport Aircraft. In: Proc. ECCOMAS'96 Conference, Paris (1996)
2. Baalbergen, E.H., van der Ven, H.: SPINEware – A Framework for User-Oriented and Tailorable Metacomputers, Future Generation Computer Systems 15 (1999) 549-558
3. Bishop. Chr. M: Neural Networks for Pattern Recognition. Clarendon Press, Oxford (1997)
4. Fonseca, C.M., Fleming, P.J.: Genetic Algorithms for Multi-Objective Optimization: Formulation, Discussion and Generalization. In: Genetic Algorithms: Proc. of the Fifth Int. Conf., San Mateo, CA, USA (1993) 416-423
5. Fonseca, C.M., Fleming, P.J.: An Overview of Evolutionary Algorithms in Multi-Objective Optimization. Evolutionary Computing 3 (1995) 1-16

6. The MathWorks: Matlab Optimization Toolbox. http://www.mathworks.com/products/optimization.
7. The MathWorks: Matlab Neural Network Toolbox. http://www.mathworks.com/products/neural.
8. Pareto, V.: Manuale di Economia Politica. Sociata Editrice Libraria, Milan, Italy (1906).
9. Sawaraki, Y., and Nakayama, H.: The Theory of Multiobjective Optimization. Academic Press, Inc, (1985)
10. Schultheiss, B.C., Baalbergen, E.H.: Utilizing Supercomputer Power from your Desktop. HPCN 2001 Conference, Amsterdam (2001)
11. Tan, K.C.,: MOEA - Multi-objective Evolutionary Algorithm. http://vlab.ee.nus.edu.sg/~kctan.
12. Vogels M.E.S., Arendsen, P., Krol, R.J., Laban M., and Pruis, G.W.: From a Mono-disciplinary to a Multidisciplinary Approach in Aerospace: As Seen from Information and Communication Technology Perspective. In: ICAS 98, Melbourne, Australia (1998)
13. "MOB: A Computational Design Engine Incorporating Multi-disciplinary Design and Optimisation for Blended-Wing-Body Configuration". EC 5th Framework Programme, Contract nr. GRD1-1999-11162 (2000).
14. Shyy W., Nilay P., Vaidyanathan R., Tucker K.: Global design optimization for aerodynamics and rocket proulsion components. Progress in Aerospace Sciences 37 (2001) 59-118.

A Web-Based Problem Solving Environment for Solution of Option Pricing Problems and Comparison of Methods

Minas D. Koulisianis, George K. Tsolis, and Theodore S. Papatheodorou

High Performance Information Systems Laboratory,
Computer Engineering & Informatics Department,
University of Patras, 26500, Patras, Greece
{mik,gkt,tsp}@hpclab.ceid.upatras.gr

Abstract. In this paper we present a Problem Solving Environment (PSE) for solving option pricing problems and comparing methods used for this purpose. An open underlying library of methods has been developed to support the functionality of the proposed PSE. PHP, a server-side, cross-platform, HTML embedded scripting language, is exploited to make the proposed environment available through the World Wide Web. The PSE is not addressed to expert users only. It is simple in use, fast and interactive, proposing dynamically selections depending on user's input. The output is returned in "real time", in either simple or graphical representation.

1 Introduction

In recent years a very important progress in the area of option pricing methodologies and implementations has been observed [3]. Large financial institutions and many researchers have developed their own methods to solve a wide range of option pricing problems, as fast and accurately as possible. However, availability of these methods for the single user/investor is in question, as there are obviously great benefits for those institutions/investors using the best methods.

In the work presented in this paper, a significant number of option pricing algorithms have been implemented, providing a library of solution methods. In more detail, this library contains popular, widely applied methods [5, 17], as well as new state-of-the-art methods, for some categories of options, proposed recently by the authors [11–13]. To make such a library useful to a wide community of users, we need tools to provide remote access. For this reason, a Problem Solving Environment (PSE) has been developed, providing services for remote submission and monitoring of results from a Web-browser. The proposed PSE is not addressed to expert users only. The end-user poses the problem through a series of high-level definitions. The PSE facilitates this procedure providing the necessary guidance. After processing the input data, the server returns the results in simple arithmetic form (e.g. option prices) or through graphical representations (e.g. charts). A PHP platform [1] is used to enable communication between the PSE and the underlying option pricing library.

P.M.A. Sloot et al. (Eds.): ICCS 2002, LNCS 2329, pp. 673–682, 2002.

The structure of the paper is as follows. In Section 2 we provide some option pricing basics. The main characteristics of the proposed PSE are discussed in Section 3, while the underlying library of methods supporting the PSE is described in Section 4. The detailed definition and solution process is given in Section 5. Finally, conclusions and future work are stated in the last Section.

2   Option Pricing Basics

Options are kind of financial derivative products, which are based on one or more underlying assets [17]. The underlying assets include stocks, indices, foreign currencies, commodities and other. Stock options were first traded on an organized exchange in 1973. Since then there has been a dramatic growth in options markets and they are now traded on many exchanges throughout the world. Huge volumes of options are also traded by banks and financial institutions.

There are two basic types of option contracts, calls and puts. A call (put) option gives the holder the right, not the obligation, to buy (sell) the underlying asset, by a certain date, for a certain price. The price of the contract is known as the exercise price; the date in the contract is known as the expiration or exercise date. American options can be exercised at any time up to the expiration date, while European options can be exercised only on the expiration date itself. Most of the options that are traded on exchanges are American.

The process of calculating the fair price for which an option should be offered in the financial markets is known as the option valuation or option pricing problem. A number of input parameters are required and play significant role in this calculation, namely risk-free interest rate, volatility of the option, dividend yield, exercise price and expiration date.

In the current work we concentrate on stock options. A mathematical formulation commonly used for this class of options, is based on the well-known Black Scholes model [2]. Most of the times it is difficult to calculate closed form solutions [4, 17], (e.g. for American options) which are of course favorable whenever they exist. In case that closed form solutions can not be obtained, numerical methods comprise an efficient approach [3, 17, 5, 8]. Other approaches include binomial or multinomial trees and simulation methods [5, 17].

3   Proposed Problem Solving Environment

Generally speaking, a PSE must satisfy a number of basic requirements [9]:

1. *Ease of use even by non-expert end-users.*
2. *Wide applicability.*
3. *Guided and user-friendly problem definition, including on-line help, validity checks, dynamic PSE-user interaction and error handling.*
4. *Presentation and analysis of results in a meaningful and understandable way.*

A significant number of environments exist for solving option pricing problems, each of which imposes its own restrictions, has its own advantages and uses its own underlying methods. Many of them are intended for professional use by stock-exchange companies, financial investment houses, etc. and as a result they are not offered for use by the single user/investor.

However, individual researchers, and a small number of financial companies, have developed PSEs that can be accessed through the World Wide Web. Their use and applicability are most of the times very restricted, because they deal with a small number of cases. Morever, the methods used in some cases may become slow and inefficient or even unable to achieve the desirable accuracy. Representative underlying platforms supporting existing environments include Java, JavaScript, or even Plain Html.

The PSE presented in this paper, first of all satisfies all basic requirements mentioned above. More than that it has a number of very important advantages over the existing environments just mentioned:

1. *Underlying library of methods:* An underlying library of algorithms constitutes the basic component of the proposed PSE. A wide range of methods from traditional to new, state-of-the-art, methods, have been implemented and comprise the basis of this library. The user is given the ability of selecting methods and defining not only financial but also, "numerical" parameters, interfering in this way directly with the underlying library.

2. *Availability through the Web:* An efficient web technology is used to support implementation of the proposed PSE under the World Wide Web. More specifically, a PHP platform enables communication between the underlying option pricing library and the end-user working on a Web-browser. The availability of the PSE through the web, facilitates access from different places and long distances. At the same time, due to the PHP platform used, user submissions are satisfied in "real time". It should be mentioned that the proposed PSE is independent of the browser platform used and has been tested and works properly with the most common browsers.

3. *Simplicity / Guidance / Functionality:* The user defines his problem through a series of simple steps. The PSE provides the necessary guidance, interacts dynamically with him/her and proposes selections that depend on input. Morever, mechanisms like on-line help or a variety of validity checks ensure for the proper functionality of the PSE.

4. *Graphical Representation of the Output:* The results returned to the user are not exclusively of numeric type. Most of the times, a graphical representation of the output is displayed.

5. *Open Architecture:* The proposed PSE is open. The underlying library of methods can be enriched by addition of new ones. The web front-end can be easily extended to support new methods.

As a drawback, one can mention the absence of an underlying expert system. The PSE proposes a list of methods, not the best one. To overcome this deficiency, a comparison menu is provided through which the end-user has the option to compare methods of the same type for a given problem. The resulting

charts of execution times lead directly to conclusions about the efficiency of the
compared methods.

The proposed PSE can be accessed through a typical Web-browser. Its architecture is apparent in Figure 1. First the user is guided to define the problem
and choose from a variety of proposed solution methods. At the same time,
parameters that are critical for the solution process must be entered. After all
necessary input has been provided, the problem is directed to the server for solution. There, a Web-server takes all responsibility to call the required methods
from the underlying library and pass the necessary input parameters to them.
The output produced is finally returned to the browser. Typical output consists
of option prices and other valuable information like execution times. As already
mentioned, communication between the browser and the server is based on an
underlying PHP platform the graphical capabilities [1] of which are exploited to
produce chart representations.

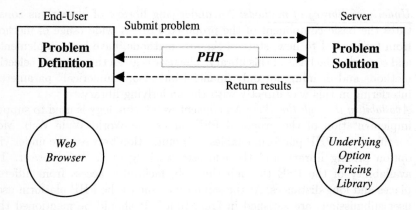

Fig. 1. Proposed Problem Solving Environment Architecture

4   Underlying Option Pricing Library

A plethora of stock option pricing methods have been developed to support the
proposed PSE. Generally speaking, they can be enlisted in two basic categories:
popular, widely used methods that can be found in the literature of the past
three decades, and new methods, proposed recently by the authors. Up to this
time the library contains methods for pricing European and American options
on stocks. A categorization of these methods is given in Figure 2.

For European options, three basic categories of methods have been developed, namely closed form, grid and tree. Closed form solutions come directly
from the Black-Scholes Partial Differential Equation [2]. The other categories
are essentially based on numerical approximations. Tree methods (binomial and
multinomial) are very easy to implement but they are generally inefficient. Grid

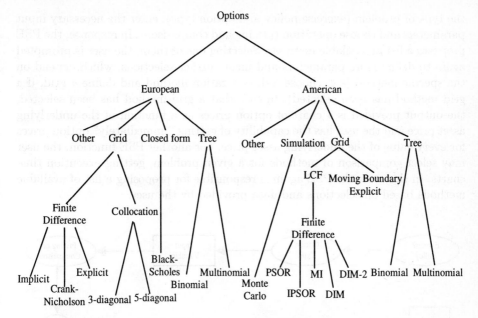

Fig. 2. Categorization of methods supported by the underlying option pricing library

methods have the advantage of providing option prices for a range of underlying asset prices based on a preselected grid; a discretization method plays significant role in the efficiency and performance of these methods. Discretizations supported by the underlying library up to now include Finite differences [17] (explicit, implicit and Crank Nicholson) and Collocation [15] (3-diagonal and 5-diagonal).

For American options, again grid methods, either based on Linear Complementarity Formulation (implicit) [6, 10] or making explicit use of the moving boundary (explicit) [14, 16] as well as tree methods (binomial and multinomial) [17, 5] have been implemented. The library contains five implicit grid methods: two iterative techniques, namely Projected Successive OverRelaxation (PSOR) [7] and Improved PSOR (IPSOR)[12], and three direct techniques, namely Moving Index (MI) [11], Direct Inverse Multiplication (DIM) and Stable DIM (DIM-2) [13]. Monte Carlo simulation can be also used, although it is not an efficient approach [17].

As already mentioned, the underlying library is open, and new methods can be added at any time to cover other kinds of financial derivatives (e.g. options on indices, currencies or commodities with simple or exotic payoff functions [5]).

5   Problem Definition and Solution Process

In this Section, we describe the sequence of the steps followed for problem definition and solution. As described in Figure 3, the user is first prompted to select the type of problem (exercise policy and option type), enter the necessary input parameters and choose operation (pricing and comparison). In response, the PSE proposes a list of available methods. Selecting one of them, the user is prompted again to define more parameters and make further selections, which depend on the specific method (e.g. choose a discretization method and define a grid, if a grid method has been selected). In case that a grid method has been selected, the output provided is a chart of option prices as a function of the underlying asset price and the user has the capability of getting "immediately" option prices for every value of the underlying asset price. For another PSE function, the user may select comparison of methods for a given problem, getting execution time charts. In this case, the PSE again is responsible for proposing a list of available methods based on selections and data provided by the user.

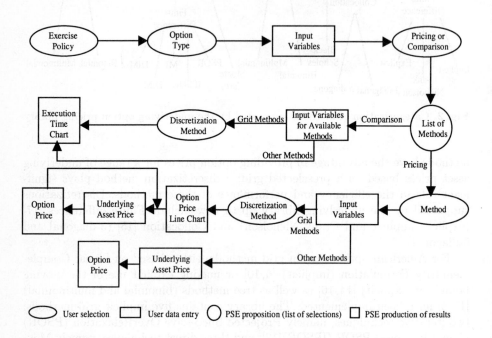

Fig. 3. Problem definition and solution process

For reasons of clarity, we now describe in more detail an indicative problem solution, giving exactly the steps followed through a web-browser. First of all, the end-user must determine if the option is European or American, call or put, if he/she is going to execute simple option pricing or comparison and finally enter the necessary input parameters, i.e. exercise price, interest rate, volatility, dividend and time to expiration. This is done, by making the suitable selections and completing the empty fields onto the form provided (see Figure 4).

The next step depends on the operation chosen in the first step. If "pricing" operation has been asked then the user can select only one of the available

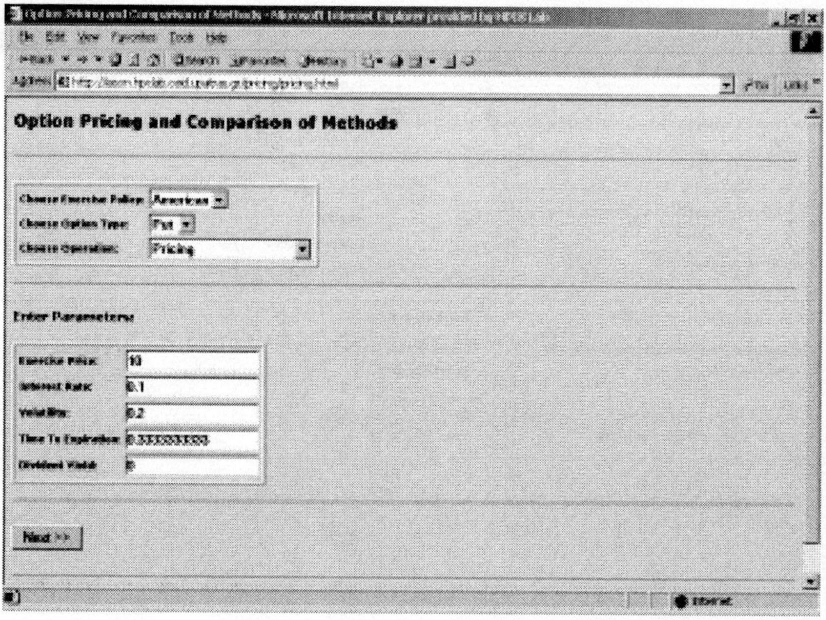

Fig. 4. First stage of problem definition

methods and define the required parameters. For example, if a grid method was selected a discretization method as well as a number of grid points and a number of time steps are necessary. Morever, if the method is iterative, a termination error should be entered. For the IPSOR method, an increment factor (see [12] for details), must also be defined. For the binomial method, only the number of time steps and the underlying asset price must be entered (other fields become non-input fields). If operation "comparison" was selected in the first step, then the user has the capability of comparing methods of the same type for the solution of the same problem. In this case he/she must select the methods to be compared and define the necessary input parameters. Only methods of the same type can be compared, otherwise the comparison does not make sense. For example, it is meaningful to compare MI with PSOR and DIM, because they are all grid methods, but none of them can be compared with the binomial method because it is a tree method. For this reason the PSE proposes available methods for comparison, depending on the problem which is currently considered for solution. An instance of the second stage of problem definition, if a pricing operation has been selected, is given in Figure 5.

For reasons of reporting, in all cases, the output has the form of a summary containing both information of problem definition and the results obtained. The form of the output primarily depends on the operation chosen in the first step and the method selected in the second step. For example, for grid methods, it has the form of a chart and the user has the capability of getting option prices

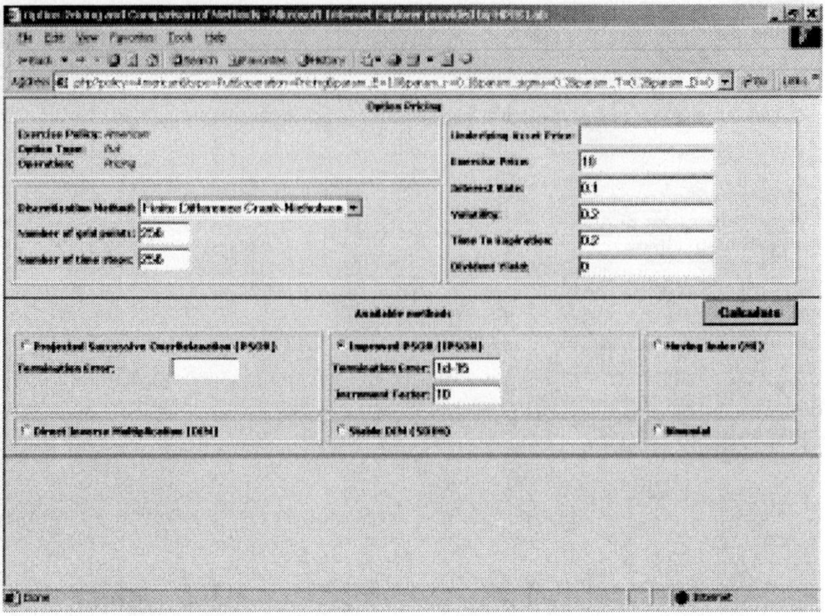

Fig. 5. Sample screen of the second stage of problem definition

for various values of the underlying asset. Figure 6 shows a sample screen for this case.

If a comparison were executed, then a chart of execution times for the selected methods is returned to the user (see Figure 7).

Finally, we should mention that a significant number of validity checks are performed for every method and special care is taken when wrong input is provided. For example, if DIM method is selected, then a stability check takes place before execution. In case that a stability problem exists, the user is informed that he can not execute this method for the current input.

The proposed PSE can be accessed through the following URL:
http://hydra.hpclab.ceid.upatras.gr/pricing/pricing.html

6   Conclusions and Future Work

In this paper we have presented a PSE for the solution of the option pricing problem. The proposed PSE has a number of very desirable properties. First, it is very simple and easy to use even by non-expert users. Second, it is available through the World Wide Web and provides immediate answers to the user in "real time". Third, although an expert system does not support the current environment, guidance is provided to the user and response of the PSE depends on user selections; morever, a comparison function is provided in order to help the end-user decide for the "best" method in terms of speed. However, the main

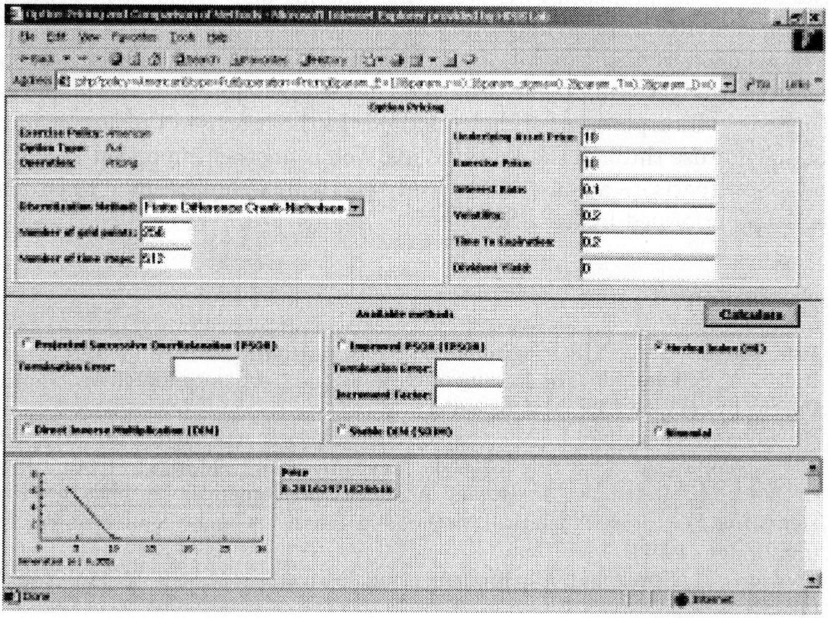

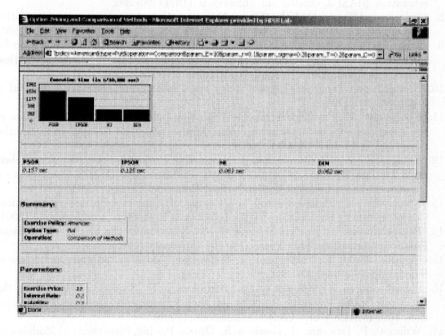

Fig. 6. Sample screen of the output for the case of pricing

Fig. 7. Sample screen of the output for case of comparison of methods

advantage of the proposed PSE over other existing ones is that it is based on an open, expandable, underlying library of methods.

Implementation and addition of new methods, for the solution of option (and other derivatives) pricing problems, in the open library, is the main task of future work. Incorporation of these methods in the proposed PSE to make them available for use through the World Wide Web is another important task. Other types of comparisons are also under study and will be implemented to enrich the functions performed by the PSE.

References

1. Atkinson, L.: Core PHP Programming. Prentice Hall, California (1999)
2. Black, F., Scholes, M.: The Pricing of Options and Corporate Liabilities. Journal of Political Economy **81** (1973) 637–654
3. Broadie, M., Detemple, J.: American option valuation: new bounds, approximations and a comparison of existing methods. Rev. Financial Stud. **9(4)** (1996)
4. Broadie, M., Detemple, J.: Recent Advances in Numerical Methods for Pricing Derivative Securities. Numerical Methods in Finance, Cambridge University Press, Cambridge (1997)
5. Clewlow, L., Strickland, C.: Implementing Derivatives Models. Wiley, New York (1998)
6. Cottle, R., Pang, J., Stone, R.: The Linear Complementarity Problem. Academic Press, New York (1993)
7. Cryer, C.: The solution of a quadratic programming problem using systematic over-relaxation. SIAM J. Control vol. /bfseries 9, (1971) 385–392
8. Geske R. and Shastri K.: Valuation by Approximation: A Comparison of Alternative Option Valuation Techniques. Journal of Financial and Quantitative Analysis (1985)
9. Houstis, E., Papatheodorou T., Rice, J.: Parallel ELLPACK: An Expert System for the Parallel Processing of Partial Differential Equations. Intelligent Mathematical Software Systems, North Holland, Amsterdam (1990) 63–73
10. Huang J. and Pang, J.: Option Pricing and Linear Complementarity. Journal of Finance **2(3)** (1998)
11. Koulisianis, M., Papatheodorou, T.: A 'Moving Index' method for the solution of the American options valuation problem. Mathematics and Computers in Simulation, Elsevier Science **54(4-5)** (2000)
12. Koulisianis, M., Papatheodorou, T.: Improving Projected Successive OverRelaxation Method for Linear Complementarity Problems. 5th IMACS Conference on Iterative Methods in Scientific Computing, Heraclion, Greece (2001)
13. Koulisianis, M., Papatheodorou, T.: Pricing of American Options using Linear Complementarity Formulation: Methods and their Evaluation. working paper (2001)
14. Pantazopoulos, K.: Numerical Methods and Software for the Pricing of American Financial Derivatives. phD Thesis, Purdue University (1998)
15. Papatheodorou, T.: Tridiagonal C^1-Collocation. Mathematics and Computers in Simulation **30** (1988) 299–309
16. Papatheodorou, T., Koulisianis, M., Hadjidoukas, P.: Numerical Methods for the American Option Valuation Problem and their Experimental Comparative Evaluation. 16th IMACS World Congress, Lausanne (2000)
17. Willmott, P., Dewynne, J., Howison, S.: Option Pricing, Mathematical Models and Computation. Oxford Financial Press, London (1993)

Cognitive Computer Graphics for Information Interpretation in Real Time Intelligence Systems

Yu.I. Nechaev, A.B. Degtyarev, A.V. Boukhanovsky

Institute for High Performance Computing and Data Bases
120 Fontanka, St. Petersburg, 198005, Russia
E-mail: {int,deg,avb}@fn.csa.ru

Abstract. Application of cognitive computer graphics (CCG) in problems of information analysis and interpretation in real-time intelligence systems (IS) is considered. Using of cognitive structures at investigation of complex dynamic processes determining interaction of floating dynamic object (DO) with environment in extreme situations is in the focus of attention.

1. Introduction and problem to solve

CCG is connected with development of representation of symbolical and graphic objects by which geometrical thinking operates. Scientific basis of CCG is founded by A. Zenkin [16,17] and D. Pospelov. These ideas have developed in papers [1,2,5,6,7,10,12,14,15]. Despite of several publications devoted to CCG problem, practical applications of obtained theoretical solutions in onboard real-time IS are not numerous, e.g. [6,7,14], probably, due to complexity of cognitive modeling of DO interaction with environment.

Let us present knowledge process as IS functioning [5,10,15]. Let us consider dynamically varied subject domain. Theoretical bases of cognitive structures formation in this area we shall develop using system of associated concepts determined on conceptual carcass. System of axioms as a finite set of statements is considered also. Due to large volume of processable information model of cognitive structure is preferable than isomorphic model. Basic principle of such model functioning is based on elimination of conflicts between observed data (results of dynamic measurements) and abstract theoretical knowledge [7,12].

One of the possible approaches to such modeling is using of integrated systems of Image Understanding (IUE – Image Understanding Environment) [3,11], based on definition of "standard" image objects and reflections between them. In the paper figurative representation of visual information as base component of intelligent block for complex onboard IS is considered. Figurative analysis as distinct from visual analysis does not use only visual human perception. It uses also special expert systems of figurative information arrays processing. These arrays represent ordered vector-matrix information containing characteristics of considered DO behaviour.

P.M.A. Sloot et al. (Eds.): ICCS 2002, LNCS 2329, pp. 683–692, 2002.

Image models as cognitive structures are formed on the basis of such information [14].

Cognitive paradigm promotes to effective dialogue between operator and computer. It is important approach to knowledge formalisation in various IS stimulating finding of original technical decisions and generation of new knowledge. CCG is especially actual in analysis and interpretation of extreme situations in real-time IS. Reliable estimation of situation and acceptance of reasoned decisions in such systems substantially depend on effective solution of situation identification problem.

Problems of CCG are connected with solution of the following problems [7]:

- development of knowledge representation models ensuring uniform objects representation of algebraic and geometrical thinking;
- visualisation of knowledge, whose textual descriptions is complicate;
- search of transition ways from observable images to hypothesis about latent processes formulation.

In the paper different ways of problems solution and construction of cognitive images are considered by the example of functioning onboard intelligent systems of navigation safety and aircraft landing [6,7]. Thus three applications of cognitive structures are shown:

- visualisation of multivariate information at interpretation of dynamic measurements;
- representation of information for operative decisions acceptance in extreme situations;
- information compression for construction and training of artificial neural networks (ANN).

Applications of carried out researches are IS using parallel algorithms of measuring information processing on the basis of supercomputer technologies.

2.   General Approach

CCG is a synthetic scientific field. It uses tools of functional analysis, both analytical and algebraic geometry and statistics from the one side, and discrete mathematics, theory of graphs and scientific visualisation from the other side. Identification and decoding of cognitive structures are carried out both on stage of knowledge system formalisation during IS synthesis, and on stage of testing by methods of imitating modelling.

Algorithm of cognitive model construction can be realised by sequence of procedures reflecting technology of knowledge representation about DO functioning [7,8]:

$$(q, U) \xrightarrow{E} J(q \in Q);$$
$$(J, U) \xrightarrow{S} \Omega; (\Omega, D) \xrightarrow{P} R;$$
$$(R, U) \xrightarrow{AD} W$$

(1)

Here q is the subregion of a subject domain Q; U is the problem of DO control; J is the visualised image; W is the program; E is the heuristic procedure of mapping q in J; S is the procedure of programs synthesis; D is the data array which is the basis for solution search during modelling; P is the solution searching procedure; R is the result; AD is the procedure of solution acceptance; W is the cognitive structure.

The general scheme of cognitive model, is characterised by the system of functional equations:

$$\Phi_j = t_i\left(F_i, f_j\right), \; i = 1, \ldots, k, \; j = 1, \ldots, n \tag{2}$$

where t_i are the terms constructed with the help of composition and recursion; F_i are the functional variables of corresponding processes in system (1); f_j are the functional constants corresponding to elementary processes.

Procedure W is most labour consuming and poorly known. Construction of W is connected with specificity of considered initial information in real-time onboard IS. Due to complexity of "exact" physical–mathematical model of DO–environmental interaction Φ_j let us consider metaphorical representations of W on the basis of initial data analysis.

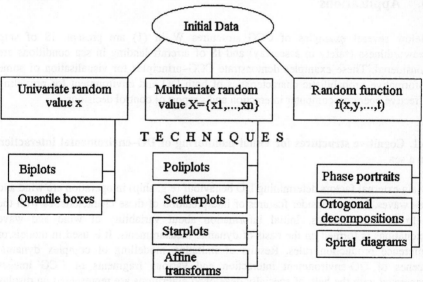

Fig. 1. Models of data and techniques of cognitive images generation

Probabilistic measure could be considered for majority of environmental processes. In this case power tools of multivariate statistics [9,13] permit to construct procedure W. Multivariate statistics operates by canonical variables (principal components, factor loadings, canonical correlations), uses analyses of regression, variance and covariance, employs classification and discrimination tools. This methodology was already developed well in terms of random value (RV) and multivariate random value (MRV) models. However, application of such methods for random function (RF) or time series model (that characteristic for real time IS) demands updating of these procedures.

In this paper basic approaches to identification and construction of cognitive structures are presented in terms of RV, MRV and RF models. Structure of used models and corresponded techniques is shown in the fig. 1.

Techniques shown in fig.1 are generally considered in paper [4]. One set of them (biplots, polyplots, affine transform, spiral diagram) is based on the idea of special functional transformation of initial data to easily recognised image (e.g., straight line). Second set contains all methods (quantile boxes, orthogonal decompositions, phase portraits) which are based on the elimination of low–informative variables. And the third set of methods (scatterplots, starplots) results in plane representation of non–planal (multivariate) images.

From the other side such measure could be fuzzy for processes which connected with human actions (driver, pilot, navigator et al.).

3.   Applications

Below several examples of CCG structures W in (1) are present. IS of ship seaworthiness (safety in a seaway) and IS of aircraft landing in sea conditions are considered. These examples demonstrate CCG–principles for visualisation of some informative probabilistic characteristics of both DO and environment for providing effective operator–computer interaction and support of control decisions.

3.1. Cognitive structures for visual monitoring of DO–enviromental interaction in a sea

Main external factors determining DO behaviour (e.g. ship) in operation are wind and sea waves. Let us consider features of interpretation of these characteristics with the help of CCG images. Initial information about variability of wind and wave excitations is formed on the basis of dynamic measurements. It is used in models of inference production rules. Results of imitating modelling of complex dynamic scenes of DO-environment interaction and various fragments of CCG images generated with the help of specially developed algorithms are represented on display together with numerical information.

Wind waves representation. Wind waves have complicated spatial structure. Geometrical characteristics of waves are determined by spatial spectral density $E_\zeta(u,v)$, or correlation surface:

$$K_\zeta(r,\theta) = D_\zeta \, exp\!\left(-\alpha(\theta)r^{1+\varepsilon}\right) cos\!\left(\beta(\theta)r\right) \tag{3}$$

where $r = \sqrt{x^2 + y^2}$, $\theta = arctan(y/x)$, D_ζ is the waves variance, $\alpha(\theta)$ is the decrement of decay, $\beta(\theta)$ is the fluctuation frequency in direction θ, ε is a small constant.

Specific wave size in direction θ is:

$$\lambda(\theta) = \frac{\pi}{\beta(\theta)} \tag{4}$$

Here value $\lambda(0)$ is traditional definition of wavelength. So, cyclic spatial structure of wave field could be described by curve in polar co-ordinates, as geometrical place of points $\{\theta, \lambda(\theta)\}$. In simple case (for visualising purpose) let us consider such structure as rotated ellipse with semiaxes $\{\lambda(0), \lambda(\pi/2)\}$.

One of the dangerous situations is ship movement along general direction of waves with speed closed to phase speed of waves. The length of dangerous wave is close to length of the ship. So, for visualising of such scenario, let us superpose the above mentioned ellipse of wave "hump" and stylistic ship image. Orthogonal axes of the plot mean module of ship speed V and cosine of course angle χ (in geographical co-ordinates), see fig. 2. From the fig 2(a) we can see that ship length is close to dimension of wave humps, and situation must be considered as dangerous.

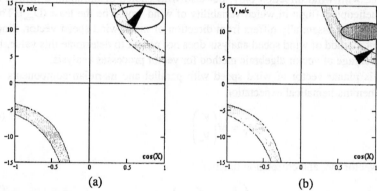

(a)   (b)

Fig. 2. Influence of environmental on ship.
(a) – danger of broaching (on quartering waves), (b) – danger of main resonance.

Resonant regimes of ship motion. Identification of ship motion resonant regimes is based on the condition (for rolling):

$$\left| \omega - \frac{V}{g} \omega^2 \cos\chi - kn_\theta \right| < \varepsilon \tag{5}$$

where n_θ is the natural frequency of main rolling resonant, k is the coefficient of overtones, v is the ship speed, ε is a small constant for comparison. After transformation u=vcosχ we obtain:

$$uv = \left(\omega - kn_{\theta} \mp \varepsilon\right)g/\omega^2 \tag{6}$$

Equation (6) defines two hyperbolic domains where resonance condition (5) is valid. Value ε of these domains approximately depends from sea state:

$$\varepsilon = 0.43h_s^{-1/2} \tag{7}$$

where h_s is the significant wave height.

If in (6) we fix $\omega = \omega_m$ (ω_m is peak spectrum frequency), than two "dangerous" domains are observes on plane (u,v). After functional transformation we will have the same on plane $(V,\cos(X))$. Dangerous resonant situation in quartering directions is shown in the fig 2(b).

3.2. Cognitive structures for landing of sea–based aircraft

Landing of aircraft in sea conditions is one of the complex extreme situations connected with flights safety. Features of landing in complicate hydrometeorological conditions are not connected with ship navigation to the certain direction in accordance with wind direction only, but also with choice of optimum direction of aircraft (helicopter) flight in which variability of wind would be the least (D_{min}). This direction usually essentially differs from direction of mean wind speed vector. Since component method of wind speed analysis does not permit to determine this value, let us use advantage of vector algebraic method for vector processes analysis.

Let $\vec{V}$ is planar vector of wind speed with parallel and meridian components V_x and V_y. Then mathematical expectation

$$\vec{m}_{\vec{V}} = M\begin{pmatrix} V_x \\ V_y \end{pmatrix} \tag{8}$$

is planar vector too, and correlation function

$$K_{\vec{V}} = \begin{pmatrix} K_{V_x} & K_{V_x V_y} \\ K_{V_x V_y} & K_{V_y} \end{pmatrix} \tag{9}$$

is dyadic tensor with the components $K_{V_i V_j}$. Variance $D_{\vec{V}} = K_{\vec{V}}(\bullet)$ when $\bullet = 0$. Hence, variability of wind speed is defined by set of five values: module $\left|\vec{m}_{\vec{V}}\right|$ and direction $\bullet$ of mean wind speed, and invariants $\bullet_1$, $\bullet_2$, $\bullet$ of variance tensor:

$$\left|\vec{m}_{\vec{V}}\right| = \sqrt{m_{V_x}^2 + m_{V_y}^2}, \tag{10}$$

$$\bullet = \arctan\frac{V_x}{V_y} \tag{11}$$

$$\bullet_{1,2} = 0.5\left(I_1 \pm \sqrt{I_1^2 - 4I_2}\right) \tag{12}$$

$$\bullet = 0.5\,arctan\left(\frac{2D_{V_x V_y}}{D_{V_y} - D_{V_x}}\right) \tag{13}$$

where I_1, I_2 are linear and quadratic invariants of tensor (9).

Tensor (9) is conveniently represented as ellipse with semiaxes λ_1, λ_2 for characterisation of V intensity in different directions,. Orientation of largest semiaxis is $\alpha \pm 180^0$. Example of wind field parameterisation in cyclone in the Mediterranean is shown in the fig. 3. Mean wind speed vectors are plotted as arrows, and standard deviations are plotted as ellipses.

Obviously that $D_{min} = \lambda_2$ and optimal landing direction is $\alpha \pm 90^0$. Thus, conventionalised graphic images of ship, mean wind speed vector, helicopter and standard deviation ellipse (fig. 4) are offered for operator (manager of landing) for acceptance of the decision of choice of optimum course angle and ship speed for helicopter landing.

Other way of CCG images using at aircraft landing is described in papers [7,14]. The approach is developed on the basis of the patent of Yu. Nechaev ensuring situation analysis and acceptance of the decisions during realisation of landing operations in sea conditions. The analysis is connected with recognition of wave field structure and of behaviour of flight deck in various external excitation conditions. Developed algorithm permits to recognise regimes of ship motion and "window of safe landing" during formation of command to the pilot about operation beginning and finalising. Display control is provided by the computer program, which carries out all necessary operations on information processing, situation estimation and generation of practical recommendations in real time.

3.3. Cognitive structures for learning ANN identifying extreme situations

In onboard IS CCG structures are not used only for analysis and interpretation of extreme situations. It is possible use it in processing of measuring information for representation of input signals for ANN. Effect of CCG application consists of "compression" of information, simplification of ANN configuration and reduction of learning time. Such approach is very useful for the problem of damaged ship state identification (flooding).

Solution of recognition problem by means of CCG was carried out with the help of transformation of information in terms of formally logic description. Search of concrete model of image was carried out with the help of cognitive spiral permiting "to compress" initial information about nonlinear ship rolling with stochastic variability of periods. Application of cognitive spiral to stochastic processes resulted in "alignment" of periods and their affine transformation to one value.

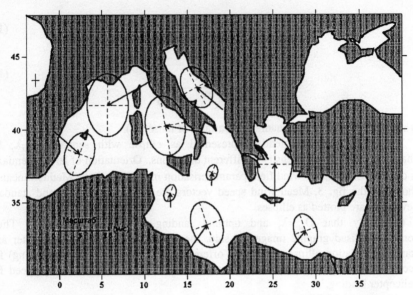

Fig. 3. Cognitive images of wind speed field. Cyclone in the Mediterranean, 1.12.1969

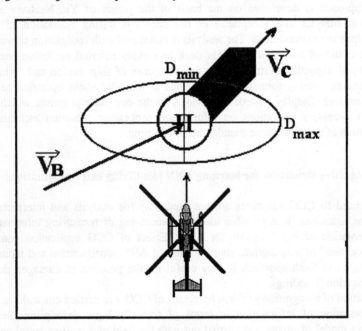

Fig. 4. Cognitive image for ship navigation to direction of wind speed $\vec{V}_B$, with account of minimal atmospheric turbulence.

Cognitive spiral, as alternative to a phase portrait, contains more information, which it is possible to select visually orienting on image structure. Breadth of bands in spiral, its saturation by colour and frequency of changing of band of one colour by band of other colour can be referred to such information. Asymmetry of upper and lower parts of cognitive spiral and distribution of colour on angles close to 0 and 180 degrees can also play important role.

Considered problem of recognition is connected with classification of typical cases of flooding. There are five classical situations of damaged ship. The first case is trivial enough: lack of static heel at symmetric flooding and positive initial metacentric height. In other emergency states ship always has heel on a side. It allows to easily pick out first case without using complicated inference procedures. As regards to the fifth case it is usually considered as subset of fourth case. Similar pictures of oscillation are typical for this case.

Analysis of CCG structures permits to obtain information for recognition. Rational decision of this problem is obtained by consideration of characteristic sections in cognitive spiral at various corners. As a result we did not exploit basic characteristics of CCG (human-computer interaction), but we used above mentioned CCG images for ANN learning. Such information permits to identify constant number of entrance neurons, to simplify structure of ANN and to have freedom in choice of length of initial time series. Cognitive image of ANN learning on the basis of two–layer map of Kohonen is shown in the fig. 5.

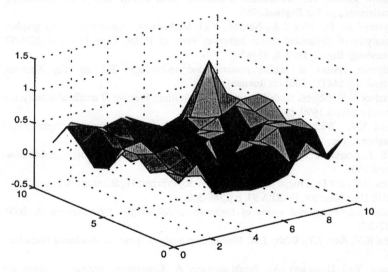

Fig. 5. Cognitive image of ANN training. Two–layer map of Kohonen (10x10 neurons)

4. Conclusions

Carried out investigation testifies to the large opportunities of CCG methods for problems of data interpretation in onboard real time IS. Methods of CCG improve efficiency of recognition and analysis of latent structures describing DO- environment interaction. Especially important role of CCG images application is for analysis and forecast of extreme situations connected with safety of ship navigation and aircraft landing in sea conditions. "Compressed" and easily interpretive cognitive representations of dynamics stimulate image thinking and allow to operate with representational objects of abstract nature.

The research is supported by grants of RFBR N 00-07-90227 and INTAS 1999 N666.

References

1. Albu V.A., Choroshevsky V.F. COG–system of cognitive graphics. Developing, realization, application. *Izv. AS USSR, Technical cybernetics*, 1990, **5**, 37–43.
2. Bashlykov A.A., Pavlova E.B. Intellectual user interface of system SPRINT–RV on the base of cognitive graphics. *Proc. of V Int. Conf.* CAI96, Kazan, Russia, 1996, **3**, 385–386.
3. Boult T.E., Fenster S.D.and Kim J.W. Dynamic attributes, code generation and the IUE, DARPA 94. vol.1,p.p.405-422.
4. Chamber J.M., Cleveland W.S., Kleiner B, Tukey P.A. Graphical methods for data analysis. Belmont, CA: Wadsworth, 1983.
5. Cognitive science. An introduction a bradford book // The MIT Press Cambridge-Messachusetts-London-England.1989.
6. DegtyarevA.B., Dmitriev S.A., Nechaev Yu.I. Methods of cognitive computer graphics for analysis of dynamical object behavior. *Proc of II Int. Conf* MORINTECH–97, St.Petersburg, Russia, 1997, **6**, 83–87
7. Intelligence system in marine research and technology /Ed. by Yu. Nechaev. St.Petersburg. SMTU. 2001 (in Russian).
8. Ivanischev V.V. Objects modelling on the visualisation. *Proc. of II artificial intelligence conference*. Minsk.1990, pp.122-124.
9. Johnson R.A., Wichern D.W. Applied multivariate statistical analysis. Prentice-Hall International, Inc., London, 1992, 642 pp.
10. Lakoff J. Cognitive Modelling. Proceedings of language and intelligence.-Moskow. Progress.1996.
11. Lowton D.T., Dai D., Frogge M.A., Gardner W,R., Pritchett H., Rathkopf A.T. and TenG. The IUE User Interface, DARPA 93, p.p.289-300.
12. Magasov S.S. Functional sceme of cognitive model. *Artificial Intelligence*. **3**, 2000, pp.257-262.
13. Mardia K.V. Kent J.T., Bibby J.M. Multivariate analysis. London, Academia Press Inc., 1979
14. Nechaev Yu., Degtyarev A., Boukhanovsky A. Coignitive computer graphic for interpretatrion of information on real time intelligence system. *Proc. of III Int. Conf.* MORINTEX-99. St.Petersburg. 2001. Vol.1, pp.297-303.
15. Shneiderman B. Perceptual and cognitive issues in the syntacsic. Semantic model of programmer behaviour. *Proc. of symp. on human factors and computer science*. Human Factor Society. Santa Monica, California,1978.
16. Zenkin A.A. Cognitive computer graphics. M., Science, 1991 (in Russian)
17. Zenkin A.A. Cognitive computer graphics – application to decision support systems. *Proc of II Int. Conf* MORINTECH–97, St.Petersburg, Russia, 1997, **8**, pp.197–203

AG-IVE: An Agent Based Solution to Constructing Interactive Simulation Systems

Z. Zhao, R.G. Belleman, G.D. van Albada, P.M.A. Sloot

Section Computational Science
Faculty of Science, Universiteit van Amsterdam
Kruislaan 403, 1098SJ Amsterdam, The Netherlands
{zhiming | robbel | dick | sloot}@science.uva.nl
http://www.wins.uva.nl/research/scs/

Abstract. An Interactive Simulation System (ISS) allows a user to interactively explore simulation results and modify the parameters of the simulation at run-time. An ISS is commonly implemented as a distributed system. Integrating distributed modules into one system requires certain control components to be added in each module. When interaction scenarios are complicated, these control components often become large and complex, and are often limited in their reusability. To make the integration more flexible and the solution more reusable, we isolated these control components out of the system's modules and implemented them as an agent framework. In this paper we will describe the architecture of this agent framework, and discuss how they flexibly integrate distributed modules and provide interaction support.

1. Introduction

An Interactive Simulation System (ISS) is a system that combines simulation and interactive visualisation and allows the users to explore the problem space and steer the execution paths of the simulator at runtime. For a complex simulation, putting a human expert in the simulation loop can reduce the parameter space. This can significantly shorten the experimental cycle and improve the efficiency of resource usage. During the last decade, interactive simulation systems have become an important research topic in the area of computational science [1,2,3,4].

In general, a minimum ISS has three basic modules: simulation, visualisation and interaction. The simulator computes and transfers data to the visualisation and interaction modules regularly; the simulation parameters can be manipulated in the interaction module. Simulators and visualisation programs often are inherently very complex. They are often developed on different platforms and with different software architectures, connected using e.g. a *Client-Server* paradigm [8]. Following the classification of paradigms for developing parallel and distributed simulation in [4], we can distinguish two principal approaches to constructing ISSs [5,6,7]. The first approach starts with building simulators using a high performance simulation engine, and then equips them with interaction capabilities. SPEEDS [5] is an example. The other uses a

P.M.A. Sloot et al. (Eds.): ICCS 2002, LNCS 2329, pp. 693–703, 2002.

run time infrastructure software to interconnect federated modules. Protocols such as Distributed Interactive Simulation (DIS) [6] and High Level Architecture (HLA) [7] support this approach. The first approach often achieves a higher efficiency because of the specific development environment for the problem domain. The second can easily interconnect modules across platforms and places fewer restrictions on the internal implementation of individual simulators. It can exhibit more flexibility and reusability of components than the first.

In an ISS, all modules have to work together under certain constraints. In our work, we use *scenarios* to specify the relation between the required behaviours of the modules. Updating and distribution of the simulation states is one of the basic actions in an ISS. Depending on the interaction context in each scenario, the simulation state needs to be updated in different ways. Often, these constraints are incorporated into the simulator and visualisation modules. This is straightforward for the system design, but makes the system modules difficult to reuse. Separately implementing them in Intelligent Agents (IA) yields a more generic and portable solution.

In earlier publications [12,13], we have described the Virtual Environment (VE), as a device for scientific visualisation and user interaction. The Intelligent Virtual Environment (IVE) integrates simulators and interactive visualisation programs in a VE, and adds interactive simulation capabilities. In the IVE, agents are used to interconnect simulators and interactive visualisation programs; they are employed to customise the inter-module communication and provide support for module co-ordination and interaction. We implement these agents as a reusable framework. In this paper, we will first introduce the architecture of this AGent based Intelligent Virtual Environment (AG-IVE), next we discuss the issues of simulation state updates and switches between system scenarios. After that, we will present experimental results on the performance of a prototype of AG-IVE.

2. Intelligent Agents and Module Integration

Intelligent Agents and Agent oriented techniques have become a modern AI approach and have been applied in many problem domains, although many researchers do not agree on apparently trivial terminology details [14]. Depending on the problem domain and system design, agents vary in their functions. A simple reflex agent is only able to respond to events from its external world, a goal-oriented agent can make plans to organise its actions.

Rather than using Agent technology for solving specific problems, we aim to use it as an integrating concept, enhancing flexibility, portability and code re-use.

2.1   Agent Paradigm: AGent Based Intelligent Virtual Environment (AG-IVE)

AG-IVE is an agent-based solution to interconnecting simulators and interactive visualisation programs that tries to combine the advantages of both development approaches mentioned in the introduction. It aims to re-use available simulation kernels and visualisation programs and to integrate them in a portable way. In AG-IVE,

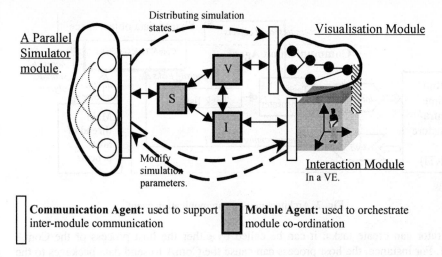

Distributing simulation
states.

Visualisation Module

A Parallel
Simulator
module.

V

S

I

Modify
simulation
parameters.

Interaction Module
In a VE.

Communication Agent: used to support inter-module communication

Module Agent: used to orchestrate module co-ordination

Fig. 1. An example of AG-IVE when it is used to interconnect a simulator and an interactive visualisation module

agents are used to federalise simulators and visualisation programs. A global run time infrastructure is employed to inter-connect them. Currently, the Run Time Infrastructure (RTI) [7] of the High Level Architecture (HLA) is used.

In AG-IVE, two types of agents are defined: "Communication Agents" (ComA) and "Module Agents" (MA). The ComAs hide the implementation details of the simulators and interactive visualisation programs, and provide them with an interface to plug into the global run time environment and interact with other modules. The MAs orchestrate the execution of modules (simulation, visualisation and interaction), and control scenario transitions of the whole system by co-operating with other agents. Fig. 1 shows how these agents can be used to integrate a system.

The example shows a parallel simulator and an interactive visualisation module. The visualisation module presents simulation results in a VE where a user can interactively study the data space. ComAs interface the modules to the RTI and transfer data between them (thick dashed lines). MAs co-operate with each other and the ComAs to control the behaviour of the system (double headed arrows).

2.2 Communication Agents

In AG-IVE, the global run time infrastructure is only used for communication between the modules; the simulators and interactive visualisation programs internally still use their original communication mechanisms. In each module, the processes needing inter-module communication are equipped with a ComA.

To minimise the influence on the execution of the host process and to make the integration easier, a ComA is implemented as a separate. The ComA is designed as an event processing architecture that contains four basic components: *a task generator, a task list, a task interpreter* and *a shared data object space* (Fig. 2). In a ComA, a task is the description of certain actions that the ComA should carry out. Only the task

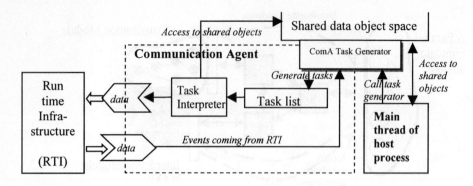

Fig. 2. Architecture of a communication agent

generator can create tasks; it can be called by either the host process or the ComA itself. For instance, the host process can cause the ComA to send data packages to the RTI, and the ComA can receive messages from other agents and react to them by creating certain tasks. Depending on the priority and the time tamp associated with each task, the task interpreter schedules tasks and executes them. This execution style hides the implementation details of the ComA from the host process, and keeps the internal functions of a ComA from being called directly.

In AG-IVE, a shared object space is used to exchange information between host process and the ComA. In the shared object space, all information is encapsulated as data objects. The structures of the data objects are defined as data classes. These are fully described in a globally accessible "Scenario Description file".

Internally, a ComA pre-defines a special kind of data class, the so called *Monitor*. During system creation, the host process can transfer access rights of its parameters to ComA by registering them as data items in the Monitor object. Therefore, the ComA can manipulate the execution of the host process via the Monitor.

Adding a ComA to a process inevitably requires certain code level adaptations. To simplify the integration and reduce the influence on the original software architecture, the interface for accessing shared objects and creating tasks of ComAs has been standardised. In order to minimise the influence on system performance, the function of a ComA is limited to transferring data and responding to external events.

The interface of each ComA, such as the set of events to which it can respond and the set of monitor parameters that it can use to manipulate its host process, is described in the Scenario Description file. From this, an MA can know how to manipulate the ComA.

2.3 Module Agents

Each module is associated with one MA, which interacts with all ComAs in that module. Depending on the type of ISS module, each Module Agent has different functions. Compared with ComAs, Module Agents are more sophisticated. An MA not only can react to the events from its external world like ComAs, but also can use its internal world model to plan its actions beforehand. For each agent (ComAs and MAs), the external world is a dynamic environment, which includes all the other live agents and the running ISS modules.

Each MA is equipped with a Reasoning Module, which is used to process the observations and make rational action plans. The Reasoning Module utilises first-order logic to represent knowledge. The Reasoning Module has following functions:

1. Maintaining the internal world model. All the information gathered from the external world will be transferred to the Reasoning Module as Prolog terms.
2. Responding to queries. The other agents can lean the current status of a module by asking the MA of that module.
3. Making plans for given goals. In each scenario, MA has a long-term goal like optimising the execution of its module. The long-term goal is normally achieved by a number of sub-goals.

RTI is the only channel for cross module and inter-agent communication. To communicate with the other agents via the RTI, each MA has a ComA. MAs perceive the dynamic external world by observing and analysing the packages transferred over the RTI. ComAs and MAs co-operate with each other to realise functions like scenario control and state update.

2.4 Interaction and Scenario Switches

ISSs allow the user to explore the computational results and to interactively manipulate the parameters of the simulation. In AG-IVE, scenarios represent the contexts of certain interactions and their required constraints on system modules. Each scenario has one or more basic interaction objectives, for instance exploring simulation results, modifying simulation parameters or probing information. In different systems, the scenarios of the interaction may be different. In AG-IVE, the following basic interaction scenarios are distinguished.

1. Obtaining data from the simulator (sample, query);
2. Explore the visualisation of the data obtained from the simulation;
3. Adjust the simulation parameters (computational steering);
4. "VCR" control of the simulation progress (re-start, rollback, pause and resume the simulation execution).

To achieve the objective of a scenario, each system module has to run under certain constraints. These are adapted when scenarios are switched. The constraints often involve issues such as module synchronisation, intermediate state storage and data distribution. In AG-IVE a Scenario Description file specifies the capability of agents and their desired behaviour in each specific scenario. At run time, each MA reads information from the description file and adjusts the behaviour of ComAs in its module to fit the requirements of a scenario. The Federation File [7] needed by the HLA RTI is also generated from the scenario description file.

Switches between scenarios are often triggered by different mechanisms such as predefined interaction routes or by the user. In AG-IVE, the interaction MA monitors the user's behaviour during interaction and tries to identify the switching of scenarios.

After a scenario has switched the interaction MA broadcasts the new scenario to all the other MAs to inform them to adapt the behaviour of their ComAs.

2.5 Optimisation of State Updates

The simulator is required to export its states selectively and to distribute them efficiently to the visualisation and interaction modules. The visualisation module also needs to update in a rational manner, so that the user can obtain clear insights into the evolution of the simulation and its problem space. The content that will be transferred for visualisation and interaction normally is a subset of the state of the simulation, which can be mapped onto subsets of the local states in each simulation process

The update of simulation states has to address issues such as speed difference between simulators and visualisation, the user's perceptual delay, and the switch between different scenarios [10]. In different interaction contexts and running conditions, the way to export and distribute simulation states will not be the same. In AG-IVE, agents are used to optimise the state updates in each module.

Each MA monitors the running status of its host module, obtaining the internal information of each individual process by communicating with the ComA. The MAs share this information with each other and estimate the communication delay over the network by using the information from the other agents. Finally, the simulation MA can decide the best update rate for the simulation to export states.

Apart from making decisions with respect to resource utilisation, optimising the update of the simulation state also involves issues like efficiently distributing data objects, and storing intermediate simulation states during interaction. In the rest of this section, we will briefly introduce the considerations on these issues in AG-IVE.

For complex parallel simulators, the vectors of the simulation states are often very large, in the order of tens of megabytes. Efficiently distributing large size data objects is an important problem in distributed interactive simulation systems. In AG-IVE, the services from the lower level RTI are employed for the basic communication support. Based on these RTI services, AG-IVE provides a number of optional techniques to decrease the size of data such as: compression, data sub-sampling and segmentation. These techniques are implemented as *filters* of data objects, they can be used to process data objects before they are sent or after they are received. For this purpose, AG-IVE explicitly defines data items such as dimensions, processing methods and mapping algorithms in the simulation state data class. These techniques can significantly shorten the data transmission delay, and improve the update rate, but as most of them are lossy, they can be activated and de-activated at runtime.

2.6 Implementation and Current Status

AG-IVE has only been tested with parallel simulators that run synchronously and in time driven mode. The interface of RTI is based on the specification of Version 1.3. The visualisation programs are executed on either VEs or normal desktops. The Reasoning Modules of the MAs are implemented by Amzi Logical Server [19].

AG-IVE provides basic templates for integrating ComA and a host program. The ComA can be added as a separate thread or in the same thread as the original process. We have used this method in a fluid flow simulator and a visualisation program, and the basic integration only took a few hours. To make the construction of the scenario description file easier, a tool that can assist a user to specify agent actions and the constraints between them is under development.

3. Test Case: Intelligent Virtual Environment (IVE)

Currently the IVE is being used in a medical application. In surgery, operations such as vascular reconstruction can often be performed in different ways. The best treatment is not always obvious, or may be hard to discern because of complicated or multi-level diseases of the patient. Verifying an operation plan is a difficult task, even for expert surgeons. Computer simulation may help a surgeon to validate his treatment, but it is almost impossible to let a computer simulate all situations, because of the huge number of possible solutions. What this application tries to do is to put a human expert into the simulation cycle, and let him apply his expertise to confine the problem space. In this system, a simulation program simulates the patient's blood flow on a parallel computer system. A visualisation module presents the simulated results together with the body's geometrical information obtained from a medical scanner (such as CT or MRI). The visualisation and interaction modules run in an immersive virtual environment (CAVE) [15] where a user can study the results of the given treatment, and modify it if necessary. These systems are connected through a high performance network.

The fluid flow simulator [16] uses MPI [9] as its internal communication interface, and runs on a PC cluster [11]. The interactive simulation program, VRE (Virtual Radiology Explorer) runs in the CAVE environment, and uses shared memory to exchange information between visualisation and interaction processes. RTI Next Generation 1.3 Version 4 is used as the global run time infrastructure.

3.1 Package Size and the State Transfer Delay

In data distribution, the size of packages significantly influences the delay in their transmission. We use the term *transfer delay of the state objects* to represent the time from when one module starts sending objects until the other one finishes receiving them. We have measured the transfer delay of ComAs between two workstations that are not in the same cluster. Both workstations use a Pentium III processor and Linux Redhat 7.2 system; they are connected via a 100 Mbits/s network. Fig. 3 shows the average of 15 different measurements. The latency of the ComA is about 0.02s, its throughput is 47 Mbits/s. The throughput is half of the bandwidth, but it is very close to 50 Mbits/s; the performance we measured with a TCP socket program which makes use of the CAVERN library [18]. Internally RTI also uses sockets as its communication mechanism and we can see that the additional services provided by RTI have no large effect on communication, except for an additional latency for small messages.

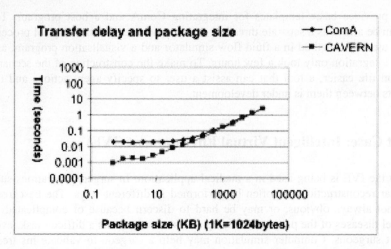

Fig. 3. Transfer delay versus package size for a ComA based method and for CAVERN

From Fig. 3 we also can see that the transfer delay increases rapidly when the package size exceeds 1 Mbyte. In order to reduce the size of packages, a number of techniques have been applied in AG-IVE, such as using general data compressing method like LZW [17] and sub-sampling simulation states. During experiments, we have seen there is not a single method that is effective for all types of simulator. For instance, for the fluid flow simulation, general compressing methods have not exhibited good performance. Normally, only less than 20% reduction of the whole simulation results can be achieved, and it needs more than 15 seconds to compress. Sub-sampling data content in simulation states will cause information loss, and requires certain code adaptation in the visualisation program. To select a proper option for distributing data objects, MAs can monitor all packages transferred between modules. MAs perform certain experiments at run time, and record the performance of each optional technique. Depending on the situation, MAs dynamically select the best option for ComA to transfer data. The user can also interactively enable or disable these functions.

3.3 ComAs and Their Penalty on System Performance

As a part of the ISS process, a ComA will require computing power from its host process. To evaluate the performance of the ComA and its influence on the original ISS processes, we have built four test configurations. The first and the second integrate ComA as separate threads but use user level scheduling vs. kernel level scheduling. In the third the ComA executes in the same thread as the simulator. The fourth one uses the CAVERN library to implement the communication between simulator and visualisation. In the first two cases, ComA and the simulation kernel run in parallel. In the last two cases, the computing of simulation kernel will be blocked during data export and distribution. All use the same simulator. Every time step, the simulation kernel exports the whole simulation state once, which is about 18 Mbytes. All these tests were executed using a single processor, two and four processors respectively. Each case has been measured 10 times, the average of each time step has been

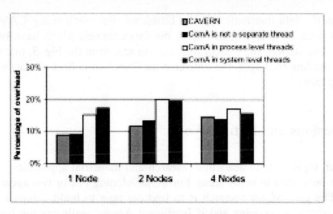

Fig. 4 Performance penalty of each test bed

compared to the original simulator. We use the ratio between the additional computing time required by these test runs and the time cost in the original simulator to obtain the penalty on system performance. Fig. 4 shows results. Using RTI, the local call back functions in each ComA only can be activated when they are ticked. When running in multithread style, in order to react on RTI events in real time, the tick function has also been called regularly even when a ComA has no task to process. That makes the system suffer higher performance penalty than the single thread integration (see Fig. 4). From the comparison with the CAVERN test case, we can say that the performance penalty of ComA is not prohibitive. As future work, we will look for more efficient scheduling strategies and improve the performance of ComAs when they are integrated as separate multithreads.

When more nodes are employed in computing, the local simulation state in each process becomes smaller, which accelerates computation. From the measurement, we found the difference between these four test beds becomes small when using more nodes. But on the other hand, the ComA in the visualisation module has to spend more time to collect and combine all sub-states. Fig. 5 shows the comparison of simulation states collection time between using ComA and using CAVERN. When using ComAs,

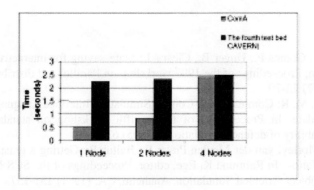

Fig. 5. Simulation sub-state collection times in the visualisation module

the RTI handles data distribution using broadcast; but when using CAVERN, data collection is done by querying data from the data channels which have been created when the system started. Because of this, we can see, from the Fig. 5, that when more ComAs are sending packages, the time cost for ComA to collect all data objects may become very large.

4. Conclusions and Future Work

As a research topic in both academic research and industrial applications, Intelligent Agents have been used in many areas. For us, developing one or two agents is not the final goal. The goal of our research is to find out how to build complex distributed interactive simulation systems and if Intelligent Agents really are the best solution. The AG-IVE project is still in progress. From the experiments that we have done, we can at least draw the following conclusions:

First, we can reuse available simulator and visualisation programs and interconnect them into one system. This is a practical way to develop interactive simulation system. Agents allow us to place essential control functionality outside the IVE modules, and thus to decrease their complexity. This makes the system more modular and increases the reusability of the modules. Second, using a Scenario Description file to explicitly represent shared object interface and interaction contexts, can make module integration more flexible and improve reusability. Third, using a run-time environment like the RTI can simplify the integration of distributed modules. The communication interface of HLA federates provides a simple and standard way to exchange information. This makes the development process of each module more independent. Finally, from the performance measurements that we have done, we can say that using Agents will not lead to large performance penalties.

In the near future, we will first finalise the basic services of AG-IVE, such as using interaction MA to handle scenario switches and providing interaction support. A user-friendly tool for specifying interaction scenarios will also be developed. After that we will also try to apply AG-IVE to the simulators like discrete event driven systems.

References

1. Franks S., Gomes F., Unger B., Cleary J.: State saving for interactive optimistic simulation; Proceedings of the 1997 workshop on Parallel and distributed simulation, (1997) 72-79
2. Salisbury, M. R: Command and Control Simulation Interface Language (CCSIL): status update. In Proceedings of the twelfth workshop on standards for the interoperability of defence simulations, (1995) 639-649
3. Ross P. Morley, van der Meulen Pieter S., Baltus P.: Getting a grasp on interactive simulation. In Raimund K. Ege, editor, Proceedings of the SCS Multiconference on Object-Oriented Simulation, Anaheim, CA, (1991) 151-157
4. Ferenci S., Perumalla K., Fujimoto R.: An Approach to Federating Parallel Simulators, in the Workshop on Parallel and Distributed Simulation, (2000)

5. Steinman, J.S.: SPEEDES: A Multiple-Synchronization Environment for Parallel Discrete Event Simulation. International Journal on Computer Simulation, (1992) 251-286
6. IEEE Std 1278.1-1995, IEEE Standard for Distributed Interactive Simulation -- Application Protocols. New York, NY: Institute of Electrical and Electronics Engineers, Inc., (1995)
7. Defence Modelling and Simulation Office (DMSO). High Level Architecture (HLA) homepage. http://hla.dmso.mil/, (2001)
8. Perumalla, K.S., Fujimoto, R.M.: Interactive Parallel Simulations with the Jane Framework. Special Issue of Future Generation Computer Systems, Elsevier Science, (2000)
9. Message Passing Interface Forum. MPI-2: A Message-Passing Interface standard. The international Journal of Supercomputer Applications and High Performance Computing, 12 (1-2), (1998)
10. Belleman R.G., Sloot P.M.A.: The Design of Dynamic Exploration Environments for Computational Steering Simulations, in the Proceedings of the SGI Users' Conference 2000, pp. 57-74. Academic Computer Centre CYFRONET AGH, Krakow, Poland, ISBN 83-902363-9-7, (2000)
11. The Beowulf cluster at SARA http://www.sara.nl/beowulf/, (2001)
12. Belleman R.G., Sloot P.M.A.: Simulated vascular reconstruction in a virtual operating theatre, in Computer Assisted Radiology and Surgery (Excerpta Medica, International Congress Series 1230), Elsevier Science B.V., Berlin, Germany, (2001) 938-944
13. Belleman R.G., Kaandorp J.A., Dijkman D., Sloot P.M.A.: GEOPROVE: Geometric Probes for Virtual Environments, in the proceedings of High-Performance Computing and Networking (HPCN Europe '99), Amsterdam, The Netherlands, in series Lecture Notes in Computer Science. Springer-Verlag, Berlin, ISBN 3-540-65821-1, (1999) 817-827
14. Wooldridge, M., Jenings, N.: Intelligent Agents: Theory and Practice, in Knowledge Engineering Review, 10(2); (1995) pp. 115-152
15. Cruz-Neira C., Sandin D.J., DeFanti T.A.: Surround-Screen Projection-Based Virtual Reality: The Design and Implementation of the CAVE. SIGGRAPH '93 Computer Graphics Conference, (1993) 135-142
16. Kandhai B.D.: Large Scale Lattice-Boltzmann Simulations (Computational Methods and Applications), PhD Thesis, Universiteit van Amsterdam, (1999)
17. Ziv J., Lempel A., A Universal Algorithm for Sequential Data Compression, IEEE Transactions on Information Theory, Vol. 23, No. 3, 337-343
18. CAVERNsoft, http://www.openchannelsoftware.org/projects/CAVERNsoft_G2/, (2001)
19. Amzi Prolog Logic Server, http://www.amzi.com/ (2002)

Computer-Assisted Learning of Chemical Experiments through a 3D Virtual Lab

Irene Luque Ruiz, Enrique López Espinosa, Gonzalo Cerruela García, and
Miguel Ángel Gómez-Nieto

Department of Computing and Numerical Analysis, University of Córdoba.
Campus Universitario de Rabanales, Edificio C2, Planta-3. E-14071 Córdoba, Spain
Phone: +34 957 212082   Fax: +34 957 218630
{mallurui, malloese, gcerruela, mangel}@uco.es

Abstract. A model for the development of virtual chemical experiments is proposed in this paper. The model is named $E(V)=M+m$ —*Experiment (Virtual)=Materials + method*— and is based on the definition of chemical experiments through the use of a set of materials which are manipulated following a method or protocol. The proposed $E(V)=M+m$ model has been implemented in a multimedia system, which is executed through a Web standard browser. The system developed allows the representation and subsequent building of chemistry experiments in virtual 3D worlds to any degree of complexity. This system integrates three main components: a complex object-relational database storing any kind of information relationship with the experiments (e.g. chemical substances, materials, apparatus, etc.), a VRML browser in charged to visualize the experiments, and a Java interface which is in charged to communicate the database with the 3D browser and the user.

1   Introduction

A great number of studies published showing the efforts investing by the chemical community to adopt the new computer technologies in the teaching of this discipline [1-7]. The published papers present teaching material like tutorials mainly aimed at secondary education [8-9], small-to-mid-sized software programs mainly directed to areas such as: periodic table information, balancing reactions, detailed composition, calculation of molecular masses, viewing molecular structures, interpretation of spectra, etc. [9-13] and laboratory software designed to simulate teaching lab experiments [14-16].

In this paper we proposed a model that permit avoid much of the existing inconveniences of the current chemical virtual labs and an environment suitable for the teaching of chemical experiments, as it offers the following features:

1. The proposed model permits a simple representation of the domain of the problem (the experiments). This model permits the inclusion of the teaching component, since content may be ordered and sequentialized for adaptation to both the student and the experiment.

P.M.A. Sloot et al. (Eds.): ICCS 2002, LNCS 2329, pp. 704–712, 2002.

2. There not exists any limitation of the experiments that can be defined. Experimental content can be reusable, allowing its partial or complete use in the building of new experiments.
3. Chemistry experiments can be performed at varying levels of complexity, in line with different educational grades.
4. The system may be executed in any platform. A standard Web browser is used as interface between the system and the user.
5. The 3D representation of the experiments in a VRML world facilitates the learning of its content by the user.
6. The integration between a complex database, a Java interface and the 3D world permit the full communication between the user and the systems during the performance of the experiments.

The paper first describes the objectives of the study, and then in Section 2 provides a description of the proposed model. The description includes specification of the structure and content of experiments in an object oriented model, allowing both the classification of knowledge in learning blocks and its sequencing in the teaching process. Section 3 describes the operating prototype, closing with a discussion of the project that mentions future projects and enhancements currently under progress.

2   Representing Virtual Chemical Experiments

The model proposed in the development of chemical virtual experiments is based on the supposition that a virtual experiment may be described by a method, which uses a set of materials. Running a virtual experiment, performed in an interactive computer environment using a system of dialogs, the user may experience a 3D presentation simulating a chemistry experiment. Each experiment takes place in a virtual environment containing all the materials with which the user might interact — the virtual laboratory. Labs may be shared by multiple experiments and may also be built using different technologies and software, so user interaction will depend on the technology employed to build them.

Each virtual experiment has an objective and is thus accompanied by descriptive content aimed at delivering knowledge for the user to learn. While the nature and extent of the experimental content may be highly varied, presentation and user interaction is independent of the experiment.

Each experiment has a unique associated method. A method is a precise description of the tasks and resources used in the experiment. Each method comprises structure and content. The structure of the method lists the processes to be carried out during the experiment, while the content includes all necessary information and the way in which this is manipulated therein.

The method's structure is made up of a sequence of phases. Each phase involves a global task or activity performed during the experiment, which may be broken down into further elementary tasks called stages.

This structural breakdown of an experiment, in which different knowledge blocks are organized chronologically in a working network, is a key factor in user learning [1,17].

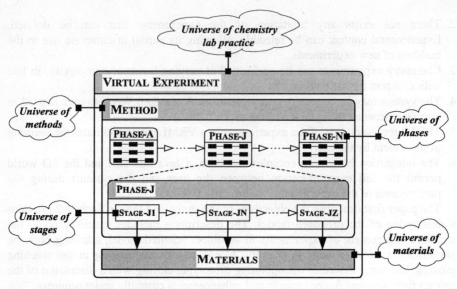

Fig. 1. Context diagram of the $E(V)=M+m$ model

Figure 1 shows a context diagram of the model $E(V)=M+m$. It may be seen that a virtual experiment represents the knowledge of a laboratory experiment through a method structured in phases, which in turn are divided into stages where the materials are manipulated. These structural elements of experiments are extracted from a general universe, enabling them to be reused in the same or indeed in a different experiment.

Let us now examine the structural elements of the model using a bottom-up approach, according to complexity.

2.1   Stages

Stages are basic tasks in the structural description of an experiment. Depending on the level of detail required for the description of an experiment, a task may vary from a simple manipulation such as "Obtain a pipette" to one like "Obtain a pipette, fill it, and empty the contents into an Erlenmeyer flask".

Stages imply manual or automated processes carried out in the lab and involving some laboratory material. They are distributed throughout the experiment, grouped in *Phases*. Within a phase, stages comprise a sequence of tasks that may be described by one or more directed graphs.

Take a *phase-k* consisting of "Pour 10 ml HCl from the stock bottle into an Erlenmeyer flask". This phase may be defined by the following stages:

PKS1: Obtain a 10 ml pipette from the rack

PKS2: Open the bottle of concentrated HCl.

PKS3: Pipette 10 ml HCl from the bottle.

PKS4: Take the Erlenmeyer flask from the shelf and place on the bench.

PKS5: Empty the pipette's contents into the Erlenmeyer flask

PKS6: Place the pipette in the sink.
PKS7: Close the bottle of concentrated HCl.

Figure 2 contains a transition diagram [18] showing the different routes that could be taken to perform *phase-k* (for a clearer description, the user is assumed to be reasonably skilled). Again in Figure 2, stages are represented by labeled boxes: shaded boxes show stages that could be the start of *phase-k* (*S1, S2, S4*), and the arrows indicate the transitions allowed between stages. If when performing *phase-k* the user chose to begin at stage *S1*, the next permissible stage would be *S2* or *S4*. If the user opted to follow the transition to *S2*, the next stage could be either *S3* or *S4*.

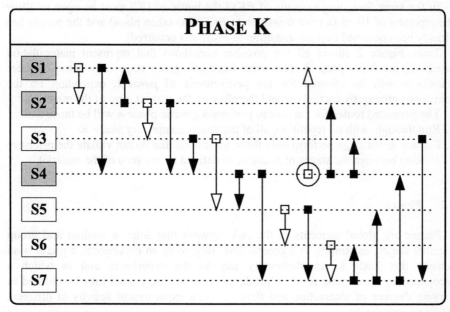

Fig. 2. Diagram showing permitted transitions for phase-k

If the transition to *S3* were chosen, the next move would have to be to *S4*, since getting to *S5* would suppose a transition prior to *S3*, that is, an Erlenmeyer flask must be present.

Alternatively, if the user had decided upon a transition to *S4*, the next move must be to *S3*, since moving on to *S5* would require a transition prior to stage *S3*, that is, the pipette has to be filled, and moving to *S7* would mean that not all the transitions to all the stages of *phase-k* could be completed by any of the existing routes.

Thus, if *phase-k* is started at *S1* it can be performed suitably or correctly by the system using any of the following routes:

R1:	S1, S2, S3, S4, S5, S6, S7	**R2**:	S1, S2, S3, S4, S5, S7, S6
R3:	S1, S2, S3, S4, S7, S5, S6	**R4**	S1, S2, S4, S3, S5, S6, S7
R5:	S1, S2, S4, S3, S5, S7, S6	**R6**:	S1, S2, S4, S3, S7, S5, S6
R7:	S1, S4, S2, S3, S5, S6, S7	**R8**:	S1, S4, S2, S3, S5, S7, S6
R9:	S1, S4, S2, S3, S7, S5, S6		

White arrows in Figure 2 show the permitted route that is considered, on defining *phase-k*, as the most recommendable and which will be taken into account by the system for the automatic execution of this phase without user intervention.

At each stage at least one *Material* is required for manipulation. The use of a material on occasion implies the presence of other materials, as shown by *PKS3* where a pipette is used —thus implying the existence of the bottle of concentrated HCl. Manipulation of material implies:

1. The presence of the material as well as those materials thereby affected.
2. A condition where the material manipulated and any other intervening material are in such a state that the manipulation may be successfully performed.

In the same foregoing example of *PKS3* the bottle of HCl must be open to allow the pipetting of 10 ml (a prior transition to *PKS2* has taken place) and the pipette has already been obtained (a prior transition to *PKS1* has occurred).

Thus, Figure 2 shows all the possible transitions that represent manipulation processes involving *phase-k*. However, from any given stage only some of these transitions will be allowed for the performance of *phase-k*, depending on the transitions previously carried out and therefore on the current state of the material.

The permitted routes for the user to perform a generic *phase-n* will be those that:

1. Run through, with no repetition, all of the stages comprising *phase-n*.
2. For any given stage perform only those transitions that do not violate the priorities existing between the stages of *phase-n*, as defined by the state of the material.

2.2 Phases

Phases are global elements in the task network that detail a method and group together stages, amounting to a general task. Regarding an experiment, a phase is an activity that leads to an elementary step in the experiment and in which a recognizable, measurable result is produced. Like stages, phases may be defined to varying degrees of abstraction and thus represent experimental activity at different levels of complexity.

2.3 Methods and Experiments

Methods describe the structure and chronological route of activities as occurring in the course of an *Experiment*. A method is an abstract sequence of procedures representing the tasks to be carried out, the material to be used and its manipulation in order to successfully perform a chemistry experiment. Since phases and stages may be described at various levels of abstraction, one laboratory experiment can be described by different methods at various levels of complexity.

The term "experiments" is used to describe practical laboratory experiments whose operational technique and underlying knowledge are to be passed on to the student through a virtual environment.

In the $E(V)=M+m$ model, an experiment is defined as the use of a method to manipulate a set of *Materials*. Thus each experiment has an associated method that is described by the phases and stages in which the materials are employed. Furthermore, an experiment has an associated virtual world containing the materials to be used and where the activities described in its associated method are carried out.

2.4   Materials

Materials represent the physical elements recognized in the virtual world or laboratory associated with an experiment and which can be used therein.

The materials category embraces a great variety of laboratory items, differentiated by their properties and associated functionalities. Hence, the context of material includes such diverse items as chemicals (e.g. existing products), furniture (e.g. lab benches), material (e.g. glassware) or instruments (e.g. pH meters).

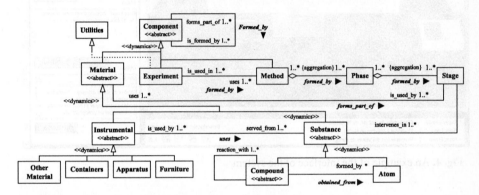

Fig. 3. Class diagram of the proposed model

2.5   Modeling Virtual Experiments

We was performed the implementation of the proposed model using the object-oriented paradigm, and complies with the UML standard [19]. Using the object model, a system can be described by a set of objects listing the properties and behavior of the system and each of its component parts.

The class diagram in Figure 3 shows the main entities of the static model representing the $E(V)=M+m$ model. A Component is considered to be any object existing in the system, and may in turn be formed by other components. The Component class is an abstract generalization class. By the term "abstract" we mean that objects belonging to the generalization class must be refined to one of the specialization classes of which it is composed.

A series of components is constructed, stored in a library and then used by a determined method, in a determined form and following a fixed timeline for the performance of an experiment, which in itself may be another component. A few examples of the objects included in this class would be: a decanting method, a precipitation flask, an experiment for the titration of NaOH with HCl, filter paper, a funnel, lab scales, etc.

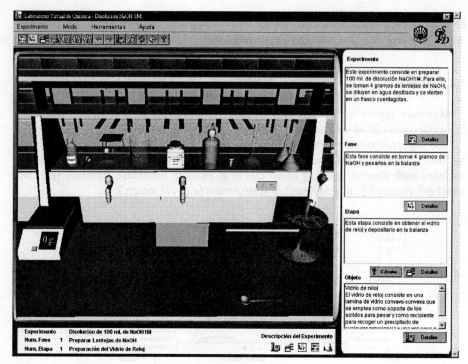

Fig. 4. An example of the interface of the system

3   Implementation of the model

An operational prototype of the proposed model has been built and is currently undergoing tests. It has been developed using the Java programming language, which is platform-independent and should run on any computer. The user interface has been created using Java applets that can be instantiated on most of the currently available browsers.

The Oracle 8i.X database manager handles information. The object-relational features of Oracle 8i.X permits the correct implementation of the class model described in the previous section. Moreover, this DBMS allows the storage of Java and VRML procedures in the very nucleus of the database alongside SQL procedure, in addition to storing media attributes like images and files in the definition of the table structure, which facilitates and enhances the development and performance of procedures [20,21].

The virtual world (the labs) and its components (materials) were developed under VRML 2.0, along with utilities to permit the graphical construction of these elements and their generation in VRML format [22,23].

The Figure 4 shows an example of the interface of the system. The Java interface coordinates the execution of the system, and it is in charged of the communication with the user, showing all information about of the evolution of the user-system interaction, which is carried out in a VRML virtual world.

Discussion

The proposal detailed in this paper is based on the use of virtual laboratories to provide extra support in the teaching of chemistry laboratory experiments. The proposal is based on $E(V)=M+m$ model that permits the structuring of knowledge in chunks of information describing basic stages in laboratory usage. The grouping of these stages in phases —both stages and phases are held in a set of permitted sequences— allows definition of the method or protocol for a lab experiment to be performed.

This structure, defined using a set of transitions that are allowed on the basis of the state of the materials employed, offers the possibility of the user carrying out an experiment in any permissible order of phases and stages and is thus not restricted to just one preset sequence, as is the case with other existing systems; moreover, the system can "advise" on the most suitable sequence and give feedback on the user's errors, a feature that adds greatly to its educational value.

The model has been implemented in an operational object-oriented prototype using the latest software technology (Oracle 8i, Java and VRML), bestowing the system with the set of properties required by authoring systems for the construction of intelligent tutorial systems; amongst these features are the reuse of software, easy integration of other components and a user-machine interface suited to the needs of assisted learning.

The authors are currently working to improve and expand several of the components of the subsystems that should form part of an ITS [24-26]. Additionally, work is underway on the definition of a suitable ontology for the construction of a dialog system between the user (student) and the system, such that through the use of an explanatory system running in parallel with the experiment, the student may both learn and be given explanations of conceptual and procedural aspects of the task in hand. In the present system this aspect is managed by static explanations held in .doc and .html files that are delivered to the user either on request or automatically, as might happen when some kind of exception arises.

References

1. Cloete, E., Electronic Education System Model., Comput. Educ., 36 (2001) 151-170
2. Lagowski, J.J. Chemical Education: Past, Present, and Future. J. Chem. Educ. 75 (1998) 425-436
3. Seal, K.C., Przasnysk, Z.H., Using The World Wide Web for Teaching Improvement., J. Chem. Inf. Comp. Sci., 36 (2001) 33-40
4. Warr, W.A., Communication and Communities of Chemistry, J. Chem. Inf. Comp. Sci., 38 (1998) 966-975
5. Borchardt, J.K., Improving the General Chemistry Laboratory, www-chenweb.com/alchem2000/news/nw-000922-edu.html

6. Jurs, P.C., Computer Software Application in Chemistry. Second Edition, John Wiley & Sons Inc., New York, (1996)
7. Rzepa, H.S., Internet-Based Computational Chemistry Tools. In Enclyclopedia of Computational Chemistry. Wiley, London, (1998)
8. (a) marian.creighton.edu/~ksmith/tutorials.html, (b) www.emory.edu/CHEMISTRY/ pointgrp, (c) www.chem.uwimona.edu.jm:1104/chemprob.html, (d) www.chemsoc. org.golbook
9. (a) chem.lapeer.org/Chem1Docs/Index.html, (b) www.geocities.com/CapeCanaveral/ 9687/index.html, (c) www.lanzadera.com/qgeum, (d) www.scripps.edu/~nwhite/ B/Download/, (e) jchemed.chem.wisc.edu/JCEsoft/Issues/ Series_SP/SP8/abs-sp8.html, (f) chem-www.mps.ohio-state.edu/~lars/ moviemol.html
10. Ivancinc, O., ChemPlus for Windows, J. Chem. Inf. Comp. Sci., 36 (1996) 919-921
11. Ivanov, A.S., Rumgantsev, A.B., Archakov, A.I., Education Program for Macromolecules Structure Examination, J. Chem. Inf. Comp. Sci., 36 (1996) 660-663
12. Yoshida, H., Mausura, H. MOLDA for Windows. A Molecular Modeling and Molecular Graphics Program using a VRML Viewer on Personal Computers. Journal of Chemical Software, 3 (1997) 147-156 (cssj.chem.sci.hiroshima-u-ac.jp/molda/download.htm)
13. Nussbaum, N., Rosas, R., Peisano, I., Cardenas, F., Development Intelligent Tutoring Systems Using Knowledge Structure. Computers and Education, 36 (2001) 15-32
14. Borman, S., Lab Systems Embrace the Web, Chem. Eng. News. (27th, January (1997) 25-27)
15. (a) www.modelscience.com/products.html, (b) www.chemnews.com, (c) website/ lineone.next/~chemie/reviews.html, (d) www2.acdlabs.com/ilabs, (e) www.chemsw.com/ 10202.htm, (f) www.compuchem.com/dldref/formdem.htm, (g) www.ir.chem.cmu.edu/ irProject/applets/virtuallab/applet_wPI.asp
16. Rzepa, H.S., Tonge, A.P., VchemLab: A Virtual Chemistry Laboratory. The Storage, Retrieval and Display of Chemical Information Using Standard Internet Tools, J. Chem. Inf. Comp. Sci., 36 (1998) 1048-1053
17. Luger, G.F., Stubblefield, W.A. Artificial Intelligence: Solutions and Strategies for Complex Problem Solving. Addison-Wesley, Reading MA (1992)
18. Jacobson, I., Booch, G., Rumbaugh, J., The Unified Software Development Process, Addison-Wesley Longman Inc, USA (1999)
19. Rumbaugh, J., Jacobson, I., Booch, G., The Unified Modeling Language. Reference Manual, Addison-Wesley Longman Inc, USA (1999)
20. Cary Anderson, J., Loy Stone, Blake Manual de Oracle JDeveloper, McGraw-Hill/Oracle Press, Madrid (1999)
21. Dorsey, P., Hudicka, J.R., Oracle8: Principios de Diseño de Base de Datos Usando UML , McGraw-Hill, Madrid (1999)
22. Roehl, Bernie, Couch, Justin, Reed-Ballreich, Cindy et al. Late Night VRML 2.0 with Java, Ziff-Davis Press, EmeryVille, California, USA (1997)
23. Lea, R., Matsuda, K. Miyashita, K. Java for 3D and VRML Worlds, New Riders, Indianápolis, USA (1996)
24. Luque Ruiz, I; Cruz Soto, J.L.; Gómez-Nieto, M.A. Error Detection, Recovery, and Repair in the Translation of Inorganic Nomenclatures. 1. A Study of the Problem. J. Chem. Inf. Comput. Sci. 36 (1996) 7-15. 2. A Proposed Strategy. J. Chem. Inf. Comput. Sci. 36 (1996) 16-24. 3. An Error Handler. J. Chem. Inf. Comput. Sci. 36 (1996) 483-490
25. Luque Ruiz, I.; Martínez Pedrajas, C.; Gómez-Nieto, M.A. Design and Development of Computer-Aided Chemical Systems: Representing and Balance of Inorganic Chemical Reactions. J. Chem. Inf. Comput. Sci. 40 (2000) 744-752
26. Luque Ruiz, I; Gómez-Nieto, M.A. Solving Incomplete Inorganic Chemical Systems through a Fuzzy Knowledge Frame. J. Chem. Inf. Comput. Sci., 41 (2001) 83-99.

Lattice-Boltzmann Based Large-Eddy Simulations Applied to Industrial Flows

Jos Derksen

Kramers Laboratorium, Applied Physics Department, Delft University of Technology, Prins Bernhardlaan6, 2628 BW Delft, The Netherlands
jos@klft.tn.tudelft.nl

Abstract. A procedure for simulating single-phase, turbulent fluid flow based on the lattice-Boltzmann method is described. In the large-eddy approach to turbulence modeling, the Smagorinsky subgrid-scale model, supplemented with wall damping, is applied exclusively. The assumptions underlying the model are discussed. Simulation results of some sample flow systems from chemical engineering practice are presented, with an emphasis on experimental validation, and grid effects.

1 Introduction

The performance of many industrial processes relies on turbulent flows. In agitated tanks, in tubular reactors, in bubble columns, and in fluidized beds, the turbulent fluctuations (that cover broad spatial and temporal spectra) enhance heat and mass transfer, control bubble and particles size distributions due to break-up and coalescence, and create contact area to facilitate chemical reactions. As a result, there definitely is a need for good insight, and accurate predictions of turbulent flows in process equipment. Computational Fluid Dynamics (CFD) is one of the tools that (to some extent) might be able to satisfy this need. In CFD, the mass and momentum balances that govern fluid flow are solved numerically. Because one of the characteristics of turbulence is its wide range of scales, a full representation (i.e. a Direct Numerical Simulation (DNS)) of the flow would require unrealistic amounts of computer resources. As an example, consider the flow with a macroscopic Reynolds number $Re = \dfrac{UL}{\nu} = 10^4$. The smallest dynamic flow scale η_K (i.e. the Kolmogorov scale) is related to the macro-scale through $\eta_K = L\,Re^{-3/4}$. A DNS would require a spatial resolution at least equal to η_K. As a result, to cover all length scales of the flow in three dimensions, a grid of $1,000^3 = 10^9$ cells has to be defined. Nowadays, these numbers start becoming feasible on massively parallel computer platforms.

There are, however, a few reasons not to pursue DNS of strongly turbulent flows in chemical engineering. In the first place, with a view to industrial applications, $Re=10^4$ is still considered a low value. In full-scale process equipment, the Reynolds number easily amounts to several millions. In the second place, the purpose of our research is

P.M.A. Sloot et al. (Eds.): ICCS 2002, LNCS 2329, pp. 713–722, 2002.
© Springer-Verlag Berlin Heidelberg 2002

to apply CFD as a scale-up, or design tool. This implies that it has to be possible to study alternative reactor designs, and process conditions within a limited amount of time, and on computer platforms widely available in academia and industry. In the third place, again with a view to the applications, the flow simulations are not the final goal of our efforts. They need to be related to the process, which next to fluid dynamics involves other physical and chemical phenomena. A good balance has to be found with respect to the computational efforts for modeling the various aspects of the process.

The alternative to DNS is turbulence modeling. Based on assumptions regarding turbulent fluctuations on the small scales (e.g. related to their supposedly universal character, their tendency to isotropy and equilibrium) turbulence models can be devised that distinguish between resolved and unresolved (i.e. modeled) phenomena. In the field of turbulence modeling, a division can be made between models based on the Reynolds Averaged Navier-Stokes (RANS) equations, and Large-Eddy Simulations (LES). For industrial (chemical engineering) applications, RANS modeling has a much longer tradition than LES. This is related to the lesser demand for computational resources of the former approach. In a RANS simulation, the grids can be coarser since only the average flow is explicitly simulated, and (in most cases) time-dependence is not taken into account. Furthermore, use can be made of symmetry in the boundary conditions. A large-eddy simulation shows a strong resemblance to a direct simulation: the three-dimensional, time-dependent Navier-Stokes equations are discretized to a grid that is (necessarily) too coarse. By viewing the flow field on the grid as a low-pass filtered representation of the true field, and by modeling the interaction between the filtered flow and the (unresolved) flow at length scales smaller than the grid spacing (the subgrid-scale (SGS) motion), the flow system is simulated.

The three-dimensionality, and time-dependence of LES make them much more computationally demanding than RANS simulations. Fine grids are required to meet the criteria for SGS modeling in LES [1]. I think, however, that it is worthwhile pursuing an LES approach to industrial flows for the following reasons. In the first place, by resolving the time-dependent flow field at high resolution, we are better able to account for the non-linearity's encountered in process modeling (e.g. chemical reactions, nucleation of crystals in flowing supersaturated liquids, transport of solid particles in a gas stream). In the second place, better flow field predictions are expected from LES in the case of inherently transient flows. By its very nature, turbulence is time-dependent in an incoherent and unpredictable way. If, on top of the turbulent fluctuations, coherent time-dependence is introduced through boundary conditions (as is the case in turbo machinery applications, or in stirred tanks), or through a macroscopic instability in the flow (such as vortex core precession in swirling flow), RANS based models might fail. In the closure models for the RANS equations, no clear spectral distinction exists between resolved and unresolved fluctuations (this in contrast to LES, where the grid makes the distinction). As a result, RANS closure models that are designed for non-coherent, turbulent fluctuations might (erroneously) affect the macroscopic, coherent fluctuations brought about by e.g. the boundary conditions.

In this paper, some examples of LES applied to industrially relevant, turbulent flows will be discussed. But first, our numerical method will be treated, which is based on the lattice-Boltzmann method. Second, a brief account of subgrid-scale modeling will be given. The examples comprise of agitated flow systems (stirred tanks), and swirling flows (with cyclone separators as the application).

2 Lattice-Boltzmann Method

Lattice-Boltzmann modeling of fluid flow has evolved from lattice-gas cellular automata (see e.g. the text book by Rothman & Zaleski [2]). The starting point of these particle-based methods is principally different from that of more traditional ways of numerical fluid flow modeling, such as finite element or finite volume methods. In the latter method, the continuous equations governing fluid flow are discretized. Lattice-gas and lattice-Boltzmann schemes start with a discrete system that can be proven to obey the Navier-Stokes equations. The discrete system that mimics fluid flow is a set of (fictitious) particles residing on a uniform lattice. Every time step, particles move to neighboring lattice sites, where they exchange momentum (i.e. collide) with particles coming in from other directions. The collision rules, and the topology of the lattice can be defined in such a way that macroscopically the Navier-Stokes equations are recovered [3]. The elegance of the method is its simplicity, and the locality of its operations. This locality (i.e. the absence of global numerical operations) is responsible for the high parallel efficiency of the method.

Lattice-Boltzmann methods have been applied to many problems involving fluid flow (see e.g. the review article by Chen & Doolen [3]). Many of the applications involve laminar flow in complexly shaped geometries (e.g. porous media modeling), interfacial phenomena, and multiphase flows. Large-eddy simulations of turbulent flows with lattice-Boltzmann schemes have been reported by Hou et al. [4], Somers [5], and Eggels [6].

We use PC based computer hardware for our lattice-Boltzmann based large-eddy simulations. By connecting the PC's (Dual-Pentium III 700 MHz machines running under Linux) through 100 BaseTX (100 Mbit/s) switched Ethernet, a Beowulf cluster is formed that can efficiently handle large jobs. A lattice-Boltzmann simulation on 360^3 (i.e. 47 million) cells, typically requires ten processors, 4 Gbyte of memory, and some 40 hours of wall-clock time to simulate one integral time scale (e.g. one impeller revolution in a stirred tank) of the flow.

3 Subgrid-scale Modeling

The most widely used model in LES is the Smagorinsky model [7]. It is an eddy-viscosity SGS model based on the assumption of local equilibrium between production and dissipation of turbulent kinetic energy. The expression for the eddy-viscosity (ν_e) reads:

$$\nu_e = \lambda_{mix}^2 \sqrt{\frac{1}{2}\left(\frac{\partial u_i}{\partial x_j} + \frac{\partial u_j}{\partial x_i}\right)^2} \tag{1}$$

with λ_{mix} a mixing length, and u_i the resolved (i.e. the grid scale) velocity in the i-th coordinate direction. In the Smagorinsky model, the mixing length is proportional to the grid spacing Δ: $\lambda_{mix} = c_s \Delta$, with c_s a constant. This constant can be estimated from turbulence theory under the assumption of local equilibrium: c_s=0.165. Note that the eddy-viscosity depends on the grid spacing. In case the grid spacing approaches zero, the eddy-viscosity vanishes, which should be, as zero grid spacing implies a direct simulation (i.e. a simulation without SGS model). In many simulations, especially of shear driven turbulence, the value of c_s is chosen between 0.06 and 1.2 [8]. This is because, especially near walls, the equilibrium assumption breaks down, and the length scales become too small to resolve. By decreasing the mixing length relative to the grid, the lack of spatial resolution is (artificially) compensated for.

The vicinity of walls poses more problems to the Smagorinsky model: the turbulence near the wall becomes anisotropic (fluctuations in the wall-normal direction are suppressed), and at the wall the SGS Reynolds stresses should become zero. These effects are not a priori accounted for in the model. The eddy-viscosity concept implies isotropic turbulence at the subgrid-scales throughout the flow domain, including the wall region. The problem of non-zero wall stresses can be repaired by applying wall-damping functions that force the eddy-viscosity (and as a result the SGS Reynolds stresses) to zero. In our studies we apply Van Driest wall damping functions [9].

In summary, the Smagorinsky model is not an obvious choice when attempting LES of flows in complexly shaped domains with intrinsic time-dependence. In the first place because of the presence of walls that pose principle problems related to anisotropy and resolution. In the second place because of the time-dependence, that will keep the flow off-equilibrium at least in parts of the flow domain. Dynamic SGS modeling (as introduced by Germano et al. [10]) might overcome some of the difficulties

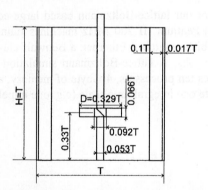

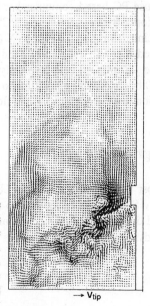

Figure 1. Left: stirred tank geometry. The tank has four, 90° spaced baffles, with a small wall clearance. The impeller has four pitched blades under a 45° angle. The top surface is closed. Right: single realization of the flow in a vertical plane midway between two baffles in terms of velocity vectors at Re=7,300. The simulation employed a 240^3 grid. The spatial resolution of the vector plot is (in axial and radial direction) twice as low as the actual resolution of the simulation.

discussed here. In dynamic modeling, a double low-pass filter operation is applied to the flow, with the aim of determining local values of the "constant" c_s. However, some problems (e.g. related to the stability of the method), even in simple flow geometries have been reported [1].

For the lack of better models, the Smagorinsky SGS model has been applied exclusively in our studies. The issues raised above with respect to the validity of this SGS model require critical evaluation and experimental validation of the simulated flow fields. In this respect, the extent to which the solution depends on the grid spacing requires special attention.

4 Agitated Flows

Blending, crystallization, suspension polymerization, particle coating, are some examples of processes carried out in stirred tanks. A revolving impeller generates a turbulent flow in a cylindrical vessel that (apart from the impeller) contains all sorts of static internals, e.g. baffles to prevent solid body rotation of the fluid, draft tubes to direct the flow, heat exchangers, nozzles for injection of reactants, etc. From experiments (e.g. [11]) it is well known that impellers generate complex vortex structures. These vortex structures are largely responsible for the mixing capacity of the flow system, and are associated with strong, anisotropic turbulent activity.

In order to standardize experimental and numerical work on stirred tank flow, a few *de facto* standard tank and impeller configurations can be identified. A standard tank has a height to diameter ratio of one, and has four, 90°-spaced baffles at its perimeter with a width of one tenth of the tank diameter. In a previous paper, large-eddy simulations on the flow driven by a Rushton turbine have been discussed [12]. In the present paper, results on flow simulations in a standard tank, equipped with a 45°-pitched blade turbine will be presented. The geometry is fully defined in Fig. 1. This

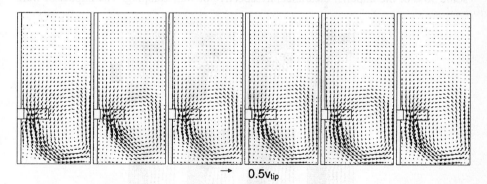

$\longrightarrow$ 0.5v_{tip}

Figure 2. Phase-averaged flow in the vertical plane midway between two baffles at Re=7,300. From left to right: experimental data [11]; LES on a 120^3 grid without wall damping functions; LES on a 240^3 grid without wall damping functions; LES on a 240^3 grid with wall damping functions; LES on a 360^3 grid without wall damping functions; LES on a 360^3 grid with wall damping functions. Per simulation, the averaging extended over at least 12 impeller revolutions. The LES results have been linearly interpolated to the experimental grid.

geometry was chosen because of the availability of a detailed experimental study by Schäfer et al. [11] on the flow field inside such a tank at $Re = \dfrac{ND^2}{v} = 7{,}300$ (with N the impeller's angular velocity in rev./s, D the tank diameter, and v the kinematic viscosity of the fluid).

In Fig. 1, a snapshot of the simulated flow field is presented. From this figure, the inhomogeneous character of the flow is obvious. A strongly turbulent, inclined impeller stream, directs fluid from the impeller towards the tank bottom. At the tank's outer wall, fluid (on average) moves in the upward direction. The stream at the tank wall is not strong enough to reach the top surface of the tank. About halfway the tank height, fluid near the outer wall is sucked into the impeller region. As a result, there is a large part of the tank volume with quiescent fluid.

In Fig. 2, experiment and simulations are compared in terms of the phase-average flow (i.e. the flow averaged over all angular positions of the impeller), in the vertical plane located midway between two baffles. The major flow features in the experimental field are the large, counter-clockwise circulation loop that approximately covers the lower half of the tank, and a clockwise rotating secondary loop in the vicinity of the bottom and the axis. The top half of the tank hardly shows coherent flow structures. In the simulations, presented in Fig. 2, the grid size has been varied, and wall damping has been switched off and on. The coarsest grid (120^3 nodes) has trouble in properly representing the primary circulation (its core being too high, and at too small radial position), as well as the secondary circulation (it extends almost to the impeller, whereas in the experiments it is limited to approximately 35% of the impeller bottom clearance). Predictions on the primary circulation significantly improve in case of the 240^3 grid. The influence of the wall damping functions is limited, and mainly related to the separation point of the primary circulation from the tank's outer wall: wall damping slightly shifts the separation point upwards, in better agreement with experiments. The secondary circulation is still too big on the 240^3 grid, its vertical size is some 50% of the impeller clearance, irrespective of wall damping. The impact of wall

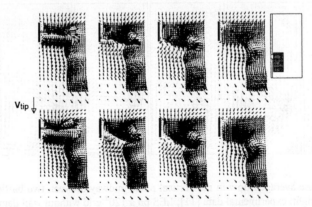

Figure 3. Phase-resolved velocity field at four different impeller angles (from left to right: 0°; 16°; 30°; 60°) in the vicinity of the turbine (the field of view is indicated in the right diagram) at $Re=7{,}300$. Top row: experiments [11]. Bottom row: LES on a 240^3 grid. The LES results have been linearly interpolated to the experimental grid.

damping functions on the simulation results is most pronounced for the 360^3 grid. This indicates that the grid has become sufficiently fine to feel the influence of details of the boundary layer. The separation point is located much too low for the 360^3 simulation without wall damping, but is almost perfectly predicted with wall damping. The size of the secondary circulation strongly benefits from the finer grid: it corresponds well with the experimental data.

Phase-averaged data obscure the details of the flow around the impeller blades. By conditionally averaging the flow at specific impeller positions, the action of the impeller on the fluid can be studied in more detail. In Fig. 3, phase-resolved LES results are confronted with experimental data [11]. Most striking flow feature is the tip vortex that is generated at the trailing edge of the inclined blade. It is a persistent phenomenon, in the sense that after two blade passages the vortex can still be distinguished in the phase-resolved flow field. The position, size, and shape (squeezed for small angular impeller positions, circular for larger angles) are well represented by the LES.

5 Swirling Flows

Swirling flows are challenging from a CFD viewpoint. They exhibit all sorts of, often counterintuitive, flow phenomena. A few examples are flow reversal, strongly anistropic turbulence, subcritical behavior, and (quasi) periodic fluctuations [13]. Our motivation for swirling flow studies lies in their application for separation of process streams containing phases with different density: Hydrocyclones are widely employed in the oil industries for separation of water and oil; gas cyclones are used for separating solid particles from a gas stream. Results on our large-eddy simulations of cyclonic flow have been published elsewhere [14]. In the present paper, I would like to focus on a different flow geometry, see Fig. 4. This so-called vortex tube was used in an experimental study by Escudier et al. [15]. In the latter article, it was demonstrated that downstream geometrical conditions had strong impact on the swirling flow in the entire geometry. The high-quality LDA data, and the visualizations presented in [15] were used to critically test the performance of our simulation procedures. One of the interesting issues here is to what extent the (isotropic) Smagorinsky SGS model [7] is capable of representing a flow system from which it is well know that isotropic RANS models (e.g. the k-ε model) are inadequate [16].

In the experimental paper, the exit diameter was varied between $D_e=D$ to $D_e=0.18D$, with D and D_e defined in Fig. 4. For the LES, three geometries were selected: $D_e=0.73D$, $D_e=0.45D$, and $D_e=0.33D$. For these three cases, the Reynolds

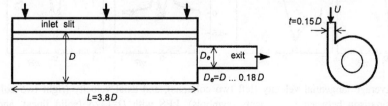

Figure 4. Vortex tube geometry as defined by Escudier et al. [15]. Fluid enters through a tangential slit, and exits through a contraction.

number was fixed at $Re = \dfrac{Ut}{\nu} = 1{,}900$, with U the (superficial) inflow velocity, and t the width of the inlet slit (see Fig. 4). This (atypical) definition of Re follows the experimental work. Reynolds numbers e.g. based on the superficial exit velocity, and diameter of the exit tube show much higher values (up to 51,000 for $D_e=0.33D$), indicating that subgrid-scale modeling is definitely necessary. The three geometrical layouts chosen were discretized on two grids: a coarse grid with $D=82\Delta$, and a fine grid with $D=132\Delta$, as a result, six simulation cases were defined. From our experience with cyclonic flow [14], we know that wall damping functions have to be applied to get a realistic representation of the swirling flow.

Velocity profiles of the tangential and axial velocity are presented in Fig. 5. The axial (i.e. z) position of the profiles, as indicated in the captions, is measured from the closed end wall of the vortex tube (located at the left side in Fig. 4). The profiles at $z=0.7D$ are close to the end wall, the profiles at $z=3.6D$ are close to the contraction into the exit tube. The impact of the exit pipe diameter on the entire flow is evident. The levels of the axial and tangential velocities strongly increase if the exit pipe diameter is made smaller (please note the different vertical scales in Fig. 5). Qualitatively, the flow features apparent from the profiles are well represented by the LES: the Rankine type of vortices as indicated by the tangential velocity profiles; the axial velocity deficit near the vortex core for the larger exit pipe configurations; the pres-

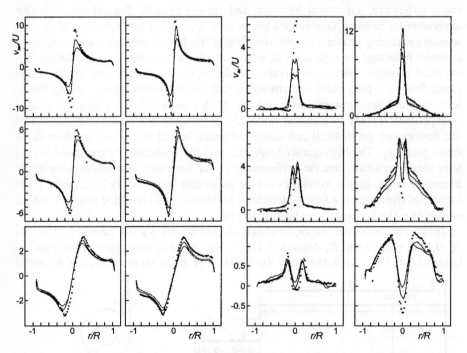

Figure 5. Average tangential velocity (left two columns), and axial velocity (right two columns). Comparison between experiments (symbols), LES with $D=132\Delta$ (solid lines), and $D=82\Delta$ (dashed lines). First and third column: $z=0.7D$; second and fourth column: $z=3.6D$. From bottom to top: $D_e/D=0.73$; $D_e/D=0.45$; $D_e/D=0.33$.

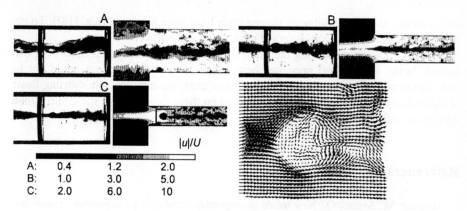

Figure 6. Visualization of the flow near the contraction. Experiment (dye injection) [15], and LES snapshots with $D=132\Delta$ (the contours indicate absolute velocity). A: $D_e/D=0.73$ ($Re_{exp}=200$, $Re_{LES}=1,900$); B: $D_e/D=0.45$ ($Re_{exp}=650$, $Re_{LES}=1,900$); C: $D_e/D=0.33$ ($Re_{exp}=1,600$, $Re_{LES}=1,900$). The rectangle in part C relates to the velocity vector field.

ence/absence of "shoulders" in the profiles of axial velocity. Quantitatively, the spatial resolution is (as it was for agitated flow systems) a key issue. The simulations with $D_e=0.33D$ that have the largest velocity gradients clearly lack spatial resolution. The trend, however, is in the right direction: the finer the grid, the better the predictions.

In the experiments, vortex breakdown was observed in the exit pipe. Vortex breakdown [17] is an abrupt change in the structure of a swirling flow. In bubble type vortex breakdown, a stagnation point is spontaneously formed at some position along the axis of the swirling flow. Vortex breakdown can trigger a laminar-turbulent transition. Figure 6 shows the experimental evidence: In cases B and C, a bubble-shaped dye structure can be observed just downstream the contraction indicating vortex breakdown. Upstream (left) of the bubble, a laminar vortex core can be observed, downstream (right) there clearly is turbulent flow. The simulations excellently represent this flow structure (a one-on-one comparison can in fact only be made for case C, where the experimental and numerical Reynolds number are close); and the laminar-turbulent transition associated with it. The stagnation point character of the breakdown is clear from the vector plot in Fig. 6. We may conclude that the (isotropic) Smagorinsky SGS model is able to resolve the major features of this swirling flow system.

6 Summary

A procedure for large-eddy simulations in complexly shaped flow geometries has been presented. The three-dimensional, time-dependent flow information that can be generated this way might be beneficial in modeling industrial processes. The lattice-Boltzmann method provided us with a means to reach high spatial, and temporal resolution. At the same time, we have limited ourselves to uniform, cubic grids. In many applications, especially if boundary layer flow is of critical importance, local grid refinement is required. In the lattice-Boltzmann community, various options for non-uniform, and even arbitrarily shaped grids have been proposed (e.g. [18]). To my best knowledge, no generally accepted method has evolved yet.

The assumptions underlying the Smagorinsky subgrid-scale model have been discussed. In the applications presented here, many of these assumptions could not be met: stirred tanks do not have equilibrium turbulence; the turbulence in swirling flows is highly anistropic. Still, the results presented on these flow systems show fairly good agreement with experimental data.

References

1. Lesieur, M., Métais, O.: New trends in large-eddy simulation of turbulence. Annual Review of Fluid Mechanics 28 (1996) 45
2. Rothman, D.H., Zaleski, S.: Lattice-gas cellular automata. Cambridge University Press, Cambridge (1997)
3. Chen, S., Doolen, G.D.: Lattice Boltzmann method for fluid flows. Annual Review of Fluid Mechanics 30 (1998) 329
4. Hou, S., Sterling, J., Chen, S., Doolen, G.D.: A lattice-Boltzmann subgrid model for high Reynolds number flows. Fields Institution Communications 6 (1996) 151
5. Somers, J.A.: Direct simulation of fluid flow with cellular automata and the lattice-Boltzmann equation. Applied Scientific Research 51 (1993) 127
6. Eggels, J.G.M.: Direct and large-eddy simulations of turbulent fluid flow using the lattice-Boltzmann scheme. International Journal of Heat and Fluid Flow 17 (1996) 307
7. Smagorinsky, J.: General circulation experiments with the primitive equations: 1. The basic experiment. Monthly Weather Review 91 (1963) 99
8. Eggels, J.G.M.: Direct and large eddy simulation of turbulent flow in a cylindrical pipe geometry. PhD thesis, Delft University of Technology (1994)
9. Van Driest, E.R.: On turbulent flow near a wall. Journal of Aerodynamic Science 23 (1956) 1007
10. Germano, M., Piomelli, U., Moin, P., Cabot, W.: A dynamic subgrid-scale eddy-viscosity model. Physics of Fluids A 3 (1991) 1760
11. Schäfer, M., Yianneskis, M., Wächter, P., Durst, F.: Trailing vortices around a 45° pitched-blade impeller. AIChE Journal 44 (1998) 1233
12. Derksen, J.J.,. Van den Akker, H.E.A.: Large eddy simulations on the flow driven by a Rushton turbine. AIChE Journal 45 (1999) 209
13. Escudier, M.P., Bornstein, J., Maxworthy, T.: The dynamics of confined vortices. Proceedings of the Royal Society London A 382 (1982) 335
14. Derksen, J.J., Van den Akker, H.E.A.: Simulation of vortex core precession in a reverse-flow cyclone. AIChE Journal 46 (2000) 1317
15. Escudier, M.P., Bornstein, J., Zehnder, N.: Observations and LDA measurements of confined turbulent vortex flow. Journal of Fluid Mechanics 98 (1980) 49
16. Hoekstra, A.J.: Gas flow field and collection efficiency of cyclone separators, PhD thesis, Delft University of Technology (2000)
17. Escudier, M.P.: Vortex breakdown: observations and explanations. Progress in Aerospace Sciences 25 (1988) 189
18. He, X., Doolen, G.: Lattice-Boltzmann method on curvilinear coordinate system: flow around a circular cylinder. Journal of Computational Physics 134 (1997) 306

Computational Study of the Pyrolysis Reactions and Coke Deposition in Industrial Naphtha Cracking

Aligholi Niaei[1], Jafar Towfighi[2], Mojtaba Sadrameli[3], Mir Esmaeil Masoumi[4]

[1]niaei@yahoo.com

[2]towfighi@modares.ac.ir

[3]sadramel@modares.ac.ir

[4]mmasoumi @modares.ac.ir

Chem. Eng. Dep., Tarbiat Modares University, P.O.Box 14155-4838, Tehran, Iran
http//www.modares.ac.ir

Abstract. The aim of this study is to develop a mechanistic reaction model for the pyrolysis of naphtha that can be used to predict the yields of the major products from a given naphtha sample with commercial indices. For this purpose, a computer control pilot plant system designed and assembled by Olefin Research Group, for studying the pyrolysis reaction kinetics, coke formation and different advance control algorithms. Experiments on the thermal cracking of naphtha were carried out in a tubular reactor under conditions as close as possible to those in the industrial operations. The reaction mechanism of thermal cracking of hydrocarbons is generally accepted as free-radical chain reactions. A complete reaction network, using a rigorous kinetic model, for the decomposition of the naphtha feed has been developed, which is used for the simulation of industrial naphtha crackers. Simultaneous simulation of the reactor and the firebox, provides a detailed understanding of the behavior and an accurate prediction of the furnace run length.

1 Introduction

Thermal cracking of hydrocarbons is one of the main processes for the production of olefins. The feed, ranging from light gaseous hydrocarbons to gas oil, is cracked in 4-8 tubular coils suspended in a fired rectangular furnace. The heat required for the endothermic reactions is provided via radiation burners in the sidewall or long flame burners in the bottom of the furnace. Mathematical models describing the simulation of the pyrolysis reactors need to be combined with complex kinetic models with important features such as coking, heat and mass transfer, firebox profiles and fluid dynamic characteristics. The central part of the model is the kinetic mechanism and related constants. Typically, the chemistry component can consist of several hundred reactions involving large numbers of species, and this together with the coupling of the kinetic rate equations to the physical features can lead to computationally calculations. The use of full kinetic mechanisms for on-line simulations such as plant optimization is therefore rarely possible in order to obtain a favorable product

P.M.A. Sloot et al. (Eds.): ICCS 2002, LNCS 2329, pp. 723–732, 2002.
© Springer-Verlag Berlin Heidelberg 2002

distribution or to reduce unwanted side effects. The present paper describes the development of a kinetic reaction network for the pyrolysis of naphtha, and with the help of an accurate coking kinetic model, the calculation of the temperature and product distribution in the reactor length and run time can be achieved. Simultaneous simulation of the reactor and the firebox provides a detailed understanding of the behavior of the cracking furnace. The experimental pilot results and simulation for the naphtha cracking are in good agreement with the industrial data.

2 Experimental Study

The aim of this study is to develop a mechanistic reaction model for the pyrolysis of naphtha that can be used to predict the yields of major products from a given naphtha sample with commercial indices. With the aid of the previous research works[1-5], pyrolysis theories, experimental data and plant output data a complete pyrolysis reaction network is generated according to the major feed component of naphtha. For this purpose experimental data from thermal cracking pilot plant system and ARPC olefin plant data is used for developing reaction model. This paper describes a computer control pilot plant system which is designed and assembled by the Olefin Research Group(ORL), for studying of the pyrolysis reaction kinetics, coke formation and different advance control algorithms. The feedstock was straight run naphtha from ARPC with the composition presented in Table 1. The standard reaction condition (T=810-860 C; residence time, 0.3-0.4 sec; total pressure 1 bar; dilution ratioH_2O /naphtha, 0.6- 0.8) are similar to those in the industrial reactors.

Table 1. Specification of Naphtha Feed (%wt)

Carbon No.	n-Paraffins	iso-Paraffins	Naphthenes	Aromatics
4	0.22	2.64		
5	25.22	17.94		
6	14.88	22.26	5.92	2.0
7	1.67	4.42	1.09	0.67
8		0.57		0.5

2.1 Experimental Setup and Procedure

The setup, used for the experiments of the naphtha thermal cracking is a computer controlled pilot plant unit, shown schematically in Figure 1. The unit is consisted to three sections:

Operating Section. The operating section includes the feeding, preheating, reaction and quench parts. Both gaseous and liquid hydrocarbons can be fed. The hydrocarbon and diluent water are pumped into the preheaters. Gas flows are metered by means of mass flow controllers. Liquid hydrocarbons and water as dilution steam are fed by

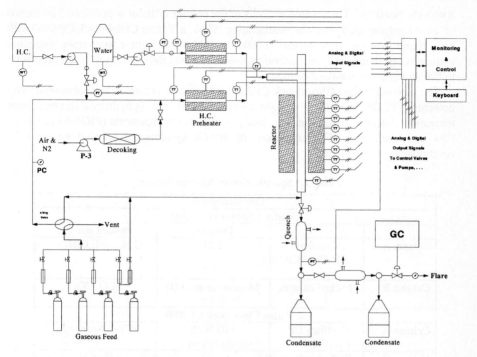

Fig.1. Flow Diagram of Thermal Cracking Pilot Plant

means of dowsing and pulsation-free pumps. The vessels containing the liquid feedstock are placed on digital balances witch is connected to the on-line computer for the feed monitoring. The setpoint of the pumps are set manually. The hydrocarbon and steam are perheated up 600 °C, and mixed.

The furnace is consisted of two electrical preheaters for the water and hydrocarbon feeds. An additional electrical heater is used for the reactor section. The preheaters are single zones and the reaction section heater is divided into eight zones, which can be heated independently to set any type of temperature profile. Each zone power can be controlled manually or by a control algorithm implemented on the process computer. The reactor is a 1000 mm long, 10 mm internal diameter tube, made of Inconel 600. There are eighteen thermocouples along the reactor, 8 inside the furnace, 8 on the external tube skin and additional 2 for measuring of XOT(Cross over temperature) and COT(Coil outlet temperature). The reactor is heated electrically and placed vertically in a cylindrical furnace. The pressure in the reactor is controlled by a back pressure from the reactor outlet. The analog signals of the thermocouples are connected to the process computer. The temperature reading is also visualized on a color digital thermometer display.

The reactor effluent is cooled in a double pipe heat exchanger by circulating ice water. Liquid products, tars and any possible solid particles cooled and separated by means of three glass condensers and cyclones. The cyclone is followed by a condenser in which steam and heavy products are condensed. The condensers is kept in a circulating ice water bath to maintain a constant temperature. A fraction of the product gas is then withdrawn for the on-line analysis on gas chromatograph, while the rest is sent directly to the flare.

Analysis Section. The on-line analysis of the reactor effluent is performed by means of two computerized gas chromatographs. First a Varian Chrompack CP3800 with two flame ionization detector (FID), analyses the light components including hydrogen, methane, oxygen, carbon monoxide, carbon dioxide, light hydrocarbons up to C4, and internal standard nitrogen.

The second Varian Chrompack CP3800 with two detectors, a thermal conductivity detectors (TCD), analyses the light components including hydrogen and one flame ionization detector (FID) for the analysis of heavy components (PIONA) including C5+ and aromatics. The conditions of the GC systems were carried out under the conditions given in Table 2.

Table 2. Specification of Analysis System

Chromatograph			
1.Varian Chrompack CP3800			
		Detector	Product analysis
Column A	Capillary CP-CIL 5CB	FID	C_2H_4,C_3H_6,C_4H_6
Column B	Packed column	Methanizer and FID in series	CO, CO_2
2.Varian Chrompack CP3800			
Column A	Capillary CP - CIL-PONA	FID With split/splittless	C_5+ , Aromatics
Column B	Packed column	TCD	H_2 ,CH_4

Computer Control Section. Computer control system consisted of hardware and software sections.

Hardware. The PIII process computer is connected on-line to the pilot plant and controls the main part of the unit. The connection with the pilot plant is done through analog to digital (A/D) converters, digital to analog (D/A) converters, and digital input-outputs.

Two Axiom A/D converters are available, one with 16-channel analog multiplexer for gas chromatographic signals and other a 32-channel analog multiplexer for the thermocouple signals. Ten D/A converters are available: two in AX-5411 system which provides the set points for water and hydrocarbon preheaters and eight in AX-5212 system which provides the set points for 8 zones of reaction section heater. An electronically kit which was made in ORL, is used for sending the control signals to the final control element of heaters (Tiristors).

Thirty tow digital input-output are available, eight in AX-5232 system which was not used, and twenty four in Ax-5411 system which has connected to a AX-755 system and used as input switches to start and stop of dowsing and cooling water pumps and main power of the electrical heaters. Two serial ports (RS-232) from the computer main board is used for the signal processing of digital balances.

Software. The on-line computer control software can be divided into two main parts of monitoring and control. In monitoring section, the process gas, tubeskin and heaters wall temperatures are displayed on a screen by means of a visual C program in 32-bit window operating system. This software was developed in ORL of T.M.U. Water and hydrocarbon flow rates are calculated from the weight change on the digital balances and the sampling rate. Monitoring of digital balances are achieved by means of a visual basic program which was developed by AND company. In control part the temperature profile of the reactor is stabilized by temperature controlling in each zone by means of a conventional PID controller. The set points for this temperature stabilizing control are included in the software. Implementation of advanced controllers and time varying set points are possible. Performances of advanced controllers can be tested experimentally by the control software in the pilot plant. All pilot plant measurements and control system information are saved in the text and graphical mode.

3 Kinetic Model Development

Many efforts have been done for the development of the reaction networks of the thermal cracking. The simplest model contains 22 molecular reactions which is developed by Kumar and Kunzru [1]. A simulation program based on the fundamental free radical reaction kinetics was developed by Dente et al.[2,3]. Pyrolysis of naphtha and lumped component such as n-paraffin, isoparaffin, naphthene and aromatics in gross molecular reactions were investigated [4]. The aim of this study is to develop a complete mechanistic reaction network that can predict the behavior of the cracking coils in different operating conditions and major feed components of naphtha. A simulation program of the reaction kinetics inside the coils of a thermal cracking unit can provide information on the effects of changing feed properties or alternative operating conditions on the thermal cracking product distribution.

The reaction mechanism of the thermal cracking is generally accepted as free-radical chain reactions. A complete reaction network, using a rigorous kinetic model, for the decomposition of naphtha feed is developed, and is used for the simulation of a naphtha cracker. The detailed mechanistic kinetic scheme in this simulation network, developed, involves over 542 of reactions and 91 molecular and radical species. As usual this chain radical mechanism consists of several radical and molecular elementary reactions, which can be briefly summarized in Table 3. On the basis of these intrinsic kinetic parameters, it was possible to describe the primary pyrolysis reactions of all the different hydrocarbon species and the generation of these reaction mechanisms allowed for development of the large kinetic schemes.

The governing mass, energy, and momentum balance equations for the cracking coil constitute the two-point boundary value problem which has a significant stiffness in the numerical simulation due to the large differences in concentration gradient between radicals and molecules. This problem can be tackled through the application of the Gear method. Details of the solution may be obtained from Towfighi et al. [5].

Table 3. Chemistry of the thermal cracking reactions

Reactions	Log (A) (l/s, 1/mol. s)	E (kcal/mol)
Radical Reactions		
Initiation		
n- $C_6H_{14} \rightarrow C_2H_5^{\cdot} + 1$- $C_4H_9^{\cdot}$	16.8	82
$C_3H_8 \rightarrow H^{\cdot} + 1$-$C_3H_7^{\cdot}$	17.3	92
H Abstraction		
$H^{\cdot} + C_2H_6 \rightarrow H_2 + C_2H_5^{\cdot}$	11.1	9.7
$CH_3^{\cdot} + C_3H_8 \rightarrow CH_4 + 2C_3H_7^{\cdot}$	8.8	10.5
Radical Decomposition		
i-$C_4H_9^{\cdot} \rightarrow CH_3^{\cdot} + C_3H_6$	14.2	32.7
n-$C_4H_9^{\cdot} \rightarrow H^{\cdot} + C_4H_8$	12.7	37.9
Radical Addition		
$CH_3^{\cdot} + C_3H_6 \rightarrow 1,n$-$C_4H_9^{\cdot}$	8.5	7.4
$CH_3^{\cdot} + C_2H_4 \rightarrow 1C_3H_7^{\cdot}$	8.1	7.7
Termination Reactions		
$C_2H_5^{\cdot} + C_2H_5^{\cdot} \rightarrow n$- C_4H_{10}	12.3	0.0
$CH_3^{\cdot} + \alpha$ - $C_3H_5^{\cdot} \rightarrow 1$- C_4H_8	12.9	0.0
Molecular Reactions		
Olefin Isomerization		
1-$C_4H_8 \rightarrow 2$-C_4H_8	12.4	14.3
2-$C_4H_8 \rightarrow 1$-C_4H_8	12.3	14.8
Olefin Decomposition		
1,3- $C_4H_6 + C_2H_2 \rightarrow C_6H_6 + H_2$	11.2	5.38
1,3- $C_4H_6 + C_2H_4 \rightarrow$ Cyclo C_6H_{10}	10.5	6.33

4 Coking Model

Coke formation in the pyrolysis of hydrocarbons is a complex phenomena due to the various free radical reactions involved. Three mechanisms contribute to the deposition of a coke layer include catalytic coking, asymptotic coking and condensation of polyaromatic. Asymptotic coking is the main mechanism occurs from the interaction between active sites on the coke layer with gas phase coke precursors. At the free radical sites unsaturated precursors from the gas phase react via addition, followed by a set of dehydrogenation and cyclization reactions finally yielding a graphitic coke layer. The present paper attends on the asymptotic coking mechanism and describes the development of a kinetic coking model in the pyrolysis of naphtha. A number of coke precursors are found to be contributed to the formation of coke. Literature survey and experimental data led to a coking model in which a number of coke precursors and the relative rates of coke deposition contribute to the formation of coke. The precursors are classified into different groups such as Olefinic (ethylene, propylene, 1-butene, 2-butene), Dienes(propadiene, butadiene), Acetylenic,

Aromatics(benzene, toluene, xylene, styrene), Naphthenes and C_5+ (unsaturate). The rate of coke formation out of olefinic group may be expressed as:

$$r_{cOlefinic} = (\alpha.C_{C2H4} + \beta.C_{C3H6} + \gamma.C_{C4H8})k_0.e^{-E/RT} \tag{1}$$

A reference component is chosen in each group (e.g. C_2H_4 in olefinic group) and α, β, γ the factors obtained from the relative coking rate derived by Kopinke et al.[6-8]. A similar approach lead to the coking model for the other groups is considered and 6 parallel reactions are achieved. The total rate of coke formation is expressed as:

$$r_c = \sum_{i=1}^{6} r_{c,Group} \tag{2}$$

Since the coking is slow, quasi steady state conditions may be assumed, so that, the deposition of coke can be shown as:

$$\frac{\partial C}{\partial t} = (d_t - 2t_c)\frac{\alpha.r_c}{4\rho_c} \tag{3}$$

Using the mathematical   model, the amount of coke deposited on the internal wall of the reactor tubes has been   calculated with a limiting value for tube skin temperature (e.g. for sample olefin plant is 1100 °C). In the following, the effect of the coke thickness on the operating parameters has been   demonstrated in the on-stream time of the furnace.

5 Results and Discussion

The results of simulation   in a sample industrial naphtha cracking furnace are illustrated in Figures 2-5. Table 4 shows the characteristics of a naphtha cracking furnace is used in present simulation. Figure 2 presents the variation of main products in the   length of the reactor. Figure 3 shows the main products yield such as ethylene, propylene, butadiene and aromatic over the run time of the reactor. The feed flow rates were also kept constant over the run length. The growth of the coke layer and the amount of heat transferred to the reacting process gas and pressure level are factors, which affect the ethylene yield and selectivity.

Figure 4 shows the simulated coke layer thickness as a function of run length. The coke formation takes   place at the temperature of the gas/coke interface. As a consequence, the coke layer grows fast there and creates an additional resistance to the heat transfer and causes a decrease of the tube cross sectional area. Increasing the heat fluxes increases the gas/ coke interface temperatures and the coking rates. The coke deposition reaches its maximum thickness in the last pass at the length of 42 m.

Figure 5   shows the evolution of the simulated process gas, coke surface, inner tube wall and external wall temperatures of coil in the length of reactor. In the first part of the reactor, the temperature profile shows a significant increase as well, but this is

Table 4. Basic Information of Industrial Cracking Furnace

Furnace characteristics	
Height (m)	11.473
Length (m)	10.488
Depth (m)	2.1
No. of side wall burners	108
Reactor configuration	
Total length (m)	45
Total length pass 1: (m)	22.5
Internal / External dia.(mm)	85 / 92
Total length pass 2: (m)	22.5
Internal / External dia.(mm)	121 / 130
Material properties	-1.257 + 0.0432 T
Tube thermal conductivity	6.46
Coke thermal conductivity	1680
Coke specific gravity (kg/m^3)	

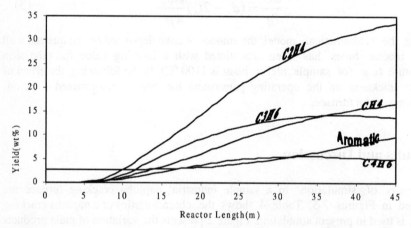

Fig. 2. Profile of main Product Yields along the Coil at End of Run of the Furnancs

mainly due to the higher heat flux. The peaks in the external tube skin temperature profile correspond to the higher temperature in the bottom of the furnace. The maximum value is reached at the end of second part. The on-stream time of cracking furnace is limited by the external tube skin temperature. In present investigation the maximum allowable temperature at the second part of coil is 1100

reached after 75 days. The simulated value is in agreement with the industrial plant data. The tube skin temperatures that are measured by the pyrometer, other thermal characteristics of the furnace such as gas and refractory temperatures and heat flux measurement are impossible or inaccurate.

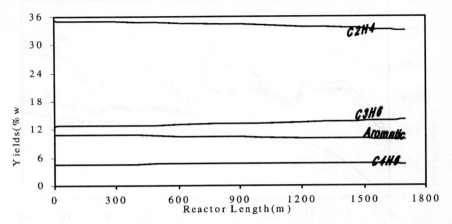

Fig. 3. Profile of main Product Yields along Run Time of the Furnace

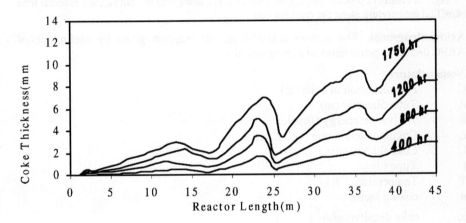

Fig. 4 Variation of Coke Thickness along the Coil at Different Run Length

6 Conclusion

Detailed and accurate information can be obtained from this simulation. The growth of a coke layer is accurately simulated, and so is the evolution of the external tube skin temperatures. The simulated and plant observation run lengths are in good agreement. Simulations of this kind can be used to optimize the furnace operation for the various feedstock and operating conditions. They can be used as a guide for the adaptation of the operating variables aiming at prolonging the run length of the furnace. The model and simulation software presented here are used as a guide for plant operation in olefin plant to control the furnace parameters.

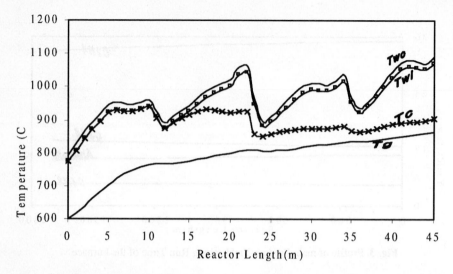

Fig. 5. Profile of process gas(T_g), coke surface(T_c), inner wall of coil(T_{wi}) and external tube skin(T_{wo}) temperature along the cracking coil

Acknowledgment. The authors acknowledge the support given by olefin units of ARPC and TPC petrochemical companies of Iran.

Nomenclature

C   Accumulation of coke (m)
d_t Tube diameter (m)
E   Activation energy (J/mole)
r_c Coking reaction rate (kg / m^3 .s)
t_c Coke thickness(m)
t   Time (hr)
T   Temperature ($^{\circ}$K)
α coking factor
ρ_c coke density(kg/m^3)

References

1. Kumar, P. and Kunzru, D.: Ind.Eng. Chem.Proc. Des. Dev., 32 (1985) 774-782
2. Dente, M., Ranzi, E., Goossens, A. G.: *Comput. Chem. Eng.*, 3 (1979) 61-72
3. Dente, M., Ranzi, E., Bozzano, G., Faravelli, T., Valkenburg, P.J.M.: AIChE Meeting, TX,75 (2001),1-20
4. Van Damme, P.S., and Froment, G.F.:   Ind. Eng. Chem. Proc. Des. Dev., 20(1981) 366-376
5. Towfighi, J., Nazari, H. Karimzadeh, R.: APCCHE / CHEMECA 93, Melbourne, Australia, 3 (1993) 337-342
6. Kopinke, F.D., Zimmermann, G., Nowak, S.: Carbon, 56,2 (1988) 117-124
7. Kopinke, F.D., Zimmermann, G., Reyners, G., Froment, G.F.: Ind. Eng. Chem. Res., 32(1993a) 56-61
8. Kopinke, F. D., Zimmermann, G., Reyners, G., Froment, G.F.: Ind. Eng. Chem. Res., 32(1993b) 2620-2625

An Accurate and Efficient Frontal Solver for Fully-Coupled Hygro-Thermo-Mechanical Problems*

Mauro Bianco[1], Gianfranco Bilardi[1], Francesco Pesavento[2],
Geppino Pucci[1], and Bernhard A. Schrefler[2]

[1] Dipartimento di Elettronica e Informatica, Università di Padova, Padova, Italy.
{bianco1,bilardi,geppo}@dei.unipd.it
[2] Dipartimento di Costruzioni e Trasporti, Università di Padova, Padova, Italy.
{pesa,bas}@caronte.dic.unipd.it

Abstract. Solving fully-coupled non-linear hygro-thermo-mechanical problems relative to the behavior of concrete at high temperatures is nowadays a very interesting and challenging computational problem. These models require an extensive use of computational resources, such as main memory and computational time, due to the great number of variables and the numerical characteristics of the coefficients of the linear systems involved.
In this paper a number of different variants of a frontal solver used within HITECOSP, an application developed within the BRITE Euram III "HITECO" EU project, to solve multiphase porous media problems, are presented, evaluated and compared with respect to their numerical accuracy and performance.
The final result of this activity is a new solver which is both much faster and more accurate than the original one. Specifically, the code runs over 5 times faster and numerical errors are reduced of up to three order of magnitude.

1 Introduction

Many successful methods exist for the solution of algebraic equations arising from the discretization of uncoupled problems. For coupled problems, especially when several fields are involved, the problem is still open. We concentrate here on a particular coupled multi-physics problem which deals with concrete under high temperature conditions [1, 2]. Such a model allows for instance to make residual lifetime predictions in concrete vessels of nuclear reactors or to predict the behavior of concrete walls in tunnel fires etc., [1–3]. The model has been implemented in the computer code HITECOSP in the framework of the BRITE Euram III "HITECO" [4] research project. This software uses a frontal technique to solve the final system resulting from the finite element (FE) implementation of the model. The aim of our work has been to improve the efficiency of HITECOSP's frontal solver in terms of performance as well as numerical accuracy, by to exploiting the various characteristics imposed by the model.

Improvements in term of performance have been obtained implementing a number of code optimizations, discussed in Section 2.2, with respect to original HITECOSP program.

In order to improve numerical accuracy in solving the linear systems produced by the Newton-Raphson like procedure used to solve the nonlinear systems arising from the FE implementation of the model, several pivoting strategies have been implemented

---

* This work was supported, in part, by MURST of Italy within the framework of the Centre for Science and Application of Advanced Computation Paradigms of the University of Padova.

P.M.A. Sloot et al. (Eds.): ICCS 2002, LNCS 2329, pp. 733–742, 2002.

and evaluated based on a modified component-wise backward error analysis (see Section 3). From our analysis it follows that the best strategy in terms of accuracy is also the best in terms of performance. In particular we have noticed errors of the same order of magnitude of the roundoff unit error and a further speed-up with respect to the optimized version of the original solver.

The rest of the paper is organized as follows. Section 2 deals with the frontal methods and the various code optimizations we have introduced and the pivoting strategies implemented. The metrics adopted to evaluate accuracy and performance are described in Section 3. Finally, the test cases used to evaluate our solvers are described in Section 4 while our results are shown in Section 5.

2   Frontal Method: Overview and Implementation

The large linear systems produced by the Newton-Raphson used in HITECOSP, are solved through the *frontal method* (see [5, 6] for a full description). The frontal method solves a linear system by working, at each step, only on a portion of the matrix (called *frontal matrix*), hence it is useful in those situations where core memory becomes the critical resource. The method works by sequentially executing two phases on each element of the finite element grid: an *assembly* phase and an *elimination* phase. During the assembly phase, the frontal matrix is augmented by the appropriate number of columns and rows relative, respectively, to the variables associated to the element and the equations containing those variables, and the matrix entries are updated to account for the new element. An entry becomes *fully-summed* if it will not receive further updates in any subsequent assembly phase. A column (resp., row) becomes fully-summed when all its entries become fully summed. A variable corresponding to a fully-summed column is also said fully-summed.

During the elimination phase, Gaussian elimination is applied to the frontal matrix, choosing the pivot in the block at the intersection of fully-summed rows and fully-summed columns. At each Gaussian elimination step, the pivot row is *eliminated*, i.e., it is stored somewhere into memory (typically onto a disk, if the problem is too large to fit in main memory). After the last elimination phase, back substitution on the reduced linear system is executed.

2.1   Pivoting Strategies

Recall that in the frontal method only a part of the matrix of the system is available at any given time, hence any pivoting strategy must be adopted to cope with this scenario. In particular, the pivot must always be chosen among those entries of the frontal matrix which reside in fully-summed rows and fully-summed columns.

Many strategies have been developed either to speed up the frontal solver or to improve its numerical stability. In this section we describe those strategies which we have implemented in order to find a solution that achieves the best trade-off between stability and performance for our particular physical problems.

Let A be the frontal matrix in a given elimination step. *Numerical pivoting* [6] entails choosing the pivot among the entries a_{ij} residing in fully-summed columns such that

$$|a_{ij}| \geq \alpha \max_k |a_{ik}|, \tag{1}$$

where $0 < \alpha \leq 1$ is a numerical constant and i is the index of a fully-summed row. Numerical pivoting was adopted to reduce the approximation error introduced by Gaussian elimination for the frontal method.

If an eligible pivot is not found, then the next element is assembled and a new search for a pivot is performed. This strategy is called *postordering for stability* [6]. The algorithm clearly terminates since eventually all the rows and all the columns of the frontal matrix will become fully-summed.

Often it is claimed in the literature that postordering for stability affects performance only slightly, while numerical stability is substantially increased. However, our experiments show that this is not the case for our physical problems. The failure of postordering in our scenario seems to be primarily due to the fact that the matrix entries in our mixed physical problems can be up to 40 orders of magnitude apart. Moreover, when a pivot is chosen, during the elimination step, fill-ins are produced in other rows, thus creating bonds among variables that do not belong to neighboring elements in the finite element mesh. Often these bonds prevent a row to be eliminated since the entries in its fully summed columns do not satisfy condition (1). In the next elimination steps this row continues to be filled with additional nonzero entries, hence the likelihood that it will not be chosen for elimination keeps on increasing, in a sort of "positive feedback" effect. Indeed we have observed extreme cases where rows entered in the frontal matrix at the very beginning of the solver's activity remains in the matrix until the very end. This phenomenon introduces two problems: not only does it cause the frontal matrix to grow inordinately, slowing down the program, but also worsens the numerical stability of the method, since a row that is present for a long time in the frontal matrix will sustain many operations on it, which is likely to amplify accumulation errors.

Another popular pivoting strategy is known as *minimum degree* [7, 6]. This strategy was proposed as a greedy way to reduce fill-ins when performing Gaussian elimination and was proved to be suited for symmetric, positive-definite matrices. Under minimum degree, the pivot is chosen as the diagonal element of a row with the minimum number of entries. Under the frontal method, the minimum degree strategy may be applied to the frontal matrix, choosing the pivot on the diagonal entry of the row with minimum number of entries in the block formed by the intersection between fully summed rows and fully summed columns. Since the full matrix of our systems has a symmetric structure, choosing pivots on the diagonal also preserves this symmetry inside the frontal matrix, allowing the data structures to be simplified.

It has to be remarked that the minimum degree strategy does not make any numerical consideration on the chosen pivot and it was originally developed for matrices that do not need such numerical precautions, e.g., positive definite symmetric matrices. Although our matrices, featuring great differences between numerical values of their entries, appear to be unsuitable for an application of minimum degree pivoting, our experiments have shown that the strategy is an effective way of reducing the accumulation error caused by postordering, perhaps due to the fact that it substantially reduces the amount of floating point operations. Indeed, a careful implementation of the minimum degree strategy has proven to feature both excellent performance and numerical accuracy for our problems.

The original application, HITECOSP, from which this work started, uses the following hybrid strategy. Before the pivot is chosen, if the absolute value of the previous pivot is less than a fixed numerical threshold value (10^{-4} in our cases), then the fully-summed rows are normalized so they will contain only values included between -1 and 1. After that, the pivot is chosen as the entry with the maximum absolute value among those in the intersection between fully-summed rows and columns. No postordering is performed. This strategy seems to work well for our physical problems. Namely, it exhibits good numerical accuracy and lends itself to an efficient implementation, which however requires a complete redesign of the relevant data structures.

2.2   Our Solutions

In this section we describe the frontal solvers which we have implemented. Each variant is characterized by a short name (in parentheses) which suggests the specific pivoting strategy adopted by the variant. Our first intervention has aimed at improving performance of the HITECOSP software (HIT) by providing an optimized implementation of its solver. In particular, the greatest improvement in performance has been obtained by the redesign of the main data structures in order to reduce at the minimum the number of linear searches inside arrays.

Another important issue that has been considered in redesigning HITECOSP's solver is the enhancement of the temporal and spatial locality of the memory accesses performed. To this purpose, extensive cache optimization has been applied, such as performing operations (e.g., row elimination, pivot search, etc.) within the solver by column rather that by row, in order to exploit the column major allocation of matrices featured by the FORTRAN compiler.

Another important source of performance enhancement has come from conditional branch optimization. As an example, consider that when computing on a sparse matrix, many operations are performed to no effect on zero entries (e.g., divisions and multiplications). However in most modern microprocessor architectures (and, in particular, on the ALPHA platform where our experiments run), keeping these operations improves performance since they take less cycles than those necessary by the processor to recover from a mispredicted branch. Indeed, conditional branch elimination in HITECOSP has improved the performance of the resulting code of up to 20%. This version of the frontal solver, implementing the same pivoting strategy as HITECOSP, named BASE, exhibits rather good performance.

The above code optimizations have also been employed to speed up the execution of the other solvers developed within our study. However, the main justification for designing new solvers mainly stems from our desire to compare the efficiency and numerical stability of the pivoting approach of HITECOSP with the other more established strategies described in the previous section. As a first step, basic numerical pivoting (as illustrated before) was implemented. Specifically we have developed a version without postordering that chooses as pivot the element that maximizes the value of α in (1) (NUMPIV), and another one implementing postordering (NUMPPO) which set $\alpha = 10^{-6}$ in (1).

Next, we have implemented the minimum degree strategy (MINDEG). This latter solver chooses the pivot on the diagonal of the frontal matrix and is endowed with recovery features when an entry in the diagonal is zero (however, this has never occurred in our experiments). A further optimization stems from the fact that, since the structure of the frontal matrix depends only on that of the finite element mesh, and, under the minimum degree strategy, the pivots depend only on the structure of the matrix, the pivotal sequence remains the same for all the executions of the frontal solver over the different iterations of the Newton-Raphson method (unless the chosen pivot is zero, which requires special care). Hence, it is possible to store the pivotal sequence during the first execution and to use it for the next ones, choosing the pivot, at each stage, with a single memory access. Our version of the minimum degree algorithm stores the pivotal sequence after the first execution of the frontal solver and uses it in the subsequent calls.

Finally, for the purpose of comparison, we have produced a further implementation (HSL) which uses the free version of the HSL (Harvell Subroutine Library) library [8]. Specifically, we have used the MA32 routine [9] that implements a frontal solver featuring a sophisticated numerical pivoting strategy. Version HSL gives us insight to compare our strategies against standard solutions available using third-party software.

3   Comparison Metrics

All the solver versions introduced in the previous section are evaluated in terms of their numerical stability (limited to the solution of the linear system) and performance. The next two sections discuss the metrics used to measure these two characteristics.

Evaluating numerical stability

Let $\tilde{x}$ be the computed (approximated) solution to one of the linear systems $Ax = b$ arising during the solution of a FEM problem. Approximation errors have been evaluated using a metric based on the *component-wise backward error* [10]

$$w = \min\{\varepsilon : (A + \Delta A)\tilde{x} = b + \Delta b, |\Delta A| \le \varepsilon|A|, |\Delta b| \le \varepsilon|b|\} \qquad (2)$$

(The absolute values and the comparisons are intended to be component-wise). It can be proved (see [10, 11]) that, setting $0/0 = 0$ and $\xi/0 = \infty$ if $\xi \ne 0$, w can be computed as

$$w = \max_i \frac{|r_i|}{(|A|\,|\tilde{x}| + |b|)_i} \qquad (3)$$

Intuitively, w measures the minimum variation that the matrix of the system and the right hand side vector should sustain to obtain the approximated $\tilde{x}$ solution. There is evidence in the literature that the component-wise backward error is more sensitive to instability than other metrics [12].

In this work, in order to have a more detailed description of the approximation errors introduced by our solvers, we have refined the component-wise backward error metric as follows. Let a_i denote the i-th row of A. Define the *ith equation error* to be

$$w_i = \min\left\{\varepsilon : (a_i + \Delta a_i)\tilde{x} = b_i + \Delta b_i, |\Delta a_i| \le \varepsilon|a_i|, |\Delta b_i| \le \varepsilon|b_i|\right\}. \qquad (4)$$

Value w_i gives a measure of "how well" vector $\tilde{x}$ satisfies the ith equation of the system and can be readily observed as follow.

Theorem 1. *Let v be a vector with $v_i = \frac{r_i}{(|A|\,|\tilde{x}|+|b|)_i}$. Then $w_i = v_i$.*

Proof. Let $\Delta\hat{a}_i$ and $\Delta\hat{b}_i$ be a pair of minimal perturbations associated to w_i. Since $|\Delta\hat{a}_i| \le w_i|a_i|$ and $|\Delta\hat{b}_i| \le w_i|b_i|$, we have that

$$|r_i| = |b_i - a_i\tilde{x}| = |\Delta\hat{a}_i\tilde{x} - \Delta\hat{b}_i| \le |\Delta\hat{a}_i||\tilde{x}| + |\Delta\hat{b}_i| \le w_i(|a_i||\tilde{x}| + |b_i|),$$

whence $w_i \ge |v_i|$. Also, from the definition of v it follows that $r_i = v_i(|a_i||\tilde{x}| + |b_i|)$. Define now $\Delta a'_i = v_i|a_i|\text{diag}(\text{sign}(\tilde{x}))$ and $\Delta b'_i = -v_i|b_i|$. It is easy to see that $(a_i + \Delta a'_i)\tilde{x} - (b_i + \Delta b'_i) = 0$. Therefore, since $|\Delta a'_i| = |v_i||a_i|$ and $|\Delta b'_i| = |v_i||b_i|$, it follows that $w_i \le |v_i|$, and the theorem follows.

It is easy to see that $\Delta\hat{A} = \text{diag}(v)|A|\text{diag}(\text{sign}(\tilde{x}))$ and $\Delta\hat{b} = -\text{diag}(v)|b|$ are such that $|\Delta\hat{A}| \le w|A|$, $|\Delta\hat{b}| \le w|b|$ and $(A + \Delta\hat{A})\tilde{x} - (b + \Delta\hat{b}) = 0$. Moreover, the above theorem proves that all the perturbations are the minimum possible, in the sense indicated by (4). Hence, vector v provides a readily obtainable indication about the minimum perturbation that each equation should sustain to obtain the approximated solution. In particular any element of v can be compared against the roundoff unit error to gain immediate appreciation of the significance of the corresponding perturbations. Finally, observe that the standard component-wise error metric w can be obtained as $w = \| v \|_\infty$.

A plot of vector v (using the equation indices as the abscissae) can be used to ascertain whether numerical errors tend to affect some groups of equations more than others. We feel that this is particularly useful in multi-physics applications as the ones treated in this paper.

All the metrics described above have been collected over several iterations of each solver. No significant variation of each metric has been observed over the different iterations. However, in what follows, we report the maximum errors encountered on each test case.

Measuring performances

Performance is measured both in terms of computational time and rate of floating point operations (Mflops) relatively to the frontal solver only.

4   Test Cases

The various solver versions have been executed on a number of test cases arising in several practical scenarios and characterized by an increasing complexity of the underlying physical system. As for the solver version, each test case is indicated by a short name (in parentheses):

1. *small column* (*smcol*): a regular 10×10 mesh of 100 elements in which all the degrees of freedom, except for the ones related to displacements, are set to zero.
2. *wall* (*wall*): 69 elements lined up in a row where the fifth degree of freedom (y-displacement) is fixed to zero;
3. *container* (*cont*): 288 elements outlining a container;
4. *column* (*col*): a square section of a column made of a 20×20 mesh of 400 elements;
5. *big column* (*bigcol*): like *column* but made of a 25×25 mesh of 625 elements;

Such a variety of test cases allows us to evaluate the behavior of the solver variants when the complexity of the physics behind the problem to be solved varies, from simpler (*smcol*) to harder (*bigcol*).

5   Results

The solver versions shown above have been tested on an Alpha workstation which uses a 21264 processor clocked at 666Mhz with two floating point units that make it capable of a 1354Mflops peak performance

The next two subsections report the results obtained by evaluating the various versions of the solvers described above. In section 5.1 we analyze the numerical stability properties exhibited by the solvers with respect to the metrics discussed in section 3. In section 5.2 we examine the performance achieved by the solvers.

5.1   Numerical Quality

Table 1 reports the component-wise backward error analysis (2) for all the solver versions and test cases described before. We note that MINDEG exhibits the least errors (in order of magnitude), scaling extremely well as the physical problems become more complex. Also, the table shows that HIT and BASE do not exhibit exactly the same errors, with BASE featuring slightly larger errors. This is explained by the fact that HIT uses some extra heuristic precautions to reduce fill-ins. We have chosen not to implement these expedients in BASE, since they complicate the code while not providing significant improvements in term of either accuracy or performance.

Table 1. Component-wise backward errors exhibited by the various solvers for each test cases.

	HIT	BASE	NEWPIV	NEWPPO	MINDEG	HSL
smcol	$3 \cdot 10^{-16}$	$3 \cdot 10^{-16}$	$3 \cdot 10^{-16}$	$3 \cdot 10^{-16}$	$3 \cdot 10^{-16}$	$4 \cdot 10^{-16}$
wall	$4 \cdot 10^{-14}$	$1 \cdot 10^{-13}$	$1 \cdot 10^{-11}$	$3 \cdot 10^{-6}$	$4 \cdot 10^{-16}$	$1 \cdot 10^{-6}$
cont	$6 \cdot 10^{-13}$	$5 \cdot 10^{-13}$	$6 \cdot 10^{-11}$	$2 \cdot 10^{-4}$	$6 \cdot 10^{-16}$	$5 \cdot 10^{-3}$
col	$4 \cdot 10^{-12}$	$2 \cdot 10^{-12}$	$2 \cdot 10^{-5}$	$2 \cdot 10^{-3}$	$9 \cdot 10^{-16}$	$4 \cdot 10^{-4}$
bigcol	$2 \cdot 10^{-12}$	$6 \cdot 10^{-13}$	$7 \cdot 10^{-5}$	$5 \cdot 10^{-3}$	$1 \cdot 10^{-15}$	$5 \cdot 10^{-5}$

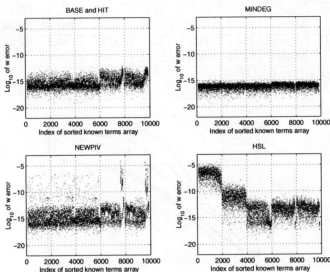

Fig. 1. Plot of the component-wise errors for the *bigcol* test case, exhibited by the various solvers. The abscissae are the equations indices and the ordinates are the base 10 logarithms of the errors.

If we compare NEWPIV with NEWPPO, implementing, respectively, numerical pivoting without and with postordering, we see that the errors exhibited by the latter (in all cases except *smcol*) are orders of magnitude worse than those exhibited by the former. Indeed, for the two largest test cases, the errors exhibited by NEWPPO in the solution of the linear systems became so large to prevent the Newton-Raphson method from converge. This provides numerical evidence that postordering does not achieve its intended purpose in our physical scenario.

To achieve a more profound understanding on the numerical behavior of the implemented variants, in Figure 1, we show all the components of the v vector (see sec. 3) for different solvers running the *bigcol* test case, plotted (in logarithmic scale) against their respective indices. For our specific physical problems, we have that indices between 1 and 2000 are relative gas pressure equations, indices between 2001 and 4000 are relative to capillary pressure, indices between 4001 and 6000 are relative to temperatures, and, finally, the remaining indices are relative to displacement equations. It is interesting to observe how errors tend to cluster according to the type of equation.

It is clear from the figure that MINDEG exhibits extremely homogeneous errors that are all close to the roundoff unit error (which is about 10^{-16} for our machine), while BASE, although still behaving quite well, tends to show a more varied range of errors, which implies that different equations (corresponding to different physical constraints) are solved with different degrees of accuracy. As for NEWPIV, we can see a rather good behavior on average, but the plot highlights a non-negligible set of outliers with high

Table 2. Times and floating point operations rates (frontal solver only). Note the $+\infty$ entries in the NEWPPO columns are related to that test cases for which the method does not converge as the errors become too large.

		HIT	BASE	NEWPIV	NEWPPO	MINDEG	HSL
wall	Time (s)	4.6	2.23	2.48	1.96	1.07	4.48
	Mflops	59.54	109.18	134.15	163.08	161.34	231.73
smcol	Time (s)	6.05	1.98	2.12	1.83	1.64	6.97
	Mflops	50.28	164.3	168.94	179.05	180.45	201.91
cont	Time (s)	77.5	27.5	30.4	790.0	22.9	207.1
	Mflops	79.63	237.63	232.06	98.4	262.02	206.47
col	Time (s)	314.1	116.8	124.5	3212	96.5	1230.5
	Mflops	54.08	224.09	222.62	7.89	256.93	154.76
bigcol	Time (s)	1276.4	260.1	274.0	$+\infty$	225.9	1609.5
	Mflops	30.58	232.53	226.6	–	261.77	137.47

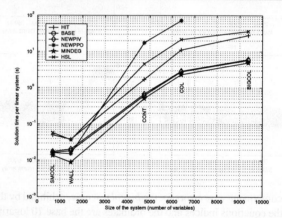

Fig. 2. Time taken by the various solvers on a single linear system as a function of the size of the system itself.

values of the error w. Finally, looking at the plot for HSL, we may note how it behaves very poorly for two entire groups of equations (especially those related to pressures) while it is only slightly worse that BASE for the other two groups.

5.2   Performances

Table 2 shows the execution times and the floating point operation rates (Mflops) exhibited by the frontal solvers. Each test case involves several time steps, with each time step in turn requiring the solution of a number of linear systems, one per Newton-Raphson iteration. More specifically, the number of linear systems solved in each test case is about 120 for *wall* and *smcol*, 45 for *cont* and *col*, and 30 for *bigcol*. For the last three test cases, the most computationally intensive ones, we have that the solver accounts for about 75% of the total time for MINDEG, and goes up to 95% for HIT. This fact justifies our focusing on the solver only, rather than other parts of the program, such as the numerical integration routines.

In Figure 2 we plot the time taken by the various solvers on a single linear system as a function the size of the system itself. We want to remark that the size of the system is not, however, the only parameter that affects performance, since, depending on the

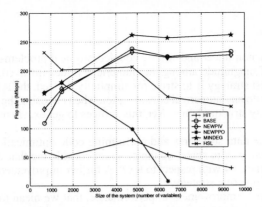

Fig. 3. Floating point operations rates exhibited by the solvers varying the size of the systems resolved.

solver used, other peculiarities may influence execution times, such as, for example, the shape of the mesh.

Looking at Table 2, we note that MINDEG exhibits the by far best performance. Together with the numerical stability data presented in the previous section, this implies that MINDEG achieves both performance and accuracy at the same time, which is somewhat surprising since an increase in accuracy often comes at the expense of a deterioration of performance.

Unlike HIT and HSL, MINDEG, BASE, and NEWPIV seem to be able to sustain high flops rate when the size of the problem increases (see fig. 3). Such scalability rewards our redesign of the data structures, which affords access times to data which are independent of the amount of the data itself. Comparing BASE against HIT, we see that redesigning the data structures, optimizing data accesses and carefully eliminating conditional branches alone made the solver more than 4 times faster. Changing the pivoting strategy yielded an extra time saving: indeed, MINDEG is more than 5.6 times faster than HIT for the *bigcol* test case.

Looking at the numerical pivoting strategies, we note that HSL exhibits a rather unsatisfactory performance. This seems to be mainly due to postordering. Comparing NEWPPO to NEWPIV, for the test cases for which both strategies converge, we note that when postordering is extensively used (as for *cont* and *col* test cases), the execution time explodes. This can be explained by the feedback effect described in section 2.1 and due to postordering. It has to be remarked that HSL behaves quite well for the *wall* and *smcol* test cases, for which it exhibits the highest flops rate. In fact, when the physical problem is simple, HSL becomes competitive. However, as the problem becomes more complex, the time performance of HSL degrades, even though its floating point operations rate remains quite high. As a bottomline we can say that HSL features a very good implementation (high flops rate), but its pivoting strategy, however, turns out to be a poor choice for our physical problems.

Going back to NEWPIV and NEWPPO, we note that for the *wall* and *smcol* test cases, the latter exhibits better execution times than the former. This is due to the fact that, while NEWPIV maximizes α in equation (1), NEWPPO simply picks the first element which satisfies (1) for a fixed $\alpha = 10^{-6}$. This strategy proves to be beneficial for performance since fewer entries of the frontal matrix need to be scanned. The gain in performance is however limited to those simple cases where postordering is rarely applied. We have chosen not to pick the best possible pivot when implementing postordering since we have observed that otherwise some rows would remain longer in the frontal matrix, which has detrimental effects on both time performance and accuracy.

6   Conclusion

When solving very non linear and strictly coupled physical problems where there may be many orders of magnitude among the numerical values involved, our experiments suggest that the best strategy is to strive for simplicity. Indeed, the MINDEG version of the solver does not do any numerical consideration when choosing the pivot in Gaussian elimination, but only structural ones. Yet, this suffices to get excellent performance and good accuracy.

Redesigning the data structures and performing code optimizations has proved to be the most effective way to speed-up the program considering that BASE achieves an improvement of a factor 4 with respect to HIT. A further improvement is then obtained by simplifying the pivoting strategy.

A possible further improvement in performance, that we mean to investigate, is to find the right tradeoff between avoiding linear searches inside the arrays and limiting indirect addressing. Specifically, observe that data structures designed to avoid linear searches make large use of indirect addressing, which, however, may disrupt temporal locality and slow down the algorithm by forcing the processor to wait for the data to become available from main memory.

References

1. D. Gawin, C. E. Majorana, F. Pesavento and B. A. Schrefler. A fully coupled multiphase FE model of hygro- thermo- mechanical behaviour of concrete at high temperature. In Computational Mechanics., Onate, E. & Idelsohn, S.R. (eds.), New Trends and Applications:, Proc. of the 4th World Congress on Computational Mechanics, Buenos Aires 1998; 1–19. Barcelona: CIMNE, 1998.
2. D. Gawin, C. E. Majorana and B. A. Schrefler. Numerical analysis of hygro-thermic behaviour and damage of concrete at high temperature. In Mech. Cohes.-Frict. Mater. 1999; 4:37–74.
3. D. Gawin, F. Pesavento and B. A. Schrefler. Modelling of hygro-thermal behaviour and damage of concrete at temperature above the critical point of water. Submitted for publication.
4. BRITE Euram III BRPR-CT95-0065 1999 "HITECO". Understanding and industrial applications of High Performance Concrete in High Temperature Environment, Final Report, 1999.
5. B. M. Irons. A frontal solution program for finite element analysis Int. J. Numer. Meth. Engng, 1970; 2:5–32.
6. I. S. Duff, A. M. Erisman and J. K. Reid. *Direct Methods for Sparse Matrices*, Clarendon Press, 1986.
7. W. F. Tinney and J. W. Walker. Direct solutions of sparse network equations by optimally ordered triangular factorization, Proc. IEEE 55, 1967; 1801–1809,
8. HSL (Formerly the Harwell Subroutine Library). http://www.cse.clrc.ac.uk/Activity/HSL
9. I. S. Duff and J. K. Reid. MA32 - a package for solving sparse unsymmetric systems using the frontal method, *Report R.10079*, HMSO, London, 1981.
10. W. Oettli and W. Prager. Compatibility of Approximate Solution of Linear Equations with Given Error Bounds for Coefficients and Right-Hand Sides, *Numerische Mathematik*, 1964; 6:405–409.
11. N. J. Higham. How accurate is Gaussian elimination? In D. F. Griffiths and G. A. Watson, editors, *Numerical Analysis 1989, Proceedings of the 13th Dundee Conference*, volume 228, pages 137–154, Essex, UK, 1990. Longman Scientific and Technical.
12. N. J. Higham. Testing Linear Algebra Software, in R. F. Boisvert, editor, *Quality of Numerical Software: Assessment and Enhancement*, pages 109-122. Chapman and Hall, London, 1997
13. I. S. Duff and J. A. Scott. The use of multiple fronts in Gaussian Elimination, *Technical Report RAL-TR-94-040*, Department for Computation and Information, Rutherford Appleton Laboratory, September 1994.

Utilising Computational Fluid Dynamics (CFD) for the Modelling of Granular Material in Large-Scale Engineering Processes

Nicholas Christakis[1], Pierre Chapelle[1], Mayur K.Patel[1], Mark Cross[1],
Ian Bridle[2], Hadi Abou-Chakra[3] and John Baxter[3]

[1] Centre for Numerical Modelling and Process Analysis, University of Greenwich, Old Royal Naval College, Park Row, Greenwich,
SE10 9LS London, UK
{N.Christakis, P.Chapelle, M.K.Patel, M.Cross}@gre.ac.uk
[2] The Wolfson Centre, University of Greenwich, Wellington Street, Woolwich,
SE18 5PF London, UK
I.Bridle@gre.ac.uk
[3] Department of Chemical and Process Engineering, University of Surrey,
6GU2 5XH 9042 Guildford, Surrey, UK
{H.Abou-Chakra, J.Baxter}@surrey.ac.uk

Abstract. In this paper, the framework is described for the modelling of granular material by employing Computational Fluid Dynamics (CFD). This is achieved through the use and implementation in the continuum theory of constitutive relations, which are derived in a granular dynamics framework and parametrise particle interactions that occur at the micro-scale level. The simulation of a process often met in bulk solids handling industrial plants involving granular matter, (i.e. filling of a flat-bottomed bin with a binary material mixture through pneumatic conveying-emptying of the bin in core flow mode-pneumatic conveying of the material coming out of a the bin) is presented. The results of the presented simulation demonstrate the capability of the numerical model to represent successfully key granular processes (i.e. segregation/degradation), the prediction of which is of great importance in the process engineering industry.

1 Introduction

In recent years significant effort has been put into the modelling of granular flows using a continuum mechanics approach ([1-2]). Although these models may be partially successful in capturing some characteristics of the flow, they do not incorporate essential information on material parameters, which are needed to model the various interactions between different particles or particles with their surrounding solid boundaries. Thus, they can not be used to simulate processes, which are of great im-

P.M.A. Sloot et al. (Eds.): ICCS 2002, LNCS 2329, pp. 743–752, 2002.

portance in the process engineering industry (i.e. hopper filling/emptying, pneumatic conveying etc.), where these interactions lead to phenomena such as particle size segregation or particle degradation/breakage.

On the other hand, micro-mechanical models are able to describe successfully the flow of granular material by accounting for all various types of interactions at the microscopic level ([3-4]). However, these models can only be applied to a small number of discrete particles, because of the complexity of the simulated processes. Therefore, such models are not suitable for large-scale process modelling, as considerable amounts of computing time are required for the simulation of processes that involve large numbers of particles.

In the present paper it is argued that the micro-mechanical behaviour of different particle species in a multi-component granular mixture, can be parametrised and employed into a continuum framework in the form of constitutive models. In this way, by embedding the necessary micro-mechanical information for the individual particle species within the continuum theory, the modelling of multi-component granular mixtures is enabled.

In this work, the continuum framework and the micro-mechanical parametrisations, employed to account for particle size segregation and breakage/degradation, are discussed. As an example, the simulation of a filling (through dilute phase pneumatic conveying) – emptying (in core flow mode) of a flat bottomed bin and further pneumatic conveying of the material coming out of the bin is presented. Conclusions are then drawn on the capability of the numerical model to realistically capture key granular processes (i.e. segregation /degradation) and its utilisation as a powerful computational tool by the process engineering community.

2 The Computational Framework

The full set of flow equations was solved using PHYSICA, a Computational Fluid Dynamics (CFD) finite-volume code developed at the University of Greenwich. The PHYSICA toolkit is a three-dimensional, fully unstructured-mesh modular suite of software for the simulation of coupled physical phenomena (see e.g. [5-7]).

The CFD code is utilised for the solution of conservation equations for mass and bulk momentum in the computational domain. Equations for energy were not solved because energy-linked flow parameters were accounted for in the micro-mechanical constitutive models, which link the granular temperature of the flow directly to the bulk velocity gradients via kinetic/theoretical considerations [8]. The effectiveness of this assumption will be demonstrated during the presentation of the numerical simulations.

It should also be noted that the computational framework made use of micro-mechanical criteria in order to predict the flow boundary and the existence of stagnant zones (core flow mode) during the flow of granular materials [9].

A scalar equation was solved for each of the individual fractions f_i of the mixture, representing the fractional volume of each of the material components in every computational cell. The summation of all individual fractions in a cell gave the total amount of material present in that cell at a certain time. This sum is only allowed to take values between 0 (cell empty of material) and the maximum allowed packing fraction (always less than unity). The maximum allowed packing fraction is a function of the individual components' shapes, sizes etc. and should be a model-input value, determined through experimental data. The scalar equation for each individual f_i may be written as

$$\partial f_i / \partial t + \nabla \cdot \{ f_i (\mathbf{u}_b + \mathbf{u}_{seg}) \} = S_i .\tag{1}$$

where $\mathbf{u}_b$ is the bulk velocity (as results from the solution of the momentum equation), $\mathbf{u}_{seg}$ is a segregation "drift" velocity and S_i is a source/sink term representing degradation in the i-th particle size class. Some details about the parametrisation of the segregation/degradation processes in the micro-mechanical framework are given in the following sections.

2.1 Parametrisation of Segregation

The segregation "drift" velocities were analysed in the micro-mechanical framework, by using principles of kinetic theory [10]. In this way, for each material component three transport processes, that effectively lead to segregation, were identified: (a) **Shear-induced segregation** (representing the flow of coarser particles in the mixture across gradients of bulk velocity), (b) **Diffusion,** (representing the flow of finer particles down a concentration gradient) and (c) **Percolation** (representing the gravity-driven motion of the finer particles through the coarse phase in a mixture). Functional forms of all three "drift" components were derived and transport coefficients were calculated for each mixture phase by using linear response theory and integrating the relevant time correlation functions in a Discrete Element Method (DEM) framework [3]. A full description and analysis of the functional forms of the derived constitutive equations for all three mechanisms, as well as validation results are given in [11].

2.2 Parametrisation of Degradation

Two distinct approaches have been incorporated in the numerical model for the modelling of degradation, according to whether the granular mixture was quite dense or quite dilute. The former case (dense-phase mixture) can be modelled in an Eulerian framework by employing population balance models to construct source terms for Equation (1), which depend on material properties and local boundary conditions (e.g. [12]). Of more engineering importance however, is the case of dilute-phase mixtures (dilute-phase pneumatic conveying systems can be found in most bulk solids handling plants), where the particles are fully dispersed in the fluid flow. The present paper concentrates on dilute-phase mixtures and their degradation modelling. An Eulerian/Lagrangian approach for the fluid and the particle phases, respectively, was adopted in this case, where a stochastic simulation procedure was included to represent particle degradation caused by particle-wall collisions.

A particle on its way through a pneumatic conveying line frequently collides with the pipe and bend walls. The force produced on the particle at the time of impact may lead to its degradation. Particle damage can occur in two principal modes according to the level of applied stresses and the material properties, namely breakage and abrasion (particle surface damage). Some of the most important parameters affecting particle impact degradation are the particle velocity, the collision angle, the properties of the particle and wall materials and the particle size and shape.

Since the main degradation usually occurs at bends [13], only collisions of particles with bend walls were assumed to give rise to particle damage and were considered in the present work. Particle degradation was simulated using a stochastic calculation procedure, based on the definition of "degradation rules", formulated from two characteristic functions, namely the probability for a particle to degrade and the size distribution of the degradation products. In the model, the probability of degradation and the size distribution of the degradation products were calibrated from experimental impact degradation studies. A statistical factor for the degradation process was introduced, since the effects of parameters such as particle shapes, particle orientation at the impact point, wall roughness etc., were not explicitly considered during the solution procedure.

3 Numerical Simulations

To demonstrate the capabilities of the presented model, a test case was chosen, which represents a full process of handling of granular material and is an example of a process commonly met in bulk solids handling operations.

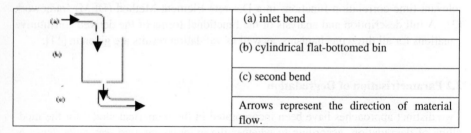

	(a) inlet bend
	(b) cylindrical flat-bottomed bin
	(c) second bend
	Arrows represent the direction of material flow.

Fig. 1. Schematic representation of the simulated system.

The process involves the modelling of the pneumatic conveying of an initially mono-sized material in dilute-phase. The material, after travelling in a pipe with a bend, degrades into various size classes. The output of the pneumatic conveyor is treated as a binary 50-50 mixture (50%/50% coarse/fine particles) of 4.6:1 particle size ratio and fills a cylindrical flat-bottomed bin. Once the bin is filled, the material is assumed to have reached its maximum packing fraction and its initial segregated distribution is set according to theoretical predictions and experimental data [14]. The modelling of the discharge of the material from the bin follows and the effects of segregation in the mixture during core flow discharge are presented. The output of the material discharge is then taken through pneumatic conveying system with a bend, and the final distribution of the mixture is recorded after its final exit, see Figure 1.

3.1 Dilute-Phase Pneumatic Conveying of Mono-sized Material

The calculation domain consists of a 90° pipe-bend with an internal diameter of 0.05 m and a ratio of the bend radius to the pipe diameter equal to 3.6, see Figure 2a. The conveyed material was mono-sized spherical particles of 931 μm diameter and solids density of 2000 kg/m³. This component simulation was performed for a conveying air velocity of 14 m/s and particle concentration of 10 kg/m³, a typical value for industrial operating conditions under which particle degradation constitutes a problem. The particle velocity at the inlet of the bend was assumed to be equal to the air velocity and homogeneously dispersed.

a   b

Fig. 2. Numerical grid of (a) the pipe bend and (b) cylindrical flat-bottomed bin

For these impact degradation calculations, the size range between 0 and 1000 μm was divided into evenly-spaced size intervals. The dependence of the particle degradation probability and the fragment size distribution on the particle velocity and the particle size were determined from experimental data, obtained from 90° impact degradation experiments on single particles, using a rotating disc accelerator type degradation tester [15]. This test facility is a bench-scale unit for assessing degradation by impact. It can control both the speed of the particles and the angle of impact. It is well known that degradation caused by solid particle impact varies with particle velocity. In the current degradation tester, the effect of particle size and/or shape is minimal on the particle travelling velocity. This is important because when testing bulk solids of various size or shape distributions, homogeneous particle velocity is essential. A restitution coefficient of 0.8 was employed for the calculations during the rebound process. In this way, particles impacting on the walls of a bend were let to degrade according to their degradation probability. In the instance of a degradation event, the original particles were split into a number of daughter particles, whose sizes were distributed according to the degradation product size distribution. Each daughter particle was given a fraction of the momentum of the original particle equal to the ratio of the daughter and original particle volumes and the new particle velocities resulting from the rebound process were calculated.

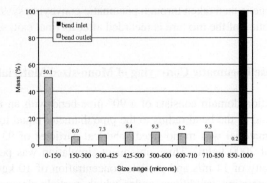

Fig. 3. Particle size distribution at the bend inlet and outlet of the filling conveying system.

The calculated particle size distribution in the bend outlet cross-section is presented in Figure 3. Almost all the original particle sample fed into the inlet was seen to have degraded (mass fraction of original 931 μm diameter particles at the outlet was 0.2 %) and a broad size distribution of the particles was predicted at the bend outlet. The proportion of debris of size below 150 μm was very high, whereas the size intervals ranging from 150 μm to 850 μm were approximately evenly populated, with a mass fraction value less than 10 % in each interval. It should be noted that the use of "degradation rules" based on 90° impact degradation tests leads most likely to over-predicting the degradation amount occurring in the bend. In reality, the smaller the impact angle, the smaller the force produced on the particle and hence, less amount of degradation should result. Refinement of the present model, by incorporating the effect of the impact angle on the degradation processes, is currently underway.

3.2 Discharge of a Flat-Bottomed Bin in Core Flow Mode

The output of the pneumatic conveyor was used to fill a cylindrical flat-bottomed bin of 65 cm diameter, 94 cm height and 6.5 cm orifice at the bottom, around its central axis. For reasons of simplification, the pneumatic conveyor output size distribution was represented as a 50-50 binary mixture of 4.6:1 particle size ratio. It consisted of fines of size below 150 μm (volume averaged diameter of 120 μm) and coarse particles of size between 150 μm and 850 μm (volume averaged diameter of 555 μm). The material was assumed to be at its maximum packing fraction (total solids fraction of 0.5) and resting in the bin at its angle of repose (34°). The initial segregated profile was fitted according to theoretical predictions and experimental data [14], so that only coarse particles could be found around the central axis of the bin, then there existed a region where the mixture composition was 50-50 and only fine particles were concentrating close to the walls of the bin. Because of the material properties and the vessel geometry, it was predicted through the micro-mechanical flow criteria [9] that the discharge of the mixture was going to occur in core flow mode (where stagnant material regions exist). Due to the axisymmetric nature of this case, a semi-3D geometry was chosen, with a slice of 5° angle being simulated. The simulated bin geometry and applied mesh are shown in Figure 2b.

During the initial stages of discharge (less than 3% of the total emptying time), the central part was observed to collapse, thus creating a channel. Once the channel reached the top surface, this began to descend and steepen until it reached the material angle of repose. Thereafter, the material emptied from the bin very slowly through avalanching from the surface region of the bulk, with the angle of repose and the central channel always being maintained (this mode of emptying is also known as "ratholing"). Eventually, a stagnant region was left in the domain, with no more material exiting the bin. The final stagnant material was also maintained at the angle of repose. The total bin emptying time until the creation of the final stagnant region was predicted by the model to be approximately 200 s, a result which is in good agreement with experimental observations of core flow bin discharges of similar dimensions (e.g. [9]). The evolution of the interface of the total material fraction and the individual components (fines and coarse) at various points in time is depicted in Figure 4.

Segregation was obvious during the intermediate stage of the discharge, when the 50-50 mixture region was discharging. As can be observed in the fourth temporal snapshot of Figure 4, the moving coarse particles were making their way towards the interface, while the finer particles of the flowing zone were moving away from the interface (this phenomenon is known as "rolling segregation"). Once the 50-50 region disappeared, then only fine particles were seen to exit the bin. This is demonstrated in Figure 5, where the mass percentage in the mixture (averaged over the outlet region) of each of the individual fractions are plotted against normalised emptying time.

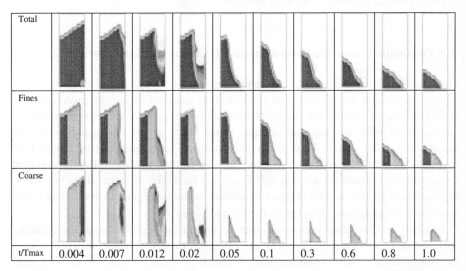

Total										
Fines										
Coarse										
t/Tmax	0.004	0.007	0.012	0.02	0.05	0.1	0.3	0.6	0.8	1.0

Fig. 4. Interface profiles for the individual/total solids fractions. Intensity of gray shade indicates increase in material.

The graph is split in three regions corresponding to the three regions of the initial vessel filling state. Region 1, initially containing coarse particles only, emptied very quickly, hence the sharp decrease (increase) in the coarse (fine) particle curve. Region 2, initially containing the 50-50 mixture, exhibited very sharp peaks, thus indicating the effects of rolling segregation with the mixture composition at the outlet alternating between mostly coarse and mostly fines. Region 3, initially containing mostly fine particles, appeared much later into the discharge, with the mixture at the outlet containing almost exclusively fines and the occasional spike appearing, indicating that some coarse (remnants of the 50-50 mixture initial distribution) exited the domain. This result was anticipated and is in agreement with theoretical predictions and experimental observations.

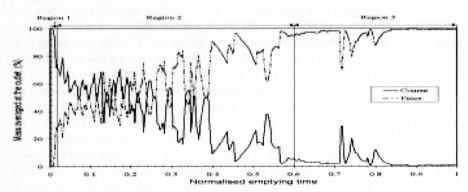

Fig. 5. Mass percentage in the mixture of the individual fractions (averaged at the outlet) plotted against normalised emptying time during core flow discharge.

3.3 Pneumatic Conveying System Located Downstream of the Storage Vessel

The final simulation was performed using the same flow conditions (pipe bend geometry, conveying air velocity and particle concentration) as for the upstream conveying system, except the orientation of the bend (see Figure 1). As it has been shown, the composition of the outgoing particles at the outlet of the bin varied during the vessel discharging due to segregation phenomena. The calculations of the particle degradation in the downstream pipe bend were performed for four different averaged compositions of coarse particles ($d = 555$ μm) and fines ($d = 120$ μm) at the bin outlet, representative of the various regimes identified on the segregation curve. The simulated input compositions are given in the table in Figure 6.

Similarly to the filling conveying system, significant degradation of the particles occurred in the bend and the particle size distributions were considerably different in the bend inlet and outlet cross-sections. Figure 6 presents for times 1 to 4 the total mass percentage of coarse particles (now defined as particles with diameter between 150 μm and 555 μm) and fines (diameter below 150 μm) in the bend outlet. For time 1 about 40 % of the particles fed into the bend were converted into fines of size below 150 μm. The same was true for about 30% of the coarse particles at time 2 and approximately 10% of the coarse particles at time 3. This result indicated that with decreasing percentage of incoming coarse particles in the bend, the percentage of them that were converted into fines also decreased. This effect could be attributed to the fact that with increasing numbers of fines coming into the pneumatic conveyor at the latter stages of the core flow bin discharge, the probability of impact between a coarse particle and the bend wall also reduced, hence leading to reduced numbers of coarse particles being allowed to collide with the wall and degrade.

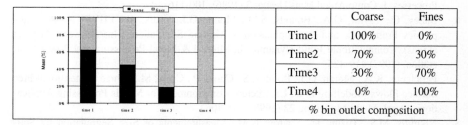

	Coarse	Fines
Time1	100%	0%
Time2	70%	30%
Time3	30%	70%
Time4	0%	100%
% bin outlet composition		

Fig. 6. Proportion of coarse particles (150 μm < d < 555 μm) and fines (d < 150 μm) at the bend outlet of the conveying system downstream the storage vessel.

4   Conclusions

In this paper, a computational framework was presented for the modelling of granular material flows, based on Computational Fluid Dynamics and implemented with micro-mechanical constitutive models for the effective representation of the interactions between particles at the microscopic level. The presented simulations demonstrated the potential capability of the model to realistically represent key granular processes (segregation/degradation), which are of great importance in the process engineering industry. The developed model represented realistically the dilute-phase pneumatic conveying, both upstream and downstream of a core flow bin, by employing "degradation rules" based on experimental studies of single-particle-impact degradation. The process of core flow bin discharge showed the correct trend for segregation, in agreement with theoretical and experimental results. It should also be noted that the presented modelling of core flow in a continuum mechanics framework is believed to be unique. Further development of the model and inclusion of parametrisations for the modelling of aggregation will make it a powerful computational tool for engineers, which will aid them in the characterisation of granular material flows and the processes involved.

References

1.   Tardos, G.I.: A fluid mechanistic approach to slow, frictional flow of powders, Powder Tech. 92 (1997) 61-74

2. Karlsson, T., Klisinski M., Runesson, K.: Finite element simulation of granular material flow in plane silos with complicated geometry, Powder Tech. 99 (1999), 29-39
3. Baxter, J., Tüzün, U., Burnell, J., Heyes, D.M.: Granular Dynamics Simulations of Two-dimensional Heap Formation, Phys. Rev. E 55 (1997), 3546-3554
4. Yang, S., Hsiau, S.: The Simulation and Experimental Study of Granular Materials Discharged from a Silo with the Placement of Inserts, Powder Tech. 120 (2001), 244-255
5. Cross, M.: Computational Issues in the Modelling of Materials Based Manufacturing Processes, J. Comp Aided Mats Design 3 (1996), 100-116
6. Bailey, C., Taylor, G.A., Bounds, S.M., Moran, G.J., Cross, M.: PHYSICA: A Multi-physics Framework and its Application to Casting Simulation. In: Schwarz, M.P. et al.(eds): Computational Fluid Dynamics in Mineral & Metal Processing and Power Generation, (1997) 419-425
7. Pericleous, K.A., Moran, G.J., Bounds, S., Chow, P., Cross, M.: Three Dimensional Free Surface Flows Modelling in an Unstructured Environment for Metals Processing Applications, Appl. Math. Modelling 22 (1998), 895-906
8. Nikitidis, M.S., Tüzün, U., Spyrou, N.M.: Measurements of Size Segregation by Self-diffusion in Slow-Shearing Binary Mixture Flows Using Dual Photon Gamma-ray Tomography, Chem. Eng. Sci. 53 (1998), 2335-2351
9. Nedderman, R.M.: The Use of the Kinematic Model to Predict the Development of the Stagnant Zone Boundary in the Batch Discharge of a Bunker, Chem. Eng. Sci. 50 (1995), 959-965
10. Zamankhan, P.: Kinetic theory of multicomponent dense mixtures of slightly inelastic spherical particles, Phys. Rev. E 52 (1995), 4877-4891
11. Christakis, N., Patel, M.K., Cross, M., Baxter, J., Abou-Chakra, H., Tüzün, U.: Continuum Modelling of Granular Flows using PHYSICA, a 3-D Unstructured, Finite-Volume Code. In: Cross, M., Evans, J.W., Bailey, C. (eds): Computatinal Modeling of Materials, Minerals and Metals Processing, San Diego, CA, TMS (2001) 129-138
12. Hogg, R.: Breakage Mechanisms and Mill Performance in Ultrafine Grinding, Powder Tech. 105 (1999) 135-140
13. Bemrose, C.R., Bridgwater, J.: A Review of Attrition and Attrition Test Methods, Powder Tech., 49 (1987) 97-126
14. Salter, G.F.: Investigations into the Segregation of Heaps of Particulate Materials with Particular reference to the Effects of Particle Size. Ph.D. Thesis, University of Greenwich, London (1999)
15. Kleis, I.R., Uuemois, H.H., Uksti, L.A., Papple, T.A.: Centrifugal Accelerators for Erosion Research and Standard Wear Testing. In: Proceedings of the International Conference on the Wear of Materials, Dearborn, MI, ASTM (1979) 212-218

Parallel Implementation of the INM Atmospheric General Circulation Model on Distributed Memory Multiprocessors

Val Gloukhov

Institute for Numerical Mathematics RAS, Moscow, Russia,
gluhoff@inm.ras.ru,
WWW home page: http://www.inm.ras.ru/~gluhoff

Abstract. The paper is devoted to a distributed memory parallel implementation of a finite difference Eulerian global atmospheric model utilizing semi-implicit time stepping algorithm. The applied two-dimensional checkerboard partitioning of data in horizontal plane necessitates boundary exchanges between the neighboring processors all over the model code and multiple transpositions in the Helmholtz equation solver. Nevertheless, quite reasonable performance has been attained on a set of cluster multiprocessors.

1 Introduction

One of the most challenging prospects requiring advances of supercomputers power is acknowledged to be the Earth climate system modeling, in particular, those studies concerning the global warming issues and related climatic changes from increasing concentration of greenhouse gases in the atmosphere. The main predictive tools here are the general circulation models run for decadal and even centennial simulation periods.

INM Atmospheric General Circulation model (AGCM) was designed at the Institute for Numerical Mathematics of the Russian Academy of Sciences and originates from earlier works of G. I. Marchuk and V. P. Dymnikov [1]. The model participated in AMIP II intercomparison project as "DNM AGCM" [2] and some other experiments [3], [4]. Though most of simulations performed with the INM AGCM are based on Monte Carlo method, allowing different trajectories to be integrated independently, the main interest of the present paper lays in the intrinsic parallelism of the model.

Mostly, the model provides substantial degree of parallelism, except the implicit part of the time-stepping scheme, that solves Helmholtz equation on a sphere, and spatial filters damping the fast harmonics of prognostic fields at the poles. In opposite to Barros and Kauranne [5], we parallelized a direct Helmholtz equation solver [7] involving the longitudinal fast Fourier transforms (FFTs) and latitudinal Gaussian elimination for tri-diagonal linear systems.

In the next section, we will outline the model structure. Then, in Sect. 3, we will proceed with details of parallelization technique. Performance of the obtained parallel version of the model is presented in Sect. 4. We have carried

P.M.A. Sloot et al. (Eds.): ICCS 2002, LNCS 2329, pp. 753–762, 2002.

out benchmarking on MBC1000M computing system, located at the Joint Supercomputer Center, and SCI-cluster, maintained by the Research Computing Center of the Moscow State University.

2   Model structure

INM AGCM solves the system of partial differential equations of hydro-thermodynamics of the atmosphere under hydrostatic approximation on the rotating globe [2]. The vertical coordinate is generally either pressure or a terrain-following sigma ($\sigma = p/\pi$, where π is the surface pressure) or a hybrid of the two. The equations are discretized on a staggered Arakawa "C" grid [6], meaning all prognostic variables are co-located, except zonal and meridional wind components that are staggered in the longitudinal and latitudinal directions, respectively. The grid has $2°$ resolution in latitude, $2.5°$ in longitude, and $K = 21$ irregularly spaced vertical levels. The size of the computational domain is thus $145 \times 73 \times 21$. All floating point data are kept in single precision (32-bit) representation.

Let u and v represent zonal and meridional wind components, correspondingly, T is the temperature, π is the surface pressure, d is the horizontal divergence; $P = \Phi + RT_0 \log \pi$, Φ is the geopotential, R is the gas constant, $T_0 = 300K$. Then, emphasizing with bold letters $K \times K$ matrices and vectors of length K, we can write the model governing equations in the discrete form as

$$\delta_t \overline{\mathbf{u}}^t + \frac{1}{a \cos \varphi} \delta_\lambda \left(\frac{1}{2} \delta_{tt} \mathbf{P} \right) = \mathbf{A}_u,$$

$$\delta_t \overline{\mathbf{v}}^t + \frac{1}{a} \delta_\varphi \left(\frac{1}{2} \delta_{tt} \mathbf{P} \right) = \mathbf{A}_v,$$

$$\delta_t \overline{\mathbf{T}}^t + \mathbf{B} \left(\frac{1}{2} \delta_{tt} \mathbf{d} \right) = \mathbf{A}_T, \tag{1}$$

$$\delta_t \overline{\log \pi}^t + \nu^T \left(\frac{1}{2} \delta_{tt} \mathbf{d} \right) = A_\pi,$$

where t stands for time, λ is longitude, φ is latitude, δ_t, δ_λ and δ_φ are the discrete analogues of the corresponding differential operators of the form

$$\delta_x \psi(x) = \left(\psi \left(x + \frac{\Delta x}{2} \right) - \psi \left(x - \frac{\Delta x}{2} \right) \right) \Big/ \Delta x,$$

$\delta_{tt} \psi(t) = \psi(t - \Delta t) - 2\psi(t) + \psi(t + \Delta t)$, $\delta_t \overline{\psi}^t(t) = (\psi(t + \Delta t) - \psi(t - \Delta t))/(2\Delta t)$, a is the mean radius of the earth, $\mathbf{B}$ is a matrix, ν is a vector.

The model parameterizes long and short wave radiations, deep and shallow convection, vertical and horizontal diffusion, large scale condensation, planetary boundary layer, and gravity wave drag.

Table 1. INM AGCM basic routines and percentage of their CPU time measured on MBC1000M computing system and SCI-cluster in uniprocessor mode

Routine	Purpose	MBC	SCI
Physics:			
FASFL	Convection processes	1%	1%
RADFL	Parameterization of radiation	33%	37%
DSP	Gravity-wave drag	8%	8%
PBL	Atmospheric boundary layer	4%	4%
Dynamics:			
ADDTEN	Add physical tendencies to the solution	1%	1%
VDIFF	Vertical diffusion	13%	16%
RHSHLM	Explicit dynamical tendencies generation	18%	10%
HHSOLV	Helmholtz equation solver	4%	4%
DYNADD	Add dynamical tendencies to the solution	5%	4%
VISAN4M	Horizontal diffusion	6%	10%
RENEW	Spatial and temporal filtration	6%	4%
Other routines		0.1%	0.1%

The fraction of time spent in various routines of INM AGCM is shown in Table 1. Physical components are computed hourly, except the radiative block (RADFL) involved every 3 model hours; dynamical block, for the grid resolution specified above, is calculated 9 times within a model hour.

3   Parallelization technique

We apply a 2-dimensional domain decomposition to the globe in both the longitude and latitude dimensions, or the so called checkerboard partitioning [8]. The resulting rectangular subdomains are assigned to the processors arranged into a $P_{lon} \times P_{lat}$ virtual grid. We will refer further this distribution of data as the basic one.

3.1   Physics

The basic distribution is highly suitable for INM AGCM's physical parameterizations that are column oriented and thus can be done concurrently by columns. Although communications are required to interpolate quantities between the staggered and non-staggered grids and some load imbalance occurs in radiative processes component (RADFL).

3.2   Helmholtz equation solver

The four equations of system (1) are reduced to a Helmholtz-like equation solved by the routine HHSOLV

$$\frac{1}{2}\delta_{tt}\mathbf{d} - (\Delta t)^2 \mathbf{G} \nabla^2 \left(\frac{1}{2}\delta_{tt}\mathbf{d} \right) = \mathbf{RHS}, \qquad (2)$$

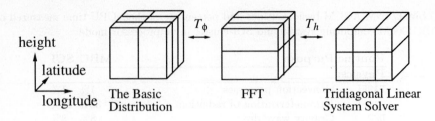

Fig. 1. Data distributions used in the Helmholtz solver ($P_{lon} = 2$, $P_{lat} = 3$)

where **G** is a matrix, ∇^2 is the discrete analogue of the horizontal Laplace operator in spherical coordinates, **RHS** is a right hand size. The periodical boundary conditions in longitude are imposed on (2) as well as on system (1). Diagonalization of matrix **G** allows to uncouple (2) and solve it independently on the vertical levels by a direct method involving Fourier transforms in longitudinal direction and Gaussian elimination in the meridional direction. Thus, HHSOLV has the following structure:

1. Transformation of the right hand side of (2) to the basis of eigenvectors of matrix **G**.
2. Forward Fourier transform.
3. Tri-diagonal matrix solver.
4. Backward Fourier transform.
5. Transformation of the solution to the original basis.

Steps 1 and 5 compute matrix vector product column-by-column, that makes them very suitable for the basic data distribution. Steps 2 and 4 require all longitudes while step 3, on the contrary, all latitudes.

To carry out step 2, we transpose the data in the height-longitude plane gathering longitudes but distributing levels over the processors (T_φ, Fig. 1). Upon completion of the FFTs each processor contains all wave-numbers but a part of levels and latitudes. To proceed with the tri-diagonal matrix solver we make use of the transposition again but in the longitude-latitude plane. Now it collects all latitudes but distribute longitudes (T_h, Fig. 1). Having obtained the solution of tri-diagonal systems, we rearrange the data back into the basic distribution performing the transpositions in the reverse order and calculating the inverse FFT.

3.3   Explicit dynamical tendencies generation

The routine RHSHLM generates the right hand side of equation (2) using the formula

$$\textbf{RHS} = \Delta t \, \text{div}\textbf{A} - \textbf{d}(t) + \textbf{d}(t - \Delta t) + \Delta t \nabla^2 \left\{ \textbf{P}(t) - \textbf{P}(t - \Delta t) - \Delta t \textbf{A}_P \right\}, \quad (3)$$

where

$$\text{div}\textbf{A} = \frac{1}{a \cos \varphi} \left\{ \delta_\lambda \textbf{A}_u + \delta_\varphi (\textbf{A}_v \cos \varphi) \right\}, \quad (4)$$

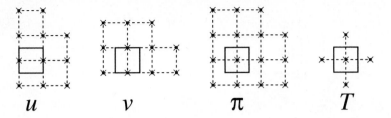

Fig. 2. Dependence stencils for the Helmholtz equation right hand side generation

$$\mathbf{A}_P = \mathbf{\Gamma} \mathbf{A}_T + RT_0 \mathbf{A}_\pi, \tag{5}$$

$\mathbf{\Gamma}$ is a matrix. Expressions of the explicit tendencies A_u, A_v, A_T and A_π are rather bulky and can be retrieved from [2]. They comprise values of known fields u, v, T, π, their derivatives and vertical integrals.

Basically, the routine RHSHLM has no global dependencies in the horizontal plane, with few exceptions occurred near the poles. In particular, to estimate vorticity ζ and surface pressure π at the North pole, u-velocity and surface pressure π are averaged along the most northern $p - 1/2$ latitude circle

$$\zeta_{i,p,k} = \frac{1}{Na\Delta\varphi} \sum_{i=1}^{N} u_{i,p-1/2,k}, \quad \pi_{i,p} = \frac{1}{N} \sum_{i=1}^{N} \pi_{i+1/2,p-1/2}, \tag{6}$$

where N is the number of points along the longitude dimension. Since both u and π are distributed in longitude, we made use of MPI_Allreduce calls [9] to evaluate the sums in (6). The South pole is processed in a similar way.

One can observe that **RHS** computation at a grid point not belonging to the polar boundaries requires values of known variables at some neighboring columns and, therefore, has to be preceded by a communication of subdomain boundaries. To investigate the dependences inherent in (3) we processed formulas (3)–(5) and the explicit tendencies expressions with a symbolic computing system, and obtained the dependence stencils depicted in Fig. 2. Boundaries of corresponding widths are interchanged beforehand. Whenever a longitudinal derivative is calculated, the very first and last processors in a processor row also interchange data to maintain the periodic boundary condition.

3.4   Filtering

The routine RENEW applies spatial and time filtering to the obtained solution of system (1). The spatial filter damping short zonal scales poleward 69° transforms the fields into Fourier space and back. We apply a transposition in the longitude-height plane to accomplish the transformation. Staying idle, the processors that don't contain any latitude poleward 69° give rise to load imbalance.

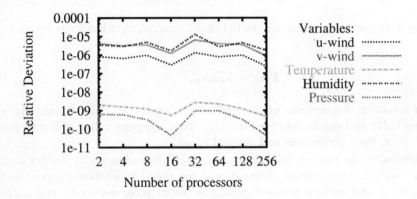

Fig. 3. Relative deviation of prognostic variables from their uniprocessor values estimated upon 6 hours' INM AGCM integrations on different number of processors of MBC1000M computing system

4   Performances

The benchmarking of the resulting parallel code of INM AGCM has been done both on MBC1000M computing system and the MSU RCC's SCI-cluster (Appendix A). For a given number of processors P we measured elapsed CPU times of 6 hours' integrations on all processor grids (P_{lon}, P_{lat}), such that $P_{lon}P_{lat} = P$ (Tables 2 and 3), and calculated speed-up as the ratio of the single processor time to the minimum time obtained on P processors (Tables 4 and 5). For instance, on 8 processors of MBC1000M we tried $(1, 8)$, $(2, 4)$, $(4, 2)$, and $(8, 1)$ grids and found $(2, 4)$ configuration to be the best.

Fig. 4 represents the elapsed CPU time of 6 hours' modeling and speed-up obtains on MBC1000M supercomputer and the SCI-cluster as a function of number of processors. On both machines the parallel version of the model by far outperforms its sequential counterpart yielding the speed-up of about 15 on 32 processors.

To detect potential bottlenecks of model we have carried out profiling of its major components. From the chart plotted in Fig. 5, one can observe that the filtering (RENEW) is becoming more and more bulky, as the number of processors increasing, while the radiative component (RADFL) relative time is reducing.

The deviation of prognostic variables from their uniprpocessor values after 6 hours of modeling is shown in Fig. 3 to confirm correctness of the parallelization.

5   Concluding remarks

INM AGCM was ported to a set of cluster multiprocessors indicating the advantage of using parallel computing in climate studies. The obtained performances, being still far from the ideal, are reasonable and improvable.

The work on optimization of communication as well as computational algorithms are in progress. It comprises fitting of transposition algorithms to particular machines, overlapping of communication and computations, also intercomparison of semi-implicit and explicit time integration schemes could be undertaken in the future.

Acknowledgments

The author would like to thank Academicians V. P. Dymnikov and V. V. Voevodin for encouragement as well as Dr. E. M. Volodin provided sequential code of the model.

The work was supported by the Russian Foundation for Basic Research under grant No. 99-07-90401.

References

1. Marchuk, G.I., Dymnikov, V.P., Zalesny, V.B.: Mathematical Models in Geophysical Hydrodynamics and Numerical Methods of their Implementation. Gidrometeoizdat, Leningrad (1987) (in Russian)
2. Alexeev, V.A., Volodin, E.M., Galin, V.YA., Dymnikov, V.P., Lykossov, V.N.: Simulation of the present-day climate using the INM RAS atmospheric model. The description of the model A5421 (1997 version) and the results of experiments on AMIP II program. Institute of Numerical mathematics RAS, Moscow (1998) (reg. VINITI 03.07.98 No. 2086-B98)
3. Volodin, E.M., Galin V.Ya.: Sensitivity of summer Indian monsoon to El Nino of 1979-1998 from the data of the atmospheric general circulation model of the Institute of Numerical Mathematics, the Russian Academia of Sciences. Meteorologia i gidrologiya **10** (2000) 10–17 (in Russian)
4. Volodin, E.M., Galin, V. Ya.: Interpretation of winter warming on Northern Hemisphere continents in 1977-1994. Journal of Climate, Vol. 12, No. 10 (1999) 2947–2955
5. Barros, S.R.M., Kauranne, T.: Spectral and multigrid spherical Helmholtz equation solvers on distributed memory parallel computers. Fourth workshop on use of parallel processors in meteorology, Workshop proceedings, ECMRWF (1990) 1–27
6. Arakawa, A. Lamb, V.R.: Computational design of the basic dynamical processes of the UCLA general circulation model, Methods Comput. Phys. **17** (1977) 173–265
7. ECMWF Forecast Model Documentation Manual, ECMWF Research Department, edited by J-F Louis, Internal Report No. 27, Vol. 1, ECMWF (1979)
8. Kumar, V., Grama, A., Gupta, A., Karypis, G.: Introduction to Parallel Computing: Design and Analysis of Algorithms. Addison-Wesley/Benjamin-Cummings Publishing Co., Redwood City, CA (1994)
9. Snir, M., Otto, S., Huss-Lederman, S., Walker, D., Dongarra, J.: MPI: The Complete Reference, The MIT Press, Cambridge, MA (1998)

Table 2. Elapsed CPU times of INM AGCM 6 hours' integrations in seconds obtained on different $P_{lon} \times P_{lat}$ processor grids of MBC1000M computing system

$P_{lon} \backslash P_{lat}$	1	2	4	8	16
1	158.3	87.76	46.22	28.50	17.65
2	92.37	49.10	27.44	16.81	10.35
4	58.75	30.75	16.80	10.63	6.61
8	37.43	20.23	12.24	7.68	5.13
16	31.47	16.78	10.74	7.85	5.09

Table 3. Elapsed CPU times of INM AGCM 6 hours' integrations in seconds obtained on different $P_{lon} \times P_{lat}$ processor grids of SCI-cluster

$P_{lon} \backslash P_{lat}$	1	2	4	8	16
1	598.05	337.36	191.56	115.19	67.83
2	340.82	190.62	105.93	65.31	42.38
4	206.26	112.82	67.38	43.07	
8	124.01	70.37	43.75		
16	84.10	52.15			

Table 4. Speed-up and efficiency of INM AGCM on MBC1000M computing system

# procs	1	2	4	8	16	32	64	128	256
(P_{lon}, P_{lat})	(1,1)	(1,2)	(1,4)	(2,4)	(4,4)	(2,16)	(4,16)	(8,16)	(16,16)
Speed-up	1.00	1.80	3.42	5.77	9.42	15.29	32.96	30.89	31.13
Efficiency	100%	90%	86%	72%	59%	48%	37%	24%	12%

Table 5. Speed-up and efficiency of INM AGCM on SCI-cluster

# procs	1	2	4	8	16	32	36
(P_{lon}, P_{lat})	(1,1)	(1,2)	(2,2)	(2,4)	(2,8)	(2,16)	(2,18)
Speed-up	1.00	1.77	3.14	5.65	9.16	14.11	14.49
Efficiency	100%	89%	78%	71%	57%	44%	40%

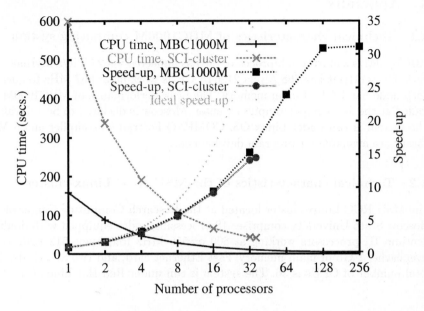

Fig. 4. Elapsed CPU time of INM AGCM 6 hours' integrations and the performance obtained on MBC1000M computing system and SCI-cluster as functions of number of processors

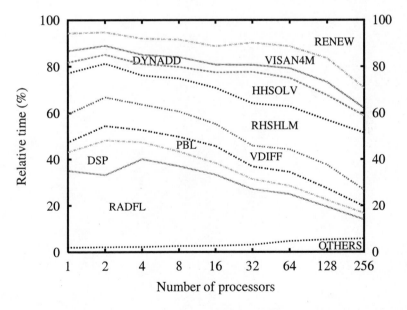

Fig. 5. Relative time spent in each part of INM AGCM on MBC1000M computing system

A   Appendix

A.1   Technical characteristics of MBC1000M computing system

MBC1000M is a cluster of biprocessor SMPs connected via 2 Gbit/s Myrinet network. Currently, it has 768 Alpha 21264A CPUs working at 667 MHz frequency. Each node has 1 Gb of main memory. The peak performance of MBC1000M installed at the Joint Supercomputer Center, Moscow is declared to be 1 Teraflop. The system is run under Linux OS. COMPAQ Fortran 90 compiler and a MPI library are available for program development.

A.2   Technical characteristics of the MSU RCC Linux-cluster

The MSU RCC Linux-cluster located at the Research Computing Center of the Moscow State University comprises 18 processing nodes equipped with doubled Pentium III processors working at 500 or 550 MHz frequency, 512 Kb second level cache, 1 Gb of main memory, Fast Ethernet card, and two SCI cards. The total number of CPUs is 36. The system is run under Red Hat Linux 6.1.

A Realistic Simulation for Highway Traffic by the Use of Cellular Automata

Enrico G. Campari[1] and Giuseppe Levi[2]

[1] Physics Department of Bologna University
and Istituto Nazionale per la Fisica della Materia
viale Berti-Pichat 6/2 I-40127, Bologna, Italy
campari@df.unibo.it
[2] Physics Department of Bologna University
and Istituto Nazionale di Fisica Nucleare
viale Berti-Pichat 6/2 I-40127, Bologna, Italy
levi@bo.infn.it

Abstract. A two lane, non-periodic highway with on-ramps and off-ramps is simulated using a simple cellular automata model. The main experimental features of highway traffic are reproduced, indicating the usefulness of this approach for the study of traffic and it is shown the need of ramps to produce congested traffic. A fractal dimension for the variables used for traffic analysis, that is car density, car speed and flow has been measured with a Box Counting algorithm. Finally, it is suggested how to perform real-time traffic forecasts.

1 Introduction

Every day an enormous number of car drivers, commuting to work or going to the countryside for the weekend, get stucked into traffic. Many of the readers of this article, as car drivers, might have noticed how sometimes, driving on an highway, the traffic suddenly slows to a crawl. For a while cars move very slowly and then are again free to accelerate. There are no accidents or roadworks to cause the jam, which apparently has no explanation.

This and other aspects of highway traffic, like the creation of car density waves, together with the great economical costs and health problems linked to traffic, spurred the scientists attention into the study of the subject. In the last decades, experimental data were collected and analysed, mainly in USA, Germany and Japan [1–5]. At the same time, models of traffic flow were built. The firsts were based on collective properties of traffic [6]. Then there were microscopic models, like the so called optimal velocity model by Bando and coworkers [7]. Both were based on the computer resolution of non-linear differential equations. Despite some succes, a real progress in the study of traffic came only with the introduction of computer simulations based on cellular automata. This method is simpler to implement on computers, provides a simple physical picture of the system and can be easily modified to deal with different aspects of traffic. For these reasons, cellular automata models are getting more and more popular.

P.M.A. Sloot et al. (Eds.): ICCS 2002, LNCS 2329, pp. 763–772, 2002.

In synthesis, a cellular automaton consists of a regular lattice with a discrete variable at each site. A set of rules specify the space and time evolution of the system, which is discrete in both variables [8, 9]. These rules are usually limited to first neighbours interactions, but this is not necessary however, and the particular application of cellular automata to traffic is one case where longer range interactions are used.

2   Cellular Automata Model

The model proposed in this article simulate a two lane highway with on and off-ramps as a long straight road divided into cells which can be occupied by *cars* moving in the direction of increasing cell number. Cars enter the highway at the road beginning and at on-ramps at a given rate by means of pseudo random number generation. Once in the road, they move according to a set of rules. These evolution rules were derived from those used by Schreckenberg, Nagel and coworkers [10–12]. These authors simulated a multilane highway with the use of few simple rules, which proved how cellular automata models are effective in simulating traffic. Although really interesting, the above cited model and other similar are unrealistic. Their main pitfalls are the use of periodic boundary conditions and the lack of ramps. Leaving aside that every highway has a beginning and an end, the use of periodic boundary conditions forbids simulating a variable car density along the highway, which is the common situation. As widely known, traffic in most highways presents two peaks, one in the morning and a second one in the afternoon [13], in addition to an absolute minimum during the night. These basic experimental facts simply cannot be reproduced in the framework of periodic models without ramps.

Ramps absence also hinders the well known fact that traffic jams usually nucleate in correspondence of ramps or other obstacles to car flow such as roadworks or junctions. Even if it is proved that above a certain car density along the road, density fluctuations naturally occur and grow into jams, as far as ramps are always present it is there that jams first nucleate [14]. Furthermore, this happens at a lower density with respect to a rampless highway.

In order to consider a more realistic road and to be able to simulate transient phenomena, we developed a cellular automata model for a two lane highway leaving periodic boundaries conditions and introducing on-ramps [15], which were modeled as a single entrance cell in lane 1. Their number could vary from simulation to simulation and were typically located 2500 cells apart. In this article we present an improved version of that model where both on and off ramps are present, data collection at a fixed point from a car detector is simulated and data analysis is performed in order to look for the fractal dimension of traffic.

In the model, the highway length can be varied from simulation to simulation and in order to compare the results of the simulations with experimental data, we considered each cell to correspond to a length of 5 m and each evolution step for the cells to occurr in 1 s. The model is asymmetric: one lane, named in the following lane 2 or *left* lane, is used for overtaking while the other, named lane 1 or *right* lane, is for normal cruise. For all on-ramps, at each time step (ts) the program checks if the cell is empty. If empty, a random number in the [0,1] interval is generated and if it is less than the threshold chosen for that simulation, a car is generated at that cell with a speed of 2 cell/ts . At the road beginning, cars are generated both on lane 1 and lane 2 with the procedure described above, with starting speed 2 cell/ts. Every car entering the highway has a given destination or off-ramp. Approaching its exit, the car slow down and eventually move into the rigth lane. When the off-ramp is reached the car is removed from the road.

For the sake of simplicity, cars can have only two values of their maximum speed: 6 or 9 cell/ts, corresponding to a maximum speed of 108 or 162 km/h. The first kind of vehicles corresponds to tracks. In this kind of simulations, cars have an unbound braking ability in order to avoid car accidents.

The evolution rules away from ramps are schematically described as follows in *pidgin algol* language:

```
MAIN:
Begin:
Init variables;
Load Highway-configuration;
Load Run-Parameters; /* RP */
Open Output-file;
   If requested by RP then
      Extract random initial state with initial car density given by RP ;
      Do Velocity-Normalization;
   Endif
/* MAIN LOOP: */
Repeat for Ntimes; /* Number of times specified in RP */
Forall on ramps:
With Propability P Do: /* Entrance Probability is given in RP for each entrance */
   If entrance cell is empty then
   generate new vehicle with velocity := 2;
   endif
   End With Propability
End Forall
Forall cars in Lane 2:
   If are verified the conditions to reenter in lane 1 then
Change Lane;
   Endif
End Forall
 Forall cars in Lane 1:
If (car distance from exit) < 2 then
remove car;
endif
If are verified the conditions to overtake then Change Lane;
```

```
    If car.velocity < car.maxv then
      car.velocity := car.velocity + 1;
    Endif
  Endif
 End Forall
 Do Velocity-normalization;
 Forall cars in Lane 1:
 With Propability P Do:
/* Random deceleration with probability P given in RP */
car.velocity := car.velocity - 1;
  End With Propability
  End Forall
    Forall cars in Lanes 1 and 2:
        Car.position := Car.position + Car.velocity;
    End Forall
Measure-and-Store physical quantities;
/* eg.: mean car density and velocity, flux etc. */
End Repeat /* End of main Loop */
Close ouput files;
End /* end of main program */
```

```
Velocity-normalization :
DEFINE  Forward.Gap as the MINIMUM between the number of free cells
from a car to the next one in the same lane along the car direction of motion
and the distance of the car fromthe exit;
  Begin:
    Forall cars in Lanes 1 and 2:
    Measure Forward.Gap;
    If car.velocity < car.maxv then
      car.velocity := car.velocity + 1;
    Endif
    If car.velocity > Forward.Gap then
/* This model (as the Schreckenberg one) avoids accidents */
    car.velocity := Forward.Gap;
    Endif
    End Forall
  Return ;
  End
```

The input parameters to the program at the beginning of each simulation are: highway length, ramp number and position, starting car density for each lane, car generation probability, random deceleration probability, fraction of slow cars, maximum speed of vehicles (maxv) and simulation length in time ticks.

At each iteration the state of the highway is computed and at selected times recorded in a file. This means computing for each cell mean car speed and density, averaging over the 100 cells before and after the given cell. From these quantities the flow is readily computed as density by mean speed. Furthermore, for each car, position, speed and gaps are recorded. The use of averages in the calculation of car speed and density is due to the use of a discrete model. Where in a real road cars can have a continuous range of speeds and distances, here there are only discrete quantities. A meaningful comparison with experimental data becomes possible only after the averaging process, which increases the number of possible values for speed and density. It is to say that in most cases experimental data from highways are averages over time and/or space [14, 16], so that the use of averages is a legitimate procedure.

In addition to the state of the road at chosen times, the traffic evolution as a function of time at selected positions is recorded. This should be particularly useful when the program will be used to simulate real highways and used to make traffic forecasts. We recently patented a system consisting of car detectors to be installed along an highway, whose output will be used to calibrate a cellular automata program implementing that particular highway. The simulations will provide a detailed knowledge of the road characteristics. As a result, it will be possible to forecast if a car jam is going to happen somewhere along the highway on the basis of the current state of traffic and try to take some action in order to prevent it. This could be done reducing car flow inside the highway or having some *ufficial* car driving at reduced speed. This second kind of action could help preventing the complete stop of cars which is the main cause of jam formation [17].

The computer program used for these simulations was written in Fortran90. In the program, software packages written at CERN (CERLIB) are used, mainly for data analysis and graphic outputs (*paw* software) [18]. Eigth hours traffic along a 100 Km long highway can be simulated in 3 to 5 minutes (depending on traffic) with a personal computer equipped with a 1 GHz processor, a 266 MHz bus, 256 Mbyte RAM memory and Linux operating system.

3   Representative results from computer simulations

Three main kind of traffic are found in highways. These will be called free flow, synchronized flow and congested flow. A wide variety of subconditions is found in synchronized and congested flow, which are sometimes named under different terms, like wide moving jams.

Free flow is a condition where fast cars can easily overtake slow ones, there are large gaps between cars and traffic flows easily. This phase occurs at low car density (below 0.1 car/cell in the computer simulations of this work, but the value is not sharply defined) and is characterised by the property that an increasing car density along the highway produce an increasing flow. Experimental data show that in terms of flow this condition lasts for car flow increasing from zero up to 2000-2500 cars per lane per hour [14].

At the higher densities (above 0.5 car/cell in the computer simulations, but again the value is not well defined) it is found the so-called congested flow. It is a condition of heavy traffic, with flow decreasing with increasing car density and easy formation of jams. Overtaking is difficult and car speed can be nearly zero.

Intermediate between the former and the latter is the so-called synchronized flow [14]. This condition is defined as a state of traffic in multilane roads in which the vehicles in different lanes move with almost the same speed. In this region flow can be high in spite of an increasing density and an average speed of cars only about 1/2 that of free flow. In synchronized flow the linear correlation between flow and density is lost and the two quantities becomes totally non-correlated, that is a density increase can be accompanied by either a decrease or an increase in flow rate. At each traffic condition corresponds a different value of the average distance between cars.

These conditions of traffic (or phases, since they resemble in some aspects phase transitions in matter [14]), correspond to three distinct regions in a flow vs. density plot, also named fundamental plot. It is obtained plotting car flow as a function of car density for all positions along the highway and for all times. Every reasonable model of traffic must be able to correctly reproduce this plot, as is the case of the cellular automata model here described. In Fig. 1 such a plot, as obtained from one of the simulations performed with this program is reported.

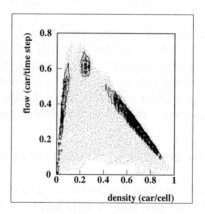

Fig. 1. Fundamental plot of traffic. The population contour lines reveal how most points (> 99%) concentrate into 3 regions. The first, on the left, corresponds to free flow and is characterized by a positive slope. The central contoured region corresponds to synchronized flow. The last contoured region, with negative slope, is due to the presence in the simulation of congested traffic.

Noteworthy, this cellular automata model is able to simulate dynamical features of traffic which were never observed with continuous and/or periodic models. Two examples are shown in Fig. 2 and in Fig. 3. In the first plot, the onset and backward propagation of a wide jam is reported. As found in real highways, jams forms when car density becomes high and are nucleated by ramps (and also by junctions or roadworks, features not introduced at present in the model).

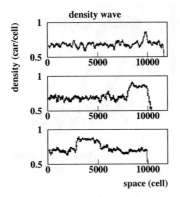

Fig. 2. In this simulation a density wave is generated at a certain time at about cell 10000 (top figure) and propagates backward (2400 and 10400 time steps later in the mid and bottom figure, respectively).

The second plot is particularly interesting to us. It shows how a single on-ramp is not enough to produce congested traffic. In the simulation relative to Fig. 3, a congested flow road with an entrance at the beginning and an exit at its end turns in time into free flow, despite the maximum possible number of cars entering the road at its beginning. In other words, only the presence of several ramps sufficiently closely spaced between them, can produce heavy traffic. This is a common experience for drivers that at the highway beginning or after a queue due to an accident, find fluid traffic and low car density.

3.1 Fractal Analysis

The results from the simple algorithm reported in the previous paragraph indicate how the complexity of traffic is not due to a great number of drivers, each behaving differently. It is rather the result of the repeted application of simple rules, which are surely not corresponding to a linear interaction between cars. Traffic is in fact a non linear system and the problem of writing down the differential equations corresponding to the evolution rules of the cellular automata problem is still open. This, together with the repeated fragmented look common to all plots obtained from the data of the various simulations performed, suggested us to look for self-similarity in traffic. To our knowledge such a research has never been done on experimental data and only once on simulated data from a unreal single lane round-about without entrance or exit [19].

Car density and flow as a function of space were analysed with the box counting algorithm [19–21]. This is probably the most widely used algorithm to

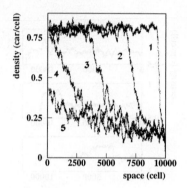

Fig. 3. Density as a function of space at different times in a simulation without ramps except at the beginning (car entrance) or the end (car exit) and a starting density of 0.8 car/cell. In this lane density decreases as a function of time: curve 1 refers to time=1000 time steps after the beginning of the simulation, curve 2 to time=5000 ts, curve 3 to time=10000 ts, curve 4 to time=15000 ts, curve 5 to time=20000 ts.

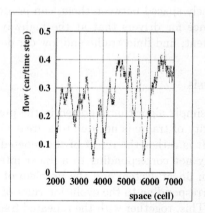

Fig. 4. Car flow as a function of space for a 5000 cell long treat of highway. The curve is repeatedly fragmented with self similar look over about 2 orders of magnification along the x-axis.

extract a non integer dimension from some data. Let consider for example Fig. 4. The plot area is divided into 25 boxes with side length 1/5 of the abscissa and ordinate range of variation. This can be done again and again using n^2 boxes of decreasing side length (1/n of the whole length). An estimate of the fractal dimention D of the curve can be obtained from a linear fit of log(S) vs. log(1/n), where S(n) is the number of squares which contain at least one point of the curve. With a simple program implementing this algorithm, the value of S is readily obtained. The capability of the program to correctly yield the fractal dimention of several point sets was tested with fractal objects taken from mathematics and with non fractal curves. A further analysis, based on the discrete Fourier transform was accomplished, revealing that our traffic data do not have periodic components.

Values of the fractal dimension D between 1.4 and 1.6 were found in the examined data set over a range of about 2 orders of magnitude in highway length. This is not much in comparison with other simulations of fractal systems, but here we are working with a discrete system, so that it is not reasonable to expect a self-similar behaviour measured over a range greater than that of the experimental data.

The spread in D values coming from simulations performed with different traffic conditions could be an indication of a different fractal dimension for the various phases of traffic. We are at the present time trying to better our data set and to use other algorithms to measure D in order to clarify this point. We also plan in the next future to accomplish a similar analysis on real data, which will be collected on italian highways, to check the simulated results.

4   Conclusions

The use of cellular automata algorithms seems to be the most effective way to simulate global and dynamic features of highway traffic. Wave nucleation and propagation, transition between phases of traffic, lane inversion (a greater car density in left than in rigth lane): all most significant features experimentally known from highways seem to be reproducible. These simulations even suggest to look for features of traffic which were never noticed before, like an eventual fractal dimension. Finally, as a result of the confidence in the goodness of the model, it would be possible to test the effect of new traffic rules or drivers behaviour. For example, as suggested in [17] the usual driver behaviour is *competitive*, that is with continuous accelerations, reduced safety distance, etc. What about the effect of a non competitive behaviour? The *experimental* results reported in [17] and our personal experience, seems to show that there should be a noticeable difference. This could be easily simulated by a modified version of the model proposed here.

References

1. B. S. Kerner, H. Rehborn, Phys. Rev. Lett. **79** (1997), 4030.
2. B. S. Kerner, H. Rehborn, Phys. Rev. E **53** (1996), 4275.
3. M. Bando, K. Hasebe, A. Nakayama, A. Shibada, Y. Sugiyama, Phys. Rev. E **51**(1995), 1035.
4. A. D. May, *Traffic flow fundamentals* (Prentice-Hall 1990, New Jersey.).
5. L. Neuber, L. Santen, A. Schadschneider, M. Schreckenberg, Phys. Rev. E **60** (1999), 6480.
6. M. J. Lighthill, G. B. Whitham, Proc. R. Soc. Lond. A **229** (1955), 281.
7. M. Bando, K. Hasebe, A. Nakayama, A. Shibada, Y. Sugiyama, Phys. Rev. E **51** (1995),1035.
8. S. Wolfram, Nature **311**, (1984), 419.
9. S. Wolfram, Theory and applications of Cellular Automata (World Scientific, Singapore, 1986).
10. K. Nagel, M. Schreckenberg, J. Physique **2** (1992), 2221.
11. K. Nagel, D. E. Wolf, P. Wagner, P. Simon, Phys. Rev. E **58**,(1998) 1425.
12. W. Knospe, L. Santen, A. Schadschneider, M. Schreckenberg, Physica A **265** (1999), 614.
13. O. Kaumann, R. Chrobok, J. Wahle, M. Schreckenberg, in: *8th Meeting of the Euro Working Group Transportation*, Editor M. Dielli (2000).
14. B. S. Kerner, Physics World **12**, (1999), 25.
15. E. G. Campari and G. Levi, Eur. Phys. J. B **17**, (2000), 159.
16. D. E. Wolf, Physica A **263**, (1999), 438.
17. http://www.eskimo.com/ billb/amateur/traffic/traffic1.html .
18. http://wwwinfo.cern.ch/asd/paw/.
19. J. Woinowski, in: *Fractal Frontieres '97* Editors M. M. Novak and T. G. Dewey (World Scientific 1997, Singapore), pp.321-327 .
20. Paul Addison, *Fractals and Chaos* (IOP 1997, London).
21. Lui Lam, *Nonlinear physics for beginners* (World Scientific 1998, Singapore), pp.237-240 .

Application of Cellular Automata Simulations to Modelling of Dynamic Recrystallization

Jiří Kroc

West Bohemia University, Research Center, Univezitní 8,
CZ 332 03, Plzeň, Czech Republic,
krocj@ntc.zcu.cz,
WWW home page: http://home.zcu.cz/~krocj/

Abstract. The cellular automaton computer simulation technique has been applied to the problem of dynamic recrystallization. The model is composed of simple microstructural mechanisms – such as evolution of dislocation density, grain boundary movement and nucleation – that acting together result in complex macrostructural response expressed by evolution of flow stress and grain size. A strong sensitivity of the flow stress curves to the initial grain size and nucleation rate is observed. Simulations lead to the conclusion that model of a dynamically recrystallizing material has a strong structural memory resulting in wide variety of experimentally observed responses.

1 Introduction

The plastic deformation of metals and alloys at homologous temperatures higher than 0.5 leads to two types of behaviour [7]. For the first type – i.e., for the high stacking fault energy materials – a dynamic equilibrium between the generation and accumulation of dislocations on one side and annihilation with rearrangement of dislocations on the other side is reached. In this way, a constant flow stress is approached after a certain amount of strain. Therefore, no dynamic recrystallization (DRX) is observed.

For the second type of behaviour – i.e., for medium and low stacking fault energy materials – a higher storage rate is reached due to lower rates of dynamic recovery. At the same time, small dislocation–free regions are produced by dislocation rearrangement at grain boundaries. These small dislocation–free regions – called nuclei – can grow into the surrounding material that contain dislocations. Such recrystallized regions are separated from the surrounding material by high angle boundaries. This process is called dynamic recrystallization where the word dynamic indicates that recrystallization operates simultaneously with deformation.

The flow stress curves of a metal subjected to DRX exhibit the *single peak* behaviour at low temperatures or high strain rates, and the *multiple peak* behaviour. Multiple peak behaviour is typically dumped towards a steady state value, at high temperatures or low strain rates. Grain growth is either deformation controlled at higher strain rates that results in fine grains, or impingement controlled at lower strain rates that results in coarse grains.

P.M.A. Sloot et al. (Eds.): ICCS 2002, LNCS 2329, pp. 773–782, 2002.

The driving force for grain boundary movement is relatively quickly decreased by the rapid production of dislocations at higher strain rates. The grain boundary velocity is not fast enough to replace the initial structure at such strain rates. Several nucleation waves operate simultaneously, and therefore the single peak on the flow stress curve is observed. There is less reduction in the driving force at lower strain rates. The synchronized recrystallization of the material results in a decrease of stress on the flow stress curve. Further deformation leads to the production of new dislocations that results in an increase of the stress on the flow stress curve until a new nucleation wave is triggered that again results in full recrystallization and a new drop on the flow stress curve. This process results in the multiple peak behaviour.

Sakai et al. [12] have shown that there exists a transition from the multiple to the single peak behaviour for materials in which nucleation operates at grain boundaries, and that this transition occurs at the ratio $D_0/D_{ss} = 2$, where D_0 is the initial grain size and D_{ss} is the steady state value. The single peak behaviour occurs when $D_0/D_{ss} > 2$ and the multiple peak behaviour when $D_0/D_{ss} \leq 2$. Sakai et al. carried out strain rate change tests in which an abrupt increase and/or decrease of the strain rate during deformation was applied, as described in [12]. It was observed that, for example, the established multiple peak behaviour could be replaced by the single peak behaviour by an abrupt increase of the strain rate. An excellent review of such DRX experiments can be found in the work of Sakai and Jonas [13].

The Monte Carlo method was applied to model DRX in work of Rollett et al. [11] and Peczak [10]. They predicted the effect of the initial grain size, the steady state grain size dependence on the steady state stress, and the dependence of the peak stress on the peak strain. Unfortunately, they were unable to reproduce the results of the test in which the initial grain size was changed under identical deformation conditions, i.e. the horizontal test.

The Cellular Automata (CA) model of DRX was proposed by Goetz and Seetharaman [4]. They focused their attention on an explanation of flow stress curve shape, the horizontal test and the hardening rate curve, but they did not explain the stress–strain and grain size curves of strain rate change test. They get rather unrealistic grain shapes using growing octagons in their simulations. General information about the CA–paradigm can be found in the the book [2]. It should be mentioned that CA was first time applied to static recrystallization (SRX) by Hesselbarth and Göbel [6].

The objective of the present work is to build a new CA–model where the attention will be focused to several following directions: the grain boundary movement, grain size curves, influence of the initial grain size, horizontal test, influence of nucleation activity, and the strain rate change test.

2   CA–model of DRX

The proposed CA–model of DRX is composed of an uniform two–dimensional lattice of cells that are updated simultaneously according to a transition rule.

The evolution of a cell is controlled by the cells that form the neighbourhood or surrounding of this cell. The Moore neighbourhood, i.e. the eight first and second nearest neighbours in a square lattice, is used with the probabilistic grain boundary movement that produce growing grains of approximately globular shape before their impingement. The lattice size used in simulations is 256×256 cells. Boundary effects of the lattice are removed by use of periodic boundary conditions. Simply, the upper with the lower edge of the lattice are attached together and the left with the right edge of the lattice are attached together as well.

Every CA–simulation of DRX can be decomposed into two steps. In the first one, the initial CA–microstructure is produced. The initial CA–microstructure is generated by use of growth of randomly generated nuclei in homogeneous medium, i.e. by simulation of SRX. The following simple transition rule is used. A cell under consideration recrystallize with 50% probability if at least one of the neighbours from the Moore neighbourhood is recrystallized. Therefore, growing grains reach approximately globular shape.

In the second step, the simulation of DRX is done by the sequential realization of the following three steps representing microstructural evolution of each cell during each time step: (a) evolution of the dislocation density, (b) recrystallization realized by the growth of grains when driving force exceeds a critical value, and (c) the nucleation of embryos of new grains. Three variables, defined for every cell separately, are used in the CA–model: the dislocation density ρ, the orientation θ, and the waiting time t_w, affecting the grain boundary migration velocity.

The dependence of the dislocation density ρ of one particular cell on the shear strain γ is expressed by the following law

$$\frac{d\rho}{d\gamma} = A\rho^{1/2} - B\rho, \tag{1}$$

which is the Mecking and Kocks law [9] at the macroscopic level, where A and B are constants. The first term represents strain hardening and the second term the recovery of dislocations. The flow stress σ is proportional to the square root of $\bar{\rho}$, i.e. $\sigma \propto (\bar{\rho})^{1/2}$, where the average dislocation density $\bar{\rho} = (\sum_{ij} \rho_{ij})/256^2$.

It is considered that 18 different orientations are present in the plane. The velocity of a moving grain boundary can be expressed by the temperature T dependent waiting time $t_w(T)$, as $v(T) = d/(t_w(T) + 2)$, where d is the cell width.

A recrystallization event will occur at a cell C under consideration with 50% probability when the following conditions are fulfilled simultaneously: (i) the cell C is situated at a grain boundary (GB), (ii) the difference in dislocation density between the cell C and neighbouring cell belonging to different grain is greater than a critical value $\Delta\rho_{cr}$, (iii) the potential new configuration of the GB is not an excluded one (e.g. one cell wide re–entrant bulges are not allowed), (iv) the waiting time t_w is zero, (v) a grain having lower dislocation density grows into a grain having a higher dislocation density.

When there is no recrystallization event for the cell C under consideration at the given time t, then the following steps are taken: the waiting time t_w is decreased by one, and the dislocation density ρ follows equation 1.

Finally, a nucleus is produced randomly at the GB with a probability of 0.1% if the dislocation density exceeds a value ρ_{nucl}. The value ρ_{nucl} is kept constant, because there is assumed no temperature the CA–model, except the case when a dynamic change of nucleation rate is applied. The production of a nucleus sets the dislocation density ρ to zero, waiting time t_w is set to its maxiaml value, and the new and different orientation θ of the nucleus is chosen randomly from all possible orientations.

3   Results and Discussion

The initial CA–microstructure of the DRX simulations (cf. Fig. 3 (a)) was produced using an improved Hesselbarth–Göbel algorithm [6], i.e. by the simulation of SRX. This leads to the globular shape of the grains at any time in a CA–simulation (cf. Fig. 3 (a)) instead of the octagonal shapes as in the work of Goetz and Seetharaman [5] or the rectangular shapes in the work of Hesselbarth and Göbel [6] and of other authors. Octagonal grain shapes were also used by Goetz and Seetharaman [4] in their CA–simulations of DRX.

3.1   Flow Stress and Grain Size Curves

DRX experiments carried out under constant strain rate and temperature are called constant strain rate tests. Those experiments displays the transition of flow stress curves from the multiple peak mode – called by some authors cyclic or oscillations – to the single peak mode of recrystallization if different temperature and/or strain rate are taken. The transition is passed by increasing the strain rate or decreasing temperature of the whole test. Sakai, Jonas and co-workers (cf. [13] and references therein) started to pay more attention to the microstructural processes in metals undergoing DRX. Beside the other results, these investigations led to the conclusion that the single peak behaviour is associated with grain *refinement* and the multiple peak behaviour is associated with several distinct cycles of grain *coarsening* [13].

Flow stress curves for two initial grain sizes $\bar{D}_0 = 2.9$ (solid lines) and 7.0 (dashed lines) simulated by the proposed CA–model of DRX are presented in Fig. 1 (a). As the complementary information, a set of grain size curves is displayed in Fig. 1 (b) for the same initial grain sizes $\bar{D}_0 = 2.9$ (solid lines) and 7.0 (dashed lines). The multiple or single peak behaviour is produced at high or low temperature, respectively (what is consistent with experimental results [14]). The peak stress σ_p is shifted upwards and to the right with an increase of the initial grain size. Two CA–microstructures are given in Fig. 3: the initial one (Fig. 3 (a)), and that corresponding to the first grain size minimum (Fig. 3 (b), curve marked by the letter m). Refinement occurs before the first minimum is reached which is replaced by coarsening until the first maximum (cf. Fig. 1 (b)).

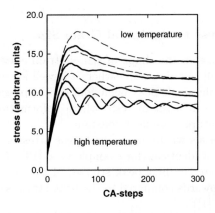
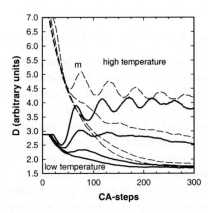

Fig. 1. (a) Flow stress curves for four different temperatures, representing deformation at a constant strain rate and for two different initial grain sizes $\bar{D}_0 = 2.9$ (solid lines) and 7.0 (dashed lines). (b) Dependence of grain size CA–step for the same deformation conditions as in figure (a) and for the same two initial grain sizes where the symbol $\bar{D}$ has been replaced by the symbol D. (The figure (a) is always on the left.)

It is observed during CA–simulations – in the CA–microstructure undergoing deformation – that the first nuclei are produced before the peak stress is reached. These nuclei grow, in the case of the multiple peak behaviour, almost in synchronism and they produce a recrystallization wave resulting in a stress decrease on the flow stress curve. The entire CA–microstructure is consumed by recrystallization after a certain amount of time, then recrystallization is terminated. Continuing accumulation of deformation gradually increases stress. The whole cycle is repeated after the production of new generations of nuclei again and again. The multiple peak behaviour is observed in this way. A decrease in the degree of synchronization of the above mentioned nucleation waves leads to the gradual disappearance of the oscillations on the flow stress curves during the multiple peak behaviour.

Several overlapping nucleation waves are produced at the beginning of recrystallization during the single peak behaviour. These overlapping waves are not synchronized, and therefore, they produce one sole peak on the flow stress curve. These characteristics of the single and the multiple peak behaviour are consistent with the experimental work of Sakui et al. [14].

Multiple peaks are observed on grain size curves as well (cf. Fig. 1 (b)). Minima and maxima of the grain size curve are phase shifted by approximately 1/4 of a period before the corresponding maxima and minima of the flow stress curve during the multiple peak behaviour. This phase shift is consistent with MC–simulations [11, 10]. On the other side, the experimental curves displaying evolution of grain size do not exhibit multiple peaks in the case of the multiple peak behaviour. The reason of this discrepancy could lay in the fact that experimental results are not in situ measurements but post mortem one (cf. details of [14]).

3.2   Sensitivity of Material Response to the Initial Grain Size

A strong influence of the initial grain size on the shape of stress–strain curves is experimentally well documented [13, 14]. As discussed earlier, it is well known that the transition from the multiple to the single peak behaviour can be passed by mere change of the initial grain size in a set of tests where deformation conditions – i.e., strain rate and temperature – are kept constant.

On one hand, there are known materials where all the first peak stresses of every curve in a horizontal test are almost identical (for example steel [13]). On the other hand, there are materials where the first peak stresses of the single peak behaviour is significantly shifted upwards compared to the peak stresses of the multiple peak behaviour (cf. copper [1]).

It should be mentioned here that the Monte Carlo simulations do not succeed to explain this type of transition [11, 10]. On the other side, CA-simulations successfully explained it [4]. In this and one subsequent section, it will be studied which other mechanisms can cause a shift of the peak stress of the single peak curve compared to the first peak stress of the multiple peak curve.

In the present simulations, it is evident from Figs. 1 and 2 that the initial grain size $\bar{D}_0$ has a strong influence on the shapes of the flow stress and grain size curves for given deformation conditions. It is observed that more coarse $\bar{D}_0$ values shift the flow stress curve upwards and result in higher peak stresses σ_p at the same time (cf. Fig. 1). The steady state value is not influenced by any change in $\bar{D}_0$ (cf. Fig. 1 (a) and 1 (b)). This means that a metal subjected to DRX reaches the same dynamically stable equilibrium for the same deformation conditions from any initial grain size D_0.

The transition from the single to the multiple peak behaviour in a horizontal test – where the strain rate, temperature and nucleation rate are kept constant – is reached for the ratio of initial grain size $\bar{D}_0$ to steady state grain size $\bar{D}_{ss}$, $\bar{D}_0/\bar{D}_{ss} = 3.5 \pm 0.5$. This is somewhat higher value than the experimental ratio $\bar{D}_0/\bar{D}_{ss} = 2$. It has to be stressed here that the experimentally derived transition ratio is just an approximate value and not the precise one (cf. details [13]).

It is well known from experiments that grain coarsening is associated with multiple peak behaviour, and that grain refinement is associated with single peak if the final grain size is less than half of the initial one. Several different inverse transition ratios $\bar{D}_{ss}/\bar{D}_0$ are observed in simulations. Inverse transition ratio of $\bar{D}_{ss}/\bar{D}_0 = 0.3$ is observed if the initial grain size is changed only. Different inverse transition ratios can be reached when nucleation rate is simultaneously changed during simulation.

3.3   Strain Rate Change Test

A special type of experiments that combine a constant strain rate test and an abrupt strain rate increase/decrease test – both during one experiment – will

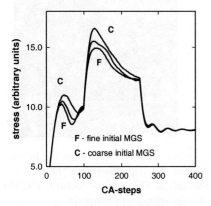
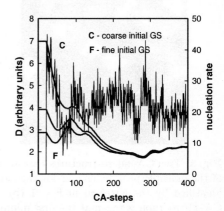

Fig. 2. (a) Flow stress curves in the case of the strain rate jump test for three different initial grain sizes $\bar{D}_0 = 2.9$ (curve F), 3.9 and 7.0 (curve C) carried out under identical deformation conditions. The sensitivity of the flow stress curves to the initial value of $\bar{D}_0$ is apparent. (b) Dependence of grain size curves on the CA–step for the same deformation conditions. The thin curve represents the nucleation rate $\dot{N}$ associated with the curve marked by the letter F.

be briefly introduced here. In this type of experiments, first more or less stable mode of behaviour for a given constant strain rate and temperature is reached, and then an abrupt change of strain rate is applied. Strain rate increase or strain rate decrease are two possible changes during such experiments. A combination of several abrupt strain rate changes is usually applied during one experiment (cf. details in [13]). Typically, the strain rate was cycled from a low value to a high value and back to the initial value.

One particular case of a simulated strain rate change test is shown in Fig. 2, where an initial strain rate was abruptly increased to a higher strain rate after 100 CA–steps, and abruptly decreased to the original strain rate after another 150 CA–steps. Three different initial grain sizes, $\bar{D}_0 = 2.9$, 3.9 and 7.0, led to three different flow stresses and grain size curves. The peak stress σ_p is shifted upwards when the initial grain size $\bar{D}_0$ is coarsened (Fig. 2 (a)). The transition from the multiple to the single peak behaviour and *vice versa* is reproduced (compare to experimental results [12]). Such transition can be interpreted as switching from one mode of behaviour to the other. This transition is consistent with experimental results of Sakai *et al.* [12]. They showed that a steel deformed at a given elevated temperature, for which the strain rate was cycled from a low value to a high value and back to the initial value, exhibits a transition from the multiple to the single peak behaviour and *vice versa*.

The evolution of grain size D_0 presented in Fig. 2 (b) is consistent with the experiments [12], in which high or low strain rate is associated with fine or coarse steady state grain size $\bar{D}_{ss}$, respectively. Details can be found in the

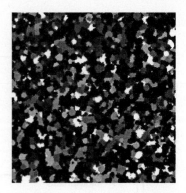

Fig. 3. Two CA–microstructures corresponding to two different CA–steps in the case of the multiple peak behaviour for the curve of grain size dependence on the CA–step identified by the letter m in Fig. 1 (b): (a) the initial CA–microstructure, (b) the CA–micrograph is taken at the first minimum. Grain refinement is apparent.

microstructural mechanism map distinguishing these two modes of DRX [12]. This and the previous simulations leads to the conclusion that a dynamically recrystallizing material has a strong structural memory resulting in a wide range of experimentally observed responses. Strain rate decrease tests or more complex tests can be easily realized by such CA–model as well.

It is worth to note that instabilities were detected in some specific cases during the simulations. These instabilities can produce "spirals" in the CA–microstructures, typically after a strain rate decrease during a strain rate test. These spirals affect the shapes of the flow stress and grain size curves. The occurrence of such spirals could be associated with the assumption of the CA–model that every cell is deformed at the very same strain rate and that there are no deformation localizations in the CA–microstructure.

The evolution of the nucleation rate is shown in Fig. 2 (b) by the thin curve. The nucleation rate $\dot{N}$ is simply defined as the number of nuclei per time step. It was recognized that the nucleation rate $\dot{N}$ is inversely correlated with the grain size curve. This correlation occurs because finer grains are associated with higher specific grain boundary areas and therefore with higher nucleation rates as long as only GB nucleation is allowed. Thus, an increase in the nucleation rate $\dot{N}$ is correlated with a decrease in the grain size $\bar{D}$ and *vice versa*.

3.4   Dynamic Change of Nucleation Rate

The influence of different initial nucleation rates on the flow stress curve for the initial grain size of $D_0 = 9.8$ can be seen in Fig. 4 where different nucleation rates are taken for the frst 50 CA-steps. Nucleation rate is changed from an initial value to the fixed value of 0.001 where the initial value is equal to 0.0001, 0.001, 0.002, 0.01 from top to bottom, i.e. A, B, C, D, respectively. The curve marked as O is computed for different initial grain size of $D_0 = 2.8$

without change of nucleation rate equal to 0.001. The ratio of the initial grain size $\bar{D}_0$ to the steady state grain size $\bar{D}_{ss}$ is $\bar{D}_0/\bar{D}_{ss} = 3.9 \pm 0.4$.

The ratio of the initial grain size $\bar{D}_0$ to the steady state grain size $\bar{D}_{ss}$ gives $\bar{D}_0/\bar{D}_{ss} = 2.8 \pm 0.4$ where the initial grain size of $D_0 = 7.0$ is taken. The lower limit of this ratio lays near of the experimentally observed value of 2 for which the inverse transition ratio $\bar{D}_{ss}/\bar{D}_0$ equal to 0.4 approaches the experimental value of 0.5. Those observations lead to the conclusion that nucleation rate can affects the shape of flow stress curves beside the initial grain size. Different thermomechanical histories of a material can lead to a different initial nucleation rates resulting in different shapes of deformation curves.

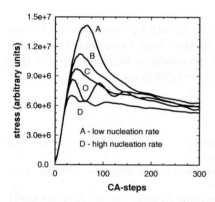
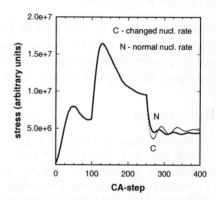

Fig. 4. The influence of different initial nucleation rates on: (a) the flow stress curves for the initial grain size of $D_0 = 9.8$ with change of the nucleation rate for the first 50 CA–steps. The nucleation rate is changed from an initial value to the fixed value of 0.001 where the initial value is equal to 0.0001, 0.001, 0.002, 0.01 from top to bottom, i.e. A, B, C, D, respectively; (b) strain rate change test at the point of strain rate decrease. Temporary increase of nucleation rate pronounce oscillations after strain rate decrease.

The influence of nucleation rate change on the shape of deformation curve during strain rate change test has been studied, see Fig. 4 (b). Solid line represents the same strain rate change test as in Fig. 2 (cf. curve C), i.e. the situation where the nucleation rate is held at constant value of 0.001. The dotted line represents the situation when the nucleation rate is increased for 20 CA–steps from 0.001 to 0.01 at the CA–step equal 250 – right at the moment of an abrupt strain rate decrease – and after this number of CA–steps decreased to its original value. It is evident that such change of nucleation rate can pronounce the maxima and minima on deformation curve.

4   Conclusions

A cellular automaton (CA) model of dynamic recrystallization (DRX) has been proposed with a probabilistic movement of the grain boundaries. The CA–model itself is simple to define and easy to handle. A strong influence of the initial grain size and nucleation rate on the shape of flow stress curves is observed. It is found that the evolution of grain size provides information that is complementary to the flow stress curve. Changes of nucleation rate shifts the maxima of the flow stress curves and transition from the multiple peak to the single peak mode. It is recognized that the material has a strong structural memory and self-organizes itself into dynamically stable modes of behaviour. Those dynamically stable modes are clearly marked on the deformation curves and grain size curves by the steady state values.

Acknowledgements

The author is greatly indebted to Prof. J.D. Eckart for his publicly available CA–environment called Cellular/Cellang [3] that have been used to compute results presented in this work. The author acknowledge financial support from the Czech Ministry of Education under project LN00B084.

References

1. Blaz, L., Sakai, T., Jonas, J.J.: Effect of initial grain size on dynamic recrystallization of cooper. Metal Sci. **17** (1983) 609–616
2. Chopard, B., Droz, M.: Cellular Atomata Modeling of Physical Systems. Cambridge University Press, Cambridge (1998)
3. Eckart, J.D.: The Cellular Automata Simulation System: Cellular/Cellang (1991–1999) http://www.cs.runet.edu/˜dana/
4. Goetz, R.L., Seetharaman, V.: Modelling Dynamic Recrystallization Using Cellular Automata. Scripta Mater. **38** (1998) 405–413
5. Goetz, R.L., Seetharaman, V.: Static Recrystallization Kinetics with Homogeneous and Heterogeneous Nucleation Using a Cellular Automata Model. Metall. Mater. Trans. **A 29** (1998) 2307–2321
6. Hesselbarth, H.W., Göbel, I.R.: Simulation of Recrystallization by Cellular Automata. Acta Metall. Mater. **39** (1991) 2135–2143
7. Humphreys, F.J., Hatherly, M.: Recrystallization and Related Annealing Phenomena. Pergamon, Oxford, New York, Tokyo, 1996
8. Kroc, J.: Simulation of Dynamic Recrystallization by Cellular Automata. PhD Thesis. Charles University, Prague, 2001
9. Mecking, H., Kocks, U.F.: Kinetics of flow and strain-hardening. Acta. Metall. **29** (1981) 1865–1875
10. Peczak, P.: A Monte Carlo study of influence of deformation temperature on dynamic recrystallization. Acta Metall. Mater. **43** (1995) 1279–1291
11. Rollett, A.D., Luton, M.J., Srolovitz, D.J.: Microstructural Simulation of Dynamic recrystallization. Acta Metall. Mater. **40** (1992) 43–55
12. Sakai, T., Akben, M.G., Jonas J.J.: Dynamic recrystallization during the transient deformation of a vanadium microalloyed steel. Acta Metall. **31** (1983) 631–642
13. Sakai, T., Jonas, J.J.: Dynamic recrystallization: mechanical and microstructural considerations. Acta. Metall. **32** (1984) 189–209
14. Sakui, S., Sakai, T., Takeishi, K.: Hot deformation of austenite in a plain carbon steel. Trans. Iron Steel Inst. Japan **17** (1977) 718–725

A Distributed Cellular Automata Simulation on Cluster of PCs

Paweł Topa

Institute of Computer Sciences, University of Mining and Metallurgy, al. Mickiewicza 30, 30-059 Kraków, Poland
email: `topa@uci.agh.edu.pl`

Abstract. Inherent parallelism of Cellular Automata as well as large size of automata systems used for simulations makes their parallel implementation indispensable. The purpose of this paper is to present a parallel implementation of two sequential models introduced for modelling evolution of anastomosing river networks. Despite the both models exploit the Cellular Automata paradigm, the nontrivial application of this approach in the second model involves defining a mixed parallel-sequential algorithm. The detailed description of the two algorithms and the results of performance test are presented.

1 Introduction

Cellular Automata (CA) [1] have gained a huge popularity as a tool for modelling problems from complex dynamics [2]. They can simulate peculiar features of systems that evolve accordingly to local interactions of their elementary parts. CA can be represented by n-dimensional mesh of cells. Each cell is described by a finished set of states. The state of the automaton can change in successive time-steps according to defined rules of interactions and the states of the nearest neighbours.

This paper introduces two Cellular Automata models implemented on cluster system. The detailed description of their sequential versions and results obtained can be found in [3]. The short review is presented below.

The presented models simulate evolution of anastomosing river networks. Anastomosing rivers develop on flat, wide areas with a small slope. The main reason of creation and existence of such the systems is a plant vegetation. Plants receive necessary resources (nutrients) from water. Products of their decay is accumulated in the interchannel areas as peatbogs and follows to rising up of the banks.

The rate of plants vegetation is controlled by the nutrients. The gradient of nutrients saturation, which appears perpendicularly to the channel axis, results in faster accumulation of peats near banks and slower accumulation on distant areas. The water plants vegetating in the channels can block the flow. In such the situation, a new channel have to be created. It starts above the jam zone and its route is determined by the local topography. Such the channel may join back

P.M.A. Sloot et al. (Eds.): ICCS 2002, LNCS 2329, pp. 783–792, 2002.

to the main river bed downstream. The evolution of the newly created branch proceeds in similar way as the evolution of the main stream. Finally, such the processes results in creating of complex, irregular network of interconnecting channels having hierarchical and fractal structure.

First model, applied for simulating anastomosing rivers, was called SCAMAN (Simple Cellular Automata Model of Anastomosing Network) [3]. It exploits the Cellular Automata paradigm in a classical way. The system is represented by a regular mesh of automata. The state of each automaton is described by the following parameters: (1) the altitudes resulting from the location of the cell on terrain mesh, (2) the amount of water, (3) the amount of nutrients and (4) the thickness of peat layer. Defined rule of water distribution simulates superficial flow. The cells containing water are the sources of nutrients, which are disseminated among surrounding cells in such a way which is able to provide expected gradient of nutrient saturation. The thickness of peat layer is updated accordingly to the amount of nutrient in the cell.

Due to limitations of SCAMAN model, another model was proposed. In MANGraCA (Model of Anastomosing Network with Graph of Cellular Automata), a network of river channels is represented by the directed graph of CA. The graph is constructed on the classical, regular mesh of Cellular Automata by establishing additional relationships between some neighbouring cells.

The state of each automaton in the regular mesh is described by three parameters: (1) the altitude, (2) the amount of nutrient and (2) the thickness of peat layer. Their meaning are the same as in SCAMAN model. Also their values are calculated by using similar algorithms.

When a new cell is added to the graph, its state is additionally described by two parameters: (a) the throughput and (b) the flow rate. These parameters describe local state of a part of the channel. The flow rate in the first cell (source) is initialized to an arbitrary value and then propagated to other cells currently belonging to the graph. Overgrowing of the channel is modelled by slowly decreasing of the throughput value. Occasionally, in randomly chosen cell, the throughput is suddenly decreased below the flow rate. This corresponds to creation of flow jam in the channel. Such the event leads to creation of a new channel, which starts before the blocked cell. The route of the new channel is determined by a local topography of the terrain.

The cells which belong to the graph are the sources of nutrients. Nutrients are spread to other cells in the mesh by using the same algorithm as in SCAMAN model. This way, the graph stimulates changes in terrain topography, which consequently influences development of the network.

Neither SCAMAN nor MANGraCA are able model entirely the anastomosing river. MANGraCA produces global pattern of anastomosing network, but without any information about local distribution of water in the terrain. SCAMAN simulates flow of water in terrain but in order to obtain more complex network a very large mesh has to be applied and several thousands of time-steps of the simulation must be performed. The models work in different spatio-temporal scales. Basing on this observation a hybrid multiresolution model have been proposed in which MANGraCA model produces global river pattern, while SCAMAN, basing on generated network, calculates the local water distribution and simulates the growth of the peat layer.

MANGraCA can be also applied to modelling other network structures. It can be useful to modelling the evolution of transportation network immersed in consuming environment such as vascular system, plant roots, internet.

Parallel computers are the most natural environment for Cellular Automata simulation [4], [5]. In fact, sequential simulation of real phenomena using CA, where a huge number of automata are employed to represent the system, are impossible in practice. For a long time, such the simulations required access to massively parallel processors (MPP) located in supercomputer centers. The breakthrough has been brought by Beowulf [6], the first low cost parallel computer built in NASA laboratory. The Beowulf was constructed using only low priced COTS (commodity of the shelf) components. Regular PC computers with Intel 486 processors were deprived of floppy disks, keyboards and monitors and connected using 10Mbit Ethernet. The system worked under Linux operating system. Some large computations were successfully performed on this machine. At present, the Beowulf name relates to the certain class of clusters, followed by the example of the machine built at NASA.

Simple recipe [7] of cluster assembling and low cost of components (software is mostly public domain) enables parallel computing for institutes with limited financial resources. Low cost of maintenance and homogeneity of the environment simplify computations. Allocating the whole or part of node for exclusive use is easy and allows to neglect the load balancing problems.

Clusters share many features with MPP architecture. Application of popular paralellel computing libraries (like MPI and PVM) makes easy to port the algorithms from clusters to MPP machines. Clusters can be used as developing and testing platforms.

The next two sections contain detailed description of Parallel-SCAMAN and Parallel-MANGraCA. Some results of performance tests on cluster of PCs are presented. Conclusions are discussed at the end.

2   Parallel-SCAMAN

In Parallel-SCAMAN the mesh of cells is geometrically decomposed onto domains which are processed independently on different nodes. Mesh can be divided on blocks or stripes. The striped partitioning is preferred due to simplified communication and load balancing implementation.

The nodes in parallel machine are ordered in one-dimensional array: $P_1, \ldots, P_N$, where N stands for number of nodes participating in computation. Processor P_1 processes columns from 1 to m, P_2 — from $m+1$ to $2m$ and so on ($m = N/M$ where M stands for total number of columns in the mesh). Each node store two additional columns, which are not processed by this node (see Fig. 1a — dark grey marked cells). These columns contain copy of the border cells (light gray marked cells in Fig. 1a) from the adjacent processors. Such the copy is necessary on P_i node to update its own border cells. After each time-step P_i exchanges the border columns to its neighbours: P_{i-1} and P_{i+1}. The exceptions are P_0 and P_N, which exchanges data with only one neighbour. Such the scheme applies only to calculating the nutrients distribution and the thickness of peat layer.

The algorithm, which calculates the water distribution consist of two stages.

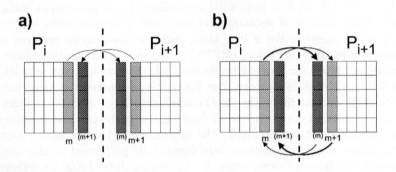

Fig. 1. Exchange of data in Parallel-SCAMAN: a) nutrients and thickness of peat layer, b) flows.

First, for each cell in the mesh the flow rate from this cell to each of its neighbours is calculated. This means, that two border columns with flow rates must be exchanged between neighbouring nodes (see Fig. 1b). On P_i node light greyed column updates the copy of water data on the adjacent processor (P_{i+1}). The second column (dark greyed) contributes to the total flow rates for the border cells on P_{i+1} processors. In the following step of the algorithm, current amount of water is calculated for each automaton basing on the flow rates.

On a single P_i processor the following operations are executed:

```
loop:
    update_water();
    update_nutrients();
    update_peats();
    exchange_data(P_{i+1}); {flow, nutrients, peats}
    exchange_data(P_{i-1});
```

In Parallel-SCAMAN the *Master* node is required only to gather current data from *worker* in order to store them or for visualization. The changes in state of cells, propagate very slowly through the mesh, especially for very large meshes. The performance can be improved if the *Workers* do not communicate with *master* in every time-step of simulation. Data can be gathered only every 10, 100 or more time-steps. The interval value depends on the size of system and quality required producing images or animation for inspection.

3   Parallel-MANGraCA

In MANGraCA model, the number of cells participating in graph computation is significantly less than their total number. The structure of the model allows for separating the computations connected with graph structure from the processing of the regular mesh.

Fig. 2 presents the general Parallel-MANGraCA architecture. The distinct processor (P_0) performs sequentially all operations concerned the graph i.e. throughput updating and flow propagation. It also handles the changes in graph topology caused by newly created branches.

The algorithm, which trace the route of a new branch requires the most up-to-date data about the terrain topography. It means that the altitude data and thickness of peat layer must be also stored on this node.

Distribution of nutrients and updating of the thickness of peat layer are

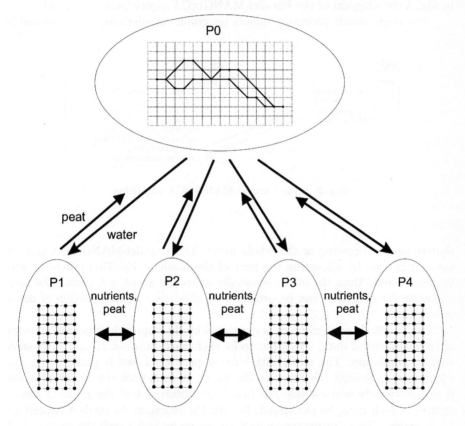

Fig. 2. General Parallel-MANGraCA architecture

performed on a regular mesh of automata. It can be easily implemented in parallel, similarly as it is in SCAMAN model. The mesh is distributed between the nodes of clusters $(P_1 \ldots P_N)$. The algorithm, which calculates the nutrients distribution, requires information, which cells are the sources of nutrients (i.e. which automata currently belongs to the graph). Therefore, the three types of communication must be provided:

1. P_0 sends information to all P_i $(i = 1, 2, \ldots, N)$, about which automata currently belongs to the graph.
2. Each P_i sends to P_0 information about the changes in the landscape, i.e., current thickness of the peat layer.
3. In every time-step, each P_i exchange the information about the state of boundary cells (the nutrient distribution and the thickness of the peat layer) with its neighbours: P_{i+1} and P_{i-1}.

In Fig. 3 the diagram of the Parallel-MANGraCA algorithm is presented.

The time, which processor spends on graph calculations, is incomparable

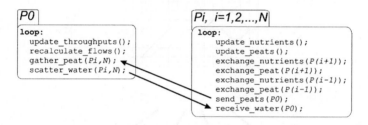

Fig. 3. Basic Parallel-MANGraCA algorithm

shorter than processing of the whole mesh. The Parallel-MANGraCA can be easily improved by allocating the part of the mesh to P_0. This processor will be more loaded than the other, especially when the graph will grow. The load balancing for this node can be provided by allocating smaller portion of mesh at startup.

A substantial problem which is generated by the approach presented consists of large amount of data, which has to be exchanged between P_0 and P_i in each step of simulation. The mesh with current peat thickness is gathered on P_0, which next broadcast the graph to the all P_i. In fact, such the communication is necessary only when a jam has just been occurred and the route of newly created branch must be calculated. In such the situation P_0 sends a request to all P_i nodes. The P_i processor, in each time-step looks for such the request and in case of such the event, it sends to P_0 their parts of mesh with the thickness of peat layer. Then P_0 node, basing on the new terrain topography, calculates, the route of the new branch. The reconfigured graph is broadcasted to all P_i nodes.

Processing of the graph structure and regular mesh can be performed more independently. In the approach presented above, P_0 node processes one part of the mesh so it is synchronized with P_i. In Asynchronous Parallel-MANGraCA, mesh and graph are processed independently. Mesh computations are executed in parallel (e.g. on cluster), while the graph is processed sequentially on a separate machine. Communication occurs only when the graph is reconfigured.

4 Results

Parallel-SCAMAN and Parallel MANGraCA have been implemented in C language, using MPI (mpich 1.2.0) library. The models implementation have been running and testing on the Beowulf-like cluster of PCs, consisting of up-to 8 uniprocessors Intel Pentium III Celeron Coppermine 600 MHz (cache L2 128 kB) computing nodes with 64 MB of memory per node plus one Intel Pentium III Katmai 500 MHz (cache L2 512 kB) as a front-end node. The nodes are connected with Fast Ethernet (100 Mbit/s).

Fig. 4 and 5 show the results of speedup measurements. Parallel algorithms have been compared with its sequential equivalents. Thus, we can see real benefits from porting models to the parallel architecture.

As shown Fig. 4 the Parallel-SCAMAN scales linearly with increasing number of processors and the job size (see Fig. 4). In this algorithm, practically there is no serial fraction. The only factor, which worsen the speedup is communication. As it was expected the best performance is obtained when the nodes are fully loaded. Otherwise, the communication will take more time than computation. The algorithm scales better for larger problems. If a small job is computed, speedup closes to linear only for 2-3 nodes. When more node are applied, the speedup is decreasing. The same result will appear if the faster processor is applied.

Fig. 5 presents the results of tests with Parallel-MANGraCA. In these tests the front-end node was exploited as P_0 machine. Basic algorithm of Parallel-MANGraCA has very poor efficiency (diamonds). Allocating the part of mesh for processor P_0 gives insignificant profits (triangles). The real improvement have just gave Parallel-MANGraCA with reduced graph-mesh communication (square and circle).

The development of anastomosing network (represented by the graph) is a stochastic process. The creation of a new branch occurs in randomly chosen moment of time. For the tests, the parameter T_b describing the mean time between the events of creation of new branches, has been introduced. Its value is dependent on parameters of graph processing i.e. the rate of throughput decrease and the probability of jam occurrence. The greater value of T_b means less frequent formation of new branches, which results in the reduction of time spent fo communication between P_0 and P_i nodes. In Fig. 6a we present the influence of T_b on the Parallel-MANGraCA efficiency. Fig. 6b shows, how the algorithm scales with the mesh size.

Unlike the Parallel SCAMAN, Parallel-MANGraCA has a significant serial fraction which influence badly the speedup. Also communication overhead worses

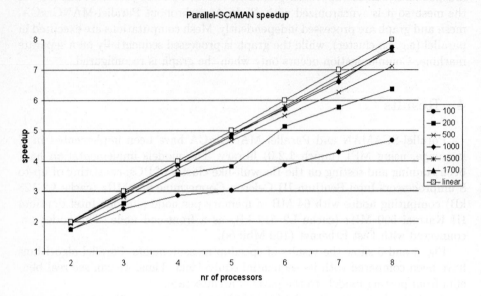

Fig. 4. Speedup of Parallel-SCAMAN

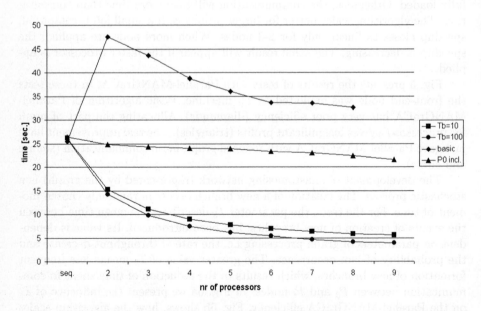

Fig. 5. Comparison of three Parallel-MANGraCA algorithms (see text for details)

the efficiency. An improvement of processor performance will result in the same issue as it is in SCAMAN model.

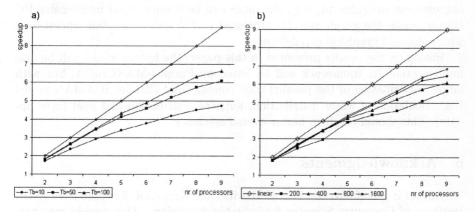

Fig. 6. Parallel-MANGraCA with reduced mesh comunication: a) speedup vs. T_b (400× 400 cells, b) speedup vs. mesh size ($T_b = 50$)

5   Conclusions

The two parallel Cellular Automata models implemented on cluster architecture has been introduced. Their conception base on sequential SCAMAN and MANGraCA model introduced for modelling anastomosing river network. This paper presents general architecture of these models and preliminary results of performance tests. Homogenous architecture of cluster environment as well as exclusive availability of necessary numbers of nodes allows to neglect some critical aspects of parallel algorithms such as load balancing. The works presented is concentrated rather on designing, implementing and testing general model structure than on maximizing performance on a specific architecture. In the future, when the Parallel-SCAMAN, Parallel-MANGraCA will be ported to high performance MPP architecture, the procedures of load balancing will be provided.

Parallel-SCAMAN was designed in a classical way. Its algorithm base on simple division of processed mesh between the nodes of clusters. The results of performance tests show clearly the benefits of parallel computing especially for large and very large tasks.

The concepts tested in Parallel-SCAMAN were applied in Parallel-MANGraCA. Due to the nontrivial application of Cellular Automata paradigm, the original parallel-sequential architecture have been implemented to obtain satisfactory efficiency. Furthermore, this approach gives an interesting issue in studies on mutual interaction of two coupled systems. Some of the future works should concentrate on Asynchronous Parallel-MANGraCA. The another area of studies on Parallel-MANGraCA may concern the choice of computer architecture, on which the model can be implemented in more efficient way. Separation of the graph and the mesh computations allows to perform the simulations in non-homogenous environment, e.g., the mesh can be computed on inexpensive PC cluster, while the graph may be processed and visualized on fast workstation with enhanced graphical capabilities.

Basing on the results presented in this paper, the hybrid model will be also implemented. Its framework will be based on Parallel-MANGraCA, but with larger participation of the parallel code coming from Parallel-SCAMAN model (i.e. calculating of water distribution, nutrient distribution and peat layer update). This should result in better overall efficiency.

6   Acknowledgments

Author is grateful to Drs Witold Dzwinel and Krzysztof Boryczko from AGH, Institute of Computer Sciences for valuable discussions. This project was partially supported by The Polish State Committee for Scientific Research (KBN) under grant 7T11C00521.

References

[1] S. Wolfram, *Cellular Automata and Complexity: Collected Papers*,1994,
[2] B. Chopard and M. Droz, *Cellular Automata Modeling of Physical Systems*, Cambridge University Press 1998,
[3] P. Topa, M. Paszkowski, Anastomosing transportation networks, In *Proceedings of PPAM'2001 Conference*, Lecture Notes in Computer Science, 2001,
[4] D. Talia, Cellular Automata + Parallel Computing = Computational Simulation, *Proc. 15th IMACS World Congress on Scientific Computation, Modelling and Applied Mathematics*, vol.6, pp.409-414, Wissenschaft&Technik Verlag, Berlin, August 1997,
[5] G. Spezzano, D. Talia, Programming cellular automata algorithms on parallel computers, *Future Generations Computers Systems*, 16(2-3):203-216, Dec. 1999,
[6] http://www.beowulf.org,
[7] T.L. Sterling, J. Salmon, D.J. Becker, D.F. Savarese *How to build a Beowulf?*, MIPT Press, 1999,

Evolving One Dimensional Cellular Automata to Perform a Non-Trivial Collective Behavior Task: One Case Study

F. Jiménez-Morales[1], M. Mitchell[2], and J.P. Crutchfield[2].

[1] Departamento de Física de la Materia Condensada. Universidad de Sevilla.
P. O. Box 1065, 41080-Sevilla, Spain.
[2] Santa Fe Institute, 1399 Hyde Park Road
Santa Fe, New Mexico, 87501, USA.

Abstract. Here we present preliminary results in which a genetic algorithm (GA) is used to evolve one-dimensional binary-state cellular automata (CA) to perform a non-trivial task requiring collective behavior. Using a fitness function that is an average area in the iterative map, the GA discovers rules that produce a period-3 oscillation in the concentration of 1s in the lattice. We study one run in which the final state reached by the best evolved rule consists of a regular pattern plus some defects. The structural organization of the CA dynamics is uncovered using the tools of computational mechanics.

PACS: 82.20Wt Computational modeling; simulation.

1 Introduction

A cellular automata (CA) is a regular array of N cells, each of whose state $s_i(t)$ is taken from a finite number of values, and which evolves in discrete time steps according to a local rule ϕ. CAs provide simple models of complex systems in which collective behavior can emerge out of the actions of simple, locally connected units. This collective behavior obeys laws that are not easily deduced from the local rule. However it is not well established how to design a CA to exhibit a specific behavior.

The application of genetic algorithms (GA) to the design of one dimensional CA that perform useful computations has both scientific and practical interest [4, 13, 14]. In the original work of Crutchfield and Mitchell [4] a GA was able to discover CAs with high performance on tasks requiring cooperative collective behavior, namely the density and the synchronization tasks. Figure 1 shows two space-time diagrams of two evolved CA. A successful CA for the density classification task decides whether or not the initial configuration contains more than half 1s. If it does, the whole lattice eventually iterates to the fixed point configuration of all cells in state 1; otherwise it eventually iterates to the fixed-point configuration of all 0s [6, 4]. For the synchronization task, a successful CA will

P.M.A. Sloot et al. (Eds.): ICCS 2002, LNCS 2329, pp. 793–802, 2002.

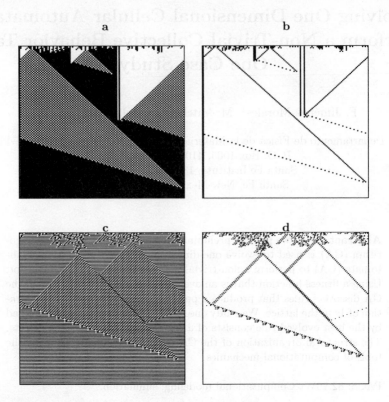

Fig. 1. Space-time diagrams illustrating the behavior of two CAs that perform the density classification task **(a)** and the synchronization task **(c)** [5, 13]. Time starts on $t = 0$ and goes from up down and space is displayed on the horizontal axis. Black represents a cell in state 1 while white is a cell in state 0. The corresponding space-time diagram after filtering out regular domains are shown in **(b)** and **(d)**.

reach a final configuration in which all cells oscillate between all 0s and all 1s on successive time steps [5, 9].

The computational task that we study in this paper is the QP3(P3) task in which the goal is to find a CA that, starting from a random initial condition, reaches one final configuration in which the concentration $c(t) = \frac{1}{N}\sum_i^N s_i(t)$ oscillates among three different values. This task was previously studied in $d = 3$ where the concentration of some CAs show many non-trivial collective behaviors (NTCB).

Figure 2 shows the iterative map, i.e. the graph of $c(t + 1)$ versus $c(t)$, for a family of totalistic CA rules [10]. Here the different collective behaviors are represented by distinct clouds of points. A few years ago a generic argument was given against the existence of collective behaviors with period larger than 2 in extended systems with local interactions [1]. Nonetheless, larger-period collective behaviors have been observed in such systems. The most interesting NTCB is the quasiperiod three behavior (QP3). Several attempts have been made to understand its phenomenology and have addressed the possible mechanisms by which this puzzling collective behavior emerges [2, 3, 8] but at the moment there is not an answer to the question of how NTCB can be predicted from the local rule. In [11] a GA was used to evolve a population of three dimensional CAs to perform a QP3(P3) task, i.e. under an appropriate fitness function the GA selected rules with P3 or QP3 collective behavior. In this work we evolve a population of one-dimensional CAs rules to perform the QP3(P3) task.

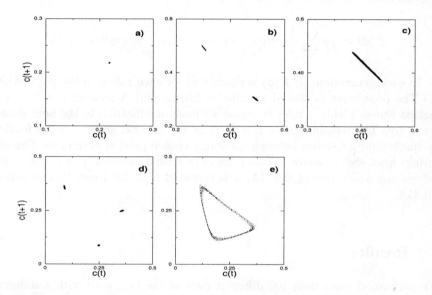

Fig. 2. Iterative map for five totalistic cellular automata rules [10]. The different collective behaviors are shown by distinct clouds of points: **(a)** period-1; **(b)** period-2; **(c)** intermittent period-2; **(d)** period-3; **(e)** quasiperiod-3.

2   The Genetic Algorithm

Our GA begins with a population of P=20 randomly generated chromosomes. Each chromosome is a bit string representing the output bits for all possible neighborhood configurations in a binary state CA of radius 3, listed in lexicographic order. ("Radius 3" means that each neighborhood of a cell consists of the cell itself plus the three nearest neighbors on either side.) Since there are $2^7 = 128$ such neighborhood configurations, the chromosomes are 128-bits long. Thus there is a huge space of 2^{128} possible rules in which the GA is to search. The fitness evaluation for each CA rule is carried out on a lattice of N cells starting from a random initial condition of concentration 0.5. After a transient time of $N/2$ time steps, we allow each rule to run for a maximum number of $M = N/2$ iterations. From the iterative map of Figure 2 we observe that the area of period-1 and period-2 behaviors are smaller than the area of period-3 and QP3. This heuristic lead us to define the fitness $F(\phi)$ of a rule ϕ as the average area covered in the iterative map corresponding to ϕ. It is worth pointing out that the final state the CA must reach for this task is not necessarily a periodic one such as it is for the synchronization and the density tasks.

The values of concentration are assembled in groups of 4 consecutive values (c_1, c_2, c_3 and c_4) and the fitness function $F(\phi)$ is defined by:

$$F(\phi) = \frac{4}{M} \sum_{i}^{M/4} \frac{1}{2} abs[(c_2 - c_1)(c_4 - c_2) - (c_3 - c_2)(c_3 - c_1)]_i$$

In each generation: (i) $F(\phi)$ is calculated for each rule ϕ in the population. (ii) The population is ranked in order of fitness. (iii) A number $E = 5$ of the highest fitness ("elite") rules is copied without modification to the next generation. (iv) The remaining $P - E = 15$ rules for the next generation are formed by single-point crossover between randomly chosen pairs of elite rules. The off-springs from each crossover are each mutated with a probability $m = 0.05$. This defines one generation of the GA; it is repeated $G = 10^3$ times for one run of the GA.

3   Results

We performed more than 100 different runs of the GA, each with a different random-number seed. In the density and in the synchronization task the usual lattice size was 149 cells, but here to define clearly the collective behavior we need a greater lattice size. We used a lattice of $N = 10^3$ cells.

Symbol	Rule Table Hexadecimal Code	NTCB	Fitness	λ
ϕ_a	21088418-01091108-41038844-10c18080	P3	0.048	0.211
ϕ_b	ffbe84bc-10874438-c6a08204-9d1b800b	P3	0.186	0.414
ϕ_c	146157d1-fbb53fec-7dfbeffc-eaf0fa28	QP3(P3)	0.066	0.625
ϕ_d	f193800-c06b0eb0-e000461c-80659c11	P3	0.031	0.336

Table 1. The best evolved rules, the rule table hexadecimal code, the type of non-trivial collective behavior, the fitness function averaged over 100 initial conditions and the lambda parameter. To recover the 128-bit string giving the output bits of the rule table, expand each hexadecimal digit to binary. The output bits are then given in lexicographic order.

Generation	Rule Table Hexadecimal Code	Fitness	λ
10	21008100-00200500-4001a000-1080c000	0.004	0.117
25	21008108-20200d20-41412000-1090521b	0.006	0.203
48	21008908-20200500-40298008-1081c082	0.012	0.180
104	21088118-20200108-41098008-1081c142	0.015	0.195
140	21088518-01091108-41018040-10c1a002	0.043	0.203
415 (ϕ_a)	21088418-01091108-41038844-10c18080	0.048	0.211

Table 2. CA look-up table output bits given in hexadecimal code, value of the fitness function and the lambda parameter for some ancestors of rule ϕ_a.

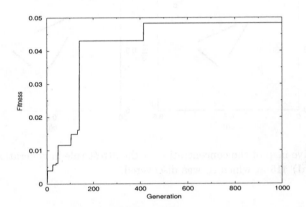

Fig. 3. Best fitness rule versus generation in the run in which rule ϕ_a was found at generation 415. Lattice size is 10^3 cells. The rules of the initial population were randomly selected.

Table 1 shows some of the best evolved rules, the rule table hexadecimal code, the kind of collective behavior observed, the fitness of the rule averaged over 100 initial configurations and the parameter λ, which is the fraction of 1s in the rule table's output bits [12]. Under the fitness function $F(\phi)$ the GA was able to find many rules with the desired behavior and about 30% of the runs ended up with a rule that showed a P3 collective behavior. Only one rule ϕ_c showed a QP3 collective behavior that after a long time decays into a P3, but this will be studied elsewhere.

The progression of a typical evolutionary run is depicted in Figure 3 which plots the fittest rule of each generation in the run in which rule ϕ_a was found. It is observed that the fitness of the best CA rules increases in jumps. Qualitatively the rise in fitness can be divided into several epochs, each one corresponding to the discovery of a new improved strategy. The generational progression of the GA can give important information about the design of CA rules with a specific behavior. Table 2 shows some ancestors of rule ϕ_a. Their iterative map and the time series of the concentration are shown in Figures 4 - 5. In the initial gener-

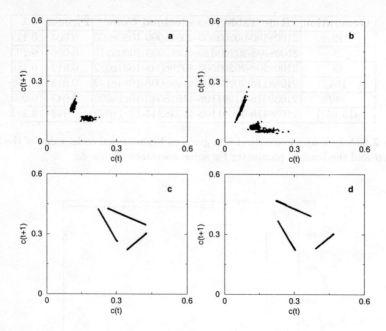

Fig. 4. Iterative map of the concentration of the fittest rule in generation: **(a)** 10; **(b)** 48; **(c)** 140; **(d)** 415 at which ϕ_a was discovered.

ations the GA discovers a rule (ϕ_{10}) that displays a P2 behavior , Figures 4a-5a, and the fitness of the rule is very small $F(\phi_{10}) = 0.039$. In generation 48 there is a jump in $F(\phi_{48}) = 0.116$ and now the iterative map, Figure 4b, shows two clouds of points with a greater size than in generation 10. In generation 140 there is a big jump in fitness $F(\phi_{140}) = 0.430$, and the iterative map now shows a triangular object, while the time series of the concentration shows clearly three branches. Finally, when rule ϕ_a is discovered at generation 415, a further improvement in the fitness is attained.

Under the fitness function $F(\phi)$ the evolutionary process has selected some rules that, starting from a random initial condition, synchronize the concentration to a three-state cycle. To see how such a synchronization is obtained we use the tools of the computational mechanics developed by Crutchfield and Hanson [7]. This point of view describes the computation embedded in the CA space-time configuration in terms of domains, defects, and defect interactions. Figure 6 shows a space-time diagram of rules ϕ_{10}, ϕ_{48}, ϕ_{140}, and ϕ_a. The space-time diagrams of ϕ_{10} and ϕ_{48}, Figures 6a-b, show irregular regions of fractal structures separated by straight lines. While for rules ϕ_{140} and ϕ_a the space-time diagrams show patterns in which there is an easily recognized spatio-temporally periodic background -a domain- on which some dislocations move. In the simplest case, a domain Λ consists of a set of cells in the space-time diagram that are always repeated. For example, the domain for rule ϕ_a and ϕ_{140} is shown in

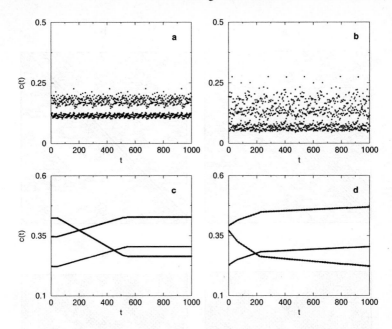

Fig. 5. Time series of the concentration of the fittest rule in generation: **(a)** 10; **(b)** 48; **(c)** 140; **(d)** 415(ϕ_a).

Table 3. The concentration of activated cells in Λ oscillates among three values $1/2$, $1/3$, and $1/6$. The computational strategy used by ϕ_a and ϕ_{140} seems to be their capacity to discover a domain that fulfills the requirements imposed by the task. In many cases, two adjacent synchronized regions left and right are out of phase i.e., there is some displacement of the right domain along the temporal axis when compared with the left domain. At the boundaries between them there are several defect cells, or particles that propagate with a given velocity. The particles, named with Greek letters, can be seen in Figure 7 which is a filtered space-time diagram of Figure 6d. The filtered diagrams reveal defect cells that interact among them and are transmitted from distant parts of the lattice until more synchronized regions are obtained.

Though the concentration of the CA can attain a P3 collective behavior, rules like ϕ_a fail to synchronize the whole lattice to a single domain: there are some particles like α and μ that appear at the boundary between two Λ domains which are in phase. Figure 8 shows in a log-log plot the decaying of the density of defects ρ_d versus the time for ϕ_a. Data points have been averaged over ten different initial condition with the system running for 5×10^4 time steps. It can be observed that the density of defects ρ_d decreases with time to a non-zero asymptotic value. This decay of the number of defects means that there are increasingly large homogeneous domains and the particles become increasingly less important to the collective behavior over time.

a   b

c   d

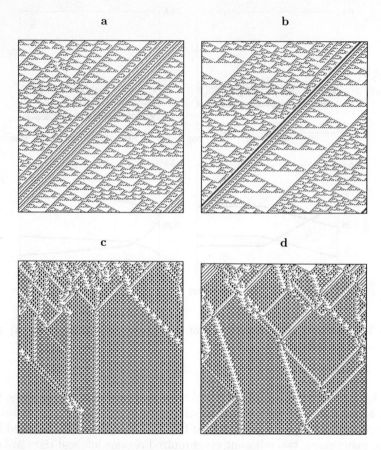

Fig. 6. Space-time diagrams of the fittest evolved rules in the run in which ϕ_a was discovered. For each space-time diagram a window of 256x256 cells is displayed. Generations: **(a)** 10. **(b)** 48. **(c)** 140. **(d)** 415 (ϕ_a).

Domain Λ	Particles (velocities)	
1 1 0 1 0 0	$\alpha \sim \Lambda\Lambda$	(-1)
1 1 0 0 0 0	$\beta \sim \Lambda\Lambda^{+1}$	(1/3)
0 1 0 0 0 0	$\delta \sim \Lambda\Lambda^{-1}$	(0)
1 0 0 1 1 0	$\epsilon \sim \Lambda\Lambda^{+2}$	(1)
0 0 0 1 1 0	$\gamma \sim \Lambda\Lambda^{+1}$	(1)
0 0 0 0 1 0	$\mu \sim \Lambda\Lambda$	(1)
Main Particle Interactions		
$\beta + \gamma \rightarrow \alpha + \delta$	$\alpha + \beta \rightarrow \delta + \epsilon$	
$\alpha + \gamma \rightarrow \mu$	$\alpha + \epsilon \rightarrow \gamma$	

Table 3. Domain Λ for rule ϕ_a, the particle catalog with their boundaries and velocities and several of the interactions.

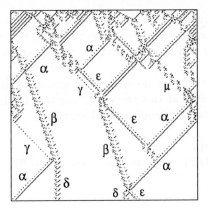

Fig. 7. Filtered space-time diagram of rule ϕ_a corresponding to Figure 6d.

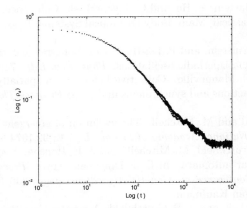

Fig. 8. The density of defects versus time for rule ϕ_a

4   Conclusion

The emergence of collective behaviors that arise from the local interaction of a collection of individuals is a phenomenon observed in many natural systems. To investigate emergent properties in spatially extended systems a theoretical framework was proposed in [4], which requires a decentralized system such as a CA, an idealized computational model of evolution such as a genetic algorithm and finally a computational task that necessitates global information processing.

The computational task that we studied was the QP3(P3) collective behavior task in spatial dimension $d = 1$, where the entire system has to cooperate in order that the concentration of activated cells is oscillating among three values. Using a fitness function that was the size of the attractor in the iterative map, the GA discovers many CA rules whose space-time dynamics displays the desired behavior. It is worth pointing out that for this task the system does not need to be globally synchronized. Here we have focused on a run in which the best evolved rule's strategy used domains and particles to produce the desired behavior. Domains and particles were also shown to be central to the GA's solutions for the density and synchronization tasks, described in [4, 5].

To our knowledge there are no systematic studies of this computational task in $d = 1$, and we expect that this work will motivate further investigations such as the study of globally synchronized P3 behavior.

Acknowledgments

This work was partially supported by Grant No.PB 97-0741 of the Spanish Government.

References

1. T. Bohr, G. Grinstein, Y. He, and C. Jayaprakash. Coherence, Chaos, and Broken Symmetry in Classical, Many-Body Dynamical Systems. *Phys. Rev. Lett.*, 58:2155-2158, 1987.
2. H. Chaté, G. Grinstein, and P. Lei-Hang Tan. Long-range correlations in systems with coherent(quasi)periodic oscillations. *Phys. Rev. Lett.*, 74:912-915, 1995.
3. H. Chaté and P. Manneville. Collective behaviors in spatially extended systems with local interactions and synchronous updating. *Progress Theor. Phys.*, 87(1):1-60, 1992.
4. J. P. Crutchfield and M. Mitchell. The evolution of emergent computation. *Proceedings of the National Academy of Science U.S.A.*, 92:10742-10746, 1995.
5. R. Das, J. P. Crutchfield, M. Mitchell, and J. E. Hanson. Evolving globally synchronized cellular automata. In L. J. Eshelman, editor, *Proceedings of the Sixth International Conference on Genetic Algorithms*, pages 336-343, San Francisco, CA, 1995. Morgan Kaufmann.
6. R. Das, M. Mitchell, and J. P. Crutchfield. A genetic algorithm discovers particle-based computation in cellular automata. In Y. Davidor, H.-P. Schwefel, and R. Männer, editors, *Parallel Problem Solving from Nature—PPSN III*, volume 866, pages 344-353, Berlin, 1994. Springer-Verlag (Lecture Notes in Computer Science).
7. J. E. Hanson and J. P. Crutchfield. Computational mechanics of cellular automata: An example. *Physica D*, 103:169-189, 1997.
8. J. Hemmingsson. A totalistic three-dimensional cellular automaton with quasiperiodic behaviour. *Physica A*, 183:225-261, 1992.
9. W. Hordijk. Dynamics, Emergent Computation, and Evolution in Cellular Automata. *Ph.D. dissertation*, Univ. New Mexico, 1999.
10. F. Jiménez-Morales and K. Hassan. Non-trivial collective behavior in three-dimensional totalistic illegal cellular automata with high connectivity. *Physics Letters A*, 240:151-159, 1998.
11. F. Jiménez-Morales. Evolving three-dimensional cellular automata to perform a quasiperiod-3(p3) collective behavior task. *Phys. Rev. E*, 60(4):4934-4940, 1999.
12. C. G.Langton. Studying Artificial Life with Cellular Automata. *Physica D*, 22:120-149, 1986.
13. M. Mitchell, J. P. Crutchfield, and P. T. Hraber. Evolving cellular automata to perform computations: Mechanisms and impediments. *Physica D*, 75:361 – 391, 1994.
14. M. Sipper. *Evolution of Parallel Cellular Machines*. Springer, Germany, 1997.

New Unconditionally Stable Algorithms to Solve the Time-Dependent Maxwell Equations

J.S. Kole, M.T. Figge, and H. De Raedt

Centre for Theoretical Physics and Materials Science Centre
University of Groningen, Nijenborgh 4,
NL-9747 AG Groningen, The Netherlands
E-mail: j.s.kole@phys.rug.nl, m.t.figge@phys.rug.nl, deraedt@phys.rug.nl
http://rugth30.phys.rug.nl/compphys

Abstract. We present a family of unconditionally stable algorithms, based on the Suzuki product-formula approach, that solve the time-dependent Maxwell equations in systems with spatially varying permittivity and permeability. Salient features of these algorithms are illustrated by computing the density of states and by simulating the propagation of light in a two-dimensional photonic material.

1 Introduction

The Maxwell equations describe the evolution of electromagnetic (EM) fields in space and time [1]. They apply to a wide range of different physical situations and play an important role in a large number of engineering applications. In many cases, numerical methods are required to solve Maxwell's equations, either in the frequency or time domain. For the time domain, a well-known class of algorithms is based on a method proposed by Yee [2] and is called the finite-difference time-domain (FDTD) method. The FDTD method has matured during past years and various algorithms implemented for specific purposes are available by now [3, 4]. These algorithms owe their popularity mainly due to their flexibility and speed while at the same time they are easy to implement. A limitation of Yee-based FDTD techniques is that their stability is conditional, depending on the mesh size of the spatial discretization and the time step of the time integration [3].

In this paper we describe the recently introduced family of unconditionally stable algorithms that solve the time-dependent Maxwell equations (TDME) [5]. From the representation of the TDME in matrix form, it follows that the time-evolution operator of the EM fields is the exponential of a skew-symmetric matrix. This time-evolution operator is orthogonal. The key to the construction of rigorously provable unconditionally stable algorithms is the observation, that orthogonal approximations to this operator automatically yield unconditionally stable algorithms. The Lie-Trotter-Suzuki product formulae [6] provide the mathematical framework to construct orthogonal approximations to the time-evolution operator of the Maxwell equations.

P.M.A. Sloot et al. (Eds.): ICCS 2002, LNCS 2329, pp. 803–812, 2002.

2 Theory

The model system we consider describes EM fields in a d-dimensional ($d = 1, 2, 3$) medium with spatially varying permittivity and/or permeability, surrounded by a perfectly conducting box. In the absence of free charges and currents, the EM fields in such a system satisfy Maxwell's equations [1]

$$\frac{\partial}{\partial t}\mathbf{H} = -\frac{1}{\mu}\nabla \times \mathbf{E} \quad \text{and} \quad \frac{\partial}{\partial t}\mathbf{E} = \frac{1}{\varepsilon}\nabla \times \mathbf{H}, \tag{1}$$

$$\operatorname{div}\varepsilon\mathbf{E} = 0 \quad \text{and} \quad \operatorname{div}\mathbf{H} = 0, \tag{2}$$

where $\mathbf{H} = (H_x(\mathbf{r}, t), H_y(\mathbf{r}, t), H_z(\mathbf{r}, t))^T$ and $\mathbf{E} = (E_x(\mathbf{r}, t), E_y(\mathbf{r}, t), E_z(\mathbf{r}, t))^T$ denote the magnetic and electric field vector respectively. The permeability and the permittivity are given by $\mu = \mu(\mathbf{r})$ and $\varepsilon = \varepsilon(\mathbf{r})$. For simplicity of notation, we will omit the spatial dependence on $\mathbf{r} = (x, y, z)^T$ unless this leads to ambiguities. On the surface of the perfectly conducting box the EM fields satisfy the boundary conditions [1]

$$\mathbf{n} \times \mathbf{E} = 0 \quad \text{and} \quad \mathbf{n} \cdot \mathbf{H} = 0, \tag{3}$$

with $\mathbf{n}$ denoting the vector normal to a boundary of the surface. The conditions Eqs. (3) assure that the normal component of the magnetic field and the tangential components of the electric field vanish at the boundary [1]. Some important symmetries of the Maxwell equations (1)-(2) can be made explicit by introducing the fields

$$\mathbf{X}(t) = \sqrt{\mu}\mathbf{H}(t) \quad \text{and} \quad \mathbf{Y}(t) = \sqrt{\varepsilon}\mathbf{E}(t). \tag{4}$$

In terms of the fields $\mathbf{X}(t)$ and $\mathbf{Y}(t)$, the TDME (1) read

$$\frac{\partial}{\partial t}\begin{pmatrix} \mathbf{X}(t) \\ \mathbf{Y}(t) \end{pmatrix} = \begin{pmatrix} 0 & -\frac{1}{\sqrt{\mu}}\nabla \times \frac{1}{\sqrt{\varepsilon}} \\ \frac{1}{\sqrt{\varepsilon}}\nabla \times \frac{1}{\sqrt{\mu}} & 0 \end{pmatrix}\begin{pmatrix} \mathbf{X}(t) \\ \mathbf{Y}(t) \end{pmatrix} \equiv \mathcal{H}\begin{pmatrix} \mathbf{X}(t) \\ \mathbf{Y}(t) \end{pmatrix}. \tag{5}$$

Writing $\Psi(t) = (\mathbf{X}(t), \mathbf{Y}(t))^T$, Eq. (5) becomes

$$\frac{\partial}{\partial t}\Psi(t) = \mathcal{H}\Psi(t). \tag{6}$$

It is easy to show that $\mathcal{H}$ is skew-symmetric, i.e. $\mathcal{H}^T = -\mathcal{H}$, with respect to the inner product $\langle\Psi|\Psi'\rangle \equiv \int_V \Psi^T \cdot \Psi'\, d\mathbf{r}$, where V denotes the volume of the enclosing box. The formal solution of Eq. (6) is given by

$$\Psi(t) = U(t)\Psi(0) = e^{t\mathcal{H}}\Psi(0), \tag{7}$$

where $\Psi(0)$ represents the initial state of the EM fields. The operator $U(t) = e^{t\mathcal{H}}$ determines the time evolution. By construction $\|\Psi(t)\|^2 = \langle\Psi(t)|\Psi(t)\rangle = \int_V \left[\varepsilon\mathbf{E}^2(t) + \mu\mathbf{H}^2(t)\right]\, d\mathbf{r}$, relating the length of $\Psi(t)$ to the energy density $w(t) \equiv$

$\varepsilon\mathbf{E}^2(t) + \mu\mathbf{H}^2(t)$ of the EM fields [1]. As $U(t)^T = U(-t) = U^{-1}(t) = e^{-t\mathcal{H}}$ it follows that $\langle U(t)\Psi(0)|U(t)\Psi(0)\rangle = \langle \Psi(t)|\Psi(t)\rangle = \langle \Psi(0)|\Psi(0)\rangle$. Hence the time-evolution operator $U(t)$ is an orthogonal transformation, rotating the vector $\Psi(t)$ without changing its length $\|\Psi\|$. In physical terms this means that the energy density of the EM fields does not change with time, as expected on physical grounds [1].

The fact that $U(t)$ is an orthogonal transformation is essential for the development of an unconditionally stable algorithm to solve the Maxwell equations. In practice, a numerical procedure solves the TDME by making use of an approximation $\tilde{U}(t)$ to the true time evolution $U(t)$. A necessary and sufficient condition for an algorithm to be unconditionally stable is that $\|\tilde{U}(t)\Psi(0)\| \leq \|\Psi(0)\|$. In other words, the length of $\Psi(t)$ should be bounded, for arbitrary initial condition $\Psi(t=0)$ and for any time t [7]. By chosing for $\Psi(0)$ the eigenvector of $\tilde{U}(t)$ that corresponds to the largest eigenvalue of $\tilde{U}(t)$, it follows that the algorithm will be unconditionally stable by construction if and only if the largest eigenvalue of $\tilde{U}(t)$ (denoted by $\|\tilde{U}(t)\|$) is less or equal than one [7]. If the approximation $\tilde{U}(t)$ is itself an orthogonal transformation, then $\|\tilde{U}(t)\| = 1$ and the numerical scheme will be unconditionally stable.

3   Unconditionally Stable Algorithms

A numerical procedure that solves the TDME necessarily starts by discretizing the spatial derivatives. This maps the continuum problem described by $\mathcal{H}$ onto a lattice problem defined by a matrix H. Ideally, this mapping should not change the basic symmetries of the original problem. The underlying symmetry of the TDME suggests to use matrices H that are real and skew-symmetric. Since formally the time evolution of the EM fields on the lattice is given by $\Psi(t+\tau) = U(\tau)\Psi(t) = e^{\tau H}\Psi(t)$, the second ingredient of the numerical procedure is to choose an approximation of the time-step operator $U(\tau)$. A systematic approach to construct orthogonal approximations to matrix exponentials is to make use of the Lie-Trotter-Suzuki formula [8, 9]

$$e^{t(H_1+\ldots+H_p)} = \lim_{m\to\infty}\left(\prod_{i=1}^{p} e^{tH_i/m}\right)^m,\tag{8}$$

and generalizations thereof [6, 10]. Applied to the case of interest here, the success of this approach relies on the basic but rather trivial premise that the matrix H can be written as $H = \sum_{i=1}^{p} H_i$, where each of the matrices H_i is real and skew-symmetric. The expression Eq. (8) suggests that

$$U_1(\tau) = e^{\tau H_1}\ldots e^{\tau H_p},\tag{9}$$

might be a good approximation to $U(\tau)$ if τ is sufficiently small. Most importantly, if all the H_i are real and skew-symmetric, $U_1(\tau)$ is orthogonal by construction. Therefore, by construction, a numerical scheme based on Eq. (9) will be unconditionally stable. Using the fact that both $U(\tau)$ and $U_1(\tau)$ are orthogonal matrices, it can be shown that the Taylor series of $U(\tau)$ and $U_1(\tau)$ are identical up to first order in τ [11]. We will call $U_1(\tau)$ a first-order approximation to $U(\tau)$. The product-formula approach provides simple, systematic procedures to improve the accuracy of the approximation to $U(\tau)$ without changing its fundamental symmetries. For example the orthogonal matrix

$$U_2(\tau) = U_1(-\tau/2)^T U_1(\tau/2) = e^{\tau H_p/2} \dots e^{\tau H_1/2} e^{\tau H_1/2} \dots e^{\tau H_p/2}, \quad (10)$$

is a second-order approximation to $U(\tau)$ [6, 10]. Suzuki's fractal decomposition approach [6] gives a general method to construct higher-order approximations based on $U_1(\tau)$ or $U_2(\tau)$. A particularly useful fourth-order approximation is given by [6]

$$U_4(\tau) = U_2(a\tau) U_2(a\tau) U_2((1 - 4a)\tau) U_2(a\tau) U_2(a\tau), \quad (11)$$

where $a = 1/(4 - 4^{1/3})$. The approximations Eqs. (9), (10), and (11) have proven to be very useful in many applications [9–19] and, as we show below, turn out to be equally useful for solving the TDME. From Eqs. (9)-(11) it follows that, in practice, an efficient implementation of the first-order scheme is all that is needed to construct the higher-order algorithms Eqs. (10) and (11).

4   Implementation

The basic ingredients of our approach will be illustrated for a one-dimensional (1D) system. A discussion of the two-dimensional (2D) and three-dimensional (3D) case are given in Ref. [5].

Maxwell's equations for a 1D system extending along the x-direction contain no partial derivatives with respect to y or z and also ε and μ do not depend on y or z. Under these conditions, the TDME reduce to two independent sets of first-order differential equations [1]. The solutions to these sets are known as the transverse electric (TE) mode and the transverse magnetic (TM) mode [1]. Restricting our considerations to the TM-mode, it follows from Eq. (5) that the magnetic field $H_y(x,t) = X_y(x,t)/\sqrt{\mu(x)}$ and the electric field $E_z(x,t) = Y_z(x,t)/\sqrt{\varepsilon(x)}$ are solutions of

$$\frac{\partial}{\partial t} X_y(x,t) = \frac{1}{\sqrt{\mu(x)}} \frac{\partial}{\partial x} \left(\frac{Y_z(x,t)}{\sqrt{\varepsilon(x)}} \right), \quad (12)$$

$$\frac{\partial}{\partial t} Y_z(x,t) = \frac{1}{\sqrt{\varepsilon(x)}} \frac{\partial}{\partial x} \left(\frac{X_y(x,t)}{\sqrt{\mu(x)}} \right). \quad (13)$$

Note that in 1D the divergence of $H_y(x,t)$ and $E_z(x,t)$ is zero, hence Eqs. (2) are automatically satisfied. Using the second-order central-difference approximation to the first derivative with respect to x, we obtain

$$\frac{\partial}{\partial t} X_y(i,t) = \frac{1}{\delta\sqrt{\mu_i}} \left(\frac{Y_z(i+1,t)}{\sqrt{\varepsilon_{i+1}}} - \frac{Y_z(i-1,t)}{\sqrt{\varepsilon_{i-1}}} \right), \quad (14)$$

$$\frac{\partial}{\partial t} Y_z(j,t) = \frac{1}{\delta\sqrt{\varepsilon_j}} \left(\frac{X_y(j+1,t)}{\sqrt{\mu_{j+1}}} - \frac{X_y(j-1,t)}{\sqrt{\mu_{j-1}}} \right), \tag{15}$$

where the integer i labels the grid points and δ denotes the distance between two next-nearest neighbor lattice points, as indicated in Fig. 1. For notational simplicity we will, from now on, specify the spatial coordinates through the lattice index i, e.g. $X_y(i,t)$ stands for $X_y(x = i\delta/2, t)$. Following Yee [2] it is convenient to assign $X_y(i,t)$ and $Y_z(j,t)$ to the odd, respectively, even numbered lattice site, as shown in Fig. 1 for a grid of n points. The equations (14) and (15) can now be combined into one equation of the form Eq. (6) by introducing the n-dimensional vector $\Psi(t)$ with elements

$$\Psi(i,t) = \begin{cases} X_y(i,t) = \sqrt{\mu_i} H_y(i,t), & i \text{ odd} \\ Y_z(i,t) = \sqrt{\varepsilon_i} E_z(i,t), & i \text{ even} \end{cases}. \tag{16}$$

The vector $\Psi(t)$ describes both the magnetic and the electric field on the lattice points $i = 1, \ldots, n$ and the i-th element of $\Psi(t)$ is given by the inner product $\Psi(i,t) = \mathbf{e}_i^T \cdot \Psi(t)$, where $\mathbf{e}_i$ denotes the i-th unit vector in the n-dimensional vector space. Using this notation, it is easy to show that $\Psi(t) = U(t)\Psi(0)$ with $U(t) = \exp(tH)$, where the matrix H is represented by $H = H_1 + H_2$ and

$$H_1 = \sum_{i=1}^{n-2}{}' \beta_{i+1,i} \left(\mathbf{e}_i \, \mathbf{e}_{i+1}^T - \mathbf{e}_{i+1}\mathbf{e}_i^T \right), \tag{17}$$

$$H_2 = \sum_{i=1}^{n-2}{}' \beta_{i+1,i+2} \left(\mathbf{e}_{i+1}\mathbf{e}_{i+2}^T - \mathbf{e}_{i+2}\mathbf{e}_{i+1}^T \right). \tag{18}$$

Here, $\beta_{i,j} = 1/(\delta\sqrt{\varepsilon_i\mu_j})$ and the prime indicates that the sum is over odd integers only. For n odd we have

$$\frac{\partial}{\partial t}\Psi(1,t) = \beta_{2,1}\Psi(2,t) \quad \text{and} \quad \frac{\partial}{\partial t}\Psi(n,t) = -\beta_{n-1,n}\Psi(n-1,t), \tag{19}$$

such that the electric field vanishes at the boundaries ($Y_z(0,t) = Y_z(n+1,t) = 0$, see also Fig. 1), as required by the boundary conditions Eqs. (3).

The representation of H as the sum of H_1 and H_2 divides the lattice into odd and even numbered cells. Most important, however, both H_1 and H_2 are skew-symmetric block-diagonal matrices, containing one 1×1 matrix and $(n-1)/2$ real, 2×2 skew-symmetric matrices. Therefore, according to the general theory

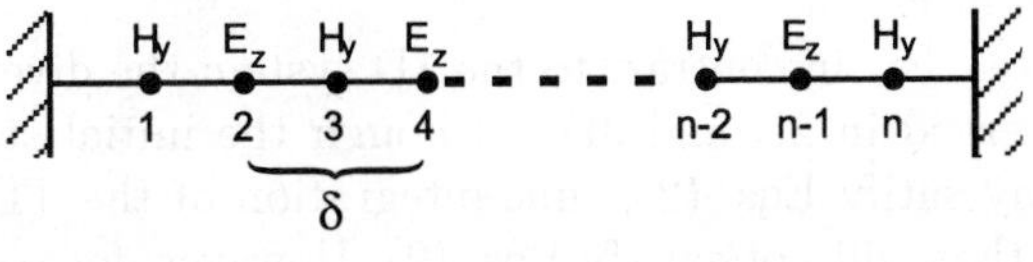

Fig. 1. Positions of the two TM-mode EM field components on the 1D grid.

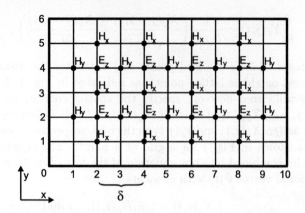

Fig. 2. Positions of the three TM-mode EM field components on the 2D grid for $n_x = 9$ and $n_y = 5$

given above, this decomposition of H is suitable to construct an orthogonal first-order approximation

$$U_1(\tau) = e^{\tau H_1} e^{\tau H_2} , \qquad (20)$$

which is all that is needed to construct unconditionally stable second and higher-order algorithms. As the matrix exponential of a block-diagonal matrix is equal to the block-diagonal matrix of the matrix exponentials of the individual blocks, the numerical calculation of $e^{\tau H_1}$ (or $e^{\tau H_2}$) reduces to the calculation of $(n-1)/2$ matrix exponentials of 2×2 matrices. The matrix exponential of a typical 2×2 matrix appearing in $e^{\tau H_1}$ or $e^{\tau H_2}$ is simply given by the rotation

$$\exp\left[\alpha \begin{pmatrix} 0 & 1 \\ -1 & 0 \end{pmatrix}\right] \begin{pmatrix} \Psi(i,t) \\ \Psi(j,t) \end{pmatrix} = \begin{pmatrix} \cos\alpha & \sin\alpha \\ -\sin\alpha & \cos\alpha \end{pmatrix} \begin{pmatrix} \Psi(i,t) \\ \Psi(j,t) \end{pmatrix} . \qquad (21)$$

The implementation for 1D can be readily extended to 2D and 3D systems [5]. In 2D, the TDME (1) separate again into two independent sets of equations and the discretization of continuum space is done by simply re-using the 1D lattice introduced above. This is shown in Fig. 2 for the case of the 2D TM-modes. The construction automatically takes care of the boundary conditions if n_x and n_y are odd and yields a real skew-symmetric matrix H. Correspondingly, in 3D the spatial coordinates are discretized by adopting the standard Yee grid [2], which also automatically satisfies the boundary conditions Eqs. (3). A unit cell of the Yee grid is shown in Fig. 3.

We finally note, that in contrast to the 1D system the divergence of the EM fields is not conserved in 2D and 3D. Although the initial state (at $t = 0$) of the EM fields may satify Eqs. (2), time-integration of the TDME using $U_k(\tau)$ yields a solution that will not satisfy Eqs. (2). However, for an algorithm based on $U_k(\tau)$ the deviations from zero vanish as τ^k, so that in practice these errors are under control and can be made sufficiently small.

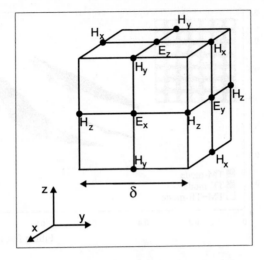

Fig. 3. Positions of the EM field components on the 3D Yee grid.

5 Simulation Results

We present simulation results for a 2D photonic bandgap (PBG) material. PBGs prohibit the propagation of EM fields in a range of frequencies that is character-istic for its structure [20]. A PBG is called absolute if it exists for any wave vector of the EM fields. The most common method used to compute a PBG employs a plane-wave expansion to solve the time-independent Maxwell equations (see e.g. [21]). This kind of calculation requires a Fourier transform of the unit cell of the dielectric structure which is for simplicity considered to be periodic. With our time-domain algorithm the existence of a PBG can be demonstrated with relative ease. It suffices to compute the spectrum of such a dielectric structure with a random initial state. If the spectrum is gapless there is no need to make additional runs. If there is a signature of a gap, it can be confirmed and refined by making more runs.

For numerical purposes we choose dimensionless quantities, where the unit of length is λ and the velocity of light in vacuum, c, is taken as the unit of velocity. Then, time and frequency are measured in units of λ/c and c/λ, respectively, while the permittivity ε and permeability μ are measured in units of their cor-responding values in vacuum. As an example we consider a system consisting of a dielectric material pierced by air-filled cylinders [22]. The geometry is taken to be a square parallelepiped of size $L = 45.1$ that is infinitely extended in the z-direction and hence is effectively two-dimensional.

In Fig. 4 we present our results for the PBGs for both the transverse magnetic (TM) and transverse electric (TE) modes as a function of the filling fraction. The data have been generated by means of the algorithm $U_4(\tau)$ with a mesh size $\delta = 0.1$ and a time step $\tau = 0.1$. To compute the density of states (DOS) $\mathcal{D}(\omega)$

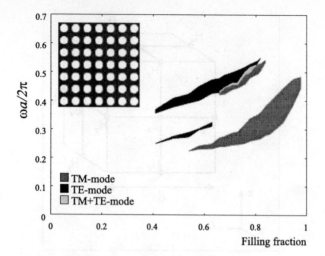

Fig. 4. Absolute photonic bandgaps of a dielectric material ($\varepsilon = 11.4$) pierced by air-filled cylinders. The largest overlap of the TM- and TE-mode gaps occurs at a filling fraction of approximately 0.77.

we used only a single random initial state for the EM fields. The results shown in Fig. 4 are in good agreement with those presented in Ref. [22].

In Fig. 5 we study the propagation of time-dependent EM fields through the above described PBG material consisting of twelve unit cells. The PBG material is placed in a cavity which contains a point source that emits radiation with frequency ω. The TDME were solved by the $U_2(\tau)$ algorithm with $\delta = 0.1$ and $\tau = 0.01$ in the presence of a current source. The snapshots show the absolute intensity E_z^2 of the TM-mode at $t = 102.4$. The DOS of the PBG material is given in Fig. 6. We used the $U_4(\tau)$ algorithm with $\delta = 0.1$ and $\tau = 0.1$ in this computation and applied a time-domain algorithm to obtain the DOS [5, 23].

The presence or absence of gaps in the DOS leads to qualitative changes in the transmitted (and reflected) intensities. Since a gap is present in the DOS at $\omega = 1.89$, radiation with this frequency does not easily propagate through the (thin slice of) PBG material. On the other hand, the DOS has no gaps at $\omega = 1.50$ and $\omega = 2.50$, so that propagation of EM fields through the PBG material should be possible, as is indeed confirmed by Fig. 5.

6   Conclusion

We have described a new family of algorithms to solve the time-dependent Maxwell equations. Salient features of these algorithms are:

- rigorously provable unconditional stability for 1D, 2D and 3D systems with spatially varying permittivity and permeability,

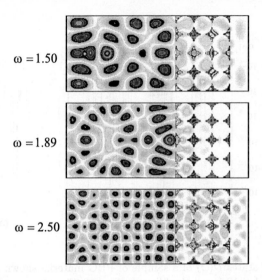

ω =1.50

ω =1.89

ω =2.50

Fig. 5. Snapshot of the intensity E_z^2 at $t = 102.4$. The size of the system is 30×12.1. A point source is located at (6,6) (see Fig. 2), emitting radiation with frequency ω.

- the use of a real-space (Yee-like) grid,
- the order of accuracy in the time step can be systematically increased without affecting the unconditional stability,
- the exact conservation of the energy density of the electromagnetic field,
- easy to implement in practice.

Although we believe there is little room to improve upon the time-integration scheme itself, for some applications it will be necessary to use a better spatial discretization than the most simple one employed in this paper. There is no fundamental problem to extend our approach in this direction and we will report on this issue and its impact on the numerical dispersion in a future publication.

We have presented numerical results for the density of states and the propagation of light in a two-dimensional photonic material. This illustrative example showed that our algorithms reproduce known results. The first of the above mentioned features opens up possibilities for applications to left-handed materials [24, 25] and we intend to report on this subject in the near future. In view of the generic character of the approach discussed in this paper, it can be used to construct unconditionally stable algorithms that solve the equations for e.g. sound, seismic and elastic waves as well.

Acknowledgements

This work is partially supported by the Dutch 'Stichting Nationale Computer Faciliteiten' (NCF).

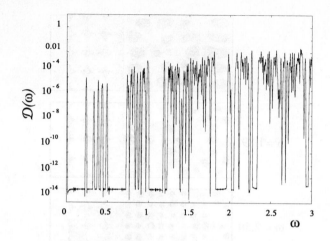

Fig. 6. Density of states $\mathcal{D}(\omega)$ of a sample of PBG material shown in Fig. 5. The size of the sample is 9.1×12.1 and the filling factor is 0.77.

References

1. M. Born and E. Wolf, *Principles of Optics*, (Pergamon, Oxford, 1964).
2. K.S. Yee, IEEE Transactions on Antennas and Propagation **14**, 302 (1966).
3. A. Taflove and S.C. Hagness, *Computational Electrodynamics - The Finite-Difference Time-Domain Method*, (Artech House, Boston, 2000).
4. See http://www.fdtd.org
5. J.S. Kole, M.T. Figge, and H. De Raedt, Phys. Rev. E **64**, 066705 (2001).
6. M. Suzuki, J. Math. Phys. **26**, 601 (1985); *ibid* **32** 400 (1991).
7. G.D. Smith, *Numerical solution of partial differential equations*, (Clarendon, Oxford, 1985).
8. H.F. Trotter, Proc. Am. Math. Soc. **10**, 545 (1959).
9. M. Suzuki, S. Miyashita, and A. Kuroda, Prog. Theor. Phys. **58**, 1377 (1977).
10. H. De Raedt and B. De Raedt, Phys. Rev. A **28**, 3575 (1983).
11. H. De Raedt, Comp. Phys. Rep. **7**, 1 (1987).
12. H. Kobayashi, N. Hatano, and M. Suzuki, Physica A **211**, 234 (1994).
13. H. De Raedt, K. Michielsen, Comp. in Phys. **8**, 600 (1994).
14. A. Rouhi, J. Wright, Comp. in Phys. **9**, 554 (1995).
15. B.A. Shadwick and W.F. Buell, Phys. Rev. Lett. **79**, 5189 (1997).
16. M. Krech, A. Bunker, and D.P. Landau, Comp. Phys. Comm. **111**, 1 (1998).
17. P. Tran, Phys. Rev. E **58**, 8049 (1998).
18. K. Michielsen, H. De Raedt, J. Przeslawski, and N. Garcia, Phys. Rep. **304**, 89 (1998).
19. H. De Raedt, A.H. Hams, K. Michielsen, and K. De Raedt, Comp. Phys. Comm. **132**, 1 (2000).
20. E. Yablonovitch, Phys. Rev. Lett. **58**, 2059 (1987).
21. K.M. Ho, C.T. Chan and C.M. Soukoulis, Phys. Rev. Lett. **65**, 3152 (1990).
22. C.M. Anderson and K.P. Giapis, Phys. Rev. Lett. **77**, 2949 (1996).
23. R. Alben, M. Blume, H. Krakauer, and L. Schwartz, Phys. Rev. B **12**, 4090 (1975).
24. V.G. Veselago, Sov. Phys. USPEKHI **10**, 509 (1968).
25. R.A. Shelby, D.R. Smith, and S. Schultz, Science **292** 77 (2001).

Coupled 3–D Finite Difference Time Domain and Finite Volume Methods for Solving Microwave Heating in Porous Media

Duško D. Dinčov[1], Kevin A. Parrott[1], and Koulis A. Pericleous[1]

University of Greenwich,
School of Computing and Mathematical Sciences,
30 Park Row
London SE10 9LS
{D.Dincov, A.K.Parrott, K.Pericleous}@gre.ac.uk
http://cms1.gre.ac.uk

Abstract. Computational results for the microwave heating of a porous material are presented in this paper. Combined finite difference time domain and finite volume methods were used to solve equations that describe the electromagnetic field and heat and mass transfer in porous media. The coupling between the two schemes is through a change in dielectric properties which were assumed to be dependent both on temperature and moisture content. The model was able to reflect the evolution of temperature and moisture fields as the moisture in the porous medium evaporates. Moisture movement results from internal pressure gradients produced by the internal heating and phase change.

1 Introduction

Microwaving is a common means of heating foodstuffs as well as an important industrial process for heating water-based materials and removing moisture from porous materials such as the drying of textiles, wood, paper and ceramics. Perhaps the largest consumer of microwave power is the food industry where it is used for cooking, thawing, pasteurization, sterilization etc. The ability of microwave radiation to penetrate and interact with materials provides a basis for obtaining controlled and precise heating. Although microwave heating is most beneficial when used for materials with a high moisture content, other materials can still be heated efficiently and quickly.

Because of dielectric losses, microwave absorption provides a volumetrically distributed source. The temperature and moisture distributions in a material during microwave heating are influenced by the interaction and absorption of radiation by the medium and the accompanying transport processes due to the dissipation of electrical energy into heat.

Dielectric properties of most biomaterials vary with temperature [1]. Torres at al. proposed in [11] a 3–D algorithm for microwave heating by coupling of the power distribution with the heat transfer equations in both frequency and

P.M.A. Sloot et al. (Eds.): ICCS 2002, LNCS 2329, pp. 813–822, 2002.

temperature dependent media. The model was able to predict the locations of hot spots within the material. The transfer of mass was assumed to be unimportant.

However, if significant drying occurs during heating, mass transfer must be accounted for too. An one dimensional multiphase porous media model for predicting moisture transport during intensive microwave heating of wet biomaterials was proposed in [8]. The transport properties varied with structure, moisture and temperature. The model successfully predicted the moisture movement. Excellent research into drying of wood was carried out by Perre [9]. Their 3–D algorithms coupled heat and mass transfer with Maxwell's equations and the dielectric properties depended both on temperature and moisture.

The present work considers microwave heating of a simple saturated porous material with temperature and moisture dependent dielectric properties. The solution of Maxwell's equations is performed in the time domain using the Finite Difference Time Domain technique. The 3–D heat and mass transfer equations are solved using the Finite Volume code PHOENICS [10]. Full coupling between the two calculations is achieved by mapping the (moisture and temperature dependent) porous media dielectric properties from the CFD mesh onto the electromagnetic solver and then mapping the microwave heating function from the finite difference mesh to the finite volume mesh.

2   Electomagnetic field in microwave enclosures

The distribution of electromagnetic field in space and time is governed by Maxwell's equations [13]. When material interfaces are present, boundary conditions must be imposed to account for discontinuities of charge and current densities. Provided both media have finite conductivity and there are no sources at the interface, the tangential electric and magnetic fields along the interface are continuous.

2.1   Dielectric Properties

The dielectric properties of materials subjected to microwave heating play a key role in designing proper microwave applicators. They are: permittivity ϵ, permeability μ and conductivity σ. Permittivity describes the interaction of the material with the high frequency electric field and is defined by the following equation :

$$\epsilon = \epsilon_0(\epsilon_r' - \epsilon_{eff}'') \tag{1}$$

where ϵ_r' is the relative dielectric constant, ϵ_{eff}'' is the effective relative loss factor and $\epsilon_0 = 8.85 \times 10^{12} F/m$ is the permittivity of air. ϵ_r' is a measure of the polarizability of a material in the electric field, and ϵ_{eff}'' includes the loss factors which are relevant to high frequency heating.

Both ϵ_r' and ϵ_{eff}'' are temperature (T) dependent and a number of investigations have been made in order to explain this behaviour [1],[5]. In most cases, their values increase as the material thaws and thereafter decrease as the temperature increases. However, these changes for the temperatures up to the boiling point are relatively small [5].

Since most foodstuffs contain an appreciable amount of water, the variation of ϵ'_r and ϵ''_{eff} with the moisture content plays an important role in the design of the microwave heating process. If the temperature dependence in known, the relative dielectric constant and the electric permittivity can be averaged using the following representation (see e.g. [9]):

$$\bar{\epsilon}(M,T) = (1 - \phi)\epsilon^{solid}(T) + \phi(M\epsilon^{liquid}(T) + (1 - M)\epsilon^{gas}(T)) \qquad (2)$$

where M is a moisture content and ϕ porosity.

Losses under the influence of the magnetic field can be described in a similar way to losses in electric materials. However, most materials used in microwave processing are magnetically transparent. The magnetic permeability in this work is assumed to have the value of the free space permeability $\mu_0 = 4\pi \times 10^{-7} H/m$.

2.2   Finite–Difference Time–Domain (FTDT) Scheme

Yee's scheme [13] is used to discretize the Maxwell's equations. The FTDT Scheme proceeds by segmenting the volume into a three–dimensional mesh composed of a number of finite volumes.

It makes use of finite difference approximations to electric and magnetic fields components, that are staggered both in time and space. E and H field components are positioned at half–step intervals around unit volumes and they are evaluated at alternate half–time steps, effectively giving centered difference expressions for both time and space derivatives.

Values for ϵ and σ are specified at cell centres as $\epsilon(i, j, k)$ and $\sigma(i, j, k)$. Continuity across the interface of the tangential field is implemented automatically.

This scheme is second order accurate in both time and space on uniform and non–uniform meshes [7], and can be locally refined [6] without significant loss of accuracy. The use of (semi)–structured meshes ensures optimally fast computations for the most time–consuming component of the overall calculations and is sufficient for our geometric modelling requirements.

3   Heat and Mass Transfer in Porous Media

3.1   Dissipated Power

The dissipated power density is the microwave energy absorbed by the material. It is eventually converted into thermal energy. The dissipated power density is influenced by the field intensity distribution and electric properties. The heating, Q, which will be included as a source term in the heat transfer equation, is computed from peak field amplitudes as:

$$Q = \frac{1}{2} \mid E_{max} \mid^2 . \qquad (3)$$

3.2   Heat Transfer Equations

There are two energy transfer equations within the processed specimen each corresponding to one "phase". The first phase consists of solid and liquid and the second one is the gas phase.

$$\overline{\rho}_1 \frac{\partial H_1(x,y,z,t)}{\partial t} + \nabla \cdot (\overline{\rho}_1 \mathbf{u}_1 H_1 - \frac{\overline{k}_1}{\overline{C}_{p1}} \nabla H_1(x,y,z,t)) = Q(x,y,z,t) + S_1^{int} \quad (4)$$

$$\overline{\rho}_2 \frac{\partial H_2(x,y,z,t)}{\partial t} + \nabla \cdot (\overline{\rho}_2 \mathbf{u}_2 H_2 - \frac{\overline{k}_2}{\overline{C}_{p2}} \nabla H_2(x,y,z,t)) = S_2^{int} \quad (5)$$

where the specific heat, C_p is averaged for each phase as follows:

$$\overline{C}_{p1} = (1 - \phi)C_{p1}^{solid} + \phi M C_{p1}^{liquid} \quad (6)$$

$$\overline{C}_{p2} = \phi(1 - M)C_{p2}^{gas} \quad (7)$$

Thermal conductivity k and density ρ are averaged in a similar way. The heating function is included in the solid-liquid heat transfer equation. There will be an interphase transfer between the two phases, represents by the interface source S_i^{int} defined as:

$$S_i^{int} = h_{ij} A_s (H_i^{int} - H_i) \quad (8)$$

where h_{ij} is a bulk–to interface heat transfer coefficient, A_s is the total interface area and H^{int} are interface enthalpies.

3.3   Mass transfer equations

Gas phase continuity equation:

$$\frac{\partial(\rho_g r_g)}{\partial t} + \nabla \cdot (\rho_g r_g \mathbf{u}_g) = \dot{m} . \quad (9)$$

Liquid phase continuity equation:

$$\frac{\partial(\rho_l r_l)}{\partial t} + \nabla \cdot (\rho_l r_l \mathbf{u}_l) + \dot{m} = 0 . \quad (10)$$

r_g and r_l are gas and liquid volume fractions respectively. The interface mass transfer rate, $\dot{m}$, is determined from the balance of heat through the interface between the phases.

3.4   Darcy's Law Equations

Darcy's law can be used to represent momentum transfer in the porous medium. For liquid it can be expressed in the form:

$$\mathbf{u}_l = -\frac{K_l}{\lambda_l}(\nabla P_g + l_r \nabla r_l + l_T \nabla T) \quad (11)$$

in which $l_r = -\partial P_c(r_l, T)/\partial r_l$ and $l_T = -\partial P_c(r_l, T)/\partial T$ are terms related to capillary pressure P_c. Also K_l and λ_l denote the permeability tensor and the liquid viscosity, respectively, and P_g gas pressure.

The appropriate version of Darcy's law for the gas phase is

$$\mathbf{u}_g = -\frac{K_g}{\lambda_g}(\nabla P_g) \tag{12}$$

where K_g and λ_g are the permeability tensor and the liquid viscosity, respectively.

3.5   Initial and Boundary Conditions

There are number of surface heat or surface loss mechanisms during the microwave heating. The general boundary condition on the material's surface can be expressed as:

$$-k\frac{\partial T}{\partial n} = h_c(T_s - T_\infty) + \sigma_{rad}\epsilon_{rad}(T_s^4 - T_\infty^4) - \dot{m}L \tag{13}$$

where T_s is the load surface temperature, T_∞ is the convective air temperature, n represents the normal to the surface, h_c is the convective heat transfer coefficient, σ_{rad} is the Stefan–Boltzmann constant, ϵ_{rad} is the radiative surface emissivity and L is the latent heat of vaporization of the evaporated liquid. The first term in the equation represent the natural convection by which the load is cooled. The second term is the surface radiation and is important as a cooling mechanism at high load temperatures, or as a heating mechanism if susceptors are used. Since materials with a high moisture content are being observed in this study, the last term, evaporative cooling will have the strongest impact on the temperature profile.

Besides, initial conditions for the example considered here also included:

$$P_g(t=0) = P_{atm}, T(t=0) = T_{ambient} \ . \tag{14}$$

4   Results and Discussion

4.1   Model Description

The microwave oven model used in this paper consists of a waveguide and a microwave cavity containing a block of porous material of rectangular cross section, Fig. 1. The input plane of the waveguide is excited in the dominant TE_{10} mode having the spacial distribution corresponding to this mode and with an amplitude of $100\frac{kV}{m}$. The excitation plane is located far away from the junction where the higher order modes are virtually non-existent. The energy which is reflected back due to a mismatch between the impedance of the waveguide and that of the located cavity passes through the excitation plane, which appears transparent, and is absorbed by the absorbing boundary.

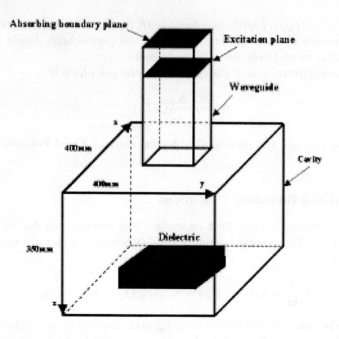

Fig. 1. Microwave oven model

4.2   Solution Technique

The coupling algorithm can be summarized as follows: the FTDT solver is run until the steady state (pure time harmonic) is sufficiently well approximated and the power distribution Q, computed from the peak field amplitudes, has converged. The temperature distribution and moisture content then evolve according to this power distribution until the electric properties of the medium have changed significantly. The electromagnetic solver is then re–run, taking into account the new electrical properties. Once the steady state is reached again, the whole procedure is repeated until the required heating time is reached.

It should be pointed out that the electromagnetic field distribution within the microwave oven is stabilized by the dielectric properties of the biomaterial on a timescale that is very small compared with the thermal process. There are many possible criteria that could be used to determine when the electromagnetic field components have converged. However the key quantity linking the two models is the microwave heating function Q. This heating function was determined at each FTDT time step and the total sum over the biomaterial calculated i.e.

$$I^n = \int_V dV = \frac{1}{2} \sum_{i,j,k} \sigma_{i,j,k} \mid E^{max}_{i,j,k} \mid^2 \Delta x_i \Delta y_j \Delta z_k .$$ (15)

The relative change in the total sum was used as a measure of the convergence of the FTDT scheme to the time–harmonic state.

The cavity and the waveguide are meshed for the FTDT computation with a tensor–product cartesian mesh, but only the biomaterial needs to be meshed for the heat and mass transfer calculations within PHOENICS. Since the two meshes are independent, a transfer function (conserving the total Q for the biomaterial) is used to map the heating function Q from the FTDT mesh to the heat and mass transfer mesh. The time step for the electromagnetic solver is limited by the stability condition

$$\Delta t < \frac{c}{\left(\frac{1}{\Delta x^2} + \frac{1}{\Delta y^2} + \frac{1}{\Delta z^2}\right)^{\frac{1}{2}}} \tag{16}$$

where $c = (\epsilon\mu)^{-1/2}$ is the local electromagnetic wave speed. The time step was set to be 95% of the minimum time step for the mesh and material.

4.3   Discussion

Results are shown for a simple biomaterial with material properties set to those of mashed potato, with density $\rho = 1050\frac{kg}{m^3}$, and porosity $\phi = 0.55$. The dielectric properties of mashed potato as a function of temperature at $10^9 Hz$ were taken from literature [4]. Values for the liquid and gas permeabilities were $K_l = 5 \times 10^{-16} m^{-2}$ and $K_g = 1 \times 10^{-16} m^{-2}$, respectively. The initial temperature was taken to be $15°C$. An ambient temperature of $25°C$ was assumed together with a convective heat transfer of $10 W m^{-2} K^{-1}$ corresponding to a flow of air of approximately $0.5 m s^{-1}$ across the surface via the action of the oven fan. The slab had dimensions $200mm \times 200mm \times 4mm$ and was positioned such that the planes of the horizontal and vertical symmetry of the load coincided with those of the waveguide. The load was placed at a distance of $17mm$ from the oven–waveguide junction. The total heating time was $120s$.

Due to the symmetry of the oven, the maximum values of the heating function on the face are in the centre, Fig. 2. The calculated temperature distributions is shown in Fig. 4. Temperature increase will be very steep in the centre corresponding to the power distribution, slowing down towards the edges due to surface cooling.

Microwave heating generates heat constantly, and as the temperature increases a change of phase will occur and the moisture content will decrease accordingly, Fig. 5. The elevated internal temperature and increasing internal vapor pressure drive the liquid from the medium quickly and efficiently. The vectors in Fig. 5 show the transfer of fluid out of the potato due to evaporation. Further increase of the temperature will further decrease liquid concentrations and vapour transport will become the dominant migration mechanism.

As the water in the food is transferred into vapour and is lost through the boundary, energy absorption is reduced because liquid water is the most active component in absorbing microwave energy. The microwave power distribution will change as shown in Fig. 3.

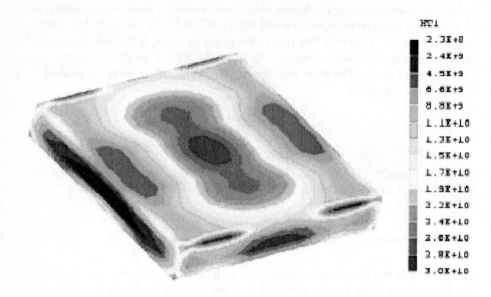

Fig. 2. Heating function,Q, inside the potato at $t = 20sec$

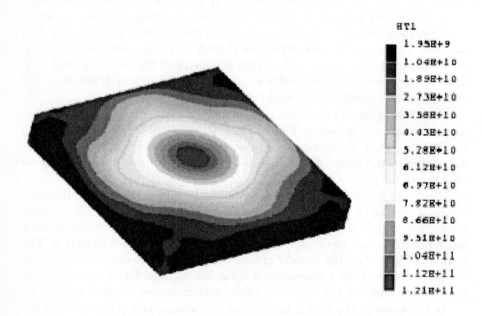

Fig. 3. Heating function,Q, inside the potato at $t = 120sec$

Fig. 4. Temperature profile inside the potato at $t = 120sec$

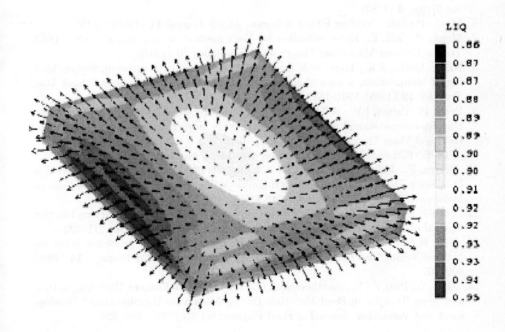

Fig. 5. Liquid concentration inside the potato at $t = 40sec$

5   Conclusion

Heat and mass processes during the microwave heating of food materials are complicated and require knowledge of several areas. The model, which combines a three–dimensional transfer code with a three–dimensional electromagnetic computational scheme, was able to predict the overall drying behaviour. The two calculations have been closely coupled in an optimally cost–efficient manner. This model can be used as a tool for studying in detail microwave heating for real cases, provided the exact dependencies of dielectric properties are measured and implemented in the code.

References

1. Ayappa, K.G., Davis, H.T., Davis, E.A.,Gordon J.: Analysis of Microwave Heating of Materials with Temperature–Dependent Properties. AEChE Journal **37** (1991) 313–322
2. Bear, J.: Dynamics of Fluids in Porous Media. Dover Publications, New York (1998)
3. Jia, X.,Jolly,P.: Simulation of microwave field and power distribution in a cavity by a three dimensional finite element method. Journal of Microwave Power and Electromagnetic Energy **27** (1992) 11–22
4. Kent,M.: Electrical and Dielectric Properties of Food Materials. Science and Technology Publishers Ltd, England,(1987)
5. Metaxas, A.C., Meredith, R.J.: Industrial Microwave Heating. IEE Power Engineering Series, **4** (1983)
6. Monk, P.: Sub–Gridding FTDT Schemes. ACES Journal **11** (1996) 37-46
7. Monk, P., Suli, E.: Error estimates for Yee's method on non–uniform grids. IEEE Transactions on Microwave Theory and Techniques **30** (1994)
8. Ni, H., Datta, A.K., Torrance,K.E.: Moisture transport in intensive microwave heating of biomaterials: a multiphase porous media model. Int. J. of Heat and Mass Transfer **42** (1999) 1501–1512
9. Perre, P., Turner, I.W.: A 3–D version of TransPore: a comprehensive heat and mass transfer computational model for simulating the drying of porous media. Int. J. of Heat and Mass Transfer **42** (1999) 4501–4521
10. PHOENICS code,CHAM ltd, Wimbledon (http://www.cham.co.uk)
11. Torres, F., Jecko,B.: Complete FTDT Analysis of Microwave Heating Process in Frequency–Dependent and Temperature–Dependent Media. IEEE Transactions on Microwave Theory and Techniques **45** (1997) 108–117
12. Turner, I.,Jolly,P..: The effect of dielectric properties on microwave drying kinetics. Journal of Microwave Power and Electromagnetic Energy **25** (1990) 211–223
13. Yee, K.S.: Numerical solution of initial boundary value problems involving Maxwell's equations in isotropic media. IEEE Trans. Antennas Propag., **14** (1996) 302–307
14. Zhao, L., Puri,V.M., Anantheswaran,G., Yeh,G.: Finite Element Modelling of Heat and Mass Transfer in Food Materials During Microwave Heating–Model Development and Validation. Journal of Food Engineering **25** (1995) 509–529

Numerical Solution of Reynolds Equations

for Forest Fire Spread

Valeri Perminov

Belovo Branch of Kemerovo State University, Sovetskay, 41
652600, Belovo, Kemerovo region, Russia
pva@belovo.kemsu.ru

Abstract. In the present paper it is planned to develop mathematical model for description of heat and mass transfer processes at crown forest fire spread. The paper suggested in the context of the general mathematical model of forest fires [1] gives a new mathematical setting and method of numerical solution of a problem of a forest fire spread. It was based on numerical solution of two dimensional Reynolds equations for the description of turbulent flow taking into account for diffusion equations chemical components and equations of energy conservation for gaseous and condensed phases. To obtain discrete analogies a method of controlled volume [2] was used. Numerical solution of this problem during surface and crown fires in exemplified heat energy release in the forest fire front was found.

1. Introduction

The research was made by means of the mathematical modeling methods of physical processes [3]. It was based on numerical solution of two dimensional Reynolds equations for the description of turbulent flow taking into account for diffusion equations chemical components and equations of energy conservation for gaseous and condensed phases. Powerful up-currents of gas occur in surface and crown forest fires, causing entrapment of heated gaseous combustion products of forest fuels in the ground layer. In this context, a study - mathematical modeling - of the conditions of forest fire spreading that would make it possible to obtain a detailed picture of the change in the velocity, temperature and component concentration fields with time, and determine as well as the limiting conditions of forest fire propagation is of interest.

2. Forest Fire Equations

Let us examine a plane problem of radiation-convection heat and mass exchange of forest fuels in all forest strata with gaseous combustion products and radiation from the tongue of flame of the surface forest fire. The surface fire source is modeled as a plane layer of burning forest fuels with known temperature and increasing area of burning. It is assumed that the forest during a forest fire can be modeled as a two-temperature multiphase non-deformable porous reactive medium [1]. Let there be a so-called "ventilated" forest massif, in which the volume of fractions of condensed forest fuel phases, consisting of dry organic matter, water in liquid state, solid pyrolysis products, and ash, can be neglected compared to the volume fraction of gas phase (components of air and gaseous pyrolysis products). To describe the transfer of energy by

P.M.A. Sloot et al. (Eds.): ICCS 2002, LNCS 2329, pp. 823–832, 2002.

radiation we use a diffusion approximation, while to describe convective transfer controlled by the wind and gravity, we use Reynolds equations.

Let the coordinate reference point x_1, x_2 $x_3= 0$ be situated at the center of the forest fire source, axis Ox_3 directed upward, axis Ox_1 and Ox_2 directed parallel to the Earth's surface(x_1 - to the right in the direction of the unperturbed wind speed, x_2 - perpendiculary of the wind direction).

Because of the horizontal sizes of forest massif more than height of forest – h, system of equations of general mathematical model of forest fire [1] was integrated between the limits from height of the roughness level - 0 to h. Besides, suppose that

$$\int_0^h \phi \; dx_3 = \overline{\phi} h,$$

$\overset{...}{\phi}$ - average value of ϕ. The problem formulated above is reduced to a solution of the following system of equations:

$$\frac{\partial \rho}{\partial t} + \frac{\partial}{\partial x_j}(\rho v_j) = Q - (\dot{m}^- - \dot{m}^+)/h, \; j=1,2, \; i=1,2; \quad (1)$$

$$\rho \frac{dv_i}{dt} = -\frac{\partial p}{\partial x_j} + \frac{\partial}{\partial x_j}(-\rho \overline{v_i' v_j'}) - \rho s c_d v_i \,|\,\vec{v}\,| - \rho g_i - Q v_i +$$
$$+ (\tau^- - \tau^+)/h; \quad (2)$$

$$\rho c_p \frac{dT}{dt} = \frac{\partial}{\partial x_j}(-\rho c_p \overline{v_j' T'}) + q_5 R_5 - \alpha_v (T - T_s) + (q_T^- - q_T^+)/h; (3)$$

$$\rho \frac{dc_\alpha}{dt} = \frac{\partial}{\partial x_j}(-\rho \overline{v_j' c_\alpha'}) + R_{5\alpha} - Q c_\alpha + (J_\alpha^- - J_\alpha^+)/h, \; \alpha = 1,3; \; (4)$$

$$\frac{\partial}{\partial x_j}\left(\frac{c}{3k}\frac{\partial U_R}{\partial x_j}\right) - k(c U_R - 4\sigma T_s^4) + (q_R^- - q_R^+)/h = 0; \quad (5)$$

$$\sum_{i=1}^4 \rho_i c_{pi} \varphi_i \frac{\partial T_s}{\partial t} = q_3 R_3 - q_2 R_2 + k(c U_R - 4\sigma T_s^4) + \alpha_v (T - T_s); (6)$$

$$\rho_1 \frac{\partial \varphi_1}{\partial t} = -R_1, \; \rho_2 \frac{\partial \varphi_2}{\partial t} = -R_2,$$

$$\rho_3 \frac{\partial \varphi_3}{\partial t} = \alpha_c R_1 - \frac{M_C}{M_1} R_3, \; \rho_4 \frac{\partial \varphi_4}{\partial t} = 0; \quad (7)$$

$$\sum_{\alpha=1}^3 c_\alpha = 1, p_e = \rho R T \sum_{\alpha=1}^3 \frac{c_\alpha}{M_\alpha}, \vec{v} = (v_1, v_2), \vec{g} = (0,0,g) ; \quad (8)$$

$$Q = (1 - \alpha_c)R_1 + R_2 + \frac{M_c}{M_1} R_3, R_{51} = -R_3 - \frac{M_1}{2M_2} R_5,$$

$$R_{52} = v(1 - \alpha_c)R_1 - R_5, R_{53} = 0. \tag{9}$$

The system of equations (1) – (9) must be solved taking into account the following initial and boundary conditions:

$$t = 0 : v_1 = 0, v_2 = 0, \ T = T_e, c_\alpha = c_{\alpha e}; \tag{10}$$

$$x_1 = x_{10} : v_1 = V_e, v_2 = 0, T = T_e, c_\alpha = c_{\alpha e}; \tag{11}$$

$$x_1 = x_{1e} : \frac{\partial v_1}{\partial x_1} = 0, \frac{\partial v_2}{\partial x_1} = 0, \ \frac{\partial c_\alpha}{\partial x_1} = 0, \frac{\partial T}{\partial x_1} = 0; \tag{12}$$

$$x_2 = x_{20} : \frac{\partial v_1}{\partial x_2} = 0, \frac{\partial v_2}{\partial x_2} = 0, \ \frac{\partial c_\alpha}{\partial x_2} = 0, \frac{\partial T}{\partial x_2} = 0; \tag{13}$$

$$x_2 = x_{2e} : \frac{\partial v_1}{\partial x_2} = 0, \frac{\partial v_2}{\partial x_2} = 0, \ \frac{\partial c_\alpha}{\partial x_2} = 0, \frac{\partial T}{\partial x_2} = 0; \tag{14}$$

It should be noted that the condition of symmetry is used in this statement instead of (13)

$$x_2 = x_{20} : \frac{\partial v_1}{\partial x_2} = 0, v_2 = 0, \ \frac{\partial c_\alpha}{\partial x_2} = 0, \frac{\partial U_R}{\partial x_2} = 0, \frac{\partial T}{\partial x_2} = 0, \tag{15}$$

Because of the patterns of flow and distribution of all scalar functions are symmetrical relative to the axis $0x_2$.

Here and above $\dfrac{d}{dt}$ is the symbol of the total (substantial) derivative; α_v is the coefficient of heat and mass exchange; t is time; x_i, v_i (i = 1, 2) are the Cartesian coordinates (x and y) and the velocity components; index α=1,2,3, where 1 corresponds to the density of oxygen, 2 - to carbon monoxide CO, 3 – concentration of inert components of air; V_e – equilibrium wind velocity.

The thermodynamic, thermophysical and structural characteristics correspond to the forest fuels in the canopy of a pine forest [1]. To define source terms, which characterize inflow (outflow of mass) in a volume unit of the gas-dispersed phase, the following formulae (9) were used for the rate of forming of the gas-dispersed mixture Q, outflow of oxygen R_{51} and changing carbon monoxide R_{52}. The source of ignition is defined as a function of time and turned off after the forest fire initiation.

Thus, the solution of the system of equations (1) - (9) with initial and boundary conditions (10) - (15) may result in defining the fields of velocity, temperature, component concentrations and radiation density.

The system of equations (1)–(8) contains terms associated with turbulent diffusion, thermal conduction, and convection, and needs to be closed. The components of the tensor of turbulent stresses, $\overline{\rho v_i' v_j'}$ as well as the turbulent fluxes of heat and mass $\overline{\rho c_p v_i' T'}$, $\overline{\rho v_i' c_\alpha'}$ are written in terms of the gradients of the average flow properties using the formulas:

$$- \overline{\rho v_i' v_j'} = \mu_t \left(\frac{\partial v_i}{\partial x_j} + \frac{\partial v_j}{\partial x_i} \right) - \frac{2}{3} K \delta_{ij},$$

$$- \overline{\rho c_p v_j' T'} = \lambda_t \frac{\partial T}{\partial x_j}, - \overline{\rho v_i' c_\alpha'} = \rho D_t \frac{\partial c_\alpha}{\partial x_i}, \tag{16}$$

$$\lambda_t = \mu_t c_p / \mathrm{Pr}_T, \rho D_t = \mu_t / \mathrm{Sc}_T, \mu_t = c_\mu \rho K^2 / \varepsilon,$$

where μ_t, λ_t, D_t are the coefficients of turbulent viscosity, thermal conductivity, and diffusion, respectively; Pr_t, Sc_t are the turbulent Prandtl and Schmidt numbers, which were assumed to be equal to 1. The coefficient of dynamic turbulent viscosity is determined using the local-equilibrium model of turbulence [1].

3. The Method of Solution

The boundary-value problem (1) – (15) was solved numerically using the method of splitting according to physical processes. In the first stage, the hydrodynamic pattern of flow and distribution of scalar functions was calculated. The system of ordinary differential equations of chemical kinetics obtained as a result of splitting was then integrated. A discrete analog for equations (1) – (9) was obtained by means of the control volume method using the SIMPLE algorithm [2,3].

The accuracy of the program was checked by the method of inserted analytical solutions. Analytical expressions for the unknown functions were substituted in (1)–(9) and the closure of the equations were calculated. This was then treated as the source in each equation. Next, with the aid of the algorithm described above, the values of the functions used were inferred with an accuracy of not less than 1%. The effect of the dimensions of the control volumes on the solution was studied by diminishing them. The time step for chemical stage was selected automatically.

4. Typical Calculated Results and Discussion

Fields of velocity, temperature, component mass fractions, volume fractions of phases were obtained numerically. The distribution of basic functions shows that the process goes through the some stages. As a result of heating of forest fuel elements, moisture evaporates, and pyrolysis occurs accompanied by the release of gaseous products, which then ignite. and burn away in the forest canopy.

In the vicinity of the source of heat and mass release, heated air masses and products of pyrolysis and combustion float up. At $V_e \neq 0$, the wind field in the forest canopy interacts with the gas-jet obstacle that forms from the surface forest fire source and from the ignited forest canopy base. On the windward side the movement of the air flowing past the ignition region accelerates (Fig. 1).

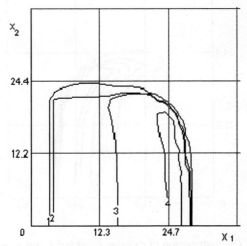

Fig. 1. . Isotherms of the forest fire for t=5 s and $V_e = 5$ m/s:
$1 - \overline{T} = 1.5, 2 - \overline{T} = 2.6., 3 - \overline{T} = 3.5, 4 - T = 5.$

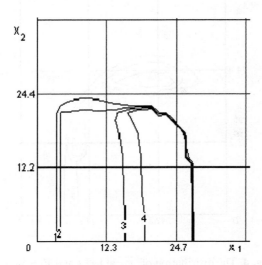

Fig. 2. . Isotherms of the solid phase for t=5 s and $V_e = 5$ m/s:
$1 - \overline{T}_s = 2.6, 2 - \overline{T}_s = 3.5, 3 - T_s = 5.$

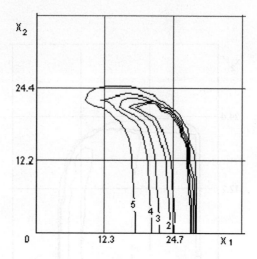

Fig.3. The distribution of oxygen $\bar{c}_1$ at t=5 s and $V_e = 5$ m/s:
$1-\bar{c}_1 = 0.5,\ 2-\ \bar{c}_1 = 0.6,\ 3-\bar{c}_1 = 0.75,\ 4-\bar{c}_1 = 0.85,\ 5-\bar{c}_1 = 0.9.$

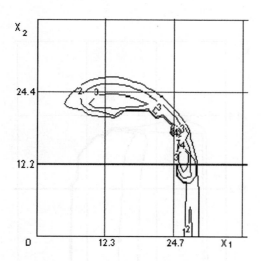

Fig. 4. The distribution of $\bar{c}_2$ at t=5 s and $V_e = 5$ m/s:
$1-\bar{c}_2 = 0.01,\ 2-\ \bar{c}_2 = 0.03,\ 3-\bar{c}_2 = 0.12,\ 4-\bar{c}_2 = 0.4.$

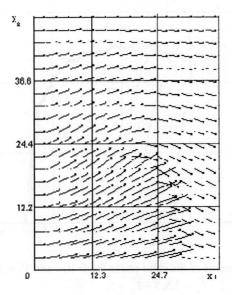

Fig. 5.

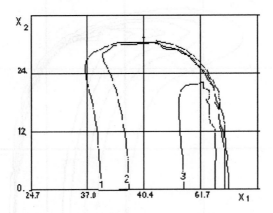

Fig. 6. Isotherms of the forest fire for t=10 s and $V_e = 5$ m/s:
$1-\overline{T} = 2.6, 2-\overline{T} = 3.5., 3-\overline{T} = 5.$

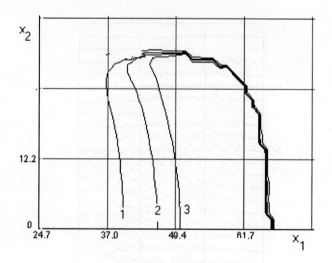

Fig. 7. Isotherms of the solid phase for t=10 s and $V_e = 5$ m/s: $1 - \overline{T}_s = 2.6, 2 - \overline{T}_s = 3.5, 3 - \overline{T}_s = 5$.

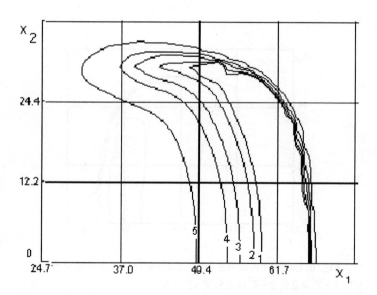

Fig. 8. The distribution of oxygen $\overline{c}_1$ at t=10 s and $V_e = 5$ m/s: $1 - \overline{c}_1 = 0.5, 2 - \overline{c}_1 = 0.6, 3 - \overline{c}_1 = 0.75, 4 - \overline{c}_1 = 0.85, 5 - \overline{c}_1 = 0.9$.

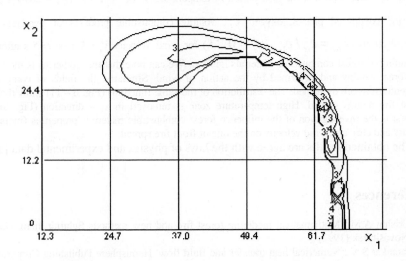

Fig. 9. The distribution of $\bar{c}_2$ at t=10 s and V_e = 5 m/s:
$1-\bar{c}_2$ =0.01, 2− $\bar{c}_2$ =0.03, 3−$\bar{c}_2$ =0.12, 4−$\bar{c}_2$ =0.4.

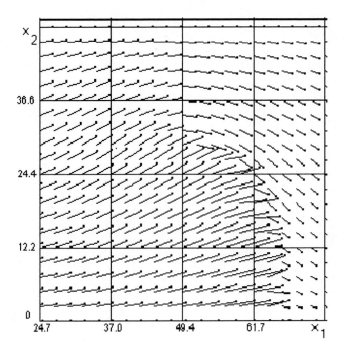

Fig. 10.

Fig. 2, 3 and 4 present the distribution of temperature $\overline{T}$ ($\overline{T} = T/T_e, T_e = 300K$) for gas and condensed phases, oxygen $\overline{c}_1$, volatile combustible products of pyrolysis $\overline{c}_2$ concentrations ($\overline{c}_\alpha = c_\alpha / c_{1e}, c_{1e} = 0.23$) for wind velocity $V_e= 5$ m/s: $t=5$ s after the beginning of forest combustible materials ignition. We can note that the isotherms is moved in the forest canopy and deformed by the action of wind. Similarly, the fields of component concentrations are deformed. The distribution of the same functions (Fig. 6 – 11) show that the forest fire begins spread. High temperature zone is removed in x_1 – direction (Fig. 6). Of interest is the investigation of the influence forest combustible materials properties (moisture, density and etc) and wind velocity on the rate of front fire spread.

The obtained results are agree with the laws of physics and experimental data [1].

References

1. Grishin A.M.: Mathematical modeling forest fire and new methods fighting them, Nauka, Novosibirsk (1992)
2. Patankar S.V.: Numerical heat transfer and fluid flow. Hemisphere Publishing Corporation, New York (1984)
3. Perminov V.A., Mathematical Modeling of Crown and Mass Forest Fires Initiation With the Allowance for the Radiative - Convective Heat and Mass Transfer and Two Temperatures of Medium, Ph.D Thesis, Tomsk State University, Tomsk, (1995)

FEM-based Structural Optimization with Respect to Shakedown Constraints

Michael Heitzer

Central Institute for Applied Mathematics (ZAM),
Forschungszentrum Jülich, D–52425 Jülich, Germany
m.heitzer@fz-juelich.de
http://www.fz-juelich.de/zam/ZAMPeople/heitzer.html

Abstract. In this paper, a mathematical programming formulation is presented for the structural optimization with respect to the shakedown analysis of 3-D perfectly plastic structures on basis of a finite element discretization. A new direct algorithm using plastic sensitivities is employed in solving this optimization formulation. The numerical procedure has been applied to carry out the shakedown analysis of pipe junctions under multi-loading systems. The computational effort of the new approach is much lower compared to so called *derivative-free* search methods.

1 Introduction

In many technically meaningful problems of structures (e.g. in the plant manufacturing) under variable loads the nonlinear behaviour of the material must be considered. Inelastic analyses of the plastic (time-independent) or viscous (time-dependent) behaviour are increasingly used to optimize industrial structures for safety and for an economic operation. Incremental analyses of the path-dependent plastic component behaviour are very time-consuming. The shakedown analysis belongs to the so-called direct or simplified methods which do not achieve the full details of plastic structural behaviour. The objective of shakedown analysis is the determination of an operation regime (i.e. safety margin, load carrying capacities) in which no early failure by plasticity effects has to be expected. Depending on the magnitude of loading, a structure can show the following structural failure modes:

- plastic collapse by unrestricted plastic flow at limit load,
- incremental collapse by accumulation of plastic strains over subsequent load cycles (ratchetting),
- plastic fatigue by alternating plasticity in few cycles, (low cycle fatigue),
- plastic instability of slender compression members (not considered here).

Within the Brite-EuRam Project LISA [17] a procedure is developed using the finite element discretization for direct calculation of the limit and shakedown load of structures made of ductile material. The shakedown analysis is formulated as optimization problem, such that it is easily reformulated for use in a structural

P.M.A. Sloot et al. (Eds.): ICCS 2002, LNCS 2329, pp. 833–842, 2002.

optimization process. Determining optimal values for relevant design variables characterizing the geometrical shape as well as the material behaviour requires an efficient strategy to perform sensitivity analyses with respect to the design variables. Heyman [9] was the first to study the problem of optimal shakedown design. Recently, optimal plastic design of plates with holes under multi-loading systems [16] have been performed. The presented approach is suitable to general 3-D structures which can be analyzed by a Finite Element code. In the new approach the sensitivity analysis is integrated in the formulation of the shakedown analysis. This permits an integrated treatment of structural and sensitivity analysis and results into easily applicable and efficient numerical algorithms. For a general review of sensitivity methods in nonlinear mechanics see [12]. Different types of problems may be considered in structural optimization:

- maximum shakedown load for a given structure (shape).
- optimum shape (e.g. minimum weight) for given shakedown load.

In this contribution we maximize the shakedown range of pipe junctions of variable thickness of the pipe and the junction.

2   Concepts of Shakedown Analysis

Static shakedown theorems are formulated in terms of stress and define safe structural states leading to an optimization problem for safe load domains. The maximum safe load domain is the load domain avoiding plastic failure (with the exception of plastic buckling). We restrict our presentation to perfectly plastic material and no elastic failure modes are considered (i.e. no elastic buckling or high cycle fatigue).

2.1   Static or Lower Bound Shakedown Analysis

The shakedown analysis starts from Melan's lower bound theorem for time variant loading for perfectly plastic material. It is assumed that the loads vary in a convex load domain $\mathcal{L}_0$ such that every load $P(t) = (\mathbf{b}(t), \mathbf{p}(t))$ which lays in $\mathcal{L}_0$ is generated by NV non-degenerated load vertices P_j. The equilibrium conditions of the shakedown analysis and the yield criterion for the actual stresses have to be fulfilled at every instant of the load history. For the following considerations the VON MISES function F is preferred. The maximum enlargement of $\mathcal{L}_0$ is searched for which the structure is safe. The structure is safe against low cycle fatigue and against ratchetting if there is a stress field $\boldsymbol{\sigma}(t)$ such that the equilibrium equations are satisfied and the yield condition (with yield stress σ_y) is nowhere and at no instant t violated.

$$\begin{aligned}
\max \ & \alpha \\
\text{s.t.} \ & F(\boldsymbol{\sigma}(t)) \leq \sigma_y \quad \text{in } V \\
& \operatorname{div}\boldsymbol{\sigma}(t) = -\alpha\mathbf{b}_0(t) \quad \text{in } V \\
& \boldsymbol{\sigma}(t)\,\mathbf{n} = \alpha\mathbf{p}_0(t) \quad \text{on } \partial V_\sigma
\end{aligned} \tag{1}$$

for body forces $\alpha \mathbf{b}_0(t)$, surface loads $\alpha \mathbf{p}_0(t)$. By convexity of $\mathcal{L}_0$ the constraints need to be satisfied only in the load vertices P_j. This makes the problem time invariant for any deterministic or stochastic load history.

Problem (1) can be transformed into a finite optimization problem by FEM discretization. For structures with NG Gaussian points in the FEM model one has to handle $O(NG)$ unknowns and $O(NG)$ constraints. The number of Gaussian points becomes huge for realistic discretizations of industrial structures (several 100000 points) and no effective solution algorithms for discretizations of the nonlinear optimization problem (1) are available. A method for handling such large–scale optimization problems for perfect plasticity is called *basis reduction technique* or *subspace iteration* [7], [18], [15]. This reduction technique generalizes the line search technique, well–known in optimization theory. Instead of searching the whole feasible region for the optimum a sequence of subspaces with a smaller dimension is chosen and one searches for the best value in these subspaces.

3   Optimization Techniques

Hooke and Jeeves coined the phrase *direct search* in a paper that appeared in 1961 [10]. It describes *direct search* by the sequential examination of trial solutions involving comparison of each trial solution with the *best* obtained up to that time together with a strategy for determining (as a function of earlier results) what the next trial solution will be. Many of the direct search methods are based on heuristics and recent analyses guarantee global convergence behavior analogous to the results known for globalized quasi-Newton techniques [14]. Direct search methods succeed because many of them can be shown to rely on techniques of classical analysis like bisection or golden section search algorithms. For simplicity, we restrict our attention here to unconstrained maximization of function $f : I\!\!R^n \to I\!\!R$. We assume that f is continuously differentiable, but that information about the gradient of f is either unavailable or unreliable. Because direct search methods neither compute nor approximate derivatives, they are often described as *derivative-free*. For a recent survey on direct search methods and genetic algorithms see [14] and [6], respectively. A classification of the most methods for numerical optimization can be done according to how many terms of the expansion they need [14], e.g.:

- **Newton's method** (*second order*)
  assumes the availability of first and second derivatives and uses the second-order Taylor polynomial to construct local quadratic approximations of f.
- **Steepest descent** (*first order*)
  assumes the availability of first derivatives and uses the first-order Taylor polynomial to construct local linear approximations of f.

In this classification, *zero-order methods* do not require derivative information and do not construct approximations of f, such that they rely only on values of the objective function. For comparison with the proposed new method using plastic sensitivities the optimization codes PDS2 [14], SIMANN [5] and PIKAIA [1] are used.

3.1   Quasi-Newton Methods

A technique used for iteratively solving unconstrained optimization problems is the line search method. The method determines the optimal point on a given line (*search direction*). A back tracking algorithm is used which starts from an initial step length and decreases the step length until it is sufficient. The algorithm of Dennis/Schnabel is used to omit the exact solution of the one-dimension optimization on the search line [3].

The IMSL routines BCONF and BCONG are used for the maximization [11] if analytic gradients are available or not, respectively. BCONF/BCONG use a quasi-Newton method and an active set strategy to solve maximization problems subject to simple bounds l, u on the variables. From a given starting point x^c, an active set IA, which contains the indices of the variables at their bounds, is built. A variable is called a *free variable* if it is not in the active set. The routine then computes the search direction for the free variables from a positive definite approximation of the Hessian and the gradient evaluated at x^c; both are computed with respect to the free variables. Routine BCONF calculates the gradient by a finite-difference method evaluated at x^c. The search direction for the variables in IA is set to zero. A line search is used to find a better point $x^n = x^c + \lambda d$, $\lambda \in (0, 1]$. Finally, the optimality conditions are checked for a suitable gradient tolerance. Another search direction is then computed to begin the next iteration. The active set is changed only when a free variable hits its bounds during an iteration or the optimality condition is met for the free variables but not for all variables in IA, the active set. In the latter case, a variable that violates the optimality condition will be dropped out of IA. For more details on the quasi-Newton method and line search, see [3].

The line search method needs for solving optimization problems a search direction d. Here the search line is given by the gradients of the shakedown factor with respect to the design parameters. The shakedown factor is a solution of a nonlinear optimization problem and therefore the gradients are given by the sensitivities with respect to the design parameters. Using the chain rule the problem of the plastic structural behaviour can be reduced to the sensitivity analysis of the elastic structural response, which is a significant reduction of computational effort (see [8],[4]). The sensitivity analysis of the elastic response is performed by a finite-difference method for a small number of parameters, see [12] for alternative techniques.

3.2   Pattern Search Method

Pattern search methods are characterized by a series of *exploratory moves* that consider the behavior of the objective function at a pattern of points, all of which lie on a rational lattice. The exploratory moves consist of a systematic strategy for visiting the points in the lattice in the immediate vicinity of the current iterate [14]. For each move the parameter is varied and it is decided if there is an improvement, such that the procedure is a direct search. The used multi-directional search algorithm proceeds by reflecting a simplex through the centroid of one of the faces [14].

If one replaces a vertex by reflecting it through the centroid of the opposite face, then the result is also a simplex. The first single move is that of reflection which identifies the *worst* vertex in the simplex (i.e. the one with the least desirable objective value) and then reflects the worst simplex through the centroid of the opposite face. If the reflected vertex is still the worst vertex, then next choose the *next worst* vertex and repeat the process. The ultimate goals are either to replace the *best* vertex or to ascertain that the best vertex is a candidate for a maximizer. Until then, the algorithm keeps moving the simplex by flipping some vertex (other than the best vertex) through the centroid of the opposite face. An *expansion step* allows for a more progressive move by doubling the length of the step from the centroid to the reflection point, whereas a *contraction steps* allow for more conservative moves by halving the length of the step from the centroid to either the reflection point or the worst vertex. These steps allow a deformation of the shape of the original simplex.

3.3   Simulated Annealing

Simulated annealing is a global optimization method that distinguishes between different local maxima. Starting from an initial point, it randomly chooses a trial point within the step length of the user selected starting point. The function is evaluated at this trial point and its value is compared to its value at the initial point. When maximizing a function, any uphill step is accepted and the process repeats from this new point. An downhill step may be accepted, such that it can escape from local maxima. Downhill moves may be accepted; the decision is made by the Metropolis criteria. It uses T (temperature) and the size of the downhill move in a probabilistic manner. The smaller T and the size of the downhill move are, the more likely that move will be accepted. If the trial is accepted, the algorithm moves on from that point. If it is rejected, another point is chosen instead for a trial evaluation. As the optimization process proceeds, the length of the steps decline and the algorithm closes in on the global optimum. Since the algorithm makes very few assumptions regarding the function to be optimized, it is quite robust with respect to non-quadratic surfaces.

The simulated annealing algorithm of Corana et al. [2] was implemented and modified in [5] and is obtained from net NETLIB collection of mathematical software.

3.4   Genetic Algorithms

Genetic algorithms are a class of search techniques inspired from the biological process of evolution by means of natural selection. A top-level view of a canonical genetic algorithm could be as follows: Start by generating a set (*population*) of trial solutions, usually by choosing random values for all model parameters; then:

1. Evaluate the (*fitness*) of each member of the current population.
2. Select pairs of solutions (*parents*) from the current population, with the probability of a given solution being selected made proportional to that solution's fitness.
3. Produce two new solutions (*offspring*) from the two solutions selected in (2).
4. Repeat steps (2)-(3) until the number of offspring produced equals the number of individuals in the current population.
5. Use the new population of offspring to replace the old population.
6. Repeat steps (1) through (5) until some termination criterion is satisfied.

The probability of a given solution being selected for breeding is proportional to the fitness, such that better trial solutions breed more often, the computational equivalent of natural selection. The production of new trial solutions from existing ones occurs through breeding. This involves encoding the parameters defining each solution as a string-like structure (*chromosome*), and performing genetically inspired operations of *crossover* and *mutation* to the pair of chromosomes encoding the two parents. The end result of these operations are two new chromosomes defining the two offsprings that incorporate information from both parents. No derivatives of the goodness of fit function with respect to model parameters need to be computed. In most real applications, the model will need to be evaluated a great number of times, such that the evaluation is computationally expensive.

The used PIKAIA subroutine[1] maximizes a user-supplied FORTRAN function using uniform one-point crossover, and uniform one-point mutation.

4   Pipe-junction Subjected to Internal Pressure and Temperature Loads

A pipe-junction subjected to internal pressure and temperature loads is analyzed as a simple example. The shakedown analyses are performed for perfectly plastic material with a yield stress $\sigma_y = 250N/mm^2$. The inner diameters $D = 39mm$ and $d = 15mm$ of the pipe and of the junction are fixed, respectively. The length of the pipe and of the junction are $L = 81.9mm$ and $l = 17.1mm$ fixed, respectively. The variable dimensions are the wall-thickness s and t of the pipe and the junction, respectively. The meshes of the pipe-junction are generated my an automatic mesh-generator. The different models are discretized with 125 solid 20-node hexahedron elements (HEXEC20). The dimensions of the model are based on a pipe benchmark problem of PERMAS [15]. The FE-mesh and the essential dimensions of the different pipe-junctions are represented in Fig. 1. The pipe junction is subjected to two-parameter loading. Pressure P and temperature difference T vary independentaly

$$0 \le P \le \alpha\mu_1 P_0,$$
$$0 \le T_i \le \alpha\mu_2 T_0, \qquad 0 \le \mu_1, \mu_2 \le 1.$$

P_0 and T_0 are a reference pressure and temperature difference, respectively.

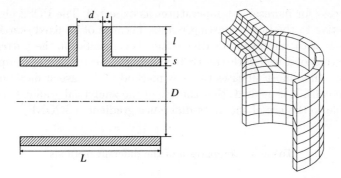

Fig. 1. FE-mesh and dimensions of the pipe-junction

The goal of the structural optimization in this example is to maximize the shakedown factor α_s for the wall-thickness s and t varying in given bounds:

$$\max \ \alpha_s$$
$$\text{s.t.} \ \ 0 < s, t \le 7.5. \tag{2}$$
$$\alpha_s \ \text{solution of problem (1)}$$

For pure pressure variation the optimal wall thickness of the pipe and of the junction will tend to infinity because of the decreasing elastic stresses. On the other hand for pure temperature variation the wall thickness of the pipe and of the junction will tend to zero. Therefore, different finite positive ratios between the initial pressure P_0 and the initial temperature T_0 are chosen. The design variables vary between bounds $0 < s, t \le 7.5mm$ to guarantee that the chosen mesh of the pipe-junction is not degenerated, otherwise the chosen automatic mesh-generator leads to meshes with degenerated elements. The shakedown factor α_s is the solution of the corresponding shakedown optimization problem with the load domain $\mathcal{L}_0$ defined by P_0 and T_0.

Five different mathematical optimization codes using the FEM-based shakedown analysis are compared in the example. The direct methods PDS2, PIKAIA, SIMANN without the use of gradients and the IMSL routines BCONF and BCONG, using finite-difference method and analytical gradients, respectively. All codes use the same subroutine to calculate the shakedown load factor α_s. The default parameters are used for all methods. PDS2 uses a fixed search pattern. BCONF performs a finite-difference method to estimate the gradient, such that for each gradient at least two additional shakedown analyses have to be performed. BCONG uses the given analytical gradients. The sensitivities are calculated using a finite-difference method for the elastic stresses and the algorithm described above.

A comparison of the different methods is shown in Table 1. In addition to the optimal values s^*, t^* and the corresponding shakedown load magnitudes $P_s = \alpha_s P_0$ and $T_s = \alpha_s T_0$ the number of function calls (i.e. shakedown analyses) are given. The results for the methods are comparatively close. The values

s^*, t^* decrease for increasing temperatures as expected. The PDS2 algorithm is quite sensitive to the initial starting points, because of its fixed search scheme. Nevertheless with improved starting values the results for the pattern search method correspond well with the BCONG routine using the new implemented sensitivities. In allmost all cases the new method is the fastest method in terms of function calls. It is evident, that the use of the analytical gradients (BCONG) is preferable to the use of the finite-difference gradients (BCONF).

Table 1. Comparison of the different methods

T_i^0/P^0 [K/MPa]	Method	s^* [mm]	t^* [mm]	P_s [MPa]	T_s [K]	function calls
	PDS2	7.4665	6.8200	19.90	796	17
	GA	6.3342	5.1185	20.72	829	17
40	SIMANN	5.5894	3.8753	19.56	782	59
	BCONF	5.5654	3.7504	18.08	720	52
	BCONG	7.5000	6.3041	19.75	790	12
	PDS2	6.9828	6.6913	13.52	1082	50
	GA	6.6958	5.4724	14.44	1155	10
80	SIMANN	5.9183	4.5799	13.56	1085	16
	BCONF	4.1465	5.7304	11.65	932	45
	BCONG	7.1101	4.9480	13.15	1052	4
	PDS2	5.1344	4.0902	10.07	1007	22
	GA	6.3168	4.5315	11.97	1197	13
100	SIMANN	6.8760	4.7475	12.21	1221	13
	BCONF	4.5733	3.4209	10.71	1072	41
	BCONG	4.1404	4.2517	10.10	1011	9
	PDS2	6.9835	6.6906	10.35	1242	7
	GA	7.3553	5.0235	10.47	1256	4
120	SIMANN	6.6813	5.0391	10.61	1273	19
	BCONF	7.5000	4.4788	9.13	1096	27
	BCONG	7.5000	6.7115	10.57	1268	6
	PDS2	5.2488	4.6217	7.21	1154	41
	GA	5.5639	5.0309	8.35	1336	11
160	SIMANN	4.4377	3.2308	7.50	1200	21
	BCONF	4.9742	4.0640	7.81	1249	75
	BCONG	3.8007	2.9375	6.96	1114	12
	PDS2	6.0650	5.7721	6.60	1320	15
	GA	5.3168	3.5315	6.26	1252	13
200	SIMANN	4.5928	3.7583	6.52	1304	12
	BCONF	4.4038	3.9833	6.43	1286	68
	BCONG	3.9576	3.5771	6.03	1207	5

It has to be noticed that the function $f(s,t) = \alpha_s$ is not convex, such that probably local maxima exist in the region $0 < s, t \leq 7.5mm$. For instance in load level $T_i^0/P^0 = 100$ [K/MPa] the resulting shakedown load magnitudes P_s

and T_s are close for BCONG and PDS2, but for GA and SIMANN the values s^*, t^* are fairly different. This may indicate that the global maximum in the region $0 < s, t \leq 7.5mm$ has not yet reached by BCONG and PDS2 but with the global methods GA and SIMANN. Additional computations suggest, that the temperature difference $T_s \approx 1400$ K is the highest allowable temperature load in the region $0 < s, t \leq 7.5mm$.

5 Conclusion

Shakedown theorems are exact theories of classical plasticity for the direct computation of safety factors or of the load carrying capacity under varying loads. This method can be based on static and kinematic theorems for lower and upper bound analysis. Using Finite Element Methods more realistic modeling can be used for a more rational design. A mathematical programming formulation is presented for the structural optimization with respect to the shakedown analysis of 3-D perfectly plastic structures. A new direct algorithm using plastic sensitivities is employed in solving this optimization formulation. The numerical procedure has been applied to carry out the shakedown analysis of pipe junctions under multi-loading systems. The computational effort of the proposed method is lower compared to so-called *derivative-free* direct search methods.

6 Acknowledgements

Parts of this research have been funded by the Brite-EuRam III project LISA: FEM-Based Limit and Shakedown Analysis for Design and Integrity Assessment in European Industry (Project N°: BE 97-4547, Contract N°: BRPR-CT97-0595).

References

1. Charbonneau, P.: Genetic Algorithms in Astronomy and Astrophysics, *The Astrophysical Journal (Supplements)*, **101** (1995) 309-334
2. Corana A., Marchesi M., Martini C., Ridella S.,: Minimizing Multimodal Functions of Continuous Variables with the "Simulated Annealing" Algorithm, *ACM Transactions on Mathematical Software* **13** (1987) 262-280
3. Dennis, J.E., Schnabel, R.B.: *Numerical methods for unconstrained optimization*, Engelwood Cliffs, N.J., Prentice Hall (1983)
4. Fiacco, A.: *Introduction to Sensitivity and Stability Analysis in Nonlinear Programming*, New York, Academic Press (1983)
5. Goffe, W.L., Ferrier,G., Rogers, J.: Global Optimization of Statistical Functions with Simulated Annealing, *Journal of Econometrics* **60** (1994) 65-99
6. Goldberg, D.E.: *Genetic Algorithms in Search, Optimization, and Machine Learning* Addison Wesley (1989)
7. Heitzer, M.: Traglast– und Einspielanalyse zur Bewertung der Sicherheit passiver Komponenten, *Berichte des Forschungszentrums Jülich* Jül–3704 (1999)

8. Heitzer, M.: Structural optimization with FEM-based shakedown analyses *Journal of Global Optimization* (to appear).
9. Heyman, J.: Minimum weight of frames under shakedown loading. *Journal of Engineering Mechanics, ASCE* **84** (1958) 1–25
10. Hooke, R., Jeeves, T.A.: Direct search solution of numerical and statistical problems. *J. Assoc. Comput. Mach.* **8** (1961) 212–229
11. IMSL: Fortran Subroutines for Mathematical Application, Math/Library Vol. 1 & 2 Visual Numerics, Inc. (1997)
12. Kleiber, M., Antúnez, H., Hien, T.D., Kowalczyk, P.: *Parameter Sensitivity in Nonlinear Mechanics: Theory and Finite Element Computations*, New York, J. Wiley & Sons (1997)
13. König, J. A.: *Shakedown of Elastic–Plastic Structures*, Amsterdam and Warsaw, Elsevier and PWN (1987)
14. Lewis, R.M., Torczon, V., Trosset, M.W.: Direct search methods: then and now, *Journal of Computational and Applied Mathematics* **124** (2000) 191–207
15. PERMAS: *User's Reference Manuals*, Stuttgart, INTES Publications No. 202, 207, 208, 302, UM 404, UM 405 (1988)
16. Schwabe, F.: Einspieluntersuchungen von Verbundwerkstoffen mit periodischer Mikrostruktur, Phd-Thesis, RWTH-Aachen (2000)
17. Staat, M., Heitzer M.: LISA a European Project for FEM-based Limit and Shakedown Analysis, *Nuclear Engineering and Design* **206** (2001) 151-166
18. Stein, E., Zhang, G., and Mahnken, R.: Shakedown analysis for perfectly plastic and kinematic hardening materials, in Stein, E. (ed.): *Progress in computational analysis of inelastic structures*, Wien, Springer, 175–244 (1993)

Tight Bounds on Capacity Misses for 3D Stencil Codes

Claudia Leopold

Friedrich-Schiller-Universität Jena
Institut für Informatik
07740 Jena, Germany
claudia@informatik.uni-jena.de

Abstract. The performance of linear relaxation codes strongly depends on an efficient usage of caches. This paper considers one time step of the Jacobi and Gauß-Seidel kernels on a 3D array, and shows that tiling reduces the number of capacity misses to almost optimum. In particular, we prove that $\Omega(N^3/(L\sqrt{C}))$ capacity misses are needed for array size $N \times N \times N$, cache size C, and line size L. If cold misses are taken into account, tiling is off the lower bound by a factor of about $1+5/\sqrt{LC}$. The exact value depends on tile size and data layout. We show analytically that rectangular tiles of shape $(N-2) \times s \times (sL/2)$ outperform square tiles, for row-major storage order.

1 Introduction

Stencil codes such as the Jacobi and Gauß-Seidel kernels are used in many scientific and engineering applications, in particular in multigrid-based solvers for partial differential equations. Stencil codes own their name to the fact that they update array elements according to some fixed pattern, called stencil. The codes perform a sequence of sweeps through a given array, and in each sweep update all array elements except the boundary. This paper considers the Gauß-Seidel and Jacobi kernels on a 3D array, which are depicted in Fig. 1. Since each update operation (*) accesses seven array elements, we speak of a 7-point stencil.

In Fig. 1b), the copy operation can alternatively be replaced by a second loop nest in which the roles of arrays A and B are reversed. Throughout the paper, we consider a single iteration of the time loop. Copy operation or second loop nest are not taken into account. Obviously, the Gauß-Seidel scheme incurs data dependencies from iterations $(i-1,j,k)$, $(i,j-1,k)$, and $(i,j,k-1)$ to iteration (i,j,k), whereas the Jacobi scheme does not incur data dependencies.

When implemented in a straightforward way, the performance of stencil codes lags far behind peak performance, chiefly because cache usage is poor [1]. On the other hand, stencil codes have a significant cache optimization potential since successive accesses refer to neighbored array elements. In 3D codes, five types of data reuse can be distinguished:

(1) Assuming row-major storage order, the cache line that holds elements $A[i,j,k]$, $A[i,j,k+1]\dots$ is reused in the k-loop,

P.M.A. Sloot et al. (Eds.): ICCS 2002, LNCS 2329, pp. 843–852, 2002.

```
a) for (t=0; t<time; t++)      /* or: while (!converged) */
     for (i=1; i<N-1; i++)
       for (j=1; j<N-1; j++)
         for (k=1; k<N-1; k++)
           A[i,j,k] += A[i-1,j,k] + A[i+1,j,k]
                     + A[i,j-1,k] + A[i,j+1,k]
                     + A[i,j,k-1] + A[i,j,k+1];           (*)

b) for (t=0; t<time; t++) {    /* or: while (!converged) */
     for (i=1; i<N-1; i++)
       for (j=1; j<N-1; j++)
         for (k=1; k<N-1; k++)
           B[i,j,k] = A[i,j,k] + A[i-1,j,k] + A[i+1,j,k]
                    + A[i,j-1,k] + A[i,j+1,k]
                    + A[i,j,k-1] + A[i,j,k+1];            (*)
     Copy(B->A);
   }
```

Fig. 1. a) Gauß-Seidel and b) Jacobi code for $N \times N \times N$ arrays A, B

(2) $A[i, j, k]$ is reused in the k-loop, in the updates of $A[i, j, k-1]$, $A[i, j, k]$, and $A[i, j, k+1]$,

(3) $A[i, j, k]$ is reused in the j-loop,

(4) $A[i, j, k]$ is reused in the i-loop, and

(5) $A[i, j, k]$ is reused in the time loop.

Reuse is exploited if data are kept in cache in-between successive accesses; exploited reuse is denoted as *locality*. Reuse types 1 and 2 are always exploited by the input codes; whether reuse types 3-5 are exploited depends on array size N and cache capacity C. As explained by Rivera and Tseng [7], reuse type 3 is exploited for $C > 2N$. This condition often holds in practice, and thus many 2D codes perform well without cache optimization. The exploitation of type 4 reuse requires $C > 2N^2$, which does not hold for typical cache and array sizes. Tiling can transform programs such that type 4 reuse is exploited, too. We will study this technique in the paper. Type 5 reuse is out of our scope.

Tiling rearranges the updates such that successive type 4 reuses are moved closer together. Fig. 2 depicts a tiled Gauß-Seidel code; tiled Jacobi code is analogous and has been omitted for brevity. Tiling groups the updates into rectangular blocks (called tiles) that are processed one after another. Note that only two loops are blocked, and the third is not. Why it is better to block the j- and k-loops instead of the i-loop will be explained in Sect. 3.

Tiling speeds up 3D stencil codes, as has been shown experimentally by Rivera and Tseng [7]. The present paper complements their work by analytically investigating the question as to how close to optimum tiling gets in terms of cache misses. In line with common notation, we distinguish cache misses into cold misses (the first access to an element), capacity misses (misses due to limited

```
for (t=0; t<time; t++)
   for (jj=1; jj<N-1; jj+=sj)
      for (kk=1; kk<N-1; kk+=sk)
         for (i=1; i<N-1; i++)
            for (j=jj; j<min(jj+sj,N-1); j++)
               for (k=kk; k<min(kk+sk,N-1); k++)
                  A[i,j,k] += A[i-1,j,k] + A[i+1,j,k]
                            + A[i,j-1,k] + A[i,j+1,k]
                            + A[i,j,k-1] + A[i,j,k+1];      (*)
```

Fig. 2. Tiled Gauß-Seidel code

cache capacity), and conflict misses (misses due to limited cache associativity). Conflict misses can be eliminated through tile size selection, array padding, and copying [6, 7], and are not considered in this paper.

This paper makes three contributions. First, we prove that $\Omega(N^3/(L\sqrt{C}))$ capacity misses are needed for any ordering of the update operations (*), where L denotes cache line size. Second, we estimate the constant factor in this bound by 0.35 under the assumptions $N \geq 1000$, $C \geq 10000$ and $N^2 \geq 100C$. Third, we analyze the tiled codes for different data layouts and tile sizes. While Rivera and Tseng [7] use square tiles, we argue that rectangular tiles are superior. For these, we prove that the number of cold and capacity misses is off the lower bound by a factor of at most $1 + 4.6/\sqrt{LC}$.

The paper is organized as follows. Section 2 proves the lower bound and estimates the constant factor. Section 3 analyzes the tiled codes. Section 4 surveys related work, and Section 5 finishes with conclusions.

2   Lower Bound

The number of cold misses is trivially equal to $N^3 - 8$ since all elements of A, except the corners, must be read. This number is independent on the order of updates, and is, thus, valid for both the input and tiled codes. In the following, we consider capacity misses only.

We do not rely on a particular cache replacement scheme, but assume user-controlled data placement. The so-derived lower bound applies to any common scheme such as LRU. Similarly, we do not take data dependencies into account, which is correct because a lower bound to an unconstrained problem is always a lower bound to a constrained problem, too. Accesses to B are ignored as they can only increase the number of cache misses. Consequently, the lower bound holds for both the Jacobi and Gauß-Seidel schemes.

An ordering of the updates is denoted as *schedule*. We allow for redundancy, that is, updates may be carried out repeatedly with same indices i, j, k. Updates must not be split into subcomputations, however, to guarantee bitwise-identical results as compared to the input codes.

Theorem 1. *Any schedule of the update operations (*) takes* $\Omega(N^3/(L\sqrt{C}))$ *capacity misses.*

Proof: Let *Sched* be any schedule. We partition *Sched* into subsequences $S_1, S_2, \ldots, S_r$ of successive updates such that $S_1, S_2, \ldots S_{r-1}$ contain exactly $C\sqrt{C}$ updates, and S_r contains up to $2C\sqrt{C}$ updates. Note that

$$r \geq \lfloor (N-2)^3/(C\sqrt{C}) \rfloor \ ,$$

where inequality indicates redundancy.

We define an array element e to be *touched-only* by S_l $(1 \leq l \leq r)$ if the number of accesses to e in S_l is at least one, but less than the total number of accesses to e in *Sched*. In other words, an array element is touched-only if it is accessed by multiple S_l.

The rest of this proof uses geometric argumentation. We model A by a large cube that is composed of small unit cubes for the $A[i,j,k]$. Hence, $A[i,j,k]$ corresponds to a cube of side length $1{\times}1{\times}1$ at position (i,j,k) of the large cube.

In the following, we consider any particular S_l. Let F denote the geometric figure (also called arrangement) that corresponds to S_l, that is, the figure that is composed of those unit cubes whose array elements are updated in S_l. Furthermore, let Q be the smallest axes-parallel cuboid that completely holds F. Obviously, the dimensions q_i, q_j, and q_k of Q (in i-, j-, and k-direction, respectively) fulfill $q_i \cdot q_j \cdot q_k \geq C\sqrt{C}$. We define a unit cube to be *i-touched* by F if

- the cube belongs to F, but its left or right neighbor in i-direction does not, or
- the cube does not belong to F, but its left or right neighbor in i-direction does.

Analogously, we define *j-touched* and *k-touched*. We say that a cube is *touched* if it is touched in at least one of the three directions. Since an array element has up to 6 neighbors, a cube can be touched up to sixfold.

The concepts of *touched* and *touched-only* are related. Consider, for instance, a cube inside F that is i-touched by F at the left boundary. Then the corresponding array element $A[i,j,k]$ is updated in S_l, but the left neighbor $A[i-1,j,k]$ is accessed and not updated. Consequently, $A[i-1,j,k]$ is either updated in another subsequence $S_{l'}$ with $l' \neq l$, or $A[i-1,j,k]$ is not updated at all. In the former case, both S_l and $S_{l'}$ access $A[i,j,k]$, and thus $A[i,j,k]$ is touched-only by S_l. In the latter case, $A[i-1,j,k]$ belongs to the boundary of A.

Lemma 1 below shows that any arrangement of $C\sqrt{C}$ cubes induces at least $4C - 6\sqrt{C}$ touched cubes. Consequently, within the whole schedule, the number of pairs (S_l, e) for which the cube that corresponds to e is touched by the figure that corresponds to S_l is at least

$$Z \geq r \cdot (4C - 6\sqrt{C}) \ .$$

We distinguish these pairs into five types:

(1) e is touched-only by S_l, and S_l is the first subsequence that accesses e.
(2) e is touched-only by S_l, and S_l is the last subsequence that accesses e.
(3) e is touched-only by S_l, and S_l is neither the first nor the last subsequence that accesses e.
(4) e is not touched-only by S_l, and e belongs to the boundary of A.
(5) e is not touched-only by S_l, and a neighbor of e belongs to the boundary of A.

The whole schedule comprises at most $6N^2+6(N-2)^2 = 12N^2-24N+24$ pairs of types 4 and 5. Let Z^{in}, Z^{out}, and Z^{inOut} denote the total number of type 1, 2, and 3 pairs, respectively. Then, $Z^{in} + Z^{out} + Z^{inOut} = Z - 12N^2 + 24N - 24$ and $Z^{in} = Z^{out}$ imply $Z^{out} + Z^{inOut} \geq Z/2 - 6N^2 + 12N - 12$.

In-between successive S_l's, at most C data are kept in cache, summing up to $W \leq (r-1) \cdot C$ data in the whole schedule. Consequently, the number of data that must be reloaded after having been replaced from cache is at least

$$Z^{out} + Z^{inOut} - W \geq r \cdot (2C - 3\sqrt{C}) - 6N^2 + 12N - 12 - rC + C$$

$$\geq \left\lfloor (N-2)^3/(C\sqrt{C}) \right\rfloor \cdot (C - 3\sqrt{C}) - 6N^2 + 12N - 12 + C .$$

Since at most L elements are loaded per memory access, the number of capacity misses is at least

$$(1/L) \cdot \left(\left\lfloor (N-2)^3/(C\sqrt{C}) \right\rfloor \cdot (C - 3\sqrt{C}) - 6N^2 + 12N - 12 + C \right)$$

$$= \Omega(N^3/(L\sqrt{C})) ,$$

provided that N is significantly larger than $\sqrt{C}$.   ∎

In the following, we estimate the constant factor in this bound under the assumptions $N \geq 1000$, $C \geq 10000$ and $N^2 \geq 100C$. Imposing stronger assumptions increases the constant, and imposing weaker assumptions decreases it. With $X \underset{D}{=} (N-2)^3/(C\sqrt{C})$, we get

$$X = \left(N^2(N-6) + 12N - 8 \right) / (C\sqrt{C})$$

$$\geq \left(100C(10\sqrt{C} - 6) + 120\sqrt{C} - 8 \right) / (C\sqrt{C})$$

$$= 1000 - 600/\sqrt{C} + 120/C - 8/(C\sqrt{C})$$

$$\geq 994 .$$

Consequently, $\lfloor X \rfloor \geq X - 1 \geq (1 - 1/994)\, X$.   Let $Y \underset{D}{=} \lfloor X \rfloor \cdot (C - 3\sqrt{C})$ and $Z \underset{D}{=} 6N^2 - 12N + 12 - C$. Then, $C - 3\sqrt{C} \geq 0.97C$ implies

$$Y \geq (1 - 1/994) \cdot 0.97 \cdot (N-2)^3/\sqrt{C} \geq 0.963\, N^3/\sqrt{C} \geq 9.63N^2 .$$

From $Z < 6N^2$ follows $Z \leq 0.63\,Y$, and thus $Y - Z \geq 0.37\,Y$. Putting it all together, the number of capacity misses is at least

$$(1/L) \cdot (Y - Z) \geq (0.37/L) \cdot 0.963 \cdot (N^3/\sqrt{C}) \geq 0.35N^3/(L\sqrt{C}) .$$

The proof of Lemma 1 is still open:

Lemma 1. *For any arrangement of $C\sqrt{C}$ cubes, at least $4C - 6\sqrt{C}$ cubes are touched.*

<u>Proof:</u> The proof is by contradiction. Let F be an arrangement with less than $4C - 6\sqrt{C}$ touched cubes. Let *column* denote a set of cubes in F that have the same i- and j-coordinates, and let *plane* denote a set of cubes in F that have the same k-coordinates. Furthermore, let w_p denote the number of cubes in plane p. We distinguish four cases and derive a contradiction in each of them.

Case 1: At least two planes p, p′ (p ≠ p′) exist such that $\mathbf{w_p \geq C}$ and $\mathbf{w_{p'} \geq C}$: The total number of k-touched cubes in F equals the sum of k-touched cubes in the individual columns (columns are independent wrt. touches in k-direction). In a column with one cube, at least three cubes are k-touched; in a column with two or more cubes, at least four cubes are k-touched. Hence, if p and p' have x columns in common, the total number of k-touched cubes is at least $4x + 3(\#\text{touched}(p)-x) + 3(\#\text{touched}(p')-x) \geq 4C$, a contradiction.

Case 2: A plane p exists with $\mathbf{w_p \geq 4C/3}$: Since there are at least $4C/3$ non-empty columns, the total number of k-touched cubes is at least $3 \cdot (4C/3) = 4C$, a contradiction.

Case 3: Exactly one plane p contains $\mathbf{C \leq w_p < 4C/3}$ cubes; all other planes contain less than C cubes: Obviously, the total number of i/j-touched cubes equals the sum of i/j-touched cubes in the individual planes (planes are independent wrt. touches in i/j-direction). In [3], it is shown that any 2D arrangement of Q unit squares induces at least $4\sqrt{Q}$ i/j-touched squares. Thus, in 3D, any plane of Q cubes i/j-touches at least $4\sqrt{Q}$ cubes. Hence, F altogether i/j-touches at least $4\sqrt{C} + \sum_p (4\sqrt{w_p})$ cubes, with $\sum_p w_p \geq C\sqrt{C} - 4C/3 \overset{D}{=} D$.

We assume $\sum_p w_p = D$, which can only underestimate the number of touched cubes. Lemma 2 below shows that, for any fixed value of $\sum_p w_p$, the value of $\sum_p \sqrt{w_p}$ is maximized if the w_p take extreme values, that is, if $\lfloor D/(C-1) \rfloor$ of the w_p take value $C-1$, one w_p takes value $D - \lfloor D/(C-1) \rfloor \cdot (C-1)$, and the other w_p take value zero. Thus, the total number of i/j-touched cubes is at least

$$4\sqrt{C} + 4 \cdot \lfloor D/(C-1) \rfloor \cdot \sqrt{C-1} \geq 4D/\sqrt{C} > 4C - 6\sqrt{C} ,$$

which contradicts the assumption.

Case 4: All planes contain less than C cubes: In analogy to Case 3, the number of i/j-touched cubes can be estimated by

$$4 \left\lceil C\sqrt{C}/(C-1) \right\rceil \cdot \sqrt{C-1} \geq 4C - 4\sqrt{C-1} > 4C - 6\sqrt{C} ,$$

which contradicts the assumption.   ∎

Lemma 2 completes the lower bound proof:

Lemma 2. *For fixed values $d, e \in \mathbb{N}$, a finite set $X \subseteq \mathbb{N}$ with $\sum_{x \in X} x = d$ and $\forall x \in X : x \leq e$ minimizes $\sum_{x \in X} \sqrt{x}$ if*

- $\lfloor d/e \rfloor$ *elements of* X *have value* e,
- *one element of* X *has value* $d - \lfloor d/e \rfloor \cdot e$, *and*
- *all other elements of* X *have value zero.*

Proof: The proof is by contradiction. Let X be defined as to the items above, and let $Y \neq X$ with $\sum_{y \in Y} y = d$ and $\forall y \in Y : y \leq e$ minimize $\sum_{y \in set} \sqrt{y}$ for all finite sets with these properties. Since $Y \neq X$, elements $y_a, y_b \in Y$ exist such that $0 < y_a, y_b < e$. Let y_a be the smallest and y_b be the largest element with this property. We define $Y' \underset{D}{=} (Y \setminus \{y_a, y_b\}) \cup \{y_a - 1, y_b + 1\}$. Then, $\sum_{y' \in Y'} y' = d$ and $\forall y' \in Y' : y' \leq e$. Since for function $f(y) = \sqrt{y}$, the derivative $f'(y) = 1/(2\sqrt{y})$ is monotonely decreasing, we observe $\sqrt{y_a - 1} < \sqrt{y_a} - 1/(2\sqrt{y_a})$ and $\sqrt{y_b + 1} < \sqrt{y_b} + 1/(2\sqrt{y_b})$, which yields $\sqrt{y_a - 1} + \sqrt{y_b + 1} < \sqrt{y_a} + \sqrt{y_b}$. Thus,

$$\sum_{y' \in Y'} \sqrt{y'} = \sum_{y \in Y, y \neq y_a, y_b} \sqrt{y} + \sqrt{y_a - 1} + \sqrt{y_b + 1} < \sum_{y \in Y} \sqrt{y} \ ,$$

in contradiction to the minimality of Y. ∎

3   Upper Bound

The tiled codes in Fig. 2 partition the updates into blocks of size $(N-2) \times s_j \times s_k$. Since the i-loop is not blocked, we say that the tiles proceed in i-direction. Alternatively, tiling can block the i- and k-loops so that tiles proceed in j-direction, or the i- and j-loops so that tiles proceed in k-direction. In either case, s_i, s_j, and s_k denote the tile size in directions i, j, and k. Below, we analyze the number of capacity misses assuming row-major storage order. For column-major storage order, the argumentation is analogous, except that i- and k-directions are interchanged.

The following analysis applies to both the Jacobi and Gauß–Seidel codes since tiling preserves the data dependencies of Gauß–Seidel, and since accesses to array B in the Jacobi code do not incur cache misses if a write-through cache (cache-bypassing) is used.

Tiling in k-direction Updates are carried out subplane-by-subplane, for increasing values of k. Here, similar to Sect. 2, a subplane is a set of array elements with same k index that are accessed in the present tile. Since an update refers to data from at most three subplanes, only three subplanes must be kept in cache instead of the whole tile [7].

Nevertheless, since cache line direction equals tile direction, it is not possible to keep exactly three planes in cache. Instead, the cache must hold up to $2L(s_i + 2)(s_j + 2)$ data, with factor 2 reflecting the possibility that $A[i, j, k]$ and $A[i, j, k-1]$ belong to different cache lines.

Capacity misses occur at tile boundaries only. If the updates of $A[i, j, k]$ and $A[i+1, j, k]$ (analogously of $A[i, j, k]$ and $A[i, j+1, k]$) belong to different tiles, then both the cache line of $A[i, j, k]$ and the cache line of $A[i+1, j, k]$ must be loaded twice. Of the four misses, two are capacity misses. Since data from the same cache line are reused in successive planes, the two misses occur only once during the processing of L planes, so that the total number of capacity misses can be estimated by

$$(2/L) \cdot \text{total area of tile boundaries} \leq (2/L) \cdot (N^3/s_i + N^3/s_j) .$$

This bound is minimized for $s_i = s_j = \lfloor \sqrt{C/(2L)} \rfloor - 2$, which implies about $(2/L) \cdot N^3 \cdot 2 \cdot (\sqrt{2L}/\sqrt{C}) = 5.7 \cdot (N^3/\sqrt{LC})$ capacity misses. Taking cold misses into account, this value is off the lower bound by a factor of at most

$$\frac{N^3 - 8 + 5.7N^3/\sqrt{LC}}{N^3 - 8 + 0.35N^3/\sqrt{LC}} = 1 + \frac{5.35N^3/\sqrt{LC}}{N^3 - 8 + 0.35N^3/\sqrt{LC}} < 1 + 5.4/\sqrt{LC} .$$

Tiling in i- or j-direction We refer to i-direction; tiling in j-direction is analogous. In both cases, the cache lines proceed orthogonal to the tiles, and thus are cut by tile boundaries.

Again, the cache must hold about three subplanes of A (here: set of array elements with same i index), although the exact value is somewhat different. On one hand, it is sufficient that the cache holds about two subplanes instead of three since after updating $A[i, j, k]$, element $A[i-1, j, k]$ can be removed, and before updating $A[i, j, k]$, element $A[i+1, j, k]$ is not yet needed. On the other hand, cache lines must be stored completely, even if they are cut by tile boundaries and part of the elements is not accessed in the present tile. Hence, $C > rs_j s_k$, for some value r with $2 < r < 3$. Below we use $r = 3$, which somewhat overestimates the real cost (for reasonably large s_j, s_k).

Capacity misses occur at tile boundaries. We distinguish three cases, which are depicted in Fig. 3. The figure shows a plane of A for any fixed i. Referring to the figure, we denote tile boundaries with fixed j ($1 \leq i, k \leq N-2$) as horizontal, and tile boundaries with fixed k ($1 \leq i, j \leq N-2$) as vertical.

Cache lines along horizontal tile boundaries (such as lines a and b) must be loaded twice during the overall computation; after loading, they are used in L iterations of the k-loop. Of the four misses, two are capacity misses. Thus, the total number of capacity misses at horizontal tile boundaries is $(2/L) \cdot N^3/s_j$.

At vertical boundaries, cache lines that are cut (such as c) incur one capacity miss since each line must be loaded twice. Cache lines that end at tile boundaries (such as d and e) incur one capacity miss each, since both lines must be loaded twice. Hence, the case that cache lines are cut induces less misses than the case that cache lines end at tile boundaries. We assume here that all cache lines are cut, which can be achieved by padding the array. Then, N^3/s_k capacity misses are taken at vertical tile boundaries.

Adding the values of horizontal and vertical boundaries, we get a total of

$$(2/L) \cdot N^3/s_j + N^3/s_k$$

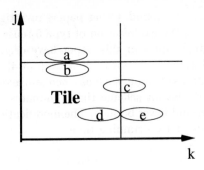

Fig. 3. Cache misses at tile boundaries

capacity misses. For $s_j = s_k \approx \sqrt{C/3}$, this value equals $\sqrt{3}(1+2/L) \cdot (N^3/\sqrt{C})$.

While Rivera and Tseng restrict consideration to square tiles [7], the following analysis suggests that rectangular tiles $(s_j \neq s_k)$ perform better. Substituting $s_j = C/(rs_k)$ into the above formula yields

$$f(s_k) = (2rs_k N^3)/(LC) + N^3/s_k$$

capacity misses. This function is minimized for

$$f'(s_k) = (2rN^3)/(LC) - N^3/s_k^2 = 0 \ ,$$

that is, for $s_k = \sqrt{(LC)/(2r)}$ and $s_j = \sqrt{(2C)/(rL)}$. Here, $s_k = (L/2) \cdot s_j$, that is, rectangular tiles of shape $(N-2) \times s \times (sL/2)$ minimize the number of capacity misses. For these tiles, $2\sqrt{2r} \cdot (N^3/\sqrt{LC}) \approx 4.9 \cdot (N^3/\sqrt{LC})$ capacity misses are taken, which is slightly less than the number of capacity misses for tiling in k-direction, and less than the number of capacity misses for square tiles. Taking cold misses into account, i-direction tiling is off the lower bound by a factor of at most $1 + 4.6/\sqrt{LC}$, which is very close to one.

4   Related Work

Tiling is a well-known compiler optimization [6, 10]. Nevertheless, Douglas et al. [1] observe that current production compilers are not able to tile stencil codes. The problem is partly due to differences in the way tiling works for stencil codes as opposed to dense linear algebra codes (the focus of compiler research). In tiled stencil codes, the data sets of neighbored tiles overlap, whereas in dense linear algebra codes, the data sets are disjoint.

Several papers deal with tiling specifically for stencil codes. Closest to ours is work by Rivera and Tseng [7] who, like us, consider one iteration of the time loop for 3D arrays. Tiling is studied experimentally, including a discussion of tile size selection and array padding to reduce conflict misses. Rivera and Tseng observe that tiling speeds up 3D stencil codes by $17 - 121\%$, but they do not relate their result to a lower bound. Other papers investigate tiling schemes for the time loop, that is, for the exploitation of type 5 reuse [1, 2, 5, 8, 9].

Part of the proof techniques in this paper have originally been developed for the simpler case of 2D stencils with $2N > C$ in [3, 4]. These references also discuss minor modifications of tiling, such as a snake-like order of tile evaluation, and a sophisticated data layout scheme that eliminates the L factor from the quotient between upper and lower bounds. The modifications can be generalized to 3D, but have little impact on running time.

5   Conclusions

This paper has analyzed the numbers of cold and capacity misses in one time step of the Jacobi and Gauß-Seidel kernels on 3D arrays. We have proven a lower bound, and have shown that the well-known technique of tiling gets very close to this bound. Moreover, we have compared different tiling schemes, and observed that tiles of shape $(N{-}2) \times s \times (sL/2)$ perform best. Future work should investigate this claim experimentally, including padding.

References

1. C. C. Douglas, U. Rüde, J. Hu, and M. Bittencourt. A guide to designing cache aware multigrid algorithms. In *Concepts of Numerical Software, Notes on Numerical Fluid Mechanics*. Vieweg-Verlag, 2001. To appear.
2. C. E. Leiserson, S. Rao, and S. Toledo. Efficient out-of-core algorithms for linear relaxation using blocking covers. *Journal of Computer and System Sciences*, 54(2):332–344, Apr. 1997.
3. C. Leopold. On optimal locality of linear relaxation. To appear in Proc. IASTED Int. Multi-Conf. on Applied Informatics, 2002.
4. C. Leopold. On optimal temporal locality of stencil codes. To appear in Proc. ACM Symp. on Applied Computing, 2002.
5. C. Leopold. An analytical evaluation of tiling for stencil codes with time loop. To appear in Workshop-Proc. of Int. Parallel and Distributed Processing Symp. (4th Workshop on Advances in Parallel and Distributed Computational Models), 2002.
6. G. Rivera and C.-W. Tseng. A comparison of compiler tiling algorithms. In *8th Int. Conf. on Compiler Construction*, pages 168–182. LNCS 1575, 1999.
7. G. Rivera and C.-W. Tseng. Tiling optimizations for 3D scientific computations. In *Proc. Supercomputing*. IEEE, 2000. Available at http://www.supercomp.org/sc2000/Proceedings/start.htm.
8. S. Sellappa. Cache-efficient multigrid algorithms. Master's thesis, University of North Carolina at Chapel Hill, Dept. of Computer Science, 2000.
9. Y. Song and Z. Li. New tiling techniques to improve cache temporal locality. In *Proc. of the ACM SIGPLAN Conf. on Programming Language Design and Implementation*, pages 215–228, 1999.
10. M. E. Wolf and M. S. Lam. A data locality optimizing algorithm. *SIGPLAN Notices*, 26(6):30–44, 1991.

A Distributed Co-operative Problem Solving Environment

Mark Walkley, Jason Wood, and Ken Brodlie

School of Computing, University of Leeds,
Leeds, LS2 9JT, UK.
{markw,jason,kwb}@comp.leeds.ac.uk
http://www.comp.leeds.ac.uk

Abstract. Scientific research is often multidisciplinary in nature and hence large projects are frequently collaborative with participants from several separate research centres. Rather than being restricted to infrequent dissemination of results and meetings a framework is described for embedding scientific computing applications within a collaborative problem solving environment. This allows users to combine their expertise in an interactive visual environment. Separate users can collaboratively steer and visualize data from a numerical simulation by embedding the simulation within the IRIS Explorer visualization system. By making use of this system the user has access to the COVISA toolkit which facilitates collaboration between separate users of IRIS Explorer. The flexibility of this system makes it straightforward to visualize and control as many aspects of the solution process as are desired.

1 Introduction

Major scientific research is today carried out by teams, often at geographically separate locations. A drawback to this typical scenario is that while results can be distributed and discussed, it is often difficult to interact constructively without physically meeting. This is particularly true of work involving numerical simulation where the collaborating researchers may include experts in the physics of the problem, the numerical model of the problem, and the visualization of the data produced by the numerical model. It is the aim of this work to provide a collaborative framework from within which the different researchers can combine their expertise and analyse the various aspects of the work.

Modern scientific computing applications typically produce far more data than can be easily handled without the use of further scientific tools. Effective visualization of this data allows the users to extract information of interest in an efficient manner, for example to quantify the performance of the algorithm or the quality of the solution produced. A number of visualization tools exists for processing of scientific data. These tools can broadly be split into two categories: turnkey systems, where the tool is generally tailored to a specific problem domain and offers a simple to use interface at the expense of being non extensible; and modular visualization environments (MVEs) which offer a more generic approach

P.M.A. Sloot et al. (Eds.): ICCS 2002, LNCS 2329, pp. 853–861, 2002.

to visualization, are more complex tools to use, but are user extensible. It is this second category of tools that is of most interest for this work since their general extensibility allows the addition of components, such as numerical simulation codes, into a single environment along with the provided visualization functionality. There are several commercial MVEs available, e.g. IRIS Explorer [1] and AVS [2], and several open source systems, e.g. OpenDX [3] and SCIRun [4]. All of these tools work along similar lines offering the user a library of modules, presented as graphical blocks, that represent a variety of functions. These modules can be placed onto a workspace and graphically connected to define the route of the data through the modules. The modules implement different algorithms changing the data from its raw state to some viewable image or geometry. In practice it is common to view this visualization as a post-processing stage in scientific computing; however this approach leads to a cycle of probably lengthy computational simulation time followed by relatively fast visualization of the data. In some cases it may have been apparent quite early on that modifications to the numerical algorithm were required but this is not possible once the simulation has begun.

Computational steering allows the user control of the simulation process from within the visualization environment and, simultaneously, access to the data produced by the simulation which can be visualized and analysed while the simulation is progressing. This allows parameters of the model to be changed during the simulation if necessary. Some previous work in this area has used a web-based interface [5] however, as previously discussed, the visual environment of IRIS Explorer, and systems such as SCIRun, naturally extend to this purpose.

Collaborative visualization allows geographically separate users to share data, and hence co-operate in its visualization and analysis. If the data directly from the simulation can be shared then users can apply different visualization techniques at each location rather than looking at a static display provided by one of the participants. This approach is obviously flexible and allows users with possibly different backgrounds and motivations to explore data in their own way. A number of projects have been undertaken to investigate collaborative visualization using a variety of systems. One of the simplest methods for collaboration is to replicate the user interface and then control access to the system through the use of a *token*. Only the user in possession of the token can make changes to the system, other collaborators must simply sit and watch until the token is released. This approach, while being simple to comprehend, limits the potential for users to visualize the data in their own way as well as taking part in the collaborative session. This mechanism is employed by the COVISE [6] visualization system to provide collaboration but in general can be implemented on top of any visualization tool through the use of application sharing technologies such as that contained within tools like NetMeeting [7]. An alternative approach is to add components to existing visualization systems that allow them to be used in a collaborative context. This is particularly applicable to the MVE class of systems due to their extensibility and open architecture. COVISA [8] was the first extension of a modular visualization environment, using IRIS Explorer, to provide collaboration. It provides modules that allow the sharing of data and control between connected IRIS Explorer sessions, as well as providing a collaborative API for module writers. A similar approach using AVS Express was used in the MANICHORAL project [9].

In this work we will combine the advantages of visualization, computational steering and collaboration for scientific computing applications into a framework within IRIS Explorer that allows the user(s) simultaneous control over every aspect of the process. This approach provides detailed visual data on the performance of the algorithm and allows adjustments to be made to aspects of the algorithm while it is progressing. The combination of computational steering and collaborative visualization allows separate users access to both the data produced and the control of the simulation process, resulting in a collaborative problem solving environment within IRIS Explorer.

2   Numerical Application

Fluid transport, or convection, is an important physical process, for example in atmospheric dispersion [10], requiring expertise in both the physical processes involved and the numerical modeling of those processes. The numerical model combines transport of scalar chemical components with reaction between those components and is a challenging problem involving physical processes with widely different timescales. A prototype for such a model is that of passive scalar convection of a single component. Here this is modeled using a standard streamline-upwind finite element method [11]. The problem is typified by a convection velocity which directs the transport process, and the solution typically contains sharp layers dividing regions of different concentration.

The collaborating researchers will have different motivations in the analysis of data from such a model. For example, the chemist may wish to monitor the concentration of the component throughout the domain whilst the numerical analyst may wish to view the errors in the numerical solution at that point. In particular by monitoring the process as it proceeds and being able to interrogate the simulation in a variety of ways, it may become apparent quite early on that problems are occurring. Decisions can then be taken as to whether the simulation should be stopped and restarted or, more usefully, if parameters of the numerical model can be adjusted and the simulation continued. These issues may be resolved by taking a more flexible view of the visualization process and integrating the simulation within the visualization environment.

3   Visualization using IRIS Explorer

IRIS Explorer is a general-purpose modular visualization environment. A visualization *map* is built by combining *modules* into a pipeline through which the simulation data is passed and converted into a rendered image. A module *librarian* is available that contains many of the standard visualization techniques, such as isosurface computation or cutting planes. Used in the simplest way the data set is supplied at the start of the dataflow pipeline and converted into a rendered image which is viewed at the other end. The disadvantage of this *standalone* approach is that visualization is a post-processing stage and that the overall work cycle is limited to successive computations followed by visualization.

IRIS Explorer is an open system and the user can also design their own visualization techniques and compile them into modules that can be included directly in the visual environment, and linked into the pipeline. This facility opens up the possibility of including modules that are not necessarily directly associated with visualization and here, in particular, modules containing numerical simulation algorithms.

4 Computational Steering

The aim of this work is to fully integrate the computation and visualization stages previously discussed and so allow synchronous computation and visualization. Here the numerical application is *wrapped* into an IRIS Explorer module and so can be included directly in the dataflow pipeline. An obvious advantage of such an approach is the immediate availability of the full visualization capabilities of IRIS Explorer for data generated during the computation. Further modules may be connected to the pipeline at any time allowing alternative, or more detailed, visualizations of the numerical model while the computation is taking place. A further advantage of including the numerical application as a module is that parameters can then be fed into the module from the IRIS Explorer environment which control its operation. The design of the simulation module allows the creation of a module interface and hence access to any parameters of interest. These parameters can be features of the numerical algorithm, e.g. error tolerance or timestep, or the physical model, e.g. density or convection velocity, as well as controlling the visualization process. This allows the users simultaneous and flexible control over the physical model, the numerical model and the data visualization.

The numerical application described in Section 2 is shown in Figure 1 together with the control panel interface used to steer the simulation and a visualization of a solution isosurface.

5 Collaborative Visualization and Steering

IRIS Explorer allows synchronous collaborative visualization through the COVISA modules [8]. These enable a team of people, at different internet locations, to link their individual IRIS Explorer dataflow pipelines. Essentially the data can flow between pipelines, as well as within each pipeline. Special 'Share' modules are wired by the researchers into their pipelines to achieve this data transfer. Different collaborative scenarios can then be programmed. For example, results generated in the pipeline of researcher A can be fed across to a (possibly different) pipeline of researcher B, as illustrated in Figure 2, for further analysis. Researchers with identical pipelines can share control of any module in the

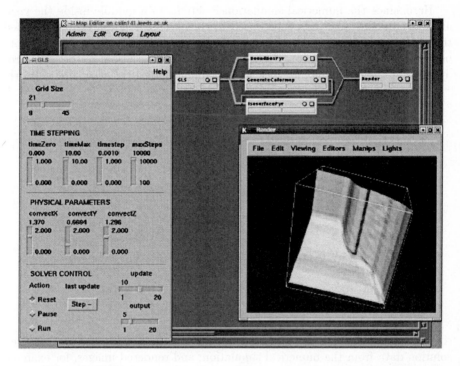

Fig. 1. IRIS Explorer map and visualization of the numerical application

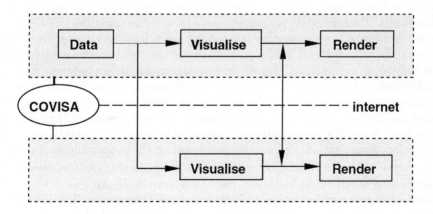

Fig. 2. Schematic of collaborative visualization

pipeline, by wiring in shared parameter modules at the appropriate point.

Here, since the numerical simulation functions as a module inside the visualization pipeline, the parameters used to steer the simulation can be supplied through the collaborative modules, as illustrated in Figure 3, allowing remote users to collaboratively steer the simulation as well as visualize the data produced by it. Several different types of data can be transferred across the network:

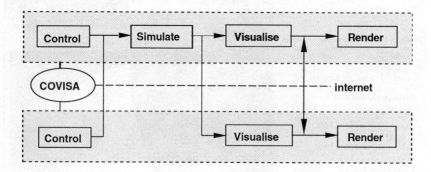

Fig. 3. Schematic of collaborative computational steering

single parameters, such as those used to steer the simulation; raw unstructured solution data from the numerical simulation; and rendered images, for example in the IRIS Explorer geometry format, or more standard formats such as gif files. Depending on the network bandwidth available, the most appropriate can be selected by the user to produce a responsive system. A useful collaborative scenario for the work considered here is that the numerical application, which is generally computationally intensive, can be run on the most powerful computer, while a *remote* user, who may have a relatively small computational resource, can control the simulation and receive data to visualize using the COVISA modules. In such a case it may be impractical for a user to monitor the process continuously. However it is also possible to use the collaborative modules to log in and out of a simulation, indeed the simulation can be run unmanned and the collaborative modules used to connect, or reconnect, a remote user. The user can monitor progress periodically and steer the simulation when necessary. This also avoids data having to be sent across the network unless it is required by the user. Note that computationally intensive visualization operations, such as isosurface computation, can also be performed on the most suitable platform.

Figure 4 shows a collaborative visualization session between two users for the application described in Section 2. The first, shown in Figure 4(a), is hosting the application and viewing an estimate of the error in the solution. Tetrahedral cells with an error exceeding a specified tolerance are visualized. The data from the simulation is shared with a second user, shown in Figure 4(b), who is viewing an isosurface of the computed solution. The parameters that control the simulation are also shared collaboratively allowing either user to steer the computation.

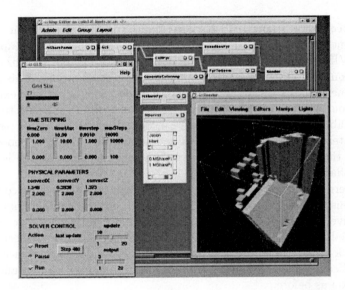

(a) User A running the simulation and visualizing the error

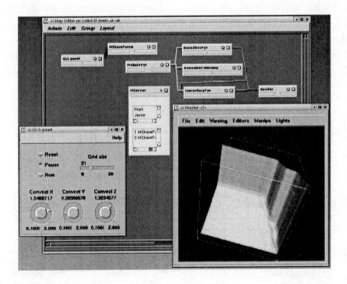

(b) User B collaboratively controlling the simulation and visualizing the solution

Fig. 4. A collaborative session using IRIS Explorer

6   Conclusions

It has been shown how IRIS Explorer can be used as a problem solving environment for numerical simulations. By using the collaborative capabilities of the COVISA modules this is extended to allow steering and visualization of the simulation by colleagues at geographically separate locations. This interactive environment allows researchers to share not only results but also experience of the various aspects of the numerical model in a simple and effective manner. The collaborative scenario developed can be tailored to the resources available, both in terms of network bandwidth and computing power, by choosing which parts of the computation are performed on each machine and the level of data that is shared collaboratively.

Modern scientific computing applications are typically highly computationally intensive and it is unrealistic to assume that they can always be run in a interactive manner. However the system developed will naturally extend to the unmanned mode described in Section 5. The collaborative connection can be used by a remote user to log in and out of the simulation and once connected the user can visualize the data produced in any manner desired and steer the simulation as before.

The computational Grid represents a new paradigm for distributed computing and recent research in computational steering and simulation is frequently concerned with allowing all, or parts, of the computation to be distributed in this way [5, 12]. Our current work in this area is focusing on extending the application to use the Globus *middleware* [13] to allow the computing to be spread over a network of computational resources. The scenario under development, depicted in Fig. 5, will use Globus to distribute the computationally intensive numerical

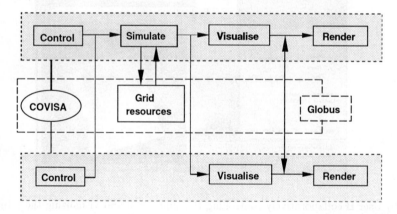

Fig. 5. Schematic of grid-based collaborative computational steering

simulation, with an interface from IRIS Explorer allowing collaborative steering and visualization of the process to be performed.

References

1. Foulser, D.: IRIS Explorer: A Framework for Investigation. Computer Graphics. 29(2) (1995) 13-16
2. Advanced Visual Systems: http://www.avs.com/
3. Open Visualization Data Explorer: http://www.opendx.org/
4. Miller, M., Hansen, C., Johnson, C.: Simulation Steering with SCIRun in a Distributed Environment. Lecture Notes in Computer Science. Springer-Verlag (1998)
5. Allen, G., Benger, W., Goodale, T., Hege, H.-C., Lanfermann, G., Merzky, A., Radke, T., Seidel, E., Shalf, J.: The Cactus Code: A Problem Solving Environment for the Grid. In: Proceedings of the Ninth IEEE International Symposium on High Performance Distributed Computing (2000)
6. COVISE: http://www.hlrs.de/organization/vis/covise/
7. NetMeeting: http://www.microsoft.com/windows/netmeeting/default.asp
8. Wood, J., Wright, H., Brodlie, K.W.: Collaborative Visualization. In: IEEE Visualization Conference (1997) 253-260
9. Duce, D.A., Gallop, J.R., Johnson, I.J., Robinson, K., Seelig, C.D., Cooper, C.S.: Distributed Cooperative Visualization - The MANICHORAL Approach. In: Eurographics-UK. Leeds (1998)
10. Hart, G., Tomlin, A., Smith, J., Berzins, M.: Multi-scale Atmospheric Dispersion Modelling by the Use of Adaptive Gridding Techniques. Environmental Monitoring and Assessment. 52 (1998) 225-228
11. Johnson, C.: The Finite Element Method. Wiley (1990)
12. Miller, M., Moulding, C., Dongarra, J., Johnson C.R.: Grid-enabling Problem Solving Environments: A Case Study of SCIRun and Netsolv. In: Proceedings of HPC 2001 (2001)
13. The Globus Project: http://www.globus.org

The Software Architecture of a Problem Solving Environment for Enterprise Computing

Xu Jun Gang, Wang Hong An, and Dai Guo Zhong

Chinese Academy of Sciences, Institute of Software,
Intelligence Engineering Lab., P.O. 8718, 100080 Beijing, China
xujungang@sina.com, wha99@sina.com,
guozhong@admin.iscas.ac.cn

Abstract. Problem Solving Environment (PSE) is a new and diverse area of research, and till now it resists simple universally accepted definitions. In brief, PSE is a complete and integrated computing environment in which new application can be constructed, compiled and executed. This paper describes the concept and software architecture of a PSE for Enterprise computing (E-PSE). The E-PSE is developed to provide transparent access to heterogeneous distributed computing resources in one enterprise, and is to improve research productivity by making it easier to construct, compile and run computer applications so that the enterprise can achieve more advanced economic benefit as soon as possible. The E-PSE includes four modules: Visual Enterprise Modeling System, Visual Application Construction Environment, Application Wrapping & Integrating Tool and Application Runtime Environment. The application of the E-PSE in an oil-refining enterprise is presented. Related work about PSE is also described.

1 Introduction

The current advances in networking protocols (including ATM and fast Ethernet), software development tools, and emerging WWW technologies have enabled the development of a cost-effective, high performance distributed computing environment, network-based computing [1]. But, the available software tools for a high-performance computing environment still require detailed understanding of the underlying architecture and components of application. Writing a parallel and distributed program overwhelms most of the users due to the complexity of communication and synchronization issues [2]. Furthermore, information technologies have almost been applied into every field, but it is impossible to require scientists and engineers in each field to learn computer knowledge and programming. In order that scientists and engineers can devote themselves to research work of their own field, and improve research efficiency and achieve advanced research production, it is necessary to develop an application developing environment for their field. This is the problem which computer experts and experts in the other field face.

This research is supported by major research issue (79931000) of National Natural Science Foundation of China.

P.M.A. Sloot et al. (Eds.): ICCS 2002, LNCS 2329, pp. 862–871, 2002.
© Springer-Verlag Berlin Heidelberg 2002

As advanced programming language replaced the original machine language owing to shielding the computer hardware, it is believed that there must exist one tool that can shield computer hardware and software to replace current complicated programming work, and can be used to develop application for any field. According to this problem, computer experts have done much research work. Now we have had the answer, which is to develop Problem Solving Environment for each field.

In engineering application field, there also exists the problem that enterprise supervisors and engineers don't know much about information technologies. How to apply information technologies into enterprise better, and how to make use of the advantages that enterprise personnel are very familiar with the enterprise business, which are the problems that computer experts and enterprise engineers need to solve. With the development of global economy, the traditional business mode of enterprise is increasingly confronted with the challenge of market globalization. Day after day people are realizing the importance that information technologies can reengineer traditional enterprise business process. And enterprise must organize and regulate production quickly, achieve the whole optimization of business process, and improve the reactivity and competition capability so that it can stand at an advantageous station in market competition, and achieve more economic benefit. So, to develop PSE for enterprise computing is very necessary.

According to some applications of PSE in engineering field, this paper puts forward the concept of a PSE for Enterprise computing (E-PSE), and describes its features and software architecture. The E-PSE includes four modules: Visual Enterprise Modeling System (VEMS), Visual Application Construction Environment (VACE), Application Wrapping & Integrating Tool (AWIT) and Application Runtime Environment (ARE).

This paper is organized as follows. Section 2 presents the concept, main purpose and features of the E-PSE. Section 3 describes the software architecture of the E-PSE and modules that the E-PSE includes. Section 4 describes the application of E-PSE in an oil-refining enterprise. Section 5 introduces some related work. Section 6 contains conclusions and prospects.

2 The Concept and Features of the E-PSE

According to some applications of PSE in engineering field, this paper brings forth the concept of an E-PSE. An E-PSE is defined as follows, "An E-PSE is one computer-integrated platform for enterprise computing, and it is a complete and integrated Problem Solving Environment in which enterprise application system can be constructed, compiled and executed. It provides various computing tools that can solve problems existed in enterprise, which make supervisors and engineers of an enterprise keep away from the ignorance of computer technologies and devote all themselves to product development, business process management and improvement."

The main purpose of the E-PSE is to provide a model-driven and component-based, application integrating and developing environment. Firstly, through enterprise business process and application modeling, the E-PSE can realize the seamless integration from enterprise business process to enterprise application, and support the integration of the existing enterprise applications and development of new enterprise applications. Secondly, the E-PSE provides a set of wrapping and integrating tools for the existing and legacy applications of an enterprise to take full advantage of the existing resources and protect the previous investment. Thirdly, the E-PSE provides a component-based application development method. One application can be represented as one application flow graph that is composed of some icons of component that can realize certain business logic. Fourthly, the E-PSE provides a set of component repository for certain applications, and the component repository can be extended at any time when it needs. Furthermore, the model-driven and component-based application integrating and developing method provided by the E-PSE can meet the requirements of continuous improvements of enterprise business process. In the E-PSE, business model describes enterprise business process, and application model describes the implementing method of an application. When business process changes, the only thing to do is to change the components or the links between the corresponding components, then the change of the application can be implemented, which diminishes the influences of changes of business process on the relative applications as possible.

The E-PSE has main features as follows: 1) It can make computer resources (including computer hardware, networks and legacy systems) in one enterprise get used sufficiently. 2) It can fit the changes of business object of the enterprise, including the changes of business process and improvement of technical implementation. 3) It can relieve supervisors and engineers from tedious computer technologies and make them concentrate on new management concepts, optimal business practice and process technology, which can improve production efficiency to the maximum extent. 4) It is not necessary to program for application developer to construct new application system, what he needs to do is just to transfer his thoughts into application flow graph in E-PSE, and to improve it in the future continuously.

3 The Software Architecture of the E-PSE

The E-PSE is composed of four modules: Visual Enterprise Modeling System (VEMS), Application Wrapping & Integrating Tool (AWIT), Visual Application Construction Environment (VACE) and Application Runtime Environment (ARE). The VEMS is used to build and modify enterprise model. The AWIT wraps third party application, standard component repository and legacy system into CORBA components, and places them into Component Repository in the VACE. The VACE parses the enterprise model and constructs the corresponding application model and then utilizes the existing software components in the Component Repository to construct new application system, or improve the existing system. Once a new application is completed by the VACE, and then it is delivered to the ARE and is scheduled to run on distributed computer resources. The software architecture of the E-PSE is illustrated in Fig. 1.

3.1 VEMS

The main function of the VEMS is to build and modify enterprise business model. The VEMS consists of two parts: Modeling Tool and Reference Model Repository. We adapt ARIS (Architecture of integrated Information Systems) Toolset as the Modeling Tool, which was published by IDS Scheer AG Corp., Germany. ARIS modeling concept keeps ahead of the world, which was first brought forth by Prof. Scheer in 1992 *(see Scheer, Architecture of integrated Information Systems, 1992)*. It provides an integrated method to represent one enterprise through 5 kinds of views: Function view, Data view, Organization view, Process view and Product/Service view.

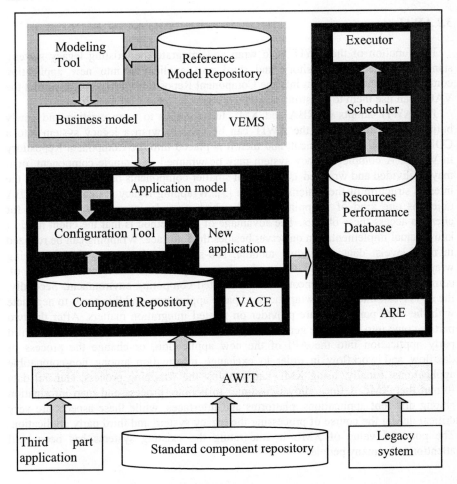

Fig. 1. The software architecture of the E-PSE

When we are building the enterprise business model, we usually look for similar model in the Reference Model Repository first, and then modify it to the required model, it is not necessary to build them from scratch. After customizing one reference model into one enterprise's own model, the enterprise can adjust the customized model according to the changes of market requirements. The Reference Model Repository is based on the analysis of a lot of successful enterprise paradigms and is a kind of knowledge base that collects the experience of many experts in the corresponding industry. The Reference Model Repository describes the specialized requirements of one industry, and so it has comprehensive instructional significance for the enterprise requirements definition and other aspects in that industry. The enterprise feedback information and customer questionnaire make clear that using reference model can decrease workload of defining enterprise requirements dramatically.

3.2 AWIT

The function of the AWIT is to wrap and integrate the existing legacy system, standard component repository and third part application into new application components, and place them into the Component Repository in the VACE so that the VACE can use them to construct new application.

The AWIT adopts CORBA and XML technologies to wrap existing and newly built components. Firstly, the AWIT can be used to wrap a legacy system into a CORBA-compliant component, and then it is placed into the Component Repository in VACE. A complete legacy system may be wrapped as a single component, or it may be divided and wrapped as a series of smaller components on assumption that the internal structure of the system is known [3]. Wrapping legacy system into CORBA object is one kind of wrapping method, which provides convenient interfaces for client to access these objects. The advantage of this method is that the client needn't know actual implementation on server besides the interface. Wrapping can be realized in multi-levels: data, individual module, sub-system and system [4]. After being wrapped into a CORBA object, the legacy system can be reused in form of component in the heterogeneous and distributed computing environment. Secondly, the AWIT can integrate or wrap the third party application, and this needs to negotiate with the third party software provider on related integration matters. After the two parties come into terms, the general adopted method is to convert the API of the third party application into the API of the new application, or change the process of dataflow and workflow in order to exchange information among non-compatible applications. Finally, using XML can simplify the wrapping process. General data format that XML defines allows customer to denote, process and map information among different applications, platforms and interfaces, while these aspects are very critical during the course of processing the legacy system and third party application. The potential value of XML in integrating the existing systems has been paid attention to by many people.

3.3 VACE

VACE includes two parts: Component Repository and Configuration Tool. The main function of the VACE is to integrate the components that correspond to the business model into a new application by parsing the enterprise business model and utilizing the Configuration Tool. At the same time, the VACE provides the interfaces to real-time database and relative database for applications to realize data accessing.

Component Repository. The Component Repository is a warehouse that stores all kinds of components, mainly including software component and sub-system. The Software component is one basic software object such as COM object, CORBA object, Java Bean and so on, which has its own interface describing file, attributes, methods and events; the sub-system is system resources that consist of some kinds of software components. Users can operate the Component Repository in the same way as operating database, such as retrieving component that has special function, adding and deleting the corresponding component and etc.

Configuration Tool. The Configuration Tool is a visual component composition tool, it makes users create and edit application through composing some kinds of component. In active editor area, a component is represented as a clickable and draggable graphical icon, each such icon includes component name and a set of markers for logical ports. The user can select one new component from the Component Repository and add it into the application sketch, then link it into the application flow graph. In a graph-based programming environment, an application is defined as a directed graph where nodes denote one component and links denote communication and synchronization between nodes. The Configuration tool mainly include key modules as follows:

– Project manager
– Drawing tool
– Control manager
– Real-time database manager interface
– Script editing environment
– Report form defining environment
– Variable manager
– Page editing interface
– Control network interface

Fig. 2 shows the windows of the Configuration Tool. The Project manager implements the operations of creating, deleting and updating of one project. The Drawing tool supports the familiar drawing and rendering methods. The Control manager calls the interface of system registration form, and it can make controls visually display in the page editing interface. If necessary, use the real-time database manager interface to connect with the real-time database manager. The Script editing environment is organized with the format of object, and it provides the method of editing control logic. The Report form defining environment can define the format and content of the real-time report form, which is managed by the Project manager, and executes in the real-time database manager. The Variable manager is an independent module, and it can be installed and published independently, we can use it to execute data gathering and physical variable configuration. The Page editing interface provides one blank template, on which we can build an application flow graph to construct a new application. The Configuration tool can define data source for data gathering and publication through calling the Control network interface.

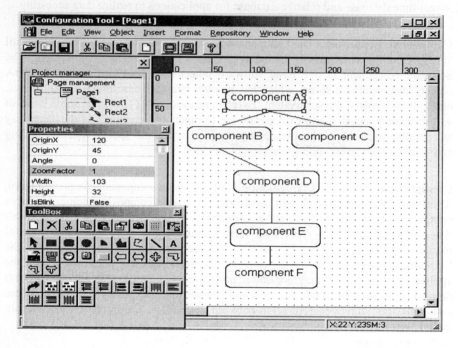

Fig. 2. The windows of the Configuration Tool

3.4 ARE

The main function of the ARE is to maintain the distributed computable resources of an enterprise, and to schedule the applications constructed by the VACE so that they can run on these resources.

The Computable resources in distributed heterogeneous system should be registered in Resources Performance Database before they are used. The Resources Performance Database maintains information about these resources. The information mainly includes resources attributes and parameters [2][3]. These attributes are grouped into two parts: 1) Static attributes, i.e. Host name, IP address, Architecture type, OS Type, CPU speed, Hard disk and Memory size, Number and type of available network interface, etc.; 2) Dynamic attributes, i.e. recent load measurement, available memory, point-to-point network latency, available network bandwidth, etc.

The Scheduler is the most important part of the ARE. It interprets the application constructed by the VACE and assigns the current best available resources for running the application in order to minimize the total execution time. The scheduling algorithm computes the predicted time of each component and selects the current best available resources within the enterprise networks. The scheduling decision is based on the software and hardware requirements, location and configuration of resources, up-to-date machine/network loads. Once the scheduling decision is completed, the application is delivered to the Executor and runs on the resources under the control of the Executor.

4 The Application of the E-PSE in an Oil-Refining Enterprise

Supported by one National Research Project of China, we utilize the model-driven and component-based method to implement the planning and development of some information systems in an oil-refining enterprise. According to process enterprise automation chain in the ISO specification and the whole framework of the oil-refining enterprise, we present the reference models of the oil-refining enterprise. We first use the Modeling Tool to build the business models of the demonstrated enterprise mentioned above, and then refine the reference business models that fit the oil-refining enterprise from these models. These reference models are helpful for the other oil-refining enterprises to build their own business models. And based on these models, the enterprises can use the Configuration Tool to improve the existing information applications or create new information application, such as the application that can monitor or distribute the process data generating during the production process, and etc. The practicability of these reference models needs proving and improving further in the other oil-refining enterprises in the future.

5 Related Work

PSEs have important significance to the research of each field, so researchers all around the world research PSE in different aspects, such as its infrastructure, reusable components, PSE for special field, or technologies used in PSEs, including distributed collaborative work, visualization, artificial intelligence and so on. The current advances in high-speed networks and WWW technologies have made network computing a cost-effective and high-performance computing environment, which provide more extensive foundation for researching on PSEs.

The current concept of a PSE has its origins in an April 1991 workshop funded by the US National Science Foundation [5][6]. The workshop found that the availability of high performance computing resources, coupled with advances in software tools and infrastructure, made the creation of PSEs for computational science a practical goal, and that these PSEs would greatly improve the productivity of scientists and engineers. This is even truer today with the advent of web-based technologies, such as Java, CORBA and XML for accessing remote resources and sharing information.

In 1994, Gallopoulos ·· Houstis and Rice presented one detailed definition for PSE as follows, "A PSE is a computer system that provides all the computational facilities needed to solve a target class of problems. These features include advanced solution methods, automatic and semiautomatic selection of solution methods, and ways to easily incorporate novel solution methods. Moreover, PSEs use the language of the target class of problems, so users can run them without specialized knowledge of the underlying computer hardware or software. By exploiting modern technologies such as interactive color graphics, powerful processors, and networks of specialized services, PSEs can track extended problem solving tasks and allow users to review them easily. Overall, they create a framework that is all things to all people: they solve simple or complex problems, support rapid prototyping or detailed analysis, and can be used in introductory education or at the frontiers of science [5]."

Since the 1991 workshop, PSE research has been mainly directed at implementing prototype PSEs and at developing the software infrastructure or "middleware" for constructing PSEs. Initially, many of the prototype PSEs that were developed focused on linear algebra computations [7] and the solution of partial differential equations [8]. More recently prototype PSEs have been developed specifically for science and engineering applications. More generic infrastructure for building PSEs is also under development ranging from fairly simple RPC-based tools for controlling remote execution [9] to more ambitious and sophisticated systems such as Legion [10] and Globus [11] for integrating geographically distributed computing and information resources. The architecture of Virtual Distributed computing Environment (VDCE) developed at Syracuse University [12] is very similar to the software architecture of the E-PSE described in section 3. However, VDCE has not modeling module for enterprise, which is more suitable for computing science than enterprise computing.

As one kind of PSE, E-PSE has many features that common PSE has, and has the similar architecture with other PSEs. However, E-PSE has also some difference from other PSEs, such as PSEs for computational science mentioned above. The difference mainly includes: 1) Common PSEs are usually designed to solve very complex computational problems, and the demands for hardware and scheduling algorithms are very high. But, as to E-PSE, these demands are not very high; 2) E-PSE can support dynamic modeling and continuous improvements of business process of enterprise, which is very important feature of E-PSE. But, many other PSEs didn't have this feature, and it is enough for users to use application modeling function of these PSEs; 3) Common PSEs are mainly used to develop new applications to solve specified problems. However, E-PSE is mainly used to integrate and improve the existing applications, And to develop new application is not major function of E-PSE.

In China, research projects on PSE aren't so many as those in other countries, especially in engineering field. We wish our research could facilitate the development of researching on PSE for engineering field.

6 Conclusions and Prospects

In this paper, we present the concept and the software architecture of an E-PSE. The chief advantage of the E-PSE is that it can relieve the supervisors and engineers of an enterprise from tedious computer application technologies and make them concentrate on new management concepts, optimal business practice and process technologies, which can improve production efficiency to the maximum extent. The E-PSE is composed of four modules: VEMS, AWIT, VACE and ARE.

We have put the E-PSE into practice in an oil-refining enterprise of China, and the effects are very remarkable. But, due to various causes, the theory and method about the E-PSE are just put into practice in enterprise in part, it is expected that the E-PSE would be further popularized and applied in enterprise in the future. And we'll improve these theories and methods further.

If one enterprise wants to keep its invincible position in the situation of more and more intense market competition and globalization, it must adjust its business process according to the changes of the market in time, it is a continuously changing course. The E-PSE adapts a model-driven and component-based software development method, which provides technical support for continuous improvements of business process of an enterprise, and can meet the requirements of high-speed developing information technologies that the enterprise faces nowadays. Consequently, researches on E-PSE are very significant and important for the development of the enterprise, and the application prospects of the E-PSE are also very bright, and it is sure that the E-PSE can bring considerable economic benefits and immeasurable value to the enterprise in the future.

References

1 Panda, D. K., Ni, L. M.: Special Issue on Workstation Clusters and Network Based Computing. Journal of Parallel and Distributed Computing, 40 (1997) 1-3.
2 Hariri, S., Topcuoglu, H., Furmanski, W., Kim, D., Kim, Y., Ra, I., Bing, X., Ye, B., Valente J.: Problem Solving Environments. IEEE Computer Society Press (1997).
3 Walker, D. W.: The Software Architecture of a Distributed Problem Solving Environment. Computer Sciences, 10 December (1999).
4 Otte, R., Patrick, P., Roy, M.: CORBA tutorial, Shixian Li., first edition. Tsinghua Press Beijing (1999) (in Chinese).
5 Gallopoulos, E., Houstis, E. N., Rice, J. R.: Computer as Thinker/Doer: Problem-Solving Environments for Computational Science. IEEE Computational Science and Engineering, Vol. 1 (2) (1994) 11-23.
6 Gallopoulos, E., Houstis, E. N., Rice, J. R.: Workshop on Problem-Solving Environments: Findings and Recommentations. ACM Computing Surveys, Vol. 27(2) (1995) 277-279.
7 Casanova, H., Dongarra, J. J.: NetSolve: A Network-Enabled Server for Solving Computational Science Problems. Int. J. Supercomputing Appl., Vol. 11(3) (1997) 212-223.
8 Houstis, E. N., et al.: Parallel ELLPACK Elliptic PDE Solvers. Intel Supercomputer Users Group Conference, Albuquerque (1995).
9 Arben, P., Sprenger, C., Luthi, H. P., Vogel, S.: SCIDDLE: A Tool for Large-Scale Distributed Computing. Technical Report 213, Institute for Scientific Computing, ETH Zurich (1994).
10 Grimshaw, A. S., Nguyen-Tuong, A., Lewis, M. J., Hyett M.: Campus-Wide Computing: Early Results Using Legion at the University of Virginia. Int. J. Supercomputing Appl., Vol. 11(2) (1997) 129-143.
11 Foster, I., Kesselman, C.: GLOBUS: A Meta-computing Infrastructure Toolkit. Int. J. Supercomputing Appl., Vol. 11(2) (1997) 115-128. See also web site at http://www.globus.org/.
12 Topcuoglu, H., Hariri, S., Furmanski, W., Valente, J., Ra, I., Kim, D., Kim, Y., Bing, X., Ye, B.: The Software Architecture of a Virtual Distributed Computing Environment. Proceedings of the Sixth IEEE International Symposium on High Performance Distributed Computing (HPDC-6), Portland (1997).

Semi-automatic Generation of Web-Based Computing Environments for Software Libraries

Pedher Johansson[1] and Daniel Kressner[2]

[1] Department of Computing Science, Umeå University, SE-901 87 Umeå, Sweden
`pedher@cs.umu.se`
[2] Institut für Mathematik, MA 4-5, Technische Universität Berlin, D-10623 Berlin, Germany
`kressner@math.tu-berlin.de`

Abstract. A set of utilities for generating web computing environments related to mathematical and engineering library software is presented. The web interface can be accessed from a standard world wide web browser with no need for additional software installations on the local machine. The environment provides a user-friendly access to computational routines, workspace management, reusable sessions and support of various data formats, including MATLAB binaries. The creation of new interfaces is a straightforward process. All necessary web pages are automatically generated from XML description files. The integration of the control and systems library SLICOT demonstrates the efficacy of this approach.

1 Introduction

Highly reliable and efficient library software is of particular importance for sophisticated engineering solutions. However, there is a gap between the number of existing mathematical routines and those actually used by engineers. A main obstacle for potential users is that, in order to benefit from new software, they typically have to go through the painful process of searching, downloading, installing and understanding. This requires a substantial amount of time; often even *before* the usefulness of the software can be evaluated. The user may moreover not have access to computing facilities and proprietary software.

We have developed a new collection of utilities, referred to as the *Web Computing Utilities* (or briefly, *Webcut*), that addresses these concerns. It provides a solution in the following way. The programmer can make new mathematical and engineering software available *on the web* where it is accessible and executable through standard web browsers. The programmer provides information about the routine parameters and standardized calling routines in so called description files. From this input, Webcut automatically generates HTML pages offering a user-friendly web computing environment. The essential prerequisites of software users are then, to know the type of the problem to be solved and to provide the input data in a convenient way. The use of web computing does not

P.M.A. Sloot et al. (Eds.): ICCS 2002, LNCS 2329, pp. 872–880, 2002.

require any installation of software on the local computer nor does it require any documentation besides that which is integrated into the user interface.

This work is related to a wide variety of projects providing easy access to mathematical software. Among those, the most popular and successful is certainly MATLAB[1] [1]. There exists a web environment for MATLAB [2], its interfacing functionalities, however, are very limited. Others are ATLAS [3], JSPICE [4], MMM [5], Paraweb [6] and WOS [7]; those facilitate Java applets. We feel that such applets unnecessarily limit the range of admissible browsers or devices and thus preclude their use in Webcut. The network-computing system PUNCH [8] is possibly the most closely related project. Here, interfaces are generated by HTML templates quite similar to the XML strategy described in Section 3. While being more mature in general, PUNCH lacks a sound workspace management, i.e., an easy way for the user to administrate and reuse input / output data among several computational tasks. It should be noted that Webcut is not aimed to be a tool for accessing and using globally distributed resources. Although it could be coupled with projects like NetSolve [9] the objectives of this work are focused on handy interfaces.

The paper is organized as follows. Section 2 gives a short description of the web computing environment, from a user's point of view. This includes presentation of the major functionalities and a step-by-step illustration of a computing session. In Section 3, we describe the way in which the programmer has to provide information about the routines to be implemented. Details about the internal design are given in Section 4. Conclusions and future work are presented in Section 5, mainly to show how the software library SLICOT [10, 11] benefited from this work and how other libraries could benefit, too.

2   A User's Point of View

The web computing environment can be accessed from a standard web browser, with no need for additional software installations on the local computer.

In the following, we illustrate how the environment can be used to solve a sample problem, namely to compute the circular convolution of two signals

$$c_k = \sum_{j=1}^{n} a_j b_{k+1-j}, \quad k = 1, \ldots, n \ . \tag{1}$$

The vectors defining the problem can be entered by the user in the web interface. However, default values are provided so that computations can be performed without the need for the user to define an own problem. Input data can also be previously used data or be uploaded from MATLAB binary or plain text files. The underlying software is the SLICOT routine DE01PD based on discrete Hartley transforms [12].

Figure 1 shows the user interface for the routine DE01PD. Here, the vectors have already been entered and the user may choose between computing the

---
[1] MATLAB is a registered trademark of The MathWorks, Inc.

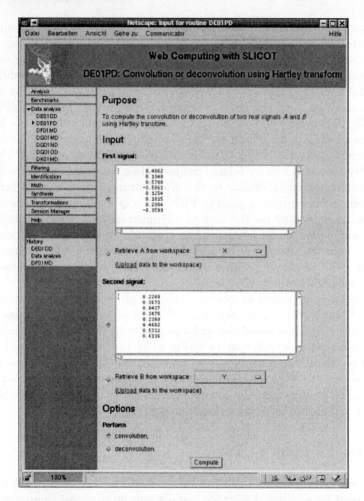

Fig. 1. Web interface for computing the convolution of real signals.

convolution or its inverse function, deconvolution. This is a simple example with only two input parameters and one option. With more complex computational tasks, the number of input parameters and options can be much larger, including scalar parameters and multiple choice lists.

After pressing the compute button in the window of Fig. 1, the convoluted signal is presented as in Fig. 2. It is also possible to download the input/output data as plain text files or MATLAB binaries. During the web computing session, matrices are stored in a workspace, making the data from previous tasks available as input in subsequent computations. For example, the discrete Fourier transform of the signal c can be computed using the web interface of the SLICOT routine DG01ND.

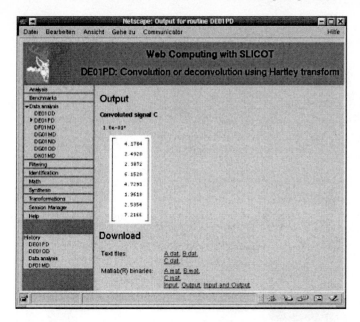

Fig. 2. Convoluted signal and possibilities for downloading data.

A convenient interface enables administrating this workspace, as illustrated in Fig. 3. One of the features of the session manager is that users can login and reuse data of previous computations. All information is stored on a remote machine so that data computed in Umeå, Sweden could easily be recovered, e.g., in an Internet cafè in Berlin.

3   A Programmers Point of View

Apart from being user-friendly it is important to keep web computing facilities as programmer-friendly as possible. The motivation for putting routines on-line will remain low if software producers have to struggle with HTML dialects and CGI techniques. Our approach only assumes the access to a web server that is able to parse the HTML-embedded scripting language PHP and to call executable binaries. The integration of new routines in the environment is then a process divided into two parts.

External routines to be included are first formalized in an XML description file. The syntax is well-defined and deals with a wide range of attributes of the routine, e.g., parameters, conditions, options, default values, description of the routine and so on. When dealing with software libraries which have very restrictive in-line documentation standards the production of the description file could be automatized.

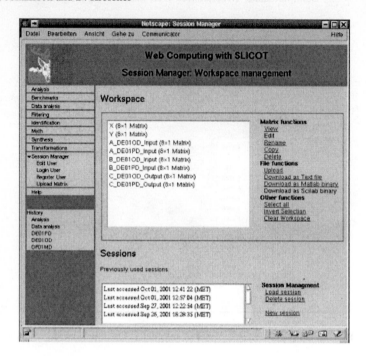

Fig. 3. The workspace manager.

The XML-format currently used in Webcut differs slightly from the format described below. Considerations concerning generality and reusability made it necessary to develop a modified version, together with improved parsers.

For the example from Section 2, the convolution of two signals, the description in the reviewed format is shown below.

```
<routine name = "DE01PD" supgroup = "DE - Covariances">
  <description>
  Convolution or deconvolution using Hartley transform
  </description>
  <parameters>
    <matrix name = "A">
      <description>First signal:</description>
      <default>
        \t0.4862\n\t0.1948\n\t0.5788\n\t-0.5861\n
        \t0.8254\n\t0.1815\n\t0.2904\n\t-0.3599
      </default>
    </matrix>
    <matrix name = "B">
      <description>Second signal:</description>
      <default>
        \t0.2288\n\t0.3671\n\t0.6417\n\t0.3875\n
```

```
      \t0.2380\n\t0.4682\n\t0.5312\n\t0.6116
    </default>
  </matrix>
  <optionlist name = "conv">
    <description>Perform:</description>
    <option name = "C" description = "convolution;"/>
    <option name = "D" description = "deconvolution."/>
    <default> C </default>
  </optionlist>
  <functional type = "length">
    <argument> A </argument>
  </functional>
</parameters>
<conditions>
  <condition type = "is_vector">
    <argument> A </argument>
  </condition>
  <condition type = "is_vector">
    <argument> B </argument>
  </condition>
  <condition type = "equal_length">
    <argument> A, B </argument>
  </condition>
</conditions>
</routine>
```

The top entry **routine**, specifies the name and description of the routine. It contains specifications of name, location and parameters to the callable routine together with possible conditions on those. Restricting parameters is convenient, especially with routines that are not designed for parameter control and may therefore crash without any further information. More about the currently used syntax and semantics of the description files can be found in the documentation accompanying Webcut [13].

The description files are then used by a parser, currently implemented in Perl, that regenerates and modifies the environment to include new routines or modify old ones. An important aspect is that each description file contains exactly the information related to one routine. There are no relations to other parts of the environment so that modifications in single routines keep local.

One weakness with our choice of platform for the server, In the current version, most added routines need an additional routine that handles the parameters in a for the Webcut compatible manner.

Data transfer follows at the moment a simple strategy; input data is supplied to the standard input stream and output data is read from the standard output stream. Hence, the input/output behavior of calling routines must satisfy a well-defined standard.

4   Implementation Details

The implementation of the Webcut environment is mainly done with PHP, a server side hypertext preprocessing language. PHP is a suitable choice for writing dynamic web content and it is intuitive and easy to integrate into existing HTML. Furthermore, a crucial aspect for our project, data bases are well supported. The overall design of the web computing environment is in a schematic way described in Fig. 4.

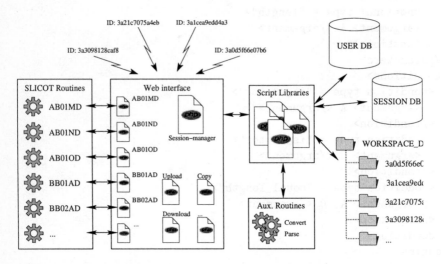

Fig. 4. A schematic overview of the Webcut design.

Within Webcut, several PHP libraries have been developed for the tasks performed by the environment. The following parts are provided.

Layout rendering. Functions that render HTML code and thereby the layout of Webcut are gathered in one library. The structure of the library makes it easy to modify parts or the entire layout.

Session Identification. Contains functions that interact with the browser and maintain the identification of the user during one session.

Invocation and interaction with external routines. Here, the execution of external routines is handled together with the passing of input parameters and catching output data.

Session, user and workspace management. The functions of this library take care of user, session and workspace data by providing high level access to the corresponding data bases.

Based on these libraries, a wide range of user-interactive pages has been developed. Those include workspace and user management, interaction with MATLAB and so on. The main part however consists of pages that call external routines.

As above described in Section 3, the PHP files are generated from the XML-description files, and included into the environment. Since they are based on the developed libraries it is easy to change their layout or functionality.

The session life time starts when a web interface is opened for the first time. The user is then provided with a unique ID that links her/him with a new session. The session itself consists of information about the user and a workspace for produced or uploaded data. This session is accessible as long as the user's web browser keeps track of the provided ID. Since there is a time limit, the ID expires after a period of inactivity. Registered users can restore previously used sessions at any time and at any place.

5   Conclusions and Future Work

In this paper we described tools which automatically generate web computing environments from given description files and calling routines. The resulting interfaces are intuitive to use and provide a sound workspace and session management.

The SLICOT library consists of about 250 user callable routines and benchmark collections in various domains of systems and control. Webcut has been used to provide large parts of this library with web interfaces. For example, Riccati benchmark collections [14] as well as solvers for the algebraic Riccati equation [15] can be tested on-line. The SLICOT web computing project can be found under `http://wc2.hpc2n.umu.se`.

Future developments will concentrate on the coupling of Webcut with grid computing environments. The web interfaces will enable users to direct computations to a heterogeneous set of computers. A major step to be taken is to improve the input/output data management. For instance, the data transfer between computational routines and the web server must be generalized. It is further planned to apply Webcut to other software libraries relevant to the engineering community. Of particular interest is software designed for parallel computations like ScaLapack [16] or PSLICOT [17].

6   Acknowledgments

The authors wish to thank Erik Elmroth, Bo Kågström and Volker Mehrmann for not only taking initiative for this project, but also for useful discussions and help throughout this work. We are also grateful to participate in the NICONET project [18], which develops and distributes the SLICOT library [10], and to the High Performance Computing Center North (HPC2N) [19] which provides the web computing server resources.

References

[1] The MathWorks, Inc.: MATLAB User's Guide. Natick, MA (1992)
[2] The MathWorks, Inc.: MATLAB web server, version 1.2.1. (2001) http://www. mathworks.com/products/webserver/
[3] Baldeschwieler, J., Blumofe, R., Brewer, E.: ATLAS: An infrastructure for global computing. Proceedings of the Seventh ACM SIGOPS European Workshop on System Support for Worldwide Applications (1996)
[4] Souder, D., Herrington, M., Garg, R. P., Deryke, D.: JSPICE: a component-based distributed Java front-end for SPICE. Concurrency: Practice and Experience **10** (1998) 1131–1141
[5] Günther, O., Müller, R., Schmidt, P., Bhargava, H. K., Krishnan, R.: MMM: A web-based system for sharing statistical computing modules. IEEE Internet Computing **1** (1997) 59–68
[6] Brecht, T., Sandhu, H., Shan, M., and Talbot, J.: ParaWeb: Towards word-wide supercomputing. Proceedings of the Seventh ACM SIGOPS European Workshop, Connemara, Ireland (1996) 181–188
[7] Kropf, P. G.: Overview of the web operating system (WOS) project. Advanced Simulation Technologies Conference, San Diego, CA (1999) 350–356
[8] Kapadia, N., Fortes, J.: PUNCH: An architecture for web-enabled wide-area network-computing. Cluster Computing **2** (1999) 153–164
[9] Arnold, D., Agrawal, S., Blackford, S., Dongarra, J., Miller, M., Sagi, K., Shi, Z., Vadhiyar, S.: Users' Guide to NetSolve V1.4. Computer Science Dept. Technical Report CS-01-467, University of Tennessee, Knoxville, TN (2001)
[10] Benner, P., Mehrmann, V., Sima, V., Van Huffel, S., Varga, A.: SLICOT—a subroutine library in systems and control theory. Applied and computational control, signals, and circuits, Vol. 1, Birkhäuser Boston, Boston, MA (1999) 499–539
[11] Elmroth, E., Johansson, P., Kågström, B. Kressner, D.: A web computing environment for the SLICOT library. The Third NICONET Workshop on Numerical Control Software (2001) 53–61
[12] Van Loan, C. F.: Computational frameworks for the fast Fourier transform. Society for Industrial and Applied Mathematics, Philadelphia, PA (1992)
[13] Johansson, P. and Kressner, D.: Webcut - a documentation. Preliminary version available from http://wc2.hpc2n.umu.se (2001)
[14] Abels, J. and Benner, P.: CAREX - a collection of benchmark examples for continuous-time algebraic Riccati equations (version 2.0). SLICOT working note 1999-14, WGS (1999)
[15] Laub, A. J.: A Schur method for solving algebraic Riccati equations. IEEE Trans. Automat. Control, **24** (1979) 913–921
[16] Blackford, S., Choi, J., Cleary, A., D'Azevedo, E., Demmel, J., Dhillon, I., Dongarra, J., Hammarling, S., Henry, G., Petitet, A., Stanley, K., Walker, D., Whaley, R. C.: ScaLAPACK Users' Guide. Society for Industrial and Applied Mathematics, Philadelphia, PA (1997)
[17] Benner, P., Quintana-Ortí, E. S., Quintana-Ortí, G.: PSLICOT routines for model reduction of stable large-scale systems. The Third NICONET Workshop on Numerical Control Software (2001) 39–44
[18] NICONET. Numerics in Control Network. http://www.win.tue.nl/wgs/ niconet.html
[19] HPC2N. High Performance Computing Center North, Umeå University, Sweden. http://www.hpc2n.umu.se.

The Development of a Grid Based Engineering Design Problem Solving Environment

A. D. Scurr[1] and A. J. Keane
School of Engineering Sciences
University of Southampton.
[1]{ads294@soton.ac.uk}

Abstract: This paper gives an overview of the grid based Engineering Design Problem Solving Environment (PSE) being developed at Southampton University. Our current PSE is based on our Options optimiser and the Cardiff VCCE and XML component model. Essentially, VCCE provides a GUI to enable a user to setup and execute a computation by creating a task graph from available components via drag and drop operations on a sketchpad display.

In order to provide an environment that more naturally meets the data-centric view of users, two major enhancements to the PSE are planned. The first concerns scheduling and task farming. The ultimate goal is to achieve within the PSE, an asynchronous computational workflow pattern where analysis tasks can seek to exploit whatever computational resources are available in various workstation clusters. The second enhancement concerns computational resource control and job control and the setting up of an Engineering Design Grid Portal.

1. Problem Solving Environments (PSEs) and Grid Portals

To enhance engineering insight, reduce development costs and improve product quality, engineering design studies are increasingly using sophisticated analysis packages together with optimisation tools. The chief characteristics of these design studies are the need to optimise a design where the analysis is very time consuming, where there are multiple methods or domains of analysis and where we need to deploy distributed, cluster based computing systems. The use of PSE's facilitates working on such problems. The accepted definition of a PSE is: "A PSE is a computer system that provides all the computational facilities needed to solve a target class of problems." (J. Rice - Purdue University). Closely aligned with PSE development is the concept of grid computing and the idea of a grid portal. A Grid Portal is an access point (usually a web browser page) designed to facilitate the use of a PSE in a particular discipline by providing seamless access to the PSE's range of computational tools, information and resource.

P.M.A. Sloot et al. (Eds.): ICCS 2002, LNCS 2329, pp. 881–889, 2002.

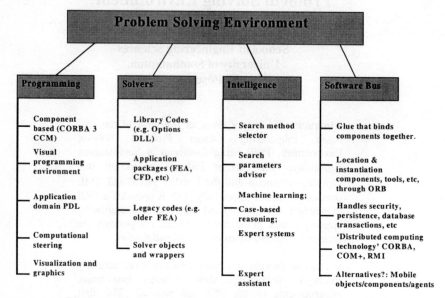

Figure 1: The main components of a generic PSE

The main sub-systems in a generic PSE (Figure 1.) include visual-programming environments for graphically composing, steering and monitoring applications, various component wrapped analysers and solvers, an integrating software-bus (here CORBA) and optionally one or more AI systems to assist the user formulate a computational strategy.

3.   Prototype Engineering Design PSE

We have developed a prototype PSE built around our Options [1] suite of search/optimisation programs and various CFD codes. Our current Engineering Design PSE incorporates the Cardiff University Visual Component Composition Environment (VCCE) as its front-end [2] and uses CORBA 2.3 as its distributed object middleware (software bus above). In Figure 2, components are dragged from the Component Repository on the left column of a sketchpad display and dropped onto the canvas of the Component Composition Tool on the right of the display. Components can then be joined together if their interfaces are compatible to form a task graph. This defines the execution order and dependencies between the various components making up a job. For a component to be available for use in VCCE, an XML definition of the component must be available in the Component repository. The component's XML definition is based on the Component Model defined in VCCE.

The Engineering Design PSE (Figure 3.) uses a component architecture based on CORBA objects (here, the terms *CORBA Object* and *component* are synonymous) whose structure and functionality are defined in CORBA IDL interfaces. Optimisation is a procedure that searches through the design domain for optimum designs. During the search, the optimiser continuously calls the analysis code and, based on the results to date, decides on the next design point to evaluate. In the Engineering Design PSE, the optimiser is contained in the Options CORBA Object and the analysis codes are contained in the OPTFUN/OPTCON CORBA Object.

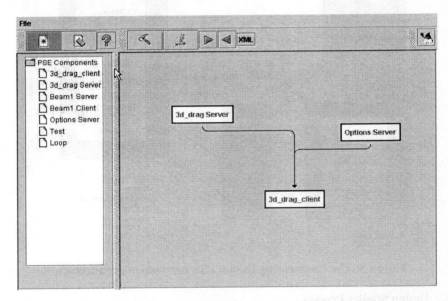

Figure 2: VCCE drag and drop interface to Engineering Design PSE

The current PSE has the following components:

- An Options component which consists of the Options search package wrapped in a C++ wrapper which uses the Options direct subroutine access facilities to set up and run a problem defined in the Client Application code. Options stores all problem data in its own database which can be output as a problem definition file for later re-input.
- A user supplied problem specific component that provides the analysis calculations on the optimisation design variable values found in the Options component solution method.
- A Client Application component, which defines the problem to be solved and specifies the Options search/optimisation method to be used, input parameter values, etc.
- A callback object created by the Client Application, which helps de-couple the Options and OPTFUN/OPTCON components.

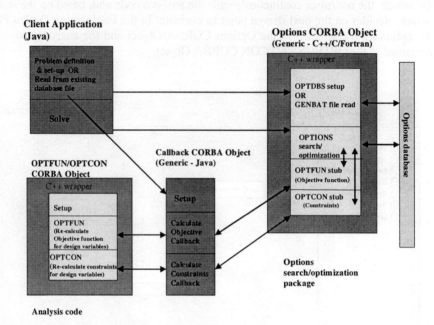

Figure 3: The Engineering Design PSE component architecture

4. Design Studies Process

Designers generally use approximation techniques to solve optimisation problems in order to reduce the otherwise enormous computational effort that would be involved if only full fidelity models were used. Depending on the size of the problem domain space, local, mid-range or global approximations are employed. For single design point optimisation, the above PSE architecture is appropriate, but for global approximation techniques such as Response Surface Modeling (RSM) [3], rather different computational workflow strategies are required.

In the DoE/RSM process (left part of Figure 4.) some technique (e.g., Latin hypercubes) is used to select a set of initial design points (typically say 100 points) that will span the design space. The analysis codes are then run for each design point, ideally in parallel on some large compute cluster, as the design point evaluations are independent of each other. From the resulting database, a response surface is built using the chosen approximation method (e.g., kriging [4]). If satisfactory, the surface is searched for an optimal design point (right part of Figure 4.) and if found, full analysis codes are run at that point. If the analysis is satisfactory and all design constraints are met, the process terminates; otherwise the new design point is added to the design set, the surface regenerated and the optimisation repeated.

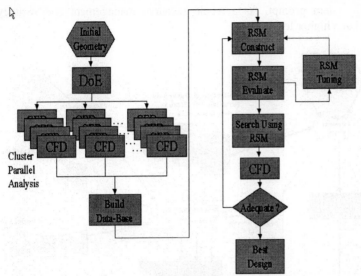

Figure 4: Overall Design of Experiments (DoE)/RSM Process

From a user perspective, the above RSM process suggests a data-centric approach to the computational workflow, where the user is primarily looking at problem design point and results data to-date (and possibly other test data such as flight trial or wind tunnel results in aerospace design problems) in order to decide where new problem analyses need to be run (see figure 5 below). From a computational perspective, the aim is to achieve the maximum amount of parallelism in the analysis computations by exploiting the various cluster resources available. This can be accomplished by decoupling the search engine computation from the analysis computations and using some task farming approach to schedule parallel running of analysis tasks over the available resources.

5. Engineering Design Grid Portal/PSE Architecture

Rapid advances in commodity computing and the emerging nation-wide computational grid are significantly influencing the development of the next generation of PSE's. The term 'commodity computing' encompasses the development of distributed computing technologies (Web services, Java, JINI, CORBA, DCOM, etc.), plus the use of standard PC hardware in networked clusters. 'Grid computing' refers to the high-performance computing community's creation of Grids; advanced infrastructures designed to enable the co-ordinated use of distributed high performance computers for scientific problem solving. Closely aligned with the concept of grid computing is the idea of portal computing [5]. A user or application portal is a web-based collection of information presented together on a browser page. In many cases a portal can replace the need for users to log on to a number of different computers to gain access to the various resources that exist. In portal computing well-defined services are delivered to the portals instead of giving users direct access to the operating system prompt. As a result, resource management and security can be exercised at a higher level.

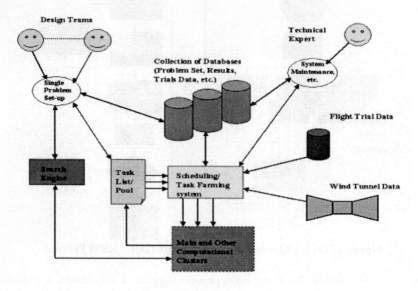

Figure 5: User view of engineering design system

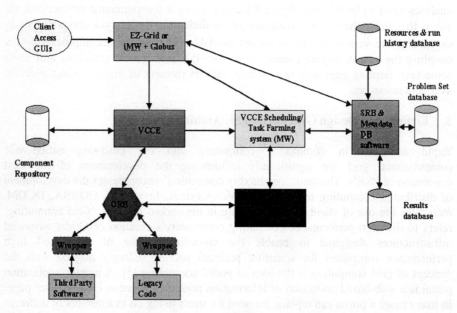

Figure 6: Grid Portal/PSE Architecture

The proposed PSE design involves integrating a number of existing PSE and Globus based software packages in a layered structure, in order to provide a fully functional Engineering Design Grid Portal (Figure 6.). The Globus project [6] aims to provide PSE developers with Grid services for resource management, security and resource discovery. The University of Houston EZ-Grid project [7] uses Globus services to make the use of the Grid easier and transparent for the user. This is achieved by developing easy-to-use interfaces coupled with brokerage systems to assist the resource selection and job execution process.

We propose using the PSE prototype described above to provide the basis for a Design Optimisation Grid portal. This involves extending the PSE functionality with the EZ-Grid/Globus software at a level above the existing VCCE system to provide for Grid user login, job submission, etc. In addition, the development of a task farming approach on the analysis side should de-couple optimisation from analysis task execution in order to exploit Grid technologies to gain more parallelism.

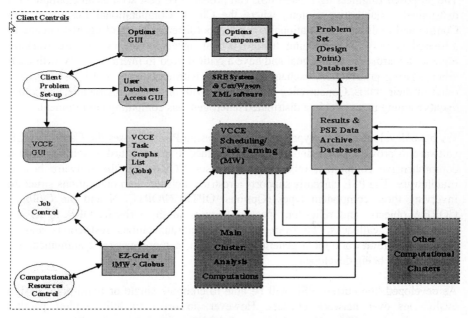

Figure 7: Grid Portal/PSE Computational Workflow

The aim of our Grid Portal/PSE system is to provide the computational workflow scheme shown in figure 7 above where analysis and search/optimisation tasks are de-coupled by the two central database systems (the Problem Set and Results databases). Thus the search and analysis tasks can compute asynchronously, with the Search Engine adding new design point data sets into the Problem Set database and large scale "task farming" on the analysis side. Here, a number of independent analysis tasks can be run in parallel over one or more linked computational clusters under some form of computational resources control. The analysis results are added to the Results

888   A.D. Scurr and A.J. Keane

database, where the Search Engine can be guided via the user or advisor (machine learning based) systems in selecting the most promising new design to analyse.

At present, the problem is set-up using the pre-defined components available to the VCCE task-graph interactive front-end. Once the problem task graph is built, the only options are to run or reset the components in the graph. The Job Control function provides facilities to control computations and monitor their progress. Essentially the Job Control will be able to dynamical start/stop/restart computations and change parameters of running processes i.e., provide interactive computational control as opposed to the process scheduling/task farming available in the Computational Resources Control. Also, for some applications, it might be useful to interactively display feedback on the progress of computations (e.g., provide on-demand graphical displays of analysis or search results to-date).

6. Conclusions:

The proposed Engineering Design PSE can probably be best seen as an example of a distributed collaborative system, where the Client, Computational Resources, Job Control and multiple computational components (different kinds of agents) engage in a shared activity. The defining feature is that agents in the system are working together towards a common goal and have a critical need to interact closely with each other: sharing information, exchanging requests with each other and informing each other of their status. Concurrency is important, with any agent interacting with other agents running in parallel in a distributed heterogeneous computing environment.

We have developed a prototype Engineering Design PSE that uses the Cardiff VCCE software to provide a drag and drop user interface to create a task graph based on a comprehensive component model and that uses CORBA as the software bus or middleware. The PSE currently supports a restricted cluster communications structure involving three component types (Options, OPTFUN/OPTCON and the Callback CORBA objects) and relies on specific features in the Orbacus ORB to handle asynchronous callbacks. To overcome these communications restrictions and to provide the framework for a generic PSE system, a more general communications sub-system is being developed.

As developed, the current PSE will be able to compute single or multi-point design evaluations over network clusters. However, in order to fully exploit parallel computation over intranet or even internet clusters and to provide an environment which more naturally meets the data-centric view of users, two additional major enhancements to the PSE are planned.

- The first concerns scheduling and task farming. The ultimate goal is to achieve within the PSE, an asynchronous computational workflow pattern where analysis tasks can seek to exploit whatever computational resources are available in various workstation clusters (i.e., task farming), running independently of search/optimisation tasks. Communication between the search and analysis tasks is through the Problem Set and Results databases. Whilst the Options package has some of this functionality available through script files, this functionality should be provided in a generic way through the development of a scheduling/task farming PSE component.

- The second enhancement concerns computational resource control and job control and the setting up of an Engineering Design Grid Portal. EZ-Grid provides a basic set of components for creating a grid portal, including a GUI for Grid authentication and login, a Job Manager for job submission and monitoring and a Broker Kernel for resource control. However, these components are likely to require considerable enhancements to work with the scheduling and task farming components above.

References:

[1]   A.J. Keane, The Options Design Exploration System Reference Manual and User Guide – Version B3.0 February 2000.

[2]   O. F. Rana, M. Li, M. S. Shields, and D. W. Walker, A Wrapper Generator for Wrapping High Performance Legacy Codes as Java/CORBA Components in Proceedings of the IEEE/ACM SC2000 Conference, held in Dallas, TX, Nov 10-12, 2000.

[3]   Myers, R. H. and Montgomery, D. C. (1995), Response Surface Methodology: Process and product optimization using designed experiments, John Wiley and Sons inc.

[4]   Jones, D. R., Schonlau, M. and Welch, W. J. (1998), Efficient global optimization of expensive black-box functions, Journal of Global Optimization, 13, 455-492.

[5]   G. von Laszewski, Ian Foster, Jarek Gawor, Peter Lane, Mike Russell. Designing Grid-based Problem Solving Environments and Portals. HICSS 2001.

[6]   The Globus Project web-site: http://www.globus.org/

[7]   The EZ-Grid Project web-site: http://www.cs.uh.edu/~ezgrid/EZ-Grid.htm

TOPAS - Parallel Programming Environment for Distributed Computing*

G. T. Nguyen, V. D. Tran, and M. Kotocova†

Institute of Informatics, SAS, Dubravska cesta 9, 84237 Bratislava, Slovakia
giang.ui@savba.sk

Abstract. In this paper, TOPAS, a new parallel programming environment for distributed systems is presented. TOPAS automatically analyzes data dependence among tasks and synchronizes data, which reduces the time needed for parallel program developments. TOPAS also provides supports for scheduling, dynamic load balancing and fault tolerance. Experiments show simplicity and efficiency of parallel programming in TOPAS environment.

1. Introduction

Nowadays, advances in information technologies have led to increased interest and use of clusters of workstations for computation-intensive applications. The main advantages of cluster systems are scalability and good price/performance ratio. One of the largest problems in cluster computing is software [1]. PVM [13] and MPI [14] are standard libraries used for parallel programming for clusters. Although these libraries allow programmers write portable high-performance applications, parallel programming still difficult. Problem decomposition, data dependence analysis, communication, synchronization, race condition, deadlock and many other problems make parallel programming much harder.

As parallel programming is difficult, there are efforts to make parallel compilers that can automatically parallelize sequential programs. However, in general, the parallel programs generated by parallel compilers are much slower than those written by experienced programmers using PVM/MPI. Parallel programs get more and more complex with many unsolved dependences at compilation time, distributed systems get heterogeneous, therefore, according to many experts, parallel programs generated by parallel compilers will not catch performance of PVM/MPI in near future [1][2].

TOPAS (Task-Oriented PArallel programming System, formerly known as Data Driven Graph - DDG [7][9]) is a new parallel programming environment for solving the problem. The objectives of TOPAS are as follows:

- to make parallel programming in TOPAS as easy as by parallel compilers, with the performance of programs in TOPAS is comparable with parallel programs written in PVM/MPI;

---

* This work is supported by the Slovak Scientific Grant Agency within Research Project No. 2/7186/20
† Department of Computer Science, FEI-STU, Ilkovicova 3, 81219 Bratislava,Slovakia

P.M.A. Sloot et al. (Eds.): ICCS 2002, LNCS 2329, pp. 890–899, 2002.

- to make parallel programs structured, easy to understand and debug, and to allow error checking at compilation time for removing frequent errors;
- to provide support for optimization techniques (scheduling and load balancing);
- to provide some facilities for Grid computing (heterogeneous architectures, task migration, fault tolerance).

The objectives are rather ambitious, but not unachievable. Section 2 shows the main features of parallel program in TOPAS, including scheduling and load balancing. Section 3 focuses on fault tolerance support in TOPAS environment, and Section 4 demonstrates real examples of parallel programs written in TOPAS.

2. Basic Ideas of TOPAS

Fig. 1 shows the steps of parallel programming development. At first, the program is divided into a set of tasks. Next, data dependence among tasks is analyzed; this is the basis for writing communication and synchronization routines in the next step. Finally, the codes of tasks are added. After these steps, programmers have a correct program; however, its performance may not be adequate. Therefore, the parallel program has to be optimized by scheduling that often requires task graphs of the parallel program.

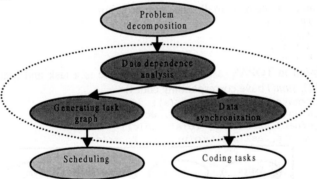

Fig. 1. Parallel program development

Except for the step of coding tasks that is similar to sequential programming, the other five steps can be divided into two groups. The first group consists of problem decomposition and scheduling, which have the following common features:
- these steps are important for performance. Every change in them may affect performance largely;
- there are many possible solutions for these steps and there is no general way to choose the best one.

Experienced programmers often do these steps better than parallel compilers. The other three steps: data dependence analysis, data synchronization and generating task graphs have the following common features:
- each of these steps has only one solution. There are simple algorithms to get it;

- these steps have little effect on performance (note that inter-processor communication is minimized by problem decomposition and scheduling);
- these steps require a lot of work and most frequent errors in parallel programming occur here.

In TOPAS, the performance critical steps (problem decomposition and problem scheduling) are done by programmers and the remaining are done automatically by TOPAS runtime library. This way, writing parallel programs written in TOPAS is as easy as by parallel compiler, while the parallel programs in TOPAS can achieve performance comparable with parallel programs in PVM/MPI.

Parallel programming in TOPAS

The basic units of parallel programs in TOPAS are tasks. Each task is a sequential program segment that can be assigned to a processor; once it starts, it can finish without interaction with other tasks. To extend, tasks can be understood as the sequential parts between two communication routines in PVM/MPI.

Each task in TOPAS consists of a function/subprogram that contains task code and a set of variables the task uses. Tasks are created by calling *create_task()* function with its code and its variables as parameters. Each variable may have read-only (*ro*), read-write (*wr*) or write-only (*wo*) access.

Here is an example of a small sequential program

```
1. a = 10
2. b = a + 10
3. c = a * 5
4. d = b + c
```

and its version in TOPAS (assuming that each line is a task and functions *assign()*, *incr()*, *mult()*, *sum()* have corresponding codes)

```
1. ddg_create_task(assign, wo(a))
2. ddg_create_task(incr, ro(a), wo(b))
3. ddg_create_task(mult, ro(a), wo(c))
4. ddg_create_task(sum, ro(b), ro(c), wr(d))
```

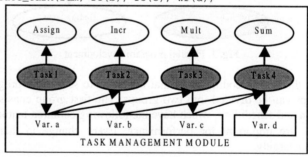

Fig. 2. Memory structures of program in TOPAS

Fig. 2 shows memory structures of the parallel program in TOPAS. Each task is represented by a task object that has pointers to its code and its variables. Using the memory structure, TOPAS runtime library can detect the data dependence among tasks and synchronize data automatically. Real programs in TOPAS are shown in

Section 4. If several tasks write to the same variable, then multi-version technique is used to solve the data dependence among tasks. Each task that writes to the variable creates a new version of the variable. Versions are created dynamically when needed, and they are removed from memory if there are non-unexecuted tasks that use the version. More details on multi-version variable are given in [7]. In any case, the complexity of data dependence analysis and creating internal memory structures of programs in TOPAS is proved to be linearly increasing with the number of tasks.

Applications in TOPAS are compiled as single executable files, which are available on every processor. When a program runs, it creates tasks, schedules, distributes and executes tasks according to the generated schedule. Tasks in TOPAS are executed in non-preemptive mode. A task is executed by calling the code of the task, which is a function in High Level Languages (C/C++, Fortran) with its parameters. When a task is finished, the output data of the task become available for other tasks. According to the memory structures, TOPAS can test all tasks, which use the output data of the executed task, if they are ready. If some of the tasks, which use the output data of the executed task, are assigned to another processor, one copy of the output data is sent to the target processor. Assume that N is the number of tasks, $T_{overhead}$ is the overhead of TOPAS runtime library, T_{total} is the total execution time and $T_{average}$ is the average task execution time. As $T_{overhead}$ is linearly increased with N, there does exist a constant k that $T_{overhead} \leq k.N$. The relative overhead of TOPAS runtime library (the ratio $T_{overhead} / T_{total}$) is

$$\frac{T_{overhead}}{T_{total}} \leq \frac{kN}{T_{total}} = \frac{k}{T_{total} / N} = \frac{k}{T_{average}}$$

In other words, the relative overhead of TOPAS runtime library does not increase with the number of task. If average task execution time is long enough, then the relative overhead is negligible. Section 4 shows the relative overhead of TOPAS runtime library of real problems.

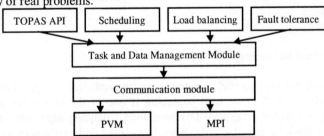

Fig. 3. TOPAS architecture

The architecture of TOPAS library is shown in **Fig. 3**. The kernel of TOPAS library is Task and data Management Module (TMM) that manages all tasks and data, executes tasks and synchronizes data. Communication module creates interface between TMM and underlying communication libraries (PVM, MPI, Nexus) and make parallel programs in TOPAS independent from communication library. TOPAS API is the application-programming interface, which allows programmers to create tasks and variables. Other modules (scheduling, load balancing and fault tolerance) use the memory structures in TMM.

Scheduling in TOPAS

Scheduling is one of the most important techniques to improve the performance of parallel programs. It distributes tasks among processors and arranges task executions in order to get the performance goals such as minimizing the inter-processor communications, balancing load of processors and/or overlapping communications by computations. There are many scheduling algorithms available however their application for programs in PVM/MPI is rather limited [3][4][6][8]. Most scheduling algorithms need the task graphs of parallel programs as input and generating the task graph from a parallel program written in PVM/MPI is difficult. TOPAS solves the problem completely by using the memory structure in TMM. If each task and its output variables are considered as a node in task graph; then the memory structures can be considered as the task graph of the program

When a parallel program in TOPAS runs, after tasks have been created, TMM calls function *ddg_schedule()*, which receives the task graph as input parameter and returns a generated schedule as output parameter. Programmers can customize scheduling by overwriting the *ddg_schedule()*. There are many choices: they can call one of the implemented scheduling algorithms in TOPAS library, write their own scheduling algorithms or save the task graph in a file, use external scheduling tools to schedule and then read the generated schedule back from a file. In any case, when programmers change the scheduling algorithms they do not have to modify the other parts of codes. This is a considerable advantage of parallel programming in TOPAS: in classical PVM/MPI programming, change task distributions and execution orders often require major modification in the parallel program source code. TOPAS runtime library also eliminates frequent errors in parallel programming, including race conditions and deadlocks. Default scheduling in TOPAS library is done by Dominant Sequence Clustering [12] algorithm. Program codes are not affected by changing task distribution or execution order of new scheduling algorithms provided by programmers.

Dynamic load balancing

If the behavior of parallel programs is unpredictable, e.g. tasks are created dynamically, then dynamic load balancing is one of the best ways to optimize the execution of parallel programs. Dynamic load balancing algorithms try to keep all processors busy by moving tasks from heavily loaded processors to lightly loaded ones. One of the largest obstacles of using dynamic load balancing is task migration [10]. Migration of a running task requires large overhead: stopping the task, saving the state of the task, sending the state and code to the target processor, restarting the task with the saved state. When a system is heterogeneous, codes of tasks generated in different platforms can differ from each other and do not support code migration. Task migration also often requires additional work from programmers or supports from operating system.

In TOPAS, moving task from a processor to another can be done easily by sending its task object to the target processor. The task code is already available in the target processor because the same executable file runs on each processor (like SPMD programs). Therefore the cost of task migration is sending 70-80 bytes (the size of the task object) to the target processor. Task migration in TOPAS can be done in systems with heterogeneous architectures due to the fact that programmers do not have to move the code. From the viewpoint of data synchronization among tasks, it is not significant if a task is placed in the source or the target one and data are sent to the target processor if necessary.

3. Fault Tolerance in TOPAS

As the size of computer systems increases unsteadily, the probabilities that faults may occur in some nodes of the system increase as well. Therefore, it may be necessary to ensure that the applications may continue in the system despite the occurrence of faults. This is very important in Grid computing, as computational nodes are not managed by one owner and communications via Internet are unreliable.

Traditional system recovery applies in fixed order: fault detection and location, FT reconfiguration, restarting system with valid data (e.g. from the last valid checkpoint). When hardware redundancy is not available, software redundancy is unavoidable.

Output data checkpoints

The basic concept of fault tolerance in TOPAS is to keep spare instances of output data in other processors as checkpoints. Assume that only one single hardware permanent fault can occur in a node; then
- one spare instance of the output data is sufficient, if the output data is required by successors that are allocated in the same node
- no spare instance is necessary, if the output data is required by successors, which are allocated in at least two different nodes. After a fault occurrence, if one node is faulty, another instance of output data is still available in another healthy node.

Each node needs information about the system and task states in its memory to maintain its duties. Such information can be created when an application starts and it is updated when each task finishes. The information can be divided to the data-flow graph of the application, states of all nodes, tasks and data, a list of data instance locations. There is a problem to keep low number of inter-node communications, otherwise the system performance decrease significantly. If a spare instance is created, its location must be announced to all nodes and this can cause a large number of broadcasts. Therefore, in TOPAS, spare instances are located in the neighbor node of the successor node. If the output data is required by at least two tasks in different nodes, locations of all data instances are in data-flow graph of the application.

Removing obsolete output data

In TOPAS, when fault-tolerant feature is not supported, removing unnecessary output data is quite simple. Each node has the data-flow graph of the application and knows about the need of all data in the node exactly. There is no requirement to know if another node will need checkpoints (spare instances of data) allocated in the node. When the fault-tolerant feature of TOPAS environment is supported, the use of multi-version variables is unavoidable and the situation is more complicated. Output data checkpoints are kept until they are unnecessary for all tasks. The moment when no task needs certain data version is not easy to determine. Broadcast messages and task synchronization are frequently mentioned in literature. They usually cause a large number of inter-node communications and/or keep large amount of obsolete data.

Because of such disadvantages, TOPAS removal method has been designed. When a task is running, the node has own information about finished and unfinished tasks in the system. The information is collected

- using the data-flow graph of the application, which is available on each node to determine all finished task predecessors
- when there is a spare instance of an output data, the node knows the task, to which the spare instance belongs. In this case, the owner of the spare instance must be finished and its predecessors are finished too.

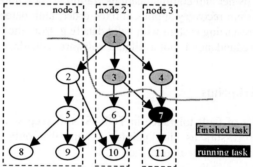

Fig. 4. Detected finished tasks from the third node viewpoint

From the viewpoint of a certain node, the set of all tasks is divided into subset of finished tasks and a subset of remaining tasks. The subset of remaining tasks contains also finished tasks that are not detected by the node. The data on the gray boundary curve (**Fig. 4**) must be kept and the remaining data can be removed. The advantage of this method is no need of broadcast and synchronization. The disadvantage is that not all obsolete data are removed. In **Fig. 4**, the multiprocessor system has three nodes. Task 7 is one of running tasks. The third node determines that task 1, 3 and 4 are finished. Task 2 had been finished on the first node, but it has not been detected by this node. The data moved from task 1 to task 4 and from task 1 to task 3 can be removed, e.g. the third node can remove such data instances, which are located on it.

The number of data that must be kept in the whole system is smaller than $O(n)$ where n is the number of application tasks and is greater than $O(p)$ where p is the level of parallelism of the parallel application. When a faulty node is detected and located, TOPAS system diagnosis reports to all remaining nodes *"i-th node is down"* and applications have to restart from the last valid checkpoint. In addition, nodes in multiprocessor systems are assumed to be fail-silent, i.e. they only send correct messages or nothing when faults occur.

Task reconfiguration

The memory structures, automatic data synchronization, runtime scheduling and task migration of TOPAS provide the basic services for implementing fault-tolerance features. When a fault occurs, TOPAS moves the unexecuted tasks on the faulty processor to a healthy one. After fault occurrence, new task locations are determined using reconfiguration algorithms with the criteria of fast recovery time and a graceful degraded performance. Of course, no solution can fulfil these criteria in the best way and an optimal solution is somewhere between them. Scheduling algorithms can be used and can provide the best performance for the system after reconfiguration, but do not give any guaranty for fast recovery time. Usually, the code of application program is small in comparison with data. In TOPAS, task codes are available on every node and there is no problem to reorganize the task-running order. The problem is in moving existing data due to task reallocations. Such data can be large and cause communication delay and long reconfiguration time. Default reconfiguration algorithm finds a suitable node to reallocate each task in the faulty node with the criterion to minimize the amount of data needed to move. It also provides possible good load balancing for the rest of the system. Similarly to like TOPAS scheduling, programmers can customize reconfiguration algorithm by overwriting the *ddg_reconfiguration()* to achieve their desired goals [5].

In healthy nodes, running tasks finish their work as normally, then refresh their states and start to realize new configuration. After reconfiguration, running tasks in the faulty node are added to waiting queues for processors on other healthy nodes and will be restarted with valid input data. Waiting tasks in the faulty node are added to waiting queues in other nodes. All remaining tasks continue from their last states.

Details of fault tolerance in TOPAS are also given in [11].

4. Case Study

In this section, Gaussian elimination algorithm is implemented as parallel program in TOPAS. This chosen problem is representative for large classes of parallel applications, so it can prove the usability of TOPAS library.

```
1.  #define N 1200
2.  main()
3.  { float a[N][N];
4.     init(a);                      // initial the values of a
5.     for (int i = 0; i < N-1; i++)
6.        for (int j = i+1; j < N; i++)
7.        { coef = a[j][i] / a[i][i];
8.           for (int k = i+1; k < N; k++)
9.              a[j][k] = a[j][k] - coef*a[i][k];
10.       }
11.       print(a);                   //print the result values of a
12. }
```
Gem01: Sequential Gaussian elimination algorithm

The sequential Gaussian elimination algorithm - GEM is described in *Gem01*. We concentrate only on GEM, the input and output functions *init()* and *print()* are not considered. The tasks are defined from the lines inside two outer loops, (line 7, 8, 9 in *Gem01*). Before defining a task, the code of the tasks has to be moved to a function. Finally, the task is created from the code (*Gem02*). It can be seen in the final version of GEM in TOPAS, that there are no communication routines in the program. All data synchronization is done by TOPAS runtime library.

```
1.  #include "ddg.h"
2.  #define N 1200
3.  typedef float vector[N];
4.  void ddg_main()
5.  { ddg_var_array<vector> arr(N);
6.    init(a);
7.    for (int i = 0; i < N - 1; i++)
8.      for (int j = i+1; j < N; j++)
9.        ddg_create_task(compute,ddg_direct(i),
                          ddg_ro(arr[i]), ddg_rw(arr[j]));
10.    print(a);
11. }
12. void compute(ddg_var<int> i,
                 ddg_var<vector> i_line, ddg_var<vector> j_line)
13. {  coef = j_line[i] / i_line[i];
14.    for (int k = i + 1; k < N; k++)
15.      j_line[k]=j_line[k] -      i_line[k]*coef;
16. }
```

Gem02: Final version in TOPAS

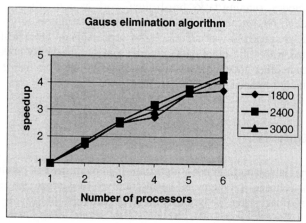

Fig. 5. Speedups of Gaussian elimination algorithm

Experiments are performed on a PC cluster of six Pentium 500 MHz connected by 100Mb Ethernet. The speedup of Gaussian elimination algorithm in TOPAS is shown in **Fig. 5**. TOPAS runtime library causes about 0.3% overhead of total execution time.

When TOPAS provides fault-tolerant feature, runtime library causes larger overhead, i.e. about 1.3% to 1.9% total non-fault execution time.

5. Conclusion

In this paper, the TOPAS parallel programming environment is described. TOPAS does not only allow programmers to write parallel programs easily but also provides facilities for scheduling, dynamic load balancing and fault tolerance features. Experiments have demonstrated the usability of TOPAS on real problems, the simplicity of parallel programming in TOPAS, the performance of TOPAS programs and the overhead of TOPAS runtime library. The latest work in TOPAS is involved in make TOPAS library to Grid computing.

References

1. H. El-Rewini, T.G. Lewis: Distributed and Parallel Computing. Manning Publication, USA, 1998.
2. Kennedy: Compilers, Languages and Libraries. The Grid: Blue Print for a New Computing Infrastructure, pp. 181-204, Morgan Kaufmann, 1999.
3. Senar M.A., Cortes A., Ripoll A., Hluchy L., Astalos J.: Dynamic Load Balancing. Parallel Program Development for Cluster Computing. Nova Science Publishers, USA, 2001.
4. H. El-Rewini, H. H. Ali, T. Lewis: Task Scheduling in Multiprocessing Systems. Manning Publication, USA, 1999.
5. L. Hluchy, M. Dobrucky, J. Astalos: Hybrid Approach to Task Allocation in Distributed Systems. Computers and Artificial Intelligence, vol.17, No.5, pp. 469-480, 1998.
6. B. A. Shirazi, A. R. Hurson, K. M. Kavi: Scheduling and Load Balancing on Parallel and Distributed Systems. IEEE Computer Society Press, 1995.
7. V. D. Tran, L. Hluchy, G. T. Nguyen: Parallel Programming Environment for Cluster Computing. CLUSTER'2000, pp. 395-396, November 2000, Germany. IEEE Computer Society Press.
8. L. Hluchy, M. Dobrucky, D. Dobrovodsky: Distributed Static and Dynamic Load Balancing Tools under PVM. First Austrian - Hungarian Workshop on Distributed and Parallel Systems, Miskolc, Hungary, 1996, pp.215-216.
9. V. D. Tran, L. Hluchy, G. T. Nguyen: Parallel Program Model for Distributed Systems. EuroPVM/MPI'2000, pp. 250-257, September 2000, Hungary. Springer Verlag.
10. M. Richmond, M. Hitchens: A New Process Migration Algorithm, Operating System Review, 31(1), 1997, 31-42.
11. G. T. Nguyen, V. D. Tran, L. Hluchy: DDG Task Recovery for Cluster Computing. PPAM'2001, Poland, September 2001. Springer Verlag. To appear.
12. T. Yang, A. Gerasoulis: DSC: Scheduling Parallel Tasks on an Unbounded Number of Processors. IEEE Transaction on Parallel and Distributed Systems, Vol. 5, No. 9, pp. 951-967, 1994.
13. PVM: Parallel Virtual Machine http://www.epm.ornl.gov/pvm/pvm-home.html
14. MPI - Message Passing Interface http://www.erc.msstate.edu/mpi/

Parallel Implementation of a Least-Squares Spectral Element Solver for Incompressible Flow Problems

Margreet Nool[1], Michael M. J. Proot[2]

[1] CWI, P.O. Box 94079, 1090 GB Amsterdam, The Netherlands
E-mail:Margreet.Nool@cwi.nl ***
[2] Delft University of Technology, Aerospace Engineering, Section Aerodynamics,
Kluyverweg 1, Delft, The Netherlands.
E-mail: m.m.j.proot@lr.tudelft.nl

Abstract. Least-squares spectral element methods are based on two important and successful numerical methods: spectral/hp element methods and least-squares finite element methods. Least-squares methods lead to symmetric and positive definite algebraic systems which circumvent the Ladyzhenskaya-Babuška-Brezzi stability condition and consequently allow the use of equal order interpolation polynomials for all variables. In this paper, we present results obtained with a parallel implementation of the least-squares spectral element solver on a distributed memory machine (Cray T3E) and on a virtual shared memory machine (SGI Origin 3800).

1 Introduction

For many engineering flow problems, the least-squares principles offer several theoretical and computational advantages in the algorithmic design and implementation [1, 2, 3, 4] of the corresponding finite element methods, advantages that are not present in standard Galerkin based discretization. In particular, the least-squares formulations lead to symmetric and positive definite algebraic systems [5] which circumvent the Ladyzhenskaya-Babuška-Brezzi stability condition irrespective of the underlying partial differential equations. Due to these advantages, least-squares finite element methods are becoming increasingly popular to solve the Stokes [6, 7, 8] and Navier-Stokes equations [9, 10, 5].

Least-squares spectral element methods (LSQSEM) seem very promising since these methods combine the generality of finite element methods with the accuracy of the spectral methods and also the theoretical and computational advantages in the algorithmic design and implementation of the least-squares methods. In [11, 12], the accuracy of a least-squares spectral discretization of the Stokes problem (cast in velocity-vorticity-pressure form) has been reported for

*** Funding for this work was provided by the National Computing Facilities Foundation (NCF), under project numbers NRG-2000.07 and MP-068. Computing time was also provided by HPαC, Centre for High Performance Applied Computing at the Delft University of Technology.

P.M.A. Sloot et al. (Eds.): ICCS 2002, LNCS 2329, pp. 900–909, 2002.

different boundary conditions. The interested reader is referred to these papers for a sound discussion regarding the least-squares spectral element formulation of the Stokes problem, the gathering procedure and the effect of the boundary conditions on the formulation. The present paper deals with efficient parallel solution strategies to solve the algebraic systems resulting from the least-squares spectral element formulation of the Stokes problem.

Parallelization of the least-squares finite element methods seems to be straightforward by using element-by-element techniques [1, 4]. However, this is not the case with least-squares spectral element methods since two different kinds of distribution of data are required and the conversion is rather complicated. The spectral element structure enables to calculate the local matrices corresponding to each spectral element, simultaneously. Obviously, if the number of available processors is much larger than the number of spectral elements, many processors become idle unless the data of a single spectral element will be computed along several processors. In the present paper, we consider a spectral element, also called a cell, as the smallest computational unit. The parallel solution of the algebraic problem, a large, global sparse system, requires a completely different data distribution.

The present paper is organized in the following way. In Sect. 2, some implementation aspects of least-squares spectral element methods are treated. The program structure and parallel implementation are discussed in Sect. 3. The results of the numerical simulations are discussed in Sect. 4. Conclusions are given in Sect. 5.

2   Implementation aspects of least-squares spectral element methods

The domain is discretized with a mesh of k non-overlapping conforming quadrilateral spectral elements of the same order. As discussed in [11, 12], each quadrilateral spectral element is first mapped on the parent spectral element and then the local systems

$$A_i z_i = f_i, \quad \text{with } i = 1, \cdots, k \tag{1}$$

are calculated. The matrix A_i represents the least-squares spectral element discretization of the governing equations of spectral element i and the vectors z_i and f_i represent the corresponding local variables and the right-hand function, respectively.

In Fig. 1 an example is given of a domain discretized with a mesh of four spectral elements. Each spectral element contains nine local nodes, numbered from 1 to 9 (small-size digits). In the same figure, also a global numbering (normal-size digits) is shown. First, the internal nodes or variables are numbered $(1, \cdots, 9)$, then the *knowns* $(10, \cdots, 25)$ given by the boundary conditions. Since each local variable corresponds to a global variable, one can establish the local-global mapping operator $gm_{\mathcal{I}}$ for each spectral element. For the given ex-

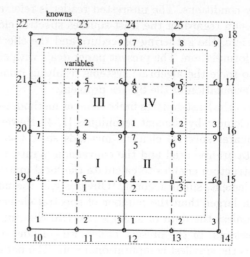

Fig. 1. Example of local and global numbering. The domain has been divided into four cells: I, II,III, IV. Each cell contains 9 nodes, denoted by a o.

ample, we have

$$
\begin{aligned}
gm_{\mathrm{I}} &= [\ 10, 11, 12, 19, 1, \ \ 2, 20, \ \ 4, \ \ 5\ ], \\
gm_{\mathrm{II}} &= [\ 12, 13, 14, \ \ 2, 3, 15, \ \ 5, \ \ 6, 16\ ], \\
gm_{\mathrm{III}} &= [\ 20, \ \ 4, \ \ 5, 21, 7, \ \ 8, 22, 23, 24\ ], \\
gm_{\mathrm{IV}} &= [\ \ 5, \ \ 6, 16, \ \ 8, 9, 17, 24, 25, 18\ ].
\end{aligned}
\tag{2}
$$

The local-global mapping operator $gm_{\mathcal{I}}$ can also be expressed by the sparse *gathering matrix* $\mathcal{G}_i$ which has nonzero entries according to $\mathcal{G}_i(i, gm_{\mathcal{I}}(i)) = 1, \mathcal{I} = \mathrm{I}, \cdots, \mathrm{IV}$. The global assembly of the k local systems (1) can now readily be obtained with:

$$
KU = F \Leftrightarrow \left[\sum_{i=1}^{k} \mathcal{G}_i^T A_i \mathcal{G}_i\right] U = \sum_{i=1}^{k} \mathcal{G}_i^T f_i.
\tag{3}
$$

where the matrix K represents the symmetrical globally gathered matrix of full bandwidth and the vectors U and F represent the global nodes (e.g., variables and knowns) and the global right-hand side function, respectively.

Since the known nodes are numbered last, one can subdivide the vector U into an unknown component U_1 and a known component U_2. Consequently, the matrix K can be factored into submatrices $K_{1,1}$, $K_{1,2}$, $K_{1,2}^T$ and $K_{2,2}$. Also the the right-hand side vector F can be factored into the submatrices F_1 and F_2. Hence, system (3) has the following matrix structure

$$
\begin{bmatrix} K_{1,1} & K_{1,2} \\ K_{1,2}^T & K_{2,2} \end{bmatrix} \begin{bmatrix} U_1 \\ U_2 \end{bmatrix} = \begin{bmatrix} F_1 \\ F_2 \end{bmatrix},
\tag{4}
$$

which readily allows "*static condensation*" of the knowns, leading to the following sparse symmetric and positive definite system

$$K_{1,1}U_1 = F_1 - K_{1,2}U_2 . \tag{5}$$

System (5) will be solved in parallel with the conjugate gradient method.

3   Program structure and parallel implementation

3.1   Redistribution of data due to renumbering

After we have built up the grid completely and after the calculation of the local systems (1), we have to switch from local numbering to global numbering as discussed in Sect. 2. As a result, we obtain a global CSR-matrix which can easily be distributed long an arbitrary number of processors. Each processor has to send data from one cell to a *few* other processors or possibly to itself, a very unbalanced task due to the chosen numbering. However, if this task is completed, each processor contains a part of the global assembled matrix (3), and the data per processor will be balanced again.

Let us return to the example grid of Fig. 1. If we consider only the internal nodes and investigate the case of four processors, then before redistribution processor p_0 corresponds to cell I, p_1 to cell II and so on. After the rearrangement of the data, p_0 contains the first three rows of matrix K of (3), p_1, p_2 and p_3 each t wo rows. The distribution is as follows:

$$
\begin{aligned}
p_0(\text{ cell I}) &\Rightarrow p_0(N_1, N_2), p_1(N_4, N_5), \\
p_1(\text{ cell II}) &\Rightarrow p_0(N_2, N_3), p_1(N_5), \quad\ p_2(N_6), \\
p_2(\text{ cell III}) &\Rightarrow \qquad\qquad\ p_1(N_4, N_5), p_2(N_7), p_3(N_8), \\
p_3(\text{ cell IV}) &\Rightarrow \qquad p_1(N_5), \quad\ p_2(N_6), p_3(N_8, N_9).
\end{aligned} \tag{6}
$$

3.2   Parallel Conjugated Gradient Performance

Since system (5) is symmetric and positive definite, the conjugate gradient (CG) method can be applied directly. The performance of this iterative solution strategy for least-squares finite element approximation of flow problems on distributed parallel computers is clearly of relevance to computational fluid dynamics. In this report, we describe results with the simple, easy to parallelize, Jacobi or diagonal preconditioning. At this moment, we test the efficiency of other preconditioning schemes for the incompressible Navier-Stokes problem: block-Jacobi, SSOR, FEM-matrix and Additive Schwarz and their parallel possibilities. The latter seems to be a good candidate.

Having assembled the system locally in parallel, solution by CG iteration involves repeated matrix-vector products, dot products and DAXPY operations. More specifically, each iteration involves one matrix-vector product, t wo dot products, t wo DAXPY and one DAYPX operations ($y = y + \alpha x, y = x + \alpha y$). Local dot products are computed in parallel on the processors and the scalar results are accumulated across the processors using global summation followed by a broadcast. The communication of the dot product will increase logarithmically with increasing number of processors. The matrix-vector products, which clearly require the greatest fraction of the computation, are computed in parallel.

Consider the matrix-vector product

$$Y = \alpha\, A\, X + \beta\, Y, \tag{7}$$

where A is stored in Compressed Row Storage mode. A fast method to parallelize this operation is to divide matrix A and vector Y into equal parts for the sake of a good load balancing. For the matrix A this means that each processor gets the same number of rows m_p, following the next distribution:

$$m_p = m/p, \tag{8}$$

if m, the number of rows of A, is a multiple of the number of processors p. If not, which will be true in most cases, some adjustment of this approximate proces partitioning will be needed and the number of grid points per processor may vary slightly. We assume that each part has a comparable number of nonzero elements. The complete vector X must be available on each processor.

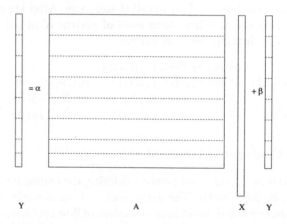

Fig. 2. Parallel distribution of matrix A and vector Y, along 8 processors

The CSR-matrix A of Fig. 2 is defined as

```
TYPE,PUBLIC :: matvec_csr
   REAL(DOUBLE), DIMENSION(:), POINTER    :: FKE
   INTEGER, DIMENSION(:), POINTER         :: JFKE
   INTEGER, DIMENSION(:), POINTER         :: IFKE
   INTEGER                                :: no_rows
END TYPE matvec_csr
```

Then on each processor the matrix A can be declared as:

```
TYPE(matvec_csr)   :: A,
```

where A%FKE contains the nonzero values at the processor involved, where the INTEGER array A%JFKE contains their column numbers and where A%IFKE(i+1)-A%IFKE(i) denotes the number of nonzeros of row i on that particular processor.

4 Numerical results

In (least-squares) spectral element applications, t wo different kinds of refinement strategies are commonly used: h-refinement and p-refinement. The purpose of the numerical simulations is to check the parallel performance for both refinement strategies. To this end, the least-squares spectral element formulation of the velocity-vorticity-pressure formulation of the Stokes problem is demonstrated by means of the smooth model problem of Gerritsma-Phillips [13] with $v = 1$. This model problem involves an exact periodic solution of the Stokes problem defined on the unit-square ($[0, 1] \times [0, 1]$). The velocity boundary condition is used for all the numerical simulations. The pressure constant is set at the point $(0, 0)$. The $h-$ and $p-$grids used in the present paper correspond to the grids in [11, 12].

4.1   The h- and p-refinement approach and its parallel performance

Six different grids are used to check the parallel performance of the h-refinement. As can be observed in Table 1, the polynomial order of all the spectral elements equals 4, which means that each direction has four Gauss-Legendre-Lobatto(GLL) collocation points, and the number of spectral elements is varied from 4 to 144. For the moment, we consider a cell as the smallest computational unit. Obviously, an increase of the number of cells allows to use more processors, and the parallel efficiency will grow. In case the number of processors is less than the number of cells, one or more processors will compute data of more than one cell.

In the middle column of Tables 1 and 2 the order of the large sparse global system is given together with the number of iterations required to solve this system using CG. The parallel solution of the systems may give a slightly different number of iteration steps. The right column in the Tables lists the L_2 norm of the different components, like the velocity (L_2 norm of $x-$ and $y-$components agree), the vorticity and pressure. Only four different grids have been used to check the parallel performance in case of the p-refinement (see Table 2). Each grid contains four spectral elements. The order of the approximating polynomial varies from 4 to 10 and is the same in all the variables. A growth of the polynomial order in the p-refinement case will increase the number of nodes per cell and so does the amount of computational effort per cell. However, the highest parallel efficiency will be achieved in case the number of cells equals the number

Table 1. The different grids used for the investigation of the h−refinements.

Spectral elements	GLL-order	size of global system	# iterations	L_2 norm		
				Velocity	Vorticity	Pressure
2×2	4	259	132	$9.2 \ 10^{-4}$	$4.8 \ 10^{-2}$	$1.8 \ 10^{-2}$
4×4	4	1027	232	$5.0 \ 10^{-5}$	$1.6 \ 10^{-3}$	$7.1 \ 10^{-4}$
6×6	4	2307	326	$5.2 \ 10^{-6}$	$2.8 \ 10^{-4}$	$6.9 \ 10^{-5}$
8×8	4	4099	431	$1.1 \ 10^{-6}$	$8.7 \ 10^{-5}$	$1.3 \ 10^{-5}$
10×10	4	6403	569	$3.2 \ 10^{-7}$	$3.5 \ 10^{-5}$	$3.6 \ 10^{-6}$
12×12	4	9219	707	$1.2 \ 10^{-7}$	$1.7 \ 10^{-5}$	$1.3 \ 10^{-6}$

Table 2. The different grids used for the investigation of the p−refinements.

Spectral elements	GLL-order	size of global system	# iterations	L_2 norm		
				Velocity	Vorticity	Pressure
2×2	4	259	132	$9.2 \ 10^{-4}$	$4.8 \ 10^{-2}$	$1.8 \ 10^{-2}$
2×2	6	579	224	$8.7 \ 10^{-6}$	$7.5 \ 10^{-4}$	$1.9 \ 10^{-3}$
2×2	8	1027	305	$6.5 \ 10^{-8}$	$7.1 \ 10^{-6}$	$1.6 \ 10^{-6}$
2×2	10	1603	388	$4.4 \ 10^{-10}$	$4.5 \ 10^{-8}$	$7.6 \ 10^{-9}$

of processors. If the number of processors is larger than the number of cells, processors will become idle and for parallel performance and scalability this result is dramatic.

We remark that four spectral elements and a GLL-order of 8 gives a higher accuracy compared to the grid with 12×12 spectral elements and a GLL-order of 4. Moreover, the systems to solve are much smaller whereas the number of iterations is halved.

4.2   Parallel platforms and implementation

The calculations have been performed on

- Cray T3E system Vermeer (named after the Dutch painter) at HPαC with 128 user PEs interconnected by the fast 3D torus interconnect network with a peak performance of 76.8 Gigaflop/s. Each PE is configured with 128 Mbytes of local memory, providing more than 16 Gbytes of globally addressable distributed memory.
- The SGI Origin 3800 Teras with 1024 500 MHz RI 14000 processors, subdivided into six partitions, t wo (interactive) 32-CPU partitions and four batch partitions of 64, 128, 256 and 512 CPU's, respectively. The theoretical peak performance is 1 Teraflop/s. The Teras is a CC-NUMA machine, Cache-Coherent, Non Uniform Memory Access. For the user the complete memory is accessible, though as a matter of fact the memory is distributed along all

processors. The memory access is not uniform, because each processor can access its own memory much faster than the memory of other processors.

To get good portable programs which may run on distributed-memory multi-processors, networks of workstations as well as shared-memory machines we use MPI, Message Passing Interface. At this moment, **standard** or blocking communication mode is used: a send call does not return until the message data have been safely stored away so that the sender is free to access and overwrite the send buffer. All routines have been implemented in FORTRAN 90.

4.3   Parallel performance and speedups

The grid creation and the calculations of the global systems can be performed completely in parallel and is very fast compared to the solution of the global systems. However, the conversion of the cell distribution to the parallel CSR-format distribution becomes more expensive in case more processors are involved. Table 3 shows wall-clock timings for the Teras of this conversion simulated on a single processor and we do not expect a high parallel speedup for this process that is mainly dominated by communication.

Table 3. Teras: Wall-clock timings in seconds for conversion of cell-wise distribution of grid with 2×2 spectral elements into global matrix in CSR-format, simulated on a single processor.

# Processors converted for	GLL-order 4	6	8	10
1	0.03	0.12	0.35	0.86
2	0.04	0.17	0.51	1.25
4	0.07	0.28	1.01	2.40
8	0.12	0.65	2.31	5.30
16	0.25	1.20	5.85	13.25
32	0.69	4.78	13.64	30.67

In Fig. 3, speedups for the solution part are given for grids with different numbers of spectral elements. The speedups, obtained at Teras and Vermeer, are achieved for 2,4,8,16 and 32 processors. The speedup S_p is defined as the quotient of the wall-clock time measured on one processor and the time measured on p processors. Obviously, the speedup on the distributed memory machine Vermeer is much higher than on the virtual shared memory Teras (cf. Fig. 3a and 3b). Since the SGI MPI-implementation on Teras takes into account that the CPUs *share* the memory, we did not expect this behaviour. The disappointing speedup may be dominated by the *slow* communication compared to its high performance. To get an indication of the performance of both machines, the solution times for grid 8×8 on 1 and 32 processors are listed in Table 4.

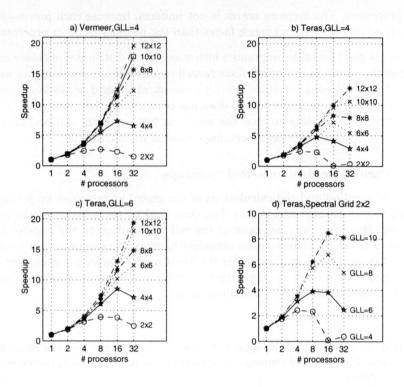

Fig. 3. Speedups achieved on both Vermeer and Teras for different kind of grids.

Table 4. Wall-clock timings in seconds for the solution part obtained for the grid of 12×12 spectral elements and GLL-order=4.

	Teras		Vermeer	
	$p = 1$	$p = 32$	$p = 1$	$p = 32$
	46.9	3.8	314.9	16.0

If we add per spectral element t wo more GLL-collocation points per direction, the computational efforts increase and the speedup on Teras is nearly t wice as much (see Fig. 3c). Finally, Fig. 3d demonstrates that the efficiency of the CG-solution method depends on the GLL-order. Actually, the model problem discussed here appears to be too small for both machines.

5   Conclusions and future plans

The LSQSEM method results in symmetric and positive definite systems of linear equations which can be solved by CG in parallel. At the moment, a Jacobi preconditioner is used that does not converge very fast. Since the total execution time is dominated by solving the linear systems it is necessary to concentrate on good parallelizable preconditioners for these systems. Obviously, we have to complete the parallelization of the conversion part and to reduce communication time by making use of nonblocking MPI-routines. The execution times listed in Fig. 4 indicate that the parallel implementation is very suitable for large-scale problems arising in scientific computing.

Acknowledgements

The authors wish to thank B. Koren for his careful reading of the paper.

References

1. G. F. Carey and B.-N. Jiang. Element-by-element linear and nonlinear solution schemes. *Communications in Appl. Numer. Meth.*, 2:145–153, 1986.
2. G. F. Carey and B.-N. Jiang. Least-squares finite element method and preconditioned conjugate gradient solution. *Int. J. Numer. Meth. Eng.*, 24:1283–1296, 1987.
3. B.-N. Jiang, T. L. Lin, and L. A. Povinelli. Large-scale computation of incompressible viscous flow by least-squares finite element method. *Comput. Methods Appl. Mech. Eng.*, 114:213–231, 1994.
4. G. F. Carey, Y. Shen, and R. T. Mclay. Parallel conjugate gradient performance for least-squares finite elements and transport problems. *Int. J. Num. Meth. Fluids*, 28:1421–1440, 1998.
5. P. B. Bochev and M. D. Gunzburger. Finite element methods of least–squares type. *SIAM Rev.*, 40(4):789–837, 1998.
6. J. M. Deang and M. D. Gunzburger. Issues related to least-squares finite element methods for the Stokes equations. *SIAM J. Sci. Comput.*, 20(3), 1998.
7. B.-N. Jiang. On the least-squares method. *Comput. Methods Appl. Mech. Eng.*, 152:239–257, 1998.
8. B.-N. Jiang and C. L. Chang. Least-squares finite elements for the Stokes problem. *Comput. Methods Appl. Mech. Eng.*, 78:297–311, 1990.
9. B.-N. Jiang and L. Povinelli. Least-squares finite element method for fluid dynamics. *Comput. Methods Appl. Mech. Eng.*, 81:13–37, 1990.
10. B.-N. Jiang. A least-squares finite element method for incompressible Navier-Stokes problems. *Int. J. Num. Meth. Fluids*, 14:843–859, 1992.
11. M. M. J. Proot and M. I. Gerritsma. A least-squares spectral element formulation for the Stokes problem. Accepted in *SIAM J. Sci. Comput.*, 2002.
12. M. M. J. Proot and M. I. Gerritsma. Least-squares spectral elements applied to the Stokes problem: best of all worlds? Submitted to *SIAM J. Sci. Comput.*
13. M. I. Gerritsma and T. N. Phillips. Discontinuous spectral element approximations for the velocity-pressure-stress formulation of the Stokes problem. *Int. J. Numer. Meth. Eng.*, 43:1401–1419, 1998.

Smooth Interfaces for Spectral Element Approximations of Navier-Stokes Equations

Sha Meng[1], Xin Kai Li[2], and Gwynne Evans[2]

[1] Department of Statistics, School of Mathematics,
University of Leeds, Leeds LS2 9JT, England
sha@amsta.leeds.ac.uk
[2] Institute of Simulation Sciences, Faculty of Computing Science and Engineering,
De Montfort University, Leicester LE1 9BH, England
xkl@dmu.ac.uk, gaevans@dmu.ac.uk

Abstract. A smoothing technique is developed to calculate the interface conditions of spectral element method for solving the incompressible Navier-Stokes equations. The first derivative of spectral element solution at the interface is calculated by using only the adjacent element information. Numerical simulations of an incompressible laminar fluid flow through a 2 : 1 planar contraction channel are presented for various Reynolds numbers.

Keywords: Spectral element method, A least square method, Navier-Stokes equations, Contraction channel flow

1 Introduction

Spectral element methods are high-order weighted residual techniques for the solution of partial differential equations typically in computational fluid dynamics [6]. Their success in the recent past in simulating complex flows derives from the flexibility of the method in representing accurately non-trivial geometries while preserving the good resolution properties of spectral method [1]. In the spectral element simulations, both the geometry and the solution are described through smooth functions so that the spectral element methods can obtain exponential accuracy by fully exploiting that regularity [6]. There are numerous fluid dynamics applications, however, where either very steep gradient of solutions or even discontinuous solutions are presented, e.g., a fluid through a channel with abrupt symmetrical contraction, interfaces in multiphase flows, or free surfaces in a die swell. A straightforward application of the spectral element methods in these situations may cause numerical instability as large errors induced by the discontinuous propagate in each element and eventually render the solution with oscillations everywhere. One reason for this instability phenomenon is that the spectral element method only enforces C^0 continuity at interfaces between each element. There have been methods proposed in which continuity of the first

P.M.A. Sloot et al. (Eds.): ICCS 2002, LNCS 2329, pp. 910–919, 2002.

derivatives at the element interfaces is maintained [4], but they have the disadvantage that knowledge of a solution is required across the entire domain. When used on a parallel computer this translates into inter-processor communication which requires an extra amount of time.

In the current work we attempt to develop a spectral element scheme to approximate the interface conditions for the Gauss-Lobatto-Legendre polynomial approximations to the solutions of Navier-Stokes equations, and a smoothing technique to calculate the interface conditions at each element. The main idea presented here is to modify the Gauss-Lobatto-Legendre polynomial basis of the spectral element formulation by using a least square reconstruct procedure implemented on the first derivatives at interfaces of each element, in which the interface values can be calculated by using only the information on the adjacent elements. As a result the proposed interfacial smooth technique is implemented on the Navier-Stokes equations based on a channel flow with a symmetric abrupt contraction.

The paper is organized as follows: in Section 2 we introduce the basic idea upon which the spectral element formulation is based. In Section 3 we describe a smooth method to examine the accuracy between the numerical and analytical solution for the first derivative of trigonometric periodic function. Finally, the numerical simulation of the flow in a symmetric contraction channel is presented. The size of the salient corner vortex and the shape of stream function contours show a good agreement with the work of Dennis *et al.* [3] and Karageorghis *et al.* [5]. The separation length and the strength of the vortex increase as Re increases after $Re > 50$, and the downstream recirculations have been seen in the streamlines.

2   Spectral element approximation

The spectral element method is high-order weighted residual technique for the approximation of partial differential equations that combines the generality of finite element method with the accuracy of spectral method. In this section we will briefly describe the spectral element method based on a simple one-dimensional Poisson equation defined by

$$-u_{xx} = f, \quad x \in I = [a, b], \tag{1}$$

with homogeneous Dirichlet boundary conditions

$$u(a) = u(b) = 0, \tag{2}$$

where f is a given function. Using the Galerkin technique, equations (1)-(2) can now be characterized by the following variational problem:

Find $u \in H_0^1(I)$, such that

$$a(u, v) = (f, v), \quad \forall v \in H_0^1(I), \tag{3}$$

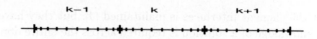

Fig. 1. Spectral element discretization in 1 D.

where

$$a(u,v) = \int_a^b u_x(x)v_x(x)dx, \quad (f,v) = \int_a^b f(x)v(x)dx,$$

$$H_0^1(I) = \{v|v \in L^2(I), v_x \in L^2(I), v(a) = v(b) = 0\},$$

$L^2(I)$ is the space of square integrable functions.

Following standard spectral element procedure, we begin, as usual, by introducing a family of partitions of I such that $I = \cup_{k=1}^K I^k$, $\forall k, l, k \neq l, I_k \cap I_l = \emptyset$, where K denotes the number of elements. In the development, we use N as the degree of the polynomial. Fig. 1 shows the spectral element discretization on one dimensional geometry. On elements $k-1$ and k the grid points are denoted $x_0^{k-1}, ..., x_N^{k-1}$ and $x_0^k, ..., x_N^k$, respectively. And then each element is mapped onto the parent element $\mathcal{I} = [-1,1]$ by using the equation

$$x = \frac{L_k}{2}\xi^k + \frac{x_L^k + x_R^k}{2}, \tag{4}$$

where x_L^k and x_R^k denote the left and right coordinates of the elemental boundaries, L_k is the element length, and $\xi \in \mathcal{I}$. The interpolant of $u(x)$ in the kth element is then represented as

$$u^k(\xi) = \sum_{i=0}^N u_i^k h_i(\xi^k). \tag{5}$$

Here u_i^k are nodal values of u, and $h_i(\xi^k)$ are the basis functions corresponding to element k and node i, with property $h_i(\xi_j^k) = \delta_{ij}$. Expressions for these interpolants in terms of Chebyshev, Lengendre and other polynomials can be found in [2]. In this paper we choose the basis functions are the Gauss-Lobatto-Lengendre polynomials defined as

$$h_i(\xi) = -\frac{1}{N(N+1)L_N(\xi_i)} \frac{(1-\xi^2)L_N'(\xi)}{\xi - \xi_i}, \quad \xi \in \mathcal{I}, \tag{6}$$

where $L_N(\xi)$ is the Legendre polynomial of degree N, $L_N'(\xi) = dL_N(\xi)/d\xi$ and the ξ_i are the Gauss-Lobatto-Legendre collocation points. Furthermore, there also exists a unique set of positive real numbers, ρ_i, corresponding with ξ_i, $(0 \leq i \leq N)$, such that the integration rule

$$\int_{-1}^1 \phi(\xi)d\xi = \sum_{i=0}^N \rho_i\phi(\xi_i), \tag{7}$$

is exact for all polynomials $\phi(x)$ of degree $\leq (2N-1)$ on the interval $[-1,1]$. We follow the standard spectral element method [6], an expansion of the function $u(x)$ can be written in terms of elements as

$$u(x) = \sum_{k=1}^{K} u^k(x) = \sum_{k=1}^{K} \sum_{i=0}^{N} u_i^k h_i(x), \tag{8}$$

where u_i^k are the point values for element k, and x refers to the local coordinate. We require that the approximation function $u(x)$ is continuous through the interface of each element, i.e.,

$$u^k(x_N) = u^{k+1}(x_0), \quad 1 \le k \le K-1, \tag{9}$$

and also satisfies the essential boundary condition

$$u^1(x_0) = u^K(x_N) = 0. \tag{10}$$

Expansion (8) together with the boundary conditions (9) and (10) are now inserted into the weak formulation (3) and the discrete equations are then generated by choosing appropriate functions v which are unity at a point ξ_i and zero at all other Gauss-Lobatto-Legendre points. In matrix form the spectral element procedure for equation (3) can be written as

$$\mathbf{Au} = \mathbf{b}, \tag{11}$$

where $\mathbf{A}$ is the discrete Laplace operator and $\mathbf{b}$ is the right hand side vector with the boundary conditions.

It is clear that the approximation by spectral element discretization $u(x)$ defined by formulation (8) is only C^0 continuity and the first derivative may not continuous across each element. There has been a method proposed by Gottlieb *et al.* [4] in which continuity of the first derivatives at the sub-domain interfaces is maintained for the domain decomposition method. Following the same idea in [4], the element interfaces in the spectral element method can be calculated by enforcing C^1 continuity across each element interface, i.e., the first derivatives are calculated directly from formulation (8)

$$\frac{d}{dx} u(x) = \sum_{k=1}^{K} \frac{d}{dx} u^k(x) = \sum_{k=1}^{K} \sum_{i=0}^{N} u_i^k \frac{d}{dx} h_i(x), \tag{12}$$

and continuity of the first derivatives is enforced at each element interface

$$\frac{d}{dx} u^k(x_N) = \frac{d}{dx} u^{k+1}(x_0), \quad \forall k \in \{1, ..., K-1\}, \tag{13}$$

Equation (13) implies that

$$\sum_{i=0}^{N} u_i^k \frac{d}{dx} h_i(x_N) = \sum_{i=0}^{N} u_i^{k+1} \frac{d}{dx} h_i(x_0). \tag{14}$$

Although the continuity of the first derivative at each element interface is satisfied, the principal disadvantage of this enforcing method is that all of the elements are coupled together, which results in a relatively large amount of inter-processor communication and requires extra computer time to solve, in particular, in the parallel computational programs. The goal of this paper is to propose a simple modification for equation (14) so that the first derivative of the function $u(x)$ is almost continuous at each element interface but only using information from the adjacent elements in the approximation.

3   Smoothing interface method

In this section, we present a smoothing interface method (SIM) based on one-dimensional spectral element approximations. Let $u^{k-1}(x)$ and $u^k(x)$ denote the approximations to the unknown variable $u(x)$ on elements $k-1$ and k, respectively, and they can be represented as

$$u^{k-1}(x^{k-1}) = \sum_{i=0}^{N} u_i^{k-1} h_i(x^{k-1}), \quad u^k(x^k) = \sum_{i=0}^{N} u_i^k h_i(x^k), \tag{15}$$

where u_i^{k-1} and u_i^k are the values of $u(x)$ in the grid points x_i^{k-1} and x_i^k, respectively. Let $d^{k-1}(x)$ and $d^k(x)$ denote the first derivatives of $u^{k-1}(x)$ and $u^k(x)$, they can be written as

$$d^{k-1}(x) = \sum_{i=0}^{N} d_i^{k-1} h_i(x^{k-1}), \quad d^k(x) = \sum_{i=0}^{N} d_i^k h_i(x^k), \tag{16}$$

where d_i^{k-1}, d_i^k are the values of the first derivatives of the variable $u(x)$ at the local nodes x_i^{k-1} and x_i^k, respectively. The idea of the smoothing method is to find an alternative approach to calculate d_i^{k-1} and d_i^k by using the information from the adjacent elements $k-1$ and k only. We wish to determine the best values for d_i^k so that the deviations among d_i^k and $\frac{d}{dx} u(x_i^k)$ are minimized. It turns out that to find a functional Φ such that

$$\min \ \Phi(d_0^{k-1}, d_1^{k-1}, ..., d_N^{k-1}, d_0^k, ..., d_N^k, \lambda), \tag{17}$$

where the functional Φ is given by

$$\Phi = \int_{k-1} \left((d^{k-1}(x) - \frac{d}{dx} u^{k-1}(x) \right)^2 dx + \int_k \left(d^k(x) - \frac{d}{dx} u^k(x) \right)^2 dx$$

$$+ \ \lambda(d_N^{k-1} - d_0^k), \tag{18}$$

and λ is a parameter. If we substitute $d^k(x)$ as defined by equation (16) and $du^k(x)/dx$ as defined by equation (14) into equation (18), we get

$$\Phi = \int_{k-1} \left(\sum_{i=0}^{N} d_i^{k-1} h_i(x) - \sum_{i=0}^{N} u_i^{k-1} \frac{d}{dx} h_i(x) \right)^2 dx$$

$$+ \int_{k} \left(\sum_{i=0}^{N} d_i^{k} h_i(x) - \sum_{i=0}^{N} u_i^{k} \frac{d}{dx} h_i(x) \right)^2 dx + \lambda(d_N^{k-1} - d_0^{k}). \qquad (19)$$

We observe that Φ is an ordinary function of the unknowns d_i^{k-1} and d_i^k as reflected in our notation. To minimize Φ, we need only to take its partial derivatives with respect to each unknown d_i^{k-1} (as well as d_i^k) and set to zero. This implies that, at the minimum, all the partial derivatives $\partial\Phi/\partial d_0^{k-1}, \dots, \partial\Phi/\partial d_N^{k-1}$ and $\partial\Phi/\partial d_0^{k}, \dots, \partial\Phi/\partial d_N^{k}$ vanish. Writing the equations for these gives $2(N+2)$ equations:

$$\frac{\partial\Phi}{\partial d_i^{k-1}} = 2 \int_{k-1} \left(\sum_{i=0}^{N} d_i^{k-1} h_i(x) - \sum_{i=0}^{N} u_i^{k-1} \frac{d}{dx} h_i(x) \right) h_i(x) dx = 0, \quad i = 0, \dots, N-1,$$

$$\frac{\partial\Phi}{\partial d_N^{k-1}} = 2 \int_{k-1} \left(\sum_{i=0}^{N} d_i^{k-1} h_i(x) - \sum_{i=0}^{N} u_i^{k-1} \frac{d}{dx} h_i(x) \right) h_N(x) dx + \lambda = 0,$$

$$\frac{\partial\Phi}{\partial d_i^{k}} = 2 \int_{k} \left(\sum_{i=0}^{N} d_i^{k} h_i(x) - \sum_{i=0}^{N} u_i^{k} \frac{d}{dx} h_i(x) \right) h_i(x) dx = 0, \quad i = 1, \dots, N,$$

$$\frac{\partial\Phi}{\partial d_0^{k}} = 2 \int_{k} \left(\sum_{i=0}^{N} d_i^{k} h_i(x) - \sum_{i=0}^{N} u_i^{k} \frac{d}{dx} h_i(x) \right) h_0(x) dx - \lambda = 0.$$

Solving these equations by using the integration formulation (7), we obtain

$$d_i^{k-1} = \frac{1}{J^{k-1}} \sum_{j=0}^{N} u_j^{k-1} \frac{d}{d\xi} h_j(\xi_i), \quad i = 0, \dots, N-1, \qquad (20)$$

$$d_i^{k} = \frac{1}{J^{k}} \sum_{j=0}^{N} u_j^{k} \frac{d}{d\xi} h_j(\xi_i), \quad i = 1, \dots, N, \qquad (21)$$

$$d_N^{k-1} = d_0^{k} = \frac{1}{J^{k-1} + J^{k}} \sum_{j=0}^{N} \left(u_j^{k-1} \frac{d}{d\xi} h_j(\xi_N) + u_j^{k} \frac{d}{d\xi} h_j(\xi_0) \right), \qquad (22)$$

where J^{k-1} and J^k are the values of Jacobian that come from mapping subdomains Ω_{k-1} and Ω_k onto the parent element $\chi^2 = [-1, 1]$. With the above equations (20)-(22), the interface of elements can be calculated in the sense of least-square approximations by using only information from the adjacent elements $k-1$ and k. It is important to realize that the deviations squared of the first derivatives should continually decrease as the degree of the polynomial is raised. If the approach is implemented on a parallel computer the only communication that is required is between adjacent processors.

Before leaving this section, we show some numerical experiments which illustrate the accuracy of the spectral element method with smoothing interface strategy. We computed the first derivatives of the test function

$$f(x) = |\sin(x)|, \quad -\pi \le x \le \pi. \tag{23}$$

Its first derivative is

$$\frac{d}{dx}f(x) = \begin{cases} -\cos(x), & -\pi \le x \le 0, \\ \cos(x), & 0 \le x \le \pi. \end{cases}$$

Clearly, the first derivative function $df(x)/dx$ is discontinuous at $x = 0$. How well does the smoothing interface method do on such a function? Fig. 2 shows the exact and numerical solutions for the first derivatives $df(x)/dx$ with different numbers of Gauss-Lobatto-Legendre collocation points as well as different numbers of the elements. As illustrated in Fig. 2, the numerical solutions agreed well with the exact solutions when the smoothing interface technique is used. It is also important to note that the smoothing solutions do not have any oscillations at $x = 0$ where the first derivative function $df(x)/dx$ is discontinuous. Table 1 shows that the numerical solutions calculated with and without smooth interface technique by using four elements and twenty-four collocation points on each element. The data show that both algorithms performed well for the discontinuous function $df(x)/dx$. It is quite self-evident that the smoothing solutions (SIM) are more accurate than the solutions without smoothing technique (SEM). This can be seen in Table 1 when $x = -\pi/2$ and $x = \pi/2$ the error for the SIM solutions is 10^{-13}, but it is only 10^{-6} for the SEM solutions.

Table 1. A comparison among the analytical, SIM and SEM solutions with 24 collocation points and 8 elements.

x	Analytical solution of $f'(x)$	SIM solution of $f'(x)$	SEM solution of $f'(x)$
$-\frac{3}{4}\pi$	$.70710670E+00$	$.70710670E+00$	$.70710680E+00$
$-\frac{1}{2}\pi$	$-.75497900E-07$	$-.75497890E-07$	$.36091110E-07$
$-\frac{1}{4}\pi$	$-.70710680E+00$	$-.70710680E+00$	$-.70710670E+00$
$\frac{1}{4}\pi$	$.70710680E+00$	$.70710680E+00$	$.70710690E+00$
$\frac{1}{2}\pi$	$.75497900E-07$	$.75497890E-07$	$.18708690E-06$
$\frac{3}{4}\pi$	$-.70710670E+00$	$-.70710670E+00$	$-.70710660E+00$

4   Numerical results

In this section, numerical results are presented for the incompressible Navier-Stokes flows in a $2:1$ contraction channel by using the spectral element method

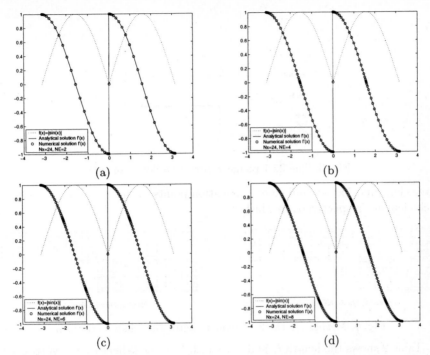

Fig. 2. Numerical solutions with the smoothing interface technique compare with the analytical solutions by using different numbers of collocation points (Nx) and elements (NE).

with smoothing interface technique. The incompressible Navier-Stokes equations are given by:

$$\frac{\partial \mathbf{u}}{\partial t} + \mathbf{u} \cdot \nabla \mathbf{u} - \frac{1}{Re}\triangle \mathbf{u} + \nabla p = \mathbf{f}, \tag{4.6}$$

$$-\nabla \cdot \mathbf{u} = 0, \tag{4.7}$$

where $\mathbf{u} = (u, v)$ is the velocity, p is the pressure, $\mathbf{f}$ is a given function, and $Re = \rho U L / \eta$ is Reynolds number defined the properties of the fluid.

4.1 Planar contraction flow

The 2 : 1 contraction channel flow geometry is shown in Fig. 3. Here the fluid enters upstream in the channel as a fully developed parabolic profile, and exits far downstream as a flat liquid sheet. The 2 : 1 planar contraction channel is chosen in order to compare with the results already published in Dennis *et al.* [3] and Karageorghis *et al.* [5]. In such a geometrical channel, the height of the inflow half channel is taken to be unit and the height of the outflow channel is $a = 1/2$, and the total length of the channel is 4.

Two non-uniform different meshes shown in Fig. 4 were used in the numerical simulations. Mesh 1 has three elements, twelve collocation points in the x direction and six collocation points in the y direction in each element, while there are

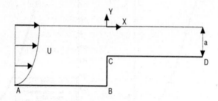

Fig. 3. The 2 : 1 planar contraction flow geometry.

three elements, sixteen and eight collocation points in the x and y directions on each element, respectively, in Mesh 2.

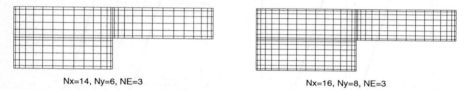

Nx=14, Ny=6, NE=3   Nx=16, Ny=8, NE=3

Fig. 4. Meshes for the 2 : 1 planar contraction channel problem.

Table 2 shows the length l_1 and the width l_2 of the salient corner vortex with different Reynolds numbers on Mesh 1 and Mesh 2. We can see that the size of corner vortex diminishes as Re increases from $Re = 0$ to $Re = 50$, and then begins to grow slowly with $Re > 50$. In Table 3, we compare the values of l_1 and l_2 for various Reynolds numbers on Mesh 2 with the results obtained by Dennis *et al.* [3] and Karageorghis *et al.* [5]. Table 3 shows that our numerical results are in good agreement with those results. Also, we find that l_1 grows more quickly than l_2 when Re increases from 50 to 200, which implies that the corner vortex grows in size along the upstream channel more quickly than up the wall at $x = 0$ as Re increases.

		$Re = 0$	$Re = 1$	$Re = 10$	$Re = 50$	$Re = 100$	$Re = 150$	$Re = 200$
l_1	Mesh 1	0.2548	0.2201	0.1477	0.1280	0.1379	0.1576	0.1870
	Mesh 2	0.2610	0.2399	0.1545	0.1077	0.1389	0.1467	0.1825
l_2	Mesh 1	0.3086	0.2793	0.1767	0.1215	0.1328	0.1474	0.1621
	Mesh 2	0.3181	0.2954	0.1706	0.1396	0.1298	0.1494	0.1706

Table 2. Values of l_1 and l_2 for various Re numbers on the Mesh 1 and Mesh 2.

Contours of the stream function for $Re = 0, 10, 50, 100, 150, 200$ on Mesh 2 are plotted in Fig. 5. The streamline plots give a qualitatively satisfactory to the flow solutions, we can see that the eddy of recirculation is happened for the Stokes flow at $Re = 0$, and as Re increases from zero, the length of corner vortex l_1 initially decreases until at $Re = 50$. For higher value Re number, the eddy length l_1 increases monotonically with Re. This phenomenon shows that the vortex develops as Re is increased.

		$Re=0$	$Re=1$	$Re=10$	$Re=50$	$Re=100$	$Re=150$	$Re=200$
l_1	(a)	0.2610	0.2399	0.1545	0.1077	0.1389	0.1467	0.1825
	(b)	0.268	0.243	0.153	0.128	0.138	–	–
	(c)	0.285	–	0.150	0.129	0.143	0.160	0.183
l_2	(a)	0.3181	0.2954	0.1706	0.1396	0.1298	–	–
	(b)	0.311	0.281	0.163	0.124	0.122	–	–

Table 3. The length and the width of the salient corner vortex for (a) smoothing interface spectral element method; (b) finite-difference scheme of Dennis *et al.* [3]; (c) spectral collocation method of Karageorghis *et al.* [5].

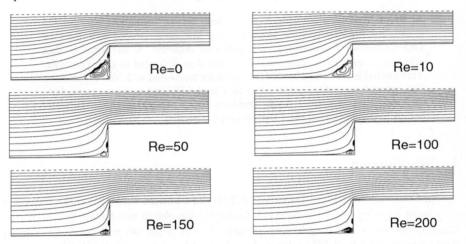

Fig. 5. Flow streamlines for various Re numbers on Mesh 2.

Acknowledgements

Sha Meng gratefully acknowledges the financial support of Ph.D scholarship of De Montfort University of England.

References

1. C. Canuto, M. Hussaini, A. Quarteroni, and T. Zang. *Spectral methods in Fluid Dynamics.* Springer-Verlag, New York, 1987.
2. S. A. Orszag D. Gottlieb. *Numerical Analysis of Spectral methods: Theory and Applications.* SIAM, Philadelphia, 1977.
3. S. C. R. Dennis and F. T. Smith. Stesdy flow through a channel with a symmetrical constriction in the form of a step. *Proc. Roy. Soc. London A*, 372:393–414, 1980.
4. D. Gottlieb and R. S. Hirsh. Parallel pseudospectral domain decomposition techniques. *J. Sci. Comput*, 4:309–325, 1989.
5. A. Karageorghis and T. N. Phillips. Conforming Chebyshev spectral methods for the solution of Laminar flow in a constricted channel. *IMA J. Numer. Anal*, 11:33–54, 1991.
6. Y. Maday and A. T. Patera. Spectral element methods for the incompressible Navier-Stokes equations. *in State of the Art Surverys in Computational Mechanics*, pages 71–143, 1989.

Simulation of a Compressible Flow by the Finite Element Method Using a General Parallel Computing Approach

André Chambarel[1] and Hervé Bolvin[1]

[1] Complex Hydrodynamics Laboratory, Faculté des Sciences, 33 rue Louis Pasteur
F-84000 Avignon, France
andre.chambarel@univ-avignon.fr

Abstract. We have developed a coherent set of techniques for parallel computing. We have used the Finite Element Method associated with the C++ Object-Oriented Programming with only one database. A technique of data selection is used in the determination of the data dedicated to each processor. This method is performed by SIMD technology associated with MPI capabilities. This parallel computing is applied to very large CPU cost problems particularly the unsteady problems or steady problems using iterative methods. Different results in Computational Fluid Dynamics are presented.

1 Introduction

In Computational Fluid Dynamics (CFD) the transient flows generally request a very large memory and CPU time [1]. Generally we obtain this results in a high cost calculus because of the step-by-step process. In this paper we will present a parallel computing method for CFD problems by a Finite Element approach [2]. We propose a coherent method for easy implementation including the following key words:
- Finite Element Method with C++ Object-Oriented Programming code,
- Selection data technique, matrix-free technique and iterative method.

We have developed an easy method for parallel computing which seems to be a natural way of performing intensive computation. Our purpose is to carry out parallel algorithms without modifying the object structure of the solvers, and the data structure [3]. To answer this requirement, we use a selected data method resulting in suitable load balancing with the choice of lists of elements. This technique is independent from the geometry, and can be applied to general cases. This new concept is a natural way for the standardization of parallel codes. In fact, parallelization is here applied to the resolution of a large sized differential system by a semi-implicit algorithm associated with a matrix-free technique.

Among different hardware concepts the SIMD — Single Instruction Multiple Data — architecture has proved to be the most promising for parallel computers. This technology is used for high performance computing especially when problems such as solving large sets of differential equations are dealt with [4]. A SIMD parallel computer consists of a set of processors connected with a fast communication network. Each processor performs the same program with different data. In our work the different data are obtained from a single file and each processor selects its dedicated data. For parallel programming we use the MPI — Message Passing Interface — library.

P.M.A. Sloot et al. (Eds.): ICCS 2002, LNCS 2329, pp. 920–929, 2002.

2 Mathematical Model

In CFD all types of cases of transient compressible flows and oscillating flows can be found. The latter case is obtained by increasing the velocity or the Reynolds number. Consequently the nozzle and its jet are often studied separately. The aim of this paper is to propose a numerical study of the transient flow in a convergent nozzle and its jet, incorporating in particular the unstable solutions.

The Navier-Stokes equations are the mathematical model. The molecular Reynolds number is approximately 10^6. The Reynolds number value Re is based on the diameter of the nozzle outlet and the sound velocity. We have here chosen a Reynolds number of 100, coherent with a zero equation turbulent model [5].

Using the usual notation, the dimensionless Navier-Stokes equations are as follows:

$$\frac{\partial \rho}{\partial t} + \frac{\partial}{\partial x_j}\left(\rho.u_j\right) = 0$$

$$\rho\left(\frac{\partial u_i}{\partial t} + u_j.\frac{\partial u_i}{\partial x_j}\right) = -\frac{1}{\gamma}.\frac{\partial}{\partial x_i}(\rho.T) + \frac{\partial}{\partial x_j}\left[\frac{1}{Re}\left(\frac{\partial u_i}{\partial x_j} + \frac{\partial u_j}{\partial x_i}\right) - \frac{2}{3.Re}.\delta_{ij}\left(\frac{\partial u_k}{\partial x_k}\right)\right]$$

$$\rho\left(\frac{\partial T}{\partial t} + u_j.\frac{\partial T}{\partial x_j}\right) = -(\gamma-1).\rho.T.\frac{\partial u_j}{\partial x_j} + \frac{\partial}{\partial x_j}\left(\frac{\gamma}{Re.Pr}.\frac{\partial T}{\partial x_j}\right) + F(u_i)$$

$$with \quad F(u_i) = \gamma.(\gamma-1).\left[\frac{1}{Re}.\left(\frac{\partial u_i}{\partial x_j} + \frac{\partial u_j}{\partial x_i}\right)^2 - \frac{2}{3.Re}.\left(\frac{\partial u_j}{\partial x_j}\right)^2\right] \tag{1}$$

The domain of integration (Ω) is presented in figure 1. The initial values are the normal thermodynamical conditions and the gas is motionless [5]. The boundary conditions are the following:
-   at the inlet of the nozzle, we simulate the start of a turbo-engine so the time-dependent pressure is as follows:

$$0 \le t \le t_0 : p = p_0 + p_{max}.\frac{t}{t_0} \quad and \quad t \ge t_0 : p = p_0 + p_{max}$$

- in the nozzle, a wall condition for the velocity and adiabatic conditions for the temperature are imposed,
- in the free space boundary, an outflow condition is used.
With the finite element formulation the following matricial form is obtained:

$$\sum_{i=1}^{ne} < \delta U_i > \left([m_i].\frac{\partial}{\partial t}\{U_i\} - \{f_i\} + [k_i].\{U_i\}\right) = 0 \tag{2}$$

After the assemblage process, we built the following differential system:

$$[M].\frac{d}{dt}\{U\}=\{F\}-[K].\{U\} \qquad (3)$$

We built a grid with 10392 nodes and 20346 elements. The differential system above is composed of approximately 40,000 equations.

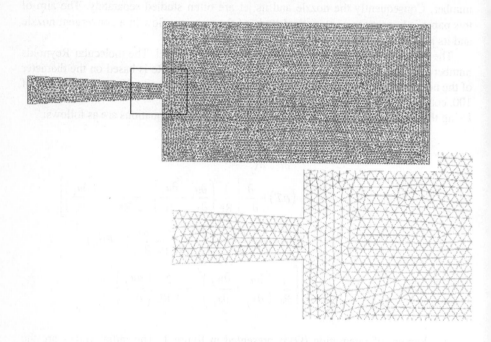

Fig. 1. Finite element grid.

3   Structure of Code

Figure 2 shows the general structure of the compact code. It is organized in three classes corresponding to the functional blocks of the Finite Element Method's different stages. With these classes we built three objects that are connected by a single heritage. So the transmission of the parameters between the objects is defined by a list technique.

We use efficient C++ Object-Oriented Programming for the Finite Element code called FAFEMO (Fast Adaptive Finite Element Modular Object) [4]. In practice, three objects compose a solver and they are separated in three source files. When we built a solver, we merge the three source files of each object. This process is performed by a software called FEOM (Finite Element Object Manager). This technology allows the implementation of very low sized solvers. In our examples their sizes are about 31 Kb — 900 C++ lines — Each solver is dedicated to a problem and can be considered as an element of an algebraic structure. In fact the set of solvers is organized into a valued graph [3].

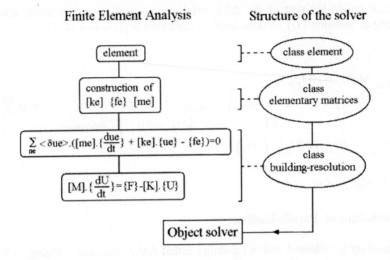

Fig. 2. Object structure of a standard solver.

4   Method of Parallel Computing

4.1 Principle of Parallelization

The principal CPU cost corresponds to the elementary matrices computing and secondarily to the   time step updating. In the case of an unsteady problem, the analytical discretization of the problem with the Finite Element Method is given by the following scalar product [6]:

$$\sum_{NE} \langle \delta ue \rangle .\left([me]\left\{ \frac{due}{dt} \right\} + [ke]\{ue\} - \{fe\} \right) = 0 \text{ with } NE = 1 .. \, ne \qquad (4)$$

If p is the number of processors, we select a list of elements Nk by an expert system called AMS (Automatic Multigrid System) described above:

$$\bigcup_{k=1}^{p} Nk = NE \quad \text{and} \quad Ni \cap Nj = 0 \quad \text{for } i \neq j \qquad (5)$$

Each elementary matrix can be assembled into a global matrix by classical Finite Element process [6]. Each processor computes its part of the global matrices:

$$\sum_{k=1}^{p}[Mk]=[M]\, mass\, matrix \qquad \sum_{k=1}^{p}\{\Psi k\}=\{\Psi\}\, residuum \qquad (6)$$

In this case the Bernstein conditions are verified [1]. So we have a correct load balancing if the lists size of elements is similar for each processor. The communications between the processors exist only at the end of the time step. Each processor builds his part of the differential system and algorithm below allows the updating of solution {U}. A semi-implicit algorithm is presented [4] [7] :

$$t_n = 0$$

$$while \quad (t_n \le t_{max})$$

$$\left\{ \begin{array}{l} for \ j = 1 \ to \ p \quad \left\{\Delta U_n^i\right\}_j = \Delta t_n . \left[M_n^i\right]^{-1} . \left\{\Psi_j \left(U_n + \alpha.\Delta U_n^{i-1}, t_n + \alpha.\Delta t_n\right)\right\} \\ \qquad i = 1, 2, ... \quad until \quad \left\|\Delta U_n^i - \Delta U_n^{i-1}\right\| \le tolerence \end{array} \right\}$$

$$\left\{U_{n+1}\right\} = \left\{U_n\right\} + \left\{\Delta U_n\right\}$$

$$t_{n+1} = t_n + \Delta t_n$$

$$end \quad while$$

4.2 Technique of Parallelization.

Each solver is endowed with a capability called AMS (Automatic Multigrid System). It is an expert system with several possibilities. The applications of this capability are very large, i.g. multiprocessor computing (in this paper), wave front, multi-domain calculus, multi-grid simulation, ...
According to the problem, the AMS expert system can choose different analytical or geometrical components (Fig. 3).
In the case of parallel computing the AMS expert system chooses here the elements dedicated to each processor for the sharing of the scalar product (2) .

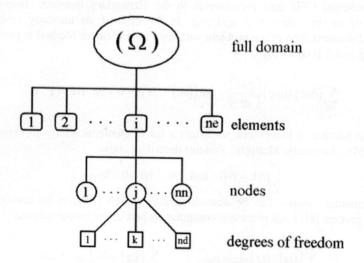

Fig. 3. Taxinomy of the Finite Element parameters.

In fact we can summarize the principal stages of parallelization:
-    A first stage consists in creating an expert system for the selection of the data for each processor.

-     In the second stage each processor calculates its elementary matrix without communication.
-     In the third stage each secondary processor sends its assembled elementary matrices to the principal processor, so that the solution may be updated.

5   Application

Nevertheless, no communications are required between the processors. Each of them performs a completely independent computation for each iteration. This is particularly well adapted to the object structure of the solver. The SIMD architecture is used for the parallel computing management [8]. The AMS capabilities select the data for each processor. The corresponding software is developed with the MPI-C++ library.

Table 1. Parallel computing efficiency.

Equations	CPU time	Nb. of processors	Speed Up (%)
40,000	22 h 28 mn	1	---
40,000	12 h 20	2	87 %

Table 1 presents the parallel efficiency for a CFD problem [9]. We notice that the parallel solver is almost the same as the one used in a sequential process. These examples are performed on a 2 processor-PC. The different examples of parallel computing are applied to unsteady CFD problems. Different cases of compressible flow are presented in figures 4, 5 and 6. Figure 5a presents a qualitative comparison between the experiment and the numerical simulation [10] .

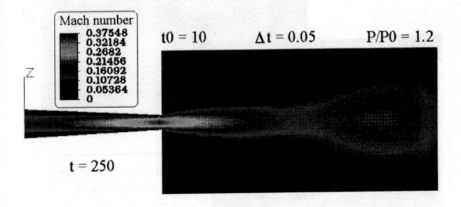

Fig. 4a. Transient compressible flow.

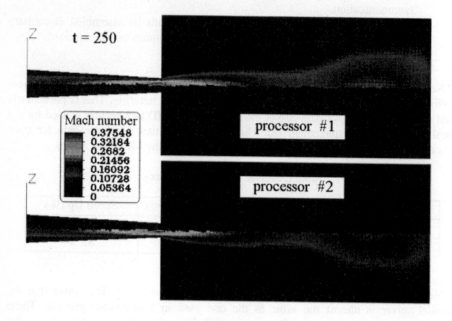

Fig. 4b. Computing with two processors.

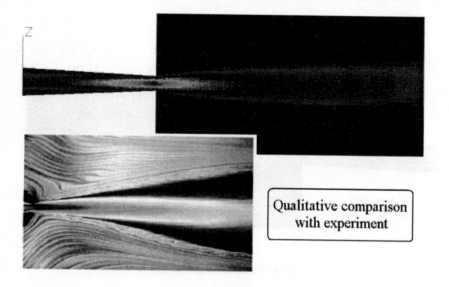

Fig. 5a. Permanent jet.

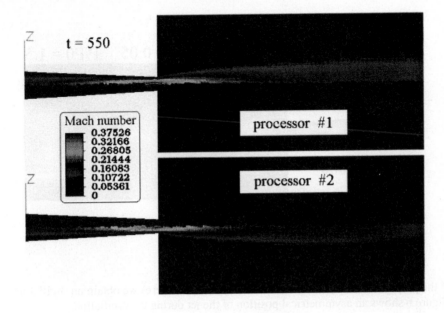

Fig. 5b. Computing with two processors.

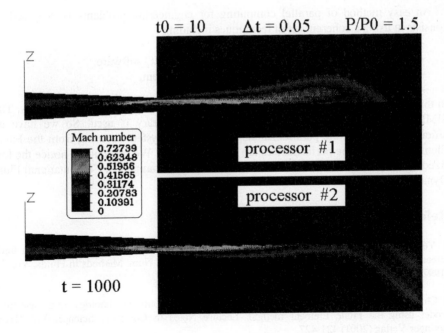

Fig. 6a. Two processor computing of an oscillating jet.

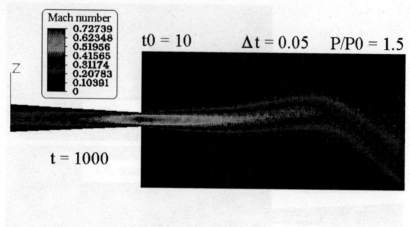

Fig. 6b. Oscillating jet.

If the pressure ratio is greater than a limit value (1.28 here) we obtain an oscillating jet. Figure 6 shows an asymmetrical position of the jet during the oscillation.

6 Conclusion

An easy method of parallel computing for engineering problems is proposed. It consists in using a coherent set of techniques including :
- The Finite Element Method,
- C++ Object-Oriented Programming by FAFEMO software,
- A selection data technique by AMS expert system,
- Matrix-free algorithms.

In this context the implementation of the low sized solvers concerned is very easy. The SIMD architecture associated with the MPI-C++ library is used. So we have an efficient method for the parallelization of differential systems coming from the Finite Element Method. The performances are interesting [11]. We particularly notice the low sized memory and the good load balancing. Different examples in Computational Fluid Dynamics are presented.

References

1. Yeckel, A., Smith, J.W., Derby, J.J.: Parallel finite element calculation of flow in a three dimensional lid-driven cavity using the CM-5 and T3D. Int. J. Num. Methods in Fluids, Vol. 24 (1997) 1449-1461.

2. Chambarel, A., Bolvin, H.: Application of the parallel computing technology to a wave front model using the Finite Element method. Lecture Notes in Computer Science, Vol. 2127, Springer-Verlag (2001) 421-427.

3. Chambarel, A., Onuphre, E.: Finite Element software based on Object Programming. International Conference of the twelfth I.A.S.T.E.D., Annecy France, May 18-20, 1994.

4. Chambarel, A., Ferry, E.: Finite Element formulation for Maxwell's equations with space dependent electric properties. Revue européenne des Eléments Finis, Vol. 9, n° 8 (2000) 941-967.

5. Gutmark, E., Schadow, K.C., Bicker, C.J.: Near acoustic field and shock structure of rectangular supersonic jet. A.I.A.A. Journal, Vol. 28 (1990) 1163-1170.

6. Dhatt, G., Touzot, G.: Une présentation de la méthode des éléments finis. Editions Maloine S.A., Paris (1981).

7. Gresho, P.M.: On the theory of semi-implicit projection methods for viscous incompressible flow and its implementation via a finite element method that also introduces a nearly consistent mass matrix. Int. J. Numer. Meth.Fluids, Vol. 11 (1990) 621-659.

8. Hempel, R., Calkin R., Hess, R., Joppich, W., Keller, U., Koike, N., Oosterlee, C.W., Ritzdorf, H., Washio, T., Wypior, P., Ziegler, W.: Real applications on the new parallel system NEC Cenju-3. Parallel Computing, Vol. 22 (1996) 131-148.

9. Chambarel, A., Fougère, D.: A general parallel computing approach using the Finite Element method and the object-oriented programming by selected data technique. 6th International Conference, PACT 2001, Novosibirsk, Russia, September 3-7, 2001.

10. Gharib, M., Derango, P.:Flow studies of a two-dimensional flow. Physics of fluids, Vol. 31, n°9 (1998) 2389-2394.

11. Laevsky, Y.M., Banushkina, P.V., Litvinenko, S.A., Zotkevich, A.A.: Parallel algorithms for non-stationary problems: survey of new generation of explicit schemes. Lecture Notes in Computer Science, Vol. 2127, Springer-Verlag (2001) 442-446.

A Class of the Relaxation Schemes for Two-Dimensional Euler Systems of Gas Dynamics

Mapundi K. Banda[1] and Mohammed Seaïd[2]

[1] Fachbereich Mathematik, TU Darmstadt, 64289 Darmstadt, Germany
banda@mathematik.tu-darmstadt.de
[2] Fachbereich Mathematik, TU Darmstadt, 64289 Darmstadt, Germany
seaid@mathematik.tu-darmstadt.de

Abstract. In the computation of approximate solutions to hyperbolic conservation laws, relaxation schemes have proven to be very useful. In this paper we present a new higher order relaxation scheme based on higher order nonoscillatory central space discretization and higher order time discretization without use of Riemann solvers. Numerical experiments with 2D Euler systems of gas dynamics are presented to demonstrate the remarkable accuracy of the relaxation scheme.

1 Introduction

In this paper, we further generalise and extend the relaxation schemes of S. Jin, and Z. Xin [4] to higher order accuracy to approximate solutions of compressible Euler system of equations for gas dynamics written in conservative form as:

$$\frac{\partial}{\partial t}\mathbf{U} + \frac{\partial}{\partial x}\mathbf{F}(\mathbf{U}) + \frac{\partial}{\partial y}\mathbf{G}(\mathbf{U}) = 0, \tag{1}$$

where

$$\mathbf{U} = \begin{pmatrix} \rho \\ \rho u \\ \rho v \\ E \end{pmatrix}, \quad \mathbf{F}(\mathbf{U}) = \begin{pmatrix} \rho u \\ \rho u^2 + p \\ \rho u v \\ u(E+p) \end{pmatrix}, \quad \mathbf{G}(\mathbf{U}) = \begin{pmatrix} \rho v \\ \rho u v \\ \rho v^2 + p \\ v(E+p) \end{pmatrix}.$$

Here ρ, u, v, p and E are the density, the x- and y-velocities, the pressure and the total energy, respectively. The equation of state $p = (\gamma - 1)\big(E - \frac{\rho}{2}(u^2 + v^2)\big)$ is required, where the specific heat ratio $\gamma = 1.4$ for an ideal gas.

The eigenvalues of the Jacobian matrix $\partial \mathbf{F}/\partial \mathbf{U}$ (or $\partial \mathbf{G}/\partial \mathbf{U}$) are $\lambda_1 = u - c$, $\lambda_2 = \lambda_3 = u$ and $\lambda_4 = u + c$ (or $\mu_1 = v - c$, $\mu_2 = \mu_3 = v$ and $\mu_4 = v + c$). These are the characteristic speeds for one-dimensional gas dynamics and are needed here only for the estimation of relaxation variables. The sound speed c is defined by $c^2 = \gamma p/\rho$.

Many numerical schemes have been developed to approximate the solutions of the system (1). For instance, the Godunov methods. All these methods are easy to formulate and to implement. What we would like to present here is an

P.M.A. Sloot et al. (Eds.): ICCS 2002, LNCS 2329, pp. 930–939, 2002.
© Springer-Verlag Berlin Heidelberg 2002

alternate scheme which provides high resolution schemes at a low cost. Here the characteristic information of the flow is included but there is no need to solve (approximate) Riemann problems.

Relaxation methods make use of the characteristic variables of the system, finite speed of propagation and do not need Riemann solvers. In the same spirit they are very similar to central schemes [5]. The relaxation system proposed by Jin and Xin [4] and used henceforth for designing schemes is:

$$\frac{\partial \mathbf{U}}{\partial t} + \frac{\partial \mathbf{V}}{\partial x} + \frac{\partial \mathbf{W}}{\partial y} = 0,$$

$$\frac{\partial \mathbf{V}}{\partial t} + \mathbf{A}^2 \frac{\partial \mathbf{U}}{\partial x} = -\frac{1}{\tau}\left(\mathbf{V} - \mathbf{F}(\mathbf{U})\right), \tag{2}$$

$$\frac{\partial \mathbf{W}}{\partial t} + \mathbf{B}^2 \frac{\partial \mathbf{U}}{\partial y} = -\frac{1}{\tau}\left(\mathbf{W} - \mathbf{G}(\mathbf{U})\right),$$

where $\tau > 0$ is the relaxation rate. The matrices $\mathbf{A} = \mathrm{diag}\{a_1, \ldots, a_4\}$ and $\mathbf{B} = \mathrm{diag}\{b_1, \ldots, b_4\}$ are appropriate diagonal matrices. In the zero relaxation limit, $\tau \longrightarrow 0$, solution of (2) approaches a solution of the original system (1) by the local equilibrium

$$\mathbf{V} = \mathbf{F}(\mathbf{U}) \qquad \text{and} \qquad \mathbf{W} = \mathbf{G}(\mathbf{U}), \tag{3}$$

provided the subcharacteristic condition [4,9,7,8], holds:

$$\frac{|\lambda_k|}{a_k} + \frac{|\mu_k|}{b_k} \leq 1, \qquad \forall \quad k = 1, \ldots, 4. \tag{4}$$

The relaxation schemes proposed in [4] are based on the finite difference discretization of (2), where, in particular, the authors consider a first order upwind and a second order MUSCL scheme, together with a second order TVD implicit-explicit time integration scheme. The resulting relaxation schemes are used in the regime $\tau << h$ or even when $\tau \longrightarrow 0$ (relaxed schemes). Here h stands for the space discretization parameter.

Notice that in (2) we construct a linear hyperbolic system with a stiff source term that approximates the original system (1) with a small dissipative correction. The main advantage of considering such a system is that one is able to solve the system (2) numerically with underresolved stable discretizations without either using Riemann solver spatially and nonlinear systems of algebraic equation solvers temporally. Moreover, the relaxation system (2) has linear characteristic variables given by

$$\mathbf{V} \pm \mathbf{A}\mathbf{U}, \qquad \text{and} \qquad \mathbf{W} \pm \mathbf{B}\mathbf{U}. \tag{5}$$

To avoid initial and boundary layer in (2), initial and boundary conditions are chosen to be consistent to the local equilibrium (3).

The paper is organized as follows: In section 2, we formulate a higher order central weighted nonoscillatory space discretization. Then, a higher order TVD time stepping procedure based on implicit-explicit Runge-Kutta methods is described in section 3. Numerical results are presented in section 5. Finally, a brief conclusion is given in section 6.

2   High Order Nonoscillatory Space Discretization

For the space discretization of the equation (2), we cover Ω with rectangular cells $C_{i,j} := [x_{i-\frac{1}{2}}, x_{i+\frac{1}{2}}] \times [y_{j-\frac{1}{2}}, y_{j+\frac{1}{2}}]$ of uniform sizes Δx and Δy with $h = \max(\Delta x, \Delta y)$. The cells, $C_{i,j}$, are centred at $(x_i = i\Delta x, y_j = j\Delta y)$. We use the notation:

$$\omega_{i\pm\frac{1}{2},j}(t) := \omega(x_{i\pm\frac{1}{2}}, y_j, t), \qquad \omega_{i,j\pm\frac{1}{2}}(t) := \omega(x_i, y_{j\pm\frac{1}{2}}, t),$$

$$\text{and} \quad \omega_{i,j}(t) := \frac{1}{\Delta x}\frac{1}{\Delta y}\int_{x_{i-\frac{1}{2}}}^{x_{i+\frac{1}{2}}}\int_{y_{j-\frac{1}{2}}}^{y_{i+\frac{1}{2}}} \omega(x, y, t)\,dx\,dy,$$

to denote the point-values and the approximate cell-average of the function ω at $(x_{i\pm\frac{1}{2}}, y_j, t)$, $(x_i, y_{j\pm\frac{1}{2}}, t)$, and (x_i, y_j, t), respectively. We define the following difference operators

$$\mathcal{D}_x\omega_{i,j} := \frac{\omega_{i+\frac{1}{2},j} - \omega_{i-\frac{1}{2},j}}{\Delta x}, \qquad \mathcal{D}_y\omega_{i,j} := \frac{\omega_{i,j+\frac{1}{2}} - \omega_{i,j-\frac{1}{2}}}{\Delta y} \tag{6}$$

Then, the semi-discrete approximation of (2) is

$$\frac{d\mathbf{U}_{i,j}}{dt} + \mathcal{D}_x\mathbf{V}_{i,j} + \mathcal{D}_y\mathbf{W}_{i,j} = 0,$$

$$\frac{d\mathbf{V}_{i,j}}{dt} + \mathbf{A}^2\mathcal{D}_x\mathbf{U}_{i,j} = -\frac{1}{\tau}\left(\mathbf{V}_{i,j} - \mathbf{F}(\mathbf{U})_{i,j}\right), \tag{7}$$

$$\frac{d\mathbf{W}_{i,j}}{dt} + \mathbf{B}^2\mathcal{D}_y\mathbf{U}_{i,j} = -\frac{1}{\tau}\left(\mathbf{W}_{i,j} - \mathbf{G}(\mathbf{U})_{i,j}\right),$$

The approximate solution is reconstructed by a piecewise polynomial over the grid points as:

$$\omega(x, y, t) = \sum_{i,j} p_{i,j}(x, y; \omega)\chi_{i,j}(x, y), \qquad \chi_{i,j} = \mathbf{1}_{C_{i,j}}, \tag{8}$$

where $p_{i,j}$ are polynomials defined in $C_{i,j}$. The degree of $p_{i,j}$ is determined by the required order of accuracy of the method. In this paper we consider the third order CWENO reconstruction, dimension by dimension [6,3]. Thus,

$$p_{i,j}(x, y; \omega) := p_i(x; \omega) + p_j(y; \omega).$$

Here $p_i(x; \omega)$ and $p_j(y; \omega)$ are x- and y-polynomials. In the following we formulate the x-direction polynomial $p_i(x; \omega)$, the formulation of $p_j(y; \omega)$ can be done analogously.

$$p_i(x; \omega) = W_L P_L(x) + W_R P_R(x) + W_C P_C(x),$$

where

$$W_l = \frac{\alpha_l}{\sum_m \alpha_m}, \quad l, m \in \{L, R, C\}, \qquad \sum_l W_l = 1, \qquad \alpha_l = \frac{c_l}{(\varepsilon + IS_l)^2},$$

$$c_L = c_R = \frac{1}{4}, \quad c_C = \frac{1}{2}, \quad IS_L = (\omega_{i,j} - \omega_{i-1,j})^2, \quad IS_R = (\omega_{i+1,j} - \omega_{i,j})^2,$$

$$IS_C = \frac{13}{3}(\omega_{i+1,j} - 2\omega_{i,j} + \omega_{i-1,j})^2 + \frac{1}{4}(\omega_{i+1,j} - \omega_{i-1,j})^2,$$

$$P_L(x) = \frac{\omega_{i,j}}{2} + \frac{\omega_{i,j} - \omega_{i-1,j}}{\Delta x}(x - x_i), \quad P_R(x) = \frac{\omega_{i,j}}{2} + \frac{\omega_{i+1,j} - \omega_{i,j}}{\Delta x}(x - x_i),$$

$$P_C(x) = \frac{\omega_{i,j}}{2} - \frac{1}{24}(\omega_{i+1,j} - 2\omega_{i,j} + \omega_{i-1,j}) + \frac{\omega_{i+1,j} - \omega_{i-1,j}}{2\Delta x}(x - x_i) + \frac{(\omega_{i+1,j} - 2\omega_{i,j} + \omega_{i-1,j})}{(\Delta x)^2}(x - x_i)^2.$$

The constant ε guarantees that the denominator does not vanish and is empirically taken to be 10^{-6}.

With this background we can now discretize the characteristic variables (5) as follows:

$$(v + a_k u)_{i+\frac{1}{2},j} = p_i(x_{i+\frac{1}{2}}; v + a_k u), \quad (v - a_k u)_{i+\frac{1}{2},j} = p_{i+1}(x_{i+\frac{1}{2}}; v - a_k u);$$

$$(w + b_k u)_{i,j+\frac{1}{2}} = p_j(y_{j+\frac{1}{2}}; w + b_k u), \quad (w - b_k u)_{i,j+\frac{1}{2}} = p_{j+1}(y_{j+\frac{1}{2}}; w - b_k u).$$

Here u, v, w, a_k and b_k are the k-th $(k = 1, \ldots, 4)$ components of $\mathbf{U}$, $\mathbf{V}$, $\mathbf{W}$, $\mathbf{A}$ and $\mathbf{B}$ respectively. Hence:

$$u_{i+\frac{1}{2},j} = \frac{1}{2a_k}\left(p_i(x_{i+\frac{1}{2}}; v + a_k u) - p_{i+1}(x_{i+\frac{1}{2}}; v - a_k u)\right),$$

$$v_{i+\frac{1}{2},j} = \frac{1}{2}\left(p_i(x_{i+\frac{1}{2}}; v + a_k u) + p_{i+1}(x_{i+\frac{1}{2}}; v - a_k u)\right);$$

$$u_{i,j+\frac{1}{2}} = \frac{1}{2b_k}\left(p_j(y_{j+\frac{1}{2}}; w + b_k u) - p_{j+1}(x_{j+\frac{1}{2}}; w - b_k u)\right),$$

$$w_{i,j+\frac{1}{2}} = \frac{1}{2}\left(p_j(x_{j+\frac{1}{2}}; w + b_k u) + p_{j+1}(x_{j+\frac{1}{2}}; w - b_k u)\right).$$

Therefore, we obtain the following expressions for the numerical fluxes in (6)

$$u_{i+\frac{1}{2},j} := \frac{u_{i,j} + u_{i+1,j}}{2} - \frac{v_{i+1,j} - v_{i,j}}{2a_k} + \frac{\sigma_{i,j}^{x,+} + \sigma_{i+1,j}^{x,-}}{4a_k},$$

$$v_{i+\frac{1}{2},j} := \frac{v_{i,j} + v_{i+1,j}}{2} - a_k\frac{u_{i+1,j} - u_{i,j}}{2} + \frac{\sigma_{i,j}^{x,+} - \sigma_{i+1,j}^{x,-}}{4};$$

$$u_{i,j+\frac{1}{2}} := \frac{u_{i,j} + u_{i,j+1}}{2} - \frac{w_{i,j+1} - w_{i,j}}{2b_k} + \frac{\sigma_{i,j}^{y,+} + \sigma_{i,j+1}^{y,-}}{4b_k},$$

$$w_{i,j+\frac{1}{2}} := \frac{w_{i,j} + w_{i+1,j}}{2} - b_k\frac{u_{i,j+1} - u_{i,j}}{2} + \frac{\sigma_{i,j}^{y,+} - \sigma_{i,j+1}^{y,-}}{4},$$

where $\sigma_{i,j}^{x,\pm}$, $\sigma_{i,j}^{y,\pm}$ are defined as:

$$\sigma_{i,j}^{x,\pm} = W_L^\pm\left((v \pm a_k u)_{i,j} - (v \pm a_k u)_{i-1,j}\right) + W_R^\pm\left((v \pm a_k u)_{i+1,j} - (v \pm a_k u)_{i,j}\right)$$

$$+\frac{W_C^\pm}{2}\left((v\pm a_k u)_{i+1,j}-(v\pm a_k u)_{i-1,j}\right)$$

$$+\frac{W_C^\pm}{3}\left((v\pm a_k u)_{i+1,j}-2(v\pm a_k u)_{i,j}+(v\pm a_k u)_{i-1,j}\right)$$

$$-\frac{W_C^\pm}{6}\left((v\pm a_k u)_{i,j+1}-2(v\pm a_k u)_{i,j}+(v\pm a_k u)_{i,j-1}\right),$$

$$\sigma_{i,j}^{y,\pm}=W_L^\pm\left((w\pm b_k u)_{i,j}-(w\pm b_k u)_{i,j-1}\right)+W_R^\pm\left((w\pm b_k u)_{i,j+1}-(w\pm b_k u)_{i,j}\right)$$

$$+\frac{W_C^\pm}{2}\left((w\pm b_k u)_{i,j+1}-(w\pm b_k u)_{i,j-1}\right)$$

$$+\frac{W_C^\pm}{3}\left((w\pm b_k u)_{i,j+1}-2(w\pm b_k u)_{i,j}+(w\pm b_k u)_{i,j-1}\right)$$

$$-\frac{W_C^\pm}{6}\left((w\pm b_k u)_{i+1,j}-2(w\pm b_k u)_{i,j}+(w\pm b_k u)_{i-1,j}\right).$$

The weight parameters $W_L^\pm$, $W_R^\pm$ and $W_C^\pm$ for $\sigma_{i,j}^{x,\pm}$ are given by

$$W_l^\pm=\frac{\alpha_l^\pm}{\sum_k \alpha_k^\pm},\quad l,k\in\{L,R,C\},\quad \sum_l W_l^\pm=1,\quad \alpha_l^\pm=\frac{c_l}{(\varepsilon\pm IS_l)^2},$$

$$c_L=c_R=\frac{1}{4},\qquad c_C=\frac{1}{2}$$

$$IS_L^\pm=\left((v\pm a_k u)_i-(v\pm a_k u)_{i-1}\right)^2,\quad IS_R^\pm=\left((v\pm a_k u)_{i+1}-(v\pm a_k u)_i\right)^2$$

$$IS_C^\pm=\frac{13}{3}\left((v\pm a_k u)_{i+1}-2(v\pm a_k u)_i+(v\pm a_k u)_{i-1}\right)^2+$$
$$\frac{1}{4}\left((v\pm a_k u)_{i+1}-(v\pm a_k u)_{i-1}\right)^2.$$

The corresponding weight parameters for $\sigma_{i,j}^{y,\pm}$ are obtained by changing $v\pm a_k u$ to $w\pm b_k u$ in the above formulas.

We close by pointing out that in this higher order scheme we approximate $\mathbf{F}(\mathbf{U})_{i,j}$ and $\mathbf{G}(\mathbf{U})_{i,j}$ in (2) using the fourth-order Simpson quadrature rule as opposed to the Midpoint Rule which was used in the first and second order cases in [4].

3   High Order TVD Time Discretization

The semi-discrete formulation (7) can be rewritten as a system of ordinary differential equations:

$$\frac{d\mathcal{Y}}{dt}=\mathcal{F}(\mathcal{Y})-\frac{1}{\tau}\mathcal{G}(\mathcal{Y}),\tag{9}$$

where the time-dependent vector functions $\mathcal{Y} := \left(\mathbf{U}_{i,j}, \mathbf{V}_{i,j}, \mathbf{W}_{i,j}\right)^T$,

$$\mathcal{F}(\mathcal{Y}) := \left(-\mathcal{D}_x \mathbf{V}_{i,j} - \mathcal{D}_y \mathbf{W}_{i,j}, -\mathbf{A}^2 \mathcal{D}_x \mathbf{U}_{i,j}, -\mathbf{B}^2 \mathcal{D}_y \mathbf{U}_{i,j}\right)^T$$

and $\mathcal{G}(\mathcal{Y}) := \left(0, \mathbf{V}_{i,j} - \mathbf{F}(\mathbf{U})_{i,j}, \mathbf{W}_{i,j} - \mathbf{G}(\mathbf{U})_{i,j}\right)^T$. Due to the presence of the stiff term in (7), one can not use explicit schemes to integrate the equations (9), particularly when $\tau \longrightarrow 0$. On the other hand, integrating the equations (9) by implicit scheme, either nonlinear or linear algebraic equations have to be solved at every time step of the computational process. To find solutions of such systems is computationally very demanding. In this paper we consider an alternative approach based on Implicit-Explicit (IMEX) Runge-Kutta splittings. The non stiff stage of the splitting for $\mathcal{F}$ is straightforwardly treated by an explicit Runge-Kutta scheme, while the stiff stage for $\mathcal{G}$ is approximated by a diagonally implicit Runge-Kutta (DIRK) scheme. Compare [2, 10] for more details.

Let Δt be the time step and $\mathcal{Y}^n$ denote the approximate solution at $t = n\Delta t$. We formulate the IMEX scheme for the system (9) as:

$$K_l = \mathcal{Y}^n + \Delta t \sum_{m=1}^{l-1} \tilde{a}_{lm} \mathcal{F}(K_m) - \frac{\Delta t}{\tau} \sum_{m=1}^{s} a_{lm} \mathcal{G}(K_m), \quad l = 1, 2, \ldots, s,$$

$$\mathcal{Y}^{n+1} = \mathcal{Y}^n + \Delta t \sum_{l=1}^{s} \tilde{b}_l \mathcal{F}(K_l) - \frac{\Delta t}{\tau} \sum_{l=1}^{s} b_l \mathcal{G}(K_l).$$

(10)

The $s \times s$ matrices $\tilde{A} = (\tilde{a}_{lm})$; $A = (a_{lm})$ and the s-vectors $\tilde{b}$; b are the standard coefficients which characterize the IMEX s-stage Runge-Kutta scheme. They are given by the usual double Butcher tables

0	0	0	0	0	0
$\tilde{c}_2$	$\tilde{a}_{21}$	0	0	0	0
$\tilde{c}_3$	$\tilde{a}_{31}$	$\tilde{a}_{32}$	0	0	0
$\vdots$	$\vdots$	$\vdots$	$\vdots$	$\vdots$	$\vdots$
$\tilde{c}_s$	$\tilde{a}_{s1}$	$\tilde{a}_{s2}$	$\cdots$	$\tilde{a}_{ss-1}$	0
	$\tilde{b}_1$	$\tilde{b}_2$	$\cdots$	$\tilde{b}_{s-1}$	$\tilde{b}_s$

0	0	0	0	0	0
c_2	a_{21}	a_{22}	0	0	0
c_3	a_{31}	a_{32}	a_{33}	0	0
$\vdots$	$\vdots$	$\vdots$	$\vdots$	$\vdots$	$\vdots$
c_s	a_{s1}	a_{s2}	$\cdots$	a_{ss-1}	a_{ss}
	b_1	b_2	$\cdots$	b_{s-1}	b_s

The left and right tables represent the explicit and the implicit Runge-Kutta methods. Then, the implementation of the IMEX algorithm to solve (9) is carried out in the two following steps:

1. For $l = 1, \ldots, s$,

   (a) Evaluate K_l^* as: $K_l^* = \mathcal{Y}^n + \Delta t \sum_{m=1}^{l-2} \tilde{a}_{lm} \mathcal{F}(K_m) + \Delta t \tilde{a}_{ll-1} \mathcal{F}(k_{l-1})$.

   (b) Solve for K_l: $K_l = K_l^* - \frac{\Delta t}{\tau} \sum_{m=1}^{l-1} a_{lm} \mathcal{G}(K_m) - \frac{\Delta t}{\tau} a_{ll} \mathcal{G}(K_l)$.

2. Update $\mathcal{Y}^{n+1}$ as: $\mathcal{Y}^{n+1} = \mathcal{Y}^n + \sum_{l=1}^{s} \tilde{b}_l \mathcal{F}(K_l) - \frac{\Delta t}{\tau} \sum_{l=1}^{s} b_l \mathcal{G}(K_l)$.

Recall that what we call high order relaxation schemes are composed of the high order reconstruction (8) augmented with the high order IMEX splitting (10).

Notice that, using the relaxation scheme neither linear algebraic equation nor nonlinear source terms can arise. In addition the high order relaxation scheme is stable independently of τ, so that the choice of Δt is based only on the usual CFL condition

$$\text{CFL} := \max_{1 \le i,j \le 4}\left(\frac{\Delta t}{h},\; a_i^2 \frac{\Delta t}{\Delta x},\; b_j^2 \frac{\Delta t}{\Delta y}\right) \le 1. \tag{11}$$

In our numerical computation we consider the third order IMEX scheme developed in [2], the associated double Butcher tables can be represented as:

$$
\begin{array}{c|ccccc}
0 & 0 & 0 & 0 & 0 & 0\\
\frac{1}{2} & \frac{1}{2} & 0 & 0 & 0 & 0\\
\frac{2}{3} & \frac{11}{18} & \frac{1}{18} & 0 & 0 & 0\\
\frac{3}{4} & \frac{5}{6} & -\frac{5}{6} & \frac{1}{2} & 0 & 0\\
\frac{1}{2} & \frac{7}{6} & -\frac{6}{6} & \frac{7}{4} & -\frac{7}{4} & 0\\
1 & \frac{1}{4} & \frac{7}{4} & \frac{3}{4} & -\frac{7}{4} & 0\\
\hline
 & \frac{1}{4} & \frac{7}{4} & \frac{3}{4} & -\frac{7}{4} & 0
\end{array}
\qquad
\begin{array}{c|ccccc}
0 & 0 & 0 & 0 & 0 & 0\\
\frac{1}{2} & 0 & \frac{1}{2} & 0 & 0 & 0\\
\frac{2}{3} & 0 & \frac{1}{6} & \frac{1}{2} & 0 & 0\\
\frac{1}{2} & 0 & -\frac{1}{2} & \frac{1}{2} & \frac{1}{2} & 0\\
1 & 0 & \frac{3}{2} & -\frac{3}{2} & \frac{1}{2} & \frac{1}{2}\\
\hline
 & 0 & \frac{3}{2} & -\frac{3}{2} & \frac{1}{2} & \frac{1}{2}
\end{array}
$$

Obviously, at the limit ($\tau \longrightarrow 0$) the time integration procedure tends to a time integration scheme of the limit equations based on the explicit scheme given by the left table in (10).

Remark. Note that the first and second order relaxation schemes studied earlier in [4] can be viewed as (8) taking

$$P_{i,j}(x,y;\omega) = \omega_{i,j} \quad \text{and} \quad P_{i,j}(x,y;\omega) = \omega + \frac{\dot\omega_{i,j}}{\Delta x}(x - x_i) + \frac{\dot\omega_{i,j}}{\Delta y}(y - y_j),$$

respectively. Here $\dot\omega_{i,j}/\Delta x$ and $\dot\omega_{i,j}/\Delta y$ are discrete slopes in the x and y directions. The time integration procedure in [4] can be represented as (10) where the explicit and implicit tables are given by

$$
\begin{array}{c|cc}
0 & 0 & 0\\
1 & 1 & 0\\
\hline
 & \frac{1}{2} & \frac{1}{2}
\end{array}
\qquad\qquad
\begin{array}{c|cc}
-1 & -1 & 0\\
2 & 1 & 1\\
\hline
 & \frac{1}{2} & \frac{1}{2}
\end{array}
$$

4   Numerical Results

We present numerical experiments for several problems (1) using the scheme introduced above. We solve three model problems. All of them have been used extensively in literature to test various numerical schemes. In all our computation the computational domain Ω is divided in $Nx \times Ny$ grid points, the relaxation rate τ is fixed to 10^{-6} and the relaxation matrices $\mathbf{A}$ and $\mathbf{B}$ are chosen locally according to (4) as:

$$a_k = 2|\lambda_k|, \quad \text{and} \quad b_k = 2|\mu_k|, \quad k = 1,\ldots 4.$$

The following test cases are selected:

The Shock Reflection Problem. This problem was solved by Jin and Xin [4] using the second order relaxation scheme. In our computation we take the same parameters as [4]. Thus, $\Omega = [0,4] \times [0,1]$; initially the domain Ω is filled by a free-stream supersonic inflow with Mach number 2.9. The Dirichlet boundary conditions are imposed at left and upper boundaries as

$$\mathbf{U}(0,y,t) = (1, 2.9, 0, 5.99071)^T,$$
$$\mathbf{U}(x,1,t) = (1.69997, 4.45279, -0.86074, 21.30317)^T.$$

The bottom boundary is a reflecting wall and the supersonic outflow condition is applied along the right boundary. The simulation is performed until $t = 5$ using $\Delta t = 0.005$. Plots of the pressure are shown in Fig. 1 using 30 equi-distributed contours. As can be seen from this figure, the reflected shock was very well captured by the relaxation scheme.

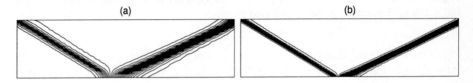

Fig. 1. The shock reflection problem. 30 equi-distributed pressure contours. (a) 60×20 grid points; (b) 120×40 grid points.

The Double-Sod Tube Shock Problem. This example is inspired by the standard 1D Sod tube shock problem [1]. Hence $\Omega = [-1,1] \times [-1,1]$ and the initial conditions are chosen as:

$$\mathbf{U}(x,y,0) = \begin{cases} (0.1, 0, 0, 0.25)^T & \text{if } xy < 0, \\ (1, 0, 0, 2.5)^T & \text{otherwise.} \end{cases}$$

Homogeneous Neumann boundary condition were used, and $\Delta t = 0.001$.

In Fig. 2 we display 30 equi-distributed contour plots of the density at time $t = 0.16$. The high resolution of the new relaxation scheme is clearly visible.

The Double-Mach Reflection Problem. This test example consist of the canonical double Mach reflection problem [11]. The domain $\Omega = [0,4] \times [0,1]$. The reflecting wall lies at the bottom of the computational domain starting from $x = \frac{1}{6}$. Initially a right-moving Mach 10 shock is positioned at $x = \frac{1}{6}$, $y = 0$ and makes a $60°$ angle with the x-axis. For the bottom boundary, the exact post-shock condition is imposed for the part from $x = 0$ to $x = \frac{1}{6}$ and a reflective boundary condition is used for the rest. At the top boundary of the domain Ω, the flow values are set to describe the exact motion of the Mach 10 shock.

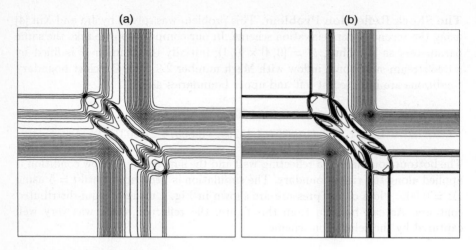

Fig. 2. The double-Sod tube shock problem. 30 equi-distributed density contours. (a) 100×100 grid points; (b) 200×200 grid points.

Fig. 3 shows 30 equi-distributed contour plots of the density at time $t = 0.2$ with $\Delta t = 0.0005$. Only part of Ω, $[0,3] \times [0,1]$, is shown. We can see the complicated structures being captured by the new relaxation scheme.

Fig. 3. The double-Mach reflection problem. 30 equi-distributed pressure contours. (a) 120×30 grid points; (b) 240×60 grid points.

5   Concluding Remarks

We have described here a third order relaxation scheme for two-dimensional Euler system of equation for inviscid flow. The system of equations is reformulated into a relaxing system. For the space discretization a generalisation of the interpolating polynomial is presented and here a third order CWENO reconstruction is used. For the time integration a third order Runge-Kutta splitting has been used. In this approach very good accuracy is achieved without using Riemann solvers nor solving nonlinear systems. The ability of the methods to handle such nonlinear systems allows for generalization to a much broader set of hyperbolic system equations.

Table 1. CPU time in minutes for the above test problems

Test problem	Grid points	Δt	N$\underline{\text{o}}$ steps	CPU
The Shock Reflection Problem	120×40	0.005	1000	16
The Double-Sod Tube Shock Problem	100×100	0.001	160	7
The Double-Mach Reflection Problem	240×60	0.0005	400	12

The multidimensional algorithms presented in this paper can be highly optimized for the vector computers, because they are explicit procedures and contain no recursive elements. Some difficulties arise from the fact that for efficient vectorization the data should be stored continuously within long vectors rather than two-dimensional arrays. For completeness, we summarize in table 1 the CPU time, measured on a PC with AMD-K6 200 processor running FORTRAN code under Linux 2.2, for the test problems presented in this paper.

Future directions for this work include the following: improvement to include unstructured grids, development of refinement strategies, and three-dimensional problems.

References

1. Aregba-Driollet, D., Natalini, R.: Discrete Kinetic Schemes for Multidimensional Systems of Conservation Laws. SIAM J. Numer. Anal. **37** (2000) 1973–2004
2. Ascher, U., Ruuth, S., Spiteri, R.: Implicit-Explicit Runge-Kutta Methods for Time-Dependent Partial Differential Equations. Appl. Numer. Math. **25** (1997) 151–167
3. Banda, M.K., Klar, A., Pareschi, L., Seaïd, M.: Lattice-Boltzmann Type Relaxation Systems and Higher Order Relaxation Schemes for The Incompressible Navier-Stokes Equation. Submitted
4. Jin S., Xin, Z.: The Relaxation Schemes for Systems of Conservation Laws in Arbitrary Space Dimensions. Comm. Pure Appl. Math. **48** (1995) 235–276
5. Kurganov, A., Tadmor, E.: New High-Resolution Central Schemes for Nonlinear Conservation Laws of Convection-Diffusion Equations. J. Comp. Phys. **160** (2000) 241–282
6. Kurganov, A., Levy, D.: A Third-Order Semi-Discrete Central Scheme for Conservation Laws and Convection-Diffusion Equations. SIAM J. Sci. Comp. **22** (2000) 1461–1488
7. Lattanzio C., Serre D.: Convergence of a relaxation scheme for hyperbolic systems of conservation laws. Numer. Math. **88** (2001) 121–134
8. Liu, H.L., WarneckeG.: Convergence rates for relaxation schemes approximating conservation laws. SIAM J. Numer. Anal. **37** (2000) 1316–1337
9. Natalini, R.: Convergence to equilibrium for relaxation approximations of conservation laws. Comm. Pure Appl. Math. **49** (1996) 795–823
10. Pareschi, L., RussoG.: Implicit-Explicit Runge-Kutta Schemes for Stiff Systems of Differential Equations. Preprint
11. Woodward, P., Colella, P.: The Numerical Simulation of Two-Dimensional Fluid Flow with Strong Shocks. J. Comp. Physics. **54** (1984) 115–173

OpenMP Parallelism for Multi-dimensional Grid-Adaptive Magnetohydrodynamic Simulations

R. Keppens[1] and G. Tóth[2]

[1] FOM-Instituut voor Plasma-Fysica Rijnhuizen, P.O. Box 1207,
3430 BE Nieuwegein, The Netherlands
[2] Department of Atomic Physics, Eötvös University,
Pázmány Péter sétány 1, 1117 Budapest, Hungary

Abstract. First results on the parallelism achieved by both automated and manually controlled OpenMP[3] programming are reported for 2D and 3D magnetohydrodynamic computations. The simulations exploit adaptive mesh refinement, capable of capturing flow features like shocks and other sharp discontinuities accurately and efficiently. Implementation details are discussed, alongside with their scaling properties on realistic plasma flow simulations.

1 Introduction

This paper focuses on the parallelization of a general-purpose software package – AMRVAC – for grid-adaptive numerical simulations. The AMRVAC package combines the versatility of the Versatile Advection Code (VAC[4]) with the advantages of Adaptive Mesh Refinement (AMR). AMRVAC [5, 4, 8] is particularly suited for time-evolving physical systems governed by (near-) conservation laws, e.g. the ideal magnetohydrodynamic (MHD) equations expressing conservation of mass, momentum, energy, and magnetic flux. The generality of the software resides in (i) a choice of the actual equations to solve for; (ii) a total independence of the implementation on the dimensionality of the problem (1D, 1.5D, 2D, 2.5D and 3D configurations are possible thanks to the LASY syntax [11]); and (iii) the availability of several high-resolution, shock-capturing spatial discretizations. Its grid-adaptivity follows an AMR technique, for which data structures and algorithmic issues handling the automated production of subgrids have been introduced by Berger [1]. Only fairly recently, this solution adaptive regridding strategy has been applied in multi-dimensional MHD simulations [10, 3, 9].

We recall that the AMR process involves the generation and destruction of hierarchically nested grids of subsequently finer resolution. In Keppens et al. [5], we concentrated on algorithmic improvements of the regridding and evaluated obtainable efficiencies by dynamic meshing for a variety of 1D, 2D and 3D problems. This 'efficiency' was characterized by the reduction in serial execution

---

[3] See http://www.openmp.org.
[4] See http://www.phys.uu.nl/~toth

P.M.A. Sloot et al. (Eds.): ICCS 2002, LNCS 2329, pp. 940–949, 2002.

times when comparing high resolution static grid simulations with corresponding AMR runs. In the latter simulations, fine meshes are triggered only when and where needed, significantly reducing computing costs. Furthermore, the memory requirements drop accordingly (by an order of magnitude for a 2D hydrodynamic shock problem as shown in [8]), allowing for much more realistic simulations of plasma behavior in regimes where both large-scale and small-scale structures are dynamically important.

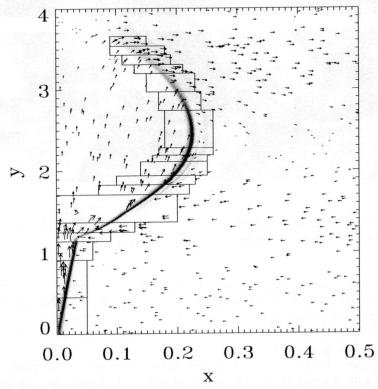

Fig. 1. A snapshot of the density (Schlieren contour plot) and the flow field at time $t = 20$ in a 2D resistive MHD reconnection simulation. The grid structure on the finest level $l = 4$ is indicated.

1.1   Grid-Adaptive Magnetofluid Simulations

Figs. 1–2 illustrate two MHD applications which greatly benefit from dynamically evolving meshes. Fig. 1 shows a snapshot of the density structure and the plasma flow field in a 2D magnetic reconnection problem. The setup is taken from [12] and considers a sideways (x-direction) mass inflow from $x = \pm 1$ at an Alfvén Mach number $M_A = 0.04$ and a plasma beta $\beta(x = 1) = 1.5$. The latter

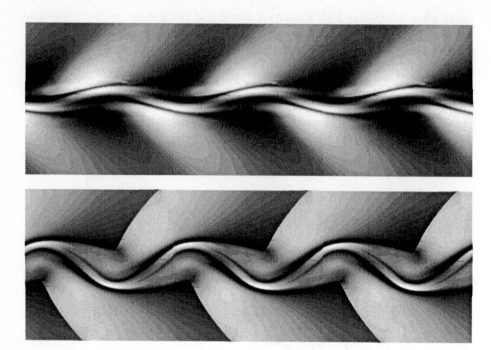

Fig. 2. The evolution of the density in a sinuously unstable, super-Alfvénic wake flow. Three periodic segments are shown at times $t = 110$ and $t = 150$.

quantity measures the ratio of thermal to magnetic pressure $\beta = 2p/B^2$. We start from a static, isothermal equilibrium state where $B_x = 0$, $B_y = \tanh(10x)$ and the pressure profile is given by $p(x) = 0.5\,B_y^2(x=1)[\beta(x=1)+1] - B_y^2(x)/2$. Due to a spatially localized anomalous resistivity which allows for topological changes in the magnetic field structure around the origin, the initial magnetic field variation changes into an X-shaped feature centered on the origin. Symmetry arguments may reduce the simulated domain to a quadrant, the central region of which is shown in Fig. 1. We allowed for 4 grid levels, such that we locally achieve a resolution where $(\Delta x, \Delta y) = (0.0025, 0.005)$. Only the highest grid level structure, which nicely traces the locations of steep gradients, is indicated in the figure. As a second example, Fig. 2 shows the temporal evolution of the density in a supersonic (Mach number $M = 3$), super-Alfvénic $M_A = 5$ wake. The problem description is given by a planar velocity profile $\mathbf{v} = [1 - sech(y)]\hat{\mathbf{e}}_x$ pervaded by a 3D magnetic field configuration with components given by $B_x = A\tanh(y)$, $B_z = A\,sech(y)$. The stability properties of such current carrying, shear flow states were analyzed in [2]. Cross-stream (vertical in the figure) perturbations lead under the chosen parameter values to transverse striations of the wake, accompanied by rightwardly traveling compressive variations reaching far out in the cross-stream direction. These ultimately lead to magnetosonic shock fronts, clearly visible in the density evolution (bottom panel of the figure). The simulation shown in Fig. 2 exploited three grid levels, with a base resolution of 100×200 per periodic segment, locally reaching 800×1600. Again, the finest level grids (not shown) only cover regions with sharp flow features, and at all times maximally covered about 20% of the computational domain.

1.2   Parallel AMR

The data structures used in AMRVAC are discussed in [8]. It should be noted that the Berger [1] approach to create higher level (i.e., finer) grids fits these grids optimally to the flow details, so that possibly many different sized grids are present on a certain grid level (see, e.g. Fig. 1). Their different sizes pose a potential load-balancing problem when parallelizing a mesh refinement code over the grids. This difficulty can be overcome by reformulating the original AMR procedure to enforce the creation of equally sized blocks, as realized in parallel MPI implementations in [9, 7]. Without switching to this block-adaptive strategy, virtually shared memory parallelization for ccNUMA[5] architectures making use of OpenMP offers a viable alternative to parallel grid-adaptive simulations. In [4], first attempts with auto-parallelization on the SGI Origin 3800 were reported: while linear scaling was demonstrated for non-adaptive, Domain Decomposition (DD) usage of AMRVAC, experiments with fully grid-adaptive simulations showed no speedup. In the remaining of this paper, we detail that a suitable manual OpenMP parallelization effort is able to achieve fair to good scaling properties for multi-dimensional, multi-level AMR calculations.

2   OpenMP Parallelization

2.1   Automated versus Manual OpenMP Parallelism

At a minimal effort, we can rely on the present version (7.3.1.2m) of the MIPSpro Fortran 90 compiler as available on the SGI Origin 3800 at Amsterdam to automatically generate parallel code. This auto-parallelization almost necessarily parallelizes individual do loops, which should contain sufficient, parallelizable operations. For AMRVAC, most computations are associated with the discretized explicit time advancing of the set of unknowns (like density ρ, momenta $\rho \mathbf{v}$, energy e, and magnetic field $\mathbf{B}$ for MHD problems) for each individual grid at a certain AMR level. Since this coincides with the lowest level subroutines in the AMRVAC calling tree, we expect the automated parallelization to scale well, as long as there is sufficient work to be done on each grid seperately.

This is indeed confirmed by timing experiments of AMRVAC shown in Fig. 3 for a full 3D MHD jet simulation. The problem is taken from Keppens and Tóth [6], and considers a cylindrical jet flow with $\mathbf{v} = V_o \tanh(r - R_{jet})/0.1 R_{jet} \hat{\mathbf{e}}_x$, in a uniform magnetic field $\mathbf{B} = B_o \hat{\mathbf{e}}_x$, when the shear flow about radius $r = R_{jet}$

---

[5] ccNUMA: cache coherent Non-Uniform Memory Access.

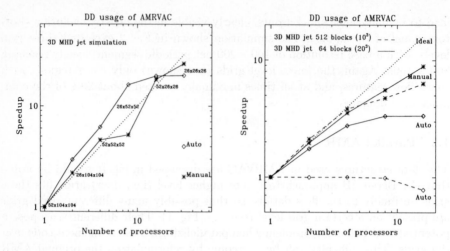

Fig. 3. Scaling behavior of both automated and manual parallelism on a 3D MHD simulation using AMRVAC as a Domain Decompositioner. *Left panel:* changing the number of blocks in accord with the number of processors. *Right panel:* running a fixed number of blocks on a varying number of processors. In both panels, the total problem size is kept constant.

is Kelvin-Helmholtz destabilized. Note that we fixed the number of refinement levels to one, so AMRVAC is only used as a Domain Decompositioner. Hence, the grid does not adjust dynamically, but is merely split into size-controlled equal blocks. The left panel of Fig. 3 shows that the automated parallelization can even achieve superlinear speedup behavior as we split a fixed size 3D problem of resolution $52 \times 104 \times 104$ in N blocks and run the N-block case on N processors. The scaling continues up to 16 processors, but no longer improves when going to 32 processors as the block size of the domains becomes 26^3. The superlinearity at lower processor numbers is due to better cache usage when the block size changes from the single processor run. Unfortunately, in our dimension-independent, general implementation the cache usage is difficult to control actively (as an indication, we probably reach at most 10% of the peak performance on 1 processor for this particular problem).

The right panel in Fig. 3 confirms that the individual block size is the critical factor for parallel efficiency in auto-parallelized code. The same 3D MHD problem was now run on a 80^3 grid, first split into 64 blocks of size 20^3 and subsequently run on 1, 2, 4, 8, and 16 processors. A second timing measurement is also shown, where the 80^3 problem is always split into 512 blocks of size 10^3. While the latter experiment shows no speedup at all, the first one reaches a speedup of 2.7 on 4 processors, with a marginal 'gain' up to 3.3 on 8 processors. Clearly, the low-level parallelism achieved by the compiler is easily lost when grids become too small.

In the following section 2.2, we outline how we made use of manually inserted OpenMP directives to bring the parallelism of the code to its natural level, namely to allow for parallel execution over the grids present at any individual level. Both panels of Fig. 3 also show the achieved speedups in this manually parallelized mode for all three DD experiments discussed above. Ideally, since the load balance is optimal in these cases, linear speedup can be obtained. In fact, we reach a speedup of 20 on 32 processors in the blocksize-adjusted experiment (left panel), while the speedup on 16 processors for the 512 and 64 block run is 6.34 and 8.95, respectively. With these scalings, we clearly outperform the automated parallel executions, although there is still room for improvement.

2.2   Implementation Details

The manual OpenMP parallelization of our adaptive mesh refinement code concentrates on the most time consuming parts of the code: denoting the set of unknowns at time t^n on grid i_g at grid level l as $U_j^n(i_g, l)$, where the subscript j indicates a cell index within the grid i_g, a partial step of an explicit conservative update can be denoted as:

$$U_j^{n+1}(i_g, l) = U_j^n(i_g, l) - \frac{\Delta t^n(l)}{\Delta x(l)} \left(f_{j+1/2}^n - f_{j-1/2}^n \right).$$

In this expression, the numerical fluxes at the cell interfaces $f_{j+1/2}^n$ and $f_{j-1/2}^n$ are calculated from the known time level U^n with a certain stencil in the cell index range j. Note that the temporal $\Delta t^n(l)$ and spatial $\Delta x(l)$ increments vary from level to level. By surrounding each grid i_g on level l with a border of ghost cells conform with the stencil of the flux evaluation, this operation can be done in parallel over the collection of grids i_g on level l. In pseudocode, this is achieved as follows

```
if(level==1)then
   !$omp parallel do &
   !$omp& shared(level,dt) private(igrid) schedule(static)
   do igrid= 1,ngrids
      call process_grid(igrid,level,dt)
   end do
else
   !$omp parallel do &
   !$omp& shared(level,dt) private(igrid) schedule(dynamic)
   do igrid= 1,ngrids
      call process_grid(igrid,level,dt)
   end do
endif
```

The number of grids on level 'level' is given by ngrids, and these are indexed by igrid. Since the number of grids on level $l = 1$ is fully controlled by setting the overall computational domain and imposing a maximal size for an individual grid at pre-processing, we enforce static scheduling there. This ensures a perfect load balance on level $l = 1$. At higher levels, the number of grids varies unpredictably and their size is only controlled in upper bound. Therefore, we resort to dynamic scheduling.

While the code segment above is responsible for most of the obtained parallelism, there are in total five instances in the code where work has to be done for all grids at a certain level. Apart from the partial advance step, they correspond to filling the ghost cells for all grids at a certain level, adding a source term, making an error estimate (needed to flag cells which need refinement), and updating all grids at a certain level by information available from overlapping finer level grids. We have inserted OpenMP directives to allow for parallel execution of all these five level-wide computations.

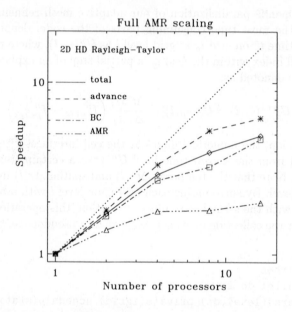

Fig. 4. Scaling of a realistic 2D hydrodynamic simulation exploiting AMR with 5 levels. The total scaling is from timings of actual simulations up to time $t = 0.5$. Individual contributions from time advancing, filling ghost cells ('BC') and error estimation plus updates ('AMR') are indicated.

2.3   Scaling Results for Full Grid-Adaptive Computations

To evaluate full grid-adaptive parallelism, beyond the DD usage as discussed in section 2.1, we perform timings for a realistic application. We set up a 2D hydrodynamic simulation of a Rayleigh-Taylor unstable configuration where a heavy compressible fluid rests on top of a lighter one (density contrast of 10) in an external gravitational field. We simulate on a $[0,1] \times [0,1]$ domain, with gravity $\mathbf{g} = -\hat{e}_y$ pointing downwards. We let a dense fluid with $\rho_{\text{dense}} = 1$ rest on top of a light fluid with $\rho_{\text{light}} = 0.1$ above the interface $y_{\text{int}} = 0.8 + 0.05 \sin 8\pi x$, so that 4 wavelengths are present in the domain at $t = 0$. The pressure field is set from a centered differenced hydrostatic balance $dp/dy = -\rho$, ensuring that the pressure about y_{int} equals unity. We allow for 5 refinement levels, with a base resolution of 48×48 and refinement ratios between consecutive levels fixed to 2. Regridding is done every sixth timestep per level, and as advocated in [5], we use a different spatial discretization on the finest level than on all underlying levels. To account for the effect of many refinements happening in a true AMR simulation, we perform timings of simulations running till time $t = 0.5$, up to which point 228 time steps have been taken on the coarsest level. To make this a tough test for the load balancing on levels $l > 1$, we enforce a maximal grid size of 24×24, ensuring the creation of many small-sized grids. Not unexpectedly, automated parallelization fails completely on this problem.

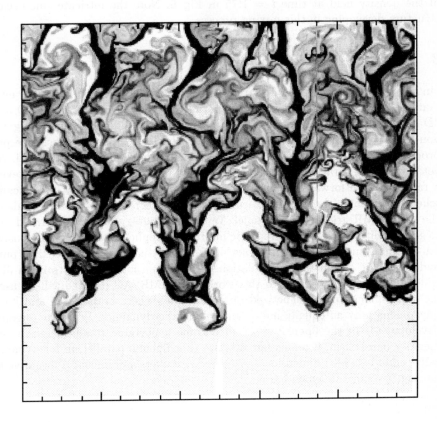

Fig. 5. A snapshot of the simulation used in the scaling experiment from Fig. 4. The density structure on the unit square is shown at time $t = 1.75$.

The result of the timings of this simulation up to 16 processors on the SGI Origin 3800 are shown in Fig. 4. We get a speedup of 2.9 on 4 processors, which only marginally improves up to 4.9 on 16 processors. If we analyze the obtained parallelism over the parallelized code segments discussed above, the most time consuming part – the partial advancing and the source term additions – scales best with a speedup of 6.2 on 16 processors. The ghost cell filling takes up a significant fraction (up to 32 % for the 4 processor run) of the computing time in this experiment due to the small maximal grid size enforced. As this is even more influenced by different grid sizes, this is responsible for the worsened scaling. The part indicated by AMR in Fig. 4 scales even less, but this combined time spent on error estimation and updating coarser from finer levels amounts to only 13 % of the total computing time on 4 processors. Overall, the obtained parallelism is in fact quite encouraging, given the severity of the test. Our timings indicate that for this particular 2D hydro simulation, it is beneficial to use up to 4 processors. We continued the calculation in that fashion and show a snapshot of the density field at time $t = 1.75$ in Fig. 5. Note the intricate small-scale structure developing in this mixing process.

3   Conclusions and Outlook

On the multi-processor ccNUMA architecture of the SGI Origin 3800, we eva-luated both automated and manually parallelized AMR simulations of 2D and 3D MHD problems. Automated parallelization can achieve linear scaling for non-adaptive, Domain Decomposition approaches as long as the block size per processor is of order 20^3 or higher for the 3D MHD simulations discussed. As yet, automated parallelization is inadequate for multi-level AMR runs. However, a rather straightforward manual parallelization strategy relies on dynamically scheduled OpenMP 'threads' to execute the time integration of multiple grids at the same AMR grid level in parallel. With this strategy, we demonstrated fair to good scaling properties for realistic multi-level applications. A more extensive set of timing measurements for a set of problems differing in complexity (pure hydrodynamic or full MHD simulations) and dimensionality (1D through 3D) will be a valuable extension to this work. The AMRVAC package will be used for multi-dimensional magnetized plasma studies where both global and local phenomena play an equally important role in the dynamics. With the ongoing developments in the OpenMP effort to become a portable standard for shared memory parallelism, it seems fair to state that OpenMP parallelism for dynami-cally regridded multi-dimensional computations offers a promising alternative to message passing implementations.

Acknowledgements.   This work was performed as part of the research programme of the 'Stichting voor Fundamenteel Onderzoek der Materie' (FOM) and Euratom, with financial support from the 'Nederlandse Organisatie voor Wetenschappelijk Onderzoek' (NWO) and computing resources by 'Nationale Computer Faciliteiten'. G.T. has been partly supported by the Education Ministry of Hungary (grant No. FKFP-0242-2000) and the Hungarian Science Foundation (OTKA, grant No. D-25519). R.K. thanks Peter Michielse (SGI Netherlands) for the OpenMP workshop (oct. 2001) which led to code improvements.

References

1. Berger, M.J.: Data structures for adaptive grid generation, SIAM J. Sci. Stat. Comput. **7**(3), 904 (1986)
2. Dahlburg, R.B., Keppens, R., Einaudi, G.: The compressible evolution of the super-Alfvénic magnetized wake, Phys. of Plasmas **8**(5), 1697-1706 (2001)
3. Friedel, H., Grauer, R., Marliani, C.: Adaptive mesh Refinement for Singular Current Sheets in Incompressible Magnetohydrodynamic Flows, J. Comput. Phys. **134**, 190-198 (1997)
4. Keppens, R., Nool, M., Goedbloed, J.P.: Zooming in on 3D magnetized plasmas with grid-adaptive simulations, in *Parallel Computational Fluid Dynamics – Recent Developments and Advances*, edited by P. Wilders et al. (Elsevier Science B.V., in press 2002)
5. Keppens, R., Nool, M., Tóth, G., Goedbloed, J.P.: Adaptive mesh refinement for conservative systems: multi-dimensional efficiency evaluation, submitted to J. Comput. Phys. (2001)
6. Keppens, R., Tóth, G.: Nonlinear dynamics of Kelvin-Helmholtz unstable magnetized jets: Three-dimensional effects, Phys. of Plasmas **6**(5), 1461-1469 (1999)
7. MacNeice, P., Olson, K.M., Mobarry, C., de Fainchtein, R., Packer, C.: PARAMESH: A parallel adaptive mesh refinement community toolkit, Comp. Phys. Comm. **126**, 330 (2000)
8. Nool, M., Keppens, R.: AMRVAC: a multidimensional grid-adaptive magnetofluid dynamics code, submitted to Comp. Methods in Applied Math. (2001)
9. Powell, K.G., Roe, P.L., Linde, T.J., Gombosi,T. I., De Zeeuw, D.L.: A Solution-Adaptive Upwind Scheme for Ideal Magnetohydrodynamics, J. Comput. Phys. **154**, 284-309 (1999)
10. Steiner, O., Knölker, M., Schüssler, M.: Dynamic interaction of convection with magnetic flux sheets: first results of a new MHD code, in *Proc. NATO advanced research workshop ASI Series C-433,* Solar Surface Magnetism, edited by R.J. Rutten and C.J. Schrijver, p. 441-470 (Kluwer Dordrecht, 1994)
11. Tóth, G.: The LASY Preprocessor and its Application to General Multi-Dimensional Codes, J. Comput. Phys. **138**, 981 (1997)
12. Tóth, G., Keppens, R., Botchev, M. A.: Implicit and semi-implicit schemes in the Versatile Advection Code: numerical tests, Astron. Astrophys. **332**, 1159-1170 (1998)

Parameter Estimation in a Three-dimensional Wind Field Model Using Genetic Algorithms*

Eduardo Rodríguez, Gustavo Montero, Rafael Montenegro,
José María Escobar, and José María González-Yuste

University Institute of Intelligent Systems and Numerical Applications in Engineering,
University of Las Palmas de Gran Canaria,
Edificio Instituto Polivalente, Campus Universitario de Tafira,
35017 Las Palmas de Gran Canaria, Spain
barrera@dma.ulpgc.es, {gustavo, rafa}@dma.ulpgc.es,
escobar@cic.teleco.ulpgc.es, josem@sinf.ulpgc.es

Abstract. The efficiency of a mass consistent model for wind field adjustment depends on several parameters that arise in various stages of the process. First, those involved in the construction of the initial wind field using horizontal interpolation and vertical extrapolation of the wind measures registered at meteorological stations. On the other hand, the Gauss precision moduli which allow from a strictly horizontal wind adjustment to a pure vertical one. In general, the values of all of these parameters are based on empirical laws. The main goal of this work is the estimation of these parameters using genetic algorithms, such that the wind velocities observed at the measurement station are regenerated as much as possible by the model.

1 Introduction

Mass consistent models are diagnostic models for constructing wind velocity fields from a few experimental measurements. In general, these models are defined by the physical laws of an incompressible fluid, by the empirical design of the wind profiles and by the values of velocities measured at the stations. This explains the existence of many parameters in the model. Some of them are clearly bounded and defined, while others are still under discussion and interpretation. Our work deals with the latter ones. There are many methods for the resolution of inverse problems involving parameter estimation and they have been largely studied in the literature. Among them, we have chosen a robust and flexible tool: genetic algorithms, which allow to solve linear and non-linear multiparameter optimisation problems.

This work has been structured as follows. First, the wind model is summarised in Sect. 2. We remark the studied parameters in Sect. 3. Next, in Sect. 4, the fitness function is established and genetic algorithms are briefly introduced, with their properties and possibilities used in this work. Numerical experiments are shown in Sect. 5 and, finally, conclusions are presented in Sect. 6.

* Partially supported by MCYT, Spain. Grant contract: REN2001-0925-C03-02/CLI

P.M.A. Sloot et al. (Eds.): ICCS 2002, LNCS 2329, pp. 950–959, 2002.

2 Mass Consistent Model in 3-D

This model [14] is based on the continuity equation for an incompressible flow with constant air density in the domain Ω and *no-flow-through* conditions on Γ_b

$$\nabla \cdot \boldsymbol{u} = 0 \qquad \text{in } \Omega . \tag{1}$$

$$\boldsymbol{n} \cdot \boldsymbol{u} = 0 \qquad \text{on } \Gamma_b . \tag{2}$$

We formulate a least-square problem in Ω with $\boldsymbol{u}(\widetilde{u}, \widetilde{v}, \widetilde{w})$ to be adjusted

$$E(\boldsymbol{u}) = \int_{\Omega} \left[\alpha_1^2 \left((\widetilde{u} - u_0)^2 + (\widetilde{v} - v_0)^2 \right) + \alpha_2^2 (\widetilde{w} - w_0)^2 \right] d\Omega . \tag{3}$$

where the interpolated wind $\boldsymbol{v}_0 = (u_0, v_0, w_0)$ is obtained from experimental measurements, and α_1, α_2 are the Gauss precision moduli. This problem is equivalent to find a saddle point $(\boldsymbol{v}, \phi)$ of the Lagrangian (see [27])

$$E(\boldsymbol{v}) = \min_{\boldsymbol{u} \in K} \left[E(\boldsymbol{u}) + \int_{\Omega} \phi \nabla \cdot \boldsymbol{u} \, d\Omega \right] . \tag{4}$$

being $\boldsymbol{v} = (u, v, w)$, ϕ the Lagrange multiplier and K the set of admissible functions. The Lagrange multipliers technique is used to minimise the problem (4), whose minimum comes to form the Euler-Lagrange equations

$$u = u_0 + T_h \frac{\partial \phi}{\partial x}, \quad v = v_0 + T_h \frac{\partial \phi}{\partial y}, \quad w = w_0 + T_v \frac{\partial \phi}{\partial z} . \tag{5}$$

where $T = (T_h, T_h, T_v)$ is the diagonal transmissivity tensor, with $T_h = \frac{1}{2\alpha_1^2}$ and $T_v = \frac{1}{2\alpha_2^2}$. Since α_1 and α_2 are constant in Ω, the variational approach results in an elliptic problem substituting (5) in (1)

$$\frac{\partial^2 \phi}{\partial x^2} + \frac{\partial^2 \phi}{\partial y^2} + \frac{T_v}{T_h} \frac{\partial^2 \phi}{\partial z^2} = -\frac{1}{T_h} \left(\frac{\partial u_0}{\partial x} + \frac{\partial v_0}{\partial y} + \frac{\partial w_0}{\partial z} \right) \quad \text{in } \Omega . \tag{6}$$

The boundary conditions result as follows (Dirichlet condition for open or *flow-through* boundaries and Neumann condition for terrain and top)

$$\phi = 0 \quad \text{on } \Gamma_a . \tag{7}$$

$$\boldsymbol{n} \cdot T \nabla \mu = -\boldsymbol{n} \cdot \boldsymbol{v}_0 \quad \text{on } \Gamma_b . \tag{8}$$

The classical formulation of the problem given by (6)-(8), is discretized using a tetrahedral mesh of finite elements (see [12]) that leads to a set of 4×4 elemental matrices and 4×1 elemental vectors, which are assembled into a global linear system of equations. A preconditioned conjugate gradient method is used for solving this symmetric linear system.

2.1   Horizontal Interpolation

The wind speeds measured at station height z_m are interpolated in function of the distance and the height difference between each point and the station [14]

$$v_0(z_m) = \varepsilon \frac{\sum\limits_{n=1}^{N} \frac{v_n}{d_n^2}}{\sum\limits_{n=1}^{N} \frac{1}{d_n^2}} + (1 - \varepsilon) \frac{\sum\limits_{n=1}^{N} \frac{v_n}{|\Delta h_n|}}{\sum\limits_{n=1}^{N} \frac{1}{|\Delta h_n|}} . \tag{9}$$

where v_n is the velocity observed at station n, N is the number of stations considered in the interpolation, d_n is the horizontal distance from station n to the point where we are computing the wind velocity, $|\Delta h_n|$ is the height difference between station n and the studied point, and ε is a weighting parameter ($0 \leq \varepsilon \leq 1$), which allows to give more importance to one of these interpolation criteria.

2.2   Vertical Profile of Wind

We have considered a logarithmic profile in the surface layer, which takes into account the previous horizontal interpolation, as well as the effect of roughness and the air stability (neutral, stable or unstable atmosphere, according to the Pasquill stability class) on the wind intensity and direction. Above the surface layer, a linear interpolation is carried out using the geostrophic wind. The logarithmic profile is given by

$$v_0(z) = \frac{v^*}{k} \left(\log \frac{z}{z_0} - \Phi_m \right) \qquad z_0 < z \leq z_{sl} . \tag{10}$$

where v^* is the friction velocity, k is von Karman constant, z_0 is the roughness length and z_{sl} is the height of the surface layer. Function Φ_m depends on the air stability

$$
\begin{aligned}
\Phi_m &= 0 & \text{(neutral)} . \\
\Phi_m &= -5\frac{z}{L} & \text{(stable)} . \\
\Phi_m &= \log\left[\left(\frac{\theta^2+1}{2}\right)\left(\frac{\theta+1}{2}\right)^2 \right] - 2\arctan\theta + \frac{\pi}{2} & \text{(unstable)} .
\end{aligned}
\tag{11}
$$

where $\theta = (1 - 16\frac{z}{L})^{1/4}$ and $\frac{1}{L} = az_0^b$, with a, b, depending on the Pasquill stability class. L is the so called Monin-Obukhov length. The friction velocity is obtained at each point from the interpolated measurements at the height of the stations (*horizontal interpolation*)

$$v^* = \frac{k\, v_0(z_m)}{\log \frac{z_m}{z_0} - \Phi_m} . \tag{12}$$

The height of the planetary boundary layer z_{pbl} above the ground is chosen such that the wind intensity and direction are constant at that height

$$z_{pbl} = \frac{\gamma \, |v^*|}{f} \, . \tag{13}$$

where $f = 2\omega \sin\varphi$ is the Coriolis parameter (ω is the earth rotation and φ the latitude), and γ is a parameter depending on the atmospheric stability. The mixing height h is considered to be equal to z_{pbl} in neutral and unstable conditions. In stable conditions, Zilitinkevich suggested (see [3])

$$h = \gamma' \sqrt{\frac{|v^*| \, L}{f}} \, . \tag{14}$$

where γ' is a constant of proportionality. The height of surface layer is $z_{sl} = \frac{h}{10}$. From z_{sl} to z_{pbl}, a linear interpolation with geostrophic wind v_g is carried out

$$v_0(z) = \rho(z) \, v_0(z_{sl}) + [1 - \rho(z)] v_g \qquad z_{sl} < z \le z_{pbl} \, . \tag{15}$$

$$\rho(z) = 1 - \left(\frac{z - z_{sl}}{z_{pbl} - z_{sl}} \right)^2 \left(3 - 2\frac{z - z_{sl}}{z_{pbl} - z_{sl}} \right) \, . \tag{16}$$

Finally, this model assumes $v_0(z) = v_g$ if $z > z_{pbl}$ and $v_0(z) = 0$ if $z \le z_0$.

3   Discussion on the Parameters to Be Estimated

First, we will consider the so called stability parameter

$$\alpha = \frac{\alpha_1}{\alpha_2} = \sqrt{\frac{T_v}{T_h}} \, . \tag{17}$$

since the minimum of the functional given by (3) is the same if we divide it by α_2^2. On the other hand, for $\alpha \gg 1$ flow adjustment in the vertical direction predominates, while for $\alpha \ll 1$ flow adjustment occurs primarily in the horizontal plane. Thus, the selection of α allows the air to go over a terrain barrier or around it, respectively [18]. Moreover, the behaviour of mass consistent models in many numerical experiments has shown that they are very sensitive to the values chosen for α. Therefore, we shall give particular attention to this problem. In the past, many authors have studied the parametrisation of stability, since the difficulty in determining the correct values of α have limited the possible wide use of mass-consistent models in complex terrain. Sherman [21], Kitada et al. [9] and Davis et al. [4], proposed to take $\alpha = 10^{-2}$, i.e., proportional to the magnitude of w/u. Other authors such as Ross et al. [20] and Moussiopoulos et al. [16] related α to the Froude number. Geai [7], Lalas et al. [10] and Tombrou et al. [24], make the α parameter vary in the vertical direction. Finally, Barnard et al. [2] proposed a procedure to obtain α for each single wind field simulation. The main idea is to use N observed wind speeds to obtain the wind field and to keep the rest, N_r, as a reference. Then, several simulations are performed with different values of α. The value which gives the best agreement with the reference observations is taken to be the final magnitude of the stability parameter. Since this method provides values of α that are only reliable for each particular case, it cannot provide an *a priori* value suitable for other simulations. Here, we follow a version of the method proposed in [2], using genetic algorithms as optimisation technique which lead to an automatic selection of α.

The second parameter to be estimate is the weighting coefficient ε $(0 \leq \varepsilon \leq 1)$ of (9). Note that $\varepsilon \to 1$ signifies more importance of the *horizontal distance* from each point to the measurement stations, while $\varepsilon \to 0$ signifies more importance of the *height difference* between each point and the measurement stations [14]. In general, the second approach has been used for complex terrains. On the other hand, the first approach has been widely used for problems with regular topography or in 2-D horizontal analysis. In realistic applications, the possibility of existing zones with complex orography and others with regular one, suggests that an intermediate value of ε should be more suitable.

The next parameter to discuss is γ, given in (13) and related to the height of the planetary boundary layer. There exist different versions of where to search this parameter. Panofsky et al. [17] proposed the interval [0.15,0.25]. On the other hand, Ratto [19] directly suggested $\gamma = 0.3$ in the *WINDS* code, while γ is located in [0.3,0.4] by de Baas [1]. Therefore, in our simulations, the search space for γ must include all these possibilities.

Finally, we are interested in obtaining suitable values of the parameter γ' involved in the computation of the mixing height for stable atmosphere, see (14). Garratt [6] proposed $\gamma' = 0.4$. Also in the *WINDS* code one may find bounds of γ' around 0.4. Thus, the value of γ' will be searched in the surroundings of 0.4.

4   Genetic Algorithms

Genetic algorithms (GAs) are optimisation tools based on the natural evolution mechanism. They produce successive trials that have an increasing probability to obtain a global optimum. This work is based on the model developed by Levine [11]. The most important aspects of GAs are the construction of an initial population, the evaluation of each individual in the fitness function, the selection of the parents of the next generation, the crossover of those parents to create the children, and the mutation to increase diversity.

Two population replacements are commonly used. The first, the generational replacement, replaces the entire population each generation [8]. The second, known as steady-state, only replaces a few individuals each generation [23, 25, 26]. Stopping criteria are iteration limit exceeded, population too similar, and no change in the best solution found in a given number of iterations. Initial population is randomly generated.

The selection phase allocates an intermediate population on the basis of the evaluation of the fitness function. We have considered four selection schemes

[11]: proportional selection (P), stochastic universal selection (SU), binary tournament selection (BT) and probabilistic binary tournament selection (PBT).

The crossover operator takes bits from each parent and combines them to create a child. One-point (OP) and uniform (U) crossover operators are used here. The first one selects randomly the place where each of the parents strings are broken in two substrings. Children will be the union of first substring of one parent and the second of the other. Uniform crossover depends on the probability of exchange between two bits of the parents [22].

The mutation operator is better used after crossover [5]. It allows to reach individuals on the search space that could not be evaluated otherwise. When part of a chromosome has been randomly selected to be mutated, the corresponding genes belonging to that part are changed. This happens with probability p. This work deals with four mutation operators. Three of them are of the form $\nu \leftarrow \nu \pm p \times \nu$, where ν is the existing allele value, and p may be a constant value (C), chosen uniformly from the interval $(0, \beta)$ with $\beta \leq 1$ (U), or selected from a Gaussian distribution (G). The fourth operator (R) simply replaces ν with a value selected uniformly random from the initialisation range of that gene.

The fitness function plays the role of the environment. It evaluates each string of a population. This is a measure, relative to the rest of the population, of how well that string satisfies a problem-specific metric. The values are mapped to a nonnegative and monotonically increasing fitness value. In the numerical experiments with this wind model, we look for suitable values of α, ε, γ and γ'. For this purpose, the average relative error of the wind velocities given by the model with respect to the measures at the reference stations is minimised

$$F(\alpha, \varepsilon, \gamma, \gamma') = \frac{\displaystyle\sum_{n=1}^{N_r} \frac{|\boldsymbol{v}_n - \boldsymbol{v}(x_n, y_n, z_n)|}{|\boldsymbol{v}_n|}}{N_r}. \tag{18}$$

where $\boldsymbol{v}(x_n, y_n, z_n)$ is the wind velocity obtained by the model at the location of station n, and N_r is the number of reference stations.

5   Numerical Experiments

We consider the same wind field problem related to the southern area of La Palma Island (Canary Islands) which was defined in [15]. A $45600 \times 31200 \times 9000 \ m^3$ domain with real data of the topography is discretized using the code developed in [12]. The maximum height in this zone of the island is $2279 \ m$. The mesh contains 11578 nodes and 52945 tetrahedra; see Fig. 1. The wind measurements were taken in four stations: MBI, MBII, MBIII and LPA. From the three cases studied in [15], we have selected case I with softly unstable conditions and case III with softly stable conditions in order to test the procedure for different stability conditions of the atmosphere. Due to the small number of available data, we have used the observed wind speeds of stations MBI, MBII and LPA to obtain the interpolated wind field (9), i.e., $N = 3$, and the measurement of MBIII is considered as reference station in the fitness function (18), i.e., $N_r = 1$.

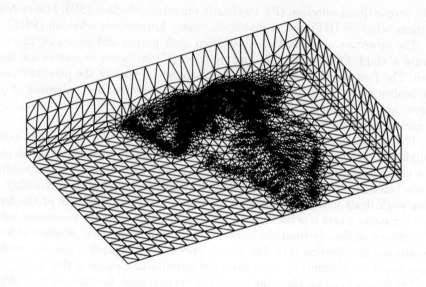

Fig. 1. Detail of the finite element mesh used for the numerical experiment. We only plot the triangulation of the terrain and two vertical wall in order to hold clarity

In the first application (case I), the parameter γ' is not involved in the modelling due to the unstable condition of the atmosphere, i.e., $h = z_{pbl}$. Thus, only α, ε and γ will be estimated in this case. The experiment has been divided in two stages. First, we fix $\gamma = 0.3$ and estimate $\alpha \in [10^{-3}, 10]$ and $\varepsilon \in [0, 1]$. The second column of Tab. 1 (*Stage 1*) shows the values obtained for α and ε, which suggest a nearly vertical wind adjustment and remark the complexity of the terrain respectively. Note that we obtain with the model an error at station MBIII about 10.7%. The strategy of GAs (*BT, U, R*) corresponds to the most efficient selection, crossover and mutation operators after several tests with different combinations. In the second stage, α, ε and $\gamma \in [0.15, 0.5]$ are estimated. The results are also shown in the third column of Tab. 1. We observe that α is near the maximum value of the space of search, ε remains quite small and γ is reduced, such that the error at station MBIII is 10.7%. We remark that in this experiment the worst evaluation of the fitness function, corresponding to values of the parameters in the search space, yields an error of 72.19%. Therefore, the knowledge of the studied parameters is essential for the efficiency of the numerical model.

For the second experiment (case III) we have followed a similar procedure. Now, $\gamma' \in [0.15, 0.5]$ must be also considered. First, a problem with two unknown parameters (α, ε) is solved. The second column of Tab. 2 (*Stage 1*) shows the values obtained for α, ε. Next, four problems arising from fixing one of the parameters each time, respectively, are studied (*Stages 2-5*). Finally, the four parameters are estimated at the same time in *Stage 6*. The atmospheric stable conditions reduce the vertical adjustment predominance arising in the previous experiment with unstable conditions, as well as augment the importance of the horizontal distance in the interpolation of the observed wind speeds. The minimum error obtained at station MBIII was about 22.2%, while the error related to the worst evaluation was 118.04%.

In both experiments, the number of individuals of the initial population was 100, except for stage 6 in case III where it was 150. Iterations and CPU timings on a 933 MHz Pentium III are shown in Tab. 1 and Tab. 2 for each stage. Evidently, the computational cost would be considerably reduced using a massive parallel machine, where the GAs become competitive with other optimisation methods.

Finally, as example, Fig. 2 shows the wind field obtained by the model, in the second experiment, at a height of 200 m using the values of the parameters corresponding to *Stage 6*. Here, the measures of the four stations have been taken into account for determining the interpolated wind field.

Table 1. First experiment corresponding to the case I analysed in [15]. Strategy of genetics algorithms, best evaluation of the fitness function and values of the parameters (*fixed values are written between brackets*)

	Stage 1	Stage 2
GAs strategy	BT, U, R	SU, U, G
Iterations	17	1
CPU time (s)	1548	108
Best Fitness	0.107	0.107
α	9.810	9.727
ε	0.010	0.029
γ	(0.300)	0.284

Table 2. Second experiment corresponding to the case III analysed in [15]. Strategy of genetics algorithms, best evaluation of the fitness function and values of the parameters (*fixed values are written between brackets*)

	Stage 1	Stage 2	Stage 3	Stage 4	Stage 5	Stage 6
GAs strategy	SU, U, G	SU, U, R	SU, U, R	SU, U, R	SU, U, R	SU, U, R
Iterations	110	147	37	16	29	61
CPU time (s)	8514	12120	2958	1362	2406	7050
Best Fitness	0.234	0.227	0.222	0.223	0.222	0.222
α	4.182	5.041	(5.041)	4.765	5.699	6.080
ε	0.003	0.272	0.281	(0.281)	0.292	0.282
γ	(0.300)	0.490	0.493	0.498	(0.498)	0.494
γ'	(0.400)	(0.400)	0.154	0.153	0.154	0.153

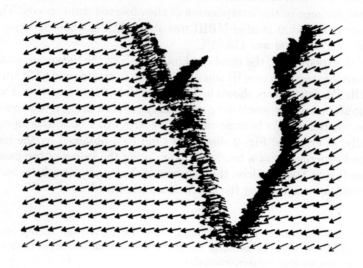

Fig. 2. Wind field solution related to the second experiment at a height of 200 m

6   Conclusions

The estimation of several parameters is essential for the efficiency of a 3-D mass consistent model for wind field adjustment. The numerical experiments have shown that these codes are very sensitive to the values chosen for α, ε, γ and γ'. A methodology for solving these parameter estimation problems is proposed. Genetic algorithms have proved to be an efficient and robust tool for these optimisation problems when several parameters are involved (see also [13]).

References

1. de Baas, A.F.: Scaling Parameters and their Estimation. In: Lalas, D.P., Ratto, C.F. (eds.): Modelling of Atmospheric Flow Fields. World Sci. Singapore (1996) 87-102
2. Barnard, J.C., Wegley, H.L., Hiester, T.R.: Improving the Performance of Mass Consistent Numerical Model Using Optimization Techniques. J. Climate Appl. Meteorol. **26** (1987) 675-686
3. Businger, J.A., Arya, S.P.S.: Heights of the Mixed Layer in the Stable, Stratified Planetary Boundary Layer. Adv. Geophys. **18A** (1974) 73-92
4. Davis, C.G., Bunker, S.S., Mutschlecner, J.P.: Atmospheric Transport Models for Complex Terrain. J. Climate Appl. Meteorol. **23** (1984) 235-238
5. Davis, L.: Handbook of Genetic Algorithms. Van Nostrand Reinhold (1991)
6. Garratt, J.R.: Observations in the Nocturnal Boundary Layer. Boundary-Layer Meteorol. **22**(1) (1982) 21-48

7. Geai, P.: Methode d'Interpolation et de Reconstitution Tridimensionelle d'un Champ de Vent: le Code d'Analyse Objective MINERVE. Technical Report DER/HE/34-87.03, EDF, Chatou, France (1985)
8. Holland, J.: Adaption in Natural and Artificial Systems. MIT Press (1992)
9. Kitada, T., Kaki, A., Ueda, H., Peters, L.K.: Estimation of Vertical Air Motion from Limited Horizontal Wind Data - A Numerical Experiment. Atmos. Environ. **17** (1983) 2181-2192
10. Lalas, D.P., Tombrou, M., Petrakis, M.: Comparison of the Performance of Some Numerical Wind Energy Siting Codes in Rough Terrain. In European Community Wind Energy Conference, Herning, Denmark (1988)
11. Levine, D.: A Parallel Genetic Algorithm for the Set Partitioning Problem. Ph. D. Thesis, Illinois Institute of Technology / Argonne National Laboratory (1994)
12. Montenegro, R., Montero, G., Escobar, J.M., Rodríguez, E., González-Yuste, J.M.: Tetrahedral Mesh Generation for Environmental Problems over Complex Terrains. Lecture Notes in Computer Science. Springer Verlag, Berlin Heidelberg New York (2002). Submitted.
13. Montero, G.: Solving Optimal Control Problems by GAs. Nonlin. Anal., Theor., Meth., & Appl. **30**(5) (1997) 2891-2902
14. Montero, G., Montenegro, R., Escobar, J.M.: A 3-D Model for Wind Field Adjustment. J. Wind Eng. & Ind. Aer. **74-76** (1998) 249-261
15. Montero, G., Sanín, N.: 3-D Modelling of Wind Field Adjustment Using Finite Differences in a Terrain Conformal Coordinate System. J. Wind Eng. & Ind. Aer. **89** (2001) 471-488
16. Moussiopoulos, N., Flassak, Th., Knittel, G.: A Refined Diagnostic Wind Model. Environ. Software **3** (1988) 85-94
17. Panofsky, H.A., Dutton, J.A.: Atmospheric Turbulence. John Wiley, New York (1984)
18. Ratto, C.F.: An Overview of Mass-consistent Models. In: Lalas, D.P., Ratto, C.F. (eds.): Modelling of Atmospheric Flow Fields. World Sci. Singapore (1996) 379-400
19. Ratto, C.F.: The AIOLOS and WINDS Codes. In: Lalas, D.P., Ratto, C.F. (eds.): Modelling of Atmospheric Flow Fields. World Sci. Singapore (1996) 421-431
20. Ross, D.G., Smith, I.N., Manins, P.C., Fox, D.G.: Diagnostic Wind Field Modelling for Complex Terrain: Model Development and Testing. J. Appl. Meteorol. **27** (1988) 785-796
21. Sherman, C.A.: A Mass-Consistent Model for Wind Fields over Complex Terrain. J. Appl. Meteorol. **17** (1978) 312-319
22. Spears, W., DeJong, K.: On the Virtues of Parametrized Uniform Crossover. Proc. of the Fourth International Conference on Genetic Algorithms (1991)
23. Syswerda, G.: Uniform Crossover in Genetic Algorithms. Proc. of the Third International Conference on Genetic Algorithms (1989)
24. Tombrou, M., Lalas, D.P.: A Telescoping Procedure for Local Wind Energy Potential Assessment. In: Palz, W. (ed.): European Community Wind Energy Conference, H.S. Stephens & Associates, Madrid (1990)
25. Whitley, D.: The GENITOR Algorithm and Selection Pressure: Why Rank-based Allocation of Reproductive Trials is Best. Proc. of the Third International Conference on Genetic Algorithms (1989)
26. Whitley, D.: GENITOR: A Different Genetic Algorithm. Rocky Mountain Conference on Artificial Intelligence (1988)
27. Winter, G., Montero, G., Ferragut, L., Montenegro, R.: Adaptive Strategies Using Standard and Mixed Finite Elements for Wind Field Adjustment. Solar Energy **54**(1) (1995) 49-56

Minimizing Interference in Mobile Communications Using Genetic Algorithms

Sa Li, Sokwei Cindy Lay, Wen Hsin Yu, and Lipo Wang[1]

School of Electrical and Electronic Engineering, Nanyang Technological University,
Block S2, 50 Nanyang Avenue, Singapore 639798
elpwang@ntu.edu.sg
[1]Corresponding author

Abstract. There is a continuously growing demand for mobile communication. With the limited frequency spectrum, the problem of channel assignment becomes increasingly important. This problem is known to belong to a class of very difficult combinatorial optimization problems. In this paper, we apply the formulation of Ngo and Li with genetic algorithms to ten benchmarking problems, for some of which interference-free solutions cannot be found but the approach is able to minimize the interference effectively.

1 Introduction

In recent years, there is a continuously growing demand for mobile communication. The rate of increase in the popularity of mobile usage has very much outpaced the availability of the usable frequencies which are necessary for the communication between mobile users and the base stations of cellular radio networks. This restriction constitutes an important bottleneck for the capacity of mobile cellular systems. Careful design of a network is necessary to provide acceptable quality of service.

An important issue on the design of a cellular radio network is to determine a spectrum-efficient and conflict-free allocation of channels among the cells while satisfying both the traffic demand and the electromagnetic compatibility (EMC) constraints. This is usually referred to as a channel assignment or frequency assignment problem (CAP1). There are basically three sources of constraints [5][21], namely, co-channel constraint (CCC), where the same channel cannot be assigned to certain pairs of radio cells simultaneously; adjacent channel constraint (ACC), where channels adjacent in the frequency spectrum cannot be assigned to adjacent radio cells simultaneously; and co-site constraint (CSC), where channels assigned in the same radio cell must have a minimal separation in frequency between each other.

This channel assignment problem is equivalent to a graph-coloring problem and is thus NP-hard. Over the recent years, several heuristic approaches have been used to solve various channel assignment problems, including simulated annealing [18], neural networks [6][7][16], tabu search, and genetic algorithms

P.M.A. Sloot et al. (Eds.): ICCS 2002, LNCS 2329, pp. 960–969, 2002.

(GAs) [11][14][19][20][22]. The existing approximative algorithm can be divided into two main groups. One group first determines ordered lists of all calls in the whole system [17] and then assign frequencies to the calls following a deterministic assignment strategy. In particular, Ngo and Li [20] developed an effective GA-base approach that obtains interference-free channel assignment by minimizing interference in a mobile network. They demonstrated that their approach efficiently converges to conflict-free solutions in some benchmarking problems.

As demand for mobile communications grows further, interference-free channel assignments often do not exist for a given set of available frequecies. Minimizing interference while satisfying demand within a given frequency spectrum is another type of channel assignment problem (CAP2) [7].

In this paper, we apply Ngo and Li's approach to several benchmarking channel assignment problems where interference-free solutions do not exist. The organization of this paper is as follows. section 2 states the channel assignment problem (CAP). In section 3 summarizes Ngo and Li's approach to solving CAP with genetic algorithms. section 4 describes the tests carried out and results obtained. Finally, section 5 concludes the paper.

2   Channel Assignment Problem

The channel assignment problem arises in cellular telephone networks where discrete frequency ranges within the available radio frequency spectrum, called channels, need to be allocated to different geographical regions in order to (CAP1) minimize the total frequency span, subject to demand and interference-free constraints, or to (CAP2) minimize the overall interference, subject to demand constraints. In this paper, we are interested in CAP2.

There are two kinds of channel allocation schemes - fixed channel allocation (FCA) and dynamic channel allocation (DCA). In FCA the channels are permanently allocated to each cell, while in DCA the channels are allocated dynamically upon request. DCA is desirable, but under heavy traffic load conditions, FCA outperforms most known DCA schemes. Since heavy traffic conditions are expected in future generations of cellular networks, efficient FCA schemes is becoming more and more important [20]. The cellular network is assumed to consist of N arbitrary cells and the number of channels available is given by M. The number of channels required (expected traffic) for cell j is D_j. Assuming that the RF propagation and the spatial density of the expected traffic have already been calculated, the non-interference constraints can be determined. The electromagnetic compatibility (EMC) constraints, specified by the minimum distance by which two channels i and j must be separated in order to guarantee an acceptably large signal-to-interference ratio S/I within the regions to which the channels are assigned, can be represented by an $N \times N$ symmetric matrix called the compatibility matrix $C = \{C_{ij}\}$.

The solution space is represented by F as an $N \times M$ binary matrix, where N is the total number of radio cells and M is the total number of available channels

[20]. Each element f_{jk} in the matrix is either one or zero such that

$$f_{jk} = \begin{cases} 1, \text{ if channel k is assigned to cell } j \text{ ;} \\ 0, \text{ otherwise .} \end{cases}$$

The cellular network is expected to meet the demand of the traffic and to minimize all forms of interference. The first requirement is the demand constraint, i.e., for cell i, a total of D_i channels are required. This implies that the total number of ones in row i of F must be D_i. If the assignment to cell i violates the demand constraint, then

$$(\sum_{q=1}^{M} f_{jq} - D_i) \neq 0 \quad . \tag{1}$$

The second requirement depends on the compatibility matrix C. It is made up of CSC, CCC and ACC. In order to satisfy the CSC, if channel p is within distance C_{ii} from an already assigned channel q in cell i, then channel p must not be assigned to cell i. This can be seen from the equation below:

$$\sum_{q=p-(C_{ii}-1),q\neq p,1\leq q\leq m}^{p+(C_{ii}-1)} f_{iq} > 0 \quad . \tag{2}$$

To satisfy the requirements for CCC and ACC, if channel p in cell i is within distance C_{ij} from an already assigned channel q in cell j, where $C_{ij} > 0$ and $i \neq j$, then channel p must not be assigned to cell i. This is represented as follows:

$$\sum_{j=1,j\neq i,C_{ij}>0}^{N} f_{jq} \sum_{q=p-(C_{ii}-1),q\neq p,1\leq q\leq m}^{p+(C_{ii}-1)} f_{iq} > 0 \quad . \tag{3}$$

Therefore, the cost function of CAP can be expressed as

$$C(F) = \sum_{i=1}^{N}\sum_{p=1}^{M}(\sum_{j=1,j\neq i,C_{ij}>0}^{N}\sum_{q=p-(C_{ii}-1),q\neq p,1\leq q\leq m}^{p+(C_{ii}-1)} f_{iq})f_{ip}$$

$$+\alpha\sum_{i=1}^{N}\sum_{p=1}^{M}(\sum_{q=p-(C_{ii}-1),q\neq p,1\leq q\leq m}^{p+(C_{ii}-1)} f_{iq})f_{ip} + \beta\sum_{i=1}^{N}\sum_{q=1}^{M}(\sum f_{jq} - D_i) \quad , \tag{4}$$

where α and β are weighting factors. It is noted that $C(F)$ achieves its minimum of zero when all constraints are satisfied. So, the objective of our problem is to find an F such that $C(F)$ is minimized.

3   Solving Channel Assignment Problem with Genetic Algorithm by Ngo and Li

In the encoding scheme by Ngo and Li [20], a p-bit binary string represents an individual with q fixed elements and the minimum separation between consecutive elements is represented by d_{min}. The concept of this scheme is to represent the solution in a way such that a one is followed by $d_{min} - 1$ zeros encoded as a new "one", denoted as **I**. Using the minimum separation scheme, the cost function can be simplified by exploiting the symmetry of the compatibility matrix C. Hence, the final cost function is represented by

$$
C(F) = \sum_{i=1}^{N-1} \sum_{j=i+1, C_{ij}>0}^{N} \left(\sum_{p=1}^{C_{ij}-1} \sum_{q=1}^{p-1} f_{jq} f_{ip} \right.
$$

$$
\left. + \sum_{p=C_{ij}}^{M} \sum_{q=p-C_{ij}+1}^{p-1} f_{jq} f_{ip} + \frac{1}{2} \sum_{p=1}^{M} f_{jq} f_{ip} \right) \quad . \tag{5}
$$

The mechanism of genetic algorithms start with randomly generating a population of chromosomes (suitable solutions for the population). The solution space represented by F, a $N \times M$ matrix, is treated as a chromosome in the population. This means that if a population is to contain n chromosomes, there will be n F solution matrixes in the population, each representing a chromosome. These n F solution matrixes or arrays are randomly generated and are all possible solutions for the channel assignment problem. The number of chromosomes in a population is stated by the population size, which is a parameter that should be manipulated to obtain an optimized solution. The setting of population size is generally quite ad hoc but nevertheless, a relatively small population size is suggested [20].

After randomly generating a population of chromosomes, the fitness of each chromosome should be evaluated. Therefore, all F solution arrays in the population are evaluated for their fitness values [20][20][20], by using the final cost function eq.(5). The lower the cost function value, the fitter the chromosome.

The next step in genetic algorithm is to generate a new population, using genetic algorithm operators, such as selection, crossover, and mutation. The selection process consists of selecting 2 parent chromosomes from a population according to their fitness (the better fitness, the bigger chance to be selected). The selected chromosomes have to undergo minimum-separation encoding, to ensure that the CSC satisfaction in them will continue to be satisfied even after crossover and mutation.

After selection and encoding, the selected parent chromosomes, or selected F solution arrays (encoded), will undergo crossover, with a probability of crossover, and mutation with a probability of mutation. Crossover probability and mutation probability are parameters that should be manipulated to obtain an optimized solution. The settings of these parameters, like the population size parameter, are generally quite ad hoc but a high crossover probability and low mutation probability are suggested [20]. Therefore, after crossover and mutation, the new offspring of the parent chromosomes are placed in the new population. For parents whereby no crossover or mutation is performed, they will be placed in the new population too. The selection, crossover and mutation processes will be repeated till the new population, which has the same size of the old population, is formed. After which, all new F solution arrays, or chromosomes, will be used for a further run of the entire genetic algorithm till an optimized solution is found.

4   Application of the Ngo-Li GA Approach to Benchmarking CAPs

The data sets used to test the performance of the approach were taken from [7]. Ten benchmarking problems were examined. Table 1 lists the problem specification and Table 2 lists the demands of those problems.   The CPU time taken

Table 1. Problem specifications of the ten benchmarking problems used. $[D_5]_i$ stands for the first i elements of matrix $[D_5]$.

Problem	No. Cell	No. Channel	Demand
EX1	4	11	D_1
EX2	5	17	D_2
HEX1	21	37	D_3
HEX2	21	91	D_3
HEX3	21	21	D_4
HEX4	21	56	D_4
KUNZ1	10	30	$[D_5]_{10}$
KUNZ2	15	44	$[D_5]_{15}$
KUNZ3	20	60	$[D_5]_{20}$
KUNZ4	25	73	D_5

Table 2. Demand constraints of the benchmarking problems.

	Demand
D_1^T	(1,1,1,3)
D_2^T	(2,2,2,4,3)
D_3^T	(2,6,2,2,4,4,13,19,7,4,4,7,4,9,14,7,2,2,4,2)
D_4^T	(1,1,1,2,3,6,7,6,10,10,11,5,7,6,4,4,7,5,5,5,6)
D_5^T	(10,11,9,5,9,4,5,7,4,8,8,9,10,7,7,6,4,5,5,7,6,4,5,7,5)

for each problem was dependent on the number of the size of the population used. A bigger population would take a much longer time for the minimum cost function to be found. Figure 1 shows the typical rate of convergence trajectory based on CPU time. Figure 2-3 shows the typical rate of convergence trajectory based on number of generations. Figures 4-7 show the sub-optimal solutions of channel assignments obtained from GA. Each dot represents a traffic demand for a particular cell and this demand would be allocated to a channel in a manner such that all the interferences are minimized.   During the simulation, several parameters, such as crossover probability, mutation probability and population size, need to be set. The number of generations for different problems also needs

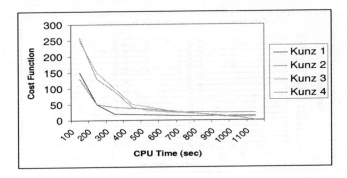

Fig. 1. A typical rate of convergence trajectory based on CPU time.

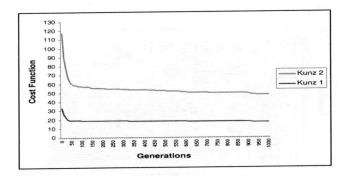

Fig. 2. A typical rate of convergence trajectory based on number of generations for KUNZ1 and KUNZ2.

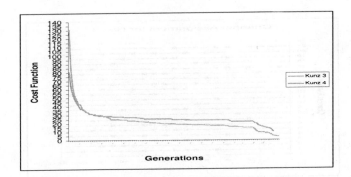

Fig. 3. A typical rate of convergence trajectory based on number of generations for KUNZ3 and KUNZ4.

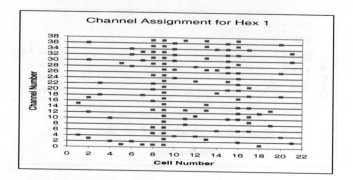

Fig. 4. Channel assignment for Hex 1 with interference value 39

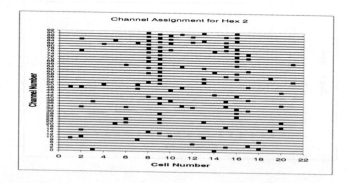

Fig. 5. Channel assignment for Hex 2 with interference value 13.5

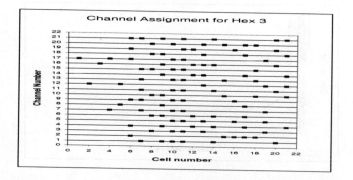

Fig. 6. Channel assignment for Hex 3 with interference value 46.5

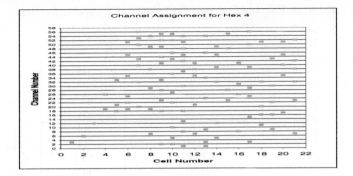

Fig. 7. Channel assignment for Hex 4 with interference value 0

to be taken into account. For example, for larger problems like Kunz 4, the number of generations needed to obtain a satisfactory result is 100000 as compared to 50000 for a smaller problem Kunz 3. For any genetic algorithms, the settings of these parameters are generally by trial-and-error. One general rule was kept throughout the simulation as suggested in [4]. That is to use relatively small population size, high crossover probability, and low mutation probability. We have systematically tested various choices of these parameters. For example, when we set the population size as 20 and the mutation probability as 0.004, the following tables show how the cost function changes with the cross-over probability:

5 Conclusion

In this paper, we applied Ngo and Li's GA-based approach to CAP2, i.e., channel assignment problems in which the total interference is minimized while traffic demands are satisfied within a given set of available channels. This approach permits the satisfaction of traffic demand requirement and co-site constraint. It is achieved by the use of a minimum-separation encoding scheme, which reduces the required number of bits for representing the solution, and with unique genetic operators that kept the traffic demand in the solution intact. This allowed the search space to be greatly reduced and hence shorten the computation time. The simulations done on benchmark problems showed that this approach could achieve desirable results.

Although we have tested a variety of choices of parameters, such as mutation rate, cross-probability, and population size, more such test with other choices of parameters should be carried out. Implementations of GAs for DCA will also be studied in future work.

References

1. Michalewicz, Z., "Genetic Algorithms + Data Structures = Evolution Programs." *Berlin; New York: Springer-Verlag*, c1994.

Table 3. Effect of Different Probability of Crossover for Hex

Problem	Probability of Crossover	Cost Function
HEX1	0.85	35
	0.90	34
	0.95	33
	0.97	33.5
	0.99	34
HEX2	0.85	14
	0.90	14
	0.95	13.5
	0.97	14.5
	0.99	14.5
HEX3	0.85	48.5
	0.90	48
	0.95	46.5
	0.97	47
	0.99	47.5
HEX4	0.85	16
	0.90	15
	0.95	14.5
	0.97	15
	0.99	15.5

2. Lai, W. K., Coghill, G. G. , "Channel assignment through evolutionary optimization", *IEEE Trans. Veh. Technol.*, vol. 45, no. 1, pp.91-96, 1996.

3. Ngo, C. Y., "Genetic algorithms for discrete optimization and their applications to radio network design", *Ph.D. thesis, Dept. Electr. Eng. Syst., Univ, Southern California, Los Angeles, CA.*, Aug.1995.

4. Goldberg, D. (Edward, D.), "1953- genetic algorithms in search, optimization, and machine learning. " *Reading, Mass.: Addison-Wesley Pub. Co.*, 1989.

5. Sivarajan, K. N., McEliece, R. J., "Channel Assignment in Cellular Radio", *Proceedings IEEE Vehicular Technology Conference*, vol. 68, pp. 1497-1514, Dec., 1980.

6. Kunz, D., "Channel assignment for cellular radio using neural networks", *IEEE Trans. Veh. Technol.*, vol. 40, pp. 188-193, Feb 1991

7. Smith, K., Palaniswami, M., "Static and dynamic channel assignment using neural network", *IEEE Journal on Selected Areas in Communications*, vol. 15, no. 2, Feb., 1997.

8. Holland, J. H., "Adaptation in Natural and Artificial Systems", *Ann Arbor, MI: Univ. Michigan Press*, 1975.

9. Falkenauer, E., "Genetic algorithms and grouping problems", *Chichester, England: Wiley*, 1998

10. Davis, L., "Genetic Algorithms and Simulated Annealing", *London, U.K.: Pitman*, 1987.

11. Lai, W. K., Coghill, G. G., "Channel Assignment for a Homogeneous Celular Network with Genetic Algorithms", *Conf. Intelligent Information Systems*, pp. 263-267, 1994.

12. Marin, J., Sole, R. V., "Macroevolutionary Algorithms: A New Optimization Method on Fitness Landscapes", *IEEE Transations on Evolutionary Computation*, vol. 3, no. 4, 1999.
13. Hata, M., "Empirical Formula for Propagation Loss in Land Mobile Radio Service", *IEEE Trans. Veh. Technol.*, vol.29, no.3, pp.317-325, 1980.
14. Cuppini, M., "A genetic algorithm for channel assignment problems", *Eur. Trans. Telecomm. Related Technol.*, vol.45, no.1, pp.91-96, 1996.
15. Kim, S., Kim, S. L., "A two-phase algorithm for frequency assignment in cellular mobile systems", *IEEE Trans. Veh. Technol.*, vol. 43, no. 3, pp. 542-548, 1994.
16. Funabiki, N., Takefuji, Y., "A neural network parallel algorithm for channel assignment problems in cellular radio networks", *IEEE Trans. Veh. Technol.*, vol. 41, no. 4, pp. 430-437, 1992.
17. Ko, T. M., "A frequency selective insertion strategy for fixed channel assignment", *Proc. 5th IEEE int, Symp.Personal, Indoor and Mobile Radio Commun*, pp. 311-314, Sept. 1994.
18. Mathar, R., Mattfeldt, J., "Channel assignment in cellular radio networks", *IEEE Trans. Veh. Technol.*, vol. 42, pp. 14-21, 1993.
19. Kim, J. S., Park, S. H., Dowd, P. W., Nasrabadi, N. M., "Channel assignment in cellular radio using genetic algorithm", *Wireless Personal Commun.* vol. 3, no. 3, pp. 273-286, Aug. 1996.
20. Ngo, Y., Li, V. O. K., "Fixed channel assignment in cellular radio networks using a modified genetic algorithm", *IEEE Trans. Veh. Technol.*, vol. 47, no. 1, 1998.
21. Ko, T. M., "A frequency selective insertion strategy for fixed channel assignment", *Proc. 5th IEEE Int. Symp. Personal, Indoor and Mobile Radio Commun.*, pp. 311-314, 1994.
22. Smith, K., "A genetic algorithm for the channel assignment problem", *IEEE Global Technology Conference.*, vol. 4, 1998.

KERNEL: A Matlab Toolbox for Knowledge Extraction and Refinement by NEural Learning

Giovanna Castellano, Ciro Castiello and Anna Maria Fanelli

Dipartimento di Informatica,
Università degli Studi di Bari
Via E.Orabona, 4 - I70126 Bari ITALY
[castellano, castiello, fanelli]@di.uniba.it

Abstract. In this paper we present KERNEL, a neuro-fuzzy system for the extraction of knowledge directly from data, and a toolbox developed in the Matlab environment for its implementation. The KERNEL system belongs to the novel approach which concerns the use and representation of explicit knowledge within the neurocomputing paradigm: the Knowledge Based Neurocomputing. A specific neural network is designed, that reflects in its topology the structure of the fuzzy inference model on which is based the KERNEL system. A well-known system identification benchmark is used as illustrative example.

1 Introduction

In recent years there has been a growing interest in combining neural networks and fuzzy systems (see for instance [6], [8], [9]). This approach belongs to the field known as Knowledge-Based Neurocomputing (KBN), which is a discipline concerning methods to address the explicit representation and processing of knowledge where a neurocomputing system is involved [5]. The KBN approach combines the powerful processing capabilities of the neural networks with the explanatory advantages of symbolic representation of knowledge, such as Boolean functions, automata, linguistic rules.

This paper proposes a toolbox for KERNEL (Knowledge Extraction and Refinement by NEural Learning) system for use with Matlab. KERNEL is a neuro-fuzzy system based on KBN approach, in which the explicit representation and the manipulation of knowledge is provided by fuzzy inference methods. As known [10], the fuzzy approach uses human-like reasoning mechanisms, thus the knowledge in the system is expressed in form of linguistic rules (IF-THEN). The final task of KERNEL is to obtain a fuzzy rule base directly from input data and to improve its efficiency and readability. The mathematical description of the neuro-fuzzy model and the presentation of the learning algorithms underlying the KERNEL system are out of the intents of this paper and can be found in our previous papers [2], [3], [4]. The focus of this paper is to describe in detail the functionalities of the Matlab user-friendly environment that has been developed to allow easy utilization of the KERNEL system.

P.M.A. Sloot et al. (Eds.): ICCS 2002, LNCS 2329, pp. 970–979, 2002.

The outline of the paper is the following: in section 2 we briefly describe the KERNEL system and its architecture. Section 3 is dedicated to the presentation of the MATLAB toolbox and its peculiarities. Finally, an example of system identification shows the utilization of the toolbox and the applicability of the proposed approach.

2 The KERNEL System and its architecture

The KERNEL system has been designed to exploit both the benefits deriving from the use of neural networks and the advantages in term of knowledge representation offered by fuzzy reasoning models. In particular, two essential characteristics should be preserved: accuracy and interpretability. To cope with these requirements, the architecture of the system is organized (see Fig. 1) into three distinct components:

- the first component extracts knowledge in form of fuzzy rules directly from data;
- the second component performs a refinement process of the fuzzy rule base in order to improve the modeling accuracy;
- the third component increases the interpretability of the fuzzy rule base through an optimization procedure.

According to the KBN approach, the fuzzy inference model is translated into a neural network: the neuro-fuzzy network (NFN). Therefore, the structure and parameters of the fuzzy rule base correspond to the structure and parameters of the NFN, respectively.

The main feature of the KERNEL system, which distinguishes it from other neuro-fuzzy systems proposed in the literature, resides in the cability of determining the structure of the rule base in an automatic fashion, through learning from data.

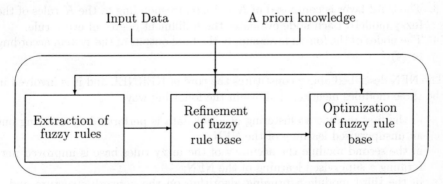

Fig. 1. Scheme of the KERNEL system

2.1   The fuzzy inference model

The KERNEL system is based on a fuzzy inference model which adopts K fuzzy rules of type:

IF x_1 is A_1^k **AND** ... **AND** x_n is A_n^k **THEN** y_1 is b_{1k} **AND** ... **AND** y_m is b_{mk},

$$(1)$$

where $x_1, \ldots, x_n$ are the input variables, $y_1, \ldots, y_m$ are the output variables, A_i^k are fuzzy sets and b_{jk} are fuzzy singletons. The membership functions related to the fuzzy sets A_i^k are Gaussian functions in the following form

$$\mu_{ik}(x_i) = \exp(-(x_i - c_{ik})^2 / \sigma_{ik}^2),\qquad (2)$$

where c_{ik}, σ_{ik} are the center and the width of the Gaussian function, respectively. For every input vector $\mathbf{x} = (x_1, \ldots, x_n)$, the output vector $\mathbf{y} = (y_1, \ldots, y_m)$ can be obtained calculating first the fulfillment of each rule by

$$\mu_k(\mathbf{x}) = \prod_{i=1}^{n} \mu_{ik} x_i, \quad k = 1, \ldots, K,\qquad (3)$$

then using the following equation:

$$y_j = \frac{\sum_{k=1}^{K} \mu_k(\mathbf{x}) b_{jk}}{\sum_{k=1}^{K} \mu_k(\mathbf{x})}, \quad j = 1, \ldots, m.\qquad (4)$$

2.2   The neuro-fuzzy network

The NFN is composed of four layers (see Fig. 2):

1. The nodes in the first layer just supply the input values $x_i (i = 1, ..., n)$.
2. In the second layer nodes are collected in K groups (corresponding to the K rules), each composed of n units (corresponding to the n fuzzy sets of every rule). These nodes estimate the values of the Gaussian membership functions.
3. The third layer is composed of K units (corresponding to the K rules of the fuzzy model). Each node evaluates the fulfillment degree of every rule.
4. The nodes of the fourth layer supply the final output of the system according to (4).

The NFN described above constitutes the core of KERNEL and it is involved in the three modules mentioned above in the following way:

- in the first module a clustering of input data is performed by carrying out an unsupervised learning of the NFN;
- in the second module the accuracy of the fuzzy rules base is improved performing a supervised learning of the NFN;
- in the third module a pruning algorithm on the network structure and a genetic algorithm on the configuration of the membership functions are performed for the benefit of the final interpretability of the fuzzy rules base.

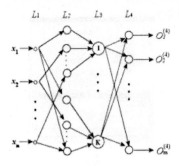

Fig. 2. The neuro-fuzzy network

3   The toolbox

The toolbox has been developed in the Matlab (ver.5.3) environment and it is designed to provide an interface to the system described in the previous section. It is composed of three modules corresponding to the three components of the KERNEL system (see Fig 3). The three modules are endowed with Graphical User Interfaces (GUIs) and have been conceived to work either sequentially (starting from the extraction of knowledge from input data to the final improvement of the fuzzy rules base) or in an unrelated way (according to the results the user needs to obtain).

The fuzzy inference model is represented by a FIS variable containing all the information related to the model and organized in a hierarchic structure. The input data can be loaded from disk or from the Matlab workspace in the form of numerical matrices.

The toolbox provides a 2D graphical representation of multidimensional data and a tool to visualize the fuzzy rules and test the accuracy of the model available at the moment.

3.1   Structure and parameters initialization

The first module of the toolbox is devoted to the automatic creation of an initial fuzzy rule base starting from the input data. The number of the rules and the parameters of the antecedents and the consequents of the rules are obtained by a clustering process. This is performed through the training of the NFN using a competitive learning algorithm [2]. A peculiarity of this clustering procedure consists in providing automatically the appropriate number of clusters, corresponding to the number of rules for the fuzzy model.

The interface realized for this first module (see Fig. 4) allows the user to load the input data and to represent them in a 2D space. An input pre-elaboration can be performed to normalized and/or shuffle the data before starting the clustering process, which can be carried out calculating two kinds of distance between the input patterns and the clusters centers: the Euclidean and the Mahalanobis

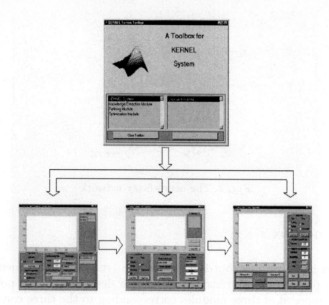

Fig. 3. The toolbox scheme

distance. The first distance performs a spherical clustering, the latter produces an elliptical one.

Once the unsupervised learning of the NFN is completed, the user can generate the initial FIS on the basis of the values obtained after the clustering. The module allows to visualize the fuzzy rule base and to test it on the training or the test set in case it is available.

3.2   Global tuning of parameters

To improve the accuracy of the fuzzy rule base, a supervised learning of the NFN must be performed. The GUI of the second module (see Fig. 5) allows the user to determine the set of parameters for the learning process.

According to the type of problem the system has to deal with (regression or classification tasks), two training algorithms can be chosen which differ in the cost functions [4]. The training of the NFN is carried out using the training set of data and, to avoid a possible overfitting, a checking set can be loaded to perform an early-stopping (see Fig. 6)

3.3   Optimization of the model

The causes of a poor readability of the fuzzy rule base can be found both in a high number of rules and in overlapping membership functions. The third module of the toolbox improves the interpretability of the fuzzy inference model while preserving the accuracy degree achieved in the previous steps. In order to realize

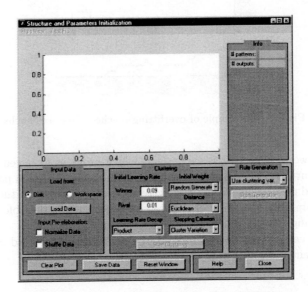

Fig. 4. GUI of the module for knowledge extraction

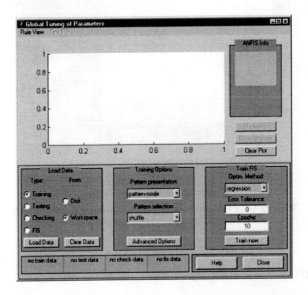

Fig. 5. GUI of the module for knowledge refinement

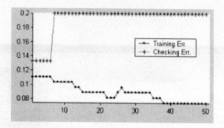

Fig. 6. An example of overfitting reached after six epochs

this task, a two step process is carried out which provides for a reduction of the model structure and a rearrangment of the membership function parameters by a pruning and genetic algorithm, respectively. The first works on the topology of the NFN to simplify it, the latter is based on population of possible membership functions configurations [3].

The user can fix the parameters for the algorithms described above by the GUI implemented for this module (see Fig. 7).

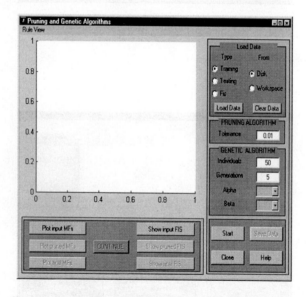

Fig. 7. GUI of the module for knowledge optimization

4   An illustrative example

To illustrate the use of the toolbox, we employ an example concerning the identification of a dynamical nonlinear system using the gas furnace data (series J) of Box and Jenkins [1].

The process is a gas furnace with a single input $u(t)$ representing the gas flow rate into the furnace, and a single output $y(t)$ representing the CO_2 concentration in the outlet gases. The data set consists of 296 pairs of input-output measurements taken with a sampling interval of 9 s. Since the process is a dynamical one, there are different values of the variables that can be considered as candidates to affect the present output $y(t)$. Typically, ten input candidates are considered, i.e. $y(t-1), \ldots, y(t-4), u(t-1), \ldots, u(t-6)$. As in [7], we considered only three input variables: $y(t-1), u(t-3)$ and $u(t-4)$. The produced CO_2 at time t, i.e. $y(t)$, is the output variable. Hence the original 296 data pairs $\langle u(t); y(t) \rangle$ were converted into 290 training points of the form $\langle y(t-1), u(t-3), u(t-4); y(t) \rangle$.

Our experiment starts with the competitive learning phase which performs a spherical clustering of the input data (see Fig. 8). The competitive learning was run with 25 initial clusters randomly generated and produced a network structure with a number of 13 nodes in the third layer (corresponding to 13 fuzzy rules for the initial knowledge base).

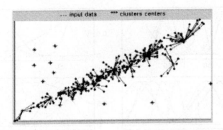

Fig. 8. Partitioning of input data with 13 clusters (the empty clusters will be discarded)

The second module realizes a global tuning of the parameters to improve the accuracy of the knowledge base. In Fig. 9 the output of the fuzzy model is compared with the one of the actual process.

In the end an optimization of the knowledge base was performed by the third module: the number of rules was reduced from 13 to 5 and the readability of the fuzzy membership functions was enhanced as can be seen in the comparison reproduced in Fig. 10.

The results obtained at each step are summed up in Table 1, in terms of numbers of rules and MSE.

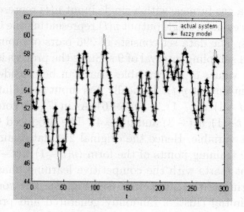

Fig. 9. Comparison between the actual output and the output of the fuzzy model

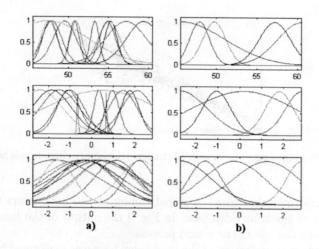

Fig. 10. Comparison of membership functions: initial (a), at the end of the third step (b).

Table 1. Results obtained at each stage of the KERNEL System

Modeling stage		rules number	MSE
stage1: structure and parameters initialization		13	1.3540
stage2: global parameter tuning		13	0.1400
stage3: first iteration	structure reduction	8	0.1459
	local parameter tuning	8	0.1421
second iteration	structure reduction	5	0.1511
	local parameter tuning	5	0.1503

References

1. Box, G. E., Jenkins, G. M.: Time Series Analysis: Forecasting and Control, 2nd ed. Holden Day, San Francisco, CA 1976.
2. Castellano, G., Fanelli, A. M.: Fuzzy inference and rule extraction using a neural network, Neural Network World Journal, 3:361-371, 2000.
3. Castellano, G., Fanelli, A. M.: Complex system modeling by a hybrid approach, International Journal of Knowledge-Based Intelligent Engineering Systems, 3(1): 42-50, 1999.
4. Castellano, G., Fanelli, A. M., Mencar, C.: A new empirical risk functional for a neuro-fuzzy classifier, In Proc. of ESIT 2000, Ahachen, Germany, pp. 429-436, 2000.
5. Cloete, I., Zurada, J. M.: Knowledge-based neurocomputing, The MIT Press, Cambridge, Massachussets, 2000.
6. Nauck, D.: Neuro-Fuzzy systems: review and prospects, Proc. of EUFIT97, pp. 1044-1053, 1997.
7. Sugeno, M., Yasukawa, T.: A Fuzzy-Logic-Based Approach to Qualitative Modeling, IEEE Trans. on Fuzzy Systems, 1:7-31, 1993.
8. Sun, C. T.: Rule-base structure identification in an adaptive-network-based fuzzy inference system, IEEE Trans. on Fuzzy Systems, 2(1):64-73, (1994).
9. Wang, L. X., Mendel, J.: Generating fuzzy rules by learning from examples, IEEE Trans. Syst., Man, and Cyb., 22: 1414-1427, (1992).
10. Zadeh, L. A.: Fuzzy sets, Information and control, 8: 338-353, (1965).

Damages Recognition on Crates of Beverages by Artificial Neural Networks Trained with Data Obtained from Numerical Simulation

Jörg Zacharias, Christoph Hartmann, and Antonio Delgado

Technische Universität München
Chair of Fluid Mechanics and Process Automation
Weihenstephaner Steig 23, 85350 Freising-Weihenstephan, Germany
Joerg.Zacharias@wzw.tum.de

Abstract. A new method to detect damages on crates of beverages is investigated. It is based on a pattern-recognition-system by an artificial neural network (ANN) with a feedforward multilayer-perceptron topology. The sorting criterion is obtained by mechanical vibration analysis which provides characteristic frequency spectra for all possible damage cases and crate models. To support the network training, a large number of numerical data-sets is calculated by finite-element-method (FEM). The combination of artificial neural networks with methods of numerical simulation is a powerful instrument to cover the broad range of possible damages. First results are discussed with respect to the influence of modelling inaccuracies of the finite-element-model and the support of ANN by training-data obtained from numerical simulation.

1 Introduction

Based on the project-idea to improve the quality requirements on deposit systems and filling lines in the beverage industry [1], a new recognition method is presented in the current contribution. The increasing variety of crate models complicates the sorting as well as added impurities and ageing of the material. Therefore, functional sorting systems are needed for the inspection of returned reusable crates of beverages which are mainly sorted by optical systems in industrial automatic filling lines. The general problem of these lines is the detection of small and hidden damages.

The scientific motivation results from some contributions in aerospace, civil engineering, seismic research and some basic mechanical engineering problems. They describe methods to detect damages and failures in structures using frequency response and transient response data of vibrating mechanical systems ([2], [3], [4]). Also numerical simulation on frequency response methods of different examples as plates, beams, bridges, buildings have been done with good agreement on experimental results ([5], [6]). Further research show the capability of artificial neural networks (ANN) in different topologies to classify some damages of structures which receive data out of different vibration based analysis ([7], [8]). This short review shows that the basic principles have been proven on buildings, aircraft-wings, beams and plates by different authors. The combination of all these principles to get finite-element (FE) supported ANN trained by numerical simulated data-sets have been

P.M.A. Sloot et al. (Eds.): ICCS 2002, LNCS 2329, pp. 980–989, 2002.

taken mainly at simple structures like plates, beams and strongly simplified models ([9], [10]), where also some problems to get fitting results with this technique are discussed [11]. This also underlines that there is not an overall solution and each single case needs its special adaptation.

The following contribution will apply the methods to the complex structure of a crate of beverage. The basic idea and the feasibility of the method have been reported in [12]. The development of a pattern-recognition-system for damaged polyethylene crates of beverages is carried out on the principle of mechanical excitation and frequency response measurement. The aim of the system is to take advantage of these reliable mechanical methods and combine them with the quick, adaptive pattern recognition of ANN which allows automatic sorting. Additionally, numerical simulation of the mechanical system provides many information about the system behaviour for the planning of the experimental device and for further use in network training. Typical damages as flaws, deformations or separated components are considered in the current contribution.

2 Analysis

In order to meet the principle objectives of the system, the damage detection of polyethylene crates of beverages, based on the pattern recognition of vibration response analysis data is established. This technique is taking advantage of the property of mechanical systems to transmit signals through the whole structure. The vibration response of damaged crates or crates of different kind vary from that of a standard (undamaged) crate. The difference is used as a criterion to select individual crates. The pattern recognition of various response spectra is done by ANN. Since a large number of possible damages occur, the training is carried out with data obtained from experimental analysis and from numerical simulation. Also the data reduction (pre-processing) of the response spectra to an appropriate amount of characteristic input data is described.

In the experimental part of the present project a mechanical excitation is applied to a crate of beverage. Therefore, a customary polyethylene crate is fixed on an electrodynamic vibration facility (see figure 1). Control by a PC-control-system provides a sinusoidal vertical movement of the shaker-piston where the crate is fixed on a central position on a specially constructed expander. The excitation is carried out over a sweep of a frequency range from 50 to 1000 Hz, controlled by an amplitude of constant power with a starting acceleration of 75 m/s² which also is the reference for the system calibration. This excitation does not cause any visible destruction or plastic deformation to the crates. The system answer is recorded separately as an acceleration signal over the whole frequency range at different locations on the crate by several acceleration-sensors. Mainly observed for the described results is one control sensor at the expander and one reference sensor on the middle of the handle of the crates. All kind of crates and damages can be processed in this way. In the present case, 20 crates with different, empirically evaluated damages are used to cover a basic range of damages. Flaws, deformations, separated components on sides, handles, compartments and also multiple damages as well as three different models are inspected.

Fig. 1. Vibration system at the laboratory

The vibrational motion can be described mathematically by the general equation of motion (1) which is the basis for the numerical analysis of the system. In a finite-element model inertia (M), damping (D), stiffness (K) of the model are determined by the geometry and the material properties. The force f(t) defines the excitation, and u represents the displacement vector of all degrees of freedom of the discretized model.

$$[M]\{\ddot{u}\}+[D]\{\dot{u}\}+[K]\{u\}=\{f\}(t) \tag{1}$$

From modal analysis it is known, that the displacement vector $\{u\}$ can be expressed as a series of superposed eigenmodes Φ_i with different amplitudes ξ_i.

$$\{u\}=\sum_i [\Phi_i]\xi_i \tag{2}$$

The natural frequencies of a structure are the frequencies at which the structure naturally tends to vibrate if it is subjected to a disturbance. The deformed shape of a structure at a specific natural frequency is called its mode shape of vibration. Natural frequencies and mode shapes are functions of the structural properties (e.g. material parameters, geometry). All the modes and frequencies are system immanent. If the excitation frequency is equal to one of the natural frequencies the response-amplitudes become very large. This is called resonance. In the case of enforced motion, the amplitude and the phase of the vibration at a distinct point are recorded. Excitation in the representative frequency range results in associated response frequency spectra which are specific for each observed measurement point as well as for each geometry, including damages, and different material properties.

Based on this knowledge, numerical simulation is carried out in order to provide both, insight into the basic vibration behaviour and frequency response spectra for the training of the ANN. The evaluation of the mode shapes ensures the optimisation of the experiment and can guide through it. In pre-test planning stages standard mode shapes can be used to indicate the best location for the accelerometers and for the position of excitation [13]. Furthermore, there is an enormous need of data-sets for the network-training and network-testing, that represent a wide range of crates of equal damages and also a wide variety of possible damages, whose experimental data collection is very costly. These data-sets can be produced in a more efficient way by numerical simulation.

A CAD-data-set is automatically meshed with support of the pre-processor MSC/PATRAN, which leads to a model of more than 1.000.000 degrees of freedom. Two different element types are deployed, 4 node (TET4) and 10 node (TET10) tetrahedral elements [14]. The finite-element-solver MSC/NASTRAN is configured to a frequency response analysis to calculate comparable spectra of the vibration behaviour of the crates, intact as well as damaged. For this purpose some simplifications are done within the configuration. Therefore, usually the first step is the calculation of mode shapes and natural frequencies. No damping is used which leads to results that characterise the basic dynamic behaviour of the structure and indicate how the structure will respond to dynamic loading. To simulate frequency-response-spectra another method is applied, which uses the technique of the large-mass-principle, where applied forces are used in conjunction with concentrated masses (see [15], [16], [17] for further information). These spectra are validated by comparison with the experimental data and by optimisation of the parameters and the simulation configuration of the system and the solver. The validated numerical database now is applicable for any possible damage case in further simulations. This data can be included into the training-data-set to enhance experimental training-data and cover a broad range of possible damages.

The third part of the system, the pattern recognition by ANN, is done with a multilayer-perceptron (MLP) network. As also reported by [18] the MLP is proved to be an appropriate topology for the desired classification. A network with 10 input nodes and 5 nodes in the hidden layer, all connected in forward direction, is built up in order to classify the signal in 2 output nodes as "intact" or "damaged" (figure 2).

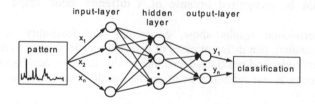

Fig. 2. Multilayer-perceptron network

The initial worths of each input are summed up and weighted in the nodes by

$$z_i = \sum_{j=1}^{n} w_{ji} x_i \tag{3}$$

leading to the activation of each node by the common type of a sigmoid transfer function

$$S_i(\vec{x}) = 1/(1 + \exp(-z_i)) . \tag{4}$$

Here x_j is the activity of the j-th input node and $S_i(\vec{x})$ is the activity of the i-th hidden layer node. This effects a binary output, which leads in the output layer to the prediction of each output node to a "1" for true and a "0" for false, as also reported by [19]. Starting with a random weight w_{ji} of each connection the supervised learning algorithm resilient propagation (RPROP) is used for the network training to adjust the weights of the initial connections to minimise errors between network output and the output target. For this feed-forward network RPROP performs a fast and robust learning algorithm, which adjusts the interconnecting weights by adapting each weight automatically in order to minimise the error function by correcting them because of the gradients algebraic sign ([20], [21]). As ANN are only able to process data in a certain format, data-reduction by splitting and integrating the spectra yields a suitable amount of data for the input nodes. Each input node processes a normalised part of response range and is not dependent on any previous knowledge about the model behaviour. A minimum of non-redundant data-sets is needed for training (20 to 50 different cases at the moment). This number depends on the variety of the damages. The implementation of more crate models will cause more training-data-sets.

3 Results and Discussion

The mode shape analysis gives insight into the preferred state of motion of the mechanical structure. Therefore, it can be evaluated in order to choose sensor positions which provide a high degree of sensitivity. For example, in figure 3, the fifth mode shape is shown. The region around the handles is strongly deformed. This means, that a sensor placed on the handle would record a high amplitude. A damaged handle would be recognised because of a different mode shape at a different frequency.

The experimental results show, concerning the feasibility of the damage recognition method, that different cases of detected spectra show different peaks with different amplitudes and positions depending on the damages. See figure 4 for typical vibration spectra at one reference point on the handle of an intact and two damaged crates.

It can be stated that damages on crates of beverages can be identified by the comparison of vibration spectra. Also the small and hidden damage in the compartment is observable. In detail, the peaks are specific in position and size, especially in two main regions from 120 to 250 Hz and from about 400 to 850 Hz. This is mainly due to the geometry modifications of every sample. The reproducibility of the spectra measured by the described method is possible in all cases. This is proved by repeated measurements. However, to obtain data-sets, which are to be expected in an industrial environment, data of up to 5% perturbation in amplitude are produced by artificial inaccuracies during the measurement process, which offers a pool of about 300 data-sets. Further investigations show that some very small damages (e.g. very small flaws) are better observable, if more than one spectra of different measurement points is taken into account.

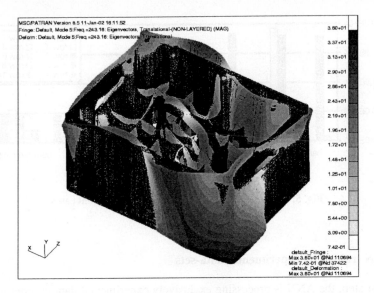

Fig. 3. Fifth mode shape at 243 Hz

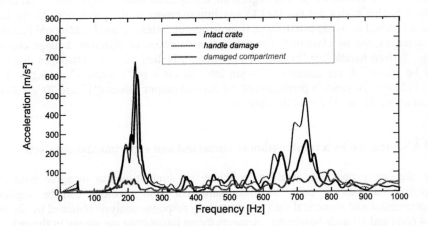

Fig. 4. Examples of vibration spectra of two different damages (handle reference point)

In order to limit the degrees of freedom of the ANN, the number of input nodes has been limited to 10. This requires a reduction of the spectrum data-sets by a factor of hundred. Different methods have been applied. Splitting the spectrum into ten parts and integration of each part leads to an appropriate data reduction. Scaling, normalising, shifting or calculating of for example inertia moments did not improve the results. Figure 5 shows four different cases, where the input values of the nodes are displayed as bars. The data-reduction for both numerically and experimentally obtained spectra is carried out in the same way.

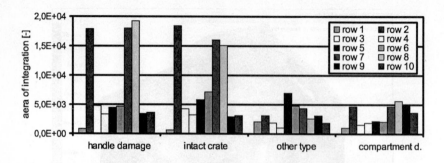

Fig. 5. Input data for ANN, bar plot of four different cases

3.1 ANN trained by experimental data-sets

In a first step, the ANN is processing exclusively experimental data in order to prove the feasibility of the pattern-recognition-system. The configuration of the network has been modified in order to analyse the sensitivity of the net topology. In all tests a MLP is used, trained by RPROP algorithm. Using more than 10 input nodes and more than 5 hidden nodes did not lead to any improvement in the prediction. Using less nodes resulted in worse predictions. If more output nodes are used, additional features of damages can be classified. For example a detection of different damage classes (e.g. "broken handle" or "broken compartment") is then possible. Damages that can not be identified, are automatically put into the class of "unknown" damages. The 10x5x2 network yields a prediction of the defined output values ("1" and "0") with an accuracy of about 5% for all the data-sets.

3.2 ANN trained by a combination of numerical and experimental data-sets

The frequency response obtained by numerical simulation has to agree with the experimental data in order to make sure that both methods can be applied. Experimental and numerical data of a frequency response analysis obtained by the use of 4 node and 10 node tetrahedral elements for an intact crate are shown in figure 6.

It can be stated, that the observable main characteristics of the curves are similar. In addition, there are two regions of interest where the numerical data are similar to those of the experimental response (around 200 Hz and above 500 Hz). It is obvious that the simulation by the 10 node element model is closer to the experimental spectra than the 4 node element model. Further improvement of the agreement of the numeric simulation with the experiment is required. Nevertheless, simulations of damaged crates have been carried out in order to investigate their accuracy. As in the measured spectra (see figure 4) it can be stated that damages can be identified by comparison of the spectra. To estimate their quality, the comparison of numerical and experimental data of a handle damage is presented in figure 7.

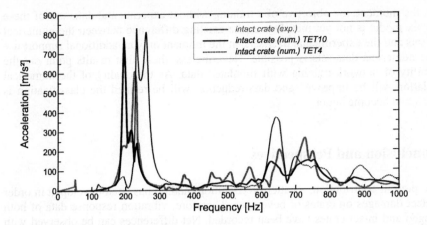

Fig. 6. Comparison of experimental and numerical spectra simulated by 4 node and 10 node tetrahedral elements (handle reference point)

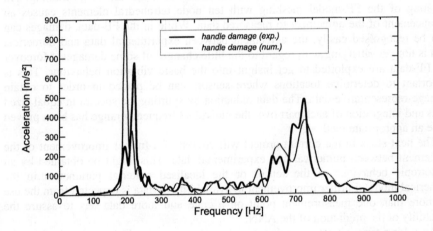

Fig. 7. Comparison of experimental and numerical handle damage (handle reference point)

The calculated response spectra as well as the experimental response spectra show net changes of peak size and location in similar frequency ranges. The difference between model and measurement of the undamaged crate in the example is, as [11] reported for a cracked beam, in the order of the change that the damage on the handle causes in measurement as well as in the simulation data. Because of this overlapping, it is important to be aware of the problem of false indication of damage in the two categories "false-positive" (indication of damage when none is present) and "false-negative" (no indication of damage when damage is present) [19]. However, other damages and other crate models cause serious changes in the spectra so that these cases differ significantly, which allows their classification.

988 J. Zacharias, C. Hartmann, and A. Delgado

At the moment, the objective of a training without experimental data-sets of these complex model is not yet met. Due to the existing difference between the numerical data-sets and the experimental data-sets, at the moment only an additional support use of the numerical data-sets is possible. Nevertheless, these first results point out the feasibility of network-training with simulated data. As the quality of the numerical simulation will be improved and data-reduction will be refined the classification is expected to become better.

4 Conclusion and Perspectives

In the present contribution, a pattern recognition method has been developed in order to detect damages on crates of beverages. Therefore, vibration response data of both damaged and intact crates have been recorded. Net differences can be observed with respect to the intact crate in all analysed damage cases. As the attempt is to use as less as possible "expert know-how", an ANN has been trained such that the detection of damaged crates is possible in all cases. Finite-element-simulation is carried out to analyse the mode shapes and in order to obtain data for the ANN-training. An updating of the FE-model meshing with ten node tetrahedral elements causes an enhancement of the agreement to measured data. While in the FE-data, damages can also be recognised easily, the agreement between experimental data and numerical data is not yet satisfying with regard to the little changes of some damages. Moreover the FE-data are exploited to get insight into the basic vibration behaviour. This is important to determine locations where sensors can be placed in order to obtain damage-representative data. The data reduction by splitting the spectra in equal sized parts and integration of each part over the individual frequency range has been proved to be an appropriate method.

The next steps in the current project will consist of a further improvement of the agreement between numerical and experimental data. This might be obtained by an anisotropic behaviour of the sensor or the idealised material parameters in the numerical model. Further improvement in availability of data is expected from the use of more than one measurement point on a crate and more data-sets to assure the reliability of the prediction of the ANN.

As a long term objective ageing or micro damages should be recognised by the system. For application of the method in an industrial context a reduction of the cycle time is required. While at the moment the cycle time is about 20 seconds, it has to be reached a level of one to two seconds. This can be obtained, if the excitation can be modified such that a shock excitation replaces the sinusoidal sweep which represents the bottle-neck of the current procedure. Also alternative processes are possible for the measurement of vibration signals, for example using contactless measurement systems or directly the signals on gripper devices.

References

1. Klein, F.: Leergutsortierung in der Getränketechnik, Getränketechnik, 8 (1992) 46-52
2. Hermans, L., Van der Auweraer, H.: Modal Testing and Analysis of Structures under Operational Conditions: Industrial Applications, Mechanical Systems and Signal Processing, 13(2) (1999) 193-216
3. Doebling, S.W., Farrar, C.R., Prime, M.B.: A Summary Review of Vibration-Based Damage Identification Methods, The Shock and Vibration Digest, 30(2) (1998) 91-105
4. Salawu, O.S.: Detection of structural damage through changes in frequency: a review , Engineering Structures, 19(9) (1997) 718-723
5. Wang, Z., Lin, R.M., Lim, M.K.: Structural damage detection using measured FRF data, Computer Methods in Applied Mechanics and Engineering, 147 (1997) 187-197
6. Alampalli, S.: Effects of Testing, Analysis, Damage, and Environment on Modal Parameters, Mechanical Systems and Signal Processing 14(1) (2000) 63-74
7. Marwala, T., Hunt, H.E.M.: Fault Identification Using Finite Element Models and Neural Networks, Mechanical Systems and Signal Processing 13(3) (1999) 475-490
8. Masri, S.F., Nakamura, M., Chassiakos, A.G., Caughey, T.K.: Neural Network Approach to Detection of Changes in Structural Parameters, Journal of Engineering Mechanics, 122(4) (1996) 350-360
9. Elkordy, M.F., Chang, K.C., Lee, G.C.: Neural Networks Trained by Analytically Simulated Damage States, Journal of Computing in Civil Engineering, 7(2) (1993) 130-145
10. Kudva, J.N., Munir, N., Tan, P.W.: Damage Detection in Smart Structures Using Neural Networks and Finite-Element-Analyses, Smart Materials and Structures, 1 (1992) 108-112
11. Fritzen, C.-P., Jennewein, D.: Damage Detection Based on Model Updating Methods, Mechanical Systems and Signal Processing, 12(1) (1998) 163-186
12. Zacharias, J., Hartmann, Ch., Delgado, A.: Recognition of Damages on Crates of Beverages by an Artificial Neural Network, Conference-Proceedings of eunite 2001, Tenerife, Spain, (2001)
13. Reynier, M.: Sensors Location for Updating Problems, Mechanical Systems and Signal Processing, 13(2) (1999) 297-314
14. Entrekin, A.: Accuracy of MSC/NASTRAN First- and Second-Order Tetrahedral Elements in Solid Modeling for Stress Analysis, MSC Aerospace Users' Conference, (1999)
15. Sitton, G.: MSC/NASTRAN Basic Dynamic Analysis User's Guide, The McNeal-Schwendler Corporation, USA, (1997)
16. Rieg, F., Hackenschmidt, R.: Finite Elemente Analyse für Ingenieure, Hanser-Verlag, München, (2000)
17. Hagedorn, P., Otterbein, S.: Technische Schwingungslehre – Lineare Schwingungen diskreter mechanischer Systeme, Springer-Verlag, (1987)
18. Rafiq, M.Y., Bugmann, G., Easterbrook, D.J.: Neural Network Design for Engineering Applications, 79 (2001) 1541-1552
19. Fugate, M.L.: Vibration-Based Damage Detection Using Statistical Process Control, Mechanical Systems and Signal Processing, 15(4) (2001) 707-721
20. http://www.lfp.blm.tu-muenchen.de/pa/Software/knn.htm, 15.01.2002
21. Zell, A.: Simulation Neuronaler Netze, Addison-Wesley, Bonn, (1994)

Simulation Monitoring System Using AVS

Tadashi Watanabe[1], Etsuo Kume[1], and Katsumi Kato[2]

[1] Center for Promotion of Computational Science and Engineering,
Japan Atomic Energy Research Institute, Tokai-mura, Ibaraki-ken,
319-1195, Japan
watanabe@sugar.tokai.jaeri.go.jp
kume@brian.tokai.jaeri.go.jp
[2] Research Organization for Information Science and Technology,
Tokai-mura, Ibaraki-ken, 319-1195, Japan
kato@sugar.tokai.jaeri.go.jp

Abstract. A simulation monitoring system has been developed to visualize ongoing numerical simulations on supercomputers or workstations. The output data for visualization are transferred from the calculation server to the visualization server and visualized automatically by the monitoring system. Visualization is performed by AVS on UNIX or WINDOWS environment. Modification of simulation program is not necessary, and the monitoring system is applied for both interactive and batch process of numerical simulations.

1 Introduction

Visualization or animation technique has been used in computational science as the post processing technology. Numerical simulations are performed on supercomputers in a computer center, and visualization is performed on graphics workstations or personal computers after the simulations. This is because supercomputers are composed of front and back end systems generally and the network queuing system is adopted. The start and end of batch jobs are not known. The numerical simulations in computational science need long computational time. The efficiency of research is not improved if the numerical results are visualized only after the simulations, even though the efficient simulations and visualization are performed. Visualization of ongoing simulations is desirable for large-scale simulations using supercomputers in a computer center.

The visualization systems, in which results of ongoing simulations are transferred from supercomputers or workstations to a visualization server, have been developed and commercialized by computer makers such as NEC[1] and Fujitsu[2]. Similar systems for research and development purposes have been proposed [3,4,5,6]. These visualization systems are adapted to specified computer systems, and it is necessary to modify simulation programs to make the display data or to use libraries for data transfer. Visualization and numerical simulations are closely linked together and applications of these systems are limited. These visualization systems might be useful for simulations using fixed computer codes and computer systems. It is, however, not

P.M.A. Sloot et al. (Eds.): ICCS 2002, LNCS 2329, pp. 990–999, 2002.

convenient to researchers who make and modify their own simulation programs frequently using several types of computer systems.

In this paper, a simulation monitoring system is described. This monitoring system is developed for researchers in a small research group where many kinds of simulation programs are developed and modified frequently, and several computer systems including supercomputers in a computer center and workstations in a laboratory are used for numerical simulations. The dependency of monitoring system on computer environments and the modification of simulation program for visualization are, thus, not preferable. UNIX-based workstations or WINDOWS PCs are used for visualization, and AVS5, AVS/Express, and MicroAVS are used as the visualization software.

2 Simulation Monitoring System

Numerical simulations, for which visualization is performed during calculations, are performed on the calculation server such as workstations in a research laboratory or supercomputer systems in a computer center. Simulation monitoring is performed on the visualization server such as graphics workstations or PCs in the laboratory. Several methods are possible in this situation to visualize results of ongoing simulations. The AVS series are used as the visualization software so that the simulation programs are not modified for visualization. The format of output data is the same for UNIX workstations and for WINDOWS PCs. Display data are not generated on the calculation server and simulation results are transferred from the calculation server to the visualization server. Time for data transfer may not be small according to the simulation data. It is, however, not significant since the computational time is generally much larger than the time for data transfer and visualization. AVS is used for visualization, and thus the change in displayed picture during monitoring is possible and easy.

The simulation monitoring system is composed of two parts: one on the calculation server and the other on the visualization server. The monitoring system on the calculation server is watching the execution and the output of numerical simulations. Once the output file of simulation results is detected by the monitoring system, the output file is transferred from the calculation server to the visualization server by FTP. On the visualization server, the arrival of output file is detected by the monitoring system and visualization is performed by AVS5, AVS/Express, or MicroAVS.

2.1 Detection of Output File on Calculation Server

The execution and the output of numerical simulations are watched by the monitoring system on the calculation server. Once the output file for visualization is generated from the simulations, the monitoring system detects the output file and send it to the visualization server. The generation of output file is easily detected on UNIX workstations, and thus the case with the computer system using batch process is described here.

The output of numerical simulations is sometimes related to the end of batch process on computer systems in a computer center, even if its OS is based on UNIX. In

992   T. Watanabe, E. Kume, and K. Kato

this case, an example of UNIX shell on the front-end of calculation server for detecting the end of batch job is shown below. This example is used in csh environment. The submission of numerical simulations is also included in this example. In this example, "qsub" in line #02 indicates a submission of calculation to the back-end computer system. Parameters for batch process are described in the file "sub.sh" in this case. The start and end of the batch process are not known since many batch jobs are submitted by many users. The output file from the back-end system is detected in line #03 in this example. In our computer system, the output file named "sub.sh.o*" is sent back from the back end to the front end, where "*" in the file name is the job number. The word "logout" in the output file indicates the normal end on the back-end system. After the normal end is detected, the job number is obtained in line #08, and the empty file "lock.*" is temporally made in line #10. The output file for visualization is transferred from the calculation server to the visualization server by FTP in line #11. After sending the data for visualization, the empty file "lock.*" is sent to the visualization server. The arrival of this file to the visualization server is corresponding to the end of data transfer, and the visualization process is started. The procedure from submission to FTP is performed again in the last line in this example. This example is used for a numerical simulation with restart calculations. In this case, the output file for visualization is made once in a batch job, and the normal end of the batch job is detected as the end of output file. In case of a job with multiple output files, it is not difficult to detect the file after the output of each file.

Example of UNIX Shell for Detecting the Output File of Numerical Simulations on Calculation Server

```
## this is submit-shell                                    #01
qsub sub.sh                                                 #02
until grep logout sub.sh.o*                                 #03
    do                                                      #04
    sleep 300                                               #05
    done                                                    #06
cat sub.sh.o* >> sub-out-file                               #07
  a1=`ls sub.sh.o* | awk '{print substr($1,9,5)}'`         #08
  /bin/rm sub.sh.o*                                         #09
  touch lock.$a1                                            #10
  ftp -n ip-address << eod                                  #11
    user user-id password                                   #12
    cd ftp-directory                                        #13
    prompt                                                  #14
    bin                                                     #15
    put avs-data avs-data.$a1                               #16
    put fld-data fld-data.$a1                               #17
    put lock.$a1                                            #18
  eod                                                       #19
  /bin/rm lock.$a1                                          #20
  nohup submit-shell &                                      #21
```

2.2 Detection of Transferred File on Visualization Server

On the visualization server, the simulation monitoring system is waiting for the arrival of the output file and the lock file. Transferred files are detected by the similar method as described above. The output file is given to AVS after the lock file is detected. If the output file itself is used for detection, AVS may start reading the data before the completion of data transfer, and thus, the lock file is used for detection. The lock file is empty and the time for data transfer is negligible.

The simulation monitoring system uses the AVS series for visualization on UNIX workstations or WINDOWS PCs. The lock file is detected by the AVS modules on UNIX workstations. The AVS modules are written in C language and registered in AVS environment after compilation. On WINDOWS PCs, the monitoring program, which detects the lock file and calls MicroAVS, is written in C language.

3   Monitoring

Examples of monitoring using AVS5, AVS/Express, and MicroAVS are shown in this section. These examples are used for numerical simulations of multi-phase flow phenomena. The monitoring of ongoing calculations is indispensable for transient analyses in many research fields such as fluid dynamics research.

3.1 AVS5

The control panel of the AVS5 module for file detection is shown in Fig.1. This module is connected to the module for reading a data file such as the read_field module in the AVS5 network. The parameters for monitoring are set by using this module. The directory for the output and lock files are selected in "search" and "select" columns. The names of the output and lock files are set in "prefix" and "lock" columns, respectively. The first step number and the increment are set in "first" and "delta" columns, respectively. The output and lock files have the same step number, which may be the simulation step number or job number. The time period for searching the lock file is set in "interval" column. Whether output files are deleted or not after visualization is selected by "delete" or "undelete" button. Monitoring is started by "run" button, and the current step number is indicated in "current" column.

An example of the AVS5 network is shown in Fig. 2. The name of the AVS5 module for file detection is "Check_Step" in this example. The Check_Step module gives the file name to the read_filed module after the lock file is detected. Scalar variables in field data are visualized in this example.

The displayed picture is shown in Fig. 3, where a snapshot of the collision process of two spherical droplets are shown. The numerical simulation is performed using the lattice Boltzmann method, which is one of the particle simulation methods for analyses of complex flow phenomena. The surface of droplets and the pressure field in a cross section are visualized using the isosurface and orthogonal_slicer modules, re-

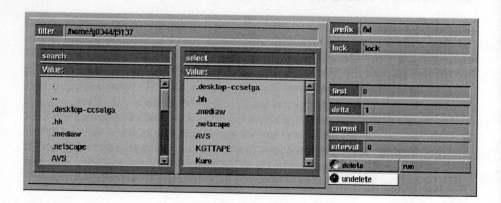

Fig. 1.   Control panel of AVS5 module for file detection

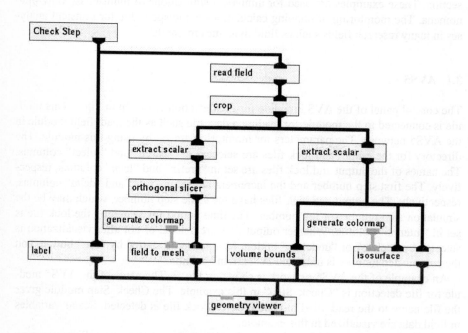

Fig. 2.   AVS5 network for monitoring

spectively. This picture is displayed until the next file arrives. The displayed picture can be changed during monitoring since AVS functions are all available.

3.2 AVS/Express

The control panel of the AVS/Express module for file detection is shown in Fig.4. As is the case with AVS5, this module is connected to the module for reading a data file in the AVS/Express application. The parameters have the same meaning as before.

An example of the AVS/Express application is shown in Fig. 5. Scalar variables in field data are visualized in this example. The displayed picture is shown in Fig. 6, where a snapshot of the two-phase flow in rectangular pipe is shown. The numerical simulation is performed using the lattice Boltzmann method, and the two-phase interfaces and the pressure field are visualized.

3.3 MicroAVS

The control panel for file detection is shown in Fig.7 for MicroAVS on WINDOWS PC. This control panel is displayed at the left corner of PC's display. The parameters are the same as before. This control panel is developed in the visual C++ environment.

An example of monitoring is shown in Fig. 8, where a snapshot of the mixing process of two fluids is shown. The numerical simulation is performed using the lattice gas method, which is one of the most simple particle simulation method for calculations of various flows. The concentration of two fluids in a mixing chamber is visualized. As is the case with AVS5 and AVS/Express, all the functions of MicroAVS are available during monitoring.

4 Summary

Computer systems and computational techniques have been developed very rapidly, and large-scale numerical simulations are performed in many science and engineering fields. A huge amount of simulation results is obtained and the efficient visualization tools, which are easy to use for researchers, are desired.

The simulation monitoring system to visualize ongoing numerical simulations on supercomputers or workstations is described in this paper. The output data for visualization are transferred from the calculation server to the visualization server and visualized automatically. Visualization is performed by AVS on UNIX or WINDOWS environment, and thus many types of visualization methods are easily used. Modification of simulation program is not necessary, and the monitoring system is applied for both interactive and batch process of numerical simulations. The monitoring system is useful for a research group, where many kinds of simulation programs are developed and modified frequently, and many types of computer systems are used. In this paper, the examples of monitoring used for numerical simulations of multi-phase flow phenomena are shown, since the monitoring of transient calculation is of importance for

996   T. Watanabe, E. Kume, and K. Kato

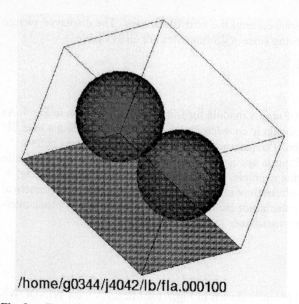

/home/g0344/j4042/lb/fla.000100

Fig. 3.   Example of monitoring using AVS5 (collision process of two droplets)

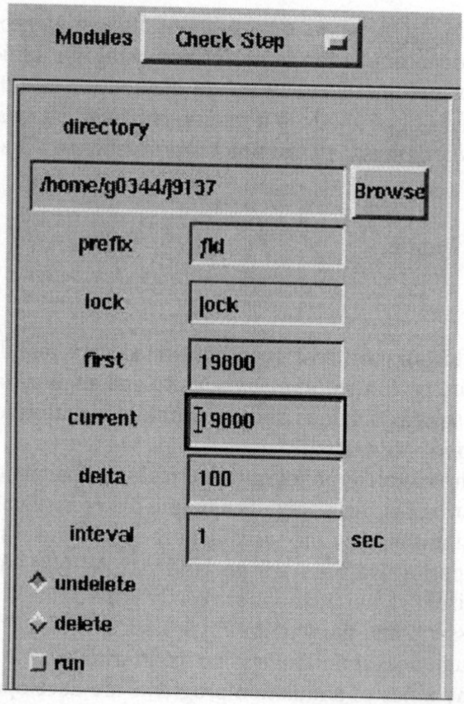

Modules	Check Step

directory

/home/g0344/j9137	Browse

prefix	fld
lock	lock
first	19800
current	19800
delta	100
inteval	1	sec

◆ undelete

◇ delete

❑ run

Fig. 4.   Control panel of AVS/Express module for file detection

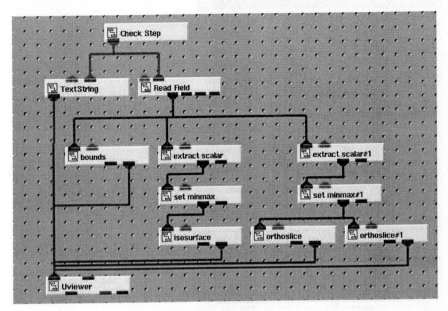

Fig. 5.   AVS/Express application for monitoring

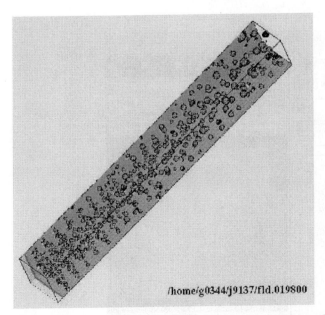

Fig. 6.   Example of monitoring using AVS/Express (two-phase flow in a pipe)

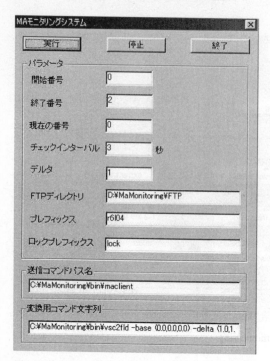

Fig. 7.   Control panel for file detection for MicroAVS

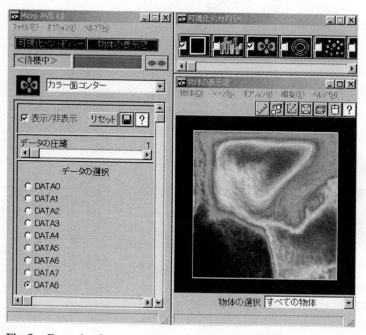

Fig. 8.   Example of monitoring using MicroAVS (mixing process of two fluids)

transient analyses in fluid dynamics research. We have a plan to use the simulation monitoring system for 3D visualization on UNIX and WINDOWS environment in the near future.

References

1. http://www.sw.nec.co.jp/APSOFT/SX/rvslib/index.
2. http://www.nagano.fujitsu.com/avs/Products/Relation/
3. Muramatsu, et al.: Development of Real Time Visualization System for Fluid Analyses on Parallel Computers, JAERI-Data/Code 98-014(1998)
4. CRIEPI-Hitachi Joint Research Group on Parallel Applications: A Study on Software Infra-structures for Computational Science and Engineering, U99033(2000)
5. http://raphael.mit.edu/visual3/visual3.html
6. http://www.icase.edu

ODEs and Redefining the Concept of Elementary Functions

Alexander Gofen

The Smith-Kettlewell Eye Research Institute, 2318 Fillmore St., San Francisco, CA 94102,
USA
galex@ski.org , www.ski.org/gofen

Abstract. The modern concept of elementary functions and the Taylor method are deeply connected. The article summarizes the remarkable (although not widely known) facts about the modern Taylor method, Ordinary Differential Equations (ODEs) and the modern notion of elementary functions - the property that actually takes place for all ODEs used in applications. Beside the typical usage of the Taylor method for integrating *initial value* problems, two new applications of the Taylor method are considered: integrating until the given *end value* of a dependent variable, and integrating the *boundary value* problem.

In the bright memory of the late Prof. Michael L. Lidov

1 Introduction

According to the tradition since the 19th century, elementary functions were defined just by a *convention* as a certain limited list of well studied functions, a desirable final format to express the solutions of different problems. They were polynomials and rational functions (of several variables), plus the exponential, logarithmic, trigonometric and the inverse trigonometric functions (of one variable), plus all finite superpositions of them (The Mathematical Encyclopedia).

Liouville adhered to a slightly different definition: he considered elementary also algebraic functions, integrals, exponential of integrals, solutions of linear differential equations with constant coefficients, and all functions which can be obtained by repeated iteration of these functions.

In any case, it was just a rather artificial convention. Probably it was Ramon Moore [1], who first suggested another approach to define the elementary functions based not on an arbitrary convention, but on a quite fundamental property, that the elementary functions are solutions of ODEs with *rational* right hand parts. Although his goal was mostly to improve and optimize the algorithm of the classical Taylor method such a way that the *modern Taylor method* would require no more than $O(N^2)$ operations, the new approach appeared to have also a profound mathematical meaning and connections. It established a fundamental dividing line between non-elementary functions and the elementary ones. The latter characterize the types of ODEs for which the modern Taylor method is applicable, and they also provide a constructive

P.M.A. Sloot et al. (Eds.): ICCS 2002, LNCS 2329, pp. 1000–1009, 2002.

way how (almost) *any* (!) non-linear source system of ODEs may be transformed to quite remarkable forms – even as simple as quadratic polynomials (Table 2).

First of all, the modern Taylor method was intended as a powerful numeric method to integrate the *initial value* problems with the ultimate accuracy. The software implementing it was known since the 80s, and the recent version of such an integrated environment for PCs (The Taylor Center) is described in these proceedings [6,7].

This paper considers also a couple of less typical tasks for the Taylor method: integration until the given *end value* of the dependent variable, and the boundary value problem.

2 Elementary Functions and ODEs

This section summarizes the fundamental facts about the modern elementary functions – the basis of the modern Taylor method.

Definition 1: a (vector-) function $f = \{f_k(x_1, ..., x_m)\}$, $(k=1,...,n)$, is called *elementary with respect to* x_1 if there exists a system of $N \geq n$ ODEs

$$\partial f_k / \partial x_1 = R_{kl}(f_1,...,f_N), \quad k = 1,..., N \tag{1}$$

with the rational right hand parts R_{kl}, for which f is a solution (f is defined by this system as a function of x_1). We say that the (vector-) function f is *elementary* if it is elementary by all its variables $x_1, ..., x_m$, thus there exist m different systems of ODEs $(l=1,...m)$ defining f as a (vector-) function of each of its arguments x_l . All those rational right hand parts R_{kl} may be organized into a matrix, called the *matrix of elementarity* of the (vector-) function f.

It was proved [2], that if we consider a similar Definition 1' (a version of Definition 1), with the *polynomial* right hand parts P_{kl} rather than the *rational* R_{kl}, both definition are equivalent.

In order to demonstrate elementarity of a (vector-) function, it may appear necessary to consider it as a component of a larger vector-function. For example, $f(x,y,z) = cos(x)e^y\zeta(z)$ is elementary together with $g(x,y,z) = sin(x)e^y\zeta(z)$ by x, $(f_x' = -g; g_x' = f)$, it is elementary "itself" by y $(f_y' = f)$, and it is probably not elementary by z (Zeta function of Riemann). Thus, the matrix of elementarity by x, y for function f (as a component of the vector-function $\{f,g\}$) is this:

$$\begin{bmatrix} -g & f \\ f & 0 \end{bmatrix}$$

The function $w=tan(t)$ is elementary together with $u=cos(t)$ and $v=sin(t)$, and the matrix of elementarity is

$$(-v \quad u \quad 1/u^2 \;)^T.$$

Table 1. Theorems and facts about the elementary functions

1	Polynomial and rational functions	Elementary
2	All traditionally elementary functions, as well as those by Liouville	Elementary
3	Superposition of elementary vector-functions	Elementary
4	Inverse to an elementary vector-function	Elementary
5	Implicit functions x_k of an equation $F(x_1,x_2,...x_n)=0$ with an elementary function F	Elementary
6	Algebraic functions	Elementary
7	Derivative of an elementary function	Elementary
8	Indefinite integral with respect to a certain variable (say t) of an elementary function $f(t,x)$	Elementary by t, but not necessarily by x
9	Vector-function $u_k(t,x)$ which is a solution of ODEs $u_k'=f_k(u_1,...u_n, x)$ with an elementary vector-function $\{f_k\}$	Elementary by t, but not necessarily by x
10	$\lim f_n(t)$ with all functions $f_n(t)$ elementary	Not necessarily elementary
11	$\sum\limits_{n=0}^{\infty} a_n t^n$ converging to $f(t)$. Coefficients a_n are arbitrary	Not necessarily elementary
12	$\sum\limits_{n=0}^{\infty} a_n t^n$ converging to $f(t)$. Coefficients a_n are obtained via the special recurrent formulas for differentiating the canonical equations	Elementary
13	A solution $x(t)$ of a finite difference equation $F(x(t), x(t+h))=0$ with an elementary non-linear function F	Not necessarily elementary
14	Euler's Gamma function defined as $\Gamma(x+1) = x\Gamma(x)$	Non-elementary

An algebraic function $x=X(y,z)$, defined by an implicit equations $P(x,y,z)=0$, where P is a polynomial of arbitrary high degree, is elementary by y and z:

$$X_y' = -P_y'(X,y,z)/P_x'(X,y,z); \quad X_z' = -P_z'(X,y,z)/P_x'(X,y,z).$$

This demonstrates existence of elementary functions of *more than one variable*, which are *not rational* (a case not covered by the definition of Moore [1]). Also, each of these ODEs gives an example, when the solution $X(y,z)$ is elementary both by its parameter and the differential variable, which is not always the case, but rather an exception.

The fundamental properties of elementary functions are summarized in Table 1. This modern definition of elementarity includes all functions, which traditionally belong to that notion, plus all solutions of ODEs with the rational right hand parts, all their possible superpositions and inverse (vector-) functions. Well, then, do non-elementary functions exist at all according to the modern definition? It is not so easy to point out even one. All types of functions in Table 1 mentioned as "not necessarily elementary" are rather "good candidates". Meanwhile the non-elementarity is known for sure only for Euler's Gamma function $\Gamma(x)$ (due to the theorem of Hödel, stating that $\Gamma(x)$ cannot be a solution of any polynomial ODE).

Table 2. The remarkable transformations possible for all ODEs with the right hand parts being elementary vector-function (practically any system of ODEs used in applications)

A standard system of ODEs whose right hand parts are an **elementary** vector-function (practically any system of ODEs used in applications) can be reduced to...		
A system of ODEs whose right hand parts are the **rational** functions [2]. It can be further reduced to...		
A **canonical** system, a mixture of the algebraic and differential equations of certain simple types, used to compute efficiently the N-order derivatives with no more than $O(N^2)$ operations.	A system of ODEs whose right hand parts are the **polynomial** functions [2]. It can be further reduced to...	
	A system of ODEs whose right hand parts are the **polynomials** of a **degree** ≤ 2 (see [4]). It can be further reduced to...	
	A system of ODEs whose right hand parts are the **polynomials** of a **degree 2** with coefficients **-1, 0** or **1** only (see [5]).	A system of ODEs whose right hand parts are the **polynomials** with **square terms only** (see [4]).

Is there any sense in defining so large a concept of the elementary functions that practically all functions used in applications belong to this class? The great reason is that for *all ODEs with elementary right hand parts* there exists a unified, constructive and efficient way of transforming them to the *rational* and several other remarkable forms, including the *canonical*, which makes possible the efficient modern Taylor method. These transformations are summarized in Table 2.

Each transformation (from the upper to lower cells in this Table) results in a certain increase in the number of equations. But the transformation required for the modern Taylor method

$$\text{Any } \textbf{Elementary } \text{ODEs} \rightarrow \textbf{Rational } \text{ODEs} \rightarrow \textbf{Canonic } \text{System}$$

adds just as little as is necessary to define the non-rational elementary functions [2]. On the contrary, transforming to the polynomial formats of a degree ≤ 2 may multiply the number of equations essentially [4, 5]. The fact itself that it is possible seems remarkable in any case.

Even with the notion of elementarity that large, it does impose certain applicability limits for the Taylor method. For example, we must be aware that the solutions of ODEs as functions of the parameters mostly are not elementary, thus we cannot apply the Taylor method to further integrate them by these parameters. Similarly, double and multiple definite integrals also cannot be computed with the Taylor method directly. On the other hand, if we know a (first, a second) integral of a certain system of ODEs, it is usually an implicit equation built of elementary functions of unknown variables and parameters. By *these* parameters, elementarity of the solution does take place.

3 Modern Taylor Method Basics

First, we assume a convention. In Modern Taylor Method, the concept of *n-th* derivative means the so called normalized derivative, which by definition is

$$u^{[n]} = u^{(n)}/n!$$

so that the Taylor expansion looks like this:

$$u(t) = \sum u^{[k]}(t - t_0)^k \; ;$$

(further on, the conventional notation $u^{(k)}$ is used, although we always mean the normalized derivatives).

What distinguishes the modern Taylor method from its classical counterpart is the special technique of computing derivatives. Let us consider a standard initial value problem for a system of ODEs with any analytical right hand parts

$$u_k' = f_k(u_1, ..., u_m), \quad u_k|_{t=a} = a_k, \qquad k=1,...m \tag{2}$$

With no additional assumptions about the right hand parts (2), to perform N-order differentiation, we have nothing simpler than the formula of Faa di-Bruno [2] for N-order derivatives of a superposition of functions – quite a complex formula with exponential growth, completely impractical.

Even for a more specific right hand part, for example like $x' = uvwy$, the Leibnitz formula for several factors is needed

$$(uvwy)^{(N)} = \sum\sum\sum u^{(i)}v^{(j)}w^{(k)}y^{(N-i-j-k)}$$

That requires as many as $O(N^4)$ operations in this specific case. What is the key issue for the modern Taylor method is a possibility to reduce whichever complex (but elementary!) right hand parts to a sequence of simple *canonical* equations [2,3], or to a sequence of formulas over two operands, for which computing N-order derivatives never requires more than $O(N^2)$ operations. Such formulas are for example, the Leibnitz formula for N-th derivative of a product of *two* functions:

$$(uv)^{(N)} = \sum_{i=0}^{N} u^{(i)}v^{(N-i)} \tag{3}$$

The similar by complexity $O(N^2)$ formulas exist for quotient, power function, exponent and logarithm, for example

$$\text{if } R=u/v, \text{ then } R^{(N)} = \left(u^{(N)} - \sum_{i=0}^{N-1} R^{(i)}u^{(N-i)} \right)/v \; ;$$

$$\text{if } R=u^a, \text{ then } R^{(N)} = \left(\sum_{i=0}^{N-1}(a(1-i/N) - i/N)R^{(i)}u^{(N-i)} \right)/u, \; (a \text{ is a constant}).$$

For the linear equations like au, $u+v$ the corresponding formulas are trivial and require $O(n)$ operations only.

Note: the canonic equations or the sequence of formulas must be computed in the given order only (not a good case for parallel computing), but the polynomial forms of a degree ≤ 2 may be used instead of the canonical equations, and unlike the latter, they may be computed in any order concurrently.

Thus, the modern Taylor method allows obtaining the numeric values of N-order derivatives in any given point for the system (2) and then summing the Taylor expansion. As soon as the source system (2) is reduced to the rational right hand parts (possibly allowing also the power, exponent and logarithm), an algorithm, similar to parsing an arithmetic expression, automatically performs further reducing to certain internal structures (like the implicit canonic system), and computes the Taylor expansion. That way the Taylor method integrates the initial value problems, applying finite steps not approaching zero and still maintaining any required high accuracy up to the ultimate – all available digits of the given binary representation [6, 7]. Practically, it is reasonable to select the order N of the Taylor method according to Moore's "rule of thumb" [1-3] – a value between 20 and 30. The heuristic radius of convergence may be obtained from the series of the derivatives [6, 7].

4 Reaching the Given End Value

The initial value problem is a "native" one for the Taylor method, but in an application it may be necessary to integrate until the moment when the given *end value* of the dependent variable is reached: that would allow determining a period of the solution (if it exists). Specifically, we are going to integrate say $u_1(t)$ until it reaches a certain value b_1. To do that, there are two approaches in the framework of the Taylor method: (1) It's always possible and easy to write the systems of ODEs for the inverse function $T(u_1)$ and to integrate it by now independent variable u_1 until b_1; (2) To apply the standard technique of numeric analysis of polynomials to determine whether there is a root of $u_1(t) = P_1(t) = b_1$ in the current domain of the Taylor expansion for $u_1(t)$.

4.1 Integrating the Inverse Function

Given the source system (2), the ODEs for the function $T(u_1)$ inverse to $u_1(t)$ are

$$\partial u_k/\partial u_1 = f_k(u_1,..., u_m)/f_1(u_1,..., u_m), \qquad k=1,...m$$
$$\partial T/\partial u_1 = 1/f_1(u_1,..., u_m) \tag{4}$$

True, when switching to system (4) with the independent variable u_1, we must be ready to run into singularities while integrating from a current point to the end value b_1. We may have to edge into the complex plane temporarily to circumvent the singularities, and to take into consideration different Riemann planes. Or we can just switch back to the system (2) and integrate it by t while near the singularity, and then return back to integration by u_1 again.

4.2 Analyzing the Polynomial

At each step of integration by t, we may obtain a polynomial expansion $P_1(t)$, representing the solution $u_1(t)$ with the ultimate accuracy in a certain finite domain. Thus, without sacrificing any accuracy, we can explore the equation $P_1(t) = b_1$ in the domain applying the well developed technique for polynomials. That would make possible either to conclude that there are no solutions (and to perform the next step of integration by t), or to obtain the solution t_1 delivering the end condition $u_1(t_1) = b_1$.

5 Boundary Value Problems

Lets us consider a boundary value problem

$$u_k'=f_k(u_1,..., u_m), \quad k=1,...m; \quad u_k|_{t=0} = a_k, \quad k=2,...m \tag{5}$$
$$u_1|_{t=0}= x, \quad u_1|_{t=1}= b_1$$

with the unknown component x of the initial value. Given the incomplete initial condition $u_1 = x$ at $t=0$, the unknown solution $u_k(t,x)$ in the general case is not elementary by x, therefore there is no hope of finding a system for the inverse function $X(t,u_1)$ and following the approach 4.1. Instead, to obtain the N-order Taylor expansion of the solution u by x at the end point $t=1$, we can adjust the approach 4.2. That will require N-times expanding the source system (5), adding ODEs defining all $\partial^n u_k/\partial x^n$ as functions of t. (Practically, the order N of the Taylor method is usually 20-30).

Here is how it is done.

$$u_k'=f_k(u_1,..., u_m), \quad u_1|_{t=0}= x, \quad u_k|_{t=0}= a_k, \; k=2,...m, \text{ (the source ODEs)}.$$

Then, we may add:

$$\partial u_k'/\partial x = \partial f_k/\partial x, \qquad \partial u_1/\partial x|_{t=0} =1, \quad \partial u_k/\partial x|_{t=0} =0, \; k=2,...m; \qquad (6)$$

$$\partial^2 u_k'/\partial x^2 = \partial^2 f_k/\partial x^2, \qquad \partial^2 u_k/\partial x^2|_{t=0} =0, \; k=1,...m;$$

$$. \; . \; . \; . \; . \; . \; . \; . \; . \; .$$

$$\partial^N u_k'/\partial x^N = \partial^N f_k/\partial x^N, \qquad \partial^N u_k/\partial x^N|_{t=0} =0, \; k=1,...m;$$

Now, if we start with an arbitrary parameter x, after integrating this system from $t=0$ until $t=1$, we will obtain the sequence in this point

$$\partial u/\partial x, \; \partial^2 u/\partial x^2 , \; . \; . \; . \; . \; , \partial^N u/\partial x^N$$

which makes available the Taylor expansion by x of $u_1(t,x)$ at $t=1$ (as though $u_1(t,x)$ were elementary by x and we had the corresponding system of ODEs):

$$u_1(t, x+\Delta x) = u_1(t,x) + (\partial u/\partial x)\Delta x + (\partial^2 u/\partial x^2)\Delta x^2 + \; . \; . \; . \; + (\partial^N u/\partial x^N)\Delta x^N$$

(the derivatives are normalized). Now, similarly to what we do with respect to t, we can estimate the domain of convergence by x of this Taylor expansion, and then analyze if there is a root of the following polynomial of Δx

$$u_1(t,x) + (\partial u/\partial x)\Delta x + (\partial^2 u/\partial x^2)\Delta x^2 + \; . \; . \; . \; + (\partial^N u/\partial x^N)\Delta x^N = b_1.$$

Here we are in a position similar to that in section 4.2 above (except that it took much more efforts to obtain this expansion). If there is no solution in the domain, we increment $x+\Delta x$ and repeat this whole step.

A question may arise how we would obtain all these right hand parts $\partial^N f/\partial x^N$ of those additional Nm equations (6). Surprising as it is, we should not care at all: the right hand parts of the source equations (5) do not depend explicitly on x: instead, the parameter x occurs in the initial values. Therefore, the same algorithm (the same postfix sequence and the instructions encoding it) [6-8], performing the automatic differentiation by t, would do that also by x, writing the corresponding results into the array (with one dimension more). It would perform it in the embedded loops: external loop by t-differentiation, the internal – by x-differentiation. (This algorithm will be added to the existing version of the Taylor Center computer program [6-8] in the future).

Thus, the automatic differentiation by the parameter x appeared to be also possible – even without the assumption about the elementarity by x of the solution. If this elementarity really took place and we knew the ODEs defining $u_i(t,x)$ with respect to x, we would be able to integrate them by x just *once* in the end point $t=1$. Otherwise, we have to do that at each t-step (integrating N times larger system). In other words, in order to have available both $\partial^n u_i/\partial t^n$ and $\partial^n u_i/\partial x^n$, $n=1,2,...,$ N, we compute N^2 derivatives of the source system (5) at each step (vs. just N derivatives for the initial value problem by t).

Note: the same approach remains valid also for integrating the source system depending on two or more parameters as soon as the parameters are in the initial values (rather than in the right hand parts). For example, for two parameters x, y, (plus t) the number of computed derivatives would be $2N^2$, i.e. it grows linearly with number of the parameters (unlike the number of the mixed partial derivatives).

Another concern indeed is that the x-convergence radius at the *end* point $t=1$ may be extremely small (depending on the instability of the problem) so that a lot of x-integration steps would be needed in order to reach the given value b_i.

6 Conclusions

The modern concept of elementary functions reveals the deep connections between seemingly unrelated issues such as the *elementarity* vs. :

(1) Existence of formulas for obtaining N-order derivatives with efficiency $O(N^2)$;

(2) Reducing of any arbitrary non-linear ODEs to surprisingly simple forms.

It also shows fundamental differences in:

(3) The nature of the dependency on t of the solutions of ODEs vs. the dependency on the parameters;

(4) The nature of the solutions of non-linear finite difference equations vs. those of ODEs.

The automatic differentiation by *parameters* of the solution of ODEs could be just as simple as switching to another system of ODEs with no increase in the per-step computation, if the dependency on the parameters of the solution were also elementary and the corresponding systems were known. Even with no elementarity by the parameters, the automatic differentiation by them is still possible, providing they are placed in the initial conditions. Then the volume of computation at each step grows proportionally with the number of parameters.

The modern Taylor method is an efficient integrator with unique features especially important when the ultimate accuracy is the main goal. Typically used for initial value problems, it may be adjusted also for boundary value problems or integration until the given end value of the unknown function. The integration by a parameter is to be added in the future to the software package for PCs [6-8], which introduces an integrating environment for exploring and graphing solutions of ODEs implemented with all the features of the Graphical User Interface.

Acknowledgement: I am very thankful to Dr. Dmitriy Sonechkin, who first pointed me out and provided the references to the remarkable transformations [4, 5] for ODEs, when I enjoyed working together with him in the Hydrometeorological Center (Moscow, Russia) in 80s.

References

1. Moore, R. E.: Interval Analysis. Prenitce-Hall, Englewood Cliffs, N.Y. (1966).
2. Gofen, A. M.: Fast Taylor-Series Expansion and the Solution of the Cauchy Problem. U.S.S.R. Comput. Maths. Math. Phys. Vol. 22, No. 5, pp. 74-88 (1982)
3. Gofen, A. M.: Taylor Method of Integrating Ordinary Differential Equations: the Problem of Steps and Singularities. Cosmic Research, Vol. 30, No. 6, pp. 581-593 (1992)
4. Charnyi, V. I.: Two Methods of integrating the equations of motion. Cosmic Research, Vol. 8, No. 5 (1970)
5. Kerner, E. H.: Universal Formats for Nonlinear Differential Systems. J. Math. Phys. Vol. 22, No. 7 (1981)
6. Gofen, A. M.: Recursive Journey to Three Bodies. Delphi Informant Magazine. Vol. 8, No. 3 (2002)
7. Gofen, A. M.: The Taylor Center for PCs: Graphing and Integrating ODEs with the Ultimate Accuracy. These Proceedings (2002)
8. Gofen, A. M.: The Taylor Center Demo for PCs. http://www.ski.org/rehab/mackeben/gofen/TaylorMethod.htm

Contour Dynamics Simulations with a Parallel Hierarchical-Element Method

R.M. Schoemaker[1,2], P.C.A. de Haas[2], H.J.H. Clercx[1], and R.M.M. Mattheij[2]

[1] Department of Applied Physics
[2] Department of Mathematics and Computing Science
Eindhoven University of Technology, The Netherlands
R.M.Schoemaker@tue.nl

Abstract. Many two-dimensional incompressible inviscid vortex flows can be simulated with high efficiency by means of the contour dynamics method. Several applications require the use of a hierarchical-element method (HEM), a modified version of the classical contour dynamics scheme by applying a fast multipole method, in order to accelerate the computations substantially. In this paper it is shown that the acceleration of contour dynamics simulations by means of the HEM can be increased further by parallelising the HEM algorithm. Speed-up, load balance and scalability are parallel performance features which are studied for several test examples. Furthermore, typical simulations are shown, including an application of vortex dynamics near the pole of a rotating sphere. The HEM has been parallelised using OpenMP and tested with up to 16 processors on an Origin 3800 cc-NUMA computer.

1 Introduction

Large-scale vortices are coherent structures that can be found in the oceans, the atmosphere of our planet as well as in the atmosphere of other planets. Examples of terrestrial nature are high and low-pressure areas, eddies in the ocean and the large polar vortex. Other planetary examples are the Great Red Spot on Jupiter, the Great Dark Spot on Neptune and huge rotating dust structures on Mars. The thickness of the layer of fluid these structures evolve in (on Earth: 1 – 10 km), is small compared to the horizontal size of the coherent structure itself (on Earth: 100 – 1000 km). This geometrical confinement, together with the planetary rotation and the density stratification in the fluid layer, implies quasi two-dimensionality of the flow. The motion of such large-scale structures is slow compared to the rotation speed of the planetary body implying that these structures are nearly non-divergent. Their dynamics can in good approximation be described by the two-dimensional incompressible inviscid variant of the Navier-Stokes equations, viz. the 2D Euler equations.

A suitable and elegant numerical technique for simulating two-dimensional (2D) vortex flows is the contour dynamics method [2] [6]. The collection of 2D vortices is approximated by nested patches of uniform vorticity. Only the evolution of the contour edges needs to be computed for determining the complete

P.M.A. Sloot et al. (Eds.): ICCS 2002, LNCS 2329, pp. 1010–1019, 2002.

dynamics of the vortices and this makes the method efficient because no grid is needed. However, when highly complicated flow patterns emerge during a simulation, the conventional numerical scheme becomes inefficient in its time-complexity. Another class of problems concerns the evolution of vortices in the presence of non-uniform background vorticity. The non-uniform background vorticity is usually a local approximation of the latitudinal variation of the Coriolis force. For example, when simulating the dynamics of vortices, located near the poles of a rotating sphere (the local approximation is denoted as the γ-plane), the initial vorticity distribution is rather intricate due to the necessity to discretise, together with the vorticity patches, the background vorticity field[1]. An acceleration of the method is in this case already appropriate from the start of the simulation. The hierarchical-element method for contour dynamics (HEM) [4] solves for the limited applicability of the conventional scheme.

The algorithmic structure of the HEM has certain features that makes parallelisation a good means for speeding-up contour dynamics simulations to an even greater extent. The HEM has been parallelised using OpenMP, which is a new industry standard for parallel programming on shared-memory architectures. The parallel HEM has been tested for speed-up, scalability and load balancing for several test cases and typical vortex configurations including one with a non-uniform background vorticity field. It is shown in this paper that the parallel HEM is a decent tool for studying flows with non-uniform background vorticity.

2   The HEM for Contour Dynamics

The spatial discretisation used in a contour dynamics method consists of three parts: The discretisation of the continuous vorticity profile into a piecewise-uniform vorticity distribution, interpolation of the bounding contours of the regions of uniform vorticity, and the redistribution of the nodes on the contours during the simulation. For the present introduction to contour dynamics and the HEM it is sufficient to focus on the first part of the spatial discretisation.

The dynamics of a 2D incompressible, inviscid fluid flow is described by the Euler equation and the equation of mass conservation. The latter implies a divergence-free velocity field or $\nabla \cdot \mathbf{u} = 0$, with $\mathbf{u}(\mathbf{r}, t)$ the velocity vector. The Euler equation, which expresses the balance of linear momentum, is then written as

$$\frac{\partial \mathbf{u}}{\partial t} + (\mathbf{u} \cdot \nabla)\mathbf{u} = -\frac{1}{\rho}\nabla p \,, \tag{1}$$

with p the pressure and ρ the density. The vorticity is defined as $\boldsymbol{\omega}(\mathbf{r}, t) = \nabla \times \mathbf{u}$. For a 2D flow, $\mathbf{u} = (u, v)^T$, $\mathbf{r} = (x, y)^T$ and $\boldsymbol{\omega} = \omega\, \mathbf{e}_z$. The non-divergence condition implies the definition of a stream function $\psi(\mathbf{r}, t)$ through $u = \frac{\partial \psi}{\partial y}$ and $v = -\frac{\partial \psi}{\partial x}$. By taking the curl of (1) we obtain an expression for conservation of

---

[1] More precisely, the potential vorticity is the conserved quantity and should thus be discretised [5].

vorticity of a fluid particle in two dimensions, viz.

$$\frac{D\omega}{Dt} = \frac{\partial \omega}{\partial t} + \frac{\partial \psi}{\partial y}\frac{\partial \omega}{\partial x} - \frac{\partial \psi}{\partial x}\frac{\partial \omega}{\partial y} = 0 \ , \tag{2}$$

whereas the vorticity ω and the stream function ψ are related through the Poisson equation

$$\nabla^2 \psi = -\omega \ . \tag{3}$$

Using the Green's function of the Laplace operator for an infinite 2D domain, $G(\mathbf{r};\mathbf{r}') = \frac{1}{2\pi}\ln|\mathbf{r}-\mathbf{r}'|$ (with $|\mathbf{r}| = \sqrt{x^2+y^2}$), the stream function can be found explicitly as

$$\psi(\mathbf{r},t) = -\frac{1}{2\pi}\iint_{\mathbb{R}^2}\omega(\mathbf{r}',t)\ln|\mathbf{r}-\mathbf{r}'|\,dx'dy' \ . \tag{4}$$

As is depicted in Figure 1, an initially continuous vorticity distribution $\omega(\mathbf{r},0)$ is approximated by a piecewise-uniform distribution $\hat{\omega}(\mathbf{r},0) = \sum_{m=0}^{M}\omega_m$. It consists in this case of a constant background vorticity ω_0 and a number of nested patches $\mathcal{P}_m$ with vorticity values ω_m (with $m \geq 1$).

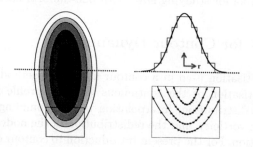

Figure 1 – *Five nested patches with uniform vorticity. Right-top: Vorticity jumps. Right-bottom: Node redistribution.*

We assume for the moment that no background vorticity is present, i.e. $\omega_0 = 0$. Due to the conservation of vorticity the piecewise-uniform distribution of vorticity remains piecewise-uniform in the course of time. The nested patches $\mathcal{P}_m(t)$ deform during the flow evolution although its area is conserved, and the bounding contours $\mathcal{C}_m(t)$ of the patches $\mathcal{P}_m(t)$ will not cut neighbouring boundary contours. Equation (4) can be reformulated as

$$\psi(\mathbf{r},t) = -\frac{1}{2\pi}\sum_{m=1}^{M}\omega_m\iint_{\mathcal{P}_m(t)}\ln|\mathbf{r}-\mathbf{r}'|\,dx'dy' \ . \tag{5}$$

The velocity field $\mathbf{u}(\mathbf{r},t)$ is obtained by taking the derivatives of $\psi(\mathbf{r},t)$ with respect to x and y and subsequently applying Stokes' theorem for a scalar field.

The following expression can be derived [2] [4] [6]:

$$\mathbf{u}(\mathbf{r},t) = -\frac{1}{2\pi} \sum_{m=1}^{M} \omega_m \oint_{C_m(t)} \ln|\mathbf{r}-\mathbf{r}'|\, d\mathbf{r}' \,, \tag{6}$$

where $d\mathbf{r}'$ denotes an infinitesimal vector tangential to the boundary. From (6) it follows that the evolution of patches of uniform vorticity is fully determined by the evolution of their bounding contours.

For contour dynamics the source distribution consists of patches of uniform vorticity. For the implementation of an acceleration scheme based on Poisson integrals, Vosbeek *et al.* [4] developed the Hierarchical-Element Method in order to reduce the $\mathcal{O}(N^2)$ operation count typical of the conventional contour dynamics approach. The HEM is an adaptation to the method of Anderson [1] and has a time-complexity of $\mathcal{O}(N)$. We present here an overview of the general structure of the HEM which is relevant for the parallelisation strategy.

In the HEM all the patches of uniform vorticity are assumed to reside in a square numerical domain, while the dynamics still involves the infinite plane. This square domain is then subdivided into a set of hierarchical levels with $2^l \times 2^l$ boxes at levels $l = 1, \ldots, l_f$ with l_f a finest level.

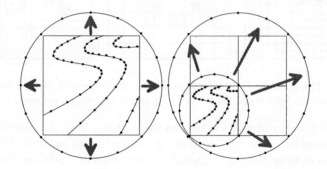

Figure 2 – *Left: Construction of an outer ring around one of the boxes at the finest level by means of direct evaluation. Right: Construction of a ring at a consecutive coarser level. Four smaller rings contribute to one coarser ring. Note that only part of the HEM domain is shown.*

Keeping in mind that our numerical approach is based on the use of rings as computational elements (and using Poisson integrals) we introduce at the finest level l_f a ring with K nodes around each box (Figure 2 left). By means of direct evaluation, the velocity is determined at each node on the ring. These rings are called the outer rings. Four outer rings at the finest level yield the input for the construction of one outer ring at the subsequent coarser level, the parent level, through a Poisson integral. In a similar way outer rings are constructed up to the level $l = 2$ (see Figure 2 right). At level $l = 2$, another kind of ring is formed via the constructed outer rings. This ring, a so-called inner ring, is to be used for the construction of inner rings at finer levels again all the way down to level

$l = l_f - 1$ by means of a similar Poisson integral. This is illustrated in Figure 3 (for $l_f = 4$), at levels $l = 3, \ldots, l_f - 1$ an inner ring is constructed from a parent inner ring at level $l - 1$ and from outer rings at the same level l. Level $l = 1$ (four boxes) is irrelevant in these hierarchical approximations.

The light grey area at each level in Figure 3 is a so-called well-separated area. It is the area with just the optimal distance for a certain box size [4]. The eight dark grey boxes at the finest level are the immediate neighbours of the box containing the evaluation points. Here, direct evaluation of velocities are needed which require $\mathcal{O}(mn)$ operations with m the number of nodes in the eight neighbouring boxes and n the number of nodes in the evaluation box. Outer rings and inner rings are both needed in the hierarchical tree of levels to calculate all the contributions from the subdivided vorticity at the finest level. The construction of the outer rings is necessary to provide information for all the approximating rings. The inner rings contain all information from the outer regions and pass this information on to the next finer level inner ring.

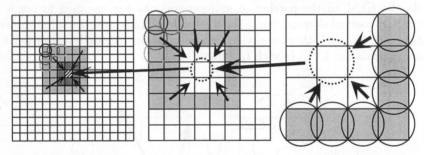

Figure 3 – *The HEM in action for $l_f = 4$. The little box with slanted lines contains evaluation points. Solid rings are outer rings and dotted rings are inner rings.*

If a small number of nodes makes up the vorticity field, it is inefficient to use a very fine meshed finest level with many boxes. In this case, the HEM is even more inefficient than the conventional contour dynamics scheme. On the other hand, however, too many nodes in a box is not efficient either because of the expensive direct interaction computations in each small domain of nine boxes at the finest level. The HEM accounts for these two restrictions by determining the hierarchical tree depth in an adaptive manner during a simulation. Optimal intervals for N are being chosen in such a way that K keeps the same order as the number of nodes per box. This leads to an $\mathcal{O}(N)$ method.

3   Parallelisation Strategy

Parallelisation of the already existing HEM method is achieved by using OpenMP. It is however important to know *a priori* if the numerical scheme can be parallelised in a convenient way. Therefore, the global structure of the HEM should be considered first. Particularly, the parallelisation of the algorithms for the velocity computations and the node redistribution should be taken into account. The numerical structure of the HEM consists of a hierarchy of levels with boxes. At the finest level these boxes contain pieces of contours, whereas coarser boxes contain the approximating outer and inner rings. Velocity computations are carried out at each level in the tree and parallelisation along all hierarchical boxes seems appropriate. Although most of the computations are at the finest level, the ring computations at each consecutive coarser level—requiring a modest amount of CPU-time—should be included for parallelisation as well. Obviously, parallelisation is most effective when implemented per level. Each level is then decomposed into several subdomains which can have one or several boxes, whereas multiple processors can be assigned to these subdomains. The part of the algorithm dealing with the redistribution of the nodes can not be skipped in the parallelisation process, because it still contributes significantly, depending on the evolving contours. Parallelisation of this part is however completely different, because it has no box-wise HEM structure. For this part, contour-wise parallelisation is adapted. This short paper however will only focus on the much more important box-wise parallelisation of the velocity computations, which is carried out independently from the node redistribution.

The assignment of processors can be scheduled in various ways. Two scheduling policies in OpenMP used for the parallel HEM are static scheduling and dynamic scheduling. When each subdomain is treated as a static subdomain, a processor assigned to this subdomain stays put until all processors have finished their own local job. After that, they proceed in unison to their next assigned subdomain. Static scheduling can be very inefficient when the computational load is non-uniform. However, when an algorithm has a predefined uniform load, processors can be assigned to steady portions of the algorithm and minimize the communication overhead showing an efficient static scheduling policy. Dynamic scheduling, on the other hand, implies that a processor can proceed with the next available subdomain when it has finished its own local job, even when other processors are still busy with their first computation.

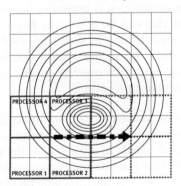 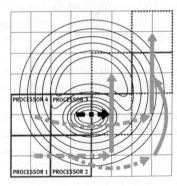

Figure 4 – *Left: Example of static scheduling for the HEM in OpenMP. Right: Dynamic scheduling for the same domain.*

Dynamic scheduling suffers more from communication overhead due to the dynamic displacements of processors. For an unevenly distributed computational load, as is the case in many contour dynamics simulations, processors taking care of less time-consuming jobs have to wait for the slower jobs and will run idle when static scheduling is applied. This so-called load imbalance is illustrated in Figure 4 for a computation with a non-uniform distribution of nodes in the domain. The left schematic shows a static scheduling which is applied to a finest level $l_f = 3$ with four processors to assign. Choosing an appropriate subdomain size—a so-called chunk size in OpenMP—can be convenient for solving a specific load imbalance problem. The chunk size in Figure 4 is four, and apparently the load is quite out of balance. Processors 1, 2, and 4 have to wait for number 3 before all can proceed in unison to the next subdomain (black arrow). The schematic on the right shows how to solve for the load imbalance when using dynamic scheduling. When processor 1 has finished its computations it will start immediately computing the first available chunk (keeping the ordering in mind). When processor 3—the one with the highest load—has finished (black arrow), processors 1, 2, and 4 have already advanced to the right-top quadrant (gray arrows). The load imbalance for this configuration has been minimized. Parallelisation of the HEM will benefit the most when using a dynamic scheduling policy with the smallest possible chunk size, i.e. as small as one box, contrary to the above illustrative example with four boxes for a chunk. The smaller the chunk size, the more balanced the load is after finishing a certain level.

In order to study the overall parallel performance of the HEM, two important features of parallel programming have been analysed, viz. processor scalability and problem scalability. Processor scalability is given through the parameter S, the speed-up of the parallelisation, and is defined as the ratio between the work done by one processor and the total execution time of a parallel programme with P processors: $S = T_1/T_P$. Problem scalability has been studied for different values of l_f and N. The number of nodes N changes continuously during a simulation and because of this, l_f changes accordingly. When l_f is large, more boxes have to be computed and more communication is necessary between processors. In a shared-memory environment this can be a limiting factor for the parallel performance because communication timing is slow compared to the crucial computations. The more computations a single processor can do in its chunk before communicating its results, the better the overall performance. This feature is called the computation-to-communication ratio.

Originally, the HEM has been introduced as a serial method for accelerating computationally expensive contour dynamics simulations. The use of OpenMP for incremental parallelisation of the method results in a fraction of the method that has been left for serial computation. This fraction depends on a heuristic pre-processing step and lies between 0.5% and 10% of the total amount of computations [3]. For the numerical experiments in the next section this has been taken into account. The serial part of the HEM however acts as a weakest link in the complete computation and slows down the method significantly. This is clearly demonstrated by Amdahl's Law, which states that for an ideal parallel machine (neglecting communication overhead) the speed-up $S_{Amdahl} = \frac{P}{Pf+(1-f)}$, where f represents the serial fraction and P the number of processors.

4   Numerical Experiments and Discussion

Three different numerical experiments have been performed. The goal of the first experiment is to investigate the scalability of the parallelised HEM, without adding additional nodes during the computation or applying any node redistribution. In the second experiment the role of static and dynamic scheduling on the speed-up S is discussed. In this numerical experiment both the number of nodes increases during the simulation (and also l_f increases) and node redistribution is included. In a third numerical experiment the parallel efficiency of the HEM for simulations of vortex evolutions in a non-uniform background vorticity field has been studied. The effort to obtain a parallelised version of the HEM is particularly aimed at solving this latter kind of problems.

EXAMPLE 1 — A first example is the simulation of concentric patches of uniform vorticity. These patches will stay circular and constant in size. The redistribution of the nodes is therefore unnecessary. During a simulation N is kept constant and the contours rotate in a clockwise direction—due to an overall negative uniform vorticity. The load is distributed symmetrically. To gain insight in the processor scalability, l_f is also kept constant during each simulation. This implies that for a certain N the optimal l_f is not changed accordingly. The following values for P, N, and l_f have been chosen. That is, the range of processors: $P = 1, 2, 4, 8, 16$, the problem size: $N = 1000, 2000, 4000, 8000, 16000$, and the choice of finest level: $l_f = 3, 4$.

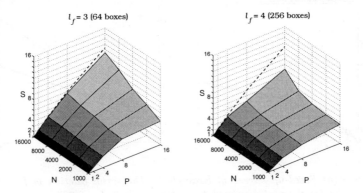

Figure 5 – $S = f(P, N)$ for $l_f = 3$ and $l_f = 4$. The speed-up for $N = 16,000$ has a dashed line indicating Amdahl's limit for a 98% parallelised algorithm ($f = 0.02$).

Each combination of P, N, and l_f represents a separate simulation. The results for the scalability S as function of P and N are given in Figure 5 as 3D graphs for different values of l_f. According to Figure 5 it is clear that for $N = 16,000$, performance is almost ideal for $l_f = 3$ in the complete processor range $P = 1, \ldots, 16$. Furthermore, an increasing computation-to-communication ratio can be observed for increasing N in both graphs. The amount of effective computations per communication step decreases for smaller boxes, resulting in a smaller S for increasing l_f. This example clearly indicates the parallel efficiency as function of P, N and l_f. For regular simulations however, a constant l_f is certainly not an option. The HEM would lose its $\mathcal{O}(N)$ behaviour.

EXAMPLE 2 — This example will highlight the difference between static scheduling and dynamic scheduling. The simulation itself requires $\mathcal{O}(10^3)$ nodes and is a relatively 'cheap' computation compared to the kind of simulations the HEM is designed for. The simulation concerns the formation of a tripole from an azimuthally perturbed isolated monopolar vortex. During the simulation the redistribution of the nodes is essential due to the formation of high-curvature segments in the contours. Node redistribution has been parallelised also. Additionally, we should keep in mind that N will increase, and as a result the tree-depth (i.e. l_f) in the HEM will automatically adapt, maintaining an $\mathcal{O}(N)$ operation count. Figure 6a shows the speed-up $S(P)$ for static and dynamic scheduling.

The static scheduling policy is clearly spoiling the parallel performance. Dynamic scheduling delivers a speed-up curve like the one in Figure 5 for $l_f = 3$. This is conform the level updates for $500 < N < 5,000$. The finest level remains relatively coarse ($l_f \leq 3$), and some parallel efficiency is lost due to the fact that no optimal benefit of load-balancing can be obtained. As a result a few processors will run idle.

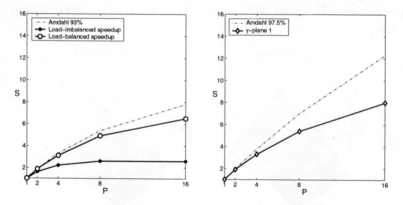

Figure 6 – (a) *Speed-up curves of Example 2. The serial fraction is $f = 0.07$.* (b) *Speed-up curve for the γ-plane simulation of Example 3, $f = 0.025$.*

EXAMPLE 3 — The last example shows the parallel efficiency of simulations the HEM had been made for originally. It concerns the evolution of an isolated monopole—like the one in Example 1—embedded in a non-uniform background vorticity field. These so-called γ-plane simulations generate complex flow patterns and require at least $\mathcal{O}(10^4)$ nodes. The simulation has 21 contours for the background field and the monopole, and a node range of $10,000 < N < 30,000$. The test starts with only a tenth of this number but arrive in the mentioned range quite early in the simulation. The simulation needs considerably more nodes than the numerical simulations for examples 1 and 2. Redistribution takes care of the node supply, particularly in the monopolar region, indicating that the load is highly out of balance in the overall domain. This implies the use of dynamic scheduling with the smallest possible chunk size.

Figure 6b shows that the speed-up for the γ-plane is on first site better than the 'cheaper' tripole test, due to a smaller serial fraction. However, the test needs more nodes and will adjust to a higher l_f, thus creating smaller boxes with on the average still the same amount of nodes in a box. The computation-to-communication ratio will unfortunately decrease because of this, although a domain with a greater amount of boxes can benefit more from the load-balancing effect of dynamic scheduling. Apparently the net effect is a lower efficiency.

5   Conclusive remarks

Despite the (inherently involved) shared-addressing complication of the HEM, the parallel HEM scales (very) good for small numbers of processors ($P \leq 8$) for every test-case discussed in this paper. For regular HEM simulations we have however non-uniform node distributions and box-size adaptations, implying a higher computation-to-communication ratio for certain boxes. The adaptations are necessary for keeping the favourable $\mathcal{O}(N)$ operation count of the HEM. For $P > 8$, all tests show that an increasing l_f demands an increasing amount of communication between (expensive) computations, whereas the dynamic scheduling solves for the non-uniform load better when l_f is actually high. A more efficient load-balancing is feasible, i.e. a locally defined tree-depth based on the requirement to keep the average number of nodes per finest level box approximately constant. This approach is not implemented yet and should be pursued in future investigations. Nonetheless, the current parallel HEM is an important improvement to run γ-plane simulations in a much shorter time for small numbers of processors.

References

1. Anderson, C.R.: *An implementation of the fast multipole method without multipoles.* SIAM J. Sci. Statist. Comput. **13** (1992) 923–947
2. Dritschel, D.: *Contour dynamics and contour surgery: Numerical algorithms for extended, high-resolution modelling of vortex dynamics in two-dimensional, inviscid, incompressible flows.* Comput. Phys. Rep. **10** (1989) 77–146
3. Schoemaker, R.M., de Haas, P.C.A., Clercx, H.J.H., Mattheij, R.M.M.: *A parallel hierarchical-element method for contour dynamics simulations.* to appear
4. Vosbeek, P.W.C., Clercx, H.J.H., Mattheij, R.M.M.: *Acceleration of contour dynamics simulations with a hierarchical-element method.* J. Comput. Phys. **161** (2000) 287–311
5. Vosbeek, P.W.C., Clercx, H.J.H., van Heijst, G.J.F., Mattheij, R.M.M.: *Contour dynamics with non-uniform background vorticity.* Int. J. Comput. Fluid Dyn. **15** (2001) 227-249
6. Zabusky, N.J.: *Contour dynamics for the Euler equations in two dimensions.* J. Comput. Phys. **30** (1979) 96–106

A Parallel Algorithm for the Dynamic Partitioning of Particle-Mesh Computational Systems*

Jing-Ru C. Cheng and Paul E. Plassmann

Department of Computer Science and Engineering
The Pennsylvania State University, University Park, PA 16802, USA,
jccheng@cse.psu.edu, and plassman@cse.psu.edu

Abstract. Particle tracking methods are a versatile computational technique central to the simulation of a wide range of scientific applications. In this paper we present a new parallel approach for the dynamic partitioning of particle-mesh computational systems. The approach uses a framework, the "in-element" particle tracking method, based on the assumption that particle trajectories are computed by problem data localized to individual elements. The parallel efficiency of such particle-mesh systems depends on the partitioning of both the mesh elements and the particles; this distribution can change dramatically because of movement of the particles and adaptive refinement of the mesh. To address this problem we introduce a combined load function that is a function of both the particle and mesh element distributions. We present experiment results that detail the performance of this parallel load balancing approach for a three-dimensional particle-mesh test problem on an unstructured, adaptive mesh.

1 Introduction

Parallel particle tracking methods have been employed to solve a variety of problems, such as Direct Simulation Monte Carlo (DSMC) methods to model the dynamics of dilute gas. Wong and Long [1] implemented parallel DSMC algorithm using a forward Euler explicit time marching scheme in the movement phase of the algorithm. They pointed out that two different levels of data parallelism (i.e., molecules and cells) cause some parallel processing difficulties. Nance et al. [2] parallelized DSMC using the runtime library CHAOS [3] on a 3-D uniform discretized mesh. Robinson and Harvey [4] parallelized DSMC by use of a spatial mesh decomposition over the processors. The domain decomposition is based on a localized "load table" computed on each processor for its neighbors, which will receive cells donated by the processor if they have load less than itself. The demonstration problem was a 2-D driven cavity with approximately 10,000 uniform cells. Another field of application of particle tracking methods

* This work was supported by NSF grants DGE–9987589 and ACI–9908057, DOE grant DG-FG02-99ER25373, and the Alfred P. Sloan Foundation.

P.M.A. Sloot et al. (Eds.): ICCS 2002, LNCS 2329, pp. 1020–1029, 2002.

is the streamline calculation, for which the most common technique is the integration of particle paths numerically or analytically. Often, for the numerical approach, a high-order ODE solver such as Runge-Kutta methods or Adam's method is required in terms of accuracy [5]. However, in 2-D and 3-D, the evaluation of velocity between grid points across processors required by the high-order ODE solver becomes costly and alternative methods should be considered. Analytic solutions for streamlines within tetrahedra were presented based on linear interpolation and therefore produces exact results for linear velocity fields [6]. However, the analytic method works for steady flow only and the parallel version has not developed yet.

The in-element particle tracking technique was developed by Cheng et al. [7] to accurately and efficiently trace fictitious particles in the velocity fields of real-world systems. Since in-element particle tracking methods are implemented with respect to the element basis, a natural approach is to develop a parallel framework based on specifying local, element-based operations. The key to the parallel algorithm is the correlated partitioning of the particle system and the unstructured element mesh. We partition particles to processors based on their element location—this approach ensures that the data required for the computation of the particle movement phase by the in-element method involves only data local to a processor. The correspondence between particles and elements is maintained through explicit references in the element and particle data structures. This correspondence is essential to ensure the correct reassignment of particles between processors. The reassignment occurs when particles move between elements owned by a processor and the ghost elements (elements with shared faces, edges, or vertices that are owned by another processor) [8].

With the goal of balancing the work load at each time step whether the explicit time-stepping method, the implicit time-stepping method, or streamline calculation, is employed, computational workload estimates need to consider the particle distribution to mesh elements. In general, the workload on each processor is a function of the number of particles and number of vertices assigned to a processor. This distribution can change significantly each time step because of the particle movement and adaptive mesh refinement. This dynamic character raises the issue of load balancing and the problem of determining representative load estimates to achieve a balancing of the assigned work. To address this problem, we define such a load function—a weighted function based on the assignment of particles and vertices to processors.

Experimental results presented in this paper demonstrate the advantage of dynamic load balancing based on this load estimation approach. For a representative parallel particle tracking problem, we present experimental results showing the performance improvement using the load function discussed above. A three-dimensional rotation cone problem is solved using the particle tracking software developed and implemented for the SUMAA3d [9] programming environment.

The remainder of this paper is organized as follows. In Sect. 2, we review in-element particle tracking technology and the extensions required for the parallel implementation are described. In Sect. 3, we discuss the encountered workload imbalance issues and present a repartitioning strategy. In Sect. 4, experimental results are presented to demonstrate the performance of our implementation. In Sect. 5, we summarize these results and discuss planned future work.

2   Particle Tracking Algorithms

Consider the following advection-dominated system as representative of many problems that are solved by particle tracking schemes. The advection equation for a scalar field $C(\mathbf{x}, t)$, for example the chemical concentration in transport equations, is represented as

$$\frac{\partial C(\mathbf{x}, t)}{\partial t} + \mathbf{u} \cdot C(\mathbf{x}, t) = 0 , \tag{1}$$

where the velocity is given by

$$\frac{d\mathbf{x}}{dt} = \mathbf{u}(\mathbf{x}, t) . \tag{2}$$

The value of $C(\mathbf{x}, t)$ is constant along the characteristic line defined by the solution to (2) (i.e., $C(\mathbf{x}, t_n) = C(\mathbf{x_o}, t_o)$) and

$$\mathbf{x} - \mathbf{x_o} = \int_{t_o}^{t_n} \mathbf{u}(\mathbf{x}, t)dt , \tag{3}$$

where $\mathbf{x_o}$ is the initial point of the particle. The accuracy of the solution to (1) depends on the accurate computation of (3). To address this problem, we have specifically developed the in-element algorithm.

2.1   The Sequential In-Element Algorithm

In Algorithm 2.1 introduced in [8], we give pseudocode describing a generic in-element particle tracking method. This algorithm takes advantage of a number of numerical features. First, the tracking follows the characteristic line element-by-element so that the computational overhead of locating the element in the mesh containing the departure point is reduced. Note that in this way the time-step size is controlled by the element size which preserves the accuracy and stability in the time integration scheme employed. Second, to increase the tracking accuracy within an element, the element can be adaptively refined into a desired number of sub-elements with the interpolated velocity computed at all vertices on sub-elements.

Algorithm 2.1 The In-Element Particle Tracking Algorithm [8]

Let P_0 be the set of particles
Set the residual time t_r to the time-step size
$n \leftarrow \|P_0\|$
Foreach ($p \in P_0$) do

> While $(t_r > 0)$ do
>> Refine the element M to the prescribed number of subelements
>> Track p subelement by subelement until time is exhausted or
>>> until the element boundary is hit
>> Compute t_r, velocity, and identify the possible neighbor element
>>> for next tracking
>> if $t_r = 0$ do
>>> Interpolate concentration
>> Endif
> Endwhile

Endfor

2.2 A Parallel Particle Tracking Algorithm

The parallel particle tracking algorithm is based on the correlated partitioning of the particle system and the unstructured element mesh. Particles are partitioned to processors based on their element location—this approach ensures that the data required for the computation of the Lagrangian step by the in-element method involves only data local to a processor. The correspondence between particles and elements is maintained through explicit references in the element and particle data structures. This correspondence is essential to ensure the correct reassignment of particles between processors. The reassignment occurs when particles move between elements owned by a processor and the ghost elements (elements with shared faces, edges, or vertices that are owned by another processor) [8]. The caching of ghost elements ensures that when the target element found by the tracking algorithm is owned by another processor, the particle is correctly moved to that processor.

Based on the element partioning resulting from the partitioning heuristic, vertex data and element data are never changed during the particle tracking process. Instead, the particles are repartitioned by inheriting their resident element's partition. This approach neither affects the coherency of element neighbor data nor destroys the coherency of vertex data. The disadvantage is the imbalance of particles as the tracking process redistributes particles among processors. Later, we present a repartitioning strategy for the partitioner to balance the workload among the processors. Algorithm 2.2 is the implementation of parallel in-element particle tracking technique neglecting repartitioning at each tracking step.

Algorithm 2.2 The Parallel In-Element Particle Tracking Algorithm[8]

> Let P_i be the set of particles on processor i
> Set the residual time t_r to the time-step size
> $n \leftarrow \sum_i \|P_i\|$
> While $(n > 0)$ do
>> $n_i \leftarrow 0$
>> Foreach $(p \in P_i)$ do
>>> While $(p \in P_i$ and $status(p) \neq$ finish$)$ do
>>>> Refine M to the prescribed number of subelements
>>>> Track p subelement by subelement until time is exhausted or
>>>>> hit the boundary of the element M
>>>> Compute t_r, velocity and identify the possible neighbor element M'

>   for next tracking
>   If $t_r > 0$ do
>   If $owner(M') \neq i$ do
>   Pack p and remove p from P_i
> $n_i \leftarrow n_i + 1$
>   Endif
>   Elseif $t_r = 0$ do
>   Interpolate concentration
> $status(p) \leftarrow$ finish
>   Endif
>   Endwhile
> Endfor
> $n = \sum_i n_i$
> Send and receive message, unpack messages to form new P_i
> Endwhile
> Update concentration on each vertex

3   A Repartitioning Strategy

For scientific simulations on high-performance parallel computers, it is essential that the partitioner is able to divide an irregular mesh into equal-sized pieces with as few interconnecting edges as possible. In general, the work performed by the simulation on each processor depends on the mesh vertices, the mesh elements, or a combination of of these objects. To balance the workload among processors, a mapping which decomposes the computational mesh onto the processors is commonly found by solving a graph partitioning problem. The graph to be partitioned is simple to construct if the computation is mainly performed on mesh vertices; and a number of adequate partitioning heuristics exist (although the graph partitioning problem itself is known to be NP-complete). However, for parallel particle tracking methods, the workload on each processor is dependent not only on mesh vertices or mesh elements, but also on particles, which move about the mesh dynamically. A different repartitioning strategy is thus required to adapt to these combined particle-mesh computational systems.

To avoid the problem of incurring unnecessary overhead in the repeated mesh partitioning in the in-element particle tracking, we first propose a balance estimator. Note that our approach is based on the assumption that the mesh is completely repartitioned, not "incrementally" repartitioned. Incremental approaches for geometric partitioning schemes have been suggested [10] and implemented for the unbalanced recursive bisection (URB) method [11]. However, the overhead in the data structure reconstructions and message-passing optimizations often makes this approach nearly as expensive (in practice) as a complete repartitioning. Incremental partitioning algorithms based on the "diffusion" of elements or vertices between processors have also been tried but suffer a similar practical downside [12].

Prior to defining the balance estimator, we define several performance metrics. At the k-th tracking step, let $N_i^{\,k}$ be the total number of particles on processor i, let A^k be the average number of particles on all processors, let M^k be the maximum number of particles, and let m^k be the minimum number of particles on any processor. These metrics can be represented as

$$
\begin{aligned}
N_i^k &= \sum_{j \in p_i} n_j \\
A^k &= \frac{1}{p} \sum_{i=1}^{p} N_i^k \\
M^k &= \max\{N_i\}_{i \in [1,p]} \\
m^k &= \min\{N_i\}_{i \in [1,p]} \; ,
\end{aligned}
\tag{4}
$$

where p_i is the processor i, n_j is number of particles in element j, and p is number of processors. After obtaining the above measures, the repartitioning is determined to be necessary if the following two criteria are met

$$
\sum_{i=1}^{p} N_i^k \geq \mathcal{N}_{threshold} \; ,
\tag{5}
$$

and

$$
\max\left\{ \left(M^k - A^k\right), \left(A^k - m^k\right) \right\} \geq \gamma \mathcal{T} \; ,
\tag{6}
$$

where $\mathcal{N}_{threshold}$ is the threshold number of particles for repartitioning, $\mathcal{T}$ is the total number of vertices of global mesh, and γ is the fraction of $\mathcal{T}$. We employ a load function which is a linear function of the total weight of particles and the total weight of vertices on each processor. Thus, the load of processor i, L_i, is

$$
L_i = \sum_{j \in i} \omega_j \; ,
\tag{7}
$$

where

$$
\omega_j = \alpha + \beta \sum_{e \ni j} n_e
\tag{8}
$$

is the weight of local vertex j in the processor i, α and β are the weighting coefficients with respect to vertex and particle, and n_e is number of particles in the element e next to vertex j.

Based on the load function $\{L_i\}_{i=1}^{i=p}$ and the distribution of $\{\omega_j\}_{j=1}^{j=\mathcal{T}}$, the partitioner repartitions the mesh based on the graph partitioning theorem and the associated heuristics. Clearly, this partitioning is based only on vertices if $\beta = 0$ and $\alpha = 1$.

4   Experimental Results

In this section we present results on the scalability and accuracy of the parallel in-element approach in solving for the advection of a sharped-peaked concentration cone in a three-dimensional domain. Backward particle tracking is performed under the following flow field:

$$
\mathbf{u} = (V_x, V_y, V_z) = (-\vartheta y, \vartheta x, 2), \quad \vartheta = \frac{\pi}{500}
\tag{9}
$$

where V_x, V_y, and V_z are the velocity components in x-, y-, and z-directions, respectively A region of $[-3000,3000] \times [-3000,3000] \times [0,3000]$ is discretized to tetrahedral elements. The fictitious particles to be backward tracked are originally the domain vertices. After a tracking time period of 500, the fictitious particles can be analytically determined by using the relationship (3) with time integration from 0 to 500. The initial conditions for this problem are specified to demonstrate the capability of parallel in-element particle tracking method to solve transport equations for a system with infinite Peclet number and for Courant numbers C_r in excess of 1 over the entire grid.

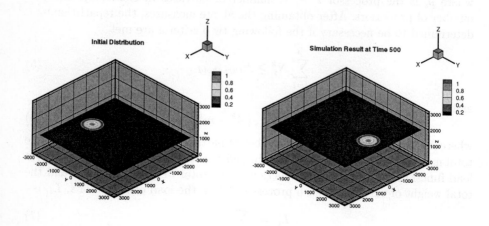

Fig. 1. The initial distribution (left) and the simulation result at time 500 (right) of the 3-D rotation cone problem

This problem is the 3-D version of "rotating cone" problem, a standard benchmark case for numerical advection algorithms [13]. The governing equation for a pure advection of concentration hills in a three-dimensional regime is given by (1). The initial and boundary conditions are given by

$$C(x,y,z,0) = C_0(x,y,z)$$
$$C(x,y,z,t) \to 0 \qquad \text{as } x^2 + y^2 + z^2 \to \infty , \tag{10}$$

in which $C_0(x,y,z)$ is given by

$$C_0(x,y,z) = \begin{cases} H(1 - \frac{|\mathbf{x}-\mathbf{x_0}|}{r_0}) & \text{if } |\mathbf{x} - \mathbf{x_0}| \le r_0, \text{ and } z = 1,000 \\ 0 & \text{otherwise} \end{cases}, \tag{11}$$

where r_0 is the radius of the initial cone-hill distribution. The exact solution for this system is

$$C(x,y,z,t) = C_0(x - V_x t, y - V_y t, z - V_z t) . \tag{12}$$

The domain is discretized to 612,821 tetrahedrons and 113,847 vertices. The initial distribution based on (11) and the simulation result from the parallel in-element tracking algorithm (i.e., Algorithm 2.2) are presented in Fig. 1. The maximum absolute pointwise error of this simulation is in the order of 10^{-6}.

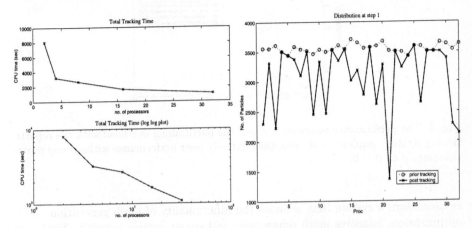

Fig. 2. On the left, the performance of the particle tracking benchmark as a function of the number of processors on an IBM SP2. On the right, we show the imbalance of particle numbers as a function of processor number resulting from a tracking step.

Concerning the performance, Fig. 2 depicts the CPU time spending in particle tracking, both in normal and log-log scale. As expected, the slope of the log-log plot in Fig. 2, i.e., speedup in tracking, close but less than 1 except a kink between 4 processors and 8 processors. To further investigate the workload across the processors, on the right we plot number of particles versus processor number and find that the distribution of particles across the processors has changed significantly after a tracking step. Thus, the presented repartitioning strategy described in Sect. 3 is applied to achiev ethe goal of load balance for parallel particle tracking methods.

The parameters required in the repartitioning strategy have been set to $\alpha = 1$, $\beta = 1, 10, 50, 100$, or 500, $N_{threshold} = 500 \times p$, where $p = 32$ in this experiment, and $\gamma = 0.02$. Fig. 3 shows the performance improvement with $\beta > 0$, i.e., the repartitioning strategy is employed. Although the partitioning overhead incurs approximately 10% of the total tracking time, the speedup due to the balanced particle tracking outperforms the required overhead. To demonstrate the partitioner does the mesh partitioning as expected, Fig. 4 shows the balanced distribution of ω_i and total number of particles among the processors as $\beta = 10$.

5   Summary and Future Plans

We have developed a dynamic repartitioning method that is appropriate for particle-mesh methods based on in-element particle tracking. We have implemented this method within the SUMAA3d unstructured mesh programming

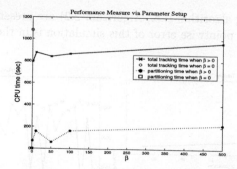

Fig. 3. The performance improvement and the partitioning overhead with the repartitioning strategy implemented. Note the relatively poor performance without any repartitioning, (i.e., $\beta = 0$).

environment, a system that includes the functionality of mesh generation, mesh optimization, adaptive mesh refinement, and sparse matrix solution. Testing of this approach for an advection-dominated test problem on 32 nodes of an IBM SP computer indicates that it significantly improves the parallel performance for this benchmark problem. From these experimental results, we observe that the proposed load function and balance estimator do improve the balance of particles to processors, in addition to the performance, for a wide range of parameter values. Additional issues, such as including an appropriate error estimator in the mesh refinement, particle tracking applications in additional areas, and experiments with more complicated flow fields are topics for further study.

References

[1] Wong, B.C., Long, L.N.: A data-parallel implementation of the dsmc method on the connection machine. Computing Systems in Engineering Journal **3** (1992)

[2] Nance, R.P., Wilmoth, R.G., Moon, B., Hassan, H.A., Saltz, J.: Parallel dsmc solution of three-dimensional flow over a finite flat plate. In: Proceedings of the ASME 6-th Joint Thermophysics and Heat Transfer Conference, Colorado Springs, Colorado, AIAA (June 20-23, 1994)

[3] Moon, B., Uysal, M., Saltz, J.: Index translation schemes for adaptive computations on distributed memory multicomputers. In: Proceedings of the 9-th International Parallel Processing Symposium, Santa Barbara, California, IEEE Computer Society Press (April 24-28, 1995) 812–819

[4] Robinson, C.D., Harvey, J.K.: A fully concurrent dsmc implementation with adaptive domain decomposition. In: Proceedings of the Parallel CFD Meeting, 1997. (1997)

[5] Malevsky, A.V., Thomas, S.J.: Parallel algorithms for semi-Lagrangian advection. International Journal for Numerical Methods in Fluids **25** (1997) 455–473

[6] Diachin, D., Herzog, J.: Analytic streamline calculations for linear tetrahedra. In: 13th AIAA Computational Fluid Dynamics Conference. (1997) 733–742

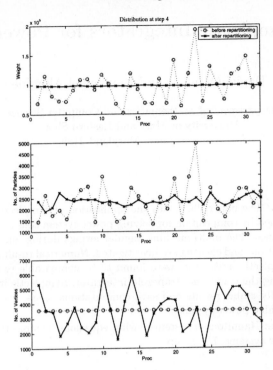

Fig. 4. The distribution of measured metrics before and after a tracking step

[7] Cheng, H.P., Cheng, J.R., Yeh, G.T.: A particle tracking technique for the Lagrangian-Eulerian finite-element method in m ulti-dimensions. International Journal for Numerical Methods in Engineering **39** (1996) 1115–1136

[8] Cheng, J.R.C., Plassmann, P.E.: The accuracy and performance of parallel in-element particle trac king methods. In: Proc. 10th SIAM Conf. on Parallel Processing, SIAM (2001) 252–261

[9] F reitag, L., Jones, M., Ollivier-Gooch, C., Plassmann, P.: SUMAA3d Web page. (http://www.mcs.anl.gov/sumaa3d/, Mathematics and Computer Science Division, Argonne National Laboratory)

[10] Berger, M., Bokhari, S.H.: A partitioning strategy for nonuniform problems on m ultiprocessors.IEEE Transactions on Computers **C-36** (1987) 570–580

[11] Jones, M.T., Plassmann, P.E.: Computational results for parallel unstructured mesh computations. Computing Systems in Engineering **5** (1994) 297–309

[12] Karypis, G., Kumar, V.: A performance study of diffusive vs. remapped load-balancing schemes. In: Proceedings of the 11th International Conference on Parallel and Distributed Computing Systems. (1998)

[13] Conv ection-Diffusion Forum: Specification of five covection-diffusion benchmark problems. 7th International Conference on Computational Methods in Water Resources, Massach usetts Institute of Technology (1988)

Stable Symplectic Integrators for Power Systems

Daniel Okunbor and Emmanuel Akinjide

Department of Mathematics and Computer Science,
University of Maryland Eastern Shore
Princess Anne, Maryland 21853

Abstract. The paper illustrates the application of symplectic integrators obtained by composition for solving power system consisting of several machines. The multi-machine angular swings during and following fault conditions and clearing are investigated. Numerical results obtained using symplectic integrators were found to be comparable to those obtained using the traditional trapezoid integrator. Symplectic integrators give globally stable results and save computational time, necessary for multi-machine power systems.

Keywords: Hamiltonian systems, swing equations, symplectic integrators, power systems, Lie group.

1 Introduction

There has been significant research in the computer solution of dynamic Hamiltonian system generally defined by scalar-valued, sufficiently smooth Hamiltonian function $H(z)$ on an even-dimenstional space, $z \in \mathcal{R}^{2d}$, see [2, 7, 11]. The corresponding Hamiltonian system of ordinary differential equations is

$$\frac{d}{dt}z = J\nabla H(z), \qquad (1)$$

where $z = \begin{pmatrix} q \\ p \end{pmatrix} \in \mathcal{R}^{2d}$, $J = \begin{pmatrix} O_d & I_d \\ -I_d & O_d \end{pmatrix}$, I_d is an identity matrix of dimension d. The increased in research interest in Hamiltonian system has been due largely to its applications in science and engineering. Two applications in which dynamical Hamiltonian system has been used extentively are bimolecular modeling and N-body simulation of planetary system [8, 9]. The general form of separable Hamiltonian for such applications is,

$$H(q(t), p(t)) = T(p) + V(q) = \frac{1}{2}p^T M^{-1}p + V(q), \qquad (2)$$

P.M.A. Sloot et al. (Eds.): ICCS 2002, LNCS 2329, pp. 1030–1039, 2002.

where $q(t), p(t) \in \mathcal{R}^d$ are respectively, the positions and momenta of the system and M is the diagonal mass matrix. The corresponding Hamiltonian system is

$$\frac{dq}{dt} = M^{-1}p, \quad \frac{dp}{dt} = -\nabla V(q), \tag{3}$$

where $-\nabla V(q)$ is the force of the system.

Symplectic integrators [7–9, 11] have been developed in favor of the conventional numerical solution techniques for Hamiltonian system of ordinary differential equations. Symplectic integrators preserve the Poincaré integrals of the t-flow of the Hamiltonian system except for truncation error. Comparison of symplectic integrators with conventional non-symplectic integrator revealed the stable characteristics of symplectic integrators making them appropriate for long-time integration.

In this paper, we investigate the application of Hamiltonian system to power system consisting of m machines. With reasonable assumption on the asychronous speed, the swing equations are modeled as a Hamiltonian system, allowing for the application of symplectic integrators. The numerical results indicate a very strong promise for symplectic integrators for power systems. The symplectic integrators tested were approximately ten times computationally faster than the conventional trapezoid-based method commonly used in power systems. The stable nature of the solution obtained utilizing symplectic integrators make them suitable for transient stability analysis as demonstrated in the paper. Section 2 describes symplectic integration schemes. The Lie group approach is used to construct symplectic integrators via composition of lower-order integrators. Section 3 focuses on the derivation of the swing equations of power systems and their relationship to the separable Hamiltonian in eq (2). This section also addresses numerical experients using a power system consisting of three machines and nine buses.

2   Symplectic Integration

Let $z(t_1)$ be the solution of Hamiltonian system (1) at time t_1 and $z(t_2)$, the solution at time t_2, where $t_2 > t_1$, we write

$$z(t_2) = \Psi_{(t,H)}(z(t_1)),$$

where Ψ_t is the transformation function from $z(t_1)$ to $z(t_2)$. A symplectic transformation is one satisfying

$$\left(\frac{\partial \Psi_t}{\partial z}\right)^{\mathrm{T}} J \left(\frac{\partial \Psi_t}{\partial z}\right) = J.$$

A symplectic integrator $z_{n+1} = \Psi_{(h,H)}(z_n)$, where z_{n+1} is the numerical solution at $t = t_{n+1}$ and z_n is the solution at $t = t_n$, is one for which $\Psi_{h,H}$ is symplectic, see [7, 11]. There are several approaches for constructing symplectic integrators [8, 11], we would use the Lie group approach. The Lie group approach involves the composition of lower-order integrators to obtain higher-order integrator. Let the time derivatives of $z(t) = (q(t))^{\mathrm{T}}, p(t)^{\mathrm{T}})^{\mathrm{T}} \in \mathcal{R}^{2d}$ be expressed as

$$\frac{d^k}{dt^k} z(t) = D_H^k z(t)$$

for the differential operator D_H defined by $D_H = \sum_{i=1}^{d} \left(\frac{\partial H}{\partial p_i} \frac{\partial}{\partial q_i} - \frac{\partial H}{\partial q_i} \frac{\partial}{\partial p_i} \right)$. Since

$$z(t) = \sum_{k=0}^{\infty} \frac{t^k}{k!} \frac{d^k}{dt^k} z(0) = e^{tD_H}(z(0)),$$

the exponential operator maps $z(0)$ to $z(t)$. For separable Hamiltonian, the differential operator is expressed as $D_H = D_T + D_V$, and the exponential operator (5) applied to some $z = (q^{\mathrm{T}}, p^{\mathrm{T}})^{\mathrm{T}}$ can be approximated by a systematic application of following operators.

$$e^{tD_T}(z) \equiv \begin{pmatrix} q + t\nabla T(p) \\ p \end{pmatrix}, \quad e^{tD_V}(z) \equiv \begin{pmatrix} q \\ p - t\nabla V(q) \end{pmatrix}.$$

These transformations are both symplectic since they satisfy the necessary and sufficient condition for a mapping to be symplectic. It is easy to show that the composition of symplectic transformations is also symplectic. A single step second-order symplectic integrator can be expressed as the composition of first-order symplectic integrators namely,

$$z_{n+1} = \Psi_{h,H}^{[2]}(z_n) = e^{\frac{h}{2}D_V} e^{hD_T} e^{\frac{h}{2}D_V}(z_n).$$

Using the Baker-Campbell-Hausdorff (BCH) formula from Lie algebra [2, 7], the product of exponentials of noncommutative operators can be expressed as

$$e^A e^B = e^{A+B+\frac{1}{2}[A,B]+\frac{1}{12}([[B,A],A]+[B,[B,A]])-\frac{1}{24}[B,[[B,A],A]]+\cdots},$$

where we have adopted the Poisson bracket notation for noncomutative operators, viz: $[A, B] = AB - BA$. Clearly, the composition method is second-order accurate, since its operator approximates e^{hD_H} up to the h^3 term:

$$e^{\frac{h}{2}D_V} e^{hD_T} e^{\frac{h}{2}D_V} = e^{hD_H + \frac{h^3}{24}(2[D_T,[D_T,D_V]]-[[D_T,D_V]D_V])+0(h^5)}$$

The composition of $\Psi_{h,H}^{[2]}$ will give a fourth-order symplectic integrator:

$$\Psi_{h,H}^{[4]} = \Psi_{x_1h,H}^{[2]} \Psi_{x_0h,H}^{[2]} \Psi_{x_1h,H}^{[2]}.$$

Yoshida [11] derived a 4-th order with the following parameters

$$x_1 = \frac{1}{3}(2 + 2^{\frac{1}{3}} + 2^{-\frac{1}{3}}), \quad x_0 = 1 - 2x_1,$$

using Lie Algebra.

3   Application of Hamiltonian System to Power Systems

In this section, we would describe the equation governing the motion of the rotors in power system consisting of synchronous m generators and buses. The equation is based on the principle of dynamics that states that accelerating torques is the product of the moment of inertia of the k-th rotor and its angular acceleration, see [1, 3, 6, 10, 12, 13].

$$G\alpha_k(t) = f_k(t) - f_e(t) = f_a(t),$$

where G is the total moment of inertia of the rotating masses in kgm^2, α_k is the k-th rotor angular acceleration in rad/sec^2, f_k is the mechanical torque supplied by the prime mover minus the retarding torque due to mechanical losses in Nm, f_e is the electrical torque that account for the total 3-phase electrical power output of the generator including the electrical losses in Nm, and f_a is the net accelerating torque in Nm. The rotor angular acceleration is given by

$$\alpha_k(t) = \frac{d\omega_k(t)}{dt} = \frac{d^2\theta_k(t)}{dt^2},$$

$$\omega_k(t) = \frac{d\theta_k(t)}{dt}, \tag{4}$$

where ω_k is the rotor angular velocity in rad/sec, and θ_k is the rotor angular postion with respect to a synchronous reference axis in radians. In steady state, f_k and f_e are equal, and the accelerating torque f_a is zero due to the synchronous speed (constant velocity). f_a and α_k are positive with an increasing rotor speed, when f_k greater than f_e. A decreasing rotor speed is obtained when f_k is less than f_e. It is easier to measure rotor angular position with respect to synchronously rotating reference.

The above equations (4) are combined to form the power generated from a typical generator. Let

$$\theta_k(t) = \omega_{\text{syn}}(t) + \delta_k(t),$$

where ω_{syn} is the synchronous angular velocity in rad/sec, δ_k is the rotor angular positon with respect to a synchronously rotating reference in radians and

$$G\frac{d^2\theta_k(t)}{dt^2} = G\frac{d^2\delta_k(t)}{dt^2} = f_k(t) - f_e(t) = f_a(t).$$

Multiplying the above equation by $\omega_{\text{syn}}(t)$ and dividing by its rated mega voltage amperes (S_{mach}) gives the per-unit asychronous speed. This is easier to work with in electrical power system and it is this case that gives rise to separable Hamiltonian. Let H represents the energy of system normalized by a rated mega voltage amperes (S_{mach})

$$H = \frac{\text{Stored kinetic energy in megajoules at synchronous speed}}{\text{Machine rating in mega voltage amperes}}.$$

In terms of the moment of inertia, $H = \frac{\frac{1}{2}G\omega_{\text{syn}}}{S_{\text{mach}}}$, where $G = \frac{2H}{\omega_{\text{syn}}}S_{\text{mach}}$. The resulting swing equations are

$$\frac{2H}{\omega_{\text{syn}}}\frac{d^2\delta_k}{dt^2} = \frac{f_a}{S_{\text{mach}}} = \frac{f_k - f_e}{S_{\text{mach}}},$$

for $k = 1, 2, \ldots, m$. This fundamental system of ordinary differential equations governs the rotational dynamics of the synchronous machine in most power systems. In Figure 1, the angle between E (the generator voltage) and V (output voltage) is called phase angle. The phase angles of the internal voltages depend on the positions of the rotors of the machines. If synchronous speed were not maintained between the generators of a power system, the phase angles of their internal voltages would be changing constantly. This will be in relation to the performance of each machine. Invariably, as a result this, a satisfactory operation would be impossible. The phase angles and the internal voltages of synchronous machines will remain constant only if the speeds of the other machines remain constant. This can happen only at the speed corresponding to the frequency of the reference phasor. If there is a change in the load profile of the entire system or in any of the generators, the current in that particular generator or the entire generator's will change to meet the demand. If the current does not effect any change, the phase angles of the internal voltages will

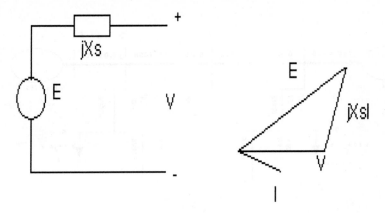

Fig. 1. Angle Diagrams

not change. There will be momentary changes in speed to demand the adjustment to the angles of the voltages to each machine. This is because the phase angles are determined by the relative positions of the rotors. This change is what is called disturbance in electrical system. When the machines have adjusted to the new positions mentioned or the cause of the change has been removed, the machines must operate at synchronous speed. If one machine does not return to synchronous state with the rest of the machines, large currents will circulate.

Stability studies are classified according to whether they involve steady state or transient conditions. An ac generator cannot deliver more than the rated power nor can synchronous motor absorb more than rated limit. Disturbances on a system that are caused by suddenly applied loads, or by the occurrence of faults, loss of excitation in the field of a generator, or switching, may cause loss of synchronous when the change in the system by the disturbance would not exceed the stability limit if the change were made gradually. The limiting value of power is called the transient stability or the steady-state stability limit, according to whether a sudden or gradual change in conditions reaches the point of instability of the system.

4   Three Machine, Nine-bus Power System

The experiments are based on a nine-bus and three-generator power system (see [5]), connected with transformers and transmission lines and associated relays as depicted in Figure 2. Transmission lines are classified

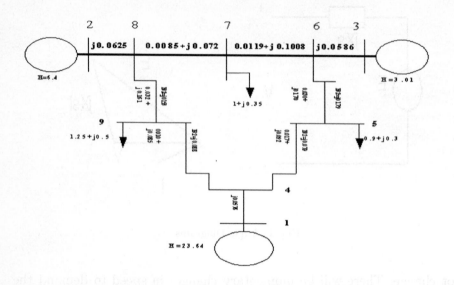

Fig. 2. Three-Machine, Nine-Bus Power System

as either single phase or three phase system. They are 60- Hz synchronous generated. They are connected and transmitted in Wye or Delta form. Usually, the transmission lines are connected to infinite bus meaning that there is no variation in the load voltage with the load voltage as a reference and at zero degrees angle. In the generating plants, paralleling of generators is necessary to supply the loads for occasional critical or peak loads. They are grouped in three to supply one-third load and for spare capacity. To parallel three generators, the output voltages of each must have the same magnitude, frequency, phase rotation, and phase angle. The same phase rotation means that the voltages of the corresponding phases of each generator must come to their peaks at the same time. In this experiment, the three generators are connected either in Wye or Delta form. The lines are represented as a single one-line diagram. In reality, there are three conductors for a three phase and two conductors for a single phase systems. The maximum power a generator can tranfer is indicated in Figure 1 and by this equation.

$$p = \frac{|E||V|}{X} \sin \delta.$$

The power transferred also depend on the variation of the angle and the generator's speed. The experiments are based on three phase faults on the transmission lines. It is also assumed that there is a three phase faults on line 1 and line 2. The moment of inertia, angular deviation, and velocity for the experiments are shown in Figure 2. The results of the experiment indicate symplectic response time are faster in CPU time than the trapezoid integration methods. A second-order simulation was performed using symplectic integration with clearing time of 0.192 seconds. Figures 3 and 4 indicate faster response time. We also extended this experiment to fourth-order using symplectic integrator. The outcomes of the experiments are displayed in Figures 5 and 6.

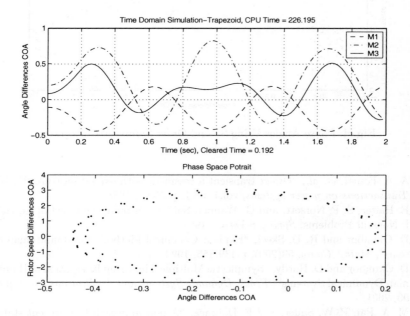

Fig. 3. Phase Angles Using Second-Order Non-Symplectic Integrator

References

1. R. Ando and S. Iwamoto, "Highly reliable transient solution method using energy function (power systems)", *Electrical Engineering in Japan*, vol 108, no.4, p.57-66, 1988.
2. V. I. Anold, Mathematical Methods of Classical Mechanics, *Springer-Verlag*, 1989.
3. J. H. Bentley, A programmed Review for Electrical Engineering, *Engineering Press, Inc.*,1994.
4. H. D. Chiang, B. Y. Ku, and J. S. Thorp, "A constructive method for direct analysis of transient stability", *IEEE*, p.3 vol.2508, 684-689, vol.1,1988.
5. K. R. Folken, The Transient Energy Function Method and Thee-Phase Unbalanced Systems, Master's Thesis, *University of Missouri-Rolla*, 1997.

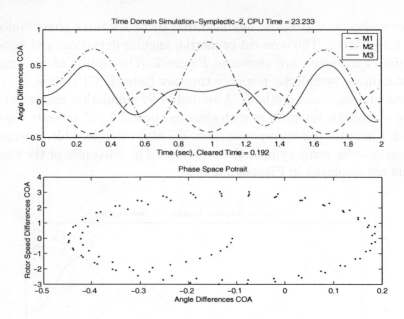

Fig. 4. Phase Angles Using Second-Order Symplectic Integrator

6. A. A. Fouad, *et. al.,* "Direct transient assessment with excitation control", *IEEE Transactions on power systems,* vol.4, no.1, p.75-82, 1988.

7. E. Hairer, S. P. Norsett, and G. Wanner, Solving Ordinary Differential Equations I: Nonstiff Problems, *Springer-Verlag,* 1991.

8. D. Okunbor and R. D. Skeel, "Explicit Canonical Methods for Hamiltonian Systems", *Math. Comp.* 59(200), p.439-455, 1992.

9. D. Okunbor and D. Hardy, "Symplectic Multiple Time Step Integration of Hamiltonian Dynamical Systems", *Procs. Dynamic Systems and Applications,* vol 3, p.483-90, 2001.

10. M. A. Pai, P. W. Sauer, and F. Dobraca, "A new approach to transient stability evaluation in power systems", *IEEE,* vol 1, 676-780, 1988.

11. J. M. Sanz-Serna and M. P. Calvo, Numerical Hamiltonian Problems, *Chapman and Hall,* 1994.

12. W. D. Stevenson, Jr., Elements of Power system analysis. *McGraw Hill,* 1982.

13. A. Seidman, H. W. Beaty, and H. Mahrous, Handbook of Electric Power Calculations, *McGraw Hill* 1984.

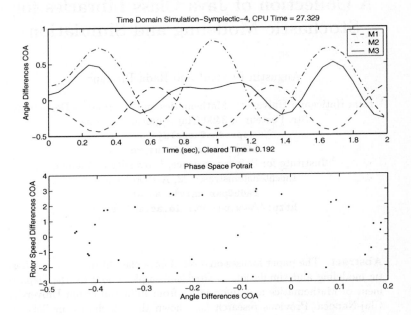

Fig. 5. Phase Angles Using Fourth-Order Symplectic Integrator

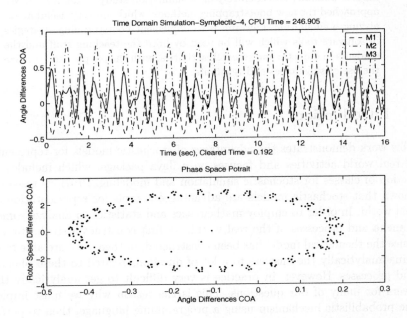

Fig. 6. Extended Time Fourth-Order Symplectic Integration Results

A Collection of Java Class Libraries for Stochastic Modeling and Simulation

Augustin Prodan[1] and Radu Prodan[2]

[1] Iuliu Haţieganu University, Mathematics and Informatics Department,
Str. Pasteur 6, 3400 Cluj-Napoca, Romania
aprodan@umfcluj.ro
http://www.umfcluj.ro
[2] Institute for Software Science, University of Vienna,
Liechtensteinstrasse 22, A-1990 Vienna
radu@par.univie.ac.at
http://www.par.univie.ac.at/~radu

Abstract. The paper focuses on a set of Java class libraries for stochastic modeling and simulation, created and implemented by the Department of Mathematics and Informatics from Iuliu Haţieganu University Cluj-Napoca. Previous research has shown that stochastic models are advantageous tools for representation of the real world activities. Due to actual spread of fast and inexpensive computational power everywhere in the world, the best approach is to model the real phenomenon as faithfully as possible, and then rely on a simulation study to analyze it. We approached the new bootstrapping methods, which are very useful in analyzing a simulation. Our aim is to implement bootstrapping strategies in software tools which will be used by both the teaching staff and the students in didactic and research activities, with the purpose to optimize the use of substances and reactants.

1   Introduction

This work demonstrates the advantages of stochastic models for representation of real world activities and focuses on a Java package, which includes a collection of classes for stochastic simulation and modeling. Previous research has shown that stochastic models are advantageous tools for representation of the real world. In order to employ mathematics and statistics to analyze some phenomena and processes of the real world, we first construct a stochastic model. Once the theoretical model has been constructed, in theory we are able to determine analytically the answers to a lot of questions related to these phenomena and processes. However, in practice is very difficult to get analytically the answers for many of our questions. This is the reason why we must implement the probabilistic mechanism using a programming language, then to perform a simulation study on a computer. Due to actual spread of fast and inexpensive computational power everywhere in the world, the best approach is to model the real phenomenon as faithfully as possible, and then rely on a simulation

P.M.A. Sloot et al. (Eds.): ICCS 2002, LNCS 2329, pp. 1040–1048, 2002.

study to analyze it. The Department of Mathematics and Informatics from Iuliu Haţieganu University Cluj-Napoca has created and implemented a collection of Java class libraries for stochastic simulation and modeling. The paper reports on the incremental development of an object-oriented Java framework, based on theoretical fundamentals in simulation and stochastic modeling defined by Ross [1] and continued by Gorunescu and Prodan [9], that supports the creation of the main elements for building and implementing stochastic models. The stochastic models constructed accurately represent real world phenomena and processes - particularly in medicine and pharmacy. The programming language we decide to use is Java, because we intend to build some software tools distributed on Internet. To build and implement such tools, we will technically use several Java technologies, particularly Enterprise Java Beans. We intend to build software tools which will be integrated in educational packages and will be used by both teaching staff and students. The software tools implemented by us will include Bayesian inference as a way of representing uncertainty, expressing beliefs and updating beliefs using experimental results. Also, we will build simulation tools for pharmaceutical research and education, with the purpose to optimize the use of substances and reactants. The basic design philosophy of our object-oriented approach to simulation of the random variables by means of classical distributions is presented in Prodan et al. [4], [5] and [8]. Some medical applications are described in [4], [7] and [8].

2   Motivation and Related Works

This research has been initiated about three years ago by A. Prodan, after his entrance as professor at Iuliu Haţieganu University, the main aim being to build software tools which will be used in didactic and research activities, particularly in medical and pharmaceutical environments. It is necessary to point out that our previous experience and preoccupations related to this domain were very poor. However, based on background of A. Prodan in programming languages (he worked more than 27 years in a research institute for computer science), we started this work from scratch, using only theoretical knowledge acquired from a book of Ross [1]. Our first objective was to build a software infrastructure, consisting of a set of Java classes for simulation of classical distributions. The idea to reinvent the wheel is not very attractive to us, but we must take the shame to say that we don't know very much about related works, mostly because of our previous isolation. What we know is that there are a lot of statistical and simulation software applications on market, but we cannot get them because of austere budget at Iuliu Haţieganu University. One difference of our approach from most others is the use of Java language and Java technologies in implementation. Based on software infrastructure already implemented, we intend to build software tools distributed on Internet, which will be used by both the teaching staff and the students in didactic and research activities.

3   Stochastic Models and Simulation

There are three levels of simulation to be considered (see Fig. 1). The *first level* consists of simulating random numbers, as they are the basis of any stochastic simulation study. The so-called pseudo-random numbers generated via Java's built in linear congruential generator are used to simulate random numbers. If Java's simple linear congruential generator is non-random in certain situations by generating undesirable regularities, it is necessary to build a class as an improvement over the one supplied by Java [2], [3]. Based on the ele-

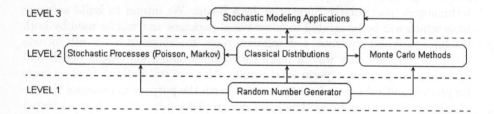

Fig. 1. The simulation levels

ments of the first level, the *second level* of simulation applied for distributional models is built. We implemented a hierarchy of Java classes, which model the classical distributions (see Fig. 2). Each distribution is determined by a distri-

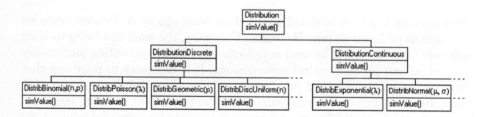

Fig. 2. The hierarchy of Java classes for distributional models

bution function and a set of parameters. Using these elements, we defined for each distribution a simulation algorithm and implemented it in a polymorphic method named **simValue()**. Each distribution class encapsulates a particular **simValue()** method. We included all distribution classes in a package named **Distrib**. To generate a single value from a particular distribution, it is necessary to import the class from **Distrib** package, then to instantiate an object of corresponding class, and finally to call the polymorphic method **simValue()** based on this object. We created Java classes for Poisson processes, renewal processes and for some particular cases of Markov processes. Also, we implemented methods for the Monte Carlo approach to solve various problems, such as approximating simple or multiple integrals, finding a solution for an equation system and finding the reverse, the eigenvectors and the eigenvalues for a matrix. The first two levels of simulation constitute the infrastructure for the *third level*, which is devoted to applications. As an application, we modeled the patient flow through chronic diseases departments [8]. Admissions are modeled as a Poisson process with parameter λ (the arrival rate) estimated by using the observed inter-arrival times. The in-patient care time is modeled as a mixed-exponential phase-type distribution [4], [5], [7], [8]. Both medical staff and hospital administrators agreed that such a model can be used to maximize the efficiency of the chronic diseases departments and to optimize the use of hospital resources in order to improve hospital care. Interesting real phenomena can be studied during simulation experiences, such as rejection at entrance due to a brimful department, the resources being limited.

4   Simulation and Visualization

In a simulation study is necessary to generate more values, a sequence of values, then to estimate some parameters and to visualize the results. For this purpose we implemented two general methods: **doSimulation(distribution)** which is called to generate the values and to compute successively actual estimators, and **doVisualization()** which is called to show the results of the simulation.

4.1   Simulation

We generate a sequence of values from a distribution, by using the general method **doSimulation(distribution)**. This method has as argument an instance of a distribution class and calls the **simValue()** method when generates each specific value for that distribution. The method generates continually additional data values, stopping when the efficiency of the simulation is good enough. Generally, we use the variance of the estimators obtained during the simulation study to decide when to stop the generation of additional values, that is when the efficiency of the simulation study is acceptable. The smaller this variance is, the smaller is the amount of simulation needed to obtain a fixed precision. For example, if our objective is to estimate the mean value $\mu = E(Xi)$, $i = 1, 2, ...,$ we should continue to generate new data until we have generated n data values for which the estimate of the standard error, $SE = \frac{s}{\sqrt{n}}$, is less than an acceptable value given by the user and named *Allowed SE* (see Fig. 3), s being the sample standard deviation. Sample means and sample variances are recursively computed. The final values of these parameters, the confidence interval and other statistics are showed as the result of the simulation study. Fig. 3 compares the results of two simulations in case of a binomial distribution with parameters n=10 and p=0.3. The number of generated values is determined by the value *Allowed SE* introduced by the user, as the simulation stops when the condition

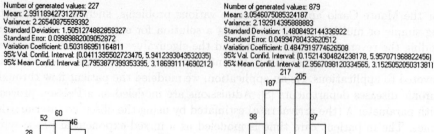

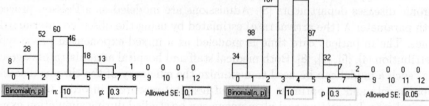

Fig. 3. A comparison between two simulations from binomial distribution Bin(10, 0.3), the first simulation with *Allowed SE* = 0.1 and the second with *Allowed SE* = 0.05

$\frac{s}{\sqrt{n}} \leq$ *Allowed SE* is true. The left side graph is obtained as a result of a simulation with *Allowed SE* = 0.1, being generated 227 values, while the right side graph is the result of a simulation with *Allowed SE* = 0.05, being generated 879 values.

4.2   Visualization

The method **doVisualization()** is used to show the results of a simulation. When we simulate a discrete random variable X, the generation of each value $x \in X$ is counted in the element $d[x]$ of a vector $d[]$. The visualization method expresses graphically in a column format the actual values contained in the elements of the vector $d[]$ (see Fig. 3). We approached in a similar way the visualization in case of a continuous distribution. When simulate from a continuous random variable X, a generated value $x \in X$ is approximated with a given *precision* expressed by the number of decimal digits to be considered. The user has the possibility to choose a precision of one, two, or even more decimal digits. If a coarse approximation is accepted, no decimals are considered and the real value x is approximated by $int(x)$, that is by integer part of the x. In this case the continuous random variable X is rudely approximated by a discrete one, and the results of a simulation can be graphically expressed in a segmented line format, each segment joining the top sides of two neighbouring columns. If a precision of one decimal digit is selected, the results of the same simulation can be more precisely visualized by a more refined segmented line. With a precision of two decimal digits, a more refined visualization is obtained. The higher this precision is, the higher is the *resolution* realized in visualization. Fig. 4 compares two visualizations for the same set of generated values from the standard normal distribution N(0, 1), the first visualization being with a precision of one decimal digit (Fig. 4, graph a), and the second with a precision of two decimal digits (Fig. 4, graph b). As can be seen in this figure, when the precision grows with one decimal digit, the resolution grows ten times. With a precision of one

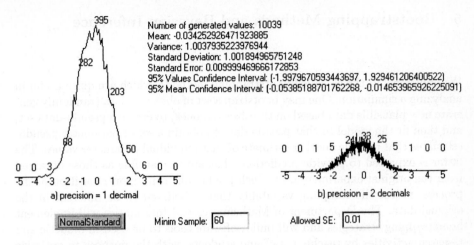

Fig. 4. A visualization with the precision of one decimal digit, versus a visualization with the precision of two decimal digits, for the same set of generated values from the standard normal distribution N(0, 1)

decimal digit, ten numbers are considered between two successive integers, while if the precision is of two decimal digits, one hundred numbers are considered between two successive integers. When necessary, intermediate resolutions can be considered. Fig. 5 shows a similar comparison for the exponential distribution with parameter $\lambda = 0.3$.

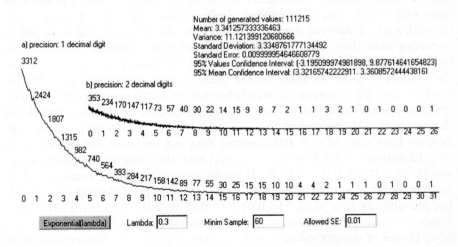

Fig. 5. A visualization with the precision of one decimal digit, versus a visualization with the precision of two decimal digits, for the same set of generated values from the exponential distribution E(0.3)

5   Bootstrapping Methods and Bayesian Inference

We implemented the so-called bootstrapping methods, which are quite useful in analyzing a simulation. One may bootstrap a set of observed data (randomly generate new plausible data based on the observed ones) to create a pseudo-data set, and then fit the model to that pseudo-data to obtain a set of parameter pseudo-estimates that provide another estimate of that individual parameter values. The latter is expected to provide predictions that are as plausible as those obtained from the original experimental fit. Such predictions can, upon repetition of the process, reflect the sampling variability that is believed to be inherent in the original data. The Department of Mathematics and Informatics will implement bootstrapping strategies and will build software tools to be used in didactic and research activities by teaching staff and students, with the purpose to optimize the utilization of substances and reactants. The bootstrapping methods will be used in medical statistics, pharmacokinetic models, bioequivalence studies and other medical and pharmaceutical applications. Also, we will implement software tools which will include Bayesian inference as a way of representing uncertainty, expressing beliefs and updating beliefs using experimental results. These software tools will assist us, as human beings, to learn from past experiments and revise our knowledge on issues around us.

We approached the bootstrap technique by considering the set of n observed data as belonging to an empirical discrete random variable X_e which is equally likely to take any of the n values. If the values $x_i, i = 1, 2, ..., n$, are not all distinct, the general discrete random variable X_e will equal the value x_i with a probability equal with $\frac{n_i}{n}$, where n_i is the number of values equal with x_i ($i = 1, 2, ...$). Starting with a set of observed data x_i, $i = 1, 2, ..., n$, the bootstrapping tools put in ascending order these data, then count the number m of distinct values and the number of occurrences for each distinct value. If the value x_i has n_i occurrences, the random variable X_e takes on the value x_i with probability $p_i = \frac{n_i}{n}$, for $i = 1, 2, ..., m$. To bootstrap the set of observed data, it is necessary to randomly generate a sequence of values from the general discrete distribution containing the values x_i with the probabilities p_i, for $i = 1, 2, ..., m$. As an example, suppose we have a set of $n = 100$ observed data and are identified the following $m = 10$ distinct values $X = (x_1, x_2, ..., x_{10})$ with the corresponding occurrences $o = (4, 15, 1, 17, 12, 9, 10, 19, 10, 3)$. It results from this that the general discrete random variable will equal the distinct values with the following corresponding probabilities $p = (0.04, 0.15, 0.01, 0.17, 0.12, 0.09, 0.1, 0.19, 0.1, 0.03)$. Fig. 6 shows the results of a simulation from this general discrete distribution, considering the set of distinct values $x_i = i - 1$, for $i = 1, 2, ..., 10$. This consideration does not diminish the generality of the algorithm. In this particular case the algorithm generates the values $x_i = i - 1$ with the corresponding probabilities p_i, but it can as well generate any values x_i with the same probabilities p_i, for $i = 1, 2, ..., 10$.

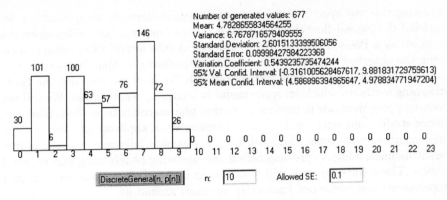

Fig. 6. The results of the bootstrapping method, approached by the simulation from a general discrete distribution

6 Future Work

We intend to use Java technologies in our activities of building and implementing models for stochastic processes (Poisson, Markov, semi-Markov, renewal) with applications in medicine and pharmacy. A continuous-time Markov model for the flow of patients around compartments of geriatric medicine will be proposed. This model will enable us to study the whole system of geriatric care and will be used to look either at the time patients spend in hospital and also the subsequent time patients spend in the community. We will implement bootstrapping strategies in software tools, which will be used by both the teaching staff and the students in didactic and research activities, with the purpose to optimize the use of substances and reactants. Also, we will implement software tools which will include Bayesian inference as a way of representing uncertainty, expressing beliefs and updating beliefs using experimental results. And finally, based on our collection of Java class libraries, we intend to build and to implement computer based learning tools to be used by students for learning statistics and stochastic processes.

7 Conclusions

The Department of Mathematics and Informatics from Iuliu Haţieganu University Cluj-Napoca has created and implemented a collection of Java class libraries for stochastic modeling and simulation. We implemented libraries for distributional models, stochastic processes and Monte Carlo methods. Due to actual spread of fast and inexpensive computational power everywhere in the world, the best approach is to model the real phenomenon as faithfully as possible, and then rely on a simulation study to analyze it. We implemented general methods for simulation and for visualization. The user has the possibility to interact with specific Java applets, to select specific values for some parameters, obtaining this way specific simulations and visualizations. As an application, we modeled the patient flow through chronic diseases departments. Admissions are modeled as a Poisson process with parameter λ (the arrival rate) estimated by using the observed inter-arrival times. The in-patient care time is modeled as a mixed-exponential phase-type distribution [8]. We implemented also the bootstrapping methods, which are quite useful in analyzing a simulation. We will use bootstrapping methods in medical statistics, pharmacokinetic models, bioequivalence studies and other medical and pharmaceutical applications. Also, we will implement software tools which will include Bayesian inference as a way of representing uncertainty, expressing beliefs and updating beliefs using experimental results. These software tools will assist us, as human beings, to learn from past experiments and revise our knowledge on issues around us.

References

1. Ross, S. M.: A Course in Simulation, Macmillan Publishing Company, New York (1990).
2. Eckel, B.: Thinking in Java, MindView Inc., Prentice Hall PTR (1998).
3. Prodan, A., Prodan, M.: Java Environment for Internet, Ed. ProMedia-plus, Cluj-Napoca, ISBN 973-9275-07-9 (1997).
4. Prodan, A., Gorunescu, F., Prodan, R.: Simulating and Modeling in Java, Proceedings of the Workshop on Parallel/High-Performance Object-Oriented Scientific Computing, June 1999, Lisbon, Technical Report FZJ-ZAM-IB-9906, Forschungszentrum Julich GmbH (1999) 55-64.
5. Prodan, A., Gorunescu, F., Prodan, R., Câmpean, R.: A Java Framework for Stochastic Modeling, Proceedings of the 16th IMACS World Congress on Scientific Computation, Applied Mathematics and Simulation, August 2000, Lausanne, CD Ed. Rutgers University, New Brunswick - NJ, USA, ISBN 3-9522075-1-9 (2000).
6. Câmpean, R., Prodan, A.: A Java Environment for Development of Stochastic Models, Clujul Medical, vol. LXXIII, No.2, (2000), 312-317.
7. Gorunescu, F., Prodan, A.: Optimizing the Chronic Healthcare Department Occupancy, Clujul Medical, vol. LXXIII, No.4, (2000), 503-507.
8. Prodan, A., Prodan, Rodica: Stochastic Simulation and Modeling, ETK-NTTS Conference (Exchange of Technology and Know-how, New Techniques and Technologies for Statistics), Crete, June 2001, Proceedings (2001) 461-466.
9. Gorunescu, F., Prodan, A.: Stochastic Modeling and Simulation, Ed. Microinformatica, Cluj-Napoca, ISBN 973-650-023-3, (2001).

Task-Oriented Petri Net Models
for Discrete Event Simulation

Ewa Ochmanska

Warsaw Technical University, Department of Transport
00-662 Warsaw, Poland
och@it.pw.edu.pl

Abstract. The paper concerns a class of task-oriented simulation models for discrete event systems. Systems are considered, in which arriving tasks are executed in alternative ways, according to the actual state of particular system resources. The models are based on Petri nets, with extensions including informative aspects of modeled processes as well as timing and decision-making rules. Model structure and behavior conform to realities of railway control. Similar models can be helpful in investigating different types of systems executing tasks by means of distributed sets of resources with dynamically changing availability. In particular they are suitable for modeling reliability behavior of control systems, built of re-configurable modules.

1 Introduction

A discrete event simulation model, presented in the paper, propagates tasks arriving to modeled system through the chain of task executors. Tasks are executed in one of possible ways or discarded, according to the current state of system resources. Such a model can be applied in investigating system reliability or performance.

General assumptions for system structure and functions, taken from realities of railway station control system, lead to the task-oriented modeling principle with dynamically changing states of resources needed for alternative task executions, formulated in section 2. The model represents processes of execution of tasks arriving to the system by occupying one of several predefined configurations of resources with limited and changing availability, corresponding to failures of technical devices [3].

Timed nets with data structures, predicates and actions [2, 4], defined upon place/transition Petri nets [1], were applied earlier and proved as effective tool for event-oriented modeling and simulation of technological processes at railway stations [4, 6] and processes of transport & distribution in logistic systems [5]. The new feature of models presented in this paper consists in task- versus event-oriented view of the system rather than in net model description. The formalism of Petri net extension adopted for this kind of models is shortly resumed in section 3.

In section 4, the composition and dynamics of net elements of simulation model are explained, as well as general structure and behavior of the model. Section 5 contains final remarks concerning some problems, which need resolving, and possible applications of the presented models.

P.M.A. Sloot et al. (Eds.): ICCS 2002, LNCS 2329, pp. 1049–1057, 2002.

2 Task-Oriented View of a System

2.1 System Resources

We assume that a control part **C** of modeled system disposes resources of a set **R** as a means to execute a stream of tasks, i.e. a sequence of tasks incoming to the system at successive points of time. All resources are divided into n disjoint subsets:

$$\mathbf{R}=\{r_f, f=1,...,p\}=\mathbf{R}_1\cup...\cup\mathbf{R}_n , \tag{1}$$

which are physically distributed and joint to the control part by communication links. The distributed subsets of resources form a family:

$$\mathbf{R}=\{\mathbf{R}_i, i=1,...,n\} . \tag{2}$$

At a given point of time, each individual resource $r \in \mathbf{R}$ may be in one of the following states: $1°$ *available*: being ready to use; $2°$ *occupied*: participating in execution of a task; $3°$ *unable*: malfunctioning; $4°$ *inaccessible* because of malfunction of its communication link (this case simultaneously concerns all resources from a set $\mathbf{R}_i$).

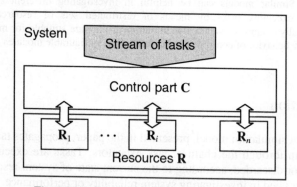

Fig. 1. Task-oriented view of the modeled system

The control part **C** of the system checks actual states of particular resources and distributes, through communication links, execution of incoming tasks to available resources proper to specific kinds of tasks, according to definitions formulated in the next subsection.

2.2 Executive Configurations of Resources

For execution of a task, some subset of system resources is needed for some period of time. Such subset of resources will be called *executive configuration* of a task. A finite family of different executive configurations, applied for execution of all system tasks, is a finished family of m resource subsets:

$$\mathbf{E}=\{\mathbf{E}_k \subseteq \mathbf{R} \mid k=1,...,m\} . \tag{3}$$

In general, an executive configuration of a task may be distributed, i.e. may contain resources of different subsets of family **R**.

A task can be alternatively executed in more than one configuration of resources, e.g. a train coming to the station can be placed on different tracks with the use of different control devices. Moreover, the modeled system executes tasks of different kinds, e.g. a station sends trains of various lengths to different directions. The kinds of tasks are distinguished by sets of possible executive configurations and their kind-specific time characteristics (duration of execution and occupation of resources).

All kinds of tasks executed by the system are described by a set $\mathbf{K}$ of pairs:

$$\mathbf{K}=\{K_j=\langle E_j, T_j \rangle \,|\, j=1,...,q\} , \tag{4}$$

where $\mathbb{E}_j \subseteq \mathbb{E}$ is a subfamily of executive configurations, and timing function T_j assigns a time period of being occupied by executing a task of j-th kind to each particular resource of each particular configuration in $\mathbb{E}_j$.

Accordingly to (4), different executive configurations may be alternatively applied to a task of given kind. On the other hand, the same executive configuration may be applicable to multiple kinds of tasks, i.e. families of configurations proper to different kinds of tasks may have common parts. However, timing functions are defined independently and the same configuration of resources may be applied for different kinds of tasks with different time characteristics.

Let us order executive configurations proper to a task of j-th kind from most to less convenient, and appropriately enumerate elements of subfamily $\mathbb{E}_j$:

$$E_j=\{E^{jd} \in E\}_{d=1,...,|\mathbb{E}_j|} , \tag{5}$$

where $|\mathbb{E}_j| \geq 1$ is cardinality of the set $\mathbb{E}_j$.

An executive configuration proper to multiple kinds of tasks may have different order for each of those kinds.

2.3 An Algorithm for System Dynamics

An executive configuration, chosen by the control part of a system for executing a task, is called its **active** configuration. During execution of a task of j-th kind, each resource in its active configuration is occupied by a period of time determined by the timing function T_j. A resource may be used for execution of one task at a time, i.e. occupied by at most one active configuration.

The tasks arrive to the input of the modeled system in some distinguished points of time. The control part of the system attempts to execute them in configurations of lowest possible order, while the states of resources change dynamically and not all of them are available. Some of resources are occupied by active configurations. Some of them have failed or became inaccessible because of damaged communication links.

Assigning an active configuration for a task of j-th kind is done by searching through all executive configurations of the system, starting from the configuration E^{j1} first in order for this kind of tasks.

Let us consider a task incoming to a system, which remains at a given moment in some current state of resources. We will distinguish five following cases with regard to possibility of executing a task of j-th kind in executive configuration $E_k \in \mathbb{E}$ as the configuration of d-th order $E^{jd} \in \mathbb{E}_j$, $d \in 1,..., |\mathbb{E}_j|$, proper to that kind of tasks:

(a) $E_k \neq E^{jd}$. Another configuration $E_l \in \mathbb{E}$, $l \neq k$ and $E_l = E^{jd}$, should be considered.

(b) $E_k = E^{jd}$ and all its resources are available.
The task is executed in active configuration E_k. The resources belonging to that configuration are occupied for a time given by the timing function $\mathcal{T}_j$.

(c) $E_k = E^{jd}$, but it is not possible to activate this configuration, because some of its resources are occupied, i.e. participating in an active configuration.
The task is waiting to be executed in the configuration E_k after releasing occupied resources.

(d) $E_k = E^{jd}$, but it is not possible to activate this configuration because of malfunctions of some elements of the system: one or more resources belonging to E_k are in states of inability or inaccessibility. Besides, $d < |\mathbb{E}_j|$, i.e. there is an executive configuration of next order in the sequence of $\mathbb{E}_j$.
Then the control part of the system attempts to activate some configuration $E_l \in \mathbb{E}$, $l \neq k$, as the next executive configuration $E^{j(d+1)}$ proper to the kind of task.

(e) $E_k = E^{jd}$, but it is not possible to activate this configuration because one or more resources belonging to E_k are in states of inability or inaccessibility, and $d = |\mathbb{E}_j|$, i.e. E^{jd} is last in the sequence of $\mathbb{E}_j$.
The task can not be executed in a regular way. For such cases we may define several treatments depending on specifics of modeled problem, e.g. refusing execution, waiting for recovering of needed resources or suspending the system normal activity for a period of time consumed by special procedures.

3.   Net Tools for Model Description

3.1   Extended Petri Nets

Our aim is to construct a task-oriented model of the system, with structure and behavior described in the previous section. To achieve it we formulate an extension of Petri nets described as follows. Place/transition Petri net is a quadruple:

$$PTN = (P, T, A, \Theta_0), \tag{6}$$

where P is a set of places, T is a set of transitions, $A \subseteq P \times T \cup T \times P$ is a set of arcs, and $\Theta . P \to \mathcal{N}$ is a state function, which assigns to each place p some nonnegative number $\Theta(p)$ of tokens. A Petri net with an initial state Θ_0 realizes a process by firing transitions. A transition t is fireable in a state Θ (it is written as $\Theta(t)$), if each place of its input $I(t) = \{ p \in P \mid \langle pt \rangle \in A \}$ contains a token:

$$\Theta(t) \Leftrightarrow \forall (p \in I(t)) \, \Theta(p) > 0 . \tag{7}$$

The fired transition t consumes one token per input place and produces one token in each place of its output $O(t) = \{ p \in P \mid \langle tp \rangle \in A \}$, moving the net into the new state Θ':

$$\Theta'(p) = \begin{cases} \Theta(p) - 1 & \text{for } p \in I(p) \backslash O(p) \\ \Theta(p) + 1 & \text{for } p \in O(p) \backslash I(p) \\ \Theta(p) & \text{otherwise} . \end{cases} \tag{8}$$

We introduce following extensions to the above definition:

- **Tokens** represent data structures defined by the state function of the net model, describing current state of the modeled system. A place **p** contains a set $\Theta(\mathbf{p})$ of some nonnegative number of data structures (in particular $\Theta(\mathbf{p})$ can be empty).

- **Predicate** of a transition **t** is a logical function which constrains its fireability, defined on values of data structures of its input $\{\Theta(\mathbf{p}) \mid \mathbf{p} \in \mathbf{I}(\mathbf{t})\}$.

- **Action** of a fired transition **t** transforms its input data to its output data, consuming one token per each input place $\{\theta(\mathbf{p}) \in \Theta(\mathbf{p}) \mid \mathbf{p} \in \mathbf{I}(\mathbf{t})\}$ and producing one token in each output place $\{\theta(\mathbf{p})' \in \Theta'(\mathbf{p}) \mid \mathbf{p} \in \mathbf{O}(\mathbf{t})\}$. Information describing actual state of realized process, transformed by actions and transported by tokens, is then used by predicates for deciding about further way of process realisation.

- **Timing** is a mechanism placing modelled processes in time by means of timestamps inserted to data structures of tokens, time conditions assigned to predicates and time updates performed by actions. A state $\Theta(\tau)$ of the net is defined by sets of tokens with data structures describing contents of places in a moment τ:

$$\Theta(\tau) = \{\Theta(\mathbf{p}, \tau) \mid \mathbf{p} \in \mathbf{P}\} . \tag{9}$$

The number of data structures $|\Theta(\mathbf{p}, \tau)| \geq 0$ for any $\mathbf{p} \in \mathbf{P}$. More detailed description of timing mechanism and other extensions, applied in somewhat different context of event-oriented models, can be found in [4].

3.2   Classes of Net Elements

Petri nets, extended by above definitions, will constitute dynamic discrete event system models, realizing processes of concurrent execution of different kinds of tasks. In order to represent their semantics, we distinguish classes of places and transitions corresponding to the structure and dynamics of the task-oriented system view described in section 2.

Places will contain tokens with class-specific data structures. Transitions with class-specific inputs and outputs will perform class-specific actions, constrained by class-specific predicates. Semantics of net elements, comprising class-specific terms, is explained together with net model description in section 4.

Task-oriented net model is built of two hierarchical levels. The model consists of two classes of places with tokens representing tasks and resources, and of three classes of "high-level" transitions, i.e. sub-nets (groups of low-level transitions) responsible for generating tasks, executing tasks and providing resources.

High-level class of task executors consists of several classes of low-level transitions, performing alternative behaviors, appropriate to particular cases of the algorithm for system dynamics described in subsection 2.3. The alternatives are realized by means of class-specific, mutually exclusive predicates.

Low-level net structures of remaining two classes of high-level transitions, representing stochastic generators for streams of tasks and states of resources, are not discussed here. Some propositions can be found in [3].

4   Task-Oriented Net Model

4.1   General Structure of the Model

The structure of task-oriented net model, presented on Fig. 2, is quite regular and has three layers of dynamic elements (high-level transitions) separated by two layers of passive elements (places). Model elements of each layer belong to one of five different classes, characterized below in order as they appear on the figure.

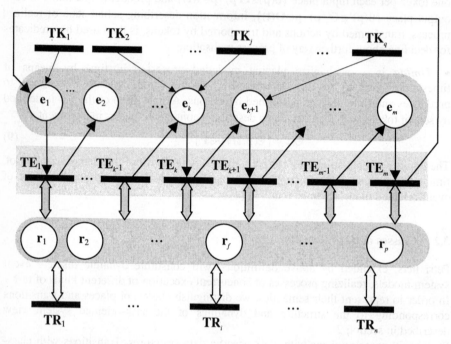

Fig. 2. General structure of the task-oriented net model

Task Generators: a class of transitions corresponding to the set **K** is responsible for generating streams of tasks of different kinds. Their actions produce tokens with data structures including descriptions of respective task kinds according to (4) and (5), completed by timestamp of their arrival to the system and by current order value set to 1. Transition TK_j, $j \in \langle 1,...,q \rangle$ is connected to the place e_k with number k such, that $E_k \in \mathbb{E}$ equals $E^{j1} \in \mathbb{E}_j$. In other words, tasks of j-th kind attempt to be executed in the configuration of the first order (the best for that kind of tasks). The connections showed on Fig. 2 by small black arrows are exemplary.

Tasks: a class of places serving as containers for tasks, waiting for execution in one of the possible configurations of resources. The places $e_1,...,e_m$, corresponding to m executive configurations used by the system, are inputs to transitions responsible for execution of tasks in respective configurations. They hold tokens representing tasks, produced by task generators.

Task Executors: high-level transitions, corresponding to the family $\mathbb{E}$ of all executive configurations used in the system. Transitions of this class consume tasks and perform actions depending on current states of needed resources. They execute tasks in particular configurations if resources are available. If some of needed resources remain in states of inability or inaccessibility, they pass tasks to the input of another task executor, next in the chain of connections showed on the figure by big black arrows. From the other side, transition $\mathbf{TE}_k$ is connected with places containing resources of executive configuration $\mathbf{E}_k$. Groups of connections $\{\mathbf{TE}_k\}\times\mathbf{E}_k\cup\mathbf{E}_k\times\{\mathbf{TE}_k\}$ between transitions $\mathbf{TE}_k$, $k=1,\ldots,m$, and their input/output places representing resources of particular executive configurations, are depicted by thick grey arrows on Fig. 2. Sets $\mathbf{E}_k$, $k=1,\ldots,m$ have in general nonempty intersections: a resource place r_f may be connected with several task executors. The detailed structure and function of this class of transitions is described in the subsection 4.2.

Resources: a class of places corresponding to the set $\mathbf{R}$ of system resources (1). These places contain tokens with data including actual states of particular resources (from $1°$ to $4°$ as enumerated in subsection 2.1), their planned durations and timestamps of last state changes. Tokens representing resources are consumed and "occupied" by task executors.

Resource Providers: a class of transitions corresponding to the family $\mathbb{R}=\mathbf{R}_1,\ldots,\mathbf{R}_n$ of resource sets (2). They are responsible for current states of resources (except of the state $2°$, caused by task executors). Acting concurrently to transitions of other classes, they consume and produce tokens representing resources and temporarily limit their accessibility, possibly in a stochastic manner. An example of detailed realisation of this class of high-level transitions, based on several stochastic generators, can be found in [3]. Groups of input/output connections $\{\mathbf{TR}_i\}\times\mathbf{R}_i\cup\mathbf{R}_i\times\{\mathbf{TR}_i\}$ between transitions $\mathbf{TR}_i$, $i=1,\ldots,n$ and places representing corresponding subsets of resources, are showed on the figure by thick white arrows. Sets $\mathbf{R}_i$, $i=1,\ldots,n$ are disjoint: each resource place r_f is connected with one particular transition $\mathbf{TR}_i$.

Active configuration for a task of j-th kind, represented by a token produced by transition $\mathbf{TK}_j$ and placed on input to transition $\mathbf{TE}_k$, is chosen by a rule similar to the "token ring" principle of communication, applied in computer nets. If k-th executive configuration can not be activated for a task waiting for execution in the place e_k (e.g. it is improper for this task, or some of resources in the set $\mathbf{E}_k$ are inaccessible), transition $\mathbf{TE}_k$ produces a token for that task in another place e_l, successive in the closed chain shown on Fig. 2 ($l=k+1$ for $k<m$ and $l=1$ for $k=m$). This mechanism is described in the following subsection.

4.2   Detailed Structure of Task Executors

A high-level transition of the task executor class forms a subnet, consisting of a group of four parallel low-level transitions of different classes **(a)**, **(b)**, **(d)** and **(e)**, as shown on Fig. 3. The predicates constraining fireability for each of these transitions are different and mutually exclusive.

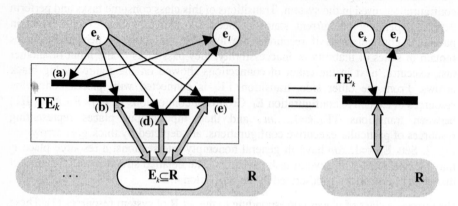

Fig. 3. Detailed structure of the task executor

The place e_k contains tokens representing tasks to be executed in some configurations of resources. Input connection from the place e_k to the high-level transition $\mathbf{TE}_k$ is branched to the four parallel low-level transitions with predicates corresponding to the cases distinguished in the algorithm of subsection 2.3. As a consequence, they alternatively consume tasks and perform actions proper to the cases (a), (b), (d) and (e) of the algorithm:

Transition (a) occurs, if an input token in place e_k represents a task of j-th kind, which attempts to be executed in some configuration $\mathbf{E}^{jd}$ of order d on the list $\mathbb{E}_j$, different from the configuration $\mathbf{E}_k$ bound to the high-level transition $\mathbf{TE}_k$. The task is moved to the input of high-level transition $\mathbf{TK}_l$, successive in the chain, by consuming the token in place e_k and producing another one with the same data structure in place e_l. This action is repeated until the proper executive configuration is reached. Note that the transition (a) has no input resource places.

Transition (b) occurs, when a token in place e_k represents a task attempting to be executed in the configuration $\mathbf{E}_k$ and all needed resources are available. It removes a task from the input place and occupies resources by consuming tokens in places of the set $\mathbf{E}_k$ and producing new ones in the same places, with states changed to "occupied" and updated values of timestamps and timing functions.

Transition (d) occurs, when a token in place e_k represents a task of j-th kind attempting to be executed in the configuration $\mathbf{E}_k = \mathbf{E}^{jd}$, which can not be activated because there is at least one token with state $3°$ or $4°$ in resource places of the set $\mathbf{E}_k$. The task is moved, with increased order value, to the input of transition $\mathbf{TK}_l$, attempting to be executed in the configuration $\mathbf{E}^{j(d+1)}$, successive on the list $\mathbb{E}_j$.

Transition (e) occurs, when a token in place e_k represents a task of j-th kind, the configuration $\mathbf{E}_k$ is last on the list $\mathbb{E}_j$ and can not be activated; the task is discarded.

The case **(c)** of the algorithm is static; waiting for releasing of occupied resources is modeled by timing condition built into the predicate of transition **(b)**.

5   Conclusions

The task-oriented approach for construction and dynamics of net simulation models bases on a set of extended Petri net modeling tools. The modeling method, with its mechanisms for timing, conditional execution and processing of data structures, has been implemented as an object library of net components and proved its usefulness for event-oriented simulation models of transport and logistic systems [4, 5, 6].

The new, task-oriented view of modeled system was inspired by the reliability behavior of control systems in transport. Such a view permits to define compact models of systems performing tasks by means of distributed resources with limited and dynamically changing accessibility. These system characteristics may be valid for various environments, e.g. for distributed computing and Web based simulation.

Using stochastic generators for streams of tasks and states of resources, this way of simulation modeling, compared with analytical models based on stochastic Petri nets, forms an alternative which is free of limitations on probability distributions.

The presented modeling rules give rise to further evolution in several topics, like multi-directional net structure instead of circular one, stochastic rules of choosing executive configurations in place of simple ordering, special treatment of "refused" tasks etc., which can increase expressive power of this modeling approach. Taking into account the easiness of object implementation of net elements, the described type of models seems promising.

References

1. Desel J., Reisig W., Place/Transition Petri Nets. In: Lectures on Petri Nets I; Basic Models, Vol. 1491 of LNCS, Springer-Verlag (1998)
2. Ghezzi C., Mandrioli D., Morasca S., Pezzè M.: A General Way to Put Time in Petri Nets. Proceedings of the 5th International Workshop on Software Specification and Design, IEEE-CS Press, Pittsburg (1989)
3. Ochmanska E., Wawrzynski W.: Simulation Model of Control System at Railway Station. Archives of Transport. Polish Academy of Science, Committee of Transport. Warsaw (to appear)
4. Ochmanska E.: System Simulating Technological Processes. ESM'97, Proceedings of the 11th European Simulation Multiconference, Istambul (1997)
5. Ochmanska E.: Object-oriented PN Models with Storage for Transport and Logistic Processes. ESS'97, Proceedings of the 9th European Simulation Symposium „Simulation in Industry", Passau (1997)
6. Ochmanska E.: Simulation of Serving Trains at Railway Freight Stations. 13th European Simulation Multiconference, Warszawa (1999)

A Subspace Semidefinite Programming for Spectral Graph Partitioning

Suely Oliveira[1], David Stewart[2], and Takako Soma[1]

[1] The Department of Computer Science
The University of Iowa, Iowa City, IA 52242, USA
{oliveira, tsoma}@cs.uiowa.edu
[2] The Department of Mathematics
The University of Iowa, Iowa City, IA 52242, USA
dstewart@math.uiowa.edu

Abstract. A semidefinite program (SDP) is an optimization problem over $n \times n$ symmetric matrices where a linear function of the entries is to be minimized subject to linear equality constraints, and the condition that the unknown matrix is positive semidefinite. Standard techniques for solving SDP's require $O(n^3)$ operations per iteration. We introduce subspace algorithms that greatly reduce the cost os solving large-scale SDP's. We apply these algorithms to SDP approximations of graph partitioning problems. We numerically compare our new algorithm with a standard semidefinite programming algorithm and show that our subspace algorithm performs better.

Keywords: semidefinite programming, subspace methods

1 Introduction

A semidefinite program (SDP) [1, 18, 22] is the problem of minimizing a linear function over symmetric matrices with linear constraints and the constraint that the matrix is positive semi-definite. That is,

$$\min_{X} C \bullet X \quad \text{subject to} \tag{1}$$

$$A_i \bullet X = a_i, \quad i = 1, 2, \dots, m, \tag{2}$$

$$X \succeq 0 \tag{3}$$

where $A \bullet B = \text{trace}(A^T B) = \sum_{i,j} a_{ij} b_{ij}$ and $A \succeq B$ means that $A - B$ is a positive semi-definite matrix. Note that A and B are symmetric, then $A \bullet B = \text{trace}(AB)$. This is essentially an ordinary linear program where the non-negativity constraint is replaced by a semidefinite constraint on matrix variables. Semidefinite programming reduces to linear program when all the matrices are diagonal. Semidefinite programs and linear programs are special instances of a more general problem class called conic linear programs, where one seeks to minimize a linear objective function subject to linear constraints and a constraint that the unknown lies in a given closed convex cone. Both the cone of semidefinite

P.M.A. Sloot et al. (Eds.): ICCS 2002, LNCS 2329, pp. 1058–1067, 2002.

matrices for semidefinite programming, and the non-negative orthant for linear program, are homogeneous, self-dual cones.

One of the main aspects in which semidefinite programming differs from linear program is that the non-negative orthant is a polyhedral cone, whereas the semidefinite cone is not. Thus, developing simplex type algorithms for semidefinite programming is a difficult task. However, it is fairly straightforward to design polynomial time primal-dual interior-point algorithms for these problems. Currently various software packages are available for solving semidefinite programs using interior-point algorithms, such as CSDP [4], SDPA [7], SDPpack [2], SDPT3 [21], SP [23], and a Matlab toolbox by Rendl [17].

Subspace methods have been developed to solve system of linear equations [11] and to compute eigenvalues [3]. In this paper we have developed a method for solving large semidefinite programming problems using subspaces. Subspace methods have been use to compute a few of the extreme eigenvalues and the corresponding eigenvectors of a large, sparse, symmetric matrices, such as the Lanczos method, which is based on Krylov subspaces [14], and the Davidson method which uses Rayleigh matrices [6]. Generalized Davidson [19] and Jacobi–Davidson type algorithms [20] have been introduced, and theoretical studies have been done in [5, 12]. Davidson type subspace methods have been applied to solve graph partitioning problem [9, 10] and extended eigenproblem [13]. Here we use subspace methods to solve semidefinite programs.

Semidefinite programming has been an active research area in mathematics and engineering. In particular, many hard optimization problems with integer constraints can be relaxed to a problem with convex quadratic constraints which can be formulated as a semidefinite program (see, for example, [15]). These semidefinite programs provide approximations to the original, hard problem which can usually be solved in polynomial time. Usually, approximations from semidefinite programming relaxations are better than those from linear programming. In particular, we will apply semidefinite programming to the graph partitioning problem.

Graph partitioning is universally employed in the parallelization of calculations on unstructured grids including finite element and finite difference techniques. In many calculations, the underlying computational structure can be conveniently modeled as a graph in which vertices correspond to computational tasks and edges reflect data dependencies. Once a graph model of a computation is constructed, graph partitioning can be used to determine how to divide the work and data for an efficient parallel computation. The goal of the graph partitioning problem is to divide the graph into equally weighted sets in such a way that the weight of the edges crossing between sets is minimized. In other words, the graph partitioning is used to evenly distribute the computations among the processors while minimizing interprocessor communication, so that the corresponding assignment of tasks to processors leads to efficient execution. In general, computing the optimal partitioning is an NP-hard problem. Therefore, heuristics need to be used to get approximate solutions for these problems. The graph partitioning problem for high performance scientific computing has been studied extensively over the past decade. While graph partitioning problems are NP-hard, in practice they are also large-scale, which can make even the application of heuristics difficult.

This paper is organized as follows. Section 2 describes semidefinite programming. The new method of solving the semidefinite programming using subspace algorithms is introduced in Section 3, and Section 4 discusses the algorithm for the new method. In Section 5, numerical results are presented.

2 Semidefinite Programming

A semidefinite program (SDP) is an optimization problem of the following form: Let C be a given symmetric $n \times n$ matrix, $a \in \Re^m$ and $\mathcal{A}$ be a linear operator that maps symmetric matrices of size n into vectors in $\Re^m$ given by $\mathcal{A}(X)_i = A_i \bullet X$. Note that the adjoint of this operator is $\mathcal{A}^T(y) = \sum_{i=1}^{m} y_i A_i$.

Primal Problem: $\min_{X} \operatorname{tr}(CX)$ subject to (a) $\mathcal{A}(X) - a = 0$,
$$\text{(b) } X \succeq 0. \tag{4}$$

The general duality theory for semidefinite programs has been studied [24, 16]. We derive the Lagrangian dual to semidefinite program directly. Introducing the Lagrange multiplier $y \in \Re^m$ for the equality constraints, we see that

$$\mathcal{L}(X, y) = C \bullet X - y^T(\mathcal{A}(X) - a) - M \bullet X$$
$$= (C - \mathcal{A}^T(y) - M) \bullet X + y^T a,$$

and the analogues of the Kuhn–Tucker conditions for optimality are

$$\nabla_X \mathcal{L}(X, y, M) = C - \mathcal{A}^T(y) - M = 0,$$
$$M, X \succeq 0, \quad M \bullet X = 0, \quad \mathcal{A}(X) = a.$$

Note that $\nabla_X f(X)$ is the matrix whose (i, j) entry is $\partial f / \partial x_{ij}(X)$. Thus, we get the dual semidefinite program

Dual Problem: $\max_{y} a^T y$ subject to (a) $C - \mathcal{A}^T(y) \succeq 0$. $\qquad$ (5)

It is assumed that there exits (X, y) such that X is positive semidefinite, $\mathcal{A}(X) = a$, and $C - \mathcal{A}^T(y)$ is positive semidefinite. These points are called feasible points. In this case it is easy to show weak duality: that is, the minimum value of primal problem (4) is greater or equal to the maximum value of the dual problem (5). If X is also positive definite, then it is a feasible interior point. In this case, Slater's theorem applies (see, e.g., [16]) and strong duality holds; that is, the minimum value of the primal problem (4) is equal to the maximum value of the Lagrange dual problem (5).

The theoretical basis of interior-point algorithms is presented in [8].

3   Subspace Semidefinite Programming

We present a subspace algorithm for solving the semidefinite program (4). The optimization problem being (4) and (5), we introduce Lagrange multipliers y and M for (4) and look for stationary points of the function

$$\mathcal{L}(X, y, M) = C \bullet X - y^T(\mathcal{A}(X) - a) - M \bullet X. \tag{6}$$

Taking the partial derivative of $\mathcal{L}$ with respect to the components of X, we obtain the residual R:

$$\nabla_X \mathcal{L}(X, y, M) = C - \mathcal{A}^T(y) - M = R.$$

At the optimal (X, y), we note that $R = 0$. Also note that the modified residual matrix $\tilde{R} = C - \mathcal{A}^T(y) = M$, is positive semidefinite at the optimal pair (X, y). Taking the partial derivative of $\mathcal{L}$ with respect to the components of y, we obtain $\nabla_y \mathcal{L}(X, y, M) = \mathcal{A}(X) - a$.

Suppose that V is an $n \times k$ matrix of orthonormal columns and W is an $m \times l$ matrix of orthonormal columns. These represent subspaces of $\Re^n$ of dimensions k and l respectively. We can represent $\hat{X}$ in subspace V with $\hat{X} = V^T X V$ and $\hat{y}$ in subspace W with $\hat{y} = W^T y$, we can rewrite (4) as

$$\min_{\hat{X}} C \bullet (V\hat{X}V^T), \qquad \text{subject to (a) } \mathcal{A}(V\hat{X}V^T) - a = 0, \tag{7}$$
$$\text{(b) } V\hat{X}V^T \succeq 0.$$

Note that $X = V\hat{X}V^T \succeq 0$ is and only if $\hat{X} \succeq 0$. This subspace SDP (7) effectively reduces the number of variables. In the dual problem (5) we can set $y = W\hat{y}$ to restrict y to a subspace. This gives the following dual subspace problem:

$$\max_{\hat{y}} a^T(W\hat{y}) \qquad \text{subject to (a) } C - \mathcal{A}^T(W\hat{y}) \succeq 0. \tag{8}$$

This dual subspace problem (8) effectively reduces the number of constraints.

Combining the subspace problems to reduce both the number of variables (in the primal problem), and the number of constraints, pre-multiply the constraint (a) in (7) by W. We introduce new operators: Let $\hat{\mathcal{A}}$ be the linear operator $\hat{\mathcal{A}}(\hat{X}) = \mathcal{A}(V\hat{X}V^T)$, that maps symmetric matrices of size $n \times n$ into vectors in $\Re^m$, and $\hat{\hat{\mathcal{A}}}$ be the linear operator $\hat{\hat{\mathcal{A}}}(\hat{X}) = W^T \hat{\mathcal{A}}(\hat{X})$ that maps symmetric matrices of size $n \times n$ into vectors in $\Re^l$.

$$\mathcal{A}(V\hat{X}V^T) = \begin{pmatrix} \operatorname{tr}(A_1 V\hat{X}V^T) \\ \vdots \\ \operatorname{tr}(A_m V\hat{X}V^T) \end{pmatrix} = \begin{pmatrix} \operatorname{tr}(V^T A_1 V\hat{X}) \\ \vdots \\ \operatorname{tr}(V^T A_m V\hat{X}) \end{pmatrix} = \begin{pmatrix} \hat{A}_1 \bullet \hat{X} \\ \vdots \\ \hat{A}_m \bullet \hat{X} \end{pmatrix} = \hat{\mathcal{A}}(\hat{X})$$

where $\hat{A}_i = V^T A_i V$ for $i = 1, 2, \dots, m$. Also,

$$W^T(\hat{\mathcal{A}}(\hat{X})) = \begin{pmatrix} \sum_j (W_{j1}\hat{A}_j) \bullet \hat{X} \\ \vdots \\ \sum_j (W_{jl}\hat{A}_j) \bullet \hat{X} \end{pmatrix} = \begin{pmatrix} \hat{\hat{A}}_1 \bullet \hat{X} \\ \vdots \\ \hat{\hat{A}}_l \bullet \hat{X} \end{pmatrix} = \hat{\hat{\mathcal{A}}}(\hat{X}).$$

Since $W^T(\mathcal{A}(V\hat{X}V^T) - a) = W^T(\hat{A}(\hat{X}) - a) = W^T(\hat{A}(\hat{X})) - W^Ta = \hat{\mathcal{A}}(\hat{X}) - W^Ta$, (7) becomes the problem with both reduced number of primal variables and reduced number of constraints:

$$\min_{\hat{X}} V^TCV \bullet \hat{X} \qquad \text{subject to (a) } \hat{\mathcal{A}}(\hat{X}) - W^Ta = 0 \tag{9}$$
$$\text{(b) } \hat{X} \succeq 0.$$

We introduce Lagrange multipliers $\hat{y}$ and N and look for stationary points of the function

$$\hat{\mathcal{L}}(\hat{X}, \hat{y}, N) = V^TCV \bullet \hat{X} - \hat{y}^T(\hat{\mathcal{A}}(\hat{X}) - W^Ta) - N \bullet \hat{X}$$

Taking the partial derivative of $\hat{\mathcal{L}}$ with respect to the components of $\hat{X}$, we obtain

$$\nabla_{\hat{X}}\hat{\mathcal{L}}(\hat{X}, \hat{y}, N) = V^T(C - \mathcal{A}^T(y) - \hat{M})V$$
$$= V^T(\nabla_X \mathcal{L}(X, y, \hat{M}))V = V^TRV$$

where $\hat{M} = VNV^T$ is an approximation of M. Taking the partial derivative of $\hat{\mathcal{L}}$ with respect to the components of $\hat{y}$, we obtain

$$\nabla_{\hat{y}}\hat{\mathcal{L}}(\hat{X}, \hat{y}, N) = \hat{\mathcal{A}}(\hat{X}) - W^Ta = W^T(W\hat{\mathcal{A}}(\hat{X}) - a)$$
$$= W^T(\hat{\mathcal{A}}(\hat{X}) - a) = W^T(\mathcal{A}(V\hat{X}V^T) - a)$$
$$= W^T(\mathcal{A}(X) - a) = W^T\nabla_y\mathcal{L}(X, y, M).$$

Projecting the constraint (a) in (8) on the subspace V, and since $V^T(\mathcal{A}^T(W\hat{y}) - C)V = V^T((W^T\mathcal{A})^T(\hat{y}) - C)V = V^T((W^T\mathcal{A})^T(\hat{y}))V - V^TCV = (W^T\hat{\mathcal{A}})^T(\hat{y}) - V^TCV = \hat{\mathcal{A}}^T(\hat{y}) - V^TCV$, we have the Lagrangian dual of (9)

$$\max_{\hat{y}}(W^Ta)^T\hat{y} \qquad \text{subject to } (\hat{\mathcal{A}})^T(\hat{y}) - V^TCV \succeq 0. \tag{10}$$

Below we present the main steps of our subspace algorithm for semidefinite programming.

Algorithm 1 – Subspace Algorithm for Semidefinite Programming

1. Define the data for the problem: C an $n \times n$ matrix, a and m-dimensional vector, and an operator $\mathcal{A}$ where $\mathcal{A}(X)_i = A_i \bullet X$.
2. Define $n \times 1$ vector V_1 and $m \times 1$ vector W_1.
3. Compute an interior feasible starting point, $\hat{X}$ and $\hat{y}$, in subspace. $\hat{X}$ is found by solving feasibility problem. $\hat{y} = W_j^Ty$, where W_j is the current orthogonal basis.
4. Compute the semidefinite programming solution $(\hat{X}, \hat{y})$ on the subspace. The projected matrix, operator and vector on the subspace

   are $\hat{C} = V_j^TCV_j$, $\hat{\mathcal{A}} = W_j^T(\hat{A}) = W_j^T \begin{pmatrix} V_j^TA_1V_j \\ \vdots \\ V_j^TA_mV_j \end{pmatrix}$ and $\hat{a} = W_j^Ta$,

   where V_j and W_j are the current orthogonal basis. smallest value.

5. Compute $\tilde{R} = C - \mathcal{A}^T(y)$ (or some representation of it), and compute the eigenvector r of $\tilde{R}$ with the minimal eigenvalue.
6. Orthonormalize r against the current orthogonal basis V_j. Append the orthonormalized vector to V_j to give V_{j+1}.
7. Estimate residual $p = \mathcal{A}(X) - a = \mathcal{A}(V_j \hat{X} V_j^T) - a$.
8. Orthonormalize p against the current orthogonal basis W_j. Append the orthonormalized vector to W_j to give W_{j+1}.

4   Implementation

The main motivation to study this kind of problem comes from applications in discrete optimization. In particular we will investigate powerful and tractable relaxation of the min-cut problem.

The min-cut problem is the problem of partitioning the node set of an edge-weighted undirected graph into two parts so as to minimize the total weight of edges cut by the partition. We assume that the graph in question is complete (if not, non existing edges can be given weight 0 to complete the graph). Mathematically, the problem can be formulated as follows. Let the graph be given by its weighted adjacency matrix A. Define the matrix $L = \text{diag}(Ae) - A$, where e is the vector of all ones. The matrix L is called the Laplacian matrix associated with the graph. If a cut S is represented by a vector x where $x_i \in \{-1, 1\}$ depending on whether or not $i \in S$, we get the following formulation for the min-cut problem, where $C = -L$.

$$\min_x \frac{1}{4} x^T C x \qquad \text{subject to (a) } x \in \{-1, 1\}^n, \qquad (11)$$
$$\text{(b) } e^T x = 0.$$

Using $X = \frac{1}{4} x x^T$, this is equivalent to

$$\min_X C \bullet X \qquad \text{subject to (a) } \text{diag}(X) = \frac{1}{4} e,$$
$$\text{(b) } \text{rank}(X) = 1, \qquad (12)$$
$$\text{(c) } X \bullet (ee^T) = 0,$$
$$\text{(d) } X \succeq 0.$$

Dropping the rank condition we obtain a problem of the form SDP with $a = \frac{1}{4} e$ and $\mathcal{A}(X) = \text{diag}(X)$.

Algorithm 1 has been implemented using CSDP developed by Borchers [4] to solve min-cut problem.

5   Numerical Results

We compared our new subspace algorithm with the original semidefinite programming. The new algorithms have been implemented and run on a HP VISU-ALIZE Model C240 workstation, with a 236MHz PA-8200 processor and 512MB

RAM. The existing semidefinite software, CSDP [4], was used for solving the problem in a subspace, and for our comparison study. Also note that the matrices in this implementation are all explicitly represented as dense matrices.

The following is the graph partitioning problem that we would like to solve:

$$\min_X C \bullet X \qquad \text{subject to (a) } \mathcal{A}(X) = \tfrac{1}{4}$$
$$\text{(b) } X \bullet (ee^T) = 0, \qquad\qquad (13)$$
$$\text{(c) } X \succeq 0$$

There are n constraints from (a) in (13). We rewrite the problem so that we can reduce the number of constraints in original problem, m. The operator $\mathcal{A}$ is defined in terms of m diagonal matrices A_i where $(A_i)_{jj}$ is one if $j \equiv i$ (mod m), and zero otherwise, $i = 1, \cdots, m$, and $m \leq n$:

$$\min_X C \bullet X \qquad \text{subject to (a) } \mathcal{A}(X) = (n/4m)e$$
$$\text{(b) } X \bullet (ee^T) = 0, \qquad\qquad (14)$$
$$\text{(c) } X \succeq 0$$

where e is the vector of ones of the appropriate size. For example, if $m = 1$ is chosen for the number of constraints in original problem, the minimization problem can be written as:

$$\min_X C \bullet X \qquad \text{subject to (a) } \mathcal{A}(X) = \tfrac{1}{4}n$$
$$\text{(b) } X \bullet (ee^T) = 0, \qquad\qquad (15)$$
$$\text{(c) } X \succeq 0$$

where $\mathcal{A}(X) = I \bullet X$. Note that this is equivalent to spectral partitioning. Therefore, (a) in (15) is same as $\text{trace}(X) = \tfrac{1}{4}n$. If the unknown is a 9×9 matrix and $m = 3$, the first constraint in (14) would be $\mathcal{A}(X) = \tfrac{3}{4}e \in \Re^3$ and $\mathcal{A}$ is given in terms of 3 symmetric matrices: $A_1 = \text{diag}(1,0,0,1,0,0,1,0,0)$, $A_2 = \text{diag}(0,1,0,0,1,0,0,1,0)$, and $A_3 = \text{diag}(0,0,1,0,0,1,0,0,1)$.

Both algorithms were run with square matrices of various sizes. Figure 1 compares the observed running timings. The vertical-axis shows the timings in seconds. The horizontal-axis is the size of the matrices, 9, 25, 100, 400, and 900. The subspace was expanded until the stopping criteria have been met, which were a constraint for the primal problem, $\text{norm}(\mathcal{A}(X)-a) = \text{norm}(p) < 10^{-3}$, and the difference between primal and dual problems (the duality gap), is less than 10^{-7}. Table 1 summarizes the size of subspace when it reached the stopping criteria and the number of constraints. From this graph we can see that the subspace algorithm takes less than the original algorithm for all the test cases.

Figure 2 shows $\text{norm}(p)$ when the subspace were expanded in the subspace method. We used a fixed 400 by 400 matrix, and the number of constraints used was 10. The horizontal-axis is number of iterations, which describes the size of the subspace. The figure shows that after three iterations $\text{norm}(p)$ is reasonably small. This means that in order to find the solution of the original problem using the subspace method, only 10 constraints were needed and the problem was solved the problem until the size of the subspace reaches 3.

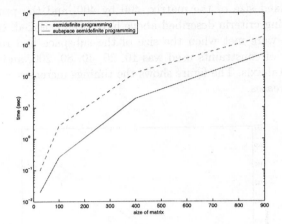

Fig. 1. Timings comparing subspace semidefinite programing against semidefinite programming.

Table 1. Timings comparing subspace semidefinite programing.

size of matrix	25	100	400	900
size of subspace	2	2	3	3
number of constraints	5	10	10	10

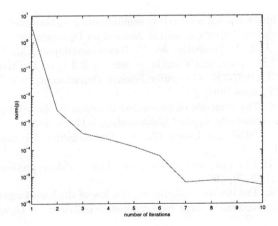

Fig. 2. norm(p) of subspace semidefinite programming.

Figure 3 shows the timings when the number of constrains were changed. We used the fixed size of the matrix, 400 by 400, and the program was run until the stopping criteria described above have been satisfied. In this example all the criteria were met when the size of the subspace had reached to 3 by 3. The number of constraints used was 10, 20, 40, 80, 200, and 400 which describes horizontal-axis. The figure shows the timings increases as the number of constraints increases.

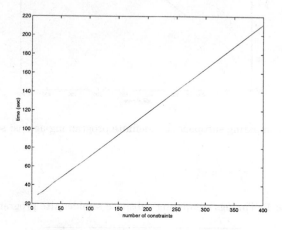

Fig. 3. Timings of subspace semidefinite programming using different number of constraints.

References

1. F. Alizadeh. Interior-point methods in semidefinite programming with applications to combinatorial optimization. *SIAM Journal on Optimization*, 5(1):13–51, 1995.
2. F. Alizadeh, J.-P. A. Haeberly, M. V. Nayakkankuppam, M. L. Overton, and S. Schmieta. SDPpack user's guide — version 0.9 beta for Matlab 5.0. Technical Report TR1997-737, Computer Science Department, New York University, New York, NY, June 1997.
3. W. E. Arnoldi. The principle of minimized iteration in the solution of the matrix eigenproblem. *Quarterly Applied Mathematics*, 9:17–29, 1951.
4. B. Borchers. CSDP, 2.3 User's Guide. *Optimization Methods and Software*, 11(1):597–611, 1999.
5. M. Crouzeix, B. Philippe, and M. Sadkane. The Davidson method. *SIAM J. Sci. Comput.*, 15(1):62–76, 1994.
6. E. R. Davidson. The iterative calculation of a few of the lowest eigenvalues and corresponding eigenvectors of large real-symmetric matrices. *J. Comp. Phys.*, 17:87–94, 1975.
7. K. Fujisawa, M. Kojima, and K. Nakata. SDPA User's Manual — Version 4.50. Technical Report B, Department of Mathematical and Computing Science, Tokyo Institute of Technology, Tokyo, Japan, July 1999.

8. C. Helmberg, F. Rendl, R. J. Vanderbei, and H. Wolkowicz. An interior-point method for semidefinite programming. *SIAM J. Optim.*, 6(2):342–361, 1996.
9. M. Holzrichter and S. Oliveira. A graph based Davidson algorithm for the graph partitioning problem. *International Journal of Foundations of Computer Science*, 10:225–246, 1999.
10. M. Holzrichter and S. Oliveira. A graph based method for generating the Fiedler vector of irregular problems. In *Lecture Notes in Computer Science*, volume 1586, pages 978–985. Springer, 1999. Proceedings of the 11th IPPS/SPDP'99 workshops.
11. C. Lanczos. Solution of systems of linear equations by minimized iterations. *J. Research Nat'l Bureau of Standards*, 49:33–53, 1952.
12. S. Oliveira. On the convergence rate of a preconditioned subspace eigensolver. *Parallel Algorithms and Applications*, 63:219–231, 1999.
13. S. Oliveira and T. Soma. A multilevel algorithm for spectral partitioning with extended eigen-models. In *Lecture Notes in Computer Science*, volume 1800, pages 477–484. Springer, 2000. Proceedings of the 15th IPDPS 2000 workshops.
14. B. N. Parlett. *The Symmetric Eigenvalue Problem*. Prentice-Hall, 1980.
15. S. Poljak, F.Rendl, and H. Wolkowicz. A recipe for semidefinite relaxation for (0,1)-quadratic programming. *J. Global Optim.*, 7(1):51–73, 1995.
16. M. V. Ramana. An exact duality theory for semidefinite programming and its complexity implications. *Mathematical Programming, Ser. B*, 77(2):129–162, 1997.
17. F. Rendl. A Matlab toolbox for semidefinite programming. Technical report, Technische Universität Graz, Institut für Mathematik, Kopernikusgasse 24, A-8010 Graz, Austria, 1994.
18. F. Rendl, R. J. Vanderbei, and H. Wolkowicz. Max-min eigenvalue problems, primal-dual interior point algorithms, and trust region subproblems. *Optimization Methods and Software*, 5:1–16, 1995.
19. Y. Saad. *Numerical Methods for Large Eigenvalue Problems*. Manchester University Press, Oxford Road, Manchester M13 9PL, UK, 1992.
20. G. L. G. Sleijpen and H. A. Van der Vorst. A Jacobi-Davidson iteration method for linear eigenvalue problems. *SIAM J. Matrix Anal. Appl.*, 17(2):401–425, 1996.
21. K. C. Toh, M. J. Todd, and P. H. Tütüncü. SDPT3 — a Matlab software package for semidefinite programming, version 2.1. Technical report, School of Operations Research and Industrial Engineering, Cornell University, Ithaca, NY, September 1999.
22. L. Vandenberghe and S. Boyd. Semidefinite programming. *SIAM Review*, 38:49–95, 1996.
23. L. Vandenberghe and S. Boyd. SP Software for semidefinite programming User's guide, version 1.0. Technical report, Information System Laboratory, Stanford University, Stanford, CA, November 1998.
24. H. Wolkowicz. Some applications of optimization in matrix theory. *Linear Algebra and its Applications*, 40:101–118, 1981.

A Study on the Pollution Error in r-h Methods Using Singular Shape Functions

Hyeong Seon Yoo and Jun-Hwan Jang

Department of Computer Science, Inha University,
Inchon, 402-751, South Korea
`hsyoo@inha.ac.kr`

Abstract: In this paper, we propose a modified pollution error in r-h methods using singular elements. The algorithm based on the element pollution error indicator concentrate on boundary nodes. The singular shape function is used where an element includes a singular point. The conventional automatic mesh generation method with this special shape function is shown to be very effective in this algorithm. The boundary node relocation phase and node insertion method are used alternatively as before. It is shown that the suggested r-h version algorithm combined with singular elements converges more quickly than the conventional one.

1 Introduction

A lot of engineering problems include geometric singularities in their problem domain and the solution near the singular points make the solution diverge to infinity. It is shown that the conventional error estimators are insufficient to estimate solution errors. Babuska pioneered works about error estimators and the effect of remote elements and introduced a concept of pollution error estimators to include this effect [1,2,3,4]. The pollution error estimator makes up for the weak points of the local error estimators. It was demonstrated that the conventional Zienkiewicz-Zhu error estimator [5,6,7,8] was insufficient and should include a pollution error indicator [1,4]. The pollution-adaptive feedback algorithm employs both local error indicators and pollution error indicators to refine the mesh outside a larger patch, which includes a patch and one to two surrounding mesh layers [2,3].

Special elements with singular shape functions were developed to overcome singularites in finite element alalysis [9]. It seems that the elements can be combined to accelerate in the pollution adaptive algorithms. We concentrate only on a problem boundary since the singularities exist on the boundary and mesh sizes change gradually regardless of the mesh generation algorithm. A mesh generation algorithm, which uses a node relocation method (r-method) as well as h-method of the finite element method for boundary elements, is proposed. The algorithm employs a boundary-node relocation at first and then does a node insertion based on the pollution error indicator.

P.M.A. Sloot et al. (Eds.): ICCS 2002, LNCS 2329, pp. 1068–1076, 2002.

2 The Model Problem

Consider a typical L-shaped polygon in two dimension, $\Omega \subseteq \mathbf{R}^2$, with mixed boundaries $\partial\Omega = \Gamma = \Gamma_D \cup \Gamma_N$, $\Gamma_D \cap \Gamma_N = \{\}$ where Γ_D is the *Dirichlet* and Γ_N is the *Newmann* boundary (Fig.1).

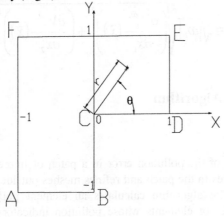

$$\Gamma_D = CD$$

$$\Gamma_N = AB \cup BC \cup DE \cup EF \cup FA$$

Fig.1 The L-shaped domain for the model problem

We will consider Laplacian with mixed boundary conditions. Let us consider the Hilbert space satisfying boundary condition $H^1_{\Gamma_D} \equiv \{u \in H^1(\Omega) \mid u = 0 \ \ on \ \ \Gamma_D\}$. Then the variational formulation of this model problem satisfies (1). And the solution space will be S^p_{h,Γ_D} combining the Hilbert and a trial function space S^p_h, [1].

Find $u_h \in S^p_{h,\Gamma_D}(\Omega) := H^1_{\Gamma_D} \cap S^p_h$ such that

$$B_\Omega(u_h, v_h) = \int_{\Gamma_N} g v_h \quad \forall v_h \in S^p_{h,\Gamma_D} \qquad (1)$$

where, $S^p_{h,\Gamma_D} \equiv \left\{ v \in C^0(\Omega) \middle| \ v \middle|_{\tau} \in p_p(\tau) \ \forall v \in T_h, v = 0 \ on \ \Gamma_D \right\}$

A patch error was expressed only by a local error, but it was demonstrated that the pollution error should include the patch error. The local error was improved by considering a mesh patch ω_h with a few surrounding mesh layers. The equilibrated residual functional is the same for the local error and the pollution error. But the pollution error was calculated by considering the outside of the larger patch, ω_h.

$$e_h \middle|_{\omega_h} = V_1^{\tilde{\omega}_h} + V_2^{\tilde{\omega}_h} \qquad (2)$$

where, $V_1^{\tilde{\omega}_h}$; local error on $\tilde{\omega}_h$

$V_2^{\tilde{\omega}_h}$; pollution error on $\tilde{\omega}_h$

$$\tilde{\omega}_h \; ; \; \omega_h + \text{a few mesh layers}$$

Let us denote $\left\|v\right\|_S = \sqrt{B_S(v,v)}$ energy norm over any domain $S \subseteq \Omega$, then the equation (3) can be a pollution estimator with $\overline{x} \in \omega_h$, [1,2,3].

$$\left\|V_2^{\tilde{\omega}_h}\right\|_{\omega_h} \cong \sqrt{|\omega_h|}\sqrt{\left(\frac{\partial V_2^{\tilde{\omega}_h}}{\partial x_1}(\overline{x})\right)^2 + \left(\frac{\partial V_2^{\tilde{\omega}_h}}{\partial x_2}(\overline{x})\right)^2} \qquad (3)$$

3 The Proposed Algorithm

3.1 The Basic Idea

For adaptive control of the pollution error in a patch of interest, the conventional algorithm fixes meshes in the patch and refines meshes outside the patch especially near singularities. The algorithm calculates an element pollution indicator and regularly divides $\gamma\%$ of elements whose pollution indicators are high [2]. This algorithm is as following Fig.2.

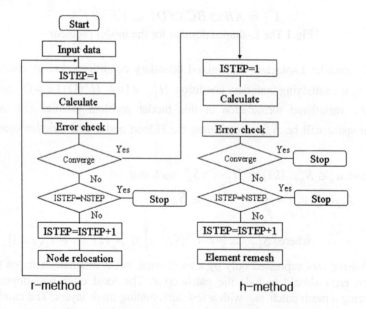

Fig. 2 Structure of the conventional algorithm

We denote the element pollution error M_{ω_h}, the local error ε_{ω_h} and the element pollution indicator μ_τ [2]. Since the conventional algorithm bisects the element length, it could be accelerated if we have smaller boundary elements near the singular points. Therefore it is natural to think about combining r and h method.

In our proposed algorithm, special elements with singular shape functions were adopted to overcome singularites. The special elements can be combined to accelerate in the pollution adaptive algorithms. Our algorithm employs two ideas. The first is to adopt special elements with singular shape functions. The other is to use r-h algorithm in which the node relocation phase and the node insertion phase are employed alternatively. In the relocation phase, the new boundary element length is calculated by using the following relationship between the pollution error estimator and the element size h, [1, 10].

$$\text{Let} \quad \left\| V_2^{\tilde{\omega}_h} \right\|_{\omega_h} \approx h^{2\lambda+1} \tag{4}$$

where λ ; the exponent for singular point.
From this expression, we can deduce old and new element length as following,

$$\left\| V_2^{\tilde{\omega}_h} \right\|_{\omega_h, old} = Ch_{old}^{2\lambda+1} \tag{4'}$$

$$\left\| V_2^{\tilde{\omega}_h} \right\|_{\omega_h, new} = Ch_{new}^{2\lambda+1} \tag{4''}$$

Combining two equations, we obtain the new element size h_{new},

$$h_{new} = h_{old} \times \left(\left\| V_2^{\tilde{\omega}_h} \right\|_{\omega_h, old} / \left\| V_2^{\tilde{\omega}_h} \right\|_{\omega_h, new} \right)^{-\frac{1}{2\lambda+1}} \tag{5}$$

In order to get the pollution error smaller than the local error we use

$t\mathcal{E}_{\omega_h} \approx t\left\| V_1^{\tilde{\omega}_h} \right\|_{\omega_h}$ instead of $\left\| V_2^{\tilde{\omega}_h} \right\|_{\omega_h, new}$. t is a user-specified constant between

0 and 1. And $\left\| V_2^{\tilde{\omega}_h} \right\|_{\omega_h, old}$ will be $\mu_\tau \approx \left\| V_2^{\tilde{\omega}_h} \right\|_{\omega_h} / \sqrt{|\omega_h|}$ since the pollution error

consists of the element pollution error indicators outside $\tilde{\omega}_h$. Finally the new element size becomes,

$$h_{new} = h_{old} \times (\zeta_\tau)^{-\frac{1}{2\lambda+1}} \tag{6}$$

where $\zeta_\tau \equiv \dfrac{\mu_\tau}{t\mathcal{E}_{\omega_h}}$

This new element size has an effect on the location of the boundary node, especially the nodes on BC and CD in Fig 1. If the ratio of the element length $(\zeta_\tau)^{-\frac{1}{2\lambda+1}}$ is less than 1, the algorithm moves the node to the singular point. But if it is greater than 1, the new length is discarded and the location of the node remains fixed to have stable solution. This relocation method is for reducing the number of iteration to get the final mesh. The boundary node insertion phase takes part in a high quality of the error estimator; this phase is the same as others [1,2,3].

3.2 Singular Shape Functions

For singular shape functions, we follow a generation scheme developed by Huges and Akin [9]. The r-directional shape function of node i, $N_i(r)$ can be expressed as the following algorithm.

Step 1. $N_{m+1}(r) \leftarrow \dfrac{N_{m+1}(r) - \sum\limits_{a=1}^{m} N_{m+1}(r_a) N_a(r)}{N_{m+1}(r_{m+1}) - \sum\limits_{a=1}^{m} N_{m+1}(r_a) N_a(r_{m+1})}$ (7)

Step 2. $N_a(r) \leftarrow N_a(r) - N_a(r_{m+1}) N_{m+1}(r)$ $a = 1, 2, \cdots, m$

Step 3. If m+1 < n (number of nodes), replace m by m+1 and repeat steps 1-3
  If m+1 = n, stop.

A binary number Flag is employed to alternate the boundary relocation and the node insertion process. If the flag is 0, the relocation phase is performed. The algorithm starts with the initial mesh and set Flag 0. The boundary node relocation is controlled by $(\zeta_\tau)^{-\frac{1}{2\lambda+1}}$. If the value is below 1, the element shrinks to singular point. In the node insertion phase, a new node is added on the middle of the boundary element. This r-h method makes fewer nodes on the boundary than the h-version.

The interior mesh generation phase is following the control of nodes on boundaries. This step is performed by the constrained Delaunay method. And the finite element analysis and error estimations are following [10].

4 Numerical Results and Discussions

We considered the mixed boundary-valued problem for the Laplacian over a L-shaped domain and applied boundary conditions consistent with the exact solution $u(r, \theta) = r^{\frac{1}{3}} \sin(\frac{1}{3}\theta)$ [1].

In order to test the effect of remote elements, an interior patch element ω_h whish is far from the singular point is chosen. A typical L-shaped domain is meshed by uniform quadratic triangles (p=2) with h=0.125 [10]. The conventional r-h method adopts r and h phase alternatively. But the node relocation phase converges rather slowly it is suggested to adopt multiple r phases until it converges under given conditions. The modified r-h method which combines multiple and single h phase shows better results than the conventional r-h methods. In table.1, we show the numerical results for the model problem. Though the local error estimate (ε_{ω_h}) is almost constant, the pollution error decreases dramatically with iterations.

The local error estimates change little and are nearly the same for all cases as we have expected. The pollution error converges after 4 or 5 iterations for the conventional shape function cases, but the singular shape function case is acceptable even for the start and changes a little.

Table 1 Results of the model problem

Iter.	ε_{ω_h} x E-05			M_{ω_h} x E-05		
	A	B	C	A	B	C
1	8.29	7.71	8.29	21.9	33.80	4.32
2	8.33	8.15	8.33	7.00	18.60	4.03
3	8.33	8.15	8.33	6.09	12.50	3.85
4	8.32	8.32	8.32	2.88	8.31	2.75
5		8.32			2.62	

A: Conventional r-h method with conventional functions
B: Modified r-h method with conventional shape functions
C: Modified r-h method with singular shape functions

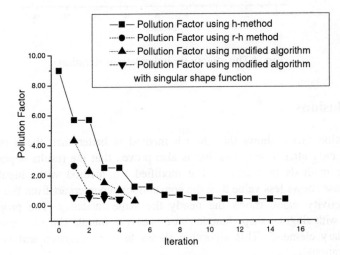

Fig. 3 The pollution factor, p_τ versus iteration

The pollution factor is defined by the ratio of a pollution error estimate and local error estimate $p_\tau = M_{\omega_h} / \varepsilon_{\omega_h}$. In Fig.3 we can see that the pollution factor decrease quickly for the conventional shape function case, but one iteration is enough for the modified shape function case. From this result, we note that the proposed algorithm with modified shape function case shows the best results and is effective to control the pollution error.

Fig.4 shows that the effectivity index has nearly the same tendency as the pollution factor. It is almost 1 for the first iteration, but in the conventional shape function case it needs 4 iterations to converge. Fig. 5 shows the final result for the proposed r-h method with conventional shape functions, which is obtained after 15 iterations. The number of element is increased to 1,461 and the node is 3,427.

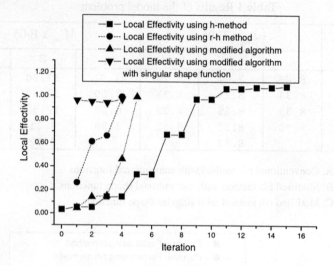

Fig. 4 The effectivity index versus iteration

5 Conclusions

The pollution factor shows that the r-h method is better than the h method. It converges only after 4 iterations, but is also proved that the results depend on the combining methods of r and h. The modified r-h method with singular shape function case shows less value than the conventional counterpart from the start. The local effectivity index shows the nearly the same tendency. The proposed r-h algorithm with singular shape function case is easy to handle since it considers only the boundary elements. This algorithm shows fast convergence and is nearly 1 during iterations.

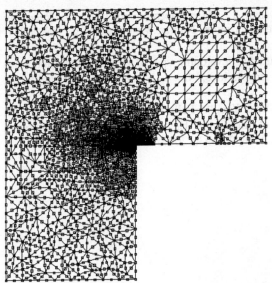

Fig.5 The final mesh after 5-th step by the proposed algorithm
(N=3427,E=1461)

References

1. I. Babŭska , T. Strouboulis, A. Mathur and C.S. Upadhyay, "Pollution error in the h-version of the finite element method and the local quality of a-posteriori error estimates", Finite Elements Anal. Des.,17,273-321(1994)
2. I. Babŭska , T. Strouboulis, C.S. Upadhyay and S.K. Gangaraj, "A posteriori estimation and adaptive control of the pollution error in the h-version of the finite element method", Int. J. Numer. Method Engrg., 38, 4207-4235(1995)
3. I. Babŭska , T. Strouboulis, S.K. Gangaraj, "Practical aspects of a-posteriori estimation and adaptive control of the pollution error for reliable finite element analysis", http://yoyodyne.tamu.edu/research/pollution/index.html(1996)
4. I. Babŭska , T. Strouboulis, S.K. Gangaraj and C.S. Upadhyay, "Pollution error in the h-version of the finite element method and the local quality of the recovered derivatives", Comput. Methods Appl. Mech. Engrg.,140,1-37(1997)
5. O.C. Zienkiewicz, and J.Z.Zhu, "The Superconvergent Patch Recovery and a posteriori estimators. Part1. The recovery techniques", Int. Numer. Methods Engrg.,33,1331-1364(1992)
6. O.C. Zienkiewicz, and J.Z.Zhu, "The Superconvergent Patch Recovery and a posteriori estimators. Part2. Error estimates and adaptivity", Int. J. Numer. Methods Engrg.,33,1365-1382(1992)
7. O.C. Zienkiewicz, and J.Z.Zhu, "The Superconvergent Patch Recovery(SPR) and adaptive finite element refinement", Comput. Methods Appl. Mech. Engrg.,101,207-224(1992)

8. O.C. Zienkiewicz, J.Z.Zhu and J. Wu, "Superconvergent Patch Recovery techniques – Some further tests", Comm. Numer. Methods Engrg., Vol. 9,251-258(1993)

9. Huges and Akin, "Techniques for developing special element shape functions with particular reference to singularities," Int. J. Numer. Methods Engrg., 33, 733-751(1980)

10. Soo Bum Pyun and Hyeong Seon Yoo, "A Pollution Adaptive Mesh Generation Algorithm in r-h Version of the Finite Element Method," Computational Science - ICCS 2001, LNCS 2073, 928-936(2001)

Device Space Design for Efficient Scale-Space Edge Detection

B.W. Scotney, S.A. Coleman, M.G. Herron

School of Information and Software Engineering, University of Ulster,
Coleraine, Northern Ireland

{bw.scotney, sa.coleman. mg.herron}@ulst.ac.uk

Abstract. We present a new approach to the computation of scalable image derivative operators, based on the finite element method, that addresses the issues of method, efficiency and scale-adaptability. The design procedure is applied to the problem of approximating scalable differential operators within the framework of Schwartz distributions. Within this framework, the finite element approach allows us to define a device space in which scalable image derivative operators are implemented using a combination of piecewise-polynomial and Gaussian basis functions.

Here we illustrate the approach in relation to the problem of scale-space edge detection, in which significant scale-space edge points are identified by maxima of existing edge-strength measures that are based on combinations of scale-normalised derivatives. We partition the image in order to locally identify approximate ranges of scales within which significant edge points may exist, thereby avoiding unnecessary computation of edge-strength measures across the entire range of scales.

1. Introduction

It is well known that the strength of a feature in an image may depend on the scale at which the appropriate detection operator is applied. It is also the case that many features in images exist significantly (in terms of operator response) over only a limited range of scales, and, moreover, that the most salient scale may vary spatially over the feature. Hence, when designing feature detection operators, it is necessary to consider the requirements for both the systematic development and efficient application of such operators adaptively over appropriately limited scale- and image-domains.

For scalable image operators, the finite element method offers a framework for systematic development and efficient implementation. In particular, scalable image derivative operators may be approximated within the framework of Schwartz distributions [4]. The general technique may be applied to a range of operators and using a variety of discrete approximations. The issue of scale (see, for example, [4,5,6]) is naturally and explicitly embraced in this approach by the introduction into the finite element method of Gaussian test functions that are controlled by a scale parameter [1]. We discuss the efficient implementation of the approach in terms of some well-established mechanisms for implementing the finite element method. It is also shown that the general finite element method naturally overcomes the 'border' problem [7]: image geometries such as corner and boundary operators may be produced with no additional effort. This is particularly important when scale-space techniques are used for problems that may require the use of very large operators.

P.M.A. Sloot et al. (Eds.): ICCS 2002, LNCS 2329, pp. 1077–1086, 2002.

2. Schwartz Distributions

An image here is represented by a rectangular $n{\times}n$ array of samples of a continuous function $u(x,y)$ of image intensity on a domain Ω. For the most refined level of finite element discretisation of the image domain Ω, nodes are placed at the pixel centres, and lines joining these form the edges of elements (see Fig. 1).

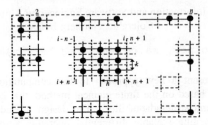

Fig. 1. Regular rectangular array of pixels and nodes for finite element discretisation

In Fig. 1 a global numbering scheme for the nodes is depicted. The dotted lines show the natural pixel edges.

In previous work [8,9] we have formulated image operators that correspond to weak forms of operators in the finite element method [2]. Operators used for smoothing may be based simply on a weak form of the image function, for which it is assumed that the image function $u \equiv u(x, y)$ belongs to the Hilbert space $H^0(\Omega)$; that is, the integral $\int_{\Omega} u^2 d\Omega$ is finite. Edge detection and enhancement operators are often based on first or second derivative approximations, for which it is necessary that the image function $u \equiv u(x, y)$ is constrained to belong to the Hilbert space $H^1(\Omega)$; i.e. the integral $\int_{\Omega} \left(\left| \underline{\nabla} u \right|^2 + u^2 \right) d\Omega$ is finite, where $\underline{\nabla} u$ is the vector $(\partial u/\partial x, \partial u/\partial y)^T$.

We now embrace a more general, and less restrictive, approach offered by Schwartz distribution theory (see [4]). For the image function $u \equiv u(x, y)$ we may obtain a distribution $F_u[v] \equiv \int_{\Omega} uv d\Omega$ for some test function v. Derivatives of order α are then defined by $\left(\nabla_\alpha F_u \right)[v] \equiv F_u[(-1)^{|\alpha|} \nabla_\alpha v]$. Hence a first or second order directional derivative may be defined by (respectively)

$$E(u) = -\int_{\Omega} u\underline{b} \cdot \underline{\nabla} v d\Omega \text{ and } Z(u) = \int_{\Omega} u\underline{\nabla} \cdot (B\underline{\nabla} v) d\Omega \qquad (1)$$

Here $\mathbf{B} = \underline{b}\, \underline{b}^T$ and $\underline{b} = (\cos\theta, \sin\theta)$ is the unit direction vector. Zero-crossing methods are often based on the isotropic form of the second order derivative, namely the Laplacian. In this formulation, this is equivalent to the general form in which the matrix $\mathbf{B}$ is the identity matrix $\mathbf{I}$.

3. Finite Element Formulation

In the finite element method a finite-dimensional space S^h is used for function approximation. We use this idea to implement an "image space" S^h. In the image space context, the superscript h corresponds to the inter-pixel distance in the image array (see Fig. 1). A finite element mesh is constructed using rectangular elements, as in Fig. 1, in which a global numbering scheme $1,...,N$ is employed. A basis for S^h may be formed by associating with each node i a basis function $\phi_i(x, y)$ which has the properties

$$\phi_i(x_j, y_j) = \begin{cases} 1 & \text{if } i = j \\ 0 & \text{if } i \neq j \end{cases} \tag{2}$$

where (x_j, y_j) are the co-ordinates of the nodal point j. $\phi_i(x, y)$ is thus a "tent-shaped" function with support restricted to a neighbourhood, centred on node i, consisting of only those elements that have node i as a vertex. We then approximately represent the image u by a function $U(x, y) = \sum_{j=1}^{N} U_j \phi_j(x, y)$ in which the parameters $\{U_1,...,U_N\}$ are mapped from the sampled image intensity values. The approximate image representation is therefore a simple function (typically a low order polynomial) on each element and has the sampled intensity value U_j at node j.

Our previous work has been based on a formulation analogous to the Galerkin finite element method [9]. However, since we now focus on the development of operators that can explicitly embrace the concept of scale, an alternative approach is preferred in which a finite-dimensional test space T_σ^h other than S^h is employed; the test space, or "device space", T_σ^h explicitly embodies a scale parameter σ. This generalisation allows sets of test functions $\psi_i^\sigma(x, y)$, $i=1, ..., N$, to be used when defining scale-space derivatives; for first and second order operators respectively, this provides the functionals

$$E_i^\sigma(U) = -\int_\Omega U \underline{b}_i \cdot \underline{\nabla} \psi_i^\sigma d\Omega_i \text{, and } Z_i^\sigma(U) = \int_\Omega U \underline{\nabla} \cdot (\mathbf{B}_i \underline{\nabla} \psi_i^\sigma) d\Omega \tag{3}$$

The scale parameter σ may now be used to explicitly control the scale of the operator, as illustrated in the example implementation in Section 4.

4. Example Implementation

We illustrate the approach applied to first derivative approximation using a regular bilinear rectangular discretisation formed from a set of rectangular elements such as that illustrated in Fig. 1. We construct a set of basis functions $\phi_i(x, y)$, $i=1,...,N$, so that the N-dimensional image subspace S^h comprises of functions that are piecewise bilinear.

For the test space T_σ^h a set of Gaussian basis functions is used that explicitly embodies the scale parameter σ. Sets of Gaussian test functions $\psi_i^\sigma(x, y)$, $i=1, \ldots, N$ of the form

$$\psi_i^\sigma(x, y) = \frac{1}{2\pi\sigma^2} e^{-\left(\frac{(x-x_i)^2+(y-y_i)^2}{2\sigma^2}\right)} \tag{4}$$

are constructed, each restricted to have support over a neighbourhood Ω_i^σ centred on the node i at (x_i, y_i). The size of the neighbourhood Ω_i^σ to which the support of $\psi_i^\sigma(x, y)$ is restricted is also explicitly related to the scale parameter σ [3]. The Gaussian is restricted so that the neighbourhood Ω_i^σ is a $2W_\sigma \times 2W_\sigma$ block of rectangular elements (where W_σ is an integer) with the nodal position (x_i, y_i) at the centre of the block. Then the relationship $1.96\sigma = W_\sigma$ ensures that along both the vertical and horizontal co-ordinate directions through the nodal point (x_i, y_i), 95% of the cross-section of the Gaussian is contained within Ω_i^σ.

On a neighbourhood Ω_i^σ we consider a locally constant unit vector $\underline{b}_i = (b_{i1}, b_{i2})^T$. Substituting the image representation $U(x, y) = \sum_{j=1}^{N} U_j \phi_j(x, y)$ into the derivative $E_i^\sigma(U) = -\int_\Omega U \underline{b}_i \cdot \nabla \psi_i^\sigma \, d\Omega_i$ gives

$$E_i^\sigma(U) = -b_{1i} \sum_{j=1}^{N} K_{ij}^\sigma U_j - b_{2i} \sum_{j=1}^{N} L_{ij}^\sigma U_j, \tag{5}$$

where K_{ij}^σ and L_{ij}^σ are respectively entries in $N \times N$ global matrices K^σ and L^σ given by

$$K_{ij}^\sigma = \int_{\Omega_i^\sigma} \frac{\partial \psi_i^\sigma}{\partial x} \phi_j \, dxdy \quad \text{and} \quad L_{ij}^\sigma = \int_{\Omega_i^\sigma} \frac{\partial \psi_i^\sigma}{\partial y} \phi_j \, dxdy, \ i,j=1,..,N. \tag{6}$$

These integrals need be computed only over the neighbourhood Ω_i^σ, rather than the entire image domain Ω, since ψ_i^σ has support restricted to Ω_i^σ.

4.1. Element-by-Element Computation

Each neighbourhood Ω_i^σ is composed of a set S_i^σ of elements. In our example implementation S_i^σ is the $2W_\sigma \times 2W_\sigma$ block of elements having nodal point (x_i, y_i) at its centre. We may thus write K_{ij}^σ as a summation $K_{ij}^\sigma = \sum_{\{m | e_m \in S_i^\sigma\}} k_{ij}^{m,\sigma}$ where $k_{ij}^{m,\sigma}$ is the element integral $k_{ij}^{m,\sigma} = \int_{e_m} \frac{\partial \phi_j}{\partial x} \psi_i^\sigma \, dxdy$.

In principle, each $E_i^\sigma(U)$ may be computed by calculating $k_{ij}^{m,\sigma}$ and $l_{ij}^{m,\sigma}$ for each element e_m in S_i^σ and then assembling these values to yield K_{ij}^σ and L_{ij}^σ respectively. In practice, much less computation than this is required. To construct $E_i^\sigma(U)$ only one quarter of the operator is computed, and via reflective symmetry in the horizontal and vertical axes the complete operator is generated. Consider the situation in Fig. 2 where the Gaussian basis function is restricted to the 4×4 block of elements (i.e. $W^\sigma = 2$). We need to compute $k_{ij}^{m,\sigma}$ and $l_{ij}^{m,\sigma}$ only for $m = a_1, b_1, c_1, d_1$, representing just one quarter of the whole support Ω_i^σ of the operator E_i^σ.

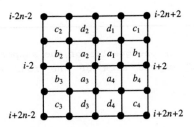

Fig. 2. Elements within neighbourhood Ω_i^σ for $W^\sigma = 2$

For each of the elements e_m, $m = a_1$, b_1, c_1 and d_1, a pair of 2×2 element operators, $k_i^{m,\sigma}$ and $l_i^{m,\sigma}$, is generated. For example, the x-directional element operator over element a_1 is given by:

$$k_i^{a_1,\sigma} = \begin{bmatrix} k_{i,i-n}^{a_1,\sigma} & k_{i,i-n+1}^{a_1,\sigma} \\ k_{i,i}^{a_1,\sigma} & k_{i,i+1}^{a_1,\sigma} \end{bmatrix} \text{ where } k_{ij}^{a_1,\sigma} = \int_{a_1} \frac{\partial \phi_j}{\partial x} \psi_i^\sigma \, dxdy. \tag{7}$$

If the element sizes are normalised, i.e. $h = k = 1$, this can readily be calculated and found to be:

$$a_1 = \begin{bmatrix} \alpha_4 & \alpha_3 \\ \alpha_1 & \alpha_2 \end{bmatrix} = \begin{bmatrix} -0.025178 & 0.025178 \\ -0.021469 & 0.021469 \end{bmatrix}. \tag{8}$$

In practice, integration over an element is done numerically using a low-order Gauss quadrature rule, typically requiring just four evaluations of the integrand (e.g. [2]). Similarly, we can compute:

$$b_1 = \begin{bmatrix} \beta_4 & \beta_3 \\ \beta_1 & \beta_2 \end{bmatrix} = \begin{bmatrix} -0.061061 & 0.061061 \\ -0.052064 & 0.052064 \end{bmatrix}, \quad c_1 = \begin{bmatrix} \gamma_4 & \gamma_3 \\ \gamma_1 & \gamma_2 \end{bmatrix} = \begin{bmatrix} -0.011802 & 0.011802 \\ -0.007433 & 0.007433 \end{bmatrix},$$

$$\text{and } d_1 = \begin{bmatrix} \delta_4 & \delta_3 \\ \delta_1 & \delta_2 \end{bmatrix} = \begin{bmatrix} -0.028621 & 0.028621 \\ -0.018025 & 0.018025 \end{bmatrix}$$

We can generate operators of any size by this means, only ever requiring four pre-computed values of the basis function derivative $DX[\cdot][g] = \dfrac{\partial \phi}{\partial x}\Big|_{(x_g, y_g)}$ evaluated at four Gauss quadrature points in a representative element. The four values of $DX[\cdot][g]$ for each node in any element can be obtained by permutation of these four pre-computed values. Further, since the corresponding four values of $DY[\cdot][g]$ required for the computation of $l_i^{m,\sigma}$ are permutations of the four values of $DX[\cdot][g]$, it is not necessary to explicitly compute $l_i^{m,\sigma}$; the operator L_{ij}^{σ} is merely a rotation of the operator K_{ij}^{σ} . The general procedure for computing the contributions required for K_{ij}^{σ} is shown below.

```
for m from 0 to Wₒ×Wₒ -1

/*for each element eₘ in one quarter of the support
of the operator Eᵢᵍ compute each of the four entries in the 2×2
operator kᵢᵐ,ᵍ */

begin
for j from 0 to 3
begin
   for g from 0 to 3
   begin

   /*compute position of Gauss point g in element eₘ*/
      compute (xᵍᵐ, yᵍᵐ)

      /*implement Gauss quadrature rule*/
      evaluate ψᵢᵍ[m] [g]

      compute DX[j] [g] × ψᵢᵍ[m] [g]

      compute kᵢ[j] [m] = kᵢ[j] [m] + DX[j] [g] × ψᵢᵍ[m] [g]
   end g
      compute kᵢ[j] [m] = kᵢ[j] [m] × h × k / 4
end j
end m
```

The assembly process to construct $E_i^{\sigma}(U)$ from the element matrices is described in Section 4.2.

4.2. Finite Element Assembly

The next stage of the finite element method consists of assembling the element matrices $k_i^{m,\sigma}$ and $l_i^{m,\sigma}$, for each element $m=1,\ldots,M$, into the global matrices K^{σ} and L^{σ} respectively. This is accomplished by distributing each entry in the element matrices into the corresponding row and column of the global matrices using the local to global node numbering mapping routinely stored in the finite element method.

col	$i\text{-}2n\text{-}2$	$i\text{-}2n\text{-}1$	$i\text{-}2n$	$i\text{-}2n\text{+}1$	$i\text{-}2n\text{+}2$
	$-\gamma_3$	$-\gamma_4\text{-}\delta_3$	$\delta_4\text{-}\delta_4$	$\gamma_4\text{+}\delta_3$	γ_3
col	$i\text{-}n\text{-}2$	$i\text{-}n\text{-}1$	$i\text{-}n$	$i\text{-}n\text{+}1$	$i\text{-}n\text{+}2$
	$-\beta_3\text{-}\gamma_2$	$-\alpha_3\text{-}\beta_4\text{-}\gamma_1\text{-}\delta_2$	$\alpha_4\text{-}\alpha_4\text{+}\delta_1\text{-}\delta_1$	$\alpha_3\text{+}\beta_4\text{+}\gamma_1\text{+}\delta_2$	$\beta_3\text{+}\gamma_2$
col	$i\text{-}2$	$i\text{-}1$	i	$i\text{+}1$	$i\text{+}2$
	$-\beta_2\text{-}\beta_2$	$-\alpha_2\text{-}\alpha_2\text{-}\beta_1\text{-}\beta_1$	$\alpha_1\text{-}\alpha_1\text{-}\alpha_1\text{+}\alpha_1$	$\alpha_2\text{+}\alpha_2\text{+}\beta_1\text{+}\beta_1$	$\beta_2\text{+}\beta_2$
col	$i\text{+}n\text{-}2$	$i\text{+}n\text{-}1$	$i\text{+}n$	$i\text{+}n\text{+}1$	$i\text{+}n\text{+}2$
	$-\beta_3\text{-}\gamma_2$	$-\alpha_3\text{-}\beta_4\text{-}\gamma_1\text{-}\delta_2$	$\alpha_4\text{-}\alpha_4\text{-}\delta_1\text{+}\delta_1$	$\alpha_3\text{+}\beta_4\text{+}\gamma_1\text{+}\delta_2$	$\beta_3\text{+}\gamma_2$
col	$i\text{+}2n\text{-}2$	$i\text{+}2n\text{-}1$	$i\text{+}2n$	$i\text{+}2n\text{+}1$	$i\text{+}2n\text{+}2$
	$-\gamma_3$	$-\gamma_4\text{-}\delta_3$	$-\delta_4\text{+}\delta_4$	$\gamma_4\text{+}\delta_3$	γ_3

Fig. 3. Element assembly to form row i of matrix $\boldsymbol{K}^\sigma$

Such a procedure is illustrated in Fig. 3 for the distribution of the values of the 16 element matrices $k_i^{m,\sigma}$ corresponding to the 16 elements depicted in Fig. 2. We see that 25 values of the global matrix K^σ are augmented in the assembly procedure for these 16 elements. Since the neighbourhood Ω_i^σ is centred on node i, these 25 values are all in row i of K^σ and in the columns of K^σ indicated in Fig. 2. Hence assembled in row i of the global matrix K^σ we have the 5×5 spatial operator centred on node i:

$$
\begin{array}{ccccc}
-\gamma_3 & -\gamma_4\text{-}\delta_3 & 0 & \gamma_4\text{+}\delta_3 & \gamma_3 \\
-\beta_3\text{-}\gamma_2 & -\alpha_3\text{-}\beta_4\text{-}\gamma_1\text{-}\delta_2 & 0 & \alpha_3\text{+}\beta_4\text{+}\gamma_1\text{+}\delta_2 & \beta_3\text{+}\gamma_2 \\
-2\beta_2 & -2\alpha_2\text{-}2\beta_1 & 0 & 2\alpha_2\text{+}2\beta_1 & 2\beta_2 \\
-\beta_3\text{-}\gamma_2 & -\alpha_3\text{-}\beta_4\text{-}\gamma_1\text{-}\delta_2 & 0 & \alpha_3\text{+}\beta_4\text{+}\gamma_1\text{+}\delta_2 & \beta_3\text{+}\gamma_2 \\
-\gamma_3 & -\gamma_4\text{-}\delta_3 & 0 & \gamma_4\text{+}\delta_3 & \gamma_3
\end{array}
$$

Assembly of all of the element matrices into K^σ thus yields a sparsely populated and highly structured $N{\times}N$ block matrix composed of $n{\times}n$ sub-matrices. Each sub-matrix is itself sparse, having non-zero values in only 5 positions in each row: the diagonal, and the first two upper and lower diagonals.

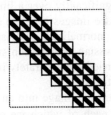

Fig. 4. The general banded block structure of $\boldsymbol{K}^\sigma$

Fig. 4 shows the general structure of K^σ; zero-values have been suppressed.

Each row of matrix K^σ and of matrix L^σ represents an operator on the vector $\mathbf{U}=\{U_j\}$, $j=1, ..., N$, by which the scale-space derivative approximations E_i^σ, $i=1, ..., N$ are computed. As an integral part of this process, appropriate operators are generated both for image corners and image borders as well as the internal regions of the image. Hence the "border problem", particularly significant for large-scale operators, is naturally accommodated.

We thus obtain the vector $E^\sigma = \{E_i^\sigma\}$, for $i=1, ..., N$, of weak first derivative approximations by using the operator $(B_1^\sigma K^\sigma + B_2^\sigma L^\sigma)$ on the vector $\mathbf{U}$ of image intensity values, namely,

$$E^\sigma = (B_1^\sigma K^\sigma + B_2^\sigma L^\sigma)\mathbf{U}, \tag{9}$$

where B_1^σ and B_2^σ are diagonal matrices of respectively the locally constant direction components b_{i1} and b_{i2} on each neighbourhood Ω_i^σ as described in Section 3.

5. Scale-Space Edge Detection

As a fundamental application of the technique for adaptive scale-space derivative approximation developed above, we choose the problem of scale-space edge detection. For this we follow closely the work on edge detection with automatic scale selection developed in [6]. In [6], significant scale-space edge points are identified by maxima of specified edge-strength measures that are located by zero-crossings of two functions $Z_1(L)$ and $Z_2(L)$ of the scale-space image $L(x, y, t)$, where t is the scale parameter. The procedure used in [6] is a pre-processing phase to tracking the zero-crossing surfaces and identifying, by use of a line integral measure, the most significant scale-space edges in the image. In [6] it is illustrated that this approach can be very successful in both identifying the most significant edges and demonstrating how the most salient scale varies along an edge.

We now illustrate an alternative approach to the pre-processing phase in [6] based on application of the adaptive scale-space derivative approximation technique developed in Sections 2 to 4. Advantages of our approach are that we exploit the fact that the image features exist significantly over a limited range of scales to avoid the computation of global scale-space images over the extensive range of scales used in [6]; moreover, our adaptive normalised scale-space derivatives naturally and systematically combine the Gaussian convolution smoothing and discrete derivative approximation steps that are carried out separately in [6], thus avoiding the use of ad hoc finite difference approximations.

As an initial step we partition the image into an array of regular rectangular sub-regions, $R_j, j = 1,..., J$, within each of which we compute the image variance σ_j. The results of numerical experimentation have established that σ_j may be used as an indicator of the appropriate device space to be used for the computation of the scale-space derivatives within R_j required for $Z_1(L)$ and $Z_2(L)$. Using the experimentally derived relationship $\sigma_j^0 = 1/\log_{10}\sigma_j$, we thus compute the operator scale σ_j^0 at which the device space response is approximately maximal. Hence a "key-scale" t_j^0 is established for each sub-region R_j. The operator size corresponding to t_j^0 is automatically derived using the procedure described in Section 4, and Fig. 5 shows an image together with these "key scales" computed for 64 image sub-regions. Here, "1" indicates 3×3, "2" indicates 5×5, etc., and it can be clearly seen that the smaller scales occur in regions representing the detail image objects, whilst larger scales are employed in the image background.

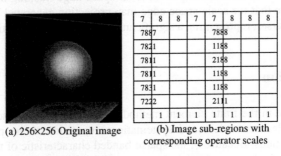

7	8	8	7	7	8	8	8
7887				7888			
7821				1188			
7811				2188			
7811				1188			
7831				1188			
7222				2111			
1	1	1	1	1	1	1	1

(a) 256×256 Original image   (b) Image sub-regions with corresponding operator scales

Fig. 5. Illustration of adaptive filters

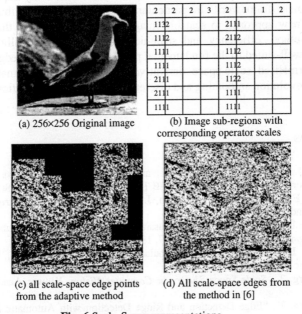

2	2	2	3	2	1	1	2
1132				2111			
1112				2112			
1111				1112			
1111				1112			
2111				1122			
2111				1111			
1111				1111			

(a) 256×256 Original image   (b) Image sub-regions with corresponding operator scales

(c) all scale-space edge points from the adaptive method   (d) All scale-space edges from the method in [6]

Fig. 6. Scale-Space representations

Fig. 6 (a) and (b) show another image and corresponding grid of sub-regional key-scales. Again the use of smaller scales for the image detail and larger scales for the background is clearly evident. Once a regional key-scale t_j^0 has been computed for each R_j, the measures $Z_1(L)$ and $Z_2(L)$ are computed over R_j for a small number of scales above and below t_j^0; in this case we have chosen five scales: $t_j^{-2}, t_j^{-1}, t_j^0, t_j^1, t_j^2$ centred on t_j^0. The inter-spatial distance has been chosen to be the same as that used in [6]; note, however, that we are not restricted to aligning the scales between sub-regions, as each key-scale t_j^0 is selected independently. In Fig. 6(c) and (d) respectively are shown all scale-space edge points identified by the method proposed in this paper and by the method described in [6]. Whilst both approaches provide suitable output for scale-space edge tracking by the method used in [6], we note that the scale-space edge map projection generated by our approach is very much sparser, representing a significant reduction in computational effort.

6. Conclusion

A key concept in the finite element method is the notion of a standard element. This concept enables convenient and systematic computation to be performed even in complex situations. Because of the sparse banded characteristic of the global matrices produced, efficiency extends to storage as well as computation. We have shown that the concept of scale for operators can be explicitly and efficiently implemented within a finite element framework. In particular, using Schwartz distributions, we have presented a new approach to scalable image derivative operators that addresses the issues of method, efficiency and scale-adaptability. We have illustrated that the approach is well-suited to the fundamental problem of scale-space edge detection: our adaptive normalised scale-space derivatives naturally and systematically combine Gaussian convolution smoothing and discrete derivative approximation.

References

[1]   Babaud,J., Witkin, A.P., Baudin, M., and Duda, R.O., "Uniqueness of the Gaussian Kernel for Scale-Space Filtering" IEEE Trans. on PAMI, Vol. PAMI-8(1), 1986, 26-33.

[2]   Becker, E.B., Carey, G.F., and Oden, J.T., *Finite Elements: An Introduction*, Prentice Hall, London, 1981.

[3]   Davies,E.R., "Design of Optimal Gaussian Operators in Small Neighbourhoods" *Image and Vision Computing*, Vol. 5(3), 1987, 199-205.

[4]   Florack, L., *Image Structure*, Computational Imaging and Vision, Vol. 10, Kluwer Academic Publishers, 1997.

[5]   Lindeberg, T., *Scale-Space Theory in Computer Vision*, Kluwer Academic Publishers, 1994.

[6]   Lindeberg, T., "Edge Detection and Ridge Detection with Automatic Scale Selection", *International Journal of Computer Vision*, Vol. 30(2), 117-154, 1998.

[7]   Prewitt, J.M.S., "Object Enhancement and Extraction", *Picture Processing & Psychopictorics*, 1970, 75-149.

[8]   Scotney, B.W., Coleman, S.A., Herron, M.G., "A Systematic Design Procedure for Scalable Near-Circular Gaussian Operators.", Proceedings of the IEEE Int. Conf. On Image Processing(ICIP 2001), Greece, 844-847, 2001.

[9]   Scotney, B.W., Herron, M.G., "Systematic and Efficient Construction of Neighbourhood Operators on Irregular and Adaptive Grids." Third Irish Machine Vision and Image Processing Conference (IMVIP'99), Dublin, 204-218, 1999.

Author Index

Lecture Notes in Computer Science

For information about Vols. 1–2248
please contact your bookseller or Springer-Verlag

Vol. 2284: T. Eiter, K.-D. Schewe (Eds.), Foundations of Information and Knowledge Systems. Proceedings, 2002. X, 289 pages. 2002.

Vol. 2285: H. Alt, A. Ferreira (Eds.), STACS 2002. Proceedings, 2002. XIV, 660 pages. 2002.

Vol. 2286: S. Rajsbaum (Ed.), LATIN 2002: Theoretical Informatics. Proceedings, 2002. XIII, 630 pages. 2002.

Vol. 2287: C.S. Jensen, K.G. Jeffery, J. Pokorny, Saltenis, E. Bertino, K. Böhm, M. Jarke (Eds.), Advances in Database Technology – EDBT 2002. Proceedings, 2002. XVI, 776 pages. 2002.

Vol. 2288: K. Kim (Ed.), Information Security and Cryptology – ICISC 2001. Proceedings, 2001. XIII, 457 pages. 2002.

Vol. 2289: C.J. Tomlin, M.R. Greenstreet (Eds.), Hybrid Systems: Computation and Control. Proceedings, 2002. XIII, 480 pages. 2002.

Vol. 2291: F. Crestani, M. Girolami, C.J. van Rijsbergen (Eds.), Advances in Information Retrieval. Proceedings, 2002. XIII, 363 pages. 2002.

Vol. 2292: G.B. Khosrovshahi, A. Shokoufandeh, A. Shokrollahi (Eds.), Theoretical Aspects of Computer Science. IX, 221 pages. 2002.

Vol. 2293: J. Renz, Qualitative Spatial Reasoning with Topological Information. XVI, 207 pages. 2002. (Subseries LNAI).

Vol. 2295: W. Kuich, G. Rozenberg, A. Salomaa (Eds.), Developments in Language Theory. Proceedings, 2001. IX, 389 pages. 2002.

Vol. 2296: B. Dunin-Kęplicz, E. Nawarecki (Eds.), From Theory to Practice in Multi-Agent Systems. Proceedings, 2001. IX, 341 pages. 2002. (Subseries LNAI).

Vol. 2297: R. Backhouse, R. Crole, J. Gibbons (Eds.), Algebraic and Coalgebraic Methods in the Mathematics of Program Construction. Proceedings, 2000. XIV, 387 pages. 2002.

Vol. 2299: H. Schmeck, T. Ungerer, L. Wolf (Eds.), Trends in Network and Pervasive Computing – ARCS 2002. Proceedings, 2002. XIV, 287 pages. 2002.

Vol. 2300: W. Brauer, H. Ehrig, J. Karhumäki, A. Salomaa (Eds.), Formal and Natural Computing. XXXVI, 431 pages. 2002.

Vol. 2301: A. Braquelaire, J.-O. Lachaud, A. Vialard (Eds.), Discrete Geometry for Computer Imagery. Proceedings, 2002. XI, 439 pages. 2002.

Vol. 2302: C. Schulte, Programming Constraint Services. XII, 176 pages. 2002. (Subseries LNAI).

Vol. 2303: M. Nielsen, U. Engberg (Eds.), Foundations of Software Science and Computation Structures. Proceedings, 2002. XIII, 435 pages. 2002.

Vol. 2304: R.N. Horspool (Ed.), Compiler Construction. Proceedings, 2002. XI, 343 pages. 2002.

Vol. 2305: D. Le Métayer (Ed.), Programming Languages and Systems. Proceedings, 2002. XII, 331 pages. 2002.

Vol. 2306: R.-D. Kutsche, H. Weber (Eds.), Fundamental Approaches to Software Engineering. Proceedings, 2002. XIII, 341 pages. 2002.

Vol. 2307: C. Zhang, S. Zhang, Association Rule Mining. XII, 238 pages. 2002. (Subseries LNAI).

Vol. 2308: I.P. Vlahavas, C.D. Spyropoulos (Eds.), Methods and Applications of Artificial Intelligence. Proceedings, 2002. XIV, 514 pages. 2002. (Subseries LNAI).

Vol. 2309: A. Armando (Ed.), Frontiers of Combining Systems. Proceedings, 2002. VIII, 255 pages. 2002. (Subseries LNAI).

Vol. 2310: P. Collet, C. Fonlupt, J.-K. Hao, E. Lutton, M. Schoenauer (Eds.), Artificial Evolution. Proceedings, 2001. XI, 375 pages. 2002.

Vol. 2311: D. Bustard, W. Liu, R. Sterritt (Eds.), Soft-Ware 2002: Computing in an Imperfect World. Proceedings, 2002. XI, 359 pages. 2002.

Vol. 2312: T. Arts, M. Mohnen (Eds.), Implementation of Functional Languages. Proceedings, 2001. VII, 187 pages. 2002.

Vol. 2313: C.A. Coello Coello, A. de Albornoz, L.E. Sucar, O.Cairó Battistutti (Eds.), MICAI 2002: Advances in Artificial Intelligence. Proceedings, 2002. XIII, 548 pages. 2002. (Subseries LNAI).

Vol. 2314: S.-K. Chang, Z. Chen, S.-Y. Lee (Eds.), Recent Advances in Visual Information Systems. Proceedings, 2002. XI, 323 pages. 2002.

Vol. 2315: F. Arhab, C. Talcott (Eds.), Coordination Models and Languages. Proceedings, 2002. XI, 406 pages. 2002.

Vol. 2316: J. Domingo-Ferrer (Ed.), Inference Control in Statistical Databases. VIII, 231 pages. 2002.

Vol. 2317: M. Hegarty, B. Meyer, N. Hari Narayanan (Eds.), Diagrammatic Representation and Inference. Proceedings, 2002. XIV, 362 pages. 2002. (Subseries LNAI).

Vol. 2318: D. Bošnački, S. Leue (Eds.), Model Checking Software. Proceedings, 2002. X, 259 pages. 2002.

Vol. 2319: C. Gacek (Ed.), Software Reuse: Methods, Techniques, and Tools. Proceedings, 2002. XI, 353 pages. 2002.

Vol. 2322: V. Mařík, O. Stěpánková, H. Krautwurmová, M. Luck (Eds.), Multi-Agent Systems and Applications II. Proceedings, 2001. XII, 377 pages. 2002. (Subseries LNAI).

Vol. 2324: T. Field, P.G. Harrison, J. Bradley, U. Harder (Eds.), Computer Performance Evaluation. Proceedings, 2002. XI, 349 pages. 2002.

Vol. 2329: P.M.A. Sloot, C.J.K. Tan, J.J. Dongarra, A.G. Hoekstra (Eds.), Computational Science – ICCS 2002. Proceedings, Part I. XLI, 1095 pages. 2002.

Vol. 2330: P.M.A. Sloot, C.J.K. Tan, J.J. Dongarra, A.G. Hoekstra (Eds.), Computational Science – ICCS 2002. Proceedings, Part II. XLI, 1115 pages. 2002.

Vol. 2331: P.M.A. Sloot, C.J.K. Tan, J.J. Dongarra, A.G. Hoekstra (Eds.), Computational Science – ICCS 2002. Proceedings, Part III. XLI, 1227 pages. 2002.

Vol. 2332: L. Knudsen (Ed.), Advances in Cryptology – EUROCRYPT 2002. Proceedings, 2002. XII, 547 pages. 2002.